Handbook of Sensor Networks: Compact Wireless and Wired Sensing Systems

Handbook of Sensor Networks: Compact Wireless and Wired Sensing Systems

TK
7872
D48H36
2005
Web

Edited by
MOHAMMAD ILYAS AND IMAD MAHGOUB

CRC PRESS

Boca Raton   London   New York   Washington, D.C.

Library of Congress Cataloging-in-Publication Data

Handbook of sensor networks : compact wireless and wired sensing systems / edited by
   Mohammad Ilyas and Imad Mahgoub.
       p.   cm.
Includes bibliographical references and index.
ISBN 0-8493-1968-4 (alk. paper)
   1. Sensor networks. 2. Wireless LANs. I. Ilyas, Mohammad, 1953- II. Mahgoub, Imad.

TK7872.D48.H36 2004
004.6′8—dc22   2004043852

This book contains information obtained from authentic and highly regarded sources. Reprinted material is quoted with permission, and sources are indicated. A wide variety of references are listed. Reasonable efforts have been made to publish reliable data and information, but the author and the publisher cannot assume responsibility for the validity of all materials or for the consequences of their use.

Neither this book nor any part may be reproduced or transmitted in any form or by any means, electronic or mechanical, including photocopying, microfilming, and recording, or by any information storage or retrieval system, without prior permission in writing from the publisher.

All rights reserved. Authorization to photocopy items for internal or personal use, or the personal or internal use of specific clients, may be granted by CRC Press LLC, provided that $1.50 per page photocopied is paid directly to Copyright Clearance Center, 222 Rosewood Drive, Danvers, MA 01923 USA. The fee code for users of the Transactional Reporting Service is ISBN 0-8493-1968-4/05/$0.00+$1.50. The fee is subject to change without notice. For organizations that have been granted a photocopy license by the CCC, a separate system of payment has been arranged.

The consent of CRC Press LLC does not extend to copying for general distribution, for promotion, for creating new works, or for resale. Specific permission must be obtained in writing from CRC Press LLC for such copying.

Direct all inquiries to CRC Press LLC, 2000 N.W. Corporate Blvd., Boca Raton, Florida 33431.

Trademark Notice: Product or corporate names may be trademarks or registered trademarks, and are used only for identification and explanation, without intent to infringe.

Visit the CRC Press Web site at www.crcpress.com

© 2005 by CRC Press LLC

No claim to original U.S. Government works
International Standard Book Number 0-8493-1968-4
Library of Congress Card Number 2004043852
Printed in the United States of America 1 2 3 4 5 6 7 8 9 0
Printed on acid-free paper

Preface

As the field of communications networks continues to evolve, a very interesting and challenging area — wireless sensor networks — is rapidly coming of age. A wireless sensor network consists of a large number of sensor nodes that may be randomly and densely deployed. Sensor nodes are small electronic components capable of sensing many types of information from the environment, including temperature; light; humidity; radiation; the presence or nature of biological organisms; geological features; seismic vibrations; specific types of computer data; and more. Recent advancements have made it possible to make these components small, powerful, and energy efficient and they can now be manufactured cost-effectively in quantity for specialized telecommunications applications. Very small in size, the sensor nodes are capable of gathering, processing, and communicating information to other nodes and to the outside world. Based on the information handling capabilities and compact size of the sensor nodes, sensor networks are often referred to as "smart dust."

Sensor networks have numerous applications, including health; agriculture; geology; retail; military; home; and emergency management. Sensor network research and development derive many concepts and protocols from distributed computer networks such as the Internet; however, several technical challenges in sensor networks need to be addressed due to the specialized nature of the sensors and the fact that many sensor network applications may involve remote mobile sensors with limited power sources that must dynamically adapt to their environment. This handbook proposes to capture the current state of sensor networks and to serve as a source of comprehensive reference material on them.

The handbook has a total of 40 chapters written by experts from around the world and is divided into the following nine sections:

1. Introduction
2. Applications
3. Architecture
4. Protocols
5. Tracking technologies
6. Data gathering and processing
7. Energy management
8. Security, reliability, and fault tolerance
9. Performance and design aspects

The targeted audience for this handbook includes professionals who are designers and/or planners for emerging telecommunication networks; researchers (faculty members and graduate students); and those who would like to learn about this field.

This handbook provides technical information about various aspects of sensor networks, networks comprising multiple compact, intercommunicating electronic sensors. The areas covered range from basic concepts to research-grade material, including future directions. This handbook should serve as a complete reference material for sensor networks.

The *Handbook of Sensor Networks* has the following specific salient features:

- It serves as a single comprehensive source of information and as reference material on wireless sensor networks.
- It deals with an important and timely topic of emerging communication technology of tomorrow.
- It presents accurate, up-to-date information on a broad range of topics related to wireless sensor networks.
- It presents material authored by experts in the field.
- It presents the information in an organized and well-structured manner.
- Although it is not precisely a textbook, it can certainly be used as one for graduate courses and research-oriented courses that deal with wireless sensor networks. Any comments from the readers will be highly appreciated.

Many people have contributed to this handbook in their unique ways. The first and the foremost group that deserves immense gratitude is the highly talented and skilled researchers who have contributed 40 chapters to this handbook. All of them have been extremely cooperative and professional. It has also been a pleasure to work with Nora Konopka and Helena Redshaw of CRC Press; we are extremely grateful for their support and professionalism. We also thank Sophie Kirkwood and Gail Renard in the CRC production department. Our families have extended their unconditional love and strong support throughout this project and they all deserve very special thanks.

Mohammad Ilyas and Imad Mahgoub
Boca Raton, Florida

Editors

Mohammad Ilyas, Ph.D., received his B.Sc. degree in electrical engineering from the University of Engineering and Technology, Lahore, Pakistan, in 1976. From March 1977 to September 1978, he worked for the Water and Power Development Authority in Pakistan. In 1978, he was awarded a scholarship for his graduate studies and he completed his M.S. degree in electrical and electronic engineering in June 1980 at Shiraz University, Shiraz, Iran. In September 1980, he joined the doctoral program at Queen's University in Kingston, Ontario, Canada; he completed his Ph.D. degree in 1983. Dr. Ilyas' doctoral research was about switching and flow control techniques in computer communication networks. Since September 1983, he has been with the College of Engineering at Florida Atlantic University, Boca Raton, Florida, where he is currently associate dean for graduate studies and research. From 1994 to 2000, he was chair of the department. During the 1993–1994 academic year, he was on his sabbatical leave with the Department of Computer Engineering, King Saud University, Riyadh, Saudi Arabia.

Dr. Ilyas has conducted successful research in various areas, including traffic management and congestion control in broadband/high-speed communication networks; traffic characterization; wireless communication networks; performance modeling; and simulation. He has published one book, three handbooks, and over 140 research articles. He has supervised 10 Ph.D. dissertations and more than 35 M.S. theses to completion. Dr. Ilyas has been a consultant to several national and international organizations; a senior member of IEEE, he is an active participant in several IEEE technical committees and activities.

Imad Mahgoub, Ph.D., received his B.Sc. degree in electrical engineering from the University of Khartoum, Khartoum, Sudan, in 1978. From 1978 to 1981, he worked for the Sudan Shipping Line Company, Port Sudan, Sudan, as an electrical and electronics engineer. He received his M.S. in applied mathematics in 1983 and his M.S. in electrical and computer engineering in 1986, both from North Carolina State University. In 1989, he received his Ph.D. in computer engineering from The Pennsylvania State University.

Since August 1989, Dr. Mahgoub has been with the College of Engineering at Florida Atlantic University, Boca Raton, Florida, where he is currently professor of computer science and engineering. He is the director of the Computer Science and Engineering Department Mobile Computing Laboratory at Florida Atlantic University.

Dr. Mahgoub has conducted successful research in various areas, including mobile computing; interconnection networks; performance evaluation of computer systems; and advanced computer architecture. He has published over 70 research articles and supervised three Ph.D. dissertations and 18 M.S. theses to completion. He has served as a consultant to industry. Dr. Mahgoub served as a member of the executive committee/program committee of the 1998, 1999, and 2000 IEEE International Performance, Computing and Communications Conferences. He has served on the program committees of several international conferences and symposia. He is currently the vice chair of the 2004 International Symposium on Performance Evaluation of Computer and Telecommunication Systems. Dr. Mahgoub is a senior member of IEEE and a member of ACM.

Contributors

T. Abdelzaher
University of Virginia
Charlottesville, Virginia

Özgür B. Akan
Georgia Institute of
 Technology
Atlanta, Georgia

Jamal N. Al-Karaki
Iowa State University
Ames, Iowa

Petr Benes
Brno University of
 Technology
Brno, Czech Republic

Jan Beutel
Swiss Federal Institute of
 Technology
Zurich, Switzerland

B. Blum
University of Virginia
Charlottesville, Virginia

Cristian Borcea
Rutgers University
Piscataway, New Jersey

Jacir L. Bordim
ATR — Adaptive
 Communications Research
 Laboratories
Kyoto, Japan

Athanassios Boulis
University of California at
 Los Angeles
Los Angeles, California

Richard R. Brooks
The Pennsylvania State
 University
State College, Pennsylvania

Mihaela Cardei
Florida Atlantic University
Boca Raton, Florida

Erdal Cayirci
Istanbul Technical University
Istanbul, Turkey

Krishnendu Chakrabarty
Duke University
Durham, North Carolina

Anantha Chandrakasan
Engim, Inc.
Acton, Massachusetts

Duminda Dewasurendra
Virginia Polytechnic Institute
 and State University
Blacksburg, Virginia

Jessica Feng
University of California at Los
 Angeles
Los Angeles, California

Kurt Fristrup
Cornell Laboratory of
 Ornithology
Ithaca, New York

Vincente González–Millán
University of Valencia
Valencia, Spain

Joel I. Goodman
MIT Lincoln Laboratory
Lexington, Massachusetts

Zygmunt J. Haas
Cornell University
Ithaca, New York

Martin Haenggi
University of Notre Dame
Notre Dame, Indiana

Hossam Hassanein
Queen's University
Kingston, Ontario, Canada

T. He
University of Virginia
Charlottesville, Virginia

Chi-Fu Huang
National Chiao-Tung University
Hsin-Chu, Taiwan

Liviu Iftode
Rutgers University
Piscataway, New Jersey

S. Sitharama Iyengar
Louisiana State University
Baton Rouge, Louisiana

Chaiporn Jaikaeo
University of Delaware
Newark, Delaware

Ram Kalidindi
Louisiana State University
Baton Rouge, Louisiana

Ahmed E. Kamal
Iowa State University
Ames, Iowa

Porlin Kang
Rutgers University
Piscataway, New Jersey

Rajgopal Kannan
Louisiana State University
Baton Rouge, Louisiana

Zdravko Karakehayov
Technical University of Sofia
Sofia, Bulgaria

Farinaz Koushanfar
University of California at
  Berkeley
Berkeley, California

Sheng-Po Kuo
National Chiao-Tung University
Hsin-Chu, Taiwan

Baohua Li
Sichuan University
Chengdu, Sichuan, China

Xiang-Yang Li
Illinois Institute of Technology
Chicago, Illinois

Alvin S. Lim
Auburn University
Auburn, Alabama

Malin Lindquist
Örebro University
Örebro, Sweden

Antonio A.F. Loureiro
Federal University of Minas
  Gerais
Belo Horizonte, Brazil

Amy Loutfi
Örebro University
Örebro, Sweden

Chenyang Lu
University of Washington at
  St. Louis
St. Louis, Missouri

David R. Martinez
MIT Lincoln Laboratory
Lexington, Massachusetts

Amitabh Mishra
Virginia Polytechnic Institute
  and State University
Blacksburg, Virginia

Koji Nakano
Hiroshima University
Higashi-Hiroshima, Japan

Eric Nettleton
The University of Sydney
New South Wales, Australia

José Marcos Nogueira
Federal University of Minas
  Gerais
Belo Horizonte, Brazil

Lee Ling (Sharon) Ong
The University of Sydney
New South Wales, Australia

Symeon Papavassiliou
New Jersey Institute of
  Technology
Newark, New Jersey

Dragan Petrovic
University of California at
  Berkeley
Berkeley, California

Miodrag Potkonjak
University of California at
  Los Angeles
Los Angeles, California

Alejandro Purgue
Cornell Laboratory of
  Ornithology
Ithaca, New York

Gang Qu
University of Maryland
College Park, Maryland

Jan M. Rabaey
University of California at
  Berkeley
Berkeley, California

Nageswara S.V. Rao
Oak Ridge National
  Laboratory
Oak Ridge, Tennessee

Lydia Ray
Louisiana State University
Baton Rouge, Louisiana

Albert I. Reuther
MIT Lincoln Laboratory
Lexington, Massachusetts

Matthew Ridley
The University of Sydney
New South Wales, Australia

Linnyer Beatrys Ruiz
Pontifical Catholic University
  of Paraná
Curitiba, Brazil
and Federal University of
  Minas Gerais
Belo Horizonte, Brazil

Ayad Salhieh
Wayne State University
Detroit, Michigan

Enrique Sanchis-Peris
University of Valencia
Valencia, Spain

Alberto Sangiovanni-Vincentelli
University of California at
  Berkeley
Berkeley, California

Loren Schwiebert
Wayne State University
Detroit, Michigan

Rahul C. Shah
University of California at
  Berkeley
Berkeley, California

Chien-Chung Shen
University of Delaware
Newark, Delaware

Amit Sinha
Engim, Inc.
Acton, Massachusetts

Sasha Slijepcevic
University of California at
  Los Angeles
Los Angeles, California

Tara Small
Cornell University
Ithaca, New York

S. Son
University of Virginia
Charlottesville, Virginia

Chavalit Srisathapornphat
University of Delaware
Newark, Delaware

John Stankovic
University of Virginia
Charlottesville, Virginia

Weilian Su
Georgia Institute of
  Technology
Atlanta, Georgia

Saleh Sukkarieh
The University of Sydney
New South Wales, Australia

Miroslav Sveda
Brno University of
  Technology
Brno, Czech Republic

Vishnu Swaminathan
Duke University
Durham, North Carolina

Yu-Chee Tseng
National Chiao-Tung
  University
Hsin-Chu, Taiwan

Radimir Vrba
Brno University of
  Technology
Czech Republic

Quanhong Wang
Queen's University
Kingston, Ontario, Canada

Yu Wang
Illinois Institute of
  Technology
Chicago, Illinois

Brett Warneke
Dust Networks
Berkeley, California

Peter Wide
Örebro University
Örebro, Sweden

Jennifer L. Wong
University of California at Los
  Angeles
Los Angeles, California

Anthony D. Wood
University of Virginia
Charlottesville, Virginia

Jie Wu
Florida Atlantic University
Boca Raton, Florida

Qishi Wu
Oak Ridge National
  Laboratory
Oak Ridge, Tennessee

Kenan Xu
Queen's University
Kingston, Ontario,
  Canada

Mark Yarvis
Intel Corporation
Hillsboro, Oregon

Wei Ye
University of Southern
 California
Los Angeles, California

Lin Yuan
University of Maryland
College Park, Maryland

Frantisek Zezulka
Brno University of Technology
Brno, Czech Republic

Jin Zhu
New Jersey Institute of
 Technology
Newark, New Jersey

Mengxia Zhu
Louisiana State University
Baton Rouge, Louisiana

Yunmin Zhu
Sichuan University
Chengdu, Sichuan, China

Yi Zou
Duke University
Durham, North Carolina

Contents

SECTION I Introduction

1 Opportunities and Challenges in Wireless Sensor Networks *Martin Haenggi*
- 1.1 Introduction 1-1
- 1.2 Opportunities 1-2
- 1.3 Technical Challenges 1-4
- 1.4 Concluding Remarks 1-11

2 Next-Generation Technologies to Enable Sensor Networks *Joel I. Goodman, Albert I. Reuther, David R. Martinez*
- 2.1 Introduction 2-1
- 2.2 Goals for Real-Time Distributed Network Computing for Sensor Data Fusion 2-5
- 2.3 The Convergence of Networking and Real-Time Computing 2-6
- 2.4 Middleware 2-11
- 2.5 Network Resource Management 2-11
- 2.6 Experimental Results 2-16

3 Sensor Network Management *Linnyer Beatrys Ruiz, José Marcos Nogueira, Antonio A. F. Loureiro*
- 3.1 Introduction 3-1
- 3.2 Management Challenges 3-2
- 3.3 Management Dimensions 3-3
- 3.4 MANNA as an Integrating Architecture 3-15
- 3.5 Putting It All Together 3-25
- 3.6 Conclusion 3-25

4 Models for Programmability in Sensor Networks *Athanassios Boulis*
- 4.1 Introduction 4-1
- 4.2 Differences between Sensor Networks and Traditional Data Networks 4-2
- 4.3 Aspects of Efficient Sensor Network Applications 4-2
- 4.4 Need for Sensor Network Programmability 4-3
- 4.5 Major Models for System-Level Programmability 4-4

	4.6	Frameworks for System-Level Programmability	4-6
	4.7	Conclusions	4-12

5 Miniaturizing Sensor Networks with MEMS *Brett Warneke*
- 5.1 Introduction 5-1
- 5.2 MEMS Basics 5-2
- 5.3 Sensors 5-4
- 5.4 Communication 5-5
- 5.5 Micropower Sources 5-10
- 5.6 Packaging 5-12
- 5.7 Systems 5-13
- 5.8 Conclusion 5-15

6 A Taxonomy of Routing Techniques in Wireless Sensor Networks
Jamal N. Al-Karaki, Ahmed E. Kamal
- 6.1 Introduction 6-1
- 6.2 Routing Protocols in WSNs 6-6
- 6.3 Routing in WSNs: Future Directions 6-21
- 6.4 Conclusions 6-22

7 Artificial Perceptual Systems *Amy Loutfi, Malin Lindquist, Peter Wide*
- 7.1 Introduction 7-1
- 7.2 Background 7-2
- 7.3 Modeling of Perceptual Systems 7-3
- 7.4 Perceptual Systems in Practice 7-8
- 7.5 Research Issues and Summary 7-12

SECTION II Applications

8 Sensor Network Architecture and Applications *Chien-Chung Shen, Chaiporn Jaikaeo, Chavalit Srisathapornphat*
- 8.1 Introduction 8-1
- 8.2 Sensor Network Applications 8-1
- 8.3 Functional Architecture for Sensor Networks 8-3
- 8.4 Sample Implementation Architectures 8-4
- 8.5 Summary 8-12

9 A Practical Perspective on Wireless Sensor Networks *Quanhong Wang, Hossam Hassanein, Kenan Xu*
- 9.1 Introduction 9-1

	9.2	WSN Applications	9-2
	9.3	Classification of WSNs	9-6
	9.4	Characteristics, Technical Challenges, and Design Directions	9-7
	9.5	Technical Approaches	9-11
	9.6	Conclusions and Considerations for Future Research	9-22

10 Introduction to Industrial Sensor Networking *Miroslav Sveda, Petr Benes, Radimir Vrba, Frantisek Zezulka*
- 10.1 Introduction ... 10-1
- 10.2 Industrial Sensor Fitting Communication Protocols ... 10-2
- 10.3 IEEE 1451 Family of Smart Transducer Interface Standards ... 10-11
- 10.4 Internet-Based Sensor Networking ... 10-13
- 10.5 Industrial Network Interconnections ... 10-15
- 10.6 Wireless Sensor Networks in Industry ... 10-21
- 10.7 Conclusions ... 10-24

11 A Sensor Network for Biological Data Acquisition *Tara Small, Zygmunt J. Haas, Alejandro Purgue, Kurt Fristrup*
- 11.1 Introduction ... 11-1
- 11.2 Tagging Whales ... 11-2
- 11.3 The Tag Sensors ... 11-3
- 11.4 The SWIM Networks ... 11-6
- 11.5 The Information Propagation Model ... 11-7
- 11.6 Simulating the Delay ... 11-9
- 11.7 Calculating Storage Requirements ... 11-12
- 11.8 Conclusions ... 11-17

SECTION III   Architecture

12 Sensor Network Architecture *Jessica Feng, Farinaz Koushanfar, Miodrag Potkonjak*
- 12.1 Overview ... 12-1
- 12.2 Motivation and Objectives ... 12-1
- 12.3 SNs — Global View and Requirements ... 12-3
- 12.4 Individual Components of SN Nodes ... 12-4
- 12.5 Sensor Network Node ... 12-8
- 12.6 Wireless SNs as Embedded Systems ... 12-13
- 12.7 Summary ... 12-16

13 Tiered Architectures in Sensor Networks *Mark Yarvis, Wei Ye*
- 13.1 Introduction ... 13-1

	13.2	Why Build Tiered Architectures?	13-3
	13.3	Spectrum of Sensor Network Hardware	13-5
	13.4	Task Decomposition and Allocation	13-8
	13.5	Forming Tiered Architectures	13-10
	13.6	Routing and Addressing in a Tiered Architecture	13-15
	13.7	Drawbacks of Tiered Architectures	13-18
	13.8	Conclusions	13-19

14 Power-Efficient Topologies for Wireless Sensor Networks *Ayad Salhieh, Loren Schwiebert*

14.1	Motivation	14-1
14.2	Background	14-2
14.3	Issues for Topology Design	14-3
14.4	Assumptions	14-8
14.5	Analysis of Power Usage	14-10
14.6	Directional Source-Aware Routing Protocol (DSAP)	14-13
14.7	DSAP Analysis	14-15
14.8	Summary	14-19

15 Architecture and Modeling of Dynamic Wireless Sensor Networks *Symeon Papavassiliou, Jin Zhu*

15.1	Introduction	15-1
15.2	Characteristics of Wireless Sensor Networks	15-2
15.3	Architecture of Sensor Networks	15-3
15.4	Modeling of Dynamic Sensor Networks	15-7
15.5	Concluding Remarks	15-13

SECTION IV Protocols

16 Overview of Communication Protocols for Sensor Networks *Weilian Su, Erdal Cayirci, Özgür B. Akan*

16.1	Introduction	16-1
16.2	Applications/Application Layer Protocols	16-2
16.3	Localization Protocols	16-4
16.4	Time Synchronization Protocols	16-5
16.5	Transport Layer Protocols	16-7
16.6	Network Layer Protocols	16-9
16.7	Data Link Layer Protocols	16-11
16.8	Conclusion	16-14

17 Communication Architecture and Programming Abstractions for Real-Time Embedded Sensor Networks *T. Abdelzaher, J. Stankovic, S. Son, B. Blum, T. He, A. Wood, Chenyang Lu*
  17.1 Introduction 17-1
  17.2 A Protocol Suite for Sensor Networks 17-2
  17.3 A Sensor-Network Programming Model 17-5
  17.4 Related Work 17-6
  17.5 Conclusions 17-8

18 A Comparative Study of Energy-Efficient (E^2) Protocols for Wireless Sensor Networks *Quanhong Wang, Hossam Hassanein*
  18.1 Introduction 18-1
  18.2 Motivations and Directions 18-2
  18.3 Cross-Layer Communication Protocol Stack for WSNs 18-5
  18.4 Energy-Efficient MAC Protocols 18-6
  18.5 Energy-Efficient Network Layer Protocols 18-10
  18.6 Concluding Remarks 18-17

SECTION V Tracking Technologies

19 Coverage in Wireless Sensor Networks *Mihaela Cardei, Jie Wu*
  19.1 Introduction 19-1
  19.2 Area Coverage 19-4
  19.3 Point Coverage 19-8
  19.4 Barrier Coverage 19-9
  19.5 Conclusion 19-10

20 Location Management in Wireless Sensor Networks *Jan Beutel*
  20.1 Introduction 20-1
  20.2 Location in Wireless Communication Systems 20-2
  20.3 Location in Wireless Sensor Networks 20-10
  20.4 Summary 20-21

21 Positioning and Location Tracking in Wireless Sensor Networks *Yu-Chee Tseng, Chi-Fu Huang, Sheng-Po Kuo*
  21.1 Introduction 21-1
  21.2 Fundamentals 21-2
  21.3 Positioning and Location Tracking Algorithms 21-4
  21.4 Experimental Location Systems 21-10
  21.5 Conclusions 21-12

22 Tracking Techniques in Air Vehicle-Based Decentralized Sensor Networks
Matthew Ridley, Lee Ling (Sharon) Ong, Eric Nettleton, Salah Sukkarieh

- 22.1 Introduction ... 22-1
- 22.2 The ANSER System and Experiment ... 22-2
- 22.3 The Decentralized Tracking Problem ... 22-2
- 22.4 Algorithmic System Design ... 22-3
- 22.5 Sensor Design ... 22-9
- 22.6 Hardware and Software Infrastructure ... 22-13
- 22.7 Conclusion ... 22-15

SECTION VI Data Gathering and Processing

23 Fundamental Protocols to Gather Information in Wireless Sensor Networks
Jacir L. Bordim, Koji Nakano

- 23.1 Introduction ... 23-1
- 23.2 Model Definition ... 23-3
- 23.3 Gathering Information in Wireless Sensor Networks ... 23-5
- 23.4 Identifying Faulty Nodes in Wireless Sensor Networks ... 23-11
- 23.5 Conclusions ... 23-15

24 Comparison of Data Processing Techniques in Sensor Networks
Vicente González-Millán, Enrique Sanchis-Peris

- 24.1 Sensor Networks: Organization and Processing ... 24-1
- 24.2 Architectures for Sensor Integration ... 24-3
- 24.3 Example of Architecture Evaluation in High-Energy Physics ... 24-18

25 Computational and Networking Problems in Distributed Sensor Networks
Qishi Wu, Nageswara S.V. Rao, Richard R. Brooks, S. Sitharama Iyengar, Mengxia Zhu

- 25.1 Introduction ... 25-1
- 25.2 Foundational Aspects of DSNs ... 25-2
- 25.3 Sensor Deployment ... 25-5
- 25.4 Routing Paradigms for DSNs ... 25-8
- 25.5 Conclusions and Future Work ... 25-16

26 Cooperative Computing in Sensor Networks *Liviu Iftode, Cristian Borcea, Porlin Kang*

- 26.1 Introduction ... 26-1
- 26.2 The Cooperative Computing Model ... 26-3
- 26.3 Node Architecture ... 26-4
- 26.4 Smart Messages ... 26-5

	26.5	Programming Interface	26-7
	26.6	Prototype Implementation and Evaluation	26-8
	26.7	Applications	26-12
	26.8	Simulation Results	26-14
	26.9	Related Work	26-15
	26.10	Conclusions	26-18

SECTION VII Energy Management

27 Dynamic Power Management in Sensor Networks *Amit Sinha, Anantha Chandrakasan*

	27.1	Introduction	27-1
	27.2	Idle Power Management	27-2
	27.3	Active Power Management	27-5
	27.4	System Implementation	27-6
	27.5	Results	27-12

28 Design Challenges in Energy-Efficient Medium Access Control for Wireless Sensor Networks *Duminda Dewasurendra, Amitabh Mishra*

	28.1	Introduction	28-1
	28.2	Unique Characteristics of Wireless Sensor Networks	28-2
	28.3	MAC Protocols for Wireless ad hoc Networks	28-4
	28.4	Design Challenges for Wireless Sensor Networks	28-10
	28.5	Medium Access Protocols for Wireless Sensor Networks	28-13
	28.6	Open Issues	28-22
	28.7	Conclusions	28-24

29 Techniques to Reduce Communication and Computation Energy in Wireless Sensor Networks *Vishnu Swaminathan, Yi Zou, Krishnendu Chakrabarty*

	29.1	Introduction	29-1
	29.2	Overview of Node-Level Energy Management	29-2
	29.3	Overview of Energy-Efficient Communication	29-4
	29.4	Node-Level Processor-Oriented Energy Management	29-4
	29.5	Node-Level I/O-Device-Oriented Energy Management	29-11
	29.6	Energy-Aware Communication	29-19
	29.7	Conclusions	29-32

30 Energy-Aware Routing and Data Funneling in Sensor Networks *Rahul C. Shah, Dragan Petrovic, Jan M. Rabaey*

	30.1	Introduction	30-1
	30.2	Protocol Stack Design	30-2

30.3	Routing Protocol Characteristics and Related Work	30-4
30.4	Routing for Maximizing Lifetime: A Linear Programming Formulation	30-5
30.5	Energy-Aware Routing	30-5
30.6	Simulations	30-7
30.7	Data Funneling	30-10
30.8	Conclusion	30-13

SECTION VIII Security, Reliability, and Fault Tolerance

31 Security and Privacy Protection in Wireless Sensor Networks *Sasha Slijepcevic, Jennifer L. Wong, Miodrag Potkonjak*

31.1	Introduction	31-1
31.2	Unique Security Challenges in Sensor Networks and Enabling Mechanisms	31-2
31.3	Security Architectures	31-4
31.4	Privacy Protection	31-11
31.5	Conclusion	31-15

32 A Taxonomy for Denial-of-Service Attacks in Wireless Sensor Networks *Anthony D. Wood, John A. Stankovic*

32.1	Introduction	32-1
32.2	Attack Taxonomy	32-5
32.3	Vulnerabilities and Defenses	32-10
32.4	Related Work	32-16
32.5	Conclusion	32-17

33 Reliability Support in Sensor Networks *Alvin S. Lim*

33.1	Introduction	33-1
33.2	Reliability Problems in Sensor Networks	33-2
33.3	Existing Work on Reliability Support	33-2
33.4	Supporting Reliability with Distributed Services	33-3
33.5	Architecture of a Distributed Sensor System	33-3
33.6	Directed Diffusion Network	33-4
33.7	Distributed Services	33-5
33.8	Mechanisms and Tools	33-8
33.9	Dynamic Adaptation of Distributed Sensor Applications	33-10
33.10	Conclusions	33-11

34 Reliable Energy-Constrained Routing in Sensor Networks *Rajgopal Kannan, Lydia Ray, S. Sitharama Iyengar, Ram Kalidindi*

| 34.1 | Introduction | 34-1 |

	34.2	Game-Theoretic Models of Reliable and Length Energy-Constrained Routing	34-2
	34.3	Distributed Length Energy-Constrained (LEC) Routing Protocol	34-5
	34.4	Performance Evaluation	34-7

35 Fault-Tolerant Interval Estimation in Sensor Networks *Yunmin Zhu, Baohua Li*

35.1	Introduction	35-1
35.2	Sensor Network Formulation	35-2
35.3	Fault-Tolerant Interval Estimation without Knowledge of Confidence Degrees	35-4
35.4	Combination Rule and Optimal Fusion for Sensor Output	35-5
35.5	Fault-Tolerant Interval Estimation with Knowledge of Confidence Degrees	35-9
35.6	Extension to Sensor Estimate with Multiple Output Intervals	35-11
35.7	Robust Fault-Tolerant Interval Estimation	35-11
35.8	Conclusion	35-16

36 Fault Tolerance in Wireless Sensor Networks *Farinaz Koushanfar, Miodrag Potkonjak, Alberto Sangiovanni-Vincentelli*

36.1	Introduction	36-1
36.2	Preliminaries	36-3
36.3	Example of Fault Tolerance in a Sensor Network System	36-3
36.4	Classical Fault Tolerance	36-4
36.5	Fault Tolerance at Different Sensor Network Levels	36-5
36.6	Case Studies	36-8
36.7	Future Research Directions	36-12
36.8	Conclusion	36-13

SECTION IX   Performance and Design Aspects

37 Low-Power Design for Smart Dust Networks *Zdravko Karakehayov*

37.1	Introduction	37-1
37.2	Location	37-1
37.3	Sensing	37-2
37.4	Computation	37-2
37.5	Hardware–Software Interaction	37-5
37.6	Communication	37-7
37.7	Orientation	37-10
37.8	Conclusion	37-10

38 Energy-Efficient Design of Distributed Sensor Networks *Lin Yuan, Gang Qu*

38.1	Introduction	38-1
38.2	Background	38-4

38.3	Preliminaries	**38**-6
38.4	DVS with Message Header	**38**-9
38.5	Simulation	**38**-11
38.6	Conclusions	**38**-17

39 Wireless Sensor Networks and Computational Geometry *Xiang-Yang Li, Yu Wang*

39.1	Introduction	**39**-1
39.2	Preliminaries	**39**-4
39.3	Topology Control	**39**-10
39.4	Localized Routing	**39**-33
39.5	Broadcasting	**39**-38
39.6	Summary and Open Questions	**39**-42

40 Localized Algorithms for Sensor Networks *Jessica Feng, Farinaz Koushanfar, Miodrag Potkonjak*

40.1	Introduction	**40**-1
40.2	Models and Abstractions	**40**-2
40.3	Centralized Algorithm	**40**-4
40.4	Case Studies	**40**-8
40.5	Analysis	**40**-12
40.6	Protocols and Distributed Localized Algorithms	**40**-13
40.7	Pending Challenges	**40**-15

Index **I**-1

Introduction

I

1
Opportunities and Challenges in Wireless Sensor Networks

Martin Haenggi
University of Notre Dame

1.1	Introduction	1-1
1.2	Opportunities	1-2
	Growing Research and Commercial Interest • Applications	
1.3	Technical Challenges	1-4
	Performance Metrics • Power Supply • Design of Energy-Efficient Protocols • Capacity/Throughput • Routing • Channel Access and Scheduling • Modeling • Connectivity • Quality of Service • Security • Implementation • Other Issues	
1.4	Concluding Remarks	1-11

1.1 Introduction

Due to advances in wireless communications and electronics over the last few years, the development of networks of low-cost, low-power, multifunctional sensors has received increasing attention. These sensors are small in size and able to sense, process data, and communicate with each other, typically over an RF (radio frequency) channel. A sensor network is designed to detect events or phenomena, collect and process data, and transmit sensed information to interested users. Basic features of sensor networks are:

- Self-organizing capabilities
- Short-range broadcast communication and multihop routing
- Dense deployment and cooperative effort of sensor nodes
- Frequently changing topology due to fading and node failures
- Limitations in energy, transmit power, memory, and computing power

These characteristics, particularly the last three, make sensor networks different from other wireless ad hoc or mesh networks.

Clearly, the idea of mesh networking is not new; it has been suggested for some time for wireless Internet access or voice communication. Similarly, small computers and sensors are not innovative per se. However, combining small sensors, low-power computers, and radios makes for a new technological platform that has numerous important uses and applications, as will be discussed in the next section.

1.2 Opportunities

1.2.1 Growing Research and Commercial Interest

Research and commercial interest in the area of wireless sensor networks are currently growing exponentially, which is manifested in many ways:

- The number of Web pages (Google: 26,000 hits for sensor networks; 8000 for wireless sensor networks in August 2003)
- The increasing number of
    - Dedicated annual workshops, such as IPSN (information processing in sensor networks); SenSys; EWSN (European workshop on wireless sensor networks); SNPA (sensor network protocols and applications); and WSNA (wireless sensor networks and applications)
    - Conference sessions on sensor networks in the communications and mobile computing communities (ISIT, ICC, Globecom, INFOCOM, VTC, MobiCom, MobiHoc)
    - Research projects funded by NSF (apart from ongoing programs, a new specific effort now focuses on sensors and sensor networks) and DARPA through its SensIT (sensor information technology), NEST (networked embedded software technology), MSET (multisensor exploitation), UGS (unattended ground sensors), NETEX (networking in extreme environments), ISP (integrated sensing and processing), and communicator programs

Special issues and sections in renowned journals are common, e.g., in the *IEEE Proceedings* [1] and signal processing, communications, and networking magazines. Commercial interest is reflected in investments by established companies as well as start-ups that offer general and specific hardware and software solutions.

Compared to the use of a few expensive (but highly accurate) sensors, the strategy of deploying a large number of inexpensive sensors has significant advantages, at smaller or comparable total system cost: much higher spatial resolution; higher robustness against failures through distributed operation; uniform coverage; small obtrusiveness; ease of deployment; reduced energy consumption; and, consequently, increased system lifetime. The main point is to position sensors close to the source of a potential problem phenomenon, where the acquired data are likely to have the greatest benefit or impact.

Pure sensing in a fine-grained manner may revolutionize the way in which complex physical systems are understood. The addition of actuators, however, opens a completely new dimension by permitting management and manipulation of the environment at a scale that offers enormous opportunities for almost every scientific discipline. Indeed, Business 2.0 (http://www.business2.com/) lists sensor robots as one of "six technologies that will change the world," and *Technology Review* at MIT and Globalfuture identify WSNs as one of the "10 emerging technologies that will change the world" (http://www.globalfuture.com/mit-trends2003.htm). The combination of sensor network technology with MEMS and nanotechnology will greatly reduce the size of the nodes and enhance the capabilities of the network.

The remainder of this chapter lists and briefly describes a number of applications for wireless sensor networks, grouped into different categories. However, because the number of areas of application is growing rapidly, every attempt at compiling an exhaustive list is bound to fail.

1.2.2 Applications

1.2.2.1 General Engineering

- *Automotive telematics.* Cars, which comprise a network of dozens of sensors and actuators, are networked into a system of systems to improve the safety and efficiency of traffic.
- *Fingertip accelerometer virtual keyboards.* These devices may replace the conventional input devices for PCs and musical instruments.
- *Sensing and maintenance in industrial plants.* Complex industrial robots are equipped with up to 200 sensors that are usually connected by cables to a main computer. Because cables are expensive

and subject to wear and tear caused by the robot's movement, companies are replacing them by wireless connections. By mounting small coils on the sensor nodes, the principle of induction is exploited to solve the power supply problem.
- *Aircraft drag reduction.* Engineers can achieve this by combining flow sensors and blowing/sucking actuators mounted on the wings of an airplane.
- *Smart office spaces.* Areas are equipped with light, temperature, and movement sensors, microphones for voice activation, and pressure sensors in chairs. Air flow and temperature can be regulated locally for one room rather than centrally.
- *Tracking of goods in retail stores.* Tagging facilitates the store and warehouse management.
- *Tracking of containers and boxes.* Shipping companies are assisted in keeping track of their goods, at least until they move out of range of other goods.
- *Social studies.* Equipping human beings with sensor nodes permits interesting studies of human interaction and social behavior.
- Commercial and residential security.

1.2.2.2 Agriculture and Environmental Monitoring

- *Precision agriculture.* Crop and livestock management and precise control of fertilizer concentrations are possible.
- *Planetary exploration.* Exploration and surveillance in inhospitable environments such as remote geographic regions or toxic locations can take place.
- *Geophysical monitoring.* Seismic activity can be detected at a much finer scale using a network of sensors equipped with accelerometers.
- *Monitoring of freshwater quality.* The field of hydrochemistry has a compelling need for sensor networks because of the complex spatiotemporal variability in hydrologic, chemical, and ecological parameters and the difficulty of labor-intensive sampling, particularly in remote locations or under adverse conditions. In addition, buoys along the coast could alert surfers, swimmers, and fishermen to dangerous levels of bacteria.
- *Zebranet.* The Zebranet project at Princeton aims at tracking the movement of zebras in Africa.
- *Habitat monitoring.* Researchers at UC Berkeley and the College of the Atlantic in Bar Harbor deployed sensors on Great Duck Island in Maine to measure humidity, pressure, temperature, infrared radiation, total solar radiation, and photosynthetically active radiation (see http://www.greatduckisland.net/).
- *Disaster detection.* Forest fire and floods can be detected early and causes can be localized precisely by densely deployed sensor networks.
- *Contaminant transport.* The assessment of exposure levels requires high spatial and temporal sampling rates, which can be provided by WSNs.

1.2.2.3 Civil Engineering

- *Monitoring of structures.* Sensors will be placed in bridges to detect and warn of structural weakness and in water reservoirs to spot hazardous materials. The reaction of tall buildings to wind and earthquakes can be studied and material fatigue can be monitored closely.
- *Urban planning.* Urban planners will track groundwater patterns and how much carbon dioxide cities are expelling, enabling them to make better land-use decisions.
- *Disaster recovery.* Buildings razed by an earthquake may be infiltrated with sensor robots to locate signs of life.

1.2.2.4 Military Applications

- *Asset monitoring and management.* Commanders can monitor the status and locations of troops, weapons, and supplies to improve military command, control, communications, and computing (C4).

- *Surveillance and battle-space monitoring.* Vibration and magnetic sensors can report vehicle and personnel movement, permitting close surveillance of opposing forces.
- *Urban warfare.* Sensors are deployed in buildings that have been cleared to prevent reoccupation; movements of friend and foe are displayed in PDA-like devices carried by soldiers. Snipers can be localized by the collaborative effort of multiple acoustic sensors.
- *Protection.* Sensitive objects such as atomic plants, bridges, retaining walls, oil and gas pipelines, communication towers, ammunition depots, and military headquarters can be protected by intelligent sensor fields able to discriminate between different classes of intruders. Biological and chemical attacks can be detected early or even prevented by a sensor network acting as a warning system.
- *Self-healing minefields.* The self-healing minefield system is designed to achieve an increased resistance to dismounted and mounted breaching by adding a novel dimension to the minefield. Instead of a static complex obstacle, the self-healing minefield is an intelligent, dynamic obstacle that senses relative positions and responds to an enemy's breaching attempt by physical reorganization.

1.2.2.5 Health Monitoring and Surgery

- *Medical sensing.* Physiological data such as body temperature, blood pressure, and pulse are sensed and automatically transmitted to a computer or physician, where it can be used for health status monitoring and medical exploration. Wireless sensing bandages may warn of infection. Tiny sensors in the blood stream, possibly powered by a weak external electromagnetic field, can continuously analyze the blood and prevent coagulation and thrombosis.
- *Micro-surgery.* A swarm of MEMS-based robots may collaborate to perform microscopic and minimally invasive surgery.

The opportunities for wireless sensor networks are ubiquitous. However, a number of formidable challenges must be solved before these exciting applications may become reality.

1.3 Technical Challenges

Populating the world with networks of sensors requires a fundamental understanding of techniques for connecting and managing sensor nodes with a communication network in scalable and resource-efficient ways. Clearly, sensor networks belong to the class of ad hoc networks, but they have specific characteristics that are not present in general ad hoc networks.

Ad hoc and sensor networks share a number of challenges such as energy constraints and routing. On the other hand, general ad hoc networks most likely induce traffic patterns different from sensor networks, have other lifetime requirements, and are often considered to consist of *mobile* nodes [2–4]. In WSNs, most nodes are static; however, the network of basic sensor nodes may be overlaid by more powerful mobile sensors (robots) that, guided by the basic sensors, can move to interesting areas or even track intruders in the case of military applications.

Network nodes are equipped with wireless transmitters and receivers using antennas that may be omnidirectional (isotropic radiation), highly directional (point-to-point), possibly steerable, or some combination thereof. At a given point in time, depending on the nodes' positions and their transmitter and receiver coverage patterns, transmission power levels, and cochannel interference levels, a wireless connectivity exists in the form of a random, multihop graph between the nodes. This ad hoc topology may change with time as the nodes move or adjust their transmission and reception parameters.

Because the most challenging issue in sensor networks is *limited and unrechargeable* energy provision, many research efforts aim at improving the energy efficiency from different aspects. In sensor networks, energy is consumed mainly for three purposes: *data transmission, signal processing,* and *hardware operation* [5]. It is desirable to develop energy-efficient processing techniques that minimize power requirements across all levels of the protocol stack and, at the same time, minimize message passing for network control and coordination.

1.3.1 Performance Metrics

To discuss the issues in more detail, it is necessary to examine a list of metrics that determine the performance of a sensor network:

- *Energy efficiency/system lifetime.* The sensors are battery operated, rendering energy a very scarce resource that must be wisely managed in order to extend the lifetime of the network [6].
- *Latency.* Many sensor applications require delay-guaranteed service. Protocols must ensure that sensed data will be delivered to the user within a certain delay. Prominent examples in this class of networks are certainly the sensor-actuator networks.
- *Accuracy.* Obtaining accurate information is the primary objective; accuracy can be improved through joint detection and estimation. Rate distortion theory is a possible tool to assess accuracy.
- *Fault tolerance.* Robustness to sensor and link failures must be achieved through redundancy and collaborative processing and communication.
- *Scalability.* Because a sensor network may contain thousands of nodes, scalability is a critical factor that guarantees that the network performance does not significantly degrade as the network size (or node density) increases.
- *Transport capacity/throughput.* Because most sensor data must be delivered to a single base station or fusion center, a *critical area* in the sensor network exists (the gray area in Figure 1.1.), whose sensor nodes must relay the data generated by virtually all nodes in the network. Thus, the traffic load at those critical nodes is heavy, even when the average traffic rate is low. Apparently, this area has a paramount influence on system lifetime, packet end-to-end delay, and scalability.

Because of the interdependence of energy consumption, delay, and throughput, all these issues and metrics are tightly coupled. Thus, the design of a WSN necessarily consists of the resolution of numerous trade-offs, which also reflects in the network protocol stack, in which a cross-layer approach is needed instead of the traditional layer-by-layer protocol design.

1.3.2 Power Supply

The most difficult constraints in the design of WSNs are those regarding the minimum energy consumption necessary to drive the circuits and possible microelectromechanical devices (MEMS) [5, 7, 8]. The energy problem is aggravated if actuators are present that may be substantially hungrier for power than the sensors. When miniaturizing the node, the energy density of the power supply is the primary issue. Current technology yields batteries with approximately 1 J/mm^3 of energy, while capacitors can achieve as much as 1 mJ/mm^3. If a node is designed to have a relatively short lifespan, for example, a few months, a battery is a logical solution. However, for nodes that can generate sensor readings for long periods of time, a charging

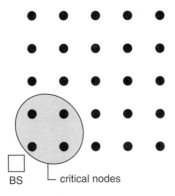

FIGURE 1.1 Sensor network with base station (or fusion center). The gray-shaded area indicates the critical area whose nodes must relay all the packets.

method for the supply is preferable. Currently, research groups are investigating the use of solar cells to charge capacitors with photocurrents from the ambient light sources. Solar flux can yield power densities of approximately 1 mW/mm^2. The energy efficiency of a solar cell ranges from 10 to 30% in current technologies, giving 300 μW in full sunlight in the best-case scenario for a 1-mm^2 solar cell operating at 1 V. Series-stacked solar cells will need to be utilized in order to provide appropriate voltages.

Sensor acquisition can be achieved at 1 nJ per sample, and modern processors can perform computations as low as 1 nJ per instruction. For wireless communications, the primary candidate technologies are based on RF and optical transmission techniques, each of which has its advantages and disadvantages. RF presents a problem because the nodes may offer very limited space for antennas, thereby demanding very short-wavelength (i.e., high-frequency) transmission, which suffers from high attenuation. Thus, communication in that regime is not currently compatible with low-power operation. Current RF transmission techniques (e.g., Bluetooth [9]) consume about 100 nJ per bit for a distance of 10 to 100 m, making communication very expensive compared to acquisition and processing.

An alternative is to employ free-space optical transmission. If a line-of-sight path is available, a well-designed free-space optical link requires significantly lower energy than its RF counterpart, currently about 1 nJ per bit. The reason for this power advantage is that optical transceivers require only simple baseband analog and digital circuitry and no modulators, active filters, and demodulators. Furthermore, the extremely short wavelength of visible light makes it possible for a millimeter-scale device to emit a narrow beam, corresponding to an antenna gain of roughly five to six orders of magnitude compared to an isotropic radiator. However, a major disadvantage is that the beam needs to be pointed very precisely at the receiver, which may be prohibitively difficult to achieve.

In WSNs, where sensor sampling, processing, data transmission, and, possibly, actuation are involved, the trade-off between these tasks plays an important role in power usage. Balancing these parameters will be the focus of the design process of WSNs.

1.3.3 Design of Energy-Efficient Protocols

It is well acknowledged that *clustering* is an efficient way to save energy for static sensor networks [10–13]. Clustering has three significant differences from conventional clustering schemes. First, data compression in the form of distributed source coding is applied within a cluster to reduce the number of packets to be transmitted [14, 15]. Second, the *data-centric* property makes an identity (e.g., an address) for a sensor node obsolete. In fact, the user is often interested in phenomena occurring in a specified area [16], rather than in an individual sensor node. Third, randomized rotation of cluster heads helps ensure a balanced energy consumption [11].

Another strategy to increase energy efficiency is to use *broadcast and multicast trees* [6, 17, 18], which take advantage of the *broadcast property* of omnidirectional antennas. The disadvantage is that the high computational complexity may offset the achievable benefit. For sensor networks, this *one-to-many* communication scheme is less important; however, because all data must be delivered to a single destination, the traffic scheme (for application traffic) is the opposite, i.e., *many to one*. In this case, clearly the *wireless multicast advantage* offers less benefit, unless path diversity or cooperative diversity schemes are implemented [19, 20].

The exploitation of *sleep modes* [21, 22] is imperative to prevent sensor nodes from wasting energy in receiving packets unintended for them. Combined with efficient medium access protocols, the "sleeping" approach could reach optimal energy efficiency without degradation in throughput (but at some penalty in delay).

1.3.4 Capacity/Throughput

Two parameters describe the network's capability to carry traffic: *transport capacity* and *throughput*. The former is a distance-weighted sum capacity that permits evaluation of network performance. Throughput is a traditional measure of how much traffic can be delivered by the network [23–30]. In a packet network, the (network-layer) throughput may be defined as the expected number of successful packet transmissions of a given node per timeslot.

The capacity of wireless networks in general is an active area of research in the information theory community. The results obtained mostly take the form of scaling laws or "order-of" results; the prefactors are difficult to determine analytically. Important results include the scaling law for point-to-point coding, which shows that the throughput decreases with $1/\sqrt{N}$ for a network with N nodes [23]. Newer results [28] permit network coding, which yields a slightly more optimistic scaling behavior, although at high complexity. Grossglauser and Tse [26] have shown that mobility may keep the per-node capacity constant as the network grows, but that benefit comes at the cost of unbounded delay.

The throughput is related to (error-free) transmission rate of each transmitter, which, in turn, is upper bounded by the channel capacity. From the pure information theoretic point of view, the capacity is computed based on the ergodic channel assumption, i.e., the code words are long compared to the coherence time of the channel. This Shannon-type capacity is also called *throughput capacity* [31]. However, in practical networks, particularly with delay-constrained applications, this capacity cannot provide a helpful indication of the channel's ability to transmit with a small probability of error.

Moreover, in the multiple-access system, the corresponding power allocation strategies for maximum achievable capacity always favor the "good" channels, thus leading to unfairness among the nodes. Therefore, for delay-constrained applications, the channel is usually assumed to be nonergodic and the capacity is a random variable, instead of a constant in the classical definition by Shannon. For a delay-bound D, the channel is often assumed to be block fading with block length D, and a *composite channel* model is appropriate when specifying the capacity. Correspondingly, given the noise power, the channel state (a random variable in the case of fading channels), and power allocation, new definitions for *delay-constrained* systems have been proposed [32–35].

1.3.5 Routing

In ad hoc networks, routing protocols are expected to implement three main functions: *determining and detecting network topology changes* (e.g., breakdown of nodes and link failures); *maintaining network connectivity*; and *calculating and finding proper routes*. In sensor networks, up-to-date, less effort has been given to routing protocols, even though it is clear that ad hoc routing protocols (such as *destination-sequenced distance vector* (DSDV), *temporally-ordered routing algorithm* (TORA), *dynamic source routing* (DSR), and *ad hoc on-demand distance vector* (AODV) [4, 36–39]) are not suited well for sensor networks since the main type of traffic in WSNs is "many to one" because all nodes typically report to a single base station or fusion center. Nonetheless, some merits of these protocols relate to the features of sensor networks, like *multihop communication* and *QoS routing* [39]. Routing may be associated with data compression [15] to enhance the scalability of the network.

1.3.6 Channel Access and Scheduling

In WSNs, scheduling must be studied at two levels: the *system level* and the *node level*. At the node level, a scheduler determines which flow among all multiplexing flows will be eligible to transmit next (the same concept as in traditional wired scheduling); at the system level, a scheme determines which nodes will be transmitting. System-level scheduling is essentially a medium access (MAC) problem, with the goal of minimum collisions and maximum spatial reuse — a topic receiving great attention from the research community because it is tightly coupled with energy efficiency and throughput.

Most of the current wireless scheduling algorithms aim at improved *fairness, delay, robustness* (with respect to network topology changes) and *energy efficiency* [62, 64, 65, 66]. Some also propose a distributed implementation, in contrast to the centralized implementation in wired or cellular networks, which originated from general fair queuing. Also, wireless (or sensor) counterparts of other wired scheduling classes, like *priority scheduling* [67, 68] and *earliest deadline first (EDF)* [69], confirm that prioritization is necessary to achieve *delay balancing* and *energy balancing*.

The main problem in WSNs is that all the sensor data must be forwarded to a base station via multihop routing. Consequently, the traffic pattern is highly nonuniform, putting a high burden on the sensor nodes close to the base station (the critical nodes in Figure 1.1). The scheduling algorithm and routing protocols must aim at *energy and delay balancing*, ensuring that packets originating close and far away from the base station experience a comparable delay, and that the critical nodes do not die prematurely due to the heavy relay traffic [40].

At this point, due to the complexity of scheduling algorithms and the wireless environment, most performance measures are given through simulation rather than analytically. Moreover, *medium access* and *scheduling* are usually considered separately. When discussing scheduling, the system is assumed to have a single user; whereas in the MAC layer, all flows multiplexing at the node are treated in the same way, i.e., a default FIFO buffer is assumed to schedule flows. It is necessary to consider them jointly to optimize performance figures such as delay, throughput, and packet loss probability.

Because of the bursty nature of the network traffic, random access methods are commonly employed in WSNs, with or without carrier sense mechanisms. For illustrative purposes, consider the simplest sensible MAC scheme possible: all nodes are transmitting packets independently in every timeslot with the same transmit probability p at equal transmitting power levels; the next-hop receiver of every packet is one of its neighbors. The packets are of equal length and fit into one timeslot. This MAC scheme was considered in Silvester and Kleinrock [41], Hu [42], and Haenggi [43]. The resulting (per-node) throughput turns out to be a polynomial in p of order N, where N is the number of nodes in the network.

A typical throughput polynomial is shown in Figure 1.2. At $p = 0$, the derivative is 1, indicating that, for small p, the throughput equals p. This is intuitive because there are few collisions for small p and the throughput $g(p)$ is approximately linear. The region in which the packet loss probability is less than 10% can be denoted as the *collisionless* region. It ranges from 0 to about $p_{max}/8$. The next region, up to p_{max}, is the practical region in which energy consumption (transmission attempts) is traded off against throughput; it is therefore called the *trade-off* region. The difference $p - g(p)$ is the *interference loss*. For small networks, all N nodes interfere with each other because spatial reuse is not possible: If more than one node is transmitting, a collision occurs and all packets are lost. Thus, the (per-node) throughput is $p(1-p)^{N-1}$, and the optimum transmit probability is $1/N$. The maximum throughput is $(1-1/N)^{N-1}/N$. With increasing N, the throughput approaches $1/(eN)$, as pointed out in Silvester and Kleinrock [41] and LaMaire et al. [44]. Therefore the difference $p_{max} - 1/N$ is the *spatial reuse gain* (see Figure 1.2). This simple example illustrates the concepts of collisions, energy-throughput trade-offs, and spatial reuse, which are present in every MAC scheme.

1.3.7 Modeling

The bases for analysis and simulations and analytical approaches are accurate and tractable models. Comprehensive network models should include the number of nodes and their relative distribution; their degree and type of mobility; the characteristics of the wireless link; the volume of traffic injected by the sources and the lifespan of their interaction; and detailed energy consumption models.

1.3.7.1 Wireless Link

An attenuation proportional to d^α, where d is the distance between two nodes and α is the so-called path loss exponent, is widely accepted as a model for path loss. Alpha ranges from 2 to 4 or even 5 [45], depending on the channel characteristics (environment, antenna position, frequency). This path loss model, together with the fact that packets are successfully transmitted if the signal-to-noise-and-interference ratio (SNIR) is bigger than some threshold [8], results in a deterministic model often used for analysis of multihop packet networks [23, 26, 41, 42, 46–48]. Thus, the radius for a successful transmission has a deterministic value, irrespective of the condition of the wireless channel. If only interferers within a certain distance of the receiver are considered, this "physical model" [23] turns into a "disk model".

The stochastic nature of the fading channel and thus the fact that the SINR is a random variable are mostly neglected. However, the volatility of the channel cannot be ignored in wireless networks [5, 8];

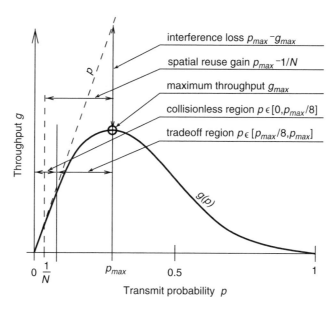

FIGURE 1.2 Generic throughput polynomial for a simple random MAC scheme.

Sousa and Silvester have also pointed out the inaccuracy of disk models [49] and it is easily demonstrated experimentally [50, 51]. In addition, this "prevalent all-or-nothing model" [52] leads to the assumption that a transmission over a multihop path fails completely or is 100% successful, ignoring the fact that end-to-end packet loss probabilities increase with the number of hops. Although fading has been considered in the context of packet networks [53, 54], its impact on the throughput of multihop networks and protocols at the MAC and higher layers is largely an open problem.

A more accurate channel model will have an impact on most of the metrics listed in Section 1.3.1. In the case of Rayleigh fading, first results show that the energy benefits of routing over many short hops may vanish completely, in particular if latency is taken into account [20, 55, 56]. The Rayleigh fading model not only is more accurate than the disk model, but also has the additional advantage of permitting separation of noise effects and interference effects due to the exponential distribution of the received power. As a consequence, the performance analysis can conveniently be split into the analysis of a zero-interference (noise-analysis) and a zero-noise (interference-analysis) network.

1.3.7.2 Energy Consumption

To model energy consumption, four basic different states of a node can be identified: transmission, reception, listening, and sleeping. They consist of the following tasks:

- *Acquisition:* sensing, A/D conversion, preprocessing, and perhaps storing
- *Transmission:* processing for address determination, packetization, encoding, framing, and maybe queuing; supply for the baseband and RF circuitry (The nonlinearity of the power amplifier must be taken into account because the power consumption is most likely not proportional to the transmit power [56].)
- *Reception:* Low-noise amplifier, downconverter oscillator, filtering, detection, decoding, error detection, and address check; reception even if a node is not the intended receiver
- *Listening:* Similar to reception except that the signal processing chain stops at the detection
- *Sleeping:* Power supply to stay alive

Reception and transmission comprise all the processing required for physical communication and networking protocols. For the physical layer, the energy consumption depends mostly on the circuitry, the error correction schemes, and the implementation of the receiver [57]. At the higher layers, the choice of protocols (e.g., routing, ARQ schemes, size of packet headers, number of beacons and other infrastructure packets) determines the energy efficiency.

1.3.7.3 Node Distribution and Mobility

Regular grids (square, triangle, hexagon) and uniformly random distributions are widely used analytically tractable models. The latter can be problematic because nodes can be arbitrarily close, leading to unrealistic received power levels if the path attenuation is assumed to be proportional to d^α. Regular grids overlaid with Gaussian variations in the positions may be more accurate. Generic mobility models for WSNs are difficult to define because they are highly application specific, so this issue must be studied on a case-by-case basis.

1.3.7.4 Traffic

Often, simulation work is based on constant bitrate traffic for convenience, but this is most probably not the typical traffic class. Models for bursty many-to-one traffic are needed, but they certainly depend strongly on the application.

1.3.8 Connectivity

Network connectivity is an important issue because it is crucial for most applications that the network is not partitioned into disjoint parts. If the nodes' positions are modeled as a Poisson point process in two dimensions (which, for all practical purposes, corresponds to a uniformly random distribution), the problem of connectivity has been studied using the tool of *continuum percolation theory* [58, 59]. For large networks, the phenomenon of a sharp phase transition can be observed: the probability that the network *percolates* jumps abruptly from almost 0 to almost 1 as soon as the density of the network is bigger than some critical value. Most such results are based on the geometric disk abstraction. It is conjectured, though, that other connectivity functions lead to better connectivity, i.e., the disk is apparently the hardest shape to connect [60]. A practical consequence of this conjecture is that fading results in improved connectivity. Recent work [61] also discusses the impact of interference. The simplifying assumptions necessary to achieve these results leave many open problems.

1.3.9 Quality of Service

Quality of service refers to the capability of a network to deliver data reliably and timely. A high *quantity of service*, i.e., throughput or transport capacity, is generally not sufficient to satisfy an application's delay requirements. Consequently, the *speed of propagation* of information may be as crucial as the throughput. Accordingly, in addition to network capacity, an important issue in many WSNs is that of quality-of-service (QoS) guarantees. Previous QoS-related work in wireless networks mostly focused on delay (see, for example, Lu et al. [62], Ju and Li [63], and Liu et al. [64]). QoS, in a broader sense, consists of the triple (R, P_e, D), where R denotes throughput; P_e denotes reliability as measured by, for example, bit error probability or packet loss probability; and D denotes delay. For a given R, the reliability of a connection as a function of the delay will follow the general curve shown in Figure 1.3

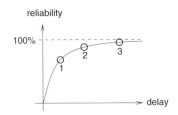

FIGURE 1.3 Reliability as a function of the delay. The circles indicate the QoS requirements of different possible traffic classes.

Note that capacity is only one point on the reliability-delay curve and therefore not always a relevant performance measure. For example, in certain sensing and control applications, the value of information quickly degrades as the latency increases. Because QoS is affected by design choices at the physical, medium-access, and network layers, an integrated approach to managing QoS is necessary.

1.3.10 Security

Depending on the application, security can be critical. The network should enable intrusion detection and tolerance as well as robust operation in the case of failure because, often, the sensor nodes are not protected against physical mishandling or attacks. Eavesdropping, jamming, and listen-and-retransmit attacks can hamper or prevent the operation; therefore, access control, message integrity, and confidentiality must be guaranteed.

1.3.11 Implementation

Companies such as Crossbow, Ember, Sensoria, and Millenial are building small sensor nodes with wireless capabilities. However, a per-node cost of $100 to $200 (not including sophisticated sensors) is prohibitive for large networks. Nodes must become an order of magnitude cheaper in order to render applications with a large number of nodes affordable. With the current pace of progress in VLSI and MEMS technology, this is bound to happen in the next few years. The fusion of MEMS and electronics onto a single chip, however, still poses difficulties. Miniaturization will make steady progress, except for two crucial components: the antenna and the battery, where it will be very challenging to find innovative solutions. Furthermore, the impact of the hardware on optimum protocol design is largely an open topic. The characteristics of the power amplifier, for example, greatly influence the energy efficiency of routing algorithms [56].

1.3.12 Other Issues

- *Distributed signal processing*. Most tasks require the combined effort of multiple network nodes, which requires protocols that provide coordination, efficient local exchange of information, and, possibly, hierarchical operation.
- *Synchronization and localization*. The notion of time is critical. Coordinated sensing and actuating in the physical world require a sense of global time that must be paired with relative or absolute knowledge of nodes' locations.
- *Wireless reprogramming*. A deployed WSN may need to be reprogrammed or updated. So far, no networking protocols are available to carry out such a task reliably in a multihop network. The main difficulty is the acknowledgment of packets in such a joint multihop/multicast communication.

1.4 Concluding Remarks

Wireless sensor networks have numerous exciting applications in virtually all fields of science and engineering, including health care, industry, military, security, environmental science, geology, agriculture, and social studies. In particular, the combination with macroscopic or MEMS-based actuators is intriguing because it permits manipulation of the environment in an unprecedented manner. Researchers and operators currently face a number of critical issues that need be resolved before these applications become reality. Wireless networking and distributed data processing of embedded sensing/actuating nodes under tight energy constraints demand new approaches to protocol design and hardware/software integration.

References

1. Sensor networks and applications, *IEEE Proc.*, 8, Aug. 2003.
2. Internet Engineering Task Force, Mobile ad-hoc networks (MANET). See http://www.ietf.org/html.charters/manet-charter.html.
3. Z.J. Haas et al., Eds., Wireless ad hoc networks, *IEEE J. Selected Areas Commun.*, 17, Aug. 1999. Special ed.
4. C.E. Perkins, Ed., *Ad Hoc Networking*. Addison Wesley, Reading, MA, 2000.
5. A.J. Goldsmith and S.B. Wicker, Design challenges for energy-constrained ad hoc wireless networks, *IEEE Wireless Commun.*, 9, 8–27, Aug. 2002.
6. A. Ephremides, Energy concerns in wireless networks, *IEEE Mag. Wireless Commun.*, 9, 48–59, Aug. 2002.
7. V. Rodoplu and T.H. Meng, Minimum energy mobile wireless networks, *IEEE J. Selected Areas Commun.*, 17(8), 1333–1344, 1999.
8. A. Ephremides, Energy concerns in wireless networks, *IEEE Wireless Commun.*, 9, 48–59, Aug. 2002.
9. Bluetooth wireless technology. Official Bluetooth site: http://www.bluetooth.com.
10. S. Tilak, N.B. Abu–Ghazaleh, and W. Heinzelman, A taxonomy of wireless micro-sensor network models, *ACM Mobile Computing Commun. Rev.*, 6(2), 28–36, 2002.
11. W.B. Heinzelman, A.P. Chandrakasan, and H. Balakrishnan, An application-specific protocol architecture for wireless microsensor networks, *IEEE Trans. Wireless Commun.*, 1, 660–670, Oct. 2002.
12. J. Kulik, W. Heinzelman, and H. Balakrishnan, Negotiation-based protocols for disseminating information in wireless sensor networks, *Wireless Networks*, 8, 169–185, March–May 2002.
13. A.B. McDonald and T.F. Znati, A mobility-based framework for adaptive clustering in wireless ad-hoc networks, *IEEE J. Selected Areas Commun.*, 17, 1466–1487, Aug. 1999.
14. S.S. Pradhan, J. Kusuma, and K. Ramchandran, Distributed compression in a dense microsensor network, *IEEE Signal Process. Mag.*, 19, 51–60, Mar. 2002.
15. A. Scaglione and S. Servetto, On the interdependence of routing and data compression in multi-hop sensor networks, in *Proc. ACM Int. Conf. Mobile Comp. Networks (MobiCom'02)*, Atlanta, GA, 140–147, Sept. 2002.
16. C. Intanagowiwat, R. Govindan, and D. Estrin, Directed diffusion: a scalable and robust communication paradigm for sensor networks, in *ACM Int. Conf. Mobile Computing Networking (MobiCom'00)*, Boston, MA, 56–67, Aug. 2000.
17. J.E. Wieselthier, G.D. Nguyen, and A. Ephremides, On the construction of energy-efficient broadcast and multicast trees in wireless networks, in *IEEE INFOCOM*, Tel Aviv, Israel, 585–594, Mar. 2000.
18. J.E. Wieselthier, G.D. Nguyen, and A. Ephremides, An insensitivity property of energy-limited wireless networks for session-based multicasting, in *IEEE ISIT*, Washington, D.C., June 2001.
19. J. Laneman, D. Tse, and G. Wornell, Cooperative diversity in wireless networks: efficient protocols and outage behavior, *IEEE Trans. Inf. Theory*. Accepted for publication. Available at: http://www.nd.edu/jnl/pubs/it2002.pdf.
20. M. Haenggi, A formalism for the analysis and design of time and path diversity schemes in wireless sensor networks, in *2nd Int. Workshop Inf. Process. Sensor Networks (IPSN'03)*, Palo Alto, CA, 417–431, Apr. 2003. Available at http://www.nd.edu/mhaenggi/ipsn03.pdf.
21. C.S. Raghavendra and S. Singh, PAMAS — power aware multi-access protocol with signaling for ad hoc networks, 1999. *ACM Computer Commun. Rev.* Available at: http://citeseer.nj.nec.com/460902.html.
22. C.-K. Toh, Maximum battery life routing to support ubiquitous mobile computing in wireless ad hoc networks, *IEEE Commun. Mag.*, 39, 138–147, June 2001.
23. P. Gupta and P.R. Kumar, The capacity of wireless networks, *IEEE Trans. Inf. Theory*, 46, 388–404, Mar. 2000.
24. P. Gupta and P.R. Kumar, Towards an information theory of large networks: an achievable rate region, in *IEEE Int. Symp. Inf. Theory*, Washington, D.C., 159, 2001.

25. L.-L. Xie and P.R. Kumar, A network information theory for wireless communication: scaling laws and optimal operation, Apr. 2002. submitted to *IEEE Trans. Inf. Theory*. Available at: http://black1.csl.uiuc.edu/prkumar/publications.html.
26. M. Grossglauser and D. Tse, Mobility increases the capacity of ad-hoc wireless networks, in *IEEE INFOCOM*, Anchorage, AL, 2001.
27. D. Tse and S. Hanly, Effective bandwidths in wireless networks with multiuser receivers, in *IEEE INFOCOM*, 35–42, 1998.
28. M. Gastpar and M. Vetterli, On the capacity of wireless networks: the relay case, in *IEEE INFOCOM*, New York, 2002.
29. G. Mergen and L. Tong, On the capacity of regular wireless networks with transceiver multipacket communication, in *IEEE Int. Symp. Inf. Theory*, Lausanne, Switzerland, 350, 2002.
30. S. Toumpis and A. Goldsmith, Capacity regions for wireless ad hoc networks, *IEEE Trans. Wireless Commun.*, 2, 736–748, July 2003.
31. D.N.C. Tse and S.V. Hanly, Multiaccess fading channels — part I: polymatroid structure, optimal resource allocation and throughput capacities, *IEEE Trans. Inf. Theory*, 44(7), 2796–2815, 1998.
32. S.V. Hanly and D.N.C. Tse, Multiaccess fading channels — part II: delay-limited capacities, *IEEE Trans. Inf. Theory*, 44(7), 2816–2831, 1998.
33. R. Negi and J.M. Cioffi, Delay-constrained capacity with causal feedback, *IEEE Trans. Inf. Theory*, 48, 2478–2494, Sept. 2002.
34. R.A. Berry and R.G. Gallager, Communication over fading channels with delay constraints, *IEEE Trans. Inf. Theory*, 48, 1135–1149, May 2002.
35. D. Tuninetti, On multiple-access block-fading channels, Mar. 2002. Ph.D. thesis, Institut EURECOM. Available at: http://www.eurecom.fr/tuninett/publication.html.
36. J. Broch, D. Maltz, D. Johnson, Y. Hu, and J. Jetcheva, A performance comparison of multi-hop wireless ad hoc network routing protocols, in *ACM Int. Conf. Mobile Computing Networking (MobiCom)*, Dallas, TX, 85–97, Oct. 1998.
37. P. Johansson, T. Larsson, and N. Hedman, Scenario-based performance analysis of routing protocols for mobile ad-hoc networks, in *ACM MobiCom*, Seattle, WA, Aug. 1999.
38. S.R. Das, C.E. Perkins, and E.M. Royer, Performance comparison of two on-demand routing protocols for ad hoc networks, in *IEEE INFOCOM*, Mar. 2000.
39. C.R. Lin and J.-S. Liu, QoS Routing in ad hoc wireless networks, *IEEE J. Selected Areas Commun.*, 17, 1426–1438, Aug. 1999.
40. M. Haenggi, Energy-balancing strategies for wireless sensor networks, in *IEEE Int. Symp. Circuits Syst. (ISCAS'03)*, Bangkok, Thailand, May 2003. Available at http://www.nd.edu/mhaenggi/iscas03.pdf.
41. J.A. Silvester and L. Kleinrock, On the capacity of multihop slotted ALOHA networks with regular structure, *IEEE Trans. Commun.*, COM-31, 974–982, Aug. 1983.
42. L. Hu, Topology control for multihop packet networks, *IEEE Trans. Commun.*, 41(10), 1474–1481, 1993.
43. M. Haenggi, Probabilistic analysis of a simple MAC scheme for ad hoc wireless networks, in *IEEE CAS Workshop on Wireless Communications and Networking*, Pasadena, CA, Sept. 2002.
44. R.O. LaMaire, A. Krishna, and H. Ahmadi, Analysis of a wireless MAC protocol with client–server traffic and capture, *IEEE J. Selected Areas Commun.*, 12(8), 1299–1313, 1994.
45. T.S. Rappaport, *Wireless Communications — Principles and Practice*, 2nd ed., Prentice Hall, Englewood Cliffs, NJ.
46. H. Takagi and L. Kleinrock, Optimal transmission ranges for randomly distributed packet radio terminals, *IEEE Trans. Commun.*, COM-32, 246–257, Mar. 1984.
47. J.L. Wang and J.A. Silvester, Maximum number of independent paths and radio connectivity, *IEEE Trans. Commun.*, 41, 1482–1493, Oct. 1993.
48. C. Schurgers, V. Tsiatsis, S. Ganeriwal, and M. Srivastava, Optimizing sensor networks in the energy–latency–density design space, *IEEE Trans. Mobile Computing*, 1(1), 70–80, 2002.

49. E.S. Sousa and J.A. Silvester, Optimum transmission ranges in a direct-sequence spread-spectrum multihop packet radio network, *IEEE J. Selected Areas Commun.*, 8, 762–771, June 1990.
50. D.A. Maltz, J. Broch, and D.B. Johnson, Lessons from a full-scale multihop wireless ad hoc network testbed, *IEEE Personal Commun.*, 8, 8–15, Feb. 2001.
51. D. Ganesan, B. Krishnamachari, A. Woo, D. Culler, D. Estrin, and S. Wicker, An empirical study of epidemic algorithms in large scale multihop wireless networks, 2002. Intel Research Report IRB-TR-02-003. Available at www.intel-research.net/Publications/Berkeley/05022002170319.pdf.
52. T.J. Shepard, A channel access scheme for large dense packet radio networks, in *ACM SIGCOMM*, Stanford, CA, Aug. 1996. Available at: http://www.acm.org/sigcomm/sigcomm96/papers/shepard.ps.
53. M. Zorzi and S. Pupolin, Optimum transmission ranges in multihop packet radio networks in the presence of fading, *IEEE Trans. Commun.*, 43, 2201–2205, July 1995.
54. Y.Y. Kim and S. Li, Modeling multipath fading channel dynamics for packet data performance analysis, *Wireless Networks*, 6, 481–492, 2000.
55. M. Haenggi, On routing in random rayleigh fading networks, *IEEE Trans. Wireless Commun.*, 2003. Submitted for publication. Available at http://www.nd.edu/mhaenggi/routing.pdf.
56. M. Haenggi, The impact of power amplifier characteristics on routing in random wireless networks, in *IEEE Global Commun. Conf. (GLOBECOM'03)*, San Francisco, CA, Dec. 2003. Available at http://www.nd.edu/mhaenggi/globecom03.pdf.
57. H. Meyr, M. Moenecleay, and S.A. Fechtel, *Digital Communication Receivers: Synchronization, Channel Estimation, and Signal Processing*. Wiley Interscience, 1998.
58. R. Meester and R. Roy, *Continuum Percolation*. Cambridge University Press, New York, 1996.
59. B. Bollobás, *Random Graphs*, 2nd ed. Cambridge University Press, New York, 2001.
60. L. Booth, J. Bruck, M. Cook, and M. Franceschetti, Ad hoc wireless networks with noisy links, in *IEEE Int. Symp. Inf. Theory*, Yokohama, Japan, 2003.
61. O. Dousse, F. Baccelli, and P. Thiran, Impact of interferences on connectivity in ad-hoc networks, in *IEEE INFOCOM*, San Francisco, CA, 2003.
62. S. Lu, V. Bharghavan, and R. Srikant, Fair scheduling in wireless packet networks, *IEEE/ACM Trans. Networking*, 7, 473–489, Aug. 1999.
63. J.-H. Ju and V.O.K. Li, TDMA scheduling design of multihop packet radio networks based on Latin squares, *IEEE J. Selected Areas Commun.*, 1345–1352, Aug. 1999.
64. H. Luo, S. Lu, and V. Bharghavan, A new model for packet scheduling in multihop wireless networks, in *ACM Int. Conf. Mobile Computing Networking (MobiCom'00)*, Boston, MA, 76–86, 2000.
65. H. Luo, P. Medvedev, J. Cheng, and S. Lu, A self-coordinating approach to distributed fair queueing in *ad hoc* wireless networks, *IEEE INFOCOM*, Anchorage, Apr. 2001.
66. A.E. Gamal, C. Nair, B. Prabhakar, E. Uysal-Biyikoglu, and S. Zahedi, Energy-efficient scheduling of packet transmissions over wireless networks, *IEEE INFOCOM*, New York, 2002, pp. 1773–1782.
67. S. Bhatnagar, B. Deb, and B. Nath, Service differentiation in sensor networks, *Fourth International Symposium on Wireless Personal Multimedia Communications*, Sept. 2001.
68. V. Kanodia, C. Li, A. Sabharwal, B. Sadeghi, and E. Knightly, Distributed multi-hop scheduling and medium access with delay and throughput constraints, *ACM MobiCom*, Rome, July 2001.
69. A. Striegel and G. Manimaran, Best-effort scheduling of (m, k)-firm real-time streams in multihop networks, *Workshop of Parallel and Distributed Real-Time Systems (WPDRTS) at IPDPS 2000*, Apr. 2000.

2
Next-Generation Technologies to Enable Sensor Networks*

Joel I. Goodman
MIT Lincoln Laboratory

Albert I. Reuther
MIT Lincoln Laboratory

David R. Martinez
MIT Lincoln Laboratory

2.1 Introduction 2-1
  Geolocation and Identification of Mobile Targets • Long-Term Architecture
2.2 Goals for Real-Time Distributed Network Computing for Sensor Data Fusion 2-5
2.3 The Convergence of Networking and Real-Time Computing 2-6
  Guaranteeing Network Resources • Guaranteeing Storage Buffer Resources • Guaranteeing Computational Resources
2.4 Middleware 2-11
  Control and Command of System • Parallel Processing
2.5 Network Resource Management 2-11
  Graph Generator • Metrics Object • Graph Search • NRM Agents • Sensor Interface • Mapping Database • Topology Database • NRM Federation • NRM Fault Tolerance
2.6 Experimental Results 2-16

2.1 Introduction

Several important technical advances make extracting more information from intelligence, surveillance, and reconnaissance (ISR) sensors very affordable and practical. As shown in Figure 2.1, for the radar application the most significant advancement is expected to come from employing collaborative and network centric sensor netting. One important application of this capability is to achieve ultrawideband multifrequency and multiaspect imaging by fusing the data from multiple sensors. In some cases, it is highly desirable to exploit multimodalities, in addition to multifrequency and multiaspect imaging.

Key enablers to fuse data from disparate sensors are the advent of high-speed fiber and wireless networks and the leveraging of distributed computing. ISR sensors need to perform enough on-board computation to match the available bandwidth; however, after some initial preprocessing, the data will be distributed across the network to be fused with other sensor data so as to maximize the information content. For example, on an experimental basis, MIT Lincoln Laboratory has demonstrated a virtual radar with ultrawideband frequency [1]. Two radars, located at the Lincoln Space Surveillance Complex

---

*This work is sponsored by the United States Air Force under Air Force contract F19628-00-C-002. Opinions, interpretations, conclusions, and recommendations are those of the authors and are not necessarily endorsed by the U.S. government.

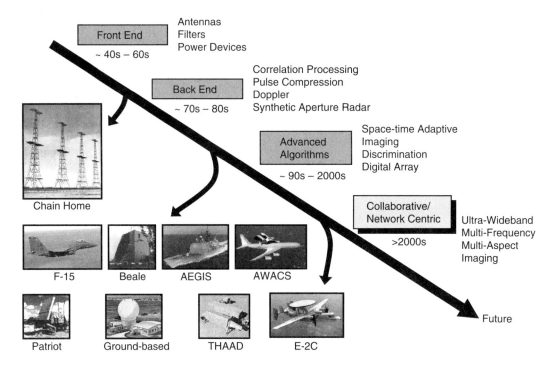

FIGURE 2.1 Radar technology evolution.

in Westford, Massachusetts, were employed; each of the two independent radars transmitted the data via a high-speed fiber network. The total bandwidth transmitted via fiber exceeded 1 Gbits/sec (billion bits per second). One radar was operating at X-band with 1-MHz bandwidth, and the second was operating at Ku-band with a 2-MHz bandwidth. A synthetic radar with an instantaneous bandwidth of 8 MHz was achieved after employing advanced ultrawideband signal processing [2].

These capabilities are now being extended to include high-speed wireless and fiber networking with distributed computing. As the Internet protocol (IP) technologies continue to advance in the commercial sector, the military can begin to leverage IP formatted sensor data to be compatible with commercial high-speed routers and switches. Sensor data from theater can be posted to high-speed networks, wireless and fiber, to request computing services as they become available on this network. The sensor data are processed in a distributed fashion across the network, thereby providing a larger pool of resources in real time to meet stringent latency requirements. The availability of distributed processing in a grid-computing architecture offers a high degree of robustness throughout the network. One important application to benefit from these advances is the ability to geolocate and identify mobile targets accurately from multiaspect sensor data.

2.1.1 Geolocation and Identification of Mobile Targets

Accurately geolocating and identifying mobile targets depends on the extraction of information from different sensor data. Typically, data from a single sensor are not sufficient to achieve a high probability of correct classification and still maintain a low probability of false alarm. This goal is challenging because mobile targets typically move at a wide range of speeds, tend to move and stop often, and can be easily mistaken for a civilian target. While the target is moving the sensor of choice is the ground moving target indication (GMTI). If the target stops, the same sensor or a different sensor working cooperatively must employ synthetic aperture radar (SAR). Before it can be declared foe, the target must often be confirmed with electro-optical or infrared (EO/IR) images. The goal of future networked systems is to have multiple sensors providing the necessary multimodality data to maximize the chances of accurately declaring a target.

A typical sensing sequence starts by a wide area surveillance platform, such as the Global Hawk unmanned aerial vehicle (UAV), covering several square kilometers until a target exceeds a detection threshold. The wide area surveillance will typically employ GMTI and SAR strip maps. Once a target has been detected, the on-board or off-board processing starts a track file to track the target carefully, using spot GMTI and spot SAR over a much smaller region than that initially covered when performing wide area surveillance. It is important to recognize that a sensor system is not merely tracking a single target; several target tracks can be going on in parallel. Therefore, future networked sensor architectures rely on sharing the information to maximize the available resources.

To date, the most advanced capability demonstrated is based on passing target detections among several sensors using the Navy cooperative engagement capability (CEC) system. Multisensor tracks are formed from the detection inputs arriving at a central location. Although this capability has provided a significant advancement, not all the information available from multimodality sensors has been exploited. The limitation is with the communication and available distributed computing. Multimodality sensor data together with multiple look angles can substantially improve the probability of correct classification vs. false alarm density. In addition to multiple modalities and multiple looks on the target, it is also desirable to send complex (amplitude and phase) radar GMTI data and SAR images to permit the use of high-definition vector imaging (HDVI) [3]. This technique permits much higher resolution on the target by suppressing noise around it, thereby enhancing the target image at the expense of using complex video data and much higher computational rates.

Another important tool to improve the probability of correct classification with minimal false alarm is high-range resolution (HRR) profiles. With this tool, the sensor bandwidth or, equivalently, the size of the resolution cell must be small resulting in a large data rate. However, it has been demonstrated that HRR can provide a significant improvement [4]. Therefore, next generation sensors depend on available communication pipes with enough bandwidth to share the individual sensor information effectively across the network. Once the data are posted on the network, the computational resources must exist to maintain low latencies from the time data become available to the time a target geoposition and identification are derived. The next subsection discusses the long-term architecture to implement netting of multiple sensor data efficiently.

2.1.2 Long-Term Architecture

In the future it will be desirable to minimize the infrastructure (foot print) forwardly deployed in the battlefield. It is most desirable to leverage high-speed satellite communication links to bring sensor data back to a combined air operations center (CAOC) established in the continental United States (CONUS).

The technology enablers for the long-term architecture shown in Figure 2.2 are high-speed, IP-based wireless and fiber communication networks, together with distributed grid computing. The in-theater commander's ability to task his organic resources to perform reconnaissance and surveillance of the opposing forces, and then to relay that information back to CONUS, allows significant reduction in the complexity, level, and cost of in-theater resources. Furthermore, this approach leverages the diverse analysis resources in CONUS, including highly trained personnel to support the rapid, accurate identification and localization of targets necessary to enable the time-critical engagement of surface mobile threats.

Space, air, and surface sensors will be deployed quickly to the battlefield. As shown in Figure 2.3, the stage in the processing chain at which the sensor data are tapped off to be sent via the network will dictate the amount of data transferred. For example, in a few applications one needs to send the data directly out of the analog-to-digital converters (A/D) to exploit coherent data combining from multiple sensors. Most commonly, it is preferable to perform on-board signal preprocessing to minimize the amount of data transferred. However, one must still be able to preserve content in the transferred data that is required to exploit features in the data not available from processing a signal sensor end to end. For example, one might be interested in transmitting wide area surveillance (WAS) data from SAR with high resolution to be followed by multiaspect SAR processing (shown in Figure 2.3 as application B). The data volume will be larger than the second example shown in Figure 2.3 as application A, in which

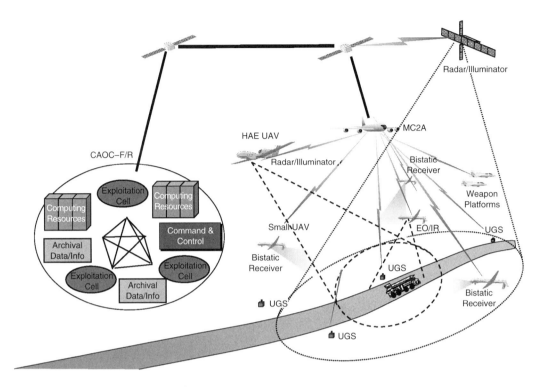

FIGURE 2.2 Postulated long-term architecture.

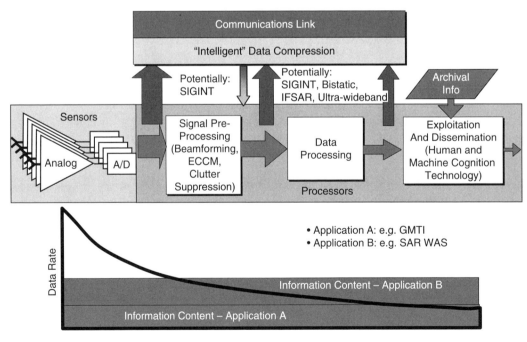

FIGURE 2.3 Sensor signal processing flow.

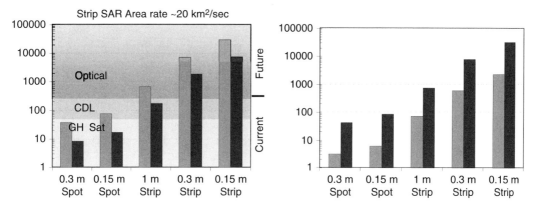

FIGURE 2.4 SAR data rate and computational throughput trade.

most of the GMTI processing is done on board. In any of these applications, it is paramount that "intelligent" data compression be done on board before data transmission to send only the necessary parts of the data requiring additional processing off board.

Each sensor will be capable of generating on-board processed data greater than 100 Mbits/sec (million bits per second). Figure 2.4 shows the trade-off between communication link data rates vs. on-board computation throughputs for different postulated levels of image resolution (for spot or strip map SAR modes). For example, for an assumed 1-m strip map SAR, one can send complex video radar data to then perform super-resolution processing off board. This approach would require sending between 100 to 1000 Mbits/sec. Another option is to perform the super-resolution processing on board, requiring between 100 billion floating-point operations per second (GFLOPS) to 1 trillion floating-point operations per second (TFLOPS).

Specialized military equipment, such as the common data link (CDL), can achieve data rates reaching 274 Mb/sec. If higher communication capacity were available, one would much prefer to send the large data volume for further processing off board to leverage information content available from multiple sensor data. As communication rates improve in the forthcoming years, it will not matter to the in-theater commander if the data are processed off board with the benefit of allowing exploitation of multiple sensor data at much rawer levels than is possible to date.

2.2 Goals for Real-Time Distributed Network Computing for Sensor Data Fusion

Several advantages can be gained by utilizing real-time distributed network computing to enable greater sensor data fusion processing. Distributed network computing potentially reduces the cost of the signal processing systems and the sensor platform because each individual sensor platform no longer needs as much processing capability as a stove-piped stand-alone system (although each platform may need higher bandwidth communications capabilities). Also, fault tolerance of the processing systems is increased because the processing and network systems are shared between sensors, thereby increasing the pool of available signal processors for all of the sensors. Furthermore, the granularity of managed resources is smaller; individual processors and network resources are managed as independent entities rather than managing an entire parallel computer and network as independent entities. This affords more flexible configuration and management of the resources.

To enable collaborative network processing of sensor signals, three technological areas are required to evolve and achieve maturity:

- Guaranteed *communication, storage buffer,* and *computation resources* must keep up with the high-throughput streams of data coming from the sensors. If any stage of the processing falls

behind due to a network problem or interruption in the processor, buffering the data will become a problem quickly as increasing volumes of data must be stored to accommodate the delayed processors. Section 2.3 addresses technological possibilities to mitigate these resource availability issues.
- *Middleware* in the network of processors must be developed to accommodate a heterogeneous mix of computer and network resources. This middleware consists of a task control interface, which facilitates the communication between network resource management agents and entities, and an application programming interface for programming applications executed on the collaborative network processors. Section 2.4 will address these middleware interfaces.
- A *network resource manager* (NRM) system is necessary for orchestrating the execution of the application components on the computation and communication resources available in the collaborative network. Section 2.5 will discuss the components and functionality of the NRM.

2.3 The Convergence of Networking and Real-Time Computing

To date, networking of sensors has been demonstrated primarily using localized- and limited-capacity data links. As a result, the data available on the network from each sensor node typically represent the product of extensive prior processing of the radar data carried at the individual sensor. For example, the Navy CEC system, a relatively advanced current system, uses detection reports from independent sensors in the network to build composite tracks of targets. Access to raw (or possibly minimally preprocessed) multisensor data opens the opportunity for more effective exploitation of these data through integrated sensor data processing. The future network-centric ISR architecture will likely employ worldwide wideband communication networks to interconnect sensors with distributed processing and fusion sites. The resulting distributed database will provide a common operational picture for deployed forces. The sensor data will return to a CONUS entry point and pass over a wideband fiber network to the various processing centers where the sensor data will be fused. The data link from the theater to CONUS is expected to be optical to achieve very high link capacity [5].

This section discusses technologies that will guarantee that wireless and terrestrial network resources, storage buffer resources, and computational resources are available for sensor signal processing.

2.3.1 Guaranteeing Network Resources

Sensor data will traverse wireless and terrestrial (e.g., optical, twisted-copper) networks in which bit errors, packet loss, and delay could adversely affect the quality and timeliness of the ultimate result. The goal then is to choose a network and processing architecture to ameliorate the deleterious effects of data loss and network delay in the data fusion process. Due to the costs associated with developing, deploying, and maintaining a fixed terrestrial infrastructure, as well as inventing wholly new modulation protocols and standards for wireless and terrestrial signaling, it is cost-effective and expedient for military technology to ride the "commercial wave" of technical investment and progress in communication technologies.

With a fixed network infrastructure consisting primarily of commercial components, combating data loss and delay in terrestrial networks involves choosing the right protocols so that the network can enforce quality of service (QoS) demands; in wireless networks, this involves aggressive coding, modulation, and "lightweight" flow control for efficient bandwidth utilization. With sufficient complexity and bandwidth, it is possible with today's IP-based protocols to differentiate high-priority data to impart the mandated QoS for time-critical applications.

2.3.1.1 Terrestrial Networks

Reserving bandwidth on an IP-based network that is uniformly recognized across administrative domains involves employing protocols like RSVP-TE [6] or CR-LDP [7]. Although having sufficient communication bandwidth is an important aspect of processing sensor data in real time on a distributed network of resources, it does not guarantee real-time performance. For example, time-critical applications mapped onto networked resources should not have processing interrupted to service unmanaged traffic or be subject to a computational resource's resident operating system switching contexts to a lower priority task. For data that originate from sensors at very high streaming rates, a storage solution, as discussed in Section 2.3.2, is needed that is capable of recording sensor data in real time as well as robust in the face of network resource failures; this insures that a high-priority application can continue processing in the presence of malfunctioning or compromised networked equipment. However, adding a buffering storage solution only alleviates part of the problem; it does not mitigate the underlying problem of losing packets during network equipment failures or periods of network traffic that exceed network capacities.

For an IP-based network, one solution to this problem is to use remote agents deployed on primary compute resources or networked terminals located at switches that can dynamically filter unmanaged traffic. This is implemented by programming computer hardware specifically tasked with packet filtering (e.g., next generation gigabit Ethernet card) or dynamically reconfiguring the switch that directly connects to the compute resource in question by supplying an access control list (ACL) to block all packets except those associated with time-critical targeting. The formation of these exclusive networks using agents has been dubbed *dynamic private networks* (DPNs) — in effect, mechanisms for virtually overlaying a circuit switch onto a packet-switched network.

2.3.1.2 Wireless Networks

Unlike terrestrial networks, flow control and routing in mobile wireless sensor networks must contend with potentially long point-to-point propagation delays (e.g., satellite to ground) as well as a constantly changing topology. In a traditional terrestrial network employing link-state routing (e.g., OSPF), each node maintains a consistent view of a (primarily) fixed network topology so that a shortest path algorithm [8] can be used to find desirable routes from source to destination. This requires that nodes gather network connectivity information from other routers.

If OSPF were employed in a mobile wireless network, the overhead of exchanging network connectivity information about a transient topology could potentially consume the majority of the available bandwidth [9]. Routing protocols have been specifically designed to address the concerns of mobile networks [10]; these protocols fall into two general categories: proactive and reactive. Proactive routing protocols keep track of routes to all destinations, while reactive protocols acquire routes on demand. Unlike OSPF, proactive protocols do not need a consistent view of connectivity; that is, they trade optimal routes for feasible routes to reduce communication overhead. Reactive routes suffer a high initial overhead in establishing a route; however, the overall overhead of maintaining network connectivity is substantially reduced. The category of routing used is highly dependent upon how the sensors communicate with one another over the network.

Traditional flow control mechanisms over terrestrial networks that deliver reliable transport (e.g., TCP) may be inappropriate for wireless networks because, unlike wireless networks, terrestrial networks generally have a very low bit error rate (BER) on the order of 10^{-10}, so errors are primarily due to packet loss. Packet loss occurs in heavily congested networks when an ingress or egress queue of a switch or router begins to fill, requiring that some packets in the queue be discarded [11]. This condition is detected when acknowledgments from the destination node are not received by the source, prompting the source's flow control to throttle back the packet transmit rate [12].

In a wireless network in which BERs are four to five orders of magnitude higher than those of terrestrial networks, packet loss due to bit errors can be mistakenly associated with network congestion, and source flow control will mistakenly reduce the transmit rate of outgoing packets. Furthermore, when the source and destination are far apart, such as the communication between a satellite and ground terminal, where propagation delays can be on the order of 240 ms, delayed acknowledgments from the destination result in source flow control inefficiently using the available bandwidth. This is due to source flow control incrementally increasing the transmit rate as destination acknowledgements are received even though the entire frame of packets may have already been transmitted before the first packet reaches the receiver [13]. Therefore, to use bandwidth efficiently in a wireless network for reliable transport, flow control must be capable of differentiating BER from packet loss and account for long-haul packet transport by more efficiently using the available bandwidth. Some work in this area is reflected in RFC 2488 [14], as well as proposals for an explicit congestion warning, where, for example, the destination site would respond to packet errors with an acknowledgment that it received the source packets with a corruption notification.

At the physical layer, high data rates for a given BER have been realized by employing low-density parity check codes, such as turbo codes, in conjunction with bandwidth efficient modulation to achieve spectral efficiencies to within 0.7 dB of the Shannon limit [15]. Furthermore, extremely high spectral efficiencies have been demonstrated using multiple input, multiple output (MIMO) antenna systems whose theoretical channel capacity increases linearly with the number of transmit/receive antenna pairs [16]. Although turbo codes are advantageous as a forward error correction mechanism in wireless systems when trying to maximize throughput, MIMO systems achieve high spectral efficiencies only when operating in rich scattering environments [17]. In environments in which little scattering occurs, such as in some air-to-air communication links, MIMO systems offer very little improvement in spectral efficiency.

2.3.2 Guaranteeing Storage Buffer Resources

For a variety of reasons, it may be very desirable to record streaming sensor data directly to storage media while simultaneously sending the data on for immediate processing. For sensor signal processing applications, this enables multimodality data fusion of archived data with real-time (perishable) data from in-theatre sensors for improved target identification and visualization [18]. Storage media could also be used for rate conversion in cases in which the transmission rate exceeds the processing rate and for time-delay buffering for real-time robust fault tolerance (discussed in the next section). The storage media buffer reuse is deterministic and periodic so that management of the buffer is straightforward.

A number of possible solutions exist:

- *Directly attached storage* is a set of hard disks connected to a computer via SCSI or IDE/EIDE/ATA; however, this technology does not scale well to the volume of streaming sensor data.
- *Storage area networks* are hard disk storage cabinets attached to a computer with a fast data link like Fibre Channel. The computer attached to the storage cabinet enjoys very fast access to data, but because the data must travel through that computer, which presents a single point of failure, to get to other computers on the network, this option is not a desirable solution.
- *Network-attached storage* connects the hard disk storage cabinet directly to the network as a file server. However, this technology offers only midrange performance, a single point of failure, and relatively high cost.

A visionary architecture in which data storage centers operate in parallel at a wide-area network (WAN) and local area network (LAN) level is described in Cooley et al. [19]. In this architecture, developed by MIT Lincoln Laboratory, high-rate streaming sensor data are stored in parallel across a partitioned network of storage arrays, which affords a highly scalable, low-cost solution that is relatively insensitive to communications or storage equipment failure. This system employs a novel and computationally efficient encoding and decoding algorithm using low-density parity check codes [20] for erasure recovery. Initial system performance measures indicate the erasure coding method described in Cooley et al. [19] has a significantly higher throughput and greater reliability when compared to Reed–Solomon, Tornado [21], and Luby [20] codes. This system offers a promising low-cost solution that scales in capability with the performance gains of commodity equipment.

2.3.3 Guaranteeing Computational Resources

The exponential growth in computing technology has contributed to making viable the implementation of advanced sensor processing in cost-effective hardware with form factors commensurate with the needs of military users. For example, several generations of embedded signal processors are shown in Figure 2.5.

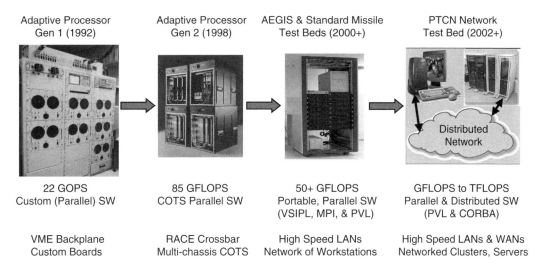

FIGURE 2.5 Embedded signal processor evolution.

In the early 1990s, embedded signal processors were built using custom hardware and software. In the late 1990s, a move occurred from custom hardware to COTS processor systems running vendor-specific software together with application-specific parallel software tuned to each specific application. Most recently, the military embedded community is beginning to demonstrate requisite performance employing parallel and portable software running on COTS hardware.

Continuing technology advances in computation and communication will permit future signal processors to be built from commodity hardware distributed across a high-speed network and employing distributed, parallel, and portable software. These computing architectures will deliver 10^9 to 10^{12} floating point operations per second (GFLOPs to TFLOPs) in computational throughput. The distributed nature of the software will apply to on-board sensor processing as well as off-board processing. Clearly, on-board embedded processor systems will need to meet the stringent platform requirements in size, weight, and power.

Wireless and terrestrial network resources are not the only areas in which delays, failures, and errors must be avoided to process sensor data in a timely fashion. The system design must also guarantee that the marshaled compute nodes will keep up with the required computational throughput of streaming data at every stage of the processing chain. This guarantee encompasses two important facets: (1) keeping the processors from being interrupted while they are processing tasks and (2) implementing fail-over that is tolerant of fault.

2.3.3.1 Avoiding Processor Interruption

It is easy to take for granted that laptop and desktop computers will process commands as fast as the hardware and software are capable of doing so. A fact not generally known is that general computers are interrupted by system task processes and the processes of other applications (one's own and possibly from others working in the background on one's system). System task processes include keyboard and mouse input; communications on the Ethernet; system I/O; file system maintenance; log file entries; etc. When the computer interrupts an application to attend to such tasks, the execution of the application is temporarily suspended until the interrupting task has finished execution. However, because such interruptions often only consume a few milliseconds of processing time, they are virtually imperceptible to the user [22].

Nevertheless, the interruptions are detrimental to the execution of real-time applications. Any delay in processing these streams of data will instigate a need for buffering the data that will grow to insurmountable size as the delays escalate. A solution for these interrupt issues is to use a real-time operating system on the computation processors.

Simply put, real-time operating systems (RTOS) give priority to computational tasks. They usually do not offer as many operating system features (virtual memory, threaded processing, etc.) because of the interrupting processing nature of these features [22]. However, an RTOS can ensure that real-time critical tasks have guaranteed success in meeting streamed processing deadlines. An RTOS does not need to be run on typical embedded processors; it can also be deployed on Intel and AMD Pentium-class or Motorola G-series processor systems. This includes Beowulf clusters of standard desktop personal computers and commodity servers. This is an important benefit, providing a wide range of candidate heterogeneous computing resources.

A great deal of press has been generated in the past several years about real-time operating systems; however, the distinction between soft real-time and hard real-time operating systems is seldom discussed. Hard real-time systems guarantee the completion of tasks in a deterministic time period, while soft real-time systems give priority to critical tasks over other tasks but do not guarantee the completion of tasks in a deterministic time period [22]. Examples of hard real-time operating systems are VxWorks (Wind River Systems, Inc. [23]); RTLinux/Pro (FSMLabs, Inc. [24]); and pSOS (Wind River Systems, Inc. [23]), as well as dedicated massively parallel embedded operating systems like MC/OS (Mercury Computer Systems, Inc. [25]). Examples of soft real-time operating systems are Microsoft Pocket PC; Palm OS; certain real-time Linux releases [24, 26]; and others.

2.3.3.2 Working through System Faults

When fault tolerance in massively parallel computers is addressed, usually the solution is parallel redundant systems for fail-over. If a power supply or fan fails, another power supply or fan that is redundant in the system takes over the workload of the failed device. If a hard disk drive fails on a redundant array of independent disks (RAID) system, it can be hot swapped with a new drive and the contents of the drive rebuilt from the contents of the other drives along with checksum error correction code information. However, if an individual processor fails on a parallel computer, it is considered a failure of the entire parallel computer, and an identical backup computer is used as a fail-over. This backup system is then used as the primary computer, while the failed parallel computer is repaired to become the backup for the new primary eventually.

If, however, it were possible to isolate the failed processor and remap and rebind the processes on other processors in that computer — in real time — it would then be possible to have only a number of redundant processors in the system rather than entire redundant parallel computers. There are two strategies for determining the remapping as well as two strategies for handling the remapping and rebinding; each has its advantages and disadvantages.

To discuss these fail-over strategies, it is necessary to define the concepts of tasks and mappings. A signal processing application can be separated into a series of pipelined stages or tasks that are executed as part of the given application. A mapping is the task-parallel assignment of a task to a set of computer and network resources. In terms of determining the fail-over remapping, it is possible to choose a single remapping for each task or to choose a completely unique secondary path — a new mapping for each task that uses a set of processors mutually exclusive from the processors in the primary mapping path. If task backup mappings are chosen for each task, the fail-over will complete faster than a full processing chain fail-over; however, the rebinding fail-over for a failed task mapping is more difficult because the mappings from the task before and the task after the failed task mapping must be reconfigured to send data to and receive data from the new mapping. Conversely, if a completely unique secondary path is chosen as a fail-over, then fail-over completion will have a longer latency than performing a single task fail-over. However, the fail-over mechanics are simpler because the completely unique secondary path could be fully initialized and ready to receive the stream of data in the event of a failure in the primary mapping path.

In terms of handling the remapping and rebinding of tasks, it is possible to choose the fail-over mappings when the application is initially launched or immediately after a fault occurs. In either case, greater latency is incurred at launch time or after the occurrence of a fault. For these advanced options, support for this fault tolerance comes mainly from the middleware support, which is discussed in the next section, and from the NRM discussed in Section 2.5.

2.4 Middleware

Middleware not only provides a standard interface for communications between network resources and sensors for plug-and-play operation, but also enables the rapid implementation of high-performance embedded signal processing.

2.4.1 Control and Command of System

Because many systems use a diverse set of hardware, operating systems, programming languages, and communication protocols for processing sensor data, the manpower and time-to-deployment associated with integration have a significant cost. A middleware component providing a uniform interface that abstracts the lower-level system implementation details from the application interface is the common object request broker architecture (CORBA) [27]. CORBA is a specification and implementation that defines a standard interface between a client and server. CORBA leverages an interface definition language (IDL) that can be compiled and linked with an object's implementation and its clients. Thus, the CORBA standard enables client and server communications that are independent of the host hardware platforms, programming language, operating systems, and so on. CORBA has specifications and implementations to interface with popular communication protocols such as TCP/IP. However, this architecture has an open specification, general interORB protocol (GIOP) that enables developers to define and plug in platform-specific communication protocols for unique hardware and software interfaces that meet application-specific performance criteria.

For real-time and parallel embedded computing, it is necessary to interface with real-time operating systems, define end-to-end QoS parameters, and enact efficient data reorganization and queuing at communication interfaces. CORBA has recently included specifications for real-time performance and parallel processing, with the expectation that emerging implementations and specification addendums will produce efficient implementations. This will enable CORBA to move out of the command and control domain and be included as a middleware component involved in real-time and parallel processing of time-critical sensor data.

2.4.2 Parallel Processing

The ability to choose one of many potential parallel configurations enables numerous applications to share the same set of resources with various performance requirements. What is needed is a method to decouple the mapping, that is, the parallel instantiation of an application on target hardware, from generic serial application development. Automating the mapping process is the only feasible way of exploring the large parameter space of parallel configurations in a timely and cost-effective manner.

MIT Lincoln Laboratory has developed a C++-based library known as the parallel vector library (PVL) [28]. This library contains objects with parameterized methods deeply rooted in linear algebraic expressions commonly found in sensor signal processing. The parameters are used to direct the object instance to process data as one constituent part of a parallel whole. The parameters that organize objects in parallel configurations are run-time parameters so that new parallel configurations can be instantiated without having to recompile a suite of software. The technology of PVL is currently being incorporated into the parallel vector, signal, and image processing library for C++ (parallel VSIPL++) standard library [29].

2.5 Network Resource Management

Given the stated goals for distributed network computing for sensor fusion as outlined in Section 2.3, the associated network communication, storage, and processing challenges in Section 2.3, and the desire for standard interfaces and libraries to enable application parallelism and plug-and-play integration in Section 2.4, an integrated solution is needed that bridges network communications, distributed storage, distributed processing, and middleware. Clearly, it is possible for a development team to implement a

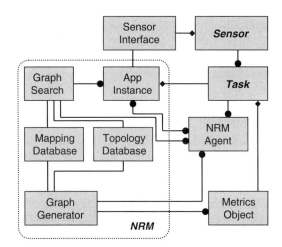

FIGURE 2.6 Object model for network resource manager (NRM).

"point" solution, but this is inherently not scalable and very difficult to maintain. Therefore an additional goal is to fully automate the process of configuring network communication, storage, and computational resources to process data for sensor fusion applications in real time, provide robust fault tolerance in the face of network resource failures, and impart this service in a highly dynamic network in the face of competing interests.

To address these needs, the network resource manager (NRM) was developed. The novelty and potency of the NRM is its capability of taking a sensor signal processing application designed and tested on single target processing element (PE) and mapping it in a task- and a data-parallel fashion across a network of computational resources to achieve real-time performance [30]. Figure 2.6 is an object-oriented model of the components that constitute the NRM. A high-level overview of the NRM follows, and details will be provided in the following subsections. The task of building a model from which the NRM launches parallel applications is broken into three distinct phases:

1. Map generation involves breaking an application into various task- and data-parallel components.
2. Map timing collects performance metric information associated with the components (or tasks) running on host resources. Using the performance metrics, the NRM creates a weighted graph-theoretic view of various permutations of an application mapped in parallel across networked resources.
3. Map selection finds the path through the graph that best meets system and application performance requirements.

The graph generator and graph search objects will heavily leverage PVL (discussed earlier) objects in the instantiation of task- and data-parallel configurations of applications on host resources. It should be noted, however, that the NRM's capabilities are fully general and independent from those of PVL and could work with other applications that are not developed using PVL to instantiate task- and data parallelism.

2.5.1 Graph Generator

As noted previously, PVL uses run-time parameters to generate new parallel configurations. This enables the NRM to launch applications in arbitrary parallel configurations using software developed for a single target PE without having to recompile the application software suite. The central challenge is to select a subset of the potentially astronomical number of permutations of parallel configurations as candidate parallel mappings. It is expected that the NRM will receive guidance in the form of performance and resource utilization bounds to help it avoid choosing undesirable configurations. It will also be given a

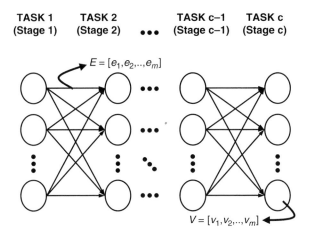

FIGURE 2.7 Sample graph with edge and vertex weights.

series of constituent tasks that comprise an application, so that its primary objective is to choose candidate data-parallel configurations for each of the individual tasks. Using a graph-theoretic model, the application space may be broken up as shown in Figure 2.7.

Each column in the graph is populated with vertices; each vertex corresponds to a mapping of the task corresponding to the given column to a potentially unique set of computational resources in the system. Each vertex has edges entering and exiting: entering edges correspond to communications with preceding tasks and exiting edges correspond to communications with succeeding tasks. Sensor signal processing applications may be represented as a stream signal processing flow, in which data move in one direction from task to task as they are processed. In this graph-theoretic model, task parallelism is represented along the horizontal axis of the graph, i.e., pipelined, overlapping execution intervals, while data parallelism is represented by the mapping of each task in the application onto one or more parallel computational resources of each vertex. The graph-theoretic representation of data- and task-parallel applications and the corresponding flow of communication enable the graph generator of the NRM to capture the potentially astronomical number of combinations of application-to-resource mappings in a concise and efficient fashion.

Finally, the graph generator is also responsible for launching the executable for each task mapping (vertex) on target resources so that performance metrics can be collected as discussed in the next subsection.

2.5.2 Metrics Object

The metrics object (MO) is responsible for collecting performance metrics of tasks launched by the graph generator. The MO works closely with the graph generator to weight the graph. Each of the resources that hosts a task is time synchronized; metric agents (see NRM agents in Subsection 2.5.4) on each of the resources will provide the MO measurements for it to formulate the following performance parameters associated with graph weights: throughput; latency; RAM memory; and PE utilization. The MO will calculate another metric known as processor cost, which is a ratio of compute horsepower used in the mapping to the overall processing horsepower available in the network.

Link utilization percentages within each mapping are also measured, as well as intertask utilization percentages. Map generation uses task column pairs to gather performance metrics in order to reduce the effort and time involved drastically. This is possible because the graph search algorithm will use a running tabulation of resource utilization percentages to ensure that simple linear superposition of path weights hold, given that these percentages remain under a given threshold. This is explained further in the next subsection. Once above the threshold, weight modifiers will be applied to subsequent stages during search. Finally, the metrics object will calculate a *network cost*, analogous to processor cost, which is a ratio of communications bandwidth used by a mapping pair with respect to the overall bandwidth available in the network.

2.5.3 Graph Search

The NRM must choose a path through the graph that determines the task mappings with which an application is launched on network resources. The choice of a path by the NRM is constrained by the time to result and the mandate to use a minimum set of networked resources. The data rate of the sensor data stream will drive required throughput for each task column in the graph; overall latency, which represents the total pipeline delay, is defined as the time period after which all data have been transmitted that a result is generated. To minimize any one application's impact on resource consumption, the path through the graph could be chosen to minimize the overall usage of computational or communication resources. This choice will depend upon whether an application is launched in a network that is compute resource or communication bandwidth limited.

The graph search problem may be formalized as a discrete and constrained optimization problem: given a set of hard constraints, minimize (or maximize) a given objective function. As described in the metrics object subsection, the NRM may choose constraints and an objective function from the set of weights shown in Table 2.1.

Scalar weights are singular — that is, only one is associated with a given vertex or edge; vector weights may include many elements in an edge or vertex association. Because each vertex and edge may represent the combination of many PE and network communication elements associated with a mapping pair, processor and network utilization may constitute weight vectors with many elements.

Although all weights tabulated previously may be chosen as constraints, memory, throughput, and network and PE utilization are not parameters that can be chosen as an objective function to optimize. This is because throughput is only a function of data rate; maximizing throughput has no impact on performance. Utilization also has no impact on performance and is only a measure of the validity of the solution. That is, subsequent stages in the graph may include resources from earlier stages, so keeping a running tabulation of utilization gives an indication of the onset of usage exceeding capacity and thereby degrading performance.

Network utilization and cost, PE utilization and cost, and memory are weights derived and constrained by the NRM, while data rate (throughput) and latency are application dependent and imposed by the sensor. The objective function that the NRM uses is chosen based on the desire to minimize an application's impact on resource usage or minimize the latency associated with an application's execution. For example, in a bandwidth-limited network, the graph search problem may be formulated as follows. While meeting application latency and throughput constraints, using less than 80% of the bandwidth available in the chosen network conduits and PEs and less than 100% of the available local PE-RAM memory, and using only a fraction of the overall processing bandwidth available network wide, select a parallel configuration for the

TABLE 2.1 Graph Weights Associated with Individual Edges and Vertices, and Corresponding Sizes (Types)

Weight	Type
Latency	Scalar
Throughput	Scalar
PE utilization	Vector
Processor cost	Scalar
Network utilization	Vector
Network cost	Scalar
Memory	Scalar

application and the associated host resources using the smallest fraction of overall network bandwidth available. Even for moderately sized graphs (e.g., 1000 vertices by 10 stages), this is a complex combinatorial optimization problem; the general problem is NP complete. The authors have developed an iterative heuristic algorithm that has shown favorable performance for this class of problem in the quality of the solution and time to solution compared to other popular combinatorial optimization algorithms [31].

2.5.4 NRM Agents

The NRM agents are information and service links between the NRM and each of the resources. Agents must first register and be authenticated (e.g., using Kerberos [32]) before an NRM will invoke their services. This registration includes a characterization of the resource capabilities and services. When registered, the NRM will use these remotely deployed agents on computational resources to download and launch parameterized executables and modify the access control list (ACL) of switches and routers under its control in the formation of DPNs. Agents also provide a mechanism for centralized software maintenance and configuration by acting as transaction managers in the download and installation of applications, databases, middleware, etc. As stated earlier, the agents also provide a measurement object that is instantiated by applications to provide the NRM's MO with performance metrics during graph generation. Finally, agents give the NRM a view of the network state, periodically sending diagnostic messages indicating its operational status.

2.5.5 Sensor Interface

Sensors can be thought of as resources much like computational and communication resources, which are served by the NRM agents; thus, the sensor interface can be thought of as another type of NRM agent. Because many different sensor platforms could be served by an NRM-managed resource network, the sensor interface provides a common, abstract mechanism for communication between the NRM and the sensor platforms.

Sensors will request services through the sensor interface from the NRM using a well-defined middleware interface such as CORBA. This request for services involves requesting the proper application for the data stream that the sensor will be delivering to the network of resources as well as a request for the required metric constraints, such as throughput and latency (discussed in Subsection 2.5.2), needed to process the sensor data stream effectively. The determination of required constraints could involve negotiations between the sensor and the NRM through the sensor interface. The NRM uses the sensor interface to direct the sensor platform to start sending a data stream once the NRM has marshaled the resources that the sensor will need to satisfy the request. Finally, the sensor interface also facilitates communications between the sensor platform and the NRM regarding flow control, application shutdown, etc.

2.5.6 Mapping Database

This mapping database is populated with data structures generated by the graph generator and metrics object; it represents the weighted graph-theoretic characterization of the various parallel permutations of an application that is mapped to networked resources. Graph search uses the mapping database to reconstitute a weighted graph for each application for which it is asked to find resources and the degree and form of parallelism needed to meet real-time constraints.

2.5.7 Topology Database

The topology database stores the current state of each of the resources; the graph generator and graph search use this database. Graph generator uses the topology database to determine which resources are available and most appropriate for candidate task-application mappings. Graph search uses this database to verify that resources are functional before a set of resources is chosen to host an application, as well as for generating and modifying weights associated with resource utilization. The topology database is generated during the discovery phase when the NRM first comes online (e.g., see Breitbart et al. [33] and Astic and Foster [34]). Alternatively, an administrator could choose to generate a topology database for the NRM that enumerates connectivity and capability among all computation and storage resources under its control. Agent reports (or lack thereof) will affect state changes in this database indicating whether the resource is online or offline.

2.5.8 NRM Federation

In a large network with a sizeable number of resources, using a single NRM may not be the most effective solution. In such a scenario, multiple NRMs are organized in a bilevel hierarchy; wide-area network (WAN) NRMs interface with sensors and administer backbone communication resources, underneath which local-area network (LAN) NRMs administer and allocate compute resources for regional compute centers (RCCs). The primary responsibility of a WAN NRM is to choose a location on the network at which distributed computing is conducted for each application and to allocate WAN bandwidth for data flow between sensors and LAN resources. The objective of the WAN NRM is to load balance WAN traffic and computational load, taking into account the relative overall processing capability of each RCC. Each LAN NRM advertises its current processing capability using standardized metrics.

Each NRM is a federated collection, using a voting mechanism to elect an executor independently at the LAN and WAN levels. Each federation monitors the health of its executor by inspecting periodic diagnostic reports that the executor broadcasts. In response to an executor's diagnostic report (or lack thereof), the federation may choose to relieve the current executor of its responsibility and elect a new one. This prevents any one NRM failure from rendering resources unusable or disabling a sensor from contracting for network services.

Earlier paragraphs have detailed the LAN NRMs graph-theoretic representation of network resources, as well as its construction, weighting, and search criteria. The WAN NRM graph-theoretic representation and weighting are somewhat different from that of a LAN NRM; however, its construction and search criteria are formulated in an identical manner. The vertices in a WAN graph represent RCCs and each column corresponds to an application, while the concatenation of applications across the columns in a WAN NRM graph spans a mission. This is in contrast to a LAN NRM, in which the concatenation of tasks in its graph spans an application.

2.5.9 NRM Fault Tolerance

The absence of a heartbeat or the delivery of an error report by an agent alerts the NRM to a system fault. The NRM's fault tolerance policy is application dependent and is derived from a mandate by the developer and/or client. The policy is a trade-off between resource usage and seamless fail-over and includes redundant processing, surgical replacement, or restart of the application. Redundant processing is the most robust fail-over mechanism; the NRM simply assigns duplicate sets of resources to process the same data. If one set of resources fails, results are obtained from one of the duplicate sets. Redundant processing has the highest resource cost of all fault tolerant policies.

Conversely, the NRM may choose to replace the failed component dynamically so that processing is able to continue. In this case, the NRM may have allocated distributed network storage to act as a time-delay buffer in the event of resource failure. This would enable the application, if so instrumented, to pick up processing at the point at which the failure occurred. Finally, the NRM could simply choose to halt execution of the application and start over with a new set of processing resources, although a certain amount of data and the corresponding results may be lost irrevocably.

2.6 Experimental Results

A proof-of-concept experiment has been conducted at MIT Lincoln Laboratory in which the NRM allocates distributed networked resources for a sensor data fusion application in various scenarios [35].

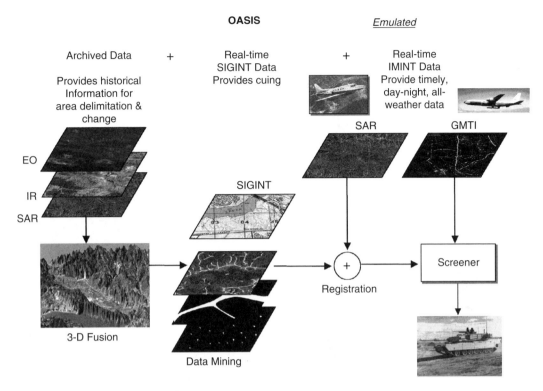

FIGURE 2.8 OASIS ATR and visualization.

TABLE 2.2 Synopsis of NRM Expected Performance

Experimental Configuration	Max Comm BW Requirement (MB/s)	Max Throughput Requirement (GFLOPS)	Processors Employed	Result Turn-Around Time
1 m data	26	0.7	1	1.6
1 m data with HDVI	26	2.2	2	2.6
1/4 m data	410	2.5	2	2.8
1/4 m data with HDVI	410	10	10	7

TABLE 2.3 Synopsis of NRM Performance

Experimental Configuration	Comm BW Measured (MB/s)	Throughput Measured (GFLOPS)	Processors Employed	Result Turn-Around Time
1 m data	26	0.7	1	1.4
1 m data with HDVI	26	2.2	2	2.5
1/4 m data	410	2.5	2	2.7
1/4 m data with HDVI	410	10	8	7.8

The sensor fusion application is OASIS (operator assisted integrated systems), which is an automatic target recognition and visualization suite (see Figure 2.8). OASIS processes real-time SAR data and archived data generated by sensors with different modalities like EO and IR [36]. A block diagram of the

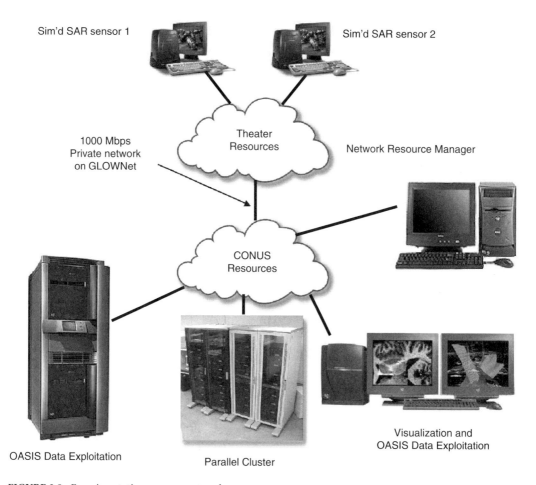

FIGURE 2.9 Experimentation resource network.

experimental test bed is shown in Figure 2.9. The experimentation resource network consisted of three SGI O2 workstations, an eight-processor SGI Origin, an eight-node, dual Pentium3 class Beowulf cluster, and a PC workstation, which hosted the NRM.

For this experiment, two SGI O2s were used as sensor surrogates to transmit unprocessed complex SAR imagery generated with range and cross-range resolutions of 1 and 1/4 m, respectively. The sensor surrogates fed data into the OASIS processing chain. To keep the complexity of the system manageable, only the most computationally intensive stage was made remappable. This stage, the HDVI processing [3] (stage 3 in Figure 2.10), had six options for the NRM ranging from a single SGI processor to six Pentium3 class cluster processors. The HDVI processing was conducted on targets detected on the two images at both resolutions, and image formation was conducted on processors in the local area network. The performance metrics for the OASIS applications were determined with a combination of actual performance measurements and modeled performance analyses. Table 2.2 is a tabulated synopsis of the expected performance of the NRM and Table 2.3 shows the actual performance of the NRM. The expected and actual performance values compared very well.

Because this network was PE resource limited, the objective of the NRM was to use the smallest fraction of PE bandwidth available across the network while meeting network conduit, PE utilization, latency, throughput, and network-wide bandwidth usage constraints. It is clear from the results that the NRM was able to tailor the communication and computation solution it delivered based on the particular

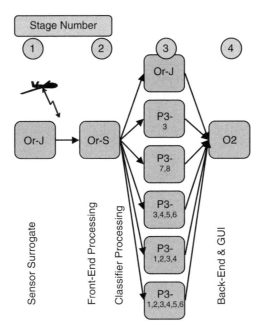

FIGURE 2.10 Graph of OASIS application onto the experimental resources.

application needs and the constraints imposed. The successful completion of this experiment has initiated further research and development to give the NRM greater functionality, automation, and flexibility.

Acknowledgments

The authors thank the members of the Precision Targeting via Collaborative Networking team at MIT Lincoln Laboratory for formulating many of the concepts discussed in this chapter. The authors also thank Dr. Mari Maeda, formerly of DARPA/ITO, and Dr. Gary Koob of DARPA/IPTO for their encouragement and support of this project.

References

1. Usoff, J., Beavers, W., and Cox, J., Wideband networked sensors processing, in *Proc. High Performance Embedded Computing Workshop*, November 2001.
2. Cuomo, K.M., Pion, J.E., and Mayhan, J.T., Ultrawide-band coherent processing, *IEEE Trans. Antenna Propagation*, 47, 1094, June 1999.
3. Benitz, G.R., High-definition vector imaging, *MIT Lincoln Lab. J.*, Special Issue Super-Resolution, 10:2, 147, 1997.
4. Nguyen, D.H. et al., Super-resolution HRR ATR Performance with HDVI, *IEEE Trans. Aerospace Electron. Syst.*, 37:4, 1267, October 2001.
5. Chan, V.W.S., Optical space communications, *IEEE J. Selected Topics Quantum Electron.*, 6:6, 959, November/December, 2000.
6. Awduche, D. et al., RSVP-TE: extensions to RSVP for LSP tunnels, RFC 3209, http://www.faqs.org/rfcs/rfc3209.html, December 2001.
7. Ash, J. et al., Applicability statement for CR-LDP, RFC 3213, http://www.faqs.org/rfcs/rfc3213.html, January 2002.
8. Cormen, T.H., Leiserson, C.E., and Rivest, R.L., *Introduction to Algorithms*. McGraw–Hill, New York, 1993.

9. Strater, J. and Wollman, B., OSPF modeling and test results and recommendations, Mitre Technical Report 96W0000017, Mitre Corporation, 1996.
10. Perkins, C., *Ad Hoc Networking*, Addison–Wesley, Boston, 2001.
11. Floyd, S. and Jacobson, V., Random early detection gateways for congestion avoidance, *IEEE/ACM Trans. Networking*, 1:4, 397, August 1993.
12. Stevens, W., TCP slow start, congestion avoidance, fast retransmit and fast recovery algorithms, RFC 2001, http://www.faqs.org/rfcs/rfc2001.html, January 1997.
13. Stadler, J.S., Performance enhancements for TCP/IP on a satellite channel, in *Proc. IEEE Military Commun. Conf. 1998 (MILCOM98)*, 1, 270, October 1998.
14. Allman, M., Glover, D., and Sanchez, L., Enhancing TCP over satellite channels using standard mechanisms, RFC 2488, http://www.faqs.org/rfcs/rfc2488.html, January 1999.
15. Berrou, C., Glavieux, A., and Thitimajshima, P., Near Shannon limit error-correcting coding and decoding: turbo codes. 1, in *Conf. Rec. IEEE Int. Conf. Commun. 1993 (ICC 93)*, 2, 1064, May 1993.
16. Foschini, G.J., Layered space-time architecture for wireless communication in a fading environment when using multiple antennas, *Bell Labs Tech. J.*, 1:2, 41, Autumn 1996.
17. Raleigh, G.G. and Cioffi, J.M., Spatio-temporal coding for wireless communications, in *Proc. IEEE Global Telecommun. Conf. 1996 (GLOBECOM 96)*, 3, 1405, November 1996.
18. Sisterson, L.K. et al., An architecture for semi-automated radar image exploitation, *Lincoln Lab. J.*, 11:2, 175–204, 1998.
19. Cooley, J.A. et al., Software-based erasure codes for scalable distributed storage, in *Proc. 20th IEEE Symp. Mass Storage Syst.*, 157–164, April 2003.
20. Luby, M.G. et al., Practical loss-resilient codes, in *Proc. 29th ACM Symp. Theory Computing*, 150–159, 1997.
21. Byers, J.W., Luby, M.G., and Mitzenmacher, M., Accessing multiple mirror sites in parallel: using tornado codes to speed up downloads, in *Proc. IEEE INFOCOM 1999*, 275–283, March 1999.
22. Silberschatz, A. and Galvin, P., *Operating System Concepts*, 5th ed., Addison–Wesley, Reading, MA, 1998.
23. Wind River Systems, Inc. http://www.windriver.com/, accessed July 2003.
24. FSMLabs (Finite State Machine Labs), Inc. http://www.fsmlabs.com/, accessed July 2003.
25. Mercury Computer Systems, Inc. http://www.mc.com/, accessed July 2003.
26. Abbott, D., *Linux for Embedded and Real-Time Applications*, Newnes, Amsterdam, 2003.
27. Object Management Group. http://www.omg.org/, accessed July 2003.
28. Hoffmann, H., Kepner, J., and Bond, R., S3P: Automatic, optimized mapping of signal processing applications to parallel architectures, in *Proc. High Performance Embedded Computing Workshop 2001*, September 2001.
29. The vector, signal, and image processing library. http://www.vsipl.org/, accessed July 2002.
30. Reuther, A.I. and Goodman, J.I., Resource management for digital signal processing via distributed parallel computing, in *Proc. High Performance Embedded Computing Workshop 2002*, September 2002.
31. Goodman, J.I. et al., Discrete optimization using decision-directed learning for distributed networked computing, in *Proc. IEEE Asilomar Conf. Signal, Syst. Computers*, 1189–1196, November 2002.
32. Neuman, B.C. and Ts'o, T., Kerberos: an authentication service for computer networks, *IEEE Commun.*, 32:9, 33, September 1994.
33. Breitbart, Y. et al., Topology discover in heterogeneous IP networks, in *Proc. IEEE INFOCOM 2000*, 265–274, March 2000.
34. Astic, I. and Foster, O., A hierarchical topology discovery service for IPv6 networks, in *Proc. 2002 Network Operations Manage. Symp.*, 497–510, April 2002.

35. Reuther, A.I. and Goodman, J.I., dynamic resource management for a sensor-fusion application via distributed parallel grid computing, in *Proc. High Performance Embedded Computing Workshop 2003*, 2003.
36. Avent, R.K., A multi-sensor architecture for detecting high-value mobile targets, in *Proc. 2002 SIAM Conf. Imaging Sci. (IS02),* March 2002.

3
Sensor Network Management

Linnyer Beatrys Ruiz
Pontifical Catholic University of Paraná and Federal University of Minas Gerais

José Marcos Nogueira
Federal University of Minas Gerais

Antonio A. F. Loureiro
Federal University of Minas Gerais

3.1 Introduction 3-1
3.2 Management Challenges... 3-2
3.3 Management Dimensions 3-3
   Dimensions for WSN Management • Management Levels • WSN Functionalities • Management Functional Areas
3.4 MANNA as an Integrating Architecture 3-15
   Management Services, Functions, and Models • Functional Architecture • Information Architecture • Physical Architecture
3.5 Putting It All Together 3-25
3.6 Conclusion 3-25

3.1 Introduction

A wireless sensor network (WSN) consists of a large number of sensor nodes deployed over an area and integrated to collaborate through a wireless network. WSNs encourage several novel and existing applications such as environmental monitoring; health care; infrastructure management; public safety; medical; home and office security; transportation; and military [1, 2, 9, 17, 18]. These have been enabled by the rapid convergence of three technologies: digital circuitry, wireless communications, and the microelectromechanical system (MEMS). These technologies have enabled very compact and autonomous sensor nodes, each containing one or more sensor devices, computation and communication capabilities, and limited power supply.

Some of the applications foreseen for WSNs will require a large number of devices in the order of tens of thousands of nodes. Traditional methods of sensor networking represent an impractical, complex, and expensive demand on cable installation. WSNs promise several advantages over traditional sensing methods in many ways: better coverage, higher resolution, fault tolerance, and robustness. The ad hoc nature and deploy-and-leave vision make it even more attractive in military applications and other risk-associated applications, such as catastrophe, toxic zones, and disasters [2, 9]. Performing the processing at the source can drastically reduce the computational burden on application, network, and management. On the other hand, any solution must take into account specific characteristics of this type of network.

WSN management must be autonomic, i.e., self-managed (self-organizing, self-healing, self-optimizing, self-protecting, self-sustaining, self-diagnostic) with a minimum of human interference, and robust to changes in network states while maintaining the quality of services [ ]. Until now, WSNs and their applications have been developed without considering an integrated management solution. The task of building and deploying management systems in environments that will contain tens of thousands of network elements with particular features and organization and that deal with the aforementioned attributes is not trivial. This task becomes more complex due to the physical restrictions of the unattended sensor nodes, in particular energy and bandwidth restrictions.

In this chapter, the focus is on WSN management, which comprises a large number of devices in the order of tens of thousands of nodes. Clearly, the mechanisms associated with traditional management paradigms must be rethought. In this sense, a new paradigm called autonomic management is explored. The rest of this chapter is organized as follows. Section 3.2 presents an overview of network management and discusses the management challenges for WSNs. In Section 3.3, management dimensions (management levels, WSN functionalities, and management functional areas) are presented and discussed. A management architecture for WSNs called MANNA is presented in Section 3.4, as well as how it works. In Section 3.5, a simple example shows the different aspects together. Finally, Section 3.6 presents conclusions.

3.2 Management Challenges

One of the major goals of network management is to promote productivity of network resources and maintain the quality of the service provided. However, the management of traditional networks and of WSNs has several significant differences. This section discusses important characteristics of WSNs that make their management different from that of other networks.

A WSN is a tool for distributed sensing of one or more phenomenon that reports the sensed data to one or more observers. A WSN provides services for observers as well as for itself. It produces and transports application data, so, in this sense, the network provides service to itself. The objective of a WSN is to monitor and, eventually, control a remote environment. Sensor nodes execute a common application in a cooperative way (i.e., a clear, common goal in the overall network), which may not be the case in a traditional network.

The traditional computer networks are designed to accommodate a diversity of applications. Network elements are installed, configured by technicians, and connected in a network in a way to provide different kinds of services. Technicians' maintenance of components or resources is a normal fact. The network tends to follow well-established planning of available resources and the location of each network element is well-known. In a WSN this is not often the case because the network is planned to have unattended operation. In fact, the initial configuration of a WSN can be quite different from what was supposed to be in cases such as throwing the nodes into an ocean, forest, or other remote regions. In unpredictable situations, a configuration error such as a planning error may cause the loss of the entire network even before it starts to operate.

Energy is a critical resource in WSNs. Thus, all operations performed in the network should be energy efficient. Topology is dynamic because sensor nodes can become out of service temporarily or permanently (nodes can be discarded, lost, destroyed, or even run out of energy). In this scenario, faults are a common fact, which is not expected in a traditional network.

Depending on the WSN application, it may be interesting to identify uniquely each node in the network. Furthermore, one may be interested in a value associated to a given region and not to a particular node — for instance, in the temperature at the top of a mountain. A WSN is typically data centric, which is not common in traditional networks.

A managed WSN is responsible for configuring and reconfiguring under varying (and, in the future, even unpredictable) conditions. System configuration ("node setup" and "network boot up") must occur automatically; dynamic adjustments need to be done to the current configuration to best handle changes in the environment and itself. A managed WSN always looks for ways to optimize its functioning; it will monitor its constituent parts and fine-tune workflow to achieve predetermined system goals. It must perform something akin to healing — it must be able to recover from routine and extraordinary events that might cause some of its parts to malfunction. The network must be able to discover problems or potential problems, such as uncovered area, and then find an alternate way of using resources or reconfiguring the system to keep it functioning smoothly. In addition, it must detect, identify, and protect itself against various types of attacks to maintain overall system security and integrity. A managed WSN must know its environment and the context surrounding its activity and act accordingly. The management entities must find and generate rules to perform the best management of the current state of the network [22].

A managed WSN with this has various characteristics can be called an autonomic system [1], which is an approach to self-managed computing systems with a minimum of human interference. This term derives from the autonomic nervous system of the human body, which controls key functions without conscious awareness or involvement. The processors in such systems use algorithms to determine the most efficient and cost-effective way to distribute tasks and store data. Along with software probes and configuration controls, computer systems will be able to monitor, tweak, and even repair themselves without requiring technology staff — at least, that is the goal [1].

WSN management must be autonomic, i.e., self-managed and robust to changes in network states while maintaining the quality of service; that is, it must be capable of self-configuration, self-organization, self-healing, and self-optimization. However, the computational cost of autonomic processes can be expensive to some WSN architectures.

Probably, the fundamental issue about the management of a WSN is concerned with how the management can promote plant and resource productivity, and how it integrates in an organized way functions of configuration, operation, administration, and maintenance of all elements and services.

The task of building and deploying autonomic management systems in environments in which tens of thousands of network elements with particular features and organization will be present is very complex. This task becomes even more involved due to the physical restrictions of the sensor nodes, in particular energy and bandwidth restrictions. The management application to be built also depends on the kind of application being monitored. A good strategy is to deal with complex management situations by using management dimensions.

3.3 Management Dimensions

In general, for traditional networks, management aspects are clearly separated from network common activities, i.e., from the services they provide to their users. It is also said that an overlap of management and network functionalities exists, although the implementation can be thought of independently. This separation can be promoted by using two traditional management dimensions: management functional areas [14] and management levels [15].

The requirements to be satisfied by systems management activities can be categorized into functional areas. These facilities have come to be known as the specific management functional areas (SMFAs): fault management; configuration management; performance management; accounting management; and security management. This has proved to be a helpful way of partitioning the network management problem from an application point of view [14].

To deal with the complexity of management, management functionality with its associated information can be decomposed into a number of logical layers: business management; service management; network management; and network element management. The architecture that describes this layering is called the logical layered architecture (LLA) [15]. Management activities can be clustered into layers and decoupled by introducing manager and agent roles. A logical layer reflects particular aspects of management and implies the clustering of management information supporting that aspect. Typically, an interaction takes place between adjacent layers, but due to operational and management considerations other interactions may also occur between nonadjacent layers.

The use of the management dimensions is a good strategy to deal with complex management situations by decomposing a problem into smaller subproblems, in successive refinements steps, and to provide a separation between application and management functionalities through a management architecture. This will make possible the integration of organizational, administrative, and maintenance activities for a given network.

WSN management must be simple, adherent to network idiosyncrasies, including its dynamic behavior, and efficient in its use of scarce resources. The adoption of a strategy based on the traditional framework of functional areas and management levels will permit management integration in the future. However, for WSN management it is necessary to go further. Using management functional areas and management levels is not enough because WSNs are application specific.

The following discussion concerns how the traditional management dimensions can be applied in WSN management. Also, new dimension for WSN management is proposed that considers the general aspects of the different types of the networks.

3.3.1 Dimensions for WSN Management

WSNs are embedded in applications to monitor the environment and act upon it. Thus, the management application should try to be "compatible" with the kind of application being monitored. In order to have better development of WSN management services and functions, it is necessary to characterize the WSN and establish a novel management dimension. Thus, looking at the characteristics of various WSN applications, five main WSN functionalities are identified: configuration; sensing; processing; communication; and maintenance. These functionalities define a novel dimension for the management, as presented in Figure 3.1[22]. Configuration is the first functionality before a network starts sensing the environment, processing, and communicating data. Maintenance treats specific characteristics of WSN applications during the entire network lifetime.

In this way, WSN management will have an organization that comes from abstractions offered by management functional areas, management levels, and WSN functionalities (configuration, sensing, processing, communication, and maintenance). The novel dimension introduced can be observed in the upper part of Figure 3.1.

The coordination among the three planes can be based on policies. Policy-based network management (PBNM) [7] is a feasible alternative because it allows the manager to set actions to be carried out by the network without worrying too much about network details. Managers can define suitable actions in due time and still have a global or local view of the network. PBNM helps to manage complex networks such as WSNs. The managers will only inform concerning what is expected, but not how it should be obtained. The agents will be intelligent to decide what to do as well as how and when to do it. Automatic services and functions can be executed toward self-management if appropriate conditions, such as residual energy level, are present.

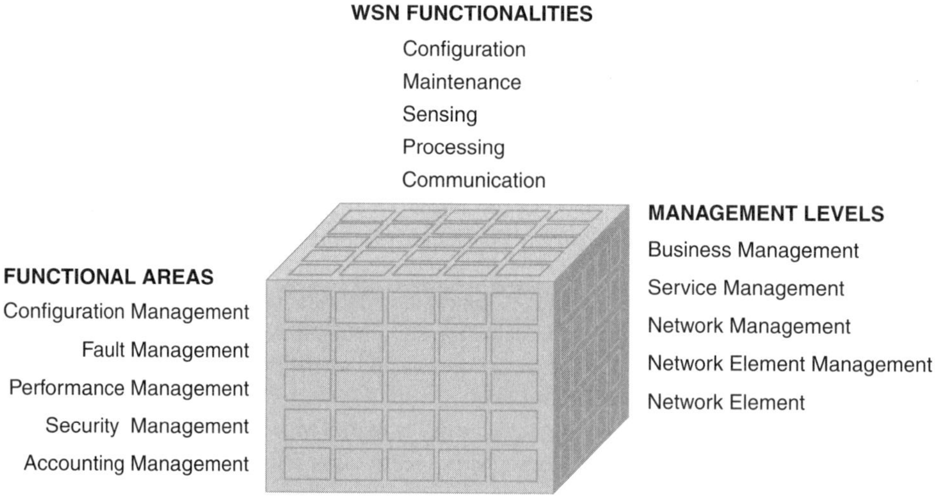

WSN FUNCTIONALITIES
Configuration
Maintenance
Sensing
Processing
Communication

FUNCTIONAL AREAS
Configuration Management
Fault Management
Performance Management
Security Management
Accounting Management

MANAGEMENT LEVELS
Business Management
Service Management
Network Management
Network Element Management
Network Element

FIGURE 3.1 Management dimensions for WSNs. (From Ruiz, L.B., Nogueira, J.M., Louriero, A.A., *IEEE Commun. Mag.*, 41(2), 116–125, 2003. With permission.)

Three management dimensions must be considered in the definition of a management function, establishment of an information model, service composition, and development of a management application. The next subsections explain WSN management from the perspective of management level, WSN functionalities, and management functional areas.

3.3.2 Management Levels

Many traditional management systems use this model in a bottom-up approach; however, in WSN management, the LLA model is used in a top-down approach. After analyzing the business level issues, the necessities of the lower levels become clear. Similarly, it is only after defining the application, including the corresponding requirements on the service layer, that one can plan the network, network element management layers, and network elements. This is a key observation when reasoning about WSN management. A brief discussion concerning WSN management from the perspective of management level is now presented.

3.3.2.1 Business Management

Requirements that allow the characterization of a sensor network come from the objectives defined for the business management layer. Because WSNs depend on applications, business management deals with service development and determination of cost functions. It represents a sensor network as a cost function associated with network setup, sensing, processing, communication, and maintenance. WSN applications have enormous potential benefits for society as a whole and represent new business opportunities. Instrumentation of environments [2, 9] with numerous networked sensor nodes can enable long-term data collection at scales and resolutions that are difficult, if not impossible, to obtain otherwise. In the future, one can expect to have Internet end-points equipped with a variety of sensors to monitor the network and their own state, as well as fairly sophisticated computing capabilities to enable them to function as decision elements and not just as repeaters. As more aspects of society are connected to networks, their sensory components become more prominent.

3.3.2.2 Service Management

A WSN is used to monitor and, sometimes to control, an environment. WSN service management introduces new challenges due to scarce network resources, dynamic topology, traffic randomness, energy restriction, and a large amount of network elements. WSN services are concerned with functionalities (see Figure 3.1) associated with application objectives. Basic WSN services are sensing, processing, and data dissemination [21]. Two main issues are associated with WSN service management: quality of service (QoS) and denial of service (DoS).

Quality of service. QoS architectures can only be effective and provide guaranteed services if QoS elements can be adequately configured and monitored; mechanisms can be defined to help managers to deal with these elements. Also, such mechanisms must allow replacement of the current device-oriented management approach by a network-oriented or cluster-oriented approach. Thus, in addition to the management of elements (physical and logical resources), management applications must also manage QoS aspects. Components involved in QoS support to WSNs include QoS models, QoS sensing, processing, and QoS dissemination [22]. The larger the number of monitored QoS parameters is, the larger the energy consumption and the lower the network lifetime are.

QoS model. A QoS model specifies an architecture in which some of the services can be provided in WSNs. All other QoS components, such as QoS sensing, QoS processing, and QoS dissemination (e.g., signaling, QoS routing, and QoS MAC), must cooperate to achieve this goal. A management application can establish the QoS model and can control the QoS signaling that coordinates behavior of the other components. QoS-related tasks must be performed by using network management functions.

QoS sensing. QoS sensing considers the sensor device calibration, environment interference monitoring, and exposure (time, distance, and angle between sensor device and phenomenon). Meguerdichian [18] defines coverage area as a measure of QoS for a WSN. In the worst-case coverage, attempts are made to quantify the quality of service by finding areas of low observability to sensor nodes and detecting breach regions. In the best-case coverage, the management application must find areas of high observability to sensors and identify the highest accuracy. A denser network will lead to more effective sensing because of the higher accuracy of the network (e.g., areas of intersection and redundant information) and better fault tolerance.

On the other hand, this will lead to a large number of collisions and potentially to congestion situations, increasing latency and reducing energy efficiency. Congestion control must be based not only on the capacity of the network, but also on the accuracy level required at the observer. The traffic in a WSN is different from conventional networks: it is a collective communication operation with redundancy. Thus, the management application has the flexibility of meeting the performance demands by controlling the reporting rate of sensors, controlling the virtual topology of the network (by turning off some sensors), or optimizing the collective reduction communication operation (by data aggregation). The provision of QoS can rely on resource reservation. When an active node goes out of service due to operational problems, the management application activates a redundant node, defining a sort of resource reservation scheme. In case of a low density of sensors, the network coverage area can be committed, thus affecting the quality of the service. Resource reservation is being applied.

QoS dissemination. Reliable data delivery is still an open issue in the context of WSNs. QoS dissemination in WSNs is a challenging task because of constraints, mainly energy and dynamic topology of WSNs. The two components for QoS dissemination are QoS routing and QoS medium access control (MAC). QoS routing finds a path that satisfies a given QoS requirement, and QoS MAC solves the problem of medium contention that supports reliable unicast communication [29]. To support QoS, a link state information such as delay, bandwidth, cost, loss rate, and error rate in network should be available and manageable. One of the objectives of the management application is to obtain and to manage link state information in WSNs for monitoring QoS. This is very difficult because the quality of a wireless link is apt to change with the circumstances, such as residual energy, node distribution, density (all change along the network lifetime), and interference. Configuration characteristics such as coverage area, density, network organization, node deployment (distribution), latency, and communication range may degrade or deny the service.

QoS processing. Processing quality depends on the robustness and complexity of the algorithms used, as well as processor and memory capacities. The computing paradigm changes from one based on computational power to one driven by data. The way to measure processing performance changes from processor speed to the immediacy and accuracy of the response and energy consumption. Individual computers become less important than lower granularity and dispersed computing attributes.

The network quality of service can be measured by the energy consumption to execute a service with a determined quality level. In most WSNs, energy consumption is one of the main metrics. However, in some situations, during certain events the network must apply the maximum of energy possible in the delivery of information — for instance, in WSNs deployed over the havoc of a cave-in where as much information as possible is needed in the shortest time period. In this kind of application, to extend the network lifetime is not that important. However, without proper management mechanisms, the network can suffer the implosion problem (a large amount of data generating congestions, collisions, and data losses in the network).

Any situation that diminishes or eliminates the capacity of the network to perform its expected job is called DoS (denial of service). Some examples of incidental threats are hardware failures, software bugs, resource exhaustion, and unexpected environmental conditions. DoS aspects will be discussed in Subsection 3.3.4.4.

3.3.2.3 Network Management

This layer aims to manage a network, which is typically distributed over an extensive geographical area, as a whole. In the network management level, relationships among sensor nodes are to be considered. It is known that individual nodes are designed to sense, process data, and communicate, thus contributing to a common objective. In this way, nodes can be involved in collaboration, connectivity, and aggregation relationships. A WSN is composed of interconnected managed objects (physical or logical) capable of exchanging information. In these cases, the WSN is basically composed of two parts: physical resources and services. Service execution depends on the physical resource capabilities.

3.3.2.4 Network Element Management

Managed network elements represent the sensor and actuators nodes or other WSN entities, which execute management functions and provide sensing, processing, and dissemination services. The basic functions of a WSN management network element are

- Power management (how a sensor node uses its power)
- Mobility management (how the movement of sensor nodes is planned, run, and registered)
- State management (how a sensor node manages the three management states defined for a node: operational, administrative, and usage)
- Task management (how a sensor node balances and schedules the sensing, processing, and dissemination tasks given to a specific network state)

Each sensor node must be autonomous and capable of organizing itself in the overall community of sensor nodes to perform coordinated activities with global objectives.

Sensor nodes have strong hardware and software restrictions in terms of processing power, memory capacity, battery lifetime, and communication throughput. These are typical characteristics of mobile and wireless devices and not of wired network elements. Thus, software designed for a sensor node must consider these limitations, whereas an element for a wired network may have other restrictions such as performance and response time. The main physical restriction of a WSN is the available energy because batteries are often not recharged during the operation of a sensor node and all activities performed by the node must take energy consumption into account.

3.3.2.5 Network Element

The network element represents physical and logical components of a managed element. Physical resources include sensor or actuator nodes; power supply; processor; memory; sensor device; and transceiver. Logical resources include communication protocols; application programs; correlation procedures; and network services. Because applications may require networks with a large number of sensor nodes, a network element can deal with a single node component or a group of nodes. In such a case, a manageable element can be a cluster of nodes or a cluster-head node, rather than an individual node. The design of a sensor node is motivated by the need to create an inexpensive device with a small form factor and low power dissipation.

Understanding node capability allows function management to be structured and fine-tuned more efficiently. The physical aspects of a network element are described in the following.

- *Power supply.* Energy consumption patterns of individual nodes and of the entire network must be characterized and profiled. This process yields a better understanding of where to apply trade-offs in the design of the management. The most widely used power supply in a WSN is the battery, which is classified into the following types [23]:
  - Linear model — the battery is considered to be a bucket of energy that is linearly drawn from this bucket by the energy consumers
  - Dependent model — considers the rate at which energy is drawn from the battery to compute the remaining battery lifetime; at high discharge rates, the capacity of the battery is reduced
  - Relaxation model — takes into account a phenomenon seen in real-life batteries in which the battery's voltage recovers if the discharge rate is decreased
- *Computational module.* This module is composed of processor and memory. It is responsible for the collaborative processing between nodes to achieve the levels of service and reliability desired by the observer.

- *Sensor element.* Sensing devices can be classified into three groups: monitors (e.g., magnetometer, light sensor, temperature, pressure, humidity); motion detectors (e.g., accelerometer); and media processing (e.g., audio, video).
- *Transceiver.* The main types of a transceiver are radio frequency (RF), infrared, and optical. RF communication is based on electromagnetic waves with frequencies ranging from tens of kilohertz to hundreds of gigahertz. Of the most important factors in the design of RF communications is the size of the antenna. To optimize transmission and reception, an antenna should be at least $\lambda/4$, where λ is the wavelength of the carrier frequency. In optical laser communication, a transmitting device uses a laser beam to send information. An optical receiver, in the form of a photodiode or charge-coupled device (CCD) array, receives the signal and decodes the data. Optical communications can be classified into two types: passive (the laser signal is generated through a secondary source) and active (the transmitting device generates its own laser signal). A few points should be noted regarding the differences between optical and RF communication. Both forms of communication are based on sending electromagnetic waves through air. To compare RF to optical communication, one must conside the receiving end of the communication system. For both, a trade-off takes place between size and receiving performance [12].
- *Software.* This is used to represent a set of programs and procedures that becomes an autonomous system capable of executing the information processing, relaying, or routing.

3.3.3 WSN Functionalities

This section presents the novel proposed dimension for the WSN management, composed by the configuration, sensing, processing, communication, and maintenance functionalities. These WSN functionalities can be observed in the upper part of Figure 3.1. This novel dimension is obtained from the functional model defined in Reference 22, which presents a scheme to characterize WSNs considering that they are application dependent. Because a management solution depends on the features of the network, this solution must also be proposed considering the type of network. For this reason, WSN functionalities are serviceable in the development of the management application [22].

3.3.3.1 Configuration

This functionality involves procedures related to planning, placement, and self-organization of a WSN. The configuration functionality (predeployment) is related to the:

- Definition of WSN application requirements
- Determination of the monitoring area (shape and dimension)
- Characteristics of the environment
- Choice of nodes
- Definition of the WSN type
- Service provided

In the deployment phase, sensor nodes can be placed by dropping them from a plane, rocket, or missile, and placed one by one by a human or a robot. Any placement approach for sensor nodes must also take into account the expense and difficulty in redeploying nodes. This is chiefly due to the limited life span of nodes and to their generally nonreplaceable power sources [19]. Another problem is the optimal location of the access point (sink node or base station). An inefficient configuration management may adversely affect overall performance.

WSNs are application specific, which means that the configuration functionality changes from one WSN to another. Next, the configuration is discussed considering the possible types of WSN and the other two management dimensions.

Considering the network management level and management functional areas based on configuration functionality, WSNs can be classified in various ways. A WSN is said to be homogeneous when all nodes have the same hardware; otherwise, it is said to be heterogeneous. A WSN is hierarchical when nodes are grouped for the purpose of communication, and flat otherwise. When nodes are stationary, a WSN is static; otherwise it is dynamic. Note that the topology may be dynamic even when nodes are stationary because new nodes can be added to the network or existing nodes can become unavailable. A WSN is symmetric concerning signal transmission when each transceiver has the same transmission range, and asymmetric otherwise. A WSN is said to be regular concerning node placement when its nodes are placed in a grid; it is called irregular when its nodes are randomly distributed, presenting different densities on the monitored area, and it is balanced when its nodes are randomly distributed and present a uniform distribution. Depending on the number of nodes per area unit, a WSN can be sparse or dense.

Considering the network element management level and the management functional areas based on the configuration functionality, the sensor nodes in a WSN are spread over a region and communicate among themselves using point-to-point wireless communication, thus forming an ad hoc network. The nodes are autonomous when they are able to execute location discovery and self-configuration tasks without human intervention, for example, the location discovery. To relay information off the network, sensor nodes are equipped with a wireless communication device (transceiver). A wireless sensor node also comprises one or more sensor elements, and a battery, memory, and processor. The size of a node is an important consideration. Nodes need to have small form factors so that they may be located unobtrusively in the environment targeted for monitoring. The restriction in size is closely related to the amount of energy available to a node. A rugged and robust construction is required if nodes are dispersed in an inhospitable terrain such as a forest.

Software developed to execute in a wireless sensor node must take into account its hardware restrictions. Because of limited energy capacity, nodes are expected to be thrown away once their energy supply is exhausted. The system can have levels of redundancy built into it to allow failures or to increase accuracy. This can be achieved by using more sensor nodes than are strictly necessary to cover an area. Also, due to environmental nature, logistics, and deploying costs, the deployment of sensors can be a one-time operation; therefore, after nodes have been distributed in the field, human intervention is not an option. The three basic different types of sensor nodes are: common nodes responsible for collecting sensing data; sink nodes (monitoring nodes) responsible for receiving, storing, and processing data from common nodes; and gateway nodes that connect sink nodes to external entities called observers. WSNs can also include actuators that enable control of or actuation in a monitored area. In a hierarchical network, it is common to have a base station (BS) that works as a bridge to external entities.

Considering the service management level and the management functional areas, the WSN comprises three entities: observer, phenomenon, and environment. The observer is a network entity or a final user that wants to have information about data collected, processed, and disseminated by sensor nodes. Depending on the type of application, the observer may send a query to the WSN, and receive a response from it. These queries can be done with or without fidelity. The translation of the query could be performed by the application software or sensor nodes. The WSN may participate in synthesizing the query (e.g., filtering some sensor data or summarizing several measurements into one value), but these procedures are related to the processing functionality. The phenomenon is the entity of interest to the observer that is sensed and optionally analyzed or filtered by the WSN. The observer is interested in monitoring a phenomenon under some latency and accuracy restrictions. A sensor element generates data about a given phenomenon such as temperature, pressure, electromagnetic field, or chemical agents because it can be comprised of different sensor elements.

3.3.3.2 Sensing

The lowest level of the sensing application is provided by the autonomous sensor nodes. An important operation in a sensor network is data gathering. Sensing functionality depends on the type of the phenomenon. Thus, WSNs can be classified in terms of data gathering required by the application as continuous (when sensor nodes collect data continuously along the time), reactive (when they answer to an observer's query or gather data referring to specific events occurring in the environment), and periodic (when nodes collect data according to conditions defined by the application). Some approaches can coexist in the same network; this model is referred to as the hybrid collect model. An example of a continuous phenomenon is temperature and an example of an application in which the phenomenon is moving is a sensor deployed for animal detection. Other examples of phenomena are video; audio; pressure; mechanical stress; humidity; soil composition; luminosity; seismic; and chemical.

Whether gathering is continuous or not, WSNs are defined based on how the data will be transmitted to the observer. The sensing encloses the exposure (time, distance, and angle of phenomenon exhibition at the sensor), calibration, and sensing coverage. Depending on the density of the phenomenon, it will be inefficient if all sensor nodes are active all the time. A model that is well-suited to this case is the Frisbee model [5]. On the other hand, redundancy (overlapping in the sensor coverage) should be utilized in such a way that fault tolerance in the communication network is avoided and better accuracy can be found [26]. Nevertheless, the sensors can be mobile. In this case, the sensors are moving with respect to each other and to the observer as well, and they have direction, orientation, and acceleration.

3.3.3.3 Processing

Memory and processor of a sensor node form the computational module, which is a programmable unit that provides computation and storage for other nodes in the system. Depending on the communication constraints of the system, algorithms must be developed that will allow individual nodes or clusters of nodes to share and process data efficiently. The computational module performs basic signal processing (e.g., simple translations based on calibrating data or threshold filters) and dispatches the data according to the application. Processing can also involve correlation procedures such as data fusion, which combines one or more data packets received from different sensors to produce a single packet (data fusion). Data fusion helps to reduce the amount of data transmitted between the sensor nodes and the observer and allows design of a network that delivers required data while meeting energy requirements. Other possible tasks are security processing and data compression.

3.3.3.4 Communication

Individual nodes communicate and coordinate among themselves. Two types of communication are proposed: infrastructure and application. Infrastructure communication refers to the communication needed to configure, maintain, and optimize operation. The configuration and topology of the sensor network may be rapidly changing in the presence of a hostile environment, a large volume of assigned work, and nodes that fail routinely. Conventional protocols may be inadequate to manage such situations; thus, new protocols are required to promote WSN productivity. In a static sensor network, an initial phase of the infrastructure communication is needed to set up the network and an additional communication is needed to perform its reconfiguration. If the sensors are mobile, additional communication is needed for path discovery/reconfiguration.

Application communication (dissemination) relates to the transfer of sensed data (or information obtained from it). The amount of energy spent in transmitting a packet has a fixed cost related to the hardware and a variable cost that depends on the distance of transmission. Receiving a data packet also has a fixed energy cost. Therefore, to conserve energy, short distance transmissions are preferred. Because the access point (sink node or the BS) may be located far away, the cost to transmit data from a given node to the access point may be high. In a homogeneous and flat WSN, the sensor nodes can form a multihop network by forwarding each other's messages, which can provide different connectivity options. In a heterogeneous and hierarchical WSN, the cluster heads can form a single-hop network for reporting aggregated data to the BS. Within a cluster, measured data are sent to the cluster head by the sensor nodes under its control. All nodes in a cluster are identical except in the heterogeneous WSN, where the cluster head has a larger transmission capacity.

In terms of the data delivery required by the application interest, WSNs can be classified as continuous, when sensor nodes collect data and send them to an observer continuously along the time, and as on demand, when they answer an observer's query. A WSN is event driven when sensor nodes send data referring to events occurring in the environment and programmed when nodes collect data according to conditions defined by the application. Some approaches can coexist in the same network; such a model is referred to as the hybrid model. The cost of sending data continuously may lead to a more rapid consumption of the scarce network resources and, thus, shorten resource lifetime.

Multihop wireless capabilities will enable communication and coordination among autonomous nodes in unplanned environments and configurations. At the same time, wireless channels present challenges of dynamic operating conditions, power constraints for autonomously-powered nodes, and complicating interactions between high level behavior and lower level channel characteristics (e.g., increased synchronized communication will significantly degrade channel characteristics).

For any of the preceding models, the communication approach can be classified as:

- Flooding (sensors broadcasting their information to their neighbors, which in turn broadcast these data until they reach the observer)
- Gossiping (sending data to one randomly selected neighbor)
- Bargaining (sending data to sensor nodes only if they are interested)
- Unicast (sensor communicating to the sink node, cluster head, or BS directly)
- Multicast (sensors forming application-directed groups and using multicast to communicate among group members)

A major advantage of flooding or broadcast is the lack of a complex network layer protocol for routing and address and location management.

In a WSN, each sensor node puts its information onto a common medium. This requires careful attention to protocols in hardware and software. In master–slave protocols, one node gives the commands and another node or a collection of nodes executes them. The cluster head is usually the master and the common nodes (sensors and actuators) are slaves. This protocol allows tight traffic control because no node is allowed to transmit unless requested by the master, and no communication is allowed between slaves except through the master (e.g., medium control access protocol using a channel fixed allocation scheme). In a peer-to-peer network, all nodes are created equal. A node can be a master one moment and then be reconfigured at another time. Peer-to-peer configurations offer the greatest flexibility, but they are the most difficult to control. Any node can communicate directly to any other node.

3.3.3.5 Maintenance

Maintenance functionality is used in the WSNs that can configure, protect, optimize and heal themselves without a lot of input from the human operators who have, until now, been required to keep traditional networks up and running. Maintenance detects failures or performance degradations, initiates diagnostic procedures, and carries out corrective actions on the network. Its ability to discover changes in the network state enables the self-management to adapt and optimize the network behavior. Beyond corrective maintenance, the other types of maintenance are: adaptive (the system should adapt to meet the changes); preventive (the system should learn to anticipate the impact of those changes); and proactive (as it gets smarter, the system should learn to intervene so as to preempt negative events). An example of maintenance concerns the density of nodes in the WSN; in case of a high node density, the maintenance can turn off some nodes temporally.

The WSN state (e.g., topology, energy, coverage area) changes frequently. In the case of static networks, changes occur because nodes may become unavailable during operation. This dynamic behavior must be observed. The maintenance depends on the knowledge of the network state. Thus, maintenance functionality is needed to keep the network operational and functional to ensure robust operation in dynamic environments, as well as optimize overall performance. Maintenance provides dependability, the main attributes of which are reliability; availability; safety; security; testability; and performability.

WSNs have important characteristics depending on the application. Some of them are:

- Planning
- Deployment
- Coverage
- Accuracy

- Fidelity
- Density
- Self-organization
- Adaptation
- Location

The points described in this subsection will play an important role in the definition of the management services and functions.

3.3.4 Management Functional Areas

WSN management considers fault, security, performance, and accounting management functional areas extremely dependent on the configuration functional area. In WSNs, all operational, administrative, and maintenance characteristics of the network elements; the network, services; and business; and the adequate execution in the activities of configuration, sensing, processing, communication, and maintenance (as shown in Figure 3.1) are dependent on the configuration of the WSN. An error in the configuration or a forgotten requisite during the planning may compromise all the functionalities of the other areas. This idea is depicted in Figure 3.2, in which the configuration functional area plays a central role. As mentioned before, there are several significant differences in the management of traditional networks and WSNs. In this sense, management functional areas must revisit considering the WSNs features.

3.3.4.1 Configuration Management

Configuration management is a functional area of high relevance in WSN management. Because the objective of a sensor network is to monitor (acquisition, processing, and delivery of data) and, eventually, to control an environment, any problem or situation not anticipated in the configuration phase can affect the offered service. The configuration management must provide basic features such as self-organization, self-configuration, self-discovery, and self-optimization. Some management functions defined for network level configuration management are:

- Requirements specification of the network operational environment
- Monitoring of environmental variations
- Size and shape definition of the region to be monitored
- Node deployment — random or deterministic
- Operational network parameters determination
- Network state discovery
- Topology discovery
- Network connectivity discovery
- Control of node density
- Synchronization
- Network energy map evaluation
- Coverage area determination
- Integration with observer

Some management functions defined for network-element level configuration management are:

- Node programming
- Node self-test
- Node location
- Node operational state
- Node administrative state
- Node usage state
- Node energy level

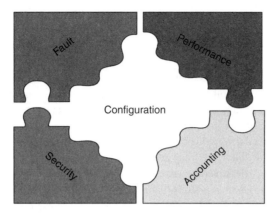

FIGURE 3.2 The role of configuration management. (From Ruiz, L.B., Nogueira, J.M., Louriero, A.A., *IEEE Commun. Mag.*, 41(2), 116–125, 2003. With permission.)

3.3.4.2 Fault Management

Faults in WSNs are not an exception and tend to occur frequently. This is one of the reasons why management of WSNs is different from the traditional network management. Faults happen all the time due to energy shortage, connectivity interruption, environmental variations, and so on. In general, sensor networks must be fault tolerant and robust and must survive despite occurrences of faults in individual nodes, in the network, or even in services provided. In addition to events caused by energy problems, other events can happen in a wireless sensor network related to communication; quality of service; data processing; physical equipment fault; environment; integrity violation; operational violation; security; and time-domain violation. Therefore, even if a node has an adequate energy level to execute its function, it may decide not to do that for other reasons. Fault management must provide basic characteristics such as self-maintenance, self-healing, and self-protection.

Failures will be frequent in a WSN, and fault management is a critical function. Several characteristics of sensor networks suggest that faults, common in traditional computer networks, will be even more common in this kind of network.

- Large-scale deployment of cheap individual nodes means that node failures from fabrication defects will not be uncommon.
- Attacks by adversaries will be likely because these networks will often be embedded in critical applications. Worse, attacks will be made easier because these networks will often be deployed in open spaces or enemy territories, where adversaries can manipulate the environment (so as to disrupt communication by jamming) and also have physical access to the nodes.
- Ad hoc wireless communication by radio frequencies means that adversaries can easily put themselves in the network and disrupt infrastructure functions (such as routing) taken by the individual nodes.

Fault management, an essential component of any network management system, will play an equally, if not more, crucial role in WSNs.

In the majority of applications, failure detection is vital not only for fault tolerance, but also for security. If, in addition to detecting a failure, one can also determine (or gather indications) that it has malicious origin, the observer can be alerted to an attack.

3.3.4.3 Performance Management

The challenge is to perform this task without adversely consuming network resources. In performance management, a trade-off must be considered: the higher the number of managed parameters, the higher the energy consumption and the lower the network lifetime are. On the other hand, if parameter values are not obtained, it may not be possible to manage the network appropriately.

The configuration (in terms of sensor capabilities, number of sensors, density, node distribution, self-organization, and data dissemination) plays a significant role in determining the performance of the network. Performance management must consider the self-service characteristic. As such, the performance of the network and provided service are best measured in terms of meeting the accuracy and delay requirements of the observer, as well as consumed energy.

The accuracy indicates the reliability or exactness of a result; it can also be defined as the fraction of valid results from all results obtained. The accuracy of a measurement at a network element (sensor) is specific to the physical transducer and the nature of the phenomenon. At the network level, accuracy depends on the delay in data delivery due to network congestion, route length, duty cycle of the sensors, or aggregation processing of data. Accuracy at the service level depends on the metric chosen by the application for establishing the coverage area and amount of energy to be spent in gathering and disseminating data. At the observer, it is likely that multiple samples will be received from different sensor nodes and with different data quality. Thus, additional performance metrics include:

- Coverage area
- Exposure
- Goodput (the ratio of the total number of packets received by the observer to the total number of packets sent by all sensors over a period of time [25])
- Sensor cost
- Scalability
- Produced data quality

In some applications, in addition to information about some features of the phenomenon, it might be necessary to know where (sensor location), when (data–time), and how (sensor calibration, exposure) to manage the WSN performance.

Regardless of the application, certain critical features can determine the efficiency and effectiveness of a sensor network [24]. These features can be categorized into quantitative features and qualitative features. Qualitative features include network settling time; network join time; network depart time; network recovery time; frequency of updates (overhead); memory requirement; and network scalability. Qualitative critical features include knowledge of nodal location; effect of topology changes; adaptation to radio communication environment; power consciousness; single- or multichannel; and preservation of network security.

3.3.4.4 Security Management

Security functionalities for WSNs are difficult to provide because of their ad hoc organization, intermittent connectivity, wireless communication, and resource limitations. A WSN is subject to different safety threats: internal, external, accidental, and malicious. Information or resources can be destroyed; information can be modified, stolen, removed, lost, or disclosed and service can be interrupted. Even if the WSN is secure, the environment can turn it insecure or vulnerable. Security management must provide self-protection, reliability, disposability, privacy, authenticity, and integrity.

Determining if a fault or collection of faults is the result of an intentional DoS attack presents a concern of its own — a point that becomes even more difficult in large-scale deployments, which may have higher nominal failure rates of individual nodes than small networks will. The robustness against physical challenges may prevent some classes of DoS attacks. Each layer of the protocol stack is vulnerable to different DoS attacks and has different options available for its defense.

3.3.4.5 Accounting Management

Accounting management includes functions related to the use of resources and corresponding reports. It establishes metrics and quotes and limits what can be used by functions of other functional areas. These functions can trace the behavior of the network and even make inferences about the behavior of a given node. Accounting management must be considered self-sustaining.

A WSN contains an energy producer (battery) and some energy consumers (transceiver, computation module, and sensor devices). Operations of the application or management can be measured or counted in terms of energy consumption. Given the node characteristics, the average sensor lifetime determines the cost of running a sensor network. One way to reduce total energy consumption is to cut down the number of high-energy operations at the cost of an increase in the number of low-energy operations. The measured cost can be amortized using prediction models [10]. Some functions related to accounting management include: discovery, counting, storing, and data reporting of a parameter; network inventory; determination of communication costs; energy consumption; and traffic checking.

3.4 MANNA as an Integrating Architecture

The MANNA architecture [22] was proposed to provide a management solution to different WSN applications. It provides a separation between both sets of functionalities, i.e., application and management, making integration of organizational, administrative, and maintenance activities possible for this kind of network.

The approach used in the MANNA architecture works with each functional area, as well as each management level, and proposes the new abstraction level of WSN functionalities (configuration, sensing, processing, communication, and maintenance) presented earlier (Figure 3.1). As a result, it provides a list of management services and functions that are independent of the technology adopted.

The MANNA architecture establishes some automatic services, which feature self-managing, self-organizing, self-healing, self-optimizing, self-protecting, self-sustaining, and self-diagnostic, with a minimum of human interference. It is robust to changes in the network state and establishes some services to maintain the quality of the provided services.

3.4.1 Management Services, Functions, and Models

The definition of management service[*] is a task that consists of finding which activities or functions must be executed, when, and with which data. Management services are executed by a set of functions, and they need to succeed to conclude a given service. Management functions represent the lowest granularity of functional portions of a management service, as perceived by users. The conditions for executing a service or function are obtained from the WSN models.

The WSN models, defined in the MANNA architecture, represent aspects of the network and serve as a reference for the management. These models provide an abstract vision of the system through which is possible to hide all nonrelevant aspects given a certain objective.

Figure 3.3 represents a scheme to construct the management, starting at the definition of management services and functions that use models to achieve their goals. A management service can use one or more management functions. Different services can use common functions that use models to retrieve a

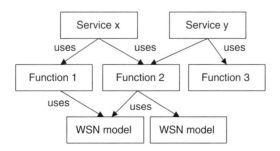

FIGURE 3.3 Services, functions, and WSN models. (From Ruiz, L.B., Nogueira, J.M., Louriero, A.A., *IEEE Commun. Mag.*, 41(2), 116–125, 2003. With permission.)

---

[*]Note that the term management service is different from the service management functional area.

network state concerning a given aspect. Therefore, the management functions use and generate management information.

MANNA architecture considers the three management dimensions in the definition of the management functions and in the development of the functional, physical, and information architectures (see Figure 3.1). A partial list of the management functions, in no particular order, follows. The complete list can be obtained from Reference 21.

- Environmental monitoring function
- Monitored area definition function
- Coverage area supervision function
- Node deployment definition function
- Node deployment function [4]
- Environmental requirements acquisition function
- Network operating parameters configuration function
- Topology map discovery function
- Network connectivity discovery function
- Aggregation function
- Data fusion function
- Node density control function
- Priority of action definition function
- Management operation schedule function
- Cooperation discovery function
- Synchronization function
- Energy map generation function
- Network coverage area definition function
- User interface function
- Self-test function
- Node localization discovery function
- Node operating-state control function
- Node administrative-state control function
- Node usage-state control function
- Node mobile function
- Navigation plan function
- Energy-level discovery function

Some functions allow one to obtain characteristics related to the efficiency and effectiveness of a WSN. Some of them are quantitative functions defined to obtain parameters presented by Subbarao [24], such as network settling time function; network join time function; network depart time function; network recovery time function; frequency of updates (overhead) function; memory requirement function; network scalability function; and energy consumption function.

The distributed management MANNA architecture is based on two paradigms: policy-based management and autonomic management. In most of the management applications, the MANNA architecture uses automatic services and functions executed by a management entity invoked as a result of information acquired from a WSN model. This is called self-management. Management functions can also be semi-automatic when executed by an observer assisted by a software system that provides a network model or invoked by a management system. They can be manual when executed outside the management system. Five possible states are defined for a function:

- Ready (when the necessary conditions to execute a function are satisfied)
- Not ready (when the necessary conditions to execute a function are not met)
- Executing (when the function is being executed)
- Done (when the function has a successful execution)

- Failed (when a failure occurs during execution of the function)

Locations for managers and agents, as well as functions that they can execute, are suggested by the functional architecture. The MANNA architecture also proposes two other architectures: physical and information.

The following discussion concerns how the MANNA architecture can cope with different kinds of network and presents the functional, information, and physical architectures.

3.4.2 Functional Architecture

The functional architecture describes the distribution of management functionalities in the network among manager, agent, and management information base (MIB). In the architecture, it is possible to have a diversity of manager and agent locations. The management choice depends on the functional areas involved, the management level considered, and the application running in the WSN, i.e., depends on the network functionalities (Figure 3.1). This architecture introduces the organizational concept of a management "domain," which is an administrative partition of a network for the purpose of network management. Domains may be useful for reasons of scale, security, or administrative autonomy. Each domain may have one or more managers monitoring and controlling agents in that domain. In addition, managers and agents may belong to more than one management domain. Domains allow the construction of strict hierarchical, fully cooperative, and distributed network management systems.

3.4.2.1 WSN Manager

WSN management can be centralized, distributed, or hierarchical. In a centralized management network, a single manager collects information from all agents and controls the entire network. A distributed management network has several managers, each responsible for a subnetwork and communicating with other managers. In a hierarchical management network, intermediate managers distribute the management tasks. The management alternative to be chosen depends on the application running on the WSN. In any solution, it may be important to have a manager entity located externally to the WSN. The external manager has a global vision of the network and can perform complex tasks (automatic services and functions) that would not be possible inside the network. However, this manager can be the only one (centralized management) or it can collaborate with another manager localized inside the network (decentralized management).

3.4.2.2 WSN Agents

The development of a functional architecture raises the question of the most adequate location for an agent, given a particular kind of WSN. A possible alternative to the agent location is to place it close to the manager, i.e., external to the network. However, this may cause isolation of the management and make it difficult to integrate it in the future and to access other management systems.

Next, some possible configurations are explored:

- *Agents in flat and homogeneous WSNs.* A flat WSN has at least one sink node to provide network access. All network nodes have the same hardware configuration. Some possible alternatives for flat and homogeneous networks considering agent location in the WSN are:
  - Agents inside the network and external manager (Figure 3.4a)
  - Agents in the sink node (Figure 3.4b)
  - Agents and manager in the network; the two possibilities for manager organization are hierarchical (Figure 3.4c) and distributed (Figure 3.4d)

  In any of these proposals, the main concern is the large amount of traffic that may be generated in response to operation requests and in sending notifications. Another alternative is to place managers inside the network and allowing them to communicate among themselves. This defines a distributed management. In case of having agents as part of common nodes, some questions

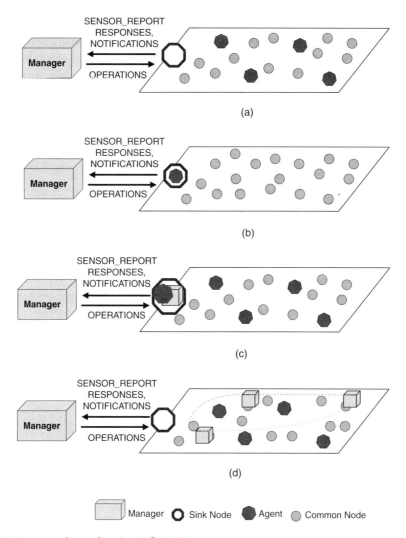

FIGURE 3.4 Manager and agent location in flat WSNs.

remain, such as how to distribute the agents, how to define domains for the agents, and how to deal with nodes with more than one agent.
- *Agents in flat and heterogeneous WSNs.* In a heterogeneous WSN, nodes differ in their physical hardware capabilities. Agents can be placed in more powerful nodes as long as they present adequate location in the network. The sink node can host an intermediate manager or even present no management function. To establish a distributed management, agents can be placed in less powerful nodes and managers in more powerful ones.
- *Agents in hierarchical homogeneous or heterogeneous WSNs.* In this kind of network, there is no sink node. A cluster-head node is responsible for sending data to a base station. It also communicates with the observer. The cluster head may also execute correlation of management data. This computation may decrease the information flow and thus energy consumption. The correlation may also allow a multiresolution in which differences are filtered and a higher precision is obtained. Some possible alternatives for a hierarchical WSN considering the agent location include:
  - Agents in cluster heads and external manager (Figure 3.5a)
  - Agent in the base station (Figure 3.5b)
  - Agents in the network and intermediate manager (Figure 3.5c)

Sensor Network Management

- Agents and distributed managers in the network (Figure 3.5d)

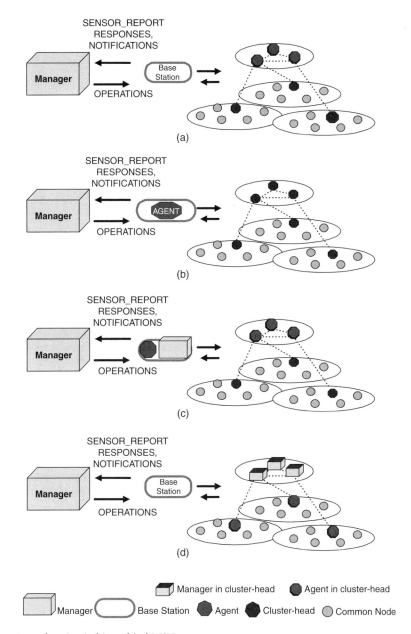

FIGURE 3.5 Agent location in hierarchical WSN.

3.4.2.3 Management Application

In the management architecture (functional, information, and physical), how the management entities receive and analyze information and react to it, which services and functions will be executed, and how the information is exchanged through the communication interface are defined. The type of management (centralized, hierarchical, or distributed) is also defined. Now, the "implosion problem" is explained and management aspects concerning WSN functionalities are addressed.

Centralized management for WSNs, as well as for traditional ad hoc networks, is not always appropriate. One main reason is the traffic concentration problem caused by a central manager that receives and originates management traffic. In addition, the response implosion problem may happen when a high volume of incoming replies is triggered by management operations or events. In case of WSNs, there will always be one access point (sometimes more than one), through which data go to the observer or to the management application. The access point represents a sink node or a base station that can make use of a gateway to communicate with the external environment.

To resolve the implosion problem for management and application, one possibility is to select only a subset of nodes sending data, known as fidelity. In the case of management, some agents are selected to send replies back. This approach may be suitable for densely populated sensor networks with a large number of sensor nodes, in which missing information from some nodes can be ignored with acceptable accuracy. The accuracy of the calculation might significantly degrade. In a sparse sensor network, or a network with a small number of nodes not collecting enough replies, however, the number of replies may not be small enough to be received without taking into account the response implosion problem. Another solution is to make a scheduled response approach [16].

A management solution depends on the features of the network. In some WSNs, only a few management functions can be implemented. In other cases, the management functions must be semiautomatic or manual because of restrictions in the computation. The MANNA architecture is built to provide a management solution to different WSN applications. Depending on the application, it may be interesting or not to use determinate management services, which also can be implemented as automatic, semiautomatic, or manual.

A management solution must also be proposed considering the type of the dissemination: continuous, on demand, programmed, or event driven (see Section 3.3.3.4). In a continuous monitoring scheme, agents are programmed to send monitoring data continuously to a manager. In an on-demand scheme, a manager sends a query to one or more agents, and it receives data back from those agent nodes. In an event-driven monitoring scheme, agents are programmed to send data to a manager only when an event happens and a local condition is satisfied.

Each one of these management solutions has pros and cons. In a continuous monitoring scheme, a management application that stops receiving data from a given node may be an indication of a problem, mainly if the previous sensor condition was normal. The cost of sending data continuously may lead to more rapid consumption of scarce network resources and thus shorten its lifetime. In an on-demand and programmed scheme, the monitoring node can become aware of a problem in the network after sending a query to the node. The cost of having this information is proportional to the number of queries sent or the number of programmed responses. Finally, the design of an event-driven monitoring scheme makes some assumptions about how events are generated. If they happen in an unpredictable way, then, again, there is the problem of consumption of network resources.

On the other extreme, if a node does not report an event, it may be an indication of a failure or of an event that did not occur. In both cases, the management application cannot differentiate them. The same is true for the on-demand network. In normal situations, an event-driven scheme only sends an event to the sink node when it happens. This is the minimum possible cost associated with an event when it must be sent to the management application.

In energy-constrained WSNs, event-driven networks represent an attractive option when compared to continuous networks because they typically send and receive far fewer messages. This translates to a significant energy saving because message transmissions are much more energy intensive when compared to sensing and (CPU) processing.

In terms of failure detection, event-driven networks present challenges not found in continuous and programmed networks. Under normal conditions, a management application of a continuous network receives sensing data at regular intervals. This stream of data not only delivers the content in which one is interested, but also works as an indication of how well the network is operating. If the management application receives data from every single node, then all is well (of course, assuming that the messages are authenticated and cannot be spoofed). If, however, the management application stops receiving data from certain nodes or entire regions of the network, a failure has occurred.

3.4.2.4 Issues Concerning Management Information Base Implementation and Usage

The description of objects present in the information model and the relationship among them are specified in the management information base. In the WSN, to update an MIB with the current network state may require measuring various parameters. In general, the collection of these parameters can have spatial and temporal errors. This is called the "uncertainty problem."

To have a higher precision in the network state, probabilistic measures should be performed with a higher granularity. As in any probing, this would take a finite amount of the system energy and could modify the network state. This is called the "probe effect"; in this way, better precision in management information requires modification of the state.

The MANNA architecture proposes the limitation in the scope as a method for reducing uncertainty and energy consumption while updating the MIB. Spatial limitation consists of defining a physical space inside which the data will be considered for management. Temporal limitation defines a time window (fixed or sliding) inside which the collected data are considered. Functional limitation selects the data of a certain functional network segment for management — for example, the data of a group of nodes or a group leader.

3.4.3 Information Architecture

To ensure common solutions for WSN management, the MANNA architecture defines an information model. WSN management has two kinds of management information: static and dynamic. Static management information describes the configuration of services, network, and network elements. Dynamic management information describes information that changes frequently.

In the MANNA architecture, static management information is based on object orientation and dynamic management information is described by WSN models (see Figure 3.3). From the management point of view, the MANNA functional architecture establishes the circumstances in which a manager will receive event notifications and how it can get its information (monitoring). It also becomes clear what kind of influence the management system has over the WSN resources and how to control them.

3.4.3.1 Static Information

Two types of object classes represent resources under the three different dimensions: managed object and support object. The managed object class directly relates with the network components and with the network. The support object classes play the role of supporting the management functions, i.e., making available to them the necessary information.

The specification of an object class is done through predefined syntactic structures called templates, based on the abstract syntax notation.1 (ASN.1) language, which is used to describe the objects and their characteristics. Object classes may be inherited or reused from standard objects; reuse allows future management integration. Some object classes and their new attributes, based on WSN characteristics, are listed next.

Support object classes. These classes can be programmed by the agent or can be present in the management application. They are mostly derived from the OSI reference model. Some support object classes include:

- Log
- State change record
- Attribute change value record
- Event record
- Event forwarding discriminator
- Management operation schedule
- Information log
- Management log
- Energy level severity assignment profile
- Current remaining energy level summary control

- Monitored object
- Current data object
- History data object
- Threshold data object
- Scanners

Managed object classes. The RFC3433 [3] describes managed objects for extending the entity MIB (RFC 2737) to provide generalized access to information related to physical sensors, which are often found in a networking equipment (such as chassis temperature, fan RPM, and power supply voltage). The RFC 3433 is used and other object classes defined. Some of the defined managed object classes follow:

- *Network* is composed by interconnected managed objects (physical or logical ones) capable of exchanging information. Examples of new attributes for this class include:
  - Network identifier
  - Composition type (homogeneous or heterogeneous)
  - Organization type (flat or hierarchical)
  - Organization period
  - Mobility (stationary, stationary nodes and mobile phenomenon, mobile node or mobile phenomenon)
  - Data delivery (continuous, event driven, on demand, programmed, or hybrid)
  - Type of access point (sink node or base station)
  - Localization type (relative or absolute)
  - Control (open or close)
  - Mission (critical or common)
  - Node distribution (regular, irregular, balanced, sparse or dense)
  - Node deployment (affected by many factors, some of which are the sensor node capabilities of individual nodes, radio propagation characteristics, and the topology of the region)

  Other constraints may include a degree of overlapping in the sensor coverage of two nodes so that they may collaborate.

- *Managed element* represents the sensor node and actuator nodes or other WSN entities that perform functions on managed elements and provide sensing, processing, and communicating services. Examples of new attributes of this class include:
  - Localization (relative or absolute)
  - Element type (common node, sink node, gateway, or cluster head)
  - Minimum energy limit
  - Mobility (direction, orientation, or acceleration)

  The problem is where to place the base station or sink node. Some approaches use a combination of computational geometry, computer-aided design, and numerical optimization methods.

- *Equipment* represents the physical components of a managed element. In this case, this class represents the physical aspects of the sensor node constitution, which is composed of memory, processor, sensor device, battery, and transceiver. The equipment class can be specialized in object classes. For instance,
  - Battery type (linear: the battery is considered to be a bucket of energy; energy is linearly drawn from this bucket by the energy consumers)
    - Discharge rate-dependent model (considers rate at which energy is drawn from the battery to compute the remaining battery life; at high discharge rates, battery capacity is reduced)
    - Relaxation model (takes into account a phenomenon seen in real-life batteries in which the battery's voltage recovers if the discharge rate is decreased)
    - Battery capacity
    - Remaining energy level
    - Energy density

- Computational module composed by processor and memory (clock; state of use; available memory; endurance; AD channel; operating voltage; IO pins)
- Sensor element (sensor type; current consumption; voltage range; min–max range; accuracy; temperature dependence; version; state current; exposure)
- Transceiver (type; modulation type; carrier frequency; operating voltage; current consumption; throughput; receiver sensitivity; transmitter power)

- *System* is used to represent hardware and software, which constitute an autonomous system capable of executing the information processing and/or transference. Examples of new attributes include:
  - Operating system type
  - Version
  - Code length
  - Complexity
  - Total MIPS per available MIPS
  - Synchronization type (mutual exclusion, synchronization of processes)

  A notification of change in an attribute value must be reported upon the event occurrence, such as a software upgrade.

- *Environment* represents the environment in which the WSN is operating. Examples of new attributes include:
  - Environment type (internal, external, and unknown)
  - Noise ratio
  - Atmospheric pressure
  - Temperature
  - Radiation
  - Electromagnetic field
  - Humidity
  - Luminosity

  The environment can present static and dynamic features.

- *Connection* represents the actual connections and is expressed as an association between particular points. The direction of connectivity can be unidirectional (asymmetric) or bidirectional (symmetric). If an instance of this class is unidirectional, the point "a" will be the origin and the terminal point "z" will be the destination. The operational state will indicate the capacity to load a signal. An example of attribute for this class is the communication direction (simplex, half duplex, full duplex). The network topology describes the connections that may exist, and it is expressed as relationships between a set of points.

- *WSN observer* represents the entity that requires WSN services. It may be a human user applying for the use of services via some human–machine communication or it may be some computer-based organizational system.

- *WSN goals* are the benefits provided to users that are obtained by carrying out WSN activities and using WSN services. They can be defined as accuracy, latency, fidelity, etc.

- *WSN management context* defines the environment in which WSN management services are carried out. The definition includes the description of the entity responsible for managing the network, what is managed, and how it can be managed. The WSN management context is described by using three dimensions: management functional areas, management levels, and WSN functionalities.

3.4.3.2 Dynamic Information

In a WSN, network conditions can vary dramatically along the time. In this case, the use of models established by MANNA is of fundamental importance for the management, although its updating cycle can be extremely dynamic and complex. Based on the information obtained with these models, services and functions are executed according to management policies. Dynamic management information is described by WSN models and needs to be obtained frequently. Because acquisition of this information has a cost in terms of energy consumption, an important aspect is to determine the adequate moment, frequency, and fidelity for updating that information. Furthermore, the information collected may not be valid at the moment at which it is processed by the management entity due to delays, omissions, and uncertainty present in WSNs. Static information is needed in order to obtain the WSN models.

In the following, some network models are presented. They always represent dynamic aspects of the network. The dynamic information represented in the network models could or could not be stored in MIBs. Some of the WSN models (map) follow:

- *Network topology map* represents the topology map and the reachability of the network.
- *Residual energy* represents the remaining energy in a node or in a network.
- *Sensing coverage area map* describes the actual sensing coverage map of the sensor elements.
- *Communication coverage area map* describes the present communication coverage map from the range of transceivers.
- *Cost map* represents the cost of energy necessary for maintaining desired performance levels.
- *Production map* represents nodes that are producing.
- *Usage standard map* represents the activity of the network. It can be delimited for a period of time, quantity of data transmitted for each sensor unit, or the number of movements made by the target.
- *Dependence model* represents the functional dependency that exists among the nodes;
- *Structural model* represents aggregation and connectivity relations among network elements.
- *Cooperational model* represents relations of interaction among network entities.

3.4.4 Physical Architecture

The physical architecture defines how management information is exchanged between management entities. It can be seen as the implementation of the functional architecture. In doing so, physical aspects such as the management protocol, physical location of agents, agent functionalities, implemented management service, and supported interfaces for WSNs are defined. The interface among management entities should use a light-weight protocol stack. The MANNA architecture does not define a protocol stack for these interfaces, but provides protocol profiles that may be adequate for each application type.

Application layer. Although the simple network management protocol (SNMP) [28], common management information protocol (CMIP) [13], Web-based management protocol (WBM) [8], and the ad hoc network management protocol (ANMP) [6] allow management in a decentralized and event-oriented way, the structure of managed components is always rather rigid. In these paradigms, management intelligence always resides in the management instance, while the information is generated in the managed instances.

An alternative method would be to delegate management functionalities to the managed systems. A solution for supporting this feature in the implementation of the physical architecture is management by delegation (MbD) [11]. Other alternatives are intelligent agents and mobile agents. In the model of mobile agents, data stay at the local place while the processing task is moved to the data locations. The management functions are performed locally and only the resulting data are sent to the manager. By transmitting the code instead of data, the mobile agent model offers several important benefits:

- Network bandwidth requirements are reduced, which is especially important for real-time applications and when communication uses low-bandwidth wireless channels.
- Agents can migrate to another node when the hosting node is compromised.
- Network scalability is supported.
- Agents can migrate to regions of interest independently of the movement of nodes, if they are mobile.
- Extensibility is supported — that is, mobile agents can be programmed to carry out task-adaptive processes, which extend the capability of the system.

- More stability is achieved because mobile agents can be sent when the network connection is alive and return results when the connection is re-established along with the network data.
- The delay in management actions is reduced.
- Managers are not required to instruct agents all the time.
- The main management part does not reside in the manager.
- Agent cloning offers means for robustness and fault tolerance.

Transport layer. For all protocols described in the application layer, the correct reception of data messages is not assured [27]. Unlike traditional networks (e.g., IP networks), reliable data delivery is still an open research question in the context of WSNs.

Network layer. This should be designed considering power efficiency, and that WSNs are mostly data centric. Data aggregation is useful only when it does not hinder the collaborative effort of sensor nodes. Energy-efficient routes can be found based on the available power in the nodes and the energy required for transmitting data in the link along the route.

Data-link layer. This is responsible for the multiplexing of data streams, data frame transmission and reception, medium access, and error control. Medium access control has two goals: (1) to create the network infrastructure to establish communication links for data transfer and give the sensor network self-organizing ability; and (2) to share communication resources fairly and efficiently between sensor nodes. Simple error control codes with low complexity encoding and decoding might present the best solutions for sensor networks. Open research issues for MAC protocols in WSNs are: determination of low bounds on the energy required for sensor network self-organization; error control coding schemes; and power-saving modes of operation [20].

Physical layer. This is responsible for frequency selection, carrier frequency generation, signal detection, modulation, and data encryption. The 915-MHz ISM band has been widely suggested for sensor networks.

3.5 Putting It All Together

Consider that a management entity has just received the topology and energy messages. It calculates the sensing and communication range area maps and detects the existence of a high node density because there are lots of intersections among the sensing range of the nodes. The management entity faces a redundancy problem of the sensing data received. On one hand, redundancy provides a mechanism for fault tolerance and multiresolution (gives better accuracy), but on the other hand, it represents a waste of resources.

This redundancy problem can be detected by the MANNA architecture using the WSN models, in particular, the "topology map," "energy map," "communication coverage area map," and "sensing coverage area map." Based on these maps, maintenance services may be executed. These services are automatic and executed by a set of functions that use and generate the management information. In this case, one of the functions invoked is the "node administrative state control function."

This function represents the intersection of the three abstraction dimensions for the configuration functional area, network element management level and sensing functionality. The function allows locking the redundant nodes in the administrative state. For this, the agent assigns the value "locked" for the administrative state attribute of the objects (present in the MIB), which represents such nodes acting over the nodes and removing them from sensing, processing, and dissemination services. Figure 3.6 shows a diagram that represents this process.

3.6 Conclusion

Monitoring applications based on wireless sensor networks represent a new important class of applications that can provide data to different kinds of observers. Furthermore, WSNs must deliver the data of interest according to different parameters, such as power efficiency and latency.

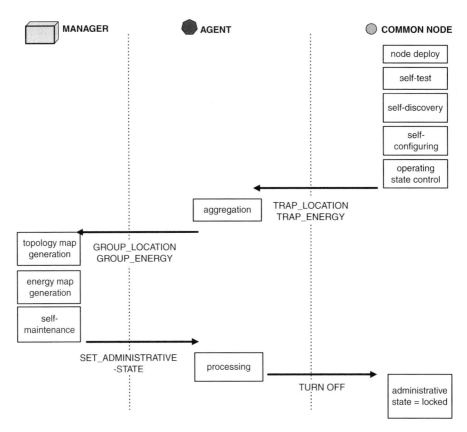

FIGURE 3.6 Applying the MANNA architecture: an example.

Management of WSNs is a new research area that only recently started to receive attention from the research community. This chapter discussed the issue of WSN management and presented autonomic management using the MANNA architecture, which is based on the traditional framework of functional areas and management levels. Adopting this strategy will permit management integration in the future. In the management architecture, the models were built that represent the network state (e.g., WSN topology map, WSN energy map, WSN coverage area map, and WSN production map). These models are important in different applications specified and designed for WSNs.

The fundamental issues about management of WSNs are concerned with how the management application promotes resource productivity and quality of services. Nevertheless, an important aspect is to verify the impact of the management services over the WSN lifetime, latency, goodput, and coverage area.

The important point to be stressed is that, although introduction of management has a cost, this must not affect the network behavior considerably. In fact, the goal is to have the benefits brought by the management solution outweighing the overhead introduced by the management application. Another interesting aspect is that the monitoring scheme to be chosen depends fundamentally on the kind of application monitored. Thus, the management requirements also change among sensor networks.

References

1. Autonomic computing. Available in http://www-3.ibm.com/autonomic/index.shtml.
2. B.R. Badrinath, M. Srivastava, K. Mills, J. Scholtz, and K. Sollins. Special issue on smart spaces and environments. *IEEE Personal Commun.*, 7(5), October 2000.

3. A. Bierman, D. Romascanu, and K.C. Norseth. RFC 3433 on entity sensor management information base (draft-ietf-entmib-sensor-mib-02.txt). Available in ftp://ftp.rfc-editor.org/in-notes/rfc3433.txt.
4. S.B.B. Deb and B. Nath. A topology discovery algorithm for sensor networks with applications to network management. Technical report DCS-TR-441, Department of Computer Science, Rutgers University, May 2002.
5. A. Cerpa, J. Elson, D. Estrin, L. Girod, M. Hamilton, and J. Zhao. Habitat monitoring: application driver for wireless communications technology. In *ACM SIGCOMM Computer Communication Review*, 31(2), 20–41, 2001.
6. W. Chen, N. Jain, and S. Singh. ANMP: ad hoc network management protocol. *IEEE J. Selected Areas Commun.*, 17(8), 1506–1531, August 1999.
7. J. Conover. Policy-based network management. *Network Computing*, November 1999.
8. Distributed Management Task Force (DMTF). Web-based management. Available in http://www.dmtg.org
9. D. Estrin, R. Govindan, and J. Heidemann. Embedding the Internet. *Commun. ACM*, 43(5), 39–41, May 2000.
10. S. Goel and T. Imieli. Prediction-based monitoring in sensor networks: taking lessons from mpeg. Technical report, Rutgers University, 2001.
11. G. Goldzmidt and Y. Yemini. Distributed management by delegation. *Proc. 15th Int. Conf. Distributed Computing Syst.*, 333–340, June 1995.
12. S.E.-A. Hollar. Cots dust. Master's thesis, University of California, Berkeley, 2000.
13. International Organization for Standardization. ISO/IEC ITU-T X.711 Information Technology — Open System Interconnection — CMIP, specification 1991.
14. International Telecommunication Union (ITU). CCITT recommendation X.700, management framework for open systems interconnection (OSI) for CCITT applications, 1992.
15. International Telecommunication Union (ITU). ITU-T M.3010 — principles for a telecommunications management network, May 1996.
16. D.B. Johnson and D.A. Maltz. Dynamic source routing in ad hoc wireless networks. In Imielinski and Korth, Eds., *Mobile Computing*, Vol. 353. Kluwer Academic Publishers, 1996, 153–181.
17. S. Lindsey, C. Raghavendra, and K. Sivalingam. Data gathering in sensor networks using the energy delay metric. In *Int. Workshop Parallel Distributed Computing: Issues Wireless Networks Mobile Computing*, San Francisco, April 2001.
18. S. Meguerdichian, F. Koushanfar, M. Potkonjak, and M.B. Srivastava. Coverage problems in wireless ad hoc sensor networks. In *INFOCOM*, 1380–1387, 2001.
19. S. Mehrotra. Distributed algorithms for tasking large sensor network. Thesis submitted to the faculty of Virginia Polytechnic Institute and State University, July 2001.
20. National Chiao Tung University Department of Computer, Information Science. Mobile computing and broadband networking lab. Wireless sensor network. Available in http://pds.cs.nctu.edu.tw.
21. L.B. Ruiz, T.R.M. Braga, F. Silva, J.M.S. Nogueira, and A.A.F. Loureiro. Service management for wireless sensor networks. *IEEE LANOMS — Latin Am. Network Operation Manage. Symp.*, 55–62, September 2003.
22. L.B. Ruiz, J.M.S. Nogueira, and A.A.F. Loureiro. MANNA: a management architecture for wireless sensor networks. *IEEE Commun. Mag.*, 41(2), 116–125, Feb. 2003.
23. A. Savvides, S. Park, and M.B. Srivastava. On modeling networks of wireless microsensors. In *Joint Int. Conf. Measurement Modeling Computer Syst.*, 318–319, Cambridge, MA, June 2001.
24. M.W. Subbarao. Ad hoc networking critical features and performance metrics. Technical report, Wireless Communications Technology Group, NIST, September 1999.
25. S. Tilak, N. Abu-Ghazaleh, and W. Heinzelman. A taxonomy of wireless microsensor network models. *ACM Mobile Computing and Commun. Rev.* (MC2R), 6(2), April 2002.
26. M.A. Vieira, L.F. Vieira, L.B. Ruiz, A.A. Loureiro, and A.O. Fernandes. Scheduling nodes in wireless sensor network: a Voronoi approach. *IEEE LCN — Local Computer Network*, October 2003.

27. C.-Y. Wan, A. Campbell, and L. Krishnamurthy. PSFQ: a reliable transport protocol for wireless sensor networks. *WSNA'02*, 1–11, September 2002.
28. W. Stallings. *SNMP, SNMPv2, SNMPv3, ROMON, and ROMON2: Practical Network Management*. Addison–Wesley, Reading, MA, 3rd ed., 1998.
29. K. Wu and J. Harms. QoS support in mobile ad hoc networks. *Crossing Boundaries — GSA J. Univ. Alberta*, 1(1), 92–106, November 2001.

4
Models for Programmability in Sensor Networks

4.1	Introduction	4-1
4.2	Differences between Sensor Networks and Traditional Data Networks..	4-2
4.3	Aspects of Efficient Sensor Network Applications	4-2
4.4	Need for Sensor Network Programmability	4-3
4.5	Major Models for System-Level Programmability..........	4-4
	Database Model • Active Sensor Model • Active Networks — Mobile Agents	
4.6	Frameworks for System-Level Programmability.............	4-6
	Directed Diffusion with In-Network Processing • Cougar • TinyDB • SQTL • Smart Messages — Spatial Programming • Maté • SensorWare • MagnetOS • DFuse	
4.7	Conclusions	4-12

Athanassios Boulis
University of California at Los Angeles

4.1 Introduction

Several aspects of the form and operation of sensor networks have been encountered in the previous chapters, as well as strong indications of the great versatility that these systems exhibit and the multiple modes of operations supported in order to achieve their diverse goals. Reading the chapters on several different applications in this book only reinforces the observation that different applications require different distributed algorithms to be handled efficiently.

Having sensor networks with long lifetimes supporting multiple transient users with different needs implies that many different distributed algorithms will run in the network — algorithms that are not known *a priori*. This fact gives rise to the following question: How does one dynamically program the network to provide the users with the needed services efficiently? This chapter examines this problem and the different models proposed by researchers to address it. The discussion begins with some background on the differences of sensor networks with traditional data networks, immediately followed by a section on the general characteristics of efficient sensor network applications. These two sections allow one to motivate the need for dynamic programmability as well as the kind of programmability desired. Description of the different models to achieve such programmability and examples supporting frameworks then follow.

4.2 Differences between Sensor Networks and Traditional Data Networks

Although sensor networks are networks of computing devices, they are considerably different from traditional data networks. The first difference of sensor networks compared to traditional data networks is that they have severe energy, computation, storage, and bandwidth constraints. For example, the wireless sensor node designed by Rockwell Scientific [24] has a 133-MHz, 32-bit, Intel StrongARM 1100 CPU, 1 MB of FLASH memory, 1 MB of RAM, and a 100-Kbps radio, and must operate on two 9-V batteries. This is considered to be toward the high end of sensor network devices. A popular, low-end node design from UC Berkeley, the mica-II [12], uses a 7.37-MHz, 8-bit Atmel CPU with 128 KB of FLASH memory, only 4 KB of RAM, and a 35-Kbps Chipcon radio. The major resource problem in such networks is energy because these are static unattended networks and the nodes cannot have renewable energy sources. Energy is so important that algorithms designed for sensor networks often sacrifice response latency, accuracy, and other user-desired qualities to save energy and prolong the operational lifetime of the network.

The second difference of sensor networks compared to traditional data networks is their overall usage scenario and the implications that this brings to the traffic and interaction with the users. Typically, in traditional networks, users are connected to a node (or group of nodes) and require a service from another node. This two-entity communication model describes the overwhelming majority of traditional network traffic. The network acts as a medium bringing the two parties together. The interaction model is also straightforward; the user interacts directly with the user or service at the other end. Certain actions from the user will produce certain data transfers to and from the other end. The most popular exceptions to these rules are free roaming mobile agents providing data mining or broker services. However, this is a small portion of today's data networks.

Sensor networks, on the other hand, are less like networks (i.e., in the sense that they loosely connect independent entities) and more like distributed systems. As stated earlier, the nodes tightly collaborate to produce information-rich results. The user will rarely be interested in the readings of one or two specific nodes, but will be interested in some parameters of a dynamic physical process. To achieve this efficiently, the nodes must form an application-specific distributed system to provide the user with the answer. This is a departure from the two-entity model: there are no clear sources and destinations based on user desires — only the users and the *whole network*. The nodes involved in the process of providing the user with information are constantly changing as the physical phenomenon is changing. In conclusion, the sensor network is not there to connect different parties together, as in the traditional networking sense, but rather to provide information services to users.

4.3 Aspects of Efficient Sensor Network Applications

The preceding remark leads to the user-interaction topic. Apart from the user input, the physical phenomena now play a central role in the actions inside the network. The actions in each individual node are affected from external physical stimuli and information from other nodes, as well as direct input from the user. Actually, it is desirable to operate in a fashion in which a node's actions are affected largely by physical stimuli detected by the node or nearby nodes. Frequent long trips to the user are undesirable because they consume time and energy. Tennenhouse [27] calls this decentralized (i.e., not all traffic flows to/from user), autonomous (i.e., user is out of the loop most of the time) way of operating "proactive computing" (as opposed to interactive). The term "proactive" is also adopted to denote an autonomous and noninteractive nature. In order for sensor networks to realize their full potential and efficiently use their limited resources, they have to be viewed as distributed proactive systems.

Another efficient design principle is to keep communications localized. Apart from the apparent benefit of saving valuable communication energy, the algorithms can be made more robust by taking advantage of the broadcast nature of the channel combined with the ability to process inputs from all neighbors

— not just selected neighboring nodes. Finally, algorithms can benefit from acknowledging and exploiting the inherent energy–accuracy–latency trade-off present in sensor networks. That is, the more energy one is willing to give, the more accuracy and less latency is achieved, or by keeping the energy consumption constant, one can trade high accuracy for lower latency. Operating in the trade-off space, an algorithm becomes more flexible in accommodating user needs.

Successful applications for sensor networks employ one or more of the preceding design aspects to achieve their goal. Some examples include target tracking algorithms [8, 28]; edge detection algorithms [9, 22]; and periodic aggregation algorithms [4]. Sensor network algorithms' diversity is interesting to those who study them. Some of these algorithms might use common services such as a wake-up protocol [25] or a geographic routing protocol [17], but in essence they are deeply different. From the communication patterns (e.g., cluster based, tree structured, nonhierarchical) to the computation tasks (e.g., custom fusion of sensing data, keeping and processing state of neighbors), these algorithms are as diverse as the problems they tackle. Even in algorithms tackling the same general problem, one can find very different solutions (e.g., edge detection tackled by Chintalapudi and Govidan [9] and by Nowak and Mitra [22]).

Efficiently designed sensor networks are application-specific distributed systems that require a different distributed proactive algorithm as an efficient solution to each different application problem. Given the nature of sensor networks (i.e., diverse solutions for diverse problems), several generic questions come to mind:

- How does one deploy different algorithms into the network?
- What is the programming model that will implement these algorithms?
- What general support does one need from a programming framework?

4.4 Need for Sensor Network Programmability

Researchers who develop sensor network algorithms have shown little concern about how to program them. Most of the time, the proposed algorithms are assumed to be hard-coded into the memory of each node. In some platforms, the application developer can use a node-level OS (e.g., TinyOS [13]) to create the application, which has the advantages of modularity, multitasking, and a hardware abstraction layer. Nevertheless, the developer must still create a single executable image to be downloaded manually into each node. However, it is widely accepted that sensor networks will have long-deployment cycles and serve multiple transient users with dynamic needs. These two features clearly point in the direction of dynamic sensor network programming.

What kind of dynamic programmability is wanted for sensor networks? Hard-coding a few algorithms into each node that are tunable through the transmission of parameters is not flexible enough for the wide variety of possible sensor network applications. An ability to download executable images into the nodes is not feasible because most of the nodes will be physically unreachable or reachable at a very high cost. An ability to use the network in order to transfer the executable images to each and every node is energy inefficient (because of the high communication costs and limited node energy) and cannot allow multiple users to share the sensor network.

Ideally, it is desirable to be able to program the sensor network dynamically as a whole — an aggregate — and not as a mere collection of individual nodes. This means that a user connected to the network at any point will be able to inject instructions into the network to perform a given (probably distributed) task. The instructions will task individual nodes according to user needs, network state, and physical phenomena, *without any intervention from the user*, other than the initial injection. Furthermore, because multiple users should be able to use the sensor network concurrently, several resources/services of the sensor node should be abstracted and made sharable by many users/applications. This kind of programmability is called "system-level programmability." The next section presents the two main models adopted by researchers who try to provide system level programmability.

4.5 Major Models for System-Level Programmability

Before delving into individual research efforts by describing several frameworks and their properties, the two major models for system level programmability will be described: (1) the database model and (2) the active sensor model. Most research efforts fall into one of these models and some frameworks can exhibit characteristics from both.

4.5.1 Database Model

One approach of programming the sensor network as an aggregate is a distributed database system. Multiple users can inject database-like queries to be distributed autonomously into the network. The sensor network is viewed as a distributed database and the query's task is to retrieve the needed information by finding the right nodes and, possibly, to process the data in predefined ways (e.g., aggregate the data) as they are routed back to the user. The strong point of the database approach is that it offers an intuitive way to extract information from a sensor network hiding the complications of *embedded* and *distributed* programming. The user simply describes the information needed. The way in which data are retrieved in nodes and the distributed algorithm needed to retrieve and process the data are not specified. The user "magically" sees the requested information in the use node.

The model's limitation is that only predefined ways to process the data exist, thus implying that only certain types of applications (i.e., applications studied by the specific researchers that are mainly aggregation applications) are addressed in the most efficient way by the database model. If a new way to process and react to the data is needed by application N&U (new and unexplored), this can only be done at the user node (assuming that the human-controlled user node is easily upgradeable). Consequently, the algorithmic pattern to address application N&U under the database model will be an iteration of the generalized steps: (1) partially processed data arriving to the user node; (2) data undergoing custom processing; and (3) based on the result, a new database query issued. In most cases, this is not the structure of the most efficient algorithm to solve an application problem. Recently researchers have tried to augment the language model (e.g., by using event triggers) to accommodate a richer variety of distributed algorithms and provide more flexibility to the user. Nevertheless, the user has no ultimate control over the distributed algorithm executed in the network; this prevents maximum efficiency in certain applications.

The database model is a good solution in the following cases: (1) used in the full-scale network for applications that are well-studied under this model and (2) used in subnetworks with small diameter (e.g., 3 to 4 hops) as a flexible local data retrieval system. For the latter case, imagine a powerful cluster head node with a few less capable nodes around it. The less capable nodes can easily run the framework to interpret and reply to database queries while the cluster head runs a more heavyweight framework (e.g., of the active sensor variety). The cluster head can use the database model to retrieve aggregated data easily from the nodes around it. These data can be further processed by the cluster head and participate in a custom, user-defined distributed algorithm among other cluster heads.

4.5.2 Active Sensor Model

The term coined in Levis and Culler [19] denotes an adaptation of the active networking idea in traditional data networks to the sensor network realm. The difference is that although active networking tasks are reacting only to reception of data packets, active sensor tasks need to react to many types of events, such as network events, sensing events, and timeouts. Active sensor frameworks abstract the run-time environment of the sensor node by installing a virtual machine or a high-level script interpreter at each node. For example, single instructions of the scripts (or bytecodes) can send packets, or read data from the sensing device. Moreover, the scripts (or bytecodes) are made mobile through special instructions, so nodes can autonomously task their peers.

Active sensor frameworks seek to remedy the limited flexibility problem found in the database model at the expense of increased responsibility for the programmer. They provide a language model powerful enough to implement any distributed algorithm while at the same time hiding unnecessary low-level details from the application programmer. Many of the frameworks also provide a way to share the resources of a node among many applications and users that might concurrently use the sensor network. The control of the distributed algorithm (which implies efficiency in any application) comes at a cost compared to the database model. The programmer must explore, define, and test the distributed algorithm for each application.

The difficulty in designing an active sensor framework lies in determining how to define the abstraction of the run-time environment properly so that one achieves compactness of code, sharing of resources for multiuser support, and portability in many platforms, while at the same time keeping a low overhead in delays and energy. Two major choices determine the run-time abstraction:

- Choice of virtual machine (interpreting machine-level bytecodes usually based around a stack architecture) or script interpreter (interpreting high-level ASCII scripts)
- Choice for number and content of native services provided

These choices affect ease of programming, mobile code compactness, time it takes to execute a task, and the memory footprint required in the sensor nodes to accommodate the framework. For example, the more services provided, the more compact the mobile code becomes but the greater the memory footprint becomes. Also, by providing more native services, the execution time of a task is reduced because it is not necessary to rely on interpreted code to implement these parts of the task. Choosing a virtual machine usually requires less memory footprint, but creates less compact code when compared to a high-level scripting language. Given the conflicting nature of the preceding "performance" criteria, it is clear that no one optimal design point exists; rather, the optimality is determined by specific implementation goals. Some of the frameworks discussed in Section 4.6, for example, make some different choices because they target different hardware platforms.

The process of populating the sensor network with viral pieces of code as the active sensor model dictates resembles the operation of multiple collaborating mobile agents, replicating/migrating to the nodes at which the distributed algorithm should be executed. For this reason, the next subsection offers a general discussion on mobile agent (MA) frameworks.

4.5.3 Active Networks — Mobile Agents

Traditional distributed applications are designed as a set of processes (mostly network unaware) cooperating within assigned execution environments. MA technology, however, promotes the design of applications made up of network-aware entities that can change their execution environment by transferring while executing. In recent years, several research groups have created mobile systems based around the notion of an agent that consists of procedures and state data that can migrate from machine to machine. Some of these, such as Agent Tcl [10], have been built on top of interpreted scripting languages; others, such as Aglets, have relied on Java, which provides code mobility via applets and object serialization. The interest in this area is propelled by the advantages agents offer in Internet applications. The advantages fall into three different categories, as reported by Cabri et al. [7], among others:

- Bandwidth and delay savings because computation is moved to the data
- Flexibility because agents do not require the availability of specific code
- Suitability for mobile computing because agents do not require continuous network connections

Thus, when considering MAs, one overwhelmingly sees them in an Internet-application environment with the possibility of mobile endpoints. Consequently, mobile agents are viewed as free-roaming entities that are mostly autonomous with no point of control and should perform well under intermittent connections and mobility. The major design issue in such systems is how the agents communicate and collaborate. Basically, four coordination models classify mobile agents in their current Internet-motivated world:

- *Client/server model.* Direct connection is that of involved agents; the main advantage is the low overhead in delay and implementation. The main disadvantage is that agents are spatially and temporally coupled.
- *Meeting-oriented model.* Agents interact by opening and joining abstract meeting points. The model achieves spatial uncoupling but preserves temporal coupling.
- *Blackboard-based model.* The agents interact by leaving messages in predefined blackboards. Temporal uncoupling is achieved, but some weak spatial coupling still exists because the agents must know each other's names.
- *Linda-like model.* The blackboard is extended by introducing associative mechanisms into the shared data space, thus making the messages' content addressable. Spatial and temporal uncoupling is achieved.

Clearly, the advantages and disadvantages coupled with these models revolve around the notion of the agent's spatial and temporal coupling with its peers or lack thereof. This is understandable, if one remembers the previous discussion on mostly autonomous agents with intermittent network connections. Spatial and temporal uncoupling is desirable, even at the cost of more complex (thus less secure and less efficient) designs.

In the realm of sensor networks, however, these concerns and classifications are becoming irrelevant. The concern is mainly with building reconfigurable and distributed applications that can be reconfigured and relocated. The pieces of mobile code in active sensor frameworks (i.e., the equivalent of mobile agents) are envisioned to perform very tight collaboration with each other, thus departing from the autonomous agent model. In addition, this kind of collaboration will happen among locally clustered nodes, making the peer-to-peer direct communication easier. Furthermore, intermittent connections and mobility are not issues that the framework should hide, but instead should let the algorithm deal with them in an application-specific manner. Remember that efficiently designed applications in sensor networks do not rely on data from specific nodes; rather, they can handle inputs from a greatly varying set of nodes. If data are not available from certain nodes due to intermittent connections or mobility, the application simply keeps on working. For these reasons, the server/client model or the more general peer-to-peer direct communication model is an acceptable choice.

In conclusion, the MA paradigm is associated with the notion of a single agent migrating from node to node, performing part of a given task in each node while sparsely communicating with *specific* remote services or other MAs. The active sensor model, on the other hand, is associated with multiple simple lightweight agents that tightly collaborate to implement a distributed algorithm; their behavior and position is influenced by physical events as well as by user needs. Most of the time, the communication is not tied to specific nodes but rather to a statistically chosen set of nodes.

4.6 Frameworks for System-Level Programmability

This section looks into individual research efforts, beginning with database model frameworks. It continues with active sensor frameworks and concludes with a framework that mixes both notions.

4.6.1 Directed Diffusion with In-Network Processing

Early sensor network research has shown the benefits of attribute-based naming (e.g., geographical information) and routing in the operation of sensor network applications. Directed diffusion [15] was the first protocol to implement such ideas. Heidemann et al. [11] incorporate data-driven, low-level naming with directed diffusion, along with in-network processing ideas, to task the sensor network. The in-network processing is limited to aggregation filters that take n stream input data and produce m stream output data. The application programmer can use simple APIs to use the directed diffusion and custom filtering mechanisms. More specifically, the commands *subscribe*, *unsubscribe*, *publish*, *unpublish*, and *send* implement the publish/subscribe mechanism of directed diffusion, while the commands *addFilter*, *removeFilter*, *sendMessage*, and *sendMessageToNext* register and utilize custom filters for in-network processing. The initial implementation of the system does not contain a method to upload filters dynamically to the nodes. Although the authors do not explicitly categorize their work in the database model, one can see most of its main notions.

4.6.2 Cougar

Other systems, such as Cougar [2], focus more on transferring the sensor querying language (SQL) semantics of traditional databases to the distributed setting of sensor networks. In this case, the naming system developed in Heidemann et al. [11] is replaced by an SQL equivalent. Each node is equipped with a fixed database query resolver. As queries arrive at a node, the local resolver decides on the best distributed plan to execute the query and distributes the query to the appropriate nodes.

4.6.3 TinyDB

The more recent and probably more advanced system that follows the database model is the TinyDB [21] developed in Berkeley. The developers' main focus is aggregate queries (e.g., min, max, average); thus, they provide special optimizations for them (e.g., exploit the shared medium, perform what they call "hypothesis testing"). A query has the following general form:

```
SELECT expr1, expr2 …
FROM sensors
WHERE pred1 [AND | OR] pred2 …
GROUP BY groupexpr1, groupexpr2 …
SAMPLE PERIOD t
```

The select clause lists the attributes or aggregates of attributes to retrieve from the sensors. Aggregates and nonaggregates cannot appear in the same select clause unless the nonaggregate fields appear in the "group by" clause. "Sensors" is the standard table containing one attribute for each type of sensor existing in the network. It is the common table on which queries are computed on the "where" clause, which filters out readings that do not satisfy the Boolean expression of predicates. The group clause is used in conjunction with aggregate expressions to specify a partitioning of readings before aggregation. For example, one might query:

```
SELECT buildingID, AVG(temp)
GROUP BY buildingID
```

to collect the average temperature from each building, instead of the average temperature over all sensor readings. Finally, the "sample period" clause specifies the time between reevaluation of the query with freshly sampled data.

TinyDB has recently added new language features to provide more flexibility to the programmers. To move beyond passive querying, clauses were added to spawn queries autonomously based on predefined events and also to create internal storage points in the network. Even with these additions, though, the declarative nature of TinyDB remains. The programmer has no ultimate control over the distributed algorithm executed in the network because its details are taken care of by the underlying TinyDB system.

4.6.4 SQTL

Jaikaeo et al. [16] developed the sensor querying and tasking language (SQTL). Starting from a database-like system, the researchers realized the limitations of a declarative language to the implementation of arbitrary distributed algorithms into the sensor network. Thus, they augmented their initial language with imperative style commands to help task the network.

SQTL fits in a more general architecture for sensor networks called sensor information networking architecture (SINA) [26], which uses SQL-like queries as well as SQTL programs. Some of its main features include: (1) hierarchical clustering; (2) attribute-based naming; and 3) a spreadsheet paradigm for organizing sensor data in the nodes. SQL-like queries use these three features to execute simple querying and monitoring tasks. When a more advanced operation is needed, SQTL plays the essential role by programming the sensor nodes and allowing proactive population of the program. In SINA, SQTL is used as an enhancement of simple SQL-like queries; thus, the framework still revolves around a database-like model.

4.6.5 Smart Messages — Spatial Programming

The Rutgers researchers have developed a mobile code platform for embedded systems called smart messages (SM) [3]. They used SM to develop their suggestion for a programmable sensor network framework, which they call spatial programming (SP) [14]. First, the characteristics of SMs will be presented and then the SP model will be discussed.

SMs are entities that carry code, data, and execution state (in order to resume execution from the same point upon migration of the SM). The code is written in Java language supporting a few extra commands relevant to the SM environment. The run-time environment consists of a KVM (Sun's Java virtual machine for embedded devices) modified to support the new commands. Apart from the mobile code entities (the smart messages), the SM environment also supports the abstraction of tags, which are essentially SM-persistent storage and are used as universal names. From naming underlying devices and OS services to naming nodes or application ports for specific data, tags do not have a specific structure. Tags can be used to access the sensor data, name the node, or leave next-hop information behind from a previously executed routing protocol.

The run-time environment also includes a manager for the tag space (essentially a name-based memory). The basic execution model of SMs is that one main agent for an application does the job by hopping from node to node, doing some portion of the work each time. Other agents (i.e., SM) perform supporting functions (e.g., routing). The new commands added to the basic Java language to create the extension of SMs are:

- Four commands to create, delete, read, and write tags
- One command to create a new SM or replicate yourself
- One command to block on a tag (used for synchronization)
- Two commands to migrate (to next hop or arbitrary)

The block command can block only on one tag thus allowing a program to wait only on a single event. Furthermore, only one smart message executes at each moment. If another is to be executed, the current active one must block or complete execution.

Based on the SM platform, researchers from Rutgers introduced a programming model for a network embedded system (a term that includes sensor networks) called spatial programming. SP is more a resource-based routing scheme than a programming model. The SM platform is augmented with a way to refer to nodes by spatial and arbitrary content properties of the node. The abstraction of spatial reference (SR) is introduced, which has the form "space:content_tag." Simple operations are defined on the space portion of an SR. For instance, one can take the difference of two spaces simply by writing space1-space2. Space can also be created with the use of the "rangeof" function, which receives a point and a radius as arguments. An SR can refer to multiple nodes (as it covers a certain space). One can reference individual nodes within an SR by using the "[i]" indexing convention. Another key point is the reference consistency; once an SR is created (and thus some nodes are referred with that name,) SR_name[i] is always the same node.

Resources in nodes (e.g., sensor modules, software services) are accessed as variable names, which can be written and read. The names do not follow a particular structure so the applications must know in advance the custom way to access them. A weak point of the SP architecture concerns resource sharing, which is absent from the system; the applications must explicitly negotiate any sharing. Obviously, this method is error prone and at times impossible to follow because applications will not always have knowledge of each other. Finally, questions are posed concerning the actual programming model in SP. How is the code distributed in the network? How is collaborative operation between agents facilitated? The examples developed by the researchers to illustrate their framework present centralized applications (executing only at one node) that access resources remotely, much like RPC calls. This kind of execution is not the most desirable one, as was discussed in the first section of this chapter.

4.6.6 Maté

An active sensor framework for sensor networks called Maté is currently being developed in Berkeley [19]. Maté is a tiny virtual machine built on top of TinyOS [13]. TinyOS is an operating system, designed specifically for the Berkeley-designed family of sensor nodes, generically named "motes" [12]. Maté's goal is to make a sensor network composed of motes dynamically programmable in an efficient manner. This includes the capability to dynamically instruct a mote to execute any program, as well as expressing this program in a concise way. This is achieved by building a virtual machine (VM) for the motes. The VM supports a very simple, assembly-like language to be used for all needs of mote tasking. Programs (called capsules) written on the VM language can be injected to any node and perform a task. Furthermore, the capsules have the ability to self-transfer by using special language commands. This model seems extremely similar to the author's in SensorWare. Indeed, Maté shares the same goals as other active sensor frameworks, as well as the same basic principles to achieve these goals. However, as discussed in Section 4.5, design choices differentiate active sensor frameworks.

Maté, like its substrate TinyOS, was built with a specific platform in mind: the extremely resource-limited mote. The main restriction for the developer of mote-targeted frameworks (such as an OS or a VM) is memory. The newest version of a mote, called mica, offers 128 Kbytes of program memory and 4 Kbytes of RAM. An older version called rene2 has 16 Kbytes of program memory and 1 Kbyte of RAM. With an ingenious architecture, Maté supports both platforms. Because it is so constrained by memory, Maté must sacrifice some features that would make programming easier and more efficient.

First, a stack-based architecture with an ultracompact instruction set (all instructions are 1 byte) reminiscent of a low-level assembly language or the byte code of the Java VM is adopted. This kind of model makes programming of even medium-sized tasks difficult. Furthermore, due to the ultracompact instruction set, many 1-byte instructions are needed to express a medium complexity algorithm, leading in turn to large programs, compared to a higher-level, more abstracted scripting language. The size of programs is important because the code is transmitted/received using the radios of the nodes spending energy for every transmitted/received bit. Second, the behavior of a program when radio packets are received is rather rigid. A handler to process such events is essentially stateless in Maté. Thus, if a new pattern of packet processing is needed, a new handler must be transferred through the network. This imposes an overhead in energy consumption and execution time. Third, because there is only one context (i.e., handler) per event (e.g., clock tick, reception of packet), multiple applications cannot run concurrently in one mote.

Other active sensor frameworks that target richer platforms (e.g., Rockwell Scientific's node [24] includes a 1-Mbyte of program memory and 128 Kbytes of RAM) have the luxury of providing much richer native services to support easy programming with a high-level scripting language, as well as concurrent multitasking of a node so that multiple applications can concurrently execute in a sensor network. One such framework is present in the next subsection.

4.6.7 SensorWare

SensorWare [5, 6] is another active sensor framework developed at UCLA. This framework uses a high-level scripting abstraction based around Tcl [23] and a highly expandable run-time environment. The run-time environment provides multiple services that achieve the sharing of the sensor node's resources among multiple applications. The programming model is event based with event handlers to react to various high-level, application-specific events that occur during a period of interest. The expandability in SensorWare is achieved through the abstraction of virtual devices.

Almost everything in SensorWare is a device (e.g., sensor modules, localization procedure, routing protocols, neighborhood discovery). All devices have a unified interface to interact with them. More specifically, the programmer can act on the device, query the device, describe and name an event the device can produce, and dispose a previously defined event name. The programmer can use the wait command to wait on any of the previously described events. The scripts are made mobile through special commands and data can be carried with the scripts in the form of parameters passed by value. SensorWare has many features to enhance efficiency, flexibility, and ease of programming, the most important of which are:

- Custom script compression based on semantic information
- Script cashing and selective script population
- Addressing tied with routing
- Ability to register scripts as dynamic devices for seamless script coordination

A small code sample of SensorWare scripts follows.

file1:
#code_id 32 small code used as a parameter to other scripts

```
send neighbor $parent "here is your packet"
```

file2:
#code_id 33 this script is an example

```
parameter total_time small_code
set neighbors_num [llength [query neighbor]]
```

#spawn to all neighbors small_code

```
spawn neighbor 0 $small_code
interest timer t1 $total_time
set index 0
while {index<neighbors_num} {
  wait packet t1
  case {$event_name} {
    packet {
      debug "received packet: $event_body"
    }
    t1 {
      debug "not all neighbors replied"
      exit.
    }
  }
  incr index
}
```

To invoke the example, do the following from a terminal (user node):

```
load small_code file1
load example_code file2
carry 5000 $small_code
spawn neighbor [id-n] $example_code
```

The preceding invocation commands simply load the code from the two files into Tcl variables, set the parameters passed to the code of the spawn command, and spawn the code in file2 in the current node. The code of file2 gets the parameters and assigns them to local names, finds out the number of neighbors by querying the neighbor device, and then spawns the small_code (which was passed as a parameter) into all its neighbors. The small code simply sends a message back to the current node. Back in the code of file2 one waits for a packet received or the timer named t1 to expire. According to which event is taking place, different messages are output. SensorWare has been used to implement complex applications such as the distributed estimation algorithm described in Boulis et al. [4] among others.

4.6.8 MagnetOS

MagnetOS [1] was developed at Cornell University and, although it is classified as an operating system for networked embedded systems, it can be seen as a method to program a sensor network dynamically. MagnetOS' key idea is a single system image. The entire network is seen as a unified Java virtual machine by the applications. The system consists of a static and a dynamic component. The static component is a partitioning service that partitions regular Java applications into objects that can be distributed into the network. The dynamic part in each node then provides services for application monitoring, object creation, and migration.

The programmer should write normal Java applications, oblivious of the distributed nature of the execution environment; MagnetOS will take care of partitioning and distribution of the application. The application is partitioned according to the objects that the programmer has defined. Thus, an object becomes a mobile application component. The objects are gradually distributed in the network following automatic object migration policies. In MagnetOS, two algorithms perform the automatic object migration: NetPull and NetCenter. NetPull watches communication at the one-hop neighborhood level and migrates components toward links with the greatest communication. NetCenter performs the same monitoring at the network level and can migrate a component several hops at a time.

Apart from the inefficiencies that the automatic code migration can create (e.g., slow convergence to a satisfactory distribution, oscillations of component placement), MagnetOS has the major drawback of completely hiding the distributed nature of the application. Despite the claim that the application can be defined with a single image in mind, the choice of object definition can greatly affect the efficiency of the distributed application because the number and type of object directly affects the partitioning of the application. The complete elimination of the distributed nature of an application from the mind of the programmer is an exciting goal, but very distant or even unattainable for a sufficiently diverse set of applications.

4.6.9 DFuse

Kumar et al. [18] aspire to generalize and facilitate the data fusion process (termed "aggregation" by other researchers) by providing a framework called DFuse. The framework consists of an API to define arbitrary fusion processing and an algorithm for automatic fusion point placement and relocation. The API allows the fusion application to be specified as a directed dataflow graph along with the definition of the fusion functions. The API hides many programming details common to fusion applications, such as buffer management, time stamping, and exception mechanism for error control. Furthermore, using the automatic placement algorithm considerably eases the deployment of such an application. The algorithm decides where the fusion points should be placed and periodically re-evaluates the placement. DFuse is evaluated in its current implementation of iPAQs + Linux + Stampede (a distributed programming system) by measuring the delay of the API's basic commands and by measuring the ability of the placement algorithm to optimize the fusion process.

DFuse seems successful because it restricts itself to a certain type of application without making overstatements on its general application. Certainly the restrictions on the dataflow graphs that the programmer can define (i.e., sources and sinks of the fusion computation are fixed) limit the type of applications that can benefit from DFuse; nevertheless, the framework presents an interesting combination of the database and active sensor models. The arbitrary definition of fusion algorithms brings an element of the imperative active sensor model, while the definition of dataflow graphs and the automatic placement of fusion points bring an element of the declarative database model.

4.7 Conclusions

Issues that concern sensor network programmability along with the major two models for dynamic system level programmability in sensor networks have been discussed. From the individual frameworks examined one can conclude that, when efficiency is the major concern in a large and diverse set of applications, the imperative active sensor model with explicit acknowledgment of the distributed nature of the applications is the solution. On the other hand, when ease of programming in a limited set of applications (e.g., aggregation) is the major concern, the declarative database model is the solution. The research community is currently moving toward a macroprogramming vision for dynamically programming the sensor network. This vision combines elements from both existing models. It will use an active sensor framework as an underlying mechanism to execute arbitrary complex distributed algorithms into the network and a declarative framework that will enable the automatic creation of these algorithms based on well-studied run-time primitives. The declarative part can include database-like queries or dataflow graphs to make the programming task easier. Such elements have already been seen in the DFuse framework for a restricted number of applications, but the generalized large-scale implementation of a macroprogramming framework is still far from realization.

References

1. Barr R. et al., On the need for system-level support for ad hoc and sensor networks, *Operating Syst. Rev., ACM*, 36(2):1–5, April 2002.
2. Bonnet P., Gehrke J., and Seshadri P., Querying the physical world, *IEEE Personal Commun.*, 7, 10–15, October 2000.
3. Borcea C. et al., Cooperative computing for distributed embedded systems, *Proc. 22nd Int. Conf. Distributed Computing Syst. (ICDCS)*, July 2002.
4. Boulis A., Ganeriwal S., and Srivastava M., Aggregation in sensor networks: an energy accuracy trade-off, *First IEEE Int. Workshop Sensor Network Protocols Applications* (SNPA 2003), Anchorage, AK, May 11, 2003.
5. Boulis A., Han C., and Srivastava M., Design and implementation of a framework for efficient and programmable sensor networks, in *Proc. MobiSys 2003*, San Francisco, CA, May 6–8 2003.
6. Boulis A. and Srivastava M.B., A framework for efficient and programmable sensor networks, in *Proc. OPENARCH 2002*, New York, June 2002.
7. Cabri G., Leonardi L., and Zamponelli F., MARS: a programmable coordination architecture for Mobile agents, *IEEE Internet Computing*, 4(4), 26–35, Jul.–Aug. 2000.
8. Chen J.C., Yao K., and Hudson R.E., Source localization and beamforming, *IEEE Signal Process. Mag.*, 19, 30–39, March 2002.
9. Chintalapudi K.K. and Govidan R., Localized edge detection in sensor fields, *First IEEE Int. Workshop Sensor Network Protocols Applications* (SNPA 2003), Anchorage, AK, May 11, 2003.
10. Gray R.S., Agent Tcl: a flexible and secure mobile-agent system, *Proc. 4th Annu. Tcl/Tk Workshop '96*, Monterey, CA, 10–13 July 1996.
11. Heidemann J. et al., Building efficient wireless sensor networks with low-level naming, *Proc. Symp. Operating Syst. Principles*, 146–159, October 2001.
12. Hill J. and Culler D., A wireless embedded sensor architecture for system-level optimization, Intel Research IRB-TR-02-00N, 2002.
13. Hill J. et al., System architecture directions for networked sensors, *Proc. ASPLOS-IX*, 93–104, Cambridge, MA, November 2000.
14. Iftode L. et al., Programming computers embedded in the physical world, *Proc. 9th IEEE Workshop Future Trends Distributed Computing Syst.* (FTDCS 2003), May 2003.
15. Intanagonwiwat C., Govindan R., and Estrin D., Directed diffusion: a scalable and robust communication paradigm for sensor networks, *Proc. 6th Ann. Intl. Conf. Mobile Computing and Networking, MobiCOM '00*, 56–67, Boston, MA, August 2000.

16. Jaikaeo C., Srisathapornphat C., and Shen C., Querying and tasking of sensor networks, *SPIE's 14th Annu. Int. Symp. Aerospace/Defense Sensing, Simulation, Control (Digitization of the Battlespace V)*, Orlando, Florida, April 26–27, 2000.
17. Karp B. and Kung H.T., GPSR: greedy perimeter stateless touting for wireless networks, in *Proc. 6th Ann. Intl. Conf. Mobile Computing Networking, MobiCom*, 243–254, Boston, MA, 2000.
18. Kumar R. et al., *DFuse: a framework for distributed data fusion*, in *Proc. ACM SenSys* 2003, Los Angeles, CA, Nov. 5–7, 2003.
19. Levis P. and Culler D., Maté: a tiny virtual machine for sensor networks, *Proc. 10th Int. Conf. Architectural Support Programming Languages Operating Syst.* (ASPLOS X), 85–95, October 5–9 2002.
20. Madden S.R. et al., TAG: a tiny aggregation service for ad-hoc sensor networks, *OSDI Conference*, 2002.
21. Madden S.R. et al., Supporting aggregate queries over ad-hoc wireless sensor networks, *Workshop on Mobile Computing and Systems Applications*, 2002.
22. Nowak R. and Mitra U., Boundary estimation in sensor networks: theory and methods, *2nd Int. Workshop Inform. Process. Sensor Networks*, Palo Alto, CA, April 22–23, 2003.
23. Ousterhout J.K., *TCL and the TK toolkit*, Addison–Wesley, Boston, 1994.
24. Rockwell WINS nodes, http://wins.rsc.rockwell.com/.
25. Schurgers C. et al., Optimizing sensor networks in the energy–latency–density design space, *IEEE Trans. Mobile Computing*, 1(1), 70–80, Jan.–March 2002.
26. Srisathapornphat C., Jaikaeo C., and Shen C., Sensor information networking architecture, *Int. Workshop Pervasive Computing (IWPC'00)*, Toronto, Canada, August 21–24, 2000.
27. Tennenhouse D., Proactive computing, *Commun. ACM*, 43(5), 43–50, May 2000.
28. Zhao F., Shin J., and Reich J., Information-driven dynamic sensor collaboration for tracking applications, *IEEE Signal Process. Mag.*, March, 61–72, 2002.

5
Miniaturizing Sensor Networks with MEMS

	5.1	Introduction 5-1
	5.2	MEMS Basics 5-2
		Micromachine Fabrication Techniques • Highly Integrated Processes
	5.3	Sensors 5-4
		Selection Criteria • Integrated Circuit Sensors • Nanosensors
	5.4	Communication 5-5
		RF Communication • Optical Communication
	5.5	Micropower Sources 5-10
		Energy Storage • Energy Harvesting
	5.6	Packaging 5-12
Brett Warneke	5.7	Systems 5-13
Dust Networks	5.8	Conclusion 5-15

5.1 Introduction

As sensor network nodes decrease in size, denser networks can be deployed and entirely new sensor network applications will be enabled. Furthermore, smaller, lighter nodes will facilitate more network deployment methods, such as microaerial vehicles (MAV) and even air-borne dispersal. An additional side effect of miniaturization techniques based on semiconductor batch fabrication is that the manufacturing cost of the sensor nodes can be reduced for large quantities, which will allow for denser and more extensive sensor networks. These factors of discrete size and large, dense networks will enable new methods of interacting with the environment and provide more information from more places in a less intrusive way than before. Application areas enabled by miniaturized sensor nodes are numerous and include defense and intelligence networks; tracking the movements of birds, small animals, and even insects; fingertip accelerometer virtual keyboards; and interfaces for the disabled.

Sensor nodes can be divided into four major components: sensors; communication; power source; and circuits for computation, data storage, and sensor signal processing. The volume of the sensor node circuits is being reduced through dramatic process scaling and greater integration of mixed signal functions into a single chip. Microelectromechanical systems (MEMS) are similarly reducing the size and cost of sensors, some communications components, and power supplies while also reducing the power consumption of the former two. Furthermore, MEMS techniques can reduce packaging size and facilitate tighter integration.

5.2 MEMS Basics

MEMS is based on microfabrication techniques developed for microelectronics. By extending these processes, micromachining techniques have been developed to fabricate micron-scale mechanical features that are often controlled or sensed electrically, forming microelectromechanical systems. Through highly integrated processes, these electromechanical components can be fabricated alongside microelectronics, yielding complex systems.

In order to provide some background on MEMS, several of the fundamental micromachining processes will first be described, followed by the highly integrated processes that are more advantageous for miniaturizing sensor network nodes. This chapter will deal only with micromachining processes based on semiconductor microfabrication techniques because they have more promise of inexpensive batch fabrication and are more easily integrated with microelectronics for a small system size. For more information, Pierret [1] provides a good introduction to microfabrication technologies; Petersen [2] has produced the seminal paper on micromachining; Muller et al. [3] and Trimmer [4] provide collections of classic papers in the field; and references 5 through 11 are reference textbooks on micromachining and MEMS.

5.2.1 Micromachine Fabrication Techniques

Most micromachining processes begin with a substrate 100 to 600 μm thick, usually composed of silicon, other crystalline semiconductors, or quartz. Upon this substrate a number of process steps are performed, such as thin film deposition; photolithography; etching; oxidation; electroplating; machining; and wafer bonding. One of the key concepts of planar micromachining is that of sacrificial and structural layers — the former refers to thin films that are etched away to allow structures patterned in the structural layers to move. Common elemental structures include cantilevers, membranes, and plates suspended on thin or narrow flexural beams.

Bulk micromachining involves removing relatively large portions of the substrate, including the entire thickness, typically with a silicon etchant such as ethylene diamine pyrochatechol (EDP); tetramethylammonium hydroxide (TMAH); sublimated XeF_2, HNA (HF + HNO_3 + acetic acid); SF_6 plasma; or deep reactive ion etch (DRIE). A simple bulk process would involve first depositing a masking material such as SiO_2, photolithographically patterning it, and then placing the wafer in the silicon etchant for a specific period of time. If the etchant etches laterally as well as vertically (such as an isotropic etchant), the mask material will be undercut, potentially releasing structures such as cantilevers. Bulk micromachining often results in structures that move vertically. Figure 5.1 illustrates this process, except with masking layers resulting from a CMOS process.

Surface micromachining consists of depositing and patterning a series of sacrificial and structural layers on top of the wafer, followed by a final release step that etches away the sacrificial layers. A basic process would start with a silicon wafer, deposit 1 μm of SiO_2, and pattern it to form places for the structural layer to be attached to the substrate. Next, 2 μm of low-stress polysilicon would be deposited and patterned to form the microstructures. In the final step, an HF etch would remove the SiO_2 and release the structures. Surface micromachining usually produces structures that move laterally, but vertical motion is also possible.

A third style of micromachining, which combines the deep etches of bulk micromachining yet yields structures more similar to surface micromachining, begins with silicon-on-insulator (SOI) wafers. These wafers contain a several-micron thick "buried oxide" that isolates a relatively thin silicon device layer from the bulk substrate. The device layer, which is where the transistors or microstructures are formed, is usually only a couple of microns thick for CMOS wafers, but for MEMS processes it can be much thicker, such as 50 μm. After patterning a photoresist mask, the device layer is etched in a DRIE that can achieve high aspect ratios — up to 100:1. This allows the formation of deep, narrow trenches. Finally, a timed oxide etch removes the buried oxide from beneath the structures to be released. Because the structural material is single crystal silicon, very flat beams and plates can be made with no residual stress,

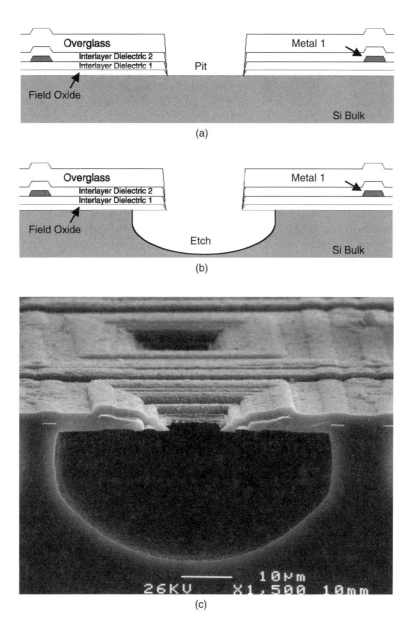

FIGURE 5.1 Cross sections of bulk micromachining in standard CMOS. (a) The wafer as it appears when it returns from the CMOS foundry with the various dielectric layers patterned so that the silicon substrate is exposed. When the wafer is then placed in an isotropic silicon etchant, such as XeF2, the silicon is dissolved and the dielectric layers become undercut, as shown in (b) and (c). (From Warneke, B. and Pister, K.S.J., *Sensors Actuators A*, 89(1–2), 142–151, 2001. With permission.)

while the thick device layer drastically reduces the compliance of beams in the vertical axis, which is advantageous for lateral structures.

5.2.2 Highly Integrated Processes

By integrating disparate components together into a single process, significant reductions in the size of the sensor node may be possible. Of particular interest are processes that combine CMOS transistors with micromachining capabilities. Analog Devices has successfully commercialized a process based on a standard BiCMOS process with a 4-μm low-stress polysilicon structural layer inserted into the flow before the interconnect metallization is deposited. An additional mask is used at the end to protect the oxide over the circuits during the sacrificial oxide release etch [12].

A number of techniques have been demonstrated to perform postprocess micromachining on foundry CMOS. One approach utilizes poly-SiGe microstructures and poly-Ge sacrificial layers on top of a CMOS wafer. These films can be deposited at temperatures low enough that the CMOS aluminum interconnects are not damaged and poly-Ge can be etched with hydrogen peroxide, which does not attack the CMOS layers [13].

One of the simplest techniques of adding micromachining to CMOS requires only a single maskless postprocess etch [14]. By stacking the contact, via, and overglass cut layers, a region of silicon will be exposed when the chip returns from the foundry. The silicon can then be sacrificially etched by bulk Si etchants such as XeF_2 [15] with the oxides and metals acting as the mask and structural layer (Figure 5.1). However, this method does not work in submicron processes that use tungsten plugs in the vias.

CMOS high aspect ratio micromachining facilitates maskless postprocessing in submicron processes by using the top metal layer as a mask for a high aspect ratio reactive ion etch that removes any oxide not protected by metal. In this way, narrow trenches down to the silicon substrate can be made. An isotropic plasma Si etch then releases the microstructures formed by the CMOS thin films.

5.3 Sensors

5.3.1 Selection Criteria

A large amount of MEMS research and product development has been in the area of sensors, so a wide variety of measurands using numerous detection techniques are available with micromachined sensors [7–9]. Examples include thermal sensors [16]; accelerometers [17]; gyroscopes [12]; pressure sensors [18]; microphones [19]; radiation detectors; magnetic sensors; flow sensors; and chemical and biological sensors. However, when selecting or designing a sensor for use in a miniature sensor network node, several criteria should be considered:

- *Volume of the complete sensor.* Although the active sensing element may be small, the complete system necessary to operate the sensor or interface it to the environment may be much larger. For example, a chemical sensor may require a sample gathering and preparation system much larger than the active region.
- *Energy consumption.* Because, for a given lifetime, energy needs directly affect the size of the power system and thus the sensor node, the energy required to make a measurement with the sensor should be minimized. The energy consumed is determined by the power consumption integrated by the time that the sensor has power applied to make a particular measurement.
- *Power consumption.* The first approach to reducing the energy consumption of a sensor is to reduce the power consumed by the device during operation, primarily by placing a high priority on minimizing the power consumption throughout the design of the sensor to guide trade-off decisions. For example, power considerations can affect the choice of detection technique — a piezoresistive sensor can have a large DC current, whereas a capacitive sensor will have no such component; however, the detection circuits are likely to have a high frequency excitation signal that consumes dynamic power. Nevertheless, power consumption cannot be considered in isolation because it is possible for the lowest power sensor to consume more energy per sample if smaller currents increase the time necessary to reach a stable measurement and thus add to the sample time and energy.
- *Suitability for power cycling.* One of the most straightforward methods of reducing the power consumed by a device is to turn it on only when necessary for as long as needed. It is therefore important that a sensor can be turned on and off relatively quickly. The gains become greater for sensors that are not sampled as frequently, such as a temperature sensor likely to be accessed once

a minute or less due to slow thermal time constants. Certain sensors, such as chemical or light, may need to integrate the measurand over a significant period of time, so their usage needs to be evaluated carefully. Additionally, low-frequency sensors such as seismometers can have long system time constants that prevent rapid power cycling. Some systems may benefit from a threshold or course sensor with a reduced energy consumption, which then triggers the sensor node to activate a higher energy consuming device with greater resolution.

- *Fabrication and assembly compatibility with the rest of the system.* A sensor that can be fabricated in the same substrate as other system components, such as the integrated circuits or communication devices, can greatly assist in building a compact node. If monolithic fabrication is not possible, assembly compatibility is also beneficial. For example, flip-chip bonding of heterogeneous substrates can yield small, integrated systems. These are all areas in which MEMS-based sensors can aid in miniaturizing a system.
- *Packaging requirements.* Some sensors may need contact with the environment, such as humidity and chemical sensors, which can limit the miniaturization potential.

5.3.2 Integrated Circuit Sensors

A number of measurands can be sensed by standard integrated circuits, which makes these sensors extremely easy to integrate with minimal additional volume. Temperature can be determined through the temperature dependence of subthreshold MOSFETs or the p–n junction of a diode or bipolar transistor. The proportional to absolute temperature (PTAT) circuit [20, 21] is most commonly used to extract the temperature signal, but other approaches that provide a digital output have been implemented, including using a counter to measure the frequency of a temperature-dependent ring oscillator [67].

Similarly, p–n junctions can also be used as photodiodes and phototransistors to measure light intensity, although a translucent window is necessary in the package. In addition, metal shields should be placed over sensitive circuits to prevent photogenerated carriers from interfering with their operation. Hall-effect sensors that detect magnetic fields can also be built from integrated circuits [22]. Sometimes ferromagnetic materials are deposited on top of the standard transistor process and patterned to form field concentrators that improve the responsivity of the sensor.

5.3.3 Nanosensors

Nanosensors can potentially provide further reductions in volume of the sensing element. The molecular scale and high relative surface area of nanowires allow precise control and sensitive detection of charged biological and chemical species [23]. In addition, nanowires can improve the responsivity of optical detectors by dramatically increasing the surface area of the detector; thermocouple-style temperature sensors are being developed with silicon nanowires. Meanwhile, carbon nanotubes have been demonstrated as chemical [24] and infrared [25] sensors.

5.4 Communication

MEMS does not impact the communication of wired sensor networks, but it can help miniaturize wireless communication. The most common form of wireless communication in use today is radio frequency (RF) radiation, including microwave and millimeter wave. However, because the relatively long wavelengths inherently limit the size of a sensor node utilizing these frequencies, free-space optical communication can be advantageous for building tiny sensor nodes.

5.4.1 RF Communication

The primary reasons to use RF communication are that it does not require line of sight and readily allows omnidirectional links. In some applications, such as asset tracking or supply chain monitoring, in which the node may be enclosed, these benefits are imperative. Nevertheless, RF does have limitations that make it less efficient for tiny, energy-constrained devices:

- Efficient antennas need to be a significant fraction of a wavelength, resulting in antennas that are many centimeters long at RF and microwave frequencies. Millimeter wave frequencies can yield more reasonably sized antennas, but the circuit efficiencies are lower and the transmission attenuation is greater.
- A small RF antenna will have very low antenna gain because beam divergence is fundamentally limited by diffraction, which is dependent on wavelength. To achieve the same milliradian collimation of an inexpensive laser pointer would require a 100-m diameter parabolic antenna at 1 GHz.
- RF transmitters have poor efficiency; a GMSK power amplifier has 50% slope efficiency (not including bias overhead), while the linear amplifiers used in CDMA systems have 10% slope efficiency. In addition, usually 1 to 100 mW of overhead is due to mixers, biasing, etc., although researchers are working to improve these efficiencies and build 100-µW radios that consume 5 nJ/(correct)b [26].
- The received power varies as the inverse of the distance raised to the second to seventh power due to multipath fading; for communication along the ground, such as cellular telephones, the average is four.

Together, these reasons make RF unattractive for tiny wireless nodes due to poor energy efficiencies and large radiators.

To illustrate these inefficiencies, the Bluetooth radio standard, which was designed for relatively low-power handheld devices, consumes about 100 nJ/b to transmit just tens of meters. Similarly, an IEEE 802.15.4 (draft) radio [27], which was actually designed for low-power wireless sensor networks, has a 100-m range; 0 dBm transmitted power (25 nJ/b); receiver sensitivity of –92 dBm; and 40 kbps data rate; it operates in the 902 to 928 MHz band. When actively communicating, it consumes 1 µJ/b on the transmit side and 2 µJ/b on the receive side, not including the power-up overhead time and idle periods.

Nevertheless, for those applications that do require RF nodes, MEMS can reduce the size of the transceiver [28, 29]. Figure 5.2 shows a block diagram of a typical wireless transceiver front end with a superheterodyne architecture. A relatively large number of high-Q, passive components are shown, including ceramic and SAW filters, discrete inductors, and discrete tunable capacitors (varactors) that cannot be fabricated with conventional integrated circuit processes. These components thus must be implemented with off-chip devices that end up dominating the size of the transceiver. Fortunately, micromachined components have been developed that may be able to replace each of these off-chip components; this will reduce the overall size of the transceiver through physically smaller components and the potential for integration with the integrated circuit chips.

Voltage-tunable high-Q capacitors can be fabricated by suspending a top aluminum plate on soft flexures over a bottom plate [30]. A DC bias on the resulting capacitor causes an electrostatic force to pull the top plate down, thus varying the capacitance. Such a structure has been demonstrated with a Q of 62 at 1 GHz.

There are a number of approaches for fabricating on-chip high-Q inductors. Two techniques improve the Q of normal planar inductors (which is 1 to 3 at 1 GHz): the first utilizes a NiFe thin film under the spiral to act as a core to increase the magnetic flux and thus the Q (6.6 at 4 MHz [31]); the second method uses a front- [32] or back-side silicon etch to remove the lossy substrate from underneath the spiral and achieve Qs of 5 at 1 GHz and 60 to 80 at 40 GHz. The latter approach is more readily integrated with circuits because it can be implemented with a postprocess etch, while the former requires adding nonstandard metal depositions into the process. More exotic fabrication techniques can be used to build three-dimensional inductors. Figure 5.3 shows a four-turn inductor fabricated on a silicon substrate with 5 µm-thick copper traces electroplated around an alumina insulating core with a 650 × 500 µm cross section. Direct-write laser lithography is used to pattern the top and sidewall photoresist. This device achieves 14 nH of inductance and a Q of 16 at 1 GHz, while a similar one-turn coil obtains an inductance of 4.8 nH and a Q of 30 at 1 GHz [33].

Miniaturizing Sensor Networks with MEMS

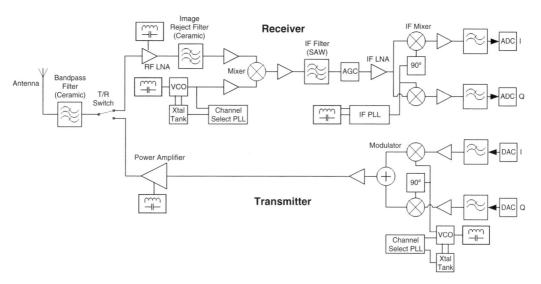

FIGURE 5.2 Diagram of a typical wireless transceiver front end showing the many off-chip, high-Q, passive components, such as filters, inductors, and capacitors, that could be replaced by micromechanical versions. Besides the component-size reduction, such components could potentially be integrated with the circuits for dramatic volume reductions.

FIGURE 5.3 Four-turn inductor fabricated on a silicon substrate with electroplated copper around an insulating core. It has an inductance of 14 nH and a Q of 16 at 1 GHz. (From Young, D.J. et al., *Tech. Dig., Int. Electron Devices Meeting*, Washington, D.C., December 1997, 67–70. With permission.)

Some applications, such as replacing quartz crystals and ceramic and SAW filters, require even higher Qs than these devices can provide. Just as the macroscopic domain utilizes vibrating mechanical structures, the microscopic domain can achieve high Qs through vibrating resonators with a second-order response. Thin film bulk acoustic resonators (FBAR) are composed of a metal–piezoelectric–metal film stack, similar to a quartz crystal, and suspended on a thin membrane to provide acoustic isolation from the substrate. Such resonators can achieve a Q of 1200 at 1.9 GHz with an area of only 100×100 µm^2 [34] and are currently in production.

An alternative method of building micromechanical resonators uses surface-micromachined polysilicon to suspend a flexural-mode beam over an electrode. The beam is electrostatically excited, resulting in the second-order resonance. Qs of 7450 have been achieved at 92 MHz [35]. These structures can also be mechanically coupled to form high-Q filters and filter-mixer structures [36] that allow multiple components of the system illustrated in Figure 5.2 to be replaced with a single passive micromechanical component. This could reduce size and power consumption of the transceiver. A similar fabrication process can also produce a ring or disk with a central anchor driven laterally through a submicron gap to obtain a Q of 9400 at 156 MHz. One of the major problems with micromechanical resonators is that they have relatively low power handling capabilities that limit their use in applications such as cellular telephones; however, these limits are high enough to be applicable to distributed wireless sensor networks that utilize short-range, multihop communication links.

Finally, the transmit/receive diplexer switch can be replaced with micromechanical relays that feature lower insertion loss ("on" impedance) and larger isolation ("off" impedance). The two major styles of switches are (1) cantilever beams with electrostatic pull-down electrodes and metal–metal contacts for DC operation; and (2) suspended membranes that are electrostatically deformed to increase the capacitive coupling through the structure dramatically [37]. Cantilever-style switches have been demonstrated with an actuation voltage of 30 V, >50 dB of isolation below 2 GHz, and <0.2 dB of insertion loss from DC to 40 GHz [38]. Besides their use as diplexers, the nearly ideal behavior of RF switches can be used to build small tunable filters, multiband antennas, true-time delay phased-array antennas, and even reconfigurable transceiver architectures [39].

It should be noted that although Figure 5.2 provides a good discussion point because of the large number of high-Q components that could be replaced by MEMS components, it is not the only transceiver architecture possible. For example, direct-conversion (zero-IF) [40] and subsampling [41] transceivers eliminate many of the filters. In addition, if the channel selectivity and other parameters of the radio band are relaxed, high-Q components may not be necessary, although the use of higher Q components can often lead to lower power consumption because of the reduced losses.

5.4.2 Optical Communication

Free-space optical communication has many advantages for miniature sensor nodes:

- Optical radiators such as mirrors and laser diodes can be made extremely tiny — 0.03-µm^3 lasers have been demonstrated [42].
- As mentioned earlier, optical transmission provides extremely high antenna gain, which yields higher transmission efficiencies.
- Although laser output slope efficiencies are only about 25%, the diode turn-on current overhead can be as low as 1 µW for vertical cavity surface emitting lasers (VCSELs), so the effective output efficiency can be much higher than RF power amplifiers.
- The received power only decays as the inverse of the distance squared, assuming line of sight.
- The high directivity of optical communication enables the use of spatial division multiple access (SDMA) [43], which is a simple network media access technique in which an imaging receiver can separately process simultaneous transmissions from different angles. SDMA thus requires no communication overhead and has the potential to be more energy efficient than RF media access methods such as frequency, time, and code division multiple access (FDMA, TDMA, CDMA).

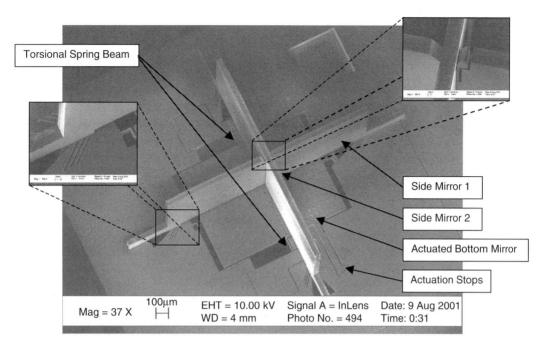

FIGURE 5.4 A quad-corner cube retroreflector (CCR) used for passive optical transmission. The electrostatically actuated bottom mirror rotates torsionally to disturb the orthogonality of the corner and switch the light reflected from the CCR from the "1" to "0" states. The insets show the spring locks that aid in assembly and maintain alignment. The device is fabricated on an SOI wafer with a 50-μm thick device layer using deep reactive ion etching. (From Zhou, L. et al., *IEEE J. Microelectromech. Syst.*, 12(3), 233–242, 2003. With permission.)

- It is extremely difficult to eavesdrop on collimated optical communication (low probability of detection and low probability of intercept), which is a significant security advantage.

The primary drawbacks of optical communication are that line of sight is necessary for all but the shortest distances and the narrow beams imply the need for accurate pointing. Fortunately, MEMS technology and clever algorithms can provide accurate pointing [44] and multihop, self-healing networking can allow messages to travel around certain obstacles.

The two primary methods of free-space optical transmissions are passive reflective systems and active-steered laser systems. The passive reflective system consists of three mutually orthogonal mirrors that form the corner of a cube (Figure 5.4) [45] — thus the name corner cube retroreflector (CCR). Light entering the CCR bounces off each of the mirrors and is reflected back to the sender parallel to the incoming beam. By electrostatically actuating the bottom mirror, the orthogonality can be disturbed, causing the reflection to no longer return to the sender. This behavior allows the CCR to communicate with an interrogator by simply modulating the reflected light and resembles the operation of a heliograph in which the operator bounces sunlight off a mirror to transmit Morse code messages to other ships. This is a concept that can be traced back to Greece in the fifth century B.C. Because the only energy consumed is that required to charge 3 pF of capacitance in the actuator, this is much more efficient than an approach that requires the generation of radiation, such as RF or lasers.

The device shown in Figure 5.4 is fabricated using deep reactive ion etching (DRIE) in an SOI wafer with a 50-μm device layer for flat, smooth mirror surfaces. It consumes 16 pJ/b transmitted, has a demonstrated range of 180 m, transmission data rates in excess of 4 kbps, and a size of $2 \times 2 \times 0.5$ mm, although it can be made smaller if less reflection is acceptable. One restriction with CCR-based communication is that it does not facilitate peer-to-peer communication, so a one-to-many network topology is required; however, distributed algorithms are under development to take advantage of such a network

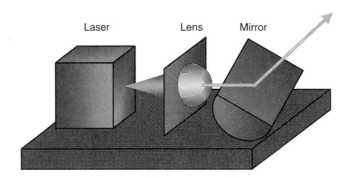

FIGURE 5.5 Conceptual diagram of a steered agile laser transmitter. A laser diode emits a beam that is collimated by a lens (may be micromachined) before bouncing off the MEMS beam steering mirror, which aims the beam toward the intended receiver.

for things such as sensor data compression. Furthermore, the communication range of a sub-mm CCR is theoretically limited to about 1 km in a practical implementation.

Active-steered laser communication utilizes a small laser diode, such as a VCSEL, a collimating lens, and MEMS beam-steering optics to transmit a tightly collimated light beam to a particular receiver (Figure 5.5). This facilitates peer-to-peer communication over a wide area, while maintaining many of the features of optical communication including high directivity and long-distance communication using little power. Because efficient lasers cannot be fabricated in silicon, monolithic integration is unlikely; however, micromachined structures can be used to aid in the alignment of a bare laser diode onto a chip [46]. On the other hand, three-dimensional micromachined collimating lenses have been demonstrated using reflowed photoresist [47]. The beam-steering optics are the most challenging part of the system because they should have close to hemispherical range, low actuation range, low cross-axis sensitivity, and be robust against shock. Current approaches use multilevel SOI MEMS for very flat mirrors, low cross-axis sensitivity, and robustness [48], but have only achieved up to 40° of optical deflection angle with a rather high actuation voltage of 90 V [49].

Finally, to illustrate the dramatic differences between the various communication schemes discussed, Figure 5.6 compares the communication range vs. energy/bit consumption of CCR, green laser, and GSM RF communication.

5.5 Micropower Sources

Miniature sensor nodes can be powered from energy storage or energy scavenging devices or a combination thereof. In addition, to allow larger peak currents or integration of charge from energy harvesters to compensate for lulls such as nighttime for a solar cell, capacitors may be used in these systems to lower the effective impedance of a battery or energy harvester. High-density capacitors, such as the Ultracapacitor [50], can store up to 10 mJ/mm^3, which is less than 1% the energy density of lithium cells.

5.5.1 Energy Storage

From the system's perspective, a good microbattery should have the following features:

- High energy density
- Large active volume to packaging volume ratio (i.e., a thin film on top of a 500-μm silicon wafer would not be desirable)
- Small cell potential (0.5 to 1.0 V) so digital circuits can take advantage of the quadratic reduction in power consumption with supply voltage
- Ability to configure efficiently into a series of cells to provide a variety of potentials for the various components of the system without requiring the overhead of voltage converters

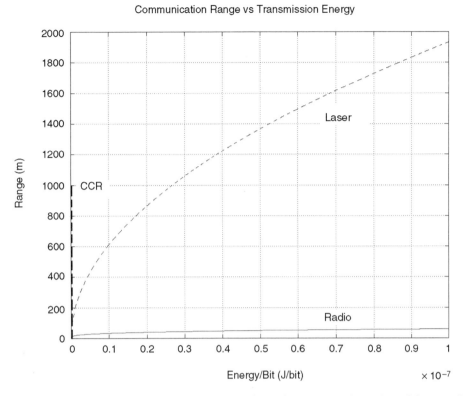

FIGURE 5.6 Communication range vs. transmission energy for RF (GSM, 1 GHz, isotropic, path loss $n = 4$); laser (532 nm green, 1 mW, 1 Mbps, 200 × 200 µm receiver aperture); and CCR (400 µm passive, 16 pJ/b independent of distance up to 1 km).

- Rechargeable in case the system has an energy harvester

A variety of tiny batteries are being developed, including thin-film vanadium oxide and molybdenum oxide [51] that are fabricated using spin-casting sol-gel techniques and micromachined cavities containing an electrolyte, although the latter devices do not have high energy densities [52]. Nickel-zinc batteries have been developed with a footprint of 2 mm^2, < 100 µm thick, and a capacity of 20 mJ/mm^2 with a discharge rate of >1 mA/mm^2 [53]. Another potential candidate chemistry is rechargeable thin-film lithium energy cells. Researchers at Oak Ridge National Laboratory have built 1 cm^2 × <15 µm Li-LiCoO$_2$ batteries with a 40,000 charge/discharge cycle life and a capacity up to 24 mJ/mm^2 [54–56]. A derivative process at the Jet Propulsion Laboratory uses microfabrication techniques to generate batteries as small as 50 × 50 µm with a 0.25-µm cathode film and capable of energy densities of 1.4 mJ/mm^2 [57].

One of the highest energy density battery chemistries available is the Zn-air cell. It is also available in the smallest button cell package: the Energizer IEC-PR63 weighs 0.2 g (including packaging); is 0.051 cm^3; V_{oc} = 1.4 V; and contains 33 mAh (160 J). TPL Inc. is using micromachining techniques to develop Zn-air volumetric batteries 2 mm in diameter and 0.5 mm thick with a capacity approaching 1 mAh (3 J/mm^3) [58]. With an areal capacity of 1.6 J/mm^2, the advantage of the volumetric approach is evident over the thin-film lithium batteries if maintaining a small footprint is a priority. To meet the demand for higher discharge current, TPL proposes to combine supercapacitors that store 30 mJ in a similar size in parallel with the batteries.

The biggest problem with current Zn-air cells is that the self-discharge is so high that, after the air terminal is opened, they have a shelf life of only a couple of weeks, although a micro Zn-air cell could potentially incorporate a micromachined air valve to control this self-discharge. The sensor node would then operate primarily off a capacitor that would be charged periodically by opening the air valve. An additional problem with Zn-air cells is that they are not rechargeable. This chemistry is thus only a candidate in short-term deployments. Even though process compatibility with the other components of the system may seem desirable, it may actually not be important due to the possibility of stacking the various components because the batteries do not need exposure to the environment.

In addition to chemical energy storage, radioactive isotopes provide another method of storing energy on a small sensor node; such techniques are already used extensively in deep space probes and satellites where long life and reliable operation are essential, just as in wireless sensor networks. Blanchard and coworkers [59] have demonstrated a micromachined radioactive battery based on a thin-film beta emitter coating a beam that performs a charge to mechanical conversion (as the beta particles leave, the beam acquires a positive charge, causing it to be attracted to the substrate). This is followed by a mechanical to electrical conversion using a piezoelectric material (the strain of the bending beam is converted to charge), also on the beam. Two companies [60, 61] have also proposed building millimeter-scale radioactive power sources based on beta emitters, the first of which is using betavoltaics — the direct conversion of beta particles to electricity by bombarding a p–n junction.

5.5.2 Energy Harvesting

Scavenging energy from the environment will allow the wireless sensor nodes to operate nearly indefinitely, without their batteries dying. Solar radiation is the most abundant energy source and yields around 1 mW/mm^2 (1 J/day/mm^2) in full sunlight or 1 μW/mm^2 under bright indoor illumination. Solar cells have conversion efficiencies up to 30%.

Vibration has been proposed as an energy source [62, 63] that can be scavenged. Vibration spectra of office windows, copy machines, microwave ovens, industrial motors, freeway traffic, and the human gait reveal that usable energy is there — typically on the order of 10 μW/g of mass of the converter. Because the mass of a cubic millimeter of silicon is about 2 mg, this energy source is only feasible at the centimeter scale and above. The basic device used to extract energy from vibrations is a mass on a spring connected to a variable capacitor. In actual implementation, a lateral or gap-closing comb resonator is typically used. A precharged reservoir, such as a capacitor or rechargeable battery, a storage capacitor, and two switches form the basic charge-constrained conversion circuit.

More exotic energy sources that have been proposed include utilizing the excess heat from microrocket engine combustion [64]; using copper and zinc electrodes to generate power from seawater; and harvesting ATP for *in vivo* applications. For applications in which duty cycling is acceptable, solar cells or other power scavenging sources can be used to trickle charge a capacitor or battery, after that the stored energy can be used at much higher power rates than the charging pace.

5.6 Packaging

As the size of the sensor node decreases, the packaging considerations become more critical to prevent the package from dominating the volume and since nonstandard packaging is necessary. Some of the requirements of the packaging include:

- The microstructure, such as a CCR or accelerometer, must be protected while still being able to move.
- Electrical connections between various chips, such as bond wires or vertical interconnects from a battery, need to be facilitated and protected.
- Solar cells require clear packaging and possibly a lens to improve the collection efficiency.
- An optical receiver photodiode may require an optical filter.
- A CCR requires an antireflective (AR)-coated cover that allows illumination along its primary axis of [111].
- The packaging must add a minimum of extra volume.

- The deployment method used in the application will place certain requirements on the packaging. For example, micro air vehicle deployment would require the packaging to protect the sensor node from being dropped 100 ft.
- Toxic battery chemistries need sufficient shielding in case a human or animal swallows the node.
- Vibration harvesting devices need a solid mechanical connection to the environment.
- Sensors may require special access to the environment, so packages may require tailoring to the application. Examples include humidity, pressure, acoustics, strain, gaseous chemical and biological sensors, and fluidic sensors.

The use of a common substrate is also a consideration because it can ease assembly, but adds volume. The die substrates can be thinned to help reduce the impact of a common substrate.

Micromachining techniques can help meet some of the packaging requirements. For instance, microstructures such as accelerometers and resonators can be fabricated in sealed vacuum cavities by defining the cavity with a sacrificial layer; depositing a structural layer; removing the sacrificial layer through a small access hole; and then sealing the cavity by depositing a CVD, sputtered, or evaporated film or by growing an oxide on a polysilicon layer until the hole is sealed. Wafer bonding can also be used to protect microstructures within a hole or cavity in the wafer. A variety of microassembly technologies [65], such as pick-and-place methods for the microdomain, batch transfer, fluidic microassembly, and flip-chip bonding, facilitate the compact assembly of heterogeneous dies.

The CCR poses some of the most difficult packaging constraints because the device must be mechanically protected, allowed to move, and have good optical properties. Three options were proposed in Hsu [66]:

- A hemispherical cover can cause lensing effects if the diameter is too small, which affects the performance of the CCR.
- A flat plate elevated on short walls eliminates the lensing effects, but the plate must be large to avoid the edge blocking the light. Because the optimum axis of the CCR is at a 45° angle to the plate and the reflectivity of the plate increases as the angle of incidence increases, this approach is not optically efficient.
- A pyramid that has surfaces normal to the body diagonals of the CCRs can be used. Because the optimum incident angles for the CCRs are closer to normal to the package, reflections will be reduced.

Steered agile laser communication also requires a package that mechanically protects the micro-optical system, allows it to move, and has good optical properties. However, because an input optical beam is not necessary, a simple hemispherical cover is the best option.

For cubic millimeter sensor nodes, such as that shown in Figure 5.9, the best proposed solution at this time involves potting the node in an optical-quality polymer with some special molds as shown in Figure 5.7. This package provides many of the necessary features detailed previously, including providing access to the environment by molding holes in the polymer. An antireflective coating can probably be placed on the polymer at the end of the process.

5.7 Systems

A number of wireless sensor nodes have been developed that take advantage of MEMS to achieve a small size. Mason and colleagues [67] at the University of Michigan created a multisensor microcluster that measures temperature, pressure, humidity, and vibration/position. It includes a microcomputer, has a 50-m RF link, is less than 10 cm^3 (Figure 5.8), operates off a single battery, and consumes 530 µW average power and 10 mW while transmitting.

The microsystem contains a variety of chips: a commercial microcontroller (Motorola 68HC11); a power management chip; a commercial transmitter (RFM HX1005); a capacitive interface chip with an

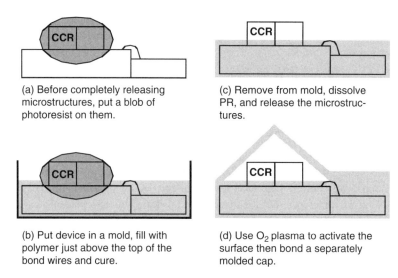

FIGURE 5.7 Polymer encapsulation process for cubic millimeter sensor node packaging.

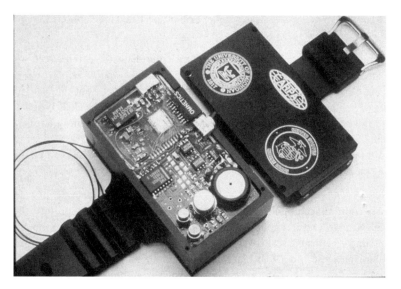

FIGURE 5.8 Multisensor microcluster containing MEMS pressure, humidity, and acceleration sensors and an RF transmitter with a 50-m range. The device is less than 10 cm^3. (Personal communication from K. Wise.)

integrated temperature sensor; a capacitive barometric pressure sensor; a capacitive relative humidity sensor; two accelerometers; a threshold accelerometer interface chip; and a lithium coin cell. The pressure sensor is fabricated using bulk micromachining and a silicon-glass dissolved-wafer process to create multiple diaphragms that segment the pressure range. The humidity sensor is fabricated with high-aspect-ratio micromolding and electroplating to form a series of interdigitated electrodes. A thin polymer film, whose dielectric constant varies as a function of moisture, fills the gaps between the electrodes and causes the capacitance to vary with humidity. A z-axis accelerometer is fabricated in a three-mask dissolved-wafer process and contains a proof mass suspended by torsional beams. At the end of the proof mass, a set of comb fingers is interdigitated with a set of fixed comb fingers that provide capacitive sensing of the movement of the proof mass. Finally, an array of threshold accelerometers, which are simply cantilever switches with varying proof masses and spring constants, is fabricated using the dissolved wafer process;

FIGURE 5.9 16-mm³ Smart Dust mote, showing a 0.25-µm CMOS ASIC with optical receiver, ambient light sensor, and controller; solar power array; accelerometer; and CCR, each on separate die. (From Warneke, B.W. et al., *Proc. IEEE Int. Conf. Sensors 2002*, Orlando. With permission.)

p++ etch stop proof masses; oxide suspension beams; and gold contacts. A second-generation microcluster system reduced the volume to less the 5 cm³, while forthcoming versions will be around 1 cm³ and even down to 0.2 cm³.

The Wireless Integrated Network Sensor (WINS) project at UCLA [68] developed a sensor node that included an infrared imager; seismometer; spectrum analyzer; RF transceiver; and lithium coin cells in a volume on the order of tens of cubic inches. The sensor integration relied on flip-chip bonding structures to a low temperature, cofired ceramic (LTCC) substrate that provided a platform for support of interface, signal processing, and communication circuits. In addition, the LTCC substrate provides small, embedded low-loss capacitors and high-Q inductors that are used by the transceiver. The infrared imager and seismometer were fabricated with bulk micromachining and flip-chip bonding. WINS also explored building a loop antenna on a CMOS die by removing the silicon substrate with a XeF_2 etch.

The PicroRadio project [26] at UC Berkeley is developing an ultralow energy transceiver for ubiquitous wireless data acquisition. The goal is to consume less than 5 nJ/(correct)b and less than 100 µW. The transceiver uses FBARs for low-phase noise oscillators [69] and filters, while vibration harvesting is being investigated for the power source [63].

The most extensive use of MEMS for miniaturizing wireless sensor nodes is the Smart Dust project [70] at UC Berkeley that seeks to push the volume of wireless sensor nodes aggressively down to a cubic millimeter. Figure 5.9 shows a 16-mm³ autonomous solar-powered sensor node [71] with bidirectional optical communication. The system consists of four die: a 0.25-µm CMOS ASIC; a trench-isolation SOI solar cell array; a micromachined four-quadrant CCR; and a capacitive accelerometer. The ASIC contains an optical receiver that consumes 69 pJ/b; an ADC that uses 180 pJ/8-b sample; a photosensor for measuring ambient light; a finite state machine to control the system; and a 1-µW, 3.9-MHz integrated oscillator. A new DRIE SOI/CMOS process has been developed to allow integration of solar cells, CCR, and accelerometer along with high-voltage FETs. Figure 5.10 shows the resulting die combined with the same ASIC as in Figure 5.9 for a total device size of 6.6 mm³.

5.8 Conclusion

Many aspects of wireless sensor network nodes can be miniaturized with MEMS technology. From the sensors to the wireless communication components and power supply, MEMS is reducing volume, improving performance, and reducing cost through batch fabrication techniques. In addition, MEMS

FIGURE 5.10 Mock-up of a 6.6-mm³ autonomous Smart Dust mote. This mote has the same functionality as the one in Figure 5.9 — the CMOS ASIC is identical but process integration allowed the devices on the other die to be fabricated on a single die, thus reducing the size.

packaging and assembly techniques can help build miniature systems out of these small components. By miniaturizing sensor networks, not only will new applications be enabled, but they can also be deployed in more places, with higher densities and less interference to the monitored area, thus allowing improved data gathering. In this way the physical world can truly be instrumented.

References

1. Pierret, R.F., *Introduction to Microelectronic Fabrication*, Addison–Wesley, Menlo Park, CA, 1990.
2. Pierret, K., Silicon as a mechanical material, *Proc. IEEE*, 70(5), 420–457, 1982.
3. Muller, R.S., Howe, R.T., Senturia, S.D., Smith, R.L., and White, R.M. (Eds.), *Microsensors*, IEEE Press, New York, 1991.
4. Trimmer, W.S., *Micromechanics and MEMS: Classic and Seminal Papers to 1990*, IEEE Press, New York, 1997.
5. Madou, M., *Fundamentals of Microfabrication*, CRC Press, Inc., Boca Raton, FL, 2002.
6. Elwenspoek, M. and Jansen, H.V., *Silicon Micromachining*, Cambridge University Press, 1999.
7. Sze, S.M., *Semiconductor Sensors*, John Wiley & Sons, Sommerset, NJ, 1994.
8. Ristic, L.J. (Ed.), *Sensor Technology and Devices*, Artech House, London, 1994.
9. Kovacs, G.T.A., *Micromachined Transducers Sourcebook*, WCB McGraw–Hill, San Francisco, 1998.
10. Senturia, S.D., *Microsystem Design*, Kluwer Academic Publishers, Norwell, MA, 2001.
11. Gad–El-Hak, M. (Ed.), *The MEMS Handbook*, CRC Press, Inc., Boca Raton, FL, 2001.
12. Geen, J.A. et al., Single-chip surface-micromachined integrated gyroscope with 50°/hour root Allan variance, *2002 IEEE Int. Solid-State Circuits Conf. Dig. Tech. Papers*, 45, 426–427, 2002.
13. Franke, A.E., King, T.-J., and Howe, R.T., Integrated MEMS technologies, *MRS Bull.*, 26(4), 291–295, *Mater. Res. Soc.*, 2001.
14. Parameswaran, M. et al., A new approach for the fabrication of micromechanical structures, *Sensors Actuators*, 19, 289–307, 1989.
15. Warneke, B. and Pister, K.S.J., *In situ* characterization of CMOS postprocess micromachining, *Sensors Actuators A (Physical)*, 89(1–2), 142–151, 2001.
16. Baltes, H. et al., Micromachined thermally based CMOS microsensors, *Proc. IEEE*, 86, 1660–1678, 1998.

17. Yazdi, N., Ayazi, F., and Najafi, K., Micromachined inertial sensors, *Proc. IEEE*, 86(8), 1640–1659, 1998.
18. Eaton, W.P. and Smith, J.H., Micromachined pressure sensors: review and recent developments, *Smart Mat. Struct.*, 6(5), 530–539, 1997.
19. Rombach, P. et al., The first low voltage, low noise differential silicon microphone, technology development and measurement results, *Sensors Actuators A (Physical)*, A95(2–3), 196–201, 2002.
20. Timko, M.P., A two-terminal IC temperature transducer, *IEEE J. Solid State Circuits*, SC-11(6), 784–788, 1976.
21. Hosticka, B.J., Fichtel, J., and Zimmer, G., Integrated monolithic temperature sensors for acquisition and regulation, *Sensors Actuators*, 6(3), 191–200, 1984.
22. Nakamura, T. and Maenaka, K., Integrated magnetic sensors, *Sensors Actuators*, A22(1–3), 762–769, 1990.
23. Cui, Y. et al., Nanowire nanosensors for highly sensitive and selective detection of biological and chemical species, *Science*, 293, 1292–1298, August 2001.
24. Kong, J. et al., Nanotube molecular wires as chemical sensors, *Science*, 287(5453), 28 622–625, 2000.
25. Xu, J.M., Highly ordered carbon nanotube arrays and IR detection, *Infrared Phys. Technol.*, 42, 485, 2001.
26. Rabaey, J. et al., PicoRadios for wireless sensor networks: the next challenge in ultra-low-power design, *2002 IEEE Int. Solid-State Circuits Conf. Dig. Tech. Papers*, San Francisco, 45, 200–201, 2002.
27. http://www.amis.com/wireless/ASTRX1.html.
28. Nguyen, C.T.-C., Katehi, L.P.B., and Rebeiz, G.M., Micromachined devices for wireless communications, *Proc. IEEE*, 86(8), 1756–1768, 1998.
29. Rebeiz, G.M., *RF MEMS: Theory, Design, and Technology*, John Wiley & Sons, Sommerset, NJ, 2002.
30. Young, D.J. and Boser, B.E., A micromachined variable capacitor for monolithic low-noise VCOs, *Tech. Dig., 1996 Solid-State Sensor Actuator Workshop*, Hilton Head Island, SC, 86–89, 1996.
31. Von Arx, J.A. and Najafi, K. On-chip coils with integrated cores for remote inductive powering of integrated microsystems, *Dig. Tech. Papers, 1997 Int. Conf. Solid-State Sensors Actuators (Transducers'97)*, Chicago, IL, June 16–19, 1997, 999–1002.
32. Rofougaran, A. et al., A 1 GHz CMOS RF front-end IC for a direct-conversion wireless receiver, *J. Solid State Circuits*, 31, 880–889, 1996.
33. Young, D.J. et al., Monolithic high-performance three-dimensional coil inductors for wireless communication applications, *Tech. Digest, Int. Electron Devices Meeting*, Washington, D.C. December 1997, 67–70.
34. Ruby, R. Micromachined cellular filters, *IEEE MTT-S Int. Microwave Symp. Dig.*, 2, 1149–1152, June 1996.
35. Nguyen, C.T.-C., Vibrating RF MEMS for low power wireless communications (invited keynote), *Proc., 2000 Int. MEMS Workshop (iMEMS'01)*, Singapore, July 4–6, 2001, 21–34.
36. Wong, A.-C., Ding, H., and Nguyen, C. T.-C., Micromechanical mixer + filters, *Tech. Dig., IEEE Int. Electron Devices Meeting (IEDM)*, San Francisco, CA, 1998, 471–474.
37. Goldsmith, C. et al., Characteristics of micromachined switches at microwave frequencies, *IEEE MTT-S Dig.*, 1141–1144, June 1996.
38. Hyman, D. et al., Surface-micromachined RF MEMs switches on GaAs substrates, *Int. J. RF Microwave Computer Aided Eng.*, 9(4), 348–361, 1999.
39. Izadpanah, H. et al., Reconfigurable low power, light weight wireless system based on the RF MEM switches, *1999 MTT-S Int. Topical Symp. Technol. Wireless Appl. Dig.*, Feb. 21–24, 1999, Vancouver, BC, 175–180.
40. Abidi, A.A., Direct-conversion radio transceivers for digital communications, *IEEE J. Solid-State Circuits*, 30(12), 1399–1410, 1995.
41. Sheng, S. et al., A low-power CMOS chipset for spread spectrum communications, *1996 Int. Solid State Circuits Conf. Dig. Tech. Papers*, 346–347, Feb. 1996.

42. Painter, O. et al., Two-dimensional photonic band-gap defect mode laser, *Science*, 284(5421), 1819–1821, 1999.
43. Kahn, J.M. et al., Imaging diversity receivers for high-speed infrared wireless communication, *IEEE Commun.*, 88–94, Dec. 1998.
44. Last, M. et al., Toward a wireless optical communication link between two small unmanned aerial vehicles, *Int. Symp. Circuits Syst. 2003*, Bangkok, Thailand, 3, III-930-3, May 2003.
45. Zhou, L., Kahn, J.M., and Pister, K.S.J., Corner-cube retroreflectors based on structure-assisted assembly for free-space optical communication, *IEEE J. Microelectromech. Syst.*, 12(3), 233–242, 2003.
46. Lin, L.Y. et al., Micromachined integrated optics for free-space interconnections, *Proc. IEEE Microelectromech. Syst. Conf.*, Amsterdam, Netherlands, Jan. 29–Feb. 2, 1995, 77–82.
47. Toshiyoshi, H. et al., A surface micromachined optical scanner array using photoresist lenses fabricated by a thermal reflow process, *J. Lightwave Technol.*, 21(7), 1700–1708, 2003.
48. Zhou, L. et al., Two-axis scanning mirror for free-space optical communication between UAVs, *IEEE Conf. Optical MEMS*, Waikoloa, HI, August 18–21, 2003.
49. Milanovic, V., Last, M., and Pister, K.S.J., Laterally actuated torsional micromirrors for large static deflection, *Photonics Technol. Lett.*, 15(2), 245–247, 2003.
50. http://www.powercache.com.
51. Harreld, J. H., Dong, W., Dunn, B. Ambient pressure synthesis of aerogel-like vanadium oxide and molybdenum oxide, *Mat. Res. Bull.*, 33(4), 561–567, 1998.
52. Lee, K.B. and Lin, L., Electrolyte-based on-demand and disposable microbattery, *Proc. 15th Annu. Int. Conf. Microelectromech. Syst. (MEMS 2002)*, Las Vegas, Nevada, 20–24 Jan. 2002, 236–239.
53. Humble, P.H., Harb, J.N., and LaFollette, R.M., Microscopic nickel-zinc batteries for use in autonomous microsystems, *J. Electrochem. Soc.*, 148(12), A1357, 2001.
54. Neudecker, B.J., Dudney, N.J., and Bates, J.B., "Lithium-free" thin-film battery with *in-situ* plated anode, *J. Electrochem. Soc.*, 147, 517–523, 2000.
55. Oak Ridge Micro-Energy, Inc., http://www.oakridgemicro.com/
56. Front Edge Technology, Inc., http://www.frontedgetechnology.com/
57. West, W.C. et al., Fabrication and testing of all solid-state microscale lithium batteries for microspacecraft applications, *J. Micromechan. Microeng.*, 12, 58–62, 2002.
58. TPL, Inc., http://www.tplinc.com/
59. Blanchard, J.P. et al., Radioisotope power for MEMS devices. *ANS Trans. Am. Nucl. Soc.*, 86, 186–187, 2002.
60. Qynergy, Corp., http://www.qynergy.com/.
61. TRACE Photonics, Inc., Charleston, IL.
62. Roundy, S., Wright, P., and Pister, K.S.J. Micro-electrostatic vibration-to-electricity converters, *Intl. Mech. Eng. Conf. Exp. 2002*, IMECE2002-39309, Nov. 17–22, 2002, New Orleans, LA.
63. Meninger, S. et al., Vibration-to-electric energy conversion, *Proc. 1999 Int. Symp. Low Power Electron. Design*, San Diego, CA, 16–17 Aug. 1999, 48–53.
64. Teasdale, D. et al., Thrust and electrical power from solid propellant microrockets, *Proc. 14th Annu. Int. Conf. Microelectromech. Syst. (MEMS 2001)*, Interlaken, Switzerland, Jan. 2001.
65. Cohn, M.B. et al., Microassembly technologies for MEMS, *Proc. SPIE*, 3511, *Micromachining Microfabrication Process Technol. IV*, Santa Clara, CA, 3511, 2–16; 21–22, 1998.
66. Hsu, V., *M.S. Report*, University of California, Berkeley.
67. Mason, A. et al., A generic multielement microsystem for portable wireless applications, *Proc. IEEE*, 86(8), 1733–1746, 1998.
68. Asada, G. et al., Wireless integrated network sensors: low power systems on a chip, *Proc. 1998 Eur. Solid State Circuits Conf.*, The Hague, 22–24 Sept. 1998, 9–16.
69. Otis, B. and Rabaey, J., A 300μW 1.9GHz CMOS oscillator utilizing micromachined resonators, *Proc. Eur. Solid-State Circuits Conf. (ESSIRC)*, Florence, Italy, Sept. 24–26, 2002.

70. Warneke, B. et al., Smart dust: communicating with a cubic-millimeter computer, *Computer Mag.*, IEEE, Piscataway, NJ, 44–51, Jan. 2001.
71. Warneke, B.A. et al., An autonomous 16-mm^3 solar-powered node for distributed wireless sensor networks, *Proc. IEEE Int. Conf. Sensors 2002*, Orlando, FL, June 12–14, 2002.

6
A Taxonomy of Routing Techniques in Wireless Sensor Networks

Jamal N. Al-Karaki
Iowa State University

Ahmed E. Kamal
Iowa State University

6.1 Introduction 6-1
   Motivation and Design Issues in WSN Routing • Routing Challenges in WSNs
6.2 Routing Protocols in WSNs.. 6-6
   Flat Routing • Hierarchical Routing • Adaptive Routing • Multipath Routing • Query-Based Routing • Negotiation-Based Protocols
6.3 Routing in WSNs: Future Directions............................. 6-21
6.4 Conclusions 6-22

6.1 Introduction

Wireless sensor networks (WSNs) contain hundreds or thousands of sensor nodes equipped with sensing, computing and communication abilities. Each node has the ability to sense elements of its environment, perform simple computations, and communicate among its peers or directly to an external base station (BS) (Figure 6.1). Deployment of a sensor network can be in random fashion (e.g., dropped from an airplane) or planted manually (e.g., fire alarm sensors in a facility). These networks promise a maintenance-free, fault-tolerant platform for gathering different kinds of data. Because a sensor node needs to operate for a long time on a tiny battery, innovative techniques to eliminate energy inefficiencies that would shorten the lifetime of the network must be used. A greater number of sensors allows for sensing over larger geographical regions with greater accuracy. The networking principles and protocols for WSNs are currently being investigated and developed [3–10]. Some application examples of WSNs include:

- Target field imaging
- Intrusion detection
- Weather monitoring
- Security and tactical surveillance
- Distributed computing
- Detecting ambient conditions such as temperature, movement, sound, light, or presence of certain objects
- Inventory control

Data sensing and reporting in sensor networks is dependent on the application and time criticality of the data reporting. As a result, sensor networks can be categorized as time-driven or event-driven

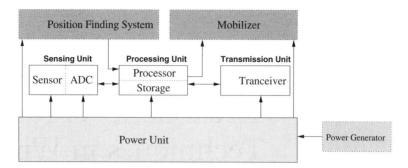

FIGURE 6.1 Components of a sensor node.

networks. The former type is suitable for applications that require periodic data monitoring. As such, sensor nodes will periodically switch on their sensors and transmitters, sense the environment, and transmit data of interest at constant periodic time intervals. Thus, they provide a snapshot of the relevant attributes at regular intervals. In the latter type, sensor nodes react immediately to sudden and drastic changes in the value of a sensed attribute due to the occurrence of a certain event. These are well suited for time critical applications.

A combination of these two types of communication is also possible. Moreover, WSNs can involve single-hop or multihop communication. In a single-hop WSN, a sensor node can directly communicate with any other sensor node or with the external base station. In multihop WSNs, however, communication between two sensor nodes may involve a sequence of hops through a chain of pairwise adjacent sensor nodes. A single-hop communication may take place between the base station and the sensor nodes, while the communication among the sensor nodes is typically multihop.

Despite the innumerable applications of WSNs, these networks have several restrictions, which should be considered when designing any protocol for these networks. Some of these limitations include:

- *Limited energy supply.* WSNs have a limited supply of energy; thus, energy-conserving communication protocols are necessary.
- *Limited computation.* Sensor nodes only have limited computing power, so WSNs cannot run a sophisticated network protocol.
- *Communication.* The bandwidth of the wireless links connecting sensor nodes is often limited, thus constraining the intersensor communication.

WSNs differ from traditional wireless networks like cellular networks in several ways. First, WSNs have severe energy constraints where the network needs to operate unattended for a long period of time. Second, in traditional wireless networks, the task of routing and mobility management is performed to optimize quality of service (QoS) and bandwidth efficiency; energy consumption is of secondary importance because the energy source can be replaced or recharged at any time. However, WSNs consist of nodes designed for unattended operation, so one task of routing is to optimize the use of energy so that the lifetime of the network is maximized. Third, nodes in WSNs are generally stationary after deployment except possibly for a few mobile nodes. Fourth, WSNs send redundant low-rate data in a many-to-one fashion.

MANETs and WSNs share some common problems. Among these are the time-varying characteristics of wireless links; limited power sources; possibility of link failures; scarce resources (e.g., bandwidth); multihop communications; and the ad hoc deployment of nodes in the network area. Although WSNs and MANETs involve multihop communications, the routing requirements are different in several ways:

- The destination in WSNs is known and communication is normally carried from multiple data sources to the BS (i.e., many to one); thus, the basic topology desired in data-gathering is a spanning tree. In MANETs, however, communication is generally on a peer–peer basis (i.e., one to one).

- Data collected by many sensors in WSNs are based on common phenomena, so there is a high probability that these data have some redundancy.
- MANETs are characterized by highly dynamic topologies due to free node mobility. In most application scenarios of WSNs, the sensors are not mobile and thus the nature of the dynamics is different.
- Mobile nodes in MANETs can have their energy sources (e.g., batteries) renewed, replaced, or recharged. The large number of sensor nodes, the necessity of unattended operation, and the long expected working lifetime of WSNs mean that the extremely limited energy resources must be managed carefully. Moreover, limited energy resources, in turn, preclude high data rate communication in WSNs.

The aforementioned reasons make the many end-to-end routing schemes proposed for MANETs in the literature inappropriate for WSNs under these conditions.

6.1.1 Motivation and Design Issues in WSN Routing

One of the main design goals of WSNs is to prolong the lifetime of the network and prevent connectivity degradation by employing aggressive energy management techniques. This is motivated by the fact that energy sources in WSNs are irreplaceable and their lifetime is limited. However, the positions of the sensor nodes are usually not engineered or predetermined and thus allow random deployment in inaccessible terrain or disaster relief operations. This implies that the nodes are expected to perform sensing and communication with no continual maintenance or human attendance and battery replenishment, which limits the amount of energy available to the sensor nodes. Therefore, extensive collaboration between sensor nodes is required to perform high-quality sensing and to behave as fault-tolerant systems. Current routing protocols designed for traditional networks cannot be used directly in a sensor network because:

- Sensor nodes should be self-organizing because the ad hoc deployment of these nodes requires the system to form connections and cope with the resultant distribution. The operation of the sensor networks is unattended, so network organization and configuration should be performed automatically.
- In most application scenarios, sensor nodes are stationary. However, in some applications, some sensor nodes may be allowed to move and change their location (though very low mobility).
- Sensor networks are application specific (i.e., design requirements of a sensor network change with application). For example, the challenging problem of low-latency precision tactical surveillance is different from that required for a periodic weather-monitoring task.
- Data collected by many sensors in WSNs are based on common phenomena; there is a high probability that these data have some redundancy (i.e., data redundancy). Therefore, in-network aggregation of data is needed to yield energy-efficient data delivery before dispatch to destinations. Data redundancy may consume sensor nodes' energy as a result of unnecessary and replicated transmissions.
- Sensor networks are data-centric networks. In traditional networks, data are requested from a specific node. In sensor networks, data are requested based on certain attributes. The sensors can remain in the sleep state, with the data reported from the few remaining sensors providing lower quality. Once an event of interest is detected, the system should be able to configure so as to obtain very high-quality results.
- WSNs have relatively large numbers of sensor nodes, potentially on the order of thousands of nodes. Therefore, sensor nodes need not have a unique ID because the overhead of ID maintenance is high. In data-centric WSNs, the data can be more important than knowing which nodes sent the data.
- WSNs use attribute-based addressing. A user issues an attribute-based address composed of a set of attribute–value pair query. For example, if the query is [temperature > 60°F], then sensor nodes that sense temperature > 60°F only need to respond and report their readings.

- Position awareness of sensor nodes is important because data collection is based on the location. Currently, it is not feasible to use global positioning system (GPS) hardware for this purpose. Methods based on triangulation [14], for example, allow sensor nodes to approximate their position using radio strength from a few known points. Bulusu and colleagues [14] have found that algorithms based on triangulation can work quite well under conditions in which only a very few nodes know their positions *a priori*, e.g., using GPS hardware. Nevertheless, it is favorable to have GPS-free solutions [15] for the location problem in WSNs.

Effective design and deployment of efficient routing protocols in WSNs still face several challenges. These are discussed briefly in the next section.

6.1.2 Routing Challenges in WSNs

The design of routing protocols in WSNs is influenced by many challenging factors that must be overcome before efficient communication can be achieved in WSNs. Some of these challenges and some design guidelines to be considered in the design process include:

- *Ad hoc deployment.* Sensor nodes are deployed randomly. This requires that the system be able to cope with the resultant distribution and form connections between the nodes. Thus, the system should be adaptive to changes in network connectivity as a result of node failure.
- *Energy consumption without losing accuracy.* Sensor nodes can use up their limited supply of energy performing computations and transmitting information in a wireless environment. As such, energy-conserving forms of communication and computation are essential. Sensor node lifetime shows a strong dependence on battery lifetime. In a multihop WSN, each node plays a dual role as data sender and data router. The malfunctioning of some sensor nodes because of power failure can cause significant topological changes and might require rerouting packets and reorganizing the network.
- *Computation capabilities.* Sensor nodes have limited computing power and therefore may not be able to run sophisticated network protocols. Therefore, new or light-weight and simple versions of traditional routing protocols are needed to fit in the WSN environment.
- *Communication range.* Intersensor communication exhibits short transmission ranges. Therefore, it is most likely that a route will generally consist of multiple wireless hops.
- *Fault tolerance.* Some sensor nodes may fail or be blocked due to lack of power, physical damage, or environmental interference. The failure of sensor nodes should not affect the overall task of the sensor network. If many nodes fail, MAC and routing protocols must accommodate formation of new links and routes to the data collection base stations. This may require actively adjusting transmit powers and signaling rates on the existing links to reduce energy consumption, or rerouting packets through regions of the network where more energy is available. Therefore, multiple levels of redundancy may be needed in a fault-tolerant sensor network.
- *Scalability.* The number of sensor nodes deployed in the sensing area may be in the order of hundreds or thousands or more. Any scheme must be able to work with this huge number of sensor nodes. Also, change in network size, node density, and topology should not affect the task and operation of the sensor network. In addition, sensor network routing protocols should be scalable enough to respond to events in the environment. Until an event occurs, most of the sensors can remain in the sleep state, with data from the few remaining sensors providing a coarse quality. Once an event of interest is detected, the system should be able to configure so as to obtain very high-quality results.
- *Hardware constraints.* Consisting of many hardware components, a sensor node may be smaller than a cubic centimeter. These components consume extremely low power and operate in an unattended mode; nonetheless, they should adapt to the environment of the sensor network and function correctly.

- *Transmission media.* In a multihop sensor network, communicating nodes are linked by a wireless medium. The traditional problems associated with a wireless channel (e.g., fading, high error rate) may also affect the operation of the sensor network. In general, the required bandwidth of sensor data will be low, on the order of 1 to 100 kb/s. Related to the transmission media is the design of medium access control (MAC). One approach of MAC design for sensor networks is to use TDMA-based protocols that conserve more energy compared to contention-based protocols like CSMA (e.g., IEEE 802.11). However, although TDMA-based protocols work fine in a flat network, they do not adapt well to clustered WSNs. Management of intercluster communication and dynamic adaptation of the TDMA protocol to variation in the number of nodes in the cluster — in terms of its frame length and time slot assignment — are key challenges for the MAC protocol in hierarchical network. In WSNs, sensors use the Bluetooth technology for transmission. Bluetooth is based upon low-cost, low-complexity, and short range radio communication of data and voice in stationary and mobile environments.
- *Connectivity.* High node density in sensor networks precludes their complete isolation from each other. Therefore, sensor nodes are expected to be highly connected. This, however, may not prevent the network topology from being variable and the network size from being changed due to sensor nodes' failures for different reasons.
- *Control overhead.* When the number of retransmissions in a wireless medium increases due to collisions, latency and energy consumption will also increase. Therefore, control packet overhead increases linearly with node density. As a result, trade-offs among energy conservation, self-configuration, per-node fairness, and latency may exist. However, fairness and throughput are of secondary importance in WSNs.
- *Quality of service.* In some applications, the data should be delivered within a certain period of time from the moment they are sensed; otherwise the data will be useless. Therefore, bounded latency for data delivery is another condition for time-constrained applications.

The communication architecture of the sensor network is shown in Figure 6.2. The sensor nodes are usually scattered in a sensor field — an area in which the sensor nodes are deployed. The nodes in these networks coordinate to produce high-quality information about the physical environment. Each sensor

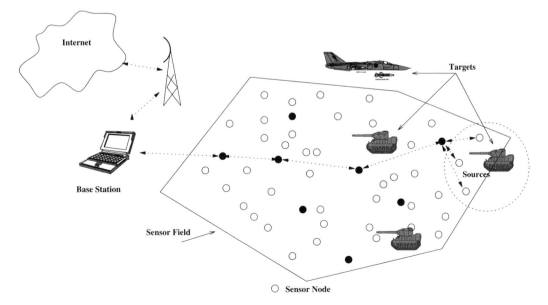

FIGURE 6.2 Communication architecture of a sensor network.

node bases its decisions on its mission, the information it currently has, and its knowledge of its computing, communication and energy resources. Each of these scattered sensor nodes has the capabilities to collect data and route data back to the base stations. A base station may be a fixed node or a mobile node capable of connecting the sensor network to an existing communications infrastructure or to the Internet where a user can have access to the reported data.

6.2 Routing Protocols in WSNs

In sensor networks, conservation of energy, which is directly related to network lifetime, is considered relatively more important than the performance of the network in terms of quality of data sent. As the energy gets depleted, the network may be required to reduce the quality of the results in order to reduce the energy dissipation in the nodes and thus lengthen total network lifetime. Therefore, conservation of energy is considered to be more important than the performance of the network.

Recently, routing protocols for WSNs have been extensively studied. In general, routing in WSNs can be divided into flat-based routing, hierarchical-based routing, and adaptive-based routing. In flat-based routing, all nodes are assigned equal roles. In hierarchical-based routing, however, nodes will play different roles in the network. In adaptive routing, certain system parameters are controlled in order to adapt to the network's current conditions and available energy levels. Furthermore, these protocols can be classified into multipath-based, query-based, or negotiation-based routing techniques depending on the protocol operation. In order to streamline this survey, classification according to the network structure and routing criteria is used. The classification is shown in Figure 6.3. Note that because the topology is static, it is preferable to have a table-driven routing protocol because a lot of energy is used in route discovery and setup of reactive protocols. Another class of routing protocols is the cooperative routing protocols in which nodes send the data to a central node at which data can be aggregated and may be subject to further processing. Therefore, reducing route cost in terms of energy use is of great importance.

Several energy-aware routing protocols have been proposed to capture this requirement. The rest of this section presents a detailed overview of the main routing paradigms in WSNs.

6.2.1 Flat Routing

The first category of routing protocols is the multihop flat routing protocols, summarized in the remainder of this subsection.

6.2.1.1 Sequential Assignment Routing (SAR)

Routing decision in SAR [11] is dependent on three factors: energy resources, QoS on each path, and the priority level of each packet. To avoid single-route failure, a multipath approach and localized path restoration schemes are used. To create multiple paths from a source node, a tree rooted at the source

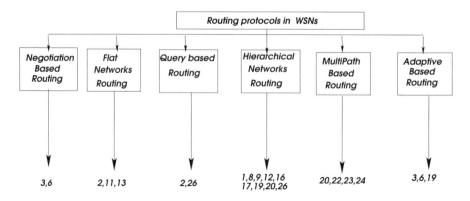

FIGURE 6.3 Routing protocols in WSNs: a taxonomy.

node to the destination nodes (i.e., the set of base stations) is built. The paths of the tree are built while avoiding nodes with low energy or QoS guarantees. At the end of this process, each sensor node will be part of the multipath tree.

For each node, two metrics are associated with each path: an additive QoS metric, i.e., delay, and a measure of the energy usage for routing on that path. The energy is measured with respect to how many packets will traverse that path. SAR will calculate a weighted QoS metric as the product of the additive QoS metric and a weight coefficient associated with the priority level of the packet. The objective of the SAR algorithm is to minimize the average weighted QoS metric throughout the lifetime of the network. If topology changes due to node failures, a path recomputation is needed. As a preventive measure, a periodic recomputation of paths is triggered by the base station to account for any changes in the topology. A handshake procedure based on a local path restoration scheme between neighboring nodes is used to recover from a failure.

6.2.1.2 Directed Diffusion

Intanagonwiwat et al. [2] have presented a data-centric and application-aware paradigm called directed diffusion. It is data centric (DC) in the sense that all the data generated by sensor nodes are named by attribute–value pairs. DC performs in-network aggregation of data to yield energy-efficient data delivery. The main idea of the DC paradigm is to combine the data coming from different sources en route — eliminating redundancy, minimizing the number of transmissions, and thus saving network energy and prolonging its lifetime.

This paradigm is different from the traditional paradigm, termed address centric (AC). In AC routing, the problem is to find short routes between pairs of addressable mobile nodes (end-to-end routing); DC finds routes from multiple sources to a single destination that allow in-network consolidation of redundant data. Figure 6.4 shows an example of the difference between address-centric and data-centric routing. In Figure 6.4(a) is an example of AC routing in which three source nodes detect a target and each uses an end-to-end path independently of the others to report data to the sink node. Using DC routing (Figure 6.4b), an aggregated form of the data received by node B is sent to the sink node, resulting in less energy expenditure.

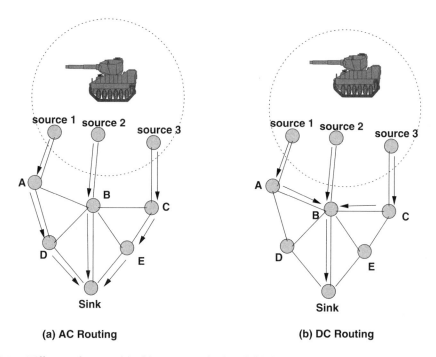

FIGURE 6.4 Differences between (a) address-centric (AC) and (b) data-centric (DC) routing.

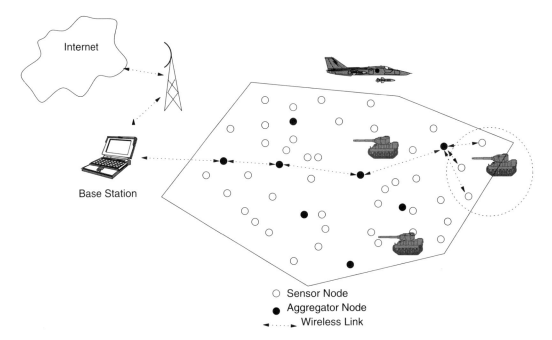

FIGURE 6.5 Sensor network used in military application and employing directed diffusion. A set of sensor nodes (black circles) are selected to work as data aggregators; through them data are sent to the external base station. If an Internet connection is available, a quality copy of the readings can be sent through the Internet to the central command, for example.

The application of this paradigm to query dissemination and processing has been demonstrated in Intanagonwiwat et al. [2]. The query is disseminated or flooded throughout the network and gradients are set up to draw data satisfying the query toward the requesting node; that is, a sink may query for data by disseminating interests and intermediate nodes propagate these interests. More generally, a gradient specifies an attribute value and a direction. Events (i.e., data) start flowing toward the requesting node from multiple paths. A small number of paths can be reinforced so as to prevent further flooding according to a local rule. Then an empirically low delay path is selected to be reinforced. The strength of the gradient may be different toward different neighbors, resulting in different amounts of information flow (see Figure 6.5, for example).

Another use of directed diffusion is to propagate an important event spontaneously to some sections of the sensor network. This type of information retrieval is well suited only for persistent queries in which requesting nodes are not expecting data that satisfy a query for duration of time. This makes it unsuitable for one-time queries because it is not worth setting up gradients, etc. for queries that employ the path only once.

Interest describes a task required to be done by the sensor net. Interest is injected at some point, normally at BS; the source is unknown at this point. Interest diffuses through the network hop by hop and is broadcast by each node to its neighbors. At this stage, loops are not checked for; they are removed at a later stage. Figure 6.6 shows an example of the working of directed diffusion (sending interests, building gradients, and data dissemination).

All sensor nodes in a directed diffusion-based network are application-aware, which enables diffusion to achieve energy savings by selecting empirically good paths and by caching and processing data in the network. In a sensor network based on directed diffusion, each sensor node names data that it generates with one or more attributes. The sink broadcasts the interest, which is a named task descriptor, to all sensors. The task descriptors are named by assigning attribute–value pairs that describe the task. Each sensor node then stores the interest entry in its cache. The interest entry contains a time stamp field and

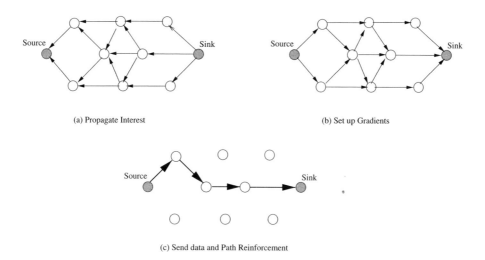

FIGURE 6.6 Interest diffusion in a sensor network.

several gradient fields. As the interest is propagated throughout the network, the gradients from the source back to the sink are set up.

Caching can increase the efficiency, robustness, and scalability of coordination between sensor nodes, which is the essence of the data diffusion paradigm. Locally cached data may be accessed by other users with lower energy consumption than if the data were to be resent end to end. When the source has data for the interest, the source sends the data along the interest's gradient path. As the data propagates, data may be transformed locally at each node. The sink periodically refreshes and resends the interest when it starts to receive data from the source. This is necessary because interests are not reliably transmitted throughout the network. The main goal of this protocol is to compute a path robustly from source to sink through the use of attribute-based naming and gradient paths.

The performance of data aggregation methods used in the directed diffusion paradigm is affected by the positions of the source nodes in the network, the number of sources, and the communication network topology. In order to investigate these factors, two models of source placement, called the event radius (ER) model and the random source (RS) model (shown in Figure 6.7), were studied. In the ER model,

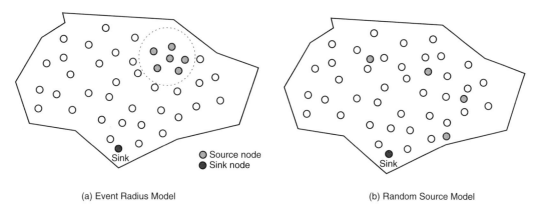

FIGURE 6.7 Two models used in data-centric routing.

a single point in the network area is defined as the location of an event. This may correspond to a vehicle or some other phenomenon tracked by the sensor nodes. All nodes within a distance S (called the sensing range) of this event that are not sinks are considered to be data sources. The average number of sources is approximately $\pi S^2 n$ for a network with n nodes.

In the RS model, k of the nodes that are not sinks are randomly selected to be sources. Unlike the ER model, in RS the sources are not necessarily clustered near each other. In both models of source placement, for a given energy budget, a greater number of sources can be connected to the sink. Thus, the energy savings with aggregation used in the directed diffusion can be transformed to provide a greater degree of robustness to dynamics in the sensed phenomena.

6.2.1.3 Minimum Cost Forwarding Algorithm

The minimum cost forwarding algorithm (MCFA) [13] exploits the fact that the direction of routing is always known (i.e., toward the fixed external base station). Thus, a sensor node need not have a unique ID or maintain a routing table. Instead, each node maintains the least cost estimate from itself to the base station. Each message to be forwarded by the sensor node is broadcast to its neighbors. When a node receives the message, it checks if it is on the least cost path between the source sensor node and the base station. If this is the case, it rebroadcasts the message to its neighbors. This process repeats until the base station is reached. In MCFA, each node should know the least cost path estimate from itself to the base station. This is obtained as follows.

The base station broadcasts a message with the cost set to zero while every node initially sets its least cost to the base station to infinity (∞). Each node, upon receiving the broadcast message originated at the base station, checks to see if the estimate in the message plus the link on which it is received are less than the current estimate. If so, the current estimate and the estimate in the broadcast message are updated. If the received broadcast message is updated, then it is resent; otherwise, it is purged and nothing further is done. However, the previous procedure may result in some nodes having multiple updates and nodes far away from the base station will get more updates from those closer to the base station. To avoid this, the MCFA was modified to run a backoff algorithm at the setup phase. The backoff algorithm dictates that a node will not send the updated message until $a^* l_c$ time units have elapsed from the time at which the message is updated, where a is a constant and l_c is the link cost from which the message was received.

6.2.1.4 Coherent and Noncoherent Processing

Data processing is a major component in the operation of wireless sensor networks. Thus, routing techniques employ different data processing techniques. In general, sensor nodes will cooperate with each other in processing different data flooded in the network area. Two examples of data processing techniques proposed in WSNs are coherent and noncoherent data processing-based routing [11]. In noncoherent data processing routing, nodes will locally process the raw data before sending them to other nodes for further processing. The nodes that perform the further processing are called the aggregators. In coherent routing, the data are forwarded to aggregators after minimum processing. The minimum processing typically includes tasks like time stamping, duplicate suppression, etc. To perform energy-efficient routing, coherent processing is normally selected.

Noncoherent functions have fairly low data traffic loading. On the other hand, because coherent processing generates long data streams, energy efficiency must be achieved by path optimality. Noncoherent cooperative processing contains three phases in the processing: (1) target detection, data collection, and preprocessing; (2) membership declaration; and (3) central node election. During phase 1, a target is detected and its data are collected and preprocessed. When a node decides to participate in a cooperative function, it will enter phase 2 and declare this intention to all neighbors. This should be done as soon as possible so that each sensor has a local understanding of the network topology. Phase 3 is the election of the central node, which is selected to perform more sophisticated information processing; therefore, it must have sufficient energy reserves and computational capability.

Sohrabi and Pottie [11] proposed single and multiple winner algorithms for noncoherent and coherent processing, respectively. In the single winner algorithm (SWE), a single aggregator node is elected for complex processing. The election of a node is based on the energy reserves and computational capability of that node. The algorithm has two components. The first computes the signaling overhead associated with the election process of the single node; the node with the least overhead will be the winner. The winner node broadcasts a message with its ID that will be stored in the node's registry. The second component of the algorithm finds a spanning tree rooted at the winner node. The building of the spanning tree follows a procedure similar to Kruskal's algorithms outlined in Sohrabi and Pottie [11]. By the end of the SWE process, a minimum-hop spanning tree will completely cover the network.

In the multiple winner algorithm (MWE), a simple extension to the SWE is proposed. When all nodes are sources and send their data to the central aggregator node, a large amount of energy will be consumed, so this process has a high cost. One way to lower the energy cost is to limit the number of sources that can send data to the central aggregator node. Instead of keeping record of only the best candidate node (master aggregator node), each node will keep a record of up to n nodes of those candidates. At the end of the MWE process, each sensor in the network has a set of minimum-energy paths to each source node (SN). After that, the SWE is used to find the node that yields the minimum energy consumption. This node can then serve as the central node for the coherent processing. In general, the MWE process has longer delay, higher overhead, and lower scalability than that for noncoherent processing networks.

6.2.2 Hierarchical Routing

Hierarchical or cluster-based routing, originally proposed in wireline networks, comprises well-known techniques with special advantages related to scalability and efficient communication. As such, the concept of hierarchical routing is also utilized to perform energy-efficient routing in WSNs. In a hierarchical architecture, higher energy nodes can be used to process and send the information while low energy nodes can be used to perform the sensing in the proximity of the target. This means that creation of clusters and assigning special tasks to cluster heads can greatly contribute to overall system scalability, lifetime, and energy efficiency.

6.2.2.1 LEACH Protocol

Heinzelman et al. [1] introduced a hierarchical clustering algorithm for sensor networks called low energy adaptive clustering hierarchy (LEACH). LEACH is a cluster-based protocol that includes distributed cluster formation. The authors allowed for a randomized rotation of the cluster head's role in the objective of reducing energy consumption (i.e., extending network lifetime) and to distribute the energy load evenly among the sensors in the network. LEACH uses localized coordination to enable scalability and robustness for dynamic networks and incorporates data fusion into the routing protocol in order to reduce the amount of information that must be transmitted to the base station. The authors also made use of a TDMA/CDMA MAC to reduce inter- and intracluster collisions.

Because data collection is centralized and performed periodically, this protocol is most appropriate when constant monitoring by the sensor network is needed. A user may not need all the data immediately. Thus, periodic data transmissions, which may drain the limited energy of the sensor nodes, are unnecessary. The authors of LEACH introduced adaptive clustering, i.e., reclustering after a given interval with a randomized rotation of the energy-constrained cluster head so that energy dissipation in the sensor network is uniform. They also found, based on their simulation model, that only 5% of the nodes need to act as cluster heads.

The operation of LEACH is separated into two phases: the setup phase and the steady state phase. In the setup phase, the clusters are organized and cluster heads are selected. In the steady state phase, the actual data transfer to the base station takes place. The duration of the steady state phase is longer than the duration of the setup phase in order to minimize overhead.

During the setup phase, a predetermined fraction of nodes, p, elect themselves as cluster heads as follows. A sensor node chooses a random number, r, between 0 and 1. If this random number is less than a threshold value, $T(n)$, the node becomes a cluster head for the current round. The threshold value is calculated based on an equation that incorporates the desired percentage to become a cluster head, the current round, and the set of nodes not selected as a cluster head in the last $(1/P)$ rounds, denoted by G. This is given by:

$$T(n) = \frac{p}{1 - p(r \bmod (1/p))} \quad \text{if } n \in G$$

where G is the set of nodes that

After the cluster heads have been elected, they broadcast an advertisement message to the rest of the nodes in the network that they are the new cluster heads. Upon receiving this advertisement, all the noncluster head nodes decide on the cluster to which they want to belong, based on the signal strength of the advertisement. The noncluster head nodes inform the appropriate cluster heads that they will be members of the cluster. Figure 6.8 shows a flowchart of the cluster head election procedure.

After receiving all the messages from the nodes that would like to be included in the cluster and based on the number of nodes in the cluster, the cluster head node creates a TDMA schedule and assigns each node a time slot when it can transmit. This schedule is broadcast to all the nodes in the cluster. During the steady state phase, the sensor nodes can begin sensing and transmitting data to the cluster heads. The cluster head node, after receiving all the data, aggregates them before sending them to the base station. After a certain time, which is determined *a priori*, the network goes back into the setup phase

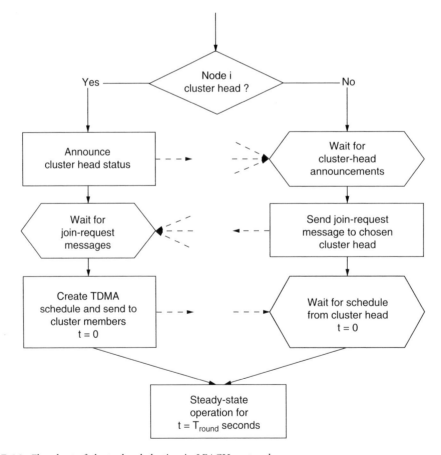

FIGURE 6.8 Flowchart of cluster head election in LEACH protocol.

again and enters another round of selecting new cluster heads. Each cluster communicates using different CDMA codes to reduce interference from nodes belonging to other clusters.

Although LEACH is able to increase the network lifetime, a number of issues about the assumptions used in this protocol remain. LEACH assumes that all nodes can transmit with enough power to reach the base station if needed and that each node has computational power to support different MAC protocols. It also assumes that nodes always have data to send, and nodes located near each other have correlated data. It is not obvious how the number of the predetermined cluster heads (p) is going to be uniformly distributed through the network. Because it is possible that the elected cluster heads will be concentrated in one part of the network, some nodes will not have any cluster heads in their vicinity. Finally, the protocol assumes that all nodes begin with the same amount of energy capacity, supposing that a cluster head removes approximately the same amount of energy for each node. The protocol should be extended to account for nonuniform energy nodes, i.e., use energy-based threshold.

Heinzelman and coworkers proposed an extension to LEACH — LEACH with negotiation [7]. The main theme of the proposed extension is that high-level negotiation using metadata descriptors (as in the SPIN protocol discussed in Section 6.2.3) precede data transfers. This ensures that only data that provide new information are transmitted to the cluster heads before being transmitted to the base station.

6.2.2.2 Power-Efficient Gathering in Sensor Information Systems (PEGASIS)

In Lindsey and Raghavendra [12], an enhancement over the LEACH protocol was proposed. This protocol, called power-efficient gathering in sensor information systems (PEGASIS), is a near optimal chain-based protocol. The basic idea of the protocol is that, in order to extend network lifetime, nodes need only communicate with their closest neighbors and take turns in communicating with the base station. When the round of all nodes communicating with the base station ends, a new round will start and so on. This reduces the power required to transmit data per round because the power draining is spread uniformly over all nodes. Thus, PEGASIS has two main objectives: (1) to increase the lifetime of each node by using collaborative techniques and thus increase network lifetime; and (2) to allow only local coordination between nodes that are close together so that the bandwidth consumed in communication is reduced.

To locate the closest neighbor node, each node uses signal strength to measure the distance to all neighboring nodes and then adjusts the strength so that only one node can be heard. The chain in PEGASIS will consist of nodes closest to each other that form a path to the base station. The aggregated form of the data will be sent to the base station by any node in the chain and the nodes in the chain will take turns sending to the base station. The authors show through simulation that PEGASIS is able to increase the lifetime of the network to twice the lifetime of the network under the LEACH protocol.

However, PEGASIS uses assumptions that may not always be realistic. First, PEGASIS assumes that each sensor node is able to communicate with the base station directly. In practical cases, sensor nodes use multihop communication to reach the base station. Second, it assumes that all nodes maintain a complete database about the location of all other nodes in the network, but the method by which the node locations are obtained is not outlined. Third, it assumes that all sensor nodes have the same level of energy and are likely to die at the same time. Fourth, although in most scenarios sensors will be fixed or immobile as assumed in PEGASIS, some sensors may be allowed to move and thus affect the protocol functions.

6.2.2.3 Threshold-Sensitive Energy-Efficient Protocols (TEEN and APTEEN)

Two hierarchical routing protocols called TEEN (threshold-sensitive energy-efficient sensor network) and APTEEN (adaptive periodic threshold-sensitive energy-efficient sensor network) have been proposed by Manjeshwar and Agarwal [8, 9] for time-critical applications. In TEEN, sensor nodes sense the medium continuously, but the data transmission is done less frequently. A cluster head sensor sends its members a hard threshold, which is the threshold value of the sensed attribute, and a soft threshold, which is a small change in the value of the sensed attribute that triggers the node to switch on its transmitter and

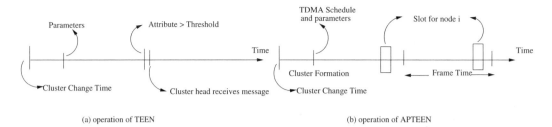

FIGURE 6.9 Time line for the operation of (a) TEEN and (b) APTEEN.

transmit. Thus, the hard threshold tries to reduce the number of transmissions by allowing the nodes to transmit only when the sensed attribute is in the range of interest.

The soft threshold further reduces the number of transmissions that might have otherwise occurred when little or no change occurs in the sensed attribute. A smaller value of the soft threshold gives a more accurate picture of the network, at the expense of increased energy consumption. Thus, the user can control the trade-off between energy efficiency and data accuracy. When cluster heads are to change (see Figure 6.9), new values for the preceding parameters are broadcast. The main drawback of this scheme is that, if the thresholds are not received, the nodes will never communicate and the user will not get any data from the network.

The nodes sense their environment continuously. The first time a parameter from the attribute set reaches its hard threshold value, the node switches on its transmitter and sends the sensed data. The sensed value is stored in an internal variable, called sensed value (SV). The nodes will transmit data in the current cluster period only when the following conditions are true: (1) the current value of the sensed attribute is greater than the hard threshold ; and (2) the current value of the sensed attribute differs from SV by an amount equal to or greater than the soft threshold.

Important features of TEEN include its suitability for time-critical sensing applications. Also, because message transmission consumes more energy than data sensing, the energy consumption in this scheme is less than the proactive networks. The soft threshold can be varied. At every cluster change time, the parameters are broadcast afresh, so the user can change them as required. The main drawback is that if the thresholds are not reached, the nodes will never communicate.

APTEEN, on the other hand, is a hybrid protocol that changes the periodicity or threshold values used in the TEEN protocol according to user needs and type of the application. In APTEEN, the cluster heads broadcast the following parameters:

- *Attributes* (A) is a set of physical parameters about which the user is interested in obtaining information.
- *Thresholds* consist of the hard threshold (HT) and the soft threshold (ST).
- *Schedule* is a TDMA schedule that assigns a slot to each node.
- *Count time* (CT) is the maximum time period between two successive reports sent by a node.

The node senses the environment continuously and only nodes that sense a data value at or beyond the hard threshold transmit. Once a node senses a value beyond HT, it transmits data only when the value of that attribute changes by an amount equal to or greater than the ST. If a node does not send data for a time period equal to the count time, it is forced to sense and retransmit the data. A TDMA schedule is used and each node in the cluster is assigned a transmission slot. Thus, APTEEN uses a modified TDMA schedule to implement the hybrid network. The main features of the APTEEN scheme include: (1) combining proactive and reactive policies; (2) offering a lot of flexibility by allowing the user to set the CT interval; and (3) controlling threshold values for the energy consumption by changing the CT as well as the threshold values. The main drawback of the scheme is the additional complexity required to implement the threshold functions and the CT. However, the authors of these two protocols showed through simulation that both protocols perform better than LEACH.

6.2.2.4 Small Minimum Energy Communication Network (SMECN)

Rodoplu and Meng [16] have proposed a protocol that computes an energy-efficient subnetwork, namely, the minimum energy communication network (MECN), for a certain sensor network. A new algorithm called small MECN (SMECN) to provide such a subnetwork has been proposed by Li and Halpern [17]. The subnetwork (i.e., subgraph G') constructed by SMECN is smaller than the one constructed by MECN if the broadcast region is circular around the broadcasting node for a given power setting. Subgraph G' of graph G, which represents the sensor network, minimizes the energy usage satisfying the following conditions: (1) the number of edges in G' is less than in G while containing all nodes in G; and (2) the energy required to transmit data from a node to all its neighbors in subgraph G' is less than the energy required to transmit to all its neighbors in graph G.

Assuming that $r = (u, u_1, \ldots, u_{k-1}, v)$ is a path between u and v, the total power consumption of one path like r is given by:

$$C(r) = \sum_{i=0}^{k-1} (p(u_i, u_{i+1}) + c)$$

where $u = u_0$; $v = u_k$; the power required to transmit data under this protocol is

$$p(u, v) = t d(u, v)^n$$

for some appropriate constant t; n is the path-loss exponent of outdoor radio propagation models $n \geq 2$ and $d(u,v)$ is the distance between u and v. A reception at the receiver takes power c.

The subnetwork computed by SMECN helps to send messages on minimum-energy paths. However, the proposed algorithm is local in the sense that it does not actually find the minimum-energy path; it just constructs a subnetwork in which the path is guaranteed to exist. Moreover, the subnetwork constructed by SMECN makes it more likely that the path used is one that requires less energy consumption.

6.2.2.5 Fixed-Size Cluster Routing

Xu and colleagues [19] have proposed a geography informed routing protocol for ad hoc networks. The network area is first divided into fixed zones; inside each zone, nodes collaborate with each other to play different roles. For example, nodes will elect one sensor node to stay awake for a certain period of time and then they go to sleep. Each sensor node is positioned randomly in a two-dimensional plane. When a sensor transmits a packet with power for a distance r, the signal will be strong enough for other sensors to hear it within the Euclidean distance r from the sensor that originates the packet. In other words, to cover a range of r, the sensor that originates the signal must transmit with enough power to cover that range.

Figure 6.10 gives an example of fixed zoning that can be used in sensor networks similar to the one proposed by Xu et al., but with an extension. The extension is to use two zones to receive signals instead of one. After the range r, the power signal starts to attenuate (i.e., fade out), so a sensor in the second zone, called the *border* zone, may or may not hear the signal depending on the signal strength. Therefore, a sensor within the *guaranteed* zone, i.e., within the distance r, is guaranteed to receive the signal, while a sensor in the border zone may or may not receive the packet. Figure 6.10 shows this situation.

Xu and colleagues' fixed clusters [19] are selected to be equal and square. The selection of the square size depends on the required transmitting power and the communication direction. One node in each cluster, called the cluster head, is elected periodically. Vertical and horizontal communication is guaranteed if the signal travels a distance of $a = \dfrac{r}{\sqrt{5}}$, chosen so that any two sensor nodes in adjacent vertical or horizontal clusters can communicate directly in the guaranteed zone. For a node in the *border* zone

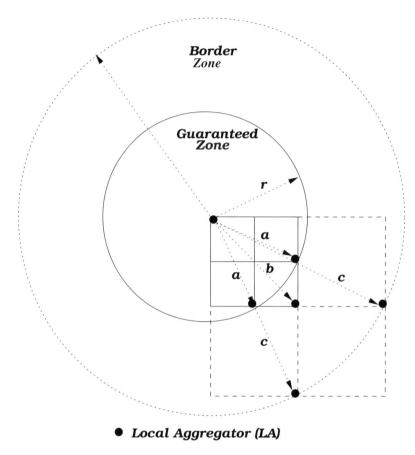

FIGURE 6.10 An example of zoning in sensor networks.

to receive the transmitted packet, the signal must travel a distance of $c = \dfrac{r}{2\sqrt{5}}$. Note also that for a diagonal communication to happen, the signal must span a distance of $b = \dfrac{r}{2\sqrt{2}}$. A cluster head is responsible for receiving raw data from other nodes in its cluster. The role of cluster head is rotated to distribute the energy draining role evenly around the network.

6.2.2.6 Virtual Grid Architecture Routing

An energy-efficient routing paradigm proposed by [26] is based on the concept of data aggregation and in-network processing. The data aggregation is performed at two levels: local and then global. A reasonable approach for WSNs is to arrange nodes in a fixed topology due to the node stationarity or extremely low mobility. Fixed, equal, adjacent, and nonoverlapping clusters with regular shapes are selected to obtain a fixed rectilinear virtual topology. Inside each zone, a node is optimally selected to act as cluster head. The set of cluster heads, also called local aggregators (LAs), performs the local aggregation. Several heuristics were formulated to allocate a subset of the cluster heads, called the master aggregators (MAs), in order to perform near optimal global data aggregation so that the total routing cost from the source nodes to the base station is minimized.

Figure 6.11 illustrates an example of fixed zoning and the resulting virtual grid architecture (VGA) used to perform two level data aggregation. Note that the location of the base station is not necessarily at the extreme corner of the grid, but rather can be located at an arbitrary place.

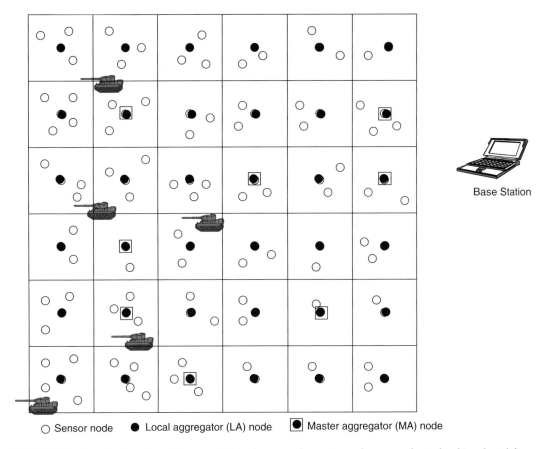

FIGURE 6.11 Regular shape tessellation applied to the network area. In each zone, a cluster head is selected for local aggregation. A subset of those cluster heads, called master nodes, are optimally selected to perform global aggregation.

All heuristics in Reference 26 start with the first node in the VGA architecture and proceed sequentially the whole topology left to right and then right to left in a top-down fashion. Although finding the optimal routes from the source nodes to the base station by using the set of MAs is an NP-complete problem, Al-Karaki and Kamal's developed dynamic program [26] is able to find the optimal values most of the time.

6.2.2.7 Hierarchical Power-Aware Routing

Li and coworkers [20] have proposed a hierarchical power-aware routing protocol that divides the network into groups of sensors. Each group of sensors in geographic proximity is clustered together as a zone and each zone is treated as an entity. To perform routing, each zone is allowed to decide how it will route a message hierarchically across the other zones.

Messages are routed along the path with the maximal–minimal of the remaining power, called the max–min path. The motivation is that using nodes with high residual power may be expensive compared to the path with the minimal power consumption. An approximation algorithm, called the *max–min zPmin* algorithm, combines the benefits of selecting the path with the minimum power consumption and the path that maximizes the minimal residual power in the nodes of the network. The algorithm finds the path with the least power consumption, P_{min}, by using the Dijkstra algorithm.

Another algorithm, called zone-based routing, that relies on *max–min zPmin* and is scalable for large scale networks has also been proposed in Reference 20. Zone-base routing is a hierarchical approach in which the area covered by the (sensor) network is divided into a small number of zones. To send a

TABLE 6.1 Hierarchical vs. Flat Topology Routing

Hierarchical Routing	Flat Routing
Reservation-based scheduling	Contention-based scheduling
Collisions avoided	Collision overhead present
Reduced duty cycle due to periodic sleeping	Variable duty cycle by controlling sleep time of nodes
Data aggregation by cluster head	Node on multihop path aggregates incoming data from neighbors
Simple but nonoptimal routing	Routing is complex but optimal
Requires global and local synchronization	Links formed on the fly without synchronization
Overhead of cluster formation throughout the network	Routes formed only in regions with data for transmission
Lower latency because multiple hops network formed by cluster heads always available	Latency in waking up intermediate nodes and setting up multipath
Energy dissipation is uniform	Energy dissipation depends on traffic patterns
Energy dissipation cannot be controlled	Energy dissipation adapts to traffic pattern
Fair channel allocation	Fairness not guaranteed

message across the entire area, a global path from zone to zone is found. The sensors in a zone autonomously direct local routing and participate in estimating the zone power level. Each message is routed across the zones using information about the zone power estimates. A global controller for message routing, which may be the node with the highest power, is assigned the role of managing the zones. If the network can be divided into a relatively small number of zones, the scale for the global routing algorithm is reduced. The global information required to send each message across is summarized by the power level estimate of each zone.

A zone graph was used to represent connected neighboring zone vertices if the current zone can go to the next neighboring zone in that direction. Each zone vertex has a power level of 1. Each zone direction vertex is labeled by its estimated power level, computed by a procedure that is a modified Bellman–Ford algorithm. Moreover, two algorithms were outlined for local and global path selection using the zone graph.

The flat and hierarchical protocols are different in many aspects. Table 6.1 outlines the major differences between the two routing approaches.

6.2.3 Adaptive Routing

Heinzelman et al. [3] and Kulik et al. [6] proposed a family of adaptive protocols, called sensor protocols for information via negotiation (SPIN). These protocols disseminate all the information at each node to every node in the network, assuming that all nodes in the network are potential base stations. This enables a user to query any node and get the required information immediately. These protocols make use of the property that nearby nodes have similar data and thus distribute only data that the other nodes do not have.

The SPIN family of protocols uses data negotiation and resource-adaptive algorithms. Nodes running SPIN assign a high-level name to describe their collected data (called metadata) completely and perform metadata negotiations before any data are transmitted. This assures that no redundant data are sent throughout the network. The format of the metadata is application specific and is not specified in SPIN. For example, sensors might use their unique IDs to report metadata if they cover a certain known region. In addition, SPIN has access to the current energy level of the node and adapts the protocol it is running based on how much energy is remaining. These protocols work in a time-driven fashion and distribute the information over the network, even when a user does not request any data.

The SPIN family is designed to address the deficiencies of classic flooding by negotiation and resource adaptation. This family of protocols is designed based on the idea that sensor nodes operate more efficiently and conserve more energy by sending data that describe the sensor data instead of sending all the data; for example, image and sensor nodes must monitor the changes in their energy resources.

SPIN protocols are motivated by the observation that conventional protocols like flooding or gossiping waste energy and bandwidth by sending extra and unnecessary copies of data by sensors covering overlapping areas. Sensor nodes use three types of messages — ADV, REQ, and DATA — to communicate. ADV advertises new data, REQ requests data, and DATA is the actual message. The protocol starts when a SPIN node obtains new data that it is willing to share. It does so by broadcasting an ADV message containing metadata. If a neighbor is interested in the data, it sends a REQ message for the DATA and the DATA is sent to this neighbor node. The neighbor sensor node then repeats this process with its neighbors. As a result, the entire sensor area will receive a copy.

The SPIN family of protocols includes two protocols, namely, SPIN-1 and SPIN-2, which incorporate negotiation before transmitting data in order to eliminate implosion and overlap by ensuring that only useful information will be transferred. Also, each node has its own resource manager, which keeps track of resource consumption, and is polled by the nodes before data transmission. The SPIN-1 protocol is a three-stage protocol, as described earlier. An extension to SPIN-1 is SPIN-2, which incorporates a threshold-based resource awareness mechanism in addition to negotiation. When energy in the nodes is abundant, SPIN-2 communicates using the three-stage protocol of SPIN-1.

However, when the energy in a node starts approaching a low energy threshold, it reduces its participation in the protocol, i.e., it participates only when it believes that it can complete all the other stages of the protocol without going below the low-energy threshold. This approach does not prevent a node from receiving, and therefore spending, energy on ADV, or REQ messages below its low-energy threshold. It does, however, prevent the node from ever handling a DATA message below this threshold.

In conclusion, SPIN-l and SPIN-2 are simple protocols that efficiently disseminate data while maintaining no per-neighbor state. These protocols are well suited for an environment in which the sensors are mobile because they base their forwarding decisions on local neighborhood information. Other protocols of the SPIN family are:

- SPIN-BC. This protocol is designed for broadcast channels. All nodes within hearing range of a sensor node will get the message. However, nodes must wait for transmission if the channel is busy. Also, nodes do not immediately send out REQ message when they hear the ADV message. Instead, each node sets a random timer and when this timer expires, the node sends the REQ message. If, waiting for their timers to expire, other nodes are able to hear this message, they will stop their timers. This prevents sending redundant copies of the same request.
- SPIN-PP. If two nodes can communicate with each other without incurring interference from other neighboring nodes, this protocol will be used. It is designed for a point-to-point communication, i.e., hop-by-hop routing, and assumes that energy is not a major constraint and that packets are never lost. Figure 6.12 shows an example of the operation of this protocol. A node will send an ADV message to advertise that it has a message to send. All nodes in the neighborhood that hear the message, if interested, will express this interest by sending REQ messages. Upon receiving the REQ message, the announcing node will send the data to the interested nodes. Once those nodes have the information, they become an information announcer and send an ADV message to their neighbors. If their neighbors are interested, they send an REQ message and the process repeats.
- SPIN-EC. This protocol works similarly to SPIN-PP, but with an energy heuristic added to it. A node will participate in the protocol if the node is able to complete all stages of the protocol without its energy dropping below a certain threshold. The energy threshold is a system parameter.
- SPIN-RL. In SPIN-PP, it is assumed that packets are not lost. When a channel is lossy, this protocol cannot be used. Instead, another protocol called SPIN-RL, in which two adjustments are added to the SPIN-PP protocol to account for the lossy channel, is used. First, each node keeps track of all ADV messages it receives. It may also ask for data to be resent if it did not get them within a specified amount of time. Second, in order to fine tune the rate of resending data, nodes will limit the frequency of this activity by having each node wait for a certain predetermined time before replying to the same REQ messages again. This procedure guarantees that data will be resent only after making sure that the reply to the previous REQ message failed.

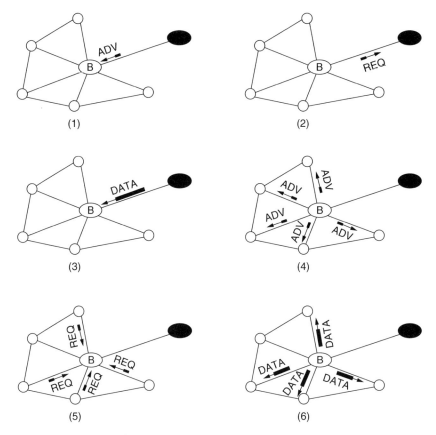

FIGURE 6.12 SPIN-PP: three-way handshake in SPIN protocol. Steps 1 through 6 show the three messages (ADV, REQ, and DATA) used in the handshaking process.

Table 6.2 compares SPIN, LEACH, and the directed diffusion routing techniques according to different parameters. The table indicates that directed diffusion shows a promising approach for energy-efficient routing in WSNS due to the use of in-network processing.

6.2.4 Multipath Routing

The resilience of a protocol is measured by the likelihood that an alternate path exists between a source and a sink when the primary path fails. This can be increased by maintaining multiple paths between the source and the sink at the expense of increased energy consumption, and keeping these alternate paths alive by sending periodic messages. Thus, the resilience of the network should be increased while keeping the maintenance overhead of these paths low. This subsection discusses routing protocols that use multiple paths rather than a single path in order to enhance network performance.

TABLE 6.2 Comparison among SPIN, LEACH, and Directed Diffusion

	SPIN	LEACH	Directed Diffusion
Optimal route	No	No	Yes
Network lifetime	Good	Very good	Good
Resource awareness	Yes	Yes	Yes
Use of metadata	Yes	No	Yes

Ganesan and coworkers [22] have proposed an energy-efficient multipath routing protocol that uses braided multipaths instead of completely disjoint multipaths so as to keep the cost of maintenance low. The costs of such alternate paths are also comparable to the primary path because they tend to be much closer to the primary path. Chang and Tassiulas [23] proposed an algorithm to route data through a path whose nodes have the largest residual energy. The path is changed whenever a better path is discovered. The primary path will be used until its energy falls below the energy of the backup path at which the backup path is used. In this way, the nodes in the primary path will not deplete their energy resources through continual use of the same route, thus achieving longer life. The path-switching cost was not quantified in the paper.

Rahul and Rabaey [24] have proposed the use of a set of suboptimal paths occasionally to increase the lifetime of the network. These paths are chosen by means of a probability that depends on how low the energy consumption of each path is.

Because the path with the largest residual energy when used to route data in a network may be very energy expensive too, a trade-off takes place between minimizing the total power consumed and the residual energy of the network. Li and colleagues [20] proposed an algorithm in which the residual energy of the route is relaxed a bit in order to select a more energy-efficient path. The operation of the algorithm is explained in Subsection 6.2.2.7.

6.2.5 Query-Based Routing

In this kind of routing, the destination nodes propagate a query for data (sensing task) from a node through the network and a node having these data sends data that match the query back to the node, which initiates the query. Usually these queries are described in natural language, or in high-level query languages. For example, client C1 may submit a query to node N1 and ask, "Are there moving vehicles in battle space region 1?"

All the nodes have tables consisting of the sensing task queries received, and hence they send data that match these queries when they receive them. Directed diffusion (described in Subsection 6.2.1.2) is an example of this type of routing. In directed diffusion, the sink node sends out interest messages to sensors. As the interest is propagated throughout the sensor network, the gradients from the source back to the sink are set up. When the source has data for the interest, the source sends the data along the interest's gradient path. To lower energy consumption, data aggregation (e.g., duplicate suppression) is performed en route.

6.2.6 Negotiation-Based Protocols

These protocols use high-level data descriptors in order to eliminate redundant data transmissions through negotiation. Communication decisions are also taken based on the resources available to them.

The SPIN family protocols discussed in Section 6.2.3 are an example of negotiation-based routing protocols. The motivation is that the use of flooding to disseminate data will produce implosion and overlap among the sent data and thus nodes will receive duplicate copies of the same data. This operation consumes more energy and more processing by sending the same data by different sensors. The SPIN protocols are designed to disseminate the data of one sensor to all other sensors assuming these sensors are potential base stations. Therefore, the main idea of negotiation-based routing in WSNs is to suppress duplicate information and prevent redundant data from being sent to the next sensor or the base station by conducting a series of negotiation messages before the real data transmission begins.

6.3 Routing in WSNs: Future Directions

The future vision of WSNs is to embed numerous distributed devices to monitor and interact with physical world phenomena and to exploit spatially and temporally dense sensing and actuation capabilities of those sensor networks. These nodes coordinate among themselves to create a network that performs higher level tasks.

Although extensive efforts have been exerted so far on the routing problem in WSNs, some challenges still confront effective solutions of the routing problem. First, there is a tight coupling between sensor nodes and the physical world. Sensors are embedded in *unattended* places or systems. This is different from traditional Internet, PDA, and mobility applications that interface primarily and directly with human users. Second, sensors are characterized by a small footprint and, as such, nodes present stringent energy constraints because they are living with small, finite, energy sources. This is also different from traditional fixed but reusable resources. Third, communications is primary consumer of energy in this environment in which sending a bit over 10 or 100 m consumes as much energy as thousands to millions of operations (known as R^4 signal energy drop-off) [27].

Future trends in routing techniques in WSNs focus on different directions, but all share the common objective of prolonging network lifetime. Some of these directions include:

- *Exploit redundancy.* Typically, a large number of sensor nodes are implanted inside or beside the phenomenon. Because sensor nodes are prone to failure, fault tolerance techniques come into the picture to keep the network operating and performing its tasks. Routing techniques that explicitly employ fault tolerance techniques in an efficient manner are still under investigation.
- *Tiered architectures (mix of form/energy factors).* Hierarchical routing is an old technique to enhance scalability and efficiency of the routing protocol. However, novel techniques to network clustering to maximize the network lifetime are also a hot area of research in WSNs.
- *Exploit spatial diversity and density of sensor/actuator nodes.* Nodes will span a network area that might be large enough to provide spatial communication between sensor nodes. Achieving energy efficient communication in this densely populated environment deserves further investigation. The dense deployment of sensor nodes should allow the network to adapt to unpredictable environments.
- *Achieve desired global behavior with adaptive localized algorithms.* That is, do not rely on global interaction or information. However, in a dynamic environment, this is hard to model.
- *Leverage data processing inside the network and exploit computation near data sources to reduce communication.* That is, perform in-network distributed processing. WSNs are organized around naming data, not node identities. Because a large collection of distributed elements is present, localized algorithms that achieve system-wide properties in terms of local processing of data before being sent to the destination are still needed. Nodes in the network will store named data and make them available for processing. The need is great to create efficient processing points in the network, e.g., duplicate suppression, aggregation, correlation of data. How to find those points efficiently and optimally is still an open research issue.
- *Time and location synchronization.* Energy-efficient techniques for associating time and spatial coordinates with data to support collaborative processing are also required.
- *Self-configuration and reconfiguration.* These are essential to the lifetime of unattended systems in dynamic, constrained-energy environments and important for keeping the network up and running. As nodes die and leave the network, update and reconfiguration mechanisms should take place. An important feature in every routing protocol is to adapt to topology changes very quickly and to maintain the network functions.

6.4 Conclusions

Routing in sensor networks is a new area of research, with a limited but rapidly growing set of research results. This chapter offered a comprehensive overview of routing techniques in wireless sensor networks that have been presented in the literature. They have the common objective of trying to extend the lifetime of the sensor network.

Overall, the routing techniques are classified based on the network structure into three categories: flat, hierarchical, and adaptive routing. Furthermore, these protocols can be classified into multipath-based, query-based, or negotiation-based routing techniques depending on the protocol operation. Design trade-offs between energy and communication overhead savings in some of the routing paradigm have been highlighted, as well as advantages and disadvantages of each routing technique. Although many of these routing techniques look promising, many challenges in the sensor networks still need to be solved; this chapter highlighted those challenges and pinpointed future research directions in this regard.

References

1. W. Heinzelman, A. Chandrakasan, and H. Balakrishnan, Energy-efficient communication protocol for wireless microsensor networks, *Proc. 33rd Hawaii Int. Conf. Syst. Sci.* (HICSS '00), January 2000.
2. C. Intanagonwiwat, R. Govindan, and D. Estrin, Directed diffusion for wireless sensor networks, *IEEE/ACM Trans. Networking*, 11(1), 2–16, 2003.
3. W. Heinzelman, J. Kulik, and H. Balakrishnan, Adaptive protocols for information dissemination in wireless sensor networks, *Proc. 5th ACM/IEEE Mobicom Conf. (MobiCom'99)*, Seattle, WA, August, 1999. 174–185.
4. I. Akyildiz, W. Su, Y. Sankarasubramaniam, and E. Cayirci, A survey on sensor networks, *IEEE Commun. Mag.*, 40(8), 102–114, August 2002.
5. A. Perrig, R. Szewzyk, J.D. Tygar, V. Wen, and D. E. Culler, SPINS: security protocols for sensor networks. *Wireless Networks*, 8, 521–534, 2000.
6. J. Kulik, W.R. Heinzelman, and H. Balakrishnan, Negotiation-based protocols for disseminating information in wireless sensor networks, *Wireless Networks*, 8, 169–185, 2002.
7. W.R. Heinzelman, A. Chandrakasan, and H. Balakrishnan, Energy-efficient communication protocol for wireless microsensor networks, *Proc. 33rd Int. Conf. Syst. Sci.*, (HICSS'00), January 2000, 1–10.
8. A. Manjeshwar and D.P. Agarwal, TEEN: a routing protocol for enhanced efficiency in wireless sensor networks, in *1st Int. Workshop Parallel Distributed Computing Issues Wireless Networks Mobile Computing*, April 2001.
9. A. Manjeshwar and D.P. Agarwal, APTEEN: a hybrid protocol for efficient routing and comprehensive information retrieval in wireless sensor networks, *Parallel Distributed Process. Symp., Proc. Int.*, IPDPS 2002, 195–202.
10. D. Ganesan, R. Govindan, S. Shenker, and D. Estrin, Highly-resilient, energy-efficient multipath routing in wireless sensor networks, *ACM SIGMOBILE Mobile Computing Commun. Rev.*, 5(4), 10–24, October 2001.
11. K. Sohrabi and J. Pottie, Protocols for self-organization of a wireless sensor network, *IEEE Personal Commun.* 7(5), 16–27, 2000.
12. S. Lindsey and C. Raghavendra, PEGASIS: power-efficient gathering in sensor information systems, *Int. Conf. Communication Protocols*, 149–155, 2001.
13. F. Ye, A. Chen, S. Liu, and L. Zhang, A scalable solution to minimum cost forwarding in large sensor networks, *Proc. 10th Int. Conf. Computer Commun. Networks (ICCCN)*, 304–309, 2001.
14. N. Bulusu, J. Heidemann, and D. Estrin, GPS-less low cost outdoor localization for very small devices, Technical report 00-729, Computer Science Department, University of Southern California, Apr. 2000.
15. A. Savvides, C.-C. Han, and M. Srivastava, Dynamic fine-grained localization in Ad-Hoc networks of sensors, *Proc. 7th ACM Annu. Int. Conf. Mobile Computing Networking (MobiCom)*, July 2001, 166–179.
16. V. Rodoplu and T.H. Meng, Minimum energy mobile wireless networks, IEEE JSAC, 17(8), Aug. 1999, 1333–1344.
17. L. Li and J.Y. Halpern, Minimum-energy mobile wireless networks revisited, *ICC '01*, Helsinki, Finland, 67–78, June 2001.
18. S. Hedetniemi, S. Hedetniemi, and A. Liestman, A survey of gossiping and broadcasting in communication networks, *Networks*, 18, 1988.

19. Y. Xu, J. Heidemann, D. Estrin, Geography-informed energy conservation for ad-hoc routing, *IEEE/ACM MobiCom*, Rome, 70–84, July 16–21, 2001.
20. Q. Li, J. Aslam, and D. Rus, Hierarchical power-aware routing in sensor networks, in *Proc. DIMACS Workshop Pervasive Networking*, May, 2001.
21. D. Braginsky and D. Estrin, Rumor routing algorithm for sensor networks, ACM First Workshop on Sensor Networks and Applications (WSNA), 2002.
22. D. Ganesan, R. Govindan, S. Shenker, and D. Estrin, Highly resilient, energy-efficient multipath routing in wireless sensor networks, *ACM Mobile Computing Commun. Rev.*, 5(4), October 2001.
23. J.-H. Chang and L. Tassiulas, Maximum lifetime routing in wireless sensor networks, *Proc. Adv. Telecommun. Inf. Distribution Res. Program (ATIRP2000)*, College Park, MD, Mar. 2000.
24. C. Rahul and J. Rabaey, Energy-aware routing for low energy ad hoc sensor networks, *IEEE Wireless Commun. Networking Conf. (WCNC)*, March 17–21, 2002, Orlando, FL.
25. W. Heinzelman, J. Kulik, and H. Balakrishnan, Adaptive protocols for information dissemination in wireless sensor networks, *Proc. 5th Annu. ACM/IEEE Int. Conf. Mobile Computing Networking*, August 1999.
26. J. Al-Karaki and A. Kamal, On the optimal data aggregation and in-network processing based routing in wireless sensor networks, technical report, Iowa State University, 2003.
27. D. Goodman, *Wireless Personal Communications Systems*. Reading, MA: Addison–Wesley, Reading, MA, 1997.

7
Artificial Perceptual Systems

Amy Loutfi
Örebro University

Malin Lindquist
Örebro University

Peter Wide
Örebro University

7.1	Introduction	7-1
7.2	Background	7-2
7.3	Modeling of Perceptual Systems	7-3
	Sensor Fusion • Time Concept • Error Handling • Reasoning • Passive and Active Perception • Memory and Knowledge Base • Human–Computer Interaction	
7.4	Perceptual Systems in Practice	7-8
	Electronic Head • Fire Indication Application	
7.5	Research Issues and Summary	7-12

7.1 Introduction

The 20th century technological revolutions in the areas of electronics, computers, and telecommunications have created a need for better techniques for interfacing, decision making, and handling of human knowledge. In general, current limitations of new technologies arise from their inadequacy and conflict with natural human behavior, mainly in three aspects related to:

- How to perceive the sensor data
- How to make intelligent decisions
- How to exchange essential information

In the first aspect, information from sensors is often imprecise and limited, with some uncertainties. A minimum component in this process is how to merge sensor data into relevant information. The second aspect concerns the way of making relevant decisions based on the dynamical sensor data and earlier experience and knowledge. An important aspect is also the way of interfacing information to humans. This crucial process often controls the effectiveness of the complete system and, if relating to human behavior, could bring about trust and understanding of system performance. Increasing the "intelligence" of perceptual machines and improving the user interfaces can strengthen the interaction between the human and the system as well as perceptual processing of activities in complex environments.

It is necessary to know the intended goals and tasks of a perceptual system in order to effectively extract information from the sensor data. To describe the benefits of a perceptual system with general abilities the following structure is presented. The perception model process shown here combines the human perspective of merging perceptual information with memory capabilities in a cyclic behavior. Figure 7.1 describes a perceptual system with general abilities. The process can be described in a human-like perspective in four subprocesses that identify the main computational activities.

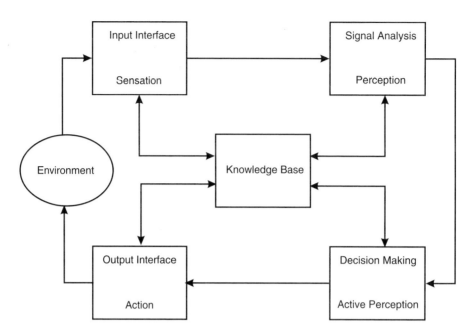

FIGURE 7.1 An overview of a perceptual system.

- *Input interface* — a sensational process with similarities to human sensation and the preprocessing activities, through a number of nuclei, of sensory information on its way to the brain. A number of different sensor capabilities ensure the ability to connect dynamical activities in the environment that correspond to the artificial system.
- *Signal analysis* — a subprocess that organizes the received data in a "structural picture." This has similarities with the functionalities in the thalamus and cerebral cortex, for example. The rich sensor data are merged and contain more information than when each of the sensors is used separately.
- *Decision making* — handles decision making in the system similarly to the motor cortex. The process of the action is then viewed and valued in order to give an appropriate qualitative result describing appropriately the activities of the system.
- *Human–computer interface* — communicates the final results of the system to a human user.

This comparison with the functionalities of the human brain is only to illustrate similar functions within a perceptual system. Although some algorithms and processes can be biologically inspired, it is not confined to such.

7.2 Background

Perception continuously gives life forms information about the relevant aspects of the surrounding environment and their own relation to it; it is necessary to have suitable perception to be able to interact with a changing surrounding. The human perceptual system can only be studied indirectly because what one experiences cannot be observed by someone else. It must be studied from the actions, descriptions, and evaluations given by the subjects of the experiment.

The study in human perception started in the 19th century with theories in psychophysics, which is the relationship between the quantitative dimensions of physical stimuli and the sensation they create, also measured quantitatively. The founder of this science was G. T. Fechner [9]. A theory had previously been developed, Weber's law, saying that if the size of the just noticeable difference in the stimulus is divided by the original stimulus it gives a constant. Fechner improved Weber's theory by saying that a sensation equals a constant multiplied by the logarithm of the stimulus; this is known as the Weber–Fechner law. That law was later replaced by Stevens' power law [19], which states that the magnitude of the psychological reaction is equal to a constant multiplied by the actual intensity of the stimulus, raised to some power. The accuracy of Stevens' law has also been questioned.

The perceptual system does not register absolute values from the sensors [14]. The sensory system always responds to relative changes and can adapt to adjust the dynamic range in order to maximize its sensitivity to changes. For example, how certain sounds in speech are perceived depends on the order of the sound and the frequency. There are differences in speech such as voice, screaming, whispering, speech rate, different dialects, and background noise. In spite of all these circumstances, the perception remains rather accurate due to its possibilities to adapt and compensate. This is also true for the human visual system: an image of a certain object can vary in quality and in viewing conditions but it is still possible to identify it. An unknown object — in the sense that it is seen under a new condition — can still be recognized by humans because the object belongs to a known family, e.g., a face or a box.

One approach to implicate the biological vision system is to compute invariance for complex patterns [20]. To gain insight into perception from a computational point of view, Fermuller and Aloimonos [10] made a working model in order to explain the abstract components of a visual system. In this case, the influences from the biological system were used to inspire what is relevant in a visual system working in an environment similar to human surroundings.

In the area of autonomous robots, a perceptual system is needed to perform tasks and interact with humans in the environment. The robot uses its perception in order to investigate the surroundings and make a decision of how to act, something referred to as the "sense–think–act" paradigm. A perceptual system can consist of a camera for recognition of signs and objects in an office environment [1]. Perception is also needed in a model where the aim is to imitate human movements and use them in a simulated humanoid [11]. There is no one general model for a perceptual system. Some of the models attempt to imitate human perception while others have not been concerned with the human model.

7.3 Modeling of Perceptual Systems

This section discusses the different components that should be considered in any model of perceptual systems. Some of these components have distinctive overlaps between perceptual systems and sensor fusion models; however, the mentioned components are focused on the contribution to the overall process of perceptual systems as defined in the introduction.

7.3.1 Sensor Fusion

Sensor fusion is an intrinsic part to perceptual systems because often more than one sensing mechanism is involved. Here a very basic general overview of different sensor fusion techniques and their contributions to the perceptual model presented in Figure 7.1 is provided.

Sensor fusion is the process of combining data so that the result provides more information compared to the handling of each source separately. Sensors can work complementarily i.e., they observe different properties to give a more complete picture of the surroundings, or they can work cooperatively, which means the sensors observe the same properties, thus making the system more stable and reliable. Sensor fusion can occur at different levels throughout the data processing. The lowest level is data fusion, which means that the raw sensor data are fused. This is often used as a example in the tracking of an object. Feature fusion occurs when features are extracted from the original quantity of data to reduce dimensionality. This makes it easier to handle fusion processes. A feature may be the mean value or edges in a picture or an output variable from a principal component analysis. Feature fusion is common in classification problems. Information fusion can be considered the highest level of fusion. An example of information fusion is threat assessment done by military-based applications in which the inputs can be tactical information and possible movement regarding enemy forces.

A large variety of different sensor fusion methods exists. Choosing the best method depends on a number of parameters, such as system requirements; accuracy; redundancy; system cost; sensor availability; and the kind of information available from the sensors. A short list of some common sensor fusion methods includes:

- Weighted averages
- Kalman filters
- Principal component analysis
- Bayesian inference
- Artificial neural networks
- Fuzzy logic
- Dempster–Shafer
- Reasoning in which the order of this list is organized from a process behavior from lowest to highest level of fusion activity (i.e., data fusion to information fusion processes)

As far as sensor fusion models are concerned, several different kinds of models and architectures have been presented over the past few decades. These models describe the system's functionality and give a simplified description of a complex entity or process. They also describe coupling between different physical components and how the components communicate together. Among the most prominent of fusion models is the joint directors of laboratories (JDL) model created in 1986 and refined in 1999 [12]. The model is a generalization of different levels of processing that may be applicable in different situations. Five levels are mentioned:

- Preprocessing (level 0)
- Single object refinement (level 1)
- Situation refinement (level 2)
- Implication refinement (level 3)
- Process refinement (level 4)

Each of these levels may consist of different elements; for example, single object refinement may consist of alignment, association, feature extraction classification and identification.

Another example of a fusion model is the observe, orient, decide, and act (OODA) model, which describes a decision making cycle [3]. The OODA is especially suitable for higher-level fusion processes. The process can also be equated to levels of the JDL model: levels 0 and 1 correspond to the observe steps; level 2 corresponds to the orient step; and levels 3 and 4 correspond to the decide and act steps, respectively. Other sensor fusion methods include the waterfall method and the omnibus model [3].

Sensor fusion is an active ingredient for most perceptual systems. Furthermore, the fusion can occur at various stages throughout the data processing. The subsequent subsections review additional components that have been addressed in different sensor fusion models and are subsequently an integral part of perceptual sensing systems.

7.3.2 Time Concept

A crucial issue to consider is how to handle data that come from sensing systems whose processing is unsynchronized. In other words, the information retrieval of the data from different sensors is processed at different instances in time. In perceptual systems, which often contain several sources of sensory input, the issue of time handling can be approached in several different ways.

One method to synchronize the incoming information is to use a process of direct perception. A reference scale is created and all incoming signals are translated onto this scale. For example Bothe and coworkers [6] explored the problem of target localization by attention control, using audio and visual perception. The goal of the work is to focus the camera onto a moving object and collect the audio information from the surroundings. The audio information is sampled at a higher rate compared to the

Artificial Perceptual Systems

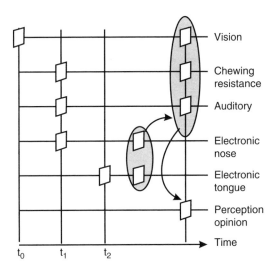

FIGURE 7.2 Sequential-based perception process of a human-related application using an electronic head. (From L. Biel, P. Wide, *IEEE Instr. Meas. Mag.*, 2000. © 2000 IEEE. With permission.)

visual information and, consequently, synchronization between the two sensor modules must be done. As shown in Figure 7.2, fusing all the audio maps in a specific time interval before the video signal is available performs the synchronization. Then the fused audio map is fused with the video map that corresponds to the same time stamp.

Another method is to fuse the information sequentially as it is generated from the sensor modules. This work was explored in Wide et al. [21]; an artificial head consisting of the five primary senses (sight, taste, smell, sound, and touch) was used to provide a quality evaluation of certain substances. In this case, two modules are fused together as the information becomes available, then the third module is merged (see Figure 7.2). This kind of time coordination is practical, especially in systems that utilize real-time implementation.

A third approach introduced by Saffiotti and Leblanc [17] is to benefit from a model that uses memory capability to strengthen or weaken the belief about a particular hypothesis. The time handling in this case is to take information that has been previously processed and fuse it with new information from the sensors, as shown in Figure 7.3. This approach was particularly useful in an example in which an unmanned flying vehicle performs a traffic surveillance task. The goal is to identify and track an object. A symbolic model is also included so that information from the vision camera is connected to different symbolic objects. The memory functioning in the time handling is used in the tracking action of the object (a moving car) in order to validate the identity of the car.

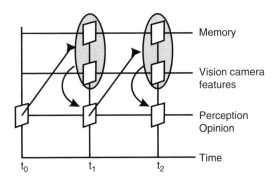

FIGURE 7.3 Direct process with memory capabilities. (From L. Biel, P. Wide, *IEEE Instr. Meas. Mag.*, 2000. © 2000 IEEE. With permission.)

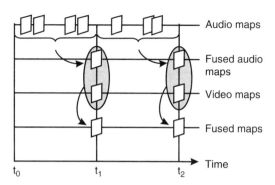

FIGURE 7.4 Time sequence using audio and video fusion. (From L. Biel, P. Wide, *IEEE Instr. Meas. Mag.*, 2000. © 2000 IEEE. With permission.)

7.3.3 Error Handling

In any multisensing system, errors are generated by missing data, corrupted data, logical errors in the software, malfunction of the sensors, etc. An intrinsic component to any sensor fusion model is to find an effective manner to treat erroneous data. In general, error handling is best conducted as close to the error source as possible in order to avoid inevitable propagation through the system. Errors that have propagated throughout the higher levels of the fusion processes are generally more difficult to rectify. Consequently, many complex fusion models attempt to compensate by including error trapping in each level of processing. Often this is a redundant process of performing cross checks, so consideration should be given to the cost of each error trap and its effect on system performance.

7.3.4 Reasoning

An active component in the modeling of a perceptual system is the ability to reason about a particular belief or hypothesis in order to make decisions. The perceptual reasoning machine (PRM) introduced by Kadar [13] provides a "governing closed loop control mechanism for intelligent adaptive information gathering, combination and monitoring, learning associative information recall and prediction as well as information assessment and interpretation." In the context of a generic information process model framework, the PRM functions between the associative systems, such as a knowledge base and the collected information from the sensing modules. The goal of the PRM is to perform a "gather and assess" task. This occurs by taking the input data and using algorithms and evidence function using Bayesian or non-Bayesian techniques; beliefs and hypotheses about observations are generated. Prior domain knowledge as well as these beliefs are sent further to a decision maker or, if required, an iterative process until the convergence of a hypothesis is achieved.

7.3.5 Passive and Active Perception

The perception process of an artificial system can be considered in two parts: the active and the passive. In a passive perception application, the incoming data are organized using a type of fusion in order to represent information about the surroundings. The information, which can be considered an "environmental picture," is then processed through the system to the various components. It is considered passive perception when no feedback component is present to readjust or redirect the environmental picture.

An active perception component introduced by Biel and Wide [5] may act as a feedback within the perception system. Active perception may initiate a redirection to specific sensing modules or may be used to adjust specific settings. A biological example of active perception is vision — the eye will compensate for luminance for the detection of objects. Another example is present in the olfactory sense when a desensitization effect occurs to adjust for odors. To determine how the active perception module interacts with the sensing components may require the use of a knowledge base. As described by Bajcsy [2], a top-down or bottom-up approach can be adopted when building an active perception system. In the top-down method, the system has no knowledge about the environment and requires a comparison with a knowledge base. In a bottom-up approach, however, the system has a predefined goal and searches for that goal in the environment.

Active perception works in cooperation with the sensor fusion or can be present within the embedded algorithm collecting the data. Basically, the use of active perception in a perceptual system is best summarized as the "intelligent goal-driven ability to make new decisions based on information feedback from past actions and consequences to the environment. It is also aimed for focusing the attention and weight of perception detectors based on internal drives and needs and considers the motivation of the system in order to generate decisions" [5].

7.3.6 Memory and Knowledge Base

Some mention of the need for a knowledge base has been introduced already, especially in the context of a sensor fusion system. In a complete perceptual system, the knowledge base can be further expanded not only to include the analogous memory components but also to contain the associative system and learning algorithms. This component is essential for any perceptual system; consequently, the purpose of the knowledge base is multifunctional. Among the most obvious tasks is the ability to store and recall knowledge based on prior information given by an expert or evolved over time in the included learning algorithms.

The knowledge base interacts with all other components and also serves in cooperation with the error trapping sequences at various stages of the data processing. The storage of knowledge may be biologically inspired and divided into different stores, such as the sensory storage, short-term storage, and long-term storage (as a reducing effect) mentioned in Best [4], or it may be based on other paradigms. What is important is that the information be double directed, i.e., that data are transmitted to as well as from the knowledge base to the other main processes. Furthermore, this component can be a means by which the human user can guide and direct the system, whether from a standpoint of directly inputting the *a priori* knowledge or acting as a supervisor in the learning process for real-time processes.

7.3.7 Human–Computer Interaction

The human–computer interaction in multisensing platforms has recently become an area of increasing interest. As a larger diversity of sensors find their way onto industrial applications as well as consumer-related domains, methods of interpreting the results from these sensors in a human friendly manner are necessary for effective and efficient operation. The communication between human and computer is a two-way process that should consider interpretation of sensor results to the human and interpretation of information from the human user to the machine. Furthermore, many more applications are considering contexts in which nonexperts may be involved; consequently, the interaction should be designed to facilitate these requirements.

The increase in sensing ability has created a particular challenge to the problem of human–computer interaction and specifically the communication process. As sensing technologies have extended our ability to perceive our environment, for example infrared, sonar, tactile and chemical sensors, there still needs to apply a method of translation from the "new" information from these complex sensors to the human perceptual domain. According to Siegel [18], this challenge has reshaped the sense–think–act paradigm generally accepted for sensing systems in robotic systems to include "communicate" as one of the essential components to robotic platforms.

The problem is to determine a means to convey information arising from sensors with more acute perception or even no counterpart in the human sensing apparatus. In some sensing systems, using scaling techniques may function as a method to translate the incoming sensor data into the human perceptual domain. For example, ultrasonic sounds can be scaled to lower frequencies and thus become detectable by the human ear. Other techniques may translate the results from one kind of sensor to another, such as using vision and color perception to view odor maps in which different colors represent various concentrations of the same odor. As the types of available sensor technologies change so do the kinds of human–machine interfaces. A new generation of interfaces is beginning to emerge that considers more advanced levels of communication, such as language and facial and body expressions, as a means of interacting with humans. Sometimes emerging technologies may require artificial sensing systems to perform higher levels of data processing, which may include categorization, conceptualization, and generalization and abstraction.

7.4 Perceptual Systems in Practice

This chapter suggests two examples of how perceptual systems could be used in a practical application. An overview of the general perceptual system is given with a focus on the components mentioned in previous chapters.

7.4.1 Electronic Head

In this example a multisensing platform presented by Wide and colleagues [21] is considered. The sensing platform is inspired by the five primary human senses and equipped with the following sensor modules:

- Vision
- Audition
- Chewing resistance (tactile sensing)
- Taste
- Olfaction

The objective is to provide qualitative estimations of different food substances based on information received from the sensors. Different motor control actions that initiate the chewing processes are also present within the platform. The entire sensing system included with the data interpretation and eventual output to the human user constitutes a perceptual system whose goals are to develop:

- A mechanism to give the system a desired degree of learning ability
- A series of perception modules to sense and analyze different features of the environment
- A fusion strategy to combine the gathered information into an overall virtual feature estimate
- An interfacing between the perceived information and the human user that exploits the ability of the learning algorithms

The perceptual system begins with the artificial sensors, which perform the sensation; from this step, feature extraction for individual sensors is applied for each of the sensors. In this case a feature fusion fuses all the results from each of the sensors together in a sequential process. A knowledge base contains storage of known substances that have been created through a training process using an artificial neural network. The task here is to create a classification of an unknown substance while using a social agent in the form of a facial expression animation to communicate the result from the classification process.

Facial expressions are generated using a facial expression driver (FED) that is an integral part of active perception [15]. The driver consists of one or more detectors for each sensory system (e.g., nose, tongue, audio, etc.) and a three-dimensional affect space mapping function as shown in Figure 7.5. The affect space described by Breazeal [7] explores how emotions can be characterized in terms of a set of discrete primary emotions such as happiness, anger, sadness, fear, etc. The model used in this work represents the discrete emotion categories by fuzzy regions around two axes denoted by arousal and pleasure (also called valence). A third dimension, called stance, is also included and represents a degree of confidence. The use of the affect space is illustrated in Figure 7.6. In the figure, two emotions of anger and fear are located on the three-dimensional affect space. These emotions are associated by a negative valence and high arousal; however, they are separated by the degree of self-confidence, where anger is represented by a higher degree.

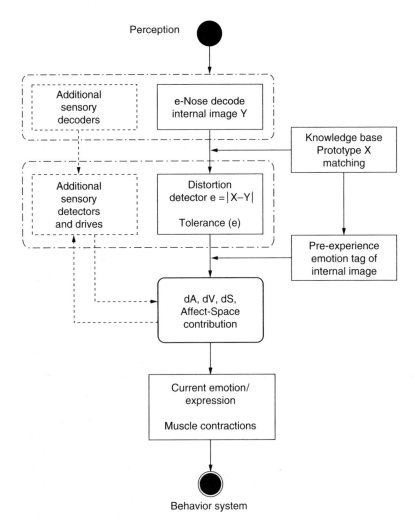

FIGURE 7.5 Details of a facial expression driver. (From A. Loutfi, et al. *Int. Symp. Virtual Environments, Human Computer Interfaces and Measurement Systems*, 2003. © 2003 IEEE. With permission.)

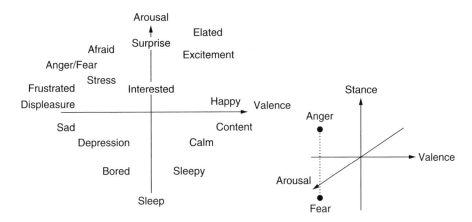

FIGURE 7.6 Motion regions in the affect space represented by arousal valence and stance. (From A. Loutfi, et al. *Int. Symp. Virtual Environments, Human Computer Interfaces and Measurement Systems*, 2003. © 2003 IEEE. With permission.)

The FED is part of a motivation system. Because the mathematical representation of goals can be extremely complex, it is difficult to build elicitors for action selection mechanisms via numerical methods. Instead, in this work, the dimensions in the affect space are treated as an additional set of internal variables or drives that can be interrelated with the motivation system. If the agent's variables are within the homeostatic regime then *pleasantness* is high. The *arousal* axis is controlled by circadian rhythm, unexpected stimuli, and rewards. *Stance* is controlled by the confidence or certainty of recognized external stimuli or statistical properties from the sensory process.

In the context of a multisensing system such as the electronic head, every prototype substance (cluster center in a classification process) is tagged with a corresponding location in the affect space called an emotion tag. This is done in a training phase guided by a human supervisor and is stored in the memory or knowledge base. In other words, it is a pivot point from which the homeostatic regions are translated to meet the expected opinion.

When new substances are detected, the FED uses a distortion detector to analyze the corresponding sensory representation, called an internal image. The distortion detector is targeted to evaluate one specific feature in that internal image. For example, different vision detectors are specialized to find specific objects, edges, or movements in a picture — one detector for the concentration of one specific compound. In addition, a secondary detector is defined to evaluate the distortion, based on the likelihood or membership value given by the electronic nose classification. The analysis of the distortion or divergence is then computed as the maximum of the absolute distance from the prototype. A tolerance drive is created that is a function of the distortion in the distance and used to produce behaviors and expressions so that the user is manipulated to regulate the system back to its expected functional balance or zero distortion. For example, in the case of an electronic nose used in quality control, the system would represent external regulation by means of social interaction.

Once the facial expression is determined, an animation sequence begins. The facial animations are based on a hybrid model of traditional geometric modeling and image-based modeling. In the traditional modeling, the governing components are the numerical description of topology, the underlying structure, and the surfaces and curves. Because the face is broken down into primitive cubes and polygons, this can be a tedious and time-consuming process. Animation is typically done by interpolation between predefined poses or so-called keyframing. Image-based modeling, on the other hand, uses photorealism, a method that uses real photos to capture shadows, lights, and depth to give three-dimensional model realism. One technique, video rewrite [8], uses modeling of social agents in user interfaces. The basic idea is to find a way to index the linear sequence of images automatically in the video format. It is then possible to produce a new arbitrary animation sequence. A drawback to image-based modeling is that its performance is limited to typically neutral expressions and neutral backgrounds.

The hybrid method takes advantage of the cost of geometry control and dynamic as well as the simplicity of photorealism. Many different kinds of hybrid models are available; however, for the work described here, the focus is not to derive necessarily a new model but rather to employ an existing model with the sensor signals. To do this, an older project introduced by Waters called SimpleFace is used [16]. This was created primarily for modeling the virtual anatomy of the muscle-based underlying structure of a face shown in Figure 7.7. This version implements only 18 of the most dominating muscles to produce a discrete set of facial expressions like happiness, sadness, surprise, anger, fear, and disgust. One of the reasons why SimpleFace was chosen was the progress in the research of EAP muscles that may motivate future implementation of the electronic head. In this perspective, the early model would serve as a good and simple reference or starting point.

This example illustrates a case of a perceptual system equipped with the ability to communicate the final result to a human user. The goals and tasks of the system used standard pattern recognition to the fused sensor data to determine how the facial expressions were to be used.

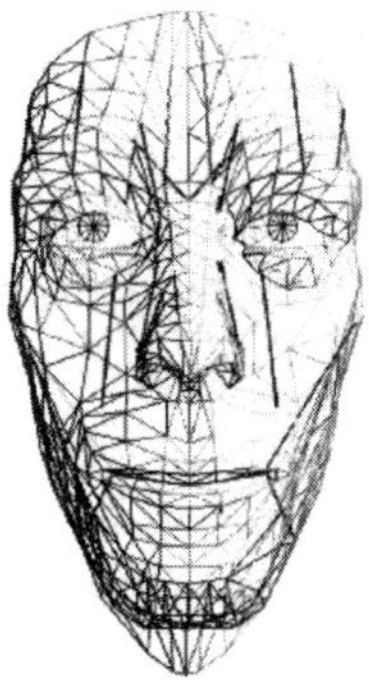

Artificial Perceptual Systems

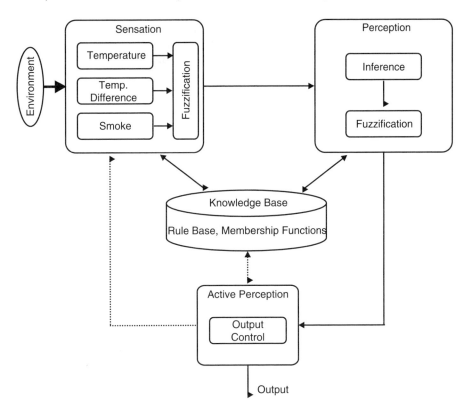

FIGURE 7.7 Model of the face topology activation muscles, (based on Parke and Waters, 1996) showing activation muscles and example values to produce a "happy" facial expression. (From A. Loutfi, et al. *Int. Symp. Virtual Environments, Human Computer Interfaces and Measurement Systems*, 2003. © 2003 IEEE. With permission.)

7.4.2 Fire Indication Application

Another example is a fire indication application as outlined by Biel and Wide [5]. In this experiment, a nonconventional multisensing fire indication is used as a platform to provide an early detection and alarm system. The entire process is summarized in Figure 7.8, which shows the sensation process occurring by using three sensors: temperature; carbon monoxide; and oxygen. Information from these sensors is immediately fed into a fuzzification algorithm that converts the crisp sensor values into a fuzzy result.

FIGURE 7.8 A fuzzy system with three inputs using active perception.

This occurs by using a set of linguistic values to represent the range of data points; for example, the value of temperature may take the values of cold, medium, or hot. For each of these linguistic values, a membership function exists that represents the "degree of belonging" that the input can correspond to a particular linguistic value.

The perception process occurs by using fuzzy inference, which combines the degrees of truth represented in the membership function by using a set of fuzzy rules. For example, *IF Temperature IS low AND CO Concentration is low THEN FIRE is NoFire.*

The set of fuzzy rules in this case is contained in the knowledge base. Also within the perception process is a defuzzification that translates the fuzzy values back into crisp output values. Of the several methods available for the defuzzification process, the center of gravity method is used in this case. (See Zadeh [22] for more detail on fuzzy logic and inference systems.) An active perception component is used here for the purpose of sensor management and sensor control. The management and control might try to focus the results to find the fire source, for which a mobile platform is required; however, the active perception component is still needed to determine the routines between robot movement and the sensing readings. Also, the active perception can be used to focus the readings further to determine the nature of the fire. Information regarding the type and source of the fire is ultimately useful to determine the best methods of extinguishment.

7.5 Research Issues and Summary

This chapter reviewed the emergence a new type of multisensing system called perceptual systems. Many components within a perceptual system, such as sensor fusion, are present in other types of data processing models; however, perceptual systems are not limited to sensor fusion techniques. One requirement is that a perceptual system encompass an important perspective of the interface between humans and an interacting system. Although case studies of perceptual systems have already been conducted, the intention of the work here is to give a general overview of some consideration in the design of any multisensing platform designed to reason and make decisions in the environment.

References

1. Adorni, G., Destri, G., and Mordonini, M. Indoor vehicle navigation by means of signs. *Proc. IEEE Intelligent Vehicles Symp., Proc.*, 76, New York, 1996.
2. Bajcsy, R., Active perception, *Proc. IEEE.*, 76, 8, 966, 1988.
3. Bedworth, M. and O'Brien, J. The omnibus model: a new model of data fusion? *Proc. 2nd Int. Conf. Inf. Fusion*, 1999.
4. Best, J.B., *Cognitive Psychology*, 4th ed., West Publishing Company, St. Paul, Minnesota, 1995.
5. Biel, L. and Wide, P., Active perception in a sensor fusion model, *Sensor Fusion: Architectures, Algorithms, Applications VI*, 4731, 164, 2002.
6. Bothe, H., Persson, M., Biel, L., and Rosenholm, M., Multivariate sensor fusion by a neural network model, *Proc. 2nd Int. Conf. Inf. Fusion*, 1094, 1999.
7. Breazeal, C., Robot in society: friend or appliance, *Agents '99 Workshop on Emotion-Based Agent Architecture*, Seattle, 18, 1999.
8. Bregler, C., Covell, M., and Stanely, M., Video rewrite: driving visual speech with audio. *Proc. 24th Annu. Conf. Computer Graphics*, 1997.
9. Fechner, G.T. *Elemente der Psychophysik.* 1860.
10. Fermuller, C. and Aloimonos Y., Vision and action, *Image Vision Computing.*, 13: 10, 725, 1995.
11. Fod, A., Mataric, M.J., and Jenkins, O.C., Automated derivation of primitives for movement classification, *Autonomous-Robots*, 12, 1, 39, 2002.
12. Hall, D.L. and Llinas, J., An introduction to multisensor data fusion. *Proc. IEEE*, 85, 6,1997.
13. Kadar, I., Adaptive prediction and off-board track and decision level fusion for enhanced surveillance, *IEEE Int. Conf. Multisensor Fusion Integration Intelligent Syst.*, 417, 1994.

14. Kluender, K.R., Coady, J.A., and Kiefte, M., Sensitivity to change in perception of speech, *Speech Commun.*, 41, 1, 59, 2003.
15. Loutfi, A., Widmark J., Wide, P., and Wikstrom, E., Social agent: expressions driven by an electronic nose, *IEEE Int. Symp. Virtual Environ., Hum.–Computer Interfaces, Measurement Syst.*, 2003.
16. Parke, I. and Waters, K., *Computer Facial Animation*, AK Peters Ltd., 1996.
17. Saffiotti, A. and Leblanc, K., Active perception anchoring of robot behavior in a dynamic environment, *IEEE Conf. Robotics Automation*, 2000.
18. Siegel, M., The sense–think–act paradigm revisited, *IEEE Workshop Robotic Sensing*, 2003.
19. Stevens, S.S., On the psychophysical law, *Psychol. Rev.*, 64, 3, 153, 1957.
20. Ullman, S. and Soloviev, S., Computation of pattern invariance in brain-like structures, *Neural Networks*, 12, 7, 1021, 1999.
21. Wide, P., Kalaykov, I., and Winquist, F., The artificial sensor head: a new approach in assessment of human-based quality, *Proc. 2nd Int. Conf. Inf. Fusion*, 1144, 1999.
22. Zadeh, L., *Fuzzy Logic and its Applications*, Academic Press, New York, 1965.

II

Applications

8
Sensor Network Architecture and Applications*

Chien-Chung Shen
University of Delaware

Chaiporn Jaikaeo
University of Delaware

Chavalit Srisathapornphat
University of Delaware

8.1 Introduction 8-1
8.2 Sensor Network Applications 8-1
    Querying Applications • Tasking Applications
8.3 Functional Architecture for Sensor Networks 8-3
8.4 Sample Implementation Architectures 8-4
    SINA (Sensor Information Networking Architecture) • TopDisc (Topology Discovery for Sensor Networks)
8.5 Summary 8-12

8.1 Introduction

The sheer number of sensor nodes and the dynamics of their operating environments (for instance, limited battery power and hostile physical environment) pose unique challenges in the design of sensor networks and their applications. Issues concerning how information collected by and stored within a sensor network can be queried and accessed are of particular importance. In this chapter, sensor network applications are categorized into two classes — querying and tasking — and a generic functional architecture, termed *sensor network architecture* (SNA), to facilitate these applications is introduced. In this architecture, functional components and their interrelationship, which should be available in sensor networks, are identified. Two existing implementation architectures, SINA [1] and TopDisc [2], are examined as a case study by describing how SNA's functional components are exploited, as well as application characteristics supported by them.

The following section describes the two categories of applications for sensor networks. Section 8.3 describes the functional architecture of SNA. Two sample implementation architectures, SINA and TopDisc, are described in Section 8.4. Section 8.5 concludes the chapter.

8.2 Sensor Network Applications

Based on the characteristics of their operations, applications of sensor networks can be divided into two classes: querying and tasking. The following subsections present sample applications for each class.

---

*Portions reprinted with permission from *IEEE Personal Communications Magazine*, 8, 4, 2001. © 2001, IEEE.

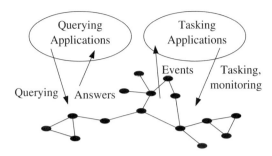

FIGURE 8.1 Querying and tasking applications in sensor networks. (From Shen, et al., *IEEE Personal Commun. Mag.*, 8(4), 52–59, 2001. With permission.)

8.2.1 Querying Applications

Querying applications concern how information collected by a sensor network can be retrieved based on specified criteria. For instance, environment sensing to extract information from the physical environments is one major application of sensor networks. Depending on its hardware capability, a sensor node can be programmed to collect temperature, humidity, light, pressure, chemical substances, or vibration information [3], and report it to the application. Applications may employ simple queries to obtain raw sensor data reported directly from each sensor node.

However, in some situations, complicated queries involving distributed data collection or aggregation become necessary. For example, to find out which region of the sensed area has the highest temperature, intelligent data collection, filtering, and aggregation could be carried out within the sensor network so that the observer will not need to obtain all raw data, thus conserving scarce system resources, such as battery energy and network bandwidth. In addition, the state of the sensor node, such as remaining energy level, operational status, or a list of neighboring sensors, can also be retrieved for management purposes [2]. The collected information could also be used to diagnose the health of sensors [4].

8.2.2 Tasking Applications

Tasking applications involve programming sensor nodes to perform specific actions upon certain events. Events can be physical environment changes, messages from nearby sensor nodes, or triggers from hardware/software modules inside a sensor node. A task can be as simple as asking individual sensor nodes to report information independently when they sense something unusual about their surrounding environments. More complex tasks may require distributed coordination, or even collaboration, among sensor nodes to achieve higher accuracy and/or efficiency. For instance, tracking a moving object in an area by simply having every single sensor node periodically and blindly monitor its surroundings can be very energy inefficient. If nodes surrounding the tracked object collaborate, more complete and accurate information can be collected with higher efficiency [5–7].

A similar idea of coordination can also be applied to reduce the number of nodes participating in data forwarding [2]. Modern equipment may have sensor modules operate in conjunction with actuator modules so that the behavior of sensor nodes can be controlled. In this case, tasking applications can utilize information obtained from sensor nodes to adapt nodes' behavior or movement pattern so as to achieve better sensing and networking performance. For environmental control applications, actuators can be controlled to affect the physical environments. An office building, for example, may have a sensor node installed in each room. These nodes then coordinate and send control signals to the air-conditioning unit, which, in turn, adjusts accordingly to achieve optimal comfort in all the rooms [8].

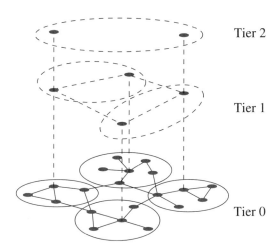

FIGURE 8.2 Clustering and a cluster hierarchy. (From Shen, et al., *IEEE Personal Commun. Mag.*, 8(4), 52–59, 2001. With permission.)

8.3 Functional Architecture for Sensor Networks

Compared to conventional distributed databases in which information is distributed across several sites, the number of sites in a sensor network equals the number of sensor nodes, and the information collected by each node (e.g., sensor readings) becomes an inherent attribute of that node [9]. To support energy-efficient and scalable operations, sensor nodes could be autonomously clustered. Furthermore, the data-centric nature of sensor information makes it more effectively accessible via an attribute-based naming approach instead of explicit addresses [10]. In addition, as these sensors are integrated into and extract information from physical environments, many applications also require the location information to be passed along with their sensor data. As a result, a generic functional architecture for sensor networks consists of the following components:

Hierarchical clustering. To facilitate scalable operations within sensor networks, sensor nodes could be aggregated to form clusters based on their energy levels and proximity. The aggregation process could also be recursively applied to form a hierarchy of clusters (Figure 8.2). Within a cluster, a cluster head will be elected to perform information filtering, fusion, and aggregation, such as periodic calculation of the average temperature of the cluster coverage area. In addition, the clustering process should be reinitiated in case the cluster head fails or runs low in battery energy. In situations in which a hierarchy of clusters is not applicable, the system of sensor nodes is perceived by applications as a one-level clustering structure in which each node is a cluster head by itself. The clustering algorithm introduced by Estrin and colleagues [10] allows sensor nodes automatically to form clusters, elect and re-elect cluster heads, and reorganize the clustering structure if necessary.

Location awareness. Because sensor nodes are operating in physical environments, knowledge about their physical locations becomes mandatory. Location information can be obtained via several methods. Global positioning system (GPS) is one of the mechanisms that provide absolute location information. For economical reasons, however, only a subset of sensor nodes may be equipped with GPS receivers and function as location references by periodically transmitting a beacon signal telling their own location information so that other sensor nodes without GPS receivers can roughly determine their position in the terrain. Other techniques for obtaining location information are also available. For example, optical trackers [11] give high-precision and -resolution location information but are only effective in a small region.

Attribute-based naming. With the large population of sensor nodes, it may be impractical to pay attention to each individual node. Users would be more interested in querying which area has temperature higher than 100°F or what the average temperature in a specific area is, rather than the temperature at sensor ID#101. To facilitate the data-centric characteristics of sensor queries, attribute-based naming is the preferred scheme [10]. For instance, the name [type=temperature, location=N-E, temperature=103] addresses all the temperature sensors located at the northeast quadrant with a temperature reading of 103°F. These sensors will reply to the query, "which area has temperature higher than 100°F?" Note that not only can physical or location attributes be part of a name, but so can logical attributes such as unique IDs, temporary variables, and clustering roles (e.g., cluster head or cluster member). Therefore, the traditional addressing scheme using node IDs becomes a special case of attribute-based naming.

With the integration of these three components, the following two sample queries may be effectively and efficiently carried out.

- *Which area has temperature higher than 100°F?* In theory, the query is broadcast to and evaluated by every node in the network. Despite possibly the best returned result, the query would suffer from long response time. In practice, each cluster head may periodically update the temperature readings of its members, and the query can now be multicast to and evaluated by cluster heads only. This results in better response time at the expense of less accurate answers. Queries under stringent timing constraints can be evaluated by cluster heads of a higher tier.
- *What is the average temperature in the southeast quadrant?* Similarly, the average temperature of each cluster can be periodically updated and cached by cluster heads. Furthermore, the query should be delivered to nodes located (named) in the southeast quadrant only.

8.4 Sample Implementation Architectures

Given the SNA functional architecture, two implementation architectures are described: SINA, which implements SNA to facilitate querying and tasking applications, and TopDisc, which is specifically designed to perform topology management of sensor networks.

8.4.1 SINA (Sensor Information Networking Architecture)

SINA [1] adopts a middleware-based approach to implementing SNA functional architecture. By modeling a sensor network as a collection of massively distributed objects, SINA modules, running on each sensor node, serve as a middleware working across all sensor nodes; provide adaptive organization of sensor information; and facilitate query, event monitoring, and tasking (Figure 8.3). SINA allows sensor applications to issue queries and command tasks into, collect replies and results from, and monitor changes within the networks. SINA provides the following mechanisms to facilitate querying and tasking of sensor networks: information abstraction; information gathering methods; sensor query and tasking language; and sensor execution environment. These mechanisms are explained in detail in the following subsections.

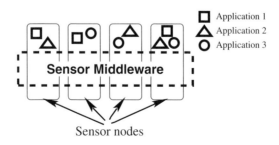

FIGURE 8.3 A model of sensor networks and SINA middleware. (From Shen, et al., *IEEE Personal Commun. Mag.*, 8(4), 52–59, 2001. With permission.)

8.4.1.1 Information Abstraction

In SINA, a sensor network is conceptually viewed as a collection of datasheets, each of which contains a collection of attributes of each sensor node. Each attribute is referred to as a cell, and the collection of datasheets of the network present the abstraction of an *associative spreadsheet*. In contrast to a conventional spreadsheet paradigm in which a data item is stored in a cell that is assigned an address according to its logical x–y coordinates, SINA refers cells via attribute-based names. Initially, a datasheet of each sensor node contains a few predefined attributes. Once these sensor nodes are deployed and form a sensor network, they can be requested by other nodes — for instance, from their cluster heads — to:

- Create new cells by evaluating valid cell construction expressions that may obtain information from other cells
- Invoke system-defined functions
- Aggregate information from other datasheets

Each newly created cell must be uniquely named and becomes a node's attribute, which can be a single value (e.g., remaining battery energy) or multiple values (e.g., history of temperature changes in the past 30 min). By incorporating a hierarchical clustering mechanism and an attribute-based naming scheme, SINA provides a set of operations to deal with data access and aggregation among sensor nodes. The mechanism of *associative broadcast* [12] has been employed to facilitate process interaction via attribute-based naming.

8.4.1.2 Information Gathering Methods

SINA provides a communication mechanism among sensor nodes to facilitate distributed applications. By providing efficient data dissemination and information-gathering supports suitable for specific application requirements, SINA abstracts low-level communications from high-level sensor applications. When users submit queries, it is not required to define how the information will be collected inside the network explicitly. SINA selects the most appropriate data distribution and collection method based on the nature of queries and current network status. Upon receiving users' queries, the *frontend* node — a special node directly connected to the user — has the responsibility to interpret and evaluate the queries by requesting information from other nodes.

With the sheer number of sensor nodes, collisions resulting from a large number of responses propagated back to the front-end node during a short period of time create the *response implosion problem* [9] depicted in Figure 8.4(a). The objective of the information-gathering mechanisms is to maximize the quality of responses in terms of their number and responsiveness while minimizing network resource consumption in conducting the query operations. Three primitive methods are provided to accomplish the information gathering task:

- *Sampling operation.* For certain types of applications (for instance, finding the average temperature over the entire network area), responses from every sensor node may cause a response implosion. To reduce the degree of the problem, some sensor nodes may not need to respond if their neighbors will. Nodes make autonomous decisions whether they should participate in this application based on a given response probability, as shown in Figure 8.4(b). This operation is also known as *Samplecast* [9]. An enhancement can be made to this approach if sensor nodes are not evenly distributed over the area. To prevent receiving more responses from dense areas, the response probability will be computed at each cluster head node based on the number of replies required from each cluster. This operation is called *adaptive probabilistic response (APR)*.
- *Self-orchestrated operation.* In a network with a small number of nodes, responses from all nodes are necessary for the accuracy of the final result. Another approach to avoiding the response implosion problem is to let each node defer sending responses for some period of time. Despite some extra delay, this method aims to improve the overall performance by reducing the chances of collision. This operation is modified from the scheduled response approach described in Johnson and Maltz [13]. Assuming that nodes are distributed uniformly within the network

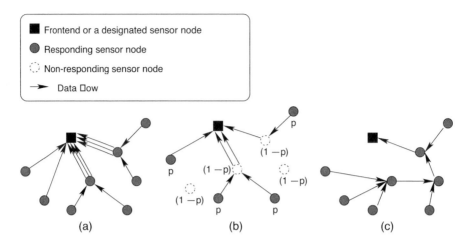

FIGURE 8.4 (a) The response implosion problem; (b) number of responses reduced by assigning sensor nodes a probability p to answer the request; (c) diffused computation operation allowing data aggregation at intermediate nodes. (From Shen, et al., *IEEE Personal Commun. Mag.*, 8(4), 52–59, 2001. With permission.)

terrain and that the number of nodes within h hops away from the front-end node proportional to h^2, the delay period at every node can be defined as

$$Delay = KH(h^2 - (2h-1)r)$$

where h is the length in number of hops away from the front end; r is a random number such that $0 < r \le 1$; and H is a constant reflecting estimated delay per hop. To incorporate potential effects from queuing and processing delays, K is used as a compensation constant. Normally, K and H are combined and used as a single adjustable parameter.

- *Diffused computation operation.* For this operation, each sensor node is assumed to have knowledge about its immediate communicating neighbors only. Algorithms used for gathering information are constrained by the capability that each node can only communicate to other nodes in its surrounding area. Information aggregation logic is programmed as a script and disseminated among sensor nodes so that they know how to aggregate information en route to the front end. The conceptual data flow is depicted in Figure 8.4(c). Because data are aggregated at intermediate nodes on the way back to the front-end node, the consumption of valuable network bandwidth is reduced and the response implosion problem alleviated considerably. However, for large sensor networks, this diffusion approach might take a longer time to deliver results back to the front end.

The hierarchical structure enabled by SINA allows different information-gathering methods to be deployed in different levels within one application in order to optimize overall performance. The effects of the integration are discussed in Shen et al. [14].

8.4.1.3 Sensor Network Programming Languages

As part of SINA, sensor querying and tasking language (SQTL) [15] plays the role of a programming interface between sensor applications and SINA middleware. This is a procedural scripting language designed to be flexible and compact, with a capability of interpreting declarative query statements. In addition to sensor hardware access (e.g., `getTemperature, turnOn`), location-aware (e.g., `isNeighbor, getPosition`), and communication primitives (e.g., `tell, execute`), it also provides an event-handling construct, which is suitable for many sensor network applications in which sensor nodes are often programmed to process asynchronous events such as receiving a message or an event triggered by a timer. By using the "upon" construct, a programmer can create an event-handling block accordingly.

TABLE 8.1 Arguments Used by Actions in SQTL Wrapper

Argument	Meaning
sender	The sender of an SQTL message wrapper
receiver	Potential receivers specify by two following subarguments
group	Subargument of receiver to specify group of receivers; its possible value can be one of ALL_NODES, or NEIGHBORS
criteria	Subargument of receiver to specify selection criteria of receivers
application-id	Unique ID for each application in the same sensor network
num-hop	Number of hops away from a gateway node
language	Specify a language used in content
content	A payload containing a program, message, or return values
with (optional)	Tuples of parameters used in the program passed from sender to receiver
parameter	Repeatable subargument of "with"
type	Data type of the parameter
name	Name of the parameter
value	Value of the parameter

Source: From Shen, et al., *IEEE Personal Commun. Mag.*, 8(4), 52–59, 2001. With permission.

Currently, three types of events are supported by SQTL: (1) events generated when a message is received by a sensor node; (2) events triggered periodically by a timer; and (3) events caused by the expiration of a timer. These types of events are defined by the SQTL keywords "receive," "every," and "expire," respectively.

An SQTL message, containing a script, is meant to be interpreted and executed by any node in the network. In order to target a script to a specific receiver, or a group of receivers, the message must be encapsulated in an *SQTL wrapper* which acts as a message header for indicating the sender, the receivers, and a particular application running on the receivers, as well as parameters for the application.

The syntax of the extensible markup language (XML) is adopted for the SQTL wrapper, which defines an application layer header capable of specifying a complicated addressing scheme for attribute-based names. Table 8.1 summarizes common SQTL wrapper fields.

For applications that collect sensor information, a user may choose to invoke the built-in query interpreter instead of explicitly writing a procedural SQTL script. The query language has been adapted from structured query language (SQL) to serve as the primary mechanism for querying sensor networks. The following sample query statement, as delivered to all cluster heads in the network (encapsulated in the SQTL wrapper), would ask every cluster head to create a new cell called *avgTemperature* that maintains the average temperature among all of its cluster members:

SELECT avg(*getTemperature*())
  AS *avgTemperature*
  FROM CLUSTER-MEMBERS

As soon as an SQTL message containing such a query statement is received by target nodes, their execution environments (explained later) will pick the most appropriate information-gathering method available to evaluate the query.

Database techniques, such as view composition, materialization, and maintenance, could be adapted to maintain consistency among associated cells. A related work on querying a sensor network modeled as a device database may be found in Bonnet et al. [16].

8.4.1.4 SEE (Sensor Execution Environment)

Running on each sensor node, a sensor execution environment (SEE) is responsible for dispatching incoming messages, examining all arrival SQTL messages, and performing the appropriate operation for each type of action specified in the messages. SEE looks inside the `receiver` argument of a message and, based on its value, decides whether to forward the message to the next hop. Messages with "ALL_NODES" in their group subarguments will be rebroadcast to every sensor node in the network and those with "NEIGHBORS" will only be forwarded to the nodes' immediate neighbors.

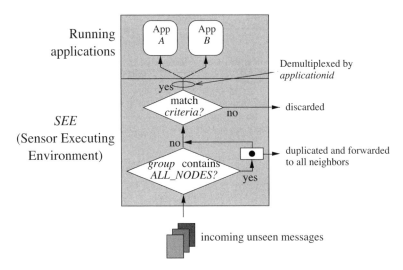

FIGURE 8.5 Dispatching of messages received by a sensor node. (From Shen, et al., *IEEE Personal Commun. Mag.*, 8(4), 52–59, 2001. With permission.)

SEE also prevents message looping by using a globally unique message ID, which is a combination of a unique node ID and message sequence number. An attribute-based name in the form of a list of attribute–value pairs indicated by the criteria field will be compared against the receiver's attributes stored in its datasheet. SEE only accepts the message if the node's attributes satisfy the criteria. This process of matching a message with its potential receivers when the message arrives at the receivers is termed *late binding* and is described by Bayerdorffer [12].

Once an SQTL script is injected from the front-end node to one or more sensor nodes, the script may push itself to other sensors in order to complete the assigned task. A `tell` message is then generated after a result is produced at each individual sensor node and is delivered back to the requesting node, which is normally the upstream node from which the script came. Figure 8.5 depicts the dispatching of incoming messages performed by SEE.

In addition to demultiplexing incoming SQTL messages, SEE also takes care of outgoing SQTL messages from all running applications. Outgoing messages will be distributed to target nodes specified in the `receiver` argument through the underlying communication mechanism. SEE may perform a translation of an attribute-based name into a unique, numeric link-layer address where applicable. Otherwise, broadcast will be used at the link layer.

8.4.1.5 Architectural View of SINA

Now the ways in which the three functional components defined in SNA are utilized and provided in SINA are examined. SINA provides an attribute-based naming mechanism by means of an associative spreadsheet in which nodes' attributes are defined in uniquely named cells. Destination groups are then determined by `criteria` fields that are part of SQTL. A mechanism for hierarchical clustering is not strictly tied to a particular algorithm and is intentionally left undefined for flexibility. A clustering algorithm such as the one described by Intanagonwiwat and colleagues [17] could be used. Once cluster heads have been elected, each node's cluster head role (i.e., whether it is a cluster member or a cluster head) will become one of its attributes. The clustering feature also allows different information-gathering methods to be used at different levels in the hierarchy in order to optimize overall performance. Similarly, mechanisms allowing nodes to obtain their location information are assumed, but not defined or used directly in SINA. It is left to the applications to target and query nodes' locations in the form of their attributes.

8.4.1.6 Sample Applications

SINA has been designed to support a wide range of sensor network applications. However, to illustrate its applicability to querying and tasking of sensor networks under this architecture, experiments were conducted on two sample applications: sensor network diagnosis and vehicle tracking; their behaviors and performance were studied using GloMoSim simulator. Results and more discussion of the two applications can be found in Shen et al. [14].

Diagnosis of sensor networks. Sensor network diagnosis is the process of querying the status of a sensor network and figuring out the problematic (group of) sensor nodes [4]. In order to monitor the status of a sensor network, one approach is to query as much information from as many sensor nodes as possible and then deliver the raw information to the manager for further processing, e.g., when a manager wants to know the remaining energy level within the network. In addition, to examine the correctness of results obtained from one sensing device, one possible method is to use the average of results obtained from other neighboring sensor nodes as a standard base to compare and diagnose the devices in doubt, given that the average has its deviation within an acceptable range. An example of using this method is to figure out which sensor node contains a faulty temperature-sensing device.

Coordinated vehicle tracking. The vehicle tracking application is to locate a specific vehicle or moving object and monitor its movement. To detect and identify an object, integrated results from more than one type of sensor, for instance, images from a camera, vibration from a seismic sensor, noise from an audio sensor, and so on, may be required. These results are to be processed and compared with the signature of the object of interest. However, the main interest is to program a coordination algorithm in the form of an SQTL script, which can be disseminated to all sensor nodes. The script controls the sensor nodes to detect the appearance of the interested object collaboratively in an effective and efficient manner. Thus, it is assumed that sensor nodes can obtain final processed results of detecting and identifying the tracked vehicle from the processing of combined sensing information.

A novice approach to tracking a moving object is to ask every sensor node to sense and detect the object's signature at the same time — an operation called the *ordinary vehicle tracking method*. However, this approach may waste sensor nodes' processing cycles, and thus inefficiently utilize a network's limited energy and shorten the overall network lifetime. In contrast, the coordinated vehicle tracking algorithm presented in Figure 8.6 is based on a suppression and reinitiation mechanism in order to achieve a better result of tracking, yet consume less network resources than the ordinary one. The main principle of the coordinated algorithm is to let the first sensor node detecting the vehicle suppress sensing activities of all other sensor nodes so that the others may stand by, which results in energy conservation. Furthermore, the node will need to reinitiate sensing activities of its neighbors in order to keep track of the moving vehicle. As long as the vehicle does not move faster than the propagation of this reinitiation message, the network can still monitor its trail. The tracking process is depicted in Figure 8.7 as well.

8.4.2 TopDisc (Topology Discovery for Sensor Networks)

TopDisc [2] provides a mechanism for data dissemination/aggregation and topology discovery in sensor networks. From an architectural point of view, TopDisc provides the same set of components specified by SNA. The following subsections describe the mechanism of TopDisc and present how its functional components are mapped to the SNA architecture. Finally, some sample applications supported by TopDisc are offered.

8.4.2.1 TopDisc Mechanism

TopDisc constructs an approximate topology of the network by collecting local topology information from *distinguished nodes* (or cluster heads) via a tree of clusters (TreC) rooted at the monitoring node. The mechanism is briefly described as follows. When TopDisc starts, all nodes are colored *white*, which means that they are undiscovered. The monitoring node initiates the topology discovery process by broadcasting a "topology discovery request." It then turns to *black*, which means that it is a distinguished

```
<execute>
   <sender> FRONTEND </sender>
   <receiver>     <group> NODE[0] </group>
                  <criteria> TRUE </criteria>
   </receiver>
   <application-id> 118 </application-id>
   <num-hop>        0        </num-hop>
   <language> SQTL </language>
   <with>
      <parameter type="clocktype" name="trackingTime"      value="600" />
      <parameter type="clocktype" name="reTrackingTime"    value="40" />
      <parameter type="clocktype" name="trackingFrequency" value="8" />
      <parameter type="object"    name="target"            value="Vehicle1" />
   </with>
   <content> <![CDATA[
      lastSensingResul = false;
      timerApplication = createTimer(trackingTime);   // instantiate a timer
      timerApplication.start();                       // turn it on
      timerReTracking  = createTimer(reTrackingTime);
      execute (ALL_NODES, "TRUE", MESSAGE["content"]);   // re-broadcast
      if ((sensor1 = getMotionSensor()).turnOn()) {   // instantiate a sensor object
         upon {                   /               /     and turn it on
            receive (msg) where msg["action"] == "tell" && msg["content"] == "suppress": {
               sensor1.standby();    break;
            }
            every (trackingFrequency): {
               if (sensor1.detect(target)) {
                  tell (ALL_NODES,       "TRUE", "suppress");
                  tell (NEIGHBORS,       "TRUE", "retrack");
                  tell (MESSAGE["sender"], "TRUE", "found");
                  lastSensingResult = true;
                  timerReTracking.start();
                  break;
               }
               else  lastSensingResult = false;
            }
            expire (timerApplication): sensor1.turnOff(); exit(0);
         }
         upon {   // After one sensor node sees the vehicle
            receive (msg) where msg["action"] == "tell" && msg["content"] == "retrack": {
               if (timerReTracking.expired()) {
                  sensor1.turnOn();
                  timerReTracking.start();
               }
            }
            receive (msg) where msg["action"] == "tell" && msg["content"] == "found":
               tell (MESSAGE["sender"], "TRUE", "found");
            every (trackingFrequency): {
               if (sensor1.detect(target)) {
                  tell (MESSAGE["sender"], "TRUE", "found");
                  if (!lastSensingResult)
                     tell (NEIGHBORS,     "TRUE", "retrack");
                  lastSensingResult = true;
                  timerReTracking.start();
               }
               else {
                  if (lastSensingResult)
                     timerReTracking.restart();
                  lastSensingResult = false;
               }
            }
            expire (timerReTracking)  : sensor1.standby();
            expire (timerApplication) : sensor1.turnOff(); exit(0);
         }
      }
      else  exit(1);
   ]]> </content>
</execute>
```

FIGURE 8.6 Complete SQTL script for the coordinated vehicle tracking algorithm. (From Shen, et al., *IEEE Personal Commun. Mag.*, 8(4), 52–59, 2001. With permission.)

node. White nodes receiving a request from a black node become *gray* and rebroadcast the request with a random delay inversely proportional to the distance between the black node and themselves.

Sensor Network Architecture and Applications

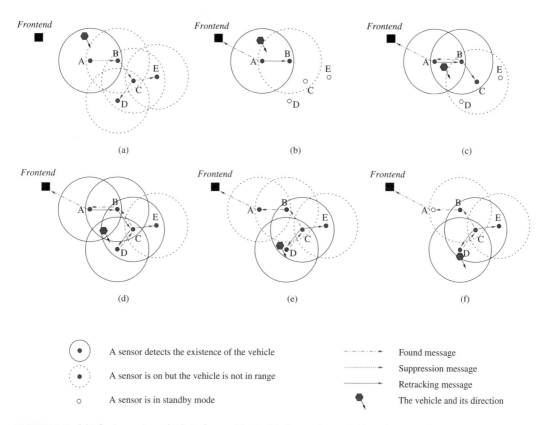

FIGURE 8.7 (a) The incoming vehicle is detected by A; (b) the sensing activities of C, D, and E are suppressed, but B starts tracking again; (c) the vehicle comes into B's area and C restarts its sensor; (d) C and D detect the vehicle and E's sensor is restarted; (e) the vehicle goes out of A's and B's ranges; (f) sensing activity at A stops. (From Shen, et al., *IEEE Personal Commun. Mag.*, 8(4), 52–59, 2001. With permission.)

However, white nodes will become black with some random delay if they receive a request from a gray node. During the delay interval, if white nodes hear any message from other black nodes, they will become gray. Note that all the black and gray nodes ignore all other incoming request messages. After the request has been propagated to the entire network, each node knows its *parent black node*, which is the last black node from which the topology discovery was forwarded to reach it. Each black node also knows the node to which it should forward packets in order to reach its parent black node. By snooping at all incoming request messages, all nodes have their neighborhood information collected.

To respond to the topology discovery message, once a node becomes black, it sets a timer, inversely proportional to the number of hops away from the monitoring node, and waits for responses from its children black nodes. A black node aggregates its own neighborhood list (obtained from snooping) together with neighborhood lists from its children and forwards the aggregated list back to the monitoring node through its default forwarding node.

8.4.2.2 Architectural View of TopDisc

Similar to SINA, TopDisc provides the same set of components described by SNA. First, TopDisc builds a TreC by selecting *distinguished nodes* to become cluster heads. Other nodes then associate with one cluster head. This process has the same functionality as the hierarchical clustering component of SNA. Nodes in TopDisc also perform information aggregation by combining messages obtained from children black nodes. The objective of a TreC and data aggregation is to reduce the number of response messages coming back to the monitoring node. TopDisc also employs attribute-based naming schemes in its data dissemination process. Subsequent requests to the network will be carried over a TreC. Recall that a TreC comprises black (cluster head) and gray (forwarding) nodes. However, only cluster heads will process the requests; gray nodes only forward the requests. This process resembles attribute-based naming. Finally, TopDisc employs location information in one of its proposed applications to schedule sensor nodes' duty cycles.

8.4.2.3 Sample Applications

By using a TreC created by TopDisc, several data dissemination/aggregation applications are possible. The following applications are described in Deb et al. [2]:

- *Retrieving network state.* Connectivity, reachability, and energy maps, as well as a usage model of sensor networks, could be obtained from data collected via TopDisc.
- *Data dissemination and aggregation.* The resulting tree created by TopDisc could also be used in data dissemination and aggregation applications.
- *Duty cycle assignment.* Each pair of closest black nodes can exchange location information of their children. After collecting the complete topology of the surrounding nodes, one of the children may decide to serve as a forwarding node. It then informs other nodes so that they can go into sleep mode. Based on the category presented in Section 8.2, this application can be considered a tasking application.

8.5 Summary

The advent of technology has facilitated development of networked systems of small, low-power devices that combine programmable computing with multiple sensing and wireless communication capability. Already, experimental applications have embedded sensor nodes in the physical environment to facilitate new information-gathering and -processing capabilities. The sheer number of sensor nodes and the dynamics of their operating environments pose unique challenges on how information collected by and stored within a sensor network can be queried and accessed, and how concurrent sensing tasks can be executed internally and programmed by external clients. This chapter described a generic functional architecture for sensor networks by identifying three required functional components: hierarchical clustering, location awareness, and attribute-based naming. Two sample implementation architectures, SINA and TopDisc, were examined in terms of their exploitation of these functional components and the application characteristics they are intended to support.

References

1. Srisathapornphat, C., Jaikaeo, C., and Shen, C.-C., Sensor information networking architecture, in *Proc. 2000 Int. Workshop on Parallel Processing*, 23–30, Toronto, Canada, August 21–24, 2000.
2. Deb, B., Bhatnagar, S., and Nath, B., A topology discovery algorithm for sensor networks with applications to network management, in *IEEE CAS Workshop Wireless Commun. Networking*, Pasadena, CA, September 5–6, 2002.
3. Akyildiz, I.F. et al. Wireless sensor networks: a survey, *Computer Networks*, 38(4), 393–422, 2002.
4. Jaikaeo, C., Srisathapornphat, C., and Shen, C.-C., Diagnosis of sensor networks, in *Proc. IEEE Int. Conf. Commun. (ICC 2001)*, 5, 1627–1632, Helsinki, Finland, June 11–15, 2001.
5. Cerpa, A. et al. Habitat monitoring: application driver for wireless communications technology, in *ACM SIGCOMM Workshop Data Commun. Latin America Caribbean*, 20–41, Costa Rica, April 3–5, 2001.
6. Huang, Q., Lu, C., and Roman, G.-C., Reliable mobicast via face-aware routing, Washington University, St. Louis, MO, Tech. Rep. WUCSE-2003-49, July 2003.
7. Zhang, W. and Cao, G., DCTC: dynamic convoy tree-based collaboration for target tracking in sensor networks, *IEEE Trans. Wireless Communications*, in press.
8. Lin, C., Federspiel, C.C., and Auslander, D.M., Multi-sensor single-actuator control of HVAC systems, in *Proc. Int. Conf. Enhanced Building Operations*, Richardson, TX, October, 2002.

9. Imielinski, T. and Goel, S., DataSpace: querying and monitoring deeply networked collections in physical space, *IEEE Personal Commun.*, 7(5), 4–9, October 2000.
10. Estrin, D., Govindan, R., and Heidemann, J., Embedding the Internet, *Commun. ACM*, 43(5), 39–41, May 2000.
11. Ward, A., Jones, A., and Hopper, A., A new location technique for the active office, *IEEE Personal Commun.*, 4(5), 42–47, October 1997.
12. Bayerdorffer, B.C., Distributed programming with associative broadcast, in *Proc. 28th Hawaii Int. Conf. System Sci.*, 2, 353–362, Hawaii, January 1995.
13. Johnson, D.B., Maltz, D.A., and Broch, J., DSR: the dynamic source routing protocol for multi-hop wireless and ad hoc networks, in *Ad Hoc Networking*, Charles E. Perkins, (Ed.), Addison-Wesley, 2001, chap. 5.
14. Shen, C.-C., Srisathapornphat, C., and Jaikaeo, C., Sensor information networking architecture and applications, *IEEE Personal Commun. Mag.*, 8(4), 52–59, August 2001.
15. Jaikaeo, C., Srisathapornphat, C., and Shen, C.-C., Querying and tasking in sensor networks, in *Proc. SPIE's 14th Annu. Int. Symp. Aerospace/Defense Sensing, Simulation, Control (Digitization of the Battlespace V)*, 4037, 184–197, Orlando, FL, April 24–28 2000.
16. Bonnet, P., Gehrke, J., and Seshadri, P., Querying the physical world, *IEEE Personal Commun.*, 7(5), 10–15, October 2000.
17. Intanagonwiwat, C. et al., Directed diffusion for wireless sensor networking, *ACM/IEEE Trans. Networking*, 11(1), 2–16, February, 2002.

9
A Practical Perspective on Wireless Sensor Networks

Quanhong Wang
Queen's University

Hossam Hassanein
Queen's University

Kenan Xu
Queen's University

9.1	Introduction	9-1
9.2	WSN Applications	9-2
	Military Applications • Environment Detection and Monitoring • Disaster Prevention and Relief • Medical Care • Home Intelligence • Scientific Exploration • Interactive Surroundings • Surveillance • Other Applications	
9.3	Classification of WSNs..	9-6
9.4	Characteristics, Technical Challenges, and Design Directions..	9-7
	Characteristics • Technical Challenges and Requirements • Design Objectives and Directions	
9.5	Technical Approaches..	9-11
	Hardware Techniques • System Architecture, Protocols, and Algorithms • Software Development	
9.6	Conclusions and Considerations for Future Research	9-22

9.1 Introduction

Rapid progress in microelectromechanical system (MEMS) and radio frequency (RF) design has enabled the development of low-power, inexpensive, and network-enabled microsensors. These sensor nodes are capable of capturing various physical information, such as temperature, pressure, motion of an object, etc., as well as mapping such physical characteristics of the environment to quantitative measurements. A typical wireless sensor network (WSN) consists of hundreds to thousands of such sensor nodes linked by a wireless medium.

WSNs have created new paradigms for reliable monitoring. They outperform conventional sensor systems, which use large, expensive macrosensors to be placed and wired accurately to an end user. Detailed discussions of such benefits can be found in the literature [1, 13, 31–33, 43]. Some of these benefits are highlighted as follows:

- *Anywhere and anytime.* The coverage of a traditional macrosensor node is narrowly limited to a certain physical area due to the constraints of cost and manual deployment. In contrast, WSNs may contain a great number of physically separated nodes that do not require human attention. Although the coverage of a single node is small, the densely distributed nodes can work simultaneously and collaboratively so that the coverage of the whole network is extended. Moreover,

sensor nodes can be dropped in hazardous regions and can operate in all seasons; thus, their sensing task can be undertaken anytime.
- *Greater fault-tolerance.* This is achieved through the dense deployment of wireless sensor nodes. The correlated data from neighboring nodes in a given area makes WSNs more fault tolerant than single macrosensor systems. If the macrosensor node fails, the system will completely lose its functionality in the given area. On the contrary in a WSN, if a small portion of microsensor nodes fails, the WSN can continue to produce acceptable information because the extracted data are redundant enough. Furthermore, alternative communication routes can be used in case of route failure.
- *Improved accuracy.* Although a single macrosensor node generates more accurate measurement than one microsensor node does, the massively collected data by a large number of tiny nodes may actually reflect more of the real world. Furthermore, after processing by appropriate algorithms, the correlated and/or aggregated data enhance the common signal and reduce uncorrelated noise.
- *Lower cost.* WSNs are expected to be less expensive than their macrosensor system counterparts because of their reduced size and lower price, as well as the ease of their deployment.

In this chapter, Section 9.2 describes diverse applications of WSNs in various domains with examples and Section 9.3 discusses the classifications of the WSNs according to different criteria. Section 9.4 presents the characteristics of WSNs, highlights how they differ from traditional wireless ad hoc networks, and reviews the technique challenges and corresponding design directions. In Section 9.5, various technical approaches with respect to hardware design, system architectures, protocols and algorithms, and software development are illustrated. Finally, Section 9.6 concludes with emphasis on several possible open issues for future research in the area of WSNs.

9.2 WSN Applications

WSNs are able to monitor a wide range of physical conditions, such as [2]:

- Temperature
- Humidity
- Light
- Pressure
- Object motion
- Soil composition
- Noise level
- Presence of a certain object
- Characteristics of an object such as weight, size, moving speed, direction, and its latest position

Due to WSNs' reliability, self-organization, flexibility, and ease of deployment, their existing and potential applications vary widely. As well, they can be applied to almost any environment, especially those in which conventional wired sensor systems are impossible or unavailable, such as in inhospitable terrains, battlefields, outer space, or deep oceans.

9.2.1 Military Applications

WSNs are becoming an integral part of military command, control, communications, computing, intelligence, surveillance, reconnaissance, and targeting (C^4ISRT) systems [2]. In the battlefield, a predictable tendency is that the targets will become smaller and less recognizable/detectable, have higher mobility, and usually move in extremely hostile terrain. To explore the position and strength of the opposing forces, a promising solution lies in dense arrays of sensors to be placed close to the intended targets. Because of their ability to be unattended by humans, ease of deployment, self-organization, and fault tolerance,

WSNs can provide highly redundant and collaborative detected data without the support of friendly forces. Also, WSNs can be mounted on unmanned robotic vehicles, tanks, fighter planes, submarines, missiles, and torpedoes to route them around obstacles, guide them to the exact position and lead them to coordinate with one another to fulfill more effective attacks or defenses. WSNs can also be deployed for remote sensing of nuclear, biological, and chemical weapons, potential terrorist attack detection, and reconnaissance [2, 37]. Obviously, WSNs will take more important roles in the military C^4ISRT tasks and make future attacks and defenses more intelligent, with less human involvement.

9.2.2 Environment Detection and Monitoring

Spreading hundreds to thousands of tiny, cheap, self-configurable wireless sensors in a given geographical region can produce a wide range of applications in collaborative monitoring or control of the environment. This encompasses complex ecosystem monitoring; flood detection; air and sewage monitoring; local climate control in large office buildings; soil composition detection and precise agriculture; wild land fire detection; and exploration of mineral reserves, geophysical studies, etc. [2, 12, 32, 64]. Some representative examples include:

- *Ecosystem monitoring.* WSNs used in ecosystem monitoring represent a class of applications with numerous potential benefits for life science study because WSNs can provide information on several environmental conditions, including soil and air chemistry as well as plant and animal species population and behaviors. It ensures the long-term automatic identification, recording, and analysis of interesting events. These long-term gathered data can help ecosystem scientists to identify, localize, track, and predict species or phenomena in areas of interest [12, 32, 64]. Compared with traditional methods of environment monitoring, WSNs have a number of unique advantages:
  - Noninvasive deployment: unattended wireless sensors can be dropped on remote islands or dangerous places where it would be unsafe, unwise, or even impossible to perform field study repeatedly.
  - Anytime deployment: wireless sensor nodes can be deployed in any selected period, for example, before the producing season of some species of animal or after frozen ground melts.
  - Minimal interference: deploying WSNs for biosystems can eliminate the disturbance impact on the measured objects. For example, some species are very sensitive to the unexpected visits necessary for large-size macrosensor equipment; this can lead to a dramatic increase of mortality in a breeding year.
  - Less cost: deployment of WSNs also leads to a more economical solution to producing long-term observations than human-attended methods do.
  - Higher level of robustness and accuracy: by integrating data aggregation and signal processing within the neighborhood sensors, WSNs become more robust to node failure. Self-configurable WSNs used for biocomplexity mapping are adaptive to the dynamic physical world.
  - Ease of networking: sensor nodes are capable of connecting to the Internet, thus enabling one remote user to control, monitor, and collect data for several different sensed spots or several remote users to gather data for the same spot.

  Mainwaring and colleagues [64] present a real-life experiment of deploying WSN in a natural area — Great Duck Island (44.09N, 68.15W), Maine — to monitor the Leach's Storm Petrel, in terms of short-term cycle (24 to 72 h) of the usage pattern of nesting burrows and long-term (7 months) changes in the burrow and surface environmental parameters. The experiment is intended to guide the reliable environmental monitoring in these previously unaccessible fields.
- *Local climate control in large buildings.* Most people who have worked in large office buildings have experienced that the temperature is seldom proper, i.e., too high or too low; the humidity level is often overly dry or overly wet; too much or too little light is present; or fresh air is lacking. Therefore, local climate monitoring and control systems are highly desirable to ensure healthy and pleasant working places. At present, traditional systems with wired sensors are dominant in such

areas. Distributed WSNs are considered a better solution than their wired counterparts in at least two respects. For one thing, the deployment of a WSN is much more flexible than a wired system. Without the restriction of wire, wireless sensors can sit wherever they are needed; they can also be moved from their original positions to more suitable places. Moreover, WSNs can produce tremendous economical gains compared to wired sensors. According to da Silva et al. [93] and Rabaey et al. [79], for sensing mission, 90% of the total installation cost of a low-cost temperature sensor is due to wiring. Obviously, installation cost can be greatly reduced if wireless sensors are used.

- *Wild land fire detection.* Although significant measures have been exerted, wild land fires still cause extensive loss of lives, property, and resources each year. According to the statistics of the National Interagency Fire Center [71], the 10-year (1992 to 2001) average of wild land fires reached 103,112 and a total of 42,150,890 acres were burned. It costs approximately $1.6 billion (U.S.) on average for fire suppression by federal agencies only. However, because fire weather conditions are predictable, wild land fire prediction is often a possible source of help to support any geographic area before and during periods of high fire danger or fire activity. Because of their ability to be deployed randomly and densely, WSNs are a good choice in wild land fire detection and reporting. By scattering massive numbers of wireless sensors in intended areas, early warning and origin of fires can be caught effectively.

9.2.3 Disaster Prevention and Relief

WSNs may also be effectively deployed in emergency situations and disaster areas [37]. The accurate and prompt location detection provided by the distributed WSNs could be critical in rescue operations, including detection of victims, potential hazards, or sources of the emergency and identification and localization of trapped personnel [83]. For example, microsensors may be embedded/enabled in large-scale buildings during construction, through strategically dropping on the spot at the rescue site, or by automatically triggering standby sensors immediately following the disaster event. The collapse of the walls or ceiling could be predicated and estimated by the stress and motion of buildings. It is also useful to deploy WSNs for long-lasting monitoring tasks, such as detecting and tracking material fatigue, so that the evidence of harmful reaction of the building can be collected continuously and effective measures can be taken before an accident happens. Another example, waterproof sensor arrays, can be automatically triggered to constantly report the location of sunken vessels in the ocean and to provide critically important information for the rescue and salvage operation. Furthermore, wireless sensors can also be used to track fuel, gas, and toxic substances leaked into the neighborhood ocean when a sunken vessel is raised.

9.2.4 Medical Care

WSNs are very helpful in providing prompt and effective health care and will lead to a healthier environment for human beings. Some uses of WSNs in this field include:

- *Remote virus monitoring.* Many widespread disease-ridden regions are impoverished and lack reliable communication. Spreading large number of wireless sensors in such regions could help to collect and transmit crucial ground-based information, such as incident of disease and characteristics of the infected population; to identify features of the area; and to monitor environmental conditions, such as the amount of rainfall and humidity, that support the proliferation of virus-carrying insects. WSNs can also be used to monitor and predict the breakout of some infectious diseases, such as malaria. A project called Health Improvements through Space Technologies and Resources (HI-STAR) proposes development of a global malaria information system [26]. Based on the gathered air and ground-based data via wireless sensors and by integrating and analyzing epidemiological information, this system can generate malaria "risk maps" and provide early warnings about malaria outbreaks. Health officials could also allocate limited disease prevention and treatment resources on a global scale.

- *Integrated patient tracking and monitoring.* Using WSNs to monitor and track possible or suspected patients is a convenient and effective measure to avoid the spread of some infectious diseases. According to a Canadian Broadcast Corporation (CBC) news report in April 2003, discussion was that some people who broke quarantine in Toronto during the period of severe acute respiratory syndrome (SARS) in Spring 2003 could be required to wear a lightweight device with a wireless sensor on their ankles. This device could monitor their movements and report them to the relevant authorities. Moreover, senior citizens without sufficient care could have wireless sensors attached to medical devices to measure their heart rates, blood pressure, etc. In abnormal conditions, an automatic alert reminds the carriers to call their doctors or an automatic notification is directly sent to emergency centers. Furthermore, WSNs can also be used for medical statistics that require data collection from a large number of people or tracing some patients for long period of time.

Schwiebert and colleagues [88] present a series of applications of WSNs in health care, such as artificial retina; glucose level monitoring for diabetes patients; organ monitoring for organ transplant purposes; and cancer detection for high-risk persons, as well as general health monitoring. WSNs can also be used in drug administration and distribution [2].

9.2.5 Home Intelligence

WSNs can take key roles in providing more convenient and intelligent living environments for human beings. Some predictable examples include:

- *Remote metering.* WSNs can be used in remote reading of utility meters, such as water, gas, or electricity, and then can transmit the readings through wireless connections [37]. Simple attachments of wireless sensors in parking meters can send out warning signals to remind users to recharge the meter remotely before the parking time expires.
- *Smart space.* With recent technological development, it becomes possible to embed various wireless sensors into individual furniture and appliances, which can be connected together to form an autonomous network. For example, a smart refrigerator can understand the family's dietary requirements or doctor's orders and take inventory of refrigerators to relay information to a shopping list on a personal digital assistant [21]. It can also create a menu according to the inventory and transmit the relevant cooking parameters to the smart stove or microwave oven, which will set the desired temperature and cooking time accordingly [46]. Moreover, contents and schedules of TV, VCR, DVD, or CD players can be monitored and operated remotely to satisfy the different requirements of family members.

9.2.6 Scientific Exploration

The effective deployment and operation of self-regulating WSNs is opening novel ways of scientific exploration in higher, further, and deeper environments such as outer space and deep oceans. Hong and colleagues [50] present an example for employing WSNs on the surface of Mars to collect measurements such as seismic, chemical, and temperature and relay the aggregated sensing results to an orbiter. Each distributed sensor node provides time- and position-dependent measurements; via energy-conserved, load-balanced, multihop communications, they can relay the information to the distant base station with prolonged network lifetime. Similarly, WSNs used for underwater exploration may also be possible in the future.

9.2.7 Interactive Surroundings

WSNs produce promising mechanisms for mining information from and reacting to the physical world. By deploying cheap and tiny wireless sensors, monitors and actuators in toys and other children's familiar objects could create "smart kindergartens" to enhance early childhood education [98]. Such a system provides a childhood learning environment with "person–physical world" interaction rather than the conventional "person–computer" or "person–person" communication. Because it allows personalized configuration to each individual child; coordinated activities of children groups; adaptation to the dynamics in children's activities; and constant and unobtrusive data collection in children's actions and learning processes, it provides effective and comprehensive problem-solving strategies in young children's education. Rabaey et al. [79] described WSNs in the real world in an interactive museum in San Francisco's Exploratorium, where children can participate actively in the experiments and get feedback to their touch and speech from the sensor-equipped objects. Yarvis and colleagues [106] present another interactive ad hoc sensor network as a voting platform in San Francisco's Moscone Convention Center.

9.2.8 Surveillance

Instant and remote surveillance inspires significant applications of WSNs. For example, a large number of networked acoustic sensors can be used to detect and track desired targets in a deterministic security area [68, 83, 109]. WSNs can be deployed in buildings, residential areas, airports, railway stations, etc. to identify intruders and report to a command center immediately so that tracking actions can be initiated promptly [62]. Similarly, installing smoke sensor nodes in strategically selected positions at homes, office buildings, or factories is critical to preventing disasters of fires and tracing the spread of fire [37, 65].

9.2.9 Other Applications

Self-configurable WSNs can be used in many other areas, such as robot control and factory instrumentation, automatic warehouse inventory tracking, chemical process control, traffic monitoring and control of smart roads, etc.

9.3 Classification of WSNs

As discussed in Section 9.2, WSNs represent a variety of applications in which environment and technical requirements may greatly differ. Therefore, the design of a WSN is usually application oriented. As a result, the architectures, protocols, and algorithms of WSNs vary case by case. However, different WSNs have some common properties in a broad point of view [100]. They can generally be classified into categories based on several important criteria.

According to the distance of sensor nodes to the base station, WSNs can be single-hop (also known as nonpropagating) or multihop (propagating) systems. In a single-hop WSN, all sensor nodes transmit the data directly to the base station, while in a multihop WSN, some nodes can only deliver their data to the base station via intermediate nodes. In these cases, the intermediate nodes execute the routing function and relay the data along the routing path. Also, data aggregation (or fusion) is an optional function for those intermediate nodes. Single-hop networks have much simpler structure and control and fit into the applications of small sensing areas; multihop networks promise wider applications at the cost of higher complexity.

Based on the sensor node density and data dependency, WSNs can be classified as aggregating and nonaggregating. In nonaggregating systems, all data from each individual node will be sent to the destination "as is." The computational load at intermediate nodes is relatively small and the system can reach high accuracy. However, the total traffic load in the entire system may increase rapidly with the enlargement of the network size, more energy will be consumed for communications, and more collisions and/or congestions will occur, leading to high latency. Therefore, the nonaggregating scheme is suitable for systems that have less node density, sufficient capacity, and/or in which extremely high accuracy is demanded by end users.

While in densely distributed networks, a sensor node is usually located close to its neighboring nodes. Thus, information from multiple sources could be highly correlated and aggregating functions may be executed at the intermediate nodes to eliminate data redundancy. In this way, the traffic load in the system could be reduced considerably, and significant energy savings due to communications can be obtained. However, the intermediate nodes will perform computational functions, which may require

TABLE 9.1 Classification of WSNs according to Different Factors

Factors	Distinct Groups
Distance to base station/processing center	Single hop vs. multihop
Data dependency	Nonaggregating vs. aggregating
Distribution of sensors	Deterministic vs. dynamic
Control scheme	Non self-configurable vs. self-configurable
Application domain	Many

the larger memory size. Therefore, the aggregating scheme is an appropriate option in large-scale systems with massively and densely distributed sensor nodes. It should be noted that end users are only interested in the collective information with moderate accuracy.

WSNs can be deterministic or dynamic according to distribution of the sensor nodes. In deterministic systems, the positions of sensor nodes are fixed or preplanned. The control of this system is simpler and its implementation is easier. However, this scheme can only be used in limited kinds of systems where the information of the sensor node placement could be obtained and planned in advance. However, in many cases, the locations of sensor nodes are not available *a priori*, such as those dropped randomly in remote areas. So, the sensor nodes must work in a distributed dynamic manner. The dynamic scheme is more scalable and flexible, but requires more complex control algorithms.

Moreover, based on the control scheme, WSNs can be non-self-configurable or self-configurable. In the former mechanism, the sensor nodes are not able to organize on their own, but rely on a central controller to offer command to and collect information from them. This scheme can only be used in small-scale networks. However, in most WSNs, the sensor nodes can autonomously establish and maintain connectivity by themselves and collaboratively fulfill sensing and control tasks. This self-configurable scheme fits better in large-scale systems to perform complicated monitoring tasks and information collection and dissemination.

The categories described here may overlap, i.e., a specific WSN may have the characteristics of different domains. For instance, WSNs in a large parking lot are self-configurable, deterministic, nonaggregating, and multihop. A classification of WSNs is shown in Table 9.1.

Although self-configurable systems are more complicated than non-self-configurable ones, they are more practical for deployment in the real world, especially when the network size becomes very large. However, they raise numerous challenges and open issues to be explored further. The remainder of this chapter concentrates mainly on self-configurable systems.

9.4 Characteristics, Technical Challenges, and Design Directions

WSNs aim to bridge the gap between the physical and computational worlds. The salient features of WSNs and their differences from other wireless networks have been discussed by a number of researchers [1, 13, 32, 33, 37, 43, 93, 97, 111, 112]. Some of these features are discussed next.

9.4.1 Characteristics

Most WSNs use the network architecture of wireless ad hoc networks, which are collections of wireless, possibly mobile, nodes that are self-configurable to form a network without the aid of any established infrastructure. The mobile nodes handle the necessary control and networking tasks in a distributed manner. The ad hoc architecture is highly appealing to sensor networks for many reasons [33]:

- Ad hoc architecture overcomes the difficulties raised by the predetermined infrastructure settings of the other families of wireless networks. WSNs can be randomly and rapidly deployed and reconfigured — new nodes can be added on demand to replace failed or powered-off ones and existing nodes can withdraw or depart from the systems without affecting the functionality of other nodes.

TABLE 9.2 Differences between WSNs and Conventional Wireless ad hoc Networks

	WSNs	Conventional Wireless Ad hoc
Number of nodes	Large; hundreds to thousands or even more	Small to moderate
Node density	High	Relatively low
Data redundancy	High	Low
Power supply	Non-rechargeable; irreplaceable batteries	Rechargeable and/or replaceable batteries
Data rate	Low; 1–100kb/s	High
Mobility of nodes	Low	Can have high mobility
Direction of flows	Predominantly unidirectional; sensor nodes $\rightarrow$ sink	Bidirectional; end-to-end flows
Packet forwarding	Many to one; data centric	End-to-end address centric
Query nature	Attribute based	Node based
Query dissemination	Broadcast	Hop by hop or broadcast
Addressing	No globally unique ID	Globally unique ID
Active duty cycle	Could be as low as 1%	High

- Ad hoc networks can be easily tailored to specific applications.
- This architecture is highly robust to single node failures and provides a high level of fault tolerance because of node redundancy and its distributed nature.
- Energy efficiency can be achieved through multihop routing communication. As reported in Rappoport [82], large-scale propagation follows as exponential law to the transmitting distance (usually with exponent 2 to 4 depending on the transmission environment). It is not difficult to show that power consumption due to signal transmission can be saved in orders of magnitude by using multihop routing with short distance of each hop instead of single-hop routing with a long range of distance for the same destination.
- Ad hoc networks have the advantage of bandwidth reuse, which also benefits from dividing the single long-range hop to multihops; each hop has a considerable short distance. In this case, the communication is local and within a small range.

It is not surprising to see that the majority of existing WSN literature is based on multihop ad hoc architectures. However, because of unique application requirements, WSNs greatly differ from conventional wireless ad hoc networks [56, 93]. As a result, existing ad hoc network architectures and protocols are not directly suitable for or extendible to WSNs. Therefore, new approaches should be developed so as to satisfy the specific requirements of WSNs; numerous research issues remain to be explored. Table 9.2 summarizes the main differences between these two types of networks. These differences raise many technical challenges on system design and implementation. Next, these technical challenges are explored in detail; the corresponding design objective and directions will follow as well.

9.4.2 Technical Challenges and Requirements

WSN design is motivated and influenced by one or more of the following technical challenges [1, 32, 69]:

- *Massive and random deployment.* Most WSNs contain a large number of sensor nodes (hundreds to thousands or even more), which might be spread randomly over the intended areas or are dropped densely in inaccessible terrains or hazardous regions. The system must execute self-configuration before the normal sensing routine can take off.
- *Data redundancy.* The dense deployment of sensor nodes leads to high correlation of the data sensed by the nodes in the neighborhood.
- *Limited resources.* WSN design and implementation are constrained by four types of resources: energy, computation, memory, and bandwidth. Constrained by the limited physical size, microsensors could only be attached with bounded battery energy supply. Moreover, WSNs usually operate in an untethered manner, so their batteries are nonrechargeable and/or irreplaceable. At the same time, their memories are limited and can perform only restricted computational functionality. The bandwidth in the wireless medium is significantly low as well.

- *Ad hoc architecture and unattended operation.* The attributes of no fixed infrastructure and human-unattended operation of such networks require the system to establish connections and maintain connectivity autonomously.
- *Dynamic topologies and environment.* On the one hand, the topology and connectivity of WSNs may frequently vary due to the unreliability of the individual wireless microsensors. For example, a node may fail to function because of exhaustion of power at any time without notification to other nodes in advance. As well, new nodes may be added randomly in an area without prior notification of existing nodes. On the other hand, the environment that the WSNs are monitoring can also change dramatically, which may cause a portion of sensor nodes to malfunction or render the information they gather obsolete.
- *Error-prone wireless medium.* Sensor nodes are linked by the wireless medium, which incurs more errors than their wired counterpart. In some applications, the communication environment is actually noisy and can cause severe signal attenuation.
- *Diverse applications.* As described in Section 9.2, WSNs could be used to perform various tasks, such as target detection and tracking, environment monitoring, remote sensing, military surveillance, etc. Requirements for the different applications may vary significantly.
- *Safety and privacy.* Safety and privacy should be an essential consideration in the design of WSNs because many of them are used for military or surveillance purposes. Denial of service attacks against these networks may cause severe damage to the function of WSNs. However, security seems to be a significantly difficult problem to solve in WSNs because of the inevitable dilemma: WSNs are resource limited and security solutions are resource hungry. Indeed, most existing communication protocols for WSNs do not address security and are susceptible to adversaries [104].
- *QoS concerns.* The quality provided by WSNs refers to the accuracy with which the data reported match what is actually occurring in their environment. Different from others, accuracy in WSNs emphasizes the characteristic of the aggregated data of all sources instead of individual flows. One way to measure accuracy is the amount of data. Another aspect of QoS is latency. Data collected by WSNs are typically time sensitive, e.g., early warning of fires. It is therefore important to receive the data at the destination/control center in a timely manner. Data with long latency due to processing or communication may be outdated and lead to wrong reactions.

9.4.3 Design Objectives and Directions

The following objectives and directions are identified in the design of WSNs so as to deal with the challenges and satisfy the various application requirements [1, 13, 32, 33, 40, 43, 55, 69, 78, 97]:

- *Small microsensor devices.* Affordable and compact sensor units are essential factors to massive and random deployment of WSNs. For a large-scale WSN application, the cost of individual sensor devices would contribute to the major part of the total expense. Besides, the smaller the sensor is, the lower interference the sensor would have on the observed objects and the easier the deployment would be.
- *Scalable and flexible architectures and protocols.* In addition to the requirement on individual sensor devices, the system should be scalable and flexible to the enlargement of the network scale. The approaches to scalability and flexibility include clustering, multihop delivery, and localization of computation and protocols.
- *Localized processing and data fusion.* To eliminate data redundancy, collaborative efforts should be made among the sensor nodes performing a variety of localized processing. Instead of sending the raw data to the destination directly, sensor nodes might locally filter the data according to the requirements, carry out simple computation, process the data, and transmit only the processed data. Some intermediate nodes may also perform data fusion in order to reach high efficiency.
- *Resource efficiency design.* In WSNs, resource efficiency is extremely critical and is desirable regardless of its complexity. Above all, energy-efficient protocols are in high demand in order to extend the lifetime of the system. Indeed, power saving should be achieved in every component of the

network by integrating the corresponding mechanisms, such as power-saving mode on MAC layer, power-aware routing on network layer, etc. In addition, efforts should be made to increase efficiency for the utilization of other resources. For example, using algorithms with low complexity will reduce the computation time and thus save power; it also decreases the latency of data delivery. Bandwidth-efficient architectures and protocols can accelerate data delivery as well.

It should be noted that it is difficult to issue a unique definition of system lifetime for all applications or cases. The system can be declared dead when the first node exhausts its energy, when a certain fraction of nodes dies, or even when all nodes die. Using one or the other definition depends on the particular application. On the other hand, system lifetime can also be measured using application-specific parameters, such as the time until the system can no longer provide acceptable results.

- *Self-configuration.* Naturally, randomly and massively deployed sensor nodes have to execute self-configuration in order to set up the network connection and commence routine operation. WSNs are highly dynamic during the lifetime of the network. Sensor nodes transit among the states of off, sleep, startup, idle, transmitting, receiving, and failure* for the purpose of energy conservation. Thus, WSN protocols should have the capability of forming connections autonomously — regardless of the condition of sensor nodes. New links should be accommodated in case of node failure or link congestion, and the transmitting power or signaling rate may be adjusted actively to reduce energy consumption based on up-to-date topology information. As well, packets could be rerouted through some subsets of the network in which nodes have more residual energy so as to realize an equal dissipation of energy among nodes over the entire network.
- *Adaptability.* To cope with dynamic/varying conditions, WSNs should adapt to changing connectivity and system stimuli over time. To detect the nondeterministic phenomena with disturbance caused by communication noise and sensor diversity, adaptive fidelity signal processing at individual sensor nodes is also desired to make trade-offs among resources, accuracy, and latency requirements.
- *Reliability and fault tolerance.* For many WSN applications, data must be delivered reliably over the noisy, error-prone, and time-varying wireless channel. In such cases, data verification and correction on each layer of the network are critical to provide accurate results. Additionally, sensor nodes are expected to perform self-testing, self-calibrating, self-repair and self-recovery procedures during their lifetime.
- *Application-specific design.* Because no unique protocol satisfies all applications of WSNs, the design of WSNs is in many cases application specific.
- *Security design.* Data privacy and safe communications are of utmost importance. Wood and Stankovic [104] argue that the best way to ensure successful network deployment is to take security issues into consideration at the design stage of WSNs.
- *QoS design with resource constraints.* As stated previously, the two measures of QoS in WSNs are accuracy and timely delivery of information. Accuracy reflects the basic value of the information. In general, the amount of data determines the level of accuracy. Data should be delivered in a timely manner. It is essential to make a trade-off between these two aspects because large amounts of data consume a large portion of bandwidth and cause more contention during transmission. As a result, the latency would be increased with higher accuracy requirement. Furthermore, it is critical to realize the trade-off between QoS and resource consumption. High accuracy requires large amounts of data delivery, thus leading to more power and bandwidth consumption. Local computation is helpful to eliminate the amount of data transmitted, but complex and memory costly computation will cause long latency. At the same time, more complex computation reduces power efficiency.

---

*Note that nodes in the same network may be in different states.

TABLE 9.3 Summary of Technical Challenges and Design Objectives in WSNs

Technical Challenges and/or Requirements	Design Objectives and Directions
Massive and random deployment	Cheap and small sensor node; scalable and flexible architecture and protocols
Data redundancy	Localized processing and data fusion
Limited resources	Resource efficiency design
Ad hoc architecture and unattended operation	Self-configuration and coordination
Dynamic surrounding	Adaptability
Error-prone medium	Reliability and fault tolerance
Diverse applications	Application-specific design
Safety and privacy	Security
QoS concerns	QoS design with resource constraint; localization; attribute-based naming and data-centric routing

- *Other attributes.* In addition to the preceding objectives and directions, WSN design should accommodate the following objectives:
  - *Locality of information.* The reported data from a sensor are only meaningful when associated with exact knowledge of the sensor's location. This can significantly simplify the network discovery and maintenance efforts. The data-centric query should be forwarded directly and efficiently to targeted areas of interest.
  - *Attribute-based naming and data centric routing.* When deploying WSNs, users are more interested in querying the property of the interested phenomenon, rather than a specific node. For example, "the temperature in room 717" or "the areas where the temperature is over 50°C" are more common than the query of "the temperature read by a certain sensor node."

It is impractical to achieve all objectives in a single network. Most WSN designs are application specific and have different stress on some of the objectives described previously. Thus, the protocols should be designed to satisfy the unique quality demands of each individual network and trade-offs should be made among the different parameters when designing protocols and algorithms for WSNs. Table 9.3 summarizes the technical challenges and corresponding design objectives and directions.

9.5 Technical Approaches

In many cases, it is very challenging to design and implement a resource-efficient and QoS-enabled WSN. This is usually constrained by many factors and has several objectives to meet at the same time; often such factors and objectives are contradictory to each other. Nevertheless, research on WSNs have achieved significant progress. Emphasizing on one or two aspects of the constrained factors or objectives, these research efforts take diverse approaches. Here, they are broadly grouped into three categories: hardware techniques; system architecture, protocols, and algorithms; and software development.

9.5.1 Hardware Techniques

9.5.1.1 Cheap, Compact, Low-Power Wireless Sensor Nodes

A WSN node integrates sensing, signal processing, data collection and storage, computation, and wireless communications, along with attached power supply on a single chip. The system architecture of a typical microsensor node is shown in Figure 9.1 [81, 95]. Generally, each node is composed of four components: (1) a power supply unit that is usually an attached battery with desirable output voltage to drive all other components in the system; (2) a sensing unit consisting of the embedded sensor and actuator as well as an analog-digital converter that links the sensor node to the physical world; (3) a computing/processing unit that is a microcontroller unit (MCU) or microprocessor with memory and provides intelligence to the sensor node (widely used MCUs include Intel's Strong ARM microprocessor and Atmel's AVR microcontroller); and (4) a communication unit consisting of a short-range RF circuit and performing

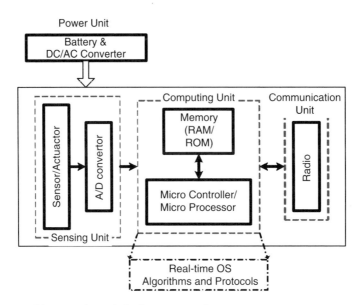

FIGURE 9.1 System architecture of a typical microsensor node.

data transmission and reception. Moreover, a real-time micro-operating system controls and operates the sensing, computing, and communication units through microdevice drivers and decides which parts to turn off and on.

Advances in microelectromechanical systems (MEMS) and continuous developments in wireless communications are spurring more intelligent, less expensive, much smaller sensor nodes to be embedded into the physical world. For example, *piconodes* in the PicoRadio project are a promising "system-on-chip" implementation to provide ubiquitous distribution of computation and communications for sensor/monitor networks. Each PicoRadio node has a small size of less than 0.10 to 0.15 in.³, consumes less than 10 mW, and costs less than $1 [79, 80, 103].

Another system, called *WINS* (wireless integrated network sensors), integrates multiple functions including sensing, signal processing, decision making, and wireless networking capability in a compact, low-power device. These intelligent sensors are tiny and powerful in establishing low-cost and robust self-organizing networks for continuous sensing and event detection and identification [4, 75, 76].

A project called μAMPS (microadaptive multidomain power-aware sensors) [67] has the objective of implementing a microsensor system on a chip of 1 cm³, with the integration of MEMS sensors, A/D, data and protocol processing, and a radio transceiver on a single die. Moreover, the *Smart Dust* project aims to explore the limits on size and power consumption of self-organizing sensor nodes that are not more than a few cubic millimeters in size, i.e., small enough to float in the air detecting and communicating for hours or days [54, 110–112]. For information on other experimental systems, refer to Hill et al. [47, 48], Mainwaring et al. [64], and Yarvis et al. [106].

9.5.1.2 Low Duty Cycle Electronics

Because the detected environment would not vary frequently or rapidly, the sensor node and its components should operate in alternating active and inactive modes for the purpose of power conservation. As the major contributors of the power consumption in a sensor node, data processing and radio subsystems have been under extensive study [13, 19, 92]. The energy consumed by the static CMOS-based microprocessor unit in a typical sensor node can be modeled as follows [92, 94]:

$$E_{total} = E_{switch} + E_{leakage} = C_{total}V_{dd}^2 + (V_{dd}t)I_0 e^{\frac{V_{dd}}{n \cdot V_T}} \tag{9.1}$$

Total power consumption is composed of two parts: switching power and leakage power. Switching power is determined by supply voltage, V_{dd}, and the total capacitance switched by executing software, C_{total}. The leakage power refers to the energy consumption while no computation is conducted. Here, V_T is the thermal voltage. An effective way to reduce the energy consumption in the processor is to minimize the power wasted while no useful work is done, i.e., the leakage power part.

For the radio module, a possible scheme of power conservation is to turn off the radio electronics (such as frequency synthesizers, mixers, etc.) during periods of inactivity and to wake them up when interesting events occur [13, 92]. The average power consumed by the radio is modeled as [92]:

$$P_{radio-ave} = N_{tx}\left[P_{tx}\left(T_{tx-on} + T_{start}\right) + P_{out}T_{tx-on}\right] + N_{rx}\left[P_{rx}\left(T_{rx-on} + T_{start}\right)\right] \qquad (9.2)$$

where $N_{tx/rx}$ is the average number of times per second that the transmitter/receiver is active; $P_{tx/rx}$ is the power consumed by the transmitter/receiver; P_{out} is the output transmit power; T_{start} is the transceiver startup time; and $T_{tx/rx-on}$ is the actual data transmitting/receiving time equal to L/R, where L is the packet length in bits and R is the data rate in bits per second. Obviously, it is natural to turn off the radio as long as no work is to be done in order to reduce power consumption. However, significant overhead in terms of time and energy dissipation will be raised when switching the electrics from the inactive to the active state. Optimal schemes are necessary to estimate the traffic dynamics and make the switching decision accordingly.

9.5.2 System Architecture, Protocols, and Algorithms

9.5.2.1 Sensor Deployment Strategies

Sensor deployment is a fundamental issue for WSNs. The objective of a sensor deployment plan is to achieve desirable coverage with a minimum number of sensor nodes while complying with constraints of QoS, cost, reliability, and scalability of a certain application.

In WSNs, coverage has a twofold meaning: range and spatial localization. Range refers to the geometric area of a designated sensing mission, while spatial localization emphasizes the relative spatial positions of sensor nodes and targets so as to extract accurate measurements. Meguerdichian and colleagues [65] interpret the coverage problem in terms of deterministic vs. statistical and worst vs. best cases in WSNs, and propose an optimal polynomial-time algorithm for coverage calculation by combining computational geometry (specifically, Voronoi diagrams) and graph search algorithm. Mehta and coworkers [66] describe several algorithms that quickly and interactively compute the optimal coverage paths in WSNs. With greatly diverse applications, sensor deployment strategies and mechanisms vary significantly from case to case. In general, four methods of sensor deployment exist: predetermined, self-regulated, randomly undetermined, or biased distribution [24, 101].

Predetermined strategy applies to two situations: (1) knowledge about the environment or the possible targets is sufficient, as described in Musman et al. [70]; (2) sensor nodes can be regularly placed in some grid-based topology in which the sensing site is spatially modeled as a grid-based distribution, i.e., the two- or three-dimensional space is represented by point coordinates. The granularity of the grid (distance between adjunctive grid points) is determined by the desired accuracy [24]. Salhieh [87] and Schwiebert and colleagues [88] illustrate several examples of placing sensor nodes in some preplanned geometric topologies for medical care purposes. Using code identification, Chakrabarty and coworkers [14] describe methods for determining the placement of sensor nodes for unique target location and provide code-theoretic bounds on the number of sensors. Chakrabarty et al. [15] developed an integer linear programming (ILP) model for optimistically minimizing the cost of sensor deployment under the constraint of complete coverage of the sensor field. In general, predetermined strategy can provide an optimal solution for desirable coverage and obtain high QoS and cost efficiency at the same time. However, the first situation is often impractical in the real world because knowledge of the environment and targets is often not available *a priori*. A regular grid-based approach has better adaptation to the variation of the conditions, although it experiences some drawbacks as well. For one thing, the computational complexity makes the schemes not scalable to large-scale networks. However, the grid coverage relies on accurate sensor detection, although, in reality, sensor detection is often uncertain.

To overcome the difficulties of the predetermined approach, self-regulated strategy is developed. Howard and colleagues [51] propose a potential field-based method to deploy sensor nodes automatically in an unknown environment. Because the sensing fields are established in a manner in which each sensor node is repelled by obstacles and by other nodes, the entire network is self-spread throughout the environment and can reach the maximum coverage. Clouqueur et al. [20] present a scheme to deploy sensor nodes sequentially in steps by introducing path exposure as a metric of goodness. With the strategy of properly choosing the number of sensors in each step, the cost of deployment can be minimized to achieve the desired detection performance. Self-regulated methods are scalable to increasing the number of sensor nodes, but the computational expense may become prohibitive.

Randomly undermined strategy is more realistic for a large-scale WSN application, such as unknown battlefields or hostile terrains. With methods of this approach, sensor nodes are generally spread uniformly in a given area [42–44, 60, 61, 101]. This strategy is preferable because of easy placement of nodes and therefore low cost. Although sensing devices can be randomly deployed in two- or three-dimensional spaces, the coverage might not be uniform due to obstacles or other sources of noise in an environment. Based on an initial random distribution, Zou and Chakrabarty [109] introduced a practical virtual force algorithm (VFA) to reposition the sensors in order to enlarge coverage to the desired optimal results, thus dealing with cases of high- and low-detection accuracy while considering energy constraints.

Furthermore, in some contexts, the uniform deployment of sensor nodes may not always satisfy the design requirements and biased deployment can then be a viable option. Willig and coworkers [103] illustrate an example of biased placement of sensors in a large-scale office building in which the density of sensor nodes close to the windows is much higher than that in the middle of the room. Some comparisons of different deployment strategies by means of simulations have been presented by Tilak et al. [101].

Most research on sensor deployment discussed here has an implicit assumption that every sensor node operates in a reliable manner; however, because this is not always true in reality, some proposals have been introduced to handle unreliable conditions. Considering the uncertainty of sensor detection, a statistical optimization framework is presented in Dhillon et al. [24]. Assuming a given set of detection probabilities in a sensor field, it optimizes the number of sensors and determines their position so as to achieve sufficient grid coverage. Guibas [116] discusses the coverage and connectivity for WSNs with unreliable sensor nodes, deriving the necessary and sufficient conditions to cover a unit square region by a random grid network and maintain connectivity. These authors also formulate the sufficient conditions for connectivity between active nodes.

The framework described in Ray et al. [83] allows the sensor coverage areas to overlap so that each resolvable position is covered by a unique set of sensors. Using novel identification codes and based on a polynomial-time algorithm, it not only requires fewer sensors than existing proximity-based schemes in order to achieve required coverage, but also is robust against sensor failure or physical damage to the system. An alternate approach to achieving desirable and reliable coverage is by means of hardware redundancy, i.e., to deploy a greater density of sensor nodes in a sensing region and exploit redundancy to extend the overall system lifetime by operating distinct subsets that are, in turn, based on local density and local demand [32]. This is effective when the cost of deploying a node during the initial placement is much smaller than the cost of adding a new node at a later time.

9.5.2.2 Dynamic Power Optimization at the Nodal Level

Energy consumption at sensor node level has been described in Raghunathan et al. [81], Shih et al. [92], and Sinha and Chandrakasan [95]. From a functionality perspective, energy is consumed for sensing, computation, and communications. Power conservation can be achieved in any of these functions.

First, it should be noted that workload in WSNs typically has the characteristic of burstiness [10, 96]. Therefore, some nodes or certain components of nodes should switch to power-saving states between consecutive bursts while the functionality and QoS are still maintained. Dynamic power management (DPM) [9, 14, 81] is an example of this approach. As listed in Table 9.4, a particular combination of

TABLE 9.4 States of the Sensor Node and its Components

No.	Node State	MCU	Memory	Sensor and A/D	Radio
S0	Transmitting	Active	Active	On	Tx
S1	Receiving	Active	Active	On	Rx
S2	Ready	Idle	Sleep	On	Rx
S3	Observing	Sleep	Sleep	On	Rx
S4	Standby	Sleep	Sleep	On	Off
S5	Sleep	Sleep	Sleep	Off	Off
S6	Off	Off	Off	Off	Off

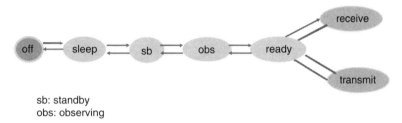

sb: standby
obs: observing

FIGURE 9.2 State transition diagram of a sensor node.

component states will determine a specific node state [92, 95]. For a sensor node, the states in decreasing order of power consumption are: transmitting, receiving, ready, observing, standby, sleep, and off. The state transition diagram of a sensor node is shown in Figure 9.2. For detailed numerical analysis of power consumption, see Raghunathan and colleagues [81]. However, transitions among states have power consumption and latency costs. Specifically, some transitions, for example, from "off" to "sleep," might cost much more energy than others, such as from "sleep" to "active." As a result, well-designed control algorithms are needed to achieve the trade-off between power saving and latency, power consumption, and state transitions.

Second, adaptively adjusting the operating voltage and frequency to meet the dynamically changing workload without degrading performance is a method of energy saving on computation. The rationale behind this technique is that the computational workload of MCU in WSNs is usually time varying and peak system performance is not always demanded. Dynamic voltage scaling (DVS) [14, 39, 73, 81] is an example of this approach. However, this scheme needs to predict the microprocessor's workload so as to adjust the power supply and operating frequency. A workload prediction strategy in WSNs is described in Chakrabarty et al. [14]. More accurate prediction can lead to higher power efficiency with less degradation to the system's performance. Nevertheless, workloads in current and future WSNs are mostly nondeterministic, so accurately modeling the workload is an open issue.

Another approach is to optimize the transmission power of sensor nodes. The change in transmission power has great impact on many aspects of WSN communication, including one-hop communication radius; network topology and hierarchy; retransmission rate; routing path selection; etc. Researches of this approach can be further divided into two types, depending on whether the node has the power control.

According to [113], an optimal transmission range, or transmission power in terms of energy efficiency, exists in certain ad hoc networks. The optimal value is mainly affected by propagation environment and device parameters. Contrary to intuition, [114] discovered that small transmission power might cause excessive power consumption due to a combined effect of increased number of hops and larger retransmission probabilities. Both researches were conducted in a flat, symmetric, multihop ad hoc network with no power control for individual nodes. Further research with various network and nodal conditions is strongly desired in the future.

Some other research assumes the power control capability on individual nodes. In such case, a large amount of communication energy can be saved through dynamically adjusting the transmission power based on the estimation of transmitting distance of each transmission. Proposed in [115], ROAD is a new MAC scheme for variable-radius multihop networks.

9.5.2.3 Optimal Schemes at System Level

9.5.2.3.1 Topology Management

As discussed earlier, dense deployment of sensors ensures the required coverage and sufficient precision of detection. Meanwhile, the redundant data generated by densely deployed nodes can be treated as backups for each other, so as to ensure the reliable function of the network. In the process of system operation, some node may operate in low duty cycles by transiting the hardware to sleep or off states to conserve energy. In these states, the sensor nodes are unable to communicate and forward packets. The nodes would then need to be awakened in certain situations, such as when it is time to collect data or neighboring nodes are depleted. Therefore, the active topology of the network changes over time. This leads to the critical issue of how to arrange sleep state transitions while ensuring robust, undegraded information collection [81].

A typical approach is to rotate the node functionality periodically to achieve balanced energy consumption among nodes. The protocol SPAN, proposed in Chen et al. [17], is an example of this approach for wireless ad hoc networks. Randomly, a limited number of nodes are self-selected as coordinators to construct the backbone in a peer-to-peer fashion within the network for traffic forwarding, while others can make local decisions to transit to a sleep state or keep active. The geographical adaptive fidelity (GAF) algorithm proposed in Xu and colleagues [105] is another way to rotate the active nodes within the network. Identified equivalent nodes, based on geographic locations on a virtual grid, can substitute each other directly and transparently without affecting the routing topology. Considering the fact that a WSN is only sensing its environment and waiting for an interesting event to happen, a new technique — sparse topology and energy management (STEM) described in Schurgers and coworkers [89, 90] — claims to improve beyond SPAN and GAF in terms of obtaining higher energy savings so as to prolong the system lifetime by trading off an increased latency to establish a multihop path.

9.5.2.3.2 Clustering and Hierarchical Architectures

It is reported that the energy consumed by communication is much higher than that for sensing and computation; in fact, this actually dominates the total energy consumption in WSNs. Experiments show that the ratio of communicating 1 bit over the wireless medium to that of processing the same bit could be in the range of 1000 to 10,000 [108]. Furthermore, in most WSNs, power for transmission contributes to a majority of the total energy consumed for communication and the required transmission power grows exponentially with the increase of transmission distance. Therefore, reducing the amount of traffic and distance of communications can greatly prolong the system's lifetime.

On the other hand, a WSN usually contains a large number of sensor nodes in a wide area, and the base station may be far from the wireless sensors. Thus, dividing the entire system into distinct clusters replaces the one-hop long-distance transmission by multihop short-distance data forwarding. This would reduce the energy consumed for data communications and also has the advantages of load balancing, and scalability when the network size grows. Challenges faced by such clustering-based approach include how to select the cluster heads and how to organize the clusters. The clustering strategy could be single-hop cluster or multihop cluster, based on the distance between the cluster heads and their members, as shown in Figure 9.3(a) and Figure 9.3(b), respectively [38]. According to the hierarchy of clusters, the clustering strategies can also be grouped into single-level or multilevel clustering. Figure 9.4 illustrates the system architecture of multilevel hierarchical clustering [7].

Various clustering approaches for wireless ad hoc and/or sensor networks have been proposed in the literature [6–8, 16, 30, 36, 38, 42–44, 59, 72, 84, 87]. Heinzelman et al. [42] propose a distributed low-energy adaptive clustering hierarchy (LEACH). At the beginning, each node self-selects itself as a cluster head with a predetermined probability; the cluster head then advertises its decision to the other nodes, which decide to join a specific cluster that requires minimum communication energy. In order to ensure the balanced energy dissipation among all nodes, LEACH invokes the rotation of the cluster head by calling the self-selection and cluster formation procedure periodically. Moreover, the analytical and

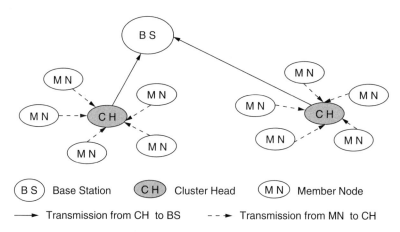

FIGURE 9.3(A) Single-hop clustering architecture.

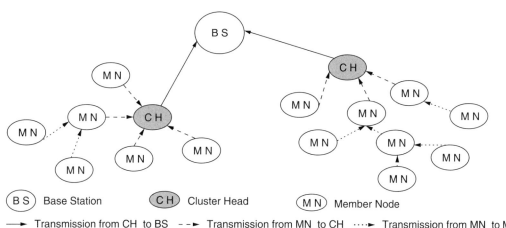

FIGURE 9.3(B) Multihop clustering architecture.

simulation results show that there is an optimal number of cluster heads that minimize the energy consumption.

Chiasserini et al. [18] attempt to solve the optimal problem of the balanced k-clustering, where k denotes the number of cluster heads in the system. Based on minimum weight matching, the algorithm attempts to realize load balancing among different clusters by partitioning the nodes into groups so that each cluster has a similar number of nodes. It achieves minimum energy consumption by optimizing the total spatial distance between the cluster members and the cluster heads. The power-aware virtual base stations (PA-VBS) protocol proposed by Safwat and colleagues [84, 86] is a first attempt to use the residual power capacity to select cluster heads in mobile ad hoc networks. It is attractive to WSNs because of its characteristics of load balancing and scalability to the growth of network size. In Gupta and Younis [38], a load-balanced clustering approach is introduced for heterogeneous sensor networks. The gateway nodes (cluster heads) with high energy manage the cluster member nodes and forward the data collected from the cluster member to a faraway base station. However, all the preceding schemes are single-hop cluster head formation algorithms, which may result in a large number of clusters. Therefore, they are only suitable for networks with a small to medium number of nodes.

In a large-scale network, multihop clusters and multilevel clustering hierarchy are highly in demand in order to decrease the communication distance further. Amis et al. [3] propose the max–min d-cluster to generate d-hop clusters, which can achieve better load balancing among clusters with fewer clusters than the

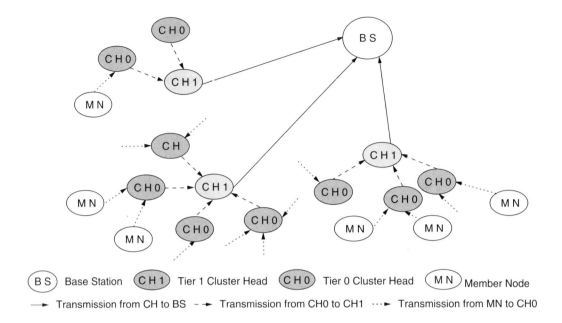

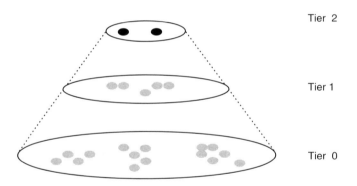

FIGURE 9.4 Multilevel hierarchical clustering architecture.

single-hop clustering schemes can [6, 30]. In Chiasserini et al. [18], a clustering algorithm is described to maximize the system lifetime through optimizing cluster size and assignment of nodes to each cluster head. However, this requires predetermining the number and locations of cluster heads, and each node should have knowledge of global network topology, which is impractical in WSNs. A chain-based protocol called power-efficient gathering in sensor information systems (PEGASIS) is presented in Lindsey and Raghavendra [60] and Lindsey et al. [61]. Instead of sending data packets directly to the cluster heads as shown in the LEACH protocol, each node forwards its packets to the destination through its closest neighbors.

Inheriting the feature of randomized creation and rotation of cluster heads as proposed in LEACH, as well as the advantages of a multihop clustering algorithm, Bandyopadhyay and Coyle [7] introduce a new energy-efficient, single-level, multihop clustering algorithm; these authors also provide the formulation for finding optimal parameter values to minimize the energy consumption. Moreover, based on the results of Foss and Zuyev [35] and Baccelli and Zuyev [5], Bandyopadhyay and Coyle [7] also provide a novel energy-efficient hierarchical clustering algorithm with a total of h levels, (i.e., some of the cluster heads in level $k-1$ select themselves as kth level cluster heads, and the remaining level $k-1$ cluster heads are cluster members in level k). They derive optimal parameters to achieve minimum energy consumption within the whole system. Experimental results for up to 10,000 nodes have been reported.

9.5.2.3.3 Traffic Distribution and System Partitioning

Due to the limited resources in WSNs, one key element of traffic forwarding is the selection of an energy-efficient path from the source to the destination. Some algorithms have been proposed to select a route that minimizes total energy consumption within the entire network. However, this is not always the case in order to maximize the overall system lifetime. Because the nodes on such route are overused, their batteries are more likely to be exhausted. This can result in discontinuity of the network, as well as unavailability of sensing in the corresponding regions. Therefore, taking the point of view of the system's overall availability and longevity, it is preferable to avoid continuously forwarding traffic through the same route, even though it always consumes the minimum energy from source to destination. Thus, it is desirable to distribute the traffic more evenly within the whole system [81].

It is also possible to introduce the concept of system partitioning [13] to reduce power dissipation in the sensor nodes by removing some intensive computation to remote base stations that are not energy constrained, or spreading some of the complex energy-consuming computation among more sensor nodes instead of overloading several centralized processing elements. Chandrakasan et al. [13], Min et al. [68], and Wang and Chandrakasan [102] describe examples of implementing system partitioning.

9.5.2.3.4 Collaborative Signal and Information Processing (CSIP) and Data Aggregation

In addition to the approaches described in previous subsections, local processing of raw data before direct forwarding will effectively reduce the amount of communication and improve the efficiency (information per bit transmitted). CSIP and data aggregation are two typical localized paradigms for the purpose of data processing in WSNs.

With the combination of interdisciplinary techniques, such as low-power communication and computation, space-time signal processing, distributed and fault-tolerant algorithms, adaptive systems, and sensor fusion and decision theory, CSIP is expected to provide solutions to many challenges, including dense spatial sampling of interested events; distributed asynchronous processing; progressive accuracy; optimized processing and communication; data fusion; and querying and routing tasks [58]. CSIP can be implemented through coherent signal processing on a small number of nodes in a cluster or through noncoherent processing across a larger number of nodes when synchronization is not a strict requirement [32]. CSIP algorithms can be classified [78] as information-driven schemes [107, 108], mobile agent-based schemes [77], or relation-based schemes [116].

Data aggregation or fusion [45, 52, 56] is another efficient data processing approach in WSNs. It tries to minimize traffic load (in terms of number and/or length of packets) through eliminating redundancy. Specifically, when an intermediate node receives data from multiple source nodes, instead of forwarding all of them directly, it checks the contents of incoming data and then combines them by eliminating redundant information under some accuracy constraints. It applies a novel data-centric approach to replace the traditional address-centric approach in data forwarding [56]. The examples depicted in Figure 9.5(a) and Figure 9.5(b) demonstrate the difference in these two approaches. In an address-centric approach, the intermediate node, M, must forward all the packets received from different source nodes, e.g., S1, S2, to the destination D. However, in a data-centric approach, node M first fuses the data from S1 and S2 by eliminating the redundant information, then relays the processed data to D. This leads to higher efficiency and more energy savings.

Several data aggregation algorithms have been reported in the literature. The most straightforward is duplicate suppression, i.e., if multiple sources send the same data, the intermediate node will only forward one of them. Maximum or minimum functions are also very simple approaches. Heinzelman and colleagues [41] and Julik and coworkers [57] propose a scheme named sensor protocols for information via negotiation (SPIN) to realize traffic reduction for information dissemination. It introduces metadata negotiations between sensors to avoid redundant and/or unnecessary data through the network. Proposed in Intanagonwiwat et al. [52], directed diffusion is a data distribution scheme that incorporates in-network data aggregation, data caching, and data-centric dissemination, while enforcing adaptation to the empirically best path. It aims to establish efficient n-way communication from single or multiple sources to sinks. Heidemann and colleagues [45] present a physical implementation of directed diffusion

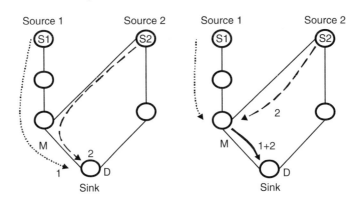

FIGURE 9.5 (A) Example of address-centric data forwarding. (B) Example of data-centric data forwarding.

with a wireless sensor test bed and shows that the traffic can be reduced by up to 42% when deploying a duplicate suppression data aggregation scheme.

The greedy aggregation approach proposed in Intanagonwiwat et al. [53] can improve path sharing and attain significant energy savings when the network has higher nodal density compared with the opportunistic approach. Krishnamachari and coworkers [56] describe the impact of source-destination placement on the energy costs and delay associated with data aggregation; they also investigate the complexity of optimal data aggregation. In [117], a polynomial-time algorithm for near-optimal maximum lifetime data aggregation (MLDA) is described for data collection in WSNs. The scheme is superior to others in terms of system lifetime, but has a high computational expense for large sensor networks. In Dasgupta et al. [22], a simple and efficient clustering-based heuristic for maximum lifetime data aggregation (CMLDA) is proposed for small- and large-scale sensor networks.

9.5.2.3.5 Cross-Layer Design

Traditional design of wireless ad hoc network protocols is mainly based on the layered stack as shown in Figure 9.6(a). This layered model makes a significant contribution to simplifying network design. Consequently, the layer structure leads to robust and scalable protocols. However, the design and operation of each layer in the stack are isolated, and the interface between layers is static and independent of the individual network constraints and applications. Therefore, inheriting such a stack will lead to poor WSN performance in which resources, especially energy, bandwidth, memory size, and CPU speed are greatly constrained. Many WSNs are dedicated for real-time data collection and strict delay bounds and high bandwidth demands could occur. Thus, new approaches are desirable to break the traditional border between the adjunct layers and create cross-layer paradigms. A possible cross-layer stack architecture is depicted in Figure 9.6(b) [37].

Goldsmith and Wicker [37] discuss not only the principles and strategies of cross-layer design in wireless ad hoc networks, but also the functionality of the individual layers and interactions among the different layers. Cross-layer design has become an attractive and active research topic in protocol designs of WSNs in recent years. Although some efforts have been made in literature, such as Heinzelman et al. [42, 43] and Safwat et al. [85], numerous open issues — how to understand and apply this design principle, how to deal with problems of information exchange across stack layers, and how to realize a specific application requirement with global system constraints — remain to be explored.

9.5.3 Software Development

Because of severe resource constraints, the software environment of WSNs is very different from those other distributed and parallel computing systems. Issues such as energy efficiency, scalability, and reliability are fundamental factors in software development for WSNs [13, 47, 49, 67, 81, 94, 99].

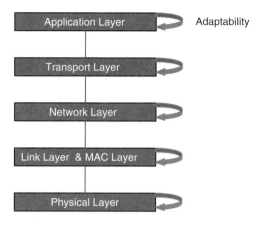

FIGURE 9.6(A) Traditional layered protocol stack for ad hoc networks.

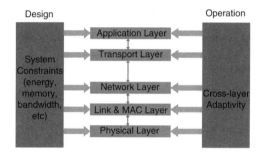

FIGURE 9.6(B) Cross-layer protocol stack for WSNs.

9.5.3.1 Single Node Level

System support on the lowest level begins at each single node. The concept of energy-aware software is introduced in Sinha and Chandrakasan [95]; who also illustrate the energy model of a typical microprocessor used for microsensors. With the proper operating systems, DPM and DVS can be deployed to reduce system power consumption at the node level. Described in Hill et al. [47, 49], TinyOS is one of the earliest operating systems dedicated for tiny sensor nodes; this system is event driven and uses only 178 bytes of memory, but supports communication, multitasking, and code modularity. Min and colleagues [67] present the concept of energy-scalable software, which is claimed to balance the trade-off between energy and quality characteristics.

9.5.3.2 Middleware

The middleware in WSNs abstracts the system as a collection of massively distributed objects and enables sensor applications to originate queries and tasks, gather responses and results, and monitor the changes within the network [91]. Sensor information networking architecture (SINA), proposed in Shen et al. [91], provides a middleware implementation of the general abstraction; these authors also describe sensor query and tasking language (SQTL), the sensor programming language used to implement such middleware architecture.

9.5.3.3 Application Programming Interface (API)

Considerable operation complexity exists in a WSN. However, with proper API implementation, the underlying system complexity can be transparent to end users who are experts in their specific application domain, but not necessarily experts in WSNs. The detailed functionalities of API in WSNs have been discussed in Shen and colleagues [91]. Stankovic et al. [99] consider other issues and advances in WSN software development.

9.6 Conclusions and Considerations for Future Research

A wireless sensor network consists of a large number of sensor nodes performing various distributed sensing and control tasks that are linked by a wireless medium. In general, a sensor is a device capable of capturing physical information, such as temperature, pressure, motion of an object, and mapping such physical characteristics of the environment to quantitative measurements. WSNs are evolving from simple networks with a small number of sensor nodes into diverse forms containing rapidly growing numbers of distributed nodes with enriched functions. These networks exhibit many benefits over their conventional wired counterparts and have been turning impossible monitoring and detection tasks into reality. Because of their ease of deployment, self-organization, reliability, versatility, scalability, and flexibility, WSNs have revealed significant potential in providing safer and healthier environments for human beings and thus have attracted much attention from academia as well as industry over the past few years.

This chapter presented an overview of WSNs and their evolution, describing numerous applications of self-configurable WSNs for target monitoring, detection, localization, and tracking in distinct military and civil domains. A discussion on technical challenges and design requirements was provided. Also highlighted were the state-of-the-art technical approaches in three aspects: hardware design; systems architectures, protocols, and algorithms; and software development.

Despite of the great progress on development of WSNs, quite a few issues still need to be explored in the future:

- *Tiny hardware components and sensor nodes with high efficiency* are still to be developed.
- *Protocols and algorithms for WSNs with heterogeneous sensor nodes.* Currently, many WSN protocols/algorithms are based on homogeneous sensor networks. However, sensors with different power capacities, sensing and transmitting range, and computing/processing abilities are usually more practical for constructing highly reliable networks [55, 63].
- *Combination of data-centric and address-centric operations.* As a long-term goal, WSNs are designated to be the first-class candidates in ubiquitous networks [118]. However, end-to-end communication fashion in traditional networks may not be suitable for the collective fashion in sensor networks. Combining WSNs' data-centric operation with the address-centric operation in traditional networks will lead to numerous open issues.
- *Security issues.* Most existing WSN communication protocols have not addressed security and are susceptible to attacks by adversaries. The issue of integrating security at the design stage in a resources-constrained WSN is a serious technical challenge.
- *Analytical modeling.* More accurate and expeditious implementation of WSNs in the real world is highly dependent on the ability to devise analytical models to evaluate and predict WSNs' performance characteristics, such as efficiency for information gathering, delay properties, granularity, and energy consumption. However, due to the diverse forms of applications and massive number of nodes in a single network, many technical problems remain to be solved in modeling the behavior of WSNs.
- *Clock synchronization.* Large numbers of sensor nodes in a WSN need to collaborate to fulfill the sensing task and the collected data are time sensitive in most cases. Thus, clock synchronization is a key requirement for algorithm and protocol design. However, due to resource and size limitation and lack of a fixed infrastructure, as well as the dynamic topology, existing time synchronization strategies designed for other traditional wired and wireless networks are not suitable for WSNs. Although Elson and Estrin [27] and Elson et al. [29] propose some synchronization proposals for WSNs, and some design principles are given in Elson and Romer [28], quite a few open issues still need to be explored in the future.

- *Other issues*. Optimal sensor node selection and allocation, discovery, localization, and network diagnoses are other open issues in this direction. Many software issues remain open as well. These include the design of distributed control and coordination algorithms to ensure balanced load assignment and energy consumption; efficient techniques for sensor data storage; and protocols with mobility consideration and dynamic group communications.

The issues discussed in this chapter are not exhaustive: many open issues remain to be explored so as to enable WSNs to achieve desirable connectivity, availability, reliability, and survivability in an energy-efficient fashion.

References

1. J. Agre and L. Clare, An integrated architecture for cooperative sensing networks, *Computer Mag.*, 33(5), 106–108, 2000.
2. I.F. Akyildiz, W. Su, Y. Sankarasubramaniam, and E. Cayirci, Wireless sensor network: a survey, *Computer Networks*, 38(4), 393–422, March 2002.
3. A.D. Amis et al., Max–min D-cluster formation in wireless ad hoc networks, IEEE *INFOCOM 2000*, 1, 32–41, Tel Aviv, March 2000.
4. G. Asada et al., Wireless integrated network sensors: low power systems on a chip, *24th IEEE Eur. Solid-State Circuits Conf.*, 9–12, The Hague, the Netherlands, 1998.
5. F. Baccelli and S. Zuyev, Poisson Voronoi spanning trees with applications to the optimization of communication networks, *Operations Res.*, 47(4), 619–631, 1999.
6. D.J. Baker, and A. Ephremides, The architectural organization of a mobile radio network via a distributed algorithm, *IEEE Trans. Commun.*, 29(11), 1694–1701, November 1981.
7. S. Bandyopadhyay and E.J. Coyle, An energy efficient hierarchical clustering algorithm for wireless sensor networks, *IEEE INFOCOM 2003*, 3, 1713–1723, San Francisco, March–April 2003.
8. S. Basagni, Distributed clustering for ad hoc networks, *Int. Symp. Parallel Architectures, Algorithms Networks*, 310–315, Freemantle, Australia, June 1999.
9. L. Benini and G.D. Micheli, *Dynamic Power Management: Design Techniques and CAD Tools*, Kluwer Academic Pub., Norwell, MA, 1998.
10. L. Benini et al., A discrete-time battery model for high-level power estimation, *Design, Automation Test in Eur. Conf.*, 35–41, Paris, March 2000.
11. D.W. Carman, P.S. Kruus, and B.J. Matt, Constraints and approaches for distributed sensor network security, NAI labs Technical Report 00-010, September 2000.
12. A. Cerpa, J. Elson, D. Estrin, L. Girod, M. Hamilton, and J. Zhao, Habitat monitoring: application driver for wireless communications technology, *1st ACM SIGCOMM Workshop Data Commun. Latin Am. Caribbean*, San Jose, Costa Rica, 31(2), 20–41, 2001.
13. A. Chandrakasan et al. Design considerations for distributed microsensor systems, *IEEE 1999 Custom Integrated Circuits Conf.*, 279–286, San Diego, May 1999.
14. K. Chakrabarty, S.S. Iyengar, H. Qi, and E. Cho, Coding theory framework for target location in distributed sensor networks, *Int. Conf. Inf. Technol.: Coding and Computing*, 130–134, Las Vegas, April 2001.
15. K. Chakrabarty, S.S. Iyengar, H. Qi, and E. Cho, Grid coverage for surveillance and target location in distributed sensor networks, *IEEE Trans. Computers*, 51, 1448–1453, 2002.
16. M. Chatterjee, S.K. Das, and D. Turgut, WCA: a weighted clustering algorithm for mobile ad hoc networks, *J. Cluster Computing*, special issue on mobile ad hoc networking, 5, 193–204, 2002.
17. B. Chen et al., SPAN: An energy-efficient coordination algorithm for topology maintenance in ad hoc wireless networks, *ACM/IEEE MOBICOM 2001*, 85–96, Rome, July 2001.
18. C.F. Chiasserini et al., Energy efficient design of wireless ad hoc networks, *Proc. Networking 2002*, 387–398, Pisa, May 2002.

19. S. Cho and A.P. Chandrakasan, Energy efficient protocols for low duty cycle wireless microsensor networks, *ICASSP'01*, 4, 2041–2044, Salt Lake City, May 2001.
20. T. Clouqueur, V. Phipatanasuphorn, P. Ramanathan, and K.K. Saluja, Sensor deployment strategy for target detection, *ACM WSNA*, 42–48, Atlanta, 2002.
21. Collaborative Sensor Networks, Internet article, http://wwwhome.cs.utwente.nl/~havinga/sensor.html, 2003.
22. K. Dasgupta, K. Kalpakis, and P. Namjoshi, An efficient clustering-based heuristic for data gathering and aggregation in sensor networks, *IEEE WCNC'03*, 3, 1948–1953, New Orleans, 2003.
23. V. De and S. Borkar, Technology and design challenges for low power and high performance, *ISLPED 1999*, 163–168, San Diego, August 1999.
24. S.S. Dhillon, K. Chakrabarty, and S.S. Iyengar, Sensor placement for grid coverage under imprecise detections, *Int. Conf. Inf. Fusion (FUSION) 2002*, 2, 1581–1587, Annapolis, 2002.
25. S. Dulman et al., Collaborative communication protocols for wireless sensor networks, *Eur. Res. Middleware Architectures for Complex Embedded Syst. Workshop*, Pisa, 2003.
26. M. Easton, Using space technology to fight malaria, *Queen's Gazette*, 13, April 7, 2003.
27. J. Elson and D. Estrin, Time synchronization for wireless sensor networks, in *2001 Int. Parallel Distributed Processing Symp.* (IPDPS), *Workshop Parallel Distributed Computing Issues Wireless Networks Mobile Computing*, 1965–1970, April 2001.
28. J. Elson and K. Romer, Wireless sensor networks: a new regime for time synchronization, *Workshop Hot Topics Networks (HotNets-I)*, Princeton, NJ, October 2002.
29. J. Elson, L. Girod, and D. Estrin, Fine-grained network synchronization using reference broadcasts, *Symp. Operating Syst. Design Implementation (OSDI 2002)*, Boston, MA. December 2002. UCLA technical report 020008.
30. A. Ephremides, J.E. Wiesethier, and D.J. Baker, A design concept for reliable mobile radio networks with frequency hopping signaling, *Proc. IEEE*, 75(1), 56–73, 1987.
31. D. Estrin, R. Govindan, J. Heidemann, and S. Kumar, Next century challenges: scalable coordination in sensor networks, *ACM/IEEE MOBICOM '99*, 263–270, Seattle, August 1999.
32. D. Estrin, L. Girod, G. Pottie, and M. Srivastava, Instrumenting the world with wireless sensor networks, *IEEE ICASSP 2001*, 4, 2033–2036, 2001.
33. J. Feng, F. Koushanfar, and M. Potkonjak, System-architectures for sensor networks issues, alternatives, and directions, *IEEE ICCD'02: VLSI Computers Processors*, 226–231, Freiburg, Germany, 2002.
34. J. Feng and M. Potkonjak, Power minimization by separation of control and data radios, short paper, *IEEE CAS Workshop Wireless Commun. Networking*, Pasadena, September 2002.
35. S.G. Foss and S.A. Zuyev, On a Voronoi aggregative process related to a bivariate Poisson process, *Adv. Appl. Probability*, 28(4), 1981, 965–981.
36. S. Ghiasi et al., Optimal energy aware clustering in sensor networks, *Sensors Mag.*, 19(2), 258–269, 2002.
37. A.J. Goldsmith and S.B. Wicker, Design challenges for energy-constrained ad hoc wireless networks, *IEEE Wireless Commun.*, 8–27, August 2002.
38. G. Gupta and M. Younis, Load-balanced clustering of wireless sensor networks, *IEEE ICC 2003*, 3, 1848–1852, May 2003.
39. V. Gutnik and A.P. Chandrakasan, An embedded power supply for low-power DSP, *IEEE Trans. VLSI Syst.*, 5(4), 425–435, December 1997.
40. P.J.M. Havinga and G.J.M. Smit, Energy-efficient wireless networking for multimedia applications, *Wireless Communications and Mobile Computing*, John Wiley & Sons, New York, 2001, 165–184.
41. W. Heinzelman, J. Kulik, and H. Balakrishnan, Adaptive protocols for information dissemination in wireless sensor networks, *ACM/IEEE MOBICOM '99*, 174–185, Seattle, August 1999.
42. W. Heinzelman, A. Chandrakasan, and H. Balakrishnan, Energy-efficient communication protocol for wireless microsensor networks, *HICSS 2000*, 8020–8029, Maui, January 2000.

43. W. Heinzelman, Application-specific protocol architecture for wireless networks, Ph.D. dissertation, Massachusetts Institute of Technology, June 2000.
44. W.B. Heinzelman, A.P. Chandrakasan, and H. Balakrishnan, An application-specific protocol architecture for wireless microsensor networks, *IEEE Trans. Wireless Commun.*, 1(4), 660–670, October 2002.
45. J. Heidemann et al., Building efficient wireless sensor networks with low-level naming, *18th ACM Symp. Operating Syst. Principles*, 146–159, October 2001.
46. C. Herring and S. Kaplan, Component-based software systems for smart environments, *IEEE Personal Commun.*, 60–61, October 2000.
47. J. Hill et al., System architecture directions for networked sensor networks, *9th Int. Conf. Architectural Support Programming Languages Operating Syst.*, 28(5), 93–104, Cambridge, MA, November 2000.
48. J. Hill and D. Culler, A wireless embedded sensor architecture for system level optimization, University of California Berkeley technical report, 2002.
49. J. Hill et al., TinyOS: operating system for sensor networks, hppt://tinyos.millennium.berkeley.edu, 2003.
50. X. Hong, M. Gerla, H. Wang, and L. Clare, Load balanced, energy-aware communications for Mars sensor networks, *IEEE Aerospace Conf.*, 3, 1109–1115, 2002.
51. A. Howard, M.J. Mataric, and G.S. Sukhatme, Mobile sensor network deployment using potential fields: a distributed, scalable, solution to the area coverage problem, *6th Int. Symp. Distributed Autonomous Robotics Syst.* (DAR02) 299–308, Fukuoka, Japan, June 2002.
52. C. Intanagonwiwat, R. Govindan, and D. Estrin, Directed diffusion: a scalable and robust communication paradigm for sensor networks, *ACM/IEEE MOBICOM 2000,* 56–67, Boston, August 2000.
53. C. Intanagonwiwat, D. Estrin, and R. Govindan, Impact of network density on data aggregation in wireless sensor networks, Technical report 01-750, University of Southern California, November 2001.
54. J.M. Kahn, R.H. Katz, and K.S.J. Pister, Next century challenges: mobile networking for Smart Dust, *ACM/IEEE MOBICOM,* 271–278, Seattle, 1999.
55. F. Koushanfar, M. Potkonjak, and A. Sangiovanni–Vincentelli, Fault-tolerance techniques for sensor networks, *IEEE Sensors Mag.*, 2, 1491–1496, 2002.
56. B. Krishnamachari, D. Estrin, and S. Wicker, Impact of data aggregation in wireless sensor networks, *Int. Workshop Data Aggregation Wireless Sensor Networks*, 575–578, Vienna, Austria, July 2002.
57. J. Julik, W. Heinzelman, and H. Balakrishnan, Negotiation-based protocols for disseminating information in wireless sensor networks, *Wireless Networks*, 8, 169–185, 2002.
58. S. Kumar, F. Zhao, and D. Shepherd, Collaborative signal and information processing in microsensor networks, *IEEE Signal Processing Mag.*, 13–14, March 2002.
59. C.R. Lin and M. Gerla, Adaptive clustering for mobile wireless networks, *J. Selected Areas Commun.*, 15(9), 1265–1275, September 1997.
60. S. Lindsey and C.S. Raghavendra, PEGASIS: power-efficient gathering in sensor information systems, *IEEE Aerospace Conf. 2002*, 3, 1125–1130, March 2002.
61. S. Lindsey, C. Raghavendra, and K.M. Sivalingam, Data gathering algorithms in sensor networks using energy metrics, *IEEE Trans. Parallel Distributed Syst.*, 13(9), 924–935, September 2002.
62. C. Lu et al., RAP: A real-time communication architecture for large-scale wireless sensor networks, *8th IEEE Real-Time Embedded Technol. Applications Symp.* (RTAS), 55–66, San Jose, CA, 2002.
63. Y. Ma et al., ROP: A resource oriented protocol for heterogeneous sensor networks, *2003 Virginia Tech Symp. Wireless Commun.*, Blacksburg, VA, 2003.
64. A. Mainwaring, J. Polastre, R. Szewczyk, D. Culler, and J. Anderson, Wireless sensor networks for habitat monitoring, *ACM WSNA'02*, 88–97, Atlanta, September, 2002.
65. S. Meguerdichian, F. Koushanfar, M. Potkonjak, and M.B. Srivastava, Coverage problems in wireless ad-hoc sensor networks, *IEEE INFOCOM*, 3, 1380–1387, Anchorage, 2001.

66. D.P. Mehta, M.A. Lopez, and L. Lin, Optimal coverage paths in ad-hoc sensor networks, *IEEE ICC*, 1, 507–511, Anchorage, 2003.
67. R. Min et al., Low-power wireless sensor networks, *IEEE VLSID 2001*, 205–210, India, 2001.
68. R. Min et al., Energy-centric enabling technologies for wireless sensor networks, *IEEE Wireless Commun.*, 28–39, August 2002.
69. J. Mirkovic, G.P. Venkataramani, S. Lu, and L. Zhang, A self-organizing approach to data forwarding in large-scale sensor networks, *IEEE ICC*, 5, 1357–1361, St. Petersburg, Russia, 2001.
70. S. Musman, P.E. Lehner, and C. Elsaesser, Sensor planning for elusive targets, *J. Computer Math. Modeling*, 25(3), 103–115, 1997.
71. National Interagency Fire Center, http://www.nifc.gov
72. A.K. Parekh, Selecting routers in ad-hoc wireless networks, *Proc. ITS*, 1994.
73. T.A. Pering, T.D. Burd, and R.W. Brodersen, The simulation and evaluation of dynamic voltage scaling algorithms, *Int. Symp. Low Power Electron. Design* (ISLPED), 1998.
74. E.M. Petriu et al., Sensor-based information appliances, *IEEE Instrumentation Measurement Mag.*, 31–35, December 2000.
75. G.J. Pottie and L.P. Clare, Wireless integrated network sensors: towards low cost and robust self-organizing security networks, *Proc. of SPIE* 1998, 3577, 86–95, 1999.
76. G.J. Pottie and W.J. Kaiser, Wireless integrated network sensors, *Commun. ACM* 2000, 43(5), 51–58, 2000.
77. H. Qi, S.S. Iyengar, and K. Chakrabarty, Multi-resolution data integration using mobile agents in distributed sensor networks, *IEEE Trans. Syst., Man Cybernetics (part C)*, 31, 383–391, August 2001.
78. H. Qi, P.T. Kuruganti, and Y. Xu, The development of localized algorithm in wireless sensor networks, *Sensors Mag.*, 2, 286–293, 2002.
79. J.M. Rabaey et al., PicoRadio supports ad hoc ultra-low power wireless networking, *IEEE Computer Mag.*, 33(7), 42–48, July 2000.
80. J.M. Rabaey, PicoRadio communication/computation piconodes for sensor networks year one report, 2001, EECS Department, University of California at Berkeley.
81. V. Raghunathan, C. Schurgers, S. Park, and M.B. Srivastava, Energy-aware wireless microsensor networks, *IEEE Signal Process. Mag.*, 40–50, March 2002.
82. T.S. Rappoport, *Wireless Communications, Principles and Practice*, Prentice Hall, Upper Saddle River, NJ, 1996.
83. S. Ray et al., Robust location detection in emergency sensor networks, *IEEE INFOCOM 2003*, 2, 1044–1053, March 2003.
84. A. Safwat, H. Hassanein, and H.T. Mouftah, Power-aware fair infrastructure formation for wireless mobile ad hoc communications, *IEEE GLOBECOM 2001*, 5, 2832–2836, November 2001.
85. A. Safwat, H. Hassanein, and H. Mouftah, Optimal cross-layer designs for energy-efficient wireless ad hoc and sensor networks, *22nd IEEE Int. Performance, Computing, Commun. Conf.*, (IPCCC 2003), 123–128, April 2003.
86. A. Safwat, H. Hassanein, and H.T. Mouftah, Power-aware virtual base stations (PW-VBS) for wireless mobile ad hoc communications, *J. Computer Networks*, 41(3), 331–346, 2003.
87. A. Salhieh et al., Power efficient topologies for wireless sensor network, *Int. Conf. Parallel Processing*, 156–163, Spain, 2001.
88. L. Schwiebert, S.K.S. Gupta, and J. Weinamann, Research challenges in wireless networks of bio-medical sensors, *ACM SIGMOBILE 2001*, 151–165, Rome, July 2001.
89. C. Schurgers, V. Tsiatsis, and M. Srivastava, STEM: topology management for energy efficient sensor networks, *2002 IEEE Aerospace Conf.*, 3, 1099–1108, March 2002.
90. C. Schurgers, V. Tsiatsis, S. Ganeriwal, and M. Srivastava, Optimizing sensor networks in the energy–latency–density design space, *IEEE Trans. Mobile Computing*, 1(1), 70–80, January–March 2002.
91. C-C. Shen, C. Srisathapornphat, and C. Jaikaeo, Sensor information networking architecture and applications, *IEEE Personal Commun.*, 52–59, August 2001.

92. E. Shih et al., Physical layer driven protocol and algorithm design for energy-efficient wireless sensor networks, *ACM/IEEE MOBICOM'01*, 272–287, Italy, July 2001.
93. J.L. da Silva Jr. et al. Design methodology for PicoRadio networks, *Design, Automation Test Eur.*, 314–325, Germany, March 2001.
94. A. Sinha and A. Chandrakasan, Energy aware software, *VLSID'00*, 50–57, Calcutta, January 2000.
95. A. Sinha and A. Chandrakasan, Dynamic power management in wireless sensor networks, *IEEE Design Test Computers*, 18(2), 62–74, 2001.
96. S. Slijepcevic and M. Potkonjak, Power efficient organization of wireless sensor networks, *IEEE ICC*, 2, 472–476, St. Petersburg, Russia, 2001.
97. K. Sohrabi, J. Gao, V. Ailawadhi, and G.J. Pottie, Protocols for self-organization of a wireless sensor network, *IEEE Personal Commun.*, 16–27, October 2000.
98. M. Srivastava, R. Muntz, and M. Potkonjak, Smart kindergarten: sensor-based wireless networks for smart developmental problem-solving environments, *ACM MOBICOM 2001*, 132–138, Italy, July 2001.
99. J.A. Stankovic et al., Real-time communication and coordination in embedded sensor networks, *Proc. IEEE*, 91(7), 1022–1032, 2003.
100. L. Subramanian and R.H. Katz, An architecture for building self-configurable systems, *MOBIHOC 2000*, 63–73, Boston, 2000.
101. S. Tilak, N.B. Abu–Ghazaleh, and W. Heinzelman, Infrastructure trade-offs for sensor networks, *ACM WSNA'02*, 49–58, September 2002.
102. A. Wang and A. Chandrakasan, Energy efficient system partitioning for distributed wireless sensor networks, *2001 IEEE Int. Conf. Acoustics, Speech, Signal Processing*, 2, 905–908, Salt Lake City, 2001.
103. A. Willig, R. Shah, J. Rabaey, and A. Wolisz, Altruists in the PicoRadio sensor network, *4th IEEE Int. Workshop Factory Commun. Syst.*, 175–184, Sweden, August 2002.
104. A.D. Wood and J.A. Stankovic, Denial of service in sensor networks, *IEEE Computer*, 35(10), 48–56, October 2002.
105. Y. Xu, J. Heidemann, and D. Estrin, Geography-informed energy conservation for ad hoc routing, *MOBICOM 2001*, 70–84, Italy, 2002.
106. M.D. Yarvis et al., Real-world experiences with an interactive ad hoc sensor network, *Int. Conf. Parallel Processing Workshops (ICPPW'02)*, 143–151, 2002.
107. F. Zhao, J. Shin, and J. Reich, Information-driven dynamic sensor collaboration for tracking applications, *IEEE Signal Process. Mag.*, 19(2), 61–72, March 2002.
108. F. Zhao et al., Collaborative signal and information processing: an information directed approach, *Proc. IEEE*, 91(8), 1199–1209, 2003.
109. Y. Zou and K. Chakrabarty, Sensor deployment and target localization based on virtual forces, *IEEE INFOCOM 2003*, 2, 1293–1303, San Francisco, 2003.
110. J.M. Kahn, R.H. Katz, and K.S.J. Pister, Emerging challenges: mobile networking for "Smart Dust," *Journal of Communications and Networks*, 2(3), 188–196, 2000.
111. K.S.J. Pister, SMART DUST — Autonomous sensing and communication in a cubic millimeter, *Internet article*, http//www-bsac.eecs.berkeley.edu/~pister/SmartDust/.
112. K.S.J. Pister, My view of sensor networks in 2010, *Internet article*, http://robotics.eecs.berkeley.edu/~pister/SmartDust/in2010.
113. P. Chen, B. O'Dea, and E. Callaway, Energy efficient system design with optimum transmission range for wireless ad hoc networks, *IEEE ICC 2002*, 2, 945–952, New York, May 2002.
114. S. Bansal et al., Energy efficiency and throughput for TCP traffic in multi-hop wireless networks, *IEEE INFOCOM 2002*, 1, 210–219, 2002.
115. C.-H. Yeh, ROAD: A variable-radius MAC protocol for ad hoc wireless networks, *IEEE VTC 2002 (Spring)*, 1, 399–403, 2002.
116. L.J. Guibas, Sensing, tracking, and reasoning with relations, *IEEE Signal Processing Magazine*, 19(2), 73–85, March 2002.

117. K. Kalpakis, K. Dasgupta, and P. Namjoshi, Efficient algorithms for maximum lifetime data gathering and aggregation in wireless sensor networks, Technical Report UMBC-TR-02-13, 2002, Computer Science and Electrical Engineering Department, University of Maryland, Baltimore, County.
118. A. Köpke, V. Handziski, and H. Karl, Making sensor networks intelligent, *7th Wireless World Research Forum (WWRF)*, the Netherlands, 2002.

10
Introduction to Industrial Sensor Networking

Miroslav Sveda
Brno University of Technology

Petr Benes
Brno University of Technology

Radimir Vrba
Brno University of Technology

Frantisek Zezulka
Brno University of Technology

10.1 Introduction 10-1
10.2 Industrial Sensor Fitting Communication Protocols ... 10-2
HART • ASI • Interbus • Measurement Bus • Controller Area Network (CAN) • LonWorks • Sercos • Bitbus (Updated as IEEE 1118) • Foundation Fieldbus • Profibus • Profibus PA • Microwire
10.3 IEEE 1451 Family of Smart Transducer Interface Standards... 10-11
IEEE 1451.1 • IEEE 1451.2 • IEEE P1451.3 • IEEE P1451.4 • IEEE P1451.5
10.4 Internet-Based Sensor Networking............................. 10-13
IEEE 1451.1 Concepts Utilized • IEEE 1451.1 Networking • Multicast Communication • Internet Coupling Architectures • Detailed Interconnecting Architectures
10.5 Industrial Network Interconnections 10-15
Interconnection Structures • Actuator-Sensor-Interface Standard • Nine-Bit Interprocessor Protocol
10.6 Wireless Sensor Networks in Industry 10-21
Problem Definition • Topology • Network Traffic • Communication Maintenance • Network Routing • Network Topology • Network Structuring Protocol
10.7 Conclusions 10-24

10.1 Introduction

The general trend in process instrumentation, including sensors and actuators directly contacting industrial processes, can be characterized by the attribute *intelligent* or *smart*. In the past decade, particularly, sensors have made the greatest progress toward being smart. At present, microcontrollers embedded in smart sensors enable signal conditioning, filtering, characteristics linearization, and other functions required to provide validity, reliability, and efficiency of measurement processes. The next important property of smart sensors is their capability to be networked.

Typical application domains for sensor networking are in automobile, aircraft, and spacecraft industries, process automation, and building/office/home automation. By means of sensor networking a large number of point-to-point connected sensors can be replaced by serial bus connections in order to achieve higher reliability, lower wiring costs, and easy set-up and maintenance. The conventional point-to-point voltage, or current loops, that have been successfully used for 30 years can be replaced by multiplexing, and, particularly, by serial networks.

The following sample of international standards also demonstrates application domains for industrial sensor networking:

- Automotive: ISO 11898
- Textile industry: ISO TC72
- Home automation: ISO/IEC-JTC1 SC25
- Trains: IEC TC9
- Shipbuilding: ISO TC8
- Mineral–oil industry: ISO TC67
- Mining industry: ISO TC82
- Medical domain and hospitals: CEN TC247

Communication systems for industrial automation, which launched industrial sensor networking initiatives, can be split into three network categories. The simplest category is the sensor/actuator network, which provide for multidrop sensor and actuator connection. Short data field and usage on the lowest hierarchical communication level in the hierarchical control and data acquisition architecture characterize this type of industrial sensor network. The second category, device buses, is characterized by a larger packet's data field and represents more powerful serial communication systems for automation. Several device buses are also efficiently used for sensor networking; thus, not only the sensor/actuator buses cover the domain of sensor networking. The third category of industrial networks, fieldbus, is applied on the higher hierarchical control and data acquisition levels and utilized for more complex measurement and data acquisition systems.

Recently, some fieldbuses have also been used for direct sensor networking. For this reason they are considered in the following comprehensive review. Actually, in addition to the use of fieldbuses for process automation as sensor networks in recent applications, local area networks (LANs) are currently used for sensor interconnections. The most popular LAN, Ethernet with TCP/IP, is increasingly employed for connection of measurement devices and systems including smart sensors.

This chapter introduces major concepts utilized in the area of industrial sensor networking. The main focus is on proper communication protocols, network interfaces, and network interconnections. Concurrently, case studies stemming from realized projects demonstrate approaches typical in this application domain.

10.2 Industrial Sensor Fitting Communication Protocols

Industrial communication networks (ICNs) can be classified into several groups: (1) industrial LANs; (2) fieldbuses; (3) device buses; and (4) sensor/actuator buses [1, 2]. LANs have emerged since the 1970s for multidrop connection of PCs, workstations, and complex electronic devices such as analyzers or PLCs. In industry, they are mostly based on TCP/IP protocol communication profiles over the industrial Ethernet. The other types of ICNs mentioned earlier can be characterized by

- Topology
- Segment length
- Bus control
- Transmission rate and timing
- Physical medium and signal modulation
- Medium access method
- Safety mechanism of data transmission
- Flexibility
- Economy
- Real time properties

Introduction to Industrial Sensor Networking

- Power supply
- Robustness
- Installation and maintenance properties
- Application areas

The following subsections review the main representatives of ICNs.

10.2.1 HART

The HART (highway addressable remote transducer) protocol is the oldest protocol and network for measurement purposes. HART supports simple star or point-to-point chain topology. It uses the 4- to 20-mA current loop for signal transfer; parameter propagation; status set-up; diagnostics and configuration by FSK modulation with 0.5 mA peak sine wave; logical 1 represented by 1200 Hz; and logical 0 represented by two cycles of 2200 Hz. The HART protocol is low cost, simple, and, because of the 4- to 20-mA physical interface, supported by many sensor producers. A low data transfer rate (10 measurements per second) can suffice for temperature, level and chemical quantity measurements, and processes control.

HART provides 13 compulsory commands and other commands are optional. Compulsory commands enable reading measured data, sensor number, measurable range, etc. Optional commands provide for calibration, setting physical values, writing a serial number, dialing one of four physical units, resetting the sensor, etc.

Technical summary:

Topology: basically, point-to-point: one field node can be connected to two higher devices (supervisor devices), analog and digital transmission modes; alternatively: bus topology (multidrop) with a maximum of 15 nodes including two supervisor devices, in this case for transmission of digital signals
Segment length: 3000 m in point–point and 300 m in multidrop topologies
Medium access control: master–slave
Data transmission rate: 1.2 kb/s (standard) and 19.2 kb/s (high-speed mode)
Response time: guaranteed; about 500 ms for one node
Medium: twisted pair: 4 to 20 mA
Modulation: frequency shift keying (FSK)
Power supply: via signal wiring
Ex mode: in special cases
State of the art: wide range of sensors and actuators of many producers on the market: Rosemount–Emerson, Siemens, Yokogawa, Krohne, ABB Automation, Endress+Hauser, Ametek, Foxboro Eckardt, etc.
Application area: temperature, pressure, flow, density, level, analytical sensors, actuators
www: http://www.hartcomm.org

10.2.2 ASI

A simple sensor/actuator bus provides for use in automation of machines, production lines, and technologies. It is available predominantly for connection of binary sensors and binary actuators. Tree network topology is available. The segment length must not exceed 100 m without repeaters. Any combination of active and passive slaves up to 256 binary slaves and actuators is permitted on a segment. The network cycle period must not exceed 128 ms. Physical layer is based on reliable alternating pulse modulation (APM) methods. The physical medium is an unshielded, untwisted pair in special mechanical shape.

Technical summary:

Topology: bus, tree
Segment length: 100 m

Medium access control: master–slave (cycle polling) with 31 active or 124 passive binary slaves on a segment; alternatively, analog nodes translate analog signals with a maximum resolution of 18 b
Data transmission rate: 156 kb/s
Network cycle: 5 ms (time for response of all active nodes)
Medium: unshielded untwisted pair
Modulation: alternate pulse modulation APM (pulse width modulation), full duplex
Power supply: by signal conduction (2 to 10 A) or by a separate two-wire connection
Physical interface: ASI
Ex mode: yes
Response time: guaranteed
Standardized: EN 50 295, IEC 62026
State of the art: more then 32 firms, e.g., Balluff, Pepperl & Fuchs, ifm electronic, Siemens, Bernstein
Application area: digital sensors, actuators, I/O modules
www: http://www.as-interface.com

10.2.3 Interbus

One of the oldest proprietary industrial sensor/actuator and device communication buses in use. Its topology is a double ring (main trunk) with short cross segments. Interbus is aimed at real-time data acquisition and control. Besides master and slave stations, there also are up to 64 data switchers (repeaters). The length of the main trunk is up to 13 km in copper wire and up to 100 km in optical fiber. The local, 10-m segments extending the ring can connect up to eight nodes each. The voltage level in the local bus segment is 0 to 5 V. The most common version, called Interbus S, supports a kind of express transmission of short process data blocks in combination with a slow message cycle for configuration, diagnostics, and other special functions.

Technical summary:

Topology: double ring (main trunk) with short cross segments
Segment length: 13 km for copper, 100 km for optical fiber, 10 m (local bus with a maximum of eight nodes with distances up to1.5 m)
Medium access control: master–slave with 256 slaves; highly effective bus access method
Data transmission rate: 500 kb/s (main trunk); 300 kb/s (local bus)
Electrical interface: EIA RS 485
Medium: unshielded twisted pair; optical fiber
Power supply: local
Ex mode: no
Response time: guaranteed
Error coding: CRC
Standardized: DIN E 19258, EN 50 254, EN 50 170 (prepared for extension)

10.2.4 Measurement Bus

The measurement bus, also known as the DIN mess bus, is designed primarily for measurement (see Figure 10.1). The maximum length of the bus is 500 m and the data transmission rate is 115.2 kb/s in free (rootless tree) topology. The control is master–slave with a maximum of 32 nodes or 961 and 4096 nodes, respectively, with extended address field and cascade sequencing. The physical medium is four-line wire for full duplex; the maximum bit rate in bus topology is optional between 1.2 kb/s and 1 Mb/s. The master uses one twisted pair of messages; the other pair is used by slaves for responses in time-division multiplex mode with polling. The method preserves basic functions of the system even in case of alarms and network reconfiguration. The measurement bus is equipped with several safety mechanisms based on parity bit control, BCC (block checksum character), and time-out.

Technical summary:

Introduction to Industrial Sensor Networking

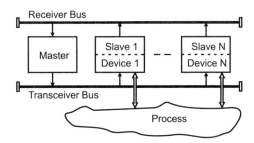

FIGURE 10.1 Measurement bus topology.

Topology: bus; root-free tree up to 115.2 kb/s
Segment length: 500 m at the maximum transmission rate of 1 Mb/s and short node connection (maximum length of 5 m)
Medium access control: master–slave with up to 32, 961, or 4096 slaves, respectively
Data transmission rate: 115.2 kb/s
Medium: two twisted pairs
Electrical interface: RS 485
Modulation: NRZ base band
Power supply: by signal conduction
Ex mode: no
Standardized: DIN 66348
Error coding: parity (HD = 4)
State of the art: emerging applications
Application area: measurement devices

10.2.5 Controller Area Network (CAN)

The CAN is one of the most popular fieldbuses. Bosch and Intel developed it at the end of the 1980s for the automobile industry. It has been applied in cars but also in manufacturing. The topology is tree or bus with maximum communication speed of 1 Mb/s. CAN is a real-time protocol with multicasting; the medium access method is CSMA/CA (carrier sense multiple access with collision avoidance) for multi-master mode. CAN is equipped with the following safety mechanisms: differential voltage for dominant and recessive levels; CRC coding with bit stuffing; message frame checking with acknowledgments; and error counters with active, passive, and off-line modes.

Technical summary:

Topology: bus; passive connection type
Segment length: 40 m up to1 Mb/s; 1000 m up to 50 kb/s
Medium access control: multimaster with CSMA/CA
Data transmission rate: 50 kb/s to 1 Mb/s
Medium: shielded pair; optical fiber
Modulation: recessive and dominant differential levels
Power supply: local
Ex mode: no
Response time: guaranteed
Robustness: high data safety grade (HD = 6)
Standardized: ISO 11898, open standard of physical and link layers
Extras: different application layers: DeviceNet, CANopen and SDS
State of the art: Bosch, Balluff, Baumer, Pepperl+Fuchs, Fraba Sensorsysteme, ifm electronic, Druck
Application area: pressure, temperature, inclinometer, actuators, encoders
www: http://can-cia.de

10.2.6 LonWorks

LonWorks technology aims at completely distributed data acquisition and control (see Figure 10.2). Layer protocols are implemented by NEURON microcontroller. The set of transceivers corresponds to the set of communication media, including twisted pair, coaxial cable, radio, optical fiber, and power line. The related communication protocol, LonTalk, provides CSMA/CA medium access control. Priority slots in the protocol frame guarantee soft real-time properties. The NEURON chip consists of three 8-b microprocessors: the first implements medium access control; the second provides higher network layer protocols; and the third one supports the user application program. The technology was originally designed for building automation with special purpose address formats respecting domains and subdomains connected by routers and bridges. The total number of nodes is up to 32,385. Principles of connecting network segments by a router are depicted in Figure 10.2. Lon technology can connect simple sensors and actuators as well as high-efficiency devices. Besides building automation, the LonWorks technology is used in data acquisition and control systems.

Technical summary:

Topology: tree
Segment length: depends on network architecture
Medium access control: peer-to-peer predictive p-persistent CSMA/CD
Data transmission rate: 79 kb/s till 1.25 Mb/s
Medium: twisted pair, coaxial cable, radio, power line, optical fiber
Electrical interface: EIA RS 485 and others
Modulation: base band with Manchester II or NRZ
Power supply: depending on physical media
Ex mode: no
Response time: soft real time, almost guaranteed for priority slots
Standardized: IEC 62026
Extras: implements all seven layers of the ISO/OSI RM
State of the art: Zellweger Analytics, Hubbell, Honeywell, Siemens

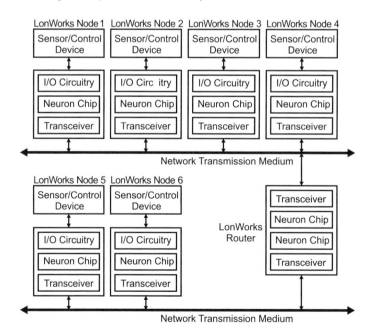

FIGURE 10.2 Connection of domains by LonWorks router.

Application area: conductivity, gas concentration, light, temperature, pressure, pH, actuators
www: http://www.lonmark.org

10.2.7 Sercos

Sercos (serial real time communication system) is developed for CNC (computer numeric control) as well as for direct connection of conventional sensors and actuators in factory automation. Due to extremely high response-time requirements, the physical communication medium is a special optical fiber.

Technical summary:

Topology: ring with active node connection
Segment length: 50 m for plastic optical fiber and 250 m for glass optical fiber, up to 254 nodes
Medium access control: master–slave
Data transmission rate: 2 to 4 Mb/s
Medium: special optical fiber
Modulation: base band with NRZI (nonreturn to zero inverted)
Network cycle: 0.062 to 65 ms; 1 ms typically (response time of all active nodes)
Power supply: local
Ex mode: no
Response time: guaranteed
Error coding: HD = 4
Standardized: IEC 61491
State of the art: instruments for CNC come from many manufacturers
Application area: CNC and motion controllers, drives, I/O modules
www: http://www.sercos.org

10.2.8 Bitbus (Updated as IEEE 1118)

Bitbus, developed by Intel, is one of the oldest serial buses for industrial use. The bitbus specification allows interconnecting 28 nodes over a distance of 30 m for synchronous mode with bit rate 2.4 Mb/s up to 250 nodes over 13.2 km in a self-clocked mode with bit rate 62.5 kb/s.

Technical summary:

Topology: one or more interconnected buses
Maximum length: 13.2 km with 62.5 kb/s and 250 nodes; maximum of 28 nodes per segment
Medium access control: master–slave with acknowledgment
Data transmission rate: 62.5 kb/s (with repeaters) up to 13.2 km (with repeaters); 375 kb/s up to 300 m; and 2.4 Mb/s up to 30 m length via twisted pair; 1.5 Mb/s via optical fiber
Medium: twisted pairs or optical fiber
Electrical interface: EIA RS 485
Modulation: base band, NRZ or NRZI
Power supply: external
Ex mode: no
Response time: guaranteed
Error Coding: CRC
Standardized: IEEE 1118
Extras: SDLC (synchronous data link control)
State of the art: many applications in the past; not available for direct sensor connection
Application area: controllers, I/O modules
www: http://www.bitbus.org

10.2.9 Foundation Fieldbus

Foundation fieldbus is the result of cooperation of the ISP (Interoperable System Project) and WorldFIP initiatives. It is the youngest and the most advanced fieldbus for industrial applications, namely, for process control in chemical, pharmaceutical, petrochemical, and other processing industries. Foundation fieldbus in the H1 variant can be used also for the explosive area because it is based on the IEC 1158-2 physical layer standard. Foundation fieldbus nodes can be classed as basic devices (BD); link master (LMD); bridge; or LAS (link active scheduler). LMD can play the role of LAS, but several LASs can cooperate in the network. Special communication modes implement broadcasting, multicasting, and distributed data transfer. The LAS compels data from a BD and the BD publishes data to all nodes programmed as subscribers to receive the data in the basic rapid cycle mode. Complementary mode makes it possible to send data in the spare time between two basic cycles. Foundation fieldbus contains a user application layer that extends the ISO/OSI communication model by application blocks accessible directly by user applications (see Figure 10.3).

Technical summary:

Topology: bus

Segment length: 1900 m with up to 32 nodes per segment

Medium access control: multimaster with the CSMA/CD and CSMA/CA medium access method for broadcasting, multicasting, and distributed data transfer

Data transmission rate: 31.25 kb/s with H1 (low-speed variant); 100 Mb/s with fast Ethernet (high-speed variant)

Electrical interface: IEC 1158-2 for H1 variant; fast Ethernet

Medium: twisted pair

Modulation: base band, Manchester II

Power supply: via double wire signal cable in explosive area

Ex mode: yes for H1 variant

Response time: guaranteed

Error coding: CRC; special coded characters in preamble, start delimiter and end delimiter

Standardized: IEC 61491

State of the art: ABB, Rosemount–Emerson, Endress+Hauser, Foxboro, Fuji, Honeywell, Krohne, Smar, Yokogawa

Application area: pressure, flow, temperature, conductivity, level, pH

www: http://www.fieldbus.org

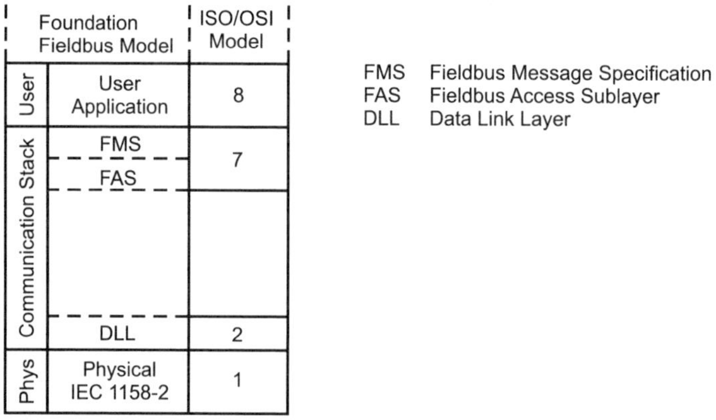

FIGURE 10.3 Foundation fieldbus communication model.

Introduction to Industrial Sensor Networking

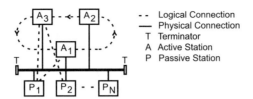

FIGURE 10.4 Medium access method by profibus DP.

10.2.10 Profibus

Profibus (process fieldbus) is an industrial communication standard of German origin (DIN 19245, EN 50 170). For sensor networking, the profibus DP (distributed periphery) and the profibus PA (process automation) variants can be employed. As depicted in Figure 10.4, the combined token passing and master–slave medium access method can be used to adapt profibus to a concrete industrial application. Two classes of nodes include active stations, which can obtain a token to control the network for a preset time, and passive stations that play the role of slaves and send data on demand of active stations. A large number of smart sensors and actuators are already equipped with the profibus DP connection.

Technical summary:

Topology: bus with passive node connection
Segment length: up to 9.6 km in copper and 90 km in optical fiber, up to 5 bus segments
Medium access control: combined token passing and multi master–slave; polling
Data transmission rate: wide range from 9.6 kb/s to 12 Mb/s (segment length up to 100 m)
Electrical interface: EIA RS 485
Medium: shielded twisted pair
Modulation: NRZ
Nodes number: 31 or 128 (with repeaters)
Power supply: external
Ex mode: no
Response time: guaranteed
Error coding: HD = 4
Standardized: DIN 19 245, EN 50 170
State of the art: FRABA, Hengstler, TWK Elektronik, Heidenhain, Siemens, AutomationX, Keller HCW, Brooks Instrument, Emerson, Barksdale Control, Mettler Toledo, Pepperl+Fuchs, IVO, SICK, Max Stegmann
Application area: flow, pressure, temperature, position, encoder
www: http://www.profibus.com

10.2.11 Profibus PA

Profibus PA is a communication system for networking of sensors and actuators in process control and data acquisition systems (see Figure 10.5). It extends the application area of profibus DP to process control and, particularly, to explosive areas. The profibus PA communication interface is embedded into several actuators and high-performance sensors.

Technical summary:

Topology: bus and tree structure with passive node connection
Segment length: up to 1900 m; maximum of 32 nodes in segment
Medium access control: combined token passing and multi master–slave method
Data transmission rate: 31.25 kb/s
Electrical interface: IEC 1158-2

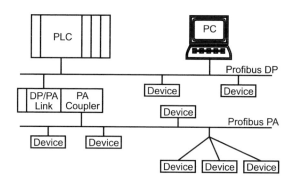

FIGURE 10.5 Profibus PA topology.

Medium: twisted pair
Modulation: base band, differential Manchester
Response time: guaranteed
Power supply: via signal wiring, or separate supply for nodes in explosive areas
Ex mode: yes
Error coding: CRC, HD = 4
Standardized: EN standard in preparation
State of the art: Foxboro, ABB Automation, Endress+Hauser, Mettler Toledo, Krohne, Emerson, Siemens, SMAR, Klay Instruments, WIKA
Application area: level, density, pressure, temperature
www: http://www.profibus.com

10.2.12 Microwire

The Dallas technology is based on the 8-b ASIC Dallas microcontroller with a 32-b unique addressing. The technology consists of three elements: PC or microcontroller-based master; wiring and connectors; and one-wire devices (slaves). Based on the usual TTL voltage UART interface, the Microwire enables connection of eight slaves to one segment with the maximum length of 100 m per segment. The network control is master–slave. The system is designed for building automation — particularly for temperature monitoring — and also for autonomous meteorological stations.

Technical summary:

Topology: bus
Segment length: up to 100 m and eight slaves
Medium access control: master–slave, time slots
Data transmission rate: 14.4 kb/s
Electrical interface: TTL voltage (log. 0 lower than 0.8 V; log. 1 higher than 2.5 V)
Medium: unshielded twisted pair (one wire for GND)
Coding: base band NRZ
Response time: 7 ms for each slave
Power supply: slave supply via signal; one wire cable
Ex mode: no
Standardized: proprietary
Extras: smart devices equipped with oscillators synchronized by master messages
State of the art: microcontrollers Dallas; IEEE 1451.4
Application area: temperature
www: www.maxim-ic.com

10.3 IEEE 1451 Family of Smart Transducer Interface Standards

The IEEE 1451 Smart Transducer Interface Standards describe open and network-independent communication architecture for smart transducers. The IEEE Instrument and Measurement Society, Technical Committee on Sensor Technology (TC-9) and the NIST Manufacturing Engineering Laboratory support the work. These standards are also being developed in cooperation with many sensor and measurement companies: for example Agilent, Analog Devices, Boeing, Telemonitor, National Instruments, PCB, Brüel & Kjaer, Sensor Synergy, Endevco, Crossbow Technology, Eaton, and EDC.

The ideas of a smart sensor communication interface standard were proposed in September 1993 at the TC9 Committee Meeting on Sensors Conference and Expo. In the following years four working groups were formed [7]. The working groups developed two accepted standards and two proposed standards:

- IEEE 1451.1 Network capable application processor information model (approved in 1999 by IEEE as a full-use standard)
- IEEE 1451.2 Transducer to microprocessor communication protocol and transducer, electronic data sheet (TEDS) formats (approved in 1997 by IEEE as a full-use standard)
- IEEE P1451.3 Digital communication and transducer electronic data sheet (TEDS) formats for distributed multidrop systems
- IEEE P1451.4 Mixed-mode communication protocols and TEDS formats

In 2002 two new working groups started work for the next standards:

- IEEE P1451.5 wireless communication protocols and TEDS formats [8]
- IEEE P1451.0 Study group with the interest area aimed at harmonizing individual standards of the 1451 family [6]

Figure 10.6 shows the basic IEEE 1451 working groups' relationship. The standards are designed to be complementary; moreover, they can be used independently or together.

10.3.1 IEEE 1451.1

The IEEE 1451.1 standard defines a common object model description for a networked smart transducer and software interface specifications for each class representing the model [3]. IEEE 1451.1 allows for flexible, modular assembly of network interface, measurement and control functions, and transducer interface modeled and/or implemented by a network capable application processor (NCAP). Thus, any control network can be connected to any transducer, or group of transducers, with an appropriately configured NCAP. The NCAP typically consists of a processor with an embedded operating system and timing capability. The IEEE 1451.1 standard provides two models for network communication between objects. The point-to-point client/server model is tightly coupled, while producer/subscriber is relatively free for one-to-many and many-to-many communications. Network software suppliers are expected to provide code libraries that contain routines for calls between the IEEE 1451.1 communication operations and the network.

10.3.2 IEEE 1451.2

The IEEE 1451.2-1997 transducer to microprocessor communication protocol and TEDS formats [4] standard defines a transducer-to-microprocessor, digital point-to-point serial communication protocol allowing any smart transducer, or group of transducers, to receive and send digital data using a common interface. Any transducer can be adapted to the P1451.2 protocol with the smart transducer interface module (STIM). Integral to this standard are the definition of the STIM; format for the TEDS; calibration and correction data engine; and 10-wire transducer independent interface (TII) — a physical interface between the STIM and the NCAP. The TEDS, stored in a nonvolatile memory, contains fields that describe

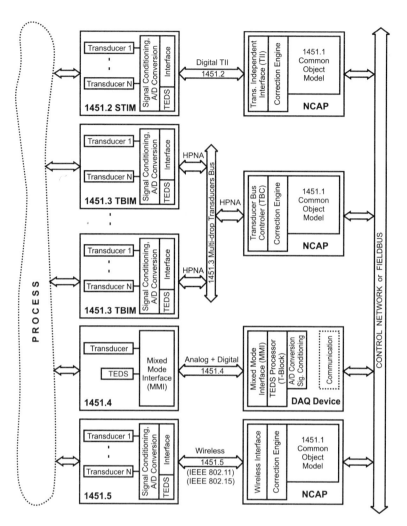

FIGURE 10.6 The IEEE 1451 family of smart transducer interfaces.

the manufacturer's name; device type; model number; revision code; serial number; transducer characteristic; and calibration constant. Eight TEDS structures are defined. Of these, two are required (meta-TEDS, channel TEDS) and six are optional (calibration TEDS; generic-extension TEDS; meta-ID TEDS; channel-ID TEDS; calibration-ID TEDS; end user application-specific TEDS).

10.3.3 IEEE P1451.3

The IEEE P1451.3 proposal defines a multidrop distributed system for interfacing smart transducers. Digital data signals are multiplexed on a common transmission medium. The wire protocol is based on the home phoneline networking alliance (HPNA) technology. A transmission line is used to supply power to the transducers and to provide communication between a single transducer bus controller (TBC) and a number of transducer bus interface modules (TBIMs). A set of TEDS fields, including many of the TEDS described in the IEEE 1451.2 standard, is based on the use of XML (extensible markup language).

10.3.4 IEEE P1451.4

IEEE P1451.4 defines a mixed-mode interface (MMI) for analog transducers with analog and digital operating modes and specific TEDS formats. For communication with the TEDS memory device, IEEE

PI451.4 uses a simple, low-cost serial transmission protocol (one-wire MicroLan protocol from Dallas Semiconductor) that provides power and data on one wire, with a second wire used for ground reference. The working group defines two classes of mixed-mode sensors, allowing analog and digital TEDS data to share the same two wires sequentially (class 1) or to be available simultaneously through separate wires (class 2).

10.3.5 IEEE P1451.5

The IEEE P1451.5 proposal defines a wireless communication protocol and TEDS formats. This proposed standard utilizes the IEEE 802 family as a basis of wireless communication protocols [8, 9]. Section 10.6 of this chapter demonstrates basic principles of wireless communication in industry using IEEE 802.15 or Bluetooth. In 2001 a new initiative started aiming at a standards review with a goal to extend some parts of the 1451 family to satisfy new industry demands. Attention was given to alternative physical layers and to enhancements of the TEDS with new features such as XML format of the TEDS, hot swapping possibilities, and physical layers information [5].

10.4 Internet-Based Sensor Networking

This section presents architectural concepts for direct sensor interconnections by Internet. It deals with the IEEE 1451.1, object-based networking model application, complemented by the Internet protocol (IP) multicast communication, that mediates unified access from Internet to sensors and vice versa.

10.4.1 IEEE 1451.1 Concepts Utilized

The 1451.1 information model deals with an object-oriented definition of an NCAP, which is the object-oriented embodiment of a smart networked device. This model includes the specification of all application-level access to network resources and transducer hardware. The object model definition encompasses a set of objects' classes, attributes, methods, and behaviors that provide a concise description of a transducer and a network environment to which it may connect. The standard brings a network and transducer hardware-neutral environment in which a concrete implementation can be developed.

The standard uses block and base classes to describe the transducer device. The 1451.1 object model defines four component classes offering patterns for (1) one physical block; (2) one or more transducer blocks; (3) function blocks; and (4) network blocks. Each block class may include specific base classes from the model. The base classes include parameters, actions, events, and files, and provide component classes.

All classes in the model have an abstract or root class from which they are derived. This abstract class includes several attributes and methods common to all classes in the model and offers a definition facility to be used for instantiation and deletion of concrete classes. In addition, methods for getting and setting attributes within each class are also provided.

10.4.2 IEEE 1451.1 Networking

Block classes form the major blocks of functionality that can be plugged into an abstract card cage to create various types of devices. One physical block is mandatory because it defines the card-cage and abstracts the hardware and software resources used by the device. All other blocks, components, and base classes can be referenced from the physical block.

The transducer block abstracts all the capabilities of each transducer physically connected to the NCAP I/O system. During the device configuration phase, the description of the kinds of sensors and actuators connected to the system is read from the hardware device. The transducer block includes an I/O device driver style interface for communication with the hardware. The I/O interface includes methods for reading and writing to the transducer from the application-based function block using a standardized interface.

The I/O device driver provides plug-and-play capability and a hot-swap feature for transducers. Of course, any application written to this interface should work interchangeably with multiple vendor transducers. In a similar fashion, the transducer vendors provide an I/O driver to the network vendors with their product that supports this interface. The driver is integrated with the transducer's application environment to enable access to its hardware. This approach is identical to the interface found in device drivers for UNIX.

The function block equips a transducer device with a skeletal area in which to place application-specific code. The interface does not define any restrictions on how an application is developed. In addition to a state variable that all block classes maintain, the function block contains several lists of parameters typically used to access network-visible data or to make internal data available remotely. This means that any application-specific algorithms or data structures are contained within these blocks to allow separately for integration of application-specific functionality using a portable approach.

The network block is used to abstract all access to the network by the block and base classes employing a network-neutral, object-based programming interface. The network model provides an application interaction mechanism based on the remote procedure call (RPC) framework for distributed computing settings. The RPC mechanism props a client–server and a publisher–subscriber paradigm for event and message generation.

In support of these two types of application interaction, a communication model that stems from the notion of a port is defined in the specification. This means that, if a block wishes to communicate with any other block in the device or across the network, it must first create a port that logically binds the block to the port name. Once enough information about addressing the port is known, the port can be bound to a network-specific block address. At this point, the logical port address has been bound to the actual destination address by underlying network technology control. Any transducer application's use of the port name is now resolved to the endpoint associated with the logical destination.

This scheme allows a late binding effect on application uses of the ports so that addresses are not hard-coded or dependent upon a specific architecture. The port capability is similar to the TCP/IP socket-programming interface in which a socket is created and bound using an application-specific port number and IP address. Once bound, the socket can be used for data transfer.

10.4.3 Multicast Communication

A traditional network computing paradigm involves communication between two network nodes. However, emerging Internet applications require simultaneous group communication based on multipoint configuration propped by multicast IP, which saves bandwidth by forcing the network to replicate packets only when necessary. Multicast improves the efficiency of multipoint data distribution by building a distribution tree from a sender to a set of receivers.

IP multicasting is the transmission of an IP datagram to a host group, which is a set of zero or more hosts identified by a single IP destination address of class D. Multicast groups are maintained by an Internet group management protocol, IGMP (IETF RFC 1112, RFC 2236). Multicast routing considers multicasting routers equipped with multicasting routing protocols such as DVMRP (IETF RFC 1075); MOSPF (IETF RFC 1584); or PIM (IETF RFC 2117). For Ethernet-based intranets, the address resolution protocol provides last-hop routing by mapping class D addresses on multicast Ethernet addresses.

10.4.4 Internet Coupling Architectures

Typically, Ethernet LANs can connect smart transducers directly to an Intranet. In this case the assigned IP address provides transducers with a unique identity not only in the system under development but also in the Internet. Such a transducer can be coupled with a client in two basic ways: direct connection and connection via a gateway. In case of the direct connection, the client communicates with the transducer directly, using some common messaging protocol. In the transducer object model, basic network block functions initialize and cover communication between a client and the transducer. In a special case, another smart transducer can provide a client. This type of connection expects a full software implementation of network functions and complete knowledge of transducer functions and architecture on the client side.

The implementation of client–server style communication software is defined by two basic network block functions: *execute* and *perform*. The standard defines a unique ID for each class function and data item. In order to call some function on the server side, the client uses the command *execute* with the following parameters: ID of requested function; enumerated arguments; and requested variables. On the server side, this request is decoded and used by the function *perform*. That function evaluates the requested function with the given arguments and, in addition, returns the resulting values to the client. Those data are delivered by requested variables sets in *execute* arguments.

The subscriber–publisher style of communication employs IP multicasting. All clients wishing to receive messages from a group of transducers defined by a common IP multicast address of class D register to this group using IGMP. After that, when any of those transducers generates a message by block function *publish*, this message is delivered to all members of this class D group.

10.4.5 Detailed Interconnecting Architectures

The term *gateway* in this subsection refers to a software process that translates messages received from the Internet into requests for the specific control network (full gateway) or provides direct communication with Internet/Ethernet-coupled device (half gateway). Generally speaking, the full gateway translates between appropriate messaging protocols while respecting the complete protocol profiles of interconnected networks. On the contrary, the half gateway resides on the same network as interconnected nodes with common lower layers and translates only between different application protocols or initializes the subsequent direct communication announcing proper node addresses.

A computer physically joining one or more transducers and clients usually implements the full gateway. For Ethernet-coupled devices that gateway can also provide a data protection barrier if the application requires security and/or efficiency support, or an application-layer protocol converter if messaging protocols of the client and the transducer differ. That full gateway evaluates and filters messages or translates them. In this case, full gateway also provides a substitute of the transducer for Internet clients when the transducers are connected to a separate local network or protected subnetwork while enabling them to communicate to the outside network. For outside computers, this gateway represents a virtual transducer. In fact, it resends an incoming command to a concrete target in the local network and, similarly, resends the reply from the responding transducer to the relevant client. This solution allows the inside architecture of the protected local network to be hidden and, concurrently, allows transparent access to transducers. In this variant, the gateway does not modify messages, but only translates them between different messaging systems.

Internet half gateway provides an efficient interconnecting architecture for Ethernet-compatible devices. A Web server usually implements the half gateway. Similarly to full gateway, it can also provide a data protection barrier if the application requires security and/or efficiency support, or a protocol converter if messaging protocols of the client and the transducer differ. Nevertheless, the specific role of the half gateway consists in initialization of the subsequent direct communication between a group of clients and a group of transducers announcing relevant addresses, including a possible multicast address.

10.5 Industrial Network Interconnections

Contemporary industrial distributed computer-based systems encompass, at their lowest level, various digital actuator/sensor-controller connections. These connections usually constitute the bottom segments of hierarchical communication systems that typically include higher level fieldbus or Intranet backbones. Thus, the systems must comprise suitable interconnections of incident higher and lower fieldbus segments, with intermediate top-down commands and bottom-up responses. Interconnecting devices for such wide-spread fieldbuses as CAN, Profibus, or WorldFIP are currently commercially available; however, some real-world applications can also demand various couplers dedicated to special-purpose protocols or fitting particular operation requirements.

10.5.1 Interconnection Structures

This subsection focuses on the domain terminology. The first part collects some relevant notions aimed originally at wide- or local-area networks that offer a natural nomenclature background; the second introduces phraseology fitting ICNs' interconnections properly.

According to the ISO open systems interconnection vocabulary, two or more subnetworks are interconnected using equipment called an intermediate system whose primary function is to relay information from one subnetwork to another selectively and to perform protocol conversion when necessary (see Jain and Agrawala [11]). A bridge or a router provides the means for interconnecting two physically distinct networks, which differ occasionally in two or three lower layers respectively. The bridge converts frames with consistent addressing schemes at the data-link layer while the router deals with packets at the network layer. Lower layers of these intermediate systems are implemented according to the proper architectures of interconnected networks. When subnetworks differ in their higher layer protocols, especially in the application layer, or when the communication functions of the bottom three layers are not sufficient for coupling, the intermediate system, called in this case gateway, contains all layers of the networks involved and converts application messages between appropriate formats.

An intermediate system represents typically a node that belongs simultaneously to two or more interconnected networks. The backbone (sub)network interconnects more intermediate systems that enable access to different subnetworks. If two segments of a network are interconnected through another network, the technique called *tunneling* enables transfer of protocol data units of the end segments nested in the proper protocol data units of the interconnecting network.

The following taxonomy of ICN interconnections covers the network topology of an interconnected system and the structure of its intermediate system, often called in the industrial domain a *coupler* or *bus coupler*. On the other hand, the term gateway sometimes denotes an accessory connecting PC or PLC to an ICN. In this chapter, the expression preserves its original meaning according to ISO-OSI terminology as discussed earlier.

The first item to be classed appears at the level ordering of interconnected networks. A peer-to-peer structure occurs when two interconnected networks interchange commands and responses through a bus coupler in both directions so that no one of the ICNs can be distinguished as a higher level. If two interconnected ICNs arise hierarchically ordered, the master–slaves configuration appears usual, at least for the lower-level network.

The second classification point of view for couplers stems from the protocol profiles involved. In this case, the standard taxonomy using the general terminology mentioned earlier can be employed: bridge, router, and gateway. Also, the tunneling and backbone networks can be distinguished in a standard manner.

The next refining items to be classed include internal logical and physical architectures of the coupler, such as routing strategy (source or adaptive) and routing and relaying algorithms (more detailed specification), as well as the number of processors and the type of their connection (direct serial or parallel, indirect through FIFO queue or through dual-port RAM). In short, the following case studies employ the source routing strategy, which demonstrates a cheap and robust solution. Of course, the complete information about a coupler can be offered only by a detailed description of the concrete implementation. The next two subsections introduce basic information about two ICNs utilized in related case studies.

10.5.2 Actuator-Sensor-Interface Standard

The ASI defines the communication and pertinent management of a controlling device with digital sensors and actuators (see Kriesel and Madelung [12] and Section 10.2). A bus topology with tree-shaped physical structure interconnects one master station and a maximum of 31 slaves with up to 124 binary

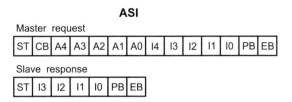

FIGURE 10.7 ASI frame formats.

actuators/sensors (a maximum of 4 binary units or 1 more complex digital unit per slave). The prescribed implementation provides a power supply through the bus and simple slave-side electronics (sensor with integrated ASI or separate ASI circuit with up to four standard sensors or actuators). An asynchronous transfer with polling mode communication at the rate of up to 167 kb/s supports 5-ms cycle time for the maximum configuration with 31 slaves.

As depicted in Figure 10.7, the request-frame format includes:

- One start bit ST
- One control bit CB (to discriminate control of internal circuits)
- Five address bits, A4…A0
- Five data bits, I4…I0 (I4 distinguishes data or parameter-control values; I3…I0 transmit a value)
- One even-parity bit PB
- One end bit EB

The response-frame format consists of one start bit ST; 4-b data I3 … I0; one even-parity bit PB; and one end bit EB. The error-detection scheme includes the following checks: power monitoring, bit coding, frame format, and parity. The master can initiate a recovery procedure by repeating the poll. Moreover, the ASI master, in the form of a card for PLC, PC, or gateway to a higher-level ICN, projects, initiates, manages, monitors, and commands the connected and active slaves in a dynamic fashion.

The ASI slave —typically an LSI circuit complemented by quartz crystal and four capacitors — carries out the communication with the ASI master and supplies the sensor or actuator with power. The ASI slave provides the connection between the ASI transmission system and the interface 1 to which the sensors and actuators are attached. Interface 1 consists of several connection points including four data input/output ports; four parameter output ports; one data strobe; and one parameter strobe. Interface 2, which joins the ASI slave to the transmission system, consists of two connection points: ASI+ and ASI–.

The ASI master yields the host interface, called interface 3, for connecting a controller, i.e., a PLC, PC, or bus coupler. Typically, the ASI master is a system board equipped with a system bus or an autonomous device with an EIA RS-232C/RS-485 serial interface. The host interface provides several functions that deliver/collect the actual user/application data, as well as set up and manage the ASI system's configuration. At the opposite side at interface 2, the ASI master is responsible for transmission control in the form of poll sequences on interface 2, accessing all the slaves.

10.5.3 Nine-Bit Interprocessor Protocol

The NBIP (9-b interprocessor protocol NBIP) [10] is a character-oriented data-link layer communication protocol for a master–slaves multidrop configuration with polling. The protocol makes full use of the so-called multiprocessor communication modes, which are based on 9-b characters. The NBIP communication procedure was designed by Intel's researchers to fit serial ports of the MCS-51 and MCS-96 families of microcontrollers interconnected by a serial bus; nevertheless, such communication modes are nowadays supported by a variety of microcontrollers and serial communication processors of miscellaneous producers.

The basic concept of the protocol can be briefly described as follows. When the master processor wants to transmit/receive a block of data to/from one of several slaves, it first sends out an address-control

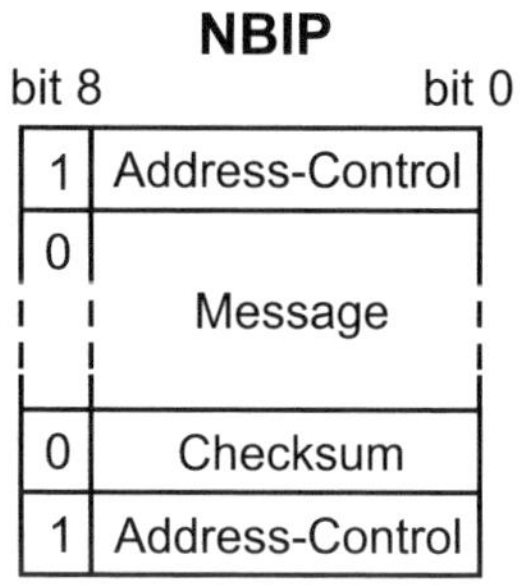

FIGURE 10.8 NBIP frame format.

character (see Figure 10.8). In this character, the ninth bit is set. The address-control character will interrupt all slaves so that each slave can examine the received byte and see if it is being addressed.

The addressed slave can find out from the control part of the byte whether the master wants to transmit or receive data. According to this information, the addressed slave changes its status to allow interruption by incoming characters, with the ninth bit cleared, or starts its own data transmission. Slaves that were not addressed leave their status unchanged, so they are not interrupted by the subsequent data bytes. The transmitting node closes the NBIP frame with a copy of the address-control character, preceded by a checksum. The NBIP definition includes also so called special functions, which are not used in the following case studies and therefore not discussed in this chapter.

10.5.3.1 Ethernet–Fieldbus Coupler: Tunneling Gateway

The first related case study deals with a regular conception of the coupler reusable for various higher level protocols, namely for Ethernet-based TCP/IP. This coupler interconnects the Ethernet backbone to a low-level fieldbus or sensor bus. This coupler is based on the tunneling conception: fieldbus messages are carried between a sensor, which is typically coupled to the fieldbus, and a client that usually resides on the Ethernet backbone, at this stage embedded in TCP/IP packets.

10.5.3.2 ASI–ASI Coupler: Fragmenting Gateway

The second case deals with an ASI–ASI coupler that enables realization of a two-level hierarchical ASI system. Such a configuration appears worthwhile when the relevant application requires more than 31 slaves and, in addition, when the employment of a higher fieldbus as the backbone interconnecting two or more ASI systems seems to be too costly. Because the ASI communication protocol does not offer a regular possibility to extend the addressing scheme, that capability must be embedded into the standard procedure of the application gateway. The technique applied can be denoted as fragmentation with multiplexing. In fact, this implementation, which can be interesting from a theoretical viewpoint, can be effectively replaced by an ASI master or an ASI gateway managing two ASI branches.

The ASI–ASI coupler can be designed as follows. Let two slaves engage in the data exchange between a backbone ASI and field ASI nets. To compact the ASI control, the address of the first slave is N, where N is an even number equal to or between 2 and 30, while the second slave's address is $N + 1$. Each of these two slaves provides four 1-b parameter outputs and four 1-b data inputs/outputs on interface 1. The total of eight output parameter pins and eight input/output data pins can submit the following information:

- Five output bits for a lower level net address
- Five output bits for a lower level net command
- Four lower level net input/output data bits
- One auxiliary output strobe bit
- One auxiliary input acknowledge bit

Introduction to Industrial Sensor Networking

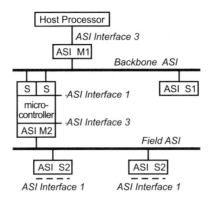

FIGURE 10.9 ASI–ASI interconnection.

This mapping fits a source routing scheme with the explicit addresses of the two gateway slaves at the backbone and with the destination field ASI address and data/parameter values carried by the data/parameter values of two subsequent backbone frames (see Figure 10.9).

In this case, the coupler consists of two backbone slaves and a microcontroller that behaves as a host processor for the field ASI net. Interface 1 defines the connection of each slave with the microcontroller. Two communication tasks of that microcontroller translate requests and responses between the backbone multiple-frame format, processed on two interfaces 1, and the standard field host message format, treated by the interface 3. The host message, composed of signal values of interface 1 and processed by field ASI master, results in a regular frame on the field ASI bus.

10.5.3.3 Bitbus–NBIP Coupler: Router

This case study presents a router conception fitting the low-level fieldbus domain. The interconnection can profit from the hierarchical model introduced in the bitbus definition [15]. In accordance with the bitbus specification, some of the nodes can be composed of two processors: device and extension. The device, which is incident with the bus, enables the extension to access the bus indirectly. The bitbus and NBIP protocols use a master–slave configuration with polling. The interconnected system shares a single global master, M1, that polls all slaves installed on both buses (see Figure 10.10).

The master M1 communicates with the slaves S2 of the NBIP bus through a coupler consisting of a bitbus slave device S1E2 and the NBIP master extension. Figure 10.11 introduces the bitbus message format, regarded also in the extended NBIP implementation.

The routing algorithm, based on an inserted network sublayer, can operate on flags that carry the following meanings: MT expresses the order/response type of message; SE represents the device/extension as the source of order message (and the destination of the response message); DE denotes the destination device/extension of order message; and TRK indicates delivery through the bus. The node address represents the address of the polled slave. Source and destination task fields identify tasks according to their roles with respect to the order message.

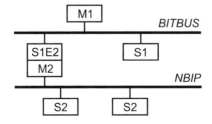

FIGURE 10.10 Bitbus-NBIP interconnection.

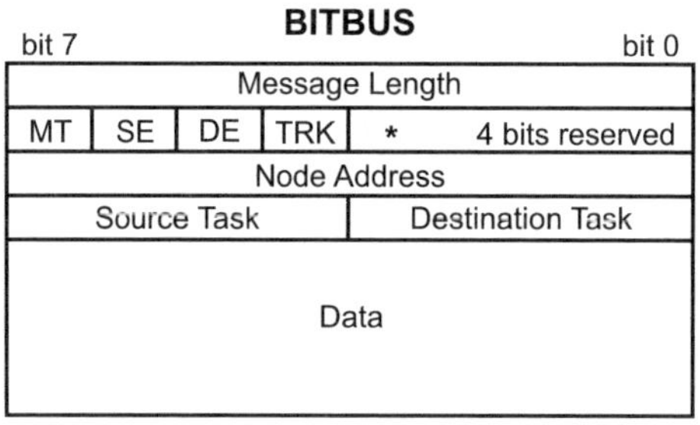

```
                    BITBUS
    bit 7                                    bit 0
    ┌─────────────────────────────────────────────┐
    │              Message Length                 │
    ├────┬────┬────┬─────┬───┬───────────────────┤
    │ MT │ SE │ DE │ TRK │ * │   4 bits reserved │
    ├────┴────┴────┴─────┴───┴───────────────────┤
    │              Node Address                   │
    ├──────────────────────┬──────────────────────┤
    │     Source Task      │   Destination Task   │
    ├──────────────────────┴──────────────────────┤
    │                                             │
    │                  Data                       │
    │                                             │
    └─────────────────────────────────────────────┘
```

FIGURE 10.11 Bitbus and NBIP message format.

The insertion of the network sublayer leads to the new format of message with the standard header containing items for the attached bus. The data for the other header, valid at the indirectly interconnected bus, are placed in the message body. The following software structure implements the routing. The source router task, which is inserted between the master source and master communication tasks in the order-message route, analyzes the original message header (see Figure 10.11). If the reserved field is not empty, it carries an address of the other bus. In this case, the source router generates new double-header items, according to Figure 10.12. The modified message passes through the bitbus and reaches the router task running on the S1E2 router processor. The router task exchanges the header items and passes the order message through the NBIP bus to the destination task. The response message goes through the inverted route, with the header items swapping in the router task.

The router implementation consists of the prototyping board connected to a standard bitbus controller board. The microcontroller 8031 of the prototyping board and the bitbus-enhanced microcontroller of the bitbus controller board communicate through their parallel ports and an FIFO emulated on the prototyping board or delivered by the bitbus controller.

10.5.3.4 Bitbus–NBIP Coupler: Bridge

The last case study describes a bridge configuration as another solution of the previous interconnection application (Figure 10.10) striving for shorter communication delay. The bridge interconnects networks on the data-link level. The frame addresses must be unambiguous in the whole interconnected system. For centralized polling, the configuration considers the global master initializing all communication transactions in the system.

The standard function of the bitbus communication task includes copying the slave-node address field of the message into the address field of the SDLC frame (Figure 10.13). Instead, at the global-master M1 communication task (see Figure 10.10), the simple copying is replaced by the routine that chooses for the frame the target node S1 address or the bridge S1E2 address instead of the target S2 address, according

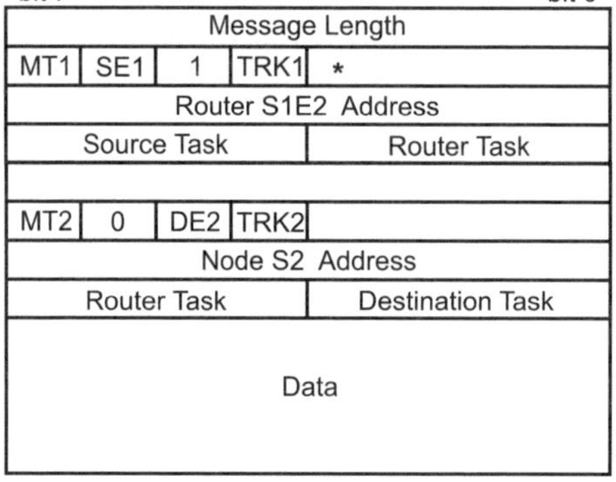

```
    bit 7                                      bit 0
    ┌─────────────────────────────────────────────┐
    │              Message Length                 │
    ├─────┬─────┬───┬──────┬──────────────────────┤
    │ MT1 │ SE1 │ 1 │ TRK1 │ *                    │
    ├─────┴─────┴───┴──────┴──────────────────────┤
    │          Router S1E2  Address               │
    ├──────────────────────┬──────────────────────┤
    │     Source Task      │     Router Task      │
    ├─────┬───┬─────┬──────┴──────────────────────┤
    │ MT2 │ 0 │ DE2 │ TRK2 │                      │
    ├─────┴───┴─────┴──────┴──────────────────────┤
    │           Node S2  Address                  │
    ├──────────────────────┬──────────────────────┤
    │     Router Task      │   Destination Task   │
    ├──────────────────────┴──────────────────────┤
    │                                             │
    │                  Data                       │
    │                                             │
    └─────────────────────────────────────────────┘
```

FIGURE 10.12 Enhanced bitbus and NBIP message format with routing information.

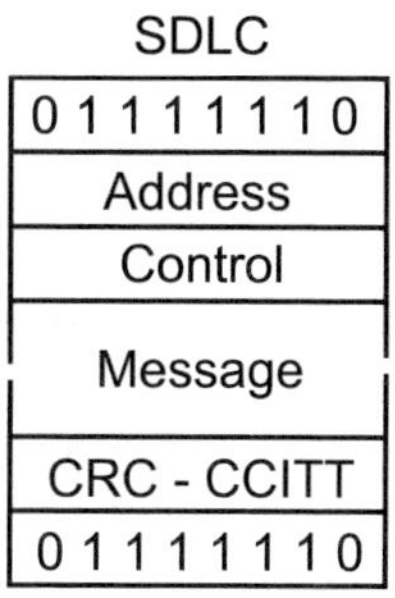

```
            SDLC
        ┌───────────┐
        │ 01111110  │
        ├───────────┤
        │  Address  │
        ├───────────┤
        │  Control  │
        ├───────────┤
        │  Message  │
        ├───────────┤
        │CRC - CCITT│
        ├───────────┤
        │ 01111110  │
        └───────────┘
```

FIGURE 10.13 SDLC frame format.

to the knowledge of address distribution along the two buses. The proper bridge now only copies the message slave-node address to the frame address field.

The implementation is similar to that in the previous case. The only difference consists in employing the 8744 microcontroller with modified software instead of the standard bitbus-enhanced microcontroller firmware. The software implementation of the bridge includes a subset of the standard bitbus and extended NBIP communication tasks, located at the bridge device and extension processors. This subset transfers only the contents of the first type of a frame to the other type across the parallel interface employing FIFO circuitry.

10.6 Wireless Sensor Networks in Industry

This recent development in communication reflects requirements for wireless systems. RF point-to-point communication evokes similar requirements in smart sensors. Wireless interfaces for smart sensor networking enable simple measurement data transmission from mobile robots and platforms, as well as from not easily accessible parts of processes or machines, e.g., rotating. The automobile industry represents another great accelerator of wireless sensor developments, and in the near future it may be the most important market for wireless sensors.

As mentioned in Aagaard et al. [17], one promising application can be a wireless version of an "electronic nose." Instead of sound and audible alarm, a wireless sensor can be mounted similarly to the way in which a dome light housing is mounted in the car. Without wire connections, the sensor would provide a central position in a car dome and send a wireless signal to a remote car receiver. Other interesting applications are in wireless on-body sensors for monitoring the health status of people with potentially debilitating conditions. The following case study covers some of the principal problems of mobile node wireless communication, namely, establishing, carrying out, and optimizing wireless messaging [16].

10.6.1 Problem Definition

For cooperative mobile robots, which, from the viewpoint of this chapter, are only moving platforms carrying wireless sensors or systems with sensors, communication is a necessity in many applications that enable them to carry out an assigned task. Each mobile node scenario has its characteristics with different communication requirements. How the communication problems should be handled depends widely on the kind of task the nodes will perform. For the communication design, the following characteristics of the robot application are relevant: topology changes; the robot group character; network traffic; and the traffic pattern. When frequent topology changes are expected, there will be high requirements on network maintenance to assure connectivity and on the routing protocol to be able to find alternate routes. The topology changes can maintain connectivity because of the mobility of robots or to optimize or adapt the structure for given needs. Requirements on the initial formation will not be so strict because the initial topology is not expected to last long. In scenarios with low rates of necessary topology changes, more demands can be made on the initial formation to prepare the network for later operation (as in Christensen and Overgaard [18]).

10.6.2 Topology

The topology is first formed in the initial phase, with some defined properties. How the robots are allocated in the environment during their mission can have significant effects on the network structure because some structures cannot be implemented in all formations (e.g., a ring). In *homogeneous groups*, all the robots are the same or have the same capabilities and requirements. *Heterogeneous groups* are composed of robots with different capabilities. The character of the robot formation implies the possible number of communication links. In *close formations*, the robots keep close together and most of them are within communication range of all others. *Loose formations* occur in situations when robots need to move far from each other (or in environments with many obstacles) so that few communication links are possible.

10.6.3 Network Traffic

Network traffic requirements can significantly affect the network structure. In different scenarios, various network traffic can be expected. Properties of the network traffic are important mostly for network optimizations. For example, *high bandwidth* data can be video data in a monitoring scenario. The data can be transferred regularly and the network should be able to reach a steady state by establishing new links where needed and canceling all unnecessary links, or the data are transferred suddenly, so the network does not have time for optimizing and must be constructed to be able to handle this kind of traffic.

Links should be available so that they can be used when needed. When *low bandwidth* data are regularly broadcast through the network — for example, periodic update messages — the main requirements are on the routing protocol to prevent unnecessary retransmissions of packets. For *a priori known* traffic, the network can be constructed to meet expected requirements. In practice, this means providing short paths between the nodes that will need to communicate; the nodes with expected high traffic loads should avoid participating in roles that carry additional communication overheads, such as a bridge. When the traffic pattern is *unknown*, the network should be able to adapt or be prepared for the worst-case situation. The listed aspects will often be combined in real applications or can change during the mission, and the communication protocol should be prepared for all of these possibilities.

10.6.4 Communication Maintenance

The communication for the mobile robots in this case study is based on Bluetooth technology. Maintaining the Bluetooth communication must primarily cover the following activities: initial network formation; routing; maintaining and optimizing the network structure; and the intra and interpiconet scheduling. The initial network formation covers the situation when a group of mobile robots is deployed and powered up. The robots have no communication links and require building up a network to enable communications as fast as possible. They have no *a priori* knowledge of their positions or any information about the others or knowledge that not all of the robots must be within communication range of all others.

The Bluetooth standard has other limits. Bluetooth specification does not define any routing mechanism. This is a task for a higher level routing protocol. The protocol should be able to handle the high mobility and changing topology and to provide a fault-tolerant message delivery. Structure of the Bluetooth network is based on the scatternet, but no specification of the topology or network formation is defined. For this purpose, a network-controlling algorithm deciding roles of the nodes and establishing the links needs to be proposed. The network-controlling algorithm must deal with mobility of the nodes and must handle link and node breakdowns. For communication inside a piconet, a polling scheme controlled by the master is used. The intrapiconet scheduling controls the polling scheme to maximize the piconet capacity. The bridging nodes between piconets can be active only one at a time; the interpiconet scheduling decides switching of the bridges between the piconets.

10.6.5 Network Routing

The routing is needed for intrapiconet routing maintained by the master to allow for communication between slaves and for interpiconet communication in the scatternet. Routing is closely related to the network structure. Some topologies have been designed to facilitate routing. This applies in particular to structures with dedicated topology (Bluetrees, BlueRings, clustered networks). However, the advantage of simple or even trivial routing is paid by increased complexity of the network structure. For the proposed network structure, a general and robust routing protocol needs to be designed. The main challenges for the routing represent mobility, causing route changes, packet losses, and potential network partitioning. All routing protocols require some kind of broadcast to discover routes and some way of storing the available route information.

Proactive routing maintains routing information for all of the nodes in the network. Protocols from this group evaluate all the routes within the network, so a route is ready immediately when a packet needs to be forwarded. The available routes are stored in tables maintained at each node. This solution requires keeping the tables consistent, which is maintained by broadcasting updates through the network. Destination-sequenced distance vector routing (DSDV) is a table-driven algorithm proposed by Perkins and Bhagwat [19]. It uses the Bellman–Ford algorithm improved to include freedom from loops. The routing loops are avoided by employing sequence numbers. Each node in the network stores routing tables listing all available destinations and number of hops to each. The number of hops is used as a metric. The proactive routing maintains routing information for all of the nodes in the network. Protocols from this group evaluate all the routes within the network, so a route is ready immediately when a packet needs to be forwarded. This solution requires keeping the routing tables consistent and is maintained by broadcasting updates through the network.

The main issue in mobile robot communication is to provide efficient and reliable message delivery. With a high rate of mobility, the DSDV can have problems with converging, which can lead to packet losses due to stalled route information. To avoid packet losses, an extension to the DSDV has been proposed that will use flooding to deliver a packet that cannot be forwarded. With the increasing mobility in the network, the routing will converge to flooding, which is the only possibility in extremely mobile situations [18].

The *update broadcast packets* (UBP) are the regularly broadcast packets. The routing protocol broadcasts the packets through the network; packets are identified through their sequence number (SN) and discarded when received again. The *directly addressed packets* (DAP) are packets sent from robot to robot. The packets carry control information for the task and network structuring. When a DAP packet cannot be delivered, it is marked as a *lost flooded packet* (LFP) and broadcast to reach its destination. The introduction of the LFP packets is an extension to the DSDV method. This kind of delivery is invoked for packets that cannot be delivered on known routes. Although the available routes would be refreshed after the next UBP broadcast, this packet would be lost. For this reason the packet is broadcast so that so it will reach its destination if it is possible.

The DSDV-based protocol introduced previously is trying to utilize the nature of messages in the given scenario. Regular broadcasts are expected as traffic background and a number of directed control messages. The directly addressed messages are not needed in the described exploration scenario because the robots do not require more information than the broadcast data. For more complicated tasks (e.g., situations in which not all of the robots have the same task), the number of directly addressed messages will increase. The extension to the DSDV with the LFP packets can handle highly mobile situations in which flooding becomes the only possible solution.

10.6.6 Network Topology

Network topologies and structuring algorithms designed particularly for Bluetooth include a large variety of different approaches and principles. The design of the topology can significantly affect network properties such as available capacity, transmission time, tolerance for link and node failures, and the frequency of necessary topology changes. However, the efficiency of a network topology differs depending on given requirements and circumstances; for some applications the network throughput is the most demanding, but for others, the reliability and robustness or the small rate of topological changes can be the most important. The proposed network structure should be built by a rule-based protocol to assure connectivity, minimize the number of piconets and the node degree, and reduce the transmission radius.

10.6.7 Network Structuring Protocol

The protocol for network structuring maintains the communication by controlling the network structure. It must cover problems of (1) initial network formation; (2) network maintenance, including node discovery, failures, and mobility; and (3) optimization of the network for needs of the given application. The most important rules that keep the network together are the *connectivity rules*, whose task is to assure that all the nodes are connected if possible. Connectivity rules must handle node mobility, link and node failures, and incoming and disappearing nodes. For basic applications these rules should be enough to maintain the network structure; with increasing mobility in the network, they can become the only rules applicable because no time will remain for optimizations.

The *structure optimizing rules* are defined to improve the network structure obtained by the connectivity rules to meet the application requirements. The requirements are given by general network structure requirements, as well as specific requirements such as traffic adaptation to enable for adapting the structure to the traffic patterns. Structuring rules are based on at least a minimal knowledge of network. Because maintaining global information about the network structure in mobile networks would be complicated and unreliable, it is proposed to use only local information for optimizations. This will cover: take-join principle; minimizing the number of piconets; avoiding multiple bridging; and reducing long links. The structure optimizations control the network constructed by the connectivity rules. The rules are designed to (1) avoid small piconets and thereby reduce the average number of piconets; (2) avoid unnecessary bridging; (3) reduce the degree of bridge nodes; and (4) reduce the connection distance by allowing slaves to select closer masters.

10.7 Conclusions

This chapter introduced major concepts utilized currently in the area of industrial sensor networking. The main focus was on proper communication protocols, network interfaces, and network interconnections dealing with common classes of ICNs in use for sensors interconnections. Industrial local area networks, fieldbuses, device buses, and sensor/actuator buses provide the platform for sensor-based industrial distributed system implementations aimed at various application domains. This chapter stems from case studies based on real projects developed by the authors enabling them to demonstrate some approaches typical in industrial application domains, such as communication systems; electrical drives; the textile industry; the chemical industry; and general machinery (see, for example, Sveda [13] and Sveda and Vrba [14]).

Acknowledgments

This work has been partly funded by the Ministry of Education of the Czech Republic in frame of the research intentions MSM 262200012: research in information and control systems; and MSM 262200022: MIKROSYT: research of microelectronic systems and technologies; by Grant Agency of the Czech Republic in frame of the projects GACR 102/02/1032: embedded control systems and their intercommunication; GACR 102/03/0619: IMAM — smart microsensors and microsystems for measurement, control and environment; and by industrial research project FD-K/104: SENSVISION — Internet access to process, supported by the Ministry of Education of the Czech Republic.

References

1. Frank, R., *Understanding Smart Sensors*, 2nd ed., Artech House Publishing, Boston, 2000.
2. Kriesel, W., *Bustechnologien fuer die Automation*, Huthing Verlag, Heidelberg, 1998.
3. Standard for a smart transducer interface for sensors and actuators — network-capable application processor (NCAP) information model, IEEE Std 1451.1-1999, IEEE, Piscataway, NJ, 1999.
4. Standard for a smart transducer interface for sensors and actuators — transducer to microprocessor communication protocols and transducer electronic data sheet (TEDS) formats, IEEE Std 1451.2-1997, IEEE, Piscataway, NJ, 1997.
5. Johnson, R.N. and Woods, S.P., Proposed enhancement to the IEEE 1451.2 standard for smart transducers, *Sensors*, 18, 74, 2001.
6. Johnson, R.N., Proposed IEEE standard P1451.0, 2003, http://www.telemonitor.com/doc/dot0vg.pdf.
7. IEEE 1451 homepage, 2003, http://ieee1451.nist.gov.
8. IEEE 1451.5 Web page at http://grouper.ieee.org/groups/1451/5.
9. Lee, K. et al., *Workshop on Wireless Sensing Proceedings*, Sensors Expo and Conference, IEEE Instrumentation and Measurement Society, Chicago, 2001.
10. Dhuse, J. and Hayek, G.R., Standard protocols are needed for distributed microcontrollers, *Data Commun.*, 15, 171, 1986.
11. Jain, J.N. and Agrawala, A.K., *Open Systems Interconnection: Its Architecture and Protocols*, Elsevier, Amsterdam, 1990.
12. Kriesel, W.R. and Madelung, O.W., *ASI: the Actuator–Sensor Interface for Automation*, Carl Hanser Verlag, Munich, 1995.
13. Sveda, M., Routers and bridges for small area network interconnection, *Computers Ind.*, 22, 25, 1993.
14. Sveda, M. and Vrba, R., Actuator–sensor interface interconnectivity, *Control Eng. Pract.*, 7, 95, 1999.
15. The BITBUS interconnect serial control bus specification, Order Number 280645-001, Intel Corp., Hillsboro, OR, 1988.
16. Hyncica, O., Autonomous mobile robot communication, diploma thesis, BUT FEEC, Brno University of Technology, Czech Republic, 2003.
17. Aagaard, M. et al., Experiments in task scheduling and distribution among Bluetooth-enabled robots, technical report, Aalborg University, Denmark, 2002.
18. Christensen, M.H. and Overgaard, E.M., Cooperative robots using Bluetooth, master thesis, Aalborg University, Denmark, 2000.
19. Perkins, C.E. and Bhagwat, P., Highly dynamic destination-sequenced distance vector routing (DSDV) for mobile computers, *Computer Commun. Rev.*, 24, 234, 1994.

11
A Sensor Network for Biological Data Acquisition[*]

Tara Small
Cornell University

Zygmunt J. Haas
Cornell University

Alejandro Purgue
Cornell Laboratory of Ornithology

Kurt Fristrup
Cornell Laboratory of Ornithology

11.1 Introduction 11-1
11.2 Tagging Whales 11-2
11.3 The Tag Sensors 11-3
11.4 The SWIM Networks 11-6
11.5 The Information Propagation Model 11-7
11.6 Simulating the Delay 11-9
11.7 Calculating Storage Requirements 11-12
  Single-Packet Storage Methods • Multiple-Packet Storage Methods
11.8 Conclusions 11-17

11.1 Introduction

Infostations offer geographically intermittent coverage at high data rates for mobile wireless networks. The Infostation model trades delay of data delivery for increased network capacity. Replication and storage of information in multiple nodes of a mobile network can also be traded for reduction in delay. Thus, augmenting the Infostation model with information replication, a new concept referred to here as the *Shared Wireless Infostation Model (SWIM)*, results in overall improved capacity–delay trade-off at the expense of modestly increased storage requirements.

In this chapter, SWIM is applied to solve a practical problem: information acquisition from radio-tagged whales; in particular, expected storage increase for the reduction in delay is calculated. Storage requirements can be further improved without affecting the delay by wisely erasing the replicated information from the network nodes. The performance of five storage/erasure techniques, which increase the computational complexity of the storage algorithm in order to further mitigate the storage increase, is studied. The results of this study will allow a network designer to implement such a system with a sufficient buffer size to ensure, with some level of confidence, that the information will be successfully carried through the mobile network.

---

[*]This work is based on an earlier work: The shared wireless Infostation model: a new ad hoc networking paradigm (or where there is a whale, there is a way), in *Proc. 4th ACM Int. Symp. Mobile Ad Hoc Networking Computing*, 233–244, June 2003, http://doi.acm.org/10.1145/778415.778443.

11.2 Tagging Whales

Large whales and marine mammals in general are keystone species in public interest and in assessing the environmental impacts of human activities. Eight species of large whales are on the endangered species list: blue whales, bowhead whales, finback whales, humpback whales, northern and southern right whales, sei whales, and sperm whales. Upon hearing noise from underwater tests, beluga whales will often flee the location at full speed for 2 to 3 days and not return to the site for weeks. Beaked whales have been stranded in association with naval exercises on several occasions. All of these species are difficult to study because of their enormous home ranges, the expense of oceanographic cruises, and the paucity of locations for fixed monitoring stations.

Wireless telemetry offers unequalled opportunities for monitoring the movements and behaviors of whales and other marine mammals. Whales are favorable subjects for radio telemetry because of their large size and their regular visits to the surface to breathe. Radio-tagged whales can provide a wealth of oceanographic information, along with data regarding their movements, because they collectively exploit a variety of resources across a wide range of oceanic habitats.

Implanting animals with miniature electronic sensing and transmitting tags provides unique opportunities to observe physiology, movements, and social behavior in a free-ranging context [1]. The addition of environmental sensors to animal tags provides the capacity to monitor ecological and oceanographic processes, which is an efficient method to monitor regions of biological interest that may be difficult to reach otherwise. Although some scenarios permit recovery of the tags, a much broader domain of applications requires implementation of a telemetry system to obtain the data from the tags, usually using radio frequency signals [2]. Designing radio tags confronts conflicting demands. Transmit power must be minimized to enable extended operations in a small form factor. On the other hand, the enormous home ranges of whales argue for substantial transmission power to maximize the distance over which the tag telemetry can be received.

The vast majority of today's radio tags are simple beacons that broadcast signals with frequency on the order of a second. To recover the data from the tags, animals are tracked with intensive operator effort by approaching and following the animal or by making coordinated measurements of bearings to the triangulate signals from two or more locations. The operator often measures the bearing by swinging a directional antenna through an arc and deciding on the direction that presents the strongest signal. This approach yields valuable data; however, it also suffers from severe limits on the number of animals that can be tracked and the area that can be monitored.

Alternatively, satellite radio tags have been also in use for some time, with the ARGOS system the primary provider of such a service.* ARGOS satellites orbit the Earth with an approximately 5000-km-diameter "footprint"; in most areas, they provide only a limited number of opportunities for offloading (recovering) data each day. Furthermore, each offloading is limited to a data packet of 256 b of data, and the system limits each tag to one message transmission every 45 to 200 s.

The seemingly irresolvable conflict between minimizing transmit power consumption and maximizing the area monitored can be successfully addressed by bringing the infrastructure for receiving the tag telemetry close to the tags, where "close" usually ranges from a few hundred meters to a few kilometers. Of course, bringing the infrastructure close to the free-roaming animals may not be a trivial matter when the size of the animals' habitats is taken into account. Many fixed receiving stations may be required, especially if the animals' movement patterns are not well specified or if it is unlikely that tagged individuals will pass close to a single receiver before exhausting the data storage capability or battery lifetime of their tags. Another option is to use mobile receiving systems, which systematically survey the animals' habitat. Data would be offloaded from each animal's tag when the receiving system reaches the vicinity of the animal. However, for large areas of habitat difficult to access (open ocean, tropical rainforest), the safety, expense, and logistical difficulties of sustaining regular surveys may be insurmountable.

---

*www.argosinc.com

Here, a different approach is advocated. In the authors' approach, the infrastructure is extended to the mobile nodes by the mobile nodes themselves; i.e., by creating a sensor network [3]. In other words, the information created in the network is replicated among the network nodes. More specifically, a piece of information is allowed to propagate among the mobile nodes in the network. When the two nodes come into communication contact due to their mobility, they exchange their stored information, saving a single copy of each packet on each whale tag. Then, when any one of the network nodes, which carries the information, reaches the vicinity of a collecting station, the information is offloaded.

To increase the probability that the information is recovered from the network, a number of collecting stations can be distributed throughout the habitat. Distribution of the collecting stations should be done in a way that maximizes the chances of information offloading.* Thus, only *one* replica of the information piece needs to reach only *one* collecting station to be successfully offloaded. Of course, this system might require each node to store and forward a substantial amount of data that originated from many other network nodes.

The idea of intermittent connectivity through a multiplicity of stations is not new; the Infostation model proposed by researchers at WINLAB** at Rutgers University offers a similar approach [4]. The novelty in the authors' design is the replication, storage, and propagation (i.e., diffusion) of the information within the Infostation environment. This system is essentially a marriage of the Infostation model with ad hoc networking technology [5]. Thus, this augmented Infostation approach is referred to as the shared wireless Infostation model (SWIM) [6]. SWIM allows delay reduction of the Infostation model, especially when the number of Infostations (SWIM stations) is relatively low.

SWIM tags and network communications protocols combine the best features of two existing marine mammal technologies: the small size and light weight of line-of-sight implantable radio tags with global coverage, similar to what the ARGOS satellite system can provide. SWIM tags exceed the capabilities of existing systems by enabling higher telemetry rates than satellite tags, with much lower power consumption and package size. The smaller package enables attachment to a wider range of organisms from greater distances. The tags are equipped with microprocessors and frequency-synthesized transmitters, so they can make measurements from a variety of sensors and implement sophisticated digital telemetry protocols; they are designed to collect sensor data continuously, and store summaries of these data in time-stamped packets for subsequent uploading to a receiving system.

Examples of desirable data regarding an animal's status are electrophysiological signals (cardiograms, myograms); body temperature; feeding activity; orientation; depth/altitude; and local movements (acceleration). Examples of desirable environmental data are ambient acoustic spectra; ambient temperature (and salinity in the ocean, humidity in the atmosphere); and light level. The value of these data increases when they are delivered relatively promptly because this enables adjustment of other observational schemes to take advantage of the unexpected opportunities or phenomena.

11.3 The Tag Sensors

The radio tag utilizes a Texas Instruments MSP430F149 microprocessor to enable field programmable operation and to schedule transmissions for power savings. The MSP430 processor provides 60 kbytes of flash memory, very low dormant power consumption (0.9 μA), an extremely small footprint, and a very low cost per unit. The MSP430 provides opportunities to monitor a variety of sensors. These include pressure sensors; light and temperature sensors; accelerometers; clinometers; microphones; and physiological electrodes. The sensor integration strategy must emphasize the following factors: minimal addition in size and weight; power shutoff capability; breadth of potential research applications; and ease of incorporating flexible logging and telemetry features in the tag software.

---

*For example, the collecting stations should be placed near areas that animals often frequent, such as water reservoirs.

**winwww.rutgers.edu/pub/docs/research/Infostations.html

Radio tags can be programmed in the field, which enables researchers to adapt the transmit schedule and operating frequency to local conditions. This embedded microprocessor is also the key to dramatic power savings. Scheduling transmissions to satisfy specific biological criteria can realize more efficient use of transmission power. Alternatively, tag sensor inputs (direct measures of activity) or an external signal (such as proximity to another whale or to a SWIM station) could be used to trigger transmissions. Scheduling is also relevant to the triggered systems because neither the sensor nor receiver systems require uninterrupted power. Accordingly, a flexible scheduling scheme is integral to the transmitter tags.

Scheduling requires accurate timekeeping; for this the tag uses a 32-kHz quartz crystal reference and achieves clock drift to less than 1 s per day. The researcher specifies the rate of regular timekeeping events and all systems are powered down between these events. The interval is defined by a 16-b integer that determines the number of 32-kHz oscillator cycles per timekeeping event (or "chronos"), but specified by the user in seconds. Thus, chronos periods range from 30.5 μs to 2 s. Total elapsed time is stored using three sixteen-bit words, where the least significant bit corresponds to a single chronos. Forty-eight bits provide a maximum tag endurance of 272 years with a resolution of 30.5 μs. A 32-b counter would impose a restrictive limit on tag endurance of just over 1.5 days at the fastest chronos rate. The cost (memory, processing time) for using three words is significant.

The scheduling algorithm is based on a repeating sequence of up to 256 tasks. The original 32-kHz clock signal is divided by a 16-b integer; the resultant clock signal constitutes the chronos that drives the timer/counter. The task list begins with a series of tasks that are executed once, followed by a series of tasks that are repeated. Each task is stored as a pair of 16-b counter values representing the durations of a pair of ON and OFF actions. These counter values are followed by a field with binary flags, specifying branching conditions, and a 16-b integer specifying the number of times to repeat the task before moving to the next task. Thus, each individual sequence of ON and OFF actions can be repeated up to 2^{16} times. Note that the ON or OFF actions can have zero duration, to enable a sequence of tasks to behave as an uninterrupted period of dormancy (or, less likely, activity) (Figure 11.2).

A continuous period of almost 4.3 billion event cycles (2^{16} counter $* 2^{16}$ repeat $= 2^{32}$ event cycles) can be scheduled with a single task (equivalent to 36.4 hours with 30.5 μs event cycles). Tasks are processed in sequence, until a task with a "repeat indefinitely" flag is encountered or the end of the task list is reached. If the end is reached, the task sequence restarts with the first task following the initialization sequence. Very complicated transmission schedules can be realized with this scheme and very rapid "schedules" can be used to implement pulse code identifiers (including Morse code).

The timekeeping system runs using interrupts, thus leaving the microprocessor in power-saving mode during the time between successive events. All microprocessor functions are implemented using interrupts, so the default state of the processor is dormant. This strategy results in approximately 64% lower power consumption than a constantly active microprocessor, with some variation in savings dependent on the precise mix of tasks.

The MSP430 software includes a simple monitor program, which manages communications with a notebook computer or PDA through a serial interface. The schedule is specified by a series of commands paired with the corresponding tasks. A host program running on a laptop (or PDA) enables the user to specify the timer tasks in an hs:mm:sec format using a simple script language (see Figure 11.1).

```
\event{\Ontime{03:30:00}\Offtime{4:45:10}
\flags{2}\repeatN{5}}
\event{\Ontime{00}\Offtime{10:5.301}
\flags{1}\repeatN{0}}
\event{\Ontime{0:0:0.010}\Offtime{0:0:59.990}
\flags{2}\repeatN{7}}
```

FIGURE 11.1 A short sample of the timer programming script language depicting three events. (From Small, T. and Haas, Z.J., *Proc. 4th ACM Int. Symp. Mobile Ad Hoc Networking Computing*, 233–234, Annapolis, 2003. With permission.)

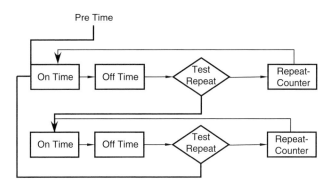

FIGURE 11.2 Timer algorithm flowchart. (From Small, T. and Haas, Z.J., *Proc. 4th ACM Int. Symp. Mobile Ad Hoc Networking Computing*, 233-234, Annapolis, 2003. With permission.)

Once the schedule has been uploaded, the timer schedule is stored in flash "eeprom" and a flag is set internally to indicate that a valid schedule is in memory. If the tag remains powered up, then the first scheduled task is executed immediately. When power is applied to the tag, the processor checks for a valid schedule and proceeds to the first task if a schedule is present. Otherwise, the processor goes into a low-power state and waits for scheduling information. The monitor and schedule execution software are stored in flash memory and can be updated from a personal computer through a serial link.

The 150-MHz tag (see Figure 11.3) uses the Silicon Labs Si4112 phase-locked loop (PLL) RF synthesizer to generate the RF signal. The PLL is controlled by the MSP430 to produce the operating frequency specified by the user. The frequency reference is provided by the microprocessor's 4-MHz oscillator. A single external inductor determines the band of frequencies that can be programmed in the field. With appropriate inductors, this part can generate frequencies between 62 MHz and 1 GHz. With a given inductor, the operating frequency is controlled by writing to registers that specify the operating frequency as a multiple of the 4-MHz clock. At the center frequency of 150 MHz, the microcontroller can tune from 147 to 153 MHz, with a resolution of 600 Hz. This allows the user to select the operating frequency at the time of deployment. The tag can also be programmed to produce ranges of frequencies in higher RF bands with this inductor, although operation in other bands would probably require changes to the matching network and antenna.

The microprocessor and PLL can generate CW or FM signals and thus implement pulse interval coding or frequency shift keying (FSK) telemetry protocols with no additional parts. For FSK, the settling time of the Si4112 (typically 40 µs) allows for modulation that supports data rates on the order of 25 kb per second. This is 250% of the typical voice telephony data rate, and about half the data rate of the fastest telephone modems. Output power is typically 0.4 mW into a 50 Ω load for the low power configuration and 20 mW into an 11 Ω load for the high power version. Current consumption for the tag will be 0.9 µA dormant, 2.5 µA processing, 8 mA transmitting for the low-power configuration, and 35 mA for the high-power configuration. Power to the Si4112 is cut when the tag is dormant.

The RF output of the PLL was boosted by a simple cascode RF amplifier to deliver a total of 20 mW of RF power. The prototype antenna for this system was a normal mode radial helix, to satisfy mechanical and hydrodynamic constraints while achieving desirable radiation efficiency. The radio tag is controlled by a microprocessor, based on parameter settings specified by the researcher at the time of deployment. The researcher specifies these operating parameters using a host program running on a laptop or PDA. This program accepts a simple text script containing the researcher's specifications, and it performs consistency checks and displays a visual summary of these settings to enable the researcher to scan for errors. This provides the capacity for last-minute changes in operating characteristics, including broadcast frequency.

Flexible support for complex sensor measurement and RF broadcast schedules is crucial to efficient use of battery power. Power will be supplied to all systems for very brief intervals as needed. In fact, the microprocessor will spend most of the time in a dormant state, with brief intervals of processing activity

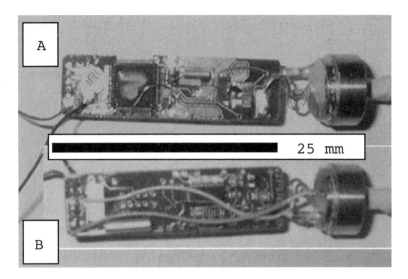

FIGURE 11.3 VHF-FM 148-MHz radio tag transmitter, showing serial port connectors and the base of the normal mode helical antenna mounted on a partially assembled Titanium housing. A: Microprocessor view; B: PLL side. Scale bar represents 25 mm. (From Small, T. and Haas, Z.J., *Proc. 4th ACM Int. Symp. Mobile Ad Hoc Networking Computing*, 233-234, Annapolis, 2003. With permission.)

triggered by a combination of regular interrupts and sensor readings. The processor runs on a 4-MHz clock to enable rapid processing of interrupts. All the timekeeping functions are based on a more power-efficient 32.768-kHz (32 kHz) clock (with the 4-MHz clock shut down between interrupts).

The electronics and battery are protected from the external environment by a custom-made housing machined from 100% grade 2 Titanium with a pressure rating in excess of 1700 m. This choice of materials serves two purposes: to prevent corrosion and to reduce tissue reaction during implantation to a minimum.

Here, the transmitting portions of the tags have been presented. To utilize this SWIM scheme, it is necessary to receive functionality on the tags as well. In order to implement this full transceiving capability, the RF chip needs to be replaced with one that is a transceiver, like the RF Monolithics TR1000 or the ChipCon CC1000. All of these chips are programmed through a serial connection, so the changes to the electronic circuit layout would be minimal. The commands that control the transmitter would need to be changed and physical and MAC layer protocols implemented. One approach would be to work with TinyOS, the operating system developed at U. C. Berkeley to support wireless sensor networks.

11.4 The SWIM Networks

In the Infostation model, users can connect to the network in the vicinity of ports (or Infostations), which are geographically distributed throughout the area of network coverage. The Infostation architecture includes low-power base stations,* which collectively provide strong radio signal reception in small and disjoint geographical areas and, as a result, offer very high rates to users in those areas. However, due to the lack of continuous coverage, this high data rate comes at the expense of providing intermittent connectivity only. Consequently, the Infostation network architecture should be used for applications that can tolerate significant delay because a node that wishes to transmit data may be located outside the Infostations' coverage areas for an extended period of time. Thus, the Infostation model trades delay for capacity by varying the degree of connectivity and by exploiting the mobility of the nodes.

Although significant delays can be tolerated in the whale tag application, if the delays are too long the data will likely be lost. The tags are foreign objects injected into the whales, and they are typically expelled

---

*The information collecting stations.

from the host's body within 3 to $3^1/_2$ months. Therefore, data retrieval must occur through transmissions from the tag while it remains attached to the whale. In the original Infostation model, a user was required physically to travel to the vicinity of an Infostation to communicate, which could lead to a significant delay in the whale tag application. Thus, to address the requirements of this application, the Shared Wireless Infostation Model has been developed as a more timely method for data retrieval. It is proposed to allow information to travel through the network by sharing (replicating, storing, and diffusing) itself as well, using the mobile nodes as physical carriers.

Clearly, allowing the packet to spread throughout the mobile nodes can significantly reduce the delay until one of the replicas reaches an Infostation. However, this comes at a price: spreading of the packets to other nodes consumes network capacity and storage space. Thus, again, the capacity–delay trade-off occurs. A new way to control this trade-off has been developed in which parameters of the spread are controlled — for example, by controlling the probability of packet transmission between two adjacent nodes, the transmission range of each node, or the number and distribution of the Infostations. In this chapter, the trade-off between the amount of storage required and the delay experienced in the system is examined.

First, methods to calculate delays of packets in the system are developed. Then we examine the increase in the required storage of the SWIM system, compared with the traditional Infostation model, for a particular reduction in delay. Because the delay in these systems is a random variable and is unbounded, a probabilistic metric is defined to describe the reduction in delay of the models.

Let P_{thresh} be some chosen threshold probability with which the packet will be offloaded (reach an Infostation) from the network. Compare the time necessary for the packet to be offloaded with probability P_{thresh} for the different network models. In general, the storage capacity necessary for the SWIM model is expected to increase, relative to the traditional Infostation model, because in the SWIM model packets are copied on many nodes; however, the time necessary to store packets (before they are offloaded with probability P_{thresh}) is also smaller. Therefore, the overall and relative storage requirements of the two schemes are subject of the study here.

The case in which the packets are shared between nodes with probability 1 each time two whales are "close" to each other represents the largest delay reduction and the highest increase in storage of the system. Sharing with probability 0 represents the pure Infostation architecture. Thus, by sharing packets with probabilities between 0 and 1, SWIM can achieve many different instantiations of the trade-off between network capacity and network delay.

11.5 The Information Propagation Model

Although this framework has a broad range of applications, the prime application addressed here is the support of biological data acquisition and animal tracking systems, such as the whale tags. Data collected on a whale tag like the tag shown in Figure 11.4 are stored locally. As a whale comes in close proximity to another whale, the stored information is transmitted, with some "probability of packet transmission," p, and is stored in the recipient whale tag's memory as well. As the whales migrate throughout the system, when a whale surfaces and comes in close contact with one of the SWIM stations, its tag offloads all the data in its memory (whether its own data or data from other whale tags) onto the SWIM station. Thus, as the whales feed and socialize near the surface of the water, the devices upload the packets of data at high data rates to the appropriately placed SWIM stations.

Typically, the SWIM stations are placed on buoys floating in the water. Because moving information from a whale tag to a SWIM station may be time-consuming, several SWIM stations are placed along the whales' paths. After receiving and storing the information from the whales, the SWIM stations transmit the information to shore, by coordination with other SWIM stations, or directly to a satellite, whenever the next satellite passes overhead. SWIM stations could alternatively be placed on seabirds high above the water. These stations would then be mobile and the data would be collected at known seabird roosting grounds.

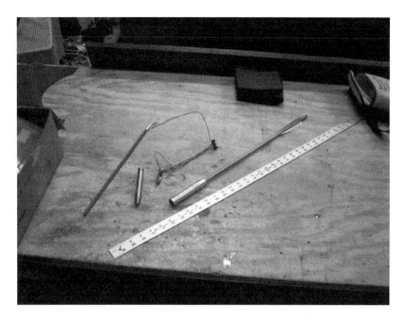

FIGURE 11.4 Whale tag prototype, to be delivered using a crossbow. (From Small, T. and Haas, Z.J., *Proc. 4th ACM Int. Symp. Mobile Ad Hoc Networking Computing,* 233-234, Annapolis, 2003. With permission.)

If a whale does not come in contact with any SWIM station for a long time, the tag may selectively discard the information in its memory when there is high probability that the information has already been offloaded to one of the SWIM stations by another whale tag. On the other hand, if a whale tag has been able to offload its stored information, its memory could be cleared immediately. The whale tag might also retain the identifier of the packet that it offloaded so that, in the future, it would not store (or even accept) information stored previously. These different methods of erasure of the stored packets will be addressed in more detail later in this chapter.

Delay experienced in the network (the time for a whale to reach a SWIM station) varies considerably depending on the mobility patterns of the whales, which are specific to the species of whales under consideration. One might expect daily surfacing near SWIM stations for humpback whales off the coast of the Hawaiian Islands, leading to delays on the order of hours. In contrast, some migratory whales visit known feeding grounds once a year, so delays may be on the order of months.

In order to study the delay of packets, the propagation of *each packet* of data information generated by a whale is modeled as the spread of *one* infectious disease. First, the propagation of a unique packet is considered. A whale tag is "infected" if it has the data packet stored in its memory. A whale tag is "susceptible" (to infection) if it does not yet have the packet stored in memory, but could potentially acquire the packet from another whale tag. A whale tag is "recovered" (healed from the disease) if it has offloaded the packet to a SWIM station. A packet is stored only once on each tag (one cannot be infected multiple times with the same disease); that is, by storing the unique identifiers of the previously received packets, a whale tag may become "immune" to receiving the same packet again. When modeling the sharing of the packet in this way, formulae from epidemiology can be used to find the probability that a packet is offloaded ("healed") as a function of the time it has spent in the system.

In Figure 11.5, the $S(t)$ represents the state of "susceptible" whale tags at time t; $I(t)$ represents the state of "infected" whale tags; and $R(t)$ represents the state of "recovered" whale tags. β is the average contact rate between two whales. Suppose that N whales are in the system and then a whale tag contacts $\beta(N-1)$ other whale tags per unit time, of which $\dfrac{S}{(N-1)}$ do not yet have the disease. Therefore, the transition rate from state S to state I becomes

FIGURE 11.5 Markov chain model of an "infectious disease" with susceptible, infected, and recovered states. (From Small, T. and Haas, Z.J., *Proc. 4th ACM Int. Symp. Mobile Ad Hoc Networking Computing*, 233-234, Annapolis, 2003. With permission.)

total infection rate

= (#infected) (contact rate) (#susceptible whales)

$$= I[\beta(N-1)]\left[\frac{S}{(N-1)}\right] = \beta SI$$

The recovery rate is labeled γ; it is the rate of contact between a whale and a SWIM station.

total recovery rate

= (whale – station contact rate) (# infected whales)

$$= \gamma I$$

Recall that if multiple SWIM stations are present, then γ represents the contact rate per station; e.g., γ will double if the number of SWIM stations is doubled.

Let T be a random variable representing the amount of time a packet has spent in the system, i.e., the time from packet creation until it is offloaded to a SWIM station. Once one packet reaches state R (meaning it has been offloaded), the rates will change, so the model is deemed invalid. Because only one packet in the model is considered, at time $t = 0$ only one whale tag carries the packet. All the N whales are in the state S or in the state I while the model is valid, so this means

$$S(0) = N - 1, \quad I(0) = 1, \quad S + I = N,$$

$$R(t) = 0 \text{ for } t < T \text{ and } R(T) = 1$$

By solving the differential equations defined by the rates of the Markov chain in Figure 11.5, it is possible to arrive at the cumulative distribution function $F(T)$, which represents the probability that the packet is offloaded after spending time T in the system. For example, if $F(300) = .5$, this means there is probability .5 that a packet is offloaded in 300 time steps or less. When the inverse of this function is used, a desired probability P_{thresh} can be chosen and the value T_p, for which $T_p = F^{-1}(P_{thresh})$, found. This means that with probability P_{thresh}, by time T_p, the packet will be offloaded. The formula for this function $F(t)$ is given by

$$F(t) = 1 - K\left(\frac{N-1}{e^{\beta Nt} + N - 1}\right)^{\frac{\gamma}{\beta}}, \text{ for K constant} \tag{11.1}$$

11.6 Simulating the Delay

Many possible mobility patterns exist for the whales. Each of these mobility patterns is represented in Equation 11.1 through the values of the contact parameters β and γ. A simple mobility pattern is random linear mobility. This pattern will be used to examine some common $F(T)$ properties. In the simulation, the whales swim in straight lines for a fixed number of time steps, s, with a randomly chosen velocity, and in a random direction. At the beginning of each group of s time steps, a new velocity and a new direction are chosen for the whales to swim in a rectangular area. The area has edges that wrap around, so a whale that swims off the right edge reenters at the left edge; similar wrap exists for the top and bottom edges.

At the beginning of the simulation, one whale carries the only replica of the packet. At every iteration, if a whale carrying a packet is within the infection range of another whale, the packet is replicated at the other whale. If any whale carrying the packet is within infection range of a SWIM station, then the simulation is stopped and the time, T, is recorded from the creation of the packet until the termination of the simulation. The simulation has been run multiple times, and the data have been compiled, representing an empirical probability function, $F(T)$. As one would expect, the $F(T)$ curves are steeper (representing shorter delay) as the number of whales increases and as the number of SWIM stations increases.

Figure 11.6 shows the empirical $F(T)$ curves with different numbers of SWIM stations, $M = 1, 2, 3, 4$. In this example, swimming speeds of the whales were chosen from 0 to 6 units per time step on a 300 × 300 toroidal area, and the reception radius of each station was 15 units. The curves are also steeper, due to increased sharing, as the number of whales increases. In order to validate the empirical $F(T)$, the corresponding theoretical $F(T)$ was found using the simulation to find β and γ. Through the use of the χ^2 goodness-of-fit test, very good agreement between the theoretical and empirical solutions was observed.

A more realistic mobility model captures the physical whale behavior by incorporating feeding grounds. In this enhanced model, three issues govern the direction of the whales' positions at any time: migration in a specified direction, grouping of whales, and direction of the nearest feeding ground. Females tend to group together with other females, while grown males tend to be more solitary in their behavior and group with females, but not with other males. All whales are attracted to feeding grounds when they are hungry. Inside the feeding grounds, whales move slowly and sometimes stop. When a whale becomes less hungry, it can leave the grounds for a significant time before returning. Direction for the whales' mobility is determined by a weighted vector sum of the direction of migration, the direction to the nearest female, and the direction to the nearest feeding area.

Because the whales are attracted to the centers of the feeding grounds, they are likely to swim close enough to a SWIM station inside the feeding grounds to offload their packet. Thus, when SWIM stations are placed inside the feeding grounds, delays can be significantly reduced. This is shown in Figure 11.7

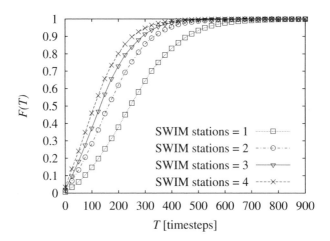

FIGURE 11.6 Probability functions of T, the time from packet creation until offloading, for different numbers of SWIM stations in the system. (From Small, T. and Haas, Z.J., *Proc. 4th ACM Int. Symp. Mobile Ad Hoc Networking Computing*, 233-234, Annapolis, 2003. With permission.)

A Sensor Network for Biological Data Acquisition

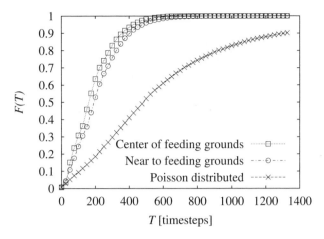

FIGURE 11.7 Effect of different SWIM station arrangements on the cumulative distribution $F(T)$. (From Small, T. and Haas, Z.J., *Proc. 4th ACM Int. Symp. Mobile Ad Hoc Networking Computing,* 233-234, Annapolis, 2003. With permission.)

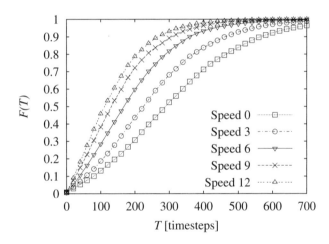

FIGURE 11.8 Cumulative distribution curves with varying speeds of the mobile SWIM stations. (From Small, T. and Haas, Z.J., *Proc. 4th ACM Int. Symp. Mobile Ad Hoc Networking Computing,* 233-234, Annapolis, 2003. With permission.)

by the "center of feeding grounds" and "near to feeding grounds" curves. If the SWIM stations are sometimes placed outside the feeding grounds, delays increase because the whales are attracted to regions far from the SWIM stations, as shown by the Poisson distributed curve in Figure 11.7. Obviously, location of the SWIM stations is a very significant parameter. The grouping of whales can also significantly affect the delay because more grouping promotes more packet sharing.

Up to this point, it has been assumed that the collection points (i.e., the SWIM stations) are fixed in their locations. Another possible model for the biological information acquisition system considers *mobile* collection points as well as mobile nodes — for example, SWIM stations mounted on seabirds that glide above the ocean along the turbulent air above the waves. Figure 11.8 shows that increasing the speed of the mobile SWIM stations has a positive effect on the packet offload time when both whales and SWIM stations use random linear mobility. Larger speeds of the SWIM stations allow them to pass through groups of whales more often, although they stay near the groups for shorter periods each time. This larger frequency of visiting allows the packets to be offloaded more often and at more regular intervals.

11.7 Calculating Storage Requirements

Equipped with these $F(T)$ curves, the information about the contact rate between the whales, and the contact rate between the whales and the SWIM stations, the expected storage requirement for all the copies of one packet can be calculated, given a desired confidence level of the packet delivery, P_{thresh}. As an example, suppose that the designer specified a confidence level of .9, then $T_{.9} = F^{-1}(.9)$ is the time necessary to wait to achieve the probability of .9 of packet offloading. This is the "expiration time" of the packet and its replicas. If any replica of the packet remains in the system for this maximum delay, it is erased, even if it has not yet been offloaded.

A quick, though naïve, approach to calculating the required storage for the system involves the average number of packets in the system at the time of offloading and applying Little's formula. Suppose that 10 adult whales are tagged and, at each time step, placed randomly* in an area of 900 km² with 1 SWIM station. The transmitting range of the radio tags is 1.4 km and reception range of the stations is 3 km. This can be modeled as a system with $N = 10$ whales, $M = 1$ SWIM station, and the probability of transmission $p = 1$. From the corresponding $F(T)$ curve, $F^{-1}(.9) \approx 78$. The "expiration" time of the packets is therefore 78 time steps.

Now suppose that each whale generates a packet every 30 time steps. Using Little's formula with generation rate $\lambda = 1/30$ time steps per packet per whale, the expected number of all the packet replicas in the system is:

$$EP = (\text{number of whales}) \lambda T_{0.9}$$

$$= 10 \frac{1}{30} [\text{packets/time} - \text{step}] (78[\text{time} - \text{steps}])$$

$$= 26 [\text{packets}]$$

An estimate of the expected number of copies of each packet in the system, EI, is the average number of whales infected with the packet at the time of offloading. It can be shown from the simulation that $EI = 2.5523$ in this case. This number assists in calculating a global storage requirement for the radio devices:

$$\text{storage requirement}$$

$$= (\text{duplicates})(\text{different packets})(\text{bytes/packet})$$

$$= EI * EP * (330 \text{ bytes/packet})$$

$$= (2.5523) * (26 \text{ packets}) * (330 \text{ bytes/packet})$$

$$= 21898 \text{ bytes}$$

$$= 21.4 \text{ kB} = 2.14 \text{ kB/whale}$$

Recall that, in this example, the probability of sharing packets between close-by whales is 1, so the results correspond to the largest delay decrease and the largest storage requirement of the SWIM model. Figure 11.9 shows that the increase in storage is very reasonable for the achieved large decrease in delay. The advantage of SWIM is even more pronounced as the number of whales increases.

In the nonsharing case, the per-whale storage requirement remains the same as the number of whales increases; however, the storage requirement in the SWIM case grows slightly due to the replication of packets. If the density of whales is larger, then the packet is shared with more whales and more storage

---

*With a Poisson distribution

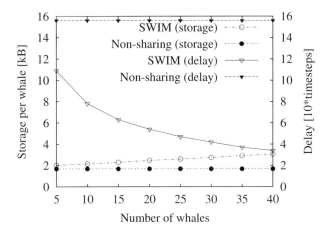

FIGURE 11.9 Necessary storage requirements and expected delays using SWIM vs. the nonsharing model. (From Small, T. and Haas, Z.J., *Proc. 4th ACM Int. Symp. Mobile Ad Hoc Networking Computing*, 233-234, Annapolis, 2003. With permission.)

is required. This increased storage requirement is mitigated by the fact that the delay is reduced; i.e., in SWIM, more packet copies are in the network, but they remain for a shorter time.

The expected delay for the nonsharing system is constant over the different numbers of whales because more whales in the system offer no advantage in this case. In other words, every whale must reach the SWIM station for the whale tag to offload its packets. On the other hand, SWIM replicates packets among the network nodes, so if there is a larger density of whales, more copies of a packet will be present in the network. Thus, SWIM achieves smaller delays as the number of whales increases.

In practice, one would want to include an extra safety factor in the memory calculation to protect against statistical variability in the number of packets stored in a tag.[*] This safety factor is not included in the simple approach presented in this section. However, storage evaluation in the more precise calculations of storage requirements will be reexamined in the following subsections.

11.7.1 Single-Packet Storage Methods

Numerous methods can be used to model packet generation, storage, and erasure. Here, five possible methods are considered: JUST_TTL; FULL_ERASE; IMMUNE; IMMUNE_TX; and VACCINE. These methods progressively extend one another. In all of them, the original packet and all of its copies are erased $T_p = F^{-1}(P_{thresh})$ time steps after the original packet was created.

- JUST_TTL is the simplest method. All packets remain in the system until $T_p = F^{-1}(P_{thresh})$ time steps have elapsed from the original packet creation.
- FULL_ERASE erases the copy of the packet completely from the offloading node just after it has been offloaded to a SWIM station. Once a copy of the packet has reached a SWIM station, there is no need for it to be stored on any of the whale tags. It is, however, possible that other whale tags still carry the packet once it has been erased from the offloading whale tag, so a whale tag might get infected with the same packet multiple times.
- IMMUNE erases the packet when it is offloaded like FULL_ERASE does, but keeps an identifier of the offloaded packet so that the whale tag will not receive that packet again. This identifier is referred to as an "antipacket"[**] because it prevents reinfection of packets.

---

[*] Not including this factor would assume that the loss due to buffer overflow is negligible.
[**] Similar to *antibody* of a biological agent.

- IMMUNE_TX erases the packet when it is offloaded, keeping the antipacket, like IMMUNE does. It also shares this antipacket with other whale tags that carry copies of the offloaded packet. This means a whale tag may receive an identifier antipacket from a transmitting whale tag only if a copy of the offloaded packet is already stored. At that point, the copy would be erased and the antipacket identifier kept.
- VACCINE erases the packet when it is offloaded, like previous methods do. It also shares all packets and antipackets between whale tags. In this case, a whale tag may receive an antipacket from a transmitting whale tag even if the receiving whale tag does not have a copy of the packet stored.

The average number of copies of the packet in the system can be calculated for a given time using each of these five methods. Call this average number $EI(T)$. Because the $F(T)$ curve expresses the desired confidence level as a function of time, it is possible to plot parametrically the average storage requirement as a function of the desired confidence level, $(F(T), EI(T))$, as shown in Figure 11.10. Notice that, for methods IMMUNE, IMMUNE_TX, and VACCINE, the average storage begins to decrease at a high confidence level due to the confidence level's dependence on T. As the confidence level approaches 1, the necessary time T for the packet to remain in the system becomes higher and higher; eventually, $T \to \infty$ as $F(T) \to 1$. In the methods IMMUNE, IMMUNE_TX, and VACCINE, the packet identifiers prohibit the whale tags from storing a copy of the packet again; thus, eventually, as T gets large, nearly all the whale tags refuse storage of the packet, and the average required storage is thereby reduced.

The storage–delay trade-off can also be depicted using SWIM. The desired $P_{thresh} = .9$ is fixed; then, to reduce the delay, the sharing of the packets is increased by increasing the density of whale tags in the system. Figure 11.11 exhibits the storage–delay trade-off due to this increased sharing; clearly, to achieve shorter delay, one must invest more storage in the system.

11.7.2 Multiple-Packet Storage Methods

For each of these five methods, an average time of the number of replicas of a packet in the system, EI, can be obtained. With the help of the Little's formula, the mean storage requirement per whale is:

$$EI * \lambda * T_p * (330 \text{ bytes/packet})$$

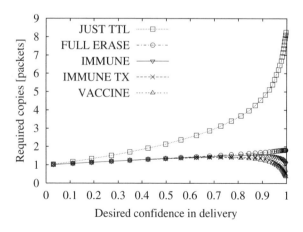

FIGURE 11.10 Expected storage required for 10 whales, assuming 4-byte identifier and packet contents of 326 bytes. (From Small, T. and Haas, Z.J., *Proc. 4th ACM Int. Symp. Mobile Ad Hoc Networking Computing*, 233-234, Annapolis, 2003. With permission.)

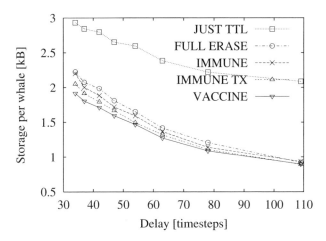

FIGURE 11.11 Storage–delay trade-off of SWIM using the different methods of erasure. (From Small, T. and Haas, Z.J., *Proc. 4th ACM Int. Symp. Mobile Ad Hoc Networking Computing*, 233-234, Annapolis, 2003. With permission.)

This does not, however, provide an indication of the variance of the number of packets stored on each whale tag. The variance is important to calculate the "safety factor" in evaluation of the necessary buffer size so as to ensure that, due to the statistical behavior of the packet arrival process at the nodes, at most only a small fraction of the packets would be lost.

In order to learn more about the probability distribution of the number of packets on each whale tag, Q_i, the system is modeled as an imaginary global queue that, at each point in time, contains all the packets present in the system. In particular, let $Q = \sum_{i=1}^{N} Q_i$ represent the number of all the copies of all the different packets in the system; i.e., the number of packets in the global queue.

However, due to the complex nature of the global queue, an approximation is employed; it is assumed that the arrival of all the copies of a packet to the global queue occurs at the time of the original packet creation, rather than when the packet is replicated from one whale to another. This is a conservative approximation for the purpose of evaluation of the variance of Q_i; in reality the arrival of the copies of a packet will be spread in time, reducing the variance due to the aggregation of such arrival processes of many other packets. It is further assumed that the number of replicas of the packet to arrive at the global queue is equal to the maximum number of the packet copies that will ever be present in the system. For the JUST_TTL case, packets are replicated when they are shared between whales, but are never removed from the system. Thus, at time $T_p = F^{-1}(P_{thresh})$, the number of copies of a particular packet in the system will be a maximum. Using other methods, the maximum number of packets may occur at a value smaller than T_p.

This global queue, Q, can be simulated. The simulation generates packets periodically for every whale in the system, given the set of periods and their offsets in time. Let $I(t_{max})$ be a random variable representing the distribution of the number of packets in the system when the expected number of packets in the system is a maximum. When a new packet is generated, the maximum number of its copies that will ever be present in the network is drawn from the distribution $I(t_{max})$. Those copies are then added to the global queue, Q, as soon as they are generated and they are removed after time T_p. After the simulation ends, the sample mean and sample variance of the number of packets in the global queue at steady state are calculated.

The global queue can also be solved analytically. When the number of whales is moderately large and the arrival processes of new packets at different whales have slightly different periods, the arrivals of groups of packets act like a Poisson queue with batch arrivals. The system is said to have infinitely many servers because all the packets are "served" at the same time. A Poisson queue with batch arrivals involves

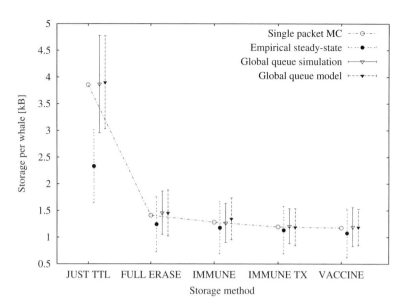

FIGURE 11.12 Mean number of packets with error bars indicating one standard deviation for a 10-whale system. (From Small, T. and Haas, Z.J., *Proc. 4th ACM Int. Symp. Mobile Ad Hoc Networking Computing*, 233-234, Annapolis, 2003. With permission.)

groups of customers that reach servers with *i.i.d.* exponentially distributed interarrival times. The numbers of customers in these groups is determined by the distribution function $I(t_{max})$. The service times in this case are deterministic, meaning that any customer leaves the system after a constant time T_p. Finally, because infinitely many servers are available, customers never need to wait in the global queue; i.e., the only delay is due to the deterministic service time.

By using the global queue with deterministic service times described previously and by assuming that all of the whale tags are *i.i.d* with respect to the number of packets they carry, one can simply divide the global queue by the number of whales to find the distribution of the number of packets on each individual tag. This provides not only the mean number of packets, which is already known from the single-packet methods, but also any quantile that the designer wishes to use in order to provide the "safety factor" in packet buffering.

Figure 11.12 compares the numbers of packets in the 10-whale system for four different metrics, exemplifying the methods described earlier, using a random mobility pattern for 10 whales. The first metric uses the single packet Markov chain to find $EI(t_{max})$ and uses the Little's formula N ∗ $EI(t_{max})$ ∗ λ ∗ T_p ∗ (330 bytes/packet) described earlier to give a conservative estimate of the mean number of packets in the system at a given time. This method does not provide error bars because the variation in numbers of different packets on a tag cannot be measured.

The second metric is the average number of packets in the system measured in the multiple packet whale simulation at steady state. This is an empirical measurement of the actual number of packets in the system, rather than the conservative estimate used in the other methods. For this reason, the curve of the second metric is lower than the other curves. The third metric uses the simulation of the global queue with batch arrivals with distribution $I(t_{max})$, and the fourth metric calculates the probabilities analytically. These metrics all provide error bars for the variance because they supply the entire distribution of the number of packets stored in the system.

As shown in Figure 11.12, the single packet Markov chain gives a reasonably conservative estimate of the packets in the system. Adding one standard deviation of the number of packets in the queue to the mean ensures even less packet loss in the system. Using this estimate, the storage requirement per whale for JUST_TTL is 4.77 kB; for FULL_ERASE, it is 1.89 kB; for IMMUNE, it is 1.73 kB; for IMMUNE_TX, it is 1.54 kB; and, for VACCINE, it is 1.53 kB. Compare these values to the average requirement of 2.14

kB per whale calculated at the beginning of this section. One can conclude that even accounting for variability in the tag queues, the storage requirements remain reasonable for practical implementation.

11.8 Conclusions

SWIM, an augmented Infostation model, has been proposed and applied as an efficient method to solve the problem of data retrieval from animal tags. In this model, users disseminate information packets throughout the system, sharing them with other users, and because only one of the replicas needs to reach a collection point, the overall delay in offloading the data is reduced. Using a probabilistic metric for the delay and comparing a system with 5 whales to one with 40 whales — each with packet generation every 30 time steps — the delay could be reduced by 320% for a 50% increase in storage compared to traditional Infostation networks.

This chapter has shown a number of methods for storing and erasing packets, using single- and multiple-packet models. By using the single-packet model to find the mean storage per whale and the multiple-packet model to find the variance, one can efficiently design a system with reasonably sized storage requirements at each node and low packet loss. This model is well suited to design and evaluate moderately delay-tolerant applications such as the preceding biological information acquisition system; however, the same methodology can also be used to model and evaluate other systems that use this augmented Infostation model.

References

1. H. Wang, J. Elson, L. Girod, D. Estrin, and K. Yao, Target classification and localization in habitat monitoring, *IEEE Int. Conf. Acoustics, Speech, Signal Process. (ICASSP 2003)*, pp. IV-844–IV-847, Hong Kong, April 2003.
2. I.G. Priede, Wildlife telemetry: an introduction, in *Wildlife Telemetry: Remote Monitoring and Tracking of Animals*, E. Horwood, I.G. Priede, and S.M. Swift, Eds., Chichester, UK, 1992, 3–25.
3. I.F. Akyildiz, W. Su, Y. Sankarasubramaniam, and E. Cayirci, Wireless sensor networks: a survey, *Computer Networks*, 38(4), 393–442, March 2002.
4. A. Iacono and C. Rose, Infostations: new perspectives on wireless data networks, WINLAB technical document, Rutgers University, 2000.
5. Z.J. Haas, J. Deng, P. Papadimitratos, and S. Sajama, Wireless ad hoc networks, in *Encyclopedia of Telecommunications*, J. Proakis, Ed., John Wiley & Sons, New York, 2002.
6. T. Small and Z.J. Haas, The shared wireless Infostation model — a new ad hoc networking paradigm (or where there is a whale, there is a way), *ACM MOBIHOC'03*, Annapolis, pp. 233–244, MD, June 2003.
7. L. Kleinrock, *Queueing Systems Volume I: Theory*, John Wiley & Sons, Inc., New York, 1975.

III

Architecture

12
Sensor Network Architecture

Jessica Feng
University of California at Los Angeles

Farinaz Koushanfar
University of California at Berkeley

Miodrag Potkonjak
University of California at Los Angeles

12.1	Overview	12-1
12.2	Motivation and Objectives	12-1
12.3	SNs — Global View and Requirements.........................	12-3
12.4	Individual Components of SN Nodes	12-4
	Processor • Storage • Power Supply • Sensors • Radio	
12.5	Sensor Network Node...	12-8
	Berkeley Mote Node • UCLA Medusa MK-2 Node • BWRC Piconode • Sensor-Centric Design: Light Compass	
12.6	Wireless SNs as Embedded Systems	12-13
12.7	Summary...	12-16

12.1 Overview

Emergence of the concept of multihop ad hoc wireless networks, low-power electronics, low-power, short-range wireless communication radios, and intelligent sensors is considered the major technological enabler for deployment of sensor networks (SNs). The goal in this survey is to identify key architectural and design issues related to sensor networks, critically evaluate the proposed solutions, and outline the most challenging research directions. The evaluation has three levels of abstraction:

- Individual components on SN nodes (processor, communication, storage, sensors and/or actuators, and power supply)
- Node level
- Distributed networked system level

Special emphasis is placed on architecture, system software, to some extent, and new challenges related to using new types of components in networked systems. The evaluation is guided by anticipated technology trends and current and future applications. The main conclusion of the analysis is that the architectural and synthesis emphasis will be shifted from computation and, to some extent communication, components to sensors, actuators, and different types of sensors and applications that require distinctly different architectures at all three levels of abstraction.

12.2 Motivation and Objectives

Embedded wireless SNs are systems consisting of a large number of nodes, each equipped with a certain amount of computational, communication, storage, sensing, and actuation resources [20]. SNs aim to provide efficient and effective connection between physical and computational worlds and are also widely considered the new big frontier for the Internet. Furthermore, they have high potential economic impact in many fields, including military, education, monitoring, retail, and science. At the same time, SNs pose numerous new research and development challenges, including the need for the next generation of low power; low cost; small size; error and fault resiliency; flexibility; conceptually new security and privacy; and a need for new types of input/output (I/O) operations.

However, before any of these challenges can be properly addressed, one must have the sensor network in place; the network must be designed and implemented and the need for flexible mechanisms and means for efficient and convenient use must be realized. In addition to algorithms, hardware and software architecture will decide to a significant extent the effectiveness of SNs. Furthermore, SN design methodology will have primary impact on the cost and performance of SNs. The third aspect with major potential impact — algorithms and modeling techniques for SN — is mainly out of the scope of this survey. Comprehensive surveys on SNs include Estrin et al. [20], Pottie and Kaiser [50], and Akyildiz et al. [2].

The overall strategic goal is to summarize current state of the art with respect to architecture and synthesis techniques for SNs and to provide a starting point and impetus for research and development of new architectures and synthesis tools for SNs. More specifically, the emphases are on:

- *Identifying requirements for typical SN application.* Traditionally, design of new computer architectures has been based on comprehensive and representative benchmark suites for typical target applications. It is of exceptional importance to create such benchmarks for sensor networks. In addition, it is important to predict the nature of future SN applications. However, even before the benchmarks are available, qualitative analysis of representative application can greatly facilitate identification of more accurate design goals.
- *Identifying relevant technological trends.* It is well known that many electronics and optical systems follow exponential performance growth rates. SN systems are heterogeneous and complex; therefore, it is important to anticipate which design and cost bottlenecks are intrinsic and which will be resolved due to technological progress. Importance of technological trends is well illustrated during power optimization. Depending on future ratios of computation, communication, and storage cost, very different types of algorithms will be best suited for SNs.
- *Balanced design.* In order to achieve a balanced design, the first instinct could be to optimize each and every component to the maximum extent. From a research and economic point of view, it is important to identify where to put the main optimization effort. In addition, new computational models are needed, but one must keep in mind that they are not the ultimate goal per se.
- *Techniques for design and the use of the design components.* The six components of SN node can be grouped in two categories according to their maturity. Power supplies, and in particular storage and power supply, are considered mature technologies. On the other hand, ultralow power wireless communication, sensors, and actuators are technologies waiting for major technological revolutions. It is important to identify which techniques, architectures, and tools can be reused, and where the new design effort is required.
- *Overall node architecture and trade-offs.* One can envision a number of possible trade-offs. For example, the TinyOS approach [27] advocates aggressive communication strategy in order to reduce complexity of computation and storage at local sensor nodes. On the other hand, the sensor-centered approach [22] advocates aggressive sensor data processing, filtering, and compression in order to reduce communication.
- *Survey of state-of-the-art technology, components, and sensor network nodes.* Special emphasis is placed on providing qualitative and quantitative analysis. In addition, several state-of-the-art sensor nodes are surveyed and decisions that influenced their structure are critically evaluated.

12.3 SNs — Global View and Requirements

It is well known that characteristics of computing or communication systems are direct consequences of targeted applications. A number of characteristics of sensor networks that have direct impact on architectural and design decisions have been identified. These characteristics rise naturally from a confluence of typical application requirements and technology limitations. Typical SN applications include contaminant transport monitoring; marine microorganisms analysis; habitat sensing; and seismic and home monitoring [9]. These applications show a great deal of diversity. Nevertheless, a number of general characteristics are shared among the majority of SN applications, regardless of the specific types of sensors and application objectives. These characteristics include low cost; small size; low power consumption; robustness; flexibility; resiliency on errors and faults; autonomous mode of operation; and privacy and security.

Sensor network nodes typically consist of six components: processor; radio; local storage; sensors and/or actuators; and power supply. A number of relevant technology trends need to be considered. For example, a huge variety of powerful low-power, low-cost processors, and low-cost memory technologies are widely accessible. Also, memory and processor technologies are growing more and more powerfully according to Moore's law, and wireless bandwidth has increased by a factor of more than 100 in the last 7 years; the capacity of batteries is growing at a rate as low as 3% per year. The cost of application-specific designs is growing rapidly: only masks cost $1 million and keep increasing by the factor of two every 2 years. Sensors and actuators are relatively young industrial fields and predictions are still uncertain.

Because of these application requirements and technology constraints, the following architectural and design objectives are most relevant:

- *Small physical size*. Reducing physical size has always been one of the key design issues. Therefore, the goal is to provide powerful processor, memory, radio, and other components while keeping a reasonably small size, dictated by a specific application.
- *Low power consumption*. The capability, lifetime, and performance of the sensors are all constrained by energy. The sensors should be able to be active for a reasonably long time without recharging the battery because maintenance is expensive.
- *Concurrency-intensive operation*. In order to achieve the overall performance, the sensor data must be captured from the sensor, processed, compressed, and then sent to the network simultaneously in pipelined processing mode, instead of sequential action. Two conceptual approaches address this requirement: (1) partitioning the processor into multiple units in which each is assigned responsibility for a specific task; and (2) reduction of the context switching time.
- *Diversity in design and usage*. Because each node should be small in size, low on power consumption, and have limited physical parallelism, the sensor nodes tend to be application specific. However, different sensors have different requirements. For example, cameras and simple thermometers are two extremes in terms of functionality and complexity. Therefore, the design should facilitate trade-offs among reuse, cost, and efficiency.
- *Robust operations*. Because sensors will be deployed over a large and sometimes hostile environment (forests, military usage, human body), they must be tolerant of fault and error. Therefore, sensor nodes need abilities to self-test, self-calibrate, and self-repair [33].
- *Security and privacy*. Each sensor node should have sufficient security mechanisms in order to prevent unauthorized access, attacks, and unintentional damage of the information inside the SN node. Furthermore, additional privacy mechanisms must be included.
- *Compatibility*. The cost to develop software dominates the cost of the overall system. In particular, it is important to be able to reuse the legacy code through binary compatibility or binary translation.
- *Flexibility*. It is necessary to accommodate functional and timing changes. Flexibility can be achieved through two means: (i) programmability (by employing programmable processors such as microprocessors, DSP processors, and microcontrollers); and (2) reconfiguration (by using FPGA-based platforms). Flexibility will be mainly achieved by programmability and use of specialized ASIC and coprocessors due to low power consumption.

12.4 Individual Components of SN Nodes

SN nodes generally are composed of six components: processor; storage unit; power supply; sensors and/or actuators; and, finally, communication (radio) subsystems. It is apparent that standard processors, possibly augmented with DSP, and other coprocessors and some ASIC units will provide adequate processing capabilities at acceptable low-energy rates. Also the state of the art of the actuators is such that they are still not used in the current generation of SN nodes. Therefore, the focus is on the other five components. For the sake of completeness, the discussion begins by presenting a processor specifically designed for sensor networks.

12.4.1 Processor

Berkeley BWRC research group has designed and implemented a prototype processor; its main target areas include voice processing and related applications for wireless devices. For example, the processor can be used in museums to provide better interaction between visitors and displayed items. The Maia processor [63] is built around an ARM8 core with 21 coprocessors. These 21 processors include: two MACs; two ALUs; eight address generators; eight embedded memories; and an embedded low-energy FPGA [24]. The goal is to provide enough parallelism at low energy levels. ARM8 core configures the memory-mapped satellites using a 32b configurable bus and also communicates data with the satellite coprocessors using two pairs of I/O interface ports by applying direct memory reads/writes. The interactions between the ARM8 and coprocessor satellites are carried out through an interface control unit.

A two-level, hierarchical, mesh-structured, reconfigurable interconnect network is used to establish the connections between all satellites. This network provides a favorable trade-off between bandwidth and low area (cost) and low power requirements. This 210-pin chip contains 1.2 M transistors and measures 5.2×6.7 mm^2 in 0.25-µm, six-metal CMOS. In order to minimize the overall energy consumption, the embedded ARM8 core is additionally optimized and can operate under variable supply voltages [8]. In addition, the dualstage pipelined media access control (MAC) and the ALU are configurable. The address generators and embedded memories provide multiple concurrent data streams to the computational components. The embedded FPGA has a 4×8 array of five-input, three-output CLBs. It can be optimized for tasks such as arithmetic operations and data-flow control functions. The interface control unit interacts and coordinates the synchronization and communication between the synchronous ARM8 core and the asynchronous reconfigurable data paths. It also enables the ARM8 core to reconfigure the satellites. The overall targeted computation model is globally asynchronous, locally synchronous computation and supports multirate operation.

12.4.2 Storage

Depending on the overall sensor network structure, the requirements for storage in terms of fast and nonvolatile memory at each node can be sharply different. For example, if one follows the architecture model in which all information is instantaneously sent to the central node, there is very little need for local storage on individual nodes. However, in a more likely scenario in which the goal is to minimize the amount of communication and conduct a significant part of computation at each individual node, there will be significant requirement for local storage. At least two alternatives exist for storing data in a local node. In addition, in the case in which the node is physically larger, one can store the data in microdisks [17].

The first option is to use flash memory, which is very attractive in terms of cost and storage capacity. However, it has relatively severe limitations in terms of how many times it can be used for storing different data in the same physical locations [28]. The second option is to use nanoelectronics-based MRAM [56]. It is expected that MRAM will soon be able to support significant numbers of applications in a number of areas.

It is important to note that historically, nonvolatile semiconductor and disk storage capacity has been growing at a rate higher than that indicated in Moore's law. At least two major challenges for the use of nonvolatile memory in sensor nodes are: (1) partitioning for power reduction and (2) developing memory structures that will fit short, word-length data produced by sensors. Note that a significant percentage of network control and sensor data will have low entropy. Therefore, it is likely that aggressive compression techniques will be used to reduce the amount of data that must be stored or transferred [14].

12.4.3 Power Supply

A wide consensus is that energy will be one of the main technological constraints for SN nodes [46, 57]. For example, the current generation of smart badges and motes enables continuous operations for only a few hours. Energy supply can be addressed in at least two conceptually different ways. The first is to equip each sensor node with a (rechargeable) source of energy. Two main options for this approach exist. Currently, the dominant option is to use high-density battery cells [23, 37]; the other alternative is to use full cells. Full cells provide exceptionally high density and a clean source of energy. However, they are not currently available in a physical format appropriate for SN nodes.

The second conceptual alternative is to harvest energy available in the environment [52]. In addition to solar cells, which are already widely used for mobile appliances such as calculators, a number of proposals concern converting vibration to electric energy [45]. An interesting solution for a power source is introduced in Douseki et al. [18]. A battery-less wireless system that harvests ambient heat is used instead of adopting traditional batteries as the power source. The main component of the system is a switched-capacitor DC–DC converter; a microthermoelectric module makes such a system possible. The chip is fabricated in a 0.8-μm fully depleted SOI process and its effectiveness has been demonstrated.

12.4.4 Sensors

The importance of sensors cannot be overstated. The purpose of SN nodes is not to compute or to communicate, but rather to sense. The sensing component of SN nodes is the current technology bottleneck; these technologies currently are not progressing as fast as semiconductors. Conceptual limitations are significantly stricter for sensors than for processors or storage. For example, sensors interface to the real physical world, while computing and communicating units are dealing with a greatly controlled environment of a single chip. Transducers are front-end components in sensor nodes that are being used to transform one form of energy into another. Design of transducers is considered out of a system architect's scope. In addition, sensors may have four other components: analog, A/D, digital, and microcontroller. The simplest design option includes only the transducer; however, because the current trend is to put more "smartness" into sensor network nodes, significant processing and computing abilities are being added to sensor nodes [41].

One of the main challenges of SNs is to select the type and quantity of sensors and determine their placement. This task is difficult because of the numerous types of sensors with different properties such as resolution, cost, accuracy, size, and power consumption. In addition, often more than one sensor type is needed to ensure the correctness of operation and data from different sensors that can be combined. For example, in the Cricket Compass [51, 65], the orientation and the movement of the studying object can be obtained by measuring the distance between several fixed-location referencing sensors; therefore the location of the sensor is crucial to minimize error [65].

Another challenge is to select the correct types of sensors and the way to operate them. The source of difficulty is sensor interactions. For example, consider determining distance using audio sensors. Because the speed of sound depends greatly on temperature and humidity of the environment, it is necessary to take both measurements into account in order to get the accurate distance.

Several other design tasks are associated with sensors, including fault tolerance, error control, calibration, and time synchronization [33]. There are a large number of different sensor technologies [46, 60]; as an example, consider Kulah et al. [35] and Luo et al. [39]. The accelerometer is one of the most popular

MEMS-based sensors. A state-of-the-art capacitive accelerometer was recently reported by the MEMS group at the University of Michigan. It uses a two-element sensor array in two Σ (sigma-delta) loops to improve accuracy by a factor more than two times in comparison with a traditional second-order Σ modulator. The design is clocked at 1 MHz and provides 1 V/pF sensitivity. It has dynamic range of more than 120 dB and consumes less than 12 mW. Another state-of-the-art accelerometer has been designed at Carnegie Mellon University. The design combines lateral accelerometer and vertical gyroscopes with signal processing circuits.

12.4.5 Radio

Short-range radios as communication components are exceptionally important because the part of the energy budget dedicated to sending and receiving messages usually dominates the overall energy budget [52]. During the design and the selection of radios, one must considers at least three different abstraction layers: physical, MAC, and network. The physical layer is responsible for establishing physical links between a transceiver and one or more receivers. The main tasks at this level involve signal modulation and encoding of data in order to maintain communication in the presence of channel noise and signal interferences. In order to use the bandwidth efficiently and reduce the development cost to some extent, the standard practice is that several radios share the same interconnect medium. The sharing of media (e.g., time or frequency) is facilitated by the MAC layer. Finally, the network layer is responsible for establishing the path that a message must travel through the network in order to be transferred from its source to the destination.

Design of power and bandwidth efficient radios is one of the main research and development tasks. It is important to realize that radio architecture is a function of the employed network structure and protocols. The main trade-off is between the relative energy cost of transmission and reception; the key observation is that listening to the channel is expensive. Therefore, it is necessary to develop schemes that will enable long periods of sleep mode for receivers. For example, one option is to use coordinated policy for deciding which node will go to sleep while the connectivity in the node is maintained [53]. The other option is to use two radios; one of them is responsible for data reception and is power hungry. It is used only when the other ultralow power radio invokes it. The ultralow power radio is only used to detect if one wants to transmit data to this node.

Table 12.1 surveys the state-of-the-art radio design alternatives from ISSCC 2001 [29] and ISSCC 2002 [30]; several notable radio designs are briefly outlined. One radio design alternative is the fully integrated GPS radio described in Behbahani et al. [4]. The low-IF architecture of the radio enables a high level of integration and low power consumption simultaneously. The integrated radio measures a 9.5-mm^2 chip area. It can operate under a various range of voltage and temperature, namely, from 2.2 to 3.6 V and from −40 to +85°C and consumes 27 mW from a 2.2-V supply.

Another notable design is the IEEE 802.11a wireless local area network (WLAN) transceivers presented in Xargari et al. [62]. A 0.25-µm CMOS technology is used to integrate a 5-GHz transceiver compressing the RF and analog circuits of an IEEE 802.11a-compliant wireless local area network (WLAN). The integrated circuit has 22 dBm maximum transmitted power; 8 dB overall receive-chain noise figure; and −112 dBc/Hz synthesizer phase noise at 1 MHz frequency offset.

Other state-of-the-art radio designs have been developed [7, 11, 32, 64]. Chien and colleagues [11] introduced a fully integrated 2.4-GHz transceiver in 0.25-µm CMOS and its associated baseband processor in 0.15-µm CMOS. Kluge and coworkers [32] have recently designed advanced microdevices — a 2.4-GHz CMOS radio for 802.11b wireless LAN. They used 0.25-µm feature size to design 10-mm^2 integrated circuits that consume 86 mA in receiver mode and 73 mA in transceiver mode from a 2.5-V supply. The receiver has a short settling time and is equipped with a separate receiver channel filter and transceiver pulse-shaping filter. In addition, it provides filter calibration circuitry. Bouros and colleagues [7] introduced a digitally calibrated transceiver in 0.18-µm CMOS that occupies 18.5 mm^2. The integrated phase noise can be minimized to less than −37.4 dBc using the fully integrated VCO and synthesizer.

TABLE 12.1 Comparison of State-of-the-Art Radio Design Alternatives

	Technology	Silicon Area (mm²)	ICC_RX (mA)	ICC_TX (mA)	VCC (V)
Alcatel (RF+BB) ISSCC 2001-13.1	0.25-μm CMOS	40	41	52	2.5
IME + OKI (RF) ISSCC 2001-13.2	0.35-μm CMOS	18	66	47	2.7–3.3
Broadcom (RF) ISSCC 2001-13.3	0.35-μm CMOS	20?	46	47	2.7–3.3
Conexant (RF) ISSCC 2001-13.4	0.35-μm SiGE BiCMOS	12?	16	12	1.6–3.0
SiliconWave (RF) ISSCC 2002-5.2	0.35-μm SOI BiCMOS	19.5	39	37	2.7
Transilica (RF) ISSCC 2002-5.3	0.25-μm CMOS	13.3	45	36	3.0
Hitachi (RF) ISSCC 2002-5.5	0.35-μm BiCMOS	11.2	45	35	2.7
Bluetooth (RF) ISSCC 2002-5.1	0.18-μm CMOS	5.5 (4.0)	30	35	2.5–3.0

Source: Zeijl, P. et al., *IEEE J. Solid-State Circuits*, 37, Dec. 2002. With permission.

TABLE 12.2 MCU Comparison

	AT91FR4081	ATMega128L
Datapath	16/32 b	8 b
Clock speed (MHz)	40	4
MIPS/MHz	(ARM 0.9); (THUMB 0.7)	1
Power @ 3 V (mW)	75	15
MIPS/W	480	242

Source: Savvides, A. and Srivastava, M.B., in *Proc. Int. Conf. Computer Design*, 2002. With permission.

TABLE 12.3 Current Drawn by Node Components

Component	Active (mA)	Sleep (mA)
ATMega128L	5.5	1
RFM	2.9	5
AT91FR4081	25	10
RS-485	3	1
RS-232	3	10
Total	39.4	27

Source: Savvides, A. and Srivastava, M.B., in *Proc. Int. Conf. Computer Design*, 2002. With permission.

Chien et al. [11] have developed a 2.4-GHz radio for 802.15.4 WPANs using 0.18-μm CMOS technology that consumes 21 and 30 mW at 1.8-V supply in receiving and transmitting mode, respectively. It incorporates a poly-phase filter and applies transistor linearization techniques to achieve a low-IF architecture. Other alternatives are also available [13, 16].

12.5 Sensor Network Node

This section addresses the key issues related to the architecture and synthesis of an individual SN node. Architecture aspects are discussed along three lines: hardware, software, and middleware; design issues are presented from synthesis and analysis points of view.

The architecture of SN nodes has been addressed in at least three main directions. The first group of initial efforts comprises a number of designs of individual sensor nodes and badges [1, 3, 38, 40, 45, 50, 59]. The emphasis in this class has been placed on ensuring creation of working prototypes and, in some cases, pushing the state of the art of an individual component (e.g., radio, low power, energy harvesting). The second group was represented by the Mote/TinyOS development team at UC. Berkeley [15, 27], who made the first effort to address the trade-offs between various components of the node by developing a new architecture and operating system (OS). The main characteristic of the last effort is sensor centered. The emphasis is on exploiting relatively inexpensive off-the-shelf components in terms of cost and energy as a basis for exploring qualitative and quantitative trade-offs between node components and, in particular, sensors.

It is difficult to anticipate technological trends, but one can easily identify at least some high-impact trends and required solutions. For example, it is apparent that overall energy consumption-balanced architectures are needed. Another high-impact research topic concerns sensor organization and development of the interface between components. Finally, due to privacy, security, and authentication needs, techniques such as unique ID for CPU and other components that facilitate privacy will be in high demand.

In the software domain, main emphasis will be on RTOS (real time operating system) [36]. Ultra-aggressive low-power management is needed because of energy constraints and comprehensive resource accounting is desired due to demands for privacy and security. In a number of cases, support for mobility functions (e.g., location discovery) is also needed. Middleware will be in even stronger demand in order to enable rapid development and deployment of new applications. Tasks such as sensor data filtering; compression; sensor data fusion, sensor data searching and profiling; exposure coverage; and tracking will be ubiquitous.

Synthesis of sensor nodes will pose a number of new problems in the CAD world. It is obvious that new types of models, abstractions, and tasks will be defined and solved. Sensor allocation and selection, sensor positioning, sensor assignment, and efficient techniques for sensor data storage are typical examples of pending synthesis tasks. Development of conceptually simple and clean, yet inexpressive, models of computation is of prime importance as a starting point for synthesis of modern computing systems. Sensor nodes will require new models of computations as well as new models of the physical world. One such example is standard Euclidian space with classical physical laws (e.g., Newton's law, thermodynamics law).

It is well known that modern design flow, debugging, and verification are the most expensive and time-consuming components. Due to the heterogeneous nature of and complex interactions between components, the same scenario is expected in the case of sensor nodes. In particular, techniques for error and fault discovery, testing, and calibration will be of prime importance. In the rest of this section, four representative SN nodes designs are described: Berkeley mote; piconode node; UCLA Medusa II; and light compass node.

12.5.1 Berkeley Mote Node

The starting point for designing modern computer systems is a comprehensive set of benchmarks that are representative for common users. Unfortunately, such a set of benchmarks currently is not available to designers of SN nodes. The starting point for designing mote wireless sensor network nodes was the set of qualitative observations about the requirements of wireless sensor networks. Special emphasis was placed on small physical size and low energy consumption. In addition, attempts were made to facilitate concurrency intensive operations to provide control hierarchy and take advantage of limited physical parallelism. Furthermore, the design decisions were driven by robust operations' ability to be retargeted, at least at the network level.

The design went though several iterations and until recently was leveraging on the availability of standard off-the-shelf components. Generally speaking, the design is radio centric in the sense that all main decisions are made in order to facilitate low-energy communications. The main processor is Atmel 90LS8535 microcontroller that has 8-b Harvard architecture with 16-b addresses. It achieves a speed of 4 MHz at 3 W. The system has a rather minimal amount of memory that consists of 8 kbytes of flash for program memory and 512 bytes SRAM for data memory. Therefore, the system can be integrated only with low-frequency sampling sensors and must communicate frequently.

The processor integrates a system of timers and counters and can be placed in four energy modes: active, idle, power down, and power save. In the idle mode, the processor is completely shut off. In the power-down mode, only the watchdog and asynchronous interrupt logic are awake. Finally, in the power-save mode, in addition to watchdog and interrupt logic, the asynchronous timer is also active. The system also has a coprocessor Atmel 90LS2343 microcontroller that has 2 kbytes flash instruction memory and 128 bytes of SRAM and EEPROM memory. The coprocessor can be used to reprogram the main microcontroller.

The authors consider the RF Monolithic 916.50 transceiver as the central part of the design. The radio is equipped with an antenna and a system of discrete components that can be used to alter characteristics of the physical layer such as signal strength. The radio operates at a speed of 19.2 kbytes/sec. The transceiver can operate in three modes: transmission, reception, and power off. The system can have up to eight sensors; the two most widely used are photoelectric optical sensor and temperature sensor. Each sensor is placed on the bus that is controlled using software.

It is instructive to consider power characteristics of the design. MCU core consumes between 2.5 to 6.5 mA; radio consumes between 5 to 12 mA. Optical sensor and temperature sensor consume 0.3 and 1 mA, respectively, and the coprocessor consumes 1 to 2.4 mA. Finally, EEPROM consumes 1 to 3 mA. In particular, it is instructive to compare energy spent for bit transmission and bit processing. The system spends about 1 mJ to send, and 0.5 mJ to receive, 1 b. At the same time, the system can execute approximately 120 instructions for each millijoule spent. The system does provide for energy reduction using variable voltage; therefore, energy is saved mainly by turning the system off. The core of the system software for the design is an exceptionally compact microthreading operating system (TinyOS).

The Berkeley design team concluded that the new application domain requires a new OS; therefore, they decided not to adopt any great variety of RTOS 8-b controllers. Although this decision certainly resulted in higher power efficiency and more interesting system software architecture, it also created additional demands and constraints in programming already highly constrained hardware. Nevertheless, the system has been highly popular in the research community. Several thousand copies of the motes in several versions have been used by more than 200 research teams. The greatest strength in the system is its small size and low power. Probably the most serious disadvantages are related to the development of real applications. Although motes have been tremendously popular in research communities, it is still unclear how well they are suited for applications in which more complex systems of sensors are needed.

12.5.2 UCLA Medusa MK-2 Node

The Medusa MK-2 node is a representative of the state-of-the-art design of more powerful sensor nodes [55]. The computational unit of Medusa MK-2 nodes consists of two microcontrollers. The first is an 8-b Atmel STMega128L MCU with 4 MHz that has 32 K of flash memory and 4 KB of RAM. This processor serves as an interface between sensors and radio base band processing. The second microcontroller is an ATMEL ARM THUMB processor enclosed within a 120-ball BGA package. It has significantly more processing power and 40 MHz. It includes 136 KB of RAM and 1 MB of on-chip FLASH memory.

The communication unit of Medusa MK-2 nodes is a combination of a TR 1000 low-power radio from RF Monolithics for wireless and an RS-485 serial bus transceiver for wireline communication. The sensing unit has two components: a MEMS accelerometer and a temperature sensor. It can also be augmented with other types of sensors. Medusa nodes also incorporate a variety of interfaces, including eight 10-b ADC inputs, serial ports, and numerous general purpose I/O ports. An ultrasonic ranging unit is implemented on an accessory board using 40-kHz transducers. Ultrasonic measurements are coordinated with RF measurements in order to calculate internode distances and therefore enable localization of nodes. Localization is conducted using iterative linearized multilateration.

The nodes also have two external connectors. The first is used to communicate with a PC to download and debug software. It also provides the necessary wiring requirements for connecting to an external GPS module. The second connector serves as an expansion slot for attaching add-on boards carrying different sensors because it has a set of ADC and GPIO. Finally, Medusa nodes also have two pushbuttons that serve as a user interface. They are mainly used for triggering events and executing different tests during experiments.

It is interesting to take a closer look at the computational unit of Medusa Mk-2 nodes. According to the computation requirements, the computational tasks are classified into two broad categories: low-demand tasks and high-demand, low-frequency processes. The low-demand tasks are the periodic processes such as base band processing for the radio while listening for new packets, sensor samplings, handling of sensor events, and power management. Even though these tasks usually require a high concurrency, they are not particularly demanding in terms of computational resource requirements and therefore can be easily handled by an 8-b microcontroller. The Medusa-MK-2 nodes use a low-power AVRMega128L microcontroller.

The second category — the low-frequency, high-demanding tasks — is related to the processing of acquired sensor data in order to produce user-requested information. For example, in the case of a fine-grained localization problem, a sensor node is expected to compute an estimate of its location based on a set of distance measurements to known beacons or neighbors. In order to avoid error propagation, a node must perform a set of high-precision operations. If an 8-b processor were used to conduct this type of computation, it would result in high latencies and lower precision. Therefore, a high-end processor is a more adequate solution. More specifically, Medusa adopts the 40-MHz ARM THUMB processor to perform this type of operation.

Another advantage is that the node can use existing standard applications and libraries. The THUMB microcontroller also has sufficient resources to support shelf-embedded operating systems such as Red Hat eCos and uCLinux. The inclusion of the THUMB processor is also justified by a comparison of the two processors made from a power/latency perspective conducted by the UCLA group. The THUMB processor executes instructions at the rate of 0.9 MIPS per megahertz at 40 MHz while consuming 25 mA with a 3-V supply, which has a performance of 480 MIPS/W. On the other hand, the ATMega128L only provides a 242-MIPS/W performance when operating at 4 MHz and consumes 5 mA at 3-V supply.

Communication between the two processors is handled by a pair of interrupt lines — one for each microcontroller — and an SPI bus. The two nodes remain in sleep mode until an interrupt indicating the need for data exchange occurs. The communication takes place over the 1-Mbs SPI bus.

Medusa MK-2 nodes are capable of two types of communications: wired and wireless. All nodes are equipped with a wired and a wireless link. The wireless link is a low-power TR1000 radio from RF Monolithics. This radio has transmitting power of 0.75 mW at maximum and has an approximate transmission range of 20 m. Two modulation schemes are supported by this radio: of-off keying (OOK) and amplitude shift keying (ASK). Selection of the appropriate modulation can be done in software. On a Medusa MK-2 node, the base band processing for the radio is done by an ATMega128L microcontroller. This also allows the node to run the low power S-MAC [61] protocol on the ATMega128L processor. In addition to the wireless link, Medusa nodes also incorporate an RS-485 serial bus interface for wireline communication. Attaching a low-power RS-485 transceiver to one of the RS-232 ports of the THUMB processor allows the node to connect to an RS-485 network using an RJ-11 connector and regular telephone wire. A single RS-485 has occupancy up to 32 nodes that span over a total wire length distance of 1000 ft.

The power unit of Medusa MK-2 nodes consists of two main components: the power supply and the power management and tracking unit (PMTU) [12]. The power supply consists of a 540-mAh lithium-ion rechargeable battery and an up–down DC–DC converter with a 3.3-V output that can reach up to 300 mA of current from the battery. The power supply is designed in such a way that power-additional sensors can be attached later on as accessory boards because the node only requires less than 50 mA with no sensors attached. In a typical SN setting, putting the ARM THUMB processor together with the RS-485 and RS-232 transceivers in sleep mode most of the time, yields an 80% reduction of the overall node power consumption. Comprehensive energy consumption comparisons between Medusa MK-2 nodes and other SN nodes designs can be found in Savvides and Srivastava [55].

12.5.3 BWRC Piconode

Another communication-centered sensor node design is the PicoNode [52]. The main overall objective of this design is to provide flexibility and low energy consumption simultaneously. It consists of four main modules. The first two units are processors: an embedded processor unit and configurable satellite units. The embedded processor is dedicated mainly for application and protocol-stack layers that require higher flexibility but have relatively low computational complexity and are infrequently requested. Configurable processing modules are targeted for the more frequent tasks with higher computational requirements. Two other modules are dedicated to communication tasks — a parameterized and configurable digital physical layer and a simple direct-down conversion RF front end.

These modules are interconnected by a flexible and low-power consumption interconnect scheme. The authors claim that a dynamic matching between application and architecture leads to a significant energy savings for signal-processing applications while maintaining implementation flexibility. One of the main premises of the design is the observation that the processor implementation is three orders of magnitude more expensive in terms of energy consumption than the implementations of the dedicated hardware. However, a trade-off occurs between flexibility and programmability (software on programmable platforms) and energy consumption (ASIC hardware).

The traditional approach is to design the wireless transceiver using only RF and analog circuit modules. More recently, a primarily digitalized design approach has emerged. This is inspired by the insight that digital circuits can improve exponentially with the scaling of technology, while analog circuits get linearly worse because of reduction of the supply voltage. Therefore, it is beneficial to incorporate a small, noncritical analog front end and use digital back-end processing to balance the limitations.

Many design challenges are related to the physical layer. They are mostly related to the low-energy targets and variable demands from the network. Therefore, in order to satisfy various demands from the network, the PicoNode physical layer can be made into parameters. These parameters include power control modes, modulation scheme, and bit rate.

In order to meet the low-energy requirement, the physical layer must meet two mutually exclusive criteria: fast signal acquisition and low standby power. The first criterion refers to the process of requiring least amount of time to wake up, receive bursts of data, and immediately go back to sleeping mode after data acquisition. The second criterion emphasizes consuming the least amount of energy while sleeping. The reason that they are usually mutually exclusive is that an inverse proportional relationship exists between the depth of sleep (i.e., energy consumed) during standby and the time required to wake up.

PicoNode is designed so that it does not require an interval power supply. It is self-constrained and self-powered using energy extracted from the environment. The two major constraints for harvesting ambient energy from the environment are: applicability within the environment and the size of the node (Berkeley group targets the 1-cm^3 design). PicoNodes harvest energy from light and vibrations [52].

12.5.4 Sensor-Centric Design: Light Compass

The final sensor node design alternative for overviewed is the light compass node [66]. The emphasis in this approach is completely shifted from computation, communication, and storage to sensors. The first three functions are provided by a standard laptop or PDA. The rationale is that this type of design will progress on its own to become a viable platform for SN nodes. Even the interface toward sensor is built using off-the-shelf components. The focus is placed on sensors and how to select and place them in such a way that sensor data fusion is facilitated. In addition, special emphasis is placed on how to rapidly develop sensor data fusion software that can be retargeted and how to develop systematic procedures for design of sensor nodes.

Figure 12.1 shows the used light sensor components. The smallest device (on the left) is a miniature silicon solar cell used for converting light impulses directly to electrical charges (photovoltaic). It generates its own power and therefore does not require any external bias. This silicon cell is further mounted on a 0.78 × 0.58 × 0.18 cm thick plastic carrier that generates roughly 400 mV in moderate light (most typical rooms). A significantly larger sensor (on the right), measuring 2.54 × 2.15 cm, also can be referred to as a photoconductor and can be surrounded by a 0.18-cm thick plastic encapsulated ceramic package. In strong light, its resistance measures 20 Ω and 5 kΩ in complete darkness. These components are very economically viable (roughly $0.30 each) and they can be easily purchased in large quantities.

These sensors can be used in multiple prototypes, such as the ones shown in Figure 12.2. On the left side of Figure 12.2, the six-sided cut-pyramid structure has a base length of 3 cm and a top edge length of 1 cm with a 60° slope. Sensors can be attached to each side of the structure depending on the application and purpose. The structure on the right is a cube with 2-cm edges; therefore, it can incorporate up to six sensors with one on each surface.

In this light sensor platform, the heart component is an eight-channel analog to digital converter (ADC) module. It is used to read the sensor values through the parallel port of a standard PC laptop. This ADC component comprises a Maxim MAX186 ADC, which has an internal analog multiplexer that can be configured for eight single-ended, or four differential, inputs at a 12-b resolution; the conversion time is under 10 s. This component is pictured on the left in Figure 12.3. In addition, some of the other components of the circuit include: several resistors to protect the analog inputs; capacitors to filter noise; an external 4.096-V voltage regulator, and an 8-b digital latch required for parallel port communications. The overall design flow of a sensor appliance is presented in detail in Figure 12.4.

FIGURE 12.1 Light sensor components. (From Wang, J., et al., *40th IEEE/ACM Design Automation Conf.*, pp. 66–71, 2003. With permission.)

FIGURE 12.2 Light appliance prototypes: 60° six-sided, cut pyramid and cube. (From Wang, J., et al., *40th IEEE/ACM Design Automation Conf.*, pp. 66–71, 2003. With permission.)

Sensor Network Architecture

FIGURE 12.3 Light appliance platform. (From Wang, J., et al., *40th IEEE/ACM Design Automation Conf.*, pp. 66–71, 2003. With permission.)

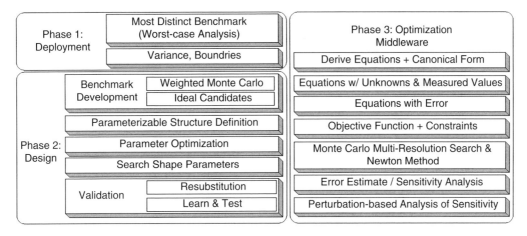

FIGURE 12.4 Overall design flow of a sensor appliance. (From Wang, J., et al., *40th IEEE/ACM Design Automation Conf.*, pp. 66–71, 2003. With permission.)

The main goal of this design was to achieve low power consumption while maintaining a tolerable level of coverage. Figure 12.5 through Figure 12.7 depict the results obtained from four different sensor structures: a four-sensor pyramid (square base); a four-sensor cut pyramid (triangular-based pyramid with a flat sensor on top); a five-sensor pyramid (pentagonal base); and a five-sensor cut pyramid (square-based pyramid with a flat sensor on top). In all cases, the objective was to estimate the positions of 5000 randomly placed light instances.

12.6 Wireless SNs as Embedded Systems

The architecture of wireless SNs at the network level is briefly surveyed in this section. For the networking of the wireless devices and appliances, several communication schemes have been proposed, such as satellite, WLAN, cellular, and ad hoc multihop architectures [25, 26, 48, 49, 58]. Based on the different architectures, the communication between the nodes can be all low power (ranges in meters), high power (ranges in megameters), or medium power (ranges in kilometers).

For example, wireless SNs are the widely used cellular wireless networks. In this architecture, a number of base stations are already deployed within the field. Each base station forms a cell around itself that covers part of the area. Mobile wireless nodes and other appliances can communicate wirelessly as long as they are at least within the area covered by one cell. An example of such a network is shown in Figure 12.8. The communication requires medium power, although the fixed and immobile base stations are consuming a large amount of power to cover a large area and to communicate to and from the lower power mobile wireless nodes. However, cellular wireless architecture has the drawback that it must be

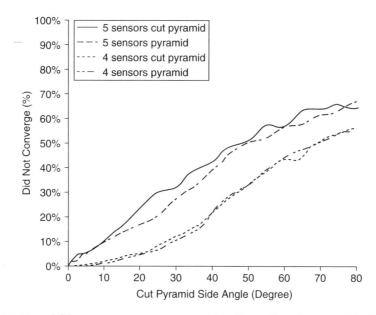

FIGURE 12.5 Fraction of failure convergence vs. sensor angles. (From Wang, J., et al., *40th IEEE/ACM Design Automation Conf.*, pp. 66–71, 2003. With permission.)

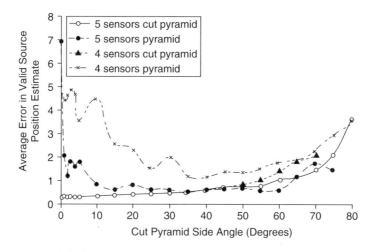

FIGURE 12.6 Fraction of valid solutions vs. sensor angles. (From Wang, J., et al., *40th IEEE/ACM Design Automation Conf.*, pp. 66–71, 2003. With permission.)

implanted in the field; also, cells should be carefully designed to have full coverage and transparency with respect to the cells.

The WLAN is built for high-frequency radio waves. The WLAN also needs its own infrastructure within the designated local area. It is very well suited for local private areas, such as offices, campuses, and buildings. In some of the applications of the sensor network, such as smart buildings, connecting the sensor networks to the WLAN implanted within the area is very suitable. The power consumption in LAN is also medium, although the fixed part of the infrastructure is naturally higher powered.

In order to overcome the difficulties caused by the infrastructure settings for wireless satellites, WLAN, and cellular networks, a new generation of wireless networks architecture has emerged — the wireless multihop ad hoc networks. In such networks, the infrastructure architecture is not needed and the nodes

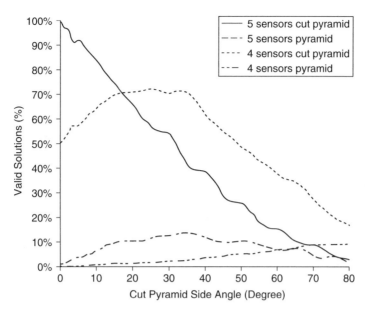

FIGURE 12.7 Average error in positions vs. sensor angles. (From Wang, J., et al., *40th IEEE/ACM Design Automation Conf.*, pp. 66–71, 2003. With permission.)

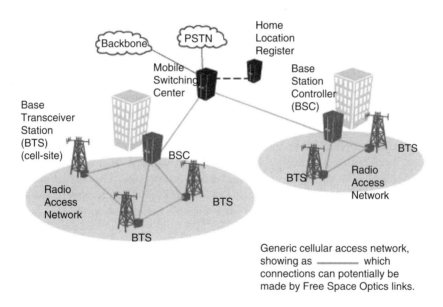

FIGURE 12.8 Wireless cellular network architecture. (From: http://w.w.w.holoplex.com/technology_backhaul.html.)

can configure to communicate to other nodes within their communication range on the fly. The nodes are short range and therefore all of the communications are low power. If two nodes that are not within each other's range need to communicate to each other, they use the intermediate nodes as the relays. The multihop ad hoc wireless SN architecture appears as an attractive alternative to the WLAN and cellular technologies for at least four reasons:

- On-demand formation of the network does not require predeployed architecture.

- Multihop routing can save orders of magnitude of power consumption when compared to long-range routing for the same distance [52].
- Because communications between the nodes are short range and local, the bandwidth is reusable, as opposed to that in long-range communications.
- The fourth reason is the fault tolerance [10]. SNs are envisioned to have a lot of inexpensive nodes embedded in the environment. The ad hoc multihop architecture supports the advent of the new nodes and departure or failure of the old ones.

Most of the current SN literature has been advocating ad hoc multihop architecture [2, 20, 27, 34, 52, 61]. Nevertheless, there are no indications that this architecture would be the best architecture for all of the sensor network applications. Because of the quantity of the radios and the number of the packets flowing in the network, a natural asymmetry is present in the multihop ad hoc implementation. In fact, for some applications, such as smart buildings or scientific experiments in which the network does not change over the space, having a number of static components in the network is a natural solution. The static parts would be connected to the constant power supply, so wireless parts could use low power to communicate to them and nodes could go into the standby mode from time to time.

Another important issue related to sensor networks is the topology of the network [10]. The question is how to distribute the nodes within the field to achieve the best range and coverage from the sensors. This question is a variation of the well-known art gallery problem [47], in which the new constraints on the nodes are that they are short communication range. The other big issue in topology consideration is that not all of the nodes should be uniformly distributed, as is the assumption in the current literature and simulations for SNs. Furthermore, network architecture should address the concerns of various layers of the network.

Better components are still needed in the physical layer [31], power control, and MAC layer [61]; routing protocols [20] are needed at the network layer. The only proposed OS for the sensor network is TinyOS, which is an operating system at the node level [27]. There is a need for a more complex network operating system (NOS) that can (1) facilitate the autonomous mode for ad hoc multihop architecture; (2) address privacy and security concerns; and (3) provide efficient execution of localized algorithms.

This section concludes with a very brief overview of three industrial wireless networks standards: IEEE 802.11b; Bluetooth; and HomeRF. IEEE 802.11b, or WiFi, primarily targets computer communication. Although its main target is indoor connectivity at speeds of 11 Mbps within 150 m, it is expected that it will provide the same level of service outdoors within a 300-m range. With specially equipped radios (amplifiers and special antenna) it may establish connectivity within a range of 30+ km. It can operate in several modes, including peer–peer and infrastructure access point. The wired equivalent privacy (WEP) standard ensures data protection using 40- and 128-b RC4-based encryption. Bluetooth mainly targets personal area networks on very short distances and applications such as audio, video, and multimedia. IEEE802.11b and Bluetooth use 2.4-GHz ISM band for unlicensed radio communication. HomeRF provides inexpensive residential-oriented wireless connectivity.

12.7 Summary

This chapter surveyed the architectural and synthesis issues related to SNs. The analysis has been conducted at three levels of abstraction: subsystem, individual node, and network. The main design objectives and current trends, as well as their relative advantages and limitations, were identified. Furthermore, several architecture and design case studies have been conducted. Special emphasis was placed on formulating the highest impact architectural and synthesis challenges.

Acknowledgment

This material is based upon work supported in part by the National Science Foundation under Grant No. ANI-0085773 and NSF CENS Grant.

References

1. Agre, J.R. et al., Development platform for self-organizing wireless sensor networks, in *Proc. SPIE Int. Soc. Optical Eng.*, 3713, 257, 1999.
2. Akyildiz, I.F. et al., Wireless sensor networks: a survey, *Computer Networks*, 38, 393, 2002.
3. Asada, G. el at., Wireless integrated network sensors: low power systems on a chip, in *Proc. 24th Eur. Solid-State Circuits Conf.*, 9, 1998.
4. Behbahani, F. et al., A fully integrated low-IF CMOS GPS radio with on-chip analog image rejection, *IEEE J. Solid-State Circuits*, 37, Dec 2002.
5. Beneden, B.V., Examining Windows CE 3.0 real time capabilities, *Dr. Dobb's J.*, 26, 66, 2001.
6. Bridges, S., The R380s — the first smartphone from the Ericsson–Symbian partnership, *Ericsson Rev.*, 78, 44, 2001.
7. Bouras, I. et al., A digitally calibrated 5.15 — 5.825-GHz transceiver for 802.11a wireless LANs in 0.18-μm CMOS, in *Proc. ISSCC*, 2003.
8. Burd, T. et al., A dynamic voltage scaled microprocessor system, *Dig. Tech. Papers ISSCC*, 2000.
9. Cerpa, A. et al., Habitat monitoring: application driver for wireless communications technology, in *Proc. ACM SIGCOMM Workshop Data Commun. Latin Am. Caribbean*, 2001.
10. Cerpa, A. and Estrin. D., ASCENT: adaptive self-configuring sensor networks topologies, in *Proc. INFOCOM 2002*, 2002.
11. Chien, G. et al., A 2.4G-Hz CMOS transceiver and base band processor chipset for 802.11b wireless LAN application, in *Proc. ISSCC*, 2003.
12. Chen, A. et al., A support infrastructure for the smart kindergarten, *IEEE Pervasive Computing Mag/*, 1, 49, 2002.
13. Cojocaru, C. et al., A 43-mW Bluetooth transceiver with −91 dBm sensitivity, in *Proc. ISSCC*, 2003.
14. Drinic, M., Kirovski, D. and Potkonjak, M. *Model-based compression in wireless ad hoc networks*, in *Proc. of Sensys*, 2003.
15. Culler, D.E. et al., EMSOFT 2001: network-centric approach to embedded software for tiny devices, in *Proc. Workshop Embedded Software*, 2001.
16. Darabi, H. et al., A dual-mode 802.11b/Bluetooth radio in 0.35-μm CMOS, in *Proc. ISSCC*, 2003.
17. Dietzel, A. and Berger, R., Trends in hard disk drive technology, in *Proc. VDE World Microtechnol. Cong.*, 1, 2000.
18. Douseki, T. et al., A batteryless wireless system uses ambient heat with a reversible-power-source compatible CMOS/SOI DC-DC converter, in *Proc. ISSCC*, 2003.
19. Doyle, M., Fuller, T.F., and Newman, J. Modeling of galvanostatic charge and discharge of the lithium/polymer/insertion cell, *J. Electrochem. Soc.*, 140, 1526, June 1993.
20. Estrin, D. et al., Next century challenges: scalable coordination in sensor networks, in *Proc. MOBICOM*, 262, 1999.
21. Faroque, M. and Maru, H.C., Fuel cells — the clean and efficient power generators, *in Proc. IEEE*, 89, 1819, Dec 2001.
22. Feng, J. et al., Sensor networks: quantitative approach to architecture and synthesis, UCLA, technical Report, 2002.
23. Fuller, T.F., Doyle, M., and Newman, J. Simulation and optimization of the dual lithium ion insertion cell, *J. Electrochem. Soc.*, 141, 1, Jan 1994.
24. George, V. et al., The design of a low-energy FPGA, in *Proc. ISLPED*, 1999.
25. Gupta, P. and Kumar, P.R., Internets in the sky: capacity of 3D wireless networks, in *Proc. IEEE Conf. Decision Control*, 3, 2290, 2000.
26. Hamburgen, W.R. et al., Itsy: stretching the bounds of mobile computing, *Computer*, 34, 28, April 2001.
27. Hill, J. et al., System architecture directions for networked sensors, *ASPLOS*, 93, 2000.
28. Ishii, T. el al., A 126.6-mm/sup 2/AND-type 512-Mb flash memory with 1.8-V power supply, *IEEE J. Solid-State Circuits*, 36, 1707, Nov 2001.

29. Session 13 on Bluetooth transceivers, *ISSCC Dig. Tech. Papers*, Feb 2001.
30. Session 5 on Bluetooth transceivers, *ISSCC Dig. Tech. Papers*, Feb 2002.
31. Kahn, J.M., Katz, R.H. and Pister, K.S. Next century challenges: mobile networking for "Smart Dust," in *Proc. MobiCom*, 271, 1999.
32. Kluge, W. et al., A 2.4GHz CMOS transceiver for 802.11b wireless LANs, in *Proc. ISSCC*, 2003.
33. Koushanfar, F., Potkonjak, M. and Sangiovanni–Vincentelli, A., Fault-tolerance techniques for sensor networks, in *Proc. IEEE Sensors*, 49, 2002.
34. Koushanfar, F. et al., Processors for mobile applications, in *Proc. Int. Conf. Computer Design*, 603, 2000.
35. Kulah, H., Yazdi, N. and Najafi, K., A multi-step electromechanical $\Sigma\Delta$ converter for micro-g capacitive accelerometers, in *Proc. ISSCC*, 2003.
36. Li, Y., Potkonjak, M. and Wolf, W., Real-time operating systems for embedded computing, in *Proc. Int. Conf. Computer Design*, 388, 1997.
37. Linden, D., *Handbook of Batteries*, 2nd ed., New York: McGraw–Hill, 1995.
38. Locher, I. et al., System design of iBadge for smart kindergarten, unpublished manuscript.
39. Luo, H., Fedder, G. and Carley, L., Integrated multiple-device IMU systems with continuous-time sensing circuitry, in *Proc. ISSCC*, 2003.
40. Maguire, G.Q. et al., Smartbadges: a wearable computer and communication system, in *Proc. 6th Int. Workshop Hardware/Software Codesign*, 1998.
41. Mason, A. et al., A generic multielement microsystem for portable wireless applications, *IEEE*, 86, 1733, Aug 1998.
42. Meguerdichian, S. et al., Coverage problems in wireless ad hoc sensor networks, in *Proc. IEEE INFOCOM*, 3, 1380, 2001.
43. Meguerdichian, S. et al., Localized algorithms in wireless ad hoc networks: location discovery and sensor exposure, in *Proc. MOBIHOC*, 106, 2001.
44. Meng, T.H. and McFarland, B., Wireless LAN revolution: from silicon to systems, in *Proc. IEEE Radio Frequency Integrated Circuits (RFIC) Symp.*, 3, 2001.
45. Meninger, S. et al., Vibration-to-electric energy conversion, in *Proc. IEEE Trans. VLSI Syst.*, 9, 64, Feb 2001.
46. Min, R. et al., An architecture for a power-aware distributed microsensor node, in *Proc. IEEE Workshop Signal Process. Syst.*, 581, 2000.
47. O'Rourke, J., *Art Gallery Theorems and Algorithms*, Oxford University Press, 1987.
48. Pehrson, S., WAP — the catalyst of the mobile Internet, *Ericsson Rev.*, 77, 14, 2000.
49. Perkins, C.E., *Ad Hoc Networking*, Boston: Addison–Wesley, 2001.
50. Pottie, G.J. and Kaiser, W.J. Wireless integrated network sensors, in *Proc. Commun. ACM*, 43, 51, May 2000.
51. Priyantha, N.B., Chakraborty, A. and Balakrishnan, H., The Cricket location-support system, in *Proc. MobiCom*, 32, 2000.
52. Rabaey, J.M. et al., PicoRodio supports ad hoc ultra-low power wireless networking, *Computer*, 33, 42, July 2000.
53. Rozovsky, R. and Kumar, P.R., SEEDEX: a MAC protocol for ad hoc networks, in *Proc. MOBIHOC*, 67, 2001.
54. Savvides, A., Han, C.C. and Srivastava, M., Dynamic fine-grained localization in ad-hoc networks of sensors, in *Proc. ACM SIGMOBILE 7th Annu. Int. Conf. Mobile Computing Networking*, 2001.
55. Savvides, A. and Srivastava, M.B., A distributed computation platform for wireless embedded sensing, *Proc. Int. Conf. Computer Design*, 2002.
56. Slaughter, J.M. et al., Fundamentals of MRAM technology, *J. Superconductivity*, 15, 19, Feb 2002.
57. Slijepcevic, S. and Potkonjak, M., Power efficient organization of wireless sensor networks, in *Proc. IEEE Int. Conf. Commun.*, 472, 2001.
58. Sohrabi, K. et al., Protocols for self-organization of a wireless sensor network, *IEEE Personal Commun.*, 16, Oct 2000.

59. Want, R. et al., The active badge location system, *ACM Trans. Inf. Syst.*, 10, 91, 1992.
60. Yazdi, N., Ayazi, F. and Najafi, K., Micromachined inertial sensors, *IEEE*, 86, 1640, Aug 1998.
61. Ye, W., Heidemann, J. and Estrin, D., An energy-efficient MAC protocol for wireless sensor networks, in *Proc. INFOCOM*, 2002.
62. Xargari, M. et al., A 5-GHz CMOS transceiver for IEEE 802.11a wireless LAN system, *IEEE J. Solid-State Circuits*, 37, Dec 2002.
63. Zhang, H. et al., A 1-V Heterogeneous reconfigurable processor IC for base band wireless applications, in *IEEE Int. Solid-State Circuits Conf.*, 2000.
64. Zeijl, P. et al., A Bluetooth radio in 0.18-µm CMOS, *IEEE J. Solid-State Circuits*, 37, Dec 2002.
65. Nissanka, P.B., Min, A.K.L., Balakrishnan, H., and Teller, S., The Cricket Compass for context-aware mobile applications, *Proc. 7th Ann. Intl. Conf. Mobile Computing and Networking (MobiComm 2001)*, pp. 1, 2001.
66. Wong, J., Megerian, S., and Potkonjak, M., Design techniques for sensor appliances: foundations and light compass case study, *40th IEEE/ACM Design Automation Conf.*, pp. 66–71, June 2003.

13
Tiered Architectures in Sensor Networks[*]

13.1	Introduction	13-1
13.2	Why Build Tiered Architectures?	13-3
	Cost-Effectiveness • Longevity • Scalability	
13.3	Spectrum of Sensor Network Hardware	13-5
	Small Sensor Nodes • Large Sensor Nodes	
13.4	Task Decomposition and Allocation.............................	13-8
	Sensing • Processing • Communication	
13.5	Forming Tiered Architectures	13-10
	Engineered Networks • Routing Mechanisms • Clustering Mechanisms	
13.6	Routing and Addressing in a Tiered Architecture	13-15
	Routing in a Hierarchy • Hierarchical Addressing	
13.7	Drawbacks of Tiered Architectures..............................	13-18
13.8	Conclusions	13-19

Mark Yarvis
Intel Corporation

Wei Ye
University of Southern California

13.1 Introduction

A wireless sensor network is a collection of nodes that self-organize to perform sensing, computation, and data delivery in the execution of a common data acquisition task. In a flat architecture, all nodes are peers and are homogeneous in form and function. In a tiered architecture, on the other hand, nodes form a hierarchy in which a node at a given level performs a specific set of tasks on behalf of a subset of nodes in the level below.

Although the notion of a flat network of completely interchangeable nodes is appealing, very few sensor networks are entirely flat. Typically, a sensor network connects to a more general-purpose network via a small number of "gateway" nodes, which can provide duplicate data removal, complex computations, buffering, and final delivery. In addition, sensor networks are often not physically homogeneous. A network may become heterogeneous from use (e.g., uneven battery drain across nodes). Phased deployment of a network and node upgrades also contribute to heterogeneity because the processing and storage capabilities of a given node technology will increase over time at a fixed cost. Finally, sensor networks are often purposely heterogeneous, due to cost and energy considerations. Tiered architectures can be employed to take advantage of unevenly distributed resources by assigning resource-intensive roles to resource-rich nodes.

In a tiered network, the functions of sensing, computation, and data delivery are divided unequally among nodes. These functions may be divided across the tiers, with the lowest tier performing all sensing,

---

[*]Other names and brands may be claimed as the property of others.

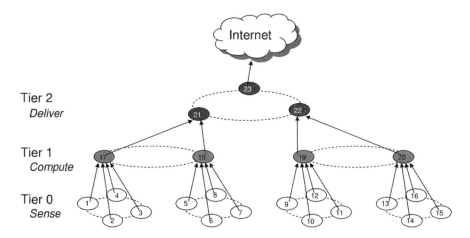

FIGURE 13.1 A partition of sensor network application functions across the levels of a tiered architecture.

the middle tier performing all computation, and the top tier performing all data delivery (Figure 13.1). Alternatively, a particular function may be divided unequally among layers; for instance, each layer could perform a specialized role in computation. In this case, the lowest level sensors might provide a simple band-pass filter or pattern recognition filter to cull interesting data from noise, while nodes at a higher tier might fuse the filtered data received from multiple sensors, characterizing a single event using multimodal sensor data. A wide variety of architectures is possible.

Functional decomposition of a sensor network can reflect physical characteristics of nodes, or it can simply be a logical distinction. For instance, a subset of nodes with a long-range communication capability may form a physically hierarchical overlay network topology (Figure 13.2). On the other hand, a subset of nodes in the network might be logically distinct in that they perform a service on behalf of the other nodes. Such services might include data aggregation, communication over a backbone, or route aggregation on behalf of a cluster of nodes. These logical role assignments can form a logically hierarchical network (Figure 13.3). Logical roles can be periodically rotated for fairness. When nodes with more computational capacity are available, computation tasks can be migrated from less capable sensor nodes. Without such "compute servers," a cluster of sensors may need to elect one node to perform tasks such as data fusion. In some cases, however, only nodes with particular physical resources are suited for a given task. For instance, a node with a global positioning system (GPS) receiver may be required to perform a lead role in localization or time synchronization.

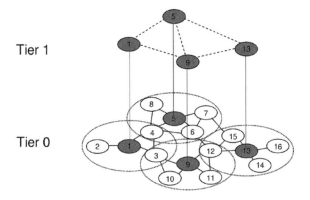

FIGURE 13.2 A physically tiered network, using the upper tier as a backbone network.

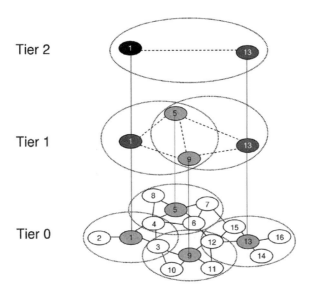

FIGURE 13.3 A logically tiered network.

It is no accident that many sensor networks are designed and built in tiered architectures. The next section explores the factors that can make tiered sensor networks more effective than flat sensor networks. The sections that follow describe characteristics of the array of hardware available for sensor networks; details of functional decomposition and role assignment; mechanisms for establishing a tiered topology in an ad hoc network; mechanisms for routing and addressing in hierarchical networks; and advantages and pitfalls of tiered architectures.

13.2 Why Build Tiered Architectures?

To understand the importance of tiered architectures, one must first consider three key characteristics of sensor networks: cost-effectiveness, longevity, and scalability. These characteristics help determine whether a sensor network is an appropriate choice for a given application.

13.2.1 Cost-Effectiveness

In the case of scientific applications, the cost of purchasing, installing, and maintaining sensor network hardware and software must be less than alternative approaches with similar application performance. In commercial applications, sensor networks must demonstrate a return on investment to be considered cost-effective. In other words, within a specified time period, the monetary benefit provided by the sensor network (e.g., reduced heating and cooling costs) must offset the cost of purchasing and installing the network.

Tiered architectures can reduce the cost of a sensor network by allocating resources where they can be most effectively utilized. Sensing typically requires a large number of nodes but relatively few resources at each node. Data analysis typically requires more processing and storage resources than sensing. Because the delay introduced by data analysis will be inversely proportional to the speed of the processor, the minimum resources required for data analysis depends on the latency that can be tolerated. Similarly, the per-node storage requirement will be, at best, inversely proportional to the number of nodes involved in data analysis and limited to the degree to which the algorithm can be distributed. Tasks such as localization and time synchronization may also require specialized hardware, such as a GPS receiver.

Clearly, if homogeneous hardware were deployed, each node would need to meet the minimum resource requirements for all tasks. Because the number of nodes required will be determined by the desired sensor coverage, the overall cost of the network will be unnecessarily high. If, instead, a large number of inexpensive nodes were allocated for sensing, and a smaller number of more expensive nodes were allocated to data analysis, localization, and time synchronization, the overall cost of the network would be reduced.

13.2.2 Longevity

Whether indoors or outdoors, sensor nodes cannot always be placed near a source of line power and must instead be powered by battery. Prolonging sensor network lifetime is a critical issue because of the limits of slowly improving battery technology, physical size requirements, and cost. The lifetime requirements of scientific applications can vary greatly; however, habitat monitoring applications typically require a lifetime of 6 to 9 months [34]. In commercial applications, the maintenance cycle must be on par with existing maintenance tasks, such as light bulb replacement, which is typically on the order of 6 months to a year.

Kumar et al. consider the suitability of two hardware platforms for various tasks, given their respective power consumption [30]. They consider the Mica mote, which uses very little power but performs complex calculations slowly, and the iPAQ, which consumes significantly more power but performs computations relatively quickly. Their results indicate that when significant computation is required, a faster processor can be more energy efficient than a slower one, due to the short time required to perform the calculation. However, for sensing tasks that require operation over a long period of time, a low-power node that meets the minimum processing requirements is more effective. Thus, a tiered architecture that partitions network functions among hardware designed for each function may increase network lifetime.

13.2.3 Scalability

A sensor network must scale with the required number of nodes in terms of bandwidth and lifetime. However, it is well known that bandwidth in a flat ad hoc network does not scale. It has been derived analytically that optimal per-node throughput in an ad hoc network of n nodes is given as $\Theta\left(\frac{W}{\sqrt{n}}\right)$, where W is the bandwidth of the shared channel [19]. Thus, as the size of the network increases, per-node throughput decreases toward 0. Moreover, experimental results have shown per-node throughput to decay as fast as $c/n^{1.68}$ [20] — even faster than the analytical result.

Analytical studies of tiered architectures are promising. One approach is to use a single channel in a hierarchical communication structure, in which nodes on the lower tier form clusters around regularly deployed base stations. Each base station acts as a bridge to the upper tier, which provides intercluster communication across a wired infrastructure. In this case, the network capacity grows linearly with the number of clusters, but only if the number of clusters grows at least as fast $\sqrt{n}$ [32]. Other researchers have explored the notion of using different channels at different levels of the network hierarchy [62]. In this case, the capacity of each layer in the tiered architecture and the capacity of each cluster in a given layer scale independently.

Scaling ad hoc wireless networks in the physical dimension leads to low density and poor connectivity. In such networks, it may make sense to introduce an overlay of nodes capable of long-distance, or even fully connected, communication [34]. Analytical results and simulations of real topologies have shown that this architecture can improve connectivity in a linear or strip topology, but has a lesser effect in more general two-dimensional networks [15].

Finally, scaling of services in an ad hoc network can be affected by tiered architectures. In particular, scalable address-lookup services have received significant study [11, 40, 46]. Such services can be fully distributed to all nodes, partially distributed to a subset of nodes, or centralized. Assuming that nodes are mobile and must change their addresses periodically, the balance between these choices depends on the relative frequency of update and lookup operations. Several mechanisms for address lookup in tiered architectures will be explored in Section 13.6.

13.3 Spectrum of Sensor Network Hardware

Although sensor network research is still in an early stage, various hardware platforms are available today that make the tiered architecture a practical choice. This section reviews the entire spectrum of hardware platforms for sensor networks. At one end of the spectrum are small nodes that have slow processors, small memory space, and short-range radios. These nodes consume very little power and normally operate on batteries. An example of the small nodes is Berkeley motes [23]. The other end of the spectrum is occupied by big nodes that have fast processors, large memories, and significantly greater energy requirements. Some of them are simply powerful PCs in a very compact form factor, such as the PC/104 [37]. Others are custom designed nodes with integrated radios and specialized sensing channels, like the Sensoria WINS NG 2.0 [36].

13.3.1 Small Sensor Nodes

An important design goal of sensor networks is the ability to embed deeply into the physical world large numbers of sensor nodes that ubiquitously perform sensing, processing, and actuation tasks over long time spans. To meet this goal, sensor nodes must be small and have low power consumption.

A first example is the Berkeley mote [23]. A mote tightly integrates an 8-bit microcontroller with a low-power radio and various sensors. The Smart Dust project first developed the mote concept [54]; the TinyOS group expanded the original hardware design [56] and developed an efficient event-driven operating system called TinyOS [23, 56]. With this important step, the mote has become one of the most widely used sensor-net platforms in the research community today.

Of the several subsequent generations of motes, current and widely used versions are the Mica mote and the Mica2 mote. Figure 13.4 shows a picture of a Mica2 manufactured by Crossbow Technology, Inc. [14]. Mica and Mica2 have an ATMega128L microcontroller from Atmel [14], which has an 8-bit RISC processor core with 128-KB flash memory, 4-KB SRAM, and a throughput of up to 1 MIPS per MHz. The CPU clock is 4 MHz on Mica [14] and 7.37 MHz on Mica2 [56].

Mica and Mica2 use different radios. Mica uses the RFM TR1000 [47] or TR3000 [48] transceiver module, while Mica2 uses the Chipcon CC1000 [12]. The RFM radios are narrow band and only operate in fixed frequency bands: 916 MHz for TR1000 and 433 MHz for TR3000. The Chipcon CC1000 is able to tune to different frequency bands from 300 MHz to 1 GHz, and can be used as a frequency hopping radio. However, on Mica2, the radio is pretuned to a specific frequency band. Table 13.1 compares some features of the RFM TR3000 and the Chipcon CC1000 at 433 MHz.

The connector on Mica and Mica2 is used to connect to extension boards with various sensors. Current supported sensors include light, temperature, humidity, pressure, infrared, acoustic, accelerometer, magnetometer, wind speed, and wind direction [14, 34]. Motes also support simple actuators such as color LEDs and buzzers.

Although Mica and Mica2 motes are small and energy efficient, they are still far from the targeted lifetime goal of operating for years on batteries. Therefore, researchers are striving to reduce the size, cost, and power consumption of sensor nodes. The latest "spec" mote is the smallest version of motes developed by UC Berkeley [22]. Its size is only about 2 × 2.5 mm. Spec has a RISC core, 3 KB of memory, 8-bit on-chip A/D converters, 4-bit I/O ports, and an integrated radio. Spec dramatically reduces the size, cost, and power consumption on motes, but it provides a reduced capability.

Motes are designed to be general-purpose platforms that are easy to use in sensor network research. A similar platform was developed by the MANTIS project [35] at the University of Colorado, called Nymph [1]. It also uses an Atmel ATMega128L microcontroller with the Chipcon CC1000 radio, as on Mica2 motes. Nymph aims to provide more flexibility, fast prototyping with multimodal sensors, and reduced hardware complexity. It is the first tiny sensor node that directly supports the GPS. The MANTIS

FIGURE 13.4 Mica2 mote, manufactured by Crossbow Technology, Inc.

TABLE 13.1 Comparison of RFM TR3000 and Chipcon CC1000 Radios

Radio module	Modulation	Date Rate (kbps)	Tx Power (dBm)	Power consumption		
				Sleep	Idle/Rx	Tx (0 dBm)
RFM TR3000	ASK	Max. 115.2	Max. 0	2.1 μW	9.3 mW	22.5 mW
Chipcon CC1000	FSK	Max. 76.8	−20 to 10	0.6 μW	22.2 mW	31.2 mW

project also developed a small multitasking operating system on Nymph that provides a programming environment similar to UNIX.

Some industry developers use similar small platforms in their products, with more focus on real world applications. For example, the EM900 and EM2400 modules developed by Ember Corporation provide a direct sequence spread spectrum radio with an 8-bit RISC processor and hardware-based advanced encryption standard (AES) [17]. With low-level network protocols implemented, these nodes are designed to act as a radio front end to other bigger nodes.

Intel Corporation developed an enhanced version of the mote called Intel® mote (Imote) [24, 28]. It utilizes a more powerful ARM processor core, a 32-bit architecture. To reduce size, Imote integrates the CPU, flash memory, SRAM, and a Bluetooth radio onto a single chip. In the current specification, the CPU clock is 12 MHz, and 512 KB of flash memory and 64 KB of SRAM are on the chip. The size of Imote is 3 × 3 cm. The main board can be extended by stackable module boards that provide sensing, actuation, and debugging capabilities, as well as different power supply options. The Imote also runs TinyOS, ported from Berkeley motes, so that most applications available for Berkeley motes are able to run on Imote without modifications.

Another enhanced mote-like platform is the Medusa MK-2 node, developed at UCLA [49], which augments the computing power of the Mica mote by integrating a second microcontroller with an ARM THUMB core. It is a 32-bit RISC processor running at 40 MHz with 1-MB flash memory and 136 KB of SRAM. Similar to the Mica mote, Medusa MK-2 also uses an Atmel ATMega128L microcontroller and the RFM TR1000 radio. The ATMega128L supports low-level radio communication and monitors simple sensors; the ARM THUMB processor is used for more extensive computations.

13.3.2 Large Sensor Nodes

Despite their advantages, small sensor nodes like motes are sometimes not capable of performing certain sensing and processing tasks on their own. For example, acoustic beamforming and localization require a fast sampling rate at high accuracy and extensive computing such as fast Fourier transform (FFT) [58]. To meet the requirement of more computing power, larger nodes have also been developed and used in sensor networks. These nodes have significant computing power, large memories, and more I/O peripherals, such as Ethernet or PCMCIA connectors. On the other hand, larger nodes consume more power and many are not easy to deploy with a battery power supply.

Nodes are roughly classified into this group if they have a high-speed 32-bit microprocessor, large memories, and high power consumption. Although the Imote has a 32-bit ARM core, it only runs at 12 MHz and has very low power consumption; thus, it is considered to be a small node. The Medusa MK-2 node has a faster ARM THUMB core at 40 MHz and consumes more power. It falls between the Imote and the large nodes described in this subsection.

The Intel StrongARM RISC processor is a popular choice in large nodes for sensor networks. Examples include AWAIR I from Rockwell Science Center and UCLA [3], the µAMPS node from MIT [51], the TCP/IP gateway node from Ember Corporation [17], PDAs like Compaq iPAQ [27], and embedded systems like Cerfcube [26]. The AWAIR I and µAMPS nodes have integrated sensors and radios. The Ember gateway node, Cerfcube, and iPAQ are more general computing platforms and use the Linux operating system for sensor network applications [17, 30, 34].

Some new nodes are based on Intel XScale™ microarchitecture, which is ARM architecture-compliant and application code-compatible with Intel StrongARM processors. Examples include the Stayton board [25] and the Stargate board [14] designed by the Intel Corporation as research platforms. Stayton and Stargate have similar components and capability, but different form factors. Stargate has a 400 MHz Intel XScale processor (PXA255), 32-MB flash memory, and 64-MB SDRAM. Standard I/O includes a PCMCIA slot, a compact flash slot, and a connector for a Mica or Mica2 mote. Stargate can be further expanded by a daughter board that provides Ethernet, serial, and USB ports.

Specialized sensor nodes have also been developed to meet the requirement of applications with highly extensive computing and sensing tasks, such as the WINS NG 2.0 [36]. WINS NG 2.0 employs the Hitachi SH-4 processor, which is a 32-bit RISC architecture with 300 MIPS CPU and 1.1 GFLOPS FPU. The node provides 15 general-purpose I/O (GPIO) lines, 4 analog input channels with a sampling frequency of 20 kHz, and 16-bit A/D converters. The node also provides an integrated GPS receiver, sensor connectors, Ethernet, and 2 PCMCIA slots. Another impressive feature is that WINS NG 2.0 provides two radios, which are useful for protocols that require two radio channels, such as those described in Subsection 13.5.3.2.

Embedded PCs are at the highest end of the sensor node spectrum. PC/104 and PC/104-plus are examples of the embedded PC architecture that support Intel microprocessors from i386™ through Pentium® III [2]. They offer full architecture, hardware, and software compatibility with the PC bus, but in ultracompact (90 × 96 mm) stackable modules [37]. Although PC/104 only supports the ISA bus, PC/104-plus supports the ISA bus and the PCI bus. Another example of embedded PCs is the system-on-modules and carrier boards (e.g., Netcard II) from PFU Systems, Inc. [43]. Its Plug-N-Run product line features 32-bit PCI with a single processor from Pentium to Pentium III in an ultracompact form factor. The carrier board provides standard peripheral connectors, such as IDE, USB, Ethernet, parallel, serial, keyboard, mouse, and CRT.

These embedded PCs function like ordinary desktop PCs, but with much smaller sizes. Some of the platforms provide GPIO lines that can be used to attach sensors. However, these PCs do not have integrated radios for wireless sensor networks. One solution is to add off-the-shelf radios, such as the RPC modules from Radiometrix Ltd. [9, 45] or IEEE 802.11 PCMCIA cards. On the other hand, to work

FIGURE 13.5 PC/104 attached to a Mica mote and a PIR sensor.

in a tiered architecture, these large nodes must be able to communicate with small nodes in the network. Therefore, they must have a radio compatible with that of the small nodes. For example, in a tiered network with motes and PC/104s, a mote can be attached to a PC/104 through its serial port and act as a radio interface with only low-level networking protocols [16]. Figure 13.5 shows a PC/104 attached with a Mica mote as the radio interface and a passive infrared (PIR) motion detector.

This wide variety of hardware allows network designers to allocate different node capabilities to different tiers of the network. The next step is to decompose the application into separate tasks and then assign each task the most appropriate hardware.

13.4 Task Decomposition and Allocation

A major characteristic of sensor networks is that all nodes in the network collaborate toward a common application. An important design issue is how to achieve good application performance in a cost-effective and energy-efficient way. Leveraging the wide hardware spectrum, a designer should decompose a complex application into different tasks and assign them to appropriate hardware in the tiered network. The goal is to match different task requirements with different node capabilities.

Although each application has a different set of tasks to be carried out, three basic types of tasks exist in a sensor network: sensing, processing, and communication. Sensing is the process of collecting data from the physical world. Data from different sensors are processed inside or outside the network to obtain a better understanding of the environment. Communication enables collaborative signal and data processing from multiple sensors and delivery of results to interested users.

13.4.1 Sensing

The sensing task uses different types of sensors to capture different signals from the physical world, such as temperature, light, acoustic, and seismic. All signals decay as they travel away from the source. As a result, the signal-to-noise ratio (SNR) decreases with distance. SNR is one of the fundamental factors that decide the quality of signal processing, such as detection and estimation. A dense deployment places sensors as close to the target as possible, thus improving the quality of sensed values. Dense deployment also increases the number of opportunities for the line-of-sight observations essential for accurate range estimation. Another way to improve sensing reliability is to deploy sensors with enough density for multiple sensed values to be combined together and thus provide higher confidence.

In order to deploy sensors with high density and close range to each target, it is most cost-effective to utilize small nodes like motes. They are also easy to deploy because they have small form factors and their own power source. Small nodes are also more power efficient than large nodes, thus enabling long network lifetime. There are various examples of utilizing motes with sensors, which send their sensing results to a large node for further processing [34, 59]. For example, in Wang et al. [59], a two-tiered network is formed with PC/104s and Mica motes. The motes record bird calls using an acoustic sensor and forward appropriate signals to PC/104s for recognition and localization.

Different sensing tasks have different hardware requirements, based on the sampling rate and accuracy. In an environmental monitoring application, ambient temperature may change slowly, potentially allowing a sampling rate as low as one sample every 10 minutes (1.67×10^{-3} Hz) [34]. Small nodes are well suited for such tasks. However, some sensing tasks have high demands on CPU and memory resources. In an application that recognizes bird calls, the acoustic sampling rate could be as high as 22 kHz [59]. Most small nodes are not capable of performing such sensing tasks.

In some cases, special sensors are only available on certain nodes. For example, in the tiered architecture described by Wang et al. [59], only PC/104s are equipped with GPS receivers. It is obvious that the task of providing location and time information should be assigned to these PC/104s. As cluster heads, they can provide such information to small nodes within their clusters.

In summary, it is desirable to allocate most sensing tasks to small nodes to take advantage of low cost, high density, and physical proximity to the target. When a sensing task exceeds the capability or resource of small nodes, it can be allocated to large nodes.

13.4.2 Processing

Processing is another basic task in sensor networks. This task can be as simple as detecting abnormal temperature changes in a fire alarm system or as complex as tracking a target moving through the network or estimating the direction of a bird call, which require extensive computation. In sensor networks, processing often combines multiple sensor outputs from local neighboring nodes, and it is thus referred to as collaborative signal and data processing. Collaborative processing has two major advantages. First, by combining multiple sensor outputs, the processing result is more reliable and accurate. Second, only the aggregate result needs to be sent to a user across the network and through gateway nodes, which can save a significant amount of energy.

In general, small nodes are only suitable for lightweight processing due to their limited computing power. An example task for small nodes is computing simple aggregates such as the average, minimum, and maximum value from different sensor readings [33]. A small node might act as a front end in a computing hierarchy and perform preprocessing for later stages. In Wang et al. [59], besides acoustic sensing, motes also perform simple filtering to reduce irrelevant events that would result from recorded sounds not produced by birds. After sampling a desired signal, motes perform data reduction by extracting the most important features in the data set and sending them to a large node after compression. Such preprocessing largely reduces the computation load on large nodes and the communication overhead between small and large nodes.

Large nodes should perform processing tasks that demand extensive computations, such as beamforming, target recognition, and classification. Sensor networks are able to take advantage of the strong computing power of these large nodes by performing most processing within the network. Compared to sending all raw data to a base station, in-network processing saves a significant amount of energy by reducing the communication cost [44].

13.4.3 Communication

Communication is perhaps the most complex task in a sensor network due to its ad hoc nature and resource constraints. The communication task can be further divided into subtasks roughly represented by different layers, as in traditional computer networks. A common decomposition includes a medium access control (MAC) layer and a routing layer. Communication enables not only collaborative processing, but also the interactions between a user and the sensor network.

To enable collaborative processing, nodes must be able to communicate with each other. In a tiered network, nodes are often organized into clusters. If a large node exists in a cluster, it is normally selected as a cluster head. No matter what size they are, these nodes must use the same radio to communicate. They also need to run the same low-level protocols, such as the link and MAC protocols. An example is LEACH [21], in which a cluster runs a TDMA protocol. Within a cluster, nodes only send their data to the cluster head. The cluster head sends aggregate data to a base station using a long-range radio. The role of cluster head will typically rotate among cluster members in order to distribute energy consumption evenly.

On the other hand, it is sometimes possible to place most of the communication burden on a subset of nodes. For example, in a TDMA cluster like LEACH, nodes only communicate with their cluster head. Therefore, only cluster heads need to participate in a routing protocol. If a cluster allows peer-to-peer communications, cluster members need only participate in intracluster routing; however, the cluster head must participate in inter- and intracluster routing. Finally, with the various hardware choices described in Section 13.3, some large nodes may have special communication capabilities, such as a long-range radio or multiple radios. These are suitable to form a communication backbone to carry more traffic than other small nodes. The interaction between routing and clustering is discussed in detail in Section 13.6.

Task decomposition and allocation are important issues in designing a tiered network. Appropriate task allocation is able to improve sensing reliability, reduce network cost, reduce energy consumption in computation and communication, and utilize special resources better.

13.5 Forming Tiered Architectures

With the hardware described in Section 13.3 and the set of tasks assigned to that hardware described in Section 13.4, the network can now be organized into a tiered architecture. A wide variety of mechanisms have been proposed to create tiered networks. Some are limited to forming two-tier hierarchies, while others can be extended to an arbitrary number of levels. Some mechanisms are designed to identify and exploit physical heterogeneity; others create small logical groups of nodes to improve scalability. The following subsections break down the approaches into three categories, describing each in more detail: engineered networks, routing mechanisms, and clustering mechanisms.

13.5.1 Engineered Networks

A simple way to organize a network into tiers is to engineer the network by hand. The network designer must specify which nodes participate at each tier and how the nodes in each tier will be organized. A tiered architecture can be created by manually configuring a routing topology, by specializing the software loaded on each node, or by providing specialized hardware on particular nodes.

A common use of this approach is in a sparse sensor network, as described in Mainwaring et al. [34]. In this case, several dense pockets of sensor networks are deployed relatively far apart, to form a single network. Within each pocket, short-range communication is possible, allowing low transmission power and simple omnidirectional antennas to be used. The spacing between sensor clusters is such that no two nodes in different clusters can communicate. Instead, one or more nodes with long-distance communication capabilities are deployed in each pocket. These nodes may include a more sensitive or powerful radio or a directional antenna, thus creating a communication backbone for the network.

Tiered architectures can include varying degrees of manual organization. Automatic organization of nodes into tiers is desirable for the same reasons that ad hoc deployment of wireless sensor networks is desirable. Manual configuration of large numbers of nodes is too time-consuming and expensive, particularly given the time-varying conditions of the interconnecting wireless links.

13.5.2 Routing Mechanisms

One way to use the resources of a subset of nodes automatically to benefit the entire network is to bias routing in favor of resource-rich nodes. Route biasing can be used to increase the packet forwarding load on nodes with more remaining energy (or that are wall powered), thus increasing the lifetime of the network. Route biasing can also be used to attract more data to nodes with greater processing power, increasing the amount of in-network processing.

Resource-biased path selection [10] introduces a delay in forwarding route-selection packets at nodes with lower than average remaining energy. Because on-demand ad hoc routing protocols like AODV [42] typically identify the path with the lowest latency, this approach tends to avoid paths containing nodes with little remaining energy, thus increasing the overall network lifetime. A small modification to the routing protocol is required, but backward compatibility with other AODV nodes is maintained. This approach works best in environments with many resource-rich nodes, in which case the latency to find a route will be low. However, because the added delay reflects the relative cost of routing through a resource-constrained node compared with the resource-rich nodes in a given network, the delay value can be difficult to determine and may need to change as the average remaining energy of nodes changes.

Energy-aware routing (EAR) is similar to the preceding approach, except that it uses a different metric to select appropriate routes. In one implementation described in Shah and Rabaey [50], each node maintains a list of neighbors and the cost of transmitting through those neighbors to a given destination. The cost is computed using the metric advertised by that neighbor, plus a hop metric consisting of a weighted multiple of the cost of transmitting a packet and the fraction of energy remaining. The average cost of forwarding through each neighbor is advertised to other nodes. The paths then selected tend to be those that include the least expensive links and the nodes with the most remaining capacity.

Although the preceding protocol was originally intended to distribute the cost of packet forwarding evenly across a homogeneous network, an extension to this protocol called EAR+A [60] allows resource-rich nodes to be altruistic and accept a disproportionate load. In EAR+A, the hop metric is inversely proportional to the remaining energy on the forwarding node. Resource-rich nodes periodically announce their altruistic nature to their neighbors. When making a forwarding decision, a node biases the metrics received from each altruistic neighbor by a *cost reduction factor*. As a result, packet routing will tend toward altruistic nodes in a greedy manner.

In all of these protocols, biasing route selection in favor of nodes with more resources allows the network as a whole to take advantage of the resources on a subset of nodes. These routing mechanisms do not form a hierarchical structure; however, they do allow resource-poor nodes to become aware of and benefit from resource-rich nodes. The benefits tend to be modest, but the overheads are low, beyond the overhead of the underlying routing protocol.

13.5.3 Clustering Mechanisms

An alternative to the routing protocols described in the preceding section is to divide the network into clusters of nodes led by a cluster head. Cluster members can utilize resources or services available at the cluster head. Because cluster heads can form clusters, clustering can be hierarchical. Clusters can be used to form a physical hierarchy, organize a logical hierarchy in a flat topology, or simply identify the set of nodes that will use a particular specialized resource, such as a GPS receiver.

Clustering algorithms can be judged on the properties of the clusters they form. Although many algorithms are designed to form one-hop clusters, others limit the size or diameter of a cluster. The size of a cluster controls the load on the cluster head and its diameter controls the cost of communication between each node and the cluster head. A balance between cluster size and cluster diameter will typically be desirable.

Cluster stability is also important. Clustering must be dynamic — adapting to mobility and changes in network connectivity. In addition, clusters must be stable in the face of small changes, or the cost of periodic cluster reformation will reduce the potential benefit. Cluster stability can also have application-specific benefits. For instance, the computed communication delay between each cluster member and the cluster head, which is required for some beamforming algorithms, can be reused until cluster membership changes [30].

The following subsections break down clustering algorithms into two classes. The first class is used to create a connected backbone in a flat network, where cluster heads are resource-rich. In a second class of algorithms, a hierarchical structure is used to constrain network communications and organizes in-network computation.

13.5.3.1 Forming a Connected Backbone

A very early example of a clustering algorithm used to form a connected backbone is the linked-cluster algorithm (LCA) [5]. LCA first selects cluster heads to form a dominating set. A distributed algorithm selects cluster heads based on node ID so that every node in the network is one hop from a cluster head. Nodes in the dominating set (the cluster heads) will now be separated by no more than three hops (two intermediary nodes) as shown in Figure 13.6. Nodes then exchange information about their two-hop neighborhood with their neighbors. Nodes that can bridge the gap between adjacent clusters become cluster gateways. Together, the cluster heads and gateway nodes form a connected dominating set. Every node in the network is part of the connected dominating set or it is one hop away from a node on the connected dominating set. The connected dominating set can be used as a communication backbone so that only nodes in the connected dominating set need to forward packets or participate in route discovery.

CEDAR [53] uses an algorithm similar to LCA to create a communication backbone called the *core*. Like the backbone created by LCA, the core is not a minimum dominating set, the creation of which is known to be NP-hard. However, because the overhead of backbone creation must be balanced against backbone optimality, a nearly optimal backbone may be a more appropriate goal. CEDAR uses a multihop beacon to identify paths between neighboring core nodes, but an extension reported in Sinha et al. [52] provides a more efficient mechanism based on local neighbor information exchange. CEDAR provides an efficient network flooding service by constraining the network topology to the backbone and single-hop links from nonbackbone nodes to the backbone. Using the constrained topology, CEDAR reduces packet transmissions required by flooding and allows flooding to replace MAC-level broadcasts with unicasts, which can utilize an RTS-CTS-DATA-ACK exchange [8].

Relay organization (ReOrg) [13] and CEC [64] are two very similar algorithms that form a backbone of nodes that have the maximal remaining energy. As with LCA, both algorithms form a connected dominating set; however, they use remaining energy as the metric for electing cluster heads and gateway nodes. In CEC, the goal is to identify network redundancy. Nodes not in the connected dominating set are considered redundant and put into a low-power state. The remaining nodes perform the sensing and communication tasks. In ReOrg, the topology of the network is artificially constrained to consist of the backbone and links from nonbackbone nodes to their elected cluster head. All nodes in the network

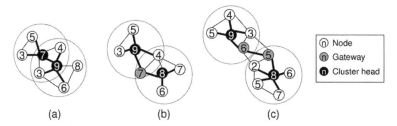

FIGURE 13.6 Gateways turn a dominating set into a connected dominating set by filling in the one- to three-hop path between cluster heads. Each node is labeled with its metric.

utilize a low duty cycle and synchronize with their neighbors periodically to allow communication. Nonbackbone nodes consume significantly less power because they do not forward packets on behalf of other nodes and they do not need to synchronize with any neighbor except their elected cluster head.

In both algorithms, clustering allows some nodes to sleep more than others. Because cluster head and backbone selection are based on remaining energy, periodic reclustering balances the load across the network, increasing network lifetime. ReOrg has also been shown to be able to leverage wall-powered nodes when they are available, thus further increasing network lifetime.

GAF [64] has similar goals to CEC, except that clusters are formed geographically, rather than according to network topology. GAF is designed for very dense sensor networks. The network is divided into a geographic grid designed so that any node in one grid square is within nominal communication range of every node in each adjacent grid square. GAF elects one node in each grid square to act on behalf of the grid square in all sensing and communication functions. All other nodes in the grid square can enter a low-power sleeping state while still maintaining full sensor coverage and a fully connected topology. By rotating the role of active node, GAF can extend the lifetime of the network in proportion to its density.

13.5.3.2 Forming a Hierarchical Communication and Processing Structure

Clustering can be used to impose a hierarchical organization in an otherwise flat ad hoc network; the hierarchy can lend structure to in-network computation. Each cluster member forwards sensor data to the cluster head, which fuses data from multiple sensors (and potentially different types of sensors) into a single observation. The resulting observation can typically be transmitted more efficiently across the network to a consumer than the individually sensed values can be. Such fusion can occur at multiple levels of a tiered architecture, allowing multiple observations to be fused together further.

Clustering can also introduce a hierarchy of data transmission. A hierarchy of clusters forms a tree structure. Nodes in each tier of the tree are divided into clusters in which the cluster head represents the cluster at the next higher tier of the tree. When a single data sink is present, it will typically be the root of the tree. When a node wishes to forward a sensed value, it can send a packet to its cluster head, which, in turn, forwards the packet to its cluster head, until the packet reaches the sink node. Control information can be flooded in the reverse direction, from cluster heads to cluster members. More general patterns of communication are also possible on a tree structure. When a node wishes to send a packet to a node outside its own cluster, it does so through the cluster head. Routing and addressing schemes for tiered architectures are discussed in Section 13.6.

The simplest clustering algorithms create one-hop clusters and use simple metrics to select cluster heads. For example the lowest ID (LID) and highest degree (HD) algorithms elect cluster heads based on a node's ID or the number of its neighbors [18, 33]. A node becomes a cluster head if it has the best metric between itself and all of its neighbors. A node relinquishes its cluster head status if a node with a better metric becomes a neighbor. Although simple, these approaches tend to be unstable in the presence of mobility because the IDs present in a neighborhood and the degree of a node will constantly change.

An alternative proposed in Basagni et al. [7] is to use node velocity as the metric. Because nodes with lower velocity tend to be chosen as cluster heads, the cluster heads tend not to change. However, nodes with high velocity may change clusters frequently, leading to rapidly changing cluster membership. Random competition-based clustering (RCC) [62] also aims to reduce the impact of mobility on cluster stability. In RCC, a node not in a cluster broadcasts a beacon after a random timeout. The first node to send a beacon becomes a cluster head (with node ID used to break ties); all other nodes hearing the beacon become cluster members. A cluster head periodically resends the beacon to retain its cluster head status. RCC allows nodes traveling together to form a stable cluster regardless of their absolute velocity. Both of these approaches have been shown to produce stable clusters with a reduced number of cluster reconfigurations, when compared with LID and HD.

Random clustering is also proposed in the dual network clustering (DNC) algorithm [55], but in this case each node is assumed to have two independent radios. DNC uses two fixed channels for communication across the entire network, and each node tunes one radio to each channel. At power-up, each node turns on both radios and each radio listens for a random time period. If no message from a cluster head is heard, the node becomes a *cluster head* on this radio; otherwise, the node is a *remote* on this radio and joins the existing cluster. A node is not allowed to have two radios acting as cluster head at the same time. As a cluster head, a node can provide a service for its remotes, such as acting as TDMA controller.

Although the preceding discussion focuses on one-hop clusters, cluster size can greatly affect performance. Researchers at UCLA have focused on creating a mobile backbone network (MBN) [62], which is a two-tier architecture forming a physically hierarchical network. The lower network tier uses a short-range radio, while the upper tier uses a long-range radio. Most nodes have only a short-range radio; however, a subset of the nodes possesses both. Nodes with two radios are able to act as cluster heads and form a mobile backbone that connects neighboring clusters. In general, because nodes are mobile as well as prone to failure, the network contains more two-radio nodes (and thus potential cluster heads) than are necessary. The number of active cluster heads (and thus cluster size) is important in an MBN because it strikes a balance between local-cluster capacity and backbone capacity. To maximize available bandwidth as the network grows (and thus scalability), the optimal number of clusters in such a network is $\frac{W_1}{W_2}\sqrt{N}$, where W_1 and W_2 are the respective bandwidths of the short- and long-range channels and N is the total number of nodes in the network. With fewer clusters, the intracluster traffic is the limiting factor, while with more clusters the backbone is the limiting factor. To achieve the optimal cluster size, the beacons used in the RCC algorithm (above) are propagated K-hops. K must be chosen to achieve the optimal number of clusters.

Cluster size is also important in a flat network topology. Flat network topologies allow collaborative processing, scalable routing solutions, and scalable service discovery. In such cases, cluster size affects the load on these service providers. The clustering algorithm described in Banerjee and Khuller [6] creates a multitiered structure of fully connected clusters with low overlap and bounded size. First, a spanning tree on the network graph is identified. Next a node is selected whose subtree has more than k nodes and whose children are each the root of a subtree with less than k nodes. Subtrees are then combined into clusters of sizes between $k - 1$ and $2k - 2$, leaving at most one subtree remaining. To ensure connectivity, the originally selected node can be included in each cluster. Finally, all clustered nodes are removed from the spanning tree and the algorithm repeats until all nodes are in clusters. The algorithm is described here as a centralized algorithm; however, Banerjee and Khuller also describe a distributed version of the algorithm, complete with cluster maintenance procedures.

The rendezvous clustering algorithm (RCA) [55] also provides a mechanism for limiting the member size of one-hop clusters in a network of nodes with two radios. This algorithm was proposed as an alternative to DNC (described previously), which does not limit cluster size. In RCA, one of the two radios on each node is initially tuned to the rendezvous channel (R-channel) used for cluster formation. Each cluster head periodically advertises its existence, as well as a metric describing the number of nodes in its cluster, on the R-channel. After a short listening period during which metrics of the existing clusters are gathered, a node can choose to join a small or moderately sized existing cluster, create a new cluster, or steal members from a large existing cluster in order to create a new cluster. Once a properly sized cluster has formed, it is moved from the R-channel to another channel. When one radio becomes a cluster head, the other is tuned to the R-channel, periodically broadcasting the existence of that cluster head. Using the cluster-size metric, RCA has been shown to control cluster size effectively in simulation.

This section has outlined a variety of mechanisms for forming a hierarchical network. The simplest, but least automatic, approach is to engineer a physically hierarchical network. Routing mechanisms provide an automatic alternative that can identify and utilize a hierarchy of services available in the network. Finally, clustering techniques can be used to organize an ad hoc network; clustering can be used to form a backbone in a two-tiered logical hierarchy. Alternatively, clustering can be used to form a logically or physically multitiered organization. Each clustering algorithm creates topologies with different degrees of connectedness, cluster size, and cluster stability. Each approach introduces a certain amount of overhead and complexity, and some algorithms assume specialized hardware, such as multiple radios.

As a result, none of the approaches presented is entirely superior. Selection of the best approach depends on the application.

13.6 Routing and Addressing in a Tiered Architecture

The efficiency of network routing can be increased by organizing a network into a tiered architecture. In a flat sensor network, route discovery usually requires packets to be flooded across the entire network (although alternatives such as geographic routing [63] are sometimes possible). The backbone creation protocols described in Subsection 13.5.3.1 reduce the cost of flood-based route discovery by constraining the set of network paths over which packets can flow. The hierarchical clustering mechanisms described in Subsection 13.5.3.2 allow a hierarchical approach to route discovery.

Hierarchical routing can take two forms. The process of discovering a route to a destination node can be tailored to take advantage of the hierarchical nature of the network. Alternatively, the hierarchical location of a node can be encoded in the node's address. The latter approach simplifies route discovery by introducing the problem of ever-changing node addresses. The following two subsections explore these approaches in more detail.

13.6.1 Routing in a Hierarchy

In a hierarchical network composed of clusters, route discovery can be simplified by splitting the problem into two cases: routes to nodes inside the local cluster and routes to nodes outside the local cluster. Route discovery within a cluster will have low overhead if the cluster size is small. In many cases, it is reasonable to assume a certain amount of communication locality. Thus, a more expensive global route discovery may only be required occasionally. Several hierarchical route discovery mechanisms have been proposed.

H-AODV provides a simple form of hierarchical routing on a physically tiered network [61] (Figure 13.2). In this approach, AODV [42] is modified to forward route request messages across the topology of each tier as well as across tiers at cluster heads. For instance, in a two-tiered network, the lower layer would consist of clusters that elect a set of backbone nodes. These backbone nodes use an independent, long-range radio to form a backbone on the second tier. While route request messages flood the lower tier, they are also forwarded by gateway nodes onto the upper tier. As route messages flood the upper tier, they are also forwarded down to the lower tier by other gateway nodes. As a result, a route can be discovered that utilizes a few hops across the backbone network as a short-cut path in place of many hops across the underlying network (Figure 13.7). This approach extends naturally to multiple tiers, but would require an additional channel for every tier.

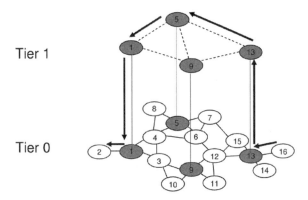

FIGURE 13.7 A backbone network on the upper tier of a two-tier network provides a shortcut, turning a six-hop route into a four-hop route.

An alternative approach, also reported in Xu and Gerla [61], uses independent routing layers at each network tier. For instance, DSDV [41] could be used for proactive routing within a cluster, while AODV [42] is used for reactive routing on the backbone. Because DSDV is proactive, each node knows the set of nodes within the cluster. If a node wishes to communicate with another node in the cluster, it already knows the route. If a node wishes to communicate with a node outside its cluster, it can forward the packet to the cluster head. The cluster head can use AODV to find a route to the cluster head of the destination's cluster. Because every cluster runs DSDV, every cluster head knows which nodes are present in that cluster. The packet is then forwarded to the destination's cluster head, which forwards the packet to the destination via DSDV. This approach can be applied to other combinations of protocols, but it is particularly attractive in this proactive–reactive combination, which reduces the amount of routing information that must be stored and maintained at each node.

A third approach for physically hierarchical networks uses reactive routing in a two-tiered architecture and breaks route discovery into two parts [29]. In the "front part," a node floods a route request on the lower tier, attempting to find a route to the destination node or the cluster head. If the route request reaches the destination node, the node sends a response containing the path and route discovery is complete. If a path to the cluster head is found first, a second message is then sent to the cluster head initiating the "rear part" of the discovery protocol. The cluster head now attempts to find a route from itself to the destination node, in its own cluster or through the upper-tier network to other clusters. When such a route is found, the cluster head sends a final route response to the originating node, establishing a route to the destination through the cluster head. This protocol, like the combined proactive–reactive discovery protocol, is more efficient at discovering local routes. Like both of the previous protocols, this protocol is able to identify efficient routes across a physically hierarchical network, utilizing the upper tier as a shortcut.

13.6.2 Hierarchical Addressing

Tiered architectures can also reduce the overhead of proactive route discovery without the latency of reactive route discovery. Each node is given a unique identifier and a logical address that designates its position in a hierarchical network. Because the node's address indicates its location in the hierarchy, packets can be directed toward their destination without reactive route discovery and with limited table maintenance. However, this approach also requires a mechanism to map unique identifiers into hierarchical addresses.

13.6.2.1 Routing with Hierarchical Addresses

The hierarchical state routing (HSR) protocol [40] uses logical clustering (Figure 13.3) to form a multitiered hierarchy. Ordinary nodes are in a cluster at the lowest level of the hierarchy. The cluster head is a member of the next level of the hierarchy, and so on. Each node has a unique address and a hierarchical address of the form $CH_n.CH_{n-1}...CH_1.ID$, where ID is the node's ID and the CH_is are the cluster head IDs from the node's cluster head to the root cluster head. Link state routing is performed in each cluster, requiring $O(N*M)$ storage, where N is the average number of nodes in a cluster and M is the number of hierarchical levels (because the root cluster head belongs to all M cluster levels). Using these link state tables, each packet can be routed using its hierarchical address alone. From the local node, a packet is forwarded up to a common point on the cluster-head tree and then down to the destination.

Landmark routing [57] also uses multilevel hierarchical addressing but with different route table management. In landmark routing, a packet is forwarded toward a successively closer sequence of landmarks until it arrives at the destination node. A landmark is a node to which packets can be routed from nodes in a neighborhood of a given radius. Landmarks form a hierarchy equivalent to a logically tiered cluster architecture so that each node in a cluster can route packets to the cluster head. A small number of landmarks have a radius larger than the network radius and act as landmarks for the entire network. Each lower level of the hierarchy has a larger number of nodes with a smaller radius. Each node receives a logical address that is the concatenation of landmarks $LM_n, LM_{n-1}, ... LM_1, LM_0$ so that LM_i is a landmark for node LM_{i-1}. LM_n is a landmark for all nodes.

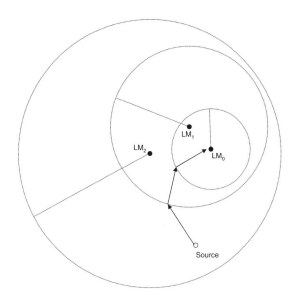

FIGURE 13.8 Routing with landmarks.

Thus, given an address, a packet can be routed toward LM_n. As it nears LM_n, a route will be available to LM_{n-1}, until the packet reaches LM_0 (Figure 13.8). Although landmark routing does not provide shortest path routing, it only requires $O(\log N)$ storage space for routing tables. The cost of route maintenance is also low because most landmarks have a small scope and only a few have a large scope.

LANMAR [39] combines the notion of landmarks with fisheye state routing (FSR) [38]; it forms a two-tiered logical hierarchy consisting of landmarks and regular nodes. Each landmark heads a cluster that represents a subnet. Each node has a logical address consisting of its subnet address and a host ID that is unique to that subnet (and can optionally be the node's globally unique ID). FSR is a link state routing mechanism that provides a variable route update interval proportional to distance. Routes within the fisheye scope (a predefined distance) are accurate, while routes to more distant nodes are updated less frequently. LANMAR uses a modified version of FSR that maintains routes only within the fisheye scope and to all landmark nodes. Thus, a packet to be delivered is first forwarded toward the destination node's landmark (identified by the subnet portion of the node's address). As the packet nears the landmark, it will enter the fisheye scope of the destination (as long as the fisheye scope is larger than the maximum subnet size), and the packet can be forwarded to the destination.

LANMAR can also be used in a physical hierarchy [62]. In this case, landmark nodes use a second, longer range radio channel to form a backbone on the upper tier. FSR is again used to forward packets on the lower tier; however, in this case, routes are only maintained within the fisheye scope. An independent routing protocol (in this case DSDV [41]) is used on the backbone network to route between landmarks. Packets destined within the subnet (and thus inside the fisheye scope) can leverage FSR. Packets destined outside the subnet are first delivered to the local landmark, which then forwards the packet to the landmark for the destination subnet over the backbone. Finally, the landmark in the destination subnet delivers the packet using FSR. This mechanism is similar to the proactive–reactive routing approach described in Subsection 13.6.1, except that both routing protocols are proactive and hierarchical addressing is used.

13.6.2.2 Mapping Unique IDs to Hierarchical Addresses

Each of the protocols described in the previous section uses a hierarchical address reflecting a node's position in the tiered network architecture to reduce the cost of route table management and packet forwarding. The drawback to this approach is that when a node changes its location in the tiered architecture, this hierarchical address must change. Although each node has a unique ID that never changes, it must be mapped to a logical address in order to be useful. Thus, each of these schemes must maintain a mapping mechanism. The mapping mechanism must allow each node to update its hierarchical address whenever its location in the tiered architecture changes, and it must allow a node to look up another node's current hierarchical address, given the node's unique ID.

The original landmark routing [57] scheme provides a single centralized database that maps unique IDs to hierarchical addresses. Unfortunately, the centralized approach does not scale for lookups, nor does it scale for updates in a mobile network. Hierarchical state routing [40] provides a semidistributed alternative. In HSR, each node's address consists of a group identifier (which is a logical distinction and does not reflect clustering) and a host identifier. An independent lookup service is provided for each group, allowing the load of lookups and updates to be balanced across several nodes. Unfortunately, because groups can be geographically spread throughout the network, no locality to look up or update operations exists; this can potentially produce significant network traffic.

L^+ [11] provides a distributed lookup mechanism for landmark routing that reduces latency for local lookups. L^+ uses a hashing function to distribute the load of lookups and updates evenly across mapping servers. When a node changes its hierarchical address, it sends an update to the level-1 landmark that it knows for which hash (*landmark-ID*) is numerically closest to hash (*node-ID*). Upon receiving an update, each landmark forwards the update up the hierarchy in the same manner, until the update reaches the root level. In addition, each landmark forwards the update down the hierarchy by selecting the child landmark with the closest hash value, until the update reaches a level-0 landmark. When a node wishes to send to a destination node x, it first sends a query to a level-1 landmark whose hash is closest to hash(x). The query is in turn forwarded to the level-0 child whose hash is closest to hash(x). If the hierarchical address of x is not found, the query is sent to a level-2 landmark, and so on, until the search succeeds. Although it increases the cost of updates, this scheme has been shown to decrease the cost of lookups in terms of the number of hops that a query must traverse.

MMWN [46] also provides a distributed address mapping service designed to improve lookup locality. However, MMWN allows a flexible trade-off between update performance and lookup performance on a per-node basis, depending on mobility. Nodes in MMWN are organized into multitiered clusters, with hierarchical addressing that reflects each node's location in the hierarchy. Each cluster has a location manager, elected from the nodes within the cluster (i.e., the node with the lowest ID in the cluster). Each location manager maintains for each node a pointer to the child cluster containing that node. Each node has an associated *roaming cluster* at some level in the cluster hierarchy.

If a node exits its roaming cluster, it will get a new roaming cluster and an update is required. An update propagates up the cluster hierarchy to the location manager in each cluster, installing new pointers until it reaches a cluster common to the old and new locations. The update then propagates down the tree until it reaches the node's previous location manager, clearing the old pointers. When a lookup is required, the query follows the pointers in the tree of location managers until it reaches the expected location of the node. If the node is not present, it must have roamed within its roaming cluster. In this case, a paging mechanism is used to locate the node within the roaming cluster. By adjusting the level of a node's roaming cluster, the frequency of required updates can be balanced against the cost of paging to find a node within its cluster. Thus, this protocol can be tuned depending on the amount of mobility vs. the number of expected lookups.

This section has presented two approaches to routing in a tiered architecture. Routing techniques can be applied directly, thus taking advantage of the hierarchical structure of tiered architectures to reduce route update traffic. Alternatively, hierarchical addressing can be used to reduce the cost of route management. The second approach requires an appropriate address management scheme to map from a node's unique ID to its hierarchical address.

13.7 Drawbacks of Tiered Architectures

Despite the advantages of tiered architectures, some drawbacks are also present. First, organizing a network into tiers has a tendency to introduce hot spots near cluster heads, where one tier connects to another. Willig et al. note that altruistic routing tends to concentrate the packet-forwarding load on nodes adjacent to altruistic nodes [60]. In sparse networks, wall-powered nodes that are altruistically attracting packets to offload the packet-forwarding burden from battery-powered nodes may inadvertently decrease the lifetime of their battery-powered neighbors. As a result, the network lifetime, too, could be decreased rather than increased. Liu et al. note that base stations that provide access to an upper tier backbone may cause hot spots in their neighborhood, thus decreasing spatial concurrency [32]. Thus, although a tiered architecture can allow resource-poor nodes to take advantage of resource-rich nodes, there may be an additional cost to some resource-poor nodes.

A second drawback to tiered architectures is the potential inefficiency of imposing a logical structure on an existing flat network. Some of the protocols described earlier require intercluster communication to pass through the cluster heads. Such a restriction means that adjacent nodes that fall into different clusters cannot communicate directly. This is particularly true of the backbone creation protocols described in Subsection 13.5.3.1, in which nonbackbone nodes may only communicate directly with their elected backbone node.

Finally, organizing nodes into a hierarchy typically introduces overhead into the network. This is particularly true of clustering algorithms. For instance, the algorithm described in Banerjee and Khuller [6] first requires that a spanning tree be identified, followed by the execution of the clustering algorithm. If node mobility results in frequent reclustering, the overhead may outweigh the benefit.

13.8 Conclusions

Applications of sensor networks include tasks such as sensing, data transport, and data fusion and processing. A wide variety of hardware is available, each with varying characteristics in terms of processing and storage capacity, sensor interfaces, communication capabilities, and specialized hardware. Constructing a heterogeneous sensor network allows the right components to be brought to bear on the individual application tasks. In particular, a node must meet the processing requirements as well as the energy consumption requirements of its given tasks. Meeting the requirements of each task individually is more efficient than meeting the minimum requirements of all tasks at every node.

Physical heterogeneity alone is not sufficient. First, the application must be broken down into its respective tasks and mapped onto a network; a tiered network provides a convenient architecture for deploying such applications. Generating a tiered organization in an ad hoc network is nontrivial. Manual engineering of the network, while simple and requiring little overhead, reduces the advantages of an otherwise ad hoc network.

Several routing-based approaches have been proposed for automatically identifying and utilizing heterogeneous resources within an ad hoc network. These approaches provide modest gains with little overhead beyond that of a typical flat routing protocol. Backbone creation algorithms provide a similar two-tiered hierarchy in which nodes with greater available resources provide service on behalf of other nodes. Backbones can help control the overhead of flooding and route discovery and allow nodes to sleep more often. Finally, more general hierarchical clustering divides the network into logical or physical clusters, defining a set of nodes that will utilize a particular service or resource. Cluster creation and address management have a relatively high overhead, but such clustering allows routing that scales with cluster size rather than the overall network size.

Taken together, the protocols and techniques described here provide a cookbook for leveraging heterogeneity in wireless sensor networks through tiered architectures. Tiered architectures will help meet the cost, lifetime, and scalability requirements of real applications of sensor networking.

References

1. H. Abrach, J. Carlson, H. Dai, J. Rose, A. Sheth, B. Shucker, and R. Han, MANTIS: system support for multimodal networks of in-situ sensors, Technical Report CU-CS-950-03, Department of Computer Science, University of Colorado, April 2003.

2. Advanced Digital-Logic, Inc., http://www.adlogic-pc104.com/.
3. J.R. Agre, L.P. Clare, G.J. Pottie, and N.P. Romanov, Development platform for self-organizing wireless sensor networks, *Proc. SPIE, Unattended Ground Sensor Technol. Applications*, 3713, April 1999.
4. Atmel Corporation, *AVR Microcontroller ATmega128L Reference Manual*, http://www.atmel.com/.
5. D.J. Baker and A. Ephremides, The architectural organization of a mobile radio network via a distributed algorithm, *IEEE Trans. Commun.*, COM-29(11), November 1981.
6. S. Banerjee and S. Khuller, A clustering scheme for hierarchical control in multi-hop wireless networks, *IEEE INFOCOM 2001*, Anchorage, AK, April 2001.
7. S. Basagni, I. Chlamtac, and A. Faragó, A generalized clustering algorithm for peer-to-peer networks, *Workshop on Algorithm Aspects of Commun.*, Bologna, Italy, July 1997.
8. V. Bharghavan, A. Demers, S. Shenker, and L. Zhang, MACAW: a media access protocol for wireless LAN's, *Proc. ACM SIGCOMM '94 Conf. Commun. Architectures, Protocols Applications*, London, August 1994.
9. A. Cerpa, J. Elson, D. Estrin, L. Girod, M. Hamilton, and J. Zhao, Habitat monitoring: application driver for wireless communications technology, *ACM SIGCOMM Workshop Data Commun. Latin Am. Caribbean*, Costa Rica, April 2001.
10. I.D. Chakeres and E.M. Belding–Royer, Resource biased path selection in heterogeneous mobile networks, University of California, Santa Barbara Computer Science Department technical report 2003-18, July 2003.
11. B. Chen and R. Morris, L[+]: scalable landmark routing and address lookup for multi-hop wireless networks, MIT LCS Technical Report 837, March 2002.
12. Chipcon AS, SmartRF CC1000 preliminary data sheet, http://www.chipcon.com/.
13. W.S. Conner, J. Chhabra, M. Yarvis, L. Krishnamurthy, Experimental evaluation of synchronization and topology control for in-building sensor network applications, *2nd ACM Int. Workshop Wireless Sensor Networks Applications (WSNA '03)*, San Diego, CA, September 2003.
14. Crossbow Technology, Inc., Wireless Sensor Networks (product data sheet), http://www.xbow.com/Products/Wireless_Sensor_Networks.htm.
15. O. Dousse, P. Thiran, and M. Hasler, Connectivity in ad-hoc and hybrid networks, *Proc. IEEE INFOCOM*, New York, June 2002.
16. J. Elson, S. Bien, N. Busek, V. Bychkovskiy, A. Cerpa, D. Ganesan, L. Girod, B. Greenstein, T. Schoellhammer, T. Stathopoulos, and D. Estrin, EmStar: an environment for developing wireless embedded systems software, CENS technical report 0009, March 24, 2003.
17. Ember Corporation, http://www.ember.com/.
18. M. Gerla and J. Tzu-Cheih Tsai, Multicluster, mobile, multimedia radio network, *Wireless Networks*, 1(3), *ACM-Baltzer*, 1995.
19. P. Gupta and P.R. Kumar, The capacity of wireless networks, *IEEE Trans. Inf. Theory*, IT-46(2), March 2000.
20. P. Gupta, R. Gray, and P.R. Kumar, An experimental scaling law for ad hoc networks, May 16, 2001. hppt://black1.csl.uiuc.edu/~prkumar/
21. W.R. Heinzelman, A. Chandrakasan, and H. Balakrishnan, Energy-efficient communication protocols for wireless microsensor networks, in *Proc. 33rd Hawaii Int. Conf. Syst. Sci.*, Maui, HI, January 2000.
22. J. Hill, Spec mote, http://www.cs.berkeley.edu/~jhill/spec/index.htm.
23. J. Hill, R. Szewczyk, A. Woo, S. Hollar, D. Culler, and K. Pister, System architecture directions for networked sensors, *Proc. 9th Int. Conf. Architectural Support Programming Languages Operating Syst.*, Cambridge, MA, November 2000.
24. Intel Corporation, Intel® Mote: Development of an enhanced universal embedded node platform for wireless sensor networks, http://www.intel.com/research/exploratory/motes.htm.
25. Intel Corporation, heterogeneous sensor networks: exploring ways to improve network performance, http://www.intel.com/research/exploratory/heterogeneous.htm.

26. Intrinsyc Corporation, Cerfcube, http://www.intrinsyc.com/products/cerfcube/.
27. iPAQ handheld pocket PC, http://www.compaq.com/.
28. R. Kling, Intel Research Mote, *Network Embedded Systems Technology* Winter 2003 Retreat, January 15-17, 2003.
29. Y.-B. Ko and N.H. Vaidya, A routing protocol for physically hierarchical ad hoc networks, technical report 97-010, Department of Computer Science, Texas A&M University, September 1997.
30. R. Kumar, V. Tsiatsis, and M.B. Srivastava, Computation hierarchy for in-network processing, *Proc. 2nd Int. Workshop Wireless Networks Applications (WSNA'03)*, San Diego, CA, September 2003.
31. C.R. Lin and M. Gerla, Adaptive clustering for mobile networks, *IEEE J. Selected Areas Commun.*, 15(7), September 1997.
32. B. Liu, Z. Liu, and D. Towsley, On the capacity of hybrid wireless networks, *IEEE INFOCOM 2003*, San Francisco, CA, April 2003.
33. S.R. Madden, M.J. Franklin, J.M. Hellerstein, and W. Hong, TAG: a tiny aggregation service for ad-hoc sensor networks, *USENIX 5th Symp. Operating Syst. Design Implementation (OSDI'02)*, Boston, December 2002.
34. A. Mainwaring, J. Polastre, R. Szewczyk, D. Culler, and J. Anderson, Wireless sensor networks for habitat monitoring, *ACM Int. Workshop Wireless Sensor Networks Applications,* Atlanta, September 2002.
35. MANTIS project, http://mantis.cs.colorado.edu/.
36. W. Merrill, K. Sohrabi, L. Girod, J. Elson, F. Newberg, and W. Kaiser, Open standard development platforms for distributed sensor networks, *Proc. SPIE, Unattended Ground Sensor Technol. Applications IV*, Orlando, FL, April 2002.
37. PC/104 Consortium, PC/104 Specification, http://www.pc104.org/.
38. G. Pei, M. Gerla, and T.-W. Chen, Fisheye state routing in mobile ad hoc networks, *Proc. 2000 ICDCS Workshops*, Taipei, Taiwan, April 2000.
39. G. Pei, M. Gerla, and X. Hong, LANMAR: landmark routing for large scale wireless ad hoc networks with group mobility, *Proc. IEEE/ACM MobiHOC 2000*, Boston, August 2000.
40. G. Pei, M. Gerla, X. Hong, and C.-C. Chiang, A wireless hierarchical routing protocol with group mobility, *IEEE WCNC'99*, New Orleans, LA, September 1999
41. C.E. Perkins and P. Bhagwat, Highly dynamic destination-sequenced distance-vector routing (dsdv) for mobile computers, *Proc. ACM SiGCOMM*, London, August 1994.
42. C.E. Perkins and E.M. Royer, Ad-hoc on-demand distance vector routing, *Proc. IEEE WMCSA 1999*, New Orleans, LA, February 1999.
43. PFU Systems, Inc., http://www.pfusystems.com/.
44. G.J. Pottie and W.J. Kaiser, Wireless integrated network sensors, *Commun. ACM*, 43(5), May 2000.
45. Radiometrix, Ltd., http://www.radiometrix.co.uk/.
46. R. Ramanathan and M. Steenstrup, Hierarchically organized, multihop mobile wireless networks for quality-of-service support, *Mobile Networks Applications*, 3(1), June 1998.
47. RF Monolithics Inc., ASH Transceiver TR1000 Data Sheet, http://www.rfm.com/.
48. RF Monolithics Inc., ASH Transceiver TR3000 Data Sheet, http://www.rfm.com/.
49. A. Savvides and M.B. Srivastava, A distributed computation platform for wireless embedded sensing, *Proc. ICCD 2002*, Freiburg, Germany, September 2002.
50. R.C. Shah and J.M. Rabaey, Energy aware routing for low energy ad hoc sensor networks, *Proc. IEEE Wireless Commun. Networking Conf. (WCNC)*, Orlando, FL, March 2002.
51. E. Shih, S.-H. Cho, N. Ickes, R. Min, A. Sinha, A. Wang, and A. Chandrakasan, Physical layer driven algorithm and protocol design for energy-efficient wireless sensor networks, *Proc. MOBICOM 2001*, Rome, July 2001.
52. P. Sinha, R. Sivakumar, and V. Bharghavan, Enhancing ad hoc routing with dynamic virtual infrastructures, *IEEE INFOCOM 2001*, Anchorage, AK, April 2001.
53. R. Sivakumar, P. Sinha, and V. Bharghavan, CEDAR: a core-extraction distributed ad hoc routing algorithm, *IEEE J. Selected Areas Commun.*, 17(8), August 1999.

54. Smart Dust project, http://robotics.eecs.berkeley.edu/~pister/SmartDust/.
55. K. Sohrabi, W. Merrill, J. Elson, L. Girod, F. Newberg, and W. Kaiser, Scalable self-assembly for ad hoc wireless sensor networks, *Proc. IEEE CAS Workshop Wireless Commun. Networking*, Pasadena, CA, September 2002.
56. TinyOS, http://webs.cs.berkeley.edu/tos.
57. P.F. Tsuchiya, The landmark hierarchy: a new hierarchy for routing in very large networks, *Proc. ACM SIGCOMM*, Stanford, CA, August 1999.
58. H. Wang, J. Elson, L. Girod, D. Estrin, and K. Yao, Target classification and localization in habitat monitoring, *Proc. IEEE Int. Conf. Acoustics, Speech, Signal Process. (ICASSP 2003)*, Hong Kong, April 2003.
59. H. Wang, D. Estrin, and L. Girod, Preprocessing in a tiered sensor network for habitat monitoring, *EURASIP J. Appl. Signal Process.*, 2003(4), March 2003.
60. A. Willig, R. Shah, J. Rabaey, and A. Wolisz, Altruists in the PicoRadio sensor network, *Int. Workshop Factory Commun. Syst. (WFCS)*, Vasteras, Sweden, August 2002.
61. K. Xu and M. Gerla, A heterogeneous routing protocol based on a new stable clustering scheme, *MILCOM'02*, Anaheim, CA, October 2002.
62. K. Xu, X. Hong, and M. Gerla, An ad hoc network with mobile backbones, *IEEE ICC 2002*, New York, April 2002.
63. Y. Xu, J. Heidemann, and D. Estrin, Geography-informed energy conservation for ad hoc routing, *Proc. ACM/IEEE Int. Conf. Mobile Computing Networking*, Rome, July 2001.
64. Y. Xu, S. Bien, Y. Mori, J. Heidemann, and D. Estrin, Topology control protocols to conserve energy in wireless ad hoc networks, technical report 6, University of California, Los Angeles, Center for Embedded Networked Computing, January 2003.

14
Power-Efficient Topologies for Wireless Sensor Networks[*]

14.1	Motivation	14-1
14.2	Background	14-2
14.3	Issues for Topology Design	14-3
	Three-Neighbors WSN • Four-Neighbors WSN • Five-Neighbors WSN • Six-Neighbors WSN • Seven-Neighbors WSN • Eight-Neighbors WSN • Six-Neighbors for Three Dimensions	
14.4	Assumptions	14-8
	Calculation of Power Usage for Each Path	
14.5	Analysis of Power Usage	14-10
	Two-Dimensional Analysis • Three-Dimensional Analysis	
14.6	Directional Source-Aware Routing Protocol (DSAP)	14-13
14.7	DSAP Analysis	14-15
	Two-Dimension Analysis • Three-Dimension Analysis	
14.8	Summary	14-19

Ayad Salhieh
Wayne State University

Loren Schwiebert
Wayne State University

14.1 Motivation

This chapter examines the relationship between power usage and the number of neighbors in a wireless sensor network. The study of wireless network topology must be approached from a point of view different from that for wired networks. In a wired network, one examines how nodes are physically connected and the resulting available routing paths. In a wireless sensor network (WSN), the definition of the network topology is derived from the physical neighborhood and transmission power, so it is necessary to determine which topology gives the optimal number of neighbors that a node can handle to transmit or receive. Many of the topologies proposed for wired networks cannot be used for wireless networks because, in wired networks, a higher dimension can be implemented by connecting the nodes in some fashion to simulate higher dimensions. In WSNs, however, one is dealing with three dimensions in the physical world and thus restricted in choice of topologies. Therefore, this chapter concentrates on two-dimensional and three-dimensional mesh topologies.

In this chapter, performance issues associated with different network topologies are analyzed. The question to answer concerns the best topology for a wireless network of sensors, assuming that one can control the placement of these sensors and the sensor locations are fixed relative to each other. Because

---

[*]This research was supported in part by National Science Foundation Grants DGE-9870720 and ANI-0086020.

control over the placement of these sensing nodes is assumed and mobility of the sensors relative to each other is not required, the research problem changes. Instead of considering self-organization of the sensor nodes into a network, efficient placement of fixed nodes is addressed.

Some of these networks can be installed in a building to monitor the building or in an assembly, where the use of regular topology will have an advantage over mobile. In a fixed topology, nodes can be placed so that they can give better coverage. Also, in the use of regular topology or mesh topologies, a node can also function as a router and can relay messages for its neighbors. These networks offer multiple redundant communication paths throughout the network. If one node dies or fails, other nodes can be used to reroute the message. Also, regular topologies enhance the overall reliability of the network.

This chapter does not consider the effects of communication with a base station. Because the topology is fixed and known, it is assumed that the base station can be placed at an appropriate place for each topology. Thus, the power requirements for communicating with the base station should be essentially independent of the topology. This enables one to concentrate on the effects of the topology on the communication among the network nodes only.

14.2 Background

Much of the related research addresses WSNs that are mobile and battery powered. Because of these requirements, most of the literature is concentrated on finding solutions at various levels of the communication protocol, including being extremely energy efficient. Energy efficiency is often gained by accepting a reduction in network performance [7]. Although one does not wish to waste energy, this system does have a constant, renewable energy source. However, a very low-power dissipation allowance offers constraint, which fits nicely with an energy-efficient scheme. Popular power-saving ideas include specialized nodes, negotiation, and data fusion.

Low-energy adaptive clustering hierarchy (LEACH) [2, 13] is a new communication protocol that tries to distribute the energy load evenly among the network nodes by randomly rotating the cluster head among the sensors. This assumes a finite amount of power and aims at conserving as much as possible despite a dynamic network. LEACH uses localized coordination to enable scalability and robustness for dynamic networks, as well as data compression to reduce the amount of data that must be transmitted to a base station. Performing some calculations and using data fusion locally conserves much energy at each node.

Sensor protocols for information via negotiation (SPIN) [3, 5] is a unique set of protocols for energy-efficient communication among wireless sensors. The authors propose solutions to traditional wireless communication issues such as network implosion caused by flooding, overlapping transmission ranges, and power conservation. The SPIN protocols incorporate two key ideas to overcome implosion, overlap, and resource blindness: negotiation and resource adaptation. Using very small metadata packets to negotiate, SPIN efficiently communicates with fewer redundancies than in traditional approaches, dealing with implosion and overlap. The metadata are application specific — they could be used to describe the amount of power dissipated, for instance. To solve the resource blindness issue, each node has an individual resource manager, allowing the node to limit activity when power is low.

Pottie has studied design issues and trade-offs that need to be considered for power-constrained WSNs with low data-rate links [8] and advocates "aggressive power management at all levels," noting that the communication protocol is more helpful in reducing the power consumption than is optimizing the hardware. Local processing of information is key to reducing the amount of communication between nodes and thus reducing the amount of power consumed by the network.

Chen and colleagues have also provided a useful comparison of multiple protocols used for WSNs [1]. Although the authors' main focus is on energy efficiency due to battery power, they provide very useful guidelines for designing access protocols for wireless networks. Specifically, they recommend that "protocols should reduce the number of contentions to improve power conservation," as well as using shorter packet lengths. The receiver usage time, however, tends to be higher for protocols that require the mobile nodes to sense the medium before attempting transmission.

Limited research has been conducted on topology's effect on wireless networking [4, 9, 12]. The concentration, however, has been on mobile networks rather than ones with fixed node placement. Although novel approaches have been devised, none of them would be appropriate, for example, in the biomedical arena, in which a surgeon places the nodes, giving a nominally fixed topology. Although much research has been completed in the area of WSN, nothing has sufficiently answered the question of fixed topology's impact on low-power requirements.

14.3 Issues for Topology Design

This section analyzes the performance issues associated with different network topologies. Unlike previous studies, mobility is not an issue. The question concerns what the best topology for a wireless network of sensors is, assuming placement of these sensors can be controlled and the sensor locations fixed relative to each other. One factor in the choice of topology is the amount of contention for the wireless media. The level of contention will vary with the application because the message pattern and overall message generation rate are functions of the application. However, this study should provide some insights that can be used, along with knowledge of the application, to select an appropriate topology. Again, the goal is not to find a single topology appropriate for all applications, but rather to provide a structured analysis of the options and give guidance on the best choices so that a more informed decision is possible.

Each of the different topologies used in this chapter will be considered as a grid on nodes in two or three dimensions. The vertices of this grid are the nodes that will transmit the packets, and the edges are the neighbors of each node that will receive the transmission. According to the mesh topologies that will be used in this section, the optimal path will be found between a source (S) and a destination (D) or the shortest path between them. We will introduce this optimal path and use it later to show how much power is used in the network using each topology to send a packet from S to D.

The WSN, WSN(m,n), is an $m \times n$ grid, where $m \times n$ represents the number of nodes in the network. Each node is represented as (y,x) for $0 \le y \le m - 1$ and $0 \le x \le n - 1$. For each of the topologies, the following will be assumed:

- $S = (y_s, x_s)$
- $D = (y_d, x_d)$
- $\Delta y = \|y_s - y_d\|$
- $\Delta x = \|x_s - x_d\|$

Each network will be defined by identifying the neighbors of each node according to the different number of neighbors (as shown in Figure 14.1) and presenting the optimal number of hops from a source to a destination. Next, identifying whether two nodes are neighbors and the optimal number of hops between a source and a destination will be discussed.

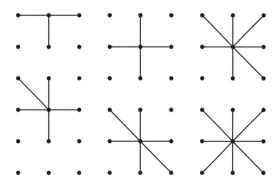

FIGURE 14.1 Possible number of neighbors.

14.3.1 Three-Neighbors WSN

According to Figure 14.2,

- Two nodes are neighbors if:
  - $\langle (y, x), (y, x + 1) \rangle$ for $x < n - 1$
  - $\langle (y, x), (y + 1, x) \rangle$ for even (y, x) and $y < m - 1$
- Two nodes are not neighbors if $\langle (y, x), (y + 1, x) \rangle$ for odd (y, x) and $y < m - 1$
- Optimal number of hops $(s, d) = \begin{cases} \Delta x + \Delta y & \text{if } \Delta x \geq \Delta y \\ 2\Delta y \pm 1 & \text{if } \Delta x < \Delta y \end{cases}$

14.3.2 Four-Neighbors WSN

According to Figure 14.3 note the following:

- Two nodes are neighbors if:
  - $\langle (y, x), (y, x + 1) \rangle$ for $x < n - 1$
  - $\langle (y, x), (y + 1, x) \rangle$ for $y < m - 1$
- Optimal number of hops $(s, d) = \Delta z + \Delta y$.

14.3.3 Five-Neighbors WSN

According to Figure 14.4,

- Two nodes are neighbors if:
  - $\langle (y, x), (y, x + 1) \rangle$ for $x < n - 1$
  - $\langle (y, x), (y + 1, x) \rangle$ for $y < m - 1$
  - $\langle (y, x), (y + 1, x + 1) \rangle$ for even x.
  - $\langle (y, x), (y - 1, x - 1) \rangle$ for odd x.

- Optimal number of hops $(s, d) = \begin{cases} \Delta x + 2 & \text{if } x_s \geq x_d \text{ and } y_s > y_d \\ & \text{or } x_s \leq x_d \text{ and } y_s < y_d \\ \Delta x + \Delta y & \text{Otherwise} \end{cases}$

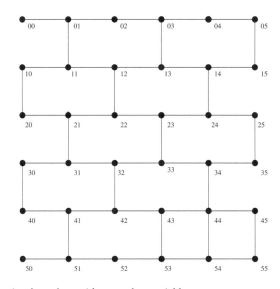

FIGURE 14.2 Two-dimensional topology with up to three neighbors.

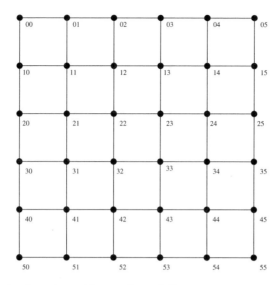

FIGURE 14.3 Two-dimensional topology with up to four neighbors.

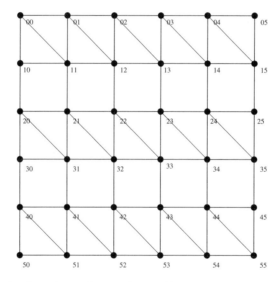

FIGURE 14.4 Two-dimensional topology with up to five neighbors.

14.3.4 Six-Neighbors WSN

According to Figure 14.5,

- Two nodes are neighbors if:
  - $\langle (y, x), (y, x + 1) \rangle$ for $x < n - 1$
  - $\langle (y, x), (y + 1, x) \rangle$ for $y < m - 1$
  - $\langle (y, x), (y + 1, x + 1) \rangle$ for every $y < y + 1$ and $x < x + 1$
  - $\langle (y, x), (y - 1, x - 1) \rangle$ for every $y < y - 1$ and $x < x - 1$
- Two nodes are not neighbors if
- Optimal number of hops $(s, d) = \begin{cases} \Delta x + \Delta y & \text{if } x_s > x_d \text{ and } y_s < y_d \\ & \text{or } x_s < x_d \text{ and } y_s > y_d \\ \max(\Delta x, \Delta y) & \text{Otherwise} \end{cases}$

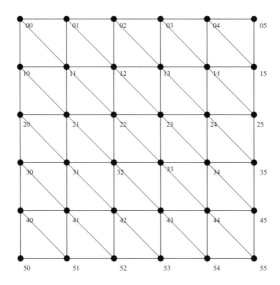

FIGURE 14.5 Two-dimensional topology with up to six neighbors.

14.3.5 Seven-Neighbors WSN

According to Figure 14.6,

- Two nodes are neighbors if:
  - $\langle(y, x), (y, x + 1)\rangle$ for $x < n - 1$
  - $\langle(y, x), (y + 1, x)\rangle$ for $y < m - 1$
  - $\langle(y, x), (y + 1, x - 1)\rangle$ for $x = 0$ or x is even.
  - $\langle(y, x), (y - 1, x + 1)\rangle$ for $x = 1$ or x is odd.
  - $\langle(y, x), (y + 1, x + 1)\rangle$ for every $y < y + 1$ and $x < x + 1$
  - $\langle(y, x), (y - 1, x - 1)\rangle$ for every $y < y - 1$ and $x < x - 1$
- Optimal number of hops $(s, d) = \begin{cases} \Delta x + 2 & \text{if } x_s > x_d \text{ and } y_s < y_d \\ & \text{or } x_s < x_d \text{ and } y_s > y_d \\ \max(\Delta x, \Delta y) & \text{Otherwise} \end{cases}$

14.3.6 Eight-Neighbors WSN

According to Figure 14.7,

- Two nodes are neighbors if:
  - $\langle(y, x), (y, x + 1)\rangle$
  - $\langle(y, x), (y + 1, x)\rangle$
  - $\langle(y, x), (y + 1, x - 1)\rangle$
  - $\langle(y, x), (y - 1, x + 1)\rangle$
  - $\langle(y, x), (y + 1, x + 1)\rangle$
  - $\langle(y, x), (y - 1, x - 1)\rangle$
- Optimal number of hops $(S, D) = \max(\Delta x, \Delta y)$.

14.3.7 Six-Neighbors for Three Dimensions

The WSN (m, n, k) is an $m \times n \times k$ grid where a node is represented as (y, x, z) for $0 \le y \le m - 1$, $0 \le x \le n - 1$, and $0 \le z \le k - 1$. For three-dimensional topology, assume the following:

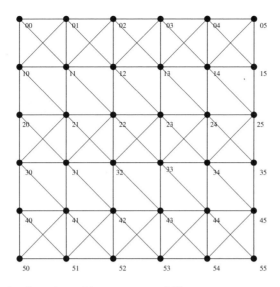

FIGURE 14.6 Two-dimensional topology with up to seven neighbors.

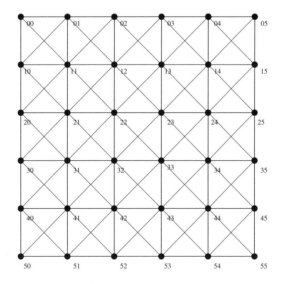

FIGURE 14.7 Two-dimensional topology with up to eight neighbors.

- $S_{3D} = (y_s, x_s, z_s)$
- $D_{3D} = (y_d, x_d, z_d)$
- $\Delta y = \|y_s - y_d\|$
- $\Delta x = \|x_s - x_d\|$
- $\Delta z = \|z_s - z_d\|$

According to Figure 14.8, two nodes are neighbors if:

- $\langle (y, x, z), (y, x+1, z) \rangle$ for $x < n-1$
- $\langle (y, x, z), (y+1, x, z) \rangle$ for $y < m-1$
- $\langle (y, x, z), (y, x, z+1) \rangle$ for $z < k-1$
- Optimal number of hops $(S_{3D}, D_{3D}) = \Delta x + \Delta y + \Delta z$.

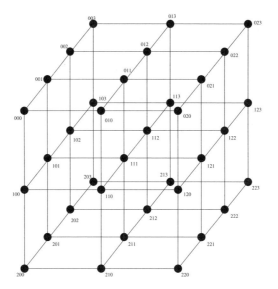

FIGURE 14.8 Three-dimensional topology with up to six neighbors.

14.4 Assumptions

In this work, a simple model is assumed in which the radio dissipates $E_{elec} = 50$ nJ/b to run the transmitter or receiver circuitry and $E_{amp} = 100$ pJ/b/m² for the transmit amplifier to achieve an acceptable E_b/N_0 (see Figure 14.9 and Table 14.1) [2]. To transmit a k-b message a distance of d meters using this radio model, the radio expends:

$$E_{Tx}(k,d) = E_{Tx-elec}(k) + E_{Tx-amp}(k,d)$$
$$= E_{elec} * k + E_{amp} * k * d^2 \tag{14.1}$$

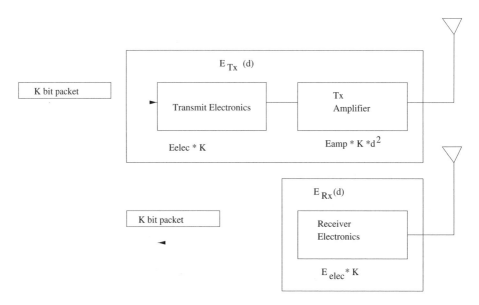

FIGURE 14.9 First-order radio model.

TABLE 14.1 Radio Characteristics

Operation	Energy Dissipated
Transmitter electronics ($E_{Tx-elec}$)	50 nJ/b
Receiver electronics ($E_{Rx-elec}$)	
($E_{Tx-elec} = E_{Rx-elec} = E_{elec}$)	
Transmit amplifier (E_{amp})	100 pJ/b/m²

Source: W.R. Heinzelman, A. Chandrakasan, and H. Balakrishnan. In *Hawaii Int. Conf. Syst. Sci.*, 2000.

To receive this message, the radio expends:

$$E_{Rx}(k) = E_{Rx-elec}(k)$$
$$= E_{elec} * k \qquad (14.2)$$

For simplicity of calculation, assume that the transmission range of each node is equal to each other on one condition: that the value of this transmission range should reach the number of neighbors allowed for each network (maximum number of neighbors). Also, assume that all data packets contain the same number of bits. Thus, a maximum distance $d = 15$ m and number of bits transmitted $k = 512$ bs are assumed. The number of nodes N was chosen to be 36 because it works nicely for two-dimensional and three-dimensional networks with the different topologies considered. This also represents an intermediate value between 16 and 64 node networks that has been used in other studies [7]. For these parameter values, receiving a message is not a low-cost operation; the protocol should thus try to minimize not only the transmit distance but also the number of transmit and receive operations for each message. Next general equations that can be used to estimate the total power used to transmit a message from source to destination will be presented.

14.4.1 Calculation of Power Usage for Each Path

In order to derive the general equations for transmitting a message from a source S to a destination D, two things must be considered for each path: (1) number of transmissions; and (2) number of receptions.

Number of transmissions can be measured as the number of hops a packet will travel through a certain path. Number of receptions is the total number of neighbors of each hop taken. Minimizing the number of transmissions and number of receptions will be the mission of any protocol designed. In general, the total power dissipated in the network for one packet to travel from a source to a destination is the sum of total power used for transmission plus the total power used for receiving the packet at each neighbor of each transmitting source.

The next equation presents an estimate for the total power used to transmit a packet over a number of hops from a source S to a destination D;

$$\text{Total power used} = \text{total power transmitted} + \text{total power received} \qquad (14.3)$$

Equation 14.3 can be written as:

$$\text{Total power transmitted} = \text{number of hops} \times \text{power transmitted}$$
$$= \text{number of hops} \times E_{Tx}(k,d) \qquad (14.4)$$

$$\text{Total power received} = \text{number of hops} \times \text{number of neighbors} \times \text{power received}$$
$$= \text{number of hops} \times \text{number of neighbors} \times E_{rx}(k) \qquad (14.5)$$

Substituting Equation 14.4 and Equation 14.5 in Equation 14.3 yields:

$$\text{Total power used} = \text{number of hops} \times (E_{Tx}(k,d)) + \text{number of neighbors} \times (E_{Rx}(k)) \quad (14.6)$$

These equations only estimate the power that will be used for a certain number of hops with a fixed number of neighbors. The idea here is to try to minimize Equation 14.3 by minimizing the total power transmitted; this can be done by minimizing the number of hops by finding the shortest path. Also, Equation 14.3 can be minimized by minimizing the total power received, which can be done by taking the paths that have the least number of neighbors. The next section presents and analyzes the effect of choosing different paths on Equation 14.3.

14.5 Analysis of Power Usage

Various network topologies are studied in this section. First, the routing is considered over the diameter of the network and two possible routes are used along the edge and through the interior. These results show that different paths consume different amounts of power. Next shortest-path routing for the various topologies for a message spanning the diameter of the network is considered. Finally, directional source-aware routing protocol (DSAP) is simulated with and without power-aware routing of arbitrary source–destination pairs and the relative performance of each is shown.

The power dissipated with respect to the network topology will be analyzed with a variable number of neighbors. First, two-dimensional networks with three, four, five, six, seven, and eight neighbors are examined. Then, three-dimensional networks with six neighbors are considered. Two kinds of routing are considered for each of the topologies: (1) edge routing; and (2) interior routing.

Edge routing consists of moving messages to the outer edges of the network where there are fewer neighbors. Interior routing keeps the messages in the middle of the network, where there is a consistent number of neighbors for each node. In some cases, longer paths were chosen for some topologies to give a similar number of transmissions. The use of these two methods of routing is only to show the effect of using topologies with different numbers of neighbors. It also shows how useful it is to increase the number of neighbors. Then, shortest-path routing will be studied to see which topology will give the most savings in power. The shortest path will be considered by using the DSAP routing protocol; and also to study the benefit of using a power-aware routing metric by using aware–DSAP will also be studied.

14.5.1 Two-Dimensional Analysis

The degree of routing freedom is the number of alternative paths that a routing protocol can select. Figure 14.2 through Figure 14.7 show that as the number of neighbors increases, the degree of routing freedom increases. For comparison purposes, the source, destination, and number of nodes were fixed to be the same (36 nodes) for all the networks under investigation. An analysis of these networks requires one to classify the routing paths into edge routes and interior routes.

14.5.1.1 Interior Routing

As defined before, interior routing keeps the messages in the middle of the network, where the number of neighbors for each node is consistent. Table 14.2 shows that as the number of neighbors increases, the number of transmissions decreases; however, the number of receptions depends on the topology. This is because, as the number of neighbors increases, the routing protocol has more freedom to choose the shortest path to the destination; by doing so the protocol will dissipate less power to route a packet from source to destination.

14.5.1.2 Edge Routing

Using edge routing is to route the packet using only the edge nodes. This strategy of routing is impossible to use at all times, of course. Here it is used to study the effect of increasing the number of neighbors

TABLE 14.2 Two-Dimensional Interior Routing

Neighbors	T_x	R_x	Energy Used
3	10	27	10.624×10^{-4}
4	10	36	12.928×10^{-4}
5	7	36	11.172×10^{-4}
6	5	27	8.768×10^{-4}
7	5	31	9.792×10^{-4}
8	5	36	10.720×10^{-4}

TABLE 14.3 Two-Dimensional Edge Routing

Neighbors	T_x	R_x	Energy Used
3	14	33	13.645×10^{-4}
4	10	28	10.880×10^{-4}
5	10	37	13.184×10^{-4}
6	10	39	13.696×10^{-4}
7	10	44	14.976×10^{-4}
8	10	46	15.488×10^{-4}

with respect to the edge nodes. As shown in Table 14.3, as the number of neighbors increases, the number of neighbors that receive the packet increases, which will increase the energy used in the network.

14.5.1.3 Edge Routing vs. Interior Routing

From Table 14.2 and Table 14.3, edge routing dissipates more power than interior routing in all cases except for four neighbors. This is because, although the path from the source to the destination in a four-neighbor case is the same, the difference is that taking the edge results in fewer neighbors and interior paths have more neighbors. With either routing strategy, as the number of neighbors increases the power dissipated increases for the same number of transmissions.

14.5.1.4 Fixed Number of Transmissions

This subsection studies the effect of increasing the number of neighbors. In order to do that it is necessary to fix the number of transmissions that a certain path can have and also certain nodes through which a path must pass. These fixed nodes are the nodes that fall on the diagonal of the network, such as nodes (1,1), (2,2), (3,3), (4,4), (5,5), (6,6), (7,7), and (8,8). By using this path, one can control the path and study the effect of increasing the number of neighbors. As shown in Table 14.4, as the number of neighbors increases, the number of receptions increases also. This yields to an increase in the energy used in the network.

TABLE 14.4 Two-Dimensional Fixed Number of Hops

Neighbors	T_x	R_x	Energy Used
3	10	27	10.624×10^{-4}
4	10	36	12.928×10^{-4}
5	10	45	15.232×10^{-4}
6	10	53	17.280×10^{-4}
7	10	61	19.328×10^{-4}
8	10	69	21.376×10^{-4}

TABLE 14.5 Routing Freedom and Power Dissipation Three and Six Neighbors

Neighbors	T_x	R_x	Energy Used
3	10	27	10.624×10^{-4}
6	5	27	8.768×10^{-4}

TABLE 14.6 Routing Freedom and Power Dissipation Four and Eight Neighbors

Neighbors	Tx	Rx	Energy Used
4	10	36	12.928×10^{-4}
8	5	36	10.720×10^{-4}

14.5.1.5 Routing Freedom

Routing freedom means that the routing protocol has the freedom to choose the optimal path. This subsection studies the effect of doubling the number of neighbors, between three and six neighbors and four and eight neighbors, to study the effect of increasing the number of neighbors and the impact it will have on routing freedom.

Table 14.5 considers the power dissipated between the source and destination for a message spanning the diameter of the network for topologies with three and six neighbors as shown in Figure 14.2 and Figure 14.5. As Table 14.5 shows, increasing the number of neighbors decreases the number of transmissions and the total power dissipated in the system. This result can only be attributed to the availability of a shorter path between the source and destination. A similar conclusion can be reached from Table 14.6.

In summary, a trade-off occurs between the number of neighbors and the total power dissipated in the system. However, this trade-off breaks in special cases in which the availability of alternative shortest paths can be used as an advantage for the power budget calculations.

14.5.2 Three-Dimensional Analysis

A three-dimensional network can be constructed from a two-dimensional network with four neighbors just by adding another dimension, which will create a three-dimensional network with six neighbors. The same thing can be done for two-dimensional networks with six neighbors, but implementing such a network with a regular structure is not possible. Figure 14.8 shows a three-dimensional network with six neighbors that has some advantages due to its inherent symmetry.

In a three-dimensional network, the routing paths between any given source and destination without misrouting would always result in the same number of transmissions but a different number of receptions. For example, from source (0,0,0) to destination (2,2,3), the number of transmissions using interior or edge routing is constant and equals seven in Figure 14.8.

From Table 14.7, the following can be concluded:

- Edge routing in the case of the three-dimensional network has lower power dissipation than interior routing does.
- The number of transmissions and receptions as well as the total power dissipated in a three-dimensional network is less than in a two-dimensional network for edge routing as well as interior routing.

For Table 14.8, the number of neighbors was fixed to study the effect of using two different dimensions on the number of transmissions each path will require using edge routing and interior routing. Using interior routing, two dimensions with six neighbors have fewer transmissions than the three dimensions with six neighbors. Also, from the nature of the two-dimensional topology, using edge routing takes

TABLE 14.7 Edge and Interior Routing Power Dissipation

Network	Path	T_x	R_x	Energy Used × 10^{-4}
2D	Interior	10	36	12.928
4 Neighbor	Edge	10	28	10.880
3D	Interior	7	33	11.046
6 Neighbor	Edge	7	25	8.998

TABLE 14.8 Six Neighbors for 2-D and 3-D Routing Power Dissipation

Network	Path	T_x	R_x	Energy Used × 10^{-4}
2D	Interior	5	27	8.768
6 Neighbor	Edge	10	39	13.696
3D	Interior	7	33	11.046
6 Neighbor	Edge	7	25	8.998

longer paths than three dimensions because the three-dimensional topology makes the edges closer than the two-dimensional one. Thus, a trade-off occurs between using edge routing and using interior routing for the two different dimensions.

14.6 Directional Source-Aware Routing Protocol (DSAP)

In order to resolve the problems of power efficiency, a unique identification system has been developed for the networks used. The idea behind this identification system is to identify the location of each node in the network that will help in routing the packets. The system has the following properties:

- Each node has unique ID.
- Each value represents how far the node is from a certain direction.
- Each ID gives how far the node is from the nodes in each direction.
- Each node can compute the direction of other nodes from its ID.

To help in studying the effect of using different numbers of neighbors, a routing scheme based on the identification system has been developed. This identification system is referred to as the *directional value* (DV). To construct the DV, each node in each topology that has been used has a fixed number of neighbors. Each neighbor represents a direction that the node can route through it, as shown in Figure 14.10. How far the node is from the edge of the network in each direction represents the directional value of each node. This number is unique for each node and can be used as the ID number for each node for the purpose of routing.

Each topology was constructed from Figure 14.10 by eliminating the directions that will make that topology. For example, constructing a seven-neighbor topology from an eight-neighbor one is done by eliminating D-7 in one node and also eliminating the corresponding direction from the other node. Each direction has a corresponding or an associate direction. D-7 has D-3, D-6 has D-2, D-5 has D-1, D-4 has D-0, and vice versa.

From this DV, a DSAP [11] was developed. DSAP incorporates the DV and power into routing protocols. For instance, in the four-neighbor case of Figure 14.3, node 31 would have an identifier of (1, 0, 3, 0, 4, 0, 2). This means that there is one node to the edge in direction 0 (left); three in direction 2 (up); four in direction 4 (right); and two in direction 6 (down). Because placement of the nodes is controlled and topology is fixed, this information can be hard-coded into each node with relative ease. However, for a random topology, it is necessary to discover the directional values of each node in the network.

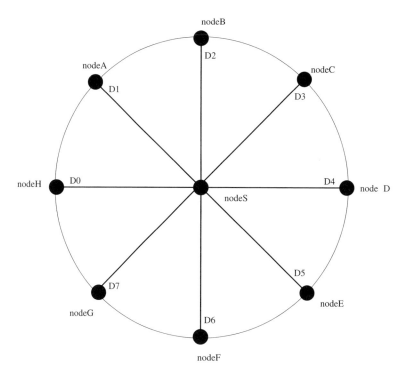

FIGURE 14.10 Directional eight-neighbor node.

In Figure 14.10, node S would have an identifier of $(DV_0, DV_1, DV_2, DV_3, DV_4, DV_5, DV_6, DV_7)$. This means that DV_0 nodes are to the edge in direction D-0; DV_1 in D-1; DV_2 in D-2; and so on. When transmitting a message, the destination node identifier is subtracted from the source node identifier. This yields at most five positive numbers (for a two-dimensional topology with eight neighbors) that describe in which direction the message needs to move. Negative numbers are ignored. The decision to move in any positive direction is determined by the DV of the nodes in question. Taking each of the neighbor's identifiers and subtracting them from the destination node's identifier computes the DV. These eight numbers are added together and the one with the smaller number is chosen. If both nodes have the same DV, then one is randomly picked. This is the basic scheme developed for routing the messages.

For example, in Figure 14.7 consider the source node $S_{1,1}$ with $DV_{1,1} = (1, 1, 1, 1, 4, 4, 4, 1)$ and destination node $D_{4,4}$ with $DV_{4,4} = (4, 4, 4, 1, 1, 1, 1, 1)$. According to the algorithm of DSAP [11], $S - D = (-3, -3, -3, 0, 3, 3, 3, 0)$, which produces D-3, D-4, D-5, D-6, and D-7 as possible positive directions to which the message can be forwarded and then computes the directional value of each positive direction to find which route to take. By doing so, the following values for each direction are obtained: 20, 17, 14, 16, and 20, respectively. By choosing the minimum directional value, the message is forwarded in direction D-5, which is obvious from Figure 14.7. Then the protocol repeats until reaching the final destination, which will have a DV of 0.

This is the basic scheme developed for routing messages. However, the objective is to incorporate energy efficiency as well. This is achieved by considering the maximum available power and minimal directional value when picking which node route to take. Instead of simply picking the node with the lowest directional value, the directional value is divided by the power available at that node. The smaller value of this power-constrained directional value is the path chosen. This allows for a least-transmission path that is also cognizant of power resources, although in some cases a longer path may be chosen if the available power dictates that choice. Salhieh and Schwiebert [10] have presented several power-aware metrics that can be incorporated with DSAP. The idea here is to show that using power-aware methods will extend the life of the network and have a fair load balance between the nodes. The method used here was only to show the effect of using power-aware rather than shortest-path metrics.

14.7 DSAP Analysis

To study the relationship between the number of neighbors and the power dissipated in the network, a controlled environment is used. This has been done to study the effect on the power dissipated in the network when the number of neighbors is increased. The effect of increasing or decreasing the number of neighbors is studied from two viewpoints: (1) power usage in the network; and (2) which topology or number of neighbors will extend the life of the network because extending the life of the network is one of the main objectives of designing WSNs.

In the simulation, two different methods for routing are used: (1) DSAP without the power aware, which is based on the shortest number of hops between a source and a destination; and (2) DSAP with power aware, which incorporates the power available at the next neighbor and tries to balance the load between the neighbors of a source. The simulation has two runs: (1) a fixed run from $S(0, 0)$ to $D(5, 5)$; and (2) a run that each node sends a message to every node in the network. Both of these should help in studying the relationship between the power usage in the system and the number of neighbors.

14.7.1 Two-Dimension Analysis

In Table 14.9, a message is sent from source $(0, 0)$ to destination $(5, 5)$ for 10,000 times. Note that:

- Increasing the number of neighbors, for DSAP in general, results in decreasing the number of transmissions that the network performs because having more neighbors creates shorter paths or alternative routes that are shorter to the destination. This is also reflected in the total power transmitted (TPT) in the network, which is decreased from a sparse topology to a more dense topology.
- Looking at the power used for both protocols, note that DSAP with power aware uses more power, which is reflected throughout Table 14.9. However, looking at Figure 14.11 and Figure 14.12, note that DSAP with power aware has a better power distribution than DSAP without power aware. This means that the life of the network can be extended using the power-aware concept.

Table 14.10 and Table 14.11 concern when the first node dies in the network. Note that:

- In Table 14.10, more than one node died in the network. This is because using DSAP without power aware uses the concept of shortest path, so every message takes the same path and thus these nodes will lose power faster than other nodes.
- In Table 14.11, the first node died at different rounds and even at a higher number of rounds than in Table 14.10 because DSAP with power aware was used in Table 14.11. This gives the routing protocol more alternative paths to use and also balances the load in the network.
- Also notice that in Table 14.11, as the number of neighbors is increased, the number of rounds when the first node dies decreases because more neighbors are hearing the transmission of each source.
- In Table 14.12 through Table 14.14, each node sends a message to every other node in the network. This will be considered as one complete run and is repeated until a fixed round or until the death of the first node. In these tables we ran the simulation for the DSAP without power aware and also for the power-aware protocol.

In Table 14.12:

- As the number of neighbors is increased, the first node dies at a lower number of rounds in both protocols because more nodes will be reached during each transmission, so more nodes will lose power.

TABLE 14.9 Round 10000 from S(0,0) to D(5,5)

		DSAP Routing				
	Neighbors	TR	TT	TPA (J)	TPR (J)	TPT (J)
	4	280,000	100,000	25.12	7.16	3.71
	5	370,000	90,000	23.19	9.47	3.34
2D	6	270,000	50,000	27.23	6.91	1.86
	7	310,000	50,000	26.20	7.94	1.86
	8	350,000	50,000	25.18	8.96	1.86

		Aware–DSAP Routing				
	Neighbors	TR	TT	TPA (J)	TPR (J)	TPT (J)
	4	314,787	100,000	24.23	8.06	3.71
	5	359,428	87,861	23.54	9.20	3.26
2D	6	301,852	65,926	25.83	7.73	2.45
	7	388,748	73,624	23.32	9.95	2.73
	8	396,424	73,212	23.13	10.15	2.72

Notes:
TR = total number of packets received by the neighbors of a source.
TT = total number of transmissions in the networks.
TPA = total power available for the network.
TPR = total power received by the neighbors of a transmitting source.
TPT = total power used for transmitting these packets.

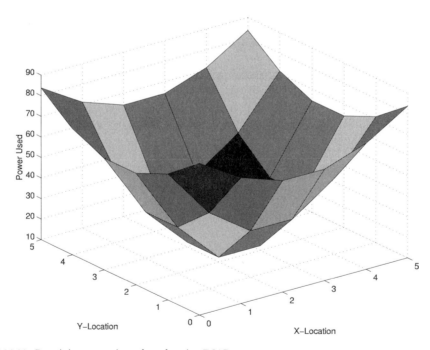

FIGURE 14.11 Remaining power in each node using DSAP.

- The number of rounds in the DSAP with power aware is higher than the DSAP without power aware. This is because alternative paths have been used, resulting in a better load balance than in the DSAP without the power aware.
- Notice that the standard deviation for the DSAP with power aware is less than that of the DSAP without power aware because DSAP with power aware has a better distribution of power usage

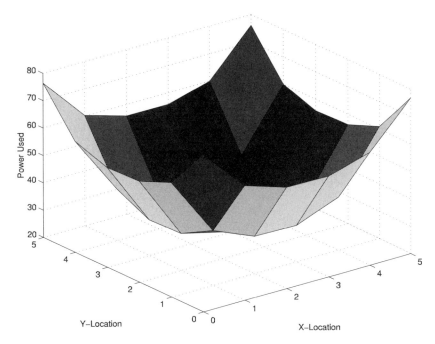

FIGURE 14.12 Remaining power in each node using aware–DSAP.

TABLE 14.10 First Node Dead for DSAP at Round 10191 from S(0,0) to D(5,5)

	Neighbors	Dead Nodes	GeoMean
	4	8	51.89
	5	7	48.20
2D	6	3	64.55
	7	3	62.42
	8	3	60.36

Note: GeoMean = geometric mean.

TABLE 14.11 First Node Dead Aware–DSAP from S(0,0) to D(5,5)

	Neighbors	Round	GeoMean
	4	14,350	49.58
	5	13,563	47.76
2D	6	14,350	52.71
	7	13,060	48.52
	8	11,456	54.82

Note: GeoMean = geometric mean.

than does DSAP without power aware. Also the geometric mean is less in the DSAP with power aware than the DSAP without power aware because DSAP with power aware balances the load among all the nodes.

In Table 14.13 and Table 14.14, the two protocols are compared at round 28,512 to study the geometric mean, the standard deviation, and different power parameters:

TABLE 14.12 First Node Dead for Fixed All Routing

Neighbors		DSAP Routing		
		GeoMean	STDEV	Number of Rounds
2D	4	39.69	21.33	39,605
	5	39.99	21.82	34,001
	6	44.33	22.04	31,715
	7	42.09	21.34	29,485
	8	45.07	22.94	29,120

Neighbors		Aware–DSAP		
		T_x	R_x	Total Power Used
2D	4	20.75	15.24	56,084
	5	31.04	18.66	30,934
	6	27.50	14.31	39,512
	7	28.76	15.71	29,485
	8	24.48	18.17	37,915

Notes:
GeoMean = geometric mean.
STDEV = standard deviation.

TABLE 14.13 Topology at Round 28512 for Fixed All Routing

Neighbors		DSAP Routing		Aware–DSAP Routing	
		GeoMean	STDEV	GeoMean	STDEV
2D	4	58.79	15.42	61.34	7.81
	5	51.75	18.38	44.66	15.40
	6	51.31	19.84	51.96	11.60
	7	44.67	20.59	43.98	13.98
	8	46.74	22.45	47.11	15.61

Notes:
GeoMean = geometric mean.
STDEV = standard deviation.

- In Table 14.13, DSAP aware has a lower standard deviation than the DSAP, but in some cases has a higher geometric mean.
- In Table 14.13, the topology with four neighbors has a lower standard deviation in both protocols.
- In Table 14.14, the number of neighbors increases, the number of transmissions decreases, as noted in Table 14.9.

In general, for the two-dimensional topologies, a trade-off occurs between increasing the number of neighbors and the power dissipated in the networks. As the number of neighbors increases, the protocol will have alternative routes; however, more power will be dissipated in the network. Also, using a power-aware routing protocol will help in extending the life of the network.

14.7.2 Three-Dimension Analysis

In Table 14.15, different runs were done for the three-dimensional topology to try to see how the power dissipated in the network would be affected by using the two different protocols. For the first 1000 rounds, there is only a difference in the number of reception in the network. This is because when the network is used more, the DSAP with power aware tries to find alternative paths with more power. If one looks at 10,000 and 100,000, it is seen that the power used is less in the DSAP with power aware than the DSAP without power aware for the same reasons mentioned before.

TABLE 14.14 Power Values at Round 28512 for Fixed All Routing

	Neighbors	TR	TT	TPA (J)	TPR (J)	TPT (J)
		\multicolumn{5}{c}{DSAP Routing}				
2D	4	390,720	110,880	21.88	10.0	4.12
	5	478,522	105,292	19.84	12.25	3.91
	6	490,776	94,556	19.93	12.56	3.51
	7	570,768	91,718	17.98	14.61	3.4
	8	544,456	78,232	19.16	13.94	2.90

	Neighbors	TR	TT	TPA (J)	TPR (J)	TPT (J)
		\multicolumn{5}{c}{Aware–DSAP Routing}				
2D	4	376,541	110,880	22.24	9.64	4.12
	5	558,634	127,596	16.96	14.30	4.74
	6	507,003	104,465	19.14	12.98	3.88
	7	608,627	103,897	16.56	15.58	3.86
	8	578,045	90,638	17.83	14.79	3.36

Notes:
TR = total number of packets received by the neighbors of a source.
TT = total number of transmissions in the networks.
TPA = total power available for the network.
TPR = total power received by the neighbors of a transmitting source.
TPT = total power used for transmitting these packets.

TABLE 14.15 Power Assessment for 3D Topology

Protocol	DSAP Routing			Aware–DSAP Routing		
Number of rounds	1000	10,000	100,000	1000	10,000	100,000
Total power used (J)	0.416	4.126	41.354	0.4	3.937	39.469
Total transmissions	3051	30,131	302,160	3051	30,131	302,160
Total reception	13,228	131,043	1,312,998	12,573	123,656	1,239,477

14.8 Summary

This chapter has looked at the WSN network topology from a different perspective: a neighborhood point of view. In these topologies, the number of neighboring nodes determines the number of receivers and therefore may result in more overall power usage, even though the number of transmissions decreases. Thus, a fundamental trade-off takes place between decreasing the number of transmissions and increasing the number of receptions. This chapter has presented a variety of topologies and examined this trade-off.

Because the number of neighbors differs with different topologies, one expects different topologies to have different power usage rates. Even simulations of the contention-free case show that different topologies have different levels of power efficiency. The results show that the total power consumption is reduced for topologies with fewer neighbors; although the topologies with more neighbors require fewer hops, the power expended by many nodes to receive these messages increases the power usage. Among the two-dimensional topologies, the best power efficiency is achieved with two dimensions with four neighbors. The three-dimensional topology performs even better, although this topology may not be feasible for some applications.

Many areas remain to be explored within this research topic. This initial set of experiments serves to demonstrate the marked difference between basic and power-aware DSAP routing. These differences are significant enough to warrant further research.

One option would be to rerun the large simulations with each node beginning with a randomly chosen power amount. This would allow for a simulation of a network that has been in use for some time. DSAP can also be extended to include a more efficient power management scheme. Because the message knows in which direction to head, it is not necessary to broadcast to all neighbors. Rather, the nodes in the wrong direction can be put to sleep. This will reduce the power used because it takes more power to transmit the large message than to poll the neighboring nodes.

Contention is also an issue that needs to be addressed in future studies because it is not realistic to have a system that sends only one message at a time. Although previous work has also ignored this issue, it is important to find a solution to give a more accurate comparison of the relative performance of the networks.

References

1. J. Chen, K.M. Sivalingam, and P. Agrawal. Performance comparison of battery power consumption in wireless multiple access protocols. *Wireless Networks*, 5(6):445–460, 1999.
2. W.R. Heinzelman, A. Chandrakasan, and H. Balakrishnan. Energy-efficient communication protocols for wireless microsensor networks. In *33rd Ann. Hawaii Int. Conf. Syst. Sci.*, 2000.
3. W.R. Heinzelman, J. Kulik, and H. Balakrishnan. Adaptive protocols for information dissemination in wireless sensor networks. In *Proc. 5th Annu. ACM/IEEE Int. Conf. Mobile Computing Networking (MobiCom'99)*, 174–185, August 1999.
4. L. Hu. Topology control for multihop packet radio networks. *IEEE Trans. Commun.*, 41(10):1474–1481, October 1993.
5. J. Kulik, W.R. Heinzelman, and H. Balakrishnan. Negotiation-based protocols for disseminating information in wireless sensor networks. In *ACM MOBICOM*, 99.
6. B. Nath and D. Niculescu. Routing on a curve. *ACM SIGCOMM Comp. Commun. Rev.*, 33(1), 155–160, 2003.
7. C. Patel, S.M. Chai, S. Yalamanchili, and D.E. Schimmel. Power/performance trade-offs for direct networks. In *Parallel Computer Routing Commun. Workshop*, 193–206, July 1997.
8. G. Pottie. Wireless sensor networks. In *Inf. Theory Workshop*, 2, 139–140, 1998.
9. R. Ramanathan and R. Rosales–Hain. Topology control of multihop wireless networks using transmit power adjustment. In *INFOCOM*, 404–413, 2000.
10. A. Salhieh and L. Schwiebert. Power-aware metrics for wireless sensor networks. In *14th IASTED Conf. Parallel Distributed Computing Syst. (PDCS 2002) Symp.*, 326–331, November 2002.
11. A. Salhieh, J. Weinmann, M. Kochhal, and L. Schwiebert. Power-efficient topologies for wireless sensor networks. In *Int. Conf. Parallel Process.*, 156–163, Sep. 2001.
12. Z. Tang and J.J. Garcia–Luna–Aceves. A protocol for topology-dependent transmission scheduling in wireless networks. In *IEEE Wireless Commun. Networking Conf.*, 3, 1333–1337, 1999.
13. A. Wang, W.R. Heinzelman, and A. Chandrakasan. Energy-scalable protocols for battery-operated microsensor networks. In *IEEE Workshop Signal Process. Syst.*, 483–492, Oct. 1999.

15
Architecture and Modeling of Dynamic Wireless Sensor Networks

Symeon Papavassiliou
New Jersey Institute of Technology

Jin Zhu
New Jersey Institute of Technology

15.1 Introduction 15-1
15.2 Characteristics of Wireless Sensor Networks 15-2
15.3 Architecture of Sensor Networks 15-3
   Functional Layers of Wireless Sensor Networks • Homogeneous vs. Heterogeneous Architectures • Communication Mode-Based Sensor Network Classification • Data Fusion Architectures
15.4 Modeling of Dynamic Sensor Networks 15-7
   Performance Metrics of Dynamic Wireless Sensor Network • Modeling Sensor Networks
15.5 Concluding Remarks 15-13

15.1 Introduction

With the development of the information society, sensors are facing ever more new challenges. Detection and monitoring requirements are becoming more complicated and difficult. They trend from single variable to multiple variables; from one point to a plane; from one sensor to a set of sensors; from simple to complex and cooperative. Networking the sensors to empower them with the ability to coordinate on a larger sensing task will revolutionize information gathering and processing in many situations. Networks of sensors can greatly improve environment monitoring for many civil and military applications. Furthermore, many environments may be unsuitable for humans and thus the use of sensors is the only solution; in some places, although accessible, in general it is more effective to place small autonomous sensors than to use humans for collection of data.

By integrating sensing, signal processing, and communications functions, a sensor network provides a natural platform for hierarchical and efficient information processing. It allows information to be processed on different levels of abstraction, ranging from detailed microscopic examination of specific targets to a macroscopic view of the aggregate behavior of targets. With focus on applications requiring tight coupling with the physical world, as opposed to the personal communication focus of conventional wireless networks, wireless sensor networks pose significantly different design, implementation, and deployment challenges.

Usually, the sensors are used to measure and/or monitor parameters that may vary with place and time. Therefore, a large number of sensors is required in order to obtain samples of these parameters at different locations and times. Moreover, these sensors should be networked in order to facilitate the transmission and dissemination of the measured/monitored parameters to some collector sites where the information is further processed for decision-making purposes. As a result, wireless sensor networks are complex systems in which the system behavior involves a large number of individual cooperating sensor nodes. A self-organized wireless sensor network provides the ability to adapt to diverse environments and unforeseeable situations. The self-organization feature is critical to achieve the wide applicability of sensor networks; however, it also makes modeling and prediction of the system behavior more difficult.

Modeling, designing, and verifying the architecture and organization of a distributed wireless sensor network with such a complicated nature require sophisticated system analysis methods and tools. There is a tremendous need for effective modeling techniques and tools to describe the large-scale sensor networks as time-varying composition of dynamically changing components and/or entities. These present additional features such as uncertainty, complexity, interaction, and collaboration.

The rest of this chapter is structured as follows. In Section 15.2, the main characteristics and most common features of wireless sensor networks involved in the development of the appropriate network architecture and modeling process are summarized. Section 15.3 provides a brief description of the architecture of sensor nodes and illustrates the general structure of sensor networks; the communication organization architecture of sensor networks is discussed, as well as the corresponding data dissemination architectures. Section 15.4 introduces and highlights the performance metrics involved in the modeling process of dynamic sensor networks; then the modeling of sensor networks from various aspects such as the sensing coverage, nodes placement, connectivity and energy consumption, etc. is addressed.

15.2 Characteristics of Wireless Sensor Networks

The progress of hardware technology in low-cost, low-power, small-sized processors, transceivers, and sensors has facilitated the development of wireless sensor networks. A distributed sensor network is a self-organized system composed of a large number (hundreds or thousands) of low-cost sensor nodes. Self-organization means that the system can achieve the necessary organizational structures without requiring human intervention, i.e., the sensor network should be able to carry out functional operations through cooperation among individual nodes rather than set up and operated by human operators. Sensor nodes are usually battery based, with limited energy resources and capabilities; it is difficult or unpractical to recharge each node. The far-ranging potential applications of sensor networks include: (1) system and space monitoring; (2) habitat monitoring [1, 2]; (3) target detection and tracking [3, 4]; and (4) biomedical applications [5–7].

In order to achieve cost-effectiveness and small sensor size, in general the individual sensor nodes present several limitations, such as limited energy and memory resources, small antennae, and limited processing capability. Although the sensor nodes and communication links are apt to fail due to these limitations and hostile operational environments, networking a large number of sensors together to form a distributed sensor network can overcome the weakness and bring great benefits and applicability. Although the organization of a distributed wireless sensor network is tightly related to the specified application, the following provides a summary of the most common features of wireless sensor networks involved in development of the appropriate network architecture and modeling process.

Extended coverage and easier deployment. The sensor network is large scaled and, in many cases, the number of sensors may be several orders of magnitude higher than the nodes in traditional ad hoc networks. Therefore, the coverage provided may be much larger compared to that provided by a single-sited sensor system. The overall coverage of a sensor network is the union of many small coverage areas of low-cost sensors, so the coverage is more flexible and can be adjusted conveniently by adding new nodes or moving nodes. Moreover, wireless sensor networks can also cover unfriendly terrains (such as battlefields, swamps, etc.) where infrastructures are not available and/or traditional deployment fashion is not feasible.

Reliability and flexibility. Although the capability and reliability of a single sensor node is restricted, multiple sensors provide fault tolerance, thus making the whole system more robust. When a sensor dies, its neighbor nodes can provide the same or similar information. Multiple routing alternatives are also available to protect the system against communication link failures. The self-organization feature of sensor networks provides the agility to adapt to unforeseeable situations, diverse environments, and dynamic changes. The flexibility refers to several aspects: sensing coverage can be adjusted by moving or replenishing nodes; trade-off between delay and information accuracy can be made via collaboration among sensors; balance of power consumption between nodes can also be achieved by cooperation.

Improved monitoring capabilities and information quality. Sensor networks can provide better monitoring capabilities about parameters that present spatial and temporal variances through the aggregation of data from plenty of nodes, and they can provide more valuable inferences about the physical world to the end user. It has been argued [8] that the gain offered by having more sensors exceeds the benefits of getting detailed information from each sensor. Thus, a network of low-cost sensors, each one with fewer capabilities, may substitute a high-accuracy but high-cost single-sited sensor and provide more accurate information about the interested conditions or track low-observable objects, while providing improved robustness.

Mobility. Sensor nodes can be fixed or mobile. Although currently most sensors are static and most existing work focuses on networks of static nodes, it is expected that in the near future mobility will be introduced into the sensor networks, because sensor movements may help to improve monitoring and tracking capabilities, achieve effective communication, and accommodate new applications. Providing movement capabilities to sensors allows them to account for initial bad positioning or potential poor propagation paths and environments so that operation of the whole system can be improved. In many environments, sensors are deployed randomly rather than located precisely. In this case, if the desired object or target area cannot be well observed based on the current location of the sensor, the sensor node may adjust its position to improve its monitoring capabilities. Moreover, in order to improve its communication quality, the sensor node may move and rearrange its connectivity with other nodes and also reduce the required transmission power for communication. Additional management and maintenance functions (such as recharging and maintenance) may benefit from sensor movement as well.

15.3 Architecture of Sensor Networks

This section provides a brief description of the architecture of sensor nodes and illustrates the general structure of sensor networks. It also discusses the communication organization architecture of sensor networks as well as the corresponding data dissemination architectures. In order to complete their task, sensor nodes need to perform the functions of sensing, processing, and communicating; Figure 15.1 demonstrates the typical architecture of a sensor node. Several experimental sensor nodes and networks have been developed, including Smart Dust mote developed by UC Berkeley [9]; WINS (wireless integrated network sensors) NG (next-generation) node by UCLA [10], and µAMPS node (microadaptive, multidomain power-aware sensors) developed by MIT [11].

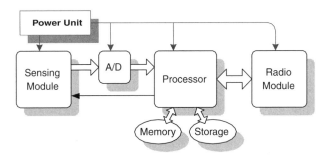

FIGURE 15.1 Architecture of a sensor node.

15.3.1 Functional Layers of Wireless Sensor Networks

The sensor network is more application specific than traditional networks designed to accommodate various applications. The organization and architecture of a sensor network should be designed or adapted to suit a special task so as to optimize the system performance, maximize the operation lifetime, and minimize the cost. Figure 15.2 depicts the various layers of functions of a distributed wireless sensor network:

- The *sensing layer* performs the work of data acquisition from the detected objects.
- The *communication layer* performs the tasks of data correlation, data compression, data dissemination, and routing. The function of this layer is to deliver the statistical observation results to the collecting center (the sink). Due to energy constraints of the wireless sensor networks and terrain characteristics, the MAC (media access control) protocols and network protocols adopted should be energy aware. The data dissemination mechanism determines which part or which kind of the information should be transmitted, while the routing mechanism makes the decision how to transmit the data and which routes should be followed. The routing and data dissemination mechanisms may affect each other to achieve maximum energy efficiency. A security layer may also be inside the communication layer that deals with security and authentication problems for some applications.
- The *data fusion layer* processes data received form the communication layer and combines them using various signal processing, data fusion, artificial intelligence, and other decision-making techniques as well as the prior knowledge of sensor performance and object characteristics. After the appropriate calculation and analysis, the data fusion layer produces the final detection results of a sensor network.
- The uppermost layer is the *user layer*, which provides a man–machine interface with displaying and interaction functions and presents the final results to human and/or computer systems in the different required forms.

Additional functional blocks provide several other supporting processes and operations such as resource management and coverage/topology monitoring and control. The resource management module monitors the available resources (such as energy, memory, and storage units) and balances the energy consumption between sensor nodes. The topology/coverage control module monitors the coverage, adjusts the network topology, and harmonizes the sensing operations among the various sensors.

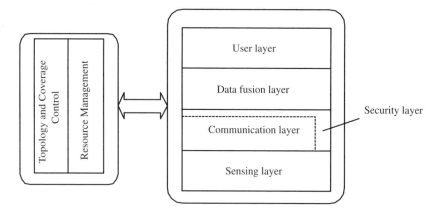

FIGURE 15.2 The function layers of distributed wireless sensor networks.

15.3.2 Homogeneous vs. Heterogeneous Architectures

In terms of the component nodes, the sensor networks can be classified in general into two categories: *homogeneous sensor network* and *heterogeneous sensor network*. In a homogeneous sensor network, the sensor nodes have identical capabilities and functionality with respect to the various aspects of sensing, communication, and resource constraints. In a heterogeneous sensor network, each node may have different capabilities and execute different functions. For example, some nodes may have larger battery capacity and more powerful processing capability and some may aggregate and relay data; other nodes may only execute the sensing function and not relay data for other nodes. A homogenous sensor network is simpler and easier to deploy, while a heterogeneous network is more complex and its deployment more complicated because different types of nodes must be dispensed in specified areas.

15.3.3 Communication Mode-Based Sensor Network Classification

With respect to the communication mechanism adopted, four basic architectures of sensor networks exist: *direct connected, flat ad hoc, peer-to-peer multihop,* and *cluster-based multihop*, as shown in Figure 15.3. Because the number of sensor nodes is usually large and the transmit range of sensor nodes may be limited due to the battery capacity limitations, in general it is cost inefficient and, in many cases, impossible, for each small sensor to communicate directly with the collector. Thus, the direct connected mode is not suitable for large-scale deployed sensor networks.

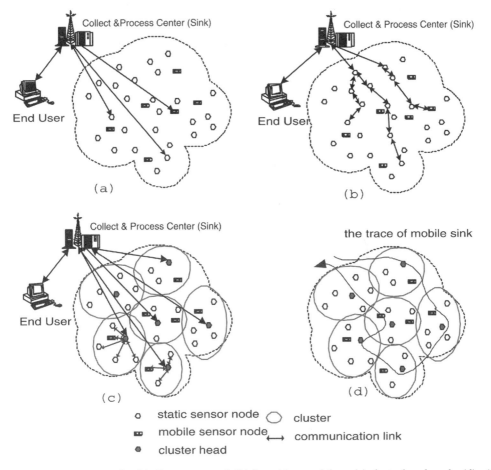

FIGURE 15.3 Sensor networks: (a) direct connected; (b) flat ad hoc multihop; (c) cluster-based mode; (d) with mobile sink.

Multihop mode is an apt alternative mainly because of its energy efficiency considerations. In addition to solving problems associated with the limited direct transmission range of nodes, multihop short-range transmission usually consumes less power than the power required by one large-hop transmission for a given pair of source and destination because, in general, the average received signal power is inversely proportional to the nth power of the distance, (usually $2 < n \leq 4$). In a flat ad hoc multi-hop network, as shown in Figure 15.3(b), some sensor nodes have routing capabilities, thus playing the role of relaying packets in addition to sensing and sending out their data. Although this mode is flexible and energy efficient, scalability is still a problem. The nodes closer to the collection and processing center will be primarily used to route data packets from other nodes to the processing center; if the network size is large, these nodes will relay a large number of data and their energy will be exhausted very fast, resulting finally in disconnection of the network.

Cluster-based multihop sensor networks attempt to address the scalability issues associated with the flat ad hoc multihop networks. In a cluster-based system, sensor nodes form clusters; a cluster head for each cluster is selected according to some negotiated rules [12]. Sensor nodes only transmit their data to their immediate local cluster head. In Figure 15.3(c), only one-level clustering is depicted; however, in general a hierarchical clustering scheme may be used, i.e., lower level cluster heads communicate with their high-level cluster heads. Local data fusion and classification at cluster heads may be used to reduce the amount of information that must be transmitted to the collection center, thereby reducing the overall energy consumed for transmission. The main disadvantage of this mode of operation is that the communication relies highly on the cluster head, thus placing a burden on the higher level cluster heads; also, the energy depletion of cluster heads is faster than that of other nodes. These issues can be addressed through rotation of the roles of various nodes.

The cases illustrated in Figure 15.3(a) through Figure 15.3(c) assume that the sink is immobile. However, in some scenarios the sink could be mobile, e.g., on a battlefield. Another scenario is to use a group of cooperating unmanned air vehicles as communication hubs for the sensors over a region of interest to collect the data.

15.3.4 Data Fusion Architectures

A sensor network is more data oriented than traditional wireless ad hoc networks are. The data fusion strategy plays an important role in the network design. Generally, the data dissemination/fusion architectures of sensor networks can be classified into the following three broad categories: *centralized, localized, or hybrid*. Figure 15.4 depicts these possibilities on a continuum.

If all sensor reports are transmitted to a collection and processing center without significant delay, it is called *centralized data fusion* [13]. For centralized fusion, all observation results are received and will be processed by the processing center at discrete instances of time; thus it could take into account all the relevant information in order to provide the optimal output. However, the realization of centralized

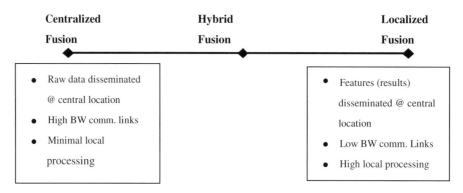

FIGURE 15.4 Sensor dissemination/fusion archiecture comparisons.

fusion architecture may face difficulties for reasons such as the limited capacity of the data links and synchronization problems. Furthermore, data transmission when fusing data from nonlocalized battery-powered sensors is a significant additional cost to a sensor. Therefore, *localized fusion architecture* has been proposed.

Unlike traditional wireless cellular networks in which the communication is person–person and the contents of conversations are irrelevant to each other, in sensor networks, the data in the neighboring nodes are considered highly correlated because observed objects in the physical world are highly correlated. Thus, localized data processing and aggregation might dramatically decrease the amount of information to be transmitted. Determining the appropriate architecture involves trading the costs of data transport vs. localized processing. Data that must be transmitted have a cost per byte and processor power used to reduce the raw data to a feature set and/or a fused result has a cost per millions of instructions per second (MIP). The processing power needed to generate feature vectors usually consumes less energy than transmitting the sensor raw data sets.

Therefore, in order to maximize the sensor network lifetime, the sensor network architecture will most likely tip toward a localized approach. At the same time, a claim may be made that optimum target detection and tracking would occur from the centralized fusion. As a result *hybrid solutions* are required in order to consider and balance the corresponding trade-offs, depending on the overall objective of the developed strategy.

15.4 Modeling of Dynamic Sensor Networks

The definition and development of models in order to analyze and evaluate sensor networks can help not only to study the network behavior and predict the evolvement of the system systematically, but also to direct deployment and implementation of these networks. This section introduces and highlights the performance metrics involved in the modeling process of dynamic sensor networks. Then the modeling of sensor networks is addressed from various aspects, such as sensing coverage, node placement, connectivity, energy consumption, etc.

15.4.1 Performance Metrics of Dynamic Wireless Sensor Network

Although traditional wireless cellular networks are mature and mobile ad hoc networking technology has been developed, the corresponding architectures and protocols still need to be tailored to the unique features of distributed wireless sensor networks. The behavior and evolution of a sensor network depend on many system parameters that are tightly correlated with the corresponding organizations and architecture forms. These parameters include:

- Total number of sensors, which indicates the size of a system
- Density, which is related to the deployment pattern
- Connectivity, which describes the communication link arrangements and related reliability
- Sensing coverage range and transmit range (radius) of sensor nodes
- Power consumption of each unit and energy availability
- Movement pattern, such as speed and direction

Before building and evaluating a sensor network, the communication mechanism, data storage scheme, and data fusion mode must be designed and the corresponding parameters determined. The goal is to obtain a balance among the various design elements in order to achieve the optimal architecture. The design of a dynamic sensor network can be evaluated by the following performance metrics:

- *Lifetime/energy efficiency.* The wireless sensor nodes can only be equipped with very limited energy resources (usually battery operated). Thus, the lifetime of a sensor is a critical issue, and energy-efficient protocols and algorithms must be designed to prolong the network lifetime. The definition of lifetime may vary for different types of applications. For non-mission-critical applications, the lifetime can be defined as the cumulative operation time of the network; for mission-critical applications, lifetime may be defined as the cumulative active time of the network until the first loss of coverage or quality failure [14]. In other works the network lifetime has been defined as the time interval from the point that the sensor network starts its operation until the point that some attributes (such as number of active nodes, the sensing coverage, information accuracy) fall off some certain threshold. Taking into consideration that the main objective of energy-efficient organizations and power conservation policies is to extend the network lifetime as long as possible, the network lifetime may be defined as the time interval from the point at which the sensor network starts its operation until the point at which loss of communication to the collector site by all sensor nodes occurs.

- *Quality.* The quality includes two aspects: accuracy and latency. The sensor network must provide sufficiently accurate information to end users and collector sites while at the same time satisfying the delay requirements. There is a trade-off between quality and energy efficiency. Usually the higher the accuracy and the smaller the delay are, the larger the power consumption. Both definitions of latency and accuracy are application dependent. For example, in target detection applications, accuracy may involve the missing-detection probability and false-alarm probability, while the latency might be defined as the interval from the time that a query is sent out to the time that the proper information is received by an end user.
- *Robustness.* Sensors and links are prone to fail. The network organization must be fault tolerant in order to avoid system failures when one or more nodes/links fail. Redundant deployment of sensors and replication of information between sensor nodes can be adopted to overcome some of the related problems. A trade-off always exists between minimizing the cost in order to keep the system affordable and improving system reliability by adding system components for redundancy and management purposes.
- *Scalability.* A scalable architecture should be able to support the growth of the system to an arbitrarily large size. Scalability in sensor networks is an important factor because the number of sensor nodes deployed in many situations may be on the order of hundreds or thousands, or even reach an extreme value of millions for some applications [15]. A scalable design requires scalable routing protocols, naming/addressing strategies, and data fusion methods so that the system can bootstrap a functional operation without significant influence from the large size of a self-organized sensor network. Hierarchical cluster-based architectures may also achieve better scalability [16].
- *Flexibility.* Although sensor networks usually are designed to accommodate certain applications, these networks should be able to adapt to possible functional and timing changes. In general, these systems should be flexible in two aspects: node configuration and network organization. For the sensor nodes, programmable devices such as programmable microprocessors and digital signal processors (DSPs) can be used to meet the flexibility requirement, while the self-organization nature of the sensor networks allows the system to adjust to suit the various environmental changes.
- *Throughput.* Because the available communication bandwidth is limited and the high node density of sensor networks may result in generation of large amounts of data, the end-to-end transmission throughput needs to be maximized in addition to providing fairness, power efficiency, and low complexity of implementation. In applications such as forest fire or nuclear power plant monitoring [17, 18], the information disseminated may increase abruptly when an emergency occurs and, as a result, the achievable peak throughput should satisfy the application requirements under this scenario.

In addition to the preceding metrics mentioned, the low cost and low complexity of implementation are also crucial parameters and may affect the transition from lab prototype networks to practical realizations and applications. The cost includes several elements such as hardware cost, deployment cost, maintenance cost, etc.

15.4.2 Modeling Sensor Networks

The following subsections describe traffic models as well as energy and battery models of the sensor nodes, and then discuss the modeling approaches of the sensor network from various viewpoints such as connectivity, sensing and coverage, power (energy) considerations, etc.

15.4.2.1 Traffic Models

Traffic characteristics of sensor networks vary and may be very different from those of conventional wireless networks. They mainly depend on operational modes that indicate the characteristics and patterns of the measurement and information to be transmitted. In general, the operational modes can be divided into three categories: *steady mode, ad hoc request/respond mode*, and *ad hoc threshold-based mode*.

The first, steady mode, assumes a steady flow of data from sensors to the collector. In this case, an accurate and current estimate of the field measured at the collector site is the goal. For instance, the communication pattern of the biosensor networks [6, 7] belongs to the steady mode; each node must transmit its data once every 250 ms and the traffic is deterministic and periodic. A field with high temporal resolution requires more frequent measurements and transmissions, while a field with low resolutions may require transmissions less often to obtain the same degree of accuracy. The second mode, ad hoc request/respond mode, corresponds to cases in which the sensors respond to requests generated by the collector site, which may be targeted to a specific set of sensors and/or for a specific time interval. The third category, ad hoc threshold-based mode, corresponds to cases in which transmission of information is triggered by an event during which a monitored/measured field exceeds some threshold. In applications of object detection or environment and system monitoring (such as forest fire monitoring), usually this mode is used. In this case the measurements and transmission frequencies may be different.

In general, in some applications, the operation may involve one or more modes. These different models generate different traffic in the network and different load conditions that affect performance of the routing strategies. For instance the ad hoc request/respond mode creates a two-way communication flow between sensors and collector site, while the other two modes mainly generate a one-way communication flow. The latter two modes may generate more bursty traffic. In general for these cases, the traffic model used in the literature is the Poisson process. Nordman and Kozlowski [19] argue that the maximum number of potential sensors accessing a wireless channel is low due to the small radio radius of sensor nodes and, in this case, the corresponding arrival process can be described as quasirandom with Engset approach.

15.4.2.2 Energy and Battery Models

In order to predict the lifetime of a sensor network and compare the quality of different algorithms and protocols, energy models for the computation and communication energy dissipation at nodes, as well as battery models used to depict battery capacity and behavior, should be specified.

As discussed earlier, the main components of a sensor node include sensing, processing, and communication units, so the *energy dissipation* comprises the energy consumed for sensing, processing, and transmitting data from source to sink. In general, the energy needed to sense and process a bit is assumed to be a constant. The energy dissipation for a radio unit includes the energy needed to receive a bit and the energy needed to transmit a bit. The former accounts for the power dissipation of the receiver electronics and the latter can be divided into two parts: transmitter electronics energy dissipation and the radio frequency (RF) transmit power. The transmit power is related to the transmission distance and the path loss exponential functions. Heinzelman and colleagues [16] provide some typical values for these parameters, while in Savvides et al. [20] the measurement values of power consumption for experimental WINS node are given. For a more comprehensive model, the effect of start-up transient behavior on energy dissipation may also be considered if sensors have short-range transmissions [21].

Battery models vary with the constituent material. In Amre El-Hoiydi [22], a constant leakage model for alkaline battery is used, where a constant leakage power equal to 10% of the full energy during 1 year is assumed. Park and coworkers have proposed three battery models for sensor networks [23]: *linear model, discharge rate-dependent model,* and *relaxation model*. In the linear model, the battery is treated as linear storage of current. The maximum capacity of the battery is achieved regardless of the discharge rate. The discharge rate dependent-model considers the effect of battery discharge rate on the maximum battery capacity. Because the battery capacity is reduced as the discharge rate increases, a battery capacity efficiency rate is introduced that varies with the current and is close to one when the discharge rate is low; it approaches zero when the rate becomes high. The relaxation model is more complicated because it also takes into account the relaxation phenomenon of real-life batteries.

15.4.2.3 Connectivity Modeling and Topology Optimization

Connectivity is a fundamental property of wireless networks. In wireless sensor networks, the connectivity relies on the actual physical conditions, such as transmit power range, network density, and node positions; it provides a good indication of network status. In-depth study and modeling of the connectivity distribution facilitates development of guidelines regarding several processes involved in design and operation of sensor networks, such as the deployment pattern and density of sensors; communication strategies among individual sensors; distributed information-processing algorithms; and, finally, routing and/or information dissemination strategies. For example, an algorithm based on multidimensional scaling that uses connectivity information to derive the locations of nodes in the network has been proposed in Shang et al. [24].

Zhu and colleagues introduced a model that gives a realistic description of the various processes and their effects as the mobile sensor-based network evolves [25]. They provide an analytical approach that describes the dynamics of the network and facilitates understanding of the effect of various events on the large-scale topology of a wireless self-organizing sensor network. The motivation of the model stems from the commonality encountered in the mobile sensor wireless networks, their self-organizing and random nature, and some concepts developed by the continuum theory [26]. Given a certain coverage and a constant number N of nodes (i.e., neither new nodes coming nor existing nodes leaving), a model is presented next that gives a realistic description of the local processes involved in network evolution, incorporating link removal, link rewiring, etc.

One of the following three operations may be executed at each time-step t:

- With probability p ($0 \leq p < 1$), m1 new links are added (m1 $\leq N$). This could happen when a node begins to contact other nodes and build new links, or when a node moves to the coverage of another node and would like to establish a new link. A node is randomly selected as the starting point of the new link while the end point is selected with probability $Q1(k_i)$, where $Q1(k_i)$ is the probability that a node i currently with k_i links is selected. This process is repeated m1 times.
- With probability q ($0 \leq q < 1$), m2 links are rewired (m2 $\leq$ m1). This will happen when a node finds that one or more new links are better than the existing ones for routing or data gathering. For this case, one node i and one link l_{ij} between node i and node j are randomly selected, and the link is rewired to another node j', where j' is selected with probability $Q2(k_i)$, defined similarly with $Q1(k_i)$. This process is repeated m2 times.
- With probability r ($0 \leq r < 1$), m3 existing links are deleted (m3 $\leq$ m1). This could happen when a node finds that it has a large number of links or its energy is being depleted faster than its schedule. One node i with probability $Q3(k_i)$ is selected, and then one of its links is randomly selected to release. This process is repeated m3 times.
- With probability 1–p–q–r, nothing happens, i.e., no connection changes.

Based on these operations, three different scenarios that represent the most common realistic situations can be defined and evaluated:

- Scenario 1: new links preferentially point to popular nodes, while the more links with which the node is associated, the higher the probability that the node removes a link.
- Scenario 2: new links preferentially are deployed evenly, while the more links with which the node is associated, the higher the probability that the node may remove a link.

- Scenario 3: the probability of removing links is relative to the connectivity conditions of the system.

For scenario 1, the connectivity approximately increases linearly with t (time-step) at the beginning and, as t becomes large, the connectivity distribution approaches a Gaussian distribution. In scenario 2, the corresponding distributions are also Gaussian, but the variances are much smaller than those in scenario 1. In scenario 3, the system will increase rapidly first and approach a dynamic balance at the same point after the transient period. The mean connectivity value depends on the ratio between adding a new link and deleting an existing link and the parameter, implying that the probability of removing links is relative to the connectivity conditions of the system.

For a given deployment of sensors, the topology can be optimized to achieve sufficient reliability, energy efficiency, or throughput by rearranging the connectivity. The topology considered here is from the viewpoint of communication, so it actually represents the logical topology of a sensor network. A distributed topology-control algorithm has been developed for a multihop packet radio network to control each node's transmission power and logical neighbors in order to construct a reliable high-throughput topology [27].

Different from the traditional cellular systems in which each mobile has at least a wireless link to the base station, the situation in multihop packet radio networks is usually more sophisticated and complicated. Philips and coworkers [28] showed that, to ensure network connectivity, the expected number of nearest neighbors of a transmitter must grow logarithmically with the area of the network. Furthermore, the critical ranges of transmitters for area coverage and connectivity purposes are discussed in Piret [29].

Recently, the connectivity problem of multihop ad hoc networks has been studied extensively. In ad hoc wireless networks, the nodes in the network are assumed to cooperate in a decentralized fashion, routing packets from other nodes; thus, each node should transmit with enough power to guarantee connectivity of the overall network. Gupta and Kumar [30] determined the critical power at which a node in the network needs to transmit in order to ensure that the network is asymptotically connected with probability one as the number of nodes in the network goes to infinity. For a one-dimensional network, Desai and Manjunath [31] obtained the exact formula for the probability that the network is connected under the assumption of uniform distribution of nodes in $[0, \pi]$ and extended this result to obtain the upper bound of the connected probability for a two-dimensional network.

The connectivity of wireless multihop networks with uniformly randomly distributed nodes was investigated by Bettstetter [32] under homogeneous and inhomogeneous transmit range assignments. A free-space radio link model and bidirectional links were considered. For the scenario without border effects, the required transmit ranges to achieve a connected or two-connected network with high probability (the probability must be close to one) for the homogeneous case were obtained as a function of the number of nodes and the system area. Considering the border effects, the threshold ranges were obtained by performing simulations and results showed that the required values are higher than those of networks without border effects. Bettstetter also gave the approximate k-connected probability of a wireless multihop network consisting of nodes with different transmit ranges [32].

Zhu and Papavassiliou [33] addressed the connectivity distribution and related power conservation issues for large-scale multihop sensor networks. It was demonstrated that, when the total number of nodes is very large and the transmit range of each node is limited and much smaller than the whole coverage, the connectivity distribution approaches a Poisson distribution with parameter depending on the density and the transmit range. Furthermore, several trade-offs among node connectivity, power consumption, and data rate were discussed. The more practical log-distance path loss model is considered for radio link instead of the free-space radio model. Utilizing the proposed model, the transmit power can be minimized by minimizing the transmit range under a specific connectivity requirement determined by the reliability requirement for a fixed data rate system. Conversely, given the transmit power and the minimum required receiver power level, the connectivity distribution of the sensor network can be obtained. Furthermore, the variable data rate was introduced in the proposed method as another adjustable parameter for the analysis of power and reliability trade-offs.

15.4.2.4 Deployment and Sensing Coverage Models

The deployment pattern of the sensors, the sensor density, and the achievable sensing coverage are critical factors that influence the overall design and effectiveness of a sensor network. The density depends on many factors, such as desired accuracy; temporal and spatial resolution; evolution of the information to be gathered and disseminated; mobility of the sensors; efficiency; and fault tolerance. Moreover, the deployment pattern and achievable sensing coverage depend not only on the previous factors, but also on other factors, such as restrictions (limitations) on the locations where sensors should be placed. Therefore, different deployment patterns and coverage models need to be considered, depending on the application of interest.

For example, for a sensor network designed to perform vehicle tracking, the user would like to maximize the probability of detection if a vehicle is in the sensor field and obtain a relatively accurate estimation of the vehicle position and speed. In this case, the deployment pattern of the sensors to achieve that objective is of interest. For an already deployed sensor network, it is often required to characterize the well-covered areas, weakly covered regions, or blind spots in order to adjust the configuration or deploy additional nodes. The sensing coverage relies on the sensing or detecting models and the placement of nodes.

In order to optimize the deployment of sensor nodes and ensure that the mandated requirement of sensing coverage is met, the sensing models of sensors must be determined. Sensing models vary with sensing devices, which generally have widely different theoretical and physical characteristics. For instance, Dhillon and colleagues [34] assumed that the probability of detection of a target by a sensor declines exponentially with the distance between the target and the sensor. In this case, a sensor detects a target at distance d from it with probability $e^{-\alpha d}$, where parameter α is used to reflect the rate at which its detection probability diminishes with distance. Other works have assumed that the sensing ability diminishes as distance increases; a sensing model at an arbitrary point is expressed as the inverse of the k-power of the distance between the sensor and the point, where k is a sensor-dependent parameter [35]. Liu and coworkers [36] used two types of sensing models for acoustic amplitude sensor and direction-of-arrival sensors; the measurement noise is considered in these two models.

The traditional sensors are characterized by specifications such as range resolution, range accuracy, bearing resolution, and accuracy. These specifications provide a good measure of the ability of a sensor. Similar specifications for sensor networks do not exist currently. Thus, Liu and colleagues [36] introduce the concept of a sensing field to be a measure of how well a sensor network can sense the phenomenon at that point. Based on this definition, a Cramer–Rao upper bound of the estimation accuracy for a target localization and tracking system is derived.

As mentioned earlier, sensor nodes can be fixed or mobile. Providing mobility to sensors allows them to account for initial bad positioning or potential poor propagation paths. However, currently most of the sensors are static and therefore the placement of sensors has significant impact on many factors, such as the desired accuracy, temporal and spatial resolution, evolution of the information to be gathered and disseminated, efficient routing, fault tolerance, etc.

Two different types of coverage exist: *deterministic coverage and stochastic coverage. Deterministic coverage* means that the placement is well controlled so that each node can be deployed at a specific position. The predefined deployment patterns could be uniform in different areas of the sensor field or can be weighted to compensate for the more critically monitored areas. This case is similar to the art gallery problem [37] and usually the suboptimum solution can be obtained by heuristics methods.

Grid-based sensor deployment is an instance of uniform patterns in which nodes are located on the intersection points of a grid. Chakrabarty et al. [38] presented different grid coverage strategies for effective surveillance and target location. The sensor field is considered as a two- or three-dimensional grid of points and the sensor placement problem is formulated in terms of cost minimization under coverage constraint. Furthermore, the authors determined the sensor placement for unique target location using the theory of identifying codes and showed that grid-based sensor placement for single targets provides asymptotically complete location of multiple targets. Although this kind of controlled node placement can provide good coverage for a given condition, it is unfeasible in many situations, especially when the sensors are deployed in unfriendly terrain or the cost is too high for determining placement of a large number of nodes.

Alternatively, the sensor field can be covered with sensors randomly distributed in the environment (e.g., dispersed by air vehicles). This type of node placement is called *stochastic coverage* [39]. The stochastic coverage scheme can be uniform, Gaussian, or Poisson, or may follow other distributions, depending on the application under consideration. A centralized polynomial time algorithm of computation of worst-case coverage and best-case coverage for random deployment using Voronoi diagram and graph search algorithms has been proposed and discussed in Meguerdichian and coworkers [39]. Furthermore, distributed algorithms to solve the best-coverage problem have been introduced in Li et al. [40].

With the assumption that the energy required to support a link is proportional to the α power ($2 < \alpha < 5$) of the Euclidean distance between the sensor and the observation spot, a solution of how to find a path with the best-coverage distance while the total energy consumed by this path is minimized among all optimum best-coverage paths is also provided [40]. In order to achieve energy efficiency, several algorithms have been proposed to address how to elect subsets of nodes from all nodes in the network to complete a specific sensing task at each moment [41, 42]. Some active nodes may stay "awake" all the time and perform multihop packet routing, while the rest of the nodes remain "passive" and periodically check whether they should become active. This coordination between nodes exploits the redundancy provided by the high sensor density in order to extend the overall system lifetime.

In most applications of sensor networks, the data transmission is relatively small compared to the Internet or other types of networks; therefore, letting the sensors go to "sleep" mode periodically can help to extend the lifetime of a sensor, especially when the traffic is low and delay constraint is not rigid. Zhu and Papavassiliou studied the power/energy consumption problem under sleeping and sleepless scenarios [33]. It has been observed that sleeping strategy is in general beneficial when traffic is low; if the traffic is high, however, it is beneficial under certain conditions only. The reason is that when the traffic is low, the node is in reception state for most of time and therefore the conserved power due to the sleeping strategies is much larger than the increase in the transmission power due to the need for increased transmission range (and power) to maintain the prespecified connectivity requirements. When traffic is high, the increased power required to keep the same connectivity during transmission due to the density decrease of active nodes is greater, in many cases, than the power conservation due to the use of the sleeping strategy.

15.5 Concluding Remarks

Because sensors are usually used to measure and/or monitor some parameters that may vary with place and time, a large number of sensors is typically required in order to obtain samples of these parameters at different locations and times. A certain set of applications requires that sensor nodes collectively form an ad hoc distributed processing network and provide information about the environment that they monitor. Without doubt, a mobile sensor-based communications and processing infrastructure will significantly enhance and facilitate the information-based detection, prevention, and response processes under several scenarios. Networking the sensors will facilitate the transmission/dissemination of the measured/monitored parameters to some collector sites at which the information is further processed for decision-making purposes. Multiple sensors can provide the end user with fault tolerance, better monitoring capabilities about parameters that present spatial and temporal variances, and valuable inferences about the physical world.

The different types of sensors may differ in size; computational and power/energy capabilities; functions to be performed; parameters to be measured; and/or mobility patterns. One of the important required features of such an infrastructure is the ability of the mobile sensor network to create the infrastructure in an ad hoc fashion and a self-organizing mode, thus allowing addition and deletion of individual sensors without any manual or centralized intervention.

This chapter examined and analyzed sensor network architecture and organization, as well as the modeling process involved in the sensor network design phase, by providing principles and guidelines that facilitate understanding of the properties of large scale sensor networks and by presenting and discussing several recent efforts and developments. Specifically, the most common features of wireless sensor networks involved in development of the appropriate network architecture and modeling process were identified and highlighted.

Along with the inherent characteristics of actual environments in which sensor networks are usually deployed and in combination with the fact that their focus is mainly on applications requiring tight coupling with the physical world, these features make modeling and design of sensor networks a very complex and sophisticated process. In order to gain some insight about these processes, the architecture of sensor nodes was presented and the general structure of sensor networks was illustrated; the communication organization architecture of sensor networks was discussed as well as the corresponding data dissemination architectures.

Due to its energy-efficiency considerations, multihop communication is used as the main communication mode in sensor networks, while the hierarchical, cluster-based multihop networking mode is described as the operational mode to address issues associated with scalability problems, especially in large-scale sensor systems. Because a sensor network is more data oriented than traditional wireless ad hoc networks are, the data fusion strategy plays an important role in the network design. Several data fusion/dissemination strategies were discussed that ranged from centralized to local/distributed methods and provide various trade-offs among accuracy, communication cost, and computing/processing cost.

As mentioned earlier, large-scale wireless mobile sensor networks correspond to the time-varying compositions of dynamically changing components and/or entities that present additional features and limitations such as uncertainty, resource constraints, complexity, interaction, and collaboration. On the one hand, the self-organization feature of sensor networks provides the ability for them to adapt to unforeseeable situations, diverse environments, and dynamic changes. On the other hand, it complicates the overall modeling process by introducing a certain degree of uncertainty and dynamic evolution. As part of the modeling process, several traffic models corresponding to different operational modes (i.e., steady mode, ad hoc request/respond mode, and ad hoc threshold-based mode) were discussed.

Furthermore, because in most cases sensors have limited available energy (usually battery operated), both energy models for the computation and communication energy dissipation at nodes, as well as battery models used to depict battery capacity and behavior, are important for evaluation and extension of the lifetime of sensor networks. Finally, connectivity is an important property of distributed wireless sensor networks that facilitates development of guidelines regarding several processes involved in design and operation of sensor networks, such as the deployment pattern and density of sensors; communication strategies among individual sensors; distributed information processing algorithms; and routing and/or information dissemination strategies. Therefore, this chapter has described several models that analyze the connectivity distribution as the network evolves.

References

1. Mainwaring, A. et al., Applications and OS: wireless sensor networks for habitat monitoring, in *Proc. ACM Int. Workshop Wireless Sensor Networks Applications*, 88, 2002.
2. Juang, P. et al., Energy-efficient computing for wildlife tracking: design trade-offs and early experiences with ZebraNet, in *Proc. 10th Int. Conf. Architectural Support Programming Languages Operating Syst. (ASPLOS-X)*, 96, 2002.
3. Shih, E. et al., Physical layer driven protocol and algorithm design for energy-efficient wireless sensor networks, in *Proc. 7th Annu. Int. conf. Mobile Computing Networking*, 272, 2001.
4. Estrin, D. et al., Next century challenges: scalable coordination in sensor networks, in *Proc. 5th ACM/IEEE Int. Conf. Mobile Computing Networking (Mobicom)*, ACM Press, 263, 1999.
5. Bauer, P. et al., The mobile patient: wireless distributed sensor networks for patient monitoring and care, in *Proc. 2000 IEEE EMBS Int. Conf. Inf. Tech. Appl. Biomed.*, 17, Nov. 2000.

6. Shankar, V. et al., Energy-efficient protocols for wireless communication in biosensor networks, in *Proc. 12th IEEE Int. Symp. Personal, Indoor Mobile Radio Commun.*, 1, D-114, 2001.
7. Schwiebert, L., Gupta, S.K.S. and Weinmann, J., Research challenges in wireless networks of biomedical sensors, in *Proc. ACM 7th Annu. Int. Conf. Mobile Computing Networking*, 151, July 2001.
8. Chamberland, J.-F. and Veeravalli, V.V., Decentralized detection in sensor networks, *IEEE Trans. Signal Process.*, 51, 407, 2003.
9. Kahn, J.M., Katz, R.H. and Pister, K.S.J., Next century challenges: mobile networking for "Smart Dust," in *Proc. 5th ACM/IEEE Int. Conf. Mobile Computing Networking (Mobicom)*, 271, 1991.
10. Pottie, G.J. and Kaiser, W.J., Wireless integrated network sensors, *Commun. ACM*, 43, 51, 2000.
11. Min, R. et al., Low-power wireless sensor networks, in *Proc. 14th Int. Conf. VLSI Design*, 205, 2000.
12. Heinzelman, W.R., Chandrakasan, A. and Balakrishnan, H., Energy-efficient communication protocol for wireless microsensor networks, *in Proc. 33rd Hawaii Int. Conf. Syst. Sci.*, 3005, 2000.
13. Koch, W., Overview of problems and techniques in target tracking, *IEEE Colloquium Target Tracking: Algorithms Applications*, 1/1–1/4, 1999.
14. Bhardwaj, M. and Chandrakasan, A.P., Bounding the lifetime of sensor networks via optimal role assignments, in *Proc. 21st Joint Conf. of IEEE Computer Commun. Soc.*, INFOCOM 2002, 3, 1587, 2002.
15. Akyildiz, I.F. et al., A survey on sensor networks, *IEEE Commun. Mag.*, 40, 102, 2002.
16. Heinzelman, W.R. et al., Energy-scalable algorithms and protocols for wireless microsensor networks, in *Proc. 2000 IEEE Int. Conf. Acoustics, Speech, Signal Process.*, 6, 3722, 2000.
17. Mladineo, N. and Knezic, S., Optimization of forest fire sensor network using GIS technology, in *Proc. IEEE 22nd Int. Conf. Inf. Tech. Interfaces*, 2000, 391.
18. Satoh, K. et al., Autonomous mobile patrol system for nuclear power plants: field test report of vehicle navigation and sensor positioning, in *Proc., 1996 IEEE/RSJ Int. Conf. Intelligent Robots Syst.*, 2, 743, 1996.
19. Nordman M.M. and Kozlowski, W.E., Modeling data transactions with standard protocols for low power wireless sensor links, in *Proc. 1st ISA/IEEE Conf. Sensor Ind.*, 51, 2001.
20. Savvides, A., Park, S. and Srivastava, M., On modeling networks of wireless microsensors, in *Proc. 2001 ACM SIGMETRICS Int. Conf. Measurement Modeling Computer Syst.*, 318, 2001.
21. Wang, A.Y. et al., Energy efficient modulation and MAC for asymmetric RF microsensor systems, in *Proc. , 2001 Int. Symp. Low Power Electron. Design*, ACM Press, 106, 2001.
22. Amre El-Hoiydi, Spatial TDMA and CSMA with preamble sampling for low power ad hoc wireless sensor networks, in *Proc. 7th Int. Symp. Computers Commun. (ISCC'02)*, 685, 2002.
23. Park, S., Savvides, A. and Srivastava, M., Simulating networks of wireless sensors, in *Proc. 2001 Winter Simulation Conf.*, 1330, 2001.
24. Shang, Y. et al., Sensor networks: localization from mere connectivity, in *Proc. 4th ACM Int. Symp. Mobile ad hoc Networking Computing*, 201, 2003.
25. Zhu, J., Papavassiliou, S. and Xu, S., Modeling and analyzing the dynamics of mobile wireless sensor networking infrastructures, in *Proc. 56th IEEE Vehicle Technol. Conf.*, 3, 1550, 2002.
26. Barabasi, A.-L., Albert, R. and Jeong, H., Mean-field theory for scale-free random networks, *Physica A*, 272, 173–187, 1999.
27. Hu, L., Topology control for multihop packet radio networks, *IEEE Trans. Commun.*, 41, 1474, 1993.
28. Philips, T.K., Panwar, S.S. and Tantawi, A.N., Connectivity properties of a packet radio network model, *IEEE Trans. Inf. Theory*, 35, 1044, 1989.
29. Piret, P., On the connectivity of radio networks, *IEEE Trans. Inf. Theory*, 37, 1490, 1991.
30. Gupta, P. and Kumar, P.R., Critical power for asymptotic connectivity, in *Proc. 37th IEEE Conf. Decision Control*, 1, 1106, 1998.
31. Desai, M. and Manjunath D., On the connectivity in finite ad hoc networks, *IEEE Commun. Lett.*, 6, 437, 2002.
32. Bettstetter, C., On the connectivity of wireless multihop networks with homogeneous and inhomogeneous range assignment, in *Proc. IEEE 56th Vehicular Tech. Conf.*, 3, 1706, 2002.

33. Zhu, J. and Papavassiliou, S., On the connectivity modeling and the trade-offs between reliability and energy efficiency in large scale wireless sensor networks, in *Proc. IEEE Wireless Commun. Networking Conf.*, 2, 1260, 2003.
34. Dhillon, S.S., Chakrabarty, K. and Iyengar, S.S., Sensor placement for grid coverage under imprecise detections, in *Proc. 5th Int. Conf. Inf. Fusion*, 2, 1581, 2002.
35. Meguerdichian, S. et al., Exposure in wireless ad-hoc sensor networks, in P*roc. ACM 7th Annu. Int. Conf. Mobile Computing Networking*, 139, 2001.
36. Liu, J. et al., Sensing field: coverage characterization in distributed sensor networks, in *IEEE Proc. Int. Conf. Acoustics, Speech, Signal Process.*, 5, 173, 2003.
37. Marengoni, M. et al., Placing observers to cover a polyhedral terrain in polynomial time, in *Proc. 3rd IEEE Workshop Appli. Computer Vision*, 77, 1996.
38. Chakrabarty, K. et al., Grid coverage for surveillance and target location in distributed sensor networks, *IEEE Trans. Computers*, 51, 1448, 2002.
39. Meguerdichian, S. et al., Coverage problems in wireless ad-hoc sensor networks, INFOCOM'01, in *Proc. 20th Annu. Joint Conf. IEEE Computer Commun. Soc.*, 3, 1380, 2001.
40. Li, X.-Y., Wan, P.-J. and Frieder, O., Coverage in wireless ad hoc sensor networks, *IEEE Trans. Computers*, 52(6), 753, 2003.
41. Slijepcevic, S. and Potkonjak, M., Power efficient organization of wireless sensor networks, in *Proc. IEEE Int. Conf. Commun.*, 2, 472, 2001.
42. Cerpa, A. and Estrin, D., ASCENT: adaptive self-configuring sensor networks topologies, in *Proc. 21st Annu. Joint Conf. IEEE Computer Commun. Soc., IEEE INFOCOM'02*, 3, 1278, 2002.

IV

Protocols

16
Overview of Communication Protocols for Sensor Networks

Weilian Su
Georgia Institute of Technology

Erdal Cayirci
Istanbul Technical University

Özgür B. Akan
Georgia Institute of Technology

16.1 Introduction 16-1
16.2 Applications/Application Layer Protocols 16-2
  Sensor Network Applications • Application Layer Protocols
16.3 Localization Protocols 16-4
16.4 Time Synchronization Protocols 16-5
16.5 Transport Layer Protocols 16-7
  Event-to-Sink Transport • Sink-to-Sensors Transport
16.6 Network Layer Protocols 16-9
16.7 Data Link Layer Protocols 16-11
  Medium Access Control • Error Control
16.8 Conclusion 16-14

16.1 Introduction

As the technology for wireless communications advances and the cost of manufacturing a sensor node continues to decrease, a low-cost but yet powerful sensor network may be deployed for various applications that can be envisioned for daily life. Although each sensor node may seem to be much less capable than a traditional stationary sensor, a collective effort of the sensor nodes may provide sensing capabilities in space and time that surpass the stationary sensor.

The communication protocols for sensor networks may leverage the capabilities of collective efforts to provide users with specialized applications. These protocols may fuse, extract, or aggregate data from the sensor field. In addition, they may self-organize the sensor nodes into clusters to complete a task or overcome certain obstacles, e.g., hills. In essence, sensor networks may provide end users with intelligence and details that traditional stationary sensors may not be able to do.

Although the sensor nodes communicate through the wireless medium, protocols and algorithms proposed for traditional wireless ad hoc networks may not be well suited for sensor networks. As previously explained, sensor networks are application specific, and the sensor nodes work collaboratively together. In addition, the sensor nodes are very energy constrained compared to traditional wireless ad hoc devices. The differences between sensor networks and ad hoc networks [29] are:

- The number of sensor nodes in a sensor network can be several orders of magnitude higher than the nodes in an ad hoc network.

- Sensor nodes are densely deployed.
- Sensor nodes are prone to failures.
- The topology of a sensor network changes very frequently.
- Sensor nodes mainly use a broadcast communication paradigm whereas most ad hoc networks are based on point-to-point communications.
- Sensor nodes are limited in power, computational capacities, and memory.
- Sensor nodes may not have global identification (ID) because of the large amount of overhead and large number of sensor nodes.
- Sensor networks are deployed with a specific sensing application in mind; ad hoc networks are mostly constructed for communication purposes.

With these differences, the design of communication protocols for sensor networks requires specific attention. Some of the potential applications as well as some application layer protocols for sensor networks are presented in Section 16.2. Next, because many of the communication protocols require the knowledge of location and time in order to function properly, localization and time synchronization protocols are described in Section 16.3 and Section 16.4. Furthermore, protocols and challenges for the transport, network, and data-link layers are consecutively explained in Section 16.5 through Section 16.7, respectively.

16.2 Applications/Application Layer Protocols

Sensor nodes can be used for continuous sensing, event detection, event identification, location sensing, and local control of actuators. The concept of microsensing and wireless connection of these nodes promise many new application areas, e.g., military, environment, health, home, commercial, space exploration, chemical processing, and disaster relief, etc. Some of these application areas are described in the next subsection. In addition, Subsection 16.2.2 introduces some application layer protocols used to realize these applications.

16.2.1 Sensor Network Applications

The number of potential applications for sensor networks is huge. Actuators may also be included in the sensor networks, thus making the number of applications that can be developed much higher. In this section, some example applications are given to provide the reader with a better insight about the potentials of sensor networks.

Military applications. Sensor networks can be an integral part of military command, control, communications, computers, intelligence, surveillance, reconnaissance and tracking (C4ISRT) systems. The rapid deployment, self-organization, and fault tolerance characteristics of sensor networks make them a very promising sensing technique for military C4ISRT. Because sensor networks are based on dense deployment of disposable and low-cost sensor nodes, destruction of some nodes by hostile actions does not affect a military operation as much as the destruction of a traditional sensor does. Military applications include: monitoring friendly forces, equipment, and ammunition; battlefield surveillance; reconnaissance of opposing forces and terrain; targeting; battle damage assessment; and nuclear, biological, and chemical attack detection and reconnaissance.

Environmental applications. Some environmental applications of sensor networks include tracking the movements of species, i.e., habitat monitoring; monitoring environmental conditions that affect crops and livestock; irrigation; macroinstruments for large-scale Earth monitoring and planetary exploration; and chemical/biological detection [1, 3, 4, 6, 15, 17, 19, 20, 39, 45].

Commercial Applications: The sensor networks are also applied in many commercial applications, including building virtual keyboards; managing inventory control; monitoring product quality; constructing smart office spaces; and environmental control in office buildings [1, 6, 11, 12, 20, 31, 33, 34, 38, 45].

16.2.2 Application Layer Protocols

Although many application areas for sensor networks are defined and proposed, potential application layer protocols for sensor networks remain largely unexplored. Three possible application layer protocols are introduced in this section: sensor management protocol; task assignment and data advertisement protocol; and sensor query and data dissemination protocol. These protocols may require protocols at other stack layers (explained in the remaining sections of this chapter).

16.2.2.1 Sensor Management Protocol (SMP)

Designing an application layer management protocol has several advantages. Sensor networks have many different application areas; accessing them through networks such as the Internet is the aim in some current projects [31]. An application layer management protocol makes the hardware and software of the lower layers transparent to the sensor network management applications.

System administrators interact with sensor networks by using sensor management protocol (SMP). Unlike many other networks, sensor networks consist of nodes that do not have global ID, and they are usually infrastructureless. Therefore, SMP needs to access the nodes by using attribute-based naming and location-based addressing, which are explained in detail in Section 16.6. SMP is a management protocol that provides software operations needed to perform the following administrative tasks:

- Introducing rules related to data aggregation, attribute-based naming, and clustering to the sensor nodes
- Exchanging data related to location-finding algorithms
- Time synchronization of the sensor nodes
- Moving sensor nodes
- Turning sensor nodes on and off
- Querying the sensor network configuration and the status of nodes, and reconfiguring the sensor network
- Authentication, key distribution, and security in data communications

Descriptions of some of these tasks are given in references 8, 11, 30, 36, and 37.

16.2.2.2 Task Assignment and Data Advertisement Protocol (TADAP)

Another important operation in the sensor networks is interest dissemination. Users send their interest to a sensor node, a subset of the nodes, or the whole network. This interest may be about a certain attribute of the phenomenon or a triggering event. Another approach is the advertisement of available data in which the sensor nodes advertise the available data to the users and the users query the data in which they are interested. An application layer protocol that provides the user software with efficient interfaces for interest dissemination is useful for lower layer operations, such as routing.

16.2.2.3 Sensor Query and Data Dissemination Protocol (SQDDP)

The sensor query and data dissemination protocol (SQDDP) provides user applications with interfaces to issue queries, respond to queries, and collect incoming replies. These queries are generally not issued to particular nodes; instead, attribute-based or location-based naming is preferred. For instance, "the locations of the nodes that sense temperature higher than 70°F" is an attribute-based query. Similarly, "temperatures read by the nodes in Region A" is an example of location-based naming.

Similarly, sensor query and tasking language (SQTL) [37] is proposed as an application that provides even a larger set of services. SQTL supports three types of events, which are defined by keywords *receive*, *every*, and *expire*. The *receive* keyword defines events generated by a sensor node when it receives a message; *every* keyword defines events occurring periodically due to a timer time-out; and *expire* keyword defines events occurring when a timer is expired. If a sensor node receives a message intended for it that contains a script, it then executes the script. Although SQTL is proposed, different types of SQDDP can be developed for various applications. The use of SQDDPs may be unique to each application.

SQDDP provides interfaces to issue queries, responds to queries, and collects incoming replies. Other types of protocols are also essential to sensor network applications: the localization and time synchronization protocols. The localization protocol enables sensor nodes to determine their locations; the time synchronization protocol provides sensor nodes with a common view of time throughout the sensor network. Because many communication protocols require knowledge of location and time, it is important to describe the localization and time synchronization techniques in detail in the following sections before transport, network, and data link protocols are discussed later.

16.3 Localization Protocols

Because sensor nodes may be randomly deployed in any area, they must be aware of their locations in order to provide meaningful data to the users. In addition, location information may be required by the network and data-link layer protocols described in Section16.6 and Section 16.7, respectively. In order to meet design challenges, a localization protocol must be:

- Robust to node failures
- Less sensitive to measurement noise
- Low error in location estimation
- Flexible in any terrain

Currently, two types of localization techniques address these challenges: (1) beacon based and (2) relative location based. Both techniques may use range and angle estimations for sensor node localization via received signal strength (RSS) [23, 42]; time of arrival (TOA) [13, 41]; time difference of arrival (TDOA); and angle of arrival (AOA).

Current localization methods [27, 36] are based on beacons with position known. The ad hoc localization system (AHLoS) [36] requires few nodes to have known location through GPS or through manual configuration. This allows nodes to discover their location through a two-phase process: ranging and estimation. During the ranging phase, each node estimates the range of its neighbors. The estimation phase then allows neighbors that do not have location to use the range estimated in the ranging phase and the known location of the beacons to estimate their locations.

Also, some methods [5, 6] assume beacon signals at known locations. This assumption may be fine for some applications, but sensor nodes may be deployed in regions in which known location is not possible. As a result, Moses and colleagues are investigating self-localization using sources at unknown locations [27]. Although these authors relax the assumption that beacons require fixed locations, the beacons still need a number of signal sources. These signal sources are deployed in the same region as the sensor nodes and used as references by the neighbor nodes to estimate the unknown locations and orientations from the signal sources.

The work of Moses et al. [27] and Savvides et al. [36] is based on signal sources. Other work [7] estimates locations of the sensor nodes by viewing the location estimation problem as a convex optimization problem because a proximity constraint exists between two nodes, i.e., the range of broadcast. In addition to these localization methods, Patwari and coworkers [28] provide the Cramer–Rao bound of sensor location accuracy based on fixed base stations capable of peer-to-peer time of arrival or received signal strength measurements.

Although beacon-based localization protocols are sufficient for certain sensor network applications, some sensor networks may be deployed in areas unreachable by beacons or GPS; they may be frequently jammed by environmental or manually induced noise. In addition, low-end sensor nodes may exhibit nonlinear device behavior and non-Gaussian measurement noise. To overcome these challenges, the location information is relayed hop by hop from the source to the sink. In order to obtain precise relative location information, the sensor nodes must collaboratively work together to assist each other. Furthermore, energy may be additionally conserved by enabling sensor nodes to track the locations of their neighbor nodes.

This relative localization technique is further explored by the perceptive localization framework (PLF) [43]. In this framework, a node is able to detect and track the location of the neighboring node by using a collaborative estimation technique and a particle filter applied to an array of sensors. To increase the accuracy of the location estimation, the sink may request all the nodes along the path to the sources to increase the number of samples (particles) for particle filtering. This process of local interaction does not require any beacon in place. In addition, a central processing unit is not required in order to determine the locations of the sources.

Whether the beacon- or relative location-based localization protocol is used, the location information is required by the protocols in the transport, network, and data-link layers. Each type of localization protocols offers different capabilities. Future sensor network applications may utilize a combination of localization techniques.

16.4 Time Synchronization Protocols

Instead of time synchronization between the sender and receiver during an application, such as in the Internet, the sensor nodes in the sensor field must maintain a similar time within a certain tolerance throughout the lifetime of the network. Combining with the criteria that sensor nodes must be energy efficient, low cost, and small in a multihop environment as described in Section 16.1, this requirement offers a challenging problem. In addition, the sensor nodes may be left unattended for a long period of time, e.g., in deep space or on an ocean floor. For short-distance multihop broadcast, data processing time and the variation of data processing time may contribute the most in time fluctuations and differences in path delays. Also, the time difference between two sensor nodes is significant over time due to the wandering effect of the local clocks.

Small and low-end sensor nodes may exhibit device behaviors much worse than those of large systems such as personal computers (PCs). Some of the factors influencing time synchronization in large systems also apply to sensor networks [21]:

- *Temperature.* Because sensor nodes are deployed in various places, the temperature variation throughout the day may cause the clock to speed up or slow down. For a typical PC, the clock drifts few parts per million during the day [25]. For low-end sensor nodes, the drifting may be even worse.
- *Phase noise.* Some of the causes of phase noise are due to access fluctuation at the hardware interface, response variation of the operating system to interrupts, and jitter in the network delay. The latter may be due to medium access and queueing delays.
- *Frequency noise.* The frequency noise is due to the instability of the clock crystal. A low-end crystal may experience large frequency fluctuation because the frequency spectrum of the crystal has large sidebands on adjacent frequencies.
- *Asymmetric delay.* Because sensor nodes communicate with each other through the wireless medium, the delay of the path from one node to another may be different from that of the return path. As a result, an asymmetric delay may cause an offset to the clock that cannot be detected by a variance type method [21]. If the asymmetric delay is static, the time offset between any two nodes is also static. The asymmetric delay is bounded by one half the round trip time between the two nodes [21].
- *Clock glitches.* Clock glitches are sudden jumps in time that may be caused by hardware or software anomalies such as frequency and time steps.

Table 16.1 shows three types of timing techniques, each of which must address the challenges mentioned earlier. In addition, the timing techniques must be energy aware because the batteries of the sensor nodes are limited. Also, they must address the mapping between the sensor network time and the Internet time, e.g., universal coordinated time. Next, examples of these types of timing techniques are described, namely, the network time protocol (NTP) [24]; the reference-broadcast synchronization (RBS) [9]; and the time-diffusion synchronization protocol (TDP) [44].

TABLE 16.1 Three Types of Timing Techniques

Type	Description
(1) Relies on fixed time servers to synchronize the network	The nodes are synchronized to time servers that are readily available. These time servers are expected to be robust and highly precise.
(2) Translates time throughout the network	The time is translated hop-by-hop from the source to the sink. In essence, it is a time translation service.
(3) Self-organizes to synchronize the network	The protocol does not depend on specialized time servers. It automatically organizes and determines the master nodes as the temporary time-servers.

In the Internet, the NTP is used to discipline the frequency of each node's oscillator. It may be useful to use NTP to discipline the oscillators of the sensor nodes, but connection to the time servers may not be possible because of frequent sensor node failures. In addition, disciplining all the sensor nodes in the sensor field may be a problem because of interference from the environment and large variation of delay between different parts of the sensor field. The interference can temporarily disjoint the sensor field into multiple smaller fields, causing undisciplined clocks among these smaller fields. The NTP protocol may be considered type 1 of the timing techniques; in addition, it must be refined to address timing challenges in the sensor networks.

The RBS, type 2 of the timing techniques, provides instantaneous time synchronization among a set of receivers within the reference broadcast of the transmitter. The transmitter broadcasts m reference packets. Each of the receivers within the broadcast range records the time of arrival of the reference packets. Afterwards, the receivers communicate with each other to determine the offsets. To provide multihop synchronization, it is proposed to use nodes receiving two or more reference broadcasts from different transmitters as translation nodes. These translation nodes are used to translate the time between different broadcast domains. As shown in Figure 16.1, nodes A, B, and C are the transmitter, receiver, and translation nodes, respectively.

Another emerging timing technique is the TDP, which is used to maintain the time throughout the network within a certain tolerance. The tolerance level can be adjusted based on the purpose of the sensor networks. The TDP automatically self-configures by electing master nodes to synchronize the sensor network. In addition, the election process is sensitive to energy requirement as well as the quality of the

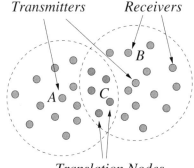

FIGURE 16.1 The RBS.

clocks. The sensor network may be deployed in unattended areas, and the TDP still synchronizes the unattended network to a common time. It is considered type 3 of the timing techniques.

In summary, these timing techniques may be used for different types of applications as discussed in Section 16.2; each has its benefits. A time-sensitive application must choose not only the type of timing techniques but also the type of transport, network, and data-link schemes described in the following sections. This is because different protocols provide different features and services to the time-sensitive application.

16.5 Transport Layer Protocols

The collaborative nature of the sensor network paradigm brings several advantages over traditional sensing, including greater accuracy, larger coverage area, and extraction of localized features. The realization of these potential gains, however, directly depends on efficient, reliable communication between the sensor network entities, i.e., the sensor nodes and the sink. To accomplish this, a reliable transport mechanism is imperative.

In general, the main objectives of the transport layer are (1) to bridge application and network layers by application multiplexing and demultiplexing; (2) to provide data delivery service between the source and the sink with an error control mechanism tailored according to the specific reliability requirement of the application layer; and (3) to regulate the amount of traffic injected into the network via flow and congestion control mechanisms. Nevertheless, the required transport layer functionalities to achieve these objectives in the sensor networks are subject to significant modifications in order to accommodate unique characteristics of the sensor network paradigm. Energy, processing, and hardware limitations of the sensor nodes bring further constraints on the transport layer protocol design. For example, conventional end-to-end, retransmission-based error control mechanisms and window-based, additive-increase, multiplicative-decrease congestion control mechanisms adopted by the vastly used transport control protocol (TCP) may not be feasible for the sensor network domain and thus may lead to waste of scarce resources.

On the other hand, unlike other conventional networking paradigms, the sensor networks are deployed with a specific sensing application objective, such as event detection, event identification, location sensing, and local control of actuators, for a wide range of applications (e.g., military, environment, health, space exploration, and disaster relief). The specific objective of the sensor network also influences the design requirements of the transport layer protocols. For example, the sensor networks deployed for different applications may require different reliability levels as well as different congestion control approaches. Consequently, development of transport layer protocols is a challenge because the limitations of the sensor nodes and the specific application requirements primarily determine design principles of transport layer protocols.

Due to the application-oriented and collaborative nature of the sensor networks, the main data flow takes place in the forward path, where the source nodes transmit their data to the sink. The reverse path, on the other hand, carries the data originated from the sink, such as programming/retasking binaries, queries, and commands to the source nodes. Therefore, different functionalities are required to handle the transport needs of the forward and reverse paths. Transport layer issues pertaining to these distinct cases are investigated separately in the following subsections.

16.5.1 Event-to-Sink Transport

Under the premise that data flows from source to sink are generally loss tolerant, Wan and coworkers questioned the need for a transport layer for data delivery in the sensor networks [32]. Although the need for end-to-end reliability may not exist because of the sheer amount of correlated data flows, an event in the sensor field needs to be tracked with a certain amount of accuracy at the sink. Therefore, unlike traditional communication networks, the sensor network paradigm necessitates an event-to-sink reliability notion at the transport layer [35]. This involves a reliable communication of the event features

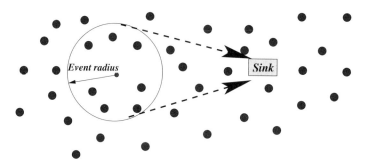

FIGURE 16.2 Typical sensor network topology with event and sink. (The sink is only interested in collective information of sensor nodes within the even radius and not in their individual data.)

to the sink rather than conventional packet-based reliable delivery of the individual sensing reports/packets generated by each sensor node in the field. Figure 16.2 illustrates an event-to-sink reliable transport notion based on collective identification of data flows from the event to the sink.

In order to provide reliable event detection at the sink, possible congestion in the forward path should also be addressed by the transport layer. Once the event is sensed by a number of sensor nodes within the coverage of the phenomenon, i.e., event radius, a significant amount of traffic is triggered by these sensor nodes; this may easily lead to congestion in the forward path. The need for transport layer congestion control to assure reliable event detection at the sink is revealed by the results of Tilak and colleagues [18], who have shown that exceeding network capacity can be detrimental to the observed goodput at the sink. Moreover, although the event-to-sink reliability may be attained even in the presence of packet loss due to network congestion (thanks to the correlated data flows), a suitable congestion control mechanism can also help conserve energy while maintaining desired accuracy levels at the sink.

On the other hand, although the transport layer solutions in conventional wireless networks are relevant, they are simply inapplicable for event-to-sink reliable transport in the sensor networks. These solutions mainly focus on reliable data transport following end-to-end TCP semantics and are proposed to address challenges posed by wireless link errors and mobility [2]. The primary reason for their inapplicability is their notion of end-to-end reliability, which is based on acknowledgments and end-to-end retransmissions. Because of inherent correlation in the data flows generated by the sensor nodes, however, these mechanisms for strict end-to-end reliability are superfluous and drain significant amounts of energy.

In contrast to the transport layer protocols for conventional end-to-end reliability, the event-to-sink reliable transport (ESRT) protocol [35] is based on the event-to-sink reliability notion and provides reliable event detection without any intermediate caching requirements. ESRT is a novel transport solution developed to achieve reliable event detection in the sensor networks with minimum energy expenditure. It includes a congestion control component that serves the dual purpose of achieving reliability and conserving energy. ESRT also does not require individual sensor identification, i.e., an event ID suffices. Importantly, the algorithms of ESRT mainly run on the sink, with minimal functionality required at resource-constrained sensor nodes.

16.5.2 Sink-to-Sensors Transport

Although data flows in the forward path carry correlated sensed/detected event features, the flows in the reverse path mainly contain data transmitted by the sink for an operational or application-specific purpose. This may include operating system binaries; programming/retasking configuration files; and application-specific queries and commands. Dissemination of this type of data mostly requires 100% reliable delivery. Therefore, the event-to-sink reliability approach introduced before would not suffice to address the tighter reliability requirements of flows in the reverse paths.

This strict reliability requirement for the sink-to-sensors transport of operational binaries and application-specific queries and commands involves a certain level of retransmission as well as acknowledgment mechanisms. However, these mechanisms should be incorporated into the transport layer protocols cautiously in order not to compromise scarce sensor network resources totally. In this respect, local retransmissions and negative acknowledgment approaches would be preferable over end-to-end retransmissions and acknowledgments to maintain minimum energy expenditure.

On the other hand, the sink is involved more in the sink-to-sensor data transport on the reverse path, so a sink with plentiful energy and communication resources can broadcast data with its powerful antenna. This helps to reduce the amount of traffic forwarded in the multihop sensor network infrastructure and thus helps sensor nodes conserve energy. Therefore, data flows in the reverse path may experience less congestion compared to the forward path, which is totally based on multihop communication. This calls for less aggressive congestion control mechanisms for the reverse path compared to the forward path in the sensor networks.

Wan and colleagues [32] propose the pump slowly, fetch quickly (PSFQ) mechanism for reliable retasking/reprogramming in the sensor networks. PSFQ is based on slowly injecting packets into the network but performing aggressive hop-by-hop recovery in case of packet loss. The pump operation in PSFQ simply performs controlled flooding and requires each intermediate node to create and maintain a data cache to be used for local loss recovery and in-sequence data delivery. Although this is an important transport layer solution for the sensor networks, PSFQ does not address packet loss due to congestion.

In summary, the transport layer mechanisms that can address the unique challenges posed by the sensor network paradigm are essential to realize the potential gains of the collective effort of sensor nodes. As discussed in the preceding two subsections, promising solutions exist for event-to-sink and sink-to-sensors reliable transports. These solutions and those currently under development, however, need to be exhaustively evaluated under real sensor network deployment scenarios to reveal their shortcomings. Therefore, necessary modifications may be required to provide a complete transport layer solution for the sensor networks.

16.6 Network Layer Protocols

Sensor nodes may be scattered densely in an area to observe a phenomenon. As a result, they may be very close to each other. In such a scenario, multihop communication may be a good choice for sensor networks with strict requirements on power consumption and transmission power levels. As compared to long distance wireless communication, multihop communication may be an effective way to overcome some of the signal propagation and degradation effects. In addition, the sensor nodes consume much less energy when transmitting a message because the distances between sensor nodes are shorter.

As discussed in Section 16.1, ad hoc routing techniques already proposed in the literature [29] do not usually fit requirements of the sensor networks. As a result, the network layer of the sensor networks is usually designed according to the following principles:

- Energy efficiency is always an important consideration.
- Sensor networks are mostly data centric.
- An ideal sensor network has attribute-based addressing and location awareness.
- Data aggregation is useful only when it does not hinder the collaborative effort of the sensor nodes.
- The routing protocol is easily integrated with other networks, e.g., Internet.

These design principles serve as a guideline when designing a routing protocol for sensor networks. Each of them is further explained to emphasize its importance. As described in the preceding section, a transport layer protocol must be energy efficient. This requirement also applies to a routing protocol because the network lifetime depends on the nodes' energy consumption when relaying messages. As a result, energy efficiency plays an important role in various protocol stack layers in addition to the network layer.

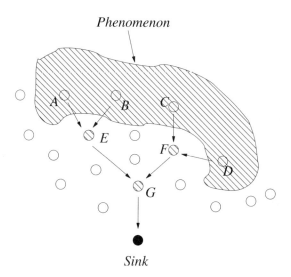

FIGURE 16.3 Data aggregation.

In sensor networks, information or data may be described by using attributes. In order to integrate tightly with the information or data, a routing protocol may be designed according to data-centric techniques. A data-centric routing protocol requires attribute-based naming [8, 10, 26, 37], which is used to carry out queries by using the attributes of the phenomenon. In essence, the users are more interested in the data gathered by the sensor networks in the phenomenon rather than by an individual node. They query the sensor networks by using attributes of the phenomenon that they want to observe. For example, the users may send out a query such as, "find the locations of areas where the temperature is over 70°F."

Furthermore, a data-centric routing protocol should also utilize the design principle of data aggregation — a technique used to solve the implosion and overlap problems in data-centric routing [15]. As shown in Figure 16.3, the sink queries the sensor network to observe the ambient condition of the phenomenon. The sensor network used to gather the information can be perceived as a reverse multicast tree, where the nodes within the area of the phenomenon send the collected data toward the sink. Data coming from multiple sensor nodes are aggregated as if they are about the same attribute of the phenomenon when they reach the same routing node on the way back to the sink. For example, sensor node E aggregates the data from sensor nodes A and B while sensor node F aggregates the data from sensor nodes C and D in Figure 16.3.

Data aggregation can be perceived as a set of automated methods of combining data from many sensor nodes into a set of meaningful information [16]. In this respect, data aggregation is known as data fusion [15]. Also, care must be taken when aggregating data because the specifics of the data, e.g., the locations of reporting sensor nodes, should not be left out. Such specifics may be needed by certain applications.

One of the design principles for the network layer is to allow easy integration with other networks such as the satellite network and the Internet. As shown in Figure 16.4, the sinks are the basis of a communication backbone that serves as a gateway to other networks. The users may query the sensor networks through the Internet or the satellite network, depending on the purpose of the query or the type of application the users are running.

A brief summary of the state of the art in the networking area is shown in Table 16.2. The schemes listed in the table utilize some of the design principles previously discussed. For example, the SMECN [22] creates an energy-efficient subgraph of the sensor networks. It tries to minimize the energy consumption while maintaining connectivity of the nodes in the network. In addition, the directed diffusion protocol [17] is a data-centric dissemination protocol in which the queries and collected data use attribute-based naming schemes.

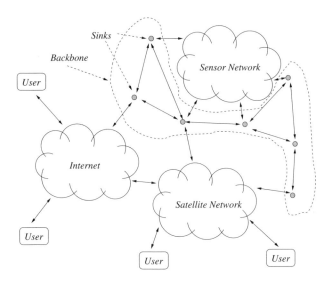

FIGURE 16.4 Internetworking between sensor nodes and user through Internet or satellite network.

TABLE 16.2 Overview of Network Layer Schemes

Network Layer Scheme	Description
SMECN [22]	Creates a sub graph of the sensor network that contains the minimum-energy path.
LEACH [16]	Forms clusters to minimize energy dissipation.
SAR [40]	Creates multiple trees where the root of each tree is one hop neighbor from the sink; select a tree for data to be routed back to the sink according to the energy resources and additive QoS Metric.
Flooding	Broadcasts data to all neighbor nodes regardless if they receive it before or not.
Gossiping [14]	Sends data to one randomly selected neighbor.
SPIN [15]	Sends data to sensor nodes only if they are interested; has three types of messages, i.e., ADV, REQ, and DATA.
Directed Diffusion [17]	Sets up gradients for data to flow from source to sink during interest dissemination.

Because different applications may require different types of network layer protocols, more advanced data-centric routing protocols are needed. In essence, application-specific requirements are part of the driving forces urging for new transport layer protocols, as described in the previous section. In addition, they push for the new data-link schemes described in the following section.

16.7 Data Link Layer Protocols

Although the transport layer mechanisms discussed in Section 16.5 are essential to achieving higher level error and congestion control, it is still imperative to have data-link layer functionalities in the sensor networks. In general, the data link layer is primarily responsible for multiplexing data streams, data frame detection, medium access, and error control; it ensures reliable point-to-point and point-to-multipoint connections in a communication network. Nevertheless, the collaborative and application-oriented nature of the sensor networks and the physical constraints of the sensor nodes, such as energy and processing limitations, determine the way in which these responsibilities are fulfilled. In the following two subsections, data-link layer issues are explored within the discussion of medium access and error control strategies in the sensor networks.

16.7.1 Medium Access Control

The medium access control (MAC) layer protocols in a multihop self-organizing sensor network must achieve two objectives:

- Establish data communication links for creating a basic network infrastructure needed for multihop wireless communication in a densely scattered sensor field.
- Regulate access to shared media so that communication resources are fairly and efficiently shared among the sensor nodes.

Due to the unique resource constraints and application requirements of sensor networks, however, the MAC protocols for the conventional wireless networks are inapplicable to the sensor network paradigm. For example, the primary goal of a MAC protocol in an infrastructure-based cellular system is to provide high QoS and bandwidth efficiency, mainly with dedicated resource assignment strategy. Such an access scheme is impractical for sensor networks because there is no central controlling agent like the base station. Moreover, power efficiency directly influences network lifetime in a sensor network and thus is of prime importance.

Although Bluetooth and the *mobile ad hoc network* (MANET) show similarities to the sensor networks in terms of communication infrastructure, both consist of nodes with portable battery-powered devices that can be replaced by the user. Therefore, unlike in the sensor networks, power consumption is only of secondary importance in these systems. Therefore, none of the existing Bluetooth or MANET MAC protocols can be directly used in the sensor networks because of network lifetime concerns.

It is evident that the MAC protocol for sensor networks must have built-in power conservation, mobility management, and failure recovery strategies. Thus far, *fixed allocation* and *random access* versions of medium access have been proposed [40, 46]. *Demand-based* MAC schemes may be unsuitable for sensor networks due to their large messaging overhead and link setup delay. Furthermore, contention-based channel access is deemed unsuitable because of the requirement to monitor the channel at all times — an energy-draining task.

The applicability of the fundamental MAC schemes in the sensor networks is discussed along with some proposed MAC solutions using that access method as follows:

- *TDMA-based medium access.* Time-division multiple-access (TDMA) access schemes inherently conserve more energy compared to contention-based schemes because the duty cycle of the radio is reduced and no contention-introduced overhead and collisions are present. Pottie and Kaiser [31] have reasoned that a MAC scheme for energy-constrained sensor networks should include a variant of TDMA because radios must be turned off during idling for precious power savings. The self-organizing medium access control for sensor networks (SMACS) [40] is such a time slot-based scheme; each sensor node maintains a TDMA-like super frame in which the node schedules different time slots to communicate with its known neighbors. SMACS achieves power conservation by using a random wake-up schedule during the connection phase and by turning the radio off during idle time slots. However, although a TDMA-based access scheme minimizes the transmit on time, it is not always preferred because of associated time synchronization costs.
- *Hybrid TDMA/FDMA-based medium access.* A pure TDMA-based access scheme dedicates the entire channel to a single sensor node; however, a pure frequency-division multiple access (FDMA) scheme allocates minimum signal bandwidth per node. Such contrast brings the trade-off between the access capacity and the energy consumption. An analytical formula is derived in Shih et al. [38] to find the optimum number of channels, which gives the minimum system power consumption. This determines the hybrid TDMA/FDMA scheme to be used. The optimum number of

channels depends on the ratio of the power consumption of the transmitter to that of the receiver. If the transmitter consumes more power, a TDMA scheme is favored, while the scheme leans toward FDMA when the receiver consumes greater power [38].

- *CSMA-based medium access.* Based on carrier sensing and backoff mechanism, traditional carrier-sense multiple access (CSMA)-based schemes are inappropriate because they make the fundamental assumption of stochastically distributed traffic and tend to support independent point-to-point flows. On the other hand, the MAC protocol for sensor networks must be able to support variable, but highly correlated and dominantly periodic traffic. Any CSMA-based medium access scheme has two important components: the listening mechanism and the backoff scheme. Woo and Culler [46] present a CSMA-based MAC scheme for sensor networks and observe from the simulations that the constant listen periods are energy efficient and the introduction of random delay provides robustness against repeated collisions.

16.7.2 Error Control

In addition to medium access control, error control of the transmitted data in the sensor networks is another extremely important function of the data-link layer. Error control is critical, especially in some sensor network applications such as mobile tracking and machine monitoring. In general, the error control mechanisms in communication networks can be categorized into two main approaches: forward error correction (FEC) and automatic repeat request (ARQ).

ARQ-based error control mainly depends on retransmission for the recovery of lost data packets/frames. It is clear that this ARQ-based error control mechanism incurs significant additional retransmission cost and overhead. Although ARQ-based error control schemes are utilized at the data-link layer for the other wireless networks, the usefulness of ARQ in sensor network applications is limited due to the scarcity of the energy and processing resources of the sensor nodes. On the other hand, FEC schemes have inherent decoding complexity that require relatively considerable processing resources in the sensor nodes. In this respect, simple error control codes with low-complexity encoding and decoding might present the best solutions for error control in the sensor networks.

On the other hand, for the design of efficient FEC schemes, it is important to have good knowledge of channel characteristics and implementation techniques. Channel bit error rate (BER) is a good indicator of link reliability. In fact, a good choice of the error correcting code can result in several orders of magnitude reduction in BER and an overall gain. The coding gain is generally expressed in terms of the additional transmit power needed to obtain the same BER without coding.

Therefore, the link reliability can be achieved by increasing the output transmit power or the use of suitable FEC scheme. Due to energy constraints of the sensor nodes, increasing the transmit power is not a feasible option. Therefore, using FEC is still the most efficient solution, given the constraints of the sensor nodes. Although the FEC can achieve significant reduction in the BER for any given value of the transmit power, the additional processing power consumed during encoding and decoding must be considered when designing an FEC scheme. If this additional power is greater than the coding gain, the whole process is not energy efficient and thus the system is better without coding. On the other hand, the FEC is a valuable asset to the sensor networks if the additional processing power is less than the transmission power savings. Thus, the trade-off between this additional processing power and the associated coding gain should be optimized in order to have powerful, energy-efficient, and low-complexity FEC schemes for error control in the sensor networks.

As researchers continue to investigate new FEC schemes for sensor networks, designers must bear in mind that the new schemes may be application specific. The data-link layer remains a challenging area in which to work because sensor nodes are inherently low end. Combining the low-end characteristic of the sensor nodes with harsh deployed terrains calls for new medium-access as well as error-control schemes.

16.8 Conclusion

An overview of the communication protocols for sensor networks is given in this chapter. Challenges and design guidelines for localization, time synchronization, application layer, transport layer, network layer, and data-link layer protocols are explored. As technology advances in the sensor network area, sensor network technologies may become an integral part of our lives.

Acknowledgments

The authors thank Dr. Ian F. Akyildiz for his encouragement and support.

References

1. Agre, J. and Clare, L., An integrated architecture for cooperative sensing networks, *IEEE Computer Mag.*, 106–108, May 2000.
2. Balakrishnan, H. et al., A comparison of mechanisms for improving TCP performance over wireless links, *IEEE/ACM Trans. Networking*, 5(6), 756–769, December 1997.
3. Bhardwaj, M., Garnett, T., and Chandrakasan, A.P., Upper bounds on the lifetime of sensor networks, *IEEE Int. Conf. Commun. '01*, Helsinki, Finland, June 2001.
4. Bonnet, P., Gehrke J., and Seshadri, P., Querying the physical world, *IEEE Personal Commun.*, 10–15, October 2000.
5. Bulusu, N., Heidemann, J., and Estrin, D., GPS-less low-cost outdoor localization for very small devices, *IEEE Personal Commun.*, 7, 28–34, October 2000.
6. Bulusu, N. et al., Scalable coordination for wireless sensor networks: self-configuring localization systems, *Int. Symp. Commun. Theory Applications (ISCTA 2001)*, Ambleside, U.K., July 2001.
7. Doherty, L., Pister, K.S.J., and Ghaoui, L.E., Convex position estimation in wireless sensor networks, *INFOCOM'01*, Anchorage, AK, April 2001.
8. Elson, J. and Estrin, D., Random, ephemeral transaction identifiers in dynamic sensor networks, *Proc. 21st Int. Conf. Distributed Computing Syst.*, 459–468, Phoenix, AZ, April 2001.
9. Elson, J., Girod, L., and Estrin, D., Fine-grained network time synchronization using reference broadcasts, *Proc. 5th Symp. Operating Syst. Design Implementation (OSDI 2002)*, Boston, MA, December 2002.
10. Estrin, D. et al., Instrumenting the world with wireless sensor networks, *Int. Conf. Acoustics, Speech, Signal Process. (ICASSP 2001)*, Salt Lake City, UT, May 2001.
11. Estrin, D. et al., Next century challenges: scalable coordination in sensor networks, *ACM Mobicom '99*, ,263–270, Seattle, WA, August 1999.
12. Estrin, D., Govindan R., and Heidemann J., Embedding the Internet, *Commun. ACM*, 43, 38–41, May 2000.
13. Fischer, S. et al., System performance evaluation of mobile positioning methods, *Proc. IEEE Vehicular Technol. Conf.*, Houston, TX, May 1999.
14. Hedetniemi, S., Hedetniemi, S., and Liestman, A., A survey of gossiping and broadcasting in communication networks, *Networks*, 18(4), 319–349, 1988.
15. Heinzelman, W.R., Kulik, J., and Balakrishnan, H., Adaptive protocols for information dissemination in wireless sensor networks, *ACM Mobicom '99*, 174–185, Seattle, WA, August 1999.
16. Heinzelman, W.R., Chandrakasan, A., and Balakrishnan, H., Energy-efficient communication protocol for wireless microsensor networks, *IEEE Proc. Hawaii Int. Conf. Syst. Sci.*, 1–10, Maui, HI, January 2000.
17. Intanagonwiwat, C., Govindan, R., and Estrin, D., Directed diffusion: a scalable and robust communication paradigm for sensor networks, *ACM Mobicom '00*, 56–67, Boston, MA, August 2000.
18. Tilak, S., Abu–Ghazaleh, N.B., and Heinzelman, W., Infrastructure trade-offs for sensor networks, *Proc. WSNA 2002*, Atlanta, GA, September 2002.

19. Jaikaeo, C., Srisathapornphat, C., and Shen, C., Diagnosis of sensor networks, *IEEE Int. Conf. Commun. '01,* Helsinki, Finland, June 2001.
20. Kahn, J.M., Katz, R.H., and Pister, K.S.J., Next century challenges: mobile networking for Smart Dust, *ACM Mobicom '99,* 271–278, Seattle, WA, August 1999.
21. Levine, J., Time synchronization over the Internet using an adaptive frequency-locked loop, *IEEE Trans. Ultrasonics, Ferroelectrics, Frequency Control,* 46(4), 888–896, July 1999.
22. Li, L. and Halpern, J.Y., Minimum-energy mobile wireless networks revisited, *IEEE Int. Conf. Commun. ICC '01,* Helsinki, Finland, June 2001.
23. Mark, B. and Zaidi, Z., Robust mobility tracking for cellular networks, *Proc. IEEE Int. Commun. Conf.,* New York, 2002.
24. Mills, D.L., Internet time synchronization: the network time protocol, *Global States and Time in Distributed Systems,* IEEE Computer Society Press, 1994.
25. Mills, D.L., Adaptive hybrid clock discipline algorithm for the network time protocol, *IEEE/ACM Trans. Networking,* 6(5), 505–514, October 1998.
26. Mirkovic, J. et al., A self-organizing approach to data forwarding in large-scale sensor networks, *IEEE Int. Conf. Commun. ICC '01,* Helsinki, Finland, June 2001.
27. Moses, R., Krishnamurthy, D., and Patterson, R., A self-localization method for wireless sensor networks, *Eurasip J. Appl. Signal Process.,* 4, 348–358, 2003.
28. Patwari, N. et al., Relative location estimation in wireless sensor networks, *IEEE Trans. Signal Process.,* August 2003.
29. Perkins, C., *Ad Hoc Networks,* Addison–Wesley, Reading, MA, 2000.
30. Perrig, A. et al., SPINS: security protocols for sensor networks, *Proc. ACM MobiCom '01,* 189–199, Rome, July 2001.
31. Pottie, G.J. and Kaiser, W.J., Wireless integrated network sensors, *Commun. ACM,* 43(5), 551–558, May 2000.
32. Wan, C.Y., Campbell, A.T., and Krishnamurthy, L., PSFQ: a reliable transport protocol for wireless sensor networks, *Proc. WSNA 2002,* Atlanta, GA, September 2002.
33. Rabaey, J. et al., PicoRadio: ad hoc wireless networking of ubiquitous low-energy sensor/monitor nodes, *Proc. IEEE Computer Soc. Annu. Workshop VLSI (WVLSI '00),* 9–12, Orlando, FL, April 2000.
34. Rabaey, J.M. et al., PicoRadio supports ad hoc ultra-low power wireless networking, *IEEE Computer Mag.,* 33, 42–48, July 2000.
35. Sankarasubramaniam, Y., Akan, O.B., and Akyildiz, I.F., ESRT: event-to-sink reliable transport for wireless sensor networks, *Proc. ACM MOBIHOC 2003,* 177–188, Annapolis, MD, June 2003.
36. Savvides, A., Han, C., and Srivastava, M., Dynamic fine-grained localization in ad hoc networks of sensors, *Proc. ACM MobiCom '01,* 166–179, Rome, July 2001.
37. Shen, C., Srisathapornphat, C., and Jaikaeo, C., Sensor information networking architecture and applications, *IEEE Personal Commun.,* 52–59, August 2001.
38. Shih, E. et al., Physical layer driven protocol and algorithm design for energy-efficient wireless sensor networks, *ACM Mobicom '01,* 272–286, Rome, July 2001.
39. Slijepcevic, S. and Potkonjak, M., Power efficient organization of wireless sensor networks, *IEEE Int. Conf. Commun. '01,* Helsinki, Finland, June 2001.
40. Sohrabi, K. et al., Protocols for self-organization of a wireless sensor network, *IEEE Personal Commun.,* 16–27, October 2000.
41. Spirito, M.A. and Mattioli, A.G., On the hyperbolic positioning of GSM mobile stations, *Proc. Int. Symp. Signals, Syst. Electron.,* September 1998.
42. Spirito, M.A., Further results on GSM mobile station location, *IEEE Electron. Lett.,* 35(22), 1999.
43. Su, W. and Akyildiz, I.F., Perceptive localization framework for sensor networks, *Georgia Tech Technical Report,* 2003.
44. Su, W. and Akyildiz, I.F., Time-diffusion synchronization protocol for sensor networks, *Georgia Tech Technical Report,* 2003.

45. Warneke, B., Liebowitz, B., and Pister, K.S.J., Smart Dust: communicating with a cubic-millimeter computer, *IEEE Computer Mag.*, 2–9, January 2001.
46. Woo, A. and Culler, D., A transmission control scheme for media access in sensor networks, *ACM Mobicom '01*, 221–235, Rome, July 2001.

17
Communication Architecture and Programming Abstractions for Real-Time Embedded Sensor Networks[*]

T. Abdelzaher
University of Virginia

J. Stankovic
University of Virginia

S. Son
University of Virginia

B. Blum
University of Virginia

T. He
University of Virginia

A. Wood
University of Virginia

Chenyang Lu
University of Washington at St. Louis

17.1 Introduction 17-1
17.2 A Protocol Suite for Sensor Networks 17-2
    Real-Time Distance-Aware Scheduling • Enforcement of Velocity Constraints • Entity-Aware Transport
17.3 A Sensor-Network Programming Model 17-5
17.4 Related Work 17-6
17.5 Conclusions 17-8

17.1 Introduction

Made possible by advances in communication technology and hardware miniaturization [11], ad hoc wireless sensor networks raise the need for a new suite of communication protocols and new programming abstractions for distributed deeply embedded computing. Such sensor networks are especially useful when an inhospitable, poorly accessible, or delicate environment prevents the installation of needed computing infrastructure; an example would be the site of a natural disaster or a target behind enemy lines. Instead, myriads of tiny, computationally equipped wireless sensor devices may be dropped to form an ad hoc network that operates autonomously to monitor its surroundings, react to distributed events, or alert appropriate authorities when specific activities are observed.

Sensor networks offer new challenges from the perspective of building communication protocols and from the perspective of developing appropriate programming models. These challenges arise due to their large scale, autonomous operation, massively parallel interactions with a spatially distributed physical environment, and a more stringent set of resource constraints.

---

[*]The work reported in this paper was supported in part by the National Science Foundation under grants CCR-02-05327 and CCR-00-92945; by DARPA under grant F33615-01-C-1905; and by MURI under grant N00014-01-1-0576.

Communication protocols for sensor networks must provide real-time assurances. Although ensuring proper timing behavior of systems has been a topic of real-time research for decades, sensor network applications offer physical *space*, in addition to time, as a new dimension for interaction with the environment. Thus, while traditional real-time computing research has been concerned with meeting time constraints, a new branch of theory is needed to analyze systems that interact with their surroundings in real time and in the real dimensions of physical space. For example, in a network that tracks vehicles through the sensor field, the application must collect sensory measurements in real time from the actual changing locale in which the vehicle is detected. Message communication must therefore be sensitive to time and distance constraints, which may depend on external factors such as the physical speed of the monitored vehicle. This chapter describes a protocol suite in which time and distance constraints are addressed.

A new programming paradigm is needed to facilitate the task of sensor network application development. Due to the large scale of sensor networks, programmers should not need to be concerned with low-level abstractions and functions such as creating and destroying individual connections between pairs of nodes. Instead, the programming environment must offer a conceptual view in which global tasks can be defined in an abstract manner, leaving it for the underlying system to translate them into computational and communication activities on individual sensor nodes.

This chapter reports on the design of a programming system developed on top of a communication protocol suite that provides the required high-level abstractions. The language allows external events in the environment to be represented as objects in the computing system facilitating the monitoring of such events by the application. The reported architecture is a part of an ongoing research effort to develop a sensor network virtual machine for future distributed deeply embedded applications.

Section 17.2 describes a protocol suite that takes into account time and space constraints and exports a useful transport-layer abstraction in which logical communication end-points can be associated with tracked objects in the external environment. Section 17.3 describes a new programming model for sensor networks that builds upon the aforementioned transport protocol to elevate environmental objects into first-class programming abstractions. Related work is summarized in Section 17.4. The chapter concludes with Section 17.5, which discusses some of the remaining challenges and directions for future research.

17.2 A Protocol Suite for Sensor Networks

Communication protocols in sensor networks are the fundamental cornerstone that glues distributed applications together. The deeply embedded nature of sensor networks presents some of the most interesting challenges in the design of their communication protocols. New research topics span all protocol stack layers, primarily motivated by a tighter interaction between the network and its physical environment. At the MAC layer, new protocols are needed that enforce message priorities consistently with time and distance constraints that arise from environmental interactions [22]. Awareness of the physical environment must also be incorporated into the network layer; for example, location should be an essential attribute of addressable networked objects [15]. Location-assisted routing protocols such as LAR [19] and DREAM [4], as well as location services [21], have been described for ad hoc wireless networks.

More generally, routing algorithms are needed in which destinations are described implicitly by their environmental attributes. For example, directed diffusion [14, 18] and the intentional naming system [3] provide addressing and routing based on data interests. A fundamental rethinking of basic protocols is required at the transport layer as well. Individual socket-style connections between nodes are too low level to be a useful abstraction for the programmer. They must be replaced with higher level alternatives more suitable for the main purpose of sensor networks, namely, monitoring the external surroundings in which they are embedded.

This section describes an answer to the challenge of incorporating environmental awareness into the design of sensor network communication protocols. This protocol stack features two important contributions. First, it implements new real-time message scheduling algorithms in which time and physical distance requirements are observed. Second, it exports a transport-layer address space that associates unique network addresses with external environmental objects. The new addresses serve as connection end-points, thereby raising the level of connection abstraction to entities of direct interest to the application. The layers of this protocol stack are described in the following subsections.

17.2.1 Real-Time Distance-Aware Scheduling

Message communication in sensor networks must occur in bounded time — for example, to prevent delivery of stale data on the status of detected events or intruders. In general, a sensor network may simultaneously carry multiple messages of different urgency communicated among destinations that are different distances apart. The network has the responsibility of ordering these messages on the communication medium in a way that respects time and distance constraints.

A protocol that achieves this goal in this architecture is called RAP [22]. It supports a notion of packet velocity and implements velocity monotonic scheduling (VMS) as the default packet scheduling policy on the wireless medium. Observe that for a desired end-to-end latency bound to be met, an in-transit packet must approach its destination at an average velocity given by the ratio of the total distance to be traversed to the requested end-to-end latency bound. RAP prioritizes messages by their required velocity so that higher velocities imply higher priorities.

Two flavors of this algorithm are implemented. The first, called *static velocity-monotonic scheduling*, computes packet priority at the source and keeps it fixed thereafter regardless of the packet's actual progression rate toward the destination. The second, called *dynamic velocity-monotonic scheduling*, adjusts packet priority *en route* based on the remaining time and the remaining distance to destination. Thus, a packet's priority will increase if it suffers higher delays on its path and decrease if it is ahead of schedule.

To achieve consistent prioritization in the wireless network, it is necessary to have priority queues at nodes, as well as a MAC layer that resolves contention on the wireless medium in a manner consistent with message priorities. A scheme similar to that of Aad and Castelluccia [1] is adopted to prioritize access to the wireless medium. The scheme is based on modifying two 802.11 parameters — the DIFS counter and the back-off window — so that they are aware of priorities. The DIFS counter determines the maximum time a node waits, after the communication channel becomes idle, prior to transmitting an RTS packet. The actual waiting time is randomly chosen between 0 and DIFS. An approximate prioritization effect is achieved by letting the DIFS value depend on the priority of the outgoing packet at the head of the transmission queue.

Because a larger value is given to packets of lower priority, more urgent packets tend to contend on the medium more aggressively. The back-off window of 802.11 increases the maximum waiting time when collisions occur. To give preferential treatment to higher priority packets, this increase is made dependent on the priority of the head of the queue. A higher increase is incurred for packets of lower priority, so collisions tend to be resolved in favor of higher priority packets.

A detailed performance evaluation of this scheme can be found in Lu et al. [22]. It is shown that VMS substantially increases the fraction of packets that meet their deadlines, taking into consideration distance constraints. More accurate schemes for medium access prioritization remain an open research topic. An interesting related topic is that of analysis of VMS. Ideally, such an analysis should allow a source node to determine whether a particular desired velocity is attainable between a source–destination pair, given current network conditions. Although an analytic expression for velocity feasibility is still an open problem, the following subsection describes a feedback-based technique that enforces velocity constraints dynamically by applying back-pressure to slow the sources when such constraints are violated.

17.2.2 Enforcement of Velocity Constraints

Consider a network that supports multiple predefined velocities. An application can choose a velocity level for each message. The network guarantees that the chosen message velocity is observed with a very high probability as long as the message is accepted from the application. A network-layer protocol with the preceding property, called SPEED [13], has recently been developed by the authors. The protocol defines the velocity of an in-transit message as the rate of decrease of its straight-line distance to its final destination. Therefore, for example, if the message is forwarded away from the destination, its velocity at that hop is negative.

The main idea of SPEED is as follows. Each node i in the sensor network maintains a neighborhood table that enumerates the set of its one-hop neighbors. For each neighbor, j, and each priority level, P, the node keeps a history of the average recently recorded local packet delay, $D_{ij}(P)$. Delay $D_{ij}(P)$ is defined as the average time that a packet of priority P spends on the local hop i before it is successfully forwarded to the next-hop neighbor j.

Given a packet with some velocity constraint, V, node i determines the subset of all its neighbors that are closer to the packet's destination. If L_{ij} is the distance by which neighbor j is closer to the destination than i, the velocity constraint of the packet is satisfied at node i if some priority level P and neighbor j exist so that $L_{ij}/D_{ij}(P) \geq V$. The packet is forwarded to one such neighbor nondeterministically. If the condition is satisfied at multiple priority levels, the lowest priority level is chosen. If no neighbor satisfies the velocity constraint, a local deadline miss occurs.

A table at node i keeps track of the number of local deadline misses observed for each velocity level V; this table is exchanged between neighboring nodes. Nodes use this information in their forwarding decisions to favor more appropriate downstream hops among all options that satisfy the velocity constraint of a given packet. No messages are forwarded in the direction of nodes with a high miss ratio. The mechanism exerts back-pressure on nodes upstream from congested areas. Congestion increases the local miss ratio in its vicinity, preventing messages from being forwarded in that direction. Messages that cannot be forwarded are dropped, thus increasing the local miss ratio upstream.

The effect percolates towards the source until a node is found with an alternative (noncongested) path toward the destination, or the source is reached and informed to slow down. The mentioned scheme is therefore effective in exerting congestion control and performing packet rerouting that guarantee the satisfaction of all velocity constraints in the network at steady state [13]. The protocol is of great value to real-time applications in which different latency bounds must be associated with messages of different priority.

17.2.3 Entity-Aware Transport

Although RAP and SPEED allow velocity constraints to be met, the abstractions provided by them are too low level for application programmers. A transport layer is developed whose main responsibility is to elevate the degree of abstraction to a level suitable for the application. In particular, a transport layer is proposed in which connection end-points are directly associated with events in the physical environment.

Events represent continuous external activities, such as the passage of a vehicle or the progress of a fire, in which an application might be interested. By virtue of this layer, the programmer can describe events of interest and logically assign "virtual hosts" to them. Such hosts export communication ports and execute programs at the locations of the corresponding events. The programmer is isolated from the details of how these hosts and ports are implemented. When an external event (e.g., a vehicle) moves, the corresponding virtual host migrates with it transparently to the programmer.

This virtual host associated with an external event of interest is called an *entity*. Sensor nodes that can sense the event are called *entity members*. Members elect an *entity leader* that uniquely represents the entity and manages its state. Thus, an entity appears indivisible to the rest of the network. The fact that it is composed of multiple nodes with a changing membership is abstracted away.

When the external event moves outside the sensing horizon of the current entity leader, the leader hands off leadership to another member. Connection state is handed off as well, allowing communication with the entity to remain uninterrupted. To ensure unique representation of external events within the computational environment, a unique entity must be associated with each event. The transport protocol meets this constraint by announcing the existence of the entity to nearby nodes that cannot yet sense

A Communication Architecture and Programming Abstractions

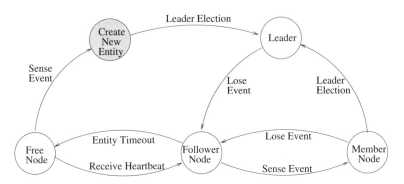

FIGURE 17.1 Node state transition.

the event. These announcements are sent periodically by the entity leader and are called *heartbeats*. Nodes that hear a heartbeat but cannot sense the event are called *entity followers* and are said to be within the *awareness horizon* of the named entity. Upon receiving a heartbeat, such nodes set an *entity timeout timer*; upon timer expiration, their status as followers expires. The timer is reset to zero every time a new heartbeat is received.

When the event enters the sensing horizon of a follower node, the node becomes a member of the entity it is following. If the node is not a follower, it recognizes that a new entity must be created. The node sets a random timer, upon expiration of which it claims leadership of the new entity. If it receives a leadership claim message from another node prior to timer expiration, it clears the timer and becomes an entity member. The algorithm ensures that a newly sensed event is represented by a single entity and that current events do not spawn spurious entities as they move from one location to another. Figure 17.1 depicts the node state transition diagram among follower, member, and leader states, as well as the free state in which a node is not cognizant of any entities.

An evaluation of this architecture reveals that entity uniqueness is maintained as long as the target event moves in the environment at a speed slower than half the nodes' communication radius per second [7]. For example, if sensor nodes can communicate within a 200-m radius, the transport layer can correctly maintain endpoints attached to targets that move as fast as 100 m/s (i.e., 360 km/h). The combination of this transport layer and the guaranteed velocity protocols described earlier provides invaluable support to real-time applications. For example, communication regarding moving targets can be made to proceed in the network at a velocity that depends on target velocity. Thus, positions of faster targets, for example, can be reported more quickly than those of slower ones. To the authors' knowledge, no other protocols in sensor networks have explicitly addressed message timing constraints.

17.3 A Sensor-Network Programming Model

The transport layer described previously gives rise to a programming model that elevates tracked activities in the physical environment into first-class programming abstractions. In this model, the application developer specifies events to be monitored. The system automatically detects such events and instantiates a so-called *context* every time an instance of an event is detected in the environment. From the programmer's perspective, the application is composed of a dynamic set of contexts, each representing a particular event. Objects can be attached to contexts and objects will logically execute in the locale of the monitored event. Contexts have unique identifiers called *context labels*. Objects attached to a context can be addressed using the context label and object name. They can communicate remotely by remote method invocation. The programmer's view of the application is depicted in Figure 17.2.

A context label around some event, e, is completely defined by two elements: (1) the function $sense_e()$, which specifies an environmental condition that spawns the context label; and (2) the function $state_e()$, which describes the environmental state to be encapsulated in the context label. The former function,

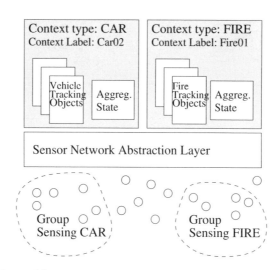

FIGURE 17.2 Programming model.

for example, might dictate that a label is to be created if magnetic distortion (e.g., the presence of a vehicle) is sensed. The state function returns a set of aggregate variables, each computed using outputs of at least N_e nodes for which $sense_e()$ was true in the last L_e time units.

N_e and L_e are called the critical mass and freshness constraints, respectively. For example, to obtain the approximate position of a vehicle we may define $state_e()$ may be defined to be the average coordinates of at least five nodes that have sensed the vehicle within the last 2 sec. Environmental tracking of event e is defined as the process of maintaining the state of this event subject to given freshness and critical mass constraints.

Syntactically, an application consists of a list of context declarations, each specifying an activation condition $sense_e()$; a set of state variables $state_e()$; and a list of attached objects. An example declaration is shown in Figure 17.3. The example defines a context of type *tracker* and specifies its activation condition, $sense_e()$, as an appropriate magnetometer reading (presumably caused by a nearby vehicle); it defines $state_e()$ as the average *location* of the tracked target. It specifies that *location* must represent the average of at least two sensor readings measured no earlier than 1 sec ago.

The attached object is invoked periodically to report the current location of the vehicle to a virtual base station object. It passes the originating context label as the identity of the reported vehicle. If several vehicles are in the field, multiple reporter objects will be automatically instantiated. The programmer does not need to worry about instantiating these objects; object execution and maintenance of aggregate state occur automatically. Because details of the underlying communication, group membership management, leader handoff, and mobility are handled transparently, the programmer's interaction with the sensor network is significantly simplified.

This subsection has described real-time communication protocols and programming abstractions motivated by a tighter interaction between sensor networks and their physical environment. This architecture might be a first step toward a comprehensive vision for next-generation programming systems supporting future real-time deeply embedded distributed sensor network applications.

17.4 Related Work

Classical distributed programming paradigms and middleware such as CORBA [27]; group communication (e.g., ISIS [5]); remote procedure calls (RPC [6]); and distributed shared memory (e.g., MUNIN [9]) share in common the fact that their programming abstractions exist in a logical space that does not represent or interact with objects and activities in the physical world. Their main goal is to abstract distributed communication rather than facilitate distributed sensory interactions with an external

```
(1)   begin context tracker
(2)       activation: MAGNETOMETER == ON
(3)       location : avg (position) mass=2, freshness=1s
(4)       begin object reporter
(5)           invocation: PERIOD(0.5s)
(6)           report_function() {
(7)               BaseStation.reportLocation (self.label, location);
(8)           }
(9)       end
(10)  end context
```

FIGURE 17.3 Sample code.

physical environment. In contrast, sensor network applications call for a paradigm that revolves around "environmentally inspired" abstractions aimed at simplifying the coding of interactions with the physical world that arise in distributed deeply embedded systems.

The work reported in this chapter is closely related to several recent projects, such as Cricket [23], Sentient Computing [2], and Cooltown [10], that propose high-level paradigms in which an embedded distributed computing system is able to share humans' perceptions of the physical world. These systems allow the location of entities in the external environment to be tracked. One major difference is that they assume cooperative users who, for example, can wear beaconing devices that interact with location services in the infrastructure for purposes of localization and tracking [2, 23]. The authors' interest, in contrast, concerns situations in which no cooperation is assumed from the tracked entity.

In the absence of cooperation, several research efforts have proposed alternative addressing schemes that do not rely on having destinations with specific identities, but rather contact sensor nodes in the vicinity of a phenomenon of interest based on the attributes of data they sense. For example, DataSpace [17] exports abstractions of physical volumes addressable by their locations. Similarly, directed diffusion [14, 18] and the intentional naming system [3] provide addressing and routing based on data interests [14, 18]. Attributed-based naming is also related to the notion of content-addressable networks [24] proposed for an Internet environment, which allow queries to be routed depending on the requested content rather than on the identity of the target machine. Context labels, a form of attribute-based naming, are adopted. In this architecture, however, context labels are *active* elements. Not only do they provide a mechanism for *addressing* nodes that sense specific environmental conditions, but also they can *host context-specific computation* that tracks a target in the environment.

Recent research on system software for sensor networks has seen the introduction of distributed virtual machines designed to provide convenient high-level abstractions to application programmers, while implementing low-level distributed protocols transparently in an efficient manner [26]. This approach is taken in MagnetOS [12], which exports the illusion of a single Java virtual machine on top of a distributed sensor network. The application programmer writes a single Java program; the run-time system is responsible for code partitioning, placement, and automatic migration so that total energy consumption is minimized. Maté [20] is another example of a virtual machine developed for sensor networks. It implements its own bytecode interpreter, built on top of TinyOS [16].

A somewhat different approach to providing high-level programming abstractions is to view the sensor network as a distributed database in which sensors produce series of data values and signal processing functions generate abstract data types. The database management engine replaces the virtual machine in that it accepts a query language that allows applications to perform arbitrarily complex monitoring functions. This approach is implemented in the COUGAR sensor network database [8]. A middleware implementation of the same general abstraction is also found in SINA [25], a sensor information networking architecture that abstracts the sensor network into a collection of distributed objects.

This system is different in that it is geared for real-time environmental tracking. To the authors' knowledge, the first programming language for sensor networks that explicitly facilitates the coding of tracking applications and the first sensor network communication protocols that consider real-time constraints are described. These novel abstractions and underlying mechanisms are well suited for monitoring targets that move in the physical world. They can therefore have a major impact on application development for sensor networks.

17.5 Conclusions

This chapter reviewed a new protocol suite and programming system for sensor network applications that may considerably improve real-time behavior and reduce the development cost of deeply embedded systems. This reduction comes from off-loading the details of managing low-level abstractions from the application developer.

Future work of the authors will involve refinement of the real-time protocols and the environmental tracking problem so that more precise semantics and failure models are achieved. It is the hope that, with such refinements, a predictable sensor network "virtual machine" can be built that exports timely, reliable behavior and well-defined semantics, implemented on the unreliable, unpredictable, and resource-constrained hardware and communication infrastructure typical of sensor networks. Such a virtual machine would hide the complexity of sensor network programming from the application developer; it would make possible a new, more robust and dynamic realm of sensor network applications to affect future defense, surveillance, habitat monitoring, and disaster management systems.

References

1. I. Aad and C. Castelluccia. Differentiation mechanisms for IEEE 802.11. In *IEEE Infocom*, Anchorage, AK, April 2001.
2. M. Addlesee, R. Curwen, S. Hodges, J. Newman, P. Steggles, A. Ward, and A. Hopper. Implementing a sentient computing system. *IEEE Computer*, 34(8):50–56, August 2001.
3. W. Adjie-Winoto, E. Schwartz, H. Balakrishnan, and J. Lilley. The design and implementation of an intentional naming system. In *ACM Symp. Operating Syst. Principles*, Kiawah Island, SC, December 1999.
4. S. Basagni, I. Chlamtac, V.R. Syrotiuk, and B.A. Woodward. A distance routing effect algorithm for mobility (dream). In *ACM MOBICOM*, 76–84, October 1998.
5. K. Birman, A. Schiper, and P. Stephenson. Lightweight causal and atomic group multicast. *ACM Trans. Computer Syst.*, 9(3):272–314, August 1991.
6. A. Birrel and B. Nelson. Implementing remote procedure calls. *ACM Trans. Computer Syst.*, 2(1), February 1984.
7. B. Blum, P. Nagaraddi, A. Wood, T. Abdelzaher, J. Stankovic, and S. Son. An entity maintenance and connection service for sensor networks. In *1st Int. Conf. Mobile Syst., Applications, Services (MobiSys)*, San Francisco, CA, May 2003.
8. P. Bonnet, J. Gehrke, and P. Seshardi. Towards sensor database systems. In *2nd Int. Conf. Mobile Data Manage.*, 3–14, Hong Kong, January 2001.
9. J. Carter, J. Bennet, and W. Zwaenepoel. Implementation and performance of munin. In *ACM Symp. Operating Syst. Principles*, 151–164, October 1991.
10. P. Debaty and D. Caswell. Uniform Web presence architecture for people, places, and things. *IEEE Personal Commun.*, 8(4):46–51, August 2001.
11. D. Estrin, R. Govindan, J. Heidemann, and S. Kumar. Next century challenges: mobile networking for smart dust. In *ACM MOBICOM*, Seattle, WA, August 1999.
12. R.B. et al. On the need for system-level support for ad hoc and sensor networks. *Operating Syst. Rev.*, 36(2):1–5, April 2002.
13. T. He, J. Stankovic, C. Lu, and T. Abdelzaher. Speed: a stateless protocol for real-time communication in sensor networks. In *Int. Conf. Distributed Computing Syst.*, Providence, RI, May 2003.

14. J. Heideman, F. Silva, C. Intanagonwiwat, R. Govindan, D. Estrin, and D. Ganesan. Building efficient wireless sensor networks with low-level naming. *Operating Syst. Rev.*, 35(5):146–159, December 2001.
15. J. Hightower and G. Borriello. Location systems for ubiquitous computing. *IEEE Computer*, 34(8), August 2001.
16. J. Hill, R. Szewczyk, A. Woo, S. Hollar, D. Culler, and K. Pister. System architecture directions for network sensors. In *ASPLOS*, Cambridge, MA, November 2000.
17. T. Imielinski and S. Goel. Dataspace — querying and monitoring deeply networked collections in physical space. *IEEE Personal Commun.*, 7(5):4–9, October 2000.
18. C. Intanagonwiwat, R. Govindan, and D. Estrin. Directed diffusion: a scalable and robust communication paradigm for sensor networks. In *ACM MOBICOM*, Boston, MA, August 2000.
19. Y.-B. Ko and V. Nitin. Location-aided routing (LAR) in mobile ad hoc networks. In *ACM MOBICOM*, 66–75, October 1998.
20. P. Levis and D. Culler. Mate: a tiny virtual machine for sensor networks. In *ASPLOS*, San Jose, CA, October 2002.
21. J. Li, J. Jannotti, D.D. Couto, D. Karger, and R. Morris. A scalable location service for geographic ad hoc routing. In *ACM MOBICOM*, Boston, MA, August 2000.
22. C. Lu, B. Blum, T. Abdelzaher, J. Stankovic, and T. He. Rap: A real-time communication architecture for large-scale wireless sensor networks. In *Real-Time Technol. Applications Symp.*, San Jose, CA, September 2002.
23. N.B. Priyantha, A. Chakraborty, and H. Balakrishnan. The cricket location-support system. In *ACM MOBICOM*, Boston, MA, August 2000.
24. S. Ratnasamy, P. Francis, M. Handley, R. Karp, and S. Shenker. A scalable content-addressable network. In *Sigcomm*, San Diego, CA, August 2001.
25. C.-C. Shen, C. Srisathapornphat, and C. Jaikeo. Sensor information networking architecture and applications. *IEEE Personal Commun.*, 8(4):52–59, August 2001.
26. E. Sirer, R. Grimm, A. Gregory, and B. Bershad. Design and implementation of a distributed virtual machine for networked computers. In *ACM Symp. Operating Syst. Principles*, 202–216, Kiawah Island, SC, December 1999.
27. S. Vinoski. Corba: integrating diverse applications within distributed heterogeneous environments. *IEEE Commun. Mag.*, 14(2), February 1997.

18

A Comparative Study of Energy-Efficient (E²) Protocols for Wireless Sensor Networks

Quanhong Wang
Queen's University

Hossam Hassanein
Queen's University

18.1 Introduction 18-1
18.2 Motivations and Directions.. 18-2
 Necessity of Resource Efficiency • QoS with Energy Efficiency Constraints • Energy Consumption in WSNs
18.3 Cross-Layer Communication Protocol Stack for WSNs... 18-5
18.4 Energy-Efficient MAC Protocols.................................... 18-6
 Sources of Energy Consumption at the MAC layer • Classification and Comparison of MAC Protocols
18.5 Energy-Efficient Network Layer Protocols................... 18-10
 Classification of Network Layer Protocols • Energy-Efficient Data Delivery Protocols • Signal and Data Processing
18.6 Concluding Remarks.. 18-17

18.1 Introduction

A typical wireless sensor network (WSN) may contain hundreds to thousands of microsensor nodes, which are connected by a wireless medium. These sensor nodes are capable of capturing various physical properties, such as temperature, humidity, or pressure, and mapping the physical characteristics of the environment to quantitative measurements. Rapid progress in microelectromechanical system (MEMS) and radio frequency (RF) design, as well as advances in communication protocols and algorithms, have made WSNs more intelligent and led them to ubiquitous deployment.

WSNs exhibit revolutionary approaches to providing reliable, time-critical, and constant environment sensing, event detecting and reporting, target localization, and tracking. Due to their ease of deployment, reliability, scalability, flexibility, and self-organization, WSNs can be deployed in almost any environment, especially those in which conventional wired sensor systems are impossible, unavailable, or inaccessible, such as in inhospitable terrain, dangerous battlefields, outer space, or deep oceans. Therefore, the existing and potential applications of WSNs span a wide spectrum in various domains such as [2, 10, 16, 21–23, 27, 39, 53, 68, 71, 76, 86, 98, 103]:

- Control, communications, computing, intelligence, surveillance, reconnaissance and targeting (C⁴ISRT) for military purposes
- Environmental detection and monitoring

- Disaster prevention and relief
- Medical care
- Home automation
- Scientific exploration
- Interactive surrounding

From a networking architecture perspective, WSNs can be classified as belonging to the family of wireless ad hoc networks, which are collections of wireless, possibly mobile, nodes that are self-configurable to form a network without the aid of any fixed infrastructure. Nodes in the system autonomously handle the necessary control and networking tasks in a distributed manner. The ad hoc architecture overcomes the difficulties raised by the predetermined infrastructure settings of other wireless networks, so a WSN can be randomly and rapidly deployed and reconfigured and easily tailored to specific applications as well. Moreover, ad hoc architecture is highly robust to single node failure and can provide a high level of fault tolerance due to node redundancy and the distributed nature. Furthermore, energy efficiency can be achieved through multihop routing communication. Bandwidth reuse can also benefit from dividing the single long-range hop to multiple short hops; each hop has a considerably short distance [25].

However, because of their unique application requirements, WSNs differ greatly from conventional wireless ad hoc networks [25, 47, 81]. For instance, a WSN usually has a considerably larger number of sensor nodes (hundred to thousands or even more), which is several orders of magnitude greater than in a conventional ad hoc network. The heavy density of nodes leads to high redundancy of data among neighboring nodes. Moreover, a WSN often encounters severe resource constraints, such as power supply, memory, computation speed, etc. Similarly, due to application diversity, the design of a WSN is normally application specific, i.e., it is difficult to devise a unified WSN architecture or deployment strategy to meet the requirements of various applications.

Furthermore, the active duty cycle of a sensor node is fairly low (possibly as low as 1%) and end users generally focus on the collective information, so the data flow is usually unidirectional, i.e., from the sensor nodes to a common processing center. As a result, many existing architectures and protocols for other wireless networks are not suitable for WSNs, and new performance metrics (e.g., system lifetime), in addition to throughput and delay characteristics, should be considered in WSN design. Therefore, novel approaches supporting resource efficiency, scalability, and reliability should be developed to satisfy the specific requirements of WSNs, and numerous research issues remain to be explored.

Section 18.2 of this chapter covers the motivation and directions of energy-efficient protocols with a discussion of QoS metrics and analysis of energy-consuming sources in WSNs. In Section 18.3, the concept of a cross-layer protocol stack dedicated for WSNs is introduced. Section 18.4 classifies and compares various MAC layer protocols targeting energy-efficient and reliable packet transmission. In Section 18.5, a comparative study is carried out on a number of energy-efficient network layer protocols. Section 18.6 concludes the chapter.

18.2 Motivations and Directions

A typical sensor node is compact, tiny, and inexpensive, but it integrates the functionalities of sensing, data processing and computation, and communication. It is normally operated by an attached power supply that is usually a nonrechargeable or nonreplaceable battery [1, 23, 60].

18.2.1 Necessity of Resource Efficiency

The limited physical size of sensor nodes has the inherent problem of severe resource limitation. Therefore, in WSNs, resource efficiency is extremely critical despite its complexity. Above all, energy-efficient protocols are in high demand in order to extend the lifetime of the system. Because a WSN often operates in a human-unattended manner, the power supply (which is usually an attached battery) cannot be replenished in most cases. In addition, efforts should be made to increase efficiency for the utilization of other resources. For example, using algorithms with low complexity will reduce computation time and thus save power. It also decreases the latency of data delivery. Bandwidth-efficient architectures and protocols can accelerate data delivery as well.

It should be noted that it is difficult to issue a unique definition of the system lifetime for all application scenarios. On one hand, a system lifetime can be measured by the time when the first node exhausts its energy, or a system can be declared dead when a certain fraction of nodes die, or even when all nodes die. Using one definition or another depends on the particular application. On the other hand, the system lifetime can also be measured by application-specific parameters, such as the time until the system can no longer provide acceptable results.

18.2.2 QoS with Energy Efficiency Constraints

Quality of service in WSNs can be evaluated by the following metrics [1, 11, 23, 25, 30, 32, 46, 60, 67, 85, 95, 105]:

- *Energy efficiency.* This determines the system lifetime and is a crucial issue in WSNs. It is clear that for the same sensing task, the higher the energy efficiency is, the longer the system will survive.
- *Accuracy.* This reflects the basic value of gathered information because the amount of received data determines the level of accuracy. In general, the more data received, the higher the accuracy should be.
- *Latency.* In most cases, information collected from the monitoring environment is time critical, so it should be delivered in a timely fashion.
- *Security.* Because many WSNs are used for military or surveillance purposes, denial of service attacks against these networks may cause severe damage to their operation. Therefore, data privacy and safe communications are of utmost importance.
- *Fault tolerance.* Although the wireless communication channel is usually noisy, prone to errors, and time varying, data must be delivered reliably. In such cases, data verification and correction on each layer of the network are critical to provide accurate results. Moreover, some sensor nodes may fail due to energy exhaustion or physical obstacles in the environment, so sensor nodes are expected to perform self-testing, self-calibration, self-repair, and self-recovery procedures.
- *Scalability and flexibility.* The system should be scalable and flexible to the enlargement of the network scale. The approaches to scalability and flexibility include clustering, multihop delivery, localization of computation, and data processing.

However, it is impossible to achieve all of these objectives at the same time because some of them conflict with each other. In terms of resource consumption, it is necessary to make a trade-off between energy efficiency and other metrics. Essentially, any of the preceding objectives except the first one is resource hungry. For example, high accuracy requires the delivery of large amounts of data, which leads to more power and bandwidth consumption. Similarly, approaches aiming for timely delivery, security, and reliability are bound to cost extra energy. Local computation is helpful to eliminate the amount of data transmitted, but complex and memory costly computation may cause long latency, and increased power consumption.

18.2.3 Energy Consumption in WSNs

As a microelectronic device, the main task of a sensor node is to detect phenomena, carry out data processing timely and locally, and transmit or receive data. A typical sensor node is generally composed of four components [4, 37, 45, 58, 64, 65, 68, 83, 94, 106]: a power supply unit; a sensing unit; a computing/processing unit; and a communicating unit. The sensing node is powered by a limited battery, which is impossible to replace or recharge in most application scenarios. Except for the power unit, all other components will consume energy when fulfilling their tasks. Extensive study and analysis of energy consumption in WSNs are available [69, 80, 83, 87].

18.2.3.1 Sensing Energy

The sensing unit in a sensor node includes the embedded sensor and/or actuator and the analog–digital converter. It is responsible for capturing the physical characteristics of the sensed environment and converts its measurements to digital signals, which can be processed by a computing/processing unit.

Energy consumed for sensing includes: (1) physical signal sampling and conversion to electrical signal; (2) signal conditioning; and (3) analog to digital conversion. It varies with the nature of hardware as well as applications. For example, interval sensing consumes less energy than continuous monitoring; therefore, in addition to designing low-power hardware, interval sensing can be used as a power-saving approach to reduce unnecessary sensing by turning the nodes off in the inactive duty cycles. However, there is an added overhead whenever transiting from an inactive state to the active state. This leads to undesirable latency as well as extra energy consumption. However, sensing energy represents only a small percentage of the total power consumption in a WSN. The majority of the consumed power is in computing and communication, as discussed next.

18.2.3.2 Computing Energy

The computing/processing unit is a microcontroller unit (MCU) or microprocessor with memory. It carries out data processing and provides intelligence to the sensor node. A real-time micro-operating system running in the computing unit controls and operates the sensing, computing, and communication units through microdevice drivers and decides which parts to turn off and on [11, 36, 38, 58, 69, 79, 82, 87].

Total computing energy consists of two parts: switching energy and leakage energy. The switching energy is determined by supply voltage and the total capacitance switched by executing software. The pattern of draining the energy from the battery affects the total computing energy expense. For example, a scheme of energy saving on computation is dynamic voltage scaling (DVS) [12, 29, 63, 69], which can adaptively adjust operating voltage and frequency to meet the dynamically changing workload without degrading performance. The leakage energy refers to the energy consumption while no computation is carried out. Some researchers have reported that it can reach 50% of the total computing energy. Therefore, it is critical to minimize leakage energy [8, 12, 29, 63, 69].

The concept of system partitioning [11, 59, 93] can also be used to reduce computing energy in sensor nodes. Two practical approaches include removing the intensive computation to a remote processing center that is not energy constrained, or spreading some of the complex computation among more sensors instead of overloading several centralized processing elements.

Energy expenditure for computing is much less compared to that for data communication. Experiments show that the ratio of communicating 1 bit over the wireless medium to that of processing the same bit could be in the range of 1000 and 10,000 [102]. Therefore, trading complex computation/data processing for reducing communication amount is effective in minimizing energy consumption in a multihop sensor network.

18.2.3.3 Communicating Energy

The communicating unit in a sensing node mainly consists of a short-range RF circuit that performs data transmission and reception. The communicating energy is the major contributor to the total energy expenditure and is determined by the total amount of communication and the transmission distance. As reported in Pottie and Kaiser [65], processing data locally to reduce the traffic amount may achieve significant energy savings. Moreover, according to Rappoport [70], signal propagation follows as exponential law to the transmitting distance (usually with exponent 2 to 4 depending on the transmission environment). It is not hard to show that the power consumption due to signal transmission can be saved in orders of magnitude by using multihop routing with a short distance of each hop instead of single-hop routing with a long-distance range for the same destination.

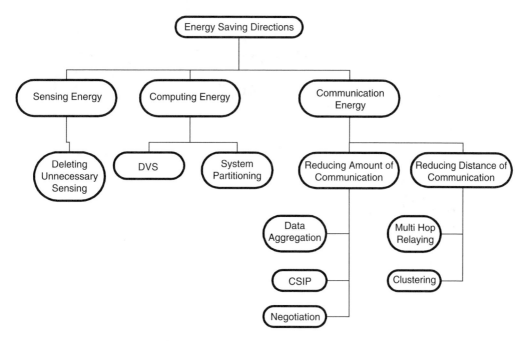

FIGURE 18.1 Energy-conserving directions in WSNs.

Therefore, minimizing the amount of data communicated among sensors and reducing the long transmitting distance into a number of short ones are key elements to optimizing the communicating energy; numerous efforts have focused on these objectives. Several approaches have been devised in order to reduce data communication. For instance,

- Data aggregation has been applied to eliminate redundancy in neighboring nodes [42, 44, 107].
- Collaborative signal and information processing (CSIP) has been used to fulfill local data processing [20, 23, 49, 66, 67, 102, 108].
- Negotiation-based protocols have been introduced to reduce unnecessary replicated data [31, 48].

Similarly, in order to decrease signal transmission distance, multihop communication and clustering-based hierarchies have been proposed to forward data in the network [14, 34, 54, 56, 96].

Figure 18.1 summarizes energy-conserving directions with respect to optimizing sensing, computing, and communication energy consumption. Such approaches exhibit a high degree of dependency on one another. For example, eliminating unnecessary sensing could reduce data communication; in turn, communication energy consumption is reduced. However, this requires more sophisticated control schemes, which are supported by higher complexity computation, and may result in higher energy use for computation. Therefore, trade-offs should be made and some specific direction may take greater importance based on the nature of the application scenario. The remainder of this chapter introduces a number of energy-efficient protocols, which concentrate in one or more of the three directions.

18.3 Cross-Layer Communication Protocol Stack for WSNs

The conventional wireless ad hoc network protocol design is mainly based on a layered stack in which each layer is designed and operated in isolation. The interfaces between layers are static and independent of the individual network constraints and applications. By using this paradigm, network design can be greatly simplified. However, this approach's lack of flexibility and optimality may result in poor performance in large-scale WSNs in which resource limitation is severe, but timely delivery is required [27].

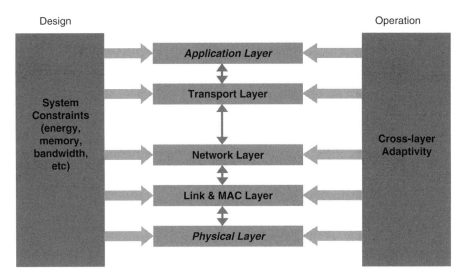

FIGURE 18.2 Cross-layer protocol stack in WSNs.

Therefore, an active theme — cross-layer design — has been recently proposed; this supports optimization and adaptability across multiple layers [27, 33, 73]. A possible cross-layer architecture is depicted in Figure 18.2.

In the concept of cross-layer design, each layer is not developed in isolation, but in an integrated and hierarchical framework. Therefore, the strict border between different layers is loosened. Some control messages as well as information concerning a layer's status will be exchanged among different layers so that the system can take advantage of the interdependencies between them. For example, the link layer can adjust rate, power, and coding to satisfy application requirements based on current channel and network conditions; MAC layer can be adaptive to underlying link and interference conditions, delay constraints, and bit priorities; Routing protocols can be developed according to up-to-date link, network, and traffic conditions; The application layer can adopt the concept of soft QoS, which is adaptive to the underlying network conditions to deliver the highest possible application quality [27].

In practice, cross-layer design may be exercised in some, rather than all, layers in the protocol stack. Discussion will focus on protocols with cross-layer design on network and MAC layers. However, many open problems exist concerning how to understand and implement this concept, what kind of information should be exchanged between layers, and what kinds of internal and external constraints should be taken into consideration.

18.4 Energy-Efficient MAC Protocols

The functions of the data link layer include framing and link access, reliable delivery, flow control, error detection, and retransmission. Because nodes share a common wireless medium for communication, MAC sublayer protocols are critical to providing coordination among nodes. These protocols attempt to provide reliable communication and achieve high throughput with bounded latency, while at the same time minimizing collisions and energy dissipation [50, 89, 90]. The following discussion covers sources influencing energy consumption at the MAC layer, which may lead to directions to improve energy efficiency. Different kinds of energy-efficient MAC layer approaches will also be discussed and some comparisons made in terms of energy efficiency.

18.4.1 Sources of Energy Consumption at the MAC layer

From the point view of energy dissipation, four major sources of energy waste are caused by MAC layer problems [100]:

- *Retransmission due to collision or congestion.* In WSNs, all nodes are capable of transmitting data through the same broadcast channel. As a tiny communication device, each sensor node may have only one receiving antenna; therefore, if two or more transmissions from multiple sources arrive at the same time, a collision will happen, and none of transmitted packets can be received correctly. To ensure reliable transmission, after source nodes detect data collision, they must retransmit, which causes extra energy expenditure. On the other hand, because of limited capacity of the wireless channel, data losses take place when traffic is heavy and the network encounters congestion. This case also requires retransmissions.
- *Idle channel sensing.* In order to eliminate or reduce collisions, nodes must sense the channel continuously to obtain scheduling information or wait before sending data until the channel is detected idle. In either case, extra sensing energy is needed. Indeed, in ad hoc networks, idle channel sensing energy is not negligible compared to data receiving and transmitting. According to Chen and colleagues [14], the ratios of $E_{idle}:E_{receiving}:E_{transmitting}$ are 1:2:2.5.
- *Overhearing.* When sharing a common wireless medium, the data transmitted by one node can reach all the other nodes within their transmission range. A node then may receive packets not destined for it. This is referred to as overhearing and it also wastes energy.
- *Overhead due to control messages.* A lot of MAC protocols operate by exchanging control messages for signaling, scheduling, and collision avoidance, which will consume extra energy.

Therefore, in order to design an energy-efficient MAC protocol, collisions must be avoided as much as possible. Moreover, energy dissipation due to idle channel sensing, overhearing, and overhead should also be reduced to a minimum. Many approaches have been proposed, but it is difficult to achieve all energy-conserving objectives at the same time.

18.4.2 Classification and Comparison of MAC Protocols

In general, wireless communication has a variety of MAC protocols, which can be classified into distinct groups according to different criteria. Based on whether a central controller is involved in coordination, WSNs' MAC protocols can be categorized as centralized, distributed (decentralized), and hybrid. Actually, hybrid protocols attempt to combine the advantages of centralized and distributed schemes, but can be more complex. Figure 18.3 shows such classification [97].

18.4.2.1 Centralized MAC Protocols

Centralized MAC protocols include polling algorithms and controlled multiplexing (or channel partitioning) algorithms [50, 70]. A centralized controller is needed to coordinate channel access among the different nodes and collision-free operation can be achieved. Thus, energy wasted due to collisions can be eliminated. However, because of the high overhead and long delay, pure polling mechanisms are not suitable in large-scale WSNs. Depending on how bandwidth is assigned, controlled multiplexing mechanisms can be frequency division multiplexing access (FDMA), code division multiplexing (CDMA), or time division multiplexing access (TDMA). This class of protocols is preferable in WSNs [65], not only because it is collision free, but also because nodes can be turned off in unassigned slots, thus saving energy expenditure due to idle sensing and overhearing.

However, drawbacks exist in channel partitioning schemes. When using TDMA, the central controller will consume more energy than other nodes, and scheduling tends to be dynamic, which will lead to a more complex mechanism. Moreover, it also requires clock synchronization among all nodes, which will

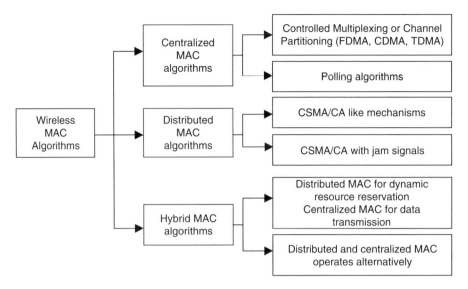

FIGURE 18.3 Classification of wireless MAC protocols.

also dissipate some extra energy. For FDMA, due to the limited bandwidth in the system, it is not realistic to assign a unique frequency for each individual node. Furthermore, bandwidth wastage will occur due to the low duty cycle. Similarly, when using CDMA, although all nodes can transmit at will, some overhead will result because each node must encode its data bits with its uniquely assigned code.

Centralized multiplexing access, therefore, lacks flexibility and scalability to adapt to the variation of WSN applications. Some efforts have been made to improve the performance in terms of energy efficiency. One way is to combine TDMA with other controlled multiplexing, such as self-organizing medium access control for sensor networks (SMACS) [85], which is a combination of TDMA/FDMA MAC protocol. Low-energy adaptive clustering hierarchical (LEACH) [32–34], on the other hand, combines TDMA and CDMA protocols, i.e., it uses TDMA protocol to prevent intracluster[*] collisions and CDMA to avoid intercluster collisions. Adaptive periodic threshold-sensitive energy-efficient sensor network protocol (APTEEN), described in Manjeshwar and Agrawal [55] and Manjeshwar et al. [56], also uses TDMA with CDMA; however, it adopts a modified TDMA in which the length of time slots assigned to idle nodes and sleeping nodes is different, and all the idle node slots are ordered to precede sleeping nodes. Another alternative is to apply dynamic reservation TDMA (DR-TDMA) [109], which is actually a hybrid approach combining TDMA and carrier sense multiple access (CSMA) mechanisms.

18.4.2.2 Distributed MAC Protocols

Distributed MAC protocols usually provide random multiple access to a wireless medium. Most prevailing MAC protocols in this category adapt carrier sensing and collision avoidance, i.e., based on CSMA/CA. Through carrier sensing, significant transmission collisions can be eliminated by deferring transmission when the channel is detected busy. To further decrease the probability of collision, some collision avoidance measures can be taken, such as a random back-off procedure; a representative example of CSMA/CA based MAC protocol is specified in IEEE 802.11 distributed coordination function (DCF) [41].

However, in some cases, location-dependent carrier sensing results in "hidden" and "exposed" terminal problems, which have a great impact on efficiency. A hidden terminal refers to the node within the range of the intended destination but out of range of the sender, so the hidden terminal cannot be aware of the ongoing transmission. An exposed terminal is the node within the range of the sender but out of range of the destination, so the exposed terminal will be improperly precluded from sending in order to avoid collision. Two types of CSMA/CA-based schemes have been proposed to solve these problems. In

---

[*] The concept of clusters will be explained later in the chapter.

DCF, the exchange of "request to send–clear to send" (RTS-CTS) control messages reserves the transmission space for subsequent data exchange, thereby eliminating hidden terminal transmission. Deng and Hass [18] and Hass and Deng [104] propose a scheme, called dual busy tone multiple access (DBTMA), that separates control and data channels to relieve the problems raised by hidden and exposed terminals by indicating the transmission or receiving status explicitly.

Other distributed MAC protocols use a jamming signal, such as elimination yield-non-preemptive priority multiple access (EY-NPMA) [3, 24], which is used in the HIPERLAN system (being developed in Europe), and black burst (BB) [84, 110], which is proposed to support prioritized data transmission in ad hoc networks.

Using distributed MAC protocols, nodes operate in a decentralized manner, so it is easy to implement and perform more flexible and scalable control mechanisms, which may fit well with the requirements of WSNs. However, they are not collision-free protocols, and the listen-before-talk scheme calls for all nodes to keep sensing the channel. This results in high energy wastage due to collisions, idle listening, overhearing, and control message overhead. In Ye et al. [100], a novel MAC protocol called sensor-MAC (S-MAC) is proposed and attempts to reduce all four types of energy wastage.

18.4.2.3 Hybrid MAC Protocols

As discussed in previous subsections, conventional centralized and distributed MAC layer protocols cannot provide optimal results in terms of energy efficiency in WSNs. Hybrid MAC protocols attempt to integrate the controllability of centralized protocols with the flexibility of distributed protocols. A number of these protocols are discussed here.

DR-TDMA [109] was originally proposed for wireless ATM networks, but can be extended to WSNs. Figure 18.4 demonstrates the frame structure of the DR-TDMA. Specifically, the fixed-length frame is divided into uplink and downlink time intervals. During the contention phase of the uplink interval, a distributed collision-based MAC scheme — the framed pseudo-Bayesian priority aloha protocol — is used for nodes to transmit temporary reservation requests for next frame to the base station. Uplink data are transmitted in TDMA mode based on the time slots assigned by the base station in the preceding downlink interval. In the downlink interval, a centralized MAC protocol is used to carry out slot assignments, as well as data transmission from base station to nodes. The resource reservation and assignment are adjusted dynamically based on work load. The nodes can turn off during periods outside assigned transmission slots or contention slots. Therefore, energy wastage due to data collision, idle sensing, and overhearing can be reduced. Dynamic slot assignment can also provide flexibility to WSNs.

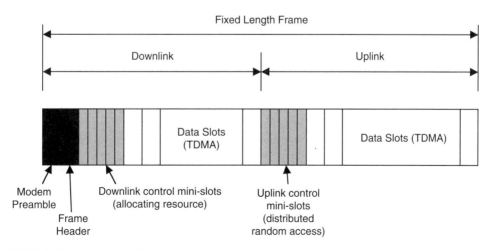

FIGURE 18.4 Frame structure of DR-TDMA.

Two other hybrid TDMA-based protocols are time reservation using adaptive control for energy efficiency (TRACE) [90] and multihop TRACE (MH-TRACE) [91]. Similarly to DR-TDMA, a central controller is in charge of arranging the TDMA transmission schedule according to continuing reservations and new reservation requests. Data are transmitted based on the transmission schedule, which is updated dynamically. The reservation requests are transmitted through a contention-based distributed MAC protocol. Because TRACE and MH-TRACE are dedicated for energy efficiency, these two approaches have a dynamic central controller as opposed to using a fixed-base station as the central controller in DR-TDMA.

A cluster formation scheme is used to manage the nodes. Each cluster head also plays the role of a TDMA scheduler within its cluster. By dynamically choosing cluster heads, balanced energy consumption among the nodes can be achieved. Moreover, these protocols introduce a novel control message — information summarization (IS) — to obtain information on future data transmission within the transmission range of nodes. This way energy wastage due to idle channel sensing and overhearing can be avoided. Due to their characteristics of energy efficiency, flexibility, and self-configuration, these two hybrid MAC protocols look promising for future WSNs. The issues of how to select central controllers dynamically and how to create data transmission schedules are major challenges for such protocols.

18.5 Energy-Efficient Network Layer Protocols

The network layer in WSNs is responsible for data delivery from source to destination via well-selected routes [50, 89]. Due to the unique characteristics of WSNs, many of the network layer protocols designed for conventional ad hoc networks may not fit with the requirements of WSNs. The following principles must be considered in WSN network layer protocols:

- Energy efficiency is always a dominant consideration.
- Routing is often data centric.
- Data aggregation/fusion is desirable, but only useful if it does not affect the collaborative efforts among sensor nodes.
- An ideal sensor network has attribute-based addressing and location awareness.
- Protocols are most likely application specific.

18.5.1 Classification of Network Layer Protocols

To reduce communication's energy consumption, network layer protocols have drawn considerable attention. Many factors influence the design of network layer protocols, and a wide range of schemes have been proposed. Table 18.1 presents a classification of energy-efficient (E^2) network layer protocols. Note that the purpose of such classification is to aid with the study of energy-efficient network layer protocols. Other researchers may opt to use different, and possibly more elaborate, classifications.

In general, the base station, which plays the role of data gathering and processing, may be located far from the sensing field, or can be placed within the network. Sensor nodes can be deployed in the sensing field according to various strategies [19, 92]. This can be predetermined [19, 75, 76] — sensor nodes are placed in preplanned positions, or fixed regular topology or self-regulated [15, 40] — sensor nodes can be spread automatically into the sensing area in sequential steps. Sensor node deployment may also follow a random [32–34, 92] or a biased distribution [94]. The sensor nodes within one system may be identical or heterogeneous in terms of functionality, and resources. The status of the network can be stationary or may change dynamically. Dynamic sensor networks may have mobile sensor nodes, end-users who are collecting the information, or sensed targets [92].

From the data perspective, WSN protocols are designed for global data delivery or local data processing. A detailed classification of the actions on data is shown in Figure 18.5. From the application perspective, the communication schemes can be grouped into three categories [54–56]. The first kind is proactive (also known as source initiated): the sensor nodes keep sensing the environment and continuously report

TABLE 18.1 Classifications of Network Layer Protocols in WSNs

Criteria	Classification		
Position of base station	Far from WSN		
	Within WSN		
Sensor deployment	Predeterministic		
	Self-regulated		
	Random distribution		
	Biased distribution		
Node properties	Homogenous		
	Heterogeneous		
Network dynamics	Static		
	Dynamic	Sensors	
		End-users	
		Targets	
Actions on data	Delivering	Information collection	
		Information dissemination	
		Hybrid	
	Processing	Data Aggregation	
		Collaborative signal and information Processing (CSIP)	Event driven
			Mobile agent based
			Relation based
Effective range of the protocols	Globalization (data forwarding)		
	Localization (data processing)		
System architecture	Flat		
	Hierarchical		
Routing approaches	Flooding		
	Unicast		
	Multicast		
Application scenarios	Proactive (source initiated, continuously)		
	Reactive	Phenomenon/event-driven	
		Query-driven (end-user initiated, request-reply on demand)	
	Hybrid		
Energy-efficiency objectives	Minimizing energy consumption in forwarding each individual packet		
	Minimizing in-network total energy consumption		
	Balancing in-network power consumption		

the sensed values. The second kind is reactive: transmission of sensed values is triggered by some specific conditions. They can be driven by a phenomenon or query. The last kind is hybrid, which is a combination of the proactive and reactive methods. In order to realize energy efficiency, different communication protocols have been proposed to fit each scenario.

The architecture of WSNs can be flat or hierarchical. Hierarchical structures can be cluster based, tree based and hierarchical chain based. Routing protocols for efficient data delivery can be flooding based, unicast, or multicast. Furthermore, different kinds of energy efficiency objectives exist, such as minimizing energy dissipation to deliver each individual packet; minimizing in-network total energy dissipation; and balancing in-network energy dissipation.

Based on these classifications, comparisons of some existing network layer protocols that claim to offer energy efficiency will be provided. These protocols are designed for WSNs or were originally designed for wireless ad hoc networks, but can scale to WSNs.

18.5.2 Energy-Efficient Data Delivery Protocols

One of the critical responsibilities of network layer protocols is to provide data delivery between desired source and destination. In WSNs, data delivery protocols should take energy efficiency into consideration. A number of protocols target E^2 data forwarding. These can be classified into distinct groups according

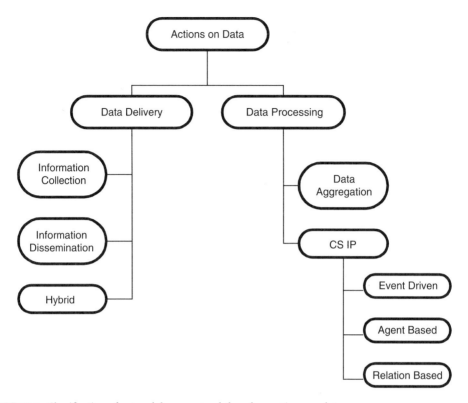

FIGURE 18.5 Classification of network layer protocols based on actions on data.

to several criteria, such as the purpose of flow delivery, application scenario, routing approach, and E^2 objectives. Figure 18.6 demonstrates such classification. Later in this chapter, a case study of data delivery protocols for the purpose of information collection and information dissemination will be provided.

18.5.2.1 Energy-Efficient Information Collection (E²IC) Protocols

Information gathering is one of the essential tasks for all WSNs. A single data-collecting and -processing center is usually assumed, so data are delivered in a unidirectional manner. Therefore, the design of information collection protocols with energy constraints may differ greatly from conventional ad hoc routing protocols.

From the perspective of system architecture, E²IC protocols can be flat or hierarchical, depending on whether the system is divided into space division groups. Figure 18.7 shows this classification of E²IC protocols.

Flat multihop E²IC protocols. The geographical adaptive fidelity (GAF) algorithm [96] and a protocol called SPAN [14] are proposed for wireless ad hoc networks. Due to their scalability, they are also applicable in WSNs. Taking advantage of redundant deployment of sensor nodes and low duty cycle, both protocols designate to rotate switching nodes between active and inactive states without losing the connectivity of the system. In GAF, equivalent nodes are identified based on geographic locations on a virtual grid, so they can substitute each other directly and transparently without affecting the routing topology. Therefore, only one node in a virtual grid needs to be on duty at any time, while all others can go to sleep. In this case, little energy is used, so energy consumption can be reduced.

In SPAN, a limited number of nodes are randomly self-selected as coordinators to construct a backbone in a peer-to-peer fashion within the network for traffic forwarding, while others can make local decisions to transit to a sleep state or keep active. Because a WSN is only sensing its environment and waiting for an interesting event to happen, a new technique described in Schurgers et al. [77, 78] — sparse topology and energy management (STEM) — claims to improve beyond SPAN and GAF in terms of obtaining

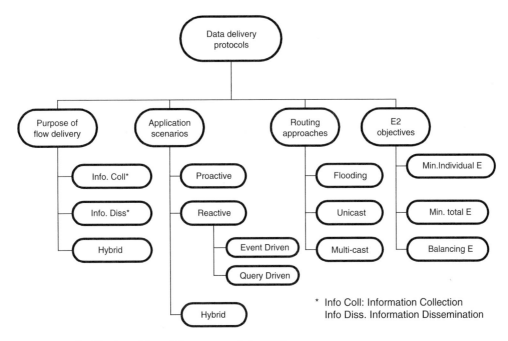

FIGURE 18.6 Classification of data delivery protocols in WSNs.

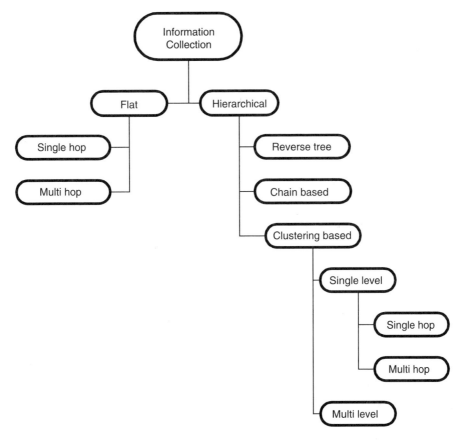

FIGURE 18.7 Classification of energy-efficient information collection protocols in WSNs.

TABLE 18.2 Comparison of Cellular Networks and Clustering-Based WSNs

	Cellular	WSNs
Peer unit	Cell	Cluster
Data processing center (DPC)	Base station	Cluster head
Location of DPC	Fixed	Randomized
Location and form of peer unit	Static	Dynamic
Node positions	Mobile	Static and/or mobile
Interpeer unit communication	Through MSC	Self-configurable

higher energy savings so as to prolong system lifetime by trading off an increased latency to establish a multihop path.

Hierarchical E^2IC protocols. Depending on how the hierarchical structure is formed, hierarchical E^2IC protocols can be grouped as reserved tree based, chain based, or clustering based. Among these, the clustering-based approach has received increased attention because of its effectiveness, lower complexity, and flexibility.

From the perspective of space division, cluster-based WSNs are similar to cellular networks, although many differences exist between them. For example, in cellular networks, a base station has no resource constraints and is placed at a fixed position, so the cell is static and the nodes there have high mobility. In WSNs, a cluster head (the peer unit of a base station in cellular networks) is generally a sensor node, which has severe resource limitations, and cluster heads are selected dynamically; therefore, clusters are not static within the network, but sensor nodes are often in stationary position. Table 18.2 shows the differences between cellular networks and cluster-based WSNs.

Dividing the entire system into distinct clusters replaces the one-hop long distance transmission by multihop short-distance data forwarding. This would reduce the energy consumed for data communications. Clustering-based E^2IC schemes also have the advantages of load balancing, and scalability when the network size grows. Challenges faced by such clustering-based approaches include how to select the cluster heads and how to organize the clusters. The clustering strategy could be single-hop cluster or multihop cluster, based on the distance between the cluster heads and their members. According to the hierarchy of clusters, the clustering strategies can also be grouped into single-level or multilevel clustering.

Various clustering approaches for wireless ad hoc and/or sensor networks have been proposed in the literature. In Heinzelman et al. [32, 33] and Heinzelman [34], the authors propose a distributed LEACH. Initially, each node self-selects itself as a cluster head with a predetermined probability; the cluster head then advertises its decision to other nodes that would make the decision to join a specific cluster that requires minimum communication energy. In order to ensure balanced energy dissipation among all nodes, LEACH invokes the rotation of cluster heads by periodically calling the self-selection and cluster formation procedure. LEACH is a well developed clustering-based protocol dedicated for continuous E^2IC in WSNs. However, it is used for proactive application scenarios and does not take the energy consumption for idle sensing of the channel into account; the formation of clusters is not energy aware. Therefore, some efforts have been made to improve its performance further.

Manjeshwar and Agrawal [54] propose the threshold-sensitive energy-efficient sensor network (TEEN) protocol. TEEN adopts the cluster formation method of LEACH, but uses thresholds to achieve enhanced control on sensor nodes. This scheme can also save energy consumption due to idle sensing. It is suitable for time-critical data delivery in reactive application scenarios. Adaptive periodic TEEN (APTEEN) is proposed in Manjeshwar and Agrawal [55] and Manjeshwar et al. [56] to fit in the requirements of hybrid application scenarios using enhanced query management and a modified TDMA MAC protocol. In TEEN and APTEEN, the concept of multilevel clustering is used.

A chain-based protocol called power-efficient gathering in sensor information systems (PEGASIS) is presented in Lindsey and Raghavendra [51] and Lindsey et al. [52]. Instead of sending data packets directly to cluster heads, as is done in the LEACH protocol, each node forwards its packets to the

TABLE 18.3 Comparison of Several Hierarchical Information Collection Protocols

	LEACH	LEACH-C	TEEN	APTEEN	PEGASIS	PA-VBS
Centralized/distributed	D	C	D	C	D	D
Cluster head selection	Self-selection	Nominated by controller	Same as LEACH	Same as LEACH-C	N/A = Not Applicable	Self-selection
Energy awareness in cluster head selection	No	Yes	No	Yes	N/A	Yes
Scalability to large heterogeneous networks	No	Yes	No	Yes	No	Yes
Location awareness	Yes	Yes	Yes	Yes	Yes	Yes
Balanced energy dissemination	Yes	Yes	Yes	Yes	Yes	Yes
Adoption of data aggregation	Yes	Yes	Yes	Yes	Yes	No
Cross-layer design	Yes	Yes	Yes	Yes	Yes	Yes

destination through its closest neighbors. Utilizing the feature of randomized creation and rotation of cluster heads as proposed in LEACH, as well as the advantages of multihop clustering algorithms, Bandyopadhyay and Coyle [7] introduced a new energy-efficient, single-level, multihop clustering algorithm.

These authors [7] also provide a formulation for finding the optimal parameter values to minimize energy consumption, as well as a novel energy-efficient hierarchical clustering algorithm with a total of h levels, i.e., some of the cluster heads in level $k-1$ select themselves as kth level cluster heads, and the remaining level $k-1$ cluster heads are cluster members in level k. Based on the results of Foss and Zuyev [26] and Baccelli and Zuyev [5], they derive the optimal parameters to achieve minimum energy consumption. Experimental results for up to 10,000 nodes have been reported.

Power-aware virtual base stations (PA-VBS) [72, 74] is a first attempt to use residual power capacity to select the cluster heads in mobile ad hoc networks. It is attractive to WSNs because of its characteristics of load balancing and scalability to the growth of network size. A load-balanced clustering approach for heterogeneous sensor networks is introduced in Gupta and Younis [28]. The gateway nodes (cluster heads) with high energy manage the cluster member nodes and forward the data collected from the cluster member to the base station, which may be far from member nodes.

Table 18.3 provides a comparison of several hierarchical protocols used for E^2IC in WSNs.

18.5.2.2 Energy-Efficient Information Dissemination (E^2ID) Protocols

Information dissemination plays a critical role in WSNs. This is particularly the case in reactive and hybrid application scenarios, in which time-sensitive information should reach other nodes as soon as serious phenomenon, e.g., early warning of a fire, is detected by some nodes, or when a query with certain attribute values should spread in the system in a timely manner. In general, information dissemination is conducted similarly to flooding, but conventional flooding schemes will cause problems of redundancy and overlap that lead to significant energy waste. In order to prolong the system lifetime, E^2ID protocols are in great demand. Most E^2ID apply the data aggregation function (discussed later) to eliminate redundant information.

A family of E^2ID protocols named sensor protocols for information via negotiation (SPIN) have been proposed in Heinzelman et al. [31] and Julik et al. [48]. They employ a new type of control message — metadata to allow negotiation between neighboring nodes — so that a node only forwards a packet to a neighbor that wants to receive the data. In such a way, the energy waste caused by classical flooding schemes can be reduced. However, overhead of control messages is created for negotiation, which will lead to long latency. Moreover, each individual node must constantly maintain a neighbor list and update it periodically. This not only requires memory space, but will also cost extra energy. Therefore, generating and controlling metadata are critical to the success of the SPIN protocols family.

Directed diffusion [42, 44] incorporates in-network data aggregation, data caching, and data-centric dissemination while enforcing adaptation to the empirically best path. It aims to establish efficient *n*-way communication from single or multiple sources to sinks. Heidemann and colleagues [35] present a physical implementation of directed diffusion with a wireless sensor test bed and show that the traffic can be reduced by up to 42% when deploying a duplicate suppression data aggregation method (see Section 18.5.3.1).

Another E^{2}ID scheme is proposed in Ye et al. [99]. Instead of generating a control message to implement data aggregation as other E^{2}ID proposals do, it attempts to relieve redundant information by allowing immediate nodes to conduct a random back-off procedure to delay the packet delivery. During this deferral period, if an intermediate node receives new data from other source nodes, it will combine them with the previous ones then transmit the processed data. Therefore, it trades latency for energy saving.

18.5.3 Signal and Data Processing

As has been mentioned in previous subsections, most E^2 communication protocols incorporate localized signal and data processing to reduce the amount of traffic in the network. Two main categories of signal and data processing applied in WSNs are (1) data aggregation and (2) collaborative signal and information processing (CSIP).

18.5.3.1 Data Aggregation

The principle of data aggregation or data fusion is to minimize traffic load (in terms of number and/or length of packets) by eliminating redundancy. It applies a novel data-centric approach to replace the traditional address-centric approach in data forwarding [47]. Specifically, when an intermediate node receives data from multiple source nodes, instead of forwarding all of them directly, it checks the contents of incoming data and then combines them by eliminating redundant information under the constraints of acceptable accuracy.

Several data aggregation algorithms have been reported in the literature. The most straightforward is duplicate suppression, i.e., if multiple sources send the same data, the intermediate node will only forward one of them. Using a maximum or minimum function is also possible. Heinzelman and colleagues [31] and Julik and colleagues [48] proposed SPIN (see Section 18.5.2.2) to realize traffic reduction for information dissemination using metadata negotiations between sensors to avoid redundant and/or unnecessary data propagation through the network. The greedy aggregation approach [43] can improve path sharing and attain significant energy savings when the network has higher node densities compared with the opportunistic approach.

Krishnamachari and colleagues [47] described the impact of source–destination placement on the energy costs and delay associated with data aggregation. They also investigated the complexity of optimal data aggregation. In Reference 111, a polynomial-time algorithm for near-optimal maximum lifetime data aggregation (MLDA) is described for data collection in WSNs. The scheme is superior to others in terms of systems lifetime, but has a high computational expense for large sensor networks. A simple and efficient clustering-based heuristic for maximum lifetime data aggregation (CMLDA) has been proposed by Dasgupta and coworkers [17] for small- and large-scale sensor networks.

18.5.3.2 Collaborative Signal and Information Processing (CSIP)

CSIP schemes are also powerful in reducing the amount of traffic transmitted and thus result in energy efficiency in WSNs. With the combination of interdisciplinary techniques, such as low-power communication and computation; space–time signal processing; distributed and fault-tolerant algorithms; adaptive systems; and sensor fusion and decision theory, CSIP is expected to provide solutions to many challenges. These include dense spatial sampling of interested events; distributed asynchronous processing; progressive accuracy; optimized processing and communication; data fusion; querying; and routing tasks [49].

CSIP can be implemented through coherent signal processing on a small number of nodes in a cluster or through noncoherent processing across a larger number of nodes when synchronization is not a strict requirement [23]. CSIP algorithms can be classified [67] as information-driven schemes [101, 102];

mobile agent-based schemes [66], which attempt to reduce the system traffic by employing an agent, thus transmitting the integration process (code) to the data sites instead of moving original data directly; and relation-based schemes [108], which use a top-down approach to select the sensor nodes to sense and communicate based on a high-level description of the task.

18.6 Concluding Remarks

Wireless sensor networks, which incorporate the functions of sensing; date collection and storage; computation and processing; communication through wireless medium; and/or actuating, have been envisioned for a wide spectrum of applications in various military and civil domains. Because of their great potential for providing safer and healthier environments for human beings through ubiquitous monitoring, objective localization, and target tracking, they have attracted extensive interest from industry and academia.

However, due to the tiny size of individual sensor nodes and human-unattended operation manners, WSNs often encounter severe resource constraints — especially, limited power supply. Therefore, from a networking perspective, energy efficiency is a critical objective in the design of communication protocols. Several proposals have been made in this direction, including physical layer approaches; data link and MAC layer protocols; network layer protocols; and transport layer and application layer strategies, as well as energy-efficient software development. This chapter has provided a classification and comparative study of such proposals.

Extending the lifetime of the network is the major concern of all energy-efficiency protocols. In order to gain a constant power supply, various proposals have been put forward to substitute conventional batteries in WSNs. With advances in power techniques, more energy-efficient, environmentally friendly alternative energy sources are raising research interest. The most familiar is called green power — solar technologies. During full and bright sunlight hours, solar cells can provide up to 45 mW/in.^{2} [112]. However, because of the variable nature of solar energy and the lack of cost-effective electricity storage techniques, solar power has been unable to become the sole power supply and is unlikely to grab a large share of the energy generation market.

Fuel cells offer a promising solution to WSNs' energy challenges. It is estimated that fuel cell use will grow by a factor of 250 over the next decade [113]. These "cells operate much like batteries, converting chemical energy into electrical power supply; however, unlike batteries, they never run down or require recharging" [114]. For example, the recent invention of regenerative fuel cells and zinc air fuel cells can be operated as a "closed loop" system in which no additional external fuel is necessary. Although they are currently just in the research and development stage, fuel cells are a potential low-cost, high-power replacement to the existing battery in WSNs. Another interesting scheme is to make sensor nodes self-contained and self-powered — able to harvest energy supply from the environment such as vibrations caused by motor vehicles driving down nearby streets or people walking on raised floors. Power output between tens and hundreds of microwatts per cubic centimeter is possible from vibrations in a normal office building with current MEMS technology [11, 81, 115].

Certainly, future WSNs can be equipped with more than one power supply. For example, a solar cell and battery can work together. The battery is used to initiate the solar cell's activity, or act as a standby power resource when there is not enough sunlight to make the solar cell create the required power. Alternatively, two power supplies can be assigned dynamically to drive distinct components of the sensor node respectively according to their residual energy.

In summary, new energy sources can have a significant impact on the development of WSNs, including [113]:

- Improving the efficiency of energy systems
- Ensuring longevity against energy disruption
- Expanding future energy choices
- Promoting energy production and use in ways that respect health and environmental values

References

1. J. Agre and L. Clare, An integrated architecture for cooperative sensing networks, *Computer Mag.*, 33(5), 106–108, 2000.
2. I.F. Akyildiz, W. Su, Y. Sankarasubramaniam, and E. Cayirci, Wireless sensor network: a survey, *Computer Networks*, 38(4), 393–422, March 2002.
3. G. Anastasi, L. Lenzini, and E. Mingozzi, Stability and performance analysis of HIPERLAN, *IEEE INFOCOM 1998*, 1, 134–141, San Francisco, 1998.
4. G. Asada et al., Wireless integrated network sensors: low power systems on a chip, *24th IEEE Eur. Solid-State Circuits Conf.*, 9–12, The Hague, the Netherlands, 1998.
5. F. Baccelli and S. Zuyev, Poisson Voronoi spanning trees with applications to the optimization of communication networks, *Operations Res.*, 47(4), 619–631, 1999.
6. D.J. Baker, and A. Ephremides, The architectural organization of a mobile radio network via a distributed algorithm, *IEEE Trans. Commun.*, 29(11), 1694–1701, November 1981.
7. S. Bandyopadhyay and E.J. Coyle, An energy efficient hierarchical clustering algorithm for wireless sensor networks, *IEEE INFOCOM 2003*, 3, 1713–1723, San Francisco, 2003.
8. L. Benini and G.D. Micheli, *Dynamic Power Management: Design Techniques and CAD Tools*, Kluwer Academic Pub., Norwell, MA, 1998.
9. D.W. Carman, P.S. Kruus, and B.J. Matt, Constraints and approaches for distributed sensor network security, NAI labs technical report 00-010, September 2000.
10. A. Cerpa, J. Elson, D. Estrin, L. Girod, M. Hamilton, and J. Zhao, Habitat monitoring: application driver for wireless communications technology, *1st ACM SIGCOMM Workshop Data Commun. Latin Am. Caribbean*, 31(2), 20–41, San Jose, Costa Rica, April 2001.
11. L. Chandrakasan et al. Design considerations for distributed microsensor systems, *IEEE 1999 Custom Integrated Circuits Conf.*, 279–286, San Diego, CA, May 1999.
12. K. Chakrabarty, S.S. Iyengar, H. Qi, and E. Cho, Coding theory framework for target location in distributed sensor networks, *Int. Symp. Inf. Technol.: Coding Computing*, 130–134, Las Vegas, 2001.
13. K. Chakrabarty, S.S. Iyengar, H. Qi, and E. Cho, Grid coverage for sureillance and target location in distributed sensor networks, *IEEE Trans. Computers*, 51, 1448–1453, 2002.
14. B. Chen et al., SPAN: An energy-efficient coordination algorithm for topology maintenance in ad hoc wireless networks, *ACM/IEEE MOBICOM 2001*, 85–96, Rome, 2001.
15. T. Clouqueur, V. Phipatanasuphorn, P. Ramanathan, and K.K. Saluja, Sensor deployment strategy for target detection, *ACM WSNA*, 42–48, Atlanta, 2002.
16. Collaborative Sensor Networks, Internet article, http://wwwhome.cs.utwente.nl/~havinga/sensor.html, 2003.
17. K. Dasgupta, K. Kalpakis, and P. Namjoshi, An efficient clustering-based heuristic for data gathering and aggregation in sensor networks, *IEEE WCNC'03*, 3, 1948–1953, New Orleans, 2003.
18. J. Deng, and Z.J. Hass, Dual busy tone multiple access (DBTMA): a new medium access control for packet radio networks, *IEEE ICUPC'98*, 2, 973–977, Florence, 1998.
19. S.S. Dhillon, K. Chakrabarty, and S.S. Iyengar, Sensor placement for grid coverage under imprecise detections, *Int. Conf. Inf. Fusion (FUSION) 2002*, 2, 1581–1587, Annapolis, 2002.
20. S. Dulman et al., Collaborative communication protocols for wireless sensor networks, *Eur. Res. Middleware Architectures for Complex Embedded Syst. Workshop*, Pisa, 2003.
21. M. Easton, Using space technology to fight malaria, *Queen's Gazette*, April 7, 2003, 13.
22. D. Estrin, R. Govindan, J. Heidemann, and S. Kumar, Next century challenges: scalable coordination in sensor networks, *IEEE MOBICOM 1999*, 263–270, August 1999.
23. D. Estrin, L. Girod, G. Pottie, and M. Srivastava, Instrumenting the world with wireless sensor networks, *IEEE ICASSP 2001*, 4, 2033–2036, 2001.
24. ETSI, HIPERLAN Functional Specification, ETSI draft standard, July 1995, http://www.etsi.org.
25. J. Feng, F. Koushanfar, and M. Potkonjak, System architectures for sensor networks issues, alternatives, and directions, *IEEE ICCD'02: VLSI Computers Processors*, 226–231, Freiburg, Germany, 2002.

26. S.G. Foss and S.A. Zuyev, On a Voronoi aggregative process related to a bivariate poisson process, *Adv. Appl. Probability*, 28(4), 965–981, 1981.
27. A.J. Goldsmith and S.B. Wicker, Design challenges for energy-constrained ad hoc wireless networks, *IEEE Wireless Commun.*, 8–27, August 2002.
28. G. Gupta and M. Younis, Load-balanced clustering of wireless sensor networks, *IEEE ICC 2003*, 3, 1848–1852, 2003.
29. V. Gutnik and A.P. Chandrakasan, An embedded power supply for low-power DSP, *IEEE Trans. VLSI Syst.*, 5(4), 425–435, December 1997.
30. P.J.M. Havinga and G.J.M. Smit, Energy-efficient wireless networking for multimedia applications, *Wireless Communications and Mobile Computing*, Wiley, 165–184, 2001.
31. W. Heinzelman, J. Kulik, and H. Balakrishnan, Adaptive protocols for information dissemination in wireless sensor networks, *ACM MOBICOM 1999*, 174–185, Seattle, 1999.
32. W. Heinzelman, A. Chandrakasan, and H. Balakrishnan, Energy-efficient communication protocol for wireless microsensor networks, *HICSS 2000*, Maui, 8020–8029, January 2000.
33. W. Heinzelman, Application-specific protocol architecture for wireless networks, Ph.D. dissertation, Massachusetts Institute of Technology, June 2000.
34. W.B. Heinzelman, A.P. Chandrakasan, and H. Balakrishnan, An application-specific protocol architecture for wireless microsensor networks, *IEEE Trans. Wireless Commun.*, 1(4), October 2002, 660–670.
35. J. Heidemann et al., Building efficient wireless sensor networks with low-level naming, *18th ACM Symp. Operating Syst. Principles*, 146–159, Banff, October 2001.
36. J. Hill et al., System architecture directions for networked sensor networks, *9th Int. Conf. Architectural Support Programming Languages Operating Syst.*, 28(5), 93–104, Cambridge, MA, November 2000.
37. J. Hill and D. Culler, A wireless embedded sensor architecture for system level optimization, University of California Berkeley Technical Report, 2002.
38. J. Hill et al., TinyOS: operating system for sensor networks, hppt://tinyos.millennium.berkeley.edu, 2003.
39. X. Hong, M. Gerla, H. Wang, and L. Clare, Load balanced, energy-aware communications for Mars sensor networks, *IEEE Aerospace Conf.*, 3, 1109–1115, 2002.
40. A. Howard, M.J. Mataric, and G.S. Sukhatme, Mobile sensor network deployment using potential fields: a distributed, scalable, solution to the area coverage problem, *6th Int. Symp. Distributed Autonomous Robotics Syst.* (DAR02) Fukuoka, Japan, 299–308, June 2002.
41. IEEE 802.11 WG, Information technology — telecommunications and information exchange between systems. Local and metropolitan area networks — specific requirements. Part 11: wireless LAN medium access control (MAC) and physical layer (PHY) specifications, 1997.
42. C. Intanagonwiwat, R. Govindan, and D. Estrin, Directed diffusion: a scalable and robust communication paradigm for sensor networks, *ACM/IEEE MOBICOM 2000*, 56–67, Boston, August 2000.
43. C. Intanagonwiwat, D. Estrin, and R. Govindan, Impact of network density on data aggregation in wireless sensor networks, technical report 01-750, University of Southern California, November 2001.
44. C. Intanagonwiwat et al. Directed diffusion for wireless sensor networking, IEEE/ACM Trans. Networking, 11(1), 2–16, February 2003.
45. J.M. Kahn, R.H. Katz, and K.S.J. Pister, Next century challenges: mobile networking for Smart Dust, *ACM MOBICOM*, 271–278, Seattle, 1999.
46. F. Koushanfar, M. Potkonjak, and A. Sangiovanni–Vincentelli, Fault-tolerance techniques for sensor networks, *IEEE Sensors*, 2, 1491–1496, 2002.
47. B. Krishnamachari, D. Estrin, and S. Wicker, Impact of data aggregation in wireless sensor networks, *Int. Workshop Data Aggregation Wireless Sensor Networks*, 575–578, Vienna, Austria, July 2002.

48. J. Julik, W. Heinzelman, and H. Balakrishnan, Negotiation-based protocols for disseminating information in wireless sensor networks, *Wireless Networks*, 8, 169–185, 2002.
49. S. Kumar, F. Zhao, and D. Shepherd, Collaborative signal and information processing in microsensor networks, *IEEE Signal Processing Mag.*, 13–14, March 2002.
50. J. F. Kurose, and K. W. Ross, *Computer Networking, a Top-Down Approach Featuring the Internet*, 1st ed., Addison–Wesley, Longman, MA, 2000.
51. S. Lindsey and C.S. Raghavendra, PEGASIS: power-efficient gathering in sensor information systems, *IEEE Aerospace Conf. 2002*, 3, 1125–1130, March 2002.
52. S. Lindsey, C. Raghavendra, and K.M. Sivalingam, Data gathering algorithms in sensor networks using energy metrics, *IEEE Trans. Parallel Distributed Syst.*, 13(9), 924–935, September 2002.
53. A. Mainwaring, J. Polastre, R. Szewczyk, D. Culler, and J. Anderson, Wireless sensor networks for habitat monitoring, *ACM WSNA'02*, 88–97, Atlanta, September, 2002.
54. A. Manjeshwar, and D. P. Agrawal, TEEN: a routing protocol for enhanced efficiency in wireless sensor networks, *IEEE IPDPS 2002 Workshop*, 2009–2015, April 2002.
55. A. Manjeshwar and D.P. Agrawal, APTEEN: a hybrid protocol for efficient routing and comprehensive information retrieval in wireless sensor networks, *IEEE IPDPS 2002 Workshop*, 195–202, April 2002.
56. A. Manjeshwar, Q. Zeng, and D.P. Agrawal, An analytical model for information retrieval in wireless sensor networks using enhanced APTEEN protocol, *IEEE Trans. Parallel Distributed Syst.*, 13(12), 1290–1302, December 2002.
57. S. Meguerdichian, F. Koushanfar, M. Potkonjak, and M.B. Srivastava, Coverage problems in wireless ad-hoc sensor networks, *IEEE INFOCOM*, 3, 1380–1387, Anchorage, 2001.
58. R. Min et al., Low-power wireless sensor networks, *IEEE VLSID 2001*, 205–210, India, 2001.
59. R. Min et al., Energy-centric enabling technologies for wireless sensor networks, *IEEE Wireless Commun.*, 28–39, August 2002.
60. J. Mirkovic, G.P. Venkataramani, S. Lu, and L. Zhang, A self-organizing approach to data forwarding in large-scale sensor networks, *IEEE ICC*, 5, 1357–1361, St. Petersburg, Russia, 2001.
61. S. Musman, P.E. Lehner, and C. Elsaesser, Sensor planning for elusive targets, *J. Computer Math. Modeling*, 25(3), 103–115, 1997.
62. A.K. Parekh, Selecting routers in ad-hoc wireless networks, *Proc. ITS*, 1994.
63. T.A. Pering, T.D. Burd, and R.W. Brodersen, The simulation and evaluation of dynamic voltage scaling algorithms, *Int. Symp. Low Power Electron. Design* (ISLPED), 1998.
64. G.J. Pottie and L.P. Clare, Wireless integrated network sensors: towards low cost and robust self-organizing security networks, *Proc. of SPIE 1998*, 3577, 86–95, 1999.
65. G.J. Pottie and W.J. Kaiser, Wireless integrated network sensors, *Commun. ACM*, 43(5), 51–58, 2000.
66. H. Qi, S.S. Iyengar, and K. Chakrabarty, Multi-resolution data integration using mobile agents in distributed sensor networks, *IEEE Trans. Syst., Man Cybernetics (part C)*, 31, 383–391, August 2001.
67. H. Qi, P.T. Kuruganti, and Y. Xu, The development of localized algorithm in wireless sensor networks, *Sensors Mag.*, 2, 286–293, 2002.
68. J.M. Rabaey et al., PicoRadio supports ad hoc ultra-low power wireless networking, *IEEE Computer Mag.*, 33(7), 42–48, 2000.
69. V. Raghunathan, C. Schurgers, S. Park, and M.B. Srivastava, Energy-aware wireless microsensor networks, *IEEE Signal Process. Mag.*, 40–50, 2000.
70. T.S. Rappoport, *Wireless Communications, Principles and Practice*, Prentice Hall, Upper Saddle River, NJ, 1996.
71. S. Ray et al., Robust location detection in emergency sensor networks, *IEEE INFOCOM 2003*, 2, 1044–1053, March 2003.
72. A. Safwat, H. Hassanein, and H.T. Mouftah, Power-aware fair infrastructure formation for wireless mobile ad hoc communications, *IEEE GLOBECOM 2001*, 5, 2822–2836, November 2001.

73. A. Safwat, H. Hassanein, and H. Mouftah, Optimal cross-layer designs for energy-efficient wireless ad hoc and sensor networks, *22nd IEEE Int. Performance, Computing, Commun. Conf.*, (IPCCC 2003), 123–128, April 2003.
74. A. Safwat, H. Hassanein, and H.T. Mouftah, Power-aware virtual base stations (PW-VBS) for wireless mobile ad hoc communications, *J. Computer Networks*, 41(3), 331–346, 2003.
75. A. Salhieh et al., Power efficient topologies for wireless sensor network, *Int. Conf. Parallel Process.*, 156–163, Spain, 2001.
76. L. Schwiebert, S.K.S. Gupta, and J. Weinamann, Research challenges in wireless networks of biomedical sensors, *ACM MOBICOM 2001*, 151–165, Rome, July 2001.
77. C. Schurgers, V. Tsiatsis, and M. Srivastava, STEM: topology management for energy efficient sensor networks, *2002 IEEE Aerospace Conf.*, 3, 1099–1108, March 2002.
78. C. Schurgers, V. Tsiatsis, S. Ganeriwal, and M. Srivastava, Optimizing sensor networks in the energy–latency–density design space, *IEEE Trans. Mobile Computing*, 1(1), 70–80, January–March 2002.
79. C-C. Shen, C. Srisathapornphat, and C. Jaikaeo, Sensor information networking architecture and applications, *IEEE Personal Commun.*, 52–59, August 2001.
80. E. Shih et al., Physical layer driven protocol and algorithm design for energy-efficient wireless sensor networks, *ACM/IEEE MOBICOM'01*, 272–287, July 2001.
81. J.L. da Silva Jr. et al. Design methodology for PicoRadio networks, *Design, Automation Test Eur.*, DATE'2001, March 2001, Germany.
82. A. Sinha and A. Chandrakasan, Energy aware software, *VLSID'00*, January 2000.
83. A. Sinha and A. Chandrakasan, Dymamic power management in wireless sensor networks, *IEEE Design Test Computers*, 18(2), 62–74, 2001.
84. J.L. Sobrinho and A.S. Krishnakumar, Real-time traffic over the IEEE802.11 medium access control layer, *Bell Labs Tech. J.*, 172–187, Autumn 1996.
85. K. Sohrabi, J. Gao, V. Ailawadhi, and G.J. Pottie, Protocols for self-organization of a wireless sensor network, *IEEE Personal Commun.*, 16–27, October 2000.
86. M. Srivastava, R. Muntz, and M. Potkonjak, Smart kindergarten: sensor-based wireless networks for smart developmental problem-solving environments, *ACM MOBICOM 2001*, 132–138, Italy, 2001.
87. J.A. Stankovic et al., Real-time communication and coordination in embedded sensor networks, *Proc. IEEE*, 91(7), 1002–1022, 2003.
88. L. Subramanian and R.H. Katz, An architecture for building self-confirgurable systems, *MOBIHOC 2000*, 63–67, Boston, 2000.
89. A.S. Tanenbaum, *Computer Networks*, Prentice Hall, Upper Saddle River, NJ, 1996.
90. B. Tavil, and W.B. Heinzelman, TRACE: time reservation using adaptive control for energy efficiency, *IEEE JSAC*, 21(10), 1506–1515, 2003.
91. B. Tavil, and W.B. Heinzelman, MH-TRACE: multi-hop time reservation using adaptive control for energy efficiency, *MILCOM 2003*, Boston, 2003.
92. S. Tilak, N.B. Abu–Ghazaleh, and W. Heinzelman, Infrastructure trade-offs for sensor networks, *ACM WSNA'02*, 49–58, September 2002.
93. A. Wang and A. Chandrakasan, Energy efficient system partitioning for distributed wireless sensor networks, 2001 *IEEE Int. Conf. Acoustics, Speech, Signal Processing*, 2, 905–908, Salt Lake City, 2001.
94. A. Willig, R. Shah, J. Rabaey, and A. Wolisz, Altruists in the PicoRadio sensor network, *4th IEEE Int. Workshop Factory Commun. Syst.*, 175–184, Sweden, August 2002.
95. A.D. Wood and J.A. Stankovic, Denial of service in sensor networks, *IEEE Computer*, 35(10), 48–56, October 2002.
96. Y. Xu, J. Heidemann, and D. Estrin, Geography-informed energy conservation for ad hoc routing, *ACM MOBICOM 2001*, 70–84, Italy, 2001.

97. K. Xu, Performance analysis of differentiated QoS MAC in wireless area networks (WLANs), master degree thesis, Queen's University, 2003.
98. M.D. Yarvis et al., Real-world experiences with an interactive ad hoc sensor network, *Int. Conf. Parallel Processing Workshops (ICPPW'02)*, 2002.
99. F. Ye et al., A scalable solution to minimum cost forwarding in large sensor networks, *IEEE ICCCN*, 304–309, 2001.
100. W. Ye, J. Heidemann, and D. Estrin, An energy-efficient MAC protocol for wireless sensor networks, *IEEE INFOCOM*, 3, 1567–1576, New York, 2002.
101. F. Zhao, J. Shin, and J. Reich, Information-driven dynamic sensor collaboration for tracking applications, *IEEE Signal Process. Mag.*, 19(2), 61–72, March 2002.
102. F. Zhao et al., Collaborative signal and information processing: an information directed approach, *Proc. IEEE*, 91(8), 1199–1209, 2003.
103. Y. Zou and K. Chakrabarty, Sensor deployment and target localization based on virtual forces, *IEEE INFOCOM 2003*, 2, 1293–1303, March 2003.
104. Z.J. Hass, and J. Deng, Dual busy tone multiple access (DBTMA) — performance evaluation, *IEEE Vehicular Technol. Conf. (VTC) 1999*, May 1999.
105. H. Qi, S.S. Iyengar, and K. Chakrabarty, Distributed sensor networks: a review of recent research, *J. Franklin Inst.*, 338, 655–668, 2001.
106. K.S.J. Pister, SMART DUST: autonomous sensing and communication in a cubic millimeter, internet article, http://www-bsac.eecs.berkeley.edu/~pister/Smart/Dust/.
107. H.S. Carvalho, et al., A general data fusion architecture, *6th Intl. Conf. on Information Fusion (Fusion 2003)*, July 2003, Australia, 1465–1472.
108. L.J. Guibas, Sensing, tracking, and reasoning with relations, *IEEE Signal Processing Magazine*, 19(2), 73–85, March 2002.
109. J.-F. Frigon, V.C.M. Leung, and H.C.B. Chan, Dynamic reservation TDMA protocol for wireless ATM networks, *IEEE JSAC*, 19(2), 370–383, February 2001.
110. J.L. Sobrinho and A.S. Krishnakumar, Quality-of-service in ad hoc carrier sense multiple access networks, *IEEE JSAC*, 17(8), 1353–1968, August 1999.
111. K. Kalpakis, K. Dasgupta, and P. Namjoshi, Efficient algorithms for maximum lifetime data gathering and aggregation in wireless sensor networks, Technical Report UMBC-TR-02-13, Computer Science and Electrical Engineering Department, University of Maryland, Baltimore County, 2002.
112. How many solar cells would I need in order to provide all of the electricity that my house needs? internet article, http://science.howstuffworks.com/question418.htm.
113. Allied business intelligence, 2001, internet article, http://www.eyeforfuelcells.com.
114. Advancing fuel cells for clean and efficient power, internet article, Pacific Northwest National Laboratory, PNNL-SA-35945, February 2002, http://www.pnl.gov.
115. K.S.J. Pister, My view of sensor networks in 2010, internet article, http://robotics.eecs.berkeley.edu/~pister/SmartDust/in2010.

V

Tracking Technologies

19
Coverage in Wireless Sensor Networks

Mihaela Cardei
Florida Atlantic University

Jie Wu
Florida Atlantic University

19.1	Introduction	**19**-1
	Sensor Coverage Problem • Design Choices • Coverage Problem in Other Fields • Overview	
19.2	Area Coverage...	**19**-4
	Energy-Efficient Random Coverage • Connected Random Coverage • Deterministic Coverage • Node Coverage as Approximation	
19.3	Point Coverage	**19**-8
	Random Point Coverage • Deterministic Point Coverage	
19.4	Barrier Coverage...	**19**-9
	Barrier Coverage Model 1 • Barrier Coverage Model 2	
19.5	Conclusion	**19**-10

19.1 Introduction

Wireless sensor networks (WSNs) have attracted a great deal of research attention due to their wide range of potential applications. A WSN provides a new class of computer systems and expands people's ability to interact remotely with the physical world. In a broad sense, WSNs will transform the way people manage their homes, factories, and environment. Applications of WSNs [11] include battlefield surveillance, biological detection, home appliance, smart spaces, and inventory tracking.

The purpose of deploying a WSN is to collect relevant data for processing and reporting. The two types of reporting are [4] *event driven* and *on demand*. Consider a WSN with a sink (also called monitoring station) and a set of sensor nodes. In event-driven reporting, the reporting process is triggered by one or more sensor nodes in the vicinity that detect an event and report it to the monitoring station. In the on-demand report, the reporting process is initiated from the monitoring station and sensor nodes send their data in response to an explicit request. A forest fire monitoring system is event driven, whereas an inventory control system is on demand. A more flexible system can be a hybrid of the two types.

19.1.1 Sensor Coverage Problem

An important problem addressed in literature is the *sensor coverage problem*. This problem is centered around a fundamental question: "How well do the sensors observe physical space?" As Meguerdichian and colleagues have pointed out, the coverage concept is a measure of the quality of service (QoS) of the sensing function and is subject to a wide range of interpretations due to a large variety of sensors and applications [17]. The goal is to have each location in the targeted physical space within sensing range of at least one sensor.

19.1.2 Design Choices

Sensor nodes, also called wireless transceivers, are tiny devices equipped with one or more sensors; one or more transceivers; processing; storage resources; and, possibly, actuators. Sensor nodes organize in networks and collaborate to accomplish a larger sensing task. One important class of WSNs is wireless ad hoc sensor networks (WASN), characterized by an ad hoc or *random* sensor deployment method in which the sensor location is not known *a priori*. This feature is required when individual sensor placement is infeasible — for example, battlefield or disaster areas.

The characteristics of a WASN include limited resources, large and dense networks, and dynamic topology. Generally, more sensors than required (compared with the optimal placement) are deployed to perform the proposed task; this compensates for the lack of exact positioning and improves the fault tolerance. The size of a WASN may reach hundreds or even thousands of sensor nodes. If the sensors can be placed exactly where they are needed, the corresponding deployment method is *deterministic* [13].

Sensors have size, weight, and cost restrictions, which impact resource availability. They have limited battery resources and processing and communication capabilities. Because replacing the battery is not feasible in many applications, low power consumption is a critical factor to be considered, not only in the hardware and architectural design, but also in the design of algorithms and network protocols at all layers of the network architecture. Therefore, an important network design objective is to maximize network lifetime. Another clear objective, especially in a deterministic node deployment, is to use a minimum number of sensors.

A sensor can be in one of the following four states: transmit, receive, idle, or sleep. The idle state is when the transceiver is neither transmitting nor receiving, and the sleep mode is when the radio is turned off. As presented by Raghunathan and coworkers [19], the power usage for WINS Rockwell seismic sensor for transmit:receive:idle:sleep operational modes is $0.38 \div 0.7$ W:0.36 W: 0.34 W:0.03 W while the sensing power is 0.02 W. An interesting observation is that the receive and idle modes may require as much energy as transmitting, whereas in the traditional ad hoc wireless networks, transmitting may use as much as twice the power of receiving.

Another observation concerns the communication/computation power usage ratio, which can be higher than 1000 (e.g., for Rockwell WINS it is $1500 \div 2700$); therefore, local data processing, data fusion, and data compression are highly desirable. Judiciously selecting the activity state of each sensor is accomplished through a *scheduling* mechanism. When the goal is to reduce the number of active sensors performing the coverage, this constitutes an important method for decreasing network energy consumption. Sometimes, the scheduling mechanism also has the objective of maintaining connectivity among active sensors.

The coverage algorithms proposed are *centralized*, or *distributed* and *localized*. In distributed algorithms, the decision process is decentralized. "Distributed and localized" algorithms refer to a distributed decision process at each node that makes use of only neighborhood information (within a constant number of hops). Because the WSN has a dynamic topology and needs to accommodate a large number of sensors, the algorithms and protocols designed should be distributed and localized in order to accommodate a scalable architecture better.

Considering the coverage concept, different problems can be formulated, based on the subject to be covered (*area* vs. *discrete points*) and on the following design choices:

- *Sensor deployment method*: deterministic vs. random. A deterministic sensor placement may be feasible in friendly and accessible environments. Random sensor distribution is generally considered in military applications and for remote or inhospitable areas.
- *Sensing & communication ranges*: WASN scenarios consider sensor nodes with same or different sensing ranges. Another factor that relates to connectivity is communication range, which can be equal or not equal to the sensing range.
- *Additional critical requirements*: energy efficiency and connectivity, referred to as energy-efficient coverage and connected coverage.
- *Algorithm characteristics*: centralized vs. distributed/localized.
- *Objective of the problem*: maximize network lifetime or minimum number of sensors.

TABLE 19.1 Coverage Approaches

Coverage Approach	Coverage Type	Problem Objectives	Ref.
Most constrained minimally constraining heuristic	Area coverage	Energy efficiency; maximize network lifetime by reducing number of working nodes	21
Disjoint dominating sets heuristic	Area coverage	-//-	2
Node self-scheduling algorithm	Area coverage	-//-	22
Probing-based density control algorithm	Area coverage	Energy efficiency; maximize network lifetime by controlling working node density	26
Optimal geographical density control (OGDC) algorithm	Area coverage	Energy efficiency; connectivity; maximize network lifetime by reducing number of working nodes	27
Coverage configuration protocol (CCP)]	Area coverage	-//-	23
Connected dominating set based coverage	Area coverage	-//-	24
Node placement algorithms	Area coverage	Energy efficiency; connectivity	13
	Point coverage	Deployment of a minimum number of sensors	
Disjoint set cover heuristic	Point coverage	Energy-efficiency; maximize network lifetime by reducing number of working nodes	3
Maximal breach path and maximal support path algorithms	Barrier coverage	Worst- and best-case coverage paths	17
Maximal support path algorithm	Barrier coverage	Best-case coverage path	14
Node density-based coverage	Barrier coverage	Find critical node density, above which a penetrating path will be detected almost surely	15
Minimum exposure path algorithm	Barrier coverage	Find path of minimum exposure	18
Critical node density for complete coverage in exposure model	Barrier coverage	Find critical density for a high probability of target detection	1

Table 19.1 and Table 19.2 summarize the methods covered in this chapter according to the preceding five choices.

TABLE 19.2 Characteristics of Approaches Listed in Table 19.1

Sensor Deployment Method	Sensing Range R_s, Communication Range R_c		Algorithm Characteristics	Coverage Approach (ref.)
	Same R_s for All Sensors?	Is $Rs==Rc$ for Each Sensor?		
Random	YES	NA	Centralized	21
Random	YES	NA	Centralized	2
Random	YES + NO (both)	NA	Distributed, localized	22
Random	YES	NA	Distributed, localized	26
Random	YES	NO	Distributed, localized	27
Random	YES	NO	Distributed, localized	23
Random	YES	YES	Distributed, localized	24
Deterministic	YES	YES	Centralized	13
Random	YES	NA	Centralized	3
Random	NA	NA	Centralized	17
Random	NA	NA	Distributed, localized	14
Random	NA	NA	Centralized	15
Random	NA	NA	Centralized	18
Random	NA	NA	Centralized	1

19.1.3 Coverage Problem in Other Fields

Coverage problems have been formulated in other fields, such as the Art Gallery problem, ocean coverage, and coverage in robotic systems.

The Art Gallery problem [20] seeks to determine the number of observers and their placement, necessary to cover an art gallery room so that every point is seen by at least one observer. This problem has a linear time solution for the two-dimensional case. The three-dimensional version is NP-hard and an approximation algorithm is presented in Marengoni et al. [16]. This visibility problem has many real-world applications, such as placement of antennas for cellular telephone companies, and placement of cameras for security purposes in banks and supermarkets.

Gregg and colleagues [8] address the ocean area coverage problem. Here, the authors are interested in satellite-based monitoring of ocean phytoplankton abundance. Given the orbit and sensor characteristics of each mission, numerical analysis results show that merging data from three satellites can increase ocean coverage by 58% for one day and 45% for four days. Additional satellites produce diminishing returns.

The coverage concept with regard to the many-robot systems was introduced by Gage [7], who defined three types of coverage: blanket coverage, barrier coverage, and sweep coverage. In blanket coverage, the goal is to achieve a static arrangement of sensors that maximizes the total detection area. Barrier coverage seeks to achieve a static arrangement of nodes that minimizes the probability of undetected penetration through the barrier; sweep coverage is more or less equivalent to a moving barrier. New applications arise in the context of mobile WSNs, in which sensors have locomotion capabilities. Thus, the nodes can spread out so that the area covered by the network is maximized (for example, see Howard et al. [10]) and can relocate to handle sensor failures.

19.1.4 Overview

This chapter surveys recent contributions addressing coverage problems in the context of static WSNs; that is, the sensor nodes do not move once they are deployed. Sensors have omnidirectional antennae and can monitor a disk whose radius is referred to as sensing range. Various coverage formulations and their assumptions are presented, as well as an overview of proposed solutions. The most discussed problems from the literature can be classified as the following types: area coverage, point coverage, and barrier coverage. This chapter continues with a discussion of these coverage problems, followed by conclusions.

19.2 Area Coverage

The most studied coverage problem is the area coverage problem in which the main objective of the sensor network is to cover (monitor) an area (also referred to sometimes as a region). Figure 19.1(a) shows an example of random deployment of sensors to cover a given square-shaped area. The connected black nodes form the set of active sensors as the result of a scheduling mechanism. Next, recent area coverage problem formulations, their models and assumptions, and proposed solutions are surveyed.

19.2.1 Energy-Efficient Random Coverage

This subsection presents several energy-efficient coverage mechanisms because energy efficiency, caused by limited battery resources, is an important issue in WASN. Mechanisms that conserve energy resources are highly desirable because they have a direct impact on network lifetime. Network lifetime is in general defined as the time interval in which the network can perform the sensing functions and transmit data to the sink. During the network lifetime, some nodes may become unavailable (e.g., physical damage, lack of power resources) or additional nodes might be deployed. An efficient, frequently used mechanism is to schedule the sensor node activity and allow redundant nodes to enter the *sleep* mode as often and for as long as possible. To design such a mechanism, the following questions must be answered:

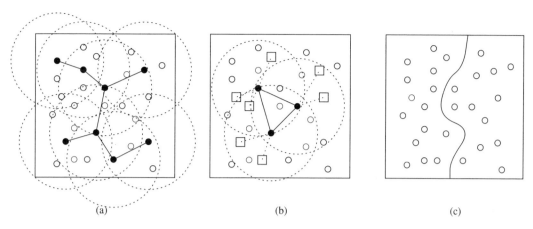

FIGURE 19.1 (a) Random sensor deployment for square-shaped area; (b) random sensor deployment to cover set of points; (c) general barrier coverage problem.

- Which rule should each node follow to determine whether to enter sleep mode?
- When should nodes make such a decision?
- How long should a sensor remain in the sleep mode?

These node scheduling mechanisms are illustrated next.

Slijepcevic and Potkonjak [21] and Cardei et al. [2] consider a large population of sensors, deployed randomly for area monitoring. The goal is to achieve an energy-efficient design that maintains area coverage. Because the number of sensors deployed is greater than the optimum required to perform the monitoring task, the solution proposed is to divide the sensor nodes into disjoint sets so that every set can individually perform the area monitoring tasks. These sets are then activated successively and, while the current sensor set is active, all other nodes are in a low-energy sleep mode. The goal of this approach is to determine a maximum number of disjoint sets because this has a direct impact on the network lifetime. The solutions proposed are centralized.

Slijepcevic and Potkonjak [21] model the area as a collection of fields in which every field has the property that any enclosed point is covered by the same set of sensors. The most constrained, least constraining algorithm [21] computes the disjoint covers successively, selecting sensors that cover the critical element (field covered by a minimal number of sensors) and giving priority to sensors that cover a high number of uncovered fields, cover sparsely covered fields, and do not cover fields redundantly. Cardei et al. [2] model the disjoint sets as disjoint dominating sets. The maximum disjoint dominating sets computation is NP complete, and an algorithm based on graph coloring is proposed. Simulations have shown that the number of sets computed is between 1.5 and 2 times greater than using Slijepcevic and Potkonjak's algorithm, with lapses in area coverage less than 5%, on average.

Tian and Georganas have proposed another energy-efficient mechanism based on node scheduling coverage [22]; the protocol is distributed and localized. The off-duty eligibility rule determines whether a node's sensing area is included in its neighbors' sensing areas. Solutions for determining whether a node's coverage can be sponsored by its neighbors (sponsored coverage calculation) is provided for several different cases in which:

- Nodes have the same sensing range and know their location.
- Nodes have the same sensing range and can obtain a neighboring node's directional information.
- Nodes have different sensing ranges in particular scenarios.

The node scheduling scheme is divided into rounds in which each round has a self-scheduling phase followed by a sensing phase. In the self-scheduling phase, the nodes investigate the off-duty eligibility rule. Eligible nodes turn off their communication and sensing units, while all other nodes will perform sensing tasks in the sensing phase. In order to obtain neighboring information, each node broadcasts a position advertisement message, at the beginning of each round, that contains node ID and node location. If the off-duty eligibility rule is tested simultaneously by neighboring nodes, a node and its sponsor may decide to turn off simultaneously, triggering the occurrence of *blind points*.

To avoid this, a back-off scheme is used in which every node starts the evaluation rule after a random time and then broadcasts a status advertisement message to announce whether it is available for turning off. Before turning off, a node waits another T_w time to listen for neighboring nodes' updates. This work does not specify synchronization mechanisms in detail. It is implemented as an extension of the data-gathering LEACH protocol [9] and simulation results show an increase of 1.7 on average in system lifetime.

A probing-based, node-scheduling solution for the energy-efficient coverage problem has been proposed by Ye and colleagues [26]. Here, all sensors are characterized by the same sensing range and coverage is seen as the ratio between the area under monitoring and total size of the network field. The off-duty eligibility rule is based on a probing mechanism. Basically, a sensor broadcasts a probing message *PRB* within a probing range *r*. Any working node that hears this message responds with a *PRB_RPY*. If at least one reply is received, the node enters the sleep mode. Probing range is selected based on the desired working node density (number of sensors per unit area) or based on the desired coverage redundancy; the wake-up time is based on the tolerable sensing intermittence. This protocol is distributed and localized, with low complexity; however, it still does not preserve the original coverage area.

19.2.2 Connected Random Coverage

An important issue in WSNs is connectivity. A network is connected if any active node can communicate with any other active node, possibly using intermediate nodes as relays. Once the sensors are deployed, they organize into a network that must be connected so that the information collected by sensor nodes can be relayed back to data sinks or controllers. An important, frequently addressed objective is to determine a minimal number of working sensors required to maintain the initial coverage area as well as connectivity. Selecting a minimal set of working nodes reduces power consumption and prolongs network lifetime. Next, several connected coverage mechanisms will be presented.

An important but intuitive result proved by Zhang and Hou [27] states that, if the communication range R_c is at least twice the sensing range R_s, a complete coverage of a convex area implies connectivity of the working nodes. If the communication range is set up too large, radio communication may be subject to excessive interference. Therefore, if the communication range can be adjusted, a good approach to assure connectivity is to set the transmission range at twice the size of the sensing range.

Based on this result, Zhang and Hou [27] further discussed the case for $R_c \geq R_s$. An important observation is that an area is completely covered if at least two disks intersect and all crossings are covered. Here a disk refers to a node's sensing area and a crossing is an intersection point of the circle boundaries of two disks. In the ideal case, in which node density is sufficiently high, the subset of working nodes can be optimally chosen for full coverage.

Based on these results, the authors proposed a distributed, localized algorithm called optimal geographical density control (OGDC). At any time, a node can be in one of the tree states: UNDECIDED, ON, and OFF. The algorithm runs in rounds, and at the beginning of each round a set of one or more starting nodes is selected as working nodes. After a back-off time, a starting node broadcasts a power-on message and changes its state to ON. The power-on message contains: (1) the position of the sender; and (2) the direction along which a working node should be located. The direction indicated by the power-on message of a starting node is randomly distributed. Selecting starting nodes randomly at the beginning of each round ensures uniform power consumption across the network. Also, the back-off mechanism avoids packet collisions.

At the beginning of each round, all nodes are UNDECIDED and will change to ON or OFF state until the beginning of the next round. This decision is based on the power-on messages received. Every node keeps a list with neighbor information. When a node receives a power-on message, it checks whether its neighbors cover its sensing area; if so, it will change to the OFF state. A node decides to change into the

ON state if it is the closest node to the optimal location of an ideal working node selected to cover the crossing points of the coverage areas of two working neighbors. Simulation based on NS-2 shows good results in terms of percentage of coverage, number of working nodes, and system lifetime.

Some applications may require different degrees of coverage while still maintaining working node connectivity. A network has a coverage degree k (k-coverage) if every location is within the sensing range of at least k sensors. Networks with a higher coverage degree can obtain higher sensing accuracy and be more robust to sensor failure. Wang et al. [23] generalized the result in Zhang and Hou [27] by showing that, when the communication range R_c is at least twice the sensing range R_s, a k-covered network will result in a k-connected network. A k-connected network has the property that removing any $k - 1$ nodes will still maintain network connectivity. The following discussion addresses the case when $R_c \geq 2R_s$.

To define the k-coverage eligibility mechanism, the problem of determining the coverage degree of a region is reduced to a simpler problem of determining the coverage degrees of all the intersection points. Given a coverage region A, a point p is called an *intersection point* if: (1) p is an intersection point of the sensing circles of any two nodes u and v, e.g., $p \in u \cap v$; and (2) for any node v, $p \in v \cap A$ and $|pv| = R_s$, where R_s is the sensing range. The authors proved that a convex region is k-covered if it contains intersection points and all these intersection points are k-covered. Based on this, a sensor is ineligible to turn active if all the intersection points inside its sensing circle are at least k-covered.

The work in Wang and coworkers [23] introduces coverage configuration protocol (CCP), which can dynamically configure the network to provide different coverage degrees requested by applications. To facilitate the computation of intersection points, every node maintains a table with neighbor information (location, status: active/inactive) and periodically broadcasts a HELLO beacon with its current location and status. A node can be in one of the three states: SLEEP, LISTEN, and ACTIVE. All nodes start in the SLEEP state for a random time. When a node wakes up, it enters the LISTEN state and, based on the outcome of the eligibility rule over a time interval, it will enter the SLEEP or ACTIVE state. Once a node is in the ACTIVE state, it will re-evaluate the coverage eligibility every time it receives a HELLO message and decide whether to go into the SLEEP state or remain in the ACTIVE state.

For the case in which $R_c < 2R_s$, CCP does not guarantee network connectivity. The solution adopted by Wang and colleagues [23] is to integrate CCP with SPAN [5] to provide sensing coverage and network connectivity. SPAN is a distributed algorithm that conserves energy by turning off unnecessary nodes while maintaining connectivity. The combined eligibility rule is as follows: (1) an inactive node goes into the active state if it satisfies the eligibility rules of SPAN and CCP; and (2) an active node withdraws if it satisfies neither the eligibility rule of SPAN or of CCP. With this combined mechanism, k-coverage can be obtained through CCP and 1-connectivity through SPAN. The algorithm was implemented and tested using NS-2 and showed good performance results in terms of coverage, active nodes, and system lifetime.

19.2.3 Deterministic Coverage

Kar and Banerjee [13] consider the problem of deterministically placing a minimum number of sensor nodes to cover a given region. The sensing area is a disk with radius r called *r-disk*, and the sensing and communication radii are equal. The basic pattern is an *r-strip*, a string of *r*-disks placed along a line so that the distance between two adjacent disks is r. This forms a connected component. To cover a given area, it is first filled with horizontal *r*-strips with distance $r(1 + \sqrt{3}/2)$ between them. Then another strip is added so as to intersect all other parallel *r*-strips. This results in a connected sensor network that covers the given area. For covering of the two dimensional plane, the performance ratio is 1.026, while for a bounded convex region with perimeter L and area A the performance ratio is $2.693(1 + 2.243Lr/A)$.

19.2.4 Node Coverage as Approximation

When a large and dense sensor network is randomly deployed for area monitoring, the area coverage can be approximated by the coverage of the sensor locations. One method to assure coverage and connectivity is to design the set of active sensors as a connected dominating set (CDS). A distributed and localized protocol for constructing the CDS was proposed by Wu and Li, using a *marking process* [24]. A node is a coverage node if two neighbors are not connected (i.e., not within the transmission range of each other). Coverage nodes (also called gateway nodes) form a CDS. A pruning process can be used to reduce the size of a coverage node set while keeping the CDS property.

Dai and Wu [6] provide a generalized pruning rule called *pruning rule k*. Basically, a coverage node can be withdrawn if its neighbor set can be collectively covered by those of k coverage nodes. In addition, these k coverage nodes have higher priority and are connected. Pruning rule k ensures a constant approximation ratio. The CDS derived from the marking process with rule k can be locally maintained when sensors switch on or off.

The node coverage problem can be related to the broadcasting problem, in which a small set of forwarding nodes is selected [25]. Forwarding set selection in broadcasting is similar to the point coverage problem, where both try to find a small coverage set. Note that directed diffusion [12] also uses this platform to collect information through broadcasting. As the interest is propagated through the network, sensor nodes set up reverse gradients to the sink in a decentralized way. The difference is that in direct diffusion all sensors forward the data. One difference between node coverage and area coverage is that neighbor set information is sufficient for node coverage, but in area coverage, geometric/directional information is needed.

19.3 Point Coverage

In the point coverage problem, the objective is to cover a set of points. Figure 19.1(b) shows an example of a set of sensors randomly deployed to cover a set of points (small square nodes). The connected black nodes form the set of active sensors, the result of a scheduling mechanism. Next, a coverage approach is presented for each sensor deployment method: random and deterministic.

19.3.1 Random Point Coverage

The point coverage scenario addressed in Cardei and Du [3] has military applicability. It considers a limited number of points (targets) with a known location that need to be monitored. A large number of sensors are dispersed randomly in close proximity to the targets; the sensors send the monitored information to a central processing node. The requirement is that every target must be monitored at all times by at least one sensor, assuming that every sensor is able to monitor all targets within its sensing range.

One method for extending the sensor network lifetime through energy resource preservation is to divide the set of sensors into disjoint sets so that every set completely covers all targets. These disjoint sets are activated successively, so at any moment in time only one set is active. Because all targets are monitored by every sensor set, the goal of this approach is to determine a maximum number of disjoint sets so that the time interval between two activations for any given sensor is longer. By decreasing the fraction of time a sensor is active, the overall time until power runs out for all sensors is increased and the application lifetime is extended proportionally by a factor equal to the number of disjoint sets. Cardei and Du have proposed a solution for this application [3]: the disjoint sets are modeled as disjoint set covers and every cover completely monitors all the target points. The authors prove that the disjoint set coverage problem is NP complete and propose an efficient heuristic for set cover computation using a mixed integer programming formulation.

19.3.2 Deterministic Point Coverage

Kar and Banerjee consider the scenario in which it is possible to explicitly place a set of sensor nodes [13]. This is feasible in friendly and accessible environments. Given a set of n points, the objective is to determine a minimum number of sensor nodes and their location so that the given points are covered and the sensors deployed are connected. For the case in which all sensors have the same sensing range and the sensing range equals the communication range, the authors propose an approximation algorithm with a performance ratio of 7.256. The algorithm begins by constructing the minimum spanning tree over the targeted points, and then successively selects sensor node locations on the tree (vertices or along the edges) so that the coverage and connectivity are maintained at every step.

19.4 Barrier Coverage

From Gage's classification, the barrier coverage can be considered as the coverage with the goal of minimizing the probability of undetected penetration through the barrier (sensor network). Figure 19.1(c) shows a general barrier coverage problem where start and end points of the path are selected from bottom and top boundary lines of the area. The selection of the path depends on the objective (discussed next).

19.4.1 Barrier Coverage Model 1

Two types of barrier coverage models are proposed in the literature. The first model has been proposed by Meguerdichian et al. [17], who address the following problem: given a field instrumented with sensors and the initial and final locations of an agent that needs to move through the field, determine a maximal breach path (MBP) and the maximal support path (MSP) of the agent. The MBP (MSP) corresponds to the worst (best) case coverage and has the property that, for any point on the path, the distance to the closest sensor is maximized (minimized). The model assumes homogeneous sensor nodes, known sensor locations (e.g., through the GPS), with sensing effectiveness decreasing as the distance increases.

The authors proposed a centralized solution, based on the observation that MBP lies on the Voronoi diagram lines and MSP lies on Delaunay triangulation lines. The proposed algorithm starts by generating the Voronoi diagram (or Delaunay triangulation diagram), assigns every segment a weight equal with the distance to the closest sensor (or equal with the segment length), and then uses binary search and breadth first search for path computation.

The best coverage problem is further explored and formalized by Li and colleagues [14], who proposed a distributed algorithm for MSP computation using the relative neighborhood graph. The authors also considered two extensions: MSP with least energy consumption and MSP with smallest path distance.

Also important is the determination of the number of sensor nodes to be deployed randomly in the field so that the probability of a penetration path is close to zero. Liu and Towsley [15] address this coverage and detectability problem in the context of grid-based sensor networks and random sensor networks. The authors propose a critical density for a given sensor network based on percolation theory so that, for nodes deployed with a lower density, a penetrating path that will not be detected almost surely exists; above this critical density any crossing object is almost surely detected.

19.4.2 Barrier Coverage Model 2

The second barrier coverage problem is the *exposure-based model*, introduced by Meguerdichian et al. [18], in which the sensing abilities of the sensors diminish as the distance increases. However, another important factor is the sensing time (exposure). The longer the exposure time, the greater the sensing ability. The two-dimensional sensing model is defined as

$$S(s,p) = \frac{\lambda}{[d(s,k)]^k}$$

where $d(s,p)$ is the Euclidean distance between the sensor s and the point p; λ and k are sensor technology-dependent parameters.

Another characteristic is the intensity of the sensor field. For example, all-sensor field intensity for a point p, field F, and n sensors $s_1, s_2, \ldots, s_n$ is defined as

$$I(F,p) = \sum_{1}^{n} S(S_i, p)$$

The exposure of an object moving in the sensor field during the interval $[t_1, t_2]$ along the path $p(t)$ is defined as

$$E(p(t), t_1, t_2) = \int_{t_1}^{t_2} I(F, p(t)) \left| \frac{dp(t)}{dt} \right| dt .$$

Given a field instrumented with n sensors and the initial and final points of the object, the authors consider the problem of determining the minimal exposure path, which corresponds to the worst-case scenario. Because the exposure analytical computation is intractable, the solution proposed uses a grid-based approach to transform the problem domain to a tractable discrete domain. The minimal exposure path in each grid square is then restricted to line segments connecting any two vertices. In the next step, the grid is transformed into a weighted graph in which the weight (exposure) of an edge is approximated using numerical techniques. Finally, the Dijkstra's single-source shortest-path algorithm is used to find the minimal exposure path between any arbitrary starting and ending points on the grid. The approximation quality improves by increasing the grid divisions — at a cost of higher storage and run time.

Adlakha and Srivastava point out another aspect of the exposure-based model [1]: to estimate sensor node deployment density, one should consider the sensor characteristics as well as target specifications. For example, detection of an enemy tank requires fewer nodes due to the strong the acoustic signal compared with soldier detection that might require more sensors. Their paper assumes the target moves in a straight line, with constant speed, between two given points. Two radii are associated for a given sensor:

- *Radius of complete influence*, defined as the distance from the sensor so that all targets originating within this radius are detected
- *Radius of no influence*, with the property that any target originating beyond it cannot be detected

Using the preceding sensing and exposure model and knowing the threshold energy, $E_{threshold}$, required to detect a target, this paper proposes a solution to calculate the influence radii as well as the sensor nodes' deployment density. Thus, to cover an area A, computing the number of nodes to be deployed as $O(\frac{A}{r^2})$, where r is the radius of no influence, achieves a probability of detection of 98% or above.

19.5 Conclusion

This chapter categorized and described recent coverage problems proposed in the literature, and their formulations, assumptions, and proposed solutions. Sensor coverage is an important element for QoS in applications with WSNs. Coverage is generally associated with energy efficiency and network connectivity — two important properties of a WSN. To accommodate a large WSN with limited resources and a dynamic topology, coverage control algorithms and protocols perform best if they are distributed and localized. Various interesting formulations for sensor coverage have been proposed recently in the literature. To meet the intended objective, these problems aim at deterministically placing sensors nodes; determining the sensor deployment density; or, more generally, designing mechanisms that efficiently organize or schedule the sensors after deployment.

Acknowledgment

The work of Jie Wu was supported in part by NSF grants CCR 0329741, CCR 9900646, ANI 0073736, and EIA 0130806.

References

1. S. Adlakha and M. Srivastava, Critical density thresholds for coverage in wireless sensor networks, *IEEE Wireless Commun. Networking*, 3, 16–20, March 2003.
2. M. Cardei, D. MacCallum, X. Cheng, M. Min, X. Jia, D. Li, and D.-Z. Du, Wireless sensor networks with energy efficient organization, *J. Interconnection Networks*, 3(3–4), 213–229, Dec 2002.
3. M. Cardei and D.-Z. Du, Improving wireless sensor network lifetime through power aware organization, accepted to appear in *ACM Wireless Networks*.
4. J. Carle and D. Simplot, Energy efficient area monitoring by sensor networks, accepted to appear in *IEEE Computer*, 37(2), 40–46, 2004.
5. B. Chen, K. Jamieson, H. Balakrishnan, and R. Morris, Span: an energy-efficient coordination algorithm for topology maintenance in ad hoc wireless networks, *ACM/IEEE Int. Conf. Mobile Computing Networking* (MobiCom 2001), 85–96, Italy, 2001.
6. F. Dai and J. Wu, Distributed dominant pruning in ad hoc wireless networks, *Proc. IEEE Int. Conf. Commun.* (ICC), (CD-ROM), 2003.
7. D.W. Gage, Command control for many-robot systems, *Proc. 19th Annu. AUVS Tech. Symp.*, AUVS-92, 22–24, Hunstville AL, June 1992.
8. W.W. Gregg, W.E. Esaias, G.C. Feldman, R. Frouin, S.B. Hooker, C.R. McClain, and R.H. Woodward, Coverage opportunities for global ocean color in a multimission era, *IEEE Trans. Geosci. Remote Sensing*, 36(5), 1620–1627, 1998.
9. W.R. Heinzelman, A. Chandrakasan, and H. Balakrishnan, Energy-efficient communication protocols for wireless microsensor networks, *Proc. HICSS*, Jan 2000, (CD-ROM).
10. A. Howard, M.J. Mataric, and G.S Sukhatme, Mobile sensor network deployment using potential fields: a distributed, scalable solution to the area coverage problem, *Proc. 6th Int. Symp. Distributed Autonomous Robotics Syst.* (DARS02), 299–308, Japan, 2002.
11. G.T. Huang, Casting the wireless sensor net, *Technol. Rev.*, 50–56, July/August 2003.
12. C. Intanagonwiwat, R. Govindan, and D. Estrin, Directed diffusion: a scalable and robust communication paradigm for sensor networks, *Proc. ACM MobiCom'00*, 56–67, 2000.
13. K. Kar and S. Banerjee, Node placement for connected coverage in sensor networks, *Proc. WiOpt 2003: Modeling Optimization Mobile, ad hoc Wireless Networks*, March 2003.
14. X.-Y. Li, P.-J. Wan, and O. Frieder, Coverage in wireless ad-hoc sensor networks, *IEEE Trans. Computers*, 52, 753–763, 2002.
15. B. Liu and D. Towsley, On the coverage and detectability of wireless sensor networks, *Proc. of WiOpt '03: Modeling Optimization Mobile, ad hoc Wireless Networks*, 2003.
16. M. Marengoni, B.A. Draper, A. Hanson, and R. Sitaraman, A system to place observers on a polyhedral terrain in polynomial time, *Image Vision Computing*, 18, 773–780, 2000.
17. S. Meguerdichian, F. Koushanfar, M. Potkonjak, and M. Srivastava, Coverage problems in wireless ad-hoc sensor networks, *IEEE Infocom 2001*, 3, 1380–1387, April 2001.
18. S. Meguerdichian, F. Koushanfar, G. Qu, and M. Potkonjak, Exposure in wireless ad hoc sensor networks, *Proc. 7th Annu. Int. Conference Mobile Computing Networking* (MobiCom'01), 139–150, July 2001.
19. V. Raghunathan, C. Schurgers, S. Park, and M.B. Srivastava, Energy-aware wireless microsensor networks, *IEEE Signal Process. Mag.*, 19, 40–50, March 2002.
20. J. O'Rourke, *Art Gallery Theorems and Algorithms*, Oxford University Press, Oxford, 1987.
21. S. Slijepcevic and M. Potkonjak, Power efficient organization of wireless sensor networks, *Proc. IEEE Int. Conf. Commun.*, 2, 472–476, Helsinki, Finland, June 2001.
22. D. Tian and N.D. Georganas, A coverage-preserving node scheduling scheme for large wireless sensor networks, *Proc. 1st ACM Workshop Wireless Sensor Networks and Applications*, pp. 32–41, 2002.

23. X. Wang, G. Xing, Y. Zhang, C. Lu, R. Pless, and C.D. Gill, Integrated coverage and connectivity configuration in wireless sensor networks, *Proc. First ACM Conf. Embedded Networked Sensor Syst.* (SenSys'03), pp. 28–39, Nov. 2003.
24. J. Wu and H. Li, On calculating connected dominating set for efficient routing in ad hoc wireless networks, *Proc. 3rd Int. Workshop Discrete Algorithms Methods Mobile Computing Commun.* (in conjunction with MobiCom 1999), 7–14, Aug 1999.
25. J. Wu and F. Dai, Broadcasting in ad hoc networks based on self-pruning, *Proc. 22nd Annu. Joint Conf. IEEE Commun. Computer Soc.* (INFOCOM), (CD-ROM), 2003.
26. F. Ye, G. Zhong, S. Lu, and L. Zhang, Energy efficient robust sensing coverage in large sensor networks, technical report, UCLA, 2002.
27. H. Zhang and J.C. Hou, Maintaining sensing coverage and connectivity in large sensor networks, technical report UIUC, UIUCDCS-R-2003-2351, June 2003.

20
Location Management in Wireless Sensor Networks

Jan Beutel
Swiss Federal Institute of Technology

20.1 Introduction 20-1
20.2 Location in Wireless Communication Systems 20-2
  Abstractions of Location • Metrics of Positioning Systems • Navigation Techniques to Derive Location • General Navigation Solutions Using Trilateration • Example Systems
20.3 Location in Wireless Sensor Networks 20-10
  The Wireless Sensor Network Difference • Locationing Characteristics in Wireless Sensor Networks • Locationing Schemes for Wireless Sensor Networks • Emerging and Open Issues
20.4 Summary 20-21

20.1 Introduction

The most profound technologies are those that disappear. They weave themselves into the fabric of everyday life until they are indistinguishable from it.

Marc Weiser, *The Computer for the 21st Century*

Since Weiser's visionary statement in 1991 [1], many technologies in the mobile and ad hoc arena have evolved. Advances in wireless communications and microsystems integration are enabling ever tighter and finer grained integration of electronic communication devices into the physical world. Wireless sensor networks (WSNs) are at one extreme of the design space, embracing such paradigms as immense scale, self-containment, self-organization, deep embedment, and therefore limited access of and to the devices involved [2]. In this new regime, Moore's law is paving the way for two directions of thrust: (1) the traditional view in which devices become ever more powerful while maintaining their size; and (2) an unprecedented possibility to shrink whole functional systems into tiny scales never anticipated before, e.g., Smart Dust [3].

Under the main prerequisite of WSNs — namely, their stringent energy budget — the first "classical" WSN approaches have focused on efficient, symmetrical ad hoc network configurations, meaning that the devices at both ends of a communication channel were essentially of the same architecture. These early prototypical "sensor nodes" are now constantly maturing and shift the focus from the fabrication of the single device [4–6] toward management of large, heterogeneous systems and architectures and the services embedded into them. This shift in paradigm brings about a second requirement: robustness. In order to be able to scale to large networks consisting of the most heterogeneous clustered devices, mechanisms and services need to be tailored specifically for interoperability and the optimal management of the limited resources available in such nodes. Location in time and space has been identified as a key technology for the successful deployment and operation of context-aware sensor network services [2, 3, 5, 7, 8].

The benefit of location technology is not limited to the subscriber of a network or to network operations like geographic routing, but will enable every wireless enabled device to become a meaningful instrumentation probe. Today, technicians gather distributed environmental information by driving to specific sample sites and making measurements, a time-consuming and inefficient solution, or by installing costly fixed infrastructure in often inaccessible target areas. Sensor data without a complete set of coordinates (this is a time stamp, t, and (x, y, z) location) are almost useless. Although the global positioning system (GPS) offers a solution for localization in an outdoor environment, no such option exists for an indoor setting.

This chapter gives an overview of the issues connected with location management in WSNs. Apart from some general observations applicable to many wireless communication systems, it will concentrate on the peculiarities of the WSN case: simplicity, robustness, and energy awareness.

20.2 Location in Wireless Communication Systems

Wireless communication systems have long been mainstream and are available on many different scales and a multitude of applications. Common to most systems is that they are standardized for control and interoperability reasons and that today they rely heavily on an infrastructure laid out, deployed, and maintained in a designated region of the objective. This infrastructure then supports up- and downlink wireless communication channels usually organized as point-to-point links or, in the case of multiple mobile units, in star topologies to and from base stations. Mobility is supported through automatic channel handoff procedures between different base stations. In the case of handoff, the topology of the network changes and routing to and from mobile devices needs to be reorganized on every change. Apart from standard address-based routing, so-called location-based [9] or geometric [10] routing schemes have been developed.

20.2.1 Abstractions of Location

The concept of location is not limited to the geographic representation of *physical location* with sets of latitude, longitude, and altitude; it is also applicable to *symbolic location* in a nongeographic sense such as location in time or in a virtual information space such as a data structure or the graph of a network. Postal zipcodes and telephone numbers are a good example of abstractions containing designated location information.

Common to all notions of location is the concept that the individual locations are all relative to each other, meaning that they depend on a predefined frame of reference. This leads to a differentiation of the *relative* and *absolute* positioning cases that will be described later in this section.

For the geometric abstractions to be used, assume the network to be a set of vertices $V = \{v_q, v_2, ..., v_n\}$ and edges $E = \{e_q, e_2, ..., e_n\}$ that describe the topology of the network by means of a graph $G(V,E)$. In the case of ranges associated with a certain edge, this can be done by assigning a weight $w(e)$ to every edge $e = (v_p, v_q)$ connecting the vertices v_p and v_q.

20.2.1.1 Location in Space and Time

When talking about physical location in the traditional way, points are usually viewed as three-dimensional coordinates (x, y, z) in a Cartesian reference coordinate system. Of course, many other transformations to other coordinate systems like polar coordinates are equivalent, but the Cartesian system will be considered here. In a three-dimensional system, the Euclidean distance between two points v_p and v_p is defined by

$$\text{dist}_G(v_p, v_q) = \sqrt{(x_p - x_q)^2 + (y_p - y_q)^2 + (z_p - z_q)^2} \tag{20.1}$$

Although other metrics exist [11], e.g., the Manhattan distance, this measure will be used as a basis for the trigonometry introduced in later sections.

Usually, mere (x, y, z) coordinates by themselves are not meaningful for context-aware system services and other information needs to be associated with these *position fixes*. The most straightforward extension is to introduce the fourth dimension, *time*, to be able to specify where and when a certain event took place resulting in sets of (x, y, z, t) for each position fix.

This four-dimensional fix can then be used to put subsequent events into a context frame. For this, a reference frame is necessary because time and location information is useless by itself. In a relative reference frame, different parts of a network would compare their data relative to each other. This enables one to discriminate orientation, distance, time difference, speed, and acceleration between nodes of a network but not to an external reference frame. Depending on the granularity required for the resolution of such data, different requirements on the accuracy of such position fixes can be defined. This allows context-aware applications to behave differently when evaluating not only the position fix but also the desired and achieved accuracy.

When position information is used in reference to a geographic map or a global time reference, the context information can be extended. Here already a single position fix can be put in context of this reference frame vs. a minimum of two position fixes necessary for the relative case. For an Earth-centered view, different reference ellipsoids and geodetic data account for the specific shape of the Earth [12]. In addition, every observer of a geometric system can act as an absolute reference point in an inertial system — namely, its own inertial system as is common on inertial platforms.

20.2.1.2 Relative Positioning

Vectors linking two points give information on their position relative to each other. When no reference points are given, the solution can be rotated and mirrored through an arbitrary axis. Each position added to a system solution reduces the problem one degree of freedom at a time. A minimum of four nodes is necessary to be able to orient a geometric position unambiguously in three-dimensional space (three nodes in two-dimensional space) because when one reference is used per axis of the coordinate system, two mirrored solutions are possible (see Figure 20.2a).

A system of many known positions that has no reference location and/or orientation is only fixed in itself, not in its position in space: If only the position of one point and no orientation is given, the system can be mirrored and rotated through any axis leading through this point. The translatory movement is prohibited by this first known position. When a second position is introduced into the system, the rotation of the system is further restricted to the axis through these two points. A two-dimensional problem would thus still have two possible solutions. A third known position finally fixes the system in two- and three-dimensional space.

A *free node*, v_f, has no *a priori* knowledge of location, but seeks to obtain a position estimate by the positioning methods described later in this subsection. A *settled node*, v_s, initially was a free node, but has calculated a position estimate and can thus serve as an additional position reference to other nodes in the network. The *scope of a node* is defined by the ρ-neighborhood of a vertex $v \in V$ as $\Gamma_\rho(v, G)$ where ρ denotes the maximal hop distance from v on the graph G.

$$\Gamma_\rho(v, G) = \{w \mid \text{dist}_G(w, v) \leq \rho\} \tag{20.2}$$

A relative position can only be given in respect to other points resolving the distances and the geometric configuration, e.g., the topology (see Figure 20.1c). The minimum requirement for relative topology discovery is that all nodes to be considered in such an algorithm must be connected and able to identify each other. Ranging and the exchange of data between nodes further allow one to weight an originally unweighted topology graph.

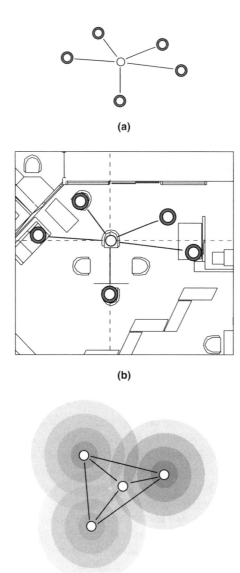

FIGURE 20.1 Absolute positioning data are used with respect to the absolute position of other nodes or to a reference position or a map. Relative positioning can discriminate topology information only in a node's local reference system. (a) Absolute positioning; (b) orienting absolute position within a map context; (c) relative positioning.

20.2.1.3 Absolute Positioning

An absolute position is given in respect to an inertial system and a reference point in this inertial system. It allows one to determine positioning information of disjoint systems independently, in reference to the same point in the inertial system. Points that know their position before the application of a navigation technique are referred to as *anchor nodes*. An anchor or beacon node, v_a, has knowledge of its location through prior configuration or an external reference source such as a GPS receiver. It is important to note that anchors do not derive position through means offered by a network positioning mechanism. They can thus serve as position references.

Absolute positioning allows one to orient the nodes of a network on a map that can be any variable or set of variables representing or assigning values to a geographic location or region, from a single point to an entire planet. This is not necessarily a two-dimensional rendition, such as a paper road map, but can be any data structure that defines a reference frame for more than one node.

20.2.2 Metrics of Positioning Systems

Different metrics can be used to describe quality aspects of positioning systems. Most important are parameters that describe the network setting and the environment. Second in importance are those that relate directly to location and the algorithms used.

The environment is mainly characterized by its signal propagation properties, the node densities, distribution, degree of connectivity, and mobility. Of course the availability of a map is also a property of the environment that can enable absolute positioning while relative positioning focuses on topology discovery only. In certain environments, the accessibility of information might be restricted to certain users or situations.

Navigation solutions can be classified based on accuracy, availability, and cost (hardware, computation, storage and communication requirements, latency, and consumed energy). The size of the database kept up-to-date in every node directly, update rates, and required mathematical accuracy influence computational complexity and therefore position accuracy. In situations in which one must start from scratch, the initial positioning error and the time needed to first fix, e.g., the duration for a first estimate to be available to an application, are the most crucial.

20.2.3 Navigation Techniques to Derive Location

Navigation techniques that use positioning generally consist of three components:

- Identification and data exchange
- Measurement and data acquisition
- Computation to derive location

The various approaches partition these tasks differently across their system components and in some cases no, or only unidirectional, data exchange between nodes is used. Measurements can be made of the distance or the angle of an incoming signal. Methods used include: received signal strength indicator (RSSI); time of arrival (ToA); time–distance of arrival (TDoA); carrier phase and code measurements; ultra wide-band (UWB); ultrasound; and even visible light pulses or the angle of arrival (AoA) of a radio signal. The important thing to note is that *these measurements always have errors* and that *individual measurements are not independent* of each other and are *strongly influenced by the surrounding environment and the transmission system used*. The availability of any one of these physical variables depends largely on the transceiver and antenna architecture available and is beyond this chapter's scope. Therefore, these nonlinear measurements will be referred to as range estimates $\hat{r}$ independent of their source.

20.2.3.1 Hyperbolic Trilateration

Using any one of the range estimation techniques listed earlier, the geometric position can be computed using hyperbolic trilateration. Here three independent range measurements with respect to globally referenced anchor nodes are used to compute the intersection of three circles (see Figure 20.2a). The inputs are the coordinates of the reference nodes $v_i = (x_i, y_i, z_i)$ and the respective range estimates $\hat{r}_i$.

Hyperbolic trilateration is essentially the core idea behind most methods to calculate geometric position and is used in variations in most systems, for example, in GPS (see Subsection 20.2.5.1), or described in detail by Capkun et al. [13]. As described in Subsection 20.2.1, at least three range estimates are necessary to solve for all ambiguities in the two-dimensional and three-dimensional cases. The mathematics of the overdetermined problem are explained in detail in Subsection 20.2.4.

20.2.3.2 Triangulation

Using the trigonometry laws of sines and cosines, the angles of an incoming signal α can be used to compute a triangulation solution (see Figure 20.2b) similar to the method used for hyperbolic trilat-

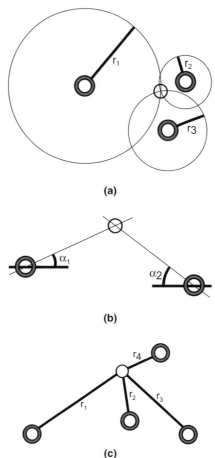

FIGURE 20.2 Three different navigation techniques and their respective inputs, $\hat{r}_i$ and $\hat{\alpha}_i$, as well as the reference nodes v_i. Inputs are drawn in bold lines; the unknown node whose position is to be computed is located in the middle of each figure. (a) Hyperbolic trilateration; (b) triangulation; (c) multilateration.

eration. These angles can be measured at the unknown node or at all reference locations. In the latter case, they need to be oriented correctly, implicating a precisely aligned infrastructure or supplemental measurements between adjacent references. Apart from this alignment problem, AoA measurements require extensive hardware, usually with multiple sectored antennae, making it currently unsuitable for most WSN applications.

20.2.3.3 Multilateration

In the case of dense anchor node populations, a maximum likelihood (ML) estimation can be performed. The prerequisite for this multilateration technique is that at least three reference nodes are visible at every unknown node when performing the position calculation. Using this multilateration method, the position of the unknown node $v_u = (x_u, y_u, z_u)$ is estimated from three or more reference nodes so that the difference between measured and estimated range is minimized for every range estimate $\hat{r}_{i,0}$ incorporated into the solution (see Figure 20.2c).

$$\min \sqrt{(x_i - x_u)^2 + (y_i - y_u)^2 + (z_i - z_u)^2} \quad \forall \hat{r}_{i,0}; i = 0\ldots n \tag{20.3}$$

An iterative algorithm applicable for the multihop case of this technique is explored in Savvides et al. [14, 15]. The geometric constraints involved in multilateration can also be formulated as a linear program

(LP) such as that proposed by Doherty and colleagues [16], in which angular and radial constraints are combined in one model. Common to all these techniques is that they rely on distinct range or angular estimates at different distributed locations. Their approaches are compatible and transformation of data from one technique to another is straightforward.

20.2.4 General Navigation Solutions Using Trilateration

In the case of three of more independent range estimates to known reference nodes, a three-dimensional trilateration problem is to be solved. If these references reside at a known location (anchors or settled nodes), the absolute position can be given in reference to their inertial system, such as is done for the GPS system [17].

For three references, a geometric approximation can be given [18]. Usually, the mathematics in a trilateration problem are overdetermined, meaning that more range estimates than necessary are incorporated into a solution. Furthermore, the errors in the range estimates make it difficult to solve the resulting set of linear equations suggesting techniques that use all available inputs to compute an approximation of position. The starting point here is *least square* methods (MMSE), to find an approximate solution $\hat{x}$ that best satisfies $Ax = b$ with agreement to the range estimates given by b. This means that the length of the residual error, $e = b - A\hat{x}$, is to be minimized.

In general, the trilateration problem can be formulated as follows: given a set of n range estimates $\hat{r}_i$ from the ith reference at position $v_i = (x_i, y_i, z_i)$ to the unknown position v_u, the n nonlinear navigation equations for the true ranges r_i are defined by

$$r_i = dist(v_i, v_u) = \sqrt{(x_i - x_u)^2 + (y_i - y_u)^2 + (z_i - z_u)^2} \quad (20.4)$$

with $r_i = \hat{r}_i + e_i$ and the estimation error e_i. This can be linearized by subtracting the last row, resulting in a system of

$$Ax = b \quad (20.5)$$

with

$$A = -2 \begin{bmatrix} (x_1 - x_n) & (y_1 - y_n) & (z_1 - z_n) \\ (x_2 - x_n) & (y_2 - y_n) & (z_2 - z_n) \\ & \vdots & \\ (x_{n-1} - x_n) & (y_{n-1} - y_n) & (z_{n-1} - z_n) \end{bmatrix} \quad (20.6)$$

$$x = \begin{bmatrix} x_u \\ y_u \\ z_u \end{bmatrix} \quad (20.7)$$

$$b = \begin{bmatrix} r_1^2 - r_n^2 - x_1^2 + x_n^2 - y_1^2 + y_n^2 - z_1^2 + z_n^2 \\ r_2^2 - r_n^2 - x_2^2 + x_n^2 - y_2^2 + y_n^2 - z_2^2 + z_n^2 \\ \vdots \\ r_{n-1}^2 - r_n^2 - x_{n-1}^2 + x_n^2 - y_{n-1}^2 + y_n^2 - z_{n-1}^2 + z_n^2 \end{bmatrix} \quad (20.8)$$

There are different ways to solve the fundamental linear problem in calculus, geometry, and linear algebra. The classical way to proceed is to solve the normal equation

$$A^T X \hat{x} = A^T b \qquad (20.9)$$

with the methods of linear algebra.

However, in order to account for system dynamics, it is common to weight input data according to their reliability using a covariance matrix Σ and to use recursive and sequential filtering techniques to reduce the computational complexity in the event of system changes and the availability of new information. The weighting matrix C is a diagonal matrix that can be derived from the covariance matrix and extends Equation 20.9 to the weighted least squares form of the normal equation

$$A^T C A \hat{x} = A^T C b \qquad (20.10)$$

that can be solved in the form $\hat{x} = \left(A^T C A\right)^{-1} A^T C b$ using QR decomposition or Choleski factorization [12]. Other methods to solve such MMSE problems are estimation using Taylor series or the householder transform.

Systems of equations like the one introduced here can be set up for any neighborhood in a graph. When applied globally, care needs to be taken to choose mathematical methods that scale well enough and can approximate all parts of a network. These are then usually iterative methods.

20.2.5 Example Systems

20.2.5.1 Global Positioning System

The GPS [17] was developed by the U.S. Department of Defense to be used to determine one's exact location and precise time anywhere on Earth at any time. It relies on 28 satellites orbiting on six different planes so that a minimum amount of four can be seen from any point on the planet. Each satellite vehicle (SV) transmits its exact position and precisely synchronized on-board clock time (see Figure 20.3) in a spread spectrum signal. A GPS receiver measures the signal transit times between its point of observation and at least four different satellites whose positions are known to be able to solve for the four unknowns: longitude x, latitude y, altitude z, and time deviation Δt [12]. No reverse uplink communication is necessary between the transceiver and the satellites; the GPS system references the Earth-centered geoid.

Today GPS receivers are highly specialized, high-performance, integrated computing devices that can be integrated into handheld devices and mobile phones. Typical performance figures of a civil receiver for continuous operation are less than 160-mW power consumption; 3-m accuracy; 20-ns timing precision; 41-sec cold start; 3.5-sec hot start; and a 4-Hz update rate at a size (without antenna) of about

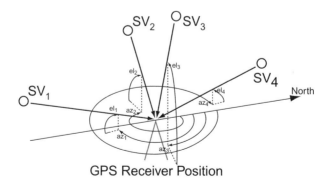

FIGURE 20.3 The position in the GPS system is computed by measuring the signal transit time from a space vehicle (SV) to the receiver and combining these data with the geometry matrix containing azimuth, *az*, and elevation, *el*, of the current SV constellation.

1.9 cm^3. GPS can be augmented with differential navigation data, postprocessing, or other sensor data to achieve even higher performance; this is common in vehicle navigation, land surveying, and civil engineering applications. In differential GPS, the error of a stationary receiver is transmitted to a mobile receiver and used to correct its result under the assumption of similar signal perturbation at both receivers. Consequently, the resulting positioning solution is improved, the closer the two receivers are colocated.

The application of GPS to sensor networks is limited because of the considerable bulk and high power consumption, but mainly because of the requirement for direct line of sight to satellites incorporated in the navigation solution. Furthermore, the cost (economic and in resources utilized) of integrating a GPS receiver is prohibitive for large-scale WSN applications.

20.2.5.2 Cell-Based Positioning

Mainly driven by the U.S. enhanced 911 mandate that requires an ability to locate mobile phones up to about 50 to 100 m, but also by emerging commercial location-based services, cell-based positioning techniques are under detailed investigation [19–21].

Common to all these techniques is that they rely on infrastructure put in place and configured to be used for positioning. Similarly to GPS, the wireless link to base stations can be estimated and, given a sufficient number of base stations, a triangulation solution can be computed. Key issues here are the interoperation of multiple base stations, inefficient surveying prior to deployment, and scalability problems in centralized services because, in most cases, mobile cellular handsets cannot be modified.

The approach used by RADAR [22] goes even further in mapping out the characteristic signal propagation in spaces with wireless LAN coverage to be used for positioning of mobile users. Individual maps per base station are then overlaid to determine a mobile unit's position. The Active Bat system [23] depends heavily on dense ceiling-mounted infrastructure that uses combined RF and ultrasound ranging to achieve 3-cm position accuracy on custom mobile devices — the bats. In contrast, the MIT Cricket location system [24] is highly decentralized and does not rely on central control and computation. The use of combined radio frequency (RF) and ultrasound ranging and identification is similar.

From a sensor network perspective, all methods used in cell-based systems are lacking in terms of available connectivity. In the case of competing cellular operators or regions with scarce base station coverage, only a few references can be incorporated into individual navigation solutions. Here, opportunities offered by dense node populations, and therefore redundant overlay triangulations, cannot be employed.

20.2.5.3 Tagging with RF-ID

Initial radio frequency identification systems (RF-ID) were meant to be passive systems confined to a very short radio range. Powered by the signal emitted by a transmitter, RF-IDs reflect a signal that can be individually identified by a usually colocated receiver. The system employed is essentially cell-based localization, but on a very local level, e.g., well under a meter [25, 26]. Cheap and easy to embed, e.g., in printable product labels or tags, this technology is ideal for simple high-volume applications. More complex active tags allow for data storage and simple computing operations like authentication and sensor data aggregation.

20.2.5.4 The Lighthouse Location System

The Lighthouse location system for Smart Dust [27] is based on direct line of sight between fixed infrastructure laser transmitters and the mobile unit. Each transmitter emits a laser beam that is rotated in two perpendicular axes, thus scanning a whole room. The mobile unit registers the phase and duration of the light flashes and uses this information to intersect three hyperboloids, each in reference to the transmitter's position. This approach is unique because high precision can be achieved with relatively low system, communication, and computational complexity on the sensor nodes. This, the line of sight requirement, and extensive calibration necessary prior to usage make the Lighthouse location system an elegant but rarely applicable solution.

20.3 Location in Wireless Sensor Networks

20.3.1 The Wireless Sensor Network Difference

Usually characterized by themes such as high overall node mobility, considerable power and resource consumption at the nodes, and moderate network sizes, mobile ad hoc networks (MANETs) have brought about quite a change in the traditional, connection-oriented, infrastructure-dominated telecommunications world [28, 29]. Wireless sensor networks have departed from this traditional viewpoint in an even more radical way as they are envisioned to be deeply embedded in very large quantities into the physical world. Myriads of smart devices will be in constant interaction with each other, the surrounding ubiquitous digital infrastructure, and human users. These will range from naive identity tags and simple sensors broadcasting singular values, such as temperature readings, to much more sophisticated computing devices with multiple sensor/actor subsystems, advanced signal processing, data aggregation, storage, and communication capabilities

When becoming ubiquitous and pervasive, such networks will inject highly distributed computation with an immense spatial density into the physical world and require clusters of devices to interoperate seamlessly, invisibly, and autonomously [2]. Characteristic to these networks is the vast number of devices and scales to be considered in one context; these are often densely clustered, but also incur occasional depletion of whole regions — at times completely disconnected from infrastructure — with nodes leaving and joining the network frequently and the possibility of every node talking to everyone else on a local, neighborhood scale. Unpredictable dynamics due to failures and changes in nodes and the environment, as well as deployment in uncontrolled areas with high dynamics and possibly hostile to radio signal propagation, require adaptable networking mechanisms.

Other characteristics of typical WSN nodes are the limited resources available on these ultralow power embedded systems, most notably the limited transmission range and low duty cycle operation of the radio transceivers. Nodes might be reactive and able to wake up on demand [30]; however, in general, in order to save energy on an ultralow duty cycle, the communication links will be offline most of the time.

Targeted for a very long lifespan, integrated into all kinds of everyday objects and building materials, deployed once, and in many cases never collected again or decommissioned [8], the vast majority of nodes will form a quasistatic, multihop network topology (see Figure 20.4) that can be best described by a locally clustered graph, with considerable variations in the local degree. Such a graph will exhibit few to no edges spanning the diameter of the whole graph and many overlapping edges that connect closely located nodes.

In WSNs, the placement of nodes will be mostly arbitrarily, in contrast to most planned ubiquitous computing settings in which infrastructure is carefully laid out. For this reason, grid-based approaches such as those put forward by Bulusu and Estrin [31] do not hold. With some nodes in WSNs mobile, they require autonomous in-network localization techniques that are *simple*, *robust*, and *energy aware*. Other requirements are more abstract, including but not limited to self-configuration, adaptability and scalability across different device types, and networking environments. Without the possibility of relying on infrastructure, methods and algorithms must be derived from the standard principles of distributed computing and graph theory.

The sensing domain has been well investigated over the past years and the first real systems are already starting to have an impact. The promising approach of sensor networks that can already be identified today distributes the instruments into the experiment [32], rather than the experiment residing in an instrumented laboratory setting, thus allowing larger and more realistic settings. In contrast, the world of actuation, especially in a distributed fashion, is still largely untouched; the power of these paradigms when perceived as completely reconfigurable systems with closed-loop sensing and control of the environment remains to be discovered. Distributed sensing and actuation can only be meaningful when applied in a context using position information in time and space. Although the focus here is on the geometric aspects of position in space, services for positioning in space and time have similar algorithmic and implementation specific requirements.

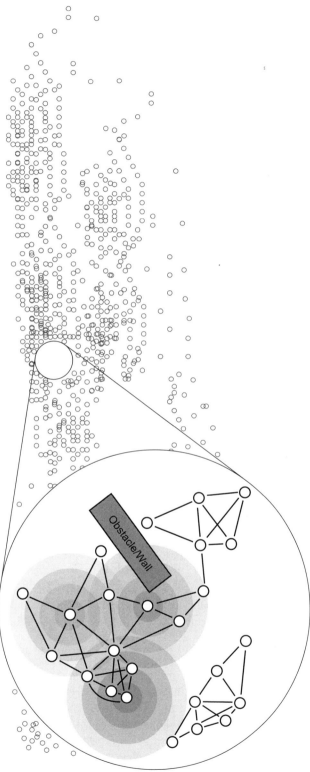

FIGURE 20.4 A view of a typical WSN topology scenario with highly clustered nodes, scarcely populated areas, and obstructed and separated regions. The limited radio range allows only a few overlapping links, thus requiring multihop communications.

Wireless front ends used in WSN nodes are designed with reduced system complexity and optimized for ultralow power duty cycles. Such front ends will not contain very fancy and high performance hardware that can be constantly operated, but rather a set of functional blocks available on demand only [31–35]. A few data transfers that seldom take place compared to traditional telecommunication applications implies that the dwell time when a certain functional block is being used, as well as the transition times from deep sleep to on and back, must be taken into account when designing operating schemes for these systems.

From this perspective, positioning techniques for WSNs must be distributed and local. The exact implementation of each location technique depends largely on the underlying communication network and its capabilities. The required abstraction of location must be defined from the viewpoint of a WSN application and the intended usage of the location information produced by such a service. It appears quite logical to try to integrate network services such as positioning with the data traffic in the network in order to reduce the number of overhead connections and amount of traffic.

20.3.2 Locationing Characteristics in Wireless Sensor Networks

The locationing problem in WSNs can be viewed as a general distributed sensor problem, with sensors that can discover other nodes, estimate ranges between nodes, etc. that serve as position references. Furthermore, the maps and datasets to be used in conjunction with the position information can also be viewed as a form of sensor information — only, in this case, the network and storage resources are the means used to provide this information to context-aware services. In densely populated WSNs, interactions between nodes are abundant. It is therefore necessary to extract and combine the appropriate information in a suitable way to make the most use out of them. To a certain degree a notion of *living with errors* must be adopted because the formal problem of distributed location is quite difficult and resources available per node in a WSN are finite. An important requirement for locationing in WSNs is a distributed approach that minimizes computational, and especially communication, overhead, and is robust enough to survive disconnection. A hybrid approach integrating networking and positioning is therefore highly applicable.

In WSNs, naming data, not nodes, and the organization around spatial and temporal coordinate systems require appropriate abstractions and programming models for location context to be developed. First approaches in this direction show that such systems can be viewed as a tuple space model, distributed databases, or even loosely coupled parallel computing structures [36].

The following subsection points to some sensor network-specific issues that can be utilized to improve performance in positioning techniques: network topology, range errors and quantization, and different filtering techniques. The subsequent subsection introduces some algorithms.

20.3.2.1 Using the Network Topology for Positioning

Sensor networks offer regions with high node densities and ad hoc networking mechanisms that fundamentally allow every node to communicate with every other node. The key idea is now to make use of the redundant network connections available when viewing the network as a fully connected graph.

Formally, a complete graph with n vertices has $\frac{1}{2}n(n-1)$ edges [37]. In an ad hoc network setting, however, not all nodes are visible to all others due to the reduced transmission range of a single node resulting in a structure such as those seen in Figure 20.4 and Figure 20.5(a). For efficient routing algorithms it is often practical to reduce the complexity of the available connections to an appropriate set of connections such as a planar graph or a dominating set [10, 38]. Geometric routing algorithms commonly use a planar graph such as the Gabriel graph shown in Figure 20.5(b). In heterogeneous network settings, the available network links might be reduced even further, to so-called backbone links, allowing fast long-haul data transfer on commonly used routes and reducing the amount of edges that can be used for positioning even further.

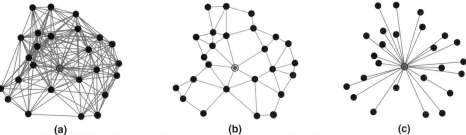

FIGURE 20.5 Using WSN specific network topology for positioning allows heavy overlaying of range estimates for individual trilaterations. (a) Complete graph for a given maximum transmission range; (b) planar Gabriel graph appropriate for geometric routing; (c) all ranges visible to the center node that can be incorporated into a combined trilateration solution.

A trilateration solution such as was discussed earlier (see subsection 20.2.3) achieves the highest position accuracy when many independent range estimates to reference nodes are used. An example scenario is shown in Figure 20.5(c). It is easy to note that the desired topology here is quite different from the routing scenario in Figure 20.5(b). A high node density in a WSN setting eases positioning because the peculiarities of the overdetermined topologies can be put to use and a redundant trilateration solution can be overlaid. This is exactly the reverse setting from a traditional cellular network, where mobile units only talk to a single base station. When more mobile units crowd into a certain area, the burden on the base station increases; the additional communication links do not help in the positioning problem because the mobile units do not communicate with each other directly.

A qualitative simulation such as the one shown in Figure 20.6 yields quite acceptable positioning errors and variances, even in the case of very high range errors. The specific improvement that can be achieved for overdetermined topologies can be seen in Figure 20.7(b) with a characteristic, nearly exponential reduction in the position error. Independent publications have agreed on a rule of thumb of using at least five to seven nodes per trilateration to achieve an acceptable accuracy for WSN applications [15, 16, 41]. From this perspective, it is desirable for a positioning service in WSNs to be able to use many more independent range estimations than are offered by the routing grid alone. This could be achieved by enforcing or switching over connections or by transceivers that can assess multiple channels (range estimates) without actually transmitting data over these channels [5, 30, 39].

20.3.2.2 Range Errors and Quantization

Range estimation based on RF propagation techniques is problematic. Especially indoors, the environment is not predictable with multipath, fading, interference, and shading effects abundant. Especially in the case of WSNs, in which the lowest power consumption and therefore the lowest transmit power is of primary concern, obstacles in the line of sight path, noise, and radio interference hinder reliable estimation using a channel model. Typical mobile radio channel models account for signal fading in the

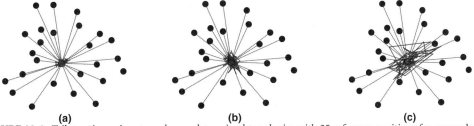

FIGURE 20.6 Trilateration using strongly overdetermined topologies with 25 reference positions for one unknown located in the center. MMSE results are shown for successive iterations on 50 sets of uncorrelated range estimates. The three figures differentiate only in the error assumed for the range estimates. (a) Range error = 20%; (b) range error = 50%; (c) range error = 80%.

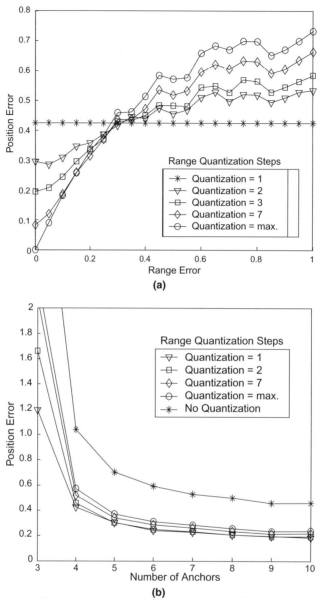

FIGURE 20.7 Quantization effects in range estimates influence local positioning results. Simulations were performed for a limited range-estimate resolution. (a) With growing uncertainty of the range estimates, their resolution becomes less important; here, four anchor nodes were used in the trilateration solution. (b) Making use of overdetermined topologies such as those in Figure 20.6 by using more anchor nodes than necessary shows improved accuracy. The independent range error used here was 50%.

order of $1/r^2$ to $1/r^4$; in WSNs it is often much worse. Sensor nodes are anticipated to be small, simple, and robust devices, down to the scale of Smart Dust [3, 40], that carry lightweight resources only and make it impossible to use complex channel estimation technology. The uncoordinated placement of nodes in environments that are often harmful to radio signals adds further to the unpredictability of range estimates.

It is therefore important to make the best out of the situation and pursue a strategy employing properties other than exact channel measurements in WSNs. The previous section suggested using as many vectors as possible for trilateration solutions. What can be done in the case of large unpredictable errors and high variance of range estimates? What if a sensor node can only discriminate very few discrete steps in a range estimate? What if it can only detect near and far? Figure 20.7 depicts the influence of reduced quantization in overdetermined trilateration using MMSE. Figure 20.7(a) shows that, for large range errors, the resulting position error actually increases with the quantization steps. This means that in the case of unreliable channel estimates, it is best not to spend too much effort on range estimation, but rather to use only topological information for positioning.

20.3.2.3 Influence of Border Effects and Filtering

A well-known influence on accuracy in GPS is the geometric dilution of precision (GDOP) [12, 17]. Simple geometry using erroneous inputs reveals that certain geometric constellations of the reference positions result in larger position errors than others. This is easy to see when the optimum constellation is considered. Assuming similar but independent perturbation on all range estimates, the best GDOP is achieved when all references are located on a circumventing sphere in equidistant spacing from the unknown location; the worst GDOP occurs in the case of all reference positions located on a straight line. In the case of the exact solution using four reference positions, this would result in a tetrahedron spanning the reference locations with the unknown in the middle.

The implication of this observation is twofold. First, edges and border regions of a network topology must be treated differently when weighting the trilateration solution than regions that are surrounded by reference points on all sides. Second, in the case of an overdetermined problem, care should be taken to select sets of equidistantly located references at a somewhat similar distance from the unknown. If multiple such sets can be selected because of a strongly overdetermined problem, the resulting position estimates can be combined in a weighted filter subsequently. Such an approach will eliminate position estimates such as those shown in Figure 20.8a through Figure 20.8c that exhibit excessively large errors, mostly due to bad GDOP.

Different bounds on the inputs and the results can be applied as well. The range estimates, $\hat{r}_i$, can be bounded by the fact that $\hat{r}_i \leq r_{max}$ with r_{max} denoting the maximum transmission range. On subsequent iterations of the position of the same node, it is also common to bound the resulting position estimates by $dist(v_i, v_{i+1}) \leq r_{max}$. Especially in the case of high mobility and dynamics in system checking, the resulting topology for infeasible results and limiting the maximum deviation of successive iteration results is an easily implemented and efficient way of improving robustness. For example, a mobile node attached to a vehicle and given a minimum speed will not exceed a certain acceleration, jump, or perform sharp turns, but rather will follow a smooth path (Figure 20.9).

Heuristics and iterations improve the resulting accuracy, but because they depend on independent sets of inputs, they add to the complexity, storage, communication, and computation requirements. Compared to storing an extensive series of inputs and computing a position estimate using heuristics, filtering over a weighted sum of individual position estimates is preferable when lightweight services for WSNs are required.

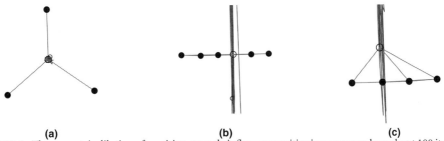

(a) (b) (c)

FIGURE 20.8 The geometric dilution of precision strongly influences positioning accuracy; here about 100 iterations were performed using a 30% range error. (a) Optimal GDOP with equally spaced references; (b) suboptimal GDOP with all nodes aligned; (c) worst-case GDOP setting. (In 20.8b and 20.8c, overshooting position errors are cut off in the illustration.)

(a)   **(b)**

FIGURE 20.9 Based on the positioning problem introduced in Figure 20.5c, a simple filtering mechanism can select nodes according to range properties (a) or (b) good GDOP properties to be used in one trilateration problem. If available, multiple position estimates can then be combined by a weighted sum or FIR filter.

20.3.3 Locationing Schemes for Wireless Sensor Networks

Subsection 20.2.3 introduced the three key components of positioning — *identification and data exchange*, *measurement and data acquisition*, and *computation to derive location* — that need to be concurrently executed at the nodes using a *cooperative ranging* scheme (see Figure 20.10). In network-based positioning, this leads to two distinct problem areas that need to be accounted for apart from the WSN-specific characteristics discussed in the previous sections. These two are the initial *start-up* problem solving the sparse anchor problem and the *convergence* on successive iterations leading to accurate positioning results.

This subsection introduces the hybrid approach of cooperative ranging integrating networking and positioning that offers *robust start-up* and *precision on-demand position updates* as a suitable means for WSNs. The start-up phase addresses the sparse anchor node problem by cooperatively spreading awareness of the anchor nodes' positions throughout the network, thus allowing all nodes to arrive at initial position estimates. These initial estimates are not expected to be very accurate, but are useful as rough approximations. The precision on-demand position update phase of the algorithm then uses the results of the start-up algorithm to improve upon these initial position estimates. Here the range error and convergence problem is addressed.

20.3.3.1 Cooperative Ranging

The cooperative ranging scheme allows one to combine the different tasks necessary for network-based positioning to operate concurrently on many nodes. Once the neighborhood information is available at a specific node, updates of *ranging*, *updating* and *positioning* (see Figure 20.10), which depend on each other to some extent, can be performed sequentially and out of order. Every node in the network is required to keep a database of this neighborhood information containing neighbors' position estimates and the range to these neighbors. The size of this database depends on the requirements of the positioning service requested as well as on the state of the network, i.e., amount and geometry of neighboring nodes.

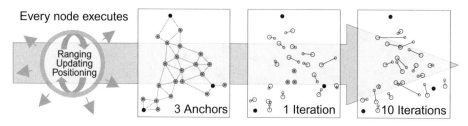

FIGURE 20.10 With the three phases of cooperative ranging data exchange, data acquisition and positioning operating at ever node in a WSN all nodes are able to compute position estimates efficiently, even when out of range of the three anchors shown here. After one iteration, the nodes in the center have not been able to acquire enough data to compute a position; this changes when first-order neighbors become settled nodes and act as position references (seen on right-hand side of figure).

This scheme constitutes the basis of all positioning schemes and can be adapted to fit specific needs or service levels as will be shown for the topology discovery, start-up, and precision on-demand update phases next.

20.3.3.2 Topology Discovery

A node discovering its topology by exchanging link pairs within its scope receives range measurements referring to a large number of neighboring nodes. This information once again can be used to construct the topology graph of the network and to derive relative location.

The *assumption-based coordinates* (ABC) algorithm [41] determines the locations of unknown nodes one at a time in the order in which they establish communication, making assumptions where necessary and compensating for errors through corrections and redundant calculations as more information becomes available. These assumptions are needed at first in order to deal with the underdetermined set of equations presented by the first few nodes. This description of the general algorithm assumes the perspective of node v_0 and can be solved successively, reducing the amount of necessary computations (Figure 20.11).

The algorithm begins with the assumption that v_0 is located at the origin (0,0,0). The first node to establish communication with v_0, v_1, is assumed to be located at $(r_{01},0,0)$, where $r_{01} = dist(v_0, v_1)$. The location of the next node, v_2, can then be explicitly solved for, given two assumptions: the square root involved in finding y_2 is assumed to yield a positive result, and z_2 is assumed to be 0.

$$x_2 = \frac{r_{01}^2 + r_{02}^2 - r_{12}^2}{2r_{01}}, \quad y_2 = \sqrt{r_{02}^2 - x_2^2}, \quad z_2 = 0 \qquad (20.11)$$

The next node, v_3, is handled much like v_2, except that only one assumption is made: the square root involved in finding v_3 is positive.

$$x_3 = \frac{r_{01}^2 + r_{03}^2 - r_{13}^2}{2r_{01}}, \quad y_3 = \frac{r_{03}^2 - r_{23}^2 + x_2^2 + y_2^2 - 2x_2 x_3}{2y_2}, \quad z_3 = \sqrt{r_{03}^2 - x_3^2 - y_3^2} \qquad (20.12)$$

From this point forth, the system of equations used to solve for further nodes is no longer underdetermined, so a standard MMSE algorithm can be employed for each new node. Under ideal conditions, this algorithm thus far will produce a topologically correct map with an orientation relative to the local node v_0. A similar approach to derive a start-up configuration with a local coordinate system (LCS) established at every node is followed by Capkun and colleagues [13]. Here a geometric transformation used to transform multiple LCSs into one oriented and networked coordinate system is described as well.

20.3.3.3 Robust Start-Up Positioning Scheme

The goal here is to have a service available at all times on every node, no matter how small the node, and independent of dedicated resources within a network node. Tight integration with network transport and efficient operation are the main aspects here. For many applications, such a simple, lightweight service that might not be able to give high accuracy, account for dynamics, and only support relative positioning is sufficient.

The purpose of the start-up phase it to solve the sparse anchor problem, which comes from the need for at least four reference points with known locations in a three-dimensional space in order to uniquely determine the location of an unknown object. Too few reference points result in ambiguities that lead to underdetermined systems of equations. For initial start-up in a multihop ad hoc network, a mandatory requirement is that a network be connected and, for generating a first unambiguous position, estimate a sufficient node degree also. In the simple case of a single node, this translates to a minimum degree of

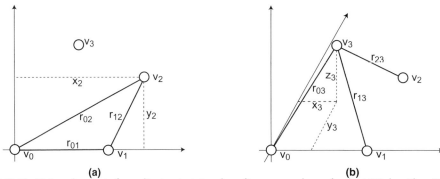

FIGURE 20.11 Fixing a local set of coordinates starts topology discovery, as shown for the ABC algorithm. By adding more nodes in the ABC algorithm, a three-dimensional reference system relative to origin node v_0 can be built up successively.

three for the unknown node (see Figure 20.8a). When assimilating a WSN topology to be a wire frame with nodes acting as hinges [42], this can be extended to the entire connected network.

However, successful start-up does not yet imply that accurate positions can be computed. As was shown earlier, different factors influence accuracy and even the convergence of the positioning problem. Especially bad GDOP and large range estimate errors in multihop scenarios can add up and cause divergence of the positioning results.

In order to support only the basics necessary for a successful start-up, different suggestions have been made by Savarese et al. [41, 43] (Hop-TERRAIN); Niculescu and Nath [42] (DV-hop); and Capkun et al. [13] (LVS, LRG). The basic idea here is to use a hop count to the nearest references that can be derived from the local topology cache of every node in order to estimate an extended range from the unknown to the reference point. Usually, most nodes will start without known locations and only a few randomly distributed anchors will exist. It is therefore highly unlikely that any randomly selected node in the network will be in direct range with a sufficient number of anchor nodes to derive its position. Hop-TERRAIN solves this problem by trading off accuracy for consistency. The start-up phase will provide rough guesses of the nodes' initial positions. Savarese and coworkers [43] have shown that this is good enough as an input to the second phase for refining the position estimates.

The Hop-TERRAIN algorithm works as follows: at large time intervals, each of the anchor nodes launches the Hop-TERRAIN algorithm by initiating a broadcast containing its known location and a hop count of 0. All of the one-hop neighbors surrounding the anchor will record the anchor's position and a hop count of 1. Then they perform another broadcast containing the anchor's position and a hop

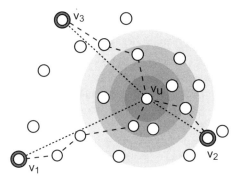

FIGURE 20.12 The Hop-TERRAIN start-up phase uses the hop-count over all intermediate nodes to the closest anchor nodes v_i to estimate extended ranges to be used in computation of initial position estimates for all unknown nodes v_u. The maximum radio range of node v_u is given by the shaded circle. Resulting range estimates $\hat{r}_1$, $\hat{r}_2$, and $\hat{r}_3$ are 4, 23, and 3, respectively.

count of 1. This process continues until each anchor's position and an associated hop count value have been spread to every node in the network, (see Figure 20.12). It is important that nodes receiving these broadcast packets only store and rebroadcast a certain anchor's position if they have not received such a packet with the same or smaller hop count before.

Algorithm 1: Hop-TERRAIN

---
for all $v_u, v_s \in G$ **do**
  while receiving position packet **do**
  if $v_a \notin \Gamma\rho\ (v, G)$ or lower hop count received **then**
  store hop count
  broadcast position packet with (hopcount + 1)
  end if
  if $|v_a \in \Gamma\rho\ (v, G)| \geq$ (dimension + 1) **then**
  estimate current position using MMSE
  end if
  end while
end for

Once a node has received data regarding at least four anchor nodes, it is able to perform a trilateration to estimate its location, (see Figure 20.2a). This will, of course, only be a very rough estimation of the actual positions.

The DV-hop algorithm proposed by Niculescu and Nath [42] works very similarly and is used to approximate location for all nodes in an isotropic environment. It allows one to orient a network on a plane with limited mobility nodes and works with few references. The claim here is that DV-hop allows distributed efficient position awareness for non-GPS-enabled nodes providing an average accuracy of less than one radio hop from the true location. The difference is that a hop count also takes place between reference locations (anchors); in combination with the true distance *a priori* known between these references, an individual weighting metric for each hop is computed rather than an average hop length as is used in Hop-TERRAIN.

20.3.3.4 Precision On-Demand Position Updates

With the initial position estimates of Hop-TERRAIN in the start-up phase, the objective of the refinement phase is to obtain more accurate positions, using the estimated ranges between nodes (see Figure 20.6). This paradigm can be abstracted as a scalable on-demand service dependent on the available infrastructure for range estimation, the environment, and the available computing resources.

In the start-up phase, the Hop-TERRAIN algorithm floods the anchor's positions through the network and nodes record the hop count of the shortest path to each anchor. Hop-TERRAIN also records the neighbor IDs on the shortest path. These IDs are collected in a set of potentially sound neighbors. When the size of this set reaches four in three dimensions (three in two dimensions), a node declares itself settled and may enter the refinement phase. The neighbors of the settled node add its ID to their sets and may in turn become settled, etc.

Refinement is an iterative algorithm in which the nodes update their positions in a number of steps. At the beginning of each step, a node broadcasts its position estimate, receives positions and corresponding range estimates from its neighbors, and computes a least squares triangulation solution to determine its new position. Often, the constraints imposed by the measured distances will force the new positions toward the true location of the node. Refinement stops and reports the final result once updates become small.

Without any prevention, the large errors induced by RSSI measurements will propagate quickly throughout the network. Therefore, a confidence metric can be included in the refinement algorithm. Instead of solving the unweighted least squares, the weighted version, as introduced in Subsection 20.2.4 is solved. Each node assigns a confidence weight between 0 and 1 to its position estimate. Anchors immediately start with a confidence value of 1; unknown nodes start with a low value (0.1) and may raise their confidence after subsequent refinement iterations. Whenever a node performs a successful triangulation, it sets its confidence level to the average of its neighbors' levels. In general, this will raise the confidence level. Savarese and colleagues [43] have shown that including confidence levels improves the refinement phase considerably. A general scheme for efficient and robust precision on-demand location awareness that can be derived from these ideas is shown in overview in Algorithm 20.2.

Algorithm 2: Precision On-Demand Positioning

for all $v_s \in G$ **do**
  while receiving position packet from $\Gamma\rho$ **do**
  establish connectivity graph, store, and exchange neighborhood information
  count hops to next references
  estimate span between hops and weight the hop count
  estimate current position using MMSE
  bound and filter position estimates based on the neighborhood topology
  check error against reference positions
  if error too large **then**
  inc(ρ)
  end if
  end while
end for

Further improvements can be made by detecting that a single node is ill connected. If the number of neighbors is less then four in three-dimensional and less than three in two-dimensional space, then the node is ill connected. However, detecting that a group of nodes is ill connected is more complicated because some global perspective is necessary. A heuristic can be employed that operates in an ad-hoc fashion, yet is able to detect most ill-connected nodes. The underlying premise for the heuristic is that a sound node has independent references, that is, the multihop routes to the anchors have no link in common.

When subsequent updates are undertaken using *cooperative ranging* schemes such as the refinement scheme described earlier, different parameters can be adapted according to the requirements of the applications using the positioning information. To achieve the goal of high local connectivity for optimal trilateration results, it is important to expand the size of the scope of each node (also termed the location reference group (LRG) by Capkun and coworkers [13]). With growing visibility of the network, more nodes as well as more references can be taken into account in each local wireframe model at a cost of higher storage and computational requirements.

20.3.4 Emerging and Open Issues

Looking over the past accomplishments in locationing techniques applicable for WSNs, many achievements have been made and the initial foundations have been laid based on graph theoretical and distributed approaches. Because transceiver architectures have not been especially designed for positioning, generation of accurate range estimates is still a problem, although several ways to circumvent this have been presented. It is unclear, however, how to detect and treat obstacles or obstructions in the line of sight path of signals, as well as mobility. Signals in this direction are certainly range-free localization schemes, as proposed by He and colleagues [44], or exact models of the environment [45].

So far most of the research in positioning has been application driven and bound; analysis of the theoretical limits of positioning are only starting [46], but the results are promising. A strong debate is also ongoing on location privacy. This is not so much an issue in WSNs as in pervasive computing, largely because single-sensor reading cannot reveal much context; however, privacy becomes an important issue when attributed with sensor data and user identities [47].

20.4 Summary

This chapter discussed the issues of location management specific to WSNs. The key concepts of network-based positioning mechanisms were introduced in comparison to other traditional wireless communication technologies. Some example algorithms were detailed and their applicability for the special WSN case were discussed. From this, the reader should be able to gain sufficient knowledge of the problems and opportunities that lie in location-based services for this class of networking systems.

Acknowledgments

The work presented in this chapter was supported (in part) by the National Competence Center in Research on Mobile Information and Communication Systems (NCCR-MICS), a center supported by the Swiss National Science Foundation under grant number 5005-67322.

References

1. M. Weiser, The computer for the 21st century, *Sci. Am.*, 265(3), 66–75, Sept. 1991.
2. D. Estrin et al., Connecting the physical world with pervasive networks, *IEEE Pervasive Computing*, 1(1), 59–69, 2002.
3. J. Kahn, R. Katz, and K. Pister, Next century challenges: mobile networking for Smart Dust, in *Proc. 5th ACM/IEEE Annu. Int. Conf. Mobile Computing Networking (MobiCom 99)*. ACM Press, New York, Aug. 1999, 271–278.
4. J. Hill et al., System architecture directions for networked sensors, in *Proc. 9th Int. Conf. Architectural Support Programming Languages Operating Syst. (ASPLOS-IX)*. ACM Press, New York, Nov. 2000, 93–104.
5. J. Rabaey et al., PicoRadio supports ad hoc ultra-low power wireless networking, *IEEE Computer*, 33(7), 42–48, July 2000.
6. J. Beutel, O. Kasten, and M. Ringwald, BTnodes — a distributed platform for sensor nodes, in *Proc. 1st ACM Conf. Embedded Networked Sensor Syst. (SenSys 2003)*. ACM Press, New York, Nov. 2003, 292–293.
7. J. Elson, Time synchronization in wireless sensor networks, Ph.D. dissertation, Dept. Computer Sciences, Univ. of California, Los Angeles, 2003.
8. Committee on Networked Systems of Embedded Computers, Computer Science and Telecommunications Board, Division on Engineering and Physical Sciences, National Research Council, Ed., *Embedded Everywhere: A Research Agenda for Networked Systems of Embedded Computers*. National Academy Press, Washington, D.C., 2001.
9. Y. Ko and N. Vaidya, Location-aided routing (LAR) in mobile ad hoc networks, in *Proc. 4th ACM/IEEE Annu. Int. Conf. Mobile Computing Networking (MobiCom 98)*. ACM Press, New York, 1998, 66–75.
10. F. Kuhn, R. Wattenhofer, and A. Zollinger, Worst-case optimal and average case efficient geometric ad hoc routing, in *Proc. 4th ACM/IEEE Annu. Int. Conf. Mobile Computing Networking (MobiHoc 2003)*. ACM Press, New York, June 2003, 267–278.
11. M. deBerg et al., *Computational Geometry: Algorithms and Applications*, 2nd ed., Springer, Berlin, 2000.
12. G. Strang and K. Borre, *Linear Algebra, Geodesy and GPS*. Wellesley–Cambridge Press, Wellesley, MA, 1997.
13. S. Capkin, M. Hamdi, and J.-P. Hubaux, GPS-free positioning in mobile ad hoc networks, *Cluster Computing*, 5(2), 157–167, 2002.

14. A. Savvides, C. Han, and M. Strivastava, Dynamic fine-grained localization in ad hoc networks of sensors, in *Proc. 7th ACM/IEEE Annu. Int. Conf. Mobile Computing Networking (MobiCom 2001)*, ACM Press, New York, July 2001, 166–179.
15. A. Savvides, H. Park, and M. Srivastava, The bits and flops of the N-hop multilateration primitive for node localization problems, in *Proc. 1st ACM Int. Workshop Wireless Sensor Networks Applications (WSNA 2002)*. ACM Press, New York, Sept. 2002, 112–121.
16. L. Doherty, K. Pister, and L. El-Ghaoui, Convex position estimation in wireless sensor networks, in *Proc. 20th Annu. Joint IEEE Conf. Computer Commun. Soc. (Infocom 2001)*, 3, 1655–1663, 2001.
17. T. Logsdon, *The Navstar Global Positioning System*. Van Nostrand Reinhold, New York, 1992.
18. J. Caffery, A new approach to the geometry of TOA location, in *Proc. 52nd IEEE Vehicular Technol. Conf. Fall 2000 (VTC 2000)*, J. Weber, J. Arnbak, and R. Prasad, Eds., 4, 1943–1949, 2001.
19. Z. Zhang, *Handbook of Wireless Networks and Mobile Computing*, John Wiley & Sons, New York, 2002, 27–50.
20. J. Caffery and G. Stuber, Overview of radiolocation in CDMA cellular systems, *IEEE Commun. Mag.*, 36(4), 38–45, Apr. 1998.
21. C. Drane, M. Macnaughtan, and C. Scott, Positioning GSM telephones, *IEEE Commun. Mag.*, 36(4), 46–54, 1998.
22. P. Bahl and V. Padmanabhan, RADAR: an in-building RF-based user location and tracking system, in *Proc. 19th Annu. Joint IEEE Conf. Computer Commun. Soc. (Infocom 2000)*, 2, 775–784.
23. M. Addlesee et al., Implementing a sentient computing system, *IEEE Computer*, 34(8), 50–56, Aug. 2001.
24. N. Priyantha, A. Chyakraborty, and H. Balakrishnan, The Cricket location-support system, in *Proc. 6th ACM/IEEE Annu. Int. Conf. Mobile Computing Networking (MobiCom 2000)*, ACM Press, New York, 2000, 32–43.
25. L. Ni et al., LANDMARC: indoor location sensing using active RFID, in *Proc. 1st IEEE Int. Conf. Pervasive Computing Commun. (PerCom 2003)*. Mar. 2003, 407–415.
26. F. Siegemund and C. Florkemeier, Interaction in pervasive computing settings using Bluetooth-enabled active tags and passive RFID technology together with mobile phones, in *Proc. 1st IEEE Int. Conf. Pervasive Computing Commun. (PerCom 2003)*, Mar. 2003, 378–387.
27. K. Rohmer, The lighthouse location system for Smart Dust, in *Proc. 1st ACM/USENIX Conf. Mobile Syst. Applications, Services (MobiSys 2003)*, ACM Press, New York, May 2003, 15–30.
28. T. Imielinski and H. Korth, Eds., *Mobile Computing*, Kluwer Academic Publishers, Norwell, MA, 1996.
29. C. Perkins, Ed., *Ad Hoc Networking*, Addison–Wesley, Boston, 2001.
30. E. Shih, P. Bahl, and M. Sinclair, Wake on wireless: an event-driven energy-saving strategy for battery-operated devices, in *Proc. 6th ACM/IEEE Annu. Int. Conf. Mobile Computing Networking (MobiCom 2001)*, ACM Press, New York, Sept. 2002, 160–171.
31. N. Bulusu, J. Heidemann, and D. Estrin, GPSless low-cost outdoor localization for very small devices, *IEEE Personal Commun.* 7(5), 28–34, Oct. 2000.
32. A. Mainwaring et al., Wireless sensor networks for habitat monitoring, in Proc. *1st ACM Int. Workshop Wireless Sensor Networks Applications (WSNA 2002)*, ACM Press, New York, Sept. 2002, 88–97.
33. M. Kubisch et al., Distributed algorithms for transmission power control in wireless sensor networks, in *Proc. 2002 IEEE Wireless Commun. Networking Conf. (WCNC 2002)*, 1, 558–563.
34. A. Willig et al., Altruists in the PicoRadio sensor network, *Proc. 4th IEEE Int. Workshop Factory Commun. Syst. (WFCS 2002)*, August 2002, 175–184.
35. M. Leopold, M. Dydensborg, and P. Bonnet, Bluetooth and sensor networks: a reality check, *Proc. 1st ACM Conf. Embedded Network Sensor Syst. (SenSys 2003)*, ACM Press, New York, Nov. 2003, 103–113.

36. P. Bonnet, J. Gehrke, and P. Seshadri, Querying the physical world, IEEE Personal Commun. 7(5), 10–15, Oct. 2000.
37. N. Biggs *Discrete Mathematics*, revised ed., Oxford University Press, New York, 1989.
38. J. Wu, *Handbook of Wireless Networks and Mobile Computing*, John Wiley & Sons, New York, 2002, 425–450.
39. C. Schurgers et al., Topology management for sensor networks: exploiting latency and density, in *Proc. 3rd ACM Int. Symp. Mobile ad hoc Networking Computing (MobiHoc 2002)*, ACM Press, New York, June 2002, 135–145.
40. L. Doherty et al., Energy and performance considerations for Smart Dust, *Int. J. Parallel Distributed Syst. Networks*, 4(3), 121–133, 2001.
41. C. Savarese, J. Rabaey, and J. Beutel, Locationing in distributed ad hoc wireless sensor networks, in *Proc. 2001 Int. Conf. Acoustics, Speech, Signal Process. (ICASSP 2001)*, 4, 2037–2040.
42. D. Niculescu and B. Nath, Ad hoc positioning system (APS), in *Proc. IEEE Global Telecommun. Conf. (GLOBECOM 2001)*, Nov. 2001, 2926–2931.
43. C. Savarese, J. Rabaey, and K. Langendoen, Robust positioning algorithms for distributed ad hoc wireless sensor networks, in *Proc. 2002 USENIX Annu. Tech. Conf.*, USENIX Assoc., Berkeley, CA, June 2002, 317–327.
44. T. He et al., Range-free localization schemes in large scale sensor networks, in *Proc. 9th ACM/IEEE Annu. Int. Conf. Mobile Computing Networking (MobiCom 2003)*, ACM Press, New York, September 2003, 81–95.
45. R. Harle, A. Ward, and a. Hopper, Single reflection spatial voting, in *Proc. 1st ACM/USENIX Conf. Mobile Syst. Applications, and Services (MobiSys 2003)*, ACM Press, New York, May 2003, 1–15.
46. R. Bischoff and R. Wattenhofer, Analyzing connectivity-based multihop ad hoc positioning, *2nd IEEE Int. Conf. Pervasive Computing and Communication, (PerCom 2004)*, March 2004.
47. J. Warrior, E. McHenry, and K. McGee, They know where you are, *IEEE Spectrum*, 20–25, July 2003.

21
Positioning and Location Tracking in Wireless Sensor Networks

Yu-Chee Tseng
National Chiao-Tung University

Chi-Fu Huang
National Chiao-Tung University

Sheng-Po Kuo
National Chiao-Tung University

21.1 Introduction 21-1
21.2 Fundamentals 21-2
    ToA, TDoA, and AoA • Positioning by Signal Strength
21.3 Positioning and Location Tracking Algorithms 21-4
    Trilateration • Multilateration • Pattern Matching • Location Tracking • Network-Based Tracking
21.4 Experimental Location Systems 21-10
    Active Badge and Bat • Cricket • RADAR and Nibble • CSIE/NCTU Indoor Tour Guide
21.5 Conclusions 21-12

21.1 Introduction

Locations of devices or objects are important information in many applications. This is particularly true for wireless sensor networks, which usually need to determine devices' context. For outdoor environments, the most well-known positioning system is the global positioning system (GPS) [5]. This positioning system uses 24 satellites set up by the U.S. Department of Defense to enable global three-dimensional positioning services; it has two levels of accuracy: stand positioning service (SPS) and precise positioning service (PPS). The accuracy provided by GPS is around 20 to 50 m.

In addition to the GPS system, positioning can also be done using some wireless networking infrastructures. Taking the PCS cellular networks as an example, the E911 emergency service requires determining the location of a phone call via the base stations of the cellular system. Several location estimation models, such as angle of arrival (AoA); time of arrival (ToA); received signal strength (RSS); phase of arrival (PoA); and assisted global positioning system (A-GPS), are widely used in cellular networks and wireless sensor networks.

Much work has been dedicated recently to positioning and location tracking in the area of wireless ad hoc and sensor networks. The purpose of this chapter is to review the recent progress in this direction. GPS is not suitable for wireless sensor networks for several reasons:

- It is not available in an indoor environment because satellite signals cannot penetrate buildings.
- For more fine-grained applications, higher accuracy is usually necessary in the positioning result.
- Sensor networks have their own battery constraint, which requires special design.

Location information can be used to improve the performance of wireless networks and provide new types of services. For example, it can facilitate routing in a wireless ad hoc network to reduce routing overhead. This is known as geographic routing [7, 9]. Through location-aware network protocols, the number of control packets can be reduced. Service providers can also use location information to provide some novel location-aware or follow-me services. The navigation system based on GPS is an example. A user can tell the system his destination and the system will guide him there. Phone systems in an enterprise can exploit locations of people to provide follow-me services. Other types of location-based services include *geocast* [6, 8], by which a user can request to send a message to a specific area, and *temporal geocast*, by which a user can request to send a message to a specific area at specific time. In contrast to traditional multicast, such messages are not targeted at a fixed group of members, but rather at members located in a specific physical area.

Section 21.2 introduces some fundamental distance estimation models; Section 21.3 discusses some positioning and location tracking algorithms. In Section 21.4, some experimental systems are reviewed and Section 21.5 gives a summary.

21.2 Fundamentals

To position an object or a device, the basic step is to use a reference point to determine the distance and angle between the device and the reference point. This has been exploited in the radar systems widely used in military applications. This section describes several such basic approaches. The next subsection discusses how to use multiple reference points jointly to estimate the location of a device.

21.2.1 ToA, TDoA, and AoA

In the ToA (time of arrival) approach, signal traveling time is used to estimate the distance between a device and the reference point. Such systems typically use signals that move at a slower speed, such as ultrasound, to measure the time of signal arrival. Figure 21.1(a) illustrates this idea. An ultrasound signal is sent from the transmitter to the receiver; in return, the receiver sends a signal back to the transmitter. After this two-way handshake, the transmitter can infer the distance from the round-trip delay of the signals:

$$\frac{((T_3 - T_0) - (T_2 - T_1)) * V}{2},$$

where V is the velocity of the ultrasound signals. The error of such measurement may come from the processing time of signals (such as computing latency and the unknown delay $T_2 - T_1$ at the receiver's side).

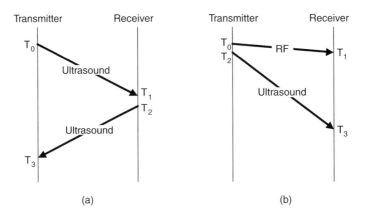

FIGURE 21.1 (a) ToA measurement; (b) TDoA measurement.

Another distance estimation technique is the time difference of arrival (TDoA). Although similar to the ToA scheme, this method uses two signals that travel at different speeds, such as the radio frequency (RF) and ultrasound. Figure 21.1(b) shows how TDoA works; transmission in one direction is sufficient. At T_0 the transmitter sends an RF signal, followed by an ultrasound signal at time T_2. The receiver can then determine its distance to the transmitter by

$$((T_3 - T_1) - (T_2 - T_0)) * (\frac{V_{RF} * V_{US}}{V_{RF} - V_{US}}),$$

where V_{RF} and V_{US} are the traveling speeds of RF and ultrasound signals, respectively. For TDoA, in addition to errors caused by processing time, the receiver also must know the precise value of $(T_2 - T_0)$ to determine the distance.

The AoA approach is another commonly used method for positioning [10, 13]. Such approaches require an antenna array or an array of ultrasound receivers, which can determine the angle and orientation of received signals.

21.2.2 Positioning by Signal Strength

Besides using the signal traveling time, another distance estimation technique is to use the property of signal degradation while traveling in a space to determine the mutual distance. Because signals traveling in a space typically reduce in strength with respect to the distance that they travel, the *received signal strength (RSS)* can be measured at the receiver's side. A mathematical propagation model can be derived to estimate the distance d between a transmitter and a receiver [14] as follows

$$PL(d) = PL(d_0) + 10n \log\left(\frac{d}{d_0}\right) \propto \left(\frac{d}{d_0}\right)^n,$$

where $PL()$ is the path loss function with respect to distance measured in decibels; n is a loss exponent that indicates the rate at which loss increases with distance; and d_0 is the reference distance determined from a measurement close to the transmitter. The path loss exponent n usually ranges from 2 to 4.

Using path loss may incur significant errors. For example, Figure 21.2 shows an experimental result based on IEEE 802.11b. As can be seen, a trend for the relation between distance and signal strength does exist; however, the curve is unstable in small ranges. The true signal strength model is complex and many uncontrollable environmental factors (such as shadows and terrain) are present.

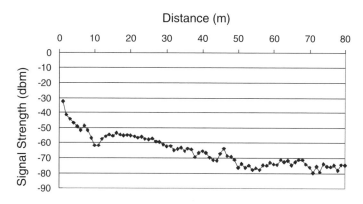

FIGURE 21.2 Signal strength vs. distance in IEEE 802.11b.

To solve the preceding problem, it is necessary to model the error for signal attenuation. One possibility is to include a random variable in the preceding path loss function as follows

$$PL(d) = PL(d_0) + 10n \log\left(\frac{d}{d_0}\right) + X_\rho,$$

where X_ρ is a zero-mean Gaussian random variable with a standard deviation ρ. Due to the existence of such errors, errors will occur as well when positioning a device based on signal strength. Assuming the similar error model in measuring distances, Slijepcevic and colleagues [16] further analyzed the location errors in a wireless sensor network and proved that the distribution of location error can be approximated by a family of Weibull distributions.

21.3 Positioning and Location Tracking Algorithms

The previous section discussed how to estimate the distance between two devices. If an object knows its distances to multiple devices at known locations, one may estimate its location. Several such methods are discussed here.

21.3.1 Trilateration

Trilateration is a well-known technique in which the positioning system has a number of *beacons* at known locations. These beacons can transmit signals so that other devices can determine their distances to these beacons based on received signals. If a device can hear at least three beacons, its location can be estimated. Figure 21.3(a) shows how trilateration works; A, B, and C are beacons with known locations. From A's signal, one can determine that the object should be located at the circle centered at A. Similarly, from B's and C's signals, it can be determined that the object should be located at the circles centered at B and C, respectively. Thus, the intersection of the three circles is the estimated location of the device.

The preceding discussion has assumed an ideal situation; however, as mentioned earlier, distance estimation always contains errors that will, in turn, lead to location errors. Figure 21.3(b) illustrates an example in practice. The three circles do not intersect in a common point. In this case, the maximum likelihood method may be used to estimate the device's location. Let the three beacons A, B, and C be located at (x_A, y_A), (x_B, y_B), and (x_C, y_C), respectively. For any point (x, y) on the plane, a difference function is computed:

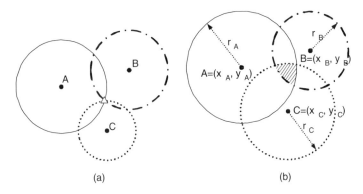

FIGURE 21.3 Trilateration method: (a) ideal situation; (b) real situation with errors.

$$\sigma_{x,y} = |\sqrt{(x-x_A)^2 + (y-y_A)^2} - r_A|$$
$$+ |\sqrt{(x-x_B)^2 + (y-y_B)^2} - r_B|$$
$$+ |\sqrt{(x-x_C)^2 + (y-y_C)^2} - r_C|$$

where r_A, r_B, and r_C are the estimated distances to A, B, and C, respectively. The location of the object can then be predicted as the point (x, y) among all points such that $\sigma_{x,y}$ is minimized.

In addition to using the ToA approach for positioning, the AoA approach can be used. For example, in Figure 21.4, the unknown node *D* measures the angle of, *ADB*, *BDC*, and *ADC* by the received signals from beacons *A*, *B*, and *C*. From this information, *D*'s location can be derived [10].

21.3.2 Multilateration

The trilateration method has its limitation in that at least three beacons are needed to determine a device's location. In a sensor network, in which nodes are randomly deployed, this may not be true. Several multilateration methods are proposed to relieve this limitation.

The AHLoS (Ad Hoc Localization System) [1] is a distributed system for location discovery. In the network, some beacons have known locations and some devices have unknown locations. The AHLoS enables nodes to discover their locations by using a set of distributed iterative algorithms. The basic one is *atomic multilateration*, which can estimate the location of a device of unknown location if at least three beacons are within its sensing range. Figure 21.5 shows an example in which, initially, beacon nodes contain only nodes marked as having a GPS. Device nodes 1, 2, 3, and 4 are at unknown locations. In the first iteration, as Figure 21.5(a) shows, the locations of nodes 1, 2, and 3 will be determined.

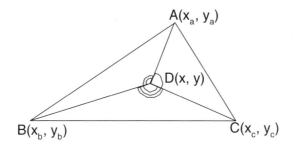

FIGURE 21.4 Angle measurement from three beacons, *A*, *B*, and *C*.

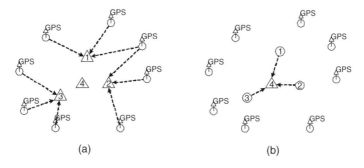

FIGURE 21.5 (a) Atomic multilateration; (b) iterative multilateration in AHLoS.

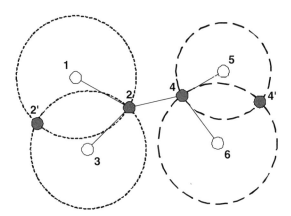

FIGURE 21.6 Collaborative multilateration in AHLoS.

The atomic multilateration is further extended to an *iterative multilateration* method. Specifically, once the location of a device is estimated, its role is changed to a beacon node so as to help determine other devices' locations. This is repeated until all hosts' locations are determined (if possible). As Figure 21.5(b) shows, in the second iteration, the location of node 4 can be determined with the help of nodes 1, 2, and 3, which are now serving as beacons.

The iterative multilateration still has its limitation. For example, as Figure 21.6 shows, it is impossible to determine node 2's and node 4's locations even if the locations of nodes 1, 3, 5, and 6 are known. The *collaborative multilateration* method may relieve this problem because it allows one to predict multiple potential locations of a node if it can hear fewer than three beacons. For example, in Figure 21.6, from beacon nodes 1 and 3, two potential locations of node 2 may be guessed (the other potential location is marked by 2′). Similarly, from beacon nodes 5 and 6, one may guess two potential locations of node 4 (the other potential location is marked by 4′). Collaborative multilateration allows estimation of the distance between nodes 2 and 4. With this information, the locations of nodes 2 and 4 can be estimated, as the figure shows.

21.3.3 Pattern Matching

Another type of location discovery is by pattern matching. Instead of estimating the distance between a beacon and a device, this approach tries to compare the received signal pattern against the training patterns in the database. Thus, this method is also known as the *fingerprinting* approach. The basic idea is that signal strength received at a fixed location is not necessarily a constant. It typically moves up and down, so it would be better to model signal strength by a random variable. This is especially true for indoor environments.

The main idea is to compare the received signals against those in the database and determine the likelihood that the device is currently located in a position. A typical solution has two phases (refer to Figure 21.7):

- *Off-line phase.* The purpose of this phase is to collect signals from all base stations at each training location. The number of training locations is decided first. Then, the received signal strengths are recorded (for a base station that is too far away, the signal strength is indicated as zero). Each entry in the database has the format: $(x, y, \langle ss_1, ss_2, \ldots, ss_n \rangle)$, where (x, y) is the coordinate of the training location, and ss_i, $i = 1 \ldots n$, is the signal strength received at the training location from the ith base station. These entries are stored in the database. Note that for higher accuracy, one may establish multiple entries in the database for the same training location. From the database, some positioning rules, which form the positioning model, will then be established.
- *Real-time phase.* With a well-trained positioning model, one can estimate a device's location given the signal strengths collected by the device from all possible base stations. The positioning model may determine a number of locations, each associated with a probability. However, the typical solution is to output only the location with the highest likelihood.

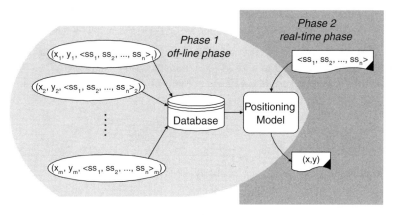

FIGURE 21.7 Pattern matching approach.

There are several similarity searching methods in the matching process; two approaches are introduced next.

21.3.3.1 Nearest Neighbor Algorithms

The simplest approach is the *nearest neighbor in signal space (NNSS)* approach [3, 11]. In the first phase, only the average signal strength of each base station at each training location is recorded. Then, in the second phase, the NNSS algorithm computes the *Euclidean distance* in signal space between the received signal and each record in the database. Euclidean distance means the square root of the summation of square of the difference between each received signal strength and the corresponding average signal strength from the access point under consideration. The training location with the minimum Euclidean distance is then chosen as the estimated location of the device. Because this algorithm only picks existing locations in the database, to improve its accuracy, it is suggested that the training set be dense enough.

One variant of the basic NNSS algorithm is *NNSS-AVG*. To take the uncertainty of a device's location into consideration, this method tries to pick a small number of training locations that closely match the received signal strengths (such as those with smaller Euclidean distances). Then, it infers the location of the device to be a function of the coordinates of the selected training locations. For example, one may take the average of the x and y coordinates of the selected training locations as the estimated result.

21.3.3.2 Probability-Based Algorithms

The probability-based positioning approach regards signal strength as a probability distribution [15]. In NNSS, because the received signal strengths are averaged out, the probability distribution would disappear. So the probability-based approach will try to maintain more complete information of signal strength distribution. The prediction result is typically more accurate.

The core of the probability-based model is the Bayes rule:

$$p(l|o) = \frac{p(o|l)p(l)}{p(o)} = \frac{p(o|l)p(l)}{\sum_{l' \in L} p(o|l')p(l')},$$

where $p(l|o)$ is the probability that the device is at location l given an observed signal strength pattern o. The prior probability that a device is resident at l is $p(l)$, which may be inferred from history or experience. For example, people may have a higher probability to appear in a hallway or lobby. If this is not available, $p(l)$ may be assumed to be a uniform distribution. L is the set of all training locations. The denominator $p(o)$ does not depend on the location variable l, so it can be treated as a normalized constant whenever only relative probabilities are required.

The term $p(o|l)$ is called the likelihood function; this represents the core of the positioning model and can be computed in the off-line phase. There are two ways to implement the likelihood function [15]:

- *Kernel method.* For each observation o_i in the training data, it is assumed that the signal strength exhibits a Gaussian distribution with mean o_i and standard deviation σ, where σ is an adjustable parameter in the model. Specifically, given o_i, the probability to observe o is

$$K(o; o_i) = \frac{1}{\sqrt{2\pi}\,\sigma} \exp\left(\frac{(o - o_i)^2}{2\sigma^2}\right).$$

Based on the kernel function, the probability $p(o|l)$ can be defined as

$$p(o|l) = \frac{1}{n_1} \sum_{l_i \in L, l_i = l} K(o; o_i),$$

where n_1 is the number of training vectors in L obtained at location l. Intuitively, the probability function is a mixture of n_1 equally weighted density functions. Also note that the preceding formulas are derived assuming that only one base station exists. With multiple base stations, the probability function will be multivariated, and the probability will become the multiplication of multiple independent probabilities, each for one base station.

- *Histogram method.* Another method to estimate the density functions is to use histogram, which is related to *discretization* of continuous values to discrete ones. A number of *bins* can be defined as a set of nonoverlapping intervals that cover the whole random variables. The advantage of this method is in its ease in implementation and low computational cost. Another reason is that its discrete property can smooth out the instability of signal strengths.

The probability-based methods can adapt to different environments. To further reduce the computational overhead, Youssef and colleagues [19] proposed a method by clustering training data in the database.

21.3.4 Location Tracking

Location tracking means that a device's location can be derived based on some history traces. Because the trace of a device may indicate where it may move in the next step, this information can be used to improve the accuracy of positioning results. For example, one possibility is to consider the relative distances between consecutive moves of a device in a short period of time. These distances are typically not long. Using this information can reduce errors in tracking results.

In Bahl et al. [3], a Viterbi-like tracking algorithm is proposed for location tracking. The Viterbe algorithm is typically used in communications theory for recognizing the most likely message that is transmitted over a noisy channel. In location tracking, because various environmental factors may interfere with signals, the Viterbi algorithm is also suitable for selecting the most likely location of a device. The idea behind the Viterbi-like tracking algorithm is to take the continuity of a user's track in the past into consideration so as to come up with a better guess of the user's current location.

Figure 21.8 shows the details of the Viterbi-like tracking algorithm. Each time the mobile device receives signals from the access points, it computes a set of k most likely locations. This may be obtained from the NNSS-AVG algorithm described earlier. After receiving continuous h samples, the Viterbi-like algorithm can generate an $h * k$ map, which is an h-stage graph in which each stage contains k possible locations of the device at that stage. The possible locations are modeled by vertices. Edges are established between continuous stages and a weight is assigned to each edge equal to the Euclidean distance of the two incident vertices.

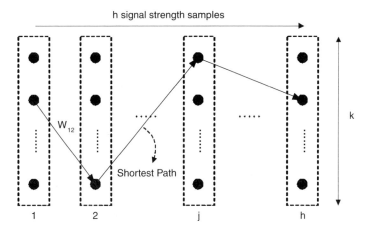

FIGURE 21.8 Viterbi-like location tracking algorithm.

Under the assumption that a user may not move too far away from his current location in a short period of time, the Viterbi-like tracking algorithm computes a shortest path in the $k * h$ map. This shortest path can be viewed as the most likely trajectory of the mobile user. Then the user's current location can be guessed to be the head of this shortest path. Note that, for this reason, the Viterbi-like algorithm may have $h - 1$ periods of delay.

The variances of environments may also complicate the problem. The radio channel condition in working hours may significantly differ from that during off hours. The positioning model may need to adapt to such factors. Recalibration is sometimes inevitable, but laborious. An environmental profile may need to be established to conquer this problem.

21.3.5 Network-Based Tracking

Special concerns— power saving, bandwidth conservation, and fault tolerance — arise when a solution is designed for a wireless sensor network. At the network level, location tracking may be done via the cooperation of sensors. Tseng and colleagues [17] addressed these issues using an agent-based paradigm. Once a new object is detected by the network, a mobile agent will be initiated to track the roaming path of the object. The agent is mobile because it will choose the sensor closest to the object to stay. The agent may invite some nearby slave sensors to cooperatively position the object and inhibit other irrelevant (i.e., farther) sensors from tracking the object. More precisely, only three agents will be used for the tracking purpose at any time and they will move as the object moves. The trilateration method is used for positioning. As a result, the communication and sensing overheads are greatly reduced. Because data transmission may consume a lot of energy, this agent-based approach tries to merge the positioning results locally before sending them to the data center. These authors also address how to conduct data fusion.

Figure 21.9 shows an example. The sensor network is deployed in a regular manner and it is assumed that each sensor's sensing distance equals the distance between two neighboring sensors. Initially, each sensor is in the *idle* state, searching for new objects. Once detecting a target, a sensor will transit to the *election* state, trying to serve as the *master agent*. The nearest sensor will win. The master agent will then dispatch two neighboring sensors as the *slave agents*; master and slave agents will cooperate to position the object. In the figure, the object is first tracked by sensors $\{S_0, S_1, S_2\}$ when resident in A_0, then by $\{S_0, S_1, S_6\}$ when in A_1, by $\{S_0, S_5, S_6\}$ when in A_2, etc.

The master agent is responsible for collecting all sensing data and performing the trilateration algorithm. It also conducts data fusion by keeping the tracking results while it moves around. At a proper time, the master agent will forward the tracking result to the data center. Two strategies are proposed for this purpose: *threshold-based (TB)* strategy, which will forward the result when the amount of data

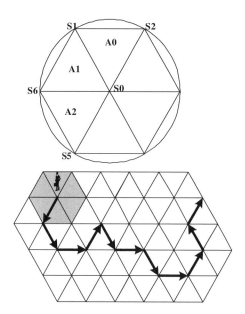

FIGURE 21.9 Roaming path of an object (dashed line) and the migration path of the corresponding master agent (arrow). Sensors that ever host a slave agent are marked by black. (From Y.-C. Tseng et al., *Int. Workshop Inf. Process. Sensor Networks (IPSN)*, 2634, 625–641, 2003. Also to be published in *The Computer Journal*. With permission.)

reaches a predefined threshold value T, and *distance-based (DB)* strategy, which will make a decision based on the routing distance from the agent's current location to the data center and the direction in which the agent is moving.

21.4 Experimental Location Systems

In this section, several location systems are introduced. Although they may not be specially designed for wireless sensor networks, these design concepts and experiences will benefit future implementations of positioning systems in wireless sensor networks.

21.4.1 Active Badge and Bat

The *Active Badge* system [18] is a cell-based location system in which objects are each attached with a badge that periodically emits infrared signals with a unique ID. Infrared receivers mounted at known positions collect these signals and relay them over a wired network. As a result, the system knows in which infrared cell a badge currently stays. The disadvantage of this badge system is that it is hard to deploy in a large-scale environment and that infrared is sensitive to external light, such as sunlight.

A successor of the Active Badge system is the *Bat* system [2], which consists of a collection of wireless transmitters, a matrix of receiver elements, and a central RF base station. The wireless transmitters, called bats, can be carried by a tagged object and/or attached to equipment. The sensor system measures the time of flight of the ultrasonic pulses emitted from a bat to receivers installed in known and fixed positions. It uses the time difference to estimate the position of each bat by trilateration.

The RF base station coordinates the activity of bats by periodically broadcasting messages to them. Upon hearing a message, a bat sends out an ultrasonic pulse. A receiver that receives the initial RF signal from the base station determines the time interval between receipt of the RF signal and receipt of the corresponding ultrasonic signal. It then estimates its distance from the bat. These distances are sent to the computer, which performs data analysis. By collecting enough distance readings, it can determine the location of the bat within 3 cm of error in a three-dimensional space at 95% accuracy. This accuracy is quite enough for most location-aware services; however, the deployment cost is high.

21.4.2 Cricket

Cricket is a system that can provide location-dependent applications [12]. Rather than explicitly tracking users' locations, Cricket helps devices learn their locations and lets them decide whether to advertise them, for preservation of privacy. Cricket does not rely on any centralized management or control and no explicit coordination occurs between beacons. To obtain information about a space, every object is attached to a *listener*, a small device that listens to messages from beacons mounted on ceilings and walls.

Similar to the Bat system, Cricket uses a combination of an RF signal and ultrasound to evaluate the distances between beacons and listeners (i.e., TDoA). A beacon sends the space information over an RF and an ultrasonic pulse at the same time. When the listener hears the RF signal, it uses the first few bits as training information and then turns on its ultrasonic receiver. It then listens for the ultrasonic pulse, which will usually arrive in a short time. The listener uses the time difference between the receipt of the first bit of RF information and the ultrasonic signal to determine its distance from the beacon.

21.4.3 RADAR and Nibble

The *RADAR* location system [11] tries to take advantage of the already existing RF data network formed by IEEE 802.11 access points. IEEE 802.11 networks are now becoming more prevalent in many office and public areas, so no extra hardware cost is incurred. In addition, users can enjoy data communications. RADAR uses the nearest neighbor technology of pattern matching discussed in Section 21.3 to infer objects' locations.

The *Nibble* [4] also adopts the IEEE 802.11 infrastructure for positioning purposes. Nibble uses the probability-based approach in Subsection 21.3.3.2. It relies on a fusion service to infer the location of an object from measured signal strengths. Data are characterized probabilistically and input into the fusion service. The output of the fusion service is a probability distribution over a random variable that represents some context.

21.4.4 CSIE/NCTU Indoor Tour Guide

The authors have also developed a prototype indoor tour guide system at the Department of Computer Science and Information Engineering, National Chiao Tung University (CSIE/NCTU), Taiwan. The hardware platforms of this project include several Compaq iPAQ PDAs and laptops. Each mobile station is equipped with a Lucent Orinoco Gold wireless card. Signal strengths are used for indoor positioning. The probability-based pattern-matching algorithm in Subsection 21.3.3.2 is used. Figure 21.10 shows the system architecture. The concept of logical areas is used to identify offices, rooms, lobbies, etc. The manager is the control center responsible for monitoring each user's movements, configuring the system, and planning logical areas and events. The location server takes care of the location discovery job and the service server is in charge of message delivery. The database can record users' profiles; the gateway can conduct location-based access control to the Internet.

One of the innovations in this project is that an event-driven messaging system has been designed. A short message can be delivered to a user when he enters or leaves a logical area. The event-driven message can also be triggered by a combination of time, location, and property of location (such as who is in the location and when the location is reserved for meetings). A user can set up a message and a corresponding event to trigger the delivery of the message. The manager will check the event list periodically and initiate messages, when necessary, with the service server. Messages can be unicast or broadcast. The expectation is that streaming multimedia can be delivered in the next stage. The system can also be applied to support a smart library. Another innovation is to provide location-based access control. In certain rooms, such as classrooms and meeting rooms, users may be prohibited from accessing certain sensitive Web pages. These rules can be organized through the manager and set up at the gateway.

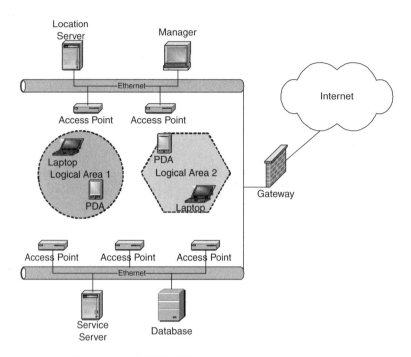

FIGURE 21.10 System architecture of the CSIE/NCTU tour guide system.

21.5 Conclusions

In this chapter, some fundamental techniques in positioning and location tracking have been discussed and several experimental systems reviewed. Location information may enable new types of services. Accuracy and deployment costs are two factors that may contradict each other, but both are important factors for the success of location-based services.

References

1. A. Savvides, C.C. Han, and M.B. Srivastava. Dynamic fine-grained localization in ad hoc networks of sensors. In *ACM/IEEE MOBICOM*, Rome, 2001, pp. 166–179.
2. M. Addlesee, R. Durwen, S. Hodges, J. Newman, P. Steggles, A. Ward, and A. Hopper. Implementing a sentient computing system. *Computer*, 34(8), 50–56, 2001.
3. P. Bahl, A. Balachandran, and V. Padmanabhan. Enhancements to the RADAR user location and tracking system. Technical report MSR-TR-00-12, Microsoft Research, 2000.
4. P. Castro, P. Chiu, T. Kremenek, and R.R. Muntz. A probabilistic room location service for wireless networked environments. In *Ubicomp*, 2001 pp. 18–34.
5. P. Enge and P. Misra. Special issue on GPS: the global positioning system. *Proc. IEEE*, 87, 3–15, 1999.
6. Y. Ko and N. Vaidya. Geocasting in mobile ad hoc networks: location-based multicast algorithms. In *IEEE Workshop Mobile Computing Syst. Applications (WMCSA)*, 101–110, 1999.
7. Y. Ko and N. Vaidya. GeoTORA: a protocol for geocasting in mobile ad hoc networks. In *8th Int. Conf. Network Protocols (ICNP)*, 2000, pp. 240–250.
8. W.-H. Liao, Y.-C. Tseng, K.-L. Lo, and J.-P. Sheu. GeoGRID: a geocasting protocol for mobile ad hoc networks based on GRID. *J. Internet Technol.*, 1(2), 23–32, 2000.
9. J.C. Navas and T. Imielinski. GeoCast — geographic addressing and routing. In *ACM/IEEE MOBICOM*, 66–76, 1997.
10. D. Niculescu and B. Nath. Ad hoc positioning system (APS) using AoA. In *IEEE INFOCOM*, San Francisco, 2003, pp. 1734–1743.

11. P. Bahl and V.N. Padmanabhan. RADAR: an in-building RF-based user location and tracking system. In *IEEE INFOCOM*, 2, 775–784, 2000.
12. N.B. Priyantha, A. Chakraborty, and H. Balakrishnan. The Cricket location-support system. In *ACM/IEEE MOBICOM*, 32–43, 2000.
13. N.B. Priyantha, A.K.L. Miu, H. Balakrishnan, and S.J. Teller. The Cricket compass for context-aware mobile applications. In *ACM/IEEE MOBICOM*, 1–14, 2001.
14. T.S. Rappaport. *Wireless Communications. Principles and Practice*, IEEE Press, 1996.
15. T. Roos, P. Myllymaki, H. Tirri, P. Misikangas, and J. Sievanen. A probabilistic approach to WLAN user location estimation. *Int. J. Wireless Inf. Networks*, 9(3), 2002.
16. S. Slijepcevic, S. Megerian, and M. Potkonjak. Characterization of location error in wireless sensor networks: analysis and applications. In *Int. Workshop Inf. Process. Sensor Networks* (IPSN), 2634, 593–608, 2003.
17. Y.-C. Tseng, S.-P. Kuo, H.-W. Lee, and C.-F. Huang. Location tracking in a wireless sensor network by mobile agents and its data fusion strategies. In *Int. Workshop Inf. Process. Sensor Networks (IPSN)*, 2634, 625–641, 2003.
18. Want, A. Hopper, V. Falcao, and J. Gibbons. The active badge location system. Technical report 92.1, Olivetti Research Ltd. (ORL), 1992.
19. M. Youssef, A. Agrawala, and U. Shankar. WLAN location determination via clustering and probability distributions. In *IEEE PerCom*, 2003.

22
Tracking Techniques in Air Vehicle-Based Decentralized Sensor Networks

Matthew Ridley
The University of Sydney

Lee Ling (Sharon) Ong
The University of Sydney

Eric Nettleton
The University of Sydney

Salah Sukkarieh
The University of Sydney

22.1 Introduction 22-1
22.2 The ANSER System and Experiment 22-2
22.3 The Decentralized Tracking Problem 22-2
   Problem Overview • System Requirements
22.4 Algorithmic System Design 22-3
   The Information Filter • Covariance Intersection • Channel Filter and Communication Management • Target Modeling • Dynamic Network Topology
22.5 Sensor Design 22-9
   Vision Sensors • Radar Sensors • Hybrid Laser/Vision Sensors
22.6 Hardware and Software Infrastructure 22-13
   Navigation Filter • Time Synchronization • Interplatform and Interprocess Communication
22.7 Conclusion 22-15

22.1 Introduction

This chapter will provide some details of developing a practical decentralized tracking system suitable for tracking ground-based targets with multiple air vehicles. Details of the design approach used to develop such a system in a multi-UAV project known as the ANSER project will be provided. This material may be seen as a justification for some of the decisions made during the development of the ANSER system, which will be presented in Section 22.2. Section 22.3 will define more specifically the problem at hand, as well as requirements of the solution.

What follows in Section 22.4 will detail the fundamental algorithms that underpin decentralized data fusion (DDF)-based tracking systems — the information filter; its operation and how it affects other aspects of a tracking system will be discussed. Section 22.5 will focus on sensor models used for the large variety of hardware available for such applications, in particular the hardware used in the ANSER experiment. An experiment of the scale of the ANSER project requires a significant amount of infrastructure; Section 22.6 will outline the numerous subsystems that enabled the experiment to be performed.

22.2 The ANSER System and Experiment

The ANSER project experiment revolves around flying four aircraft simultaneously, tracking ground-based targets and sharing these tracks using DDF [4, 6]. The aircraft are remotely piloted for take-off and landing, but autonomously maintain a preset flight path during the tracking part of the experiment. Plug and play capabilities of the sensors are desired and the air vehicles have a modular removable payload in their nose sensor substitution. There is also room for up to two internally mounted permanent sensors due to the Brumby MKIII modest payload capability (typically, there is one permanent vision sensor).

During actual operation, the four aircraft will communicate not only with each other but also with several ground-based passive sensor nodes designated as "snooper" nodes. These nodes do not actively participate in the experiment but allow those on the ground to observe it as it progresses. The scenario is more vividly depicted in Figure 22.1. The reader is encouraged to read more about the ANSER project [7, 13, 15].

22.3 The Decentralized Tracking Problem

A decentralized system should not be confused with a distributed system. Distributed systems typically contain some form of centralized resource and as a result will *never* be scalable. This is because a computational or communication bottleneck will always be associated with the central resource. This central resource is also a potential weakness because its failure will render the entire system unusable.

22.3.1 Problem Overview

The main aim is to develop a robust multisensor, multiplatform sensor system that obtains its robustness through consistently maintaining a decentralized approach to all aspects of the system. If such an approach is not maintained, a weakness is induced into the system, as mentioned earlier.

22.3.2 System Requirements

- *Target tracking* — the ability to track stationary and moving targets that may be readily discriminated from their surroundings
- *Information sharing* — each node efficiently sharing information gathered from its field of operation with other nodes in its vicinity

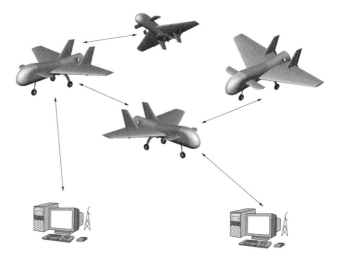

FIGURE 22.1 Operating scenario of the ANSER experiment depicting a possible network topology of four aircraft and two ground terminals.

- *Robust to failure* — entire sensing network not rendered inoperative because of single or multiple points of failure of a sensor or communication subsystem
- *Delayed communication* — support the data fusion of messages that arrive significantly late without the use of heavy buffering
- *Modularity of sensor subsystems* — system allowing for rapid removal and substitution of sensor payloads on the platform
- *Dynamic topology* — nodes capable of entering and leaving the network without significant interruption or loss of information to other nodes

22.4 Algorithmic System Design

This section will provide details of the algorithms that form the basis for decentralized data fusion: filtering methods; data fusion methods; as well as motion modeling. Algorithms suitable for dynamic network topologies are also discussed.

22.4.1 The Information Filter

Kalman filter-based methods are a common, well-tested approach for the purpose of target tracking. The information form of the Kalman filter or information filter is mathematically identical to the Kalman filter; however, it exhibits some properties that are useful for decentralized networks. A more complete derivation of the information filter may be found in Maybeck [10] and Manyika [9]

The properties of the information filter are similar to those obtained when expressing probability distributions as log likelihoods. A Bayesian update is primarily a multiplication of two distributions. When expressed in log-likelihood form, this process becomes an addition. In fact, the information filter can be derived by taking the logarithm of the Bayesian filtering equations of Gaussian distributions and differentiating them twice. Observation information is introduced as:

$$i(k) = H_k^T R_k^{-1} z(k) \tag{22.1}$$

$$I(k) = H_k^T R_k^{-1} H_k \tag{22.2}$$

Information updates are in the form of addition and therefore have lower computational costs. In a decentralized environment, multiple updates from multiple sources will occur at the same time and therefore more frequently than the prediction of target tracks. It is thus advantageous to utilize this form rather than conventional forms.

The innovation generated from each sensor is correlated because they share common information through the prediction $\hat{x}(k|k-1)$. However, in information form, estimates can be constructed from linear combinations of observation information because the information terms $i_i(k)$ from each sensor i are uncorrelated. The information update is defined as:

$$\hat{y}(k|k) = \hat{y}(k|k-1) + i(k) \tag{22.3}$$

$$Y(k|k) = Y(k|k-1) + I(k) \tag{22.4}$$

The prediction stage of the information filter is more computationally expensive than that of the Kalman filter. However, as mentioned previously, the update stage will occur more frequently in a sensor network and is therefore not a significant disadvantage. The prediction stage is defined as:

$$Y(k+1|k) = M_k - M_k G_k \Sigma_k^{-1} G_k^T M_k \tag{22.5}$$

$$\hat{y}(k+1|k) = \left[1 - M_k G_k \Sigma_k^{-1} G_k^T\right] F_k^{-T} \hat{y}(k|k) + Y(k+1|k) B_k u(k) \qquad (22.6)$$

where

$$\Sigma_k = G_k^T M_k^Y G_k + Q_k^{-1}$$

$$M_k = F_k^{-T} Y(k|k) F_k^{-1}$$

22.4.2 Covariance Intersection

The covariance intersection (CI) [8] method of data fusion provides a robust method of fusing data from two sources with unknown correlation. The approach is geometric in nature. An expression, as defined by Nettleton [12], is shown below. This provides the information form of the algorithm, which is suitable for use in combination with the information filter:

$$Y_C(k|k) = \omega Y_A(k|k) + (1-\omega) Y_B(k|k)$$

$$\hat{y}_C(k|k) = \omega \hat{y}_A(k|k) + (1-\omega) \hat{y}_B(k|k)$$

Although this approach to data fusion is conservative, it is very useful for the initialization of filters when a node previously out of contact with a group of nodes establishes contact. The common information between the nodes' common target tracks is unknown, and immediate use of the information filter would require discarding this information. However, the CI algorithm makes it possible to use some of this information. An approach to this will be described in Subsection 22.4.5

22.4.3 Channel Filter and Communication Management

Effective use of the available communication bandwidth for bandwidth-limited systems or systems with large numbers of nodes saturating a large available bandwidth is described here. The channel filter described by Grime [5] is responsible for ensuring that data from all connected nodes are shared in a coherent manner. The channel filter solves the problem of determining and maintaining common information between nodes. The upshot is that the channel filter has local information, as well as global (network) information, available. Each node communicates the full state of each target to the connected node. An indicator of what information has been received from the other node is also included (typically, a time stamp or index). With this in hand, one can achieve consistency between nodes. An update will therefore only occur when the received message contains more information than that currently held locally.

The goal of communication management is to enable the effective use of the available communication bandwidth for bandwidth-limited systems or systems with large numbers of nodes potentially saturating a large available bandwidth. Because the channel filter calculates the difference between local information and global information, it is possible to obtain a measure of the amount of information that a particular node may contribute to the global system. A system with fixed interval transmission of the target with maximum information gain (for the global system) is described by Deaves [3]. The current implementation in the ANSER experiment has no such facility because four aircraft do not impose a significant communication bandwidth constraint. It will, however, become more significant in future projects [1] as the number of nodes increases.

22.4.4 Target Modeling

The target models can be reduced to a state vector, noise matrix, and respective transition matrices. The models of the targets to be tracked should be as representative as possible for the given situation. For example, the ANSER system should be capable of tracking *ground*-based moving targets. Certain assumptions may then be made about the target's dynamics; thus, the state vector is defined as:

$$x(k) = \left[x(k), \dot{x}(k), y(k), \dot{y}(k), z(k)\right]^T \tag{22.7}$$

This model does not include vertical velocity because of the assumption that ground-based targets will never have significant vertical motion.

22.4.4.1 Time-Correlated Constant Velocity

The constant velocity model is in widespread use for typical tracking problems [2]. However, the most common applications for tracking are in situations in which the target is constantly observable, e.g., active sonar and radar. For the ANSER scenario, the flight vehicles make multiple passes with long (up to hundreds) periods in which the targets cannot be observed. A simple constant velocity model could potentially allow targets to develop velocities and translations far in excess of what is practically possible. This situation could arise if a small number of position observations result in an estimated velocity greater than a more advanced target model would typically allow. As a solution to the unbounded velocity properties of the constant velocity model, one might consider the use of the IOU (integrated Ornstein–Uhlenbeck) model.

22.4.4.2 Integrated Ornstein–Uhlenbeck Process

The target x and y position and velocity are modeled as an integrated Ornstein–Uhlenbeck process [14], which has a Brownian velocity that can be bounded by appropriate choice of the model parameter γ. The z position is modeled as a simple Brownian process because the model does not contain a velocity component. The IOU process model was selected because of the velocity-bounding property, which can be used to prevent a Brownian velocity uncertainty from increasing beyond reasonable values. For example, if one is tracking a wheeled land vehicle, the upper bound on the velocity uncertainty is the maximum speed of the vehicle. The state transition matrix for this process model is given by Equation 22.8:

$$F_k = \begin{bmatrix} 1 & \Delta T & 0 & 0 & 0 \\ 0 & F_v & 0 & 0 & 0 \\ 0 & 0 & 1 & \Delta T & 0 \\ 0 & 0 & 0 & F_v & 0 \\ 0 & 0 & 0 & 0 & 1 \end{bmatrix} \tag{22.8}$$

where

$$F_v = e^{-\Delta T \gamma} \tag{22.9}$$

The process noise is written as $G_k \, Q_k \, G_k^T$, where

$$Q_k = \begin{bmatrix} q_x & 0 & 0 \\ 0 & q_y & 0 \\ 0 & 0 & q_z \end{bmatrix} \tag{22.10}$$

and

$$G_k = \begin{bmatrix} 0 & 0 & 0 \\ \sqrt{\Delta T}(1-F_v) & 0 & 0 \\ 0 & 0 & 0 \\ 0 & \sqrt{\Delta T}(1-F_v) & 0 \\ 0 & 0 & 1 \end{bmatrix} \quad (22.11)$$

For the experiments performed to date, targets were known to be stationary, so the IOU process was tuned to decay velocity to zero.

22.4.5 Dynamic Network Topology

This subsection presents network configurations that allow for dynamic changes in the network topology. The algorithms use the same internal node structure as static topology networks, but have modified channel updates to allow for the dynamic aspect of the information. Although dynamic networks are more flexible, they cannot make optimal use of information. Because the source of information is unknown, updates must be performed in a conservative and suboptimal manner to account for unknown correlations. The unstructured network algorithms presented in this subsection use the CI algorithm to compute a conservative update.

22.4.5.1 Broadcast with CI Update

The simplest model for a dynamic network is for nodes to broadcast their full state in information form and for any receiving node to fuse this in the channel using a CI update. Nodes still use the standard information filter update for information received from locally attached sensors. This method of communication update is completely scalable and extremely simple to implement. Nodes will never do worse than if they were operating independently; however, it is possible that they will also never do better. This is the function of the inherent conservative behavior of the CI update. In practice, a few nodes with the best sensors usually end up dominating the network because they provide the best information. Information from nodes with less accurate sensors is not often used.

22.4.5.2 Broadcast with Hybrid CI/IF Update

One method of improving the conservative nature of the CI update in channels is to use a hybrid CI/full filter implementation. This aims to use the full information filter update on data known to be independent while using CI on information for which independence cannot be guaranteed.

When a node communicates, it is required to send two complete estimates in one message. The first estimate, which will be referred to as type 1 data, is the current estimate of common information at that node (i.e., the channel filter estimate). This estimate has some unknown correlation with the rest of the network because it contains information communicated previously. The second estimate, known as type 2 data, is the complete estimate at the current communication time. This estimate contains all information that was in the type 1 data, plus any new information that may have arrived from the locally attached sensor. This new sensor information is known to be independent from the rest of the network because it has not yet been communicated.

When a node receives a message from the network, it first performs a CI update with the type 1 data. This update is exactly the same as that described in Subsection 22.4.5.1 and has the same conservative and suboptimal properties. The second part of the update subtracts the type 1 from the type 2 data and performs a standard information filter update with the result. This information comes only from the sensor attached locally to the node that transmitted the data, so the information is independent and can be fused using the additive information filter update equations. When a node i communicates to node j, the complete channel update at j can be written as

$$Y_{j_{Chan}}(k|k) = \left[\omega Y_{j_{Chan}}(k|k-1) + (1-\omega)Y_{i_{Chan}}(k|k-1)\right]$$
$$+ \left[Y_i(k|k) - Y_{i_{Chan}}(k|k-1)\right] \tag{22.12}$$

$$\hat{y}_{j_{Chan}}(k|k) = \left[\omega \hat{y}_{j_{Chan}}(k|k-1) + (1-\omega)\hat{y}_{i_{Chan}}(k|k-1)\right]$$
$$+ \left[\hat{y}_i(k|k) - \hat{y}_{i_{Chan}}(k|k-1)\right] \tag{22.13}$$

This method is as scalable and as easy to implement as the broadcast with CI update of Subsection 22.4.5.1, but with the added benefit that it makes use of more information while still maintaining consistency. Because less information is thrown away, the accuracy of the estimates and performance is increased. The trade-off is that this requires twice the communication bandwidth to communicate the two estimates; however, because this is a linear increase it does not void the scalability of the network.

22.4.5.3 Dynamic Tree Structure

Although the standard tree structure for network connectivity makes the best use of communicated information, it is also susceptible to communication link failure. Conversely, the CI broadcast network structure copes well with dynamic connectivity but at the cost of poor use of communicated information. This section describes a network structure that uses a combination of both architectures. This structure operates as a tree whenever possible, but is able to reconfigure dynamically in the face of link failure. The key elements are:

- A hierarchical parent–child link structure ensuring that it is impossible to create a loop in any tree structure formed
- The use of CI to initialize a new channel filter rather than act as the nodal fusion algorithm

A node is first defined with n links. Link 1 is defined to be a parent with the remaining $n-1$ links children. The parent link can operate as a peer-to-peer or as a broadcast link, depending on the network configuration. The child links can only be used in peer-to-peer communications with other nodes. Figure 22.2 illustrates this node architecture.

The parent link on a node can operate in a broadcast mode or can be joined to a tree, but never both. When the nodes in the network form into tree structures, they do so using a hierarchical link structure. The only way that nodes may join is by joining a parent and a child link. By definition, each node can only have one parent link; this ensures that any tree structure formed can only have one single node that sits at the top of the tree and has its parent link in broadcast mode. This node is referred to as the tree master. All other nodes in the tree, by definition, must have been attached to a node using their parent links and therefore do not receive broadcast information. This ensures that the tree interacts with other nodes or trees at only one single point. In turn, this eliminates the possibility of double counting broadcast data.

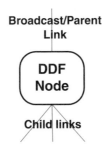

FIGURE 22.2 A single node in a dynamic tree configuration.

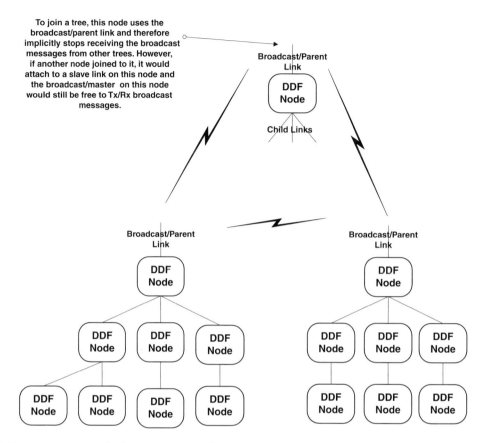

FIGURE 22.3 Structure of a dynamic tree network.

This concept is illustrated in Figure 22.3 in which a number of trees and an independent node are communicating. Note that it is only the master node in each tree that broadcasts. Once the tree master has information, it can propagate this to all other nodes within the tree in an optimal manner. An important difference in the dynamic tree architecture is that nodes know the identity of their immediate neighbors *and* the tree master node. No further local knowledge of the global topology is required, however.

The parent link on each node maintains a channel filter regardless of whether it is in a broadcast or tree topology. This is necessary because the channel mode may change with changes in network configuration. However, because all child links can only be used in a peer-to-peer mode, a channel filter is only required when a connection is made.

When a parent link is not directly connected to the child on another node, it operates in a broadcast mode. The channel filter update in broadcast mode needs to be conservative because no knowledge exists of the source of any information. The channel filter could be implemented using the conservative algorithms described in Subsection 22.4.5.1 or Subsection 22.4.5.2, although the latter would be preferable because it makes better use of the information. When configured in a tree structure, channel filters are implemented to make use of all information.

To join a tree structure, a node broadcasts a request to attach to another node (Figure 22.4). Any node with a spare child link replies and they negotiate joining. When joining a tree, it is important to ensure consistency, no double counting, between estimates in the joining nodes. The node joining the tree must send a broadcast information message to the attaching node. The child link on the attaching node must be updated with this estimate using the broadcast update algorithm. The attaching node then sends this update back down the channel and the new node need only overwrite its previous channel estimates

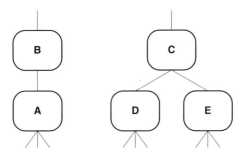

FIGURE 22.4 Joining two trees together.

with this new information. This new information is guaranteed to be at least as good as its information prior to connection.

Multiple trees can be joined in the same way as an independent node joins a tree. The node at the top of the tree uses its parent link to negotiate joining a child link on another tree. If any link in a tree connection is broken, two trees are formed. However, as soon as the link is broken, the node lower in the tree has its parent link freed again and can use this in broadcast mode or negotiate to join another tree. Importantly, the channel filter on the broadcast link must not be reset because, if a tree splits and then manages to rejoin, there will clearly be information in common up to the time of the split. The channel filter at the child link can be reset and cleared safely.

Although knowledge of the identity of the node at the top of the tree is new, it does not violate any of the criteria for a decentralized system. It is not necessary to know all nodes on the path to the top, merely the top itself. Another point of note is that it is easy and useful to implement a system for counting the number of nodes on the path to the top because this would enable a network manager to limit the size of any single tree, should that be required.

22.5 Sensor Design

A wide range of sensors can be used for these networks, including:

- Bearing-only sensors, such as vision
- Range-only sensors, such as lasers
- Sensors capable of range and bearing, such as some radar sensors and laser/vision hybrid sensors

This section details some sensor models and how errors in the sensors are modeled.

22.5.1 Vision Sensors

Vision sensors are used to measure the only bearing to a target in the sensor frame. In order to determine the angle, a pixel representing the target, in an image frame, the camera's parameters, such as focal lengths (f_u and f_v) and principal point coordinates (C_u and C_v), must be determined.

Figure 22.5 illustrates the model for a vision sensor. The relationship between the bearing of the target u and v in the captured frame with respect to the center of the sensor C_u and C_v and the bearing of the target in the actual space is defined by Equation 22.14:

$$\begin{pmatrix} u \\ v \\ 1 \end{pmatrix} = KK * \begin{pmatrix} \dfrac{x}{z} \\ \dfrac{y}{z} \\ 1 \end{pmatrix} \qquad (22.14)$$

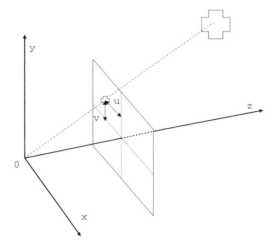

FIGURE 22.5 Vision sensor mode.

where KK is the camera matrix defined as:

$$KK = \begin{pmatrix} f_u & \alpha * f_u & -C_u \\ 0 & f_v & -C_v \\ 0 & 0 & 1 \end{pmatrix} \qquad (22.15)$$

where α is the skew coefficient, which is the angle between the x and y pixel axes.

Sources of error in the vision sensor model include:

- Radial and tangential lens distortions
- Variances in pixels
- Errors in mounting
- Changes in illumination

The camera matrix parameters, lens distortion, and pixel errors can be determined by utilizing calibration software such as the MATLAB camera calibration toolbox.

Range may be inferred from vision sensors by the size of the target. However, this method yields poor results when compared to the laser and radar sensors. Figure 22.6 shows some sample images from the camera used in a vision-sensing node in the ANSER project.

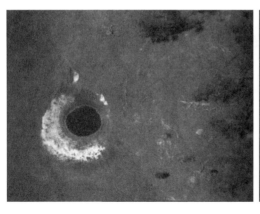

FIGURE 22.6 Sample images from video stream.

22.5.2 Radar Sensors

Radar sensors are used to measure range, azimuth, and elevation; they consist of a mirror scanner, gimbals, motor drivers, and processing capabilities (Figure 22.7 and Figure 22.8). The mirror scanner is used to reflect the radar beam toward the ground as well as to shape the beam pattern to produce a wider elevation beam width. The beam direction at the instant that a target is detected gives the bearing of that target in the radar sensor coordinate frame. The gimbals are used to remove the effects due to the change in the aircraft attitude from the direction that the antenna is pointing; this ensures that the ground area examined is consistent and continuous.

The position of the target in Earth (navigation frame), P_{et}, is:

$$P_t^e = P_b^e + C_b^e P_s^b + C_b^e C_e^b P_t^s \tag{22.16}$$

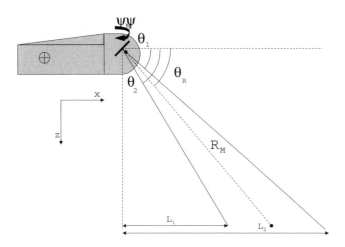

FIGURE 22.7 Radar model.

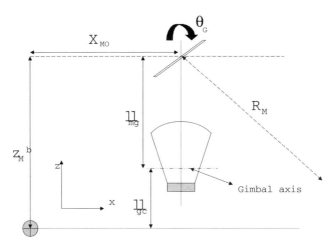

FIGURE 22.8 Radar model.

where:

$P_b^e = [x_b^e, y_b^e, z_b^e]$ is the position of the body in the Earth frame provided by the GPS/INS filter.

C_b^e is the direction cosine matrix that relates how the body frame is rotated with respect to the Earth frame.

$P_s^b = [x_s^b, y_s^b, z_s^b]$ is the position of the radar sensor to the body frame, which is determined by calibration.

C_s^b is the direction cosine matrix that relates how the sensor frame is rotated to the body frame. This is also determined via calibration from the gimbal angles.

P_t^s is the position of the target with respect to the sensor.

The cosine matrix relates one frame to another, e.g., the sensor frame (frame s) to the air vehicle body frame (frame b). It is obtained by rotating the frame by the yaw angle ψ, followed by the pitch angle θ, and the roll angle ϕ.

$$C_s^b = \begin{pmatrix} \cos\psi\cos\theta & \cos\psi\sin\theta\sin\phi - \sin\psi\cos\phi & \cos\psi\sin\theta\cos\phi + \sin\psi\sin\phi \\ \sin\psi\cos\theta & \sin\psi\sin\theta\sin\phi + \cos\psi\cos\phi & \sin\psi\sin\theta\cos\phi - \cos\psi\sin\phi \\ -\sin\theta & \cos\theta\sin\phi & \cos\theta\cos\phi \end{pmatrix} \quad (22.17)$$

$$P_s^b = \begin{pmatrix} x_s^b \\ y_s^b \\ z_s^b \end{pmatrix} = \begin{pmatrix} x_{MO} + \ell_{mg}\cos\phi_G\sin\theta_G \\ -\ell_{mg}\sin\phi_G\cos\theta_G \\ -\ell_{gc} - \ell_{mg}\cos\phi_G\cos\theta_G \end{pmatrix} \quad (22.18)$$

where:

x_{MO} is the distance from the aircraft center of mass to the center of the mirror

l_{mg} is the distance of the mirror to the gimbal center.

l_{gc} is the distance from the gimbal center to the center of mass on the y-axis.

ϕ_G is the gimbal roll angle with its positive direction on the mirror starboard.

θ_G is the gimbal pitch angle with its positive direction on the mirror aft.

$$P_t^s = \begin{pmatrix} x_t^s \\ y_t^s \\ z_t^s \end{pmatrix} = \begin{pmatrix} R_m \cos\theta_R \cos\psi_M \\ R_m \cos\theta_R \sin\psi_M \\ R_m \sin\theta_R \end{pmatrix} \quad (22.19)$$

where θ_R is the look-down angle of the target from the sensor frame on the x–y plane. This angle transforms the vector to the x–y plane; ψ_M is the mirror yaw angle.

The primary sources of radar sensor uncertainty arise from:

- Observation uncertainty
- Platform orientation uncertainty
- Platform position uncertainty

The sensor variance is defined as:

$$R_{obs} = \begin{pmatrix} \sigma_r^2 & 0 & 0 \\ 0 & r^2(\sigma_\psi^2) & 0 \\ 0 & 0 & r^2\sigma_\theta^2 \end{pmatrix} \quad (22.20)$$

where:

r is the range.

ψ is the bearing.

θ is the elevation.

22.5.3 Hybrid Laser/Vision Sensors

Laser-based systems typically use a time-of-flight measurement approach to provide the range to the nearest and furthest targets providing sufficient returns. For aerial-based target tracking, one can be assured that the range returned will be that of the ground below. On its own, a nondirected range-only sensor is difficult to use because expressing an independent observation as a Gaussian distribution is not possible.

Vision sensors return more accurate bearing measurements than radar sensors do. However, inferring range from the size of the targets is highly inaccurate. Thus, building a hybrid sensor by colocating vision and laser sensors takes advantage of the good bearing measurements from the vision sensor and range returns by the laser sensor. Figure 22.9 shows the plots of the errors in the vision, radar, and laser/vision hybrid sensors in range, bearing, and elevation. The vision sensor gives better returns in bearing and elevation compared to radar (Figure 22.10). A hybrid laser/vision sensor provides a more accurate range compared to radar.

22.6 Hardware and Software Infrastructure

This section is only concerned with the infrastructure necessary for the task of performing data fusion with multiple air vehicles. The infrastructure required to operate the flight platforms is not trivial, but out of the scope of this material.

22.6.1 Navigation Filter

Making observations of targets requires some knowledge of the sensor platform's position and orientation (pose). Most UAV applications make use of GPS and IMU devices and the ANSER project is no exception. The navigation solution of the flight path is provided by the flight control computer (FCC); a constant stream of messages provides the most up-to-date vehicle pose. The estimation filter contains nine states, position, velocity, and attitude. Linear accelerations and angular rates also made available. These messages

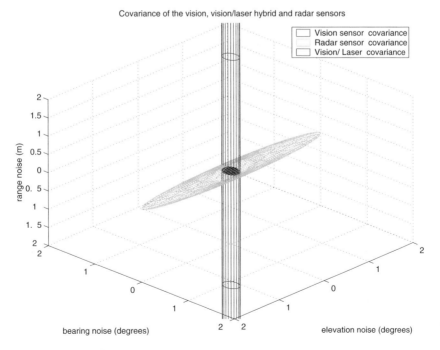

FIGURE 22.9 Covariance of the three sensors.

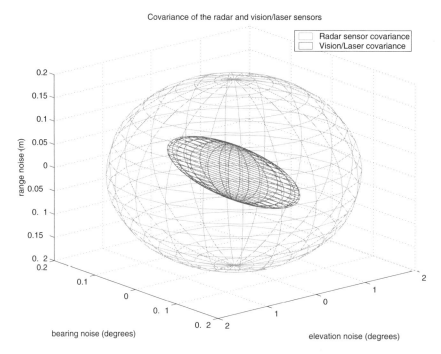

FIGURE 22.10 Covariance of the radar and laser/vision sensors.

are made available at 20 Hz and internally predicted in the sensor nodes to the time of a particular observation being made.

22.6.2 Time Synchronization

The maintenance of a real-time clock is essential for the data fusion process to work; each observation must be time stamped accurately to a time reference. This time reference may be global or local depending on the circumstances. In the case of the ANSER picture compilation system, the navigation filter operates using GPS receivers and as such can provide the necessary timing information to synchronize the local system clocks of all sensor nodes on a given platform to a global time. This does not violate the decentralized theme because the time is acquired from multiple satellites.

If an alternate method of localization was used (such as SLAM) and a GPS type clock system not available, one might choose the solution of using an NTP (network time protocol) [11] type approach to time synchronization. This approach would violate the decentralized requirements because the nominated "time server" would be a centralized component. Although some methods of redundancy may be provided by NTP, the topology of servers must be decided in advance, which also violates the decentralized "axioms."

A solution for the timing problem may be to maintain a model of the clocks of the nodes with which a particular node is in contact (using an NTP style round-trip time approach). Upon transmission of data between nodes, the time stamp of the data is corrected to match the model of the node in question. From a functional standpoint, the DDF algorithms have no idea that time stamps are being modified to correct for each conversing node's offsets. Such approaches are detailed in work by Elson and Römer [16].

22.6.3 Interplatform and Interprocess Communication

With any sensor network, one must consider the possibility of a variety of devices communicating over a variety of different media. The ANSER system utilizes wireless Ethernet, Ethernet, and CAN bus for communications between sensor nodes and sensor platforms. Because multiple sensors may exist on the same node, or on different nodes, a common communication interface needed to be developed that provided interplatform and interprocess communication. Interprocess communication was achieved using shared memory-based FIFOs to provide some buffering for tasks that may have a high latency, such as image processing.

The DDF architecture is inherently robust against communication failure and will recover from intermittent dropouts. A lightweight/nonreliable message-passing approach was therefore chosen. All Ethernet wired and wireless communication utilized the UDP/IP protocols for their simplicity, and broadcast capability. By its nature, the CAN bus is broadcast capable and was used for some internal vehicle messages.

22.7 Conclusion

This contribution has highlighted the numerous aspects associated with a decentralized tracking system involving not only air vehicles but also potentially stationary ground or mobile ground vehicles. The same principles will always apply to a variety of situations, although some subsystems may be substituted or unnecessary (e.g., navigation filter for a stationary node). The experience gained from the ANSER project will enable practical development of large-scale decentralized sensing networks in the near future.

Acknowledgments

This work is supported in part by the ARC Center of Excellence program, funded by the Australian Research Council (ARC) and the New South Wales State Government. The Autonomous Navigation and Sensing Experimental Research (ANSER) Project is funded by BAE SYSTEMS UK and BAE SYSTEMS Australia.

References

1. A. Makarenko, T. Kaupp, B. Grocholsky, and H. Durrant–Whyte. Human robot interactions in active server networks. In *IEEE Int. Symp. Computational Intelligence Robotics Automation (CIRA2003)* July 16–20, Kobe, Japan, 2003.
2. S.S. Blackman and R. Popoli. *Design and Analysis of Modern Tracking Systems*. Artech House, August 1999.
3. R.H. Deaves. The management of communication in decentralised bayesian data fusion systems. Ph.D. thesis, University of Bristol, December 1998.
4. H. Durrant–Whyte, R. Deaves, and P. Greenway. Decentralised multi-platform data fusion. In *Proc. SPIE*, 3393, 63–71, 1998.
5. S.H. Grime. Communication in decentralised sensing architectures. Ph.D. thesis, Department of Engineering Science, University of Oxford, 1992.
6. H. Durrant–Whyte, M. Stevens, and E. Nettleton. Data fusion in decentralized sensing networks. In *Proc. 4th Int. Conf. Inf. Fusion 2001*, 2001.
7. S. Sukkarieh, E. Nettleton, J.-H. Kim, M. Ridley, A. Goktogan, and H. Durrant–Whyte. The ANSER Project: data fusion across multiple uninhabited air vehicles. *Int. J. Robotics Res.*, 22(7) 505–539, July 2003.
8. S. Julier and J. Uhlmann. A nondivergent estimation algorithm in the presence of unknown correlations, *Proc. of American Control Conf.*, 4, 2369–2373, 1997.
9. J. Manyika. An information theoretic approach to data fusion and sensor management. Ph.D. thesis, The University of Oxford, 1993.
10. P. Maybeck. *Stochastic Models, Estimation and Control*, Vol. 1. Academic Press Inc, New York, 1979.
11. A. Thyagarajan Mills, D.L. Huffman, and B.C. Huffman. Internet timekeeping around the globe. In *Proc. Precision Time Time Interval (PTTI) Applications Planning* Meeting (Long Beach CA), 365–371, December 1997.

12. E. Nettleton. Decentralised architectures for tracking and navigation with multiple flight vehicles. Ph.D. thesis, Department of Aerospace, Mechanical and Mechatronic Engineering, The University of Sydney, 2003.
13. M. Ridley, E. Nettleton, S. Sukkarieh, and H. Durrant–Whyte. Tracking in decentralised air-ground sensing networks. In *Proc. 5th Int. Conf. Inf. Fusion*, Annapolis, July 2002, 616–623.
14. L.D. Stone, C.A. Barlow, and T.L. Corwin. *Bayesian Multiple Target Tracking*. Artech House, 1999.
15. S. Sukkarieh, B. Yelland, H. Durrant–Whyte, B. Belton, R. Dawkins, P. Riseborough, O. Stuart, J. Sutcliffe, J. Vethecan, S. Wishart, P. Gibbens, A. Goktogan, B. Grocholsky, R. Koch, E. Nettleton, J. Randle, K. Willis, and K.C. Wong. Decentralised data fusion using multiple UAVS — the ANSER project. In *FSR 2001, Proc. 3rd Int. Conf. on Field Service Robotics*, Helsinki, 2001, 185–192.
16. J. Elson and K. Römer. Wireless sensor networks: a new regime for time synchronization. *ACM SIGCOMM Computer Communication Review*. 33(1), 149–154, 2003.

VI

Data Gathering and Processing

23
Fundamental Protocols to Gather Information in Wireless Sensor Networks

Jacir L. Bordim
Adaptive Communications Research Laboratories

Koji Nakano
Hiroshima University

23.1 Introduction 23-1
23.2 Model Definition 23-3
23.3 Gathering Information in Wireless Sensor Networks .. 23-5
   Preliminaries • Protocols to Solve the Sum Problem in Multihop WSNs • WSNs with Dynamic Transmission Range
23.4 Identifying Faulty Nodes in Wireless Sensor Networks 23-11
   Preliminaries • Locating Faulty Sensors in Multihop WSNs
23.5 Conclusions 23-15

23.1 Introduction

Advances in microelectronic mechanical systems (MEMS) and wireless communication technologies, and the availability of sophisticated sensor-signal processing algorithms have enabled production of multifunctional and small-sized sensor nodes [25, 29]. Due to their compactness and low cost, wireless sensors can be embedded and distributed at a fraction of the cost of conventional wired sensors. A wide range of tasks can be performed by these tiny devices, such as remote object monitoring and tracking; detection of the presence or absence of certain elements; and condition-based maintenance, among other special applications.

The physical world generates a vast amount of information that can be sensed/controlled. However, bandwidth and radio frequencies are finite resources. In addition, the energy cost for communications is generally much larger than the computational cost, which exacerbates the need to process raw data at the source and carefully control the access of the wireless medium. Although sensing devices are already used in a variety of applications, interconnecting them to perform a larger task has yet to become commonplace. Designing systems involving hundreds, or even thousands, of sensors will be a challenging task.

In general, sensors are battery-driven devices that operate on a limited energy budget. Furthermore, they must have a reasonable lifespan to be worth deploying. Clearly, in a network with thousands of sensors, battery replacement is not an option, thus requiring efficient management of the energy resources. Despite these limitations, there is a promising scope for wireless sensor networks (WSNs) in the near future [1, 15]. Indeed, sensor networks have been receiving increasing attention, not only from academia, but also from government and industry [27].

This chapter focuses on fundamental protocols for a large number of sensors in the context of WSNs [4,9]. A WSN is a distributed system consisting of a base station and a number of wireless sensor nodes equipped with wireless radio communication capabilities. Energy consumption is a major factor in determining the lifespan of a sensor node. It is well known that a sensor utilizes a significant amount of energy while sending or receiving packets. Power dissipation is also expressive even when a sensor receives a packet not intended for it. State-of-the-art systems support a number of power states of the sensor nodes (see Reference 28). To simplify matters, here the assumption is that each sensor node in a WSN has two power states: *awake* and *asleep*. Although energy dissipation is negligible in the asleep mode, a significant amount of energy is utilized in the awake state. In MICA2's processor/radio transceiver, for instance, a sensor consumes up to 27 mA to send and receive packets, and less than 1 μA while asleep [28].

In this chapter, the efficiency of a protocol will be assessed by two metrics: (1) the overall amount of time required by the protocol to terminate; and (2) for each individual sensor node, the total amount of time it must be awake to transmit/receive packets. The goals of optimizing these parameters are, of course, conflicting. Sometimes, one can easily minimize the overall completion time at the expense of energy consumption and vice versa. The challenge is to strike a sensible balance between the two by designing protocols that are time and energy efficient.

This chapter first presents efficient collaborative computations and communications strategies to solve a number of fundamental problems in the context of WSNs. More specifically, it shows a number of efficient protocols to aggregate and process information among the sensor nodes and to make such information available at the base station. The fundamental problems considered in this chapter are *information gathering* and *faulty node location*.

In many applications, the sensor nodes must aggregate the sensed/monitored data. Because the sensor nodes are empowered with the ability to share their observations and coordinate among themselves to gather and process information, meaningful information can be transferred to the base station. Such information can then be retrieved and used to control the environment from remote locations.

Presented first are energy-efficient protocols that compute the sum of n numbers over any commutative and associative binary operator stored in n wireless sensor nodes arranged in a two-dimensional grid of size $\sqrt{n} \times \sqrt{n}$. This begins with a protocol that computes the sum in $O\left(r^2 + \left(\frac{n}{r^2}\right)^{\frac{1}{3}}\right)$ time slots with no sensor node awake for more than $O(1)$ time slots, where r is the transmission range of the sensor nodes. Then a fault-tolerant protocol that computes the sum in the same number of time slots with no sensor node awake for more than $O(\log r)$ time slots is presented. Finally, it is shown that, in a WSN where the sensor nodes are empowered with the ability to adjust their transmission range r dynamically during the execution of the protocol, the sum can be computed in $O(\log n)$ time slots and no sensor node needs to wake for more than $O(\log n)$ time slots.

A sensor node may cease its sensing task due to power dissipation or when affected by external events. As the number of faulty sensor nodes in the WSN increases, the accuracy of the sensed/monitored data is likely to deteriorate. If the state of the sensors in the network is known, new sensors can be added to affected areas in order to regain the desired degree of accuracy.

The second major topic in this chapter is to design efficient protocols to identify and locate the state of the sensor nodes in the WSN. As before, consider a WSN populated by n wireless sensor nodes arranged in a two-dimensional grid of size $\sqrt{n} \times \sqrt{n}$. Let q and k denote the number of fault-free and faulty nodes in the WSN, respectively. It will be shown that the task of identifying the faulty nodes and reporting their location to the base station can be completed, with high probability, in $O(\alpha + r^2)$ time slots and none of the sensors needs to wake for more than $O(\log \log \alpha)$ time slots, where $\alpha = \min(q, k)$ and r is the transmission range of the sensor nodes.

Section 23.2 in this chapter gives a formal description of the model. The main results are presented in Section 23.3 and Section 23.4. Section 23.3 begins by presenting some preliminary results that are later used in developing energy-efficient protocols to gather information on a WSN. Section 23.4 presents an energy- and time-efficient protocol to solve the detection of faulty nodes in a WSN.

23.2 Model Definition

The sensor nodes in a WSN are tiny devices operating on batteries and employing low-power radio transceivers to enable communication. The base station is equipped with a powerful antenna that enables it to monitor all the sensor nodes under consideration. It is assumed that the amount of power necessary for a sensor node to communicate with the base station does not exceed the amount of power necessary to communicate with neighboring sensor nodes. This is a reasonable assumption because one can increase the base station's receiver front-end sensitivity [24].

Another possibility is to couple the sensor nodes with *corner cube reflectors* technology to enable passive communications [14, 26, 29]. It is assumed that the base station and all the sensor nodes have a local clock that keeps synchronous time, perhaps by interfacing with the base station or through a GPS. All sensor nodes run the same protocol and can perform computations on the data being sensed. As is customary, time is assumed to be slotted and all transmissions take place at slotted boundaries [3, 6]. It is assumed that at any time slot, a sensor node can communicate with the base station and vice versa. The size of a data packet is such that its transmission can be completed within one time slot.

In a *single-hop* WSN, a sensor node can directly communicate with any other sensor node. In a *multihop* WSN, however, the communication between two sensor nodes may involve a sequence of hops through a chain of pairwise adjacent sensor nodes. There is a single-hop communication between the base station and the sensor nodes, while the communication among the sensor nodes can be single or multihop. There are several possible models for WSNs; this chapter considers WSNs in which all the sensor nodes in the network are fixed, homogeneous, and energy constrained.

A commonly accepted assumption is employed: when two or more sensor nodes are in transmission range of each other and transmitting in the same time slot, the corresponding packets *collide* and are garbled beyond recognition. Similarly, when two or more sensor nodes are broadcasting a packet in the same time slot, the base station cannot receive these packets. The computation among the sensor nodes is performed in coordination with the base station. A sensor node in a single-hop WSN can tune to a channel to send/receive a packet. At the end of a time slot, the status of the channel can be:

- NULL: no packet has been driven into the channel in the current time slot
- SINGLE: exactly one packet has been driven into the channel in the current time slot
- COLLISION: two or more packets have been driven into the channel in the current time slot

Suppose that a sensor node is positioned in a two-dimensional plane. When a sensor node transmits a packet with power r, the signal will be strong enough for other sensor nodes to hear it within the Euclidean distance r from the sensor node that originates the packet. In other words, to cover a range of r, the sensor node that originates the signal must transmit with enough power to cover that range. Every sensor node in the *intensity zone*, that is, the region within the distance r from a sensor node that originates the packet, is guaranteed to receive it. It is well known that signals are subject to fluctuations and start fading after traveling some distance [20]. Thus, sensors outside the transmission range r of a source node, e.g., $r + \delta$ for some $\delta > 0$, may or may not receive the packet.

This situation is formalized as follows: the *fading zone* of a sensor is defined as the region outside the *intensity zone* and inside the circle with radius $f(r)$, where f is an increasing function. Those sensor nodes in the *fading zone* may or may not receive the packet. The status of the channel is always SINGLE in the *intensity zone*, whereas in the *fading zone*, it is SINGLE or NULL. The sensors in the *silent zone*, that is, beyond the Euclidean distance $f(r)$ from the sensor that originated the broadcast, are guaranteed not to receive the packet, and the status of the channel is always NULL. Figure 23.1 depicts the transmission zones of a sensor node as described here.

For simplicity, assume that $f(r) = 2r$ and design the protocols under this assumption. These protocols work for any general function f as long as $f(r) = c \cdot r + o(r)$, for any fixed $c \geq 1$, by adjusting some

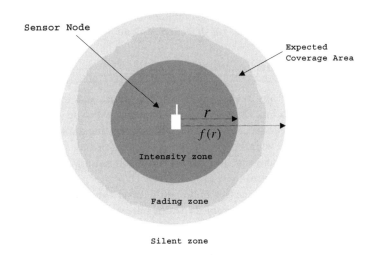

FIGURE 23.1 Transmission zones of a sensor node.

parameters used in the protocols. Although the signal may attenuate in the presence of objects, all sensor nodes in the transmission range of r are ensured to hear from the sensor that originates the signal.

Observe the channel status of a sensor node. For this purpose, let D be a sensor node in a WSN and let S be the unique sensor node broadcasting in a given time slot. The channel status of D is NULL only if D lies in the *silent zone* of S. Otherwise, the channel status is SINGLE if D lies in the *intensity zone* of S and either SINGLE or NULL if D is in the *fading zone*. Now, consider the case in which two or more sensor nodes are broadcasting at the same time. Clearly, if their transmissions do not interfere (i.e., do not overlap), the channel status of D is as discussed here.

In case of overlapping transmissions, the channel status is as follows. When D lies in the *fading zones* of two or more sensor nodes, the channel status can be NULL, SINGLE, or COLLISION. The channel status is SINGLE or COLLISION when D lies in the *intensity zone* of one sensor and in the *fading zone(s)* of other sensor(s). The channel status is COLLISION when D lies in the *intensity zones* of two or more sensors. Therefore, a sensor node is ensured to receive a packet, only if it lies in the *intensity zone* of the source node and no interference occurs from other broadcasts. Figure 23.2 illustrates the channel status in which the transmissions of two sensor nodes overlap.

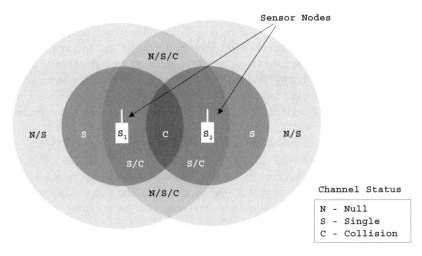

FIGURE 23.2 Status of the channel when the transmissions of two sensor nodes diverge.

In this chapter, it is assumed that the sensor nodes in the WSN are organized as a two-dimensional square plane of size $\sqrt{n} \times \sqrt{n}$ with coordinates (x,y), $(1 \leq x, y \leq \sqrt{n})$. The plane can be viewed as n cells of unit size 1×1. Let $C(x,y)$, $(1 \leq x, y \leq \sqrt{n})$, denote a cell consisting of all points (x',y'), $(x \leq x' < x + 1; y \leq y' < y + 1)$. Suppose that each cell $C(x,y)$ has a sensor denoted as $S_{x,y}$. Throughout this chapter it is assumed that each sensor node $S_{x,y}$, $(1 \leq x, y \leq \sqrt{n})$ knows its cell's location, that is, integers x and y.

If the sensors on a WSN of size $\sqrt{n} \times \sqrt{n}$ can broadcast with sufficient power to cover an area of $\sqrt{2n}$, then any pair of sensors can directly communicate. In such cases, the WSN essentially allows a single-hop communication. In other words, if the sensors with transmission range r are allocated on a WSN of size $\frac{\sqrt{2r}}{2} \times \frac{\sqrt{2r}}{2}$ ($\approx 0.71r \times 0.71r$), then a single-hop communication is ensured.

The next section discusses a fundamental protocol to collect information in WSNs and, following that, an energy-efficient protocol to identify faulty nodes in a WSN.

23.3 Gathering Information in Wireless Sensor Networks

One of the salient features of WSNs is information gathering, the ultimate goal of which is to group and collect the information sensed by the sensor nodes. This section deals with a protocol to assemble and retrieve such information. More specifically, time- and energy-efficient protocols are proposed to compute the sum over any commutative and associative binary operator for the values stored in the sensor nodes. The binary operators can be addition; multiplication; logical AND/OR; finding the maximum/minimum; etc.

Because the sensor node may fail due to power depletion or intentional or unintentional damages, a protocol has been devised that works even in the presence of faulty sensor nodes. In developing fault-tolerant protocols, it is assumed that at least one sensor remains fault free for a given group during the course of the protocol.

In the following, a formal problem definition is given. For this purpose, consider a WSN in which each sensor, $S_{i,j}$ $(1 \leq i, j \leq \sqrt{n})$, has a value $x_{i,j}$. Such value could represent temperature, humidity, gravity, seismic information, etc. Let $\otimes$ be any commutative and associative binary operator, such as addition, multiplication, or finding the maximum. The *sum* problem is to perform the $\otimes$ operation over all $x_{i,j}$, that is,

$$X = x_{1,1} \otimes x_{1,2} \otimes \cdots \otimes x_{\sqrt{n},\sqrt{n}}$$

The next subsection presents some preliminary results to solve the *sum* problem on single-hop WSNs. These results will be used later in developing protocols for multihop WSNs.

23.3.1 Preliminaries

Consider a single-hop WSN comprising m sensor nodes, in which each sensor has a unique ID in $[1,m]$. Let S_i denote a sensor node with ID i, $(1 \leq i \leq m)$, which has a value x_i stored in it. In this scenario, the sum problem can be solved in $m - 1$ time slots as follows: for each time slot i, $(1 \leq i \leq m - 1)$, the sensor node S_i broadcasts x_i on the channel and sensor node S_{i+1} monitors the channel to receive x_i. Then, S_{i+1} computes $x_{i+1} = x_i \otimes x_{i+1}$. After $m - 1$ iterations, node S_m holds the overall sum. Note that this protocol is energy efficient because each sensor node is awake for, at most, two time slots. For later reference, consider the following simple result:

Lemma 23.1. *The sum problem can be solved in a single-hop WSN comprising m sensor nodes in* $m - 1$ *time slots with each sensor node awake for, at most, two time slots.*

In case of node failure, this protocol is unable to yield the correct result. However, a way to overcome this situation is presented next.

In developing a fault-tolerant protocol, assume that faults do not occur during the execution of the protocol and at least one sensor node remains active, i.e., fault free, in the WSN. For each time slot $i\ (1 \leq i \leq m)$, sensor node S_i broadcasts x_i on the channel, and every sensor node S_k, $(k > i)$ monitors the channel to receive the value. After receiving x_i, each sensor node S_k computes $x_k = x_i \otimes x_k$. When a sensor node S_j ($j < i$) hears the broadcast of S_i, it knows that S_i exists and leaves the protocol. Consequently, this protocol terminates in m time slots. The worst case occurs when a single sensor node remains active in the WSN. In such a case, this sensor node will be awake for exactly m time slots. Thus:

Lemma 23.2. Even in the presence of faulty sensor nodes, the sum can be computed on a WSN with m sensor nodes in m time slots and no sensor node is awake for more than m time slots.

Although this protocol is fault tolerant, it is not energy efficient. Now a fault-tolerant and energy-efficient protocol is introduced to compute the sum in a single-hop WSN. When the protocol terminates, the following two conditions are satisfied:

A. The last awake sensor node, denoted as S_k, such that no sensor node S_i with $i > k$ exists, is identified and holds the final result.
B. The protocol takes $2m - 2$ time slots and no sensor node is awake for more than $2\log m$ time slots.

If $m = 1$, then S_1 knows x_1 and conditions A and B are verified. Now, assume that $m \geq 2$. The m sensor nodes are partitioned into two groups $S_1 = \left\{ S_i \middle| 1 \leq i \leq \frac{m}{2} \right\}$ and $S_2 = \left\{ S_i \middle| \frac{m}{2} + 1 \leq i \leq m \right\}$. Recursively compute the sum in P_1 and P_2. By the induction hypothesis, the A and B conditions are satisfied and therefore each of the two subproblems can be solved in $m - 2$ time slots, with no sensor node awake for more than $2\log m - 2$ time slots.

Let S_j and S_k be the last active sensor nodes in groups P_1 and P_2, respectively. In the next time slot, sensor node S_j transmits the sum $\Sigma\{x_i \mid 1 \leq i \leq j$ and S_i exists$\}$ on the channel. The last active sensor node, S_k, in P_2 monitors the channel and updates the result. In one additional time slot, the sensor node S_k reveals its identity. The reader can easily confirm that the protocol satisfies the aforementioned conditions A and B. The following lemma summarizes the preceding discussion:

Lemma 23.3. Even in the presence of faulty sensor nodes, the sum can be computed on a WSN with m sensor nodes in 2m − 2 time slots with no sensor node awake for more than 2log m time slots.

23.3.2 Protocols to Solve the Sum Problem in Multihop WSNs

Because each sensor node $S_{i,j}$, $\left(1 \leq i, j \leq \sqrt{n}\right)$ knows its cell's location within the WSN, a naive protocol can compute the sum in $O(n)$ time slots as follows: each sensor node $S_{i,j}$ broadcasts, one at a time, its value $x_{i,j}$ on the channel. The base station monitors the channel and computes the final result. Clearly, this approach is energy efficient because each sensor node is awake for only one time slot. However, it is not time efficient. The goal of this section is to present protocols that minimize the overall completion time while allowing the sensors to power-off their transceivers for the largest possible extent so as to save energy. First, an energy-efficient protocol is presented that solves the sum problem in $O\left(r^2 + \left(\frac{n}{r^2}\right)^{\frac{1}{3}}\right)$ time slots and the sensor nodes need only wake for a constant number of time slots.

23.3.2.1 Energy-Efficient Summing Protocol

The protocol begins by partitioning the n cells (nodes) into groups, blocks, and sub-blocks. Next, the sum is computed in a bottom-up fashion, starting with sub-blocks, then blocks, and finally groups. The sum of each group is later transmitted to the base station, which computes the overall sum. The parameter k used in the protocol determines the size of the groups. Later, in the description of the protocol, it will be shown how to set this parameter properly. The details of the protocol are given next.

Protocol WSN_SUM

Step 1. The n cells are divided into n/k_2 groups of size $k \times k$. Each group is then partitioned into $k^2 / \left(\frac{9r}{4}\right)^2$ blocks of size $\frac{9r}{4} \times \frac{9r}{4}$. Each block is further partitioned into 81 sub-blocks of size $\frac{r}{4} \times \frac{r}{4}$.

Step 2. Compute the sum on each block.

Step 3. Compute the sum on each group by combining the partial results of step 2.

Step 4. For each group, the sensor node that holds the sum of its group broadcasts it to the base station, one at a time. The base station monitors the channel and computes the overall sum.

The partitioning scheme in step 1 allows for single hop within a sub-block as well as between any two sensor nodes lying on adjacent sub-blocks. The partitioning scheme is illustrated in Figure 23.3. Because each sensor node $S_{i,j}$, $\left(1 \leq i,j \leq \frac{9r}{4}\right)$ knows its location, this information can be easily converted into an ID number. Because each node has a unique ID number, the sum can be computed in top-down fashion, from left to right in odd rows and from right to left in even rows, using the protocol of Lemma 23.1. In other words, the sum on each block is computed in a snake-like fashion. Because the sensor nodes in adjacent blocks lie completely outside the transmission range of each other, step 2 can be performed in parallel on all blocks. Thus, step 2 can be computed in $O(r^2)$ time slots and no sensor node needs to wake for more than two time slots. Also, the sensor node that holds the sum of its block must wake for only one time slot.

Let $S'_{i,j}$, $\left(1 \leq i,j \leq \frac{r}{4}\right)$ denote a sensor node within a sub-block and $S''_{i,j}$, $\left(1 \leq i,j \leq \frac{k}{\frac{9r}{4}}\right)$ be the sensor node on the bottom right of each block that holds the sum of its corresponding block. Note that there are eight sub-blocks of size $\frac{r}{4} \times \frac{r}{4}$ between $S''_{i,j}$ and $S''_{i,j+1}$. Clearly, these two sensor nodes cannot directly communicate because they are outside the transmission range of each other. Therefore, sensors in the sub-blocks that lie between $S''_{i,j}$ and $S''_{i,j+1}$ will be used to forward packets from $S''_{i,j}$ to $S''_{i,j+1}$. In step 3, when a sensor node, e.g., $S''_{i,j}$, broadcasts its value, the sensor node $S'_{\frac{r}{4},\frac{r}{4}}$ located in the right neighboring sub-block receives the packet and forwards it. This process is repeated until the value reaches the sensor node $S''_{i,j+1}$, which updates its value and broadcasts it in the next time slot, as illustrated in Figure 23.4.

When this process finishes, the sensor node in the rightmost column will hold the sum of its row. Next, the sum of the rightmost column is computed so that the sensor node located on the bottom-right corner of each block will hold the sum of its block. Computing the sum on the rows of all groups takes $9 \times \left(\frac{k}{\frac{9r}{4}} - 1\right)$ time slots; to compute the sum on the rightmost column requires additional $9 \times \left(\frac{k}{\frac{9r}{4}} - 1\right)$ time slots. Thus, step 3 can be computed in $O\left(\frac{k}{r}\right)$ time slots; no sensor node needs to wake for more than three time slots and the sensor that holds the sum wakes for only two time slots.

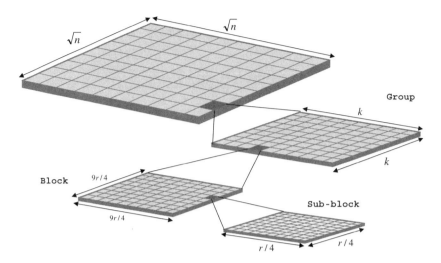

FIGURE 23.3 Grid partitioning scheme.

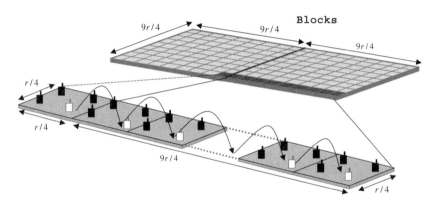

FIGURE 23.4 The computation in step 3 is performed in parallel for every row on each group.

In step 4, the sensor node located at the bottom-right corner of each group broadcasts the sum of its group to the base station. The base station monitors the channel and computes the overall sum. Because there are $\frac{n}{k^2}$ groups, this step takes $O\left(\frac{n}{k^2}\right)$ time slots to compute the sum and the sensor nodes must wake for exactly one time slot. Therefore, this protocol takes $O\left(r^2 + \frac{k}{r} + \frac{n}{k^2}\right)$ time slots to compute the sum of n numbers on a WSN with no sensor node awake for more than five time slots. The time complexity can be minimized by properly selecting the parameter k. With $k = (nr)^{\frac{1}{3}}$, the time complexity becomes $O\left(r^2 + \left(\frac{n}{r^2}\right)^{\frac{1}{3}}\right)$. Thus

Theorem 23.1. On a WSN in which the sensor nodes are arranged in cells of a grid of size $\sqrt{n} \times \sqrt{n}$, one sensor node per cell, the sum of n numbers can be computed by an energy-efficient protocol in $O\left(r^2 + \left(\frac{n}{r^2}\right)^{\frac{1}{3}}\right)$ time slots when r >> 1, and no sensor needs to wake for more than O(1) time slots.

Furthermore, if the sensor nodes were able to adjust their transmission range before the execution of the protocol, r should be selected so that $r^2 = \left(\frac{n}{r^2}\right)^{\frac{1}{3}}$ in order to minimize the running time slots. If this is the case, $r = n^{\frac{1}{8}}$, and the time complexity of the preceding protocol becomes $O\left(n^{\frac{1}{4}}\right)$. Because this protocol relies on the non-fault-tolerant approach of Lemma 23.1, it may not yield the correct result in the presence of faulty nodes. The next sub-section presents a fault-tolerant protocol for multihop WSNs.

23.3.2.2 Fault-Tolerant Energy-Efficient Summing Protocol

In developing a fault-tolerant and energy-efficient summing protocol, it is assumed that at least one sensor node remains fault free in each sub-block during the course of the protocol. Here, step 2 of the previous protocol is modified as follows:

PROTOCOL FAULT-TOLERANT WSN_SUM

Step 2.1. Compute the sum on each sub-block.
Step 2.2. Combine the partial sums of step 2.1 to obtain the sum on each block.

After partitioning the grid, step 2.1 computes the sum on each sub-block using the energy-efficient and fault-tolerant protocol of Lemma 23.3. Because step 2.1 can be computed in parallel for neighboring blocks, it takes $O(r^2)$ time slots to compute the sum on each sub-block and no sensor needs to wake for more than $O(\log r)$ time slots. Let $S'_{i,j}$ $(1 \leq i, j \leq 9)$, be the sensor node that holds the sum of sub-block i,j at the end of step 2.1. The sum of each block in step 2.2 is computed in a snake-like fashion by combining the partial results of step 2.1. Because there are 81 sub-blocks, the sum on blocks can be computed in $O(1)$ time slots and no sensor node needs to wake for more than two time slots. Step 4 and step 5 are performed as in the previous protocol.

Overall, the fault-tolerant protocol takes $O\left(r^2 + \frac{k}{r} + \frac{n}{k^2}\right)$ time slots to compute the sum of n numbers and no sensor node needs to wake for more than $O(\log r)$ time slots. By selecting $k = (nr)^{\frac{1}{3}}$, the time complexity becomes $O\left(r^2 + \left(\frac{n}{r^2}\right)^{\frac{1}{3}}\right)$. To summarize:

Theorem 23.2. On a WSN in which the sensor nodes are arranged in cells of grid size $\sqrt{n} \times \sqrt{n}$, one sensor node per cell, the sum of n numbers can be computed by a fault-tolerant and energy-efficient protocol in time slots $O\left(r^2 + \left(\frac{n}{r^2}\right)^{\frac{1}{3}}\right)$ when $r \gg 1$, and no sensor needs to wake for more than $O(\log r)$ time slots.

When the sensor nodes are empowered with the ability to change their transmission range, then with $r = n^{\frac{1}{8}}$ and the time complexity of the preceding protocol becomes $O\left(n^{\frac{1}{4}}\right)$.

23.3.3 WSNs with Dynamic Transmission Range

This section presents a protocol to solve the sum problem in WSNs in which the sensors have the ability to adjust their transmission range during the course of the protocol. Because the size of the blocks and sub-blocks dynamically changes during the execution, transmission range must change accordingly to ensure connectivity with neighboring sensors in the same sub-block. The details of the protocol are spelled out as follows:

PROTOCOL DYNAMIC WSN_SUM;

for $i = 1$ to $\lceil \log \sqrt{n} \rceil$ **odo**
   $k \leftarrow 4(2^i);$
   every sensor sets its transmission range r to $2^i \sqrt{2}$;
   divide the $\sqrt{n} \times \sqrt{n}$ grid into $\frac{n}{k^2}$ blocks of size $k \times k$;
   divide each $k \times k$ block into $\frac{k^2}{(2^i)^2}$ sub-blocks of size $2^i \times 2^i$;
      for $l = 1$ to $\frac{n}{k^2}$ **do in parallel** for each block $k \times k$
         for $j = 1$ to 16 **do**
            compute the *sum* on each sub-block j;
         endfor
      endfor
endfor

In each iteration of the outermost **for**-loop, the transmission range is properly chosen so that the communication among the sensors within a sub-block is ensured. The grid partitioning used here is similar to that of the previous protocols. In the first iteration, the size of each block is 8×8 and the size of each sub-block is 2×2. Note that each sub-block contains four cells of sensors and each block has 16 sub-blocks. When the size of a sub-block is $2^i \times 2^i$ the transmission range is set to $2^i \sqrt{2}$. This ensures communication among sensors within the same sub-block. Because the distance between any two sub-blocks, located in different blocks, is at least $3.5(2^i)$, the sum on neighboring blocks can be performed in parallel.

The communication among the sensors within the same sub-block is guaranteed, so the sum can be computed sequentially on each sub-block j, $(1 \leq j \leq 16)$ using Lemma 23.1 or Lemma 23.3. Using

Lemma 23.1, the sum on each sub-block takes exactly four time slots. Thus, computing the sum on all sub-blocks takes 64 time slots. Note that only sensors that hold the sum of its sub-block in iteration $i-1$ will participate in iteration i. Each sub-block on iteration i is formed by four sub-blocks of the previous iteration. Then, by adjusting the transmission range on iteration i, the sum on sub-blocks can be computed as discussed in the first iteration.

The outermost **for**-loop repeats for $\lceil \log \sqrt{n} \rceil$ times, and the innermost **for**-loop can be computed in constant time. Hence, our algorithm runs in $64 \cdot \frac{1}{2} \lceil \log n \rceil$ time slots using Lemma 23.1. Note that the sensor holding the sum, at the end of the protocol, is located at the bottom-right corner of the grid, and it has been awake for the largest amount of time. Specifically, the sensor that holds the sum at the end of the protocol has been awake for $2 \cdot \frac{1}{2} \lceil \log n \rceil$ time slots, according to Lemma 23.1. With Lemma 23.3, the sum on each sub-block takes six time slots; in this case, the algorithm runs in $96 \cdot \frac{1}{2} \lceil \log n \rceil$ time slots and no sensor must be awake for more than $4 \cdot \frac{1}{2} \lceil \log n \rceil$ time slots. Partitioning the grid and recursively combining the partial results results in the protocol yielding the overall sum in $O(\log n)$ time slots using Lemma 23.1 or Lemma 23.3. The following theorem summarizes these findings:

Theorem 23.3. On a WSN grid of sensor nodes, where the sensors are endowed with the ability to select the desired transmission range, the sum problem over n numbers can be computed in O(log n) time slots with no sensor node awake for more than O(log n) time slots.

23.4 Identifying Faulty Nodes in Wireless Sensor Networks

A sensor node in a WSN may cease its sensing task due to power depletion or because it was affected or destroyed by external events. In favorable circumstances, a neighboring sensor may be able to cover, even partially, the sensing task of its neighboring faulty nodes. However, the accuracy of the sensed data tends to decrease as the number of faulty sensor nodes increases — to a point at which the sensed data may not correctly reflect the physical events. If the state of the sensors in the network is known, then faulty sensors could be repaired or new sensors added to those affected areas. This task can be carried on until the desired degree of coverage is obtained.

The task of identifying faulty nodes is fundamental in network design and in multiprocessor systems and has been extensively studied in the past [2, 5, 23]. However, these studies focus on wired networks, where energy consumption and limited channel capacity is not an issue. Recently, Chessa and Santi [12] proposed a faulty identification protocol for wireless sensor networks. They focused on "soft" faults, in which a faulty node continues to operate with an alternated behavior. Later, they proposed a protocol that constructs a tree; the information obtained by the leaf nodes is routed to a sink node [13].

Here, the focus is on the task of identifying faulty nodes in a WSN. Consider a WSN with n sensor nodes, where a number of these sensors have been affected by some external event that prevents them from continuing their sensing tasks. Let k denote the number of faulty nodes, and q ($= n - k$) denote the number of fault-free nodes in the WSN. The *fault-location* problem is to identify the location of the k-faulty sensors in the WSN. In identifying faulty nodes, it is assumed that faults do not occur during the course of the protocol and that faults are permanent (i.e., a faulty node remains in that state until it is repaired or replaced). It is also assumed that the base station stores a $\sqrt{n} \times \sqrt{n}$ Boolean matrix, $B = (b_{ij})$, where each entry b_{ij}, ($1 \leq i, j \leq \sqrt{n}$), is associated with the state of the sensor node $S_{i,j}$.

In this scenario, one can easily solve the fault-location problem as follows: each sensor node S_i, ($1 \leq i \leq n$), broadcasts one at a time and the base station monitors the channel to check whether the sensor is faulty or fault-free. After n time slots, the base station knows the position within the grid of each fault-free node and, consequently, it can determine the position of the faulty nodes. This intuitive approach is energy efficient because each sensor node must wake for only one time slot. However, it is not time efficient.

Next, a time- and energy-efficient protocol that identifies and reports the location of faulty nodes to the base station is presented. It will be shown that the fault-location task can be performed in $O(\alpha + r^2)$ time slots and none of the sensors needs to wake for more than $O(\log \log \alpha)$ time slots, where $\alpha = \min(q,k)$ and r is the transmission range of the sensor nodes.

23.4.1 Preliminaries

This section presents some preliminary results that will be used in the subsequent section. To begin, suppose that a WSN is populated by n identical sensors that cannot be distinguished by any serial or manufacturing number. The *initialization* task is to assign a unique ID number in the range $[1,n]$ to each sensor so that no two sensors are assigned the same ID. To solve the initialization task, some protocols assume that the number n of sensors is known prior to the beginning of the protocol [22].

Of particular interest are the results obtained by Bordim and colleagues [10], in which the initialization task is carried out even if the number n of stations is unknown beforehand. These authors showed that a single-channel, single-hop ad hoc radio network (ARN) populated by n stations can be initialized with high probability in $O(n)$ time slots, with each station awake for $O(\log n)$ time slots, at most. More recently, they have shown that initialization can be done in the same number of time slots with each station awake for, at most, $O(\log \log n)$ time slots [11].

An ARN is a distributed system consisting of a number of wireless mobile stations that achieve communication without the aid of any network infrastructure or centralized administration. The main differences between WSNs and ARNs are the presence of a base station in the former, and the fact that the stations are mobile in the ARN. For later reference, the following important result is reproduced from Bordim et al. [11]:

Theorem 23.4. Even if the number n *of stations is not known beforehand, with probability exceeding* $1 - O(n^{-1.5})$, *an* n-*station, single-channel, single-hop ARN can be initialized in* $O(n)$ *time slots, with no station awake for more than* $O(\log \log n)$ *time slots.*

A single-hop ARN can be simulated in two time slots by a multihop WSN by relaying packets to the base station. To see this, suppose that sensor S wants to send a packet to sensor D, where S and D are outside radio reach of each other. In the first time slot, sensor S broadcasts its packet to the base station. In an additional time slot, the base station broadcasts the packet and sensor D wakes to receive it. Because a single-hop communication is between the base station and the sensor nodes in the WSN, a packet can be routed from a source sensor to any destination sensor in the WSN in two time slots — that is, one time slot to route to the base station and another for the base station to send the packet to its final destination. For the latter's reference, consider the following corollary:

Corollary 23.1. Let n *denote the number of sensors in a multihop WSN. Even if the number* n *of sensors is not known beforehand, with probability exceeding* $1 - O(n^{-1.5})$, *the* n *sensors, can be initialized in* $O(n)$ *time slots, with no sensor awake for more than* $O(\log \log n)$ *time slots.*

23.4.2 Locating Faulty Sensors in Multihop WSNs

The main contribution of this section is to present an energy- and time-efficient protocol to solve the faulty location problem in multihop WSNs. As a preliminary step, the multihop WSN is partitioned into groups and the position of the faulty nodes within each group is obtained. Next, each sensor node with faulty nodes in its vicinity sends a packet containing their location, along with its own location, to the base station. Upon learning the location of the faulty nodes, the base station can identify the position of each faulty node as well as the position of each fault-free node within the WSN. The details of this fault-location protocol for multihop WSNs follow:

Protocol Fault Location

Step 1. The $\sqrt{n} \times \sqrt{n}$ cells are divided into $\dfrac{n}{9r^2}$ groups, with each group containing $3r \times 3r$ cells.

Step 2. The sensor nodes in each group learn the state, faulty or fault free, of its immediate neighbors.

Step 3. Each sensor node with faulty sensors in its vicinity informs the base station of its location and the location of its neighboring faulty nodes,

Obviously, the partitioning scheme of Step 1 requires no broadcast. After partitioning, the fault-location task is carried out in each group G_l, $\left(1 \leq l \leq \dfrac{n}{9r^2}\right)$. Let $S^l_{i,j}$, $(1 \leq i, j \leq 3r)$ denote a sensor node within group G_l. Initially, each sensor $S^l_{i,j}$ learns the state, faulty or fault free, of its immediate neighboring sensors located at its north, south, east and west positions. This is achieved by having each sensor node $S^l_{i,j}$ transmitting on the channel and the sensor nodes $S^l_{i-1,j}$, $S^l_{i+1,j}$, $S^l_{i,j-1}$, and $S^l_{i,j+1}$ monitoring the channel. Because the coordinates can be easily converted into a unique ID number, each sensor knows the exact time slot in which it must broadcast. Similarly, each sensor knows when it must be awake to monitor the channel. By checking the channel status at the appropriate time, the immediate neighbors of $S^l_{i,j}$ know that $S^l_{i,j}$ is fault free, if the channel status is SINGLE, and faulty if the channel status is NULL.

It is not difficult to see that a sensor node located at the boundary of a group can learn the status of its adjacent neighboring sensors located in a neighboring group by listening to the channel at the appropriate time slot. Clearly, each sensor node must wake for, at most, five time slots (one time slot to broadcast on the channel) so that its neighbors learn its state. At most, it must wake for four time slots to record the state of its immediate neighbors.

To avoid collision within each group G_l, only one sensor is allowed to broadcast at a time. Therefore, $9r^2$ time slots are necessary for each sensor node to learn the state of its four immediate neighbors. Note that the sensor nodes in the corresponding locations of any two adjacent groups lie completely outside the transmission range of each other because the minimum distance between them is greater than $2r$. Thus, one can reuse the channel and compute Step 2 in parallel for neighboring groups without incurring collisions. Therefore, step 2 can be performed in $O(r2)$ time slots.

Let P denote the set of sensors that have identified faulty nodes in their vicinity at the end of Step 2. The next task is to have each sensor node in P communicating the location of its faulty neighbors to the base station. To avoid collision there, only one sensor is allowed to route its items to the base station at a time. Thus, each sensor node in P must learn the exact time slot in which it can wake and route its packet to the base station so that no other sensor is transmitting at the same time. The idea is to assign unique IDs to each sensor in P. Remember that these IDs are temporary and should not be confused with the sensors' permanent IDs. Once these IDs have been assigned, each sensor can route its packet to the base station in $|P|$ time slots without collision at the base station.

Let p denote the number of sensors in P, that is, $p = |P|$. As the exact number of sensors in P is unknown, we cannot rely on initialization protocols that require such information. Corollary 23.1 states that even if the number p of sensors is unknown beforehand, the p sensors can still be initialized with high probability. More precisely, according to this corollary, the task of assigning a unique ID number in the range $[1,p]$ to each sensor in P so that no two sensors have been assigned the same ID can be performed in $O(p)$ time slots with no sensor node in P awake for more than $O(\log \log p)$ time slots with high probability.

Let S_m, $(1 \leq m \leq p)$ so that $S_m \in P$. After completing the initialization task, each sensor S_m wakes at time slot m and routes its packet to the base station. Because each sensor has, at most, four faulty neighbors, this information can be sent in a single packet containing their respective locations. The base station monitors the channel and collects the packets that are routed for it. Clearly, this can be done in p time slots. Thus, step 3 can be computed in $O(p)$ time slots. The task of identifying the faulty nodes and reporting their locations to the base station can be completed in $O(p + r^2)$ time slots and none of the sensors need wake for more than $O(\log \log p)$ time slots with high probability.

Now consider some special cases:

- All the sensors have crashed (i.e., $k = n$).
- A single fault-free sensor is left in the field (i.e., $k = n - 1$).
- All the sensors are fault free (i.e., $k = 0$).

Verifying whether $k = n$ takes only one time slot and can be achieved as follows. All the sensors broadcast their locations and the base station monitors the channel. If the channel status is NULL, the base station learns that all the sensors have crashed and the protocol terminates. Similarly, if the channel status is SINGLE, the base station knows the location of the unique fault-free sensor in the WSN and the protocol finishes. In case of COLLISION, at least two fault-free nodes must be left in the field. However, the base station cannot check whether $k = 0$ or not. For this purpose, after step 2, all sensors that have faulty nodes in their vicinity broadcast and the base station monitors the channel. If the channel status is NULL, then all the sensors are fault free and the protocol terminates. The following lemma summarizes the results obtained so far.

Lemma 23.4. Let p *denote the number of sensors in a multihop WSN that have identified faulty neighbors adjacent to it. The task of identifying the faulty nodes and reporting them to the base station can be completed, with high probability, in* $O(p + r^2)$ *time slots and none of the sensors need wake for more than* $O(\log \log p)$ *time slots.*

As shown in Figure 23.5(a), the maximum number of sensors reporting to the base station occurs when $k = n/2$. Clearly, if the number of faulty sensors increases beyond $n/2$, the number of sensors reporting to the base stations decreases. On the other hand, when the number k of faulty sensors decreases, so does the number of sensors reporting to the base station because only sensors with faulty neighbors in their vicinity report to the base station (see Figure 23.5b). Thus, when the number of faulty sensors is small, at most, $4k$ sensors report to the base station. In other words, for large k, $p \leq n - k$ and, for small k, $p \leq 4k$.

Let q denote the number of fault-free nodes in the WSN. Clearly, the number of sensor nodes reporting to the base station cannot surpass q. As discussed earlier, for a small number of faulty nodes, $q > 4k$. Thus, in the presence of k faulty nodes, at most, $\min(q, 4k)$ sensors report to the base station. The following theorem summarizes this discussion.

Theorem 23.5. Let q *and* k *denote the number of fault-free and faulty nodes in the WSN, respectively. The task of identifying the faulty nodes and reporting their location to the base station can be completed, with high probability, in* $O(\alpha + r^2)$ *time slots and none of the sensors need wake for more than* $O(\log \log \alpha)$ *time slots, where* $\alpha = \min(q,k)$, *and* r *is the transmission range of the sensor nodes.*

Once the location of the fault sensors is obtained, the base station can identify the areas populated by faulty nodes and the areas populated by fault-free nodes. This is a trivial task when the location of all the faulty nodes is reported to the base station as in Figure 23.5(a). Clearly, the base station can identify "boundary" locations with the information received from the reporting nodes. As shown in Figure 23.5(b) and (c), such boundary locations consist of reporting nodes on one side and faulty nodes on the other. By checking such boundaries, the base station can determine the location of the faulty nodes; that is, the base station can identify which side is populated by faulty nodes and which is populated by fault-free nodes. It should be clear at this point that the base station can determine the status of each sensor node in the WSN from the information provided by the reporting nodes.

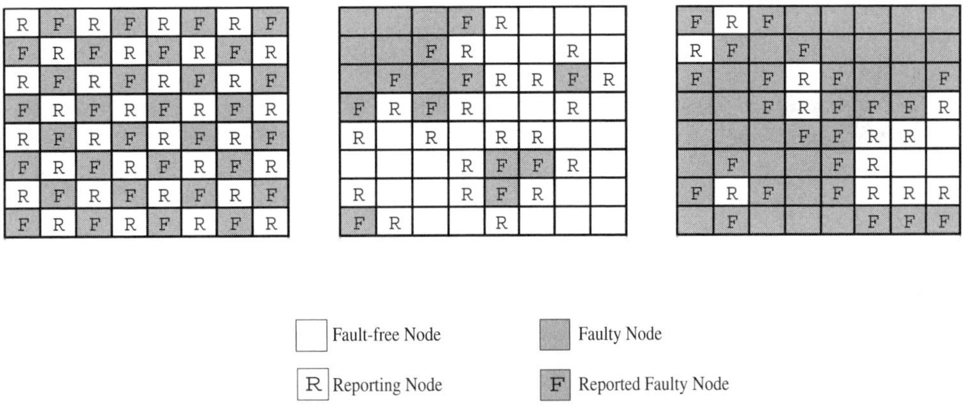

FIGURE 23.5 (a) Worst-case scenario in which half of the sensors report to the base station; (b) and (c) depict boundary formations and identify the reporting nodes and location of the faulty nodes (reported faulty nodes) as viewed by the base station.

23.5 Conclusions

WSNs can greatly augment the ability to control and supervise the environment from distant locations. For reliable monitoring, in many situations the sensor nodes must be in close proximity to the physical events. This may require deployment of large numbers of such devices, which, in turn, demands efficient distributed algorithms to enable the sensor nodes to operate in coordination with other sensors to collect and process data. By sharing observations, the sensor nodes can gather relevant data and transfer meaningful information to a sink node.

This chapter focused on the design of efficient collaborative computation and communication strategies to solve a number of fundamental problems in the context of WSNs. It began by discussing a number of energy-efficient protocols to compute the sum of n numbers over any commutative and associative binary operator stored in n wireless sensor nodes arranged in a two-dimensional grid of size $\sqrt{n} \times \sqrt{n}$. It was shown that the sum can be computed in $O\left(r^2 + \left(\frac{n}{r^2}\right)^{\frac{1}{3}}\right)$ time slots with no sensor node awake for more than $O(1)$ time slots, where r is the transmission range of the sensor nodes.

Next, discussion focused on a fault-tolerant protocol that computes the sum in the same number of time slots with no sensor node awake for more than $O(\log r)$ time slots. Then, it was demonstrated that in WSNs in which sensor nodes are empowered with the ability to adjust their transmission range, the sum can be computed in $O(\log n)$ time slots and no sensor node needs to wake for more than $O(\log n)$ time slots. Finally, a time- and energy-efficient protocol to identify the state of the sensor nodes in the WSN was presented. Here, the task of identifying faulty nodes and reporting their location to the base station can be completed, with high probability, in $O(\alpha + r^2)$ time slots and none of the sensors need wake for more than $O(\log \log \alpha)$ time slots, where q and k denote the number of fault-free and faulty nodes in the WSN, respectively; $\alpha = \min(q,k)$; and r is the transmission range of the sensor nodes.

References

1. I.F. Akyildiz, W. Su, Y. Sankarasubramaniam, and E. Cayirci, Wireless sensor networks: a survey, *Computer Networks*, 38 393–422, 2002.
2. A. Bagchi and S.L. Hakimi, An optimal algorithm for distributed system level diagnosis, *Proc. FTCS-21*, 214– 221, 1991.

3. R. Bar–Yehuda, O. Goldreich, and A. Itai, Efficient emulation of single-hop radio network with collision detection on multi-hop radio network with no collision detection, *Distributed Computing*, 5, 67–71, 1991.
4. R.S. Bhuvaneswaran, J.L. Bordim, J. Cui, and K. Nakano. Fundamental protocols for wireless sensor networks, *IEEE Trans. Fundamentals*, E-85A(11), 2479–2488, Nov. 2002.
5. D.M. Blough and H.W. Wang, The broadcast comparison model for on-line fault diagnosis in multicomputer systems: theory and implementation, *IEEE Trans. Computers*, 48(5), 470–493, May 1999.
6. D. Bertzekas and R. Gallager, *Data Networks*, 2nd ed., Prentice Hall, Upper Saddle River, NJ, 1992.
7. R. Binder, N. Abramson, F. Kuo, A. Okinaka, and D. Wax, ALOHA packet broadcasting — a retrospect, *AFIPS Conf. Proc.*, May 1975, 203–216.
8. U. Black, *Mobile and Wireless Networks*, Prentice Hall, Upper Saddle River, NJ, 1996.
9. J.L. Bordim, F. Hsu, and K. Nakano, Identifying faulty nodes in wireless sensor networks, *J. Interconnection Networks*, 3(3 & 4), 197–211, 2002.
10. J.L. Bordim, J. Cui, T. Hayashi, K. Nakano, and S. Olariu, Energy-efficient initialization protocols for ad-hoc radio networks, *IEEE*, E83-A(9), 1796–1803, Sep. 2000.
11. J.L. Bordim, J. Cui, N. Ishii, and K. Nakano, Doubly logarithmic energy-efficient initialization protocols for single-hop radio networks, *IEEE Trans. Fundamentals*, E83-A(9), 1796–1803, Sep. 2000.
12. S. Chessa and P. Santi, Comparison-based system level fault diagnosis in ad hoc networks, Proc. *IEEE 20th Symp. Reliable Distributed Syst.* (SRDS), New Orleans, 257–266, October 2001
13. S. Chessa and P. Santi, Crash faults identification in wireless sensor networks, *Computer Commun.*, 25(14), 1273–1282, Sept. 2002.
14. P.B. Chu, N.R. Lo, E. Berg, and K.S.J. Pister, Optical communication using micro corner cube reflectors, *10th IEEE Int. Micro Electro Mechanical Syst. Conf.* (MEMS 97), Nagoya, Japan, Jan. 26–30, 1997, 350–355.
15. D. Estrin, R. Govindan, J. Heidemann, and S. Kumar, Next century challenges: scalable coordination in sensor networks. In *Proc. 5th Annu. Int. Conf. Mobile Computing Networks* (MobiCOM'99), Seattle, WA, August 1999.
16. K. Feher, *Wireless Digital Communications*, Prentice Hall, Upper Saddle River, NJ, 1995.
17. W.C. Fifer and F.J. Bruno, Low-cost packet radio, *Proc. IEEE*, 75, 33–42, 1987.
18. V.K. Garg and J E. Wilkes, *Wireless and Personal Communication Systems*, Prentice Hall, Englewood Cliffs, NJ, 1996.
19. M. Gerla and T.-C. Tsai, Multicluster, mobile, multimedia radio network, *Wireless Networks*, 1, 255–265, 1995.
20. M.P.M. Hall and L.W. Barclay *Radio-Wave Propagation*, IEEE Electro Magnetic Wave Series 30, Peter Peregrinus Ltd., 1989.
21. E.P. Harris and K.W. Warren, Low-power technologies: a system perspective, *Proc. 3rd Int. Workshop Multimedia Commun.*, Princeton, 1996.
22. T. Hayashi, K. Nakano, and S. Olariu, Randomized initialization protocols for packet radio networks, *Proc. 13th Int. Parallel Process. Symp.*, (1999), 544–548, 1999.
23. S. Hosseini, J. Kuhl, and S. Reddy, *A Diagnosis Algorithm for Distributed Computing Systems with Dynamic Failure and Repair*, IEEE Trans. Computers, 33(3), 223–233, Mar. 1984.
24. J.C. Liberty and T. Rappaport, *Smart Antennas for Wireless Communications: IS-95 and Third Generation CDMA Applications*, Prentice Hall PTR, Upper Saddle River, NJ, 1999.
25. T.-H. Lin, H. Sanchez, H.O. Marcy, and W.J. Kaiser, Wireless integrated network sensors (wins) for tactical information systems, in *Proc. 1998 Government Microcircuit Applications Conf.*
26. V.S. Hsu, MEMS corner cube retro-reflectors for free-space optical communications: research project, University of California, Berkeley, 1999.

27. G.J. Pottie, Wireless integrated network sensors (WINS): the Web gets physical, *Bridge* 31(4), Winter 2001.
28. MICA2 Motes, http://www.xbow.com/.
29. B. Warneke, M. Last, B. Leibowitz, and K.S.J. Pister, Smart Dust: communicating with a cubic-millimeter computer, *IEEE Computer Mag.*, 44–51, 2001.

24
Comparison of Data Processing Techniques in Sensor Networks

Vicente González-Millán
University of Valencia

Enrique Sanchis-Peris
University of Valencia

24.1 Sensor Networks: Organization and Processing............ 24-1
   Evolution of Sensor Systems • Sensor Processing Systems
24.2 Architectures for Sensor Integration 24-3
   Problems with High Data Rate • Introduction of Preprocessing Elements
24.3 Example of Architecture Evaluation in High-Energy Physics 24-18

24.1 Sensor Networks: Organization and Processing

In the scientific study of the natural phenomena that surround us, one of the first tasks to carry out consists of detailed analysis of the physical variables of the phenomenon to obtain the maximum information about it. To perform this analysis, capable sensors are used to measure the physical variables and to transform that measure into useful information for the study. A sensor is a device made to respond to a physical variable in a predictable form. Sensors can be mechanical, electric, electromechanical, electronic, magnetic, electromagnetic, or optic, to name some. The so-called sensor transfer function assures a well-known relationship between the physical variable and the sensor output.

Sensors can be of very varied form, even those that measure the same variable. However, any sensor can be studied under two aspects: physical and functional. The physical aspect refers to how the sensor is made or to what its form is. The term *physical sensor* refers to devices that sense the physical variable of interest, for example, a barometer, radar, a thermometer, etc. The functional aspect refers to what the sensor is supposed to do or which is its abstraction. The term *abstract* or *logical sensor* is used to refer to an abstraction of the reading taken by a particular sensor. Different possible abstractions exist. The reading of a sensor can be denoted as a simple number or as an interval in the real numbers set. In most cases, sensors are always associated with a transducer element that converts the sensor variations into useful electric signal.

Because the existence of a phenomenon implies a variation in some or all of the parameters associated with it, the electric signal obtained will present a certain variation with time, directly related with the variation of the measured magnitude. Therefore, in the study of the phenomenon a change takes place from an (n + 1)-dimensional space of physical magnitudes (n magnitudes and time) to n two-dimensional spaces of electric magnitudes (amplitudes and times), each corresponding to one of the measured physical magnitudes.

The advantage gained with this transformation is that, with electrical signals, an entire series of tools and technologies allow us their analysis and treatment, something which is not always possible directly on the physical magnitudes of the phenomenon. To treat the information obtained by the sensors, a processing system is needed. This system must operate appropriately with the data of each sensor, interpreting them and obtaining the desired result. This system will be more or less complex, depending on the number of sensors used and whether they are the same or different types, which in turn depends on the phenomenon studied.

24.1.1 Evolution of Sensor Systems

The evolution of the sensor systems can be described in five stages, each represented by one different type of sensor, although a new stage does not imply the disappearance of the sensors used in the previous one [1].

- *Single-sensor systems.* An example of a single sensor system is a radar system. This equipment sends a radio signal of a determined frequency and in a given direction, and receives the signal reflected back on the objects that are on the beam way. From the time difference between the emitted and received pulses, the system calculates the object distance. Another example of a single-sensor system is the sonar used to locate objects underwater. Because of the technology that was available, single-sensor systems were used before the microelectronic era. With only one sensor, system setup and data analysis were inexpensive and easy to perform. However, its simplicity was also a disadvantage because of the limited range of applications. For example, an autonomous mobile robot needs several sensors (tactile, cameras, CCD, etc.) and therefore cannot be built as a single-sensor system. Another drawback of this type of system is its robustness and the impossibility of using it in mission-critical applications. The third disadvantage comes from the fact that a single-sensor system cannot guarantee that the reading is always correct.
- *Replied sensors.* A solution to this third drawback is the use of several sensors, each one giving a reading on the phenomenon of interest. This strategy allows for validation of the reading using different techniques such as majority voting, average, or weighted average.
- *Different sensors.* When it comes to study of a complex phenomenon, it may be necessary to gather different types of information from it. For this purpose, we may use different kinds of sensors that will get different aspects of the phenomenon. For example, an autonomous mobile robot is equipped with different sensors needed to obtain a complete apprehension of the environment. The main advantages of the integration of different sensors are [2]:
  - Reliability increase
  - Improved fault tolerance
  - Improved detection and noise reduction

  This last advantage is explained if we realize that, observing the same signal of interest, the noise picked up by different sensors tends to be uncorrelated.
- *Spatially distributed sensors.* Some applications require that observations of an object are taken simultaneously from two or more points in space. Several degrees exist in the spreading of sensors: from a limited surface area to a region or even an entire country. The type of sensor used can be any of those previously seen, even a combination of them. The peculiarity of these systems is that, now, the information varies spatially and temporarily, so the processing system will be more complex.
- *Intelligent sensors.* If a high number of sensors, replied or different, are used, the volume of information to process may grow to a point at which the problem has a difficult solution. A possibility in this case is to use intelligent sensors in the measure of the physical variables. An intelligent sensor includes certain logical circuitry to abstract information with a bigger semantic content than the one obtained with the electrical signal of the physical variable. For example, a system to detect the passing of people in an enclosure can only offer an electric pulse when a person gets in, or have the necessary logic to offer a representative numeric value of the number of people that have passed through. In this case, the sensor becomes intelligent, offering more elaborate information than the purely electric one. The abstraction of the information may come with a reduction of the information that affects the design of the processing system, reducing the computational load and the necessary bandwidth.

24.1.2 Sensor Processing Systems

Sensor processing is a crucial issue for sensor systems. For its own nature, it requires knowledge of fields like physics, electronics, and computer science. No matter how they are implemented, each sensor processing system consists of four activities: acquisition, processing, integration, and analysis. For particular systems, some of the activities may be lightly or not implemented. Single-sensor systems do not need an integration phase, whereas for a replied sensor system, processing could be minimum, but integration is crucial. In different sensor systems, processing is important to make all readings compatible for the integration phase.

However, most of the sensor processing systems will include the four activities. The physical variable will be sensed in the acquisition activity and the data obtained will be appropriately processed (for example, scaled or formatted) before passing to the integration activity. The output of this activity goes to the analysis phase, where a decision is made. The mechanism of obtaining the decision can be deterministic, stochastic, or empiric. Several options exist to organize the sensor processing system, depending on the characteristics of the problem. The main types of sensor processing systems are [1]:

- Sensor collection — referred to a group of sensors set up in series, parallel, or mixed mode in which integration takes place progressively through the different sensors.
- Hierarchical systems — applied in cases in which the data volume is high so data sent to a central processor may require high bandwidth. A hierarchical distribution may help to reduce bandwidth and increase semantic contents of data as they go down the hierarchy. An important aspect of this organization is that its size does not grow linearly with the problem.
- Tree systems —organized like trees, with sensors in the different levels of the tree. The leaf nodes basically are sensors while the intermediate and root nodes carry out the local processing of data coming from the leaf nodes and on the data read by the sensors connected to them. In this way, at the top, the root node makes the decision with the data processed. The difference between this and hierarchical systems is that, in the latter, the entirety of the sensors is processed in the first level, while in the tree system sensors are progressively integrated.
- Multisensor integration — sensors are of different types and the integration is made at multiple levels, thus implying that the information from the different sensors must be processed to assure its compatibility.
- Distributed sensor processing — the four sensor processing activities take place in a distributed form. This means that not only is acquisition of data by the sensors geographically distributed, but also the processing, integration, and decision. This kind of system gives way to distributed sensor networks (DSN) in which multiple sensors of different types are geographically distributed. Examples of DSN are robotics, particle physics experiments, medical imaging, radar tracking, and flight navigation, to name a few. These systems and others of the same characteristics constitute the logical step in the evolution of sensor processing systems. The design and implementation of these types of networks would not have been possible without advances in technology, mainly in processors and communications.

24.2 Architectures for Sensor Integration

As mentioned earlier, in cases in which the volume of information to treat is large, the sensor processing system is organized in a hierarchical way and, generally, in three levels. The question now is how to compare the goodness of this solution to other types of architectures, for example a fully parallel one. In any case, the real implementation of the sensor processing system falls within one of the well-known Flynn classes for computer architecture [3]. This taxonomy divides the systems according to their number of instruction and data flow paths, dividing them between multiple and single paths. Figure 24.1 shows the Flynn taxonomy for computer architectures.

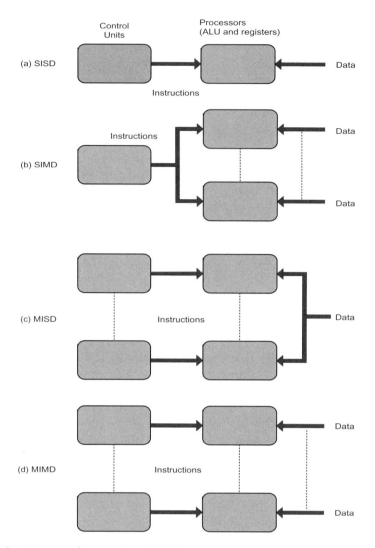

FIGURE 24.1 Flynn taxonomy for computer architectures.

- *SISD* (Figure 24.1a). This represents most of the available computers at present. The instructions are executed sequentially but they can be overlapped in the stages of execution (pipeline segmentation). An SISD computer can have more than a functional unit, but all are under the control of only a control unit
- *SIMD* (Figure 24.1b). In this class one finds the matrix processors in which several units process different data, executing the same instructions, provided by one control unit. These systems are further classified in local or global memory SIMD according to their memory organization, particularly depending on whether the memory access is local to the processor or remote through an interconnection network
- *MISD* (Figure 24.1c). This type of organization is characterized by the existence of several processors, each one executing a different instruction but on the same data flow. In this case, the output of a processor is the input of the following one
- *MIMD* (Figure 24.1d). This category includes most of the multiprocessor and multicomputer systems. An MIMD computer implies interactions between several processors because all the data flows are derived from the same data space shared by all. If the data flows come from disjoint

subspaces inside the shared memory, then one would have a multiple SISD system, or MSISD, that is really a group of independent SISD systems

24.2.1 Problems with High Data Rate

Due to improvements in technology, it is possible currently to treat large quantities of data with a reasonable cost. High-resolution image processing is now possible, even in real time, thanks to increased bandwidth and processing power of the CPUs employed; video is in the same situation. In the field of high-energy physics, trace detectors are used to observe the trajectories of the particles; the lower the processing power is, the larger is the resolution in the determination of the traces, which implies an error in identifying the particle. Because of the increased computing power of processors, it is now possible to think about the construction of systems with better resolutions.

High-resolution image processing and trace detectors are representative examples of areas of investigation in which the volume of information provided by the sensors is high; however, they are not the only ones because one can also find these problems in areas like the robotics, aerospace control, or meteorological prediction. Traditionally, in cases in which the volume of information from the sensors is high, hierarchical architectures have been used for the processing. These architectures are implemented in three or more levels, each with an MIMD structure, so the global system can be viewed as an interconnected cluster of MIMD systems.

However, the option of a hierarchical architecture may not be always the best. Indeed, one can consider other solutions different from the hierarchical system, like a full parallel MIMD system. The best choice for the architecture may not be an easy one to make because it depends on factors such as the problem itself, available technology, reliability, complexity and performance of the solution, and, inevitably, budget. The analysis that follows will be centered in system performance, defining a merit factor that would allow comparison of the systems.

24.2.1.1 Merit Factor of a System

Generally, systems to compare can be implemented in completely different ways, so one needs a parameter independent of the particular implementation; on the other hand, it should somehow indicate which of the two systems will be more complex, difficult, and expensive to implement. Therefore, the merit factor (MF) is defined as the product of the bandwidth, BW, times the processing power, PC, needed to be able to solve a certain problem. That is to say,

$$MF(Mbytes/s \cdot MIPS) = BW_r \cdot PC_r \qquad (24.1)$$

To be able to obtain the necessary expressions for the parallel and hierarchical systems, it is necessary to know the MF value of the association of a certain number of processors, each with its specific MF in serial or in parallel.

24.2.1.1.1 Merit Factor in a Parallel System

In this case, assume a number N of parallel connected processors, each with a certain MF_i value. To obtain the equivalent MF, evaluate separately the total bandwidth and computing power of the parallel system. Evidently, the bandwidth of the system is the sum of the individual bandwidths, BW_i, of each one of the processors. That is to say:

$$BW_p = \sum_{i=1}^{N} BW_i \qquad (24.2)$$

The computing power is also the sum of the capacities, PC_i, of each one of the processors:

$$PC_p = \sum_{i=1}^{N} PC_i \qquad (24.3)$$

Therefore, the merit factor of a parallel system in function of the MF of each processor is:

$$MF_p = BW_p \cdot PC_p = \sum_{i=1}^{N} BW_i \cdot \sum_{i=1}^{N} PC_i = BW_p \cdot \sum_{i=1}^{N} \frac{MF_i}{BW_i} = PC_p \cdot \sum_{i=1}^{N} \frac{MF_i}{PC_i} \qquad (24.4)$$

24.2.1.1.2 Merit Factor in a Series System

Suppose now a case in which the processors are connected sequentially, i.e., the data output of one processor is the data input of the next one. In this case, for N processors, each one with its bandwidth, BW_i, and computing power, PC_i, the equivalent system with just a single processor will have the bandwidth of the first one of the series:

$$BW_s = BW_1 \qquad (24.5)$$

On the other hand, the equivalent computing power will be determined by the total processing time, t_N, and the number of operations to carry out in this time. Therefore,

$$PC_s = \frac{D \cdot \sum_{i=1}^{N} op_i}{\sum_{j=1}^{N} t_j} = \sum_{i=1}^{N} PC_i \cdot \frac{t_i}{t_N} \qquad (24.6)$$

where D is the number of data and op_i is the number of operations carried out by processor i. From the two expressions, the required quality to the equivalent system is:

$$MF_S = BW_S \cdot PC_S = BW_1 \cdot \sum_{i=1}^{N} PC_i \cdot \frac{t_i}{t_N} = \frac{BW_1}{t_N} \cdot \sum_{i=1}^{N} PC_i \cdot t_i \cdot \frac{BW_i}{BW_i} = \frac{BW_1}{t_N} \cdot \sum_{i=1}^{N} MF_i \cdot \frac{t_i}{BW_i} \qquad (24.7)$$

24.2.1.2 Parameterization of Parallel and Hierarchical Architectures

A certain processing problem can be parameterized, indicating the required total bandwidth, BW, to read the data and the processing power necessary, PC, for their computation. Solving the problem by means of a parallel architecture as that of Figure 24.2 yields the following expressions:

$$BW_i = \frac{BW}{N} \quad 1 \leq i \leq N$$
$$CP_i = \frac{CP}{N} \quad 1 \leq i \leq N \qquad (24.8)$$

where N is the number of processors of the system.

However, it is common to have processing units (processors) with some particular processing power, $PC_i = PC_{pu}$, and bandwidth, $AB_i = AB_{pu}$. Therefore, the parameter to determine will be the number of units with those characteristics necessary to implement the system. This number, N, is:

$$N = \max\left(\frac{BW}{BW_i}, \frac{PC}{PC_i}\right) \qquad (24.9)$$

Comparison of Data Processing Techniques in Sensor Networks

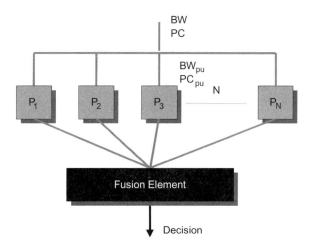

FIGURE 24.2 Parallel processing system. (From Gonzalez, V. et al., *IEEE Trans. Nucl. Sci.*, 49, 2002. With permission.)

The fusion element (fuser) picks up the decisions of the N processing elements and elaborates a final one. The ways of getting this decision are very varied. For this analysis, suppose that the system offers a yes/no binary decision elaborated performing the logical *and* of all the decisions.

The required bandwidth in the fuser depends on the size of the partial decisions of the N processing elements and on the time, T, to make the final decision. The processing power depends roughly on the number of *and* operations necessary to carry out the decision and of the time T necessary to carry them out. Using two input *and* operations, the number of necessary operations to obtain the result with N inputs is $N-1$.

In this case, suppose that the system works in a pipeline way: first the system gets the N partial decisions in T seconds; after this phase, the fuser final decision is obtained in the same T time, on time to receive the next N partial ones. In this way, final and partial decisions are overlapped in time.

If S_{dec} is the size in bytes of the partial decision; S_{dat} the number of bytes for each datum coming from the sensors; ξ the relationship between the size of the partial decision and the size of the data from each sensor; D the number of data; and op the number of operations per datum; the bandwidth and processing power of the fuser can be expressed as:

$$BW_f = N \cdot \frac{S_{dec}}{T} = \left(\begin{array}{c} S_{dec} = \xi \cdot S_{dat} \\ BW = \frac{D \cdot S_{dat}}{T} \end{array} \right) = N \cdot \frac{\xi}{D} \cdot BW = N \cdot K_1 \cdot BW$$

$$PC_f = \frac{N-1}{T} = \left(PC = \frac{D \cdot op}{T} \right) = (N-1) \cdot \frac{1}{D \cdot op} \cdot PC = (N-1) \cdot K_2 \cdot PC$$

(24.10)

If the resolution of the problem is outlined by means of the employment of a hierarchical architecture such as the one shown in Figure 24.3, it will be necessary to introduce a new parameter to be able to obtain the expressions for BW_i and PC_i that will be a function of the level in the architecture. This parameter is the reduction factor in the data volume due to the extraction of the information from the received raw data from the sensors, in the case of a measurement system, or the proportion of data discarded by not completing certain requirements if it is a detection system.

It is necessary to notice that at each level, the bandwidth, BW_i, is reduced in a factor similar to the reduction due to the extraction of information or to the elimination of data not interesting (a factor p). In this case, suppose that, from a level to the next level, all data (because a "yes" decision was taken) or no data ("no" decision) pass. This decision has a probability p, so the time for the information arrival

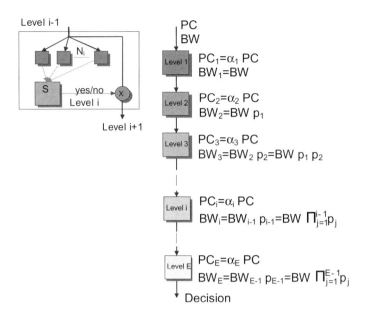

FIGURE 24.3 Hierarchical processing system. (From Gonzalez, V. et al., *IEEE Trans. Nucl. Sci.*, 49, 2002. With permission.)

between levels is increased by the inverse of that probability. Because all the data are sent, the effect is more time to send the same quantity of data, which implies a reduction of the bandwidth.

Assume that the processing power at each level i can be expressed as a proportion, α_i, of the total processing capacity PC. Each one of the levels in the hierarchical system is thought of as a parallel system with a fuser; the result of the decision makes the entirety of the data pass or not toward the next level.

24.2.1.3 Evaluation of Merit Factor for Parallel and Hierarchical Systems

With the considerations of the previous epigraph, one can now evaluate the MF for the parallel and hierarchical systems.

24.2.1.3.1 Parallel System

The MF for the parallel system has two terms. The first one, MF_{proc}, depends on the used processors while the second, MF_{fus}, is due to the fuser. According to Figure 24.2, the equivalent of the parallel system is an association of N parallel processors in series with a fuser.

Suppose that all the processors have the same characteristics; applying Equation 24.1 yields:

$$MF_{proc} = BW_{proc} \cdot PC_{proc} = \sum_{i=1}^{N} BW_i \cdot \sum_{i=1}^{N} PC_i = N \cdot BW_{pu} \cdot N \cdot PC_{pu} = N^2 \cdot MF_{pu} \qquad (24.11)$$

where N is the number of units and MF_{pu} it is the MF of each processing unit.

The second term, related to the fuser, is:

$$MF_{fus} = BW_{fus} \cdot PC_{fus} = N \cdot (N-1) \cdot K_1 \cdot K_2 \cdot BW \cdot PC = N \cdot (N-1) \cdot K \cdot BW \cdot PC \qquad (24.12)$$

The MF of the parallel system will be the series of both calculated, that is to say:

$$MF_{Paral} = \frac{MF_{proc}}{t+t} \cdot \sum_{i=1}^{2} MF_i \cdot \frac{t}{BW_i} = \frac{BW_{proc}}{2} \cdot \left[\frac{N^2 \cdot BW_{pu} \cdot PC_{pu}}{BW_{proc}} + \frac{N \cdot (N-1) \cdot K \cdot BW \cdot PC}{N \cdot K_1 \cdot BW} \right] =$$

$$= \frac{N \cdot b \cdot BW}{2} \cdot \left[\frac{N^2 \cdot b \cdot BW \cdot a \cdot PC}{N \cdot b \cdot BW} + \frac{N \cdot (N-1) \cdot K_1 \cdot K_2 \cdot BW \cdot PC}{N \cdot K_1 \cdot BW} \right] =$$

$$= BW \cdot PC \cdot \frac{N \cdot b}{2} \cdot \left[N \cdot a + (N-1) \cdot K_2 \right]$$

(24.13)

where a and b are the relationships between the processing capacity, PC_{pu}, and the bandwidth, BW_{pu}, respectively, of each processor and the total of the problem.

24.2.1.3.2 Hierarchical System

The hierarchical system is not more than a series of E sequential connected parallel systems. In each level, the system is implemented with N processors and a fuser. Therefore, the MF, MF_{Hier}, will be:

$$MF_{Hier} = \frac{BW_1}{\sum_{i=1}^{E} t_i} \sum_{i=1}^{E} MF_{Paral_i} \cdot \frac{t_i}{BW_i}$$

(24.14)

where MF_{Paral_i} is the MF of each level that is expressed as:

$$MF_{P_i} = \frac{BW_i}{t_i + t_i} \cdot \sum_{j=1}^{2} MF_{ij} \cdot \frac{t_j}{BW_j} = \frac{BW_{proc_i}}{2} \cdot \left(\frac{MF_{proc_i}}{BW_{proc_i}} + \frac{MF_{fus_i}}{BW_{fus_i}} \right)$$

(24.15)

assuming, as it was stated in the parallel association, that $t_j = t_i \ \forall i,j$ (that is, it takes the fuser the same time to get a decision as it takes the processors on the level to get theirs).

The MF for the processors on the level is the parallel association of N of them, that is:

$$MF_{proc_i} = N_i^2 \cdot a_i \cdot b_i \cdot BW \cdot PC$$

(24.16)

where a_i and b_i are defined the same way as in the parallel system, and N is defined as:

$$N_i = \max\left(\frac{BW_i}{BW_{pu_i}}, \frac{PC_i}{PC_{pu_i}} \right) = \max\left(\frac{BW \cdot \prod_{j=0}^{i-1} p_j}{BW_{pu_i}}, \frac{\alpha_i \cdot PC}{PC_{pu_i}} \right)$$

(24.17)

Looking at Figure 24.3

$$P_i = \prod_{j=1}^{i-1} P_j \quad i \geq 2 \qquad (24.18)$$

$$P_1 = 1$$

one can calculate the MF of the fuser. For this, first calculate the bandwidth of the fuser for level I, which turns out to be:

$$BW_{fus_i} = N \cdot \frac{S_{dec}}{t_i} = (t_i = T/P_i) = N \cdot \frac{S_{dec}}{T} \cdot P_i = \left[\begin{array}{c} S_{dec} = \xi \cdot S_{dat} \\ BW = \frac{D \cdot S_{dat}}{T} \end{array} \right] = \qquad (24.19)$$

$$= N \cdot \frac{\xi}{D} \cdot P_i \cdot BW = N \cdot K_1 \cdot P_i \cdot BW$$

On the other hand, the processing power necessary in the fuser of level i can be expressed as:

$$PC_{fus_i} = \frac{N-1}{t_i} = (t_i = T/P_i) = \frac{N-1}{T} \cdot P_i = \left[PC = \frac{D \cdot op}{T} \right] = (N-1) \cdot \frac{1}{D \cdot op} \cdot P_i \cdot PC = \qquad (24.20)$$

$$= (N-1) \cdot K_2 \cdot P_i \cdot PC$$

Therefore, the MF of the fuser of level i will be:

$$MF_{fus_i} = BW_{fus_i} \cdot PC_{fus_i} = N \cdot (N-1) \cdot K_1 \cdot K_2 \cdot P_i^2 \cdot BW \cdot PC \qquad (24.21)$$

Substituting in Equation 24.15 yields:

$$MF_{P_i} = \frac{BW_{proc_i}}{2} \cdot \left(\frac{N_i^2 \cdot a_i \cdot b_i \cdot BW \cdot PC}{N_i \cdot b_i \cdot BW} + \frac{N_i \cdot (N_i - 1) \cdot K \cdot P_i^2 \cdot BW \cdot PC}{N_i \cdot K_1 \cdot P_i \cdot BW} \right) =$$

$$= BW \cdot PC \cdot \frac{N_i \cdot b_i}{2} \cdot \left(N_i \cdot a_i + (N_i - 1) \cdot K_2 \cdot P_i \right) \qquad (24.22)$$

Therefore, the MF of the hierarchical system, MF_{Hier}, is:

$$MF_{Hier} = \frac{BW_1 \cdot PC}{\sum_{j=1}^{E} t_j} \cdot \sum_{i=1}^{E} \frac{N_i \cdot b_i}{2} \cdot \left(N_i \cdot a_i + (N_i - 1) \cdot K_2 \cdot P_i\right) \cdot \frac{t_i}{BW_i} =$$

$$= \frac{BW^2 \cdot PC}{\sum_{j=1}^{E} t_j} \cdot N_1 \cdot b_1 \cdot \sum_{i=1}^{E} \frac{N_i \cdot b_i}{2} \cdot \left(N_i \cdot a_i + (N_i - 1) \cdot K_2 \cdot P_i\right) \cdot \frac{t_i}{N_i \cdot b_i \cdot BW} =$$

$$= \frac{BW \cdot PC}{\sum_{j=1}^{E} t_j} \cdot N_1 \cdot b_1 \cdot \sum_{i=1}^{E} \frac{1}{2} \cdot \left(N_i \cdot a_i + (N_i - 1) \cdot K_2 \cdot P_i\right) \cdot t_i = \quad (24.23)$$

$$= \frac{BW \cdot PC}{\sum_{j=1}^{E} t_1 / P_j} \cdot N_1 \cdot b_1 \cdot \sum_{i=1}^{E} \frac{1}{2} \cdot \left(N_i \cdot a_i + (N_i - 1) \cdot K_2 \cdot P_i\right) \cdot \frac{t_1}{P_i} \Rightarrow$$

$$\Rightarrow MF_{Hier} = BW \cdot PC \cdot \frac{N_1 \cdot b_1}{2} \cdot \sum_{i=1}^{E} \frac{\left(N_i \cdot a_i + (N_i - 1) \cdot K_2 \cdot P_i\right)}{P_i \cdot \sum_{j=1}^{E} 1/P_j}$$

Evidently, Equation 24.23 becomes identical to Equation 24.13, corresponding to the parallel system, when the number of levels E is one.

The following step is to carry out the comparison between the expressions for the parallel and the hierarchical systems and to try to obtain an analytic expression that allows the decision of the better solution for a given problem. However, if the MF of the parallel case is taken as a reference, one would have one equation and $4E$ variables, which would determine infinite solutions.

A different approach to compare both systems would be the following:

1. The parallel solution is determined and its MF is calculated.
2. A hierarchical system is designed and its MF is calculated.
3. Both results are compared. If one is interested in a hierarchical solution and its MF is bigger than the one for the parallel solution, the hierarchical parameters (a, b, and P) can be adjusted and the process repeated until the MF is smaller.

However, it is possible to obtain an analytic expression if what is known is the value of the MF of the parallel system and the values of the parameters of $E - 1$ levels of the hierarchical system of E levels. In this case, imposing the condition that, for example, the MF in the hierarchical system is smaller than the one in the parallel system, the following expression is obtained to calculate the values of the parameters of the last level:

$$\frac{N_E \cdot a_E + (N_E - 1) \cdot K_2 \cdot P_E}{P_E \cdot \sum_{j=1}^{E} 1/P_E} < \frac{N \cdot b}{N_E \cdot b_E} \cdot \left[N \cdot a + (N - 1) \cdot K_2\right] - \sum_{i=1}^{E-1} \frac{N_i \cdot a_i + (N_i - 1) \cdot K_2 \cdot P_i}{P_i \cdot \sum_{j=1}^{E} 1/P_E} \quad (24.24)$$

24.2.2 Introduction of Preprocessing Elements

Up to now, the hierarchical and parallel systems for sensor processing have been studied. This subsection examines an improvement on the hierarchical system based on the introduction of preprocessing elements in the levels that will improve the system performance due to the reduction of their processing load.

24.2.2.1 Regions of Interest

When the number of sensor channels to process is very high, the required processing power and bandwidth in the levels of the system can be too high. In the field of image processing, hierarchical processing systems are used in which successive levels carry out the processing with higher resolutions [4]. Also in this field, particularly in analysis of video images, because of the great correlation that exists between frames, a technique is used based on the location of areas (*region of interest*, RoI) of the image that have changed from one frame to the next. Processing then takes place only on those regions with the consequent reduction in the processing time.

The idea is to apply the concept of RoI to a hierarchical processing sensor system with a great volume of information to release the computational load in the levels where this is possible. The RoI, Ω, can be defined as a group of sensor channels of the system. The RoI can represent a one-dimensional, two-dimensional, or three-dimensional space of measure of the physical environment. In the most general case of a three-dimensional space, the RoI is expressed as:

$$\Omega = \begin{bmatrix} c_{111} & \cdots & c_{11n} \\ \vdots & & \vdots \\ c_{1m1} & \cdots & c_{1mn} \\ c_{211} & \cdots & c_{21n} \\ \vdots & & \vdots \\ c_{2m1} & \cdots & c_{2mn} \\ \vdots & & \vdots \\ c_{p11} & \cdots & c_{p1n} \\ \vdots & & \vdots \\ c_{pm1} & \cdots & c_{pmn} \end{bmatrix} \qquad (24.25)$$

where n, m, and p are the number of channels in each one of the dimensions x, y, and z.

The size of the region of interest will be:

$$\dim(\Omega) = n \cdot m \cdot p \qquad (24.26)$$

In each data acquisition, a certain number of RoIs will be identified; call $\overline{n_{RoI}}$ the average number of RoIs of each data acquisition. In this case, the levels in the architecture will only process the channels of these regions and make the decision with only their information. In general, the average number of channels to process will be smaller than the total. The fraction of channels to process related to the total is:

$$N_{RB_i} = \max\left(\frac{\gamma_i \cdot PC}{PC_{pu_{RB}}}, \frac{BW \cdot P_i}{BW_{pu_{RB}}}\right) \qquad (24.27)$$

where N_{ch} is the total number of channels of the sensor system (normally the number of sensors).

When introducing RoIs, it is necessary to add the necessary modules for their calculation to the architecture of processing. These modules are called RoI builders, or RB, and will be placed between levels of the hierarchy. To analyze how they would affect the MF, the RB must be modeled with a bandwidth and processing power (Figure 24.4). The bandwidth of each RB module located between two

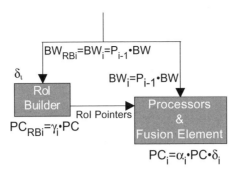

FIGURE 24.4 RoI builder placement in the hierarchical architecture. (From Gonzalez, V. et al., *IEEE Trans. Nucl. Sci.*, 49, 2002. With permission.)

levels is similar to that of the corresponding level, and its processing power can be expressed as a fraction γ_i of the total one of the problem.

Suppose that the RB in each level i is formed by a parallel system with units of bandwidth $BW_{pu_{RB}}$ and processing capacity $PC_{pu_{RB}}$, then the number of necessary units N_{RB_i} will be:

$$N_{RB_i} = \max\left(\frac{\gamma_i \cdot PC}{PC_{pu_{RB}}}, \frac{BW \cdot P_i}{BW_{pu_{RB}}}\right) \qquad (24.28)$$

The RB MF will be expressed as:

$$MF_{RB_i} = \sum_{j=1}^{N_{RB_i}} BW_{pu_{RB}} \cdot \sum_{j=1}^{N_{RB_i}} PC_{pu_{RB}} = \\ = N_{RB_i}^2 \cdot BW_{pu_{RB}} \cdot PC_{pu_{RB}} = N_{RB_i}^2 \cdot a_{RB_i} \cdot b_{RB_i} \cdot BW \cdot PC \quad 1 \leq i < E \qquad (24.29)$$

where, exactly as before, a_{RB_i} and b_{RB_i}, are the relationships between the processing power and the bandwidth of each RB unit and total of the problem. It is necessary to point out that there is no RB from level E to the $E + 1$, which explains the limits of index i in Equation 24.29.

The bandwidth of each level in the hierarchy is also modified because now it must also accept data relative to the RoI. However, this study rejects this contribution, supposing that the problem data rate is very high compared to the one due to this fact. Figure 24.5 shows the modified outline of the architecture with RoIs. Note that it is not always necessary to introduce RB between all levels of the hierarchy because it depends on the particular application.

Decreasing the number of channels to process will also decrease the processing power in the level of the hierarchy at which RoIs are used, although this does not apply to the bandwidth because each level should be capable of reading all the channels — not only those selected by the RB. The reduction factor is similar to the fraction δ, so the MF of that level is reduced by a factor $\delta < 1$, that is,

$$MF_i' = \delta \cdot MF \qquad (24.30)$$

The change in the processing power may vary the number of units needed in each level, depending on which parameter (bandwidth or processing power) determined it. In general, when introducing RoI, the number of processing units, N_i', is:

$$N_i' = m\acute{a}x\left(\frac{\delta_{i-1} \cdot PC_i}{PC_{pu_i}}, \frac{P_i \cdot BW}{BW_{pu_i}}\right) = m\acute{a}x\left(\frac{\delta_{i-1} \cdot \alpha_i \cdot PC}{PC_{pu_i}}, \frac{P_i \cdot BW}{BW_{pu_i}}\right) \qquad (24.31)$$

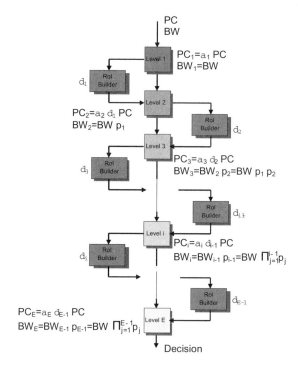

FIGURE 24.5 Hierarchical systems with RoIs. (From Gonzalez, V. et al., *IEEE Trans. Nucl. Sci.*, 49, 2002. With permission.)

for $i > 1$, with $\delta_0 = 1$ (because there is no RB from level 0 to level 1). In this calculation we supposed that the processing units used have the same characteristics as those used in the system without RoI.

The fuser MF is also modified by the use of RoI because it varies the number of processing units in the levels, although not the available time for the information processing and transmission.

Taking everything into account, the MF of each level of the hierarchy is modified and results in:

$$MF'_{P_i} = \frac{BW'_{proc_i}}{2 \cdot t_i + t_{RB_i}} \cdot \left(\frac{N^2_{RB_i} \cdot a_{RB_i} \cdot b_{RB_i} \cdot BW \cdot PC}{N_{RB_i} \cdot b_{RB_i} \cdot BW} \cdot t_{RB_i} + \frac{N'^2_i \cdot a_i \cdot b_i \cdot BW \cdot PC}{N'_i \cdot b_i \cdot BW} \cdot t_i + \frac{N'_i \cdot (N'_i - 1) \cdot K \cdot P^2_i \cdot BW \cdot PC}{N'_i \cdot K_1 \cdot P_i \cdot BW} \cdot t_i \right) =$$

$$= BW \cdot PC \cdot \frac{N'_i \cdot b_i}{2 \cdot t_i + t_{RB_i}} \cdot \left(N_{RB_i} \cdot a_{RB_i} \cdot t_{RB_i} + N'_i \cdot a_i \cdot t_i + (N'_i - 1) \cdot K_2 \cdot P_i \cdot t_i \right) = \left(K_{t_i} = t_{RB_i} / t_i \right)$$

$$= BW \cdot PC \cdot \frac{N'_i \cdot b_i}{2 + K_{t_i}} \cdot \left(N_{RB_i} \cdot a_{RB_i} \cdot K_{t_i} + N'_i \cdot a_i + (N'_i - 1) \cdot K_2 \cdot P_i \right)$$

(24.32)

where K_{t_i} is the relationship between the processing time in the RB and the one of the level, and $K_{t_E} = 0$ because there is no RB between level E and $E + 1$. As in the case without RoIs, the processing and the fusion operate in a pipelined way.

The MF of the hierarchical system with RoIs is obtained from Equation 24.32:

$$MF_{H_{RoI}} = BW \cdot PC \cdot N_1 \cdot b_1 \cdot \sum_{i=1}^{E} \frac{\left(N'_i \cdot a_i + (N'_i - 1) \cdot K_2 \cdot P_i + N_{RB_i} \cdot a_{RB_i} \cdot K_{t_i}\right)}{\left(2 + K_{t_i}\right) \cdot P_i \cdot \sum_{i=1}^{E} 1/P_i} \qquad (24.33)$$

The MF of the hierarchical systems with and without RoIs can now be compared to decide when the system with RoIs has an MF smaller than the one of the system without RoIs. Comparing Equation 24.23 with Equation 24.33 for a certain level j, as a sufficient condition for the system with RoI to have an MF lower than the one without RoI, yields:

$$MF_{H_{RoI}} < MF_{Hier} \Rightarrow$$

$$\Rightarrow \frac{N'_j \cdot a_j + N_{RB_j} \cdot a_{RB_j} \cdot K_{t_j} + (N'_j - 1) \cdot K_2 \cdot P_j}{\left(2 + K_{t_j}\right) \cdot P_j \cdot \sum_{s=1}^{E} 1/P_s} < \frac{N_j \cdot a_j + (N_j - 1) \cdot K_2 \cdot P_j}{2 \cdot P_j \cdot \sum_{s=1}^{E} 1/P_s} \qquad (24.34)$$

If $K_{t_j} \to 0$, i.e., the RB, uses very little time in identifying the channels of the RoI compared to the processing time of the corresponding level, then the previous expression can be simplified, yielding:

$$N'_i \cdot a_j + (N'_i - 1) \cdot K_2 \cdot P_j < N_j \cdot a_j + (N_j - 1) \cdot K_2 \cdot P_j \qquad (24.35)$$

Reordering,

$$(N_j - N'_j) \cdot (a_j + K_2 \cdot P_j) > 0 \qquad (24.36)$$

This condition is always true because a_j, K_2, and P_j are positive defined and $N_j > N'_j$ when using RoIs as the number of channels to process decreases. Therefore, in this case, the system with RoIs will have an MF always smaller than the one without regions.

In a case where $K_{t_j} \not\to 0$, one can obtain K_{t_j}:

$$K_{t_j} \begin{smallmatrix}\leq \\ >\end{smallmatrix} \frac{2 \cdot (N_j - N'_j) \cdot (a_j + K_2 \cdot P_j)}{2 \cdot N_{RB_j} \cdot a_{RB_j} - (N_j \cdot a_j + (N_j - 1) \cdot K_2 \cdot P_j)} \qquad (24.37)$$

where the direction of the inequality (smaller than or bigger than) depends on the sign (positive or negative) of the denominator. If the denominator is negative, the condition of bigger than is always true because, as $N_j > N'_j$, the expression on the right of the inequality will be negative and, by definition, $K_{t_j} > 0$, assuring that the system with RoIs will have a smaller MF than the system without RoIs.

The denominator can now be evaluated to see when it is positive or negative:

$$N_{RB_j} \cdot a_{RB_j} \begin{smallmatrix}\leq \\ >\end{smallmatrix} \frac{N_j \cdot a_j + (N_j - 1) \cdot K_2 \cdot P_j}{2} \qquad (24.38)$$

The expression on the left of the inequality is the relationship between the total processing capacity of the RB and the total one of the problem. The right of the expression is the half-sum of the relative processing power of the processors of the level and of the fuser referred to the total one.

Therefore, if the relative processing power of the RB is smaller than half of that of the level (considering the processors and the fuser), the denominator will be negative and it will be true, for the reason mentioned before, that the system with regions of interest has an MF smaller than the one for the system with RoIs. If, on the contrary, the relative processing power of the RB is bigger than half of that of the level, it will depend on the value of K_{t_j} whether the MF of the hierarchical system with RoIs is smaller or bigger than the one without RoIs.

If the number of processing units of the RB is limited by their processing power, Equation 24.38 transforms to offer the value of γ, the ratio of the RB processing capacity and the total one of the problem:

$$\gamma_j \lessgtr \frac{N_j \cdot a_j + (N_j - 1) \cdot K_2 \cdot P_j}{2} \qquad (24.39)$$

If the number of units of the RB is limited by the bandwidth, the relationship between the relative processing power and the relative bandwidth of each unit of the RB is obtained:

$$\frac{a_{RB_j}}{b_{RB_j}} \lessgtr \frac{1}{P_j} \cdot \frac{N_j \cdot a_j + (N_j - 1) \cdot K_2 \cdot P_j}{2} \qquad (24.40)$$

Equation 24.38 through Equation 24.40 are equivalent and the values obtained allow the calculation (Equation 24.37) of the new number of processing units at each level with the value of K_{t_j} which, in turn, can be evaluated once the processing power of the units of the RB is known. The new value for the number of units of the level is related to the reduction factor due to the employment of RoIs through Equation 24.31.

Thus, it is demonstrated that, under certain conditions, it is possible to find a hierarchical system with RoIs with lower technical requirements (lower MF) than the hierarchical system without them.

24.2.2.2 Data Clustering

Another of the improvements that can be introduced in the system is to try to avoid the dispersion of the data to process in each processing element at each level in the hierarchy. The use of an RoI suggests that all its channels should be processed in a combined way because the elaboration of the decision will be made on the basis of existing relationships among the values of the channels. If the levels of the hierarchy are implemented like parallel systems, it can happen that the channels of the RoI may be distributed among several processors, making intercommunication necessary, and thus increasing the time necessary for processing and reducing the performance of the system.

A solution to this problem is to try to gather the data so that one can maximize the probability that all the channels of an RoI are sent to only one processing unit of the parallel system inside the level. The way to do this is to study the problem to discover channels that will be part of an RoI with bigger probability. This implies that the physical process has a certain bias and is present more probably in certain subspaces of the measure space.

If this is not the case, then two options exist:

- To use the information of the RoIs to carry out a dynamic routing of the data. This solution is good but it needs the implementation of a channel multiplexing system and the use of delays to prepare the information of the RoI before routing the channels.
- To gather the channels in a static form, but following some relationship with the physical phenomenon that is being observed. This solution implies, on occasion, the necessity to exchange data among the processors, but the introduced delay will be smaller than in the case of dynamic routing.

The case in which all RoI data are supposed to be in the correct processing unit is the one studied previously. In the second case, the sharing of data will introduce delays in the processing. If one wants to keep the total processing time at the level of the hierarchy constant, the effect of the delay would be to increase the processing capacity units or its number if the processing capacity stays constant.

Taking the hierarchical system with regions of interest as a reference, the total processing power, PC_i, of the parallel system in the level i of the hierarchy can be expressed as:

$$PC_i = \delta_{i-1} \cdot \alpha_i \cdot \frac{D \cdot op}{t_i} \quad 1 \leq i \leq E \tag{24.41}$$

where D is the number of data to process; op is the number of operations to carry out per datum; t_i is the time to carry out the processing at level i; and δ is the reduction factor due to the employment of RoIs.

For each of the N_i units of this parallel system, the processing power is:

$$PC_{pu_i} = \delta_{i-1} \cdot \alpha_i \cdot \frac{D/N_i \cdot op}{t_i} \quad 1 \leq i \leq E \tag{24.42}$$

If we now have the possibility of exchanging data, this will modify the number of operations to carry out in function of the probability that this exchange will occur. However, if one wants to maintain the number of units, the processing power of each one of them will be increased. The relationship between the new processing power, PC'_{pu_i}, and the previous one is:

$$PC'_{pu_i} = \delta_{i-1} \cdot \alpha_i \cdot \frac{D/N_i \cdot op + p_{ai} \cdot D/N_i \cdot op_a}{t_i} = PC_{pu_i} \left[1 + p_{ai} \frac{op_a}{op} \right] \quad 1 \leq i \leq E \tag{24.43}$$

where op_a is the number of operations to carry out to get the data from other units of level i and p_{ai} is the probability of having to make this access; to simplify the calculation, this probability is supposed equal for all the units.

As observed, the increment in the computing power is proportional to the ratio between the number of total operations carried out when accessing other units and the number of operations when this access is not required. If one wants to maintain the processing power, then it is necessary to increase its number. To calculate this increment, we make:

$$PC'_{pu_i} = \delta_{i-1} \cdot \alpha_i \cdot \frac{D/N_i'^{(PC)} \cdot op + p_{ai} \cdot D/N_i'^{(PC)} \cdot op_a}{t_i} = \delta_{i-1} \cdot \alpha_i \cdot \frac{D \cdot op}{N_i^{(PC)} \cdot t_i} \quad 1 \leq i \leq E \tag{24.44}$$

Solving to get $N_i'^{(PC)}$, the new number of units according to the processing power,

$$N_i'^{(PC)} = N_i^{(PC)} \cdot \left[1 + p_{ai} \frac{op_a}{op} \right] \quad 1 \leq i \leq E \tag{24.45}$$

where $N_i^{(cp)} = PC_i / PC_{pu_i}$.

In either of the two cases, the required bandwidth is increased because now it must cope not only with the data of the channels but also with the data transfers between processors. To simplify the problem, suppose that the data request between units is uniformly distributed and that the request probability for data exchange is equal for all the units. If this is not the case, then the analysis would get more complicated with the introduction of the probability distribution functions of the requests, as well as that of their destination.

Assuming the simplest case, each unit carries out $D \cdot p_a$ accesses distributed among the $N-1$ remaining units, and receives the same amount from all those $N-1$ other units, where p_a is the probability of requesting external data. In this way, the required bandwidth will be:

$$BW'_{pu_i} = P_i \cdot \frac{D \cdot S_{dat}/N_i'^{(BW)} + p_{ai} \cdot D \cdot S_{dat}/N_i'^{(BW)} + \cdot p_{ai} \cdot D \cdot S_{dat}/N_i'^{(BW)}}{t_1} =$$

$$= P_i \cdot \frac{D \cdot S_{dat}/N_i'^{(BW)} + 2 \cdot p_{ai} \cdot D \cdot S_{dat}/N_i'^{(BW)}}{t_1} = \frac{D \cdot S_{dat} \cdot P_i}{t_1 \cdot N_i'^{(BW)}} \cdot (1 + 2 \cdot p_{ai}) = \frac{D \cdot S_{dat} \cdot P_i}{t_1 \cdot N_i'^{(BW)}} \quad 1 \le i \le E$$

(24.46)

From here, one can get the new number of units depending on the bandwidth, $N_i'^{(BW)}$:

$$N_i'^{(BW)} = N_i^{(BW)} \cdot (1 + 2 \cdot p_{ai}) \tag{24.47}$$

with $N_i^{(BW)} = BW_i/BW_{pu_i}$.

Therefore, the number of necessary units, N_i'', will be the biggest of the results of Equation 24.45 and Equation 24.47. That is,

$$N_i'' = \max\left(N_i'^{(PC)}, N_i'^{(BW)}\right) \tag{24.48}$$

As can be seen, when introducing a cluster of static data, we modify the number of units of the system and therefore its MF.

The new expression of the MF of the hierarchical system with data clustering is obtained by simply substituting in Equation 24.33 the number of units for the resulting value of Equation 24.48:

$$MF'_{H_{RoI}} = BW \cdot PC \cdot N_1 \cdot b_1 \cdot \sum_{i=1}^{E} \frac{\left(N_i'' \cdot a_i + (N_i'' - 1) \cdot K_2 \cdot P_i + N_{RB_i} \cdot a_{RB_i} \cdot K_{t_i}\right)}{(2 + K_{t_i}) \cdot P_i \cdot \sum_{j=1}^{E} 1/P_j} \tag{24.49}$$

If, in addition to data clustering, another type of data processing is used (formatting, detection and correction of errors, etc.), this would be reflected as a sequential element with the processors and the MF would increase. Depending on the particular case, the final result of the MF for the system with RoI could be greater than for the system without RoI.

24.3 Example of Architecture Evaluation in High-Energy Physics

High energy physics experiments try to confirm theories by detecting and measuring particle properties. For this kind of experiment, accelerators and particle detectors are used. These last are organized as a distributed sensor network with thousands, or even millions, of sensors of different types whose information must be processed in a short time, which leads to high data rates. Traditionally, hierarchical data acquisition systems have been employed for data taking because, of the total amount of data acquired, only a few are of interest. The reason is that not all the particles produced by the accelerator are of interest.

CERN, the European Laboratory for Particle Physics, is the most important laboratory in the world for the study of particle physics. It holds the biggest accelerator in construction nowadays — the LHC (large hadron collider [5]) — in which two beams of protons will collide with an energy near 14 TeV (tera electron-volts) to study the origin of mass by searching a new particle, the Higgs boson.

For analysis of the collision results, two big detectors, ATLAS [6] and CMS [7], are being constructed. ATLAS will be a huge toroid, 22 m long and 32 m high, with more than 170 million electronic channels to read, coming from sensors inside the detector. All these channels sum a total of 1.3 Mbytes to be read every 25 ns, which gives a rate of 50 TBytes/s. The total processing capacity needed to perform all the operations is estimated at $5 \cdot 10^{10}$ MIPS.

TABLE 24.1 Main Parameters of Hierarchical System Levels

	BW_{input} (MB/s)	PC (MIPS)	p_i	N_i	MF (MB/s × MIPS)
Level 1	$14.84 \cdot 10^6$	$45 \cdot 10^6$	1/400	$1.12 \cdot 10^6$	$1.24 \cdot 10^{11}$
Level 2	10^4–10^5	$32 \cdot 10^6$	1/100	$32 \cdot 10^4$	$3.5 \cdot 10^{13}$
Level 3	10^3–10^4	10^6	1/10	10^4	$1.1 \cdot 10^{14}$

TABLE 24.2 Comparison of MF for the Second Level with and without RoIs

	BW_{input} (MB/s)	PC (MIPS)	N_i	MF (MB/s × MIPS)
Without RoI	10^4–10^5	$32 \cdot 10^6$	$32 \cdot 10^4$	$3.5 \cdot 10^{13}$
With RoI	10^4–10^5	$144 \cdot 10^3$	1440	$1.548 \cdot 10^{11}$

A parallel solution for this problem would require, assuming 100 MIPS and 200 MB/s processors, 500 million processing units according to Equation 24.9. For the calculation of the MF, one needs the value of K_1 and K_2, which can be estimated [6] as $K_1 = 3'33 \cdot 10^{-9}$ and $K_2 = 6'67 \cdot 10^{-18}$. The a and b values are $a = 100/5 \cdot 10^{10} = 2 \cdot 10^{-9}$ and $b = 200/5 \cdot 10^7 = 4 \cdot 10^{-6}$. All these data compute a total MF of $2.5 \cdot 10^{21}$ MIPS × Mbytes/s.

A three-level hierarchical solution can be implemented using 40 MIPS and 200 Mbytes/s hardware processors for the first level and 100 MIPS, 200 Mbytes/s processors for the two other levels. The other parameters for this solution are, from ATLAS Collaboration [6], $a_1 = 8 \cdot 10^{-10}$; $a_2 = a_3 = 2 \cdot 10^{-9}$; and K_1 and K_2 equal to the values used in the parallel solution because they do not depend on the architecture but on the characteristics of the problem. Table 24.1 summarizes the MF for each level. The total result of the series of the three levels is $1.45 \cdot 10^{14}$.

The introduction of RoIs improved the hierarchical system. For this case, only level 2 will include RoI and RB. Simulations made [8] showed that the average number of RoIs per acquisition in level 2 would be 5, each one with 135,000 channels [9]. This leads to a reduction factor of $\delta_1 = 4.5 \cdot 10^{-3}$. The RB is estimated as a processing system of 500 units with a total processing capacity of $12 \cdot 10^3$ MIPS [10], which make the γ_1 parameter equal to $2.4 \cdot 10^{-7}$. From Brawn et al. [10], the K_{T2} parameter can be estimated to a value of $K_{T2} = 0.0875$. With these data, the new number of processing units at level two is 1440. The MF for this second level, where RoIs have been applied, is then recalculated. Table 24.2 shows the differences with and without RoIs.

References

1. Iyengar, S.S., Prasad, L. and Min H., *Advances in Distributed Sensor Integration. Application and Theory*, 1st ed., Prentice Hall, Englewood Cliffs, NJ, 1995, chap. 2.
2. Durrant–Whyte, H.F., Sensor models and multisensor integration, *Int. J. Robotics Res.*, 7(6), 97, 1988.
3. Flynn, M.J., Very high speed computing systems, *Proc. IEEE*, 54, 1901, 1966.
4. Nagin, P.A. et al., Region relaxation in a parallel hierarchical architecture, in *Real-Time Parallel Computing Image Analysis*, 1st ed., Onoe, M., Ed., Plenum Press, New York, 1981, 37.
5. LHC Study Group, *The Large Hadron Collider Accelerator Project*, CERN, Switzerland, 1993.
6. ATLAS Collaboration, *ATLAS Technical Proposal*, CERN, Switzerland, 1994.
7. CMS Collaboration, *CMS Technical Proposal*, CERN, Switzerland, 1994.
8. Kozlov, V., Functional simulation of detector, front-end and read-out parts of a LHC-like DAQ architecture, Report CERN/RD13-120, CERN, Switzerland, 1994.
9. Bock, R. and LeDu, P., Detector and readout specifications for the level-2 trigger demonstrator program, Report ATLAS/DAQ-NO-53, CERN, Switzerland, 1996.
10. Brawn, I.P. et al., The level-1 calorimeter trigger system for ATLAS, Report ATLAS-DAQ-NO-30, CERN, Switzerland, 1995.

25
Computational and Networking Problems in Distributed Sensor Networks

Qishi Wu
Oak Ridge National Laboratory

Nageswara S.V. Rao
Oak Ridge National Laboratory

Richard R. Brooks
The Pennsylvania State University

S. Sitharama Iyengar
Louisiana State University

Mengxia Zhu
Louisiana State University

25.1 Introduction 25-1
25.2 Foundational Aspects of DSNs 25-2
  Traditional Network Architectures • Mobile Agent-Based Distributed Sensor Networks • Data Integration Methods
25.3 Sensor Deployment 25-5
  Computational Complexities • Optimal Sensor Deployment Using Genetic Algorithm
25.4 Routing Paradigms for DSNs 25-8
  Mobile Agent Routing Using the Genetic Algorithm • Connectivity through Time for Mobile Wireless Networks • Adaptive Routing Using Emergent Protocols
25.5 Conclusions and Future Work 25-16

25.1 Introduction

Multisensor systems have permeated many aspects of life in various applications over the past decade. The wide variety of applications of sensor networks spans civilian services such as environment surveillance and disaster relief; industrial processes such as instrument controls and machine monitoring; and military operations such as target detection, classification, and tracking on battlefields. The ever-increasing levels of sophistication of sensor network systems continue to generate a great deal of interest in development of new computational strategies and networking paradigms.

A distributed sensor network (DSN) is a set of geographically scattered sensors designed to collect information about the environment in which they are deployed. The physical measurements (of different types, such as acoustic, seismic, or infrared) from the terminal sensor nodes are preprocessed locally into abstract and/or numerical estimates; then they are transmitted through an interconnection communication network to a processing element, where they are integrated with the information gathered from other parts of the network according to some data fusion strategy. The integrated information is then used to derive appropriate inferences about the environment for the application. A group of neighboring sensors commanded by the same processing element forms a cluster. In tracking applications, each processing element in a DSN performs a tracking function using the data from its governing cluster and possibly communicates with other processing elements in the same network to arrive at a better estimate.

Development and implementation of such spatially distributed systems involves solving a combination of many different problems in sensor deployment; network communication; data association and fusion; hypothesis testing; and other areas. In particular, design and analysis of information integration algorithms has been the focus of research since the early stages of DSN development [1, 2]. Recent advances in sensor technologies make it possible to deploy a large number of inexpensive sensors in order to achieve "quality through quantity." Exploiting useful information from an enormous amount of data collected from spatially distributed sensors in the most effective way has brought new challenges such as network architecture design, data fusion methods, sensor deployment schemes, and data routing techniques to all aspects of DSNs.

In this chapter, a broad survey of recent research efforts in the computational aspects and networking paradigms of distributed sensor networks is conducted. Section 25.2 provides a general description of the fundamental aspects of DSNs such as network architectures and multisensor data fusion methods. Two specific computational topics are covered in the next two sections. Section 25.3 discusses the computational complexities of sensor deployment problems and presents an approximate solution based on a genetic algorithm. Section 25.4 is devoted to the networking paradigms for DSNs with a focus on data routing techniques. Section 25.5 draws conclusions for all the work presented and discusses some future research directions.

25.2 Foundational Aspects of DSNs

Efficient and fault-tolerant network architectures play a very important role in successful implementations of DSNs. Apart from the timeliness and complexity of information transmission, interconnection topology has a significant impact on the computational aspects of data routing and sensor deployment schemes discussed in later sections. Therefore, the overall performance of a DSN is critically dependent on its network architecture.

Design of algorithms for data integration is one of the core tasks in the development of DSNs and has attracted a great deal of research attention during the past decades. Recent advances in sensor technology have led to better, less expensive, and smaller sensors. These advances beget a more complex tactical deployment of sensors that requires efficient and sophisticated techniques for fault-tolerant integration of sensor information. This section provides a general description of these two fundamental aspects of distributed sensor networks.

25.2.1 Traditional Network Architectures

Committees and hierarchical organizations are two basic types of network architectures [3]. In the network of a committee organization, each node is autonomous and connected to some or all of the other nodes so that the local information can be broadcast between any two connected nodes. The information collected by individual nodes in this organization is shared within the network to the fullest extent. The completely connected network is one special case of the committee organization extensively used in many practical applications. However, because $O(N^2)$ connections are required in such a network with N nodes, the network size imposes a high demand on communication resources. Moreover, the final estimate obtained in a committee organization tends to be biased because the data are shared by all participating nodes during the integration process.

A hierarchical organization arranges the nodes, each of which can only communicate with its immediate subordinate and superior nodes, in multiple levels. At each level, individual nodes receive information from the nodes at the lower level, integrate the information according to their position in the hierarchy, and report the integrated and abstracted versions of their results upward. The commander node at the highest level makes the appropriate decisions based on received information and may direct its subordinates to adjust some previous data based on the final result that it generates. In contrast to a committee organization, a hierarchical organization with N nodes only needs to create $O(N)$ links but, consequently, requires more complex communication schemes and incurs longer communication delays.

An unbiased result may be obtained because the nodes are not connected to any other nodes at the same level, but the integration errors may accumulate as the estimate moves up the hierarchy.

Due to disadvantages in committee and hierarchical organization, it is not appropriate to design the network architecture for DSNs as one of them alone. In practice, a mixed structure combining these two basic types of architectures is preferable. For example, the JIK (Jayasimha, Iyengar, Kashyap) network has such a structure, in which nodes are organized as many complete binary trees whose roots are completely connected [1, 2]. However, the JIK network still has the disadvantage of accumulated integration error, as in a hierarchical organization, thus making it difficult to identify the faulty component of the network.

Iyengar and colleagues [4] improved this by interconnecting the nodes at every level of the JIK network as a de Bruijn network, which results in a new versatile architecture referred to as the binary multilevel de Bruijn network (BMD). The BMD structure is often used as a basis for network architecture design in DSN implementation because of several promising fault-tolerant properties that make the resultant network tolerant to node or link failures. Because nodes at every level are interconnected, the BMD network is capable of eliminating estimate errors and identifying the faulty component during the process of sensor integration by comparing abstract estimates at the same level.

25.2.2 Mobile Agent-Based Distributed Sensor Networks

A novel architecture using mobile agents to meet the new challenges of the current DSNs, such as large data volume, low communication bandwidth, and unreliable environment, has been proposed by Qi and coworkers [5]. Instead of sending all the measurements collected by leaf nodes to the upper-level processing element (that performs a one-time data fusion), as occurs in a traditional hierarchical network with the server–client structure, the mobile agent-based distributed sensor network (MADSN) distributes the computation into the participating leaf nodes. Thus, this approach makes it possible to reduce the consumption of communication power and bandwidth significantly, while lowering the risk of being spied upon with hostile intent.

A MADSN is usually divided into an appropriate number of subtasks, each carried out by a mobile agent carrying the executable instructions for data integration dispatched by the processing element. The agents selectively visit the leaf sensors along a certain path to fuse the data incrementally on a sequential basis. A final data fusion is performed when all mobile agents return to the processing element. Qi et al. [5] address three technical issues associated with MADSNs: mobile agent routing, data integration, and optimum performance.

The objective of mobile agent routing is to find an optimal path for a mobile agent to visit the sensor nodes. The path quality has a significant impact on the overall performance of MADSN implementation because communication cost and detection accuracy depend on the order and the number of nodes to be visited. The NP-completeness of this problem (for a detailed proof, see Wu et al. [6]) rules out any polynomial-time solutions (unless $P = NP$). A formal description as well as an appropriate objective function of the mobile agent routing problem with certain constraints is provided in Section 25.4. That section also discusses an approximate solution based on a two-level genetic algorithm (GA) proposed by Wu and colleagues [6] and the simulation results of the GA solution are compared with those computed by two other heuristics, namely, local closest first (LCF) and global closest first (GCF).

Data integration takes into consideration problems such as the type of data processing to be conducted at the nodes and the integration results to be carried with the mobile agent. An overlap function is particularly designed to integrate the abstract estimate intervals collected from all participating nodes. In a regular DSN, the overlap function at the finest resolution is first generated at processing elements based on all readouts from the leaf sensor nodes; the multiresolution analysis procedure is then applied to find the crest at the desired resolution. In a MADSN, mobile agents migrate among sensor nodes to collect readouts and execute an overlap function of partial integration, whose results are accumulated into a final version upon the arrival of all mobile agents. The basic multiresolution integration (MRI) algorithm is adapted to MADSNs by Qi et al. [5], who applied MRI before accumulating the overlap function in order to avoid heavy data transmission.

In addition to the routing scheme and integration function, the performance of MADSNs depends on many other factors. Actually, a MADSN does not always guarantee lower data transfer time because of the overheads of time for agent creation and dispatch, and the latency of data routing. Qi and colleagues [5] make performance comparisons between DSNs and MADSNs in terms of various parameters, such as the number of agents, agent and file access overhead ratio, network transfer rate, and the number of nodes.

25.2.3 Data Integration Methods

In many military or civilian applications, sensors are typically deployed in hazardous or harsh environments where the sensor operations and data communications are not as reliable as in regular computer networks installed in structured areas. Therefore, fault tolerance is an indispensable property of data integration algorithms. The measurements collected by sensors are usually processed into interval-valued estimates serving as the inputs of an overlap function, whose redundancy may be used to provide error tolerance.

Marzullo's method yields the smallest sensor fusion interval guaranteed to contain the correct value [7]. The common sensor averaging technique by Marzullo's method combines the intervals of sensors by computing local averages. This method, however, is not stable because it exhibits an irregular behavior in the sense that a slight difference in the input may produce a quite different output. This behavior results from violation of the Lipschitz condition with respect to a certain metric on intervals [8]. Improvements can be made by combining interval estimates of sensor outputs into a best intersection estimate of outputs.

The Schimd–Schossmaier function proposed in Cho et al. [9] is a fault-tolerant interval intersection function with the same worst-case behavior as the Marzullo function but satisfying the Lipschitz condition. However, this method sacrifices integration accuracy because it produces suboptimal output intervals in some cases. Li and colleagues [10] have proposed a new fault-tolerant interval integration function based on the Dempster–Shafer theory of evidence that provides a smaller output interval than the one calculated by Marzullo function and also satisfies the local Lipschitz condition.

The Brooks–Iyengar hybrid algorithm [11] makes a weighted average of the midpoints of the regions found by the sensor fusion algorithm. The hybrid algorithm allows for increased precision, while not sacrificing accuracy in the process. A distributed system using this algorithm is truly robust and converges toward an answer within a precisely defined accuracy bound.

Most recently, Cho et al. [12] have proposed a new interval integration method that further narrows the region containing the true value of the state measured by the sensors. This proposed function satisfies the local Lipschitz condition, tolerates failures of interval-valued sensors up to a certain number, and has better performance than existing fault-tolerant interval integration functions. The detailed analysis of how this function yields a narrow interval accurately estimating the true value is given in Cho et al. [12], as well as a comparison of this new function with the existing fault-tolerant interval integration functions.

Another important formulation of the data fusion deals with combining information from multiple sensors to obtain results better than the best or best subset of sensors. Such problems are extensively studied in the target detection and tracking area [13]. Although similar problems have been studied for centuries (early work under the title Condorcet Jury models in the 18th century), recent DSNs call for a specific new formulation of data fusion problems [14]. By far a majority of these problems involve deriving a Bayesian fuser based on the joint distributions of the sensors.

However, such an approach is useful only when the joint sensor distributions are known as well as expressed in a computationally conducive form [13]. In view of the increasing sophistication of DSNs, it is particularly difficult to obtain such joint distributions; note that it is insufficient to know the individual sensor distributions because an optimal fuser must exploit the correlations between the sensors. On the other hand, it is relatively easy to collect measurements from the various sensors of a DSN by sensing known objects. Such measurements are shown to be sufficient to design fusers that can be shown to be close to optimal with a high probability [15].

25.3 Sensor Deployment

This section discusses the computational complexities of various sensor deployment schemes and presents an approximate solution to one of them based on a genetic algorithm.

25.3.1 Computational Complexities

A general sensor deployment problem and several variations have been formulated; their computational complexities are discussed by Wu et al. [16]. The sensor deployment problems can be categorized into different paradigms: (1) probabilistic deployment with investment limit, referred to as the probabilistic deployment; (2) minimum sensor set deployment for target coverage, referred to as the minimum coverage; and (3) deployment for integrity. Each has a specific application goal and certain constraint conditions. The NP-completeness proofs for the first two deployment paradigms are briefly described next.

25.3.1.1 Probabilistic Deployment

The deployment objective of the probabilistic-deployment paradigm is to place a set of sensors with probabilistic detection capability in a grid space so that the maximum detection probability is achieved under the constraint of an investment limit. Intuitively speaking, the whole surveillance region is to be covered as much as possible, while the total deployment expense does not exceed a given investment budget, where the "deployment expenses" only use abstract values incurred in purchasing available sensors.

The probabilistic-deployment problem can be shown to be NP-complete by reducing the knapsack problem to a special case of this paradigm, wherein each sensor monitors a detection area with a specified probability without overlapping with any other sensors. The knapsack problem is a well-known NP-complete problem, which is stated next for the sake of completeness:

Given a set U of n items such that, for each $u \in U$, size $s(u) \in Z^+$ and value $v(u) \in Z^+$. Does a subset $V \in U$ of exactly k items exist such that $\sum_{u \in V} s(u) \leq B$ and $\sum_{u \in V} v(u) \geq K$ for given B and K?

Note that exactly k items are required in the preceding problem statement, as opposed to unrestricted value in a traditional knapsack problem. Both problems are polynomially equivalent because $k \leq n$ and the input for either problem instance has at least n items. In the same vein, the decision version of the probabilistic-deployment problem asks for a deployment scheme consisting of exactly k sensors to be deployed.

The knapsack problem is reduced to the probabilistic-deployment problem so that only one sensor of each type is given, i.e., $q_1 = q_2 = \ldots = q_n = 1$; each sensor S_t of type t monitors a small area (compared with the whole arbitrarily large surveillance region) of size $r(t)$; and when two sensors are located in the same site, only one of them detects the target (i.e., suitable conditional probabilities are zero). For this special case, to maximize the detection probability, each deployment site is occupied by no more than one sensor. Furthermore, under the uniform prior distribution of targets in the surveillance region combined with the nonoverlapping sensor detection area, the probability of detection is simply the average of the detection probabilities of the deployed sensors. Therefore, a sensor deployment scheme $\Re$ with k deployed sensors has the detection probability calculated as $P\{\Re \mid T \in R\} = \frac{1}{k}\sum_{t=1}^{k} P\{S_t \mid T \in r(t)\}$.

Given an instance of the knapsack problem, each $u \in U$ is mapped to a sensor S_u so that its cost and value are given by $w(u) = s(u)$ and $P\{S_u | T \in r(u)\} = \dfrac{v(u)}{\sum_{a \in U} v(a)}$, and the sensor cost bound is specified as $Q = B$ and the detection probability as $A = \dfrac{K}{k \sum_{a \in U} v(a)}$.

Given a solution to the knapsack problem, a solution to the probabilistic-deployment problem exists by placing the sensors corresponding to the members of V on nonoverlapping grid points. On the other hand, given a solution to the sensor deployment problem, a solution to the knapsack problem can be obtained by choosing the items corresponding to the deployed sensors. The first condition ensures that $\sum_{u \in V} s(u) \leq B$, and the second condition ensures that $P\{\Re | T \in R\} = \dfrac{1}{k} \sum_{u \in V} P\{S_u | T \in r(u)\} = \dfrac{1}{k} \sum_{u \in V} \dfrac{v(u)}{\sum_{a \in U} v(a)} \geq A = \dfrac{K}{k \sum_{a \in U} v(a)}$ which in turn ensures that $\sum_{u \in V} v(u) \geq K$.

25.3.1.2 Minimum Coverage

In the minimum-coverage paradigm, the objective is to cover some set T of targets completely by a minimum size of set S of sensors in a surveillance region R. Its corresponding decision problem is defined as follows: given some set T of targets in a surveillance region R, determine whether some set S of sensors can completely cover all the targets. It is shown that even the restricted version of the minimum-coverage problem remains NP-complete. The proof directly follows [17].

In the restricted version of the minimum-coverage problem, a finite surveillance region R is divided into a number of uniform contiguous square cells of unit size. Any target is only located at a corner of one cell. The detection area of a sensor is a disc of some size centering at the sensor's location. In other words, each sensor has isotropic detection capability. The sensor's location can be anywhere within the surveillance region.

It is straightforward that the minimum-coverage problem belongs to NP because a successful deployment scheme can always be used as a certificate in an instance of the problem. The verifying algorithm simply checks whether every target is located within some sensor's detection area and that the number of deployed sensors does not exceed the size of the given sensor set. Obviously, this verification process can be done in polynomial time. The minimum coverage can be shown to be NP-complete by finding a polynomial-time reduction algorithm from 3-SAT to minimum coverage, i.e., $3 - SAT \leq_P optimal\ coverage$. The proof details will not be given here.

25.3.2 Optimal Sensor Deployment Using Genetic Algorithm

The NP-completeness of the probabilistic-deployment problem rules out any polynomial-time solution unless $P = NP$. A suboptimal solution is presented by Wu and colleagues [18] based on a two-dimensional genetic algorithm. This starts with a set of initial solutions and applies genetic operators to produce better solutions using random optimization techniques until a satisfactory solution is obtained. The probabilistic-deployment problem is adapted to a solution based on a genetic algorithm by reducing to a simple version where the surveillance region is restricted to a two-dimensional grid space. The method can be easily extended and applied to a three-dimensional case.

A two-dimensional surveillance region is divided into a number of uniform contiguous rectangular cells with identical dimensions. Each cell of R is labeled by a pair of indices, (i,j), and $C(i,j)$ denotes the corresponding rectangular cell. This planar surveillance region R is monitored by a set of sensors placed in it to detect a target T if located somewhere in the region. A sensor is specified by its local detection probability of detecting a target at a point within its detection region. Normally, detection is more likely as a target approaches the sensor. The cumulative detection probability of a sensor for a region is computed by integrating its local detection probability for detecting a target as the target gets close to the sensor, passes near it, and then leaves it behind.

Given the detection probability density function $p_{S_k}(x)$ for a sensor S_k of type k, its detection probability $P\{S_k | T \in C(i,j)\}$ in a cell $C(i,j)$ is defined by $P\{S_k | T \in C(i,j)\} = \int_{x \in C(i,j)} p_{S_k}(x) dx$.

To better approximate the sensor detection performance, a Gaussian function is used to formulate the measure of the continuous cumulative detection probability, which is defined by:

$$P\{S_k, \tau, \alpha_{S_k} | T \in A_{S_k,\tau}\} = e^{-\frac{\tau^2}{2 * \alpha_{S_k}^2}}, \quad \tau \in [0, d_{S_k}]$$

where $P\{S_k, \tau, \alpha_{S_k} | T \in A_{S_k,\tau}\}$ is a measure of integrated detection probability at the distance of τ to the target from the sensor location; and α_{S_k} is a coefficient parameter that determines the sensor detection quality. Distance τ is within the range between 0 and the maximum detection distance d_{S_k} of sensor S_k.

A sensor deployment is a function $\Re$ from the cells of R to $\{\varepsilon, 1, 2, \ldots, q\}$ such that $\Re(i,j)$ is the type of sensor deployed in cell (i,j); and $\Re(i,j) = \varepsilon$ indicates no sensor is deployed in cell (i,j), i.e., the deployment cost in that particular cell $w(\varepsilon) = 0$. The expense of a sensor deployment $\Re$ is the sum of the costs of all the sensors deployed in region R, defined by: $Cost(\Re) = \sum_{C(i,j) \in R} w(\Re(i,j))$.

The detection probability of deployment $\Re$, given by $P\{\Re | T \in R\}$, is the probability that a target T located somewhere in region R will be detected by at least one deployed sensor, which is evaluated by calculating the sum of all the local detection probabilities in the surveillance region as follows: $P\{\Re | T \in R\} = \sum_{i=0}^{m-1} \sum_{j=0}^{n-1} P\{\Re | T \in C(i,j)\} * P\{T \in C(i,j)\}$. According to the assumption that the location of the target has a uniform distribution in the surveillance region, the probability of target T appearing in cell $C(i,j)$ is given by: $P\{T \in C(i,j)\} = 1/(m*n)$. By plugging the occurrence probability of target T in a cell into the detection probability expression, the objective function for the genetic algorithm is obtained as: $P\{\Re | T \in R\} = \sum_{i=0}^{m-1} \sum_{j=0}^{n-1} P\{\Re | T \in C(i,j)\}/(m*n)$ with a constraint of investment limit $Cost(\Re) \leq Q$.

The genetic algorithm is a computational model that simulates the process of natural selection and adaptation in biological evolution. It has found many applications in various areas solving the combinatorial and nonlinear optimization problems with complicated constraints or nondifferentiable objective functions. The computation of the genetic algorithm is an iterative process toward achieving global optimality. On each iteration, candidate solutions are retained and ranked according to their qualities, which are indicated by their fitness values calculated based on the objective function. Any unqualified solutions are screened out of the population.

Genetic operators such as crossover, mutation, translocation, inversion, addition, and deletion are then performed on qualified solutions to produce a new generation of candidate solutions. The preceding process is carried out repeatedly until a certain convergence condition is satisfied; for example, the preset maximum generation number is reached, or the variation of fitness values between two adjacent generations is smaller than a given threshold value.

In the preceding sensor deployment problem for a surveillance region, a candidate solution can be represented by a two-dimensional matrix of sensor IDs. Thus, a two-dimensional numeric encoding scheme is adopted to make up the chromosomes instead of using the conventional linear sequence. Each element in the matrix corresponds to a cell within a surveillance region. As mentioned earlier, an empty value ε in the matrix indicates that no sensor is deployed in its corresponding cell, which must be covered by the sensors deployed in its neighborhood area.

A detailed description of the genetic algorithm implementation, including fitness function construction, genetic operator design, and candidate solution selection, can be found in Wu et al. [18]. The simulation results of different surveillance region sizes up to 1000×1000 grid points with various sensor types are also presented. Due to the difficulty of quantitatively evaluating the genetic algorithm, the performance of the solution based on genetic algorithm is compared with that of the uniform placement in terms of deployment expense and average detection probability.

25.4 Routing Paradigms for DSNs

Because the network is a critical part of a DSN, the various parts of the underlying network must be carefully designed. Various transport aspects of DSNs can be handled by suitably deployed network daemons [19]. This section discusses various routing aspects of DSNs.

25.4.1 Mobile Agent Routing Using the Genetic Algorithm

A MADSN with a simple network configuration is shown in Figure 25.1 for illustrative purposes. This sensor network contains one processing element, labeled S_0, and $N = 10$ leaf sensor nodes, labeled as S_i, $i = 1, 2, \ldots, N$, one of which is inactive or in the sleep state. The physical distances of wireless links are represented by $d_{i,j}, i \neq j, i = 0, 1, \ldots, N, j = 0, 1, \ldots, N$.

The processing element dispatches a mobile agent that visits a subset of sensors within the cluster to fuse data collected in the coverage area. Generally speaking, the more sensors that are visited, the higher the detection accuracy will be achieved using any reasonable data fusion algorithm [15]. However, visiting more sensors often incurs more communication and computing costs. The routing objective is to find a path for a mobile agent that satisfies the desired detection accuracy while minimizing energy consumption and path loss. An approximate solution based on a genetic algorithm proposed by Wu and coworkers [6] is briefly described next.

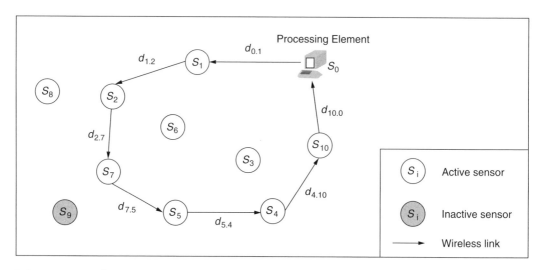

FIGURE 25.1 An illustration of a MADSN with simple configuration.

To facilitate the optimization process using genetic algorithm, an objective function of path P that considers the trade-off among energy consumption $EC(P)$, path loss $PL(P)$, and detected signal energy $SE(P)$ is defined as: $O(P) = SE(P)\left(\dfrac{1}{EC(P)} + \dfrac{1}{PL(P)}\right)$. The detailed derivation of formulas for calculating each component in $O(P)$ is presented in Wu et al. [6]. Through the punishment technique, the objective function is converted to a fitness function: $f(P) = O(P) + g$, where g represents the punishment applied for overriding the detection accuracy constraint and is defined by $g = \begin{cases} 0, & SE(P) \geq E \\ \delta \cdot (SE(P) - E)/E & SE(P) < E \end{cases}$, where E is the desired detection accuracy or signal energy level and δ is a properly selected penalty coefficient.

A two-level encoding scheme adapts the generic string-based genetic algorithm to the mobile agent routing problem in MADSN. The first level is a numerical encoding of the sensor (ID) label sequence L in the order of sensor nodes being visited by mobile agent. For the MADSN shown in Figure 25.1, the sensor label sequence L has the following contents:

The first element is always set to be 0 because a mobile agent starts from the PE S_0. The mobile agent returns to S_0 from the last visited sensor node, which is not necessarily the last element of the label sequence if inactive sensor nodes are in the network. This sequence consists of a complete set of sensor labels because it participates in the production of a new generation of solutions through the genetic operations. The new generation is required to inherit as much information as possible from the old one. For example, in Figure 25.1, although nodes 3, 6, 8, and 9 are not visited in the given solution, they or some of them may likely make up a segment of a better solution in the new generation than in the current one.

The second level is a binary encoding of the visit status sequence V in the same visiting order. For the MADSN illustrated in Figure 25.1, the visit status sequence V contains the following binary codes:

| 1 | 1 | 1 | 0 | 1 | 1 | 0 | 0 | 1 | 1 | 0 |

where "1" indicates "visited" and "0" indicates "unvisited." The first bit corresponds to the PE and is always set to be 1 because the PE is the starting point of the route. If a sensor is inactive, its corresponding bit remains 0 until it is reactivated and visited.

A candidate path P for mobile agent can be generated by masking the first level of numerical sensor label sequence L with the second level of binary visit status sequence V. In the above example, the path P is obtained as follows:

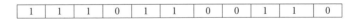

These two levels of sequences are arranged in the same visiting order for the purpose of convenient manipulations of visited/unvisited and active/inactive status in the implementation of genetic algorithm

Some common genetic operators such as crossover, mutation, inversion, and translocation, as well as a proportional selection procedure, are applied to these two levels of sequences simultaneously to create new solutions. These operators are modified from those used in the conventional genetic algorithm

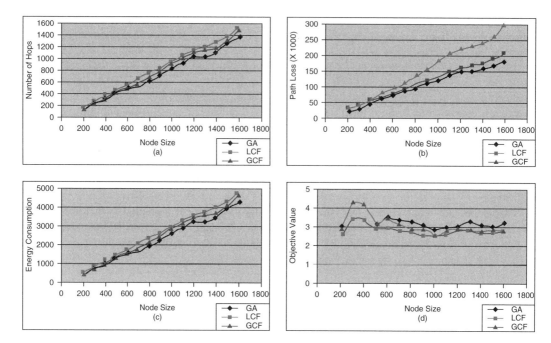

FIGURE 25.2 Performance comparison: (a) node sizes vs. hop numbers; (b) node sizes vs. path losses; (c) node sizes vs. energy consumptions; (d) node sizes vs. objective values.

solution to the traveling salesman problem in order to suit the current context of two-level string encoding. Their implementation details can be found in Wu et al. [6].

The search results computed by genetic algorithm are compared with those computed by the other two greedy heuristics, LCF and GCF, in order to demonstrate the effectiveness of the solution based on genetic algorithm. A series of sensor networks with random distribution patterns and node sizes ranging from 200 to 1600 are created for testing. An appropriate desired level of the detected signal energy for each network as well as the number of potential targets is manually selected. The sensors are randomly deployed and the targets are arbitrarily placed in the region. The comparisons of routing performance among GA, LCF, and GCF are illustrated in Figure 25.2 through Figure 25.5. Note that the quantities of path losses and energy consumption are "normalized" into reasonable ranges before they are plotted, and the objective value only serves as an indicator of the path quality according to the defined objective function, which does not bear a regular unit.

It can be observed from Figure 25.2 that, in most cases, GA is able to find a satisfying path with a smaller number of hops, lower energy consumption, and fewer path losses than LCF and GCF algorithms. The observations justify that GA has a superior overall performance over two other heuristics in terms of the defined objective function. More discussions on algorithm comparisons such as computing complexity, real-time constraint, and selection of starting point are provided in Wu et al. [6].

25.4.2 Connectivity through Time for Mobile Wireless Networks

The wireless connection is usually the only feasible means of communication between sensors in DSNs deployed in unstructured and harsh environments. Due to the lack of network infrastructures, wireless networks are always configured to operate in ad hoc mode. In some application scenarios, such as a team of robots exploring potentially radioactive areas, the moving nodes carrying sensors need to communicate effectively with other nodes in the network to coordinate their activities as well as to combine the gathered information in a timely manner. However, networking needs for this class of applications are quite specific and are not adequately addressed by existing wireless ad hoc networking technologies.

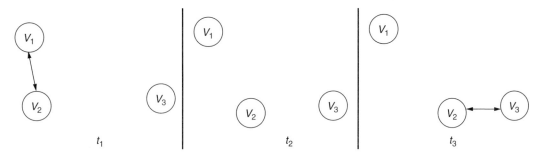

FIGURE 25.3 The CTT concept: data are delivered through intermediate node movements.

Wireless networks have very different operational characteristics from wired networks. First, packet losses in wireless networks are mostly due to physical link failures instead of network congestion. Second, the signal attenuation often causes the link to break down when environmental interferences increase or a node moves out of the maximum radio distance. Therefore, the network connectivity through wireless radio in ad hoc mobile networks can be highly dynamic, intermittent, and unpredictable. The data streams based on TCP may not meet the challenges imposed by these wireless operational characteristics because TCP is an end-to-end transport protocol that does not provide capabilities specifically accounting for connectivity constraints in wireless environments. In general, TCP needs routing support from the underlying routers and requires a continuous byte-stream connection between source and destination during the entire period of transmission.

Rao and colleagues have presented a concept of connectivity-through-time (CTT) and design a CTT protocol that utilizes node movements to enhance data transmission in ad hoc mobile networks [20]. A typical CTT example is illustrated in Figure 25.3; data are successfully delivered from a source node v_1 to a destination node v_3 even though they are never directly or indirectly connected to each other. The two-directional arrow represents a direct wireless link between two nodes located within the maximum wireless radio range.

The data delivery process is described as follows. At time t_1, the source node v_1 checks its neighbor list and notices that the destination node v_3 is unreachable at the moment. Node v_1 can wait until node v_3 comes into its radio coverage area or broadcasts the data to whomever is reachable, i.e., node v_2 in this case. Suppose node v_1 broadcasts the data as well as its destination information to node v_2, which afterwards carries the data and moves towards node v_3. At time t_2, node v_2 goes out of the radio ranges of node v_1 and node v_3 so that all three nodes are isolated. Eventually, node v_2 enters the radio area covered by node v_3 at time t_3. Once this new link is detected, node v_2 checks for the destination availability for all temporary data stored in its buffer. Because now node v_3 is reachable, node v_2 retrieves the data from its data repository and transmits them to the destination node v_3.

The CTT protocol is implemented based on user datagram protocol (UDP), which provides a connectionless data transmission service. The framework of CTT function modules is illustrated in Figure 25.4.

The connectivity computation module provides two main functions: connectivity detection and routing table construction. Each node actively broadcasts a special datagram named "IAmHere" attached with its current neighbor list to the network at a certain time interval. The receipt of such a datagram indicates that a wireless connection exists between the datagram sender and receiver. Based on the list of neighbor nodes, each node is able to construct a complete adjacency matrix of the network and compute a routing table that provides path information to the transport control module. Any changes affecting the network connectivity — for example, a link breaks down or comes back up — will be detected and reflected in the neighbor list so that the routing table can always be kept up-to-date.

The transport control module consists of two main function components: datagram receiving/sending and file table maintenance. The datagram receiving unit receives UDP datagrams from the adjacent nodes or the local host. If the received message is interpreted as a "SEND" command issued by the local host, the designated data source will be read directly from local storage devices; packed in fixed-size chunks;

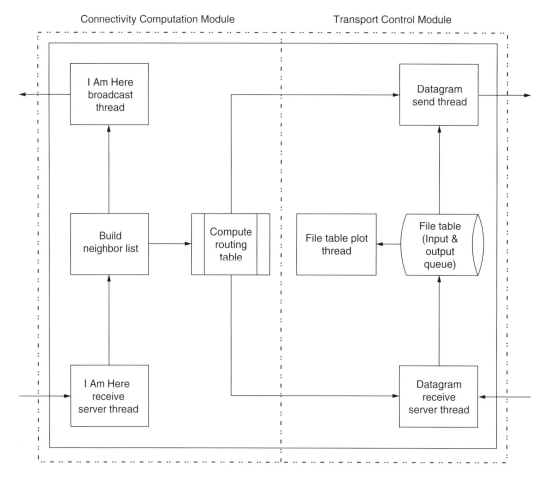

FIGURE 25.4 Framework of CTT function modules.

attached with a user-defined header of destination information; and then put in the file table. Any incoming datagrams that neither originate from nor are destined to the local host are simply placed in the datagram table of a corresponding file buffer. The file table is maintained so that each datagram is dynamically assigned a priority level per the CTT protocol, based on its waiting time and the current network connection status. The datagram sending unit repeatedly scans the whole file list on a sequential basis and sends datagrams with higher priority levels. Any incoming data destined for the local host is forwarded to the corresponding application or saved to a local storage device, while datagrams passing by are loaded into the outgoing queue for forwarding or broadcasting.

The control flow chart of the CTT protocol is illustrated in Figure 25.5. A datagram is sent one of the five modes according to the current network condition and its own status: READY, STANDBY, CTT, SENT, and ARRIVED. The policy of transition among these five modes is briefly described as follows.

A newly created datagram is set as READY mode if a direct or indirect path is found between its source and destination; otherwise, it goes to STANDBY mode. A passing-by datagram remains in READY mode if the local host is on the path and the next hop is reachable, but switches to CTT mode if the next hop is unreachable due to dynamic changes of network connectivity. A datagram received by a node that is not expecting it also enters STANDBY mode. Datagrams in READY mode or CTT mode when the next hop becomes reachable have the highest priority to be selected and put in the outgoing queue; they change to SENT mode immediately after they are successfully dispatched. A broadcast as well as a path recalculation is enforced if a certain timeout expires for a datagram in CTT, STANDBY, or SENT mode.

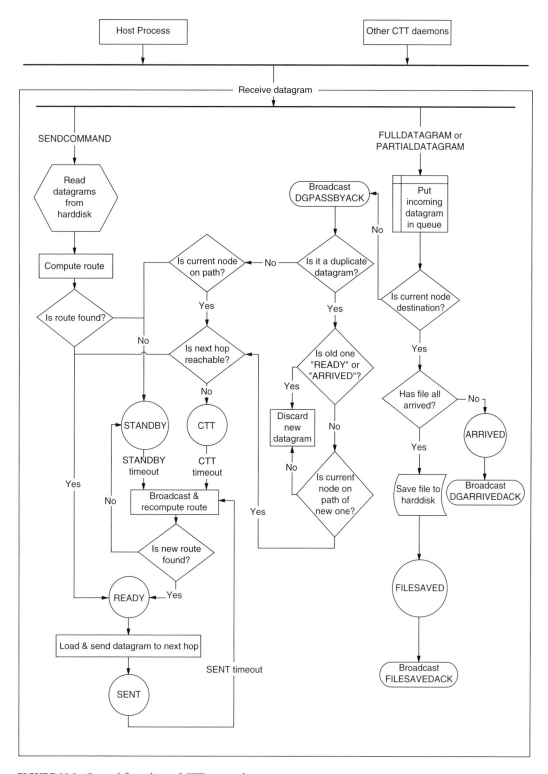

FIGURE 25.5 Control flow chart of CTT protocol.

The CTT protocol uses three different types of acknowledgments: DGARRIVEDACK, FILESAVEDACK, and DGPASSBYACK. When a datagram arrives at its destination, it is set as the ARRIVED mode and a special acknowledgment, DGARRIVEDACK, is broadcast backwards. Upon the receipt of DGARRIVEDACK, a node removes the corresponding datagram from its datagram table to release the allocated memory space and marks a special label indicating that this datagram has been received by the destination.

When the last datagram (not necessarily the one with the last sequence number) arrives at the destination, another type of acknowledgment, FILESAVEDACK, is broadcast over the network. The whole datagram table is cleaned up immediately when this acknowledgment is received, whether the table is complete or incomplete. An acknowledgment, DGPASSBYACK, is broadcast when a datagram reaches a node for which the datagram is not destined. The DGPASSBYACK carries the list of nodes that have received this datagram, which can be used effectively to reduce unnecessary flooding traffic.

The CTT protocol has been implemented and tested in various application scenarios using a small team of mini ATRV mobile robots equipped with 802.11 wireless cards. The implementation details and simulation results can be found in Rao et al. [20].

25.4.3 Adaptive Routing Using Emergent Protocols

In recent years, interest in applying classical theories in fields such as physics, chemistry, and biology to the design of new routing algorithms for DSNs has increased. The motivation behind these approaches is based on the fact that, from a microscopic perspective, the interactions between particles in a substance exhibit similarity to those between sensor nodes in a DSN to some degree. It has long been observed that a large number of identical, infinitesimal individuals interacting with each other by following simple rules are able to manifest a high-order and macroscale phenomenon that cannot be demonstrated by individuals. Such peer-to-peer interaction examples include gas molecules; fluid dynamics; sound waves; biological evolution; economics; magnetization, etc. The positive feedback in the system helps to reinforce success while the negative feedback helps to stabilize the system. Strong chaotic components behave randomly in system adaptation. Routing paradigms based on the concept of such emergent behaviors may ideally serve the routing purpose in DSNs.

The routing objective in a DSN is to find dynamic routes from sensor nodes to a data sink. The difficulty of the routing problem arises from many different factors, such as the chaotic behavior of DSNs, limited communication resources, irreplaceable power supplies, and low computation capacity. In many application scenarios, a large number of mobile sensor nodes are deployed in surveillance regions subject to unpredictable environmental disturbances. The enormous size of the network and hazardous nature of the environment always make human management infeasible. Thus, self-configuring surveillance networks that can adapt to chaotic environments are highly desired.

The following subsection briefly introduces the three most typical applications in this direction: spin glass, multifractal, and ant pheromone models proposed by Brooks et al. More technical details can be found in Brooks et al. [21].

25.4.3.1 Spin Glass Model

Spin glass is a variation of the Ising model in physics, which is one of the most important models in statistical physics. The Ising model consists of atomic magnets that can be viewed as little magnetic vectors (spins). Consider N such little magnetic spins s_i, $i = 1, 2, 3, \ldots, N$ on a two-dimensional lattice, each of which interacts with its nearest neighbors. The orientation of each spin points north ($s_i = +1$) or south ($s_i = -1$). Each of these spins interacts with its nearest neighbors and forms a magnetic field. For a ferromagnetic bond, spins with parallel directions have lower energy, while for an antiferromagnetic bond, spins with parallel directions have higher energy. The Ising model is found to be in the class of NP-complete problems.

In a two-dimensional spin glass routing model, a sensor node corresponding to a spin points to one of the eight neighboring directions. A potential field specifying the minimum energy cost to transmit data from sensor nodes to the data sink is established through local interactions. One of the most significant developments in physics in the 19th century was the discovery of the appropriate probability function to characterize the relative importance of the numerous microscopic configurations. The probabilities of taking each of the eight possible directions for each node are defined by the Boltzmann probability distribution function. This probabilistic orientation is implemented by comparing a computer-generated random number with a probability value preselected for each of the eight directions. Consequently, the spin direction with a higher probability is more likely to be selected.

Let $T[n]$ represent the energy value of node n, whose neighbor node, s, has the potential value denoted by $T[s]$. The probability $P[s]$ that node n points to neighbor node s is given as follows:

$$P[s] = e^{-E(s)/KT} \Big/ \sum_A e^{-E(A)/KT}$$

where $E(s)$ represents the energy change when node n points to neighbor s, i.e., $T[s] - T[n]$; $E(A)$ represents the energy change when node n points to all neighbors; K is a Boltzmann constant; and T is the absolute temperature.

A temperature variable is often used to tune the balance between energy minimization and entropy maximization. Intuitively speaking, sensor nodes are more likely to point to neighbors with lower energy value under low temperature. A low temperature may reduce oscillations and establish a routing mechanism with a shorter hop distance. Particularly, near freezing temperature can protect the system by refraining from erroneous action when the system is subject to harsh conditions. However, high temperature may alleviate power taxing on some hot traffic points by detouring them. The temperature is sometimes specified on a per-region basis in order to allow flexible control of the system.

25.4.3.2 Multifractal Model

This classic crystal-growing prototype for gas and fluid is referred to as diffusion limited aggregation (DLA), which was first introduced by Witten and Sander in the early 1980s [23]. Starting with some immobilized foreign seeds, wandering gas or fluid particles may become solidified in a certain way upon contact with the seeds under certain crystallization conditions. For the routing strategy in a DSN context, a data sink is always set to be a single seed, from which a routing tree is formed gradually. The sensor node has an attribute value, which defines the possibility of joining the routing tree only if the tree stretches out to its neighborhood.

However, if nodes were allowed to join the tree unequivocally, the tree structure would be out of control. In a crystallization process, the inhibiting effect of crystallization is imposed by the crystallization site on the nearby particles. This inhibition can be physically explained by interfacial surface tension and latent heat diffusion effects. When a particle becomes crystallized, its released heat will inhibit the crystallization of nearby particles. It is important to specify a set of appropriate probabilities of joining the routing tree based on the number of neighbors in the routing tree.

In general, the more neighbors a node has in the routing tree, the less likely it is to join the routing tree. Thus, a set of "stickiness" probabilities can be specified based on the number of neighboring nodes on the routing tree. Ideally, a sparse space-filling routing tree covering most of the surveillance region may be constructed after certain time steps. It is worth pointing out that DLA can be considered a self-repelling random walk, which can be modeled by Markov chain.

25.4.3.3 Ant Pheromone Model

Based on Dorigo's telecom routing work, the idea of ant-like mobile agents is utilized to tackle routing in DSNs. It has been observed that social ants coordinate with each other in accomplishing many tasks, such as food forage. Ants release a *search* pheromone when they are looking for food and a *return* pheromone when they are returning to the nest after finding food. Ant movement consists of two mechanisms: (1) they follow a random walk; and (2) they search for the opposite pheromone of the one they currently release.

Ants searching for food tend to follow the highest concentration of return pheromone. Ants returning to the nest tend to follow the highest concentration of search pheromone. Pheromone evaporates at a certain rate to accommodate topological disturbances. Such a food search behavior is very flexible because it is capable of promptly responding to any internal perturbations and outside disturbances; it is also robust because the function of the whole system is not likely to be destroyed by the failure of a single ant or a few ants.

The application of such an intelligence of distributed nature to the routing problem in DSNs results in an ant pheromone model that retains all characteristics of ant behaviors. It is straightforward to model a data source as an ant nest and a data sink as a food location. Ants are dispatched from a data source at a certain rate in search of a data sink. The pheromone gradient established by ants is used to guide their further movements. The ant pheromone model is related to the packet-driven protocols in the sense that ants are viewed as packets traversing from data sources to data sinks.

25.5 Conclusions and Future Work

A broad survey was conducted on various aspects of the design of distributed sensor networks. The issues of and approaches to the problems of multisensor systems presented in this chapter demonstrate the breadth and depth of present research efforts in this area. The successful design of multisensor systems requires solutions to various problems relating to data integration method; sensor deployment scheme; network architecture; real-time operating systems; networking paradigm; information translation cost; fault tolerance, etc. However, little work has been done so far to integrate these solutions effectively in order to achieve a systematic approach to DSN design.

Particularly, finding solutions to the fundamental mathematical problems in DSN is of great theoretical interest and practical importance. Major issues include optimal distribution of sensors; trade-off between communication bandwidth and storage; and maximization of system reliability and flexibility. Also, more attention may be paid to some research directions that are currently not in mainstream areas [22]. For example, due to the continuously increasing network size and complexity, it is very important to develop algorithms for sensor operator decomposition, subspace decomposition, function space decomposition, and domain decomposition. The techniques that transform numerical values (or measurements) to abstract estimates may improve the overall application performance; similarly, techniques that transform abstract estimates back to physical values may also improve performance. For visualization purposes, multiple source locations can be displayed as an energy intensity map using distributed image reconstruction procedures. An efficient synthesis of various methods requires the support of a distributed operating system kernel.

Acknowledgments

The work by Qishi Wu and Nageswara S.V. Rao is sponsored by Defense Advanced Research Projects Agency under MIPR No. K153, National Science Foundation, and by Engineering Research Program and High-Performance Networking Program of Office of Science, U.S. Department of Energy under Contract No. DE-AC05-00OR22725 with UT-Battelle, LLC. The work by Richard R. Brooks, S. Sitharama Iyengar, and Mengxia Zhu was supported by the Defense Advanced Research Projects Agency (DARPA), and administered by the Army Research Office under ESP MURI Award No. DAAD19-01-1-0504. Any opinions, findings, conclusions, or recommendations expressed in this publication are those of the authors and do not necessarily reflect the views of the Defense Advanced Research Projects Agency (DARPA), and Army Research Office.

References

1. D.N. Jayasimha, S.S. Iyengar, and R.L. Kashyap, Information integration and synchronization in distributed sensor networks, Tech. Rpt. 8, Dept. of Computer & Information Science, The Ohio State University, Feb. 1991, *IEEE Trans. SMC*, Sept. 1991.

2. L. Prasad, S.S. Iyengar, R.L. Kashyap, and R.N. Madan, Functional characterization of sensor integration in distributed sensor networks, *Proceedings of the Fifth International Parallel Processing Symposium*, Anaheim, California, USA, 186–193, Apr. 30–May 2, 1991.
3. R. Wesson, Network structures for distributed situation assessment, *IEEE Trans. Syst., Man Cybernetics*, Jan. 1981, 5–23.
4. S.S. Iyengar, D.N. Jayasimha, D. Nadig, and D.K. Pradhan, A versatile architecture for the distributed sensor integration problem, *IEEE Trans. Computers*, 43(2), February 1994.
5. H. Qi, S.S. Iyengar, and K. Chakrabarty, Multiresolution data integration using mobile agents in distributed sensor networks, *IEEE Trans. Syst., Man, Cybernetics Part C: Applic. Rev.*, 31(3), 383–391, August, 2001.
6. Q. Wu, S.S. Iyengar, N.S.V. Rao, J. Barhen, V.K. Vaishnavi, H. Qi, and K. Chakrabarty, On computing the route of a mobile agent for data fusion in a distributed sensor network, submitted to *IEEE Trans. Knowledge Data Eng.*, 16(6), June 2004.
7. K. Marzullo, Tolerating failures of continuous-valued sensors, *ACM Trans. Computer Syst.*, 4(4), Nov. 1990, 284–304.
8. L. Lamport, Synchronizing time servers, technical report 18, Digital System Research Center, 1987.
9. U. Schimd and K. Schossmaier, How to reconcile fault-tolerant interval intersection with the Lipschitz condition, *Distributed Computing*, 14(2), 101–111, May 2001.
10. B. Li, Y. Zhu, and X. Li, Fault-tolerant interval estimation fusion by Dempster–Shafer theory, *Proc. 5th Int. Conf. Inf. Fusion*, Annapolis, MD, July 2002, 1605–1613.
11. R. Brooks and S.S. Iyengar, *Multi-Sensor Fusion: Fundamental and Application Software*, Prentice Hall, Englewood Cliffs, NJ, 1998, 488.
12. E.C. Cho, S.S. Iyengar, K. Chakrabarty, and H. Qi, A new fault-tolerant interval integration function satisfying local Lipschitz condition, submitted to *IEEE Tran. on AES*, July 2000.
13. P.K. Varshney, *Distributed Detection*, Springer–Verlag, Heidelberg, 1996.
14. A.K. Hyder, E. Shahbazian, and E. Waltz, Eds., *Multisensor Fusion*, Kluwer Academic Publishers, 2002.
15. N.S.V. Rao, Multisensor fusion under unknown distributions: finite sample performance guarantees, in *Multisensor Fusion*, A.K. Hyder, E. Shahbazian, and E. Waltz, Eds., Kluwer Academic Publishers, 2002.
16. Q. Wu, N.S. Rao, and S.S. Iyengar, Computational complexities of sensor deployment problems, paper in preparation.
17. R.J. Fowler, M.S. Paterson, and S.L. Tanimoto, Optimal packing and covering in the plane are NP-complete, *Inf. Process. Lett.*, 12(3), 133–137, 1981.
18. Q. Wu, S.S. Iyengar, N.S.V. Rao, J. Barhen, V.K. Vaishnavi, H. Qi, and K. Chakrabarty, On efficient deployment of sensors on planar grid, in preparation.
19. N.S.V. Rao and Q. Wu, Network demons for distributed sensor networks, in *Frontiers in Distributed Sensor Networks*, S.S. Iyengar and R.B. Brooks, Eds., CRC Press Inc., Oct. 2004.
20. N.S.V. Rao, Q. Wu, S.S. Iyengar, and A. Manickam, Connectivity-through-time protocols for dynamic wireless networks to support mobile robot teams, *Proceedings of the 2003 IEEE Int. Conf. Robotics Automation*, Taiwan, 1653–1658, Sept. 14–19, 2003.
21. R. Brooks, M. Pirretti, M. Zhu, and S.S. Iyengar, Adaptive routing using emergent protocols in wireless ad hoc sensor networks, *Proc. SPIE Conf.*, 6–8 August, Vol. 5205.
22. S.S. Iyengar and Q. Wu, Computational aspects of distributed sensor networks, *Proc. Int. Symp. Parallel Architectures, Algorithms Networks*, May 22–24, 2002, Manila/Makati, Philippines, IEEE Computer Society Press (I-SPAN 2002).
23. T.A. Witten and L.M. Sander, Diffusion-limited aggregation, *Physical Review B*, 27(9), 5686–5697, May 1983.

26
Cooperative Computing in Sensor Networks

Liviu Iftode
Rutgers University

Cristian Borcea
Rutgers University

Porlin Kang
Rutgers University

26.1	Introduction	**26**-1
26.2	The Cooperative Computing Model............................	**26**-3
26.3	Node Architecture	**26**-4
	Admission Manager • Code Cache • Virtual Machine • Tag Space	
26.4	Smart Messages...	**26**-5
	Smart Message Life Cycle • Smart Message Self-Routing	
26.5	Programming Interface...	**26**-7
26.6	Prototype Implementation and Evaluation................	**26**-8
	Cost of SM Migration • Cost of Tag Space Operations	
26.7	Applications	**26**-12
	SPIN Using Smart Messages • Directed Diffusion Using Smart Messages	
26.8	Simulation Results..	**26**-14
26.9	Related Work	**26**-15
26.10	Conclusions	**26**-18

26.1 Introduction

As the cost of embedding computing becomes negligible compared to the actual cost of goods, a trend toward incorporating computing and wireless communication capabilities in most of the consumer products occurs. Therefore, the next generation of computing systems will be embedded, in a virtually unbounded number, and dynamically connected. Although these systems will penetrate every possible domain of daily life, the expectation is that they will operate outside normal cognizance, requiring far less attention from human users than today's desktop computers.

The first illustration of these systems that has received considerable interest in the last couple of years is sensor networks [11–13]. These networks have severe resource limitations in terms of processing power, amount of available memory, network bandwidth, and energy. However, during the next decade sensor networks will become part of a larger class of networks of embedded systems (NES) that have sufficient computing, communication, and energy resources to support distributed applications. For instance, already some companies propose computer systems embedded into cars or video cameras that are able to communicate with each other [1, 4].

For some of these networks, such as networks of intelligent cameras performing object tracking over a large geographical area, it might be beneficial to perform local computations and to cooperate in order to execute a global task. They may perform sophisticated filtering of data at a node that acquired an image or even distributed object tracking, rather than running a centralized algorithm at a server. The challenge is how to program NES, namely, to determine the appropriate computing model and the system support necessary to execute distributed applications in these networks.

NES pose a unique set of challenges that make traditional distributed computing models difficult to employ in programming them. The number of devices working together to achieve a common goal is orders of magnitude greater than those seen so far. These systems are heterogeneous in their hardware architectures because each embedded system is tailored to perform a specific task. Unlike the Internet, NES are typically deployed in environments void of human attention, where it is unacceptable to expect a human to hit a "reset" button to recover from a failure. NES are inherently fragile, with node and connection failures the norm rather than the exception. The availability of nodes may vary greatly over time; they can become unreachable due to mobility, depletion of energy resources, or catastrophic failures.

The nodes in NES communicate through wireless network interfaces. Thus, they can communicate directly only with nodes within their transmission range. Similarly to most ad hoc networks, the separation between hosts and routers disappears (i.e., each node must perform routing). However, the scale and heterogeneity encountered in NES as well as different application requirements preclude the existence of a common routing support. Therefore, the flexibility to use multiple routing algorithms in the same network is desirable.

The applications running in NES target specific data or properties within the network, not individual nodes. From an application point of view, nodes with the same properties are interchangeable. Fixed naming schemes, such as IP addressing, are inappropriate in most situations. The need to target specific data or properties within the network raises the issue of a different naming scheme with dynamic bindings between names and node addresses. A naming scheme based on content or properties is more appropriate for NES than a fixed naming scheme [10].

This chapter presents distributed computing model, cooperative computing, and a software architecture for NES based on execution migration. Cooperative computing applications consist of migratory execution units, called smart messages (SMs), working together to accomplish a distributed task. SMs are user-defined distributed programs (composed of code, data, and execution control state) that migrate through the network searching for nodes of interest (i.e., nodes on which the program needs to run) and execute their own routing at each node in the path. Distributed computing based on execution migration is more suitable for NES than data migration (message passing) due to the volatility and dynamic binding of names to nodes specific to these networks. Cooperative computing provides flexible support for a wide variety of applications, ranging from data collection and dissemination to content-based routing and object tracking.

Nodes in the network support SMs by providing: a name-based shared memory (tag space) for inter-SM communication and access to the host system; and an architecturally independent environment (virtual machine) for SM execution. SMs are self-routing, namely, they are responsible for determining their own paths through the network. SMs name the nodes of interest by properties and self-route to them using other nodes as "stepping stones." Applications in cooperative computing are able to adapt to adverse network conditions by changing their routing dynamically.

To validate the cooperative computing model, the authors have designed and implemented a prototype by modifying Sun Microsystem's Java KVM (kilobyte virtual machine) [3]. Microbenchmark results are reported for this prototype running over a test bed consisting of Linux-based HP iPAQs equipped with 802.11 cards for wireless communication. These results indicate that cooperative computing is a feasible solution for programming real-world applications.

For larger scale evaluation, a simulator has been developed that executes SMs and allows one to account for execution as well as communication time. In this simulator, two previously proposed applications for data collection and data dissemination in sensor networks have been implemented: directed diffusion [13] and SPIN [11]. The simulation results show that this model is able to provide high flexibility for user-defined distributed applications while limiting the increase in response time to, at most, 15% over traditional nonactive communication implementations.

The next section describes cooperative computing; Section 26.3 presents the node architecture for the model. In Section 26.4, details of smart messages are discussed, and Section 26.5 presents the API for cooperative computing. Section 26.6 shows microbenchmark results for the prototype implementation.

Section 26.7 describes the applications implemented using SMs and their simulation results are presented in Section 26.8. Section 26.9 discusses related work and the chapter concludes with Section 26.10.

26.2 The Cooperative Computing Model

Cooperative computing is a distributed computing model for large-scale, ad hoc NES. In this model, distributed applications are defined as dynamic collections of migratory execution units, called SMs, that cooperate in achieving a common goal. The execution of an SM is described in terms of computation and migration phases. The execution performed at each step is determined by the particular properties of that node. On nodes that present interest to the current computation, the SM may read and process data; on intermediate nodes, the SM executes only its routing algorithm. During migrations, SMs carry mobile data, the code missing at destination, and a lightweight execution state.

Nodes in the network cooperate by providing an architecturally independent programming environment (virtual machine) for SM execution and a name-based shared memory (tag space) for inter-SM communication and interaction with the host system. SMs, along with the system support provided by nodes, form the cooperative computing infrastructure, which allows programming user-defined distributed applications in NES.

In this model, a new distributed application can be developed without *a priori* knowledge about the scale and topology of the network or the specific functionality of each node. Placing intelligence in SMs provides this flexibility and also obviates the issue of implementing a new application or protocol in NES, which is difficult or even impossible using conventional approaches [10].

SMs are resilient to network volatility. Over time, certain nodes may become unavailable due to mobility or energy depletion, but SMs are able to adapt by controlling the routing. These messages can carry multiple routing procedures and choose the most appropriate one based on the conditions encountered in the network. Using this feature, SMs can discover routes to nodes of interest even in adverse network conditions.

Moving the execution to the source of data improves the performance for applications that need to process large amounts of data. For example, instead of transferring large size images through the network for an object tracking application, an SM can perform the analysis of the images at the nodes that acquired them. Thus, it reduces the network bandwidth and energy consumption, and in the same time, it improves the user-perceived response time. The impact of transferring code on performance can be limited by caching code at the nodes.

Figure 26.1 shows a simple application that illustrates the novel aspects of computation and communication in cooperative computing. The application performs object tracking over a large area (e.g., a campus, airport, or urban highway system) using a network of mobile robots with attached cameras [17].

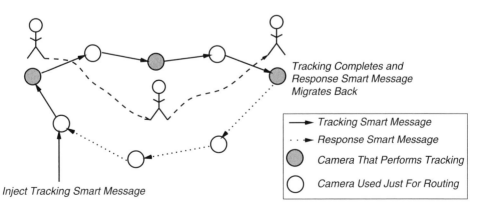

FIGURE 26.1 Distributed object tracking using cooperative computing.

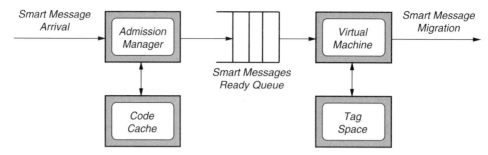

FIGURE 26.2 Node architecture.

In the figure, the target is represented by a person moving across a given geographical region. A user can inject the tracking SM into any node of the network.

The SM migrates to a node that acquired an image of a possible target object, analyzes this image, and then may decide to follow the object. The network maintains no routing infrastructure, and the SM is responsible for determining its path to cameras that detected the object. The smart message can use the direction of motion and geographical information to "chase" the object. Once the SM arrives at a new node that has a picture of the object, it generates a task to analyze the object and its motion further. The SM may migrate to neighbor nodes to obtain pictures of the object from a different angle or lighting conditions. When the tracking completes, the SM generates a response SM that will transport the gathered information back to the user node.

26.3 Node Architecture

The goal of the SM software architecture is to keep the support required from nodes in the network to the minimum, placing intelligence in SMs rather than in individual nodes. Figure 26.2 shows the common system support provided by nodes for cooperative computing. The admission manager receives incoming SMs, decides whether to accept them, and stores these messages into the SM-ready queue. The code cache stores frequently used codes to reduce the amount of traffic in the network. The virtual machine (VM) acts as a hardware abstraction layer for scheduling and executing tasks generated by incoming SMs. The tag space is a name-based shared memory that stores data objects persistent across SM executions and offers a unique interface to the host's OS and I/O system.

26.3.1 Admission Manager

To prevent excessive use of its resources (energy, memory, bandwidth), a node needs to perform admission control. Each SM presents its resource requirements within a resource table. The admission manager is responsible for receiving incoming messages and storing them in the SM ready queue, subject to admission restrictions.

26.3.2 Code Cache

Commonly, the applications executing in NES have a localized behavior, exhibiting spatial and temporal locality. Therefore, frequently used SM codes are cached in order to amortize over time the initial cost of transferring the code through the network.

26.3.3 Virtual Machine

The VM schedules, executes, and migrates SMs. To migrate an SM, the VM captures the execution state and sends it along with the code and data to the next hop. The VM at the destination will resume the

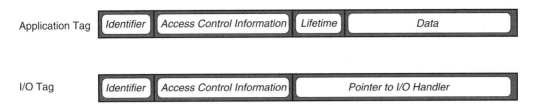

FIGURE 26.3 Structure of application and I/O tags.

SM from the instruction following the migration invocation. The VM also ensures that an SM conforms to its declared resource estimates; otherwise, the SM can be removed from the system.

26.3.4 Tag Space

Each node that supports SMs manages a name-based shared memory, called tag space, consisting of tags that are persistent across SM executions. The tag space contains two types of tags: application tags, which are created by SMs, and I/O tags that are provided by the system. The I/O tags define the basic hardware of the node and provide SMs with a unique interface to the local OS and I/O system. SMs are allowed to read and write both types of tags, but they can create or delete only application tags.

Figure 26.3 illustrates the structure of application and I/O tags. The identifier is the name of the tag and is similar to a file name in a file system; it is used by SMs for content-based node naming. The access of SMs to tags is restricted, based on the access control information associated with each tag. For application tags, the VM associates the access control information carried by the SM that created the tag (i.e., the owner of the tag). For I/O tags, the owner of the device sets the access control information.*

Application tags and I/O tags differ in terms of functionality and lifetime. Application tags offer persistent memory for a limited lifetime (i.e., application tags are still "alive" for a certain amount of time after the SMs that created them have finished the execution at the local node); after this time interval, the tags expire, and the node reclaims their memory. I/O tags, on the other hand, are permanent and provide a pointer to an I/O handler (i.e., a system call or an external process) capable of serving I/O requests. The list of all the possible utilizations of tags consists of:

- *Naming*. SMs name the nodes of interest using tag identifiers.
- *Data storage*. An SM can store data in the network by creating its own tags.
- *Data exchange and data sharing*. Exchanging data through the tag space is the only communication channel among different SMs.
- *Routing*. SMs may create routing tags at visited nodes to store routing information in the data portion of these tags.
- *Synchronization*. An SM can block on a specific tag pending a write of this tag by another SM. Once the tag is written, all SMs blocked on it are waked up and made ready for execution.
- *Interaction with the host system*. An SM can issue commands to or request data from the host OS and I/O devices using I/O tags.

26.4 Smart Messages

SMs are execution units that migrate through the network to execute on nodes of interest and route themselves at each node in the path toward a node of interest. SMs comprise code and data sections (referred to as "bricks"), a lightweight execution state, and a resource table. The code and data bricks can be dynamically used to assemble new, possibly smaller SMs. The ability to incorporate only the necessary code and data bricks in the new SMs can reduce their size and, consequently, the amount of

---

*More information about access control, protection domains, and SM security in general can be found in Xu et al. [29].

traffic in the network (i.e., the code and data carried by SMs are divided into bricks solely for this purpose). The execution state contains the execution context necessary to resume the SM after a successful migration. The resource table consists of resource estimates: execution time, tags to be accessed or created, memory requirements, and network bandwidth. These resource estimates set a bound on the expected needs of an SM at a node; they are used by the admission manager to make the admission decision.

The SM computation is embodied in tasks. During its execution, a task may modify the data bricks of the SM as well as the local tags to which it has access. It can also migrate, create new SMs, or block on tags of interest. A collection of SMs cooperating toward a common goal forms a distributed application.

26.4.1 Smart Message Life Cycle

Each SM has a well-defined life cycle at a node: (1) it is subject to admission control; (2) upon admission, a task is generated out of the SM's code and data bricks and scheduled for execution; and (3) after completion at a node, the SM may terminate or may decide to migrate to other nodes of interest.

26.4.1.1 Admission

To avoid unnecessary resource consumption, the admission manager executes a three-way handshake protocol for transferring SMs between neighbor nodes. First, only the resource table is sent to the destination for admission control. If the SM admission fails, the task will be informed, and it can decide on subsequent actions. If the SM is accepted, the admission manager checks, using the code bricks' IDs (computed off-line by applying a hash function on the code), whether the code bricks belonging to this SM are cached locally. Then, it informs the source to transfer only the missing code bricks. (These code bricks will also be cached upon arrival.)

26.4.1.2 Scheduling and Execution

Upon admission, an SM becomes a task scheduled (in FIFO order) for execution. The execution is nonpreemptive; new SMs can be accepted, but they will not be dispatched for execution until the current SM terminates. An executing SM can yield the VM, however, by blocking on a tag. The execution time is bounded by the estimated running time presented during admission (i.e., the VM may terminate an SM that does not respect the admission contract).

Nonpreemptive scheduling is used for three reasons. First, the execution time of SMs is usually short (many times a node is used merely as a "stepping stone" en route to a node of interest). Thus, context switching would incur too much overhead with respect to the total execution time of the SM. Second, it is not necessary to support multiprogramming for interactive programs (unlike traditional computer systems, embedded systems commonly operate unattended). Third, the communication always terminates the current SM (i.e., the only form of communication in cooperative computing is a migration invocation) and, consequently, the idea of using multiple threads in one application to overlap communication and computation does not make sense for SM programs. On the other hand, nonpreemptive scheduling makes inter-SM synchronization and sharing particularly simple to implement.

26.4.1.3 Migration

If the current computation does not complete at the local node, the task may continue its execution at another node. The current execution state is captured and migrated along with the code and data bricks. Because a task accesses only mobile data and tags, an efficient migration has been implemented in which only a small part of the entire execution context is saved and transferred through the network. Essentially, the instruction and stack pointers are transferred for all the stack frames corresponding to the current task. It is important to notice that migration is explicit (i.e., the programmers call a "migration" primitive when needed), and that data transferred during a migration are specified by the programmer as data bricks.

26.4.2 Smart Message Self-Routing

SMs are self-routing, i.e., they are responsible for determining their own paths through the network. SMs require no system support for routing; the entire process takes place at application level. An SM names its destinations in terms of tag identifiers and executes its routing algorithm at each node in the path. SMs may create routing tags at intermediate nodes in the network to store routing information. If routing information is not locally available, an SM may create other SMs for route discovery and block on a routing tag. A write on this tag unblocks the SM, which will resume its migration. Because tags are persistent for their lifetime, the routing information, once acquired, can be used by subsequent SMs, thus amortizing the route discovery effort.

Each SM must include at least one *routing brick* among its code bricks. A single routing algorithm, however, might not always reach a node of interest in the presence of highly dynamic network configurations. Therefore, an SM can carry multiple routing algorithms and change them during execution according to the current network conditions. For instance, an SM can use a proactive routing algorithm in a stable and relatively dense network and an on-demand algorithm in a volatile and sparse network. In this way, the SM may complete even if network conditions change significantly during its execution. Borcea and colleagues [6] offer a complete description of the self-routing mechanism.

26.5 Programming Interface

The API for the cooperative computing model is given in Table 26.1. It provides simple, yet powerful, primitives. SMs can access the tag space, dynamically create new SMs, synchronize on tags, and migrate to nodes of interest.

createTag, deleteTag, readTag, and writeTag. These operations allow SMs to create, delete, or access existing tags. As mentioned in Section 26.3, these operations are subject to access control. The same interface is used to access the I/O tags. SMs can issue commands to I/O devices by writing into I/O tags or can get I/O data by reading I/O tags.

createSMFromFiles, createSM, and spawnSM. An SM is created by injecting a program file at a node; this program calls createSMFromFiles with a list of program file names to build the new SM structure. An SM may use createSM during execution to assemble a new SM from a subset of its code and data bricks. A createSM call is commonly used to create a route discovery SM when routing information is not locally available. An SM that needs to clone itself calls spawnSM; this primitive returns true in the "parent" and false in the "child" SMs. Typically, spawnSM is invoked when the current computation needs to migrate a copy of itself to nodes of interest while continuing the execution at the local node. A newly created SM is inserted into the SM ready queue.

blockSM. This primitive implements the update-based synchronization mechanism. An SM blocks on a tag waiting for a write. To prevent deadlocks, blockSM takes a timeout as parameter. If nobody writes the tag in the timeout interval, the VM returns the control to the SM. A typical example is an SM that blocks on a routing tag while waiting for a route discovery SM to bring a new route.

TABLE 26.1 Cooperative Computing API

Category	Primitives
Tag space operations	createTag(tag_name, lifetime, data); deleteTag(tag_name); readTag(tag_name); writeTag(tag_name, value);
SM creation	createSMFromFiles(program_files); createSM(code_bricks, data_bricks); spawnSM();
SM synchronization	blockSM(tag_name, timeout);
SM migration	migrateSM(tag_names, timeout); sys_migrate(next_hop);

```
1  Typical_SM(tag){
2    do
3      migrateSM(tag, timeout);
4      <do computation>
5    until (<quality of result>);
6    migrateSM(back, timeout);
7  }
```

FIGURE 26.4 Code skeleton for typical smart message.

migrateSM and sys_migrate. The migrateSM primitive implements a high-level content-based migration, provided usually as a library function. It allows applications to name the nodes of interest by tag names and to bound the migration time. When migrateSM returns normally (no timeout), the SM is guaranteed to resume its execution at a node of interest. In case of timeout, the SM regains control at one of the intermediate nodes in the path. Figure 26.4 presents an example of a typical SM that uses migrateSM. For instance, this SM can be used in the object tracking application described in Section 26.2. The SM migrates to nodes hosting the tag of interest and executes on these nodes until a certain quality of result is achieved. When this is done, the SM migrates back to the node that injected it into the network.

The migrateSM function implements routing using routing tags, the low level primitive called sys_migrate, and possibly other SMs for route discovery. An SM can choose among multiple migrateSM functions that correspond to different routing algorithms. The sys_migrate primitive is used to migrate SMs between neighbor nodes. The entire migration protocol of capturing the execution state and sending the SM to the next hop is implemented in sys_migrate.

26.6 Prototype Implementation and Evaluation

The authors have implemented their SM prototype in Java over Linux, thus harnessing well-developed and supported Java application development tools and knowledge base.[*] Specifically, Sun Microsystem's KVM (Kilobyte Virtual Machine) [3] has been modified because it has a small memory footprint (i.e., as little as 160 KB, which makes it suitable for resource-constrained devices) and its source code is publicly available.

The SM API is encapsulated in two Java classes: *SmartMessage* and *TagSpace*; for efficiency, the API was implemented as Java native methods. The authors have also implemented their own serialization mechanism because KVM does not support serialization. In addition to the KVM interpreter thread, two additional threads have been introduced for admission control and local code injection. The design of the SM computing platform is not specific to any hardware or software environment. It can be implemented on any VM (e.g., Mate [20], Scylla [27]), language, or underlying operating system.

Next, microbenchmark results for this SM prototype are reported. Specifically, the cost of one-hop migration and the cost of tag space operations have been measured. The test bed consists of HP iPAQs 3870 running Linux 2.4.18-rmk3-hh24. Each iPAQ contains an Intel StrongARM 1110 206-Mhz RISC processor, 32-MB flash memory, and 64-MB RAM memory. For communication, Lucent Orinoco 802.11b Silver PC Cards are used in ad hoc mode. To factor out the cost of Java method call overhead (approximately 6 μs), the code for measuring costs has been inserted inside the native methods associated with the SM API.

26.6.1 Cost of SM Migration

The one-hop migration has three phases: execution capture at source, SM transfer, and execution resumption at destination. The SM is converted into a machine-independent representation to allow state capture

---

[*]The SM software distribution is freely available at http://discolab.rutgers.edu

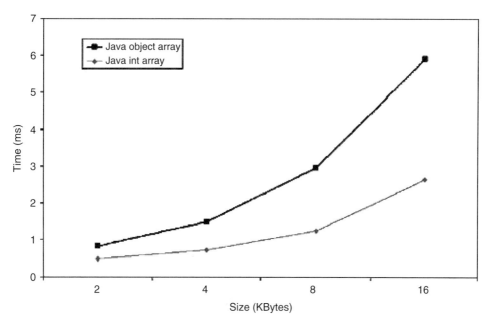

FIGURE 26.5 Cost of data brick serialization.

and resumption. Because the code bricks are already in machine-independent Java class format, only the data bricks and execution state need to be converted. This conversion is done using the authors' simple object serialization mechanism. The serialization of the execution state does not have a significant impact because only the execution control state is captured and transferred, not the local variables. Therefore, the important factors that determine the cost of one-hop migration are the data brick serialization, the SM transfer, and data brick deserialization.

26.6.1.1 Data Brick Serialization and Deserialization

To study the effect of data brick serialization, a fixed-size code brick (1197 bytes) has been used and the data brick size has been varied from 2 to 16 KB. The stack frames have also been kept constant (131 bytes for two activation records). The cost of serializing these two stack frames is 0.235 ms. Commonly, the data bricks in an SM consist of a mixture of objects and primitive types. Two types of data bricks have been used in this evaluation; they represent a practical lower and upper bound for typical data bricks: an array of integers and an array of objects. The object array represents an upper bound because each of its elements causes a call to the top-level VM serialization method, while the integer array represents a lower bound because there is only one call to the top level VM serialization method.

Figure 26.5 shows that the serialization cost is below 6 ms for data bricks as large as 16 KB. Commonly, the SMs process the data at source and therefore they carry small size data. The applications developed by the authors carry less than 2 KB, which costs less than 1 ms to serialize. Figure 26.6 presents the cost of deserialization for the same data bricks. Observe that this cost is as much as 30% larger than the cost of serialization — an increase caused by the memory allocation costs during object deserializations.

26.6.1.2 SM Transfer

To evaluate the total cost of migrating an SM (serialization, transfer, deserialization), two sets of experiments were performed. In the first, the code brick size was varied while data brick size and stack frame size were kept fixed at 53 and 131 bytes, respectively. In the second, data brick size was varied while keeping the code brick size and stack frame size fixed at 1197 and 131 bytes, respectively.

Figure 26.7 and Figure 26.8 show the results of these two experiments for two cases: when the code is not cached and when the code is cached. In Figure 26.7, the time to transfer the SM when the code is

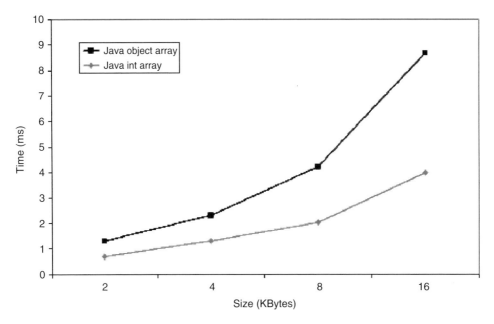

FIGURE 26.6 Cost of data brick deserialization.

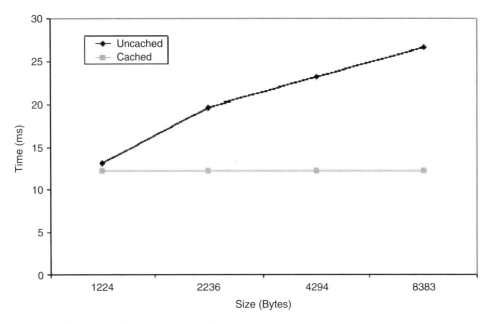

FIGURE 26.7 Effect of code brick size on single-hop migration.

cached represents, essentially, the overhead of the three-way handshake protocol because the sizes of the data bricks and stack frames are small. Figure 26.8 demonstrates that the data brick size contributes significantly to the total cost of migration. Thus, it is important to have a serialization mechanism with minimal space overhead.

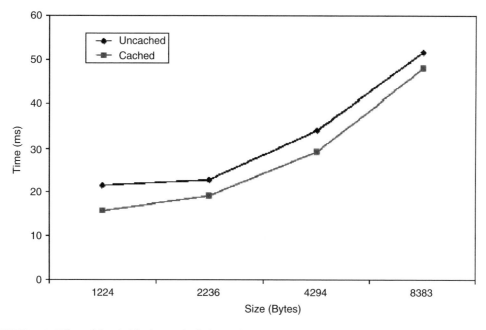

FIGURE 26.8 Effect of data brick size on single-hop migration.

26.6.2 Cost of Tag Space Operations

Table 26.2 shows the cost of the tag space operations for application tags. The *readTag* primitive has the lowest cost because it performs the least number of operations. When an SM reads a tag, the VM interpreter acquires a lock, performs a lookup in the tag space, and returns the data to the SM. The *writeTag* operation costs are slightly higher because the interpreter must check and unblock any SMs blocked on the tag. The *createTag* primitive involves an additional step to register a timer for the tag lifetime, while *blockSM* needs to append the SM to the queue and suspend the current task. The *deleteTag* primitive has the highest cost because the interpreter needs to wake up all SMs blocked on the tag, remove the timer for the tag lifetime, and remove the tag structure from the tag space.

Table 26.3 presents the access time to several I/O tags that are currently implemented in our prototype: GPS location query; neighbor discovery; camera image capture; light sensor; and system status inquiry (battery lifetime, system time, and amount of free memory). A typical node with a video camera and a GPS receiver attached is shown in Figure 26.9. The *gps_location* is updated by a user-level process that

TABLE 26.2 Time for Tag Space Operations

Tag Space Operation	Time (μs)
createTag	43.4
deleteTag	55.9
readTag	20.8
writeTag	31.7
blockSM	45.8

TABLE 26.3 Cost of Reading I/O Tags

Tag Name	Time (ms)
gps_location	0.20
neighbor_list	0.34
image_capture (32-KB)	341.23
light_sensor	0.11
battery_lifetime	25.63
system_time	0.09
free_memory	0.12

FIGURE 26.9 Prototype node with video camera and GPS receiver attached.

reads from the GPS serial interface. The location of the neighbors along with their identifiers can be returned by reading the *neighbor_list* tag, which is typically used by geographical routing algorithms carried and executed by SMs. To get the information about neighbor nodes, a neighbor discovery protocol has been implemented that maintains a cache of known neighbors. For the *image_capture* tag, the system also performs YUYV to RGB format conversion on the captured image before returning it to the tag reader. All the other tag values are obtained directly from Linux using system calls.

26.7 Applications

To prove that virtually any protocol or application can be written using SMs, two previously proposed applications — SPIN [11] and directed diffusion [13] — have been implemented. They present different paradigms for content-based communication and computation in sensor networks; SPIN is a protocol for data dissemination and directed diffusion implements data collection.

26.7.1 SPIN Using Smart Messages

SPIN is a family of adaptive protocols that disseminates information among nodes in a sensor network. The implementation of SPIN-1 is a three-stage handshake protocol for data dissemination. Each time a node obtains new data, it disseminates them in the network by sending an advertisement to its neighbors.

```
1   DisseminateSM(String tag, int timeout){
2     int timestamp;
3     Data data;
4     String tagData=tag+"data";
5     String tagTimestamp=tag+"timestamp";
6     Address src, dest;
7     while(true){ // SM at source
8       TagSpace.blockSM(TagData, timeout);
9       timestamp = TagSpace.readTag(tagTimestamp);
10      if (!SmartMessage.spawnSM()){ // child SM
11        while(true){ // SM at every node
12          src = SmartMessage.getLocalAddress();
13          SmartMessage.sys_migrate(all); // migrate to all neighbors
14          if (timestamp.CompareTo((Integer)TagSpace.readTag(tagTimestamp))<=0){
15            System.exit(0); // the same or more recent data exists at this node
16          }
17          TagSpace.writeTag(tagTimestamp, timestamp);
18          dest = SmartMessage.getLocalAddress();
19          SmartMessage.sys_migrate(src); // migrate back to source
20          data = TagSpace.readTag(tagData);
21          SmartMessage.sys_migrate(dest); // bring data to destination
22          TagSpace.writeTag(tagData, data);
23        }
24      }
25    }
26  }
```

FIGURE 26.10 SPIN with smart messages.

The node receiving the advertisement checks whether it has already received or requested those data. If not, it sends a request message to the sender asking for the advertised data. The initiator sends the requested data and then the process is executed recursively for the entire network.

As an example of a cooperative computing program, Figure 26.10 presents the code for the authors' implementation of SPIN using SMs. The tag space at each node hosts two tags: the value of the most recent data received (*tagData*) and the timestamp associated with these data (*tagTimestamp*).

The protocol is initiated by injecting a *Disseminate SM* into a node that produces data. This SM blocks on tagData (line 8) waiting for new data. Each time new data are produced, the SM reads the tagTimestamp and spawns itself (lines 9 and 10). The child SM migrates to the neighbors to advertise the new data (line 13). If a destination node does not have these or more recent data, the child SM updates the tagTimestamp and migrates back to the source to bring the data (lines 14 to 22). Upon data arrival, the child SM executes recursively the same algorithm until the data are disseminated in the entire network.

26.7.2 Directed Diffusion Using Smart Messages

In directed diffusion, a sink node requests data by sending "interests" for named data. Data matching an interest are then drawn from source nodes toward the sink node. Intermediate nodes can cache and aggregate data; they may also direct interests based on previously cached data. At the beginning, the sink may receive data from multiple paths, but after a while it will reinforce the path providing the best data rate. All future data will arrive on the reinforced path only.

For the implementation of directed diffusion using SMs, the tag space at each node hosts three tags: the most recent data value (*tagData*); the best data rate available at that node (*tagDataRate*); and the best next hop toward the source (*tagBestRoute*). Directed diffusion is initiated by injecting an SM at the sink. The execution of this SM has two main phases: (1) *exploration* starts at the sink and floods the network to find data of interest; and (2) *reinforcement* chooses the best path and brings data from source to sink.

If the information of interest is not locally available (no tagDataRate value), the *explore SM* spawns itself; the child SM migrates to all neighbors, while the parent SM blocks on tagDataRate. This operation is performed recursively at every node until an SM reaches a node containing the tagDataRate. At this point, the child SM migrates back to its parent carrying the discovered data rate. If the new data rate is better than the value stored in tagDataRate, the SM updates tagDataRate with the new value and tagBestRoute with its source as the best node in the path toward the source of data. This update unblocks the parent SM, which will carry the data rate one hop back. Eventually, the sink node is reached and the reinforcement phase begins.

During the reinforcement phase, a *collect SM* migrates to the best next hop starting from the sink. At each intermediate node, this SM spawns; the child SM migrates to the best next hop, while the parent SM blocks, waiting for data. When the SM reaches the source, it spawns new SMs to carry the data one hop back at the promised data rate. Recursively, a blocked SM is awakened by the data arrival, and it will carry the data back until it reaches the sink.

26.8 Simulation Results

For large-scale evaluation, the authors have developed an event-driven simulator, similar to ns-2 [21], extended with support for SM execution. The simulator is written in Java to allow rapid prototyping of applications. To get accurate results, the communication and the execution times must be accounted for. The simulator provides accurate measurements of the execution time by counting, at the VM level, the number of cycles per VM instruction. To account for the execution time, each node has been simulated with a Java thread, and a new mechanism has been implemented for scheduling these threads inside JVM. The communication model used in this simulator is "generic wireless," with contention solved at the message level. Before any transmission, a node "senses" the medium and backs off in case of contention.

The main goal in conducting the simulation experiments was to quantify the data convergence time for the authors' implementations of SPIN and directed diffusion using SMs and to compare these results with those for traditional message-passing implementations. Data convergence time is defined as the time when a certain percentage of the total number of nodes has received the data (SPIN), or the data rate (directed diffusion). In both cases, due to flooding, all nodes end up receiving the data and the data rate. SPIN completes after all nodes have received the data; directed diffusion will start the reinforcement phase after all nodes have received the data rate. The same network configuration is used for all experiments. The network has 256 nodes distributed uniformly over a square area, and each node has the same transmission range. The average number of neighbors per node is four.

The first set of experiments evaluated the data convergence time when only one SM is injected in the network. Figure 26.11 presents the data convergence time for a single directed diffusion SM, with the sink and source located at the diagonal corners of the square region. The data convergence time for three different cases of the same SM and a base case that uses passive communication (no SM) are plotted. The top curve shows the time when code caching is not used. The second curve shows a more than fourfold improvement in performance when code caching is activated during the first execution of the SM in the network. The code is cached when an SM visits a node for the first time and will be used by subsequent SMs during the same execution. The effects of caching are very important in this case because the SMs visit a node multiple times in directed diffusion; they travel the network forward (looking for the source) and backward (diffusion of data rate).

In the third curve, a 30% decrease can be observed in completion time when the code is already cached at all nodes. The fourth curve shows the data convergence time for a traditional implementation: the protocol is implemented at each node; only data are transferred through the network; and the execution time is not accounted for. Observe that the degradation in performance for this implementation, when the code is cached at all nodes, compared to the traditional implementation is only 5%. This is a reasonable price for the flexibility to program any user-defined distributed application in NES.

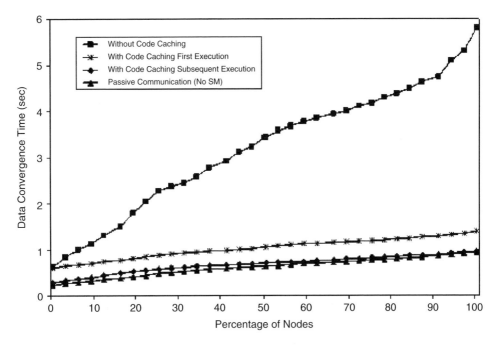

FIGURE 26.11 Directed diffusion using smart messages.

Figure 26.12 plots the same curves for a single SPIN SM launched in the network at a node located in a corner of the square area. During the first execution, code caching leads to a threefold improvement in performance (i.e., reducing the size of SMs is essential for a protocol based on flooding and three-stage communication). The third curve shows a 30% decrease in the completion time (similar to directed diffusion) when the code is already cached at all nodes. The completion time increases from 10 to 15% compared to the traditional implementation.

The second set of experiments quantified the performance of these applications when multiple SMs run simultaneously in the network. Figure 26.13 and Figure 26.14 show the data convergence time for directed diffusion and SPIN with the code already cached at nodes. For these experiments, data convergence time is the time when a certain percentage of nodes have received the data (or data rate) for all the SMs running in parallel. The nodes at which the SMs start are distributed uniformly in the network.

The results show that data convergence time increases with the number of SMs, but only during the initial flooding phase because of increased contention in the network. After that, the shapes of the curves are the same, independent of the number of SMs. The results also indicate that SPIN completes faster than directed diffusion in all cases (i.e., 2.3 s compared to 3.4 s for the top curves in the figures). The cause is that SPIN floods only the neighbors and then brings the data to them, while directed diffusion needs to flood the entire network until it finds the source and then brings the data rate back to all nodes. In the initial phase, directed diffusion generates more messages in the network leading to higher contention, but its performance will increase as soon as the reinforcement phase begins.

26.9 Related Work

SMs have been influenced by the design of mobile agents for IP-based networks [9, 18, 22, 28]. A mobile agent may be viewed as a task that explicitly migrates from node to node assuming that the underlying network assures its transport between them. SMs apply the general idea of code migration, but focus more on flexibility, scalability, reprogrammability, and the ability to perform distributed computing over unattended NES.

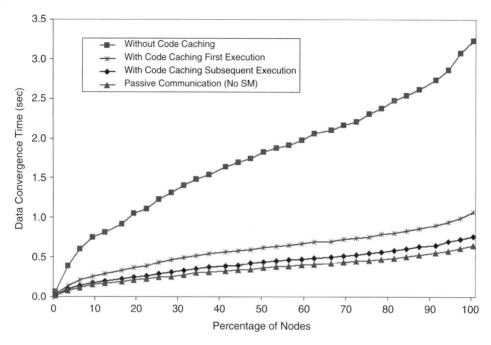

FIGURE 26.12 SPIN using smart messages.

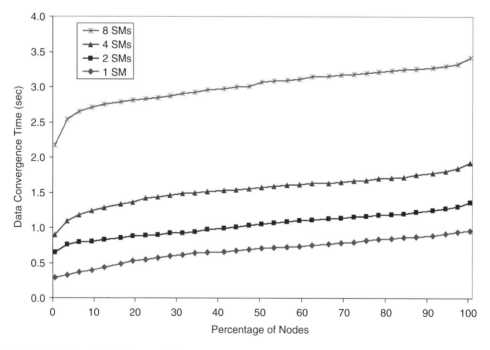

FIGURE 26.13 Directed diffusion: multiple smart messages.

Unlike mobile agents, SMs are defined to be responsible for their own routing in a network. A mobile agent names nodes by fixed addresses and commonly knows the network configuration *a priori*, while an SM names nodes by content and discovers the network configuration dynamically. Furthermore, the SM software architecture defines the common system support that each node must provide. The goal of this architecture is to reduce the support required from nodes in NES because they possess limited resources.

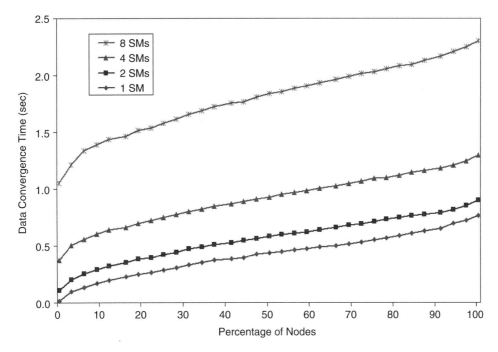

FIGURE 26.14 SPIN: multiple smart messages.

The SM self-routing mechanism shares some of the design goals and leverages work done in active networks (ANs) [8, 23, 26]; however, SMs differ from ANs in several key features. The main difference between them is in terms of programmability. Unlike ANs, which target faster communication in IP-based networks, cooperative computing defines a distributed computing model for NES whereby several SMs can cooperate, exchange data, and synchronize with each other through the tag space. Additionally, the AN model does not contain the migration of execution state as the authors' model does. The migration of execution state for SMs trades overhead for flexibility in programming sophisticated tasks that require cooperation and synchronization among several entities. For example, this execution state allows SMs to make routing decisions based on the results of computation done at previously visited nodes.

Research in mobile ad hoc networking [14, 15, 19, 24] has resulted in numerous routing protocols. These protocols have generally been designed for IP-based networks and have primarily targeted traditional mobile computing applications over networks of mobile personal computers. Some of these protocols have been leveraged into routing algorithms used by the SM self-routing mechanism.

Sensor networks represent the first step toward large networks of embedded systems. Most of the research in this area has focused on hardware [16, 25]; operating systems [12]; or network protocols [5, 11, 13]. Cooperative computing provides a solution for developing user-defined distributed applications in sensor networks, a crucial issue that has been tackled only marginally so far. As demonstrated, cooperative computing provides enough flexibility to enable implementation of previously proposed protocols over this computing platform.

SensorWare [7] is similar to cooperative computing in that both are frameworks for programmable NES based on code migration. Therefore, both are suitable to reprogram the network. However, SensorWare supports mobile control scripts and accesses the resources through virtual devices, whereas cooperative computing supports mobile Java code (i.e., Java is supported on many embedded systems today [2]), execution state migration, and uniform access to resources through tags.

Mate [20] is an efficient VM for sensor networks that can significantly simplify code development and dissemination efforts. The main difference between cooperative computing and this research is that Mate targets only the reprogrammability of the network, but the programming model is still the traditional message passing. SMs, on the other hand, are based on execution migration. An SM transfers not only the code, but also the execution state through the network.

26.10 Conclusions

This chapter has described a programming model for large-scale networks of embedded systems, in which distributed applications are implemented as collections of smart messages. The model overcomes the scale, heterogeneity, and connectivity issues encountered in these networks by using execution migration, content-based naming, and self-routing. The experimental results for this prototype implementation demonstrate the feasibility of cooperative computing. The implementation and simulation results for two sensor network applications show that this model represents a flexible, yet simple, solution for programming large networks of embedded systems.

Acknowledgments

This work was supported in part by the NSF under grant ANI-0121416. The authors thank Ulrich Kremer and Chalermek Intanagonwiwat for our frequent discussions regarding the design of cooperative computing. We also thank Philip Stanley–Marbell, Deepa Iyer, and Akhilesh Saxena for their contributions at various stages of this project.

References

1. Axis Communications. http://www.axis.com.
2. Java 2 Platform, Micro Edition (J2ME). http://java.sun.com/j2me/.
3. K Virtual Machine. http://java.sun.com/products/cldc/.
4. Sensoria Corporation. http://www.sensoria.com.
5. Blum, B., Nagaraddi, P., Wood, A., Abdelzaher, T., Son, S., and Stankovic, J. An entity maintenance and connection service for sensor networks. In *1st In. Conf. Mobile Syst., Applications, Services (MobiSys)* (May 2003), 201–214.
6. Borcea, C., Intanagonwiwat, C., Saxena, A., and Iftode, L. Self-routing in pervasive computing environments using smart messages. In *Proc. 1st IEEE Int. Conf. Pervasive Computing Commun. (PerCom)* (March 2003), 87–96.
7. Boulis, A., Han, C., and Srivasta, M. Design and Implementation of a framework for programmable and efficient sensor networks. In *Proc. 1st Int. Conf. Mobile Syst., Applications, Services (MobiSys)*. (May 2003), 187–200.
8. Wetheral, D. Active network vision reality: lessons from a capsule-based system. In *Proc. 17th ACM Symp. Operating Syst. Principles (SOSP'99)* (1999), 64–79.
9. Gray, R.S., Cybenko, G., Kotz, D., and Rus, D. Mobile agents: motivations and state of the art. In *Handbook of Agent Technology*, J. Bradshaw, Ed. AAAI/MIT Press, 2001.
10. Heideman, J., Silva, F., Intanagonwiwat, C., Govindan, R., Estrin, D., and Ganesan, D. Building efficient wireless sensor networks with low-level naming. In *Proc. 18th ACM Symp. Operating Syst. Principles (SOSP)* (October 2001), 146–159.
11. Heinzelman, W.R., Kulik, J., and Balakrishnan, H. Adaptive protocols for information dissemination in wireless sensor networks. In *Proc. 5th Annu. ACM/IEEE Int. Conf. Mobile Computing Networking (MobiCom)* (August 1999), 174–185.
12. Hill, J., Szewczyk, R., Woo, A., Hollar, S., Culler, D., and Pister, K. System architecture directions for networked sensors. In *Proc. 9th Int. Conf. Architectural Support Programming Languages Operating Syst. (ASPLOS)* (November 2000), 93–104.
13. Intanagonwiwat, C., Govindan, R., and Estrin, D. Directed diffusion: a scalable and robust communication paradigm for sensor networks. In *Proc. 6th Annu. ACM/IEEE Int. Conf. Mobile Computing Networking (MobiCom)* (August 2000), 56–67.

14. Li, J., Janotti, J., De Couto, D., Karger, D.R., and Morris, R. A scalable location service for geographic ad hoc routing. In *Proc. 6th Annu. ACM/IEEE Int. Conf. Mobile Computing Networking (MobiCom)* (August 2000), 120–130.
15. Johnson, D.B. and Maltz, D.A. *Dynamic Source Routing in Ad Hoc Wireless Networks*. In *Mobile Computing*, T. Imielinski and H. Korth (Eds.). Kluwer Academic Publishers, 1996, 153–181.
16. Juang, P., Oki, H., Wang, Y., Martonosi, M., Peh, L.-S., and Rubenstein, D. Energy-efficient computing for wildlife tracking: design trade-offs and early experiences with ZebraNet. In *Proc. 10th Int. Conf. Architectural Support Programming Languages Operating Syst. (ASPLOS)* (October 2002), 96–107.
17. Jung, B. and Sukhatme, G.S. Cooperative tracking using mobile robots and environment-embedded, networked sensors. In *2001 IEEE Int. Symp. Computational Intelligence Robotics Automation.* (July 2001), 206–211.
18. Karnik, N. and Tripathi, A. Agent server architecture for the Ajanta mobile-agent system. In *Proc. 1998 Int. Conf. Parallel Distributed Process. Tech. Applications (PDPTA'98)* (July 1998), 66–73.
19. Karp, B. and Kung, H. Greedy perimeter stateless routing for wireless networks. In *Proc. 6th Annu. ACM/IEEE Int. Conf. Mobile Computing Networking (MobiCom)* (August 2000), 243–254.
20. Levis, P. and Culler, D. Mate: a virtual machine for tiny networked sensors. In *Proc. 10th Int. Conf. Architectural Support Programming Languages Operating Syst. (ASPLOS)* (October 2002), 85–95.
21. McCanne, S. and Floyd, S. ns Network Simulator. http://www.isi.edu/nsnam/ns/.
22. Milojicic, D., LaForge, W., and Chauhan, D. Mobile objects and agents. In *Proceedings of the 4th USENIX Conf. Object-Oriented Technol. Syst.* (1998), 1–14.
23. Moore, J.T., Hicks, M., and Nettles, S. Practical programmable packets. In *Proc. 20th Annu. Joint Conf. IEEE Computer Commun. Soc. (INFOCOM'01)* (April 2001), 41–50.
24. Perkins, C.E., Royer, E., and Das, S.R. Ad hoc on demand distance vector (AODV) routing. In *2nd IEEE Workshop Mobile Computing Syst. Applications (WMCSA'99)* (February 1999), 90–100.
25. Priyantha, N.B., Miu, A.K.L., Balakrishnan, H., and Teller, S. The Cricket compass for context-aware mobile applications. In *Proc. 7th Annu. ACM/IEEE Int. Conf. Mobile Computing Networking (MobiCom)* (2001), 1–14.
26. Schwartz, B., Jackson, A.W., Strayer, W.T., Zhou, W., Rockwell, R.D., and Partridge, C. Smart packets for active networks. *ACM Trans. Computer Syst.* (February 2000), 397–413.
27. Stanley–Marbell, P. and Iftode, L. Scylla: A smart virtual machine for mobile embedded systems. In *3rd IEEE Workshop Mobile Computing Syst. Applications, WMCSA2000* (December 2000).
28. White, J. *Mobile Agents*. J.M. Bradshaw (Ed.), MIT Press, 1997.
29. Xu, G., Borcea, C., and Iftode, L. Toward a security architecture for smart messages: challenges, solutions, and open issues. In *Proc. 1st Int. Workshop Mobile Distributed Computing (MDC'03)* (May 2003).

VII
Energy Management

27
Dynamic Power Management in Sensor Networks

Amit Sinha
Engim, Inc.

Anantha Chandrakasan
Engim, Inc.

27.1 Introduction 27-1
27.2 Idle Power Management 27-2
  Multiple Shutdown States • Sensor Node Architecture • Sleep State Transition Policy
27.3 Active Power Management 27-5
  Variable Voltage Processing
27.4 System Implementation 27-6
  DVS Circuit • Idle Power Management Hooks • Processor Power Modes • OS Architecture • Sensor-Specific Application Programming Interface Extensions
27.5 Results 27-12

27.1 Introduction

Wireless distributed sensor networks have gained importance in a wide spectrum of civil and military applications [1]. Advances in microelectricalmechanical systems (MEMS) technology, combined with low-power, low-cost DSPs and RF circuits have resulted in cheap wireless sensor networks becoming feasible. A distributed, self-configuring network of adaptive sensors has significant benefits. They can be used to monitor inhospitable and toxic environments remotely. Large classes of benign environments also require the deployment of a large number of sensors such as intelligent patient monitoring, object tracking, assembly line sensing, etc. Their distributed nature provides wider resolution as well as increased fault tolerance compared to a single sensor node. Several projects, such as the MIT μAMPS project [2], that demonstrate the feasibility of sensor networks are underway.

A wireless sensor node is typically battery operated and is thus energy constrained. To maximize the lifetime of the sensor node after its deployment, all aspects, including circuits, architecture, algorithms and protocols, must be made energy efficient. Once the system has been designed, additional energy savings can be obtained by using dynamic power management concepts [3] whereby the sensor node is shut down if no interesting events occur or slowed down during periods of reduced activity. Such event-driven power consumption is critical for obtaining maximum battery life from the sensor node. In addition, it is desirable that the node has graceful energy quality scalability so that, if the application demands, the user is able to extend the mission lifetime at the cost of sensing accuracy. Energy-scalable algorithms and protocols have been proposed for such energy-constrained situations [4].

Sensing applications present a wide range of requirements in terms of data rates, computation, average transmission distance, etc. As such, protocols and algorithms will need to be tuned to each application.

Therefore, embedded operating systems and software will be critical ingredients in such sensor networks because programmability will be a necessary requirement. This chapter proposes operating system (OS)-directed dynamic power management (DPM) techniques to improve the energy efficiency of sensor nodes. DPM is an effective tool to reduce system power consumption without significantly degrading performance. The embedded OS is used to facilitate active and idle power management.

The basic idea behind idle power management is to shut down devices when they are not needed and wake them when necessary. Formulating an optimum shutdown policy, in general, is a nontrivial problem. If the energy and performance overheads in transitioning to sleep states were negligible, a simple greedy algorithm that makes the system go into the deepest sleep state as soon as it is idle would be perfect. However, in reality, transitioning to a sleep state has the overhead of storing the processor state and shutting off the power supply. Waking also takes a finite amount of time. Therefore, implementing the right policy for transitioning to various sleep states is critical for effective idle power management.

Although shutdown techniques can yield substantial energy savings in the idle states of the system, additional energy savings are possible by optimizing the performance of the sensor node in its active state. Dynamic voltage scaling (DVS) is a very effective active power management technique for reducing processor energy consumption [5]. Most microprocessor-based systems are characterized by a time-varying computational load. Simply reducing the operating frequency during periods of reduced activity results in linear decrease in power consumption but does not affect the total energy consumed per task. Reducing the operating voltage implies greater circuit delays that, in turn, mean that peak performance is compromised. Significant energy benefits can be achieved by recognizing that peak performance is not always required and therefore the operating voltage and frequency of the processor can be dynamically adapted based on instantaneous processing requirements. The goal of DVS is to adapt the power supply and operating frequency to match the workload so that the visible performance loss is negligible. Pering and colleagues have conducted an evaluation of some DVS algorithms on portable benchmarks [6].

27.2 Idle Power Management

Efficient DPM in idle mode requires power-differentiated states and optimal OS policies to transition to and from various states.

27.2.1 Multiple Shutdown States

It is not uncommon for a device to have multiple power modes. For example, the StrongARM SA-1100 processor has three power modes: "run," "idle," and "sleep" [7]. Each of these modes is associated with a progressively lower level of power consumption. The run mode is the normal operating mode of the processor; all power supplies are enabled, all clocks are running, and every on-chip resource is functional. The idle mode allows the software to halt the CPU when not in use while continuing to monitor interrupt service requests. The CPU clock is stopped and the entire processor context is preserved. When an interrupt occurs, the processor switches back to run mode and continues operating exactly at the point at which it stopped. The sleep mode offers greatest power savings and minimum functionality. The power supply is cut off to a majority of circuits and the sleep state machine watches for a preprogrammed wake-up event. Similarly, a Bluetooth radio has four different power consumption modes: "active," "hold," "sniff," and "park."

Most power-conscious devices support multiple power-down modes offering different levels of power consumption and functionality. An embedded system with multiple such devices can have a set of power states based on various combinations of device power states. In fact, an open interface specification called the advanced configuration and power management interface (ACPI), jointly promoted by Intel, Microsoft, and Toshiba [8], standardizes how the OS can interface with devices characterized by multiple power states to provide dynamic power management. ACPI supports a finite state model for system resources and specifies the hardware/software interface that should be used to control them. ACPI controls

Dynamic Power Management in Sensor Networks

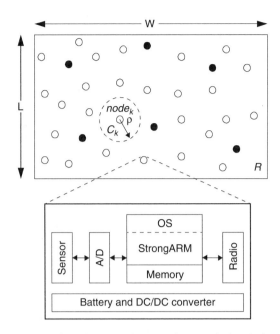

FIGURE 27.1 Sensor network and node architecture. (From Sinha, A. and Chandrakasan, A., *IEEE Des. Test Comp.* 62–74, 2001. With permission.)

the power consumption of the whole system as well as the power state of each device. An ACPI-compliant system has five global states — SystemStateS0 (working state) and SystemStateS1 to SystemStateS4 — corresponding to four different levels of sleep states. Similarly, an ACPI-compliant device has four states: PowerDeviceD0 (the working state) and PowerDeviceD1 to PowerDeviceD3. The sleep states are differentiated by the power consumed, the overhead required in going to sleep, and the wake-up time.

27.2.2 Sensor Node Architecture

Figure 27.1 illustrates the basic sensor node architecture. Each node consists of the embedded sensor, A/D converter, a processor with memory (in this case, the StrongARM SA-11x0 processor), and the RF circuits. Each of these components is controlled by the OS through primitive device drivers. An important function of the OS is to enable power management (PM). Based on event statistics, the OS decides which devices to turn off or on. The sensor network essentially consists of η homogeneous sensor nodes distributed over a rectangular region R with dimensions $W \times L$ with each node having a visibility radius ρ. There is no particular reason for the rectangular topology.

Table 27.1 enumerates the component power modes corresponding to five different useful sleep states for the sensor node. Each of these node-sleep modes corresponds to an increasingly deeper sleep state and is therefore characterized by an increasing latency and decreasing power consumption. These sleep

TABLE 27.1 Useful Sleep States for the Sensor Node

State	StrongARM	Memory	Sensor, A/D	Radio
s_0	active	active	on	tx, rx
s_1	idle	sleep	on	rx
s_2	sleep	sleep	on	rx
s_3	sleep	sleep	on	off
s_4	sleep	sleep	off	off

Source: From Sinha, A. and Chandrakasan, A., *IEEE Des. Test Comp.* 62–74, 2001. With permission.

states are chosen based on working conditions of the sensor node, e.g., it does not make sense to have the memory in the active state and everything else completely off:

- State s_0 is the completely active state of the node where it can sense, process, transmit, and receive data.
- In state s_1, the node is in sense and receive mode while the processor is in standby.
- State s_2 is similar to state s_1 except that the processor is powered down and is waked up when the sensor or the radio receives data.
- State s_3 is the sense-only mode in which everything except the sensing front-end is off.
- State s_4 represents the completely off state of the device.

The design problem is to formulate a policy of transitioning between states based on observed events so as to maximize energy efficiency. It can be seen that the power-aware sensor model is similar to the system power model in the ACPI standard. The sleep states are differentiated by the power consumed, the overhead required in going to sleep, and the wake-up time. The deeper the sleep state is, the lesser the power consumption and the longer the wake-up time.

27.2.3 Sleep State Transition Policy

Assume an event is detected by a sensor node at some time t_0; it finishes processing it at time t_1; and the next event occurs at time $t_2 = t_1 + t_i$. At time t_1, the node decides to transition to a sleep state s_k from the active state s_0 as shown in Figure 27.2. Each state s_k has a power consumption P_k, and the transition time to it from the active state and back is given by $\tau_{d,k}$ and $\tau_{u,k}$, respectively. By the definition of node-sleep states, $P_j > P_i$, $\tau_{d,i} > \tau_{d,j}$ and $\tau_{u,i} > \tau_{u,j}$ for any $i > j$. The power consumption between the sleep modes is modeled as a linear ramp between the states. When the node transitions from state s_0 to, say, state s_k, individual components such as the radio, memory, and processor are progressively powered down. This results in a stepped variation in power consumption between the states. The linear ramp is analytically simpler to handle and approximates the process reasonably well.

Now a set of sleep time thresholds $\{T_{th,k}\}$ corresponding to the states $\{s_k\}$ will be derived such that transitioning to a sleep state s_k from state s_0 will result in a net energy loss if the idle time $t_i < T_{th,k}$ because of the transition energy overhead. This assumes that no productive work can be done in the transition period, which is usually true (e.g., when a processor wakes up, the transition time is the time required

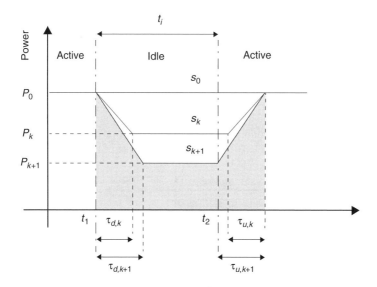

FIGURE 27.2 Dynamic voltage and frequency scaling. (From Sinha, A. and Chandrakasan, A., *IEEE Des. Test Comp.* 62–74, 2001. With permission.)

TABLE 27.2 Sleep State Power, Latency, and Threshold

State	P_k (mW)	τ_k (ms)	$T_{th,k}$
s_0	1040	–	–
s_1	400	5	8
s_2	270	15	20
s_3	200	20	25
s_4	10	50	50

Source: From Sinha, A. and Chandrakasan, A., *IEEE Des. Test Comp.* 62–74, 2001. With permission.

for the PLLs to lock, the clock to stabilize, and the processor context to be restored). The energy savings because of state transition are given by the difference in the area under the graphs shown in Figure 27.2 and are computed as follows:

$$E_{save,k} = (P_0 - P_k)t_i - \left(\frac{P_0 - P_k}{2}\right)\tau_{d,k} - \left(\frac{P_0 + P_k}{2}\right)\tau_{u,k}$$

Such a transition is only justified when $E_{save,k} > 0$. This leads to the following energy gain threshold:

$$T_{th,k} = \frac{1}{2}\left[\tau_{d,k} + \left(\frac{P_0 + P_k}{P_0 - P_k}\right)\tau_{u,k}\right]$$

This implies that the longer the delay overhead of the transition is, the higher the energy gain threshold, and the more the difference between P_0 and P_k is, the smaller the threshold.

Table 27.2 lists the power consumption of a sensor node described in Figure 27.1 designed using off-the-shelf components in different power modes and the corresponding energy gain thresholds. One can see that the thresholds are in the order of milliseconds. The OS shutdown policy is based on event interarrival statistics and energy gain thresholds and can be formulated as an optimization problem. If the events are modeled as a Poisson process, the probability of at least one event in time t_i is given by

$$p_E(t) = 1 - e^{\lambda t_i}$$

In this case, a simple algorithm that updates the average events per unit time, λ, as they happen computes the probability of an event happening within the thresholds, $T_{th,k}$, and chooses the deepest sleep state based on a minimum probability threshold that would be effective. Sinha and Chandrakasan have described the energy savings obtained from such an algorithm [9].

27.3 Active Power Management

The OS can be used to manage active power consumption in an energy-constrained sensor node. It reduces the operating frequency and voltage to a level just enough for the sensing application so that no visible loss is observed in performance while the energy consumption is reduced.

27.3.1 Variable Voltage Processing

Dynamic voltage scheduling is a very effective technique for reducing CPU energy. Several sensor systems are characterized by a time-varying computational load. Simply reducing the operating frequency during periods of reduced activity results in a linear decrease in power consumption but does not affect the total energy consumed per task, as shown in Figure 27.3(a) (the shaded area represents energy). Reduced operating frequency implies that the operating voltage can also be reduced. Because the switching power

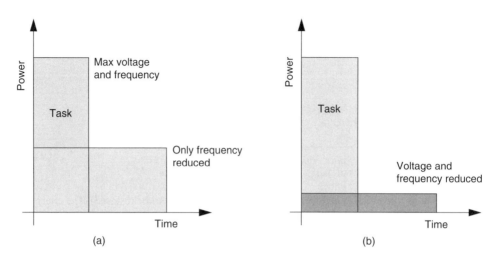

FIGURE 27.3 μAMPS processor board.

consumption scales linearly with frequency and quadratically with supply voltage, quadratic energy reduction can be obtained as shown in Figure 27.3(b). Significant system energy savings can be realized by recognizing that peak performance is not always required and therefore the operating voltage and frequency of the processor can be dynamically adapted based on instantaneous processing requirement.

27.4 System Implementation

Figure 27.4 shows the first generation μAMPS sensor node, which is based on the StrongARM SA-1110 processor and has 1 MB of on-board SRAM and flash memory. The board runs at a nominal battery (single lithium primary cell) power supply of about 4.0 V. The on-board power supply circuits generate a 3.3-V supply for all digital circuits. A separate analog power supply is also generated to isolate the digital power supply noise from the analog circuits. The 3.3-V digital power supply also powers the I/O pads of the StrongARM SA-1110 processor. The core power supply is generated through a DVS circuit that can regulate the power supply from 0.925 V to a maximum of 2.0 V with a conversion efficiency of about 85%.

The radio module is on a similarly sized board and consists of a dual power, 2.4 GHz-radio for 10- and 100-m ranges. The 16-b bus interface connector will allow the radio module to be stacked onto the processor board. In addition, the connector allows a different sensor board (e.g., a seismic sensor) to be stacked. The processor board has an RS-232 and a USB connector for remote debugging and connecting to a debug PC. The board features a built-in acoustic sensor (a microphone, some opamps, and A/D circuit) that talks to the StrongARM processor using the synchronous serial port (SSP). The opamp gains are programmable and processor controlled. An envelop detect mechanism has also been incorporated into the sensor circuit, which bypasses the A/D circuit and wakes the processor when the signal energy crosses a certain programmable threshold. Using this feature can significantly reduce power consumption in the sense mode and facilitates event-driven computation.

27.4.1 DVS Circuit

The basic variable core power supply schematic is shown in Figure 27.5. The MAX1717 step-down controller is used to regulate the core supply voltage dynamically through the 5-b digital-to-analog converter (DAC) inputs over a 0.925 to 2 V range. The converter works on the following principle. A variable duty cycle pulse width modulated (PWM) signal alternately turns on the power transistors M1 and M2. This produces a rectangular wave at the output of the transistors with duty cycle D. The LC low-pass filter passes a desired DC output equal to $DV_{battery}$ while attenuating the AC component to an acceptable ripple. The duty cycle D is controlled using the DAC pins (D0:D4), which results in 30 voltage

Dynamic Power Management in Sensor Networks

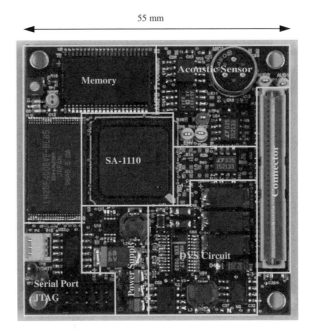

FIGURE 27.4 DVS circuit schematic.

levels (two combinations are not allowed). A two-wire remote sensing scheme compensates for voltage drops in the ground bus and output voltage rail. The StrongARM sets the DVS enable pin on the voltage regulator depending on whether DVS capability is desired or not. A feedback signal from the regulator lets the processor know if the output core voltage is stabilized. This is required for error-free operation during voltage scaling.

The processor clock frequency change involves updating the contents of the core clock configuration register (CCF) of the SA-1110 [10]. The core clock is derived by multiplying the reference crystal oscillator clocks using a phase-locked loop (PLL) based on CCF register settings, as shown in Table 27.3. The core clock (CCLK) can be driven using the fast CCLK or the memory clock (MCLK), which runs at half the frequency of the CCLK. The core clock uses CCLK normally except when waiting for fills to complete after a cache miss. Core clock switching between CCLK and MCLK can be disabled by setting a control register appropriately.

The sequence of operations during a voltage and frequency update depends on whether one is increasing or decreasing the processor clock frequency, as shown in Figure 27.6. When the clock frequency is increased, it is first necessary to increase the core supply voltage to the minimum required for that particular frequency. The optimum voltage frequency pairs are stored in a lookup table. Once the core voltage is stabilized, the frequency update can proceed. The first step involves recalibrating the memory timings. This is done by setting an appropriate value in the MSC control register. Before CCLK frequency is increased, clock switching between CCLK and MCLK is disabled to avoid an inadvertent switch of the core clock. CCLK frequency is changed by setting the CCF register. Once this is done, core clock switching between CCLK and MCLK is enabled.

The sequence of operations is somewhat reversed when reducing frequency. First, the core clock frequency is updated (following the three basic steps mentioned previously). Before one can reduce the core voltage, it is necessary to recalibrate the memory timing. This is required because, once the core clock frequency is reduced, memory read–write will result in errors unless the memory timing is adjusted (e.g., when reading the voltage–frequency lookup table). Subsequently, the core voltage is reduced and normal operation is started once it stabilizes. To ensure correct operation, the entire voltage frequency update must be done is an atomic fashion. For example, if an interrupt occurs while frequency is updated and memory has not been recalibrated, execution errors might occur.

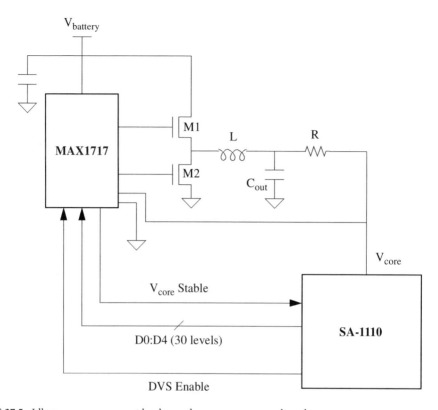

FIGURE 27.5 Idle power management hooks on the sensor processor board.

TABLE 27.3 SA-1110 Core Clock Configurations and Minimum Core Supply Voltage

CCF(4:0)	Core Clock Frequency (CCLK) in MHz		Core Voltage (V) [3.6864 MHz Osc]
	3.6864 MHz Oscillator	3.6864 MHz Oscillator	
00000	59.0	57.3	1.000
00001	73.7	71.6	1.050
00010	88.5	85.9	1.125
00011	103.2	100.2	1.150
00100	118.0	114.5	1.200
00101	132.7	128.9	1.225
00110	147.5	143.2	1.250
00111	162.2	157.5	1.350
01000	176.9	171.8	1.450
01001	191.7	186.1	1.550
01010	206.4	200.5	1.650
01011	221.2	214.8	1.750
01100-11111	–	–	

27.4.2 Idle Power Management Hooks

The sensor node has been designed specifically to allow a set of sleep states similar to the one described earlier; in addition, it has hardware support for event-driven computation. The overall schematic is shown in Figure 27.7. The general purpose I/O (GPIO) pins on the StrongARM are used to generate and receive various signals from the peripherals. The SA-1110 features 28 GPIO pins, each of which can be configured as an input or an output. In addition, the GPIO pins can be configured specifically to detect a rising or

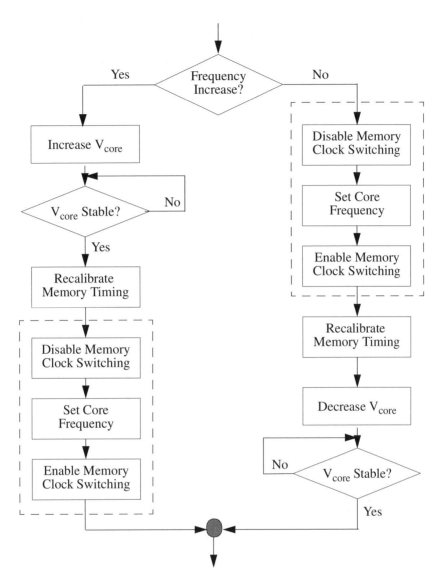

FIGURE 27.6 System-level power savings from active power management using DVS.

falling edge. In implementation, four GPIO pins are dedicated to power supply control in the system. The entire analog power supply can be switched off when no sensing is required. Alternately, only the power supply to the low pass filter (LPF) can be switched off and the envelop energy sensing circuit could be used to trigger a signal to the processor. When this happens, the processor could enable the LPF and start reading data off the A/D converter using the SSP (synchronous serial port). The signal detection threshold is also programmable using other GPIO pins and similar power supply control is available for the radio module; the processor can turn off the radio when it is not required.

27.4.3 Processor Power Modes

The SA-1110 contains power management logic that controls the transition among three different modes: run, idle, and sleep. Each of these modes corresponds to a reduced level of power consumption.

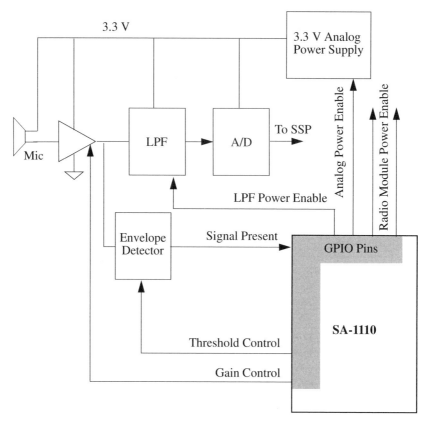

FIGURE 27.7 Degradation in DVS savings with increase in workload variance.

- *Run mode.* This is the normal mode of operation for the SA-1110. All on-chip power supplies are on, all clocks are on, and every on-chip resource is available. Under usual conditions, the processor starts up in the run mode after a power-up or reset.
- *Idle mode.* This mode allows an application to stop the CPU when not in use while continuing to monitor interrupt requests. The CPU clock is stopped and, because the SA-1110 is a fully static design, all state information is saved. When normal operation is resumed, execution is started exactly where it stopped. During idle mode, all on-chip resources (real-time clock; OS timer; interrupt controller; GPIO; power manager; DMA and LCD controllers; etc.) are on. The PLL also remains in lock so that the processor can be brought in and out of the idle mode quickly.
- *Sleep mode.* Sleep mode offers greatest power savings for the processor and, consequently, lowest functionality. When transitioning from run/idle to sleep mode, the SA-1110 performs an orderly shutdown of on-chip activity, applies an internal reset to the processor, and negates the power enable (PWR_EN) pin, thus indicating to the external system that the power supply can be turned off. Running off the 32.768 KHz crystal oscillator, the sleep state machine watches for a preprogrammed wake-up event to occur. Sleep mode is entered in one of two ways: through software control or through a power supply fault. Entry into sleep mode is accomplished by setting the force sleep bit in the power manager control register (PMCR). This bit is set by software and cleared by hardware during sleep so that, when the processor wakes, it finds the bit cleared. The entire sleep shutdown sequence takes about 90 ms.

Table 27.4 shows the power consumption in various modes of the SA-1110 processor at two different frequencies and the corresponding voltage specification [10]. Note that the minimum operating voltage

TABLE 27.4 Power Consumption of the SA-1110 Processor

		Power Consumption Modes		
Frequency	Supply Voltage (V)	Normal (mW)	Idle (mW)	Sleep (µA)
133	1.55	<240	<75	<50
206	1.75	<400	<100	<50

required (as shown in Table 27.3) at the two frequencies is slightly lower than what is shown in Table 27.4. Although the idle mode results in about 75% power reduction, the sleep mode saves almost all the power.

27.4.4 OS Architecture

The sensor OS is based on Redhat eCos, an open-source, real-time operating system for embedded applications [11]. It meets the requirements of the embedded space that Linux cannot yet reach. Linux currently scales from a minimal size of around 500 KB of kernel and 1.5 MB of RAM, all before taking into consideration application and service requirements. eCos can provide the basic runtime infrastructure necessary to support devices with memory footprints in the tens to hundreds of KB, with real-time options.

The original eCos OS is designed to be completely scalable across platforms as well as within a given platform. Essentially, source level configuration allows the user to add or remove packages from a source repository based on system requirements. For example, the user might choose to remove math libraries and the resulting kernel will be leaner. The core eCos system consists of a number of different components, such as the kernel, the C library, an infrastructure package, etc. Each of these provides a large number of configuration options, allowing application developers to build a system that matches the requirements of their particular applications.

To manage the potential complexity of multiple components and lots of configuration options, eCos has a component framework: a collection of tools specifically designed to support configuring multiple components. Furthermore, this framework is extensible, allowing additional components to be added to the system at any time. The eCos component description language (CDL) lets the configuration tools check for consistency in a given configuration and point out any dependencies that have not been satisfied.

At the core of the eCos kernel is the scheduler. This defines the way in which threads are run, and provides the mechanisms by which they may synchronize. It also controls the means by which interrupts affect thread execution. To allow threads to cooperate and compete for resources, it is necessary to provide mechanisms for synchronization and communication. The classic synchronization mechanisms are mutexes/condition variables and semaphores. These are provided in the eCos kernel, together with other synchronization/communication mechanisms that are common in real-time systems, such as event flags and message queues.

The kernel also provides exception handling. An exception is a synchronous event caused by the execution of a thread. These include the machine exceptions raised by hardware (such as divide-by-zero, memory fault, and illegal instruction) and machine exceptions raised by software (such as deadline overrun). The simplest, and most flexible, mechanism for exception handling is to call a function. This function needs context in which to work, so access to some working data is required. The function may also need to be handed some data about the exception raised — at least the exception number and some optional parameters. As opposed to exceptions, which are synchronous in nature, interrupts are asynchronous events caused by external devices. They may occur at any time and are not associated in any way with the currently running thread and are harder to deal with. The ways in which interrupt vectors are named, how interrupts are delivered to the software, and how interrupts are masked are all highly hardware specific. On the SA-1110, two kinds of interrupts are supported: fast interrupts (FIQ) and regular interrupts (IRQ); both can be masked.

TABLE 27.5 Typical Power Management Functions

Function Type	Functions Available
DVS	UAMPS_ENABLE_DVS(), UAMPS_SET_VOLTAGE(), UAMPS_SET_PROC_CLOCK(), UAMPS_CHECK_VCORE_STABLE(), UAMPS_SET_PROC_CLOCK(), uamps_set_proc_rate(), uamps_dvs_scheduler()
Shutdown	UAMPS_PERIPHERAL_POWER_ON(), UAMPS_PERIPHERAL_POWER_OFF(), UAMPS_V3_STANDBY_ON(), UAMPS_V3_STANDBY_OFF(), SA11X0_PWR_MGR_WAKEUP_ENABLE, SA11X0_PWR_MGR_GENERAL_CONFIG, SA11X0_PWR_MGR_CONTROL, uamps_set_proc_idle(), uamps_set_proc_sleep(), uamps_goto_sleep_state()

The kernel also provides a rich set of timing utilities such as counter, clocks, alarms, and timers. The counter objects provided by the kernel provide an abstraction of the clock facility that is generally provided. Application code can associate alarms with counters, where an alarm is identified by the number of ticks until it triggers, the action to be taken on triggering, and whether the alarm should be repeated. Clocks are counters associated with a stream of ticks that represent time periods. Clocks have a resolution associated with them, whereas counters do not.

27.4.5 Sensor-Specific Application Programming Interface Extensions

Table 27.5 illustrates some sample functions in the power management API. The functions are available to the µAMPS application developer to enhance the power efficiency of the sensing application. Programs written for the sensor node do not need to satisfy any unusual requirements, but some differences always exist between programs written for real-time OS platforms and those written for a time-sharing, virtual memory system like UNIX or Windows.

27.5 Results

Figure 27.8 shows the power consumption of the sensor node in the fully active state (all modules on) as a function of the operating frequency of the SA-1110. The figure shows the power consumption using DVS and only frequency scaling (done at a fixed core voltage of 1.65 V). The system power supply was 4.0 V. In the active mode, DVS is the primary source of power management. When running at the maximum operating voltage and operating frequency, the power consumption of the system is almost 1 W. Active power management using DVS results in about 53% maximum system-wide power savings. The actual savings depend on the workload requirement.

With DVS, minimum energy consumption results when the processing rate variation is minimized because of the convexity of the energy workload model. Figure 27.9 plots the relative battery life improvement as a function of the variance in workload. Each workload profile is Gaussian with a fixed average workload. Although the average workload might be constant, the battery life improvement from DVS will degrade as the fluctuations in workload increase.

Table 27.6 shows the measured power consumption of the sensor node in various modes of operation. The sensor node can be classified as a processor power-dominated architecture. The radio module follows the processor in power requirement (estimated at about 70 mA at 3.3 V). DVS can reduce system power consumption by 53%. Shutting down each of the components (analog power supply, radio module, and the processor) results in another 44% power savings, i.e., idle power management accounts for about 97% of system-wide power savings. Figure 27.10 shows overall power savings attributed to various power management hooks.

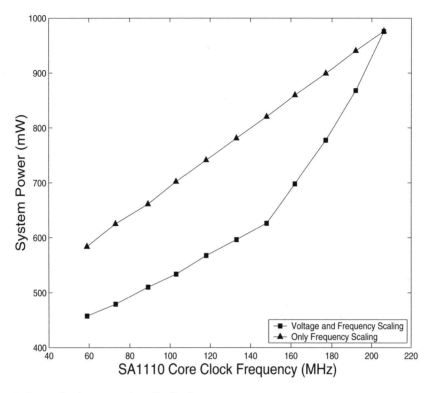

FIGURE 27.8 System-level power savings distribution.

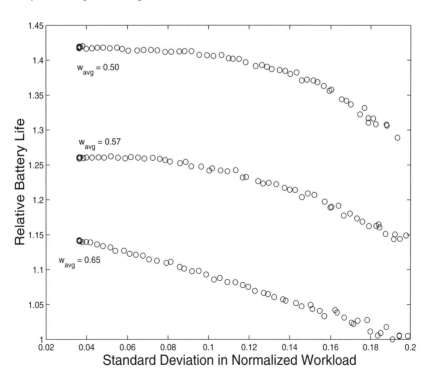

FIGURE 27.9 Battery life improvement as a function of duty cycle and active workload in a sensor node compared to a node with no power management.

TABLE 27.6 Measured Power Consumption in Various Sensor Modes

		Component Modes			Power
	System Mode	Processor	Radio	Analog	(mW)
Active	Active	Max Freq	on	on	975.6
States	Low Active	Min Freq	on	on	457.2
	Idle	idle	on	on	443.0
Sleep	Receive	idle	on	off	403.0
States	Sense	idle	off	off	103.0
	Sleep	sleep	off	off	28.0

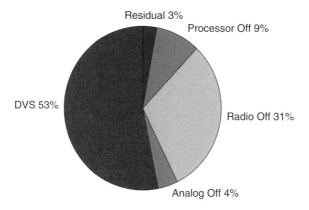

FIGURE 27.10 System level power savings distribution.

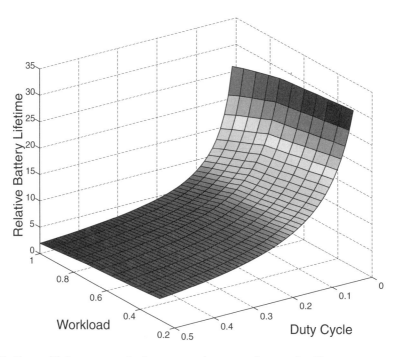

FIGURE 27.11 Battery life improvement in the sensor node compared to a node with no power management as a function of duty cycle and active workload.

Actual energy savings in the field depend significantly on processing rate requirements and event statistics. To estimate the energy savings from active mode power consumption, one would need an estimate of the workload variation on the system. If it is assumed that the average workload requirement is 50%, with slow variation, the estimated energy savings are about 30%.

Idle mode energy savings, on the other hand, can be significant. If it is assumed that the operational duty cycle is 1%, the estimated energy savings are about 96%. This implies that sensor node battery life can be improved by a factor of over 27 (i.e., a node that lasts for a day with no power management will now last for almost a month). With a 10% duty cycle, the battery life improvement is by a factor of about 10. The important point is that the system is energy scalable, i.e., it has the right hooks to tune energy consumption based on computational load and sensing requirements. Figure 27.11 shows the factor by which battery life of the sensor node can be enhanced by using power management techniques as a function of the workload and duty cycle requirement.

References

1. A.P. Chandrakasan et al., Design considerations for distributed microsensor systems, *Proc. Custom Integrated Circuits Conf.*, San Diego, CA, May 1999, 279–286.
2. The MIT μAMPS Project, http://www-mtl.mit.edu/research/icsystems/uamps/.
3. L. Benini and G.D. Micheli, *Dynamic Power Management: Design Techniques and CAD Tools*, Norwell, MA, Kluwer, 1997.
4. A. Sinha, A. Wang and A.P. Chandrakasan, Algorithmic transforms for efficient energy scalable computation, *Int. Symp. Low Power Electron. Design*, Italy, July 2000.
5. V. Gutnik and A.P. Chandrakasan, An embedded power supply for low-power DSP, *IEEE Trans. VLSI Syst.*, 5(4), Dec. 1997, 425–435.
6. T. Pering, T. Burd, and R. Broderson, The simulation and evaluation of dynamic voltage scaling algorithms, *Int. Symp. Low Power Electron. Design*, 1998, 76–81.
7. Intel StrongARM SA-1100 Microprocessor Developer's Manual, http://developer.intel.com/design/strong/manuals/278088.htm.
8. Advanced configuration and power interface specification, http://www.teleport.com/~acpi.
9. A. Sinha and A. Chandrakasan, Operating system and algorithmic techniques for energy-scalable wireless sensor networks, *Proc. 2nd Int. Conf. Mobile Data Manage.*, Hong-Kong, Jan 2001.
10. Intel Corp., Intel StrongARM SA-1110 Microprocessor — Advanced Developer's Manual, June 2000.
11. The eCos Operating System, http://www.redhat.com/ecos.

28
Design Challenges in Energy-Efficient Medium Access Control for Wireless Sensor Networks

Duminda Dewasurendra
Virginia Tech

Amitabh Mishra
Virginia Tech

28.1 Introduction 28-1
28.2 Unique Characteristics of Wireless Sensor Networks... 28-2
    Why Are MAC Layer Design Issues Important?
28.3 MAC Protocols for Wireless ad hoc Networks 28-4
    IEEE 802.11 • Bluetooth • Energy-Conserving Medium Access Control (EC-MAC) • Protocol for Wireless ATM Networks • Power-Aware Multiple Access (PAMAS) Protocol
28.4 Design Challenges for Wireless Sensor Networks........ 28-10
    Why Existing Methods for Wireless ad hoc Networks Cannot Be Used • Communication and Application Types in Sensor Networks
28.5 Medium Access Protocols for Wireless Sensor Networks... 28-13
    Sensor MAC (SMAC) • Self-Organizing MAC for Sensor Networks (SMACS) and Eavesdrop and Register (EAR) Algorithms • Traffic Adaptive Medium Access Protocol (TRAMA) • Power-Efficient and Delay-Aware Medium-Access Protocol for Sensor Networks (PEDAMACS) • Comparison
28.6 Open Issues 28-22
28.7 Conclusions 28-24

28.1 Introduction

Wireless sensor networks (WSNs) are an emerging paradigm posing new challenges for researchers in wireless communications [1]. This new class of networks closely resembles the behavior of wireless ad hoc networks. Nevertheless, they have a few unique differences; the principal one is the small size of nodes constituting a WSN. Although smaller nodes make WSNs suitable for several existing and emerging applications related to information sensing, this also implies that the nodes have limited resources, i.e., CPU speed, memory, battery, and radio interface. Because the nodes are resource constrained, they require network designs that can be customized for different types of application environments, thus placing significant demands on algorithm design, protocol specification, and technologies.

This chapter focuses on medium access control (MAC) schemes for WSNs. Unique features of these networks will be briefly discussed in order to highlight the issues demanding special attention during the design of MAC schemes. Significant research efforts currently underway in this context will be studied along with MAC schemes for generic wireless ad hoc networks (WAHNs) and wireless local area networks (WLANs). Finally, the challenges and open issues related to MAC algorithm design for the effective deployment of future WSNs will be discussed.

28.2 Unique Characteristics of Wireless Sensor Networks

WSNs consist of large numbers of distributed nodes that organize themselves to form a multihop wireless network. Each node consists of one or more types of sensors, an embedded processor, small memory, and a low-power radio transceiver. Generally, these nodes are battery powered and coordinate among themselves to achieve a common task. Compared to nodes in a generic WAHN operating under IEEE 802.11 [2] or Bluetooth [3, 4] protocols, these nodes are extremely small in size and possess limited energy resources. The transmitting power and thus the communication range are much lower, which is largely compensated by a higher density of nodes in most cases. WSNs can have distributed, hierarchical, or clustered architectures, as illustrated in Figure 28.1.

The lack of centralized control is common to WSNs as well as to other WAHNs. Nevertheless, the behavior of a WSN is largely governed by the application for which it is used. Even considering a single application, the desired role of nodes would be different from time to time. For example, in a battlefield application, it may be employed to monitor the ambient data patterns silently and generate alarms if the specified deviations are observed. The same network may be used to track the movement of a detected vehicle at another time. Such dynamic changes of network objectives and the corresponding change in node behavior are uncommon in most of the other generic WAHNs.

Furthermore, nodes of a sensor network are mostly unattended after deployment, permitting neither upgrade of energy sources nor troubleshooting. The node hardware is designed so that the overall cost is extremely low and nodes can be abandoned once the power sources are exhausted. Voids of discarded nodes may be filled with redundant nodes due to high node density, and perhaps by the deployment of additional nodes if the need arises. It is necessary that the network should accommodate such losses and new additions with least effort. These are unique issues pertaining to WSNs compared to generic WAHNs in which the nodes are mostly attended; energy sources are high capacity and can be recharged or replaced; and nodes have direct and individual interaction with users. Comparison between a WSN and a generic ad hoc network is summarized in Table 28.1.

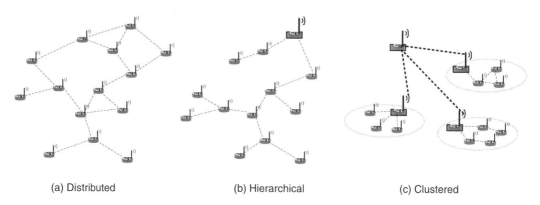

(a) Distributed   (b) Hierarchical   (c) Clustered

FIGURE 28.1 Sensor network architectures.

TABLE 28.1 Comparison of Features for WSNs and WAHNs

Wireless Sensor Network	Wireless ad hoc Network
Nodes involved in sensing the environment; events occurring in the environment can initiate certain communication in the network	No sensing behavior; network communication governed by user applications
Nodes are smaller in size	Larger nodes (e.g., PDAs, laptops)
Small and limited capacity power sources	High-capacity power sources
Inexpensive nodes	Relatively expensive nodes
Nodes unattended after deployment and designed for a prolonged lifetime with no maintenance or troubleshooting	Node troubleshooting and battery replacement possible
Node lifetime depends on the usage of attached power source	Node lifetime does not depend on energy resources because power sources are replaceable or rechargeable
Higher node density/highly redundant networks	Low node density/less redundant networks
Shorter transmission range (3 to 30 m)	Longer transmission range (10 to 500 m)
Limited processing and memory capacity	Higher processing power and memory
Nodes may stay in sleep mode for a significant amount of time	Nodes will be listening to the wireless medium most of the time
Data-centric communication; packet destination will depend on attributes of gathered data	Communication mostly occurs between specific nodes according to user requirements
Traffic profiles likely have statistically correlated properties comprising bursty traffic in case of event detection and low, continuous traffic during other times	Mostly continuous traffic, e.g., multimedia data streams
Low bandwidth (1–100 kb/s)	High bandwidth (e.g., 1 to 54 Mbps in IEEE 802.11-based WAHNs)
Network operation can be task oriented	Operation similar for all applications

28.2.1 Why Are MAC Layer Design Issues Important?

In all wireless networks, nodes must share a single medium for communication. Network performance largely depends upon how efficiently and fairly the nodes can share this common medium. Note that the packet transmission is directly handled by the MAC layer. Compared to a wired medium, a significant portion of the node's energy is spent on radio transmissions and on listening to the medium for anticipated packet reception. On the other hand, wireless networks always have restricted power sources; thus, careful design of the MAC scheme is necessary for the optimal performance and extended lifetime of the network.

In the context of WSNs, this requirement is extremely critical. According to the characteristics highlighted previously, nodes of a WSN carry extremely low energy resources and remain unattended after deployment; therefore, the node lifetime depends entirely on how energy is conserved during communication. Although some exhausted nodes could be compensated using redundant neighboring nodes, certain situations may arise rendering a part of the network completely inactive due to low connectivity and insufficient coverage, or making that part of the network inaccessible and isolated from the other parts. Such scenarios could be averted by avoiding unnecessary transmissions and longer listening periods — activities that consume the highest amount of power in nodes.

Another related issue is the high node density in WSNs. Although the transmission ranges are lower, a fairly high number of nodes can contend for the medium, at least in certain portions of the network. By the same token, transmissions from each node would increase the background noise for a large number of nodes, which may disrupt their own receptions. Thus, the MAC schemes for WSNs should be carefully designed to achieve the optimum performance toward the intended application. Previous surveys [1, 5] discuss some issues related to medium access in WSNs and WAHNs.

28.3 MAC Protocols for Wireless ad hoc Networks

The closest types of networks rendering a similar behavior to WSNs are WAHNs, although they have marked differences as highlighted in our discussion. Properly standardized MAC protocols designed to cater to the ad hoc and distributed nature of WAHNs have been developed and are in commercial use. Also, some of them focus on energy savings, mainly for mobile applications. These features are highly sought after in WSNs as well.

Currently available MAC protocols for wireless ad hoc networks are of two major types: *contention based* (CSMA) and *scheduling based* (TDMA, FDMA, or CDMA). In contention-based MAC schemes, the nodes compete among each other for channel access, whereas in scheduling-based methods, a specific schedule of channel access is used in time, frequency, or code domains. This section will briefly discuss several important medium access schemes belonging to both categories, including IEEE 802.11 [2]; Bluetooth [3, 4]; energy-conserving MAC (EC-MAC) [6]; and the power aware multiple access (PAMAS) [7]. Their merits, drawbacks, and suitability for WSNs will be highlighted.

28.3.1 IEEE 802.11

IEEE 802.11 is a standard developed for wireless LAN (WLAN) applications intended to replace conventional wired LANs so that the same applications can run seamlessly with media in 802.3 and 802.5 standards. Nodes in such networks would be mostly laptops and other typical equipment connected to a LAN. The distributed coordination function (DCF) in IEEE 802.11 is the access method used to support asynchronous data transfer on a best-effort basis when the network functions in an ad hoc mode. DCF can also coexist with an infrastructure network.

This is a contention-based protocol based on MACA [8] and MACAW [9] schemes proposed as improvements to the original CSMA scheme developed in Kleinrock and Tobagi [10]. It uses carrier sense multiple access with collision avoidance (CSMA/CA). Collision detection (CD) is not used because a node is unable to listen to the channel for collisions while transmitting. The scheme attempts to avoid the hidden terminal and exposed terminal problems in the original CSMA scheme.

28.3.1.1 Operation

Each node maintains a backoff counter controlling the channel access. Before a node starts data transmission, it senses the wireless medium. If the medium appears to be idle for a specified period of time (distributed interframe space — DIFS), it starts decrementing the backoff counter. If the carrier is detected during this time, the backoff counter is frozen; otherwise, it starts transmission once the backoff counter reaches zero. The sender and receiver exchange short request-to-send (RTS) and clear-to-send (CTS) control frames to establish a session. Data transmission is followed by an acknowledgment (ACK) frame to confirm successful reception. The gaps among RTS, CTS, DATA, and ACK frames are specified by short interframe space (SIFS). Duration of SIFS is relatively shorter than DIFS, thereby giving priority to the ongoing transmission. Contention for the channel access between two nodes, N1 and N2, is illustrated in Figure 28.2. Initially N1 is transmitting frame 1 followed by ACK reception. After waiting for the DIFS period, it starts decrementing the backoff counter in an attempt to transfer another packet. Because the backoff counter of N2 reaches zero first, it captures the medium and transmits a frame while N1 senses the medium is busy. Following the transmission of N2, N1 recaptures the medium for transmission of its second frame.

The DCF adopts a slotted binary exponential backoff mechanism to select the random backoff interval in case of unsuccessful transmission or after the completion of a successful transmission. This random number is drawn from a uniform distribution over the interval [0, CW-1], where CW is the contention window. After an unsuccessful transmission, CW is doubled; once CW reaches a maximum value (CW_{max}), it will remain there. In the case of a successful transmission, the CW value is reset to a minimum value (CW_{min}).

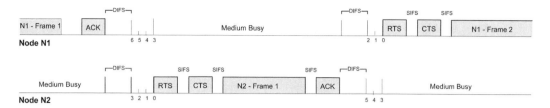

FIGURE 28.2 Contention between nodes N1 and N2 in IEEE 802.11 DCF.

The control frames RTS and CTS, as well as the data frames, include a parameter indicating the expected data transfer duration for the current session, which is used by the other nodes to update their network allocation vector (NAV). NAV is used to maintain a timer at each node, thus avoiding unnecessary transmission attempts before the current transmission is completed. This is termed a *virtual carrier sensing*. During the backoff period and the NAV timer active period, the node will be in idle mode listening to the channel with no transmission attempts.

28.3.1.2 Power-Saving Mode in IEEE 802.11

IEEE 802.11 standard also defines a power saving (PS) mode in which certain nodes can "go to sleep." Under DCF operation, PS nodes "wake up" periodically for a short interval called the Ad Hoc Traffic Indication Map (ATIM) window. It is assumed that hosts are fully connected and synchronized so that the ATIM windows of all PS hosts will start at about the same time. During this window, each node will contend to send a beacon frame. Any successful beacon serves the purpose of synchronizing node clocks and also inhibits other hosts from sending their beacons. After receiving the beacon, an active node can send a direct ATIM frame to a node in PS mode. These transmissions are also contention based and use the same DCF access procedure described earlier. On reception of the ATIM frame, the PS node will reply with an ACK and remain active for the rest of the period. Data transfer will take place after that.

28.3.1.3 Merits, Drawbacks, and Implications for WSNs

Recent work has shown that the energy consumption using IEEE 802.11 MAC protocol is significantly high because the nodes are listening to the channel most of the time. Although the 802.11 standard defines the PS mode, it provides very limited policy about when nodes should go to sleep. PS mode is designed for single-hop networks in which all nodes can hear each other. When used in multihop networks, IEEE 802.11 may have problems in clock synchronization, neighbor discovery, and network partitioning — thereby degrading the performance. Clock synchronization in a multihop WAHN is difficult because there is no centralized control; also, the synchronization packets exchanged among neighbors have variable delays due to unpredictable node mobility and radio interference.

PS mode is typically supported by letting low-power nodes wake up only at specific times. Without precise clocks, a host may not be able to know when other PS hosts will wake up to receive packets. Furthermore, a host may not be aware of a PS host at its neighborhood because a PS host will reduce its transmitting and receiving activities so that it cannot be detected. Such incorrect neighbor information may be detrimental to most routing protocols because the route discovery procedure may incorrectly report that there is no route even when routes exist with some PS hosts in the middle. Tseng et al. [11] proposed three sleep schemes to improve the PS mode in the IEEE 802.11 for its operation in multihop networks.

Requirements for clock synchronization and the suboptimal power saving makes this scheme an improper candidate for medium access in WSNs. Nevertheless, the idea of having a portion of nodes sleeping in the network while others are active may be an applicable concept to WSNs. The presence of redundant nodes in a WSN implies that all the nodes need not be active all the time because other nodes in the neighborhood can perform sensing and communication tasks covering the target area. Therefore, properly chosen redundant nodes can be put into sleep mode to achieve network-wide power savings. Open issues to be explored include the selection of redundant nodes and wake-up and connection reestablishment procedure for these nodes.

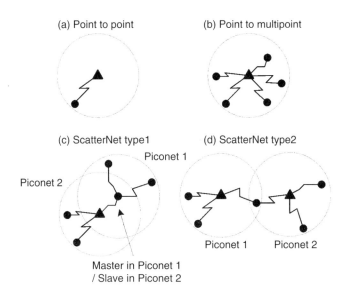

FIGURE 28.3 Piconet configurations in Bluetooth.

28.3.2 Bluetooth

Bluetooth [3, 4] is a short-range wireless networking for electronic consumer devices (mobile phones, pagers, PDAs, etc.). It uses a TDMA and CDMA hybrid scheduling based MAC scheme. The topology is a star network in which several slave nodes are attached to and synchronized with a master node to form a piconet. The number of nodes in a piconet is limited to eight in order to keep a high-capacity link among all the units and to limit the overhead required for addressing. Basic piconet configurations are shown in Figure 28.3(a) and (b). Along with the basic TDMA scheme, Bluetooth uses frequency hopping code division multiple access (FH-CDMA), which uses a large number of pseudorandom hopping sequences. Interpiconet communication is achieved by forming ScatterNets as shown in Figure 28.3(c) and (d). A single node can be a master in one piconet while it is a slave in another. Also, a node can be a slave in two piconets.

28.3.2.1 Operation

The master node determines hopping sequence, provides clock synchronization information for each slave node, and also controls the traffic in the piconet. The master/slave role is only attributed to a unit for the duration of the piconet. When a piconet is cancelled after a certain period of time, the master and slave roles are also cancelled and new piconets will be formed. Any node can become a master or slave. By definition, the unit that establishes the piconet becomes the master. Mechanisms are in place for multiple piconets to interconnect and form a multihop topology.

The time slots are alternately used for master and slave transmissions. The master transmission includes slave address of the unit for which the information is intended. In order to prevent collisions on the channel due to multiple slave transmissions, a polling technique is used: for each slave-to-master slot, the master decides which slave is allowed to transmit. This decision is performed on a per-slot basis: only the slave addressed in the master-to-slave slot directly preceding the slave-to-master slot is allowed to transmit in this slave-to-master slot. If the master has information to send to a specific slave, it is polled implicitly and the slave can return information. If the master has no information to send, it must poll the slave explicitly with a short poll packet. Because the master node schedules the traffic in the uplink and the downlink, intelligent scheduling algorithms that take into account the slave characteristics must be used. The master node control effectively prevents collisions among the participants of the piconet. Independent collocated piconets may interfere when they occasionally use the same hop carrier.

28.3.2.2 Merits, Drawbacks, and Implications for WSNs

Compared to contention-based MAC schemes, TDMA schemes have a natural advantage of energy conservation because the duty cycle of the radio is reduced and there are no contention-introduced overheads or collisions. Nodes can be put to sleep to save energy during the off intervals of the duty cycle, thereby making this an obvious candidate for WSNs.

Use of a TDMA protocol usually requires the nodes to form real communication clusters such as the piconets described here. Nodes in such clusters are restricted to communicate within the cluster, except for the master node and possible gateway nodes. Managing intercluster communication and interference is not an easy task. Moreover, when the number of nodes within a cluster changes, it is not easy for a TDMA protocol to change its frame length and time slot assignment dynamically. Thus, its scalability is normally not as good as that of a contention-based protocol.

In Bluetooth, nodes within a piconet must be synchronized to use the TDMA scheme. Achieving local synchronization within the cluster is not a difficult task. However, network-wide synchronization will be almost impractical, especially in WSNs. Thus, proper mechanisms need to be developed for intercluster communication, perhaps based on contention-based schemes.

28.3.3 Energy-Conserving Medium Access Control (EC-MAC) Protocol for Wireless ATM Networks

This particular MAC protocol is briefly described here because of its significant contribution toward minimizing the power consumption of nodes in wireless and mobile ATM networks. Goals of this access protocol are to conserve battery power; to support multiple traffic classes; and to provide different levels of service quality through bandwidth allocation. Although the IEEE 802.11 and Bluetooth standards address energy efficiency, this was not one of the central design issues in developing these protocols. The EC-MAC protocol [6], on the other hand, was developed with the issue of energy efficiency as a primary design goal.

28.3.3.1 Operation

The EC-MAC protocol is defined for an infrastructure network with a single base station serving mobiles in its coverage area. This definition can be extended to an ad hoc network by allowing the mobiles to elect a coordinator to perform the functions of a base station. Transmission in EC-MAC is organized by the base station into frames and each frame equals the basic unit of wireless data transmission.

The frame structure of EC-MAC protocol is shown in Figure 28.4. At the start of each frame, the base station transmits the frame synchronization message (FSM), which contains synchronization information and the uplink transmission order for the subsequent reservation phase. During the request and update phase, each registered mobile transmits a new connection request according to the transmission order received in the FSM. Collisions are avoided during this phase by having the base station send the explicit order of transmission using the FSM.

New mobiles that have entered the cell coverage area register with the base station during the new-user phase. Collisions during this phase are unavoidable and thus it may be operated using a variant of ALOHA. This phase also provides time for the base station to compute the data transmission schedule. The base station broadcasts a schedule message that contains the slot assignments for the subsequent uplink and downlink data transmissions (see Figure 28.4). Downlink transmission from the base station

FIGURE 28.4 Frame structure in EC-MAC.

to the mobile is scheduled considering the QoS requirements; similarly, the uplink slots are allocated using a suitable scheduling algorithm.

28.3.3.2 Merits, Drawbacks, and Implications for WSNs

Energy consumption is reduced in EC-MAC due to the use of a centralized scheduler, as in Bluetooth. Therefore, collisions over the wireless channel are avoided, thus reducing the number of retransmissions. Additionally, mobile receivers are not required to monitor the transmission channel as a result of communication schedules. The centralized scheduler may also optimize the transmission schedule so that individual mobiles transmit and receive within contiguous transmission slots. This scheme highlights the fact that scheduling algorithms that consider mobile battery power level in addition to packet priority may improve performance for low-power mobiles. Techniques used to minimize the energy consumption and performance of EC-MAC in this regard are discussed in detail in Sivalingam et al. [6].

In contrast to Bluetooth, this scheme allows new mobile nodes to join the cluster without completely disassembling it. In certain WSN applications, the network may consist of a significant portion of mobile nodes among the stationary nodes. Certain stationary nodes may act as sink nodes analogous to the base stations or cluster heads discussed here. When mobile nodes roam around these clusters, a concept similar to EC-MAC can be used to attach new mobile nodes to an existing group or cluster. After such an attachment and schedule update, the mobile nodes can communicate with the cluster head, using the set schedule, at a minimal expense of its energy.

28.3.4 Power-Aware Multiple Access (PAMAS) Protocol

PAMAS (power-aware multiple access) is a contention-based protocol [7] designed for ad hoc networks with energy efficiency as the primary design goal. It modifies the MACA protocol [8] by providing separate channels for RTS/CTS control packets and data packets (out-of-band signaling), thereby avoiding overhearing among neighboring nodes.

28.3.4.1 Operation

In PAMAS, a mobile with a packet to transmit sends an RTS message over the control channel and awaits the CTS reply message from the receiving mobile. If CTS is received, then the node transmits the packet over the data channel. This procedure is shown with nodes N3 and N4 in Figure 28.5. With the start of receipt of the data packet, the receiving mobile transmits a busy tone (BT) over the control channel with more than twice the duration of RTS/CTS packets, thus enabling users tuned to the control channel to know that the data channel is busy. Also, if it hears any other RTS packets (from node N6 in Figure 28.5), it transmits a busy tone.

If an idle node receives an RTS, it will check whether any of its neighbors is transmitting (by sensing the data channel) or receiving (by sensing BT). In either case, it will not reply with CTS (shown with nodes N2 and N5), thus causing the sender of RTS (nodes N1 and N6) to back off using a binary exponential backoff (BEB) scheme. Power conservation is achieved by requiring mobiles that are not able to receive or send packets to turn off the wireless interface. The use of a separate signaling channel allows nodes to determine when and for how long to power off. A mobile should power off when: (1) it has no packets to transmit and a neighbor begins transmitting a packet not destined for it, or (2) it has packets to transmit but at least one neighbor pair is communicating.

Once a node is powered off, two main issues arise. First, the latency due to sleeping is an issue because, if some other node wants to transmit data to a sleeping node, it must wait until this node powers up again. However, it should be noted that, even if the node was awake in this scenario, the sender must wait until the other transmissions are finished. This is a valid argument as long as the node will wake up as soon as the neighboring transmissions and receptions are complete. Therefore, the mechanisms that a node will use to decide exactly when to wake up are crucial.

The second issue is determining the length of sleep duration. It is addressed using a special probe packet. When a node wakes up after some time, it will send out a probe packet over the signaling channel

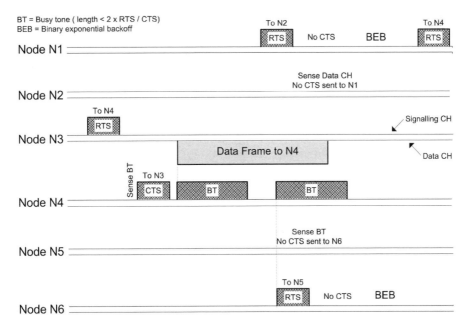

FIGURE 28.5 Operation of PAMAS.

asking any receiving nodes how much time it will take for the current transmission to end. If no collision occurs, the querying node will receive the exact time from the receiving node and will go back to sleep until this time. If the probe packet was destroyed due to a collision, the node will continue a binary search, sending more probe packets. Thus, the use of these probe packets ensures that the node sleeps no longer than necessary, thereby leaving the latency and throughput unchanged.

28.3.4.2 Merits, Drawbacks, and Implications for WSNs

Simulation results have established that this method reduces the power consumption by more than 50% in fully connected networks and at least 10% in highly loaded sparse networks. The essence of this scheme lies in the introduction of the additional signaling channel. This is also a major drawback of the scheme because employing an additional signaling channel requires additional hardware to be built into the nodes. It poses additional challenges, especially in WSNs in which node hardware is highly miniaturized.

Nevertheless, the significance of this protocol is that it can achieve high energy savings without compromising the network throughput and delay. Perhaps the concept of out-of-band signaling used here may be adaptable for intercluster communication in WSNs in which lack of synchronization does not permit using scheduling-based schemes. Such an out-of-band signaling channel can be used to set up data transfer directly between cluster heads or via gateway nodes. This will be remarkably effective for event-driven sensor networks (discussed in Section 28.4.2) in which intercluster communication occurs mostly in cases of a detected event and the regular communication is restricted primarily to nodes within the clusters.

A major contribution of the PAMAS protocol is the power savings achieved without sacrificing network throughput and latency. However, a major drawback observed here is that the power consumption of the nodes during excessive switching between the sleep and wake-up states is not given due attention. With the present WSN hardware designs, power consumption during state switching is significant. At the face of this, the PAMAS method may not perform satisfactorily without appropriate modifications for WSNs. Table 28.2 provides a brief comparison of these four medium access methods for WAHNs.

TABLE 28.2 Comparison of Media Access Protocols for Wireless ad hoc Networks

Protocol	Applications	Features	Implications to WSNs
IEEE 802.11 DCF	Wireless LAN	Optional PS mode	Redundant nodes can be sent to sleep and wake up as need arises
Bluetooth	Wireless networking for personal consumer devices	Piconets; centralized scheduling	Node clustering; local synchronization among nodes in clusters
EC-MAC	Wireless and mobile ATM networks	Scheduling for mobile nodes	Attaching mobile nodes to clusters without disassembling clusters
PAMAS	MAC protocol designed for WAHNs	Out-of-band signaling	Use of out-of-band signaling for intercluster communication

28.4 Design Challenges for Wireless Sensor Networks

As discussed in the first section, nodes in a WSN possess unique characteristics, especially the energy constraints, compact hardware, low transmission ranges, event- or task-based network behavior, and high redundancy. For a WSN, the extension of its lifetime is the most important issue. Therefore, power awareness is prominent in almost every aspect of the operation of WSNs. Currently, the research related to hardware of WSNs is focused on developing ultra low-power sensors, processors, and radio transceivers. Other drives are to reduce the form factor of batteries and improve technologies for power sources to keep nodes alive in active operation for many years. Meanwhile, software and middleware development is focused on minimizing power consumption during network operation. As highlighted in Section 28.2.1, it is extremely critical that the medium access control scheme be power optimal.

Energy consumption of a WSN occurs in three domains: sensing; data processing; and communications; among these, radio communication is the major consumer of energy. As highlighted in Pottie and Kaiser [12], energy for transmitting 1kb over a distance of 100 m is estimated as 3 J. With the same amount of energy, a general purpose processor with 100 MIPS/W power could execute 3 million instructions. The sensing circuitry consumes less power than the processor board in a typical WSN platform such as MICA [13, 14]. However, the radio consumes two to three times the power of the processor during packet transmission. Power consumption of the radio during listening to the channel for reception is also higher than the processor at full operation, but relatively lower than the transmitting power. The MICA sensor network platform defines four modes of operation, and Table 28.3 shows the typical current draw and power consumption of each node.

Thus, it is clear that the research focus should be on optimizing the medium access method in order to extend the lifetime of the network. In addition to energy conservation, the ability of the MAC scheme to adapt to network size, node density, and topology is also important. To be used in sensor networks aimed for dynamic applications, a MAC scheme should be highly scalable. Other important attributes include fairness, latency, throughput, and bandwidth utilization. However, these issues are considered secondary compared to energy considerations because they determine the entire lifetime of the network.

TABLE 28.3 Modes of Operation in MICA

Mode	Typical Current Draw	Power Consumption
Transmit (peak power)	32 mA	95 mW
Receive	18 mA	55 mW
Idle/sense	8 mA	25 mW
Sleep	20 μA	60 μW

Source: Crossbow Inc., Data sheet for MICA2 wireless measurement system, 2003.

Similar to the schemes described in Section 28.3, MAC schemes for sensor networks can be fundamentally categorized into *contention-based* or *scheduling-based* schemes. The inherent advantages of contention-based schemes in the context of WSNs include:

- No synchronization requirements
- No central scheduler required
- More robust to network dynamics
- No clustering necessary
- More suitable for event-driven WSNs

However, in terms of energy savings, contention-based schemes are not very attractive. Several sources of energy wastage in contention-based schemes during communication [15] can be identified:

- *Collision.* Usually data gathered by a node are exchanged with others using the radio. Two nodes may transfer data to each other at the same time or several nodes transfer data to the same node at the same time. When a transmitted packet is corrupted, it must be discarded and, thus, the follow-on retransmissions increase energy consumption. Collision increases latency as well.
- *Overhearing.* When a node picks up packets destined to other nodes, overhearing occurs. In an ad-hoc fashion, a transmission from one node to another is potentially overheard by all the neighbors of the transmitting node; thus, all of these nodes consume power even though the packet transmission was not directed to them.
- *Control packet overhead.* Sending and receiving control packets such as routing updates consumes energy and effectively reduces the network bandwidth for data packets.
- *Idle listening.* Nodes must listen to the channel often in order to receive possible traffic that is not sent. This is especially true in many sensor network applications because, if nothing is sensed, nodes are in idle mode for most of the time. Actual measurements have shown that idle listening consumes 50 to 100% of the energy required for receiving in such networks.

Scheduling-based schemes attempt to determine network connectivity first (i.e., discover the neighbors of each node) and assign collision-free links to each node. Links may be assigned as time slots (TDMA), frequency bands (FDMA), or spread spectrum codes (CDMA). However, the miniature hardware design of nodes in a WSN may not permit employing complex radio transceivers required for FDMA or CDMA systems. Thus, TDMA schemes are preferred as scheduling methods for WSNs. Inherently, TDMA schemes have a distinct advantage over the other methods. Except for the transmission, receiving and sensing durations, nodes can be put to sleep in order to achieve the highest amount of energy savings possible.

Nevertheless, the task of assignment of channels (i.e., TDMA slots, frequency bands, or spread spectrum codes) to links between neighbors so that packets do not collide is difficult. To ease the assignment, often a hierarchical structure is formed in the network to localize groups of nodes and make the task of channel assignment more manageable. This requires formation of node clusters and elect leaders for each cluster.

For TDMA schemes, time synchronization is a crucial factor, also. In contrast to generic WAHNs, maintaining perfect synchronization over the whole network is almost impossible, mainly because of the wide range of deployment, lower transmission ranges, and less control packet transmissions permitted due to energy constraints. Under a hierarchical clustering scheme, synchronization within each cluster can be maintained, but intercluster communication poses problems because of lack of synchronization.

28.4.1 Why Existing Methods for Wireless ad hoc Networks Cannot Be Used

The main goals of WAHNs are to provide a high throughput and low delay at a high bandwidth. In such networks, all nodes are engaged in the same type of activity and each user deserves equal opportunity in accessing the media. Thus, per-node fairness is an important issue. Network lifetime is not considered significant because the energy sources can be recharged or replaced, although power-saving schemes are recommended.

In contrast, for a WSN, extending the network lifetime is one of the priorities. To this end, it is necessary to conserve energy at each node during network operation. Toward achieving this objective, one must be ready to compromise network throughput and latency to a certain extent. Moreover, based on the type and operation of the WSN, as described in Section 28.4.2, throughput and latency requirements will depend on the application.

Also in contrast to WAHNs, nodes in WSNs are highly redundant; thus, some nodes can afford to be in sleep mode until a need arises, while others are active. Certain nodes acting as cluster heads, gateways, or sink nodes would accumulate, process, and relay larger quantities of data than ordinary, leaf-level sensor nodes. Additionally, from time to time during certain applications, the importance of data sensed by a node may vary in its importance or relevance to the current network objective. These issues call for maintaining distinct node priorities in WSNs in contrast to per-node fairness desired in WAHNs. Thus, employing MAC schemes that are developed for WAHNs would not be satisfactory in sensor networks without proper modifications.

In summary, novel MAC protocols are needed for WSNs because:

- Extending network lifetime is the primary goal in WSNs.
- Throughput and delay performance become secondary goals.
- QoS requirements may vary from time to time (e.g., in an event-driven WSN).
- Per-node fairness is not desired; instead, distinct node priorities may need to be considered for resource allocation.

28.4.2 Communication and Application Types in Sensor Networks

This section attempts to categorize the application-led behavior and possible communication types in WSNs. This is a general categorization that may help to identify relevant issues in designing an optimum medium access scheme for WSNs. Depending on application characteristics, sensor networks will behave in one of the following ways. It is possible that the same network may adopt a different role due to changes in the system objectives or firing of certain events in the observation field.

- *Centralized* data gathering and decision making. These networks are hierarchically organized, thus easier to set up and manage. At the top of the hierarchy, one or more root (sink) nodes collect all data from leaf nodes. Local processing may be performed at the sensor nodes at the bottom of the hierarchy, but the root node is responsible for gathering final data and, for the most part, governing the operation of the whole network.
- *Distributed* data gathering and decision making. Tasks in these networks are highly distributed and it is difficult to identify a particular network architecture or a hierarchy. An example would be a set of sensor nodes dropped in a harsh environment with no central control. Individual nodes must perform whatever sensing operations they can, discover their own neighbors and perhaps collectively make decisions on discovered events, and relay the decisions to the outside world via any relay nodes in reach. Instead, they would control certain actuators within their reach to perform certain reactive actions individually or by temporarily appointed leader nodes.

In both data gathering schemes, the following types of communication can occur:

- Unicast messages:
  - *Local*. When a real-world event in the network occurs, nodes are expected to perform some in-network processing. This will generally involve local messages being exchanged between neighbors. In a cluster-based scheme, this will include the messages exchanged between two nodes in the cluster or between a member node and the cluster head.

- *Multihop.* In centralized data gathering applications when a node requires sending data to the sink node (node-to-sink reporting), the sink node will not be in its direct reach most of the time. Thus, it will pass the message with the intended recipient address over multiple hops. This needs a proper addressing scheme as well as an efficient routing scheme.
- Multicast messages:
  - *Local.* These are messages originating from a node intended for several neighbors within its direct transmission range. For a clustering-based scheme, this is limited to multicast within the cluster.
  - *Multihop.* An example is a situation in which a sink node or a root node requires passing a control message to a set of nodes. In a clustering-based scheme, this may be communication from a root node to a set of cluster heads, or from one cluster head to several others. All such multihop messages need a proper addressing scheme as well as an efficient routing scheme, as mentioned earlier.
- Broadcast messages:
  - *Local.* Messages will be broadcast by a node to all the neighbors within its reach. Such messages will include anything of local importance to the neighborhood. In a clustering-based scheme, these will include messages broadcast among all nodes in a group.
  - *Multihop.* These are the messages that will impose the heaviest communication burden on the network. For instance, in a monitoring and surveillance application, a node may observe an alarming condition and may need to alert all others in the network. Unrestricted flooding may not be appropriate for such a situation, but a combination of multihop–multicast with local broadcast may be used.

Based on the application, optimal communication strategy for a sensor network would also be different. Two major categories of sensor networks dictated by their applications [16] have been identified:

- *Event driven sensor networks.* In an event-driven sensor network, sensor nodes do not send data (and are most likely asleep) until a certain event occurs. For example, in a fire-monitoring application, until a rise in temperature or smoke is detected, no data need be sent. In this way, node energy can be maximally saved. When an event occurs, how quickly the event can be reported to a central station, or how quickly other neighboring nodes can be alerted, become important issues. The main difficulty in an event-driven sensor network is to wake up the entire network, or at least the nodes along a path to the base station, when an event occurs. Moreover, the traffic pattern of the network may drastically change in case of an event.
- *Continuous monitoring sensor networks.* In a continuous monitoring sensor network, data are sampled and transmitted at regular intervals. For example, an ambient temperature monitoring station can take periodic readings and send it to a central monitoring station only at specific intervals. In these types of networks, the traffic patterns are more stationary and the routing tables (if any) remain unchanged most of the time. Scheduling-based MAC schemes can be used effectively in these networks for maximum energy savings.

The behavior and communication types identified in this section need to be considered for the optimal performance of energy-aware MAC schemes. Next, several MAC schemes proposed for WSNs will be discussed, along with their applicability to these different types of networks.

28.5 Medium Access Protocols for Wireless Sensor Networks

Recently, several authors have suggested energy-aware medium access schemes for WSNs, a number of which are modifications of existing protocols for WAHNs. This is still a growing area of research calling for attention to several open issues yet to be addressed. This section discusses four such recently proposed schemes, with their merits and drawbacks, in the context of WSNs. These include sensor MAC (SMAC) [15]; self-organizing MAC for sensor networks (SMACS) [17]; traffic adaptive medium access protocol

(TRAMA) [18]; and power-efficient and delay-aware medium access protocol for sensor networks (PEDAMACS) [19].

28.5.1 Sensor MAC (SMAC)

The main objective of SMAC [15] is to conserve energy in sensor networks; it takes into consideration that fairness and latency are less critical issues compared to energy savings. Thus, this scheme compromises fairness and latency to a certain degree. In order to save energy, SMAC establishes a low duty cycle operation in nodes. It reduces idle listening by periodically putting nodes into sleep in which the radio transceiver is completely turned off. As discussed in Section 28.2.1, a high bandwidth utilization is a goal in generic WAHNs, compelling nodes to operate in fully active mode all the time. In SMAC, the low duty cycle mode is the default operation of all nodes in the network. Nodes only become more active by changing the duty cycle when heavy traffic is present in the network, or once an event occurs in case of an event-driven WSN.

28.5.1.1 Operation

During the design of SMAC, the following assumptions have been considered:

- Short-range multihop communications will take place among a large number of nodes.
- Most communications will be between nodes as peers, rather than to a single base station.
- In-network data processing is used to reduce traffic.
- Collaborative signal processing is used to reduce traffic and improve sensing quality.
- Applications will have long idle periods and can tolerate some latency.
- Network lifetime is critical for the application.

All nodes in the network will be following a sleep-and-listen cycle called a frame, as shown in Figure 28.6. The duration of the listen period is normally fixed and the sleep interval may be changed according to application requirements, changing the duty cycle.

The same RTS/CTS/DATA/ACK procedure as that in IEEE 802.11 is adopted here for unicast packets in order to ensure collision avoidance and to avoid hidden terminal problem. Broadcast packets are sent without using RTS/CTS; NAV timer update information is included in all four types of packets. Thus, this scheme uses virtual and physical carrier sensing. After a successful exchange of RTS and CTS, the sender will start transmission and will extend it into the sleeping duration as well, if required. The nodes do not follow their sleep schedules until they finish the transmissions, thus increasing the performance.

28.5.1.2 Coordinated Sleeping

Although a node can freely choose its own active/sleep schedules in SMAC, it attempts to reduce the overhead by synchronizing schedules of neighboring nodes together. Nodes exchange their schedules by periodically broadcasting a SYNC packet to their immediate neighbors at the beginning of each listen interval. A set of nodes synchronized together will form a virtual cluster. Because the whole network cannot be synchronized together, neighboring nodes are allowed to have different schedules. However, neighboring nodes are free to talk to each other, no matter to which listen schedules they adhere. A considerable portion of the nodes will belong to more than one virtual cluster, enabling intercluster communication. Thus, this scheme is claimed to be adaptive to topology changes.

FIGURE 28.6 Periodic listen-and-sleep schedule in SMAC.

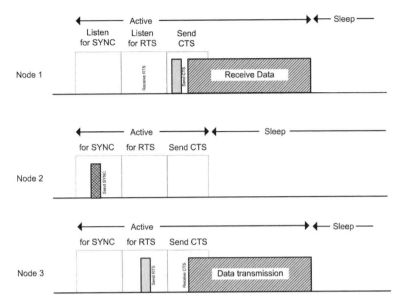

FIGURE 28.7 Timing schedules among different nodes in SMAC.

When the network is first deployed, each node tries to retrieve a sleep schedule from a neighbor first. In case of failure, it adopts one of its own and also tries to announce it to the neighbors by broadcasting a SYNC packet. Broadcasting SYNC packets must also follow the normal carrier sense and random backoff procedure. If a node receives a different schedule after it announces its own schedule, it must adopt one of the following:

- If the node detects no other neighbors, it can discard the current schedule and adopt the new.
- If it has one or more neighbors and is already a part of an existing virtual cluster, it can adopt both schedules by waking up at the listen intervals of both.

The active interval of a node is divided into three parts for SYNC, RTS, and CTS as shown in Figure 28.7. If CTS is received, data transmission will be immediately followed. Here nodes 1 and 3 are synchronized to the schedule of node 2 by receiving its SYNC packet, thus falling into the same virtual cluster. Node 3 initiates an RTS/CTS exchange with node 1 followed by a data transmission. While node 2 follows its normal sleep schedule, nodes 1 and 3 stay active until the completion of the data transfer, altering their usual schedule.

28.5.1.3 Neighbor Discovery in SMAC

When a new node powers on, it listens to the channel in anticipation of a SYNC packet. However, it is possible that a new node fails to discover an existing neighbor because of collisions or delays in sending SYNC packets by neighbor due to busy medium. To prevent a case in which two neighbors cannot find each other when they follow completely different schedules, SMAC protocol employs a simple periodic neighbor discovery procedure by requiring each node to listen periodically to the channel for the whole synchronization period. The frequency can be varied depending on the network conditions, etc.

28.5.1.4 Synchronization

Clock drift on each node can cause errors in the coordination of schedules among neighboring nodes. To minimize this problem, it uses relative timestamps in SYNC packets. Also, the listen period is made significantly longer than possible clock drift. Although this technique can tolerate relatively larger clock drifts, neighboring nodes are still required to update each other periodically with their schedules to prevent possible errors. Using experiments, authors claim that the clock drift between two nodes does not exceed 0.2 ms per second [15]; however, these figures may not be valid for certain applications of WSNs.

28.5.1.5 Adaptive Listening

An adaptive listening strategy that enables each node to adjust its schedule according to the network traffic is also used in SMAC to minimize latency. When a sensing event occurs, it is desirable that the sensing data can be passed through the network without much delay. When each node strictly follows its sleep schedule, potential delays are possible on each hop in the multihop path. The technique is that, if a node overhears its neighbor's RTS/CTS transmission during a listen period, it will receive the estimated length of that data transmission before going to sleep according to its normal schedule. However, the node will wake up for a short period of time at the end of that transmission to check whether it is the next hop in this multihop message. If so, the neighbor will immediately pass the data to it after RTS/CTS exchange, thus avoiding the neighbor's waiting for the next scheduled listen time of this node and minimizing latency.

28.5.1.6 Merits and Drawbacks

Compared to other schemes designed for the mobile ad hoc networks explained in Section 28.3, SMAC is designed particularly for use in wireless sensor networks. It attempts to combine the advantages of TDMA scheduling for power saving by periodically requiring sensing nodes to go to sleep. The sleeping patterns are coordinated in order to minimize the latency, as discussed before. Nevertheless, a solution with a fixed duty cycle does not give the optimal performance.

Authors claim that this scheme forms a flat topology and intercluster problems are absent; however, this may not be true in cases in which the application requires real clusters to be formed, at least temporarily. In such a case, the communication patterns will depend on the cluster formation and that these real clusters and the virtual clusters formed will coincide is not guaranteed. The adaptability of this scheme to such a situation should be investigated.

A significant portion of nodes will belong to two or more virtual clusters under this scheme. The energy consumption of such nodes would be higher compared to nodes within a single virtual cluster. Hence, the portion of such nodes and its effect on performance should be analyzed under real application scenarios. Also, the performance of this MAC scheme should be studied along with different routing schemes in order to assess its performance of intercluster communication, especially for multihop unicast and multicast messages. Data routing across virtual clusters needs to be studied further for its latency and throughput.

In WSN applications, it is possible for certain nodes to be exhausted with power and new nodes to be added. Performance of SMAC during times when a significant portion of nodes is discarded or added, or in cases with a higher portion of mobile nodes, should be studied. Another instance to be observed is what happens if the coordinated sleep schedules of two neighboring clusters are completely opposite. In such cases, it is not clear whether the bordering nodes could adopt both schedules.

28.5.2 Self-Organizing MAC for Sensor Networks (SMACS) and Eavesdrop and Register (EAR) Algorithms

Self-organizing MAC for sensor networks (SMACS) is designed for network startup and link layer organization in a static WSN [17]; it is one of the earliest attempts to develop MAC protocols for sensor networks. In this scheme, each node maintains a TDMA frame in which the node schedules different time slots to communicate with its known neighbors. During each time slot, it only talks to one neighbor. To avoid interference between adjacent links, the protocol uses different frequency channels (FDMA) or spread spectrum codes (CDMA). Although the frame structure is similar to a typical TDMA frame, it does not prevent two interfering nodes from accessing the medium at the same time. The actual multiple access is accomplished by FDMA or CDMA.

28.5.2.1 Operation

For the correct operation of the SMACS protocol, the following assumptions are made.

1. Nodes are able to tune the carrier frequency to different bands. It is assumed that the number of available bands is relatively large.

2. Nodes are randomly deployed. After deployment, each node wakes up at some random time according to a certain distribution.
3. The network is assumed to consist primarily of stationary nodes, with few mobile nodes.

Each node assigns links to its neighbors immediately after they are discovered. When all nodes hear all their neighbors, they have formed a connected, multihop network. Because each node is only partially aware of the radio connectivity in its vicinity, it is possible that collisions can occur if a simple TDMA scheme is used alone. To avoid this, frequency bands chosen at random from a large pool are assigned for each slot.

Length of the frame T_{frame} is fixed for all nodes in the network; however, these frames need not be synchronized and the time slots assigned inside the frame need not be aligned. This is possible because different frequency band or CDMA codes are used for communication during each slot. Such an ability to assign nonsynchronous slots is the key property that enables nodes to form links on the fly. This is illustrated in Figure 28.8, in which nodes A, B, C, and D are in the same neighborhood after deployment and they wake up at times T_1 to T_4, respectively.

Nodes A and B discover each other first and establish their own schedules for transmission and reception. Nodes C and D wake up at later times, discover each other, and establish their own schedules. However, note that all schedules are not aligned and also that the transmission slots of pair A/B and pair C/D overlap. This is made possible by using distinct frequencies f_x and f_y. After a schedule is established, a node will turn on its transceiver ahead of appropriate slots to communicate with others. Similarly, it will turn off the radio when no communication is scheduled, thereby enabling significant energy savings. In most WSN applications, mobile nodes will be present among other stationary nodes. In order to attach these mobile nodes in an energy conserving manner to the already formed network using SMACS, the eavesdrop and register algorithm (EAR) is introduced and discussed in the following section.

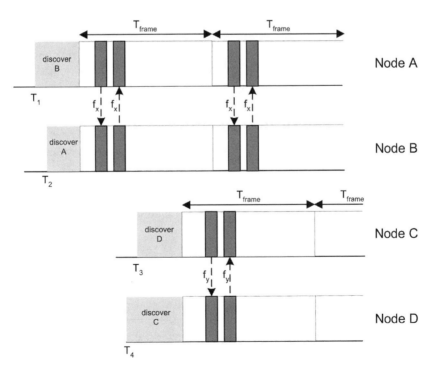

FIGURE 28.8 Nonsynchronous scheduled communication in SMACS.

28.5.2.2 EAR Algorithm

The EAR algorithm enables seamless interconnection of mobile nodes in the field of stationary wireless nodes. This protocol performs the mobility management of the network allowing mobile nodes to listen to the communication from the stationary nodes and establish connectivity with them. Because of energy limitations, the communication channels between the mobile and stationary sensors in the network must be established using as few messages as possible. This is accomplished by allowing the mobile node to decide when to invite the stationary node to establish a connection as well as when to drop a connection. In this manner, mobile nodes assume full control of the connection process to avoid the unnecessary use of power associated with lost messages.

According to the preceding third assumption, only a few stationary sensors will be within the reach of a mobile sensor at any given time. During some predetermined slot in the frame of each stationary node, it transmits an invitation message to the surrounding neighborhood with the intent of inviting new nodes to join the local network. Stationary nodes do not necessarily require a response to this message, but a mobile node with the intention of joining the network will be eavesdropping on such messages. These pilot messages will trigger the EAR algorithm in mobile nodes.

Each mobile node will maintain a list of neighbors according to the invitation messages received. It compares parameters, such as the received SNR, node ID, transmitted power, etc., and decides which node to connect. When the energy saving requirements are stringent, the decision will aim solely for minimal power connectivity. Accordingly, the mobile nodes will initiate a connection with a stationary node. Stationary nodes will also maintain a simple list of mobile nodes that have formed connections and remove the entries when the link is broken.

28.5.2.3 Merits and Drawbacks

Merits of this scheme include the ability to form links with any neighbor on the fly, with no restrictions on synchronization, which largely reduces the latency. Another advantage is that this scheme is applicable to WSNs with neither physical nor virtual clustering. This will allow the MAC scheme to function independently of any application-based clustering requirements of WSN.

However, a significant waste of resources is a trade-off to low latency. The main drawback is that the time slots are wasted if a node does not have data to send to the intended receiver. Also, the frames of a large number of nodes will mostly be vacant if the nodes are sparsely distributed in certain areas. Defining a smaller T_{frame} value will not be permitted in this case because some areas of the network may have a higher density of nodes. SMACS does not attempt to utilize these vacant time slots in order to maintain simplicity; rather, this protocol uses FDMA or CDMA and thus unnecessarily complicates the node hardware design. Bandwidth utilization would also be lower for the same reasons. Another major drawback is that the energy waste during the switching between sleep and active states is not considered. Because of assigning time slots on the fly without any synchronization, the nodes must switch between active and sleep states many times. This will drain the energy sources of nodes unnecessarily.

Apart from these drawbacks, this EAR protocol should be studied further in order to develop effective ways to manage WSNs with mobile as well as stationary nodes.

28.5.3 Traffic Adaptive Medium Access Protocol (TRAMA)

TRAMA is a recently introduced MAC protocol for energy-efficient and collision-free channel access in WSNs [18]. It uses traffic-based information to decide on schedules for individual nodes and thus is adaptive to network traffic. It is claimed that, because of this adaptability, it can deliver adequate performance and energy efficiency in both network types discussed in Section 28.4.2. TRAMA provides support for unicast, broadcast, and multicast traffic.

28.5.3.1 Operation

TRAMA assumes a single, time-slotted channel for data and signaling transmissions. The time schedule of each node is organized in two major sections, as shown in Figure 28.9. One consists of a collection of

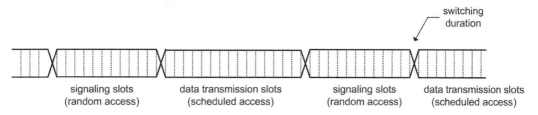

FIGURE 28.9 Time slot organization in TRAMA.

signaling slots using random access and the other of data transmission slots using schedules access. The duty cycle of switching between these states could be adjusted according to the application requirements and also according to the different network types described in Section 28.4.2. For stationary networks, the random access periods occur less frequently and vice versa for highly dynamic networks. Cycle duration is usually of the order of tens of milliseconds, making this scheme less prone to even significant clock drifts around 1 msec, which are highly unlikely in typical networks. Thus, the scheme assumes that adequate synchronization can be achieved using one of the synchronization schemes suggested for WSNs.

Communication in TRAMA consists of three major components: neighbor protocol (NP); the adaptive election algorithm (AEA); and the schedule exchange protocol (SEP). NP is used to exchange one-hop neighbor information among neighbors and to gather two-hop topology information for each node in the network. This is performed by exchanging small packets among neighbors during the random access period. Nodes always start in random access mode with NP. Also, the synchronization is performed during the random access period and the node should be in active state (transmit, receive, or listen) during this interval.

During the random access period, NP exchanges short signaling packets that include the information about connected neighbors of the sender; the goal is to provide the two-hop neighbor information to each node. Note that these include incremental information to keep the packet length small, i.e., it contains the node IDs of newly added neighbors and disconnected neighbors. These short packets are also used to maintain connectivity between neighbors.

During scheduled access, AEA selects transmitters and receivers so that collision-free transmission is achieved. AEA is based on the neighborhood-aware collision resolution protocol (NCR) proposed in Bao and Garcia–Luna–Aceves [20]. This technique claims to avoid data packet collisions among neighbors due to hidden terminals. AEA uses traffic-based information exchanged among nodes during SEP to make efficient use of the channel avoiding idle slots. The same traffic information is also used in AEA to perform receiver selection. During these selections, the node priorities in the network and two-hop neighbor information exchanged during NP are considered. By selecting the transmitter and receiver for each time slot, AEA enables nodes to switch into sleep mode whenever possible, thus achieving maximum energy savings.

SEP is used to exchange traffic schedules among neighbors during scheduled access mode. These schedules contain the set of receivers for the traffic currently originating at the node and its scheduled transmission slots. This information is periodically broadcast to the node's one-hop neighbors during scheduled access. Each node computes a schedule interval depending on how often it needs to transmit data according to its current application requirements. Following this, the node selects the highest priority slots it can acquire according to AEP; an example is illustrated in Figure 28.10. This information (schedule) is transmitted to its neighbors typically during the last slot of its schedule. Also, after emptying the current data buffers, it announces the release of its vacant time slots so that the other nodes can acquire them.

State of a node at a particular time slot is determined based on its two-hop neighborhood information and the schedules announced by its one-hop neighbors. Three possible states of a node are: transmit, receive, and sleep. At a given time slot, a node is in transmit state if it has the highest priority among its

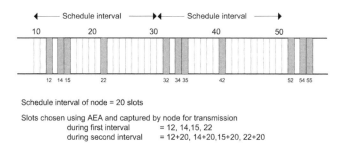

FIGURE 28.10 Example schedule of a node in TRAMA.

contending set and also if the node has data to send. A node is in the receive state when it is the intended receiver of current transmitter. If neither of these cases occurs, the node will be switched off to the sleep state in order to save energy. In other words, if a node is not the currently selected transmitter by AEP, it will consult the schedule sent by the current transmitter. If the transmitter does not have traffic destined for this node in the current slot, it can go to sleep. Under this scheme, a sleeping node is required to wake up at the schedule announcement slot (usually the last transmission slot in each schedule interval) to update itself on possible schedule changes. For this purpose, it should always be aware of the schedule of each of its one-hop neighbors.

28.5.3.2 Merits and Drawbacks

The authors provide extensive simulations to compare TRAMA with SMAC and several other comparable MAC schemes [18]. It is shown that the scheduled-based medium access protocol based on neighborhood-aware collision resolution protocol (NCR) achieves better data delivery than the contention-based protocols such as IEEE802.11, CSMA, and S-MAC. The main reason highlighted for the improvement in delivery is that the freedom from collision is guaranteed at all times during data transmission.

It is also shown that the scheduled-based medium access protocols incur higher average queuing delays. The average queuing delay for TRAMA is relatively large due to overhead involved in scheduling. Within every schedule interval, a transmission slot is used for announcing schedules in TRAMA. This decreases the effective channel access probability for data transmission and is not favorable for continuous data gathering type WSNs described in Section 28.4.2 because the traffic is homogenous across the network and all the nodes periodically generate traffic.

Simulation results also show that TRAMA exhibits high throughput compared to SMAC and IEEE 802.DCF because it avoids collisions due to hidden terminals using NCR protocol [20]. Energy savings of TRAMA depend mainly on the traffic pattern of the network as compared to duty cycle-dependent energy savings in SMAC. In TRAMA, the random access period duration plays a significant role in energy consumption. Significant features in TRAMA are the time slot reuse; using neighborhood information for collision avoidance; and use of a hybrid scheme of random and scheduled access for optimal performance.

28.5.4 Power-Efficient and Delay-Aware Medium-Access Protocol for Sensor Networks (PEDAMACS)

PEDAMACS [19] medium access protocol combines the characteristics of cellular networks with those of type 2 sensor networks for the continuous data gathering applications described in Section 28.4.2. It assumes that a single access point (AP) exists in the network and all nodes communicate with this AP. Also, it assumes that AP has no energy constraints and is capable of transmitting at higher power levels when needed so that it can reach any node in the network in a single hop. In contrast, the sensor nodes have limited transmission power and will reach the AP using multiple hops. Although it may not be possible always, in certain applications it may be possible to include a few nodes with higher energy resources to act as APs of each node cluster. The extra effort required may be compensated with optimal power savings in low-power sensor nodes.

28.5.4.1 Operation

The algorithm consists of three major phases: topology learning phase; topology collection phase; and scheduling phase. During the topology learning phase, each node identifies its interferers, neighbors, and parent node. This phase begins with a topology learning packet transmitted by AP over the longest range (highest power) in one hop to all sensors. This packet includes the current time so that each node updates its time and synchronizes with the other. Also, it includes the next anticipated incoming packet time so that every node will stop transmitting and listen for the next broadcast message of AP at this future time, as illustrated in Figure 28.11(a).

Following this, AP floods the network with a tree construction packet over a medium range (medium power). This packet contains a hop count field to avoid any retransmission loops and to facilitate choosing the parent node in the tree as shown in Figure 28.11(b). At the end of this phase, each sensor node decides the parent node to be the one with the smallest number of hops to AP, and the neighbors and interferers as the nodes with the received signal level above and below some interfering threshold, respectively. Because no prior topology information is available during this phase, the authors suggest a simple CSMA scheme with a random delay before carrier sensing.

The topology collection phase starts next with the AP transmitting a topology collection packet (with the same format as that shown in Figure 28.11(a) over the longest range (highest power). The transmission time is announced in the incoming packet time field of the topology learning packet earlier. This packet also contains current time and next incoming packet time. Following this, each node transmits its topology packet containing its parent, neighbor, and interferer information to AP as shown in Figure 28.11(c). Here again, the CSMA scheme with some random delay before the transmission is used.

During the scheduling phase, each node is explicitly scheduled by AP based on the complete topology information obtained during the previous topology collection phase. The scheduling frame is divided into time slots. At the beginning of this phase, AP performs the scheduling of the sensor nodes in the network and announces the schedule of how all the traffic will be carried during the scheduling frame by broadcasting a schedule packet over the longest range. The schedule packet includes the transmitter information corresponding to each time slot in addition to current time and next incoming packet time fields as shown in Figure 28.11(d). At the beginning of the scheduling frame, each node samples the sensor and generates one packet, which is then carried to AP according to the schedule.

| Header | Current time | Next packet transmission time | CRC |

(a) Topology learning and topology collection packet from AP

| Header | Number of hops | Parent transmission node ID | CRC |

(b) Tree construction packet from AP

| Header | Node ID | Node level | Parent ID | No. of neighbors | Neighbor IDs | No. of interferers | Interferer IDs | CRC |

(c) Topology packet from nodes

| Header | Slot seq. No. | Number of nodes scheduled for current slot | Scheduled node IDs | ...... | CRC |

(d) Schedule coordination packet form AP

FIGURE 28.11 Packet formats in PEDAMACS.

28.5.4.2 Merits and Drawbacks

Although in certain specific applications such a scheme may be able to be used for sensor networks, characteristics of this scheme are not preferred for sensor networks in general. This is mainly due to the use of a central access point that can reach all sensor nodes using high transmitting power. This assumption is not realistic in most of the sensor network applications in which the nodes are distributed over large areas or in indoor environments. Authors argue that in such cases, several APs can be used, but fail to provide details on how to establish proper coordination and synchronization among all such APs. Furthermore, it has yet to be analyzed how the large overhead associated with such a scheme affects the network performance. This overhead also makes this scheme unusable for event-driven sensor systems or dynamic systems with frequent addition and removal of nodes from the network. Authors suggest using a CSMA scheme with implicit ACKs for the topology collection phase, but this may not be an appropriate solution to avoid the huge number of possible collisions.

28.5.5 Comparison

Table 28.4 summarizes and compares the previously discussed four MAC schemes for WSNs.

28.6 Open Issues

Having discussed application and communication categories of WSNs and compared several MAC schemes for WSNs, the open research issues yet to be addressed will be discussed in this section. Designing optimal, energy-aware MAC schemes for WSNs is still an open and fast growing research area. In this context, the following issues are highlighted for consideration in future research.

TABLE 28.4 Comparison of MAC Schemes for Sensor Networks

	SMAC	SMACS/EAR	TRAMA	PEDAMACS
Features	TDMA scheduling Coordinated sleeping schedules among neighbors Adaptive listening Virtual clustering	Hybrid TDMA/FDMA scheduling Mobile node attachment	Random access (CSMA) for neighbor discovery Scheduled access (TDMA) for data transmission	Access point (AP) with high-power transmitter Centralized TDMA scheduling by AP node Hierarchical organization
Applications	WSNs with more stationary nodes	Low traffic WSN with strict latency requirements	Event-driven WSNs	Centralized data gathering WSNs
Merits	Reduced latency for multihop messages Simple hardware for TDMA	Low latency Ability to create links on the fly No clustering requirements No synchronization requirements	TDMA slot reuse No collisions due to hidden nodes Traffic adaptable	Higher energy savings in centralized WSNs
Drawbacks	Synchronization required Virtual clusters may not coincide with physical clusters	Complex hardware for FDMA or CDMA Waste of time slots Low bandwidth utilization Frequent switching can cause heavy energy losses	Synchronization required Low bandwidth utilization in periodic data gathering WSNs	Centralized control necessary AP node requires high power High overhead for scheduling

Adaptability to network objectives. How much sensor node energy to spend on a particular task entrusted on a WSN depends on how critical current application objectives are. As explained in Section 28.2 and Section 28.4.2, the same network used for a low-frequency continuous monitoring application may be employed for mission-critical tracking or emergency threat alert in the next instance. In such a scenario, less critical goals of a sensor network become highly critical and the energy saving requirements become secondary as compared to latency and throughput. A challenging and open issue is to develop medium access schemes for WSNs that have changing missions. SMAC [15] and TRAMA [18] attempt to achieve this to a certain extent; nevertheless, more work must be done in this area.

Optimal schemes depending on WSN type. Certain applications such as habitat monitoring may have stationary traffic patterns mostly over the total lifespan of the WSN employed. For these types of applications, achieving energy savings to extend the network lifetime remains the primary objective throughout the monitoring period. Medium access schemes can be optimized for energy efficiency in WSNs used for such applications. TDMA scheduling-based schemes similar to SMAC [15] may be the ideal candidate for these applications. However, the synchronization requirements and virtual clustering need to be reconsidered in this respect.

Cross-layer design. Conventional WAHNs have neatly defined protocol stacks with independent operation of each layer. For example, medium access scheme would function independent of the node connectivity, routing requirements, and application context. However, the primary goal of energy saving is tightly coupled with all these factors in a WSN and thus medium access cannot be considered alone for optimal savings. It is increasingly clear that power efficiency cannot be addressed completely at a single layer in the networking stack [21, 22]. It will often be necessary to use parameters propagated from upper layers to adapt the medium access protocol, especially in situations in which network objectives change considerably from one time to another. Another issue that must be effectively coupled with medium access is the data aggregation. No significant research efforts have been observed so far in this regard and the next generation medium access schemes for WSNs beyond SMAC and TRAMA should take these aspects into thorough consideration.

Effects of time synchronization. It is observed that higher energy savings are mostly obtained using TDMA based-scheduling schemes in WSNs. Inherently, these schemes require time synchronization of participating nodes in a single schedule. Synchronization errors always tend to degrade the end-to-end throughput performance of the network. As highlighted in Section 28.3.2.2 and Section 28.4, it may be impractical to achieve network-wide synchronization in WSNs. As an alternative, it is better to have globally asynchronous and locally synchronous architectures in which local node clusters maintain synchronization for TDMA schemes aiming for maximum energy savings, while intercluster communication is mostly contention based and asynchronous. Coupling such schemes for optimal energy saving medium access has yet to be explored.

Cluster-based hierarchy. Most WSN applications may require hierarchical architectures. This favors clustering-based systems, which is assumed in most of the research on WSNs. Highly energy-efficient medium access methods could be developed for such networks, using the cluster head as a centralized scheduler, data sink, and relay for the whole cluster. Issues arising in such contexts include ways to achieve intercluster communication, minimizing intercluster interference, and the possibility of having same application-specific clusters in the MAC layer for optimal performance. All these issues need further investigation.

Scalability. Compared to generic WAHNs, WSNs have a larger node count as well as higher density. This should be a critical consideration in designing medium access schemes. Less scalable protocols may cause unbearable overheads when applied to large networks and may cause extreme energy drains in certain nodes, even causing network failure. On the other hand, quality of service degradations can be severe. In this context, the scalability of currently available MAC schemes should be further investigated.

Mobility management. In certain scenarios, several mobile nodes may be roaming the region of deployment of a WSN among stationary nodes already organized under a certain hierarchy for communication. Sometimes the mobile nodes might serve as gateways, sinks, or localization devices, requiring their proper attachment to certain stationary points of the network. This problem is addressed partially in the SMACS/EAR algorithm described in Section 28.5.2; however, the EAR algorithm does not ensure the optimal use of resources. Thus, further investigation is required in this regard.

Hardware constraints. It is argued that energy savings can often be improved using FDMA or CDMA scheduling schemes, for example, as in SMACS [17]. However, the complexity of the required radio interface poses challenges due to compact hardware of sensor nodes. Use of such schemes must be done in conjunction with a suitable TDMA scheme for energy savings because nodes must listen to the channel all the time under pure FDMA or CDMA schemes. Nevertheless, with possible future improvements in hardware fabrication, such hybrid schemes might have a potential to play a greater role in energy savings while giving superior throughput and delay performance. The main challenge in such schemes is to optimize use of resources, for example, time slots and frequency bands. Transmission ranges of nodes are relatively lower in WSNs, so frequency reuse may be possible in hierarchical, cluster-based WSNs as in traditional cellular networks. Moreover, frequency reuse might require nodes to listen only to a limited number of channels, making such hardware more feasible.

Comparison metrics. While novel medium access schemes for optimum energy savings are developed, due attention should be paid to the metrics used in comparing these schemes. Often a trade-off takes place between energy savings and network performance. Thus, unified metrics should be used during comparisons or this will lead to unfair conclusions and probable confusion. The total energy savings of a WSN depend on percentage sleep time and average length of sleep interval. If used alone, percentage sleep time does not account for the possible higher frequency of switching that may drain a significant amount of node energy. Average sleep length is a preferred metric because it can account for the node switching. Appropriate benchmarks should be developed to facilitate accurate comparison of metrics among different MAC schemes.

28.7 Conclusions

Unique features of WSNs in comparison with generic WAHNs were identified in this chapter. Four prominent ad hoc network medium access methods were briefly discussed, as well as their merits, drawbacks, and implications toward WSNs. Design challenges in MAC for WSNs were emphasized with a classification of application and communication types in WSNs. Four medium access schemes recently proposed toward energy savings in WSNs were discussed, comparing their merits and drawbacks. In addition to these four schemes, a few other medium access schemes that have been recently proposed for WSNs [23–26] were not discussed here to preserve brevity of this chapter.

Finally, several open issues related to energy aware MAC protocol design for WSNs were emphasized. Several research efforts are already underway in this area and are being tested on open source platforms like MICA/TinyOS [27]. Significant research efforts are still required to address these open issues in order to achieve the ultimate objective of energy optimal medium access in sensor networks. Also, it is anticipated that such research efforts will soon lead to open standards for WSNs similar to the currently available, commercially deployed standards for WAHNs.

References

1. I.F. Akyildiz, W. Su, Y. Sankarasubramaniam, and E. Cayirci, Wireless sensor networks: a survey, *Computer Networks*, 38(4), 393–422, 2002.
2. The Institute of Electrical and Electronics Engineers, Wireless LAN medium access control (MAC) and physical layer (PHY) specifications, IEEE Standard 802.11, June 1997.
3. J.C. Haartsen, The Bluetooth radio system, *IEEE Personal Commun. Mag.*, 28–36, Feb. 2000.
4. Bluetooth SIG Inc., Specification of the Bluetooth system: core, http://www.bluetooth.org, 2001.
5. C.E. Jones, K.M. Sivalingam, P. Agrawal, and J.C. Chen, A survey of energy efficient network protocols for wireless networks, *Wireless Networks*, 7(4), 343–358, 2001.

6. K.M. Sivalingam, J.-C. Chen, P. Agrawal, and M. Srivastava, Design and analysis of low-power access protocols for wireless and mobile ATM networks, *Wireless Networks*, 6(1), 73–87, 2000.
7. S. Singh and C.S. Raghavendra, PAMAS: power aware multi-access protocol with signaling for ad hoc networks, *ACM Computer Commun. Rev.*, 28(3), 5–26, July 1998.
8. P. Karn, MACA — a new channel access method for packet radio networks, in *Proc. ARRL/CRRL Amateur Radio 9th Computer Networking Conf.*, 1, 134–140, 1990.
9. V. Bhargawan et al. MACAW: a media access protocol for wireless LANs, *Proc. ACM Sigcomm '94*, 24(4), 212–225, 1994.
10. L. Kleinrock and F. Tobagi, Packet switching in radio channels: carrier sense multiple access modes and their throughput delay characteristics, *IEEE Trans. Commun.*, COM-23(12), 1400–1416, Dec. 1975.
11. Y. Tseng, C. Hsu, and T. Hsieh, Power-saving protocols for IEEE 802.11-based multi-hop ad hoc networks, in *Proc. IEEE Infocom*, 1, 200–209, New York, June 2002.
12. G.J. Pottie and W.J. Kaiser, Wireless integrated network sensors, *Commun. ACM*, 43(5), 51–58, May 2000.
13. Crossbow Inc., Expected battery life vs. system current usage and duty cycle URL: www.xbow.com/Support/Support_pdf_files/PowerManagement.xls, Energy specifications for MICA motes, 2003.
14. Crossbow Inc., Data sheet for MICA2 wireless measurement system, 2003.
15. W. Ye, J. Heidemann, and D. Estrin, An energy-efficient MAC protocol for wireless sensor networks, in *Proc. IEEE Infocomm*, 3, 1567–1576, New York, June 2002.
16. S.R. Madden, M.J. Franklin, J.M. Hellerstein, and W. Hong, The design of an acquisitional query processor for sensor networks, in *Proc. (SIGMOD'03)*, 1, 491–502, San Diego, CA, June 2003.
17. K. Sohrabi, J. Gao, V. Ailawadhi and G. Pottie, Protocols for self-organization of a wireless sensor network, *IEEE Personal Commun. Mag.*, 7(5), 16–27, Oct. 2000.
18. V. Rajendran, K. Obraczka, and J.J. Garcia–Luna–Aceves, Energy-efficient, collision-free medium access control for wireless sensor networks, in *Proc. ACM SIGMOBILE Int. Conf. Embedded Networked Sensor Systems (SenSys 2003)*, 1, 181–192, Los Angeles, CA, November 2003.
19. S. Coleri, PEDAMACS: power efficient and delay aware medium access protocol for sensor networks, M.S. Thesis, Department of Electrical Engineering and Computer Science, University of California, Berkeley, December 2002.
20. L. Bao and J.J. Garcia–Luna–Aceves, A new approach to channel access scheduling for ad hoc networks, in *Proc. IEEE MOBICOMM 2001*, 1, 210–221, Rome, 2001.
21. R. Min, M. Bhardwaj, S.-H. Cho, N. Ickes, E. Shih, A. Sinha, A. Wang, and A. Chandrakasan, Energy-centric enabling technologies for wireless sensor networks, *IEEE Wireless Communications*, 9(4), 28–39, August 2002.
22. E. Shih, S.-H. Cho, N. Ickes, R. Min, A. Sinha, A. Wang, and A. Chandrakasan, Physical layer driven protocol and algorithm design for energy-efficient wireless sensor networks, in *Proc. MOBICOMM 2001*, 1, 272–287, Rome, 2001.
23. A. Woo and D. Culler. Transmission control scheme for media access in sensor networks, in *Proc. MOBICOMM 2001*, 1, 221–235, Rome, 2001.
24. R. Kannan, R. Kalidindi, S.S. Iyengar, and V. Kumar. Energy and rate based MAC protocol for wireless sensor networks, *ACM SIGMOD Record*, Special section on sensor network technology and sensor data management, 32(4), 60–65, December 2003.
25. K. Arisha, M. Youssef, and M. Younis, Energy-aware TDMA-based MAC for sensor networks, in *Proc. IEEE Integrated Manage. Power Aware Commun., Computing Networking (IMPACCT 2002)*, New York City, May 2002.
26. J.M. Van-Dam, An adaptive energy-efficient MAC protocol for wireless sensor networks, M.S. thesis, Delft University of Technology, June 2003.
27. University of California, Berkeley, TinyOS homepage. URL http://webs.cs.berkeley.edu/tos/index.html, 2003.

29
Techniques to Reduce Communication and Computation Energy in Wireless Sensor Networks

Vishnu Swaminathan
Duke University

Yi Zou
Duke University

Krishnendu Chakrabarty
Duke University

29.1 Introduction 29-1
29.2 Overview of Node-Level Energy Management 29-2
      CPU-Centric DPM • I/O-Centric DPM
29.3 Overview of Energy-Efficient Communication 29-4
29.4 Node-Level Processor-Oriented Energy
      Management 29-4
      The LEDF Algorithm • Implementation Testbed • Experimental Results
29.5 Node-Level I/O-Device-Oriented Energy
      Management 29-11
      Device Scheduling for Two-State I/O Devices • Low-Energy Device Scheduling of Multistate I/O Devices • Experimental Results
29.6 Energy-Aware Communication 29-19
      Detection Probability Table • Score-Based Ranking • Selection of Sensors to Query • Energy Evaluation Model for Target Localization in Wireless Sensor Networks • Procedural Description • Simulation Results
29.7 Conclusions 29-32

29.1 Introduction

Energy consumption is an important design consideration for wireless sensor networks. These networks are useful for a number of applications, such as environment monitoring, surveillance, and target detection and localization. The sensor nodes in such applications operate under limited battery power; they also tend to be situated at remote and/or inaccessible locations and thus the cost of replacing battery packs is high.

Figure 29.1 illustrates the basic structure of a sensor node. In sensor nodes, design for low energy entails energy-efficient sensing, computation (local information processing), and communication. Many strategies have been proposed that target the energy consumption of these functions in sensor networks. Node-level energy minimization techniques target different components, such as the CPU and I/O devices, in each sensor node. Commonly used node-level energy minimization techniques are based on

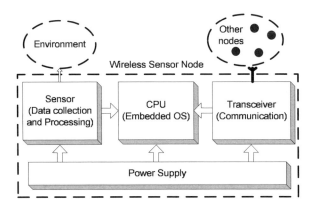

FIGURE 29.1 Typical architecture of a sensor node.

dynamic voltage scaling (DVS) for reducing processor energy and I/O-based dynamic power management (DPM) techniques for reducing the energy consumption of I/O devices.

On the other hand, techniques for energy-efficient communication address energy reduction by significantly reducing unnecessary wireless network traffic. The transmission of detailed sensed information consumes a significant amount of energy due to the large volume of sensed data. In sensor networks for target detection, localization, and classification, energy can be reduced by intelligently querying a select set of nodes for detailed target information without compromising other objectives such as coverage, reliability, etc. This reduces the traffic on the network, thus reducing the energy expended in transmitting large volumes of redundant data.

29.2 Overview of Node-Level Energy Management

One approach to reduce energy consumption is to employ low-power hardware design techniques [7, 14, 33]. These design approaches are static in that they can only be used during system design and synthesis. Thus, these optimization techniques do not fully exploit the potential for node-level power reduction under changing workload conditions and their ability to trade off performance with power reduction is thus inherently limited. An alternative and more effective approach to reducing energy in embedded systems and sensor networks is based on *dynamic power management* (DPM), in which the operating system (OS) is responsible for managing the power consumption of the system.

Many wireless sensor networks are also designed for *real-time* use. Real-time performance is defined in terms of the ability of the system to provide real-time temporal guarantees to application tasks that request such guarantees. These systems must therefore be designed to meet functional as well as timing requirements [6]. Energy minimization adds a new dimension to these design criteria. Thus, although energy minimization for sensor networks is of great importance, energy reduction must be carefully balanced against the need for real-time responsiveness.

Recent studies have shown that the CPU and the I/O subsystems are major consumers of power in an embedded system; in some cases, hard disks and network transceivers consume as much as 20% of total system power in portable devices [18, 25]. Consequently, CPU-centric and I/O-centric DPM techniques have emerged at the forefront of DPM research for wireless sensor networks.

29.2.1 CPU-Centric DPM

Designers of embedded processors used in sensor nodes now include variable-voltage power supplies in their processor designs, i.e., the supply voltages of these processors can be adjusted dynamically to trade off performance with power consumption. *Dynamic voltage scaling* (DVS) refers to the method by which quadratic savings in energy is obtained through the run-time variation of the supply voltage to the processor.

It is well known that the power consumption of a CMOS circuit exhibits a cubic dependence on the supply voltage V_{dd}. However, the execution time of an application task is proportional to the sum of the gate delays on the critical path in a CMOS processor. Because gate delay is inversely proportional to V_{dd}, the execution time of a task increases with decreasing supply voltage. The energy consumption of the CMOS circuit, which is the product of the power and the delay, therefore exhibits a quadratic dependence on V_{dd}.

In embedded sensor nodes, where peak processor performance is not always necessary, a drop in the operating speed (due to a reduction in operating voltage) can be tolerated in order to obtain quadratic reductions in energy consumption. This forms the basis for DVS; the quadratic dependence of energy on V_{dd} has made it one of the most commonly used power reduction techniques in sensor nodes and other embedded systems. When processor workload is low, the OS can reduce the supply voltage to the processor (with a tolerable drop in performance) and utilize the quadratic dependence of power on voltage to reduce energy consumption.

29.2.2 I/O-Centric DPM

Many peripheral devices possess multiple power states — usually one high-power working state and at least one low-power sleep state. Hardware-based timeout schemes for power reduction in such I/O devices have been incorporated into several device designs. These techniques shut down devices when they have been idle for a period of time specified previously. A device that has been placed in the sleep state is powered up when a new request is generated.

With the introduction of the ACPI standard in 1997, the OS was provided with the ability to switch device power states dynamically during run time, thus leading to the development of several new types of DPM techniques. Predictive schemes use various system parameters to estimate the lengths of idle periods for devices. Stochastic models with different probabilistic distributions have been used to estimate the times at which devices can be switched between power states. The goal of these methods, however, is to minimize the response times of devices. Indeed, many such probabilistic schemes see widespread use in portable and interactive systems such as laptop computers. However, their applicability in sensor systems, many of which require real-time guarantees, is limited due to a drawback inherent to probabilistic methods.

Switching between device power states incurs a time penalty, i.e., a device takes a certain amount of time to transition between its power states. In hard real-time systems in which tasks have firm deadlines, device switching must be performed with caution to avoid the potentially disastrous consequences of missed deadlines. The uncertainty inherent in probabilistic estimation methods precludes their use as effective device-switching algorithms in hard real-time systems whose behavior must be predictable with a high degree of confidence. Current-day practice consists of keeping devices in real-time systems powered up during the entirety of system operation; the critical nature of I/O devices operating in real-time prohibits shutting down devices during run time.

This chapter describes two node-level energy reduction methods for wireless sensor networks. The first algorithm focuses on reducing processor energy and was implemented on a laptop equipped with an AMD Athlon 4 processor and running the real-time Linux (RT-Linux) OS. Experiences in implementation are described and experimental power measurements provided; these validate and support simulation results. The second technique targets I/O devices in sensor nodes. The approach includes two online algorithms that schedule the shutdowns and wake-ups for I/O devices in sensor nodes that require hard real-time temporal guarantees.

29.3 Overview of Energy-Efficient Communication

Energy-efficient communication in sensor networks is crucial because most sensor nodes are battery driven and therefore severely energy constrained. Considerable research has been recently carried out in an effort to make communication in sensor networks energy efficient [5, 12, 13, 19, 23, 24, 27, 29, 31, 32, 36, 38, 39]. The focus here is on reducing energy consumption in wireless sensor networks for target localization and data communication. The transmission of detailed target information consumes a significant amount of energy because of the large volume of raw data. Contention for the limited bandwidth among the shared wireless communication channels causes additional delay in relaying detailed target information to the cluster head.

This chapter describes a technique to prolong the lifetimes of the nodes in the sensor network by adopting a new target localization procedure. It details an *a posteriori* energy-aware target localization strategy based on a two-step communication protocol between the cluster head and the sensors reporting the target detection events. In the first step, sensors detecting a target report the event to the cluster head using a very short binary yes/no message. The cluster head subsequently queries a subset of sensors in the vicinity of these likely target positions. This subset is determined through the localization procedure executed by the cluster head. Simulation results show that a large amount of energy is saved by using this procedure. These results also illustrate the built-in advantages of the proposed target localization procedure in reducing communication bandwidth and filtering out false alarms.

29.4 Node-Level Processor-Oriented Energy Management

A set $R = \{r_1, r_2, ..., r_n\}$ of n tasks is given. Associated with each task, $r_i \in R$, are the following parameters: (1) an arrival time a_i; (2) a deadline d_i; and (3) a length l_i (represented as the number of instruction cycles). Each task is placed in the ready queue at time a_i and must complete its execution by its deadline d_i. The tasks cannot be pre-empted. The CPU can operate at one of k voltages: $V_1, V_2, ..., V_k$. Depending on the voltage level, the CPU speed may take on k values: $s_1, s_2, ..., s_k$. The supply voltage to the CPU is controlled by the OS, which can dynamically switch the voltage during run time. The energy E_i consumed by task r_i is proportional to $v_i^2 l_i$. The problem is defined as follows:

P_{cpu}: Given a set R of n tasks and, for each task $r_i \in R$, (1) a release time a_i; (2) a deadline d_i; and (3) a length l_i, and a processor capable of operating at k different voltages, $V_1, V_2, ..., V_k$, with corresponding speeds $S_1, S_2, ..., S_k$, determine a sequence of voltages $v_1, v_2, ..., v_n$ and corresponding speeds $s_1, s_2, ..., s_n$ for the task set R such that the total energy consumed $\sum_{i=1}^{n} v_i^2 l_i$ by the task set is minimized, while also attempting to meet as many task deadlines as possible.

29.4.1 The LEDF Algorithm

LEDF is an extension of the well-known earliest deadline first (EDF) algorithm [15], which maintains a list of all released tasks called the ready list. These tasks have an absolute deadline associated with them that is recalculated at each release based on the absolute time of release and the relative deadline. When tasks are released, the task with the earliest deadline is selected for execution. A check is performed to see if the task deadline can be met by executing it at a lower voltage (speed). Each speed at which the processor can run is considered in order from the lowest to the highest. For a given speed, the worst-case execution time of the task is calculated based on the maximum instruction count. If this execution time is too high to meet the current absolute deadline for the task, the next higher speed is considered. Otherwise, a test is applied to verify that all ready tasks will be able to meet their deadlines when the current earliest-deadline task is run at a lower speed.

```
Procedure LEDF()
t_c: current time;
S_h > S_l1 > S_l2 > ... S_lm: Available processor speeds
schedulable = 1
1.   if ready_list ≠ NULL
2.      Sort task deadlines in ascending order;
3.      Select task τ_i with earliest deadline;
4.         for S = S_lm to S_h
```

5. $\quad\quad$ **if** $t_c + \frac{l_i}{S} \leq d_i$ **then**

6. $\quad\quad\quad t = t_c + \frac{l_i}{S}$

7. $\quad\quad\quad$ **for** each task τ_u that has not completed execution

8. $\quad\quad\quad\quad$ **if** $t + \frac{l_u}{S_h} \leq d_u$ **then**

9. $\quad\quad\quad\quad\quad t = t + \frac{l_u}{S_h}$

```
10.                  else
11.                     schedulable = 0
12.                     break
13.                  endfor
14.                  if schedulable = 0
15.                     Schedule t_i at S
16.                     break
17.                  endif
18.         endfor
```

FIGURE 29.2 LEDF algorithm.

The test consists of iterating down the ordered list of tasks and comparing the worst-case completion time for each task (at the highest speed) against its absolute deadline. If any task will miss its deadline, the selected speed is insufficient and the next higher speed for the current task is considered. If the deadlines of all tasks in the ready list can be met at the highest speed, LEDF assigns the lower voltage to the task and the task begins execution. When the task completes execution, LEDF again selects the task with the nearest deadline to be executed. As long as tasks are waiting to be executed, LEDF schedules the one with the earliest absolute deadline for execution. Figure 29.2 describes the algorithm in pseudocode form.

For a processor with two speeds, the LEDF algorithm has a computational complexity of $O(n \log n)$ where n is the total number of tasks. The worst-case scenario occurs when all n tasks are released at time $t = 0$. This involves sorting n tasks in the ready list and then selecting the task with the earliest deadline for execution. When more than two speeds are allowed, the complexity of LEDF becomes $O(n \log n + kn)$, where k is the number of speed settings allowed.

29.4.2 Implementation Testbed

29.4.2.1 Hardware Platform

The power measurement experiments were conducted on a laptop with an AMD Mobile Athlon 4 processor. AMD's PowerNow! technology offers greater flexibility in setting frequencies and core voltages [2]. The 1.1-GHz Mobile Athlon 4 processor can be set at several core voltage levels ranging from 1.2 to 1.4 V in 0.05-V increments. For each core voltage, there is a predetermined maximum clock frequency. The power states chosen to use in the scheduler and simulations are shown in

TABLE 29.1 Speed and Voltage Settings for Athlon 4 Processor

Power State	Speed (MHz)	Voltage (V)
1	1100	1.4
2	900	1.35
3	700	1.25

Source: From V. Swaminathan, et al., *Proc. IEEE Real-Time and Embedded Technology and Applications Symposium*, 229–239, 2002. With permission.

Table 29.1. Although only three speeds are used in these experiments, an extension to using all five available speeds appears to be quite straightforward.

PowerNow! technology was developed primarily to extend battery life on mobile systems. Therefore the experiments were conducted on a laptop system rather than a desktop PC. Instead of inserting a current probe into the laptop, system power was simply measured during the experiments. The laptop's system power is drawn from the power converter at approximately 18.5 V DC. Instead of using an oscilloscope or digital ammeter to take exact CPU power measurements at very high frequencies, a simpler approach used a large capacitor to average out the DC current drawn by the entire laptop. This method works primarily because of the periodic nature of these tests.

In a periodic real-time system, the power drawn over one hyperperiod is roughly the same as the power drawn over the next hyperperiod as long as no tasks are added or removed from the task set. Because a fairly large amount of energy needs to be sourced and sunk by the capacitor at the different processor speeds and activity levels, a 30-V DC 360-mF capacitance (160- and 200-mF capacitors in parallel) was used. This capacitance proved capable of averaging current loads for power state periods ranging up to hundreds of milliseconds. (When the processor power state switches at a lower rate than this, the current measurements taken between the AC/DC converter and the voltmeter readings fluctuate.) Figure 29.3 illustrates the experimental hardware setup.

29.4.2.2 Software Architecture

RT-Linux [28] was used as the OS for the experiments. In addition to providing real-time guarantees for tasks and a periodic scheduling system, RT-Linux provides a well-documented method of changing the scheduling policies. An elegant modular interface allows for easy adaptation of the scheduler module to use

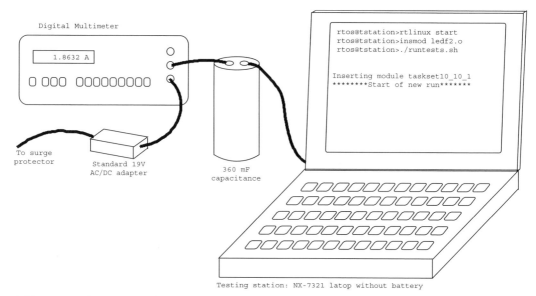

FIGURE 29.3 The experimental setup. (From V. Swaminathan, et al., *Proc. IEEE Real-Time and Embedded Technology and Applications Symposium*, 229–239, 2002. With permission.)

LEDF and then load and unload it as necessary. This feature of RT-Linux was used to swap LEDF for a regular EDF scheduler during power comparisons. Furthermore, RT-Linux uses Linux as its idle task, providing a very convenient method of control and evaluation for the execution of the real-time tasks.

LEDF sorts all tasks by their absolute deadlines and chooses the task with the earliest deadline first. If no real-time tasks are pending, the Linux/Idle task is chosen and run at the lowest available speed. A timeout is then set to preempt the Idle task at the next known release time. Once a speed is identified for a task, the switching code is invoked if the processor is not already operating at that speed.

Switching the power state of a mobile Athlon 4 processor simply consists of writing to a model-specific register (MSR). The core voltage and clock frequency at which the processor is to be set are encoded into a 32-bit word along with 3 control bits. Another 32-bit word contains the stop-grant timeout count (SGTC), which represents the number of 100-MHz system clocks during which the processor is stalled for the voltage and frequency changes. The maximum phase-locked loop (PLL) synchronization time is 50 µs and the maximum time for ramping the core voltage appears to be 100 µs. Calling the WRMSR macro then instruments the power state change. For debugging, the RDMSR macro was used with a status MSR to retrieve the processor's power state. Decoding the two 32-bit word values reveals the maximum, current, and default frequency and core voltage.

The RT-Linux high-resolution timer used for scheduling is based (in x86 systems) on the time-stamp counter (TSC), a special counter introduced by Intel that simply counts clock periods in the CPU since it was started (boot-time). The *gethrtime*() RT-Linux method (and all methods derived from it) convert the TSC value into a time value using the recorded clock frequency. Thus, a simple calculation to determine time in nanoseconds from the TSC value would be the product of TSC and clock period.

RT-Linux was initially developed without the need for dynamic frequency switching, so the speed used for the calculation of time is set at boot time and never changed. Thus, when the processor is slowed to a low-power state with a lower clock frequency, the TSC counts at a lower rate. However, the *gethrtime*() method is oblivious to this and the measurement of time slows down proportionally. It is not clear what happens to the TSC, and thus how to measure time, during a speed switch. The TSC does appear to be incremented during some part of the speed switch, but the count is not a reliable means of measuring time. Recalibrating the rate at which the TSC is incremented appears to be a nontrivial task that requires extensive rewriting of the RT-Linux timing code. Therefore, it was decided to track time from within the LEDF module.

29.4.3 Experimental Results

Data from the power measurement experiments, in which total system power consumption of the laptop was measured, are now presented. Knowledge of CPU power savings, however, is useful in generalizing the results. CPU power savings can easily be derived from a set of experiments. In order to isolate the power used by the processor and system board, one can turn off all system components except the CPU and system board, and then take a power reading when the CPU is halted. This power measurement represents the total system power excluding CPU power. This base power is then subtracted from all future power readings in order to obtain CPU power alone. However, halting a processor is far more complex than simply issuing an "HLT" instruction. Decoupling the clock from the CPU involves handshaking between the CPU and the Northbridge. Documentation was not sufficient to implement this.

As an alternative method of estimating power drawn by the system board and components, the power consumption of the CPU with maximum load can be calculated from system measurements at two power states. This can be done by devising tests to isolate power drawn by the LCD screen, hard drive, and the portion of the system beyond control. Once an estimate for system power is available, it can be eliminated from all readings to get an approximation of the fraction of CPU power saved.

Ratios for power consumption in different states can be calculated using the well-known relationship for CMOS power consumption, i.e., $P = f\, a\, C\, V_{dd}^2$, where P is the power; f is the frequency of operation; a is the average switching activity; C is the switching capacitance; and V_{dd} is the operating voltage. The switching capacitance and average switching activity are constant for the same processor and software, so only the frequency and the square of the core voltage are considered. It is also reasonable to assume that other components of the laptop (the screen and hard disk, for example) draw approximately the same current regardless of the CPU operating voltage. Therefore, power state 2 uses approximately 76% as much power as power state 1 and power state 3 uses only 50.7% as much power as the maximum power state. The minimum power configuration for this processor is 300 MHz at 1.2 V, which consumes only 20% of the power consumed in the maximum power state.

In this case, the decision was to compare a fully loaded processor operating at 700 MHz (with a core frequency of 1.25 V) and at 1100 MHz (with a core voltage of 1.4 V). The 700-MHz configuration uses $(700 * 1.25^2)/(1100 * 1.4^2)$, or 50.73% as much CPU power as the 1100-MHz configuration. For a given task running at 1100 MHz, the observed current consumption was 2.373 A. For the same task running at 700 MHz, a current reading of 1.647 A was observed. Assuming that the current consumption of the other components was approximately the same during both runs, the difference in CPU current consumption is 0.726 A. This means that:

$$I_{1100} - I_{700} = 0.726 \Rightarrow I_{1100} - 0.5073 I_{1100} = 0.726 \Rightarrow I_{1100} = 1.474 A$$

In other words, a measured difference of $(2.373 - 1.647) = 0.726$ A of current implies that the fully loaded CPU operating at 1100 MHz draws approximately 1.474 A. Knowing this, it can be deduced from the information in Table 29.2 that the system board and basic components draw approximately 0.456 A, and that under normal operation, the system (including the disk drive and display) draws about 0.976 A in addition to the load from the CPU. This estimation, although approximate, provides a useful method of isolating energy used by the CPU for various utilizations and scheduling algorithms.

Several experiments were performed with three different versions of the scheduling algorithm and different task sets at various CPU utilization levels. A pseudorandom task generator was constructed to generate the test sets. Using the task generator, several random sets of tasks were created. The release times of the tasks are set to the beginning of a period and deadlines to the end. Computation requirements for the tasks are chosen randomly and then scaled to meet the target utilization.

The test programs consist of multiple threads that execute "for" loops for specified periods of time. The time for which these threads run can be determined by examining the assembly level code for each iteration of a loop. Each loop consists of five assembly language instructions, which take one cycle each to execute. The random task set generator takes this into account when generating the task sets.

The simulator is a simple PERL program that reads in task data and generates the schedule that would be generated by the LEDF scheduler. It then takes user-supplied baseline power measurements

TABLE 29.2 Current Consumptions of Various System Components

CPU (1100 MHz)	Screen	Disk	Current Drawn (A)
Idle	Off	STBY	1.5
Idle	Off	On	1.54
Idle	On	STBY	1.91
Idle	On	Sleep	1.9
Idle	On	On	1.97
Max load	Off	STBY	1.93
Max load	On	On	2.45

Source: From V. Swaminathan, et al., *Proc. IEEE Real-Time and Embedded Technology and Applications Symposium*, 229–239, 2002. With permission.

and uses them to compute the power consumption of the task set. Summing up the fraction of the period spent in each state and multiplying it by the appropriate power consumption measurement produces the overall power consumption for the task set. As a reasonable representation of the load generated by the Linux/Idle task, the simulator assumes this task to consume a certain amount of power whose value lies between the power consumptions of a fully loaded and fully idle system running at a given speed. This power value was determined by measuring the power consumption of the laptop with regular Linux running a subset of daemon processes in the background.

A single power-state version of LEDF (in effect, EDF) was used as a comparison point. These tests show the maximum power requirements for the amount of work (computation) to be done. Two- and three-speed versions of LEDF were used to observe the effect of adding additional power states. The two-speed version used operating frequencies of 700 and 1100 MHz, and the three-speed version incorporated an intermediate 900 MHz operating frequency. The CPU utilizations ranged from 10 to 80% in increments of 10%. The maximum utilization of 80% was necessary to guarantee that the Linux/Idle task had sufficient time available for control operations. Without forcing the scheduler to leave 20% of the period open for the Linux/Idle task, the shell became unresponsive, forcing a hard reboot of the machine between each test. The cycle-conserving EDF (ccEDF) algorithm from Pillai and Shin [26] was also implemented and the algorithm compared to it. This implementation of ccEDF uses a set of discrete speeds.

The results are shown in Figure 29.4 for a 15-task task set. Each data point represents the average of three randomly generated task sets for a given utilization value and task set size. LEDF2 (LEDF3) and ccEDF2 (ccEDF3) refer to the use of two (three) processor speeds.

The power savings ranged from 9.4 W in a minimally utilized system to 2.6 W in a fully utilized system. The fully utilized system has lower power consumption under LEDF because LEDF schedules the non-real-time component at the lowest speed. Note, however, that up to the 50% mark the power

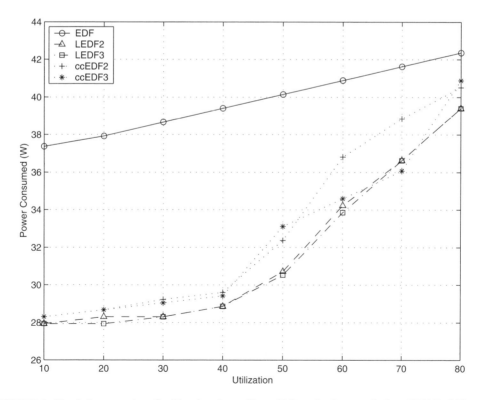

FIGURE 29.4 Heuristic comparison for 15-task task set. (From V. Swaminathan, et al., *Proc. IEEE Real-Time and Embedded Technology and Applications Symposium*, 229–239, 2002. With permission.)

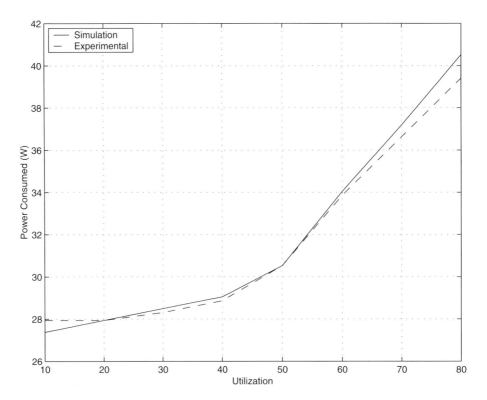

FIGURE 29.5 Comparison of experimental three-state LEDF with expected results. (From V. Swaminathan, et al., *Proc. IEEE Real-Time and Embedded Technology and Applications Symposium*, 229–239, 2002. With permission.)

savings remain over 9 W and, in most cases, remain over 7 W for 60% utilization. With a maximum utilization of 80%, the system can still save significant power with a reasonable task load.

A comparison between measured experimental results and simulation results is shown in Figure 29.5. In most cases, the simulated and measured values are the same or within 2% of each other. The simulation results thus provided a very close match to the experimental results, indicating that the simulation engine model accurately models the real hardware. Because the simulation engine does not take into account the scheduler's computation time, the fidelity of the results may degrade for very high task counts due to the extra cost of sorting the deadlines. In order to verify this, it was decided to evaluate LEDF with several randomly generated task sets with different utilizations, with the number of tasks ranging from 10 to 200, and measure the execution time of the scheduler for each task set.

Results show that the execution time of the scheduler was in the order of microseconds, while the task execution times were in the order of milliseconds. For increasing task set size, scheduler runtime increases at a very slow rate. Thus, scheduling overhead does not prove to be too costly for the power-aware version of EDF. For task sets with more than 240 tasks, the RT-Linux platform tended to become unresponsive. These results are shown in Table 29.3. The entries in the table correspond to task sets with 40% utilization, but with varying numbers of tasks. The other task sets experimented with (task set utilizations of 50 and 80%) also exhibit the same trend in scheduler runtime and are not reproduced here. The scheduler overhead in Table 29.3 indicates the time taken by the scheduler to sort the task set and to identify the active task. Even though implementation of LEDF is currently of $O(n^2)$ complexity and can be replaced by a faster $O(n \log n)$ implementation, it is obvious from the table that scheduling overhead is negligible for over a hundred tasks for utilizations ranging from 10 to 80%. For small task sets in which the task set consists of a few hundred tasks, scheduling overhead is negligible compared to task execution times.

TABLE 29.3 Measured Scheduler Overhead for Varying Task Set Sizes

Number of Tasks	Measured Scheduler Overhead (ns)
10	1739
20	1824
30	1924
60	3817
120	6621
180	10916
200	12243

29.5 Node-Level I/O-Device-Oriented Energy Management

Prior work on DPM techniques for I/O devices has focused primarily on scheduling devices in non-real-time systems. The focus of these algorithms is on minimizing user response times rather than meeting real-time task deadlines; therefore, these methods are not viable candidates for use in real-time systems. Because of their inherently probabilistic nature, the applicability of the preceding methods to real-time systems falls short in one important aspect — real-time temporal guarantees cannot be provided. Such methods perform efficiently in interactive systems in which user waiting time is an important design parameter. In real-time systems, minimizing response time of a task does not guarantee that its deadline will be met. Thus, new algorithms that operate in a deterministic manner are needed in order to ensure real-time behavior.

29.5.1 Device Scheduling for Two-State I/O Devices

This subsection describes LEDES, a deterministic device-scheduling algorithm for two-state I/O devices. It begins by defining the device scheduling problem P_{io}, and describing the assumptions in greater detail.

Take a task set $T = \{\tau_1, \tau_2, ..., \tau_n\}$ of n tasks. Each task $\tau_i \in T$ is defined by (1) an arrival time a_i; (2) a worst-case execution time c_i; (3) a period p_i; (4) a deadline d_i; and (5) a device-usage list L_i. The device-usage list L_i for a task τ_i is defined as the set of I/O devices used by τ_i. The hyperperiod H of the task set is defined as the least common multiple of the periods of all tasks. Without loss of generality, assume that the deadline of a task is equal to its period, i.e., $p_i = d_i$.

A set $K = \{k_1, k_2, ..., k_p\}$ of p I/O devices is used in the system. Each device k_i is characterized by:

- Two power states: a low-power sleep state $ps_{l,i}$ and a high-power working state $ps_{h,i}$
- A wake-up time from $ps_{l,i}$ to $ps_{h,i}$ represented by $t_{wu,i}$
- A shutdown time from $ps_{h,i}$ to $ps_{l,i}$ represented by $t_{sd,i}$
- Power consumed during wake-up $P_{wu,i}$,
- Power consumed during shutdown $P_{sd,i}$
- Power consumed in the working state $P_{w,i}$
- Power consumed in the sleep state $P_{s,i}$

Requests can be processed by the devices in the working state only. All I/O devices used by a task must be powered up before the task starts execution. Because an *online* device scheduler must be fast and efficient, assume that device scheduling decisions are made only at task starts and completions. Although software timers can potentially be used to switch device power states at *any* time instant, the processing of each timer interrupt incurs an architecture-dependent service-time penalty. This penalty, and therefore its inclusion in the task model and device scheduling algorithms, requires special handling; forcing all device switching to be performed only at task starts and completions is a simpler approach with no significant impact on energy savings.

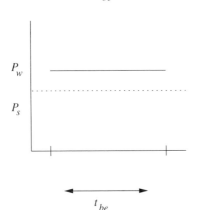 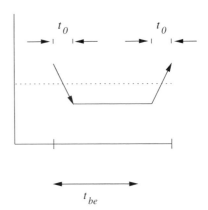

FIGURE 29.6 The time interval for which the energy consumptions are the same in (a) and (b) is called breakeven time. (From V. Swaminathan and K. Chakrabarty, *IEEE Transactions on Computer-Aided Design of Integrated Circuits & Systems*, 22, 847–858, July 2003. With permission.)

In I/O devices, the power consumed by a device in the sleep state is less than the power consumed in the working state, i.e., $P_{s,i} < P_{w,i}$. Without loss of generality, assume that for a given device, k_i, $t_{wu,i} = t_{sd,i} = t_{0,i}$ and $P_{wu,i} = P_{sd,i} = P_{0,i}$. The energy consumed by device k_i is given by:

$$E_i = P_{w,i} t_{w,i} + P_{s,i} t_{s,i} + M P_{0,i} t_{0,i},$$

where M is the total number of state transitions for k_i; $t_{w,i}$ is the total time spent by device k_i in the working state; and $t_{s,i}$ is the total time spent in the sleep state.

Incorrectly switching power states can cause increased, rather than decreased, energy consumption for an I/O device. Incorrect switching of I/O devices is eliminated using the concept of *breakeven time* [14], which is defined as the time interval for which a device in the powered-up state consumes an energy exactly equal to the energy consumed in shutting a device down, leaving it in the sleep state and then waking it up (Figure 29.6). If any idle time interval for a device is greater than the breakeven time t_{be}, energy is saved by shutting it down. For idle time periods that are less than the breakeven time, energy is saved by keeping it in the powered-up state.

Also assume that the start time for each job is known *a priori*. Several commercial RTOS support tick-driven scheduling and mission-critical embedded systems require an inherently deterministic scheduling mechanism [20]. In such types of systems, task schedules are generated offline, and the start times of all jobs are known prior to run time.

The device scheduling problem τ_2 is defined as:

- P_{io}: Given the start times $S = \{s_1, s_2, \ldots, s_n\}$ of the n tasks in a real-time task set T that uses a set K of I/O devices, determine a sequence of sleep/working states for each I/O device $v_s \in G$ such that the total energy consumed, $\Sigma_{i=1}^{p} E_i$, by K is minimized and all tasks meet their respective deadlines.

The following sections describe the conditions under which device state transitions are allowed to minimize energy and ensure the timely completion of tasks. These conditions are different for different scenarios; the scenarios are dependent on the execution times of the tasks that comprise the task set and the number of sleep states present in a device. Begin by assuming that all task execution times are greater than the maximum transition time among all devices and all devices have only one sleep state. It will then be shown that when devices have multiple power states, ensuring timeliness becomes more complex.

One notable advantage of online I/O device scheduling is that online DPM decision making can exploit underlying hardware features such as buffered reads and writes. A device schedule constructed offline and stored as a table in memory precludes the use of such features due to its inherently deterministic approach. The flexibility of online scheduling enhances the effectiveness of device scheduling.

The need for deterministic I/O device scheduling policies is motivated in detail in Swaminathan and Chakrabarty [34], who showed that it is not possible to ensure timely completion of tasks without *a priori* knowledge of future device requests. A naive, probabilistic algorithm cannot be used for real-time task sets. The determinism required to make device-scheduling decisions in hard real-time systems can be quantified through the notion of *look-ahead*, which is a bound on the number of tasks whose device-usage lists must be examined before making a state transition decision, in order to guarantee that no task deadline is missed. Next, the low-energy device scheduler (LEDES) is presented for online scheduling of I/O devices with two power states.

29.5.1.1 Online Scheduling of Two-State Devices: Algorithm LEDES

LEDES assumes that the execution times of all tasks are greater than the transition times of the devices they use. Under this assumption, the amount of look-ahead required before making wake-up decisions to ensure timeliness is easily bounded. This result is derived by presenting the following theorem from Swaminathan and Chakrabarty [34]:

Theorem 29.1. Given a task schedule for a set T *of n tasks with completion times* $c_1, c_2, ..., c_n$; *the device utilization for each task; and an I/O device* k_j, *it is necessary and sufficient to look ahead m tasks to guarantee timeliness, where m is the smallest integer such that* $\sum_{i=1}^{m} c_i \geq t_{0,j}$.

In most practical cases, the completion times of tasks are greater than the transition times $t_{0,i}$ of device k_j. This leads to the following corollary to Theorem 29.1:

Corollary 29.1. Given a task schedule for a set T *of tasks with completion times* $c_1, c_2, ..., c_n$; *the device utilization for each task; and an I/O device* k_j, *it is necessary and sufficient to look ahead one task to ensure timeliness if the completion times of all tasks in* T *are greater than the transition time* $t_{0,j}$ *of device* k_j.

The LEDES algorithm operates as follows (also see Figure 29.7). At the start of task τ_i (line 1), devices not used by the next "immediate" tasks τ_i and τ_{i+1} are put in the sleep state (lines 3 and 4). The time difference between the start of τ_{i+1} and the end of τ_i's execution is evaluated and compared with the transition time $t_{0,j}$ to determine whether k_j's wake-up can be guaranteed at τ_i's finish time. If k_j is powered down, then a wake-up decision must be made (line 8). A device must be waked up at s_i if its wake-up cannot be deferred to τ_i's finish time. This is implemented in line 12 and the device is waked up if needed.

If the scheduling instant at which LEDES is invoked is the completion time of τ_i (line 11) and if k_j is powered up (line 12), it can be shut down only if it can enter the powered-down state fully before s_{i+1} because it may be necessary for it to wake up again. If k_j is in the sleep state (line 15) and is used by τ_{i+1}, it must be waked up to ensure the timely start of τ_{i+1}. These decisions are made for each device and the entire process repeats at each scheduling instant. (Although no mention is made of the break-even time in Figure 29.7, an implicit check is made to ensure that the idle period for a given device is always greater than the breakeven time.)

A simple extension to LEDES can efficiently schedule devices that possess multiple sleep states with the ability to switch from any low-power state directly to the working state. Such a device can be viewed as a device with only two power states. Although the transition times from the sleep states to the powered-up state (and vice-versa) may be different, the correct sleep state to switch a device to is

Algorithm LEDES (k_j, τ_i, τ_{i+1})
1. **if** curr = s_i
2.   **if** k_j is powered-up
3.   **if** $k_j \notin L_i \cup L_{i+1}$
4.   **shutdown** k_j
5.   **if** $k_j \in L_{i+1}$
6.   **if** $s_{i+1} - (s_i + c_i) \geq t_0, j$
7.   **shutdown** k_j
8.   **else**
9.   **if** $k_j \in L_{i+1}$ and $s_{i+1} - (s_i + c_i) < t_0, j$
10.   **wakeup** k_j
11. **if** curr = $s_i + c_i$
12.   **if** k_j is powered-up
13.   **if** $k_j \notin L_{i+1}$ and $s_{i+1} - $ curr $\geq t_0, j$
14.   **shutdown** k_j
15.   **else**
16.   **wakeup** k_j

FIGURE 29.7 LEDES algorithm.

identified simply by performing a series of transition-time checks to verify that there is sufficient time to wake the device up if it is switched to the selected sleep state. However, LEDES cannot make full use of the available sleep states for devices which possess multiple sleep states, but do *not* possess the ability to jump to any sleep state from the powered-up state.

We next present a more general I/O-centric power management algorithm for hard real-time systems. This algorithm is called the multistate constrained low energy scheduler (MUSCLES). MUSCLES can also schedule devices without the ability to jump from the powered-up state to any sleep state. Therefore, it can be assumed that at a device scheduling instant, a device may be switched from one power state to the next higher or lower power state, i.e., only a single transition is possible at any scheduling instant. The next section describes the MUSCLES algorithm in greater detail.

29.5.2 Low-Energy Device Scheduling of Multistate I/O Devices

The properties of a real-time periodic task remain unchanged from Section 29.5.1. However, I/O device properties now include parameters to describe the different power states. These device properties are restated here for the sake of completeness. Each I/O device $k_i \in K$ is now characterized by:

- A set $PS_i = \{ps_{i,1}, ps_{i,2}, \ldots, ps_{i,m}\}$ of m sleep states
- A powered-up state $ps_{i,u}$
- Transition time from $ps_{i,j}$ to $ps_{i,j-1}$, denoted by $t_{wu}^{i,j}$
- Transition time from $ps_{i,j}$ to $ps_{i,j+1}$, denoted by $t_{sd}^{i,j}$
- Power consumed during switching up from state $ps_{i,j}$ to $ps_{i,j-1}$, denoted by $P_{wu}^{i,j}$
- Power consumed during switching down from state $ps_{i,j}$ to $ps_{i,j+1}$, denoted by $P_{sd}^{i,j}$
- Power consumed in the working state P_w^i
- Power consumed in sleep state $ps_{i,j}$, denoted by $P_s^{i,j}$,

Assume, without loss of generality, that for each device $k_j \in K$; $t_{wu+1}^{i,j} = t_{sd}^{i,j} = t_{0,i}$; and $P_{wu}^{i,j+1} = P_{sd}^{i,j} = P_{0,i}$. The total energy E_i consumed by device k_i over the entire hyperperiod is given by

$$E_i = P_w^i t_w^i + \sum_{j=1}^{m} P_s^{i,j} t_s^{i,j} + MP_{0,i} t_{0,i},$$

where M is the number of state transitions; t_w^i is the total time spent by the device in the working state; and $t_s^{i,j}$ is the total time spent by the device in sleep state $ps_{i,j}$. In order to provide conditions under which devices can be shut down and powered up, a few important terms are first defined.

Intertask time. The intertask time IT_i for task τ_i is the time interval between the start of task τ_{i+1} and completion of task τ_i. Thus, $IT_i = s_{i+1} - (s_i + c_i)$. Two scheduling instants are associated with a task τ_i. These correspond to the start and completion time of τ_i, respectively. For minimum-energy device scheduling under real-time constraints, it is not always possible to schedule devices at all scheduling instants. This is formalized using the notion of a valid scheduling instant.

Valid scheduling instant. The completion time of τ_i is defined to be a valid scheduling instant for device k_j if $s_{i+1} - (s_i + c_i) \geq t_{0,j}$. In other words, the completion time of τ_i is a valid scheduling instant if and only if $IT_i \geq t_{0,j}$. The start time of τ_i is always a valid scheduling instant. Thus, a task, τ_i, can have one or two scheduling instants, depending on the magnitude of IT_i relative to the transition time $t_{0,j}$ of a device k_j. Valid scheduling instants are important for energy minimization. Wake-ups can be scheduled at these points to minimize energy and also ensure that real-time requirements are met. Consider the example shown in Figure 29.8. This figure shows two tasks, τ_i and τ_{i+1}, with the intertask time $IT_i < t_{0,j}$. Assume that device k_1 (first used by task τ_{i+2}) is in state $ps_{1,1}$ at τ_i's completion time $(s_i + c_i)$. If a device were to be waked up at $s_i + c_i$, it would complete its transition to state $ps_{1,0}$ only in the middle of τ_{i+1}'s execution and would be in the higher powered state for the rest of τ_{i+1}'s execution (i.e., until the next scheduling instant). If, on the other hand, the device were to be waked up at s_{i+1}, one can still ensure that the device is powered-up before task τ_{i+2} starts (with the assumption that $c_{i+1} > t_{0,1}$). However, the device stays in the lower powered state until s_i, resulting in greater energy savings. Thus, wake-ups at valid scheduling instants always result in lowered energy consumption. It is always preferable to wake a device up as late as possible in order to utilize the full potential of online device scheduling.

Subsection 29.5.1 showed that a look-ahead of one task is sufficient when devices have only one sleep state. However, a look-ahead of one task is not sufficient when devices have multiple low-power sleep states. This is clarified through the example shown in Figure 29.9, which shows the execution of three tasks, τ_1, and τ_2, τ_3. Assume that the start time of τ_1 is the current scheduling instant and assume that tasks τ_1 and τ_2 do not use device k_2, which is in sleep state $ps_{2,2}$ at time s_1. An algorithm using a look-ahead of one task, i.e., looking ahead only to task τ_2, would erroneously decide that it is not necessary to wake up k_2 at time s_1. The same situation arises at scheduling instant $s_1 + c_1$. At τ_2's start time (s_2), looking ahead to task τ_3, k_2 is switched to state $ps_{2,1}$. At τ_2's completion time, again looking ahead one task to τ_3, k_2 is switched up to the powered-up state $ps_{2,u}$. However, if the intertask time IT_2 were less than $t_{0,2}$, k_2 would not have sufficient time to wake up, resulting in τ_3 missing its deadline.

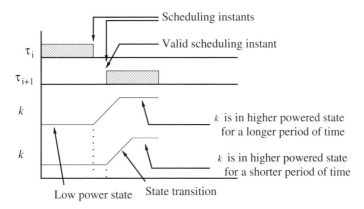

FIGURE 29.8 An invalid scheduling instant. (From V. Swaminathan and K. Chakrabarty, *IEEE Transactions on Computer-Aided Design of Integrated Circuits & Systems*, 22, 847–858, July 2003. With permission.)

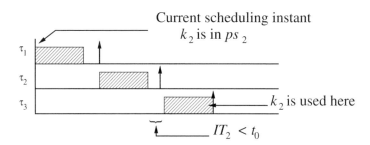

FIGURE 29.9 Look-ahead of one task is insufficient when devices have multiple sleep states. (From V. Swaminathan and K. Chakrabarty, *IEEE Transactions on Computer-Aided Design of Integrated Circuits & Systems*, 22, 847–858, July 2003. With permission.)

From this example, it is interesting to note that look-ahead represented as the number of future *tasks* is inadequate for devices with multiple low-power states. When devices have multiple states, look-ahead must be represented as the number of *valid scheduling instants* between tasks. In fact, the notion of look-ahead changes slightly when considering multiple-state I/O devices. Scheduling complexity thus increases with increasing look-ahead due to the additional computational burden of determining look-ahead; thus, minimizing look-ahead makes the scheduler more efficient. An upper bound on the look-ahead necessary to ensure timeliness while making shut-down decisions for a device [34] is now presented.

Theorem 29.2 Consider an ordered set $T = \{\tau_1, \tau_2, ..., \tau_n\}$ *of n tasks that have been scheduled a priori. Let* $c_1, c_2, ..., c_n$ *be the set of p I/O devices used by the tasks in T. In order to decide whether to switch a device* $k_i \in K$ *from state* $ps_{i,j}$ *to* $ps_{i,j+1}$ *at task* τ_{i+1}'s *start or completion time, it is necessary and sufficient to look ahead L tasks, where L is the smallest integer such that the total number of valid scheduling instants associated with the sequence of tasks* $\tau_c, \tau_{c+1}, ..., \tau_{c+L-1}$, *excluding the current scheduling instant, is at least equal to j + 1. The device* k_i *can be switched down from* $ps_{i,j}$ *to* $ps_{i,j+1}$ *if no task* τ_t, $c \le t \le c + L - 1$, *uses device* k_i.

If the intertask times of all tasks are less than the transition time $t_{0,j}$ for device k_j, Theorem 29.1 yields the following corollary.

Corollary 29.2 Suppose the intertask time IT_i *is less than the transition time* $t_{0,i}$ *for every task* $\tau_c \in T$. *In order for a device* $k_j \in K$ *to be switched down from state* $ps_{i,j}$ *to* $ps_{i,j+1}$ *at the start or completion time of task* τ_c, *it is necessary and sufficient to look ahead j + 1 tasks to ensure timeliness. Moreover, no task* τ_t, $i \le t \le j$, *must use device* k_j.

On the other hand, if the intertask times for all tasks are greater than or equal to the transition time $t_{0,j}$, Theorem 29.2 leads to the following corollary.

Corollary 29.3. Suppose the intertask time IT_i *is greater than or equal to the transition time* $t_{0,j}$ *for every task* τ_{i+1}. *In order for a device* $k_j \in K$ *to be switched down from state* $ps_{i,j}$ *to* $ps_{i,j+1}$ *at the start or completion time of task* τ_c, *it is necessary and sufficient to look ahead* $\frac{j+1}{2}$ *tasks to ensure timeliness. Moreover, device* k_j *must not be used by any task* τ_t, $i \le t \le j$.

Look-ahead increases as the depth of the sleep state increases. Next, an upper bound on look-ahead for making wake-up decisions is presented.

Theorem 29.3. Consider an ordered set $T = \{\tau_1, \tau_2, ..., \tau_n\}$ *of n tasks and a set* $K = \{k_1, k_2, ..., k_p\}$ *of p devices used by the tasks in T. Suppose the first task after* τ_c *that uses device* k_i *is* τ_{c+L}. *The device* $k_i \in K$ *must be switched up from state* $ps_{i,j+1}$ *to* $ps_{i,j}$ *at the start or completion time of task* τ_c *if and only if the total number of valid scheduling instants, including the current scheduling instant, associated with the tasks* $\tau_c, \tau_{c+1}, ..., \tau_{c+L-1}$ *is exactly equal to j + 1, where L is the look-ahead from the current scheduling instant.*

Algorithm MUSCLES (S, PS, k_i)
curr: current scheduling instant;
At s_m:
1. Find first task τ_L that uses device k_i;
2. Compute number of valid scheduling instants X between s_m and τ_L;
3. **if** $X \geq j + 1$
4.   **switchdown** k_i from $ps_{i,j}$ to $ps_{i,j+1}$;
5. **else if** $X = j$
6.   **wake up** k_i from $ps_{i,j}$ to $ps_{i,j-1}$;

At $s_m + c_m$:
7. Find first task τ_L that uses device k_i;
8. Compute number of valid scheduling instants X between s_m and τ_L;
9. **if** $X \geq j + 1$
10.   **switchdown** k_i from $ps_{i,j}$ to $ps_{i,j+1}$;
11. **else if** $X = j$ and curr is a valid scheduling instant
12.   **wake up** k_i from $ps_{i,j}$ to $ps_{i,j-1}$;
13. **else** leave k_i in $ps_{i,j}$.

FIGURE 29.10 MUSCLES algorithm.

Theorem 29.2 and Theorem 29.3 form the basis for the MUSCLES algorithm described in the next subsection.

29.5.2.1 Online Scheduling for Multistate Devices: Algorithm MUSCLES

For a precomputed task schedule, MUSCLES generates a sequence of power states for every device so that energy is minimized. It operates as follows (also see Figure 29.10): Let device ki be in state $ps_{i,j}$ at scheduling instant s_m. MUSCLES finds the next task, τ_L, that uses k_i (line 1). A check is then performed to test whether k_i can be switched down to a lower powered state. This is done by ensuring that at least $j + 1$ valid scheduling instants are between the current scheduling instant and τ_L's start time. The presence of $j + 1$ valid scheduling instants implies that device k_i can be switched down from state $ps_{i,j}$ to $ps_{i,j+1}$ (line 3). The absence of $j + 1$ valid scheduling instants precludes the shutting down of k_i to a lower powered state; a check is then performed to test whether the device must be switched up. If exactly j instants are present, then the device must be switched up in order to ensure timeliness (line 4). At the completion of a task, τ_m, the same process is repeated. However, in order to minimize energy consumption, an additional check is performed to test if the current scheduling instant is valid. If the current scheduling instant is not valid, the device is left in the same state until a valid scheduling instant (line 10) occurs. MUSCLES guarantees that no task ever misses its deadline.

LEDES and MUSCLES are polynomial-time algorithms. MUSCLES has a worst-case complexity of $O(pn^2)$, where p is the number of I/O devices used in the system and n is the number of tasks in the task set; LEDES is $O(p)$. The complexity increases in MUSCLES because the amount of look-ahead, in terms of valid scheduling instants, for each device must be computed before any state transition. Nevertheless, the relatively low complexity of MUSCLES makes online device scheduling for low-energy and real-time execution feasible.

29.5.3 Experimental Results

LEDES and MUSCLES were first evaluated with several randomly generated task sets with varying utilizations. The task sets consist of six tasks with varying hyperperiods and randomly generated device-usage lists. Because jobs may be preempted, each preempted slice of a job is considered as two jobs with identical device-usage lists. As a result, the number of jobs listed in each task set in Table 29.4 is an approximation. Each task in the task set uses one or more out of three I/O devices whose power values are shown in Table 29.5. These values pertain to real devices currently deployed

TABLE 29.4 Evaluation Task Sets for LEDES and MUSCLES

Task Set	Approximate Number of Jobs	Hyperperiod
T_1	303	1,700
T_2	68,951	567,800
T_3	36,591	341,700

Source: From V. Swaminathan and K. Chakrabarty, *IEEE Transactions on Computer-Aided Design of Integrated Circuits & Systems*, 22, 847–858, July 2003. With permission.

TABLE 29.5 Device Parameters Used in Evaluation of LEDES and MUSCLES

Device k_i	Device Type	P_w	$P_{sd}^{i,0}=P_{wu}^{i,1}$	$P_{sd}^{i,1}=P_{wu}^{i,2}$	$P_{sd}^{i,2}=P_{wu}^{i,1}$	t_0	$P_s^{i,1}$	$P_s^{i,2}$	$P_s^{i,3}$
k_1	HDD[a]	2.3 W	1.5 W	0.6 W	0.3 W	0.6 s	1.0 W	0.5 W	0.2 W
k_2	NIC[b]	0.3 W	0.2 W	0.05 W	—	0.5 s	0.1 W	3 mW	—
k_3	DSP[c]	0.63 W	0.4 W	0.1 W	—	0.5 s	0.25 W	0.05 W	—

[a] Fujitsu MHL2300AT Hard Disk Drive. http:://www.fujitsu.jp/hypertext/hdd/drive/overseas/mhl2xxx/mhl2xxx.html.
[b] AMD Am79C874 NetPHY-1LP Low-Power 10/100 Tx/Rx Ethernet Transceiver Technical Datasheet.
[c] Analog Devices Multiport Internet Gateway Processor. http://www.analog.com.

Source: From V. Swaminathan and K. Chakrabarty, *IEEE Transactions on Computer-Aided Design of Integrated Circuits & Systems*, 22, 847–858, July 2003. With permission.

in embedded systems. Each task set is scheduled using the rate-monotonic algorithm. The utilization of each task set is varied from 10 to 90% to observe the impact of slack on energy consumption of the I/O devices.

While evaluating LEDES, it was assumed that the single low-power sleep state for the devices corresponded to the highest powered sleep state of the device. The energy consumptions at different utilizations for task set T_1 are shown in Figure 29.11. Figure 29.12 illustrates the percentage energy savings for each of the task sets obtained from the LEDES algorithm.

A study of Figure 29.11 reveals that the energy consumption using LEDES and MUSCLES increases with increasing utilization because devices are kept powered up for longer periods of time within the hyperperiod. The resulting decrease in sleep time causes this increased energy consumption. However, energy savings of over 35 and 40% can be obtained for task sets with high and low utilization, respectively. No task deadlines are missed at any utilization value.

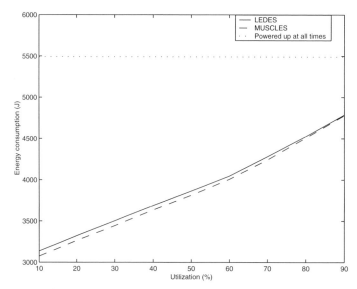

FIGURE 29.11 Comparison of LEDES and MUSCLES for task set T_1. (From V. Swaminathan and K. Chakrabarty, *IEEE Transactions on Computer-Aided Design of Integrated Circuits & Systems*, 22, 847–858, July 2003. With permission.)

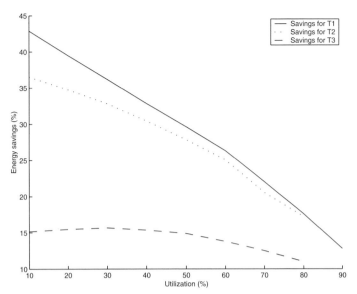

FIGURE 29.12 Energy savings using LEDES. (From V. Swaminathan and K. Chakrabarty, *IEEE Transactions on Computer-Aided Design of Integrated Circuits & Systems*, 22, 847–858, July 2003. With permission.)

One other important observation is that the savings in energy obtained from MUSCLES *over* LEDES decreases with increasing utilization because the number of valid scheduling instants decreases with increasing utilization. Thus, MUSCLES cannot place devices in deep sleep states as often in high-utilization task sets as it can in low-utilization task sets.

LEDES and MUSCLES were also evaluated with three real-life task sets. These task sets are used in an instrument navigation system (INS) [16]; a computer numerical control (CNC) system [17]; and an aviation platform (GAP) [21]. The assignment of devices to tasks in the task sets has been inferred from the functionality of the tasks. For example, task 2 in the GAP task set is a communication task that uses the NIC and task 7 is a status update task that performs occasional reads and writes and therefore uses a hard disk.

Table 29.6 presents the energy consumptions for these task sets using LEDES and MUSCLES. The energy values here are expressed in units of joules and correspond to the energy consumption of the I/O devices over the duration of a single hyperperiod. Using LEDES, an energy savings of 45% for the GAP task set are obtained. With MUSCLES, an energy savings of 80% is obtained for the INS task set. Owing to the low utilizations of real-life task sets, significant energy savings can be obtained by intelligently performing state transitions for I/O devices.

29.6 Energy-Aware Communication

This section describes a novel target localization approach based on a two-step communication protocol between the cluster head and the sensors within the cluster. Because the energy consumption in wireless sensor networks increases significantly during periods of activity, which may be triggered, for example, by a moving target [5], an energy-reduction method is proposed for target localization

TABLE 29.6 Comparison of LEDES and MUSCLES Using Real-Life Task Sets

Task Set	Energy (J)			% Savings	
	All Powered Up	LEDES	MUSCLES	(LEDES)	(MUSCLES)
CNC	403,104	197,140	117,604	51%	70%
INS	16.5×10^6	7.7×10^6	3×10^6	51%	81%
GAP	381×10^6	210×10^6	153×10^6	45%	60%

Source: From V. Swaminathan and K. Chakrabarty, *IEEE Transactions on Computer-Aided Design of Integrated Circuits & Systems*, 22, 847–858, July 2003. With permission.

in cluster-based wireless sensor networks. In the first step, sensors detecting a target report the event to the cluster head. The amount of information transmitted to the cluster head is limited. In order to conserve power and bandwidth, the message from the sensor to the cluster head is kept very small; in fact, the presence or absence of a target can be encoded in just one bit. No detailed information such as signal strength, confidence level in the detection, imagery or time series data is transmitted at this time. Based on the information received from the sensor and knowledge of the sensor deployment within the cluster, the cluster head executes a probabilistic scoring-based localization algorithm to determine the likely position of the target. It subsequently queries a subset of sensors in the vicinity of the likely target position.

29.6.1 Detection Probability Table

The cluster head first generates a detection probability table for each grid point. Consider a sensor field represented by an $m \times n$ grid. Let $\langle (y, x), (y, x + 1)\rangle$ denote the set of deployed sensor nodes and $|S| = k$. Let s be an individual sensor node on the sensor field located at grid point (x,y). Sensor detections are imprecise, so coverage is expressed in probabilistic terms. For any grid point P at (i,j), the coverage $c_{ij}(x,y)$ of P by a sensor located at $sp = (x,y)$ is expressed probabilistically in Equation 29.1, which is motivated in part by Elfes [9]:

$$c_{ij}(x,y) \begin{cases} 0, & \text{if } r + r_e \leq d_{ij}(x,y) \\ e^{-\lambda a^\beta}, & \text{if } r - r_e < d_{ij}(x,y) < r + r_e \\ 1, & \text{if } r - r_e \geq d_{ij}(x,y) \end{cases} \qquad (29.1)$$

where $r_e(r_e < r)$ is a measure of the uncertainty in sensor detection; $a = d_{ij}(x,y) - (r - r_e)$; and λ and β are parameters that measure detection probability when a target is at distance greater than r_e but within a distance from the sensor. This model reflects the behavior of range sensing devices such as infrared and ultrasound sensors. The probabilistic sensor detection model is shown in Figure 29.13. Note that distances are measured in units of grid points. This figure also illustrates the translation

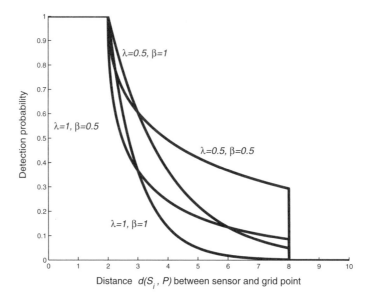

FIGURE 29.13 Probabilistic sensor detection model. (From Y. Zou and K. Chakrabarty, *ACM Transactions on Embedded Computing Systems*, 3, 61–91, February 2004. With permission.)

TABLE 29.7 Example Probability Table

l	$d_1d_2d_3$	$p\_table_{ij}(l), 0 \le l < 2^{k_{ij}}, k_{ij} = 3$
0	000	$(1 - 0.5736) \times (1 - 1) \times (1 - 0.5736) = 0.0$
1	001	$(1 - 0.5736) \times (1 - 1) \times 0.5736 = 0.0$
2	010	$(1 - 0.5736) \times 1 \times (1 - 0.5736) = 0.1819$
3	011	$(1 - 0.5736) \times 1 \times 0.5736 = 0.2446$
4	100	$0.5736 \times (1 - 1) \times (1 - 0.5736) = 0.0$
5	101	$(1 - 0.5736) \times (1 - 1) \times 0.5736 = 0.0$
6	110	$0.5736 \times 1 \times (1 - 0.5736) = 0.2446$
7	111	$0.5736 \times 1 \times 0.5736 = 0.3290$

Source: From Y. Zou and K. Chakrabarty, *ACM Transactions on Embedded Computing Systems*, 3, 61–91, February 2004. With permission.

of a distance response from a sensor to the confidence level as a probability value about this sensor response. Different values of the parameters α and β yield different translations reflected by different detection probabilities, which can be viewed as the characteristics of various types of physical sensors.

The detection probability table contains entries for all possible detection reports from sensors that can detect a target at this grid point. Assuming that the sensor field is represented by an $m \times n$ grid, and a grid point P at (i,j) is covered by a set of k_{ij} sensors denoted as S_{ij}, $|S_{ij}| = k_{ij}$, $0 \le k_{ij} \le k$ and $S_{ij} \subseteq \{s_1, s_2, \ldots, s_k\}$.

The probability table is built on the power set of S_{ij} because there are 2^{k}_{ij} possibilities for k_{ij} sensors in reporting an event. These 2^{k}_{ij} cases include the case that none of the sensors detect anything (represented by the binary string as "00...0") as well as the case that all of the sensors (represented by the binary string as "11...1") detect an event. Thus the probability table for grid point (i,j) then contains 2^{k}_{ij} entries, defined as:

$$p\_table_{ij}(l) = \prod_{s_p \in S_{ij}} p_{ij}(S_p, l) \qquad (29.2)$$

where $0 \le l \le 2^{k}_{ij}$, and $p_{ij}(sp, l) = c_{ij}(sp)$ if s_j detects a target at grid point $P(i, j)$; otherwise, $p_{ij}(sp, l) = 1 - c_{ij}(sp)$. Table 29.7 gives an example of the probability tables on a 5×5 grid with three sensors deployed.

Consider a grid point P that is covered by three sensors, s_1, s_2 and s_3, with probabilities as 0.57, 1, and 0.57, respectively.[*] For these three sensors, eight possibilities exist for their combined event detection at P. For example, the binary string 110 denotes the possibility that s_1 and s_2 report a target but s_3 does not report a target. For each such possibility $d_1d_2d_3$ (d_1, d_2, $d_3 \in \{0, 1\}$) for a grid point, the conditional probabilities that the cluster head receives $d_1d_2d_3$, given that a target is present at that grid point, are calculated. Table 29.7 lists these conditional probabilities for this example. Consider the binary string 110, the conditional probability associated with this possibility, is given by $p\_table_{24}(6) = p_{24}(s_1, 6) \, p_{24}(s_2, 6) \, p_{24}(s_3, 6) = 0.57 \times 1 \times (1 - 0.57) = 0.24$. Note that the probability table generation is only a one-time cost. Once this table is generated, there is no need to refresh it unless sensor locations are changed.

29.6.2 Score-Based Ranking

After the probability table is generated for all the grid points, localization is done by the cluster head if a target is detected by one or more sensors. An inference method based on the established probability table is used. When at time instant t the cluster head receives a positive event message from $k(t)$ sensors, it uses the grid point probability table to determine which of these sensors are most suitable to be queried for more detailed information. Detailed target reporting consumes more

---

[*]These coverage values can be obtained using the sensor detection model described in Zou and Chakrabarty [37].

energy consumption and needs more bandwidth. Therefore, the cluster head cannot afford to query all the sensors for detailed reports. Sensor detection information also has an inherent redundancy, so it is not necessary to query all sensors. The scoring approach is able to select the most suitable sensors for this purpose.

Consider the 10 × 10 grid shown in Figure 29.14. There are five sensors deployed, $k = 5$, $r = 2$, and $r_e = 1$. The zigzag shaped line is the target movement trace. The target starts to move at $t = t_{start}$ from the grid point marked as "start" and finishes at $t = t_{end}$ at the grid point marked as "end." Figure 29.15 gives the score report at the time instant t_{start} when the target is present at "start."

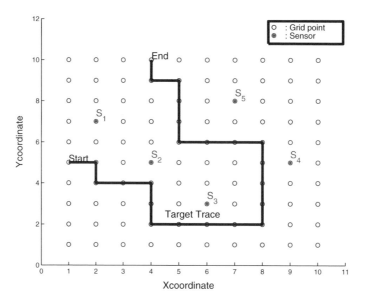

FIGURE 29.14 Sensor field with a moving target. (From Y. Zou and K. Chakrabarty, *ACM Transactions on Embedded Computing Systems*, 3, 61–91, February 2004. With permission.)

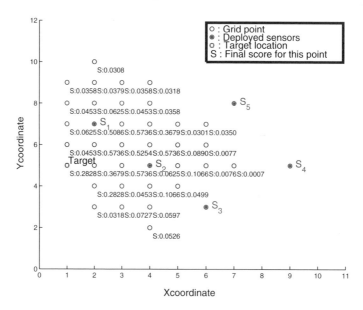

FIGURE 29.15 Scoring results for target in the sensor field at t_{start}. s_1 and s_2 have reported. (From Y. Zou and K. Chakrabarty, *ACM Transactions on Embedded Computing Systems*, 3, 61–91, February 2004. With permission.)

Assume $S_{rep}(t)$ is the set of sensors that have reported the detection of an object at time t and $S_{rep,ij}(t)$ is the set of sensors that can detect a target at point $P(i,j)$ and have also reported the detection of an object at time t. Obviously, $S_{rep,ij}(t) \subseteq S_{rep}(t)$ and $S_{rep,ij}(t) \subseteq S_{ij}$ because $S_{rep,ij}(t) = S_{rep,ij}(t) \cap S_{ij}$. The score of the grid point $P(i,j)$ at time instant t is calculated as follows:

$$SCORE_{ij}(t) = p\_table_{ij}(l) \times w_{ij}(t) \tag{29.3}$$

where l is the index of the p_table_{ij}. The parameter l is calculated from S_{ij} and $S_{rep,ij}$. The parameter p_table_{ij} ($l(t)$) corresponds to the conditional probability that the cluster head receives this event information if there was a target at $P(i, j)$. The weight $w_{ij}(t)$ reflects the confidence level in this reporting event for this particular grid point. In previous work [37], the authors have used the weight factor

$$w_{ij}(t) = \frac{k_{rep,ij}(t)}{k_{rep}(t)}$$

where $k_{rep}(t) = |S_{rep}(t)|$, and $k_{rep,ij}(t) = |S_{rep,ij}(t)|$; this is sufficient for selecting sensors in order to conserve energy. However, in order to refine the grid point scores to narrow down grid points that are most probably close to the current target location, $w_{ij}(t)$ have been redefined here to improve the accuracy for target location. The weight for the grid point $P(i, j)$ at time instant t is defined as

$$w_{ij}(t) = \begin{cases} 0 & \text{if } S_{rep,ij}(t) = \{\varnothing\} \\ 4^{-\Delta k_{rep,ij}(t)} & \text{otherwise} \end{cases} \tag{29.4}$$

where $\Delta k_{rep,ij}(t)$ measures the degree of difference in the set of sensors that reported, and those that can detect, point $P(i,j)$ at time instant t. The parameter $\Delta k_{rep,ij}(t)$ is defined as

$$\Delta k_{rep,ij}(t) = |k_{rep}(t) - k_{rep,ij}(t)| + |k_{rep}(t) - k_{ij}| \tag{29.5}$$

The parameter $w_{ij}(t)$ is therefore a decaying factor that is 1 only if $S_{rep}(t) = S_{ij}$. The number 4 in the formula for $w_{ij}(t)$ was chosen empirically after it was found to provide accurate simulation results; $w_{ij}(t)$ is used to filter out grid points not likely to be close to the actual target location. The score is based on the probability value from the probability table and the current relationship among $S_{rep}(t)$, $S_{rep,ij}(t)$, and S_{ij}. Table 29.8 gives some score calculation examples for the grid points in Figure 29.15 at the time instant t_{start}.

TABLE 29.8 Scoring Calculation Example for t at t_{start}

(x,y)	S_{ij}	$S_{rep,ij}(t)$	$w_{ij}(t)$	$p_{tableij}$ ($l(t)$)	$SCORE_{ij}(t)$
...	...	...	...	...	...
(1,6)	s_1	s_1	0.25	0.7248	0.0453
(2,6)	s_1, s_2	s_1, s_2	1.00	0.5736	0.5736
(3,6)	s_1, s_2	s_1, s_2	1.00	0.5254	0.5254
(4,6)	s_1, s_2	s_1, s_2	1.00	0.5736	0.5736
(5,6)	s_2, s_5	s_2	0.25	0.3562	0.0890
(6,6)	s_2, s_3, s_5	s_2	0.25	0.1240	0.0077
...	...	...	...	...	...

29.6.3 Selection of Sensors to Query

Assume that the maximum number of sensors allowed to report an event is k_{max}, and the set of the sensors selected by the cluster head for querying at time t is $S_q(t)$, $S_q(t) \subseteq S_{rep}(t) \subseteq \{s_1, s_2, \ldots, s_k\}$. To select the sensor to query based on the event reports and the localization procedure, first note that, for time instant t, if $k_{max} \geq k_{rep}(t)$, then all reporting sensors can be queried. Otherwise, sensors are selected on a score-based ranking. The sensors selected correspond to the ones that have the shortest distance to grid points with the highest scores. This selection rule is defined as:

$$S_q(t):d\big(S_q(t), P_{MS}\big) = min\big\{d(s_i, P_{MS})\big\} \tag{29.6}$$

where $s_i \in S_{rep}(t)$, and P_{MS} denotes the set of grid points with the highest scores.

Note that multiple grid points with the maximum score are possible. When this happens, the score concentration is calculated by averaging the scores of the current grid point and its eight neighboring grid points. The grid point with the highest score (or the score concentration) is the most likely current target location. Therefore, selecting sensors closest to this point guarantees that the selected sensors can provide the most detailed and accurate data in response to the subsequent queries. Target identification is not possible at this stage because the cluster head has no additional information other than $S_{rep}(t)$. However, the selected sensors provide enough information in the subsequent stage to facilitate target identification.

The accuracy of this target localization procedure is evaluated by calculating the distance between the grid point with the highest score and the actual target location. For the example of Figure 29.14, Table 29.9 gives some results for the selected sensor when the target is moving from "start" ($t = 1$) to "end." It is assumed that $k_{max} = 1$ and the target is moving at a constant speed. $\bar{S}_q(t)$ is the set of sensors closest to the actual location of the target at time t. The results show that $S_q(t)$ matches $\bar{S}_q(t)$ in many cases. The example does not illustrate the advantages of this proposed strategy because not many sensors are actually involved at the same time for target detection. However, in Subsection 29.6.6 the proposed algorithm performs very well when many sensors are involved in the target detection and reporting process.

29.6.4 Energy Evaluation Model for Target Localization in Wireless Sensor Networks

Consider the energy consumption for a sensor network that is actively detecting a target in the sensor field. Assume that sensor nodes are homogeneous and therefore the energy consumption for sensing is the same for each sensor node. Because the focus here is on energy minimization of communication traffic due to target activities or events, energy consumed by sensor nodes when they are in the idle

TABLE 29.9 Selected Sensors for the Example in Figure 29.14

t	$S_{rep}(t)$	$\bar{S}_q(t)$	$S_q(t)$	t	$S_{rep}(t)$	$S_q(t)$	$\bar{S}_q(t)$
...	...	...	...	...	...	...	...
3	s_1, s_2	s_1	s_2	4	s_2	s_2	s_2
5	s_2, s_3	s_3	s_2	6	s_2, s_3	s_3	s_2
7	s_2, s_3	s_3	s_3	8	s_3	s_3	s_3
...	...	...	...	...	...	...	...
16	s_4, s_5	s_4	s_4	17	s_4, s_5	s_4	s_5
18	s_2, s_3, s_5	s_2	s_2	19	s_2, s_5	s_5	s_2
20	s_1, s_2, s_5	s_2	s_2	21	s_5	s_5	s_5
...	...	...	...	...	...	...	...

state is not considered. This does not imply, however, that the energy consumption of idle sensor nodes can always be ignored.

To simplify the energy analysis, first consider a primitive sensor model that focuses on the energy consumption of the wireless sensor network due to the target activities or events. Suppose the sensor node has three basic energy consumption types — sensing, transmitting and receiving — and these power values (energy per unit time) are E_s, E_t, and E_r, respectively. If all sensors that reported the target for querying are selected, the total energy consumed for the event happening at time instant t can be evaluated using the following set of equations:

$$E_1(t) = k_{rep}(t)(E_t + E_r)T_1 \tag{29.7}$$

$$E_2(t) = (k_{rep}(t)E_r + E_t)T_2 \tag{29.8}$$

$$E_3(t) = k_{rep}(t)(E_t + E_r)T_3 \tag{29.9}$$

$$E_4(t) = E_s T_s \tag{29.10}$$

$$E(t) = E_1(t) + E_2(t) + E_3(t) + E_4(t) \tag{29.11}$$

$$E = \sum_{t=t_{start}}^{t_{end}} E(t) \tag{29.12}$$

where

E_1 is the energy required for reporting the detection of an object.

E_2 is the energy required for transmitting query information from the cluster head by broadcasting and for receiving this information at the sensor nodes.

E_3 is the energy required by sensor nodes being queried to send detailed information to the cluster head.

Parameters T_1, T_2, and T_3 denote the lengths of time involved in the transmission and reception, which are directly proportional to the sizes of data for yes/no messages, control messages to query sensors, and the detailed sensor data transmitted to the cluster head, respectively.

Parameter T_s is the time of sensing activity of sensors.

Parameter E denotes total energy — in this case for target localization from t_{start} to t_{end}.

For the proposed probabilistic localization approach, the total energy consumption E^* is calculated as follows:

$$E_1^*(t) = k_{rep}(t)(E_t + E_r)T_1 \tag{29.13}$$

$$E_2^*(t) = (k_q(t)E_r + E_t)T_2 \tag{29.14}$$

$$E_3^*(t) = k_q(t)(E_t + E_r)T_3 \tag{29.15}$$

$$E_4^*(t) = E_s T_s \tag{29.16}$$

$$E^*(t) = E_1^*(t) + E_2^*(t) + E_3^*(t) + E_4^*(t) \tag{29.17}$$

$$E^* = \sum_{t=t_{start}}^{t_{end}} E^*(t) \tag{29.18}$$

where $E_1(t)^* = E_1(t)$; $E_4^*(t) = E_4(t)$; and the total energy consumed is denoted by E^*. Therefore, the energy savings via the use of the probabilistic target localization algorithm is:

$$\Delta E = E - E^* = C \sum_{t=t_{start}}^{t_{end}} (k_{rep}(t) - k_q(t)) \tag{29.19}$$

where $C = E_r T_2 + (E_t + E_r)T_3$ is a constant. Because $k_q(t)$ is always less than or equal to $k_{rep}(t)$, $\Delta E \geq 0$. Also, ΔE is monotonically nondecreasing with time. Figure 29.16 shows the energy saved for the target trace in Figure 29.14.

29.6.4.1 Refined Energy Evaluation Model

The previous primitive energy evaluation models given by Equation 29.7 through Equation 29.19 ignore the overhead due to the two-step protocol and convey the impression that large volumes of data can greatly burden the energy consumption on sensor nodes. Therefore, the energy evaluation model is refined to incorporate the overhead introduced by this approach. The refined model is used later as the primary metric for evaluating energy consumption with parameter values from Heinzelman et al. [13] and Rappaport [27]. It is still necessary to consider a sensor node with three basic energy consumption types — sensing, transmitting, and receiving — and these power values (joules per second) are ψ_s, ψ_t, and ψ_r, respectively. Assume at time instant t, $k(t)$ sensors have detected

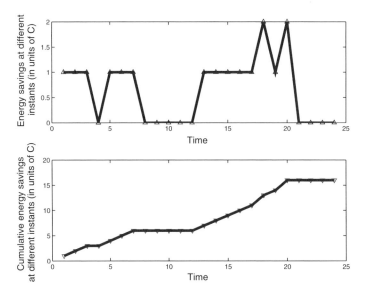

FIGURE 29.16 Energy saved for the example in Figure 29.14 using the primitive energy evaluation model. (From Y. Zou and K. Chakrabarty, *ACM Transactions on Embedded Computing Systems*, 3, 61–91, February 2004. With permission.)

the target, where $k(t) \leq k$. Therefore, the energy for sensing activities in the wireless network, denoted as $E_s(k(t))$, is

$$E_s(k(t)) = k(t)\psi_s T_s \quad (29.20)$$

where T_s is the time duration that a sensor node is involved in sensing.

For a fixed time interval, E_s is a constant if all sensor nodes are assumed to be homogenous. The energy used for communication between nodes and the cluster head can be categorized into two types, E_b and E_c. The parameter E_c is the energy consumed by a sensor node for communication with the cluster head. This includes the energy for transmitting data and the energy for receiving data. The parameter E_b is the energy needed for broadcasting data from the head to the nodes. E_b and E_c are functions of T and $k(t)$, where T is the time required for retrieving data from a sensor node or broadcasting data from the cluster head, and $k(t)$ is the number of sensors involved in this communication at time instant t. E_c and E_b are defined as:

$$E_c(k(t), T) = (\psi_t T + \psi_r T)k(t) \quad (29.21)$$

$$E_b(k(t), T) = \psi_t T + \psi_r T k(t) \quad (29.22)$$

The parameter T is directly proportional to the volume of data involved in the communication. In this work, T can be one of three values: T_d for raw target data; T_e for target event reporting; and T_q for query request. They satisfy the relationship $T_e \leq T_q \ll T_d$ because raw data collected by a sensor node can be up to hundreds of bytes in size. Assume that target detection and localization are discrete processes derived from a discrete sampling of target activities in the sensor network. Also, because the sensor network is designed to track target activities, T_s, T_e, T_q, and T_d are assumed to be less than the granularity of the time t. Thus, for the case of a target moving in the sensor field during the time interval $[t_{start}, t_{end}]$, the corresponding instantaneous energy consumption $E(t)$ and total energy consumption E in the wireless sensor network can be expressed as

$$E(t) = E_s(k(t)) + E_c(k(t), T_d) \quad (29.23)$$

$$E = \sum_{t=t_{start}}^{t_{end}} E(t). \quad (29.24)$$

From Equation 29.20 to Equation 29.24, energy consumption is evaluated using the preceding target localization method:

$$E^*(t) = E_s(k_{rep}(t)) + E_c(k_{rep}(t), T_e) + E_b(k_q(t), T_q) + E_c(k_q(t), T_d) \quad (29.25)$$

$$E^* = \sum_{t=t_{start}}^{t_{end}} E^*(t) \quad (29.26)$$

Let $k(t) = k_{rep}(t)$ in Equation 29.2 and Equation 29.3. The difference in energy consumption, $\Delta E = E - E^*$ can be expressed as:

$$\Delta E(t) = (k_{rep}(t) - k_q(t))(\psi_t + \psi_r)T_d$$
$$-(k_q(t)\psi_r + \psi_t)T_q - k_{rep}(t)(\psi_t + \psi_r)T_e \qquad (29.27)$$

$$\Delta E = \sum_{t=t_{start}}^{t_{end}} \Delta E(t) \qquad (29.28)$$

The last two terms in Equation 29.27 indicate the overhead for the proposed target localization procedure. Because $T_d \gg T_e$ and $T_d \gg T_q$, the overhead is small. Because $k_q < k_{max}$, with k_{max} properly selected, from Equation 29.27 and Equation 29.28, energy consumption is greatly reduced with the passage of time.

29.6.5 Procedural Description

Figure 29.17 shows the pseudocode of the procedure to generate the detection probability table for each grid point and Figure 29.18 shows pseudocode for simulation of the probabilistic localization algorithm. For an $n \times m$ grid with k sensors, the computational complexity involved in generating the probability table is $O(nm2k)$ because the maximum number of sensors that can detect a grid point is k for the worst case. The computational complexity of the localization procedure is $O(nmk_{max})$, $k_{max} \leq k$. Therefore, the computational complexity of the probabilistic localization algorithm is $\max\{O(nmk_{max}), O(nm2^k)\} = O(nm2^k)$. Even though the worst-case complexity of the localization procedure is exponential in k, in practice the localization procedure can execute in less time because the number of sensors that effectively detect a target at a given grid point is small.

29.6.6 Simulation Results

This subsection presents results for case studies carried out on a Pentium III 1.0GHz PC using Matlab.

29.6.6.1 Case Study

The simulation is presented on a 30×30 sensor field grid with 20 sensors randomly placed in the sensor field. The parameters of the sensor detection model are $r = 5$; $r_e = 4$; $\lambda = 0.5$; and $\beta = 0.5$.

Procedure Generate_Probability_Table (P(i, j), {s_1, ..., s_k})

```
1 /*find S_ij, the set of sensors that can detect P(i, j)*/
2 For sp ∈ {s_1, s_2, ..., s_k}
3    if d_ij (s_p) ≤ r + r_e
4       S_ij = S_ij U {s_p}
5    End
6 End
7 /*fill up the probability table */
8 For l, 0 ≤ l ≤ k_ij, k_ij = |S_ij|;
9    If s_p detects P(i, j)
10      Set p_ij (s_p, l) = cij (s_p);
11   Else
12      Set p_ij (s_p, l) = 1 − cij (s_p);
13   End
14   Set p_table_ij(l) = ∏_{s_p ∈ S_ij} p_ij (s_p, l)
15 Else
```

FIGURE 29.17 Pseudocode for generating the detection probability table. (From Y. Zou and K. Chakrabarty, *ACM Transactions on Embedded Computing Systems*, 3, 61–91, February 2004. With permission.)

Procedure *Target_Localization*(Grid, {s_1, ..., s_k}, TargetTrace)

/* k_{max} is the maximum number of sensors that are allowed for querying, p_{rep} is the threshold level for a sensor to report to the cluster head of an event. *TargetTrace* starts from t_{start} and ends at t_{end}, with time unit as 1. */

1 Set $t = t_{start}$;
2 **While** ($t \leq t_{end}$)
3   /* current target location */
4   Set *Target* = *TargetTrace*(t);
5   /* calculate the scores */
6   Calculate $S_{rep}(t)$ from {$s_1, s_2, ..., s_k$}, Target(t), p_{rep};
7   Set $k_{rep}(t) = |S_{rep}(t)|$;
8   **For** $P(i, j)$ in Grid, $i \in [1, width], j \in [1, height]$
9   Set $k_{ij} = |S_{ij}|$;
10   Calculate $S_{rep,ij}(t)$ from $S_{rep}(t)$ and $P(i, j)$;
11   Calculate the index $l(t)$ of p_table_{ij} from
             $S_{rep}(t)$, and $S_{rep,ij}(t)$;
12   Set $k_{rep,ij}(t) = |S_{rep,ij}(t)|$;
13   **If** $S_{rep,ij}(t) = \{\emptyset\}$
14             $w_{ij}(t) = 0$;
15   **Else**
16   Set $\Delta k_{rep,ij}(t) = |k_{rep}(t) - k_{rep,ij}(t)|$
                 $+ |k_{rep}(t) - k_{ij}(t)|$;
17             $w_{ij}(t) = 4^{-\Delta k_{rep,ij}(t)}$;
18   **End**
19   Set $SCORE_{ij}(t) = p\_table_{ij} (l(t) \times w_{ij}(t))$;
20   **End**
21   /* select sensors for querying */
22   Calculate $S_q(t)$ from $SCORE_{ij}(t)$ and $k_{max}, i \in [1, width], j \in [1, height]$;
23   /* next time instant */
24   Set $t = t + 1$;
25 **End**

FIGURE 29.18 Pseudocode of the target localization procedure.

Choose the energy consumption model parameters as $\psi_r \approx 400$ nJ/sec; $\psi_t \approx 400$ nJ/sec; and $\psi_s \approx 1000$ nJ/sec. These values are based on typical values given in Heinzelman et al. [13] and Rappaport [27], assuming the sensing rate for the sensor is 8 bits/sec. No physical data are available for T_d and T_e; however, because their values do not affect the target localization procedure, it is necessary only to set them manually to satisfy the relationship $T_d \gg T_e$ and $T_d \gg T_q$. In this case, $T_d = 100$ ms; $T_e = 2$ ms; and $T_q = 4$ ms.

The layout of the sensor field is given in Figure 29.19, with a target trace randomly generated in the sensor field. The target travels from the position marked "start" to the position marked "end." Assume the target locations are updated at discrete time instants in units of seconds, and the granularity of time is long enough for sampling by two neighboring locations in the target trace with negligible errors. Evaluate the algorithm for $k_{max} = 1$; $k_{max} = 2$; and $k_{max} = 3$.

Figure 29.20 illustrates the instantaneous energy savings in percentage, and Figure 29.21 shows the absolute value of the cumulative energy savings for the case study as the target moves along its trace in the sensor field. The energy savings are compiled relative to the base case when all sensors report complete target information in one step everywhere.

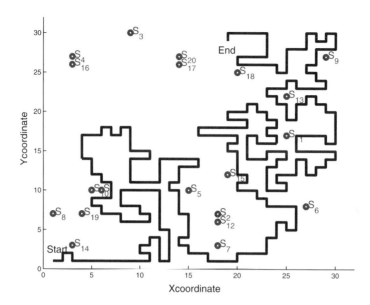

FIGURE 29.19 Sensor field layout with target trace. (From Y. Zou and K. Chakrabarty, *Proc. IEEE International Conference on Pervasive Computing and Communications*, 60–67, 2003. With permission.)

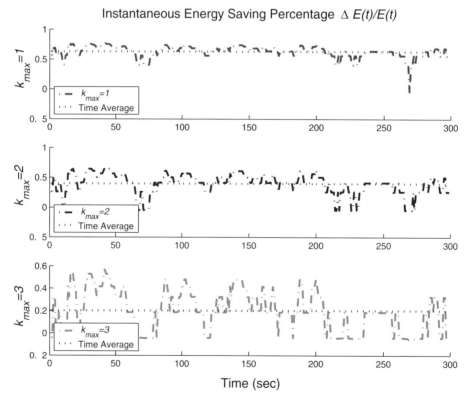

FIGURE 29.20 Instantaneous energy saving percentage during target localization relative to the "always report" one-step base case. (From Y. Zou and K. Chakrabarty, *Proc. IEEE International Conference on Pervasive Computing and Communications*, 60–67, 2003. With permission.)

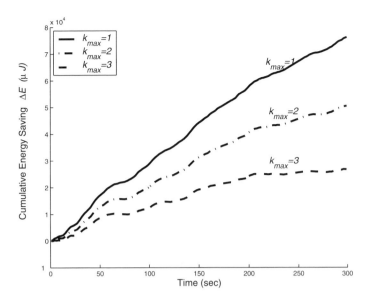

FIGURE 29.21 Cumulative energy saving during target localization relative to the "always report" one-step base case. (From Y. Zou and K. Chakrabarty, *Proc. IEEE International Conference on Pervasive Computing and Communications*, 60–67, 2003. With permission.)

It is evident from Figure 29.20 and Figure 29.21 that a large amount of energy is saved during target localization. Note that when k_{max} approaches $k_{rep}(t)$, the savings is less apparent due to the additional communication overhead of the two-stage query protocol. Nevertheless, a considerable amount of energy is saved in target localization, even when $k_{max} = 3$. With an appropriate selection of k_{max}, the proposed algorithm performs exceptionally well.

Next, consider the latency in the localization of a target by the cluster head. Latency refers to the time taken for the cluster head to collect detailed target information from sensor nodes starting from the time sensor nodes detect an event. Assume that the wireless sensor network uses the time division multiple access (TDMA) protocol [35]. The results are shown in Figure 29.22. The latency is reduced here compared to the base case using a "report once" strategy because a large amount of communication for transmitting raw data has been reduced to a smaller amount of data sent by a selected set of sensors. This is an added advantage to the proposed energy-aware target localization procedure.

Because the selection of sensors for querying is based on the detection probability table and the distance of sensors from the estimated high-score points, the *a posteriori* approach offers another important advantage: it provides a substantial amount of built-in false-alarm filtering. Figure 29.23 illustrates the false-alarm filtering ability of the proposed approach. False alarms reported by some malfunctioning sensors during $t \in [18, 22]$ by s_4; during $t \in [138, 142]$ by s_{16}; and during $t \in [239, 241]$ by s_8 were manually generated. The distance d of the target from the sensor in $S_{rep}(t)$ farthest from it was calculated, as well as the distance d^* of the target from the sensor in $S_q(t)$ farthest from it. The difference $d - d^*$ is used as a measure of the built-in filtering ability. Figure 29.23 shows the variation of $d - d^*$ with time. Note that prior to querying, the cluster head only knows which sensors have reported the detection of a target; no detailed information about the target is available to the cluster head.

The localization approach successfully narrows down the sensors closest to the real target location and selects them for detailed information querying. As shown in Figure 29.23, the three spikes represent the fact that the false alarms from the sensor (which in this case is the sensor farthest from the actual target location) have been filtered out because the proposed target localization procedure is still able to select the most appropriate sensors to be queried for detailed target information.

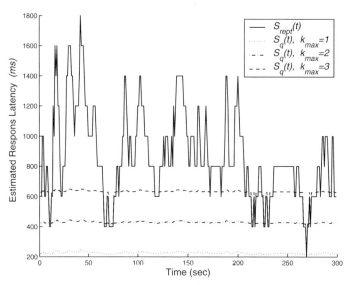

FIGURE 29.22 Latency in the localization of a target by the cluster head. (From Y. Zou and K. Chakrabarty, *Proc. IEEE International Conference on Pervasive Computing and Communications*, 60–67, 2003. With permission.)

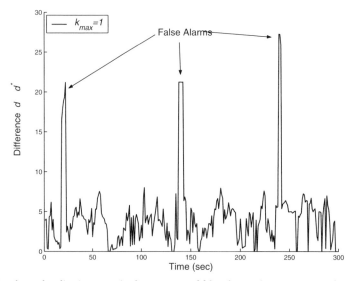

FIGURE 29.23 Results on localization error in the presence of false alarms. (From Y. Zou and K. Chakrabarty, *Proc. IEEE International Conference on Pervasive Computing and Communications*, 60–67, 2003. With permission.)

29.7 Conclusions

Energy is an important resource in battery-operated sensor systems. For such systems that operate under real-time constraints, energy consumption must be carefully balanced with real-time responsiveness. This chapter has described two approaches to energy minimization in sensor networks: node-level energy minimization and network-level energy minimization.

The two node-level energy minimization techniques described here focus on minimizing the energy consumption of the processor and I/O devices in a sensor node, respectively. Implementation of a dynamic power management scheme that uses an EDF-based scheduler to support real-time execution was described. The scheduler is efficient and can be easily integrated into the kernels of real-time operating systems on sensor nodes. The LEDF algorithm provides significant energy savings in real-time systems.

In many embedded systems, the I/O subsystem is a viable candidate to target for energy reduction. Two low-energy I/O device scheduling algorithms have also been described. The first, called LEDES, assumes that the I/O devices present in the sensor system possess two power states: a high-powered working state and a low-powered sleep state. Even under this somewhat restrictive assumption, experimental results show that energy savings of over 40% can be obtained. A generalized version of LEDES, called MUSCLES, that schedules devices with more than two low-power sleep states has also been described. Experimental case studies for real-life task sets show that energy savings of over 50% can be obtained by targeting the I/O subsystem for power reduction. The amount of energy saved decreases with increasing task-set utilization; nevertheless, energy savings of over 40% with these device scheduling algorithms in high-utilization task sets can be realized.

Finally, the chapter described an energy-aware target localization procedure for cluster-based wireless sensor networks for target localization and detection to reduce network-level energy consumption. This approach is based on the combination of a two-step communication protocol between the cluster head and the sensors in the cluster and a probabilistic localization algorithm. This approach reduces energy consumption, decreases the latency for target localization, and provides a mechanism for filtering false alarms.

References

1. AMD Am79C874 NetPHY-1LP Low-Power 10/100 Tx/Rx Ethernet Transceiver Technical Datasheet.
2. AMD PowerNow! Technology, *http://www.amd.com/us-en/Processors/ProductInformation/0,,30_118_756_807964,00.html*.
3. Analog Devices Multiport Internet Gateway Processor.*http://www.analog.com*.
4. L. Benini, A. Bogliolo, G.A. Paleologo, and G. De Micheli. Policy optimization for dynamic power management. *IEEE Trans. Computer-Aided Design*, 16(6), 813–833, June 1999.
5. M. Bhardwaj and A.P. Chandrakasan. Bounding the lifetime of sensor networks via optimal role assignments. *Proc. IEEE Infocom Conf.*, 1587–1596, 2002.
6. G.C. Buttazzo, *Hard Real-time Computing Systems: Predictable Scheduling Algorithms and Applications*, Kluwer Academic Publishers, Norwell, MA, 1997.
7. A.P. Chandrakasan and R. Broderson, *Low Power Digital CMOS Design*, Kluwer Academic Publishers, Norwell, MA, 1995.
8. E.-Y. Chung, L. Benini and G. De Micheli. Dynamic power management using adaptive learning tree. *Proc. Int. Conf. Computer-Aided Design*, 274–279, 1999.
9. A. Elfes. Occupancy grids: a stochastic spatial representation for active robot perception. *Proc. Conf. Uncertainty in AI,* 60–70, 1990.
10. Fujitsu MHL2300AT Hard Disk Drive. http:://www.fujitsu.jp/hypertext/hdd/drive/overseas/mhl2xxx/mhl2xxx.html.
11. R. Golding, P. Bosh, C. Staelin, T. Sullivan and J. Wilkes. Idleness is not sloth. *Proc. Usenix Tech. Conf. UNIX Adv. Computing Syst.*, 201–212, 1995.
12. W.B. Heinzelman, J. Kulik, and H. Balakrishnan, Adaptive protocols for information dissemination in wireless sensor networks, *Proc. IEEE/ACM MobiCom Conf.*, 174–185, 1999.
13. W.B. Heinzelman, A. Chandrakasan and H. Balakrishnan. Energy-efficient communication protocol for wireless micro sensor networks. *Proc. Annu. Hawaii Int. Conf. Syst. Sci.*, 3005–3014, 2000.
14. C. Hwang and A.C.-H. Wu. A predictive system shutdown method for energy saving of event-driven computation. *Proc. Int. Conf. Computer-Aided Design*, 28–32, 1997.
15. K. Jeffay, D.F. Stanat, and C.U. Martel. On non-preemptive scheduling of periodic and sporadic tasks with varying execution priority. *Proc. Real-Time Syst. Symp.*, 129–139, December 1991.
16. D. Katcher, H. Arakawa, and J. Strosnider. Engineering and analysis of fixed priority schedulers. *IEEE Trans. Software Eng.*, 19, 920–934, September 1993.

17. N. Kim, M. Ryu, S. Hong, M. Saksena, C. Choi, and H. Shin. Visual assessment of a real-time system design: case study on a CNC controller. *Proc. Real-Time Syst. Symp.*, 300–310, 1996.
18. K. Li, R. Kumpf, P. Horton and T. Anderson. A quantitative analysis of disk drive power management in portable computers. *Proc. Usenix Winter Conf.*, 279–292, 1994.
19. S. Lindsey and C.S. Raghavendra. PEGASIS: power-efficient gathering in sensor information systems. *Proc. IEEE Aerospace Conf.*, 3, 1125–1130, 2002.
20. J.W.S. Liu. *Real-Time Systems*. Prentice Hall, Upper Saddle River, NJ, 2000.
21. D.C. Locke, D. Vogel and T. Mesler. Building a predictable avionics platform in Ada: a case study. *Proc. Real-Time Syst. Symp.*, 181–189, 1991.
22. Y-H. Lu, L. Benini and G. De Micheli. Operating system directed power reduction. *Proc. Int. Conf. Low-Power Electron. Design*, 37–42, 2000.
23. A. Manjeshwar and D.P. Agrawal. TEEN: a routing protocol for enhanced efficiency in wireless sensor networks. *Proc. Int. Parallel Distributed Process. Symp.*, 2009–2015, 2001.
24. R. Min, M. Bhardwaj, S.H. Cho, A. Sinha, E. Shih, A. Wang, and A. Chandrakasan. Low-power wireless sensor networks. *VLSI Design*, 2001.
25. M. Newman and J. Hong. A look at power consumption and performance of the 3Com Palm Pilot. *http://guir.cs.berkeley.edu/projects/p6/finalpaper.html*.
26. P. Pillai and K.G. Shin. Real-time dynamic voltage scaling for low-power embedded operating systems. *Proc. Symp. Operating Syst. Principles*, 89–102, 2001.
27. T. Rappaport. *Wireless Communications: Principles & Practice*, Englewood Cliffs, NJ: Prentice Hall, Inc., 1996.
28. The RT-Linux Operating System, *http://www.fsmlabs.com/community/*.
29. E. Shih, B.H. Calhoun, H.C. Seong and A.P. Chandrakasan. An energy-efficient link layer for wireless micro sensor networks. *Proc. IEEE Computer Soc. Workshop VLSI*, 16–21, 2001.
30. T. Simunic, L. Benini, P. Glynn, and G. De Micheli. Event driven power management. *IEEE Trans. Computer-Aided Design*, 840–857, 2001.
31. A. Sinha and A. Chandrakasan. Dynamic power management in wireless sensor networks. *IEEE Design Test Computers*, 18, 62–74, 2001.
32. S. Slijepcevic and M. Potkonjak. Power efficient organization of wireless sensor networks. *Proc. IEEE Int. Conf. Commun.*, 472–476, 2001.
33. M.B. Srivastava, A.P. Chandrakasan, and R.W. Broderson. Predictive system shutdown and other architectural techniques for energy efficient programmable computation. *IEEE Trans. VLSI Syst.*, 4, 42–55, 1996.
34. V. Swaminathan and K. Chakrabarty. Energy-conscious, deterministic I/O device scheduling in hard real-time systems. *IEEE Trans. Computer-Aided Design Integrated Circuits Syst.*, 22, 847–858, July 2003.
35. P.K. Varshney. *Distributed Detection and Data Fusion*, Springer, New York, 1996.
36. A. Wang, W.B. Heinzelman and A.P. Chandrakasan. An energy-efficient system partitioning for distributed wireless sensor networks. *Proc. IEEE Int. Conf. Acoustics, Speech, Signal Process.*, 2, 905–908, 2001.
37. Y. Zou and K. Chakrabarty. Sensor deployment and target localization based on virtual forces. *Proc. IEEE Infocom Conf.*, vol. II, session 1, paper 1, 2003.
38. Y. Zou and K. Chakrabarty. Energy-aware target localization in wireless sensor networks. *Proc. IEEE Int. Conf. Pervasive Computing Commun.*, 60–67, 2003.
39. Y. Zou and K. Chakrabarty. Target localization based on energy considerations in distributed sensor networks. *Ad Hoc Networks*, 1, 261–272, 2003.

30
Energy-Aware Routing and Data Funneling in Sensor Networks

Rahul C. Shah
University of California at Berkeley

Dragan Petrovic
University of California at Berkeley

Jan M. Rabaey
University of California at Berkeley

30.1 Introduction 30-1
30.2 Protocol Stack Design 30-2
  Application Layer • Network Layer • Data Link Layer
30.3 Routing Protocol Characteristics and Related Work..... 30-4
30.4 Routing for Maximizing Lifetime: A Linear Programming Formulation 30-5
30.5 Energy-Aware Routing.. 30-5
  Setup Phase • Data Communication Phase
30.6 Simulations 30-7
30.7 Data Funneling... 30-10
  Setup Phase • Data Communication Phase
30.8 Conclusion 30-13

30.1 Introduction

The rising interest in ubiquitous sensor networks [12] has led to increased work on ad hoc multihop routing protocols. Unlike traditional routing protocols that minimize delay, many of these protocols try to minimize the energy required for communication because nodes in a sensor network are energy constrained. However, minimizing the energy consumption for every route can lead to undesirable effects like the creation of hotspots. These are areas that provide very good connectivity across the network and thus are used more often than other nodes. This leads to some nodes dying much earlier than others, resulting in lost sensing functionality as well as possible network partition. Many researchers have proposed ways to avoid this problem. This is typically done by using the residual energy at nodes as a routing metric rather than the energy used in communication.

Data communication in sensor networks primarily follows a pull model. In other words, most of the traffic in the network is based on a request–response model in which a request for information can set up a number of responses over a period of time. Thus, most routing protocols are reactive in nature, setting up route information during the request phase and using this information to route packets during the response phase. Based on this information, protocols use a single path or multiple paths to route the packets. Although single paths are cheaper to maintain, they run the risk of nondelivery of data if one node fails along the path. If such failure occurs, reflooding and route recomputation must be done to discover a new route.

Multipath protocols [13] get around this problem by using multiple routes so that failure of one route does not necessitate rediscovery of routes. However, using multiple routes increases the energy consumption for communication. Energy-aware routing [1] avoids this problem by discovering and using multiple routes; however, at any point in time, only one route is used. This is achieved by keeping a set of good routes and, for every packet, choosing a route in a probabilistic fashion. Thus, energy-aware routing is tolerant to failure of nodes (or, node mobility). It also uses the residual energy in its metric, thus avoiding nodes depleted of energy.

Energy-aware routing tries to optimize network lifetime — defined as the time until the first node in the network dies. Extending network lifetime translates to ensuring that energy use is equitable across the network. This is in contrast to simply minimizing the energy, which leaves the network with a wide disparity in the energy levels of the nodes and, eventually, disconnected subnets. If nodes in the network burn energy more equitably, then the nodes in the center of the network continue to provide connectivity longer, and the time to network partition increases. This leads to a more graceful degradation of the network.

Many different orthogonal techniques can be devised at the application or network level to extend the lifetime of the network. A first and obvious approach is to minimize the amount of data that must be transmitted through the network. One great opportunity arises from the observation that data transmitted by identical sensors spaced closely together tend to be spatially correlated. Thus, distributed compression schemes can be used to remove this redundancy and minimize the amount of traffic in the network [20].

A second approach to reduce the traffic volume (in terms of the number of bits an individual node must process) is to reduce transmission overhead. Packet payloads in sensor networks tend to be short (individual sensor measurements rarely need a resolution of more than 24 bits). Thus, packet headers comprise a substantial part of a packet because they contain training sequences for clock synchronization, source and destination information, and error control codes. These headers are necessary to make communication in a multihop network possible, but they do not carry any information about the phenomenon sensed. Thus substantial gains can be achieved if packets from different sources can be combined together and sent as one super-packet to the destination. This is the basic principle behind data funneling, a technique for routing with data aggregation [14]. Finally, energy-aware routing protocols can help to extend network lifetime compared to protocols that solely attempt to minimize network latency or the overall energy consumption of the network.

Section 30.2 of this chapter discusses the protocol stack design for sensor networks. Routing protocol characteristics and related work in this area are considered in Section 30.3. A linear programming formulation of routing is presented in Section 30.4. Section 30.5 introduces the energy-aware routing protocol, and Section 30.6 presents some simulation results. The data funneling protocol is described in Section 30.7 and the chapter concludes with Section 30.8.

30.2 Protocol Stack Design

This section gives a brief description of the application, network, and data link layers of the protocol stack.

30.2.1 Application Layer

Although the PicoRadio can support different types of applications, the driver application considered as an example in this chapter is environment control and monitoring. The aim is to control a typical office environment using a distributed building monitor and control approach. This is achieved by having three kinds of nodes in the system. The first are sensor nodes, which sense some environmental variable. The second node types are controllers, nodes that collect data from the sensors and, based on the data, decide the responses. They then command the third kind of nodes, actuators, to take appropriate action and affect the environment.

Based on the application, most of the nodes are expected to be static in nature, with a few low-speed mobile nodes. Furthermore, because the dominant form of data transport is from the sensors to the controller, it is important to optimize for that traffic. In addition, the total bit rate is rather low, about a few hundred bits per second per node. Also, because sensor data are inherently redundant, it is not necessary to have a transport layer that ensures reliable end-to-end delivery of every packet. If needed, the application can ensure the reliability itself. Thus, the application layer sends packets directly to the network layer.

30.2.2 Network Layer

The network layer has two primary functions: routing and addressing nodes. Although the rest of the chapter is concerned with routing, the kind of addressing used in the authors' network is briefly described in this section.

Traditional network addressing assigns fixed addresses to nodes, such as in the Internet. The advantage of such schemes is that the addresses can be made unique. However, a very high cost is associated with assigning and maintaining these kinds of addresses. This problem is exacerbated in mobile networks in which the topology information keeps changing. It is very difficult to route packets if the node address does not provide a clue as to the direction in which the packet is to be routed. Two approaches offer a solution to this problem. One is to maintain a central server that keeps up-to-date information on the position of every node. The other is to take the mobile IP approach: every node has a home agent that handles all the requests for the node and redirects them to the present position of the node.

For sensor networks, however, there is an important property of information flow that can be used advantageously. Most of the communication in sensor networks is of the form, "Give me the temperature of room 5." Thus, nodes can be addressed based on their geographical position. This information is also very useful for the routing protocol because it can direct communication in the right direction. Thus, for PicoRadio, class-based addressing is used. These addresses are triplets of the form (location, node type, node subtype). *Location* specifies a particular point or region in space that is of interest. *Node type* defines which type of node is required, such as sensor, controller, or actuator. Finally, the *node subtype* further narrows the scope of the address, such as temperature sensor, humidity sensor, etc. Thus, class-based addressing defines the type of node in the region of space that is needed.

In the rest of this chapter, class-based addressing within the network layer is assumed. Note that class-based addressing implicitly assumes that each node has knowledge of its position, which can be achieved by utilizing GPS or other distributed locationing algorithms [15]. These algorithms compute the location coordinates based on the received signal strength of neighboring nodes and the presence of certain *anchor* nodes in the network that know their exact positions. In this chapter, it is also assumed that all nodes know their position information perfectly.

30.2.3 Data Link Layer

The primary functions of the data link layer (DLL) are to provide access control, ID assignment, neighbor list management, and power control. The DLL coordinates assignments of local IDs so that each node gets a locally unique ID (up to two hops), while the IDs are reused globally. Thus a node can be identified within its neighborhood by its local ID. A common broadcast channel is used to send all DLL maintenance messages as well as for a rendezvous with another node when some data need to be sent. Once they rendezvous, the two nodes can communicate on other data channels. Using such a multichannel approach substantially reduces the number of collisions occurring during data transmission.

The link layer also keeps a list of its neighbors and metrics, such as the neighbor's position and the energy needed to reach it. This list is used by the network layer to make decisions regarding packet routing. Finally, the DLL also performs power control to ensure a power level that maintains an optimal number of neighbors.

30.3 Routing Protocol Characteristics and Related Work

Before introducing the energy-aware routing protocols, it is worthwhile to summarize the characteristics of what constitutes a good routing protocol in sensor networks. These are over and above properties traditionally required for routing protocols such as loop freedom and decentralized implementation. The additional requirements stem from the limited energy and small size of nodes in a sensor network.

- *Energy aware.* Sensor nodes may have limited battery lifetime and thus it is necessary for protocols to be aware of the residual energy at the nodes. Knowledge of the current energy use and the battery charge can enable the protocol to avoid nodes that are heavily depleted or to use energy-rich nodes.
- *Simple.* Simplicity of protocols refers to the fact that the protocol should not be too large or complex. It should minimize its memory requirement and also the amount of overhead it generates for routing. Thus, this refers to minimizing the communication and state of the protocol.
- *Adaptable.* Sensor nodes are inherently unreliable. They may fail due to lack of energy or may move to another region; thus, the routing protocol should be adaptable to such failures and be able to take appropriate action.
- *Scalable.* Because sensor networks can scale to hundreds and thousands of nodes, the routing must scale gracefully with such numbers. A key part of this is the size of the routing tables that are maintained and how they scale with the number of nodes/destinations in the network.

Now a small subset of routing protocols proposed for sensor and ad hoc networks will be discussed. The set is by no means exhaustive; however, it lists protocols that possess some of the preceding characteristics and from which energy-aware routing was derived.

Energy-aware routing is closely related to directed diffusion routing [2]. In directed diffusion, the destination sends an interest packet toward the source and sets up multiple routes in the process. The source subsequently sends data packets along the route, which has the minimum number of hops, consumes minimum energy, or something similar. Periodically, the source also sends data along all the paths to keep them alive and to check if any path has become better than the previously best route.

Both these protocols fall under the class of reactive protocols — protocols that discover routes only on demand. This is best exemplified by such ad hoc routing protocols such as ad hoc on-demand distance vector (AODV) routing [8] and dynamic source routing (DSR) [9]. AODV discovers routes on an as-needed basis by flooding the network and choosing the path with minimum delay. If the route gets broken, it discovers a new route by flooding the network again. On the other hand, DSR also discovers routes by flooding or by promiscuous listening to neighboring nodes to see what routes they have and then aggressively caching such information. It uses source routing, which means that the entire route is stored in the packet header rather than in intermediate nodes. Many reactive protocols also use the residual battery life as metrics for routing purposes [3, 4].

The other class of protocols is proactive protocols that maintain routes to all other nodes in the network. Destination sequenced distance vector (DSDV) routing [7] is one example of this kind. These protocols are not used often in sensor networks because they are better suited for high-traffic networks with a large number of source–destination pairs.

Location or geographic protocols make up one class of protocols very specific to ad hoc and, particularly, sensor networks. These protocols use the geographic position of nodes to forward packets toward the destination. This is very natural because nodes in a sensor network typically need to have a sense of their position to make sensor data meaningful. One of the first geographic routing protocols was developed in Karp and Kung [17], although there have been many others (e.g., [5, 18, 19]). Most of the work in this area until now has focused on routing around obstacles or areas in which no nodes are present.

Another class of routing protocols formulates the routing as a linear program (shown next) and tries to solve it in a decentralized fashion. The approach of Chang and Tassiulas [10] is representative of this kind of approach. Unlike most routing protocols which are based on heuristics, these protocols are based on a theoretical formulation of the problem. However, the problem is very difficult to solve in practice; to achieve a distributed implementation, it is necessary to use certain heuristics as well.

30.4 Routing for Maximizing Lifetime: A Linear Programming Formulation

As mentioned earlier, the routing protocol is designed to optimize network lifetime. Thus, it is very important to measure the performance of the protocol with respect to the optimum. The optimal protocol can be written as a linear program (as in Bhardwaj and Chandrakasan [16]). This linear program would correspond to a centralized routing scheme in which the traffic patterns are known *a priori*, in addition to the network topology and cost of communication between pairs of nodes. The cost of communication between pairs of nodes is the cost of communicating over the link by including the long-term average number of retransmissions needed for successful data delivery.

The metric to be optimized is the lifetime of the network (t), defined as the time when the first node in the network dies out. The first constraint is the non-negativity of the flow between any two nodes. Here, $r_{i,j}$ is the flow between nodes i and j. The second constraint is the flow constraint, which specifies that the amount flowing into a node is the same as the amount flowing out, except for the amount (S_i) that is absorbed or generated depending on whether the node is a sink or source, respectively. The third constraint is the energy constraint, which specifies that the energy expended in receiving and transmitting a packet should not exceed the initial amount of energy at the node. Solving this linear program would split each flow in such a fashion as to maximize the network lifetime.

$$\max t$$

$$s.t.$$

$$r_{i,j} \geq 0 \,\forall i,j$$

$$\sum_d r_{i,d} - \sum_s r_{s,i} = S(i) \cdot t$$

$$\sum_d P_{tx}(i,d) r_{i,d} + \sum_s P_{rx}(s,i) r_{s,i} \leq E_i \,\forall i$$

$$S(i) = \begin{cases} +1 & (source) \\ 0 & (relay) \\ -1 & (sink) \end{cases}$$

30.5 Energy-Aware Routing

The previous section specified optimal routing policy if centralized computation were possible; in reality the protocols need to work in a decentralized fashion. Thus, routes need to be selected based on some metric without full knowledge of the network. Even though sensor networks are energy limited, finding the lowest energy route and using that for every communication is not the best thing to do for network lifetime. The reason is that using a low-energy path frequently leads to energy depletion of the nodes along that path and, in the worst case, may lead to network partition.

To counteract this problem, a new protocol, called energy-aware routing, is proposed. The basic idea is that, to increase the survivability of networks, it may be necessary to use suboptimal paths occasionally. This ensures that the optimal path is not depleted and the network degrades gracefully as a whole rather than getting partitioned. To achieve this, multiple paths are found between source and destinations, and each path is assigned a probability of being chosen, depending on the energy metric. Every time data are to be sent from the source to destination, one of the paths is randomly chosen, depending on the probabilities. Therefore, none of the paths is used all the time, thus preventing energy depletion. Also, different paths are tried continuously, improving tolerance to nodes moving around the network.

Energy-aware routing is also a reactive routing protocol. It is a destination-initiated protocol in which the consumer of data initiates the route request and maintains the route subsequently. Thus, it is similar to diffusion in many ways. Multiple paths are maintained from source to destination. However, diffusion sends data along all the paths at regular intervals, but energy-aware routing uses only one path at all times. However, due to the probabilistic choice of routes, it can continuously evaluate different routes and choose the probabilities accordingly. The protocol has three phases:

- *Setup phase or interest propagation.* Directional flooding occurs to find all the routes from source to destination and their energy costs. This is when routing (interest) tables are built up.
- *Data propagation phase.* Data are sent from source to destination, using the information from the earlier phase. This is when paths are chosen probabilistically according to the energy costs calculated earlier.
- *Route maintenance.* Route maintenance is minimal. Directional flooding is performed infrequently from destination to source to keep all the paths alive and to collect new metrics.

30.5.1 Setup Phase

1. The destination node initiates the connection by flooding the network in the direction of the source node or region. It also sets the "cost" field to zero before sending the request.

$$Cost(N_D) = 0$$

2. Every intermediate node forwards the request only to neighbors closer to the source node (region) and farther away from the destination node than itself. Thus, at a node N_i, the request is sent only to a neighbor N_j that satisfies:

$$d(N_i, N_s) > d(N_j, N_s)$$

$$d(N_i, N_D) < d(N_j, N_D)$$

3. Upon receiving the request, the energy metric for the neighbor that sent the request is computed and is added to the total cost of the path. Thus, if the request is sent from node N_i to node N_j, N_j calculates the cost of the path as:

$$C_{N_j, N_i} = Cost(N_i) + Metric(N_j, N_i)$$

4. Paths that have a very high cost are discarded and not added to the forwarding table. Only the neighbors N_i with paths of low cost are added to the forwarding table FT_j of N_j.

$$FT_j = \left\{ i \middle| C_{N_j, N_i} \leq \kappa \cdot \left(\min_k C_{N_j, N_k} \right) \right\}$$

5. Node N_j assigns a probability to each of the neighbors N_i in the forwarding table FT_j, with the probability inversely proportional to the cost.

$$P_{N_j, N_i} = \frac{\frac{1}{C_{N_j, N_i}}}{\sum_{k \in FT_j} \frac{1}{C_{N_j, N_k}}}$$

6. Thus, each node N_j has a number of neighbors through which it can route packets to the destination. N_j then calculates the average cost of reaching the destination using the neighbors in the forwarding table.

$$Cost(N_j) = \sum_{i \in FT_j} P_{N_j,N_i} C_{N_j,N_i}$$

7. This average cost, $Cost(N_j)$ is set in the "cost" field of the request packet and forwarded along toward the source node, as in step 2.

30.5.2 Data Communication Phase

1. The source node sends the data packet to any of the neighbors in the forwarding table, with the probability of the neighbor being chosen equal to the probability in the forwarding table.
2. Each of the intermediate nodes forwards the data packet to a randomly chosen neighbor in its forwarding table, with the probability of the neighbor being chosen equal to the probability in the forwarding table.
3. This is continued until the data packet reaches the destination node.

The energy metric used in the protocol takes into account the residual energy of nodes along the path and the total energy needed for communication from the source to the destination. Thus, it was similar to the metric proposed by Chang and Tassiulas [10].

$$C_{ij} = e_{ij}^{\alpha} \cdot R_i^{-\beta}$$

Here, C_{ij} is the cost metric between two nodes i and j; e_{ij} is the energy used to transmit and receive on the link; and R_i is the residual energy at node i normalized to the initial energy at the node. The weighting factors α and β can be chosen to find the minimum energy path or the path with nodes having the most energy or a combination of these.

30.6 Simulations

Simulations were carried out in Opnet to demonstrate the increased network survivability due to energy-aware routing. The simulation consisted of 76 nodes in a typical office set up as in Figure 30.1. There were 65 sensors, seven static controllers, and four mobile nodes. Among the sensors, 47 were light sensors and 18 were temperature sensors. The controllers sent out requests for data to the sensors in their region of interest. These requests programmed the light sensors to send data every 10 sec and temperature data every 30 sec. These numbers are obtained from real application scenarios.

Every node consisted of an application and a network layer. The application layer was programmed to be a sensor or a controller, while the network layer performed the routing operations. Energy-aware routing was compared against directed diffusion routing. Both routing protocols used the same energy metrics for path selection; this was the metric function with $\alpha = 1$ and $\beta = 50$.

The MAC layer was abstracted away by providing for direct transfer of packets from the network layer of one node to the network layer of its neighbor. Thus, there was no contention for the medium when sending data. Fading effects were not considered either. Transmissions were always successful as long as a node was within the transmission range of the transmitter. The main purpose of removing the MAC was to orthogonalize the advantages of the network and media access layers and to evaluate the benefits of each separately.

Beyond its standard task of injecting, extracting, and forwarding packets, the network layer also maintained the neighbor list. An expanding ring search was used to create the list until it had the minimum number of neighbors (five) or the maximum radio range of the node was reached. Every node

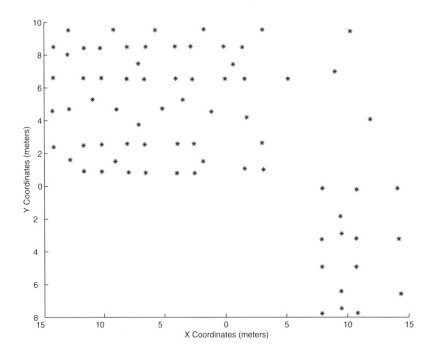

FIGURE 30.1 Layout of static nodes in the network.

was given an identical amount of initial energy at startup. Transmission used 20 nJ/bit + 1 pJ/bit/m^3 (i.e., energy drop off was r^3, which is a moderate indoor environment). The energy for reception was 30 nJ/bit. These numbers are typical values for radios of the Bluetooth class. The packets were 256 b in size.

The performance of the protocols can also be checked against the optimal routing scheme using the linear programming formulation. This gives an idea of the maximum network lifetime if it were possible to use a centralized approach.

Figure 30.2 shows the results of one of the simulation runs. It shows the energy consumed by the various nodes during a 1-h period of the network using energy-aware routing. This can be compared against the energy consumed by the directed diffusion routing protocol in Figure 30.3. As expected, energy-aware routing spreads the traffic over the network, resulting in a much "cooler" network. As a consequence, the nodes in the center of the network conserve energy longer and the time until the first node runs out of energy increases.

The simulations show that energy-aware routing reduces the average energy consumption per node from 14.99 to 11.76 mJ, an improvement of 21.5% (Table 30.1). This is primarily due to the very low overhead of the protocol. At the same time, it reduces the energy differences between different nodes. Figure 30.4 shows energy consumption for the linear programming formulation. In such an optimal scenario, the controller nodes clearly are the bottleneck because they must process all the data packets traversing the network.

In another performance run, the network was simulated till a node ran out of energy. For diffusion routing, this occurred after 150 min; it took 216 min for the energy-aware routed network to fail. This is an increase in network lifetime of 44%, which agrees with the results of the previous simulation. In that simulation, the maximum energy use among all nodes was 57.44 mJ for diffusion and 41.11 mJ for energy-aware routing. This means that diffusion had a maximum energy consumption of 1.4 times energy-aware routing and thus an increase of 40% in the network lifetime is expected. Table 30.1 shows some of the statistics of energy consumption across nodes in the network for the three routing schemes.

It is worth noting that for linear program routing, the time to failure was 234 min, while the maximum energy use was 38.46 mJ, which was very close to energy-aware routing. Thus, this shows that energy-aware routing avoids heavily depleted nodes and performs very well compared to an optimal routing

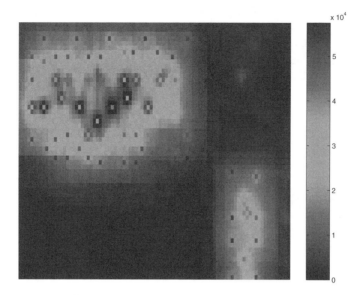

FIGURE 30.2 Energy consumption for energy-aware routing (µJ).

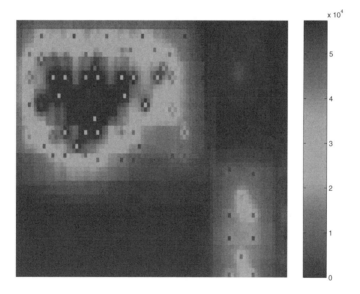

FIGURE 30.3 Energy consumption for diffusion routing (µJ).

TABLE 30.1 Energy Consumption Statistics after 1-h Simulation Time

Energy (mJ)	Avg.	Std. Dev.	Max	Min
Centralized	10.02	8.89	38.46	0.32
Diffusion	14.99	12.28	57.44	0.87
Energy-aware routing	11.76	9.67	41.11	0.98

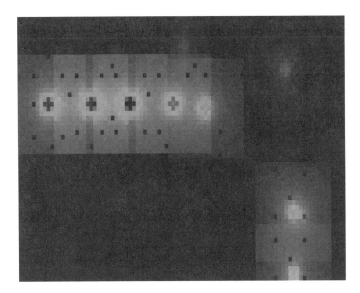

FIGURE 30.4 Energy consumption for linear program-based routing (µJ).

scheme. Also, the bit rate measured by the network is 250 b/sec, which demonstrates the extremely low data rate requirements of sensor networks. Thus, the results clearly show improved network health due to energy-aware routing.

30.7 Data Funneling

Although energy-aware routing tries to route data packets, avoiding regions in the network that are heavily depleted of energy, improvements in network lifetime are possible by reducing traffic volume in the network. To that end, this section discusses data funneling, a protocol that performs data aggregation while routing packets in the network. It is based on the energy-aware routing protocol, but instead of sending multiple packets from all nodes in a region, packets are aggregated along the way to obtain substantial energy savings.

The sensor networks envisioned by PicoRadio consist of a few controller nodes and many sensor nodes that periodically send their readings to the controllers. Because the controller nodes are required to have much greater computational and communication capabilities than the sensor nodes, the cost of controller nodes can be much greater than that of sensor nodes. Also, the controllers must decide what actions to take based upon collated readings from a large region of space. For these reasons, there are many more sensor nodes than controller nodes; each controller receives the readings of many sensors, while each sensor sends its data to only one or two controllers.

Furthermore, the amount of data in each reading is low, at most a few bytes of light, temperature, acoustic, seismic, or other measurements. However, packet headers include training sequences for clock synchronization; framing information; destination address; and error control codes and can be large relative to the packet size. Because many sensors report their data to the controller at approximately the same time and have similar headers, considerable savings can be realized by combining different packets into one large packet with a single header.

The main idea behind the algorithm, called data funneling, is that the controller breaks up the space into different regions (e.g., cuboids) and sends interest packets to each region, as shown in Figure 30.5. Upon receiving the interest packet, each node in the region will start periodically sending its readings back to the controller at an interval specified in the interest packet, usually every few minutes. Because many or all of the nodes within the region will be sending their readings to the controller at the same time, it would be much more efficient to combine these readings into a single packet so that only one

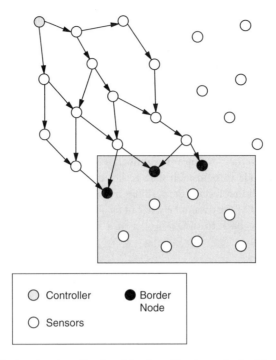

FIGURE 30.5 Control node, target region, directional flood, and border nodes in data funneling.

packet with only one header travels from the region to the controller. The question is how all these readings can be collected at a single point and combined into a single packet.

The data funneling algorithm works as follows. The interest packets are sent toward the region using directional flooding. Each node that receives the interest packet checks whether it is in the target region. If it is not, it computes its cost for communicating back to the controller, updates the cost field within the interest packet, and sends it on toward the specified region. This is the directional flooding phase.

When a node in the target region receives the interest packet from a neighbor node outside the target region, the directional flooding phase concludes. The node realizes that it is on the border of the region and designates itself to be a border node as shown in Figure 30.5. Each border node computes its cost for communicating with the controller in the same manner as was done by the nodes outside the region during the directional flooding phase. It then floods the entire region with a modified version of the interest packet. The "cost to reach the controller" field is reset to zero and becomes the "cost to reach the border node field." Within the region, each node only keeps track of its cost for communicating with the border node, not its cost for communicating with the controller. Intuitively, it is as if the border node becomes the controller of the specified region. At one of the border nodes, all the readings from within the region will be collated into a single packet.

In addition, two new fields are added to the modified interest packet. One keeps track of the number of hops that have been traversed between the border node and the node currently processing the packet. The other field specifies the border node's cost for communicating with the controller; this field, once defined by the border node, does not change as the packet travels from one node to another.

Once the nodes within the region receive the modified interest packet from the border nodes, they will then route their readings to the controller via each of the border nodes in turn. Several border nodes are within the region, so maximizing aggregation of sensor readings requires all the nodes within the region to agree to route their data via the same border node during every given round of reporting back to the controller. This is accomplished by having every node compute an identical schedule of which border node to use during each round of reporting. This is achieved by each node in the region applying the same deterministic function to the vector of costs to reach the controller seen by each border node.

Because all the nodes apply the same function to the same inputs, they will all compute the same schedule, allowing them to collect all of their data at one border node during each round of reporting. The function used to compute the schedule can be the function used to compute the probabilities for selecting different paths in energy-aware routing. This allows border nodes with a low cost for communicating to the controller to be used more frequently than the ones with a high cost.

As data flow within the region from the sensors to the border nodes, the packets can be aggregated along the way as shown in Figure 30.6. When the time comes to send a new round of observations back to the controller, the sensor nodes do not immediately start sending their packets. Instead, before sending their readings toward the border node to be used in that round of reporting, they wait an amount of time inversely proportional to their distance (in number of hops) to that border node. This allows nodes far away from the border node to send their data earlier than nodes closer to the border node. Thus, nodes close to the border will first receive the readings from upstream nodes and bundle those readings with their own. In the end, all of the data to be sent out by all the nodes within the region will be collated at one border node and sent back to the controller in a single packet, as shown in Figure 30.6. The protocol is summarized next.

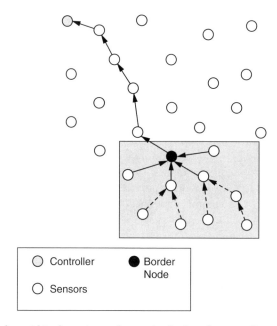

FIGURE 30.6 Funneling data within the region and reporting back to the controller.

30.7.1 Setup Phase

1. The controller divides the area it wishes to monitor into cuboids.
2. It then initiates a directional flood toward each region.
3. Each intermediate node records the cost of reaching the controller as in the routing scheme described previously.
4. When the packet reaches the region, the first node in the region that receives the packet designates itself as a border node.
5. The border node adds two new fields to the packet: the cost for reaching the border node and the number of hops to the border node. It then floods the region with this modified packet.
6. All nodes within the region receive packets from all the border nodes. Based on the energy required by each border node to reach the controller, they compute a schedule of border nodes at which data are to be aggregated.

30.7.2 Data Communication Phase

1. When a sensor has a data sample that it needs to send back to the controller, it uses the schedule to figure out the border node to use in the current round of reporting.
2. It then waits for a time inversely proportional to the number of hops from the border node before sending out the packet.
3. Along the way to the border node, the data packets are joined together until they reach the border node.
4. The border node collects all the packets and then sends one packet with all the data back to the controller, using probabilistic routing.

Data funneling creates clusters within the sensor network, but does so in a fluid fashion, which makes the approach a lot less brittle. There is no single cluster head whose failure can be devastating to the functionality of the network. Instead, the border nodes take turns acting as the cluster head, spreading out the responsibility and the load (i.e., energy consumption) among them. Also, the controller can redefine the regions into which its area of interest is divided, thereby forcing the nodes to divide into new clusters and elect new sets of border nodes. The controller can redefine the regions based on the data received from the nodes and/or the energy remaining in the nodes so as to ensure that nodes with the greatest energy reserves act as border nodes.

To demonstrate its feasibility, the data funneling algorithm was implemented in the Opnet network simulator. The simulation was performed only for the network layer; lower layers were abstracted away. Energy consumed at all the nodes for transmission, reception, and computation was measured. One sample topology is shown in Figure 30.7. The controller node, shown within the dark circle, queried a region, shown as the large rectangle, containing 15 sensor nodes. Copies of the interest packet propagated toward the region, and the four nodes shown within the squares were determined to be the border nodes. Each of the sensors sent its readings to the controller every 10 sec, and the packets were aggregated along the way. The simulation measured the number of sensor readings contained within each transmitted packet.

For the topology shown in Figure 30.7, the average number of sensor readings per transmitted packet was seven. This means that the energy expended by the network on transmitting packet headers was reduced by 86%. In general, the larger the region and the further away it is from the controller, the greater the savings are due to funneling. If n sensors are in the region and the region is far away from the controller, the energy spent on transmitting headers will be reduced by a factor of approximately n.

Let γ be the ratio of bits in a packet header to the total number of bits in a packet containing the header and a single sensor reading for a particular application; let m be the average number of sensor readings per transmitted packet when data funneling is employed. Then, the total energy expended by the network on communication is reduced by $\gamma \cdot \frac{m-1}{m} \cdot 100$ % due to data funneling if no compression of the sensor readings is done at the aggregation points. Performing compression on the sensor readings at the aggregation points within a region would result in even greater energy savings.

30.8 Conclusion

This chapter discussed the concept of the lifetime of sensor networks and presented two approaches to extend it. The first protocol, energy-aware routing, uses a very simple energy metric based on communication energy and residual battery lifetime at the nodes to figure out a set of good paths. These paths are then chosen probabilistically to route packets so that no single path is depleted. This leads to more uniform depletion of nodes across the network. Currently, the choice of the energy metric is based on heuristics and previous research; further research into the energy metric used may lead to better results.

However, upon implementing the protocol on a real-life test bed, it was found that, although the network resources depleted uniformly as expected, in some situations the link with a particular node

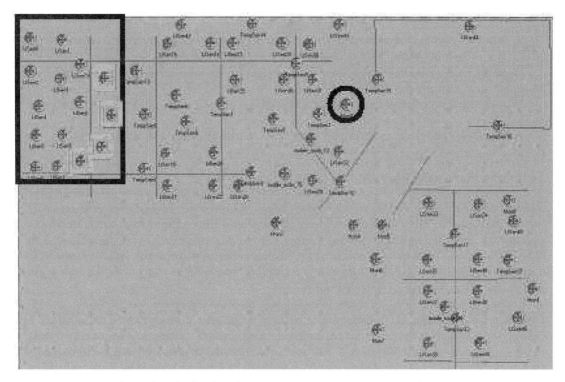

FIGURE 30.7 Sample topology for data funneling.

was bad. If, unfortunately, the node selected that link as the next hop, it would take a while for the transmission to be successful. Although this is inevitable in a wireless scenario, it still might be possible to exploit the fact that other nodes may have a good link at the same point in time. In other words, using current knowledge of the link state with nodes, either as part of the energy metric or otherwise, might lead to improved performance.

The second protocol presented uses energy-aware routing as the primary routing mechanism while performing aggregation of data. Combining data from a set of sensors in the same area can lead to substantial savings as the number of packet headers reduce. Data funneling achieves this in a completely distributed fashion and without the need for any local coordination. The algorithm may also be used in conjunction with other routing protocols and can be combined with other source-coding techniques to reduce network energy consumption further.

Acknowledgments

This work was supported by DARPA as a part of the PAC/C program and NSF as a part of the CITRIS project. Their support is greatly appreciated. The support of the Berkeley Wireless Research Center member companies is kindly appreciated also.

References

1. Shah R.C. and Rabaey J.M., Energy-aware routing for low energy ad hoc sensor networks, *IEEE WCNC*, 1, 350–355, 2002.
2. Intanagonwiwat C., Govindan R., and Estrin D., Directed diffusion: a scalable and robust communication paradigm for sensor networks, *IEEE/ACM Mobicom*, 2000, 56–67.
3. Singh S., Woo M., and Raghavendra C.S., Power-aware routing in mobile ad hoc networks, *IEEE/ACM Mobicom*, 1998, 181–190.

4. Toh C.K., Maximum battery life routing to support ubiquitous mobile computing in wireless ad hoc networks, *IEEE Commun. Mag.*, June 2001, 138–147.
5. Jain R., Puri A., and Sengupta R., Geographical routing for wireless ad hoc networks using partial information, *IEEE Personal Commun. Mag.*, Feb. 2001, 48–57.
6. Royer E. and Toh C.K., A review of current routing protocols for ad hoc mobile wireless networks, *IEEE Personal Commun. Mag.*, April 1999.
7. Perkins C.E. and Bhagwat P., Highly dynamic destination sequenced distance vector routing (DSDV) for mobile computers, *Comp. Commun. Rev.*, Oct. 1994, 234–244.
8. Perkins C.E. and Royer E., Ad hoc on demand distance vector routing, *Proc. 2nd IEEE Workshop Mobile Comp. Syst. Apps.*, Feb. 1999, 90–100.
9. Johnson D.B. and Maltz D.A., Dynamic source routing in ad hoc wireless networks, in *Mobile Computing*, Kluwer, 1996, 153–181.
10. Chang J. and Tassiulas L., Energy conserving routing in wireless ad hoc networks, *IEEE Infocom*, 2000, 22–31.
11. Rabaey J.M. et al., PicoRadio supports ad hoc ultra-low power wireless networking, *IEEE Computer Mag.*, July 2000, 42–48.
12. Akyildiz I., Wireless sensor networks: a survey, *Computer Networks*, 38(4), 2002, 393–422.
13. De S., Qiao C., and Wu H., Meshed multipath routing: an efficient strategy in sensor networks, *IEEE WCNC*, 2003.
14. Petrovic D., Shah R.C., Ramchandran K., and Rabaey J.M., Data funneling: routing with aggregation and compression for wireless sensor networks, *IEEE SNPA*, 2003.
15. Savarese C., Langendoen K., and Rabaey J.M., Robust positioning algorithms for distributed ad hoc wireless sensor networks, *Proc. 2002 USENIX Annu. Tech. Conf.*
16. Bhardwaj M. and Chandrakasan A., Bounding the lifetime of sensor networks via optimal role assignments, *IEEE Infocom*, 2002, 1587–1596.
17. Karp B. and Kung H.T., GPSR: greedy perimeter stateless routing for wireless networks, *IEEE/ACM Mobicom*, 2000, 243–254.
18. Stojmenovic I., Position-based routing in ad hoc networks, *IEEE Commun. Mag.*, 40(7), 128–134, July 2002.
19. Ko Y. and Vaidya N., Location-aided routing (LAR) in mobile ad hoc networks, *IEEE/ACM Mobicom*, 1998, 66–75.
20. Chou J., Petrovic D., and Ramchandran K., A distributed and adaptive signal processing approach to reducing energy consumption in sensor networks, *IEEE Infocom*, 2003.

VIII

Security, Reliability, and Fault Tolerance

31

Security and Privacy Protection in Wireless Sensor Networks

Sasha Slijepcevic
University of California at Los Angeles

Jennifer L. Wong
University of California at Los Angeles

Miodrag Potkonjak
University of California at Los Angeles

31.1 Introduction 31-1
31.2 Unique Security Challenges in Sensor Networks and Enabling Mechanisms 31-2
   Security-Related Properties • System-Level Security • Mobile Code • Metering
31.3 Security Architectures 31-4
   Cell-Based WSNs • Ad Hoc Sensor Networks
31.4 Privacy Protection 31-11
   Principle of Minimal Generalization • Privacy of Location Information
31.5 Conclusion 31-15

31.1 Introduction

Security and privacy protection are of extreme importance for many of the proposed applications of wireless sensor networks (WSNs). The list of potential applications that require protection mechanisms includes early target tracking and monitoring on a battlefield; law enforcement applications; automotive telemetric applications; room occupation monitoring in office buildings; measuring temperature and pressure in oil pipelines [1]; and forest fire detection. All these applications have unlimited benefits and potential; however, if the sensor information is not protected properly, possible compromises in user information, the environment, and even physical actuators could result.

The primary driving impetus for the development of sensor networks has been military applications, where security requirements are at their highest [2]. Although a WSN deployed on a battlefield can offer a reliable assessment of battlefield conditions without risking lives, an inadequately protected network could become a powerful weapon for an enemy. Strong security requirements for such applications are often combined with an inhospitable and physically unprotected environment. For commercial applications of WSNs, the issue of privacy protection is as important as secure and reliable functioning of a network. The protection of personal physiological and psychological information is expected by any user. As the applications of WSNs become more complex and widespread, the ability to protect such systems from any unauthorized access will become increasingly important.

Sensor networks operate in a variety of physical environments and under varieties of constraints. The limited resources of sensor nodes require the development of customized system architectures for each particular WSN application so that the sensor node resources are efficiently used. Because security and privacy protection mechanisms require a significant amount of computational and storage resources, such mechanisms must be tailored to the corresponding sensor system architectures and security threats specific to a given physical environment. Section 31.2 describes the unique properties of WSNs and the security challenges that they bring. Security implications and corresponding security solutions for two basic WSN system architectures, cell-based WSNs and ad hoc WSNs, are discussed in Section 31.3. The overview of the privacy protection solutions proposed for WSN, as well as the solutions originally developed for other environments but applicable in WSN, is given in Section 31.4. Section 31.5 summarizes and concludes the chapter.

31.2 Unique Security Challenges in Sensor Networks and Enabling Mechanisms

WSNs share several important properties with traditional wireless networks, most notably with mobile ad hoc networks. Both types of networks rely on wireless communication, ad hoc network deployment and setup, and constant changes in the network topology. Many security solutions proposed for wireless networks can be applied in WSNs; however, several unique characteristics of WSNs require new security mechanisms. In this section, four characteristics specific to WSNs and their resulting security challenges are discussed. Additionally, it presents work performed in the areas, system-level security, mobile code, and metering, which can be foundations for the development of security techniques in WSNs.

31.2.1 Security-Related Properties

Four properties that are specific for WSNs and require attention are hostile environment, limited resources, in-network processing, and application-specific architectures.

- *Hostile environment.* WSNs can be deployed in hostile environments such as battlefields. In these cases, the nodes cannot be protected from physical attacks. Security information potentially could be collected from compromised nodes. The development of tamper-proof nodes is one approach to security in hostile environments. However, as shown in Anderson and Kuhn [3], the development of such systems is far from simple and certainly not cheap in terms of computational and memory requirements. Because of the physical accessibility of sensor nodes, the security mechanisms for WSNs are specifically concerned with situations in which one or more nodes are compromised.
- *Limited resources.* Sensor network nodes are designed to be compact and therefore are limited by size, energy, computational power, and storage. The limited resources limit the types of security algorithms and protocols that can be implemented. Security solutions for WSNs operate in a solution space defined by the trade-off between resources spent on security and the achieved protection. Limited energy available to nodes allows for new types of attacks, such as a sleep deprivation torture attack [4].
- *In-network processing.* Communication between the nodes in a WSN consumes most of the available energy, much less than sensing and computation do. For that reason, WSNs perform localized processing [5] and data aggregation [6]. An optimal security architecture for this type of communication is one in which a group key is shared among the nodes in an immediate neighborhood. However, in an environment in which the nodes can be captured, the confidentiality offered by the shared symmetric keys is easily compromised.
- *Application-specific architectures.* As a result of the previously mentioned properties, WSN system architectures must be designed to be application specific. The flexibility of a general-purpose architecture is traded for the efficient utilization of the resources. Almost every aspect of a WSN can be adjusted to improve performance and optimize resource consumption in a network for a particular application. This allows a network designer to determine the importance of various security threats and adjust security mechanisms to these threats.

31.2.2 System-Level Security

Three types of cryptographic tools have been developed for practical security of real-life systems: firewalls, honeypots, and intrusion detection techniques. Each will be discussed briefly in order to illustrate the types of approaches that exist in the field. Although these techniques may not be well suited for WSNs as proposed, modifications of their notions may be excellent security approaches for WSN.

A firewall is a policy enforcement point (node) for a part of a network designed to restrict access from and to that subnetwork. Several classes of firewalls exist: packet filtering according to a particular set of rules; access to particular servers or ports; or application-level firewalls that protect by remembering the state of the network connection. Firewalls still face denial of service (DoS) attacks and they try to address them by filtering suspicious connections. Among the several limitations of firewalls is the fact that they do not protect the network from insider attacks and that filtering can only be done against already known attacks.

Honeypots are systems placed on networks specifically for the purpose of being attacked or compromised [7, 8]. Because they are not designed for true use, they exist only to detect and collect information about security attacks. Advantages of honeypots include low false positives; ability to capture unknown attacks; and ability to facilitate interaction with the attacker in order to gain better insights into actions and thinking. Intrusion detection techniques aim at recognizing statistical or pattern irregularities in the incoming or outgoing traffic. The most recent approach to detection of Internet attacks is probabilistic deduction of the IP traceback [9–12]. Finally, virtual private networks are logical extensions of private networks over insecure channels provided by the Internet.

31.2.3 Mobile Code

Once deployed, access to the nodes in a WSN for management and code updates poses security threats and drains resources. Despite difficulties, mechanisms that allow changes in application and system code on the nodes are necessary. One feasible solution for remote configuration and application code updates is network-wide deployment of mobile code. A legitimate mobile code is injected into the network through several nodes and then spread throughout the network [13]. This subsection surveys proposed code manipulation approaches (attacks) and techniques for secure execution of mobile code. Among mobile code intrusion techniques, four have been most popular: viruses, Trojan horses, buffer overflow, and covert communication channels.

A computer virus [14] can be defined as a small program that attaches to the host computer and co-opts its resources for the purpose of creating new copies of the virus. Detailed analysis of viruses and models of their proliferation can be found in Cohen [15] and Kephart and White [16]. Trojan horses [17, 18] disguise themselves as programs that appear to perform a function while actually performing another function. Buffer overflow has been by far the most common type of attack of computer security in the last decade [19]. These attacks use the functions of a privileged program in such a way that the attacker can take control of the program and corrupt the computer. This is commonly achieved by making suitable code available in the program address space and then inducing a program to jump to that space with suitable parameters. Recently, the first constraint-based analysis technique for automated detection of buffer overflow has been proposed [20].

Covert communication channels [21] arise from resource sharing in computer systems. For example, a process with high priority can pass information to a process with low priority by interfering or refraining from interfering with the timing of the process. The most popular and simplest is the timed Z-channel, in which the communication alphabet consists of time values [22]. Numerous generalization of the timed Z-channel have been proposed and analyzed [23–25].

Smaller mobile devices have created a strong impetus for the development of mobile code security techniques. At least three major approaches for mobile code security have emerged: code signing, sandboxes, and proof-carrying code. Code signing follows a typical client- and server-authenticated handshake protocol such as SSL or WTLS [26]. Recently, sandboxing has attracted a great deal of attention [27] as a security paradigm; Brigner [28] presented a 3-MB Java applet that implements a sandbox. In addition, Sekar and Uppuluri developed a security layer that includes a sandbox designed to protect the application against malicious users and the host from malicious applications [29]. Proof-carrying code is a mechanism that allows a host computer to determine if a program can be executed with certainty despite being provided by an untrusted source [30, 31].

31.2.4 Metering

One aspect of WSN security threat that is not often addressed as an attack is consumer access to the sensor data. As WSNs become more advanced and versatile, the notions of user access, application-specific sensor designs, and licensing of network usage will become an issue. Metering is one approach to handling these types of issues. Although many of these approaches are too computationally or memory intensive for WSNs, they provide a starting point for development of WSN techniques.

SiidTech Inc., an Oregon startup company, has proposed an approach for integrated circuit identification from random threshold mismatches in an array of addressable MOSFETs. The technique leverages on process discrepancies unavoidably formed during fabrication. This analog technique can be used in tracking semiconductor dies, authentication, and intellectual property (IP) tagging [32]. Sampling and auditing are the two main methods used for measuring the usage of media channels. Sampling conducted by Nielsen Media Research and NetRatings Inc. is based on surveys among a representative group of users [33]. Web page access metering has been addressed by a number of researchers and companies [34–36].

Licensing is the most common approach to protecting software. It provides a certain degree of control to the vendor in terms of software distribution and may prevent unauthorized duplication of software packages. The most common technique is based on the license key concept. A key is encrypted by using a string of data that contain software package ID and its usage constraints (e.g., expiration date) and the serial number of the computer where the key is installed. The invocation of the software package is done automatically when software is invoked by using one of the password schemes [37, 38]. A large number of patented licensing protocols have been proposed; for example, licenses can be used to authenticate the legal users, as well as to upgrade the products and other after-market information transmissions [39] or licensing using smart cards [40, 41].

31.3 Security Architectures

This section describes security protocols developed for two typical WSN system architectures: cell-based WSN and ad hoc WSN, with particular concentration on key establishment and distribution algorithms because they set up the necessary infrastructure for security protocols. The proposed WSN system architectures differ in many aspects, which is not surprising because WSNs operate in vastly different physical environments, supporting different applications and using different sensor nodes. The main benefit of the development of a specific architecture for each WSN application is the efficient utilization of scarce sensor node resources.

Many elements of WSN architecture, including hardware architectures of sensor nodes, routing protocols, and level of abstraction between the layers of the architecture, can be adjusted to improve performance and optimize resource consumption. One possible categorization of wireless ad hoc network systems, which includes WSN systems, is given in Law et al. [42]. Here, only the security architectures for WSN systems are discussed, while Papadimitratos and Haas [43] give an overview of security architectures for general wireless ad hoc network architectures. From the security point of view, the WSN system architectures can be broadly divided in two categories:

- Cell-based WSNs consisting of low-power low-cost sensor nodes and base stations, operating in relatively friendly environments of houses and office buildings, or in easily accessible outdoor areas
- Ad hoc WSNs consisting only of low-cost sensor nodes distributed in an ad hoc manner into remote and inhospitable environments without any wireless infrastructure

These two network architectures differ in terms of the security threats to which they are exposed and in terms of security requirements and abilities to support security architectures of various levels of complexity. The cell-based WSN allows for more sophisticated and resource-consuming protocols and algorithms because the additional computationally expensive workload can be assigned to the base stations.

31.3.1 Cell-Based WSNs

In cell-based WSN, the nodes are organized around one or more base stations that have significantly more computing and energy resources than the regular sensor nodes. These networks are most often used for user and object tracking systems in home and commercial building environments, as well as in outdoor perimeter-monitoring systems. The base stations collect information from the network and provide a link between the WSN and the outside world. Cell-based networks are often used in an environment in which it is easy to add new nodes, remove the ones that are not functioning, and even recharge the energy supplies for nodes. However, even in such an environment, the nodes can still be captured or damaged, and unauthorized nodes can be added.

The presence of base stations in a WSN offers at least two significant benefits:

- Base stations represent a trusted base that cannot be compromised. They can be used as a safe source of mobile code and configuration parameters, which enables safe bootstrapping and configuration of the network, as well as the addition of new nodes.
- Base stations offer computational resources that can be used in asymmetric security protocols in which they perform the majority of intensive computations. Such protocols allow stronger security, while not exhausting the limited resources of regular sensor nodes.

An example of a WSN organized around one or more base stations and SPINS, the security protocol suite for that network, is described by Perrig and colleagues [44]. The network consists of a trusted backbone of base stations with unlimited power supply and a large number of *motes* (low-cost, low-power sensor nodes described by Hill and colleagues [45]), distributed in the area covered by the base stations. The operation of the network is fully controlled from the base stations. A routing structure is formed as a set of routing trees; each base station is the root of one such tree. The traffic mainly consists of requests initiated at the base stations and sent down the trees to the nodes and the responses sent from the nodes back to the base stations. When the same request is sent to all nodes, the communication is most efficiently performed through broadcast messages. If a base station needs to send a unicast message to a particular node, source routing is used.

The SPINS protocol suite assumes that the base stations share a unique master key with each node in the network. The system architecture and security protocols require that the base station keep track of the route to each node and of the secret key. All other keys that the base station and a node use for communication are derived from the master key. Even though the base station is a single point of failure, it is trusted, implying no one can capture the station and recover all keys.

This security architecture efficiently uses the resources of the base stations. To keep a separate key for each node would not be possible in an architecture in which all nodes have limited resources. Also, this solution is not applicable to networks in which any two nodes are likely to communicate directly. However, because the bulk of traffic in the network is between the base station and the nodes, the inability of the nodes to communicate securely without involvement of the base station is of limited importance.

The SPINS protocols suite consists of two building blocks, sensor network encryption protocol (SNEP) and μTESLA. SNEP protects the unicast communication between the base stations and the nodes, while μTESLA provides secure broadcast communication. Each of these protocols will be discussed in more detail.

31.3.1.1 SNEP

The basic confidentiality of messages in any secure system is achieved through encryption. Encryption protects the network from adversaries who have the capability to listen to network traffic. SNEP uses

RC5 block cipher [46] for basic encryption. The original RC5 encryption algorithm is implemented with lowered functionality and generality in order to fit in the limited storage space of nodes. In addition to basic confidentiality, SNEP offers *semantic security*, which means that the encryption of the same plaintext produces a different encrypted message each time. This is achieved by keeping a shared counter on each of the two entities involved in the message exchange and incrementing the counter for each message.

Because the value of the counter is an initialization vector for the RC5 block cipher, it is guaranteed that the encrypted messages differ even if the content is the same. An additional benefit of the counters is that they ensure *freshness* of messages, i.e., a receiver can establish the partial message ordering of the messages from a particular sender. Finally, each node has its separate master key, so SNEP guarantees authentication of messages that the nodes receive from the base station.

An important property of such a solution is that it can be used in environments with relatively static forwarding structure, in which the nodes communicate with a limited number of other nodes or base stations, usually smaller than the number of neighbors. The number of the keys and counters can be estimated, and the efficiency of such a solution is known in advance. In a network with a dynamic forwarding structure in which any neighbor can be a previous or a next hop for any message, it may be prohibitively expensive to keep counters and separate keys for all possible sources and destinations. However, for a limited number of cases, two nodes that need to communicate directly can use their master keys to generate and exchange a session key through the base station.

31.3.1.2 μTESLA

The second element of the SPINS protocol suit is μTESLA. The master key shared between each node and the base station ensures confidentiality and authentication of unicast messages exchanged between the nodes and the base station. However, if the same message is sent from a base station to all nodes, it is much more efficient to broadcast the message. SNEP does not support secure broadcast because each master key is unique; allowing nodes to accept unencrypted, unauthenticated messages would allow an adversary to send arbitrary requests to nodes. Therefore, for secure broadcast communication, SPINS proposes μTESLA, the goal of which is to ensure authentication of broadcast messages sent from the base stations to the nodes.

In μTESLA, a base station generates a reverse key chain containing the keys $K_0, K_1, \ldots, K_n$. The key chain length and the key K_n are determined before the key chain is generated. Other keys are determined using one-way function F, $K_i = F(K_{i+1})$. The key K_0 is not used to authenticate any of the messages, but is distributed initially as a commitment to the key chain. The distribution of the commitment K_0 in μTESLA requires that each node and the corresponding base station share a secret key unique for that node. Then, the base station sends K_0 to all nodes as a sequence of unicast messages, before any broadcast message is transmitted.

The time is divided into the intervals $I_1, \ldots, I_n$, as shown in Figure 31.1, where each interval I_i corresponds to the key K_i. During the interval I_1, the base station sends broadcast messages with attached message authentication code (MAC) calculated using the key K_1. Because the key K_1 has not been disclosed yet, the messages could not have been forged by any of the nodes. The function F is a one-way function, so no one can determine K_1 from K_0. The nodes authenticate the messages received during the interval I_1 at the end of that interval, when the key K_1 is disclosed. At that time, the nodes compare K_0 with the value derived from $F(K_1)$. If the values match, then the messages authenticated with K_1 are sent from the base station, because only the base station could have known the value of K_1 before that key was disclosed. After K_1 is disclosed, the following broadcast messages are authenticated, using K_2, until K_2 is disclosed, and the process continues until the interval I_n expires.

Because the keys are disclosed in periodic intervals, the base stations and nodes must be at least loosely synchronized. If a node does not receive a message with a disclosed key and its clock is late, an adversary who received the disclosed key can forge and send messages with the MAC calculated using that key. A node with an unsynchronized clock would accept such messages for the interval equal to the delay of the node's clock.

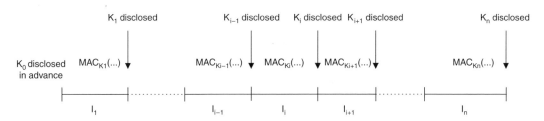

FIGURE 31.1 The key distribution timeline. The messages sent during the interval I_i have attached MAC calculated using the key K_i. The key K_i is disclosed at the end of the interval I_i.

Even if a node does not receive all the keys, it can still authenticate all messages. Once the node receives the key K_{i+1}, it can derive all previous keys $K_0, \ldots, K_i$ by successively applying the function F, and then use these keys to authenticate messages received in the corresponding intervals. However, buffering the messages for a prolonged time requires additional storage. Nodes have limited available memory, so additional mechanisms are needed to ensure that the keys are disclosed to all nodes in a timely manner.

If some nodes are deployed later and begin receiving messages during the interval I_i, they need to receive the key K_i at the end of the interval. The initialization process for those nodes is the same as for the nodes initialized when the whole network is bootstrapped; the only difference is that instead of the commitment K_0, the nodes receive the key K_{i-1} as a commitment. By doing so, new nodes save a certain amount of computation because they do not need to compute all keys from the chain $K_0, \ldots, K_{i-1}$ to compare the value of the key K_0 with the value of $F^i(K_i)$; they would need to do this if they received K_0 as a commitment. Instead, the nodes simply compare the commitment K_{i-1} with the value of $F(K_i)$.

The initial distribution of the key chain commitments requires that a base station send a separate unicast message to each node. In addition to the energy consumption of sending multiple unicast messages with the same information, the time required to initialize thousands of nodes is measured in tens of seconds, as calculated by Liu and Ning [47], who proposed that instead of costly initialization using broadcast messages, the commitment K_0 be embedded into the nodes during initialization, before deployment of the nodes. This solution brings significant savings for the majority of the nodes deployed together. For the nodes added later, there is a trade-off between computation expenses incurred when the new nodes authenticate a broadcast by comparing the embedded commitment K_0 and $F^i(K_i)$ and when the added nodes receive unicasts containing K_{i-1} and authenticating broadcast with only one calculation of the function $F(K_i)$.

The decision as to which mechanism for initialization of the added nodes is preferred could be potentially based on the number of new nodes that should be initialized. If the number is small, then the delay incurred by sending unicast messages is acceptable; however, if many nodes are added, it is more efficient to have the nodes use K_0 as a commitment. Unfortunately, the decision about the preferred commitment distribution mechanism cannot be made online because delivering the decision to the nodes would require sending authenticated unicast (which is the expense to avoid if the number of new nodes is large) or sending a nonauthenticated broadcast, which then could be a message forged by an adversary.

An implicit assumption in μTESLA is that the base stations have sufficient memory storage to hold a long key chain or that they have enough processing power to compute keys fast enough while keeping in memory only the last member of the chain. That assumption saves the nodes from buffering too many messages because the intervals can be arbitrarily short; the key chain is still long enough not to require frequent costly commitment distributions. Liu and Ning [47] propose a hierarchical organization of the keys that decreases the required memory storage at the base stations. The basic principle behind this solution is that the base stations keep only a high-level key chain in memory, while the elements of the low-level key chains are generated using the high-level keys. The keys $K_0, \ldots, K_n$ from the high-level chain are not used for message authentication. They only authenticate messages containing low-level key chain commitments $K_{<0,0>}, K_{<1,0>}, \ldots, K_{<i,0>}, \ldots, K_{<n,0>}$.

This extension of μTESLA needs to keep the property of the original scheme that even if some key disclosure messages are lost, the later key disclosures can be used to authenticate previous messages. Otherwise, the network would need to ensure that all messages are received at all nodes — an expensive proposition for WSNs. In order to enable authentication despite lost messages, the high-level key K_{i+1} is used to generate the last key for the low-level key chain for the interval I_i, $K_{<i,m>} = F_1(K_{i+1})$. Without this relation, if a message with the commitment $K_{<i,0>}$ is lost, the nodes could not authenticate the messages from the interval I_i. Even in this solution, if the message with the commitment $K_{<i,0>}$ is lost, the nodes must keep all messages from the interval I_i in a buffer until the key K_{i+1}, or some other later key, is disclosed in the interval I_{i+2}. Because the high-level intervals are intentionally kept long so that the number of keys stored in the memory of the base stations is small, the memory required to store the messages may be prohibitively large. One possible solution is to repeat messages frequently that contain key commitments.

Authentication of broadcast messages sent from the base stations to the nodes is supported by μTESLA. It may be possible to use the protocol in cell-based networks in which the nodes send broadcast messages too. Two possible solutions for this problem are: (1) a node sends a unicast to the base station using the key that the node and base station share, and then the base station broadcasts the message using the original μTESLA broadcast authentication mechanism; or (2) a node broadcasts the message and the base station handles the distribution of a key for that broadcast.

31.3.2 Ad Hoc Sensor Networks

Certain military, law enforcement, and disaster recovery WSNs are deployed in remote and inhospitable environments without any wireless infrastructure. Nodes must self-organize and bootstrap a network without any support from base stations. Such networks distributed in an ad hoc fashion must be capable of accepting requests from various points within the network because a user walking through the area may not have the capability to connect to the designated gateway. Any node in such an architecture can be a source of or a destination for messages.

Even more than in other networks, the nodes in such systems are exposed to a danger of capture or destruction. The most dangerous physical threat regarding security is physical possession of a sensor node by an adversary. Sensor nodes may contain keys that allow the adversary to decrypt the messages and even to inject false messages into the network. In circumstances in which long-term security of all nodes in a network cannot be guaranteed, the best solution is to extend the lifetime of the network as much as possible. There are two aspects of extending the lifetime of a network:

- The time period from when the network is deployed to the moment a node is compromised should be as long as possible. An adversary can determine the positions of nodes using various technologies. The easiest way is to listen to the messages exchanged between the nodes because they usually contain the locations of nodes that detected an event. In Slijepcevic et al. [48], messages are encrypted with a separate encryption algorithm for locations of nodes; this is stronger than the encryption for the rest of the message content so that the adversary has less encrypted text for cryptanalysis. If the information about the locations of nodes is adequately protected, the adversary is left using trilateration, which requires more equipment and effort than simply extracting the locations of nodes from the messages.
- Once some of the nodes are detected, the keys that these nodes contain can be extracted and used to decrypt previously exchanged messages as well as future ones. Key distribution mechanisms in WSN and secure protocols must be designed so that the security exposure is minimized when any of the cryptographic keys is compromised.

In a system architecture in which all nodes are potential senders or receivers, symmetric cryptography suits the low-power nodes better than public cryptography. Because symmetric cryptography assumes that the keys are shared, the design space between two extreme solutions remains: (1) all nodes share only one key embedded in them before the deployment; and (2) each pair of nodes shares a unique key.

The first solution is simple, does not require too much memory space, and has the broadcast primitive available. However, when one node is compromised, the adversary can decrypt all messages from the network. The second solution has a perfect security property: if a node is compromised, the recovered keys are useless because no other nodes use those keys. However, the memory space for all keys for networks of thousands of nodes is not available on most sensor node platforms. Even if only a handful out of thousands of keys is actually used when a network is deployed, the nodes must store them all because their exact physical locations are not known before the deployment, and they cannot know which nodes will be located close to each other.

Additionally, sending broadcast messages is not possible, so each broadcast message must be replaced with multiple unicast messages, and the energy consumption is multiplied accordingly. The key distribution algorithms proposed for WSNs try to find a trade-off among the various requirements. The important factors for the key distribution algorithms are:

- Impact of compromise of one or more nodes on security of the traffic in the network
- Ability of algorithm to include additionally deployed nodes into the security infrastructure
- No single point of failure
- Spatial and temporal variation in keys to reduce encrypted material for cryptanalysis
- Support for broadcast

31.3.2.1 Key Distribution Schemes

Most key distribution mechanisms shy away from key distribution after the nodes are deployed. Such schemes exist for various wired and wireless networks and they mainly include key distribution servers. They consider self-organized wireless network with no security infrastructure. Therefore, no central authority, no centralized trusted party, and no other centralized security service provider exist. The standard solutions for authenticated broadcast are not applicable. The solution used in wireless networks with more capable nodes [49, 50] employs public key cryptography to ensure authenticity of messages. However, in many WSN projects [44, 48, 51], the public key algorithms are considered too expensive in terms of memory and processing requirements to be used in WSNs, except as a one-time protocol for exchange of private keys. A thorough discussion about the energy requirements of the public key encryption algorithms and their performance on various processors is given by Yuan and Qu [52].

A possible solution for key distribution is to assign some nodes to be key distribution servers, delivering symmetric keys to nodes that need to communicate. The use of online key distribution servers in WSN has a disadvantage; because all nodes are physically exposed, key distribution servers would become single points of failure because of failures and especially because they would allow an adversary to get hold of all keys used in networks. Therefore, the key distribution algorithms presented here are based on key assignment before deployment. The addition of new nodes and the loss of previous nodes does not require an immediate key distribution process, as is the case in Internet multicast algorithms in which new nodes must not be able to read previously exchanged messages, and the old nodes must not be able to read future messages. In WSNs, new nodes are trusted and old nodes are most likely out of energy.

Eschenauer and Gligor [51] propose a probabilistic key distribution in which a node shares a key with a certain percentage of other nodes. Before the deployment, an initial pool of P keys is generated. For each node, k keys are selected from the initial pool for a key ring. After the deployment, the nodes announce and compare their key rings, looking for at least one key that belongs to both key rings. If such a key is found, those two nodes can communicate directly. When all such pairs are found, they represent the connectivity graph for the network. Now, even the pairs of neighboring nodes that do not have a direct connection can use an established secure path to generate a key, or pick a key from a set of unused keys from the key rings and exchange that key. Eventually, each pair of neighboring nodes will share the key, under the condition that the network graph was connected initially. If that was not the case, a certain number of nodes are permanently excluded from the network.

The main advantage of this scheme is its resiliency in the case of compromised keys. If a node is captured, all of its k keys are available to the adversary. The probability that a particular key is used for encryption of a link is the same for all keys, so a probability that the adversary can decrypt traffic on a particular link is k/P. As will be explained later, k is significantly smaller than P; thus, that probability is low. The scheme also achieves significant memory savings compared to a scheme in which each pair of nodes shares a key. Furthermore, when new nodes are added to the network, they announce their key rings in the same way as the nodes deployed during the initialization of the network.

Because different keys are used throughout the network, the amount of encrypted material for cryptanalysis is smaller in such a key distribution scheme than in the scheme with a shared key for all nodes. If the lifetime of the network is so long that the keys should be replaced, the nodes can revoke the keys with the expired lifetime. After some keys are revoked, the process of establishing secure connections must be run again, with fewer keys. The removal of keys decreases the probability that the network will be fully connected, so the key revocation has limited usage. Finally, the scheme does not support broadcast. However, after a node establishes secure paths to all its neighbors, it can distribute one of its keys as a broadcast key in the case of increased broadcast traffic. Obviously, the applications running on top of the WSN running this key distribution scheme need a certain amount of control over the deployment of certain mechanisms; such mechanisms are not always needed, but their deployment consumes energy.

The parameters of the scheme are determined as a trade-off between security in the case of compromised nodes and the probability that the network is connected. The parameter k, the size of a key ring, is determined by the size of the memory reserved for the keys. The size of the network and the average number of nodes within a communication range are determined by the network application and topology. The only parameter that can be changed over a large range of values is the size of the initial key pool, P. If P decreases, the probability that two key chains selected from the keys from P have one or more common keys increases. However, the value of the expression k/P, which represents the probability that an adversary can compromise a communication link when a node is compromised, also increases.

A result from graph theory, presented by Spencer [53], determines the probability, p, that an edge is between two vertices in a graph, for which the probability that the graph is connected rises from a small probability to "certainly true." Then, P is determined from the condition that the probability of two key chains with one or more common keys is equal to p. For a network of 10,000 nodes, with $d = 40$ neighbors per node on average and the size of a key ring $k = 15$, if the size of the initial pool is $P = 100,000$, the network is fully connected with the probability .99999.

Chan and colleagues [54] offer two improvements to the described scheme. The first change is that two nodes can establish a secure link only if they share q keys, instead of one as in the original scheme. The advantage of this approach is that, for a small number of captured nodes, the probability that any link in the network can be compromised is lower than if the nodes establish a link with only one shared key. However, with the increased number of captured nodes, the relation between these two probabilities changes, so with a sufficiently high number of captured nodes, an adversary has better chances of compromising the secure links than in the original scheme. This trade-off improves the protection of the network against small-scale attacks, which are easier to execute and therefore more likely, and decreases the protection against larger attacks, which are more expensive to perform.

The second improvement from Chan and colleagues [54] allows pairs of nodes that have a secure link between each other to establish new keys. That way, more keys are used, while the amount of the memory required to hold the keys is kept low, at the order of magnitude of the number of neighbors. The price paid for this improvement is increased traffic for key exchanges. It is also important to mention that the two schemes proposed here should not be used at the same time. The first scheme requires that the number of keys in the initial pool be kept low to ensure the connectivity of the network. At the same time, the second scheme tries to use different keys; however, during the exchange of the keys the probability of capture of these keys is increased.

31.4 Privacy Protection

The previous sections have examined the security architectures for two broadly defined types of WSN. The main goal of the presented architectures is to establish secure communication channels within a network in order to protect transmitted information from unauthorized access. In many WSNs, especially in military and law enforcement systems, sensor nodes and communication between them are the most exposed part of the network. In such networks, reliable and secure communication is the most important and best guarantee of uninterrupted functionality.

For another class of WSN systems — those intended for use in commercial settings — the privacy protection of individuals observed by a WSN whose living and working spaces are populated by sensor nodes is as important as the protection of applications' functionality. It is still necessary to ensure secure communication channels in commercial WSNs in order to prevent unauthorized access to the personalized information about the users of the system. However, even if the communication security architecture ensures that the personal information is well protected during transfer through the network, once such information is collected at a data collection point, the information is protected as much as the data collection hosting system is. Commercial systems tend to have lower standards for security protection of acquired information than military and law enforcement systems do. The news frequently reports about systems in which system security at data collection points was compromised and social security numbers, credit card numbers, and many other highly sensitive and personal data ended up publicly available to anyone on the Internet. These cases illustrate the need for additional mechanisms that will ensure a certain level of privacy protection without interfering with the functionality of commercial WSN systems.

31.4.1 Principle of Minimal Generalization

Sensor nodes' sensing capabilities, size, and low cost allow a large number of sensor nodes to be deployed in and a large amount of information to be acquired from physical surroundings. Except in rare cases, the larger the amount and the higher the precision of the sensing data available to a WSN, the better the performance of applications is. Although the applications may perform better if more data are available, privacy protection, by definition, strives for the minimum amount of data to be acquired about a single individual. Although the performance of applications and the need for privacy protection may seem to be two opposing goals, many applications can function effectively with their information precision at a lower level than the level of precision that WSNs are capable of delivering. That interval between the required and potential accuracy can be effectively used for privacy protection.

Samarati and Sweeney [55] have proposed a mechanism for generalization of data in databases in order to prevent matching individuals and their medical records. The medical records with the names removed, but with ZIP codes, dates of birth, and other information, are easily matched with the identities acquired from voter lists, city directories, and other publicly available sources. To prevent reidentification, nonessential information is removed, i.e., the year of the birth is kept, while the exact date is removed. The goal of the process is to depersonalize medical records so that multiple identities are equally likely to correspond to a particular medical record. This approach is called the principle of *minimal generalization*. The same principle can be applied in the context of the privacy protection in WSNs. Naturally, this may affect applications that operate on generalized data, so the principle can be applied only if the application can retain the required performance level. An informal description of the principle of minimal generalization is: accurate private information about the users of a system should be generalized so that the acquired data can be matched to no less than k identities, where k is the required level of anonymity.

Under the assumption that a data collection point and a WSN are under separate control, this principle is beneficial for both entities. The WSN offers higher privacy protection for its users, while the data collection system does not need to expend resources for privacy protection purposes and does not need to risk liability for possible breaches of privacy protection.

31.4.2 Privacy of Location Information

This principle is demonstrated on several WSN applications that rely on location information about users. Protection of the location information is highlighted for three main reasons:

- The most frequent tasks for WSN systems are concerned with detection of location of an event. Even if the goal of an application is to perform a more complex task, the location information is present as a part of the individual observations generated by sensor nodes.
- The privacy protection of location information for users observed by a WSN is a prime example of the importance of data protection because, with access to the location data for a user, an adversary can infer additional private information — for example, medical conditions, shopping habits, and patterns of social interactions between monitored users.
- Protection of location information allows the principle of minimal generalization to be demonstrated in an easily understandable case study. Generally, location discovery systems are often capable of locating users within meters indoors and within tens of meters outdoors. For many applications, that level of precision is more than required for basic functionality, so it is acceptable to reduce the precision of the information in order to achieve a required level of privacy protection.

The general system architecture for which privacy protection solutions are described is shown in Figure 31.2. The crucial part of the privacy protection framework for WSNs is the location server. The server is a part of the trusted zone, which in this context means that the server adheres to the same security policies and is controlled by the same entity as the accompanying network of sensor nodes. In fact, the location server is likely to be implemented as a service running on a gateway between the sensor network and the outside world. The assumption that sensor nodes are trusted and that they do not forward any information to an unauthorized party extends here to a location server. The responsibility of the location server is to transform the locations of users observed by the network into a representation that keeps the level of location privacy protection above a certain threshold. The main difference between various privacy protection algorithms is in the types of transformations performed by a location server. The transformed location information is then forwarded to any of the servers offering location-based services (LBS). The services are offered by various entities that do not share security trust with the WSN.

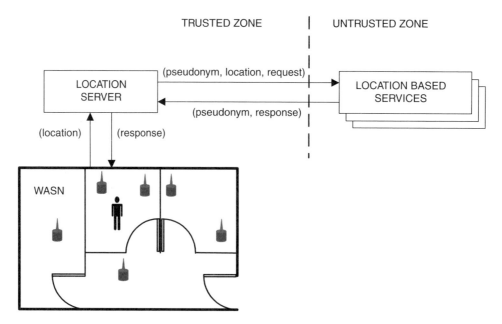

FIGURE 31.2 The architecture of the system connecting a wireless sensor network and a location-based service. The location server transforms location information, so the identities of users are hidden.

Many WSN and wireless network applications may run on top of a system architecture similar to the one shown in Figure 31.2. Two applications, which offer road maps and road condition information based on the location of a car, are proposed in several projects [56] and are commercially available [57–59]. In Beresford and Stajano [60], various proposed applications are implemented on the top of an indoor location system in an office building environment. Gruteser and coworkers examine the privacy concerns of applications that track the use of different building areas [61].

In all these applications, users send information about their locations to an LBS to update their locations or to request services offered in their vicinity. Without transformation performed in a location server, each user request or update would be accompanied with as precise location information as the location discovery technology used allows. In automotive applications, precision is defined by the precision of a GPS receiver, while in an indoor environment location precision depends on the density of the sensor network, usually precise enough to locate a room where a person is correctly located. If the information from these location discovery services is compromised, the location precision allows for easy recovery of users' movements by LBS.

The first step to protect users' privacy is to disconnect the location information from the explicit user identification. The location server performs this task by assigning an alternative identification or a *pseudonym* to each user. Some kind of identification is necessary because a response to each request among a possibly large number of requests handled by a location server must be forwarded to the original user. For many LBSs, it is not necessary for a service to be aware of a user's real identity. A road map can be generated for a user based only on the user's current location. However, two problems occur with privacy protection through anonymity of identifications. The first is the possibility that if a user uses the same pseudonym when connecting to various LBSs, the combined data from all LBSs can give a full overview of that user's activities. That problem can be solved simply by using different pseudonyms for various LBSs, similar to a solution for the same problem outside the context of WSNs, as proposed by Chaum [62].

The second problem is that a user can be easily identified despite different pseudonyms, if requests for LBSs are coming from specific locations that can be directly connected to the user. In the case of a request for a road map that an LBS issued from a location that can be identified as a private garage, the anonymous identification can be attached to the owner of the garage, and then all the movements of that ID can be personalized. In the same way, in an office building environment, an ID that spends most of the time in a particular office can be connected to the regular occupant of that office.

The solution for this problem is in the combination of the principle of minimal generalization and temporary anonymous identifications. Before details of the mechanism are described, it is necessary to include a more formal definition of privacy in order to be able to compare the benefits of different approaches to this problem. The measure of privacy in this context is the notion of *k-anonymity*. The meaning of the term *anonymity* in privacy protection research is formally defined by Pfitzmann and Koehntopp [63]; they define it as a quality of not being identifiable within an anonymity set containing a set of subjects. Then *k*-anonymity, as defined by Samarati and Sweeney [55], is anonymity within a set with the cardinality of k. This term is used to define an acceptable level of privacy protection in which a person cannot be distinguished from $k - 1$ other individuals. For the application using location information, *k*-anonymity means that the attached location information for that user comprises location information for $k - 1$ other users. If a stretch of a freeway of the length d contains k cars, a user whose location is defined with the resolution d is *k*-anonymous.

Gruteser and Grunwald [56] achieve *k*-anonymity by transforming precise, high-resolution GPS–originated location information available to a location server to low-resolution location information sent to the LBS. Additionally, the identity of a user is hidden in order to avoid continuous tracking of his location. If a user's location is defined by the intervals $[x_1,x_2]$ and $[y_1,y_2]$ in two-dimensional space and the time interval $[t_1,t_2]$ in time dimension, *k*-anonymity is achieved by extending and contracting the intervals until $k - 1$ or more objects share the resulting parallelepipe in the three-dimensional space-time coordinate system. Depending on the nature of the application, Gruteser and Grunwald have proposed two different algorithms [56] that transform resolution of location information:

- If the application requires a timely response, the spatial resolution is brought to a level at which k-anonymity is achieved. The implementation of this algorithm starts from the entire area covered by the LBS. The area is then divided into subareas, until the subarea containing the specified user also contains less than $k - 1$ other users. Examples of such applications are road map and road condition LBSs in which the delay must be on the level of minutes; otherwise, the information returned from the LBS cannot be effectively used.
- If a delay is acceptable, the application can set a threshold on the spatial resolution, requiring that the area confined with the intervals $[x_1, x_2]$ and $[y_1, y_2]$ is never above the given threshold. The property of k-anonymity for a user is then achieved by extending the temporal interval $[t_1, t_2]$ by simply waiting until k or more users pass through the space limited by the spatial intervals. Now, the LBS side of the application deals with a more precise location information, which is beneficial for the quality of service that LBSs offer.

In Gruteser and colleagues [61], the underlying application counts the number of people in various parts of a building to estimate the utilization of the rooms in the building. The nodes in the network are organized in a hierarchical structure, with sensor nodes at the lowest level detecting individuals in their vicinity, usually only in one part of a room. The nodes at the next level count the number of individuals in each room, using the sensing data from the nodes from the lowest level. The hierarchy structure assumes nodes at the floor level as the next level and then, finally, a location server that gathers the data from the floor level. Without privacy protection, the information about occupancy of the rooms would be simply transferred up to the location server. The application requirements are such that it can perform its task without identifying individuals occupying rooms. However, similar to the applications mentioned previously, if an adversary can access the data gathered at a data collection point, he can reidentify individuals from the rooms that each of them occupies most frequently. Now, identified individuals can be tracked by observing counterchanges in various parts of the building.

The solution proposed in this work leverages the hierarchical network architecture of the WSN [61]. A node determines a count of individuals in its area, by sensing, if at the lowest level of the system hierarchy, or by aggregating the counts received from the nodes one level below, if at one of the higher levels. The node then compares the count with a threshold value k. If the value is below k, that value is propagated to a higher level with decreased resolution of the location information. If the value is above k, the value sent to the upper level is the nearest multiple of k. In that case, the location information is accurate. Using this algorithm, even a location server does not need to belong to a trusted zone because the obfuscation of the location information is already performed in the network.

The work in Beresford and Stajano [60], in which anonymous users are tracked through a building, notes the same problem with permanent anonymous IDs that can be connected to a particular user based on location information from a private area. However, the applications from that work cannot allow for a location precision coarser than the level of a room, so the increased resolution is not an acceptable solution. On the other hand, at the room level, the location information can easily be used to find out which identifiers spend the most time in a particular room.

These authors propose a solution in which the concept of k-anonymity is used in special areas called *mix zones* where users change their temporary anonymous IDs [60]. The manipulation of identifiers is a responsibility of a location server. However, the authors also note an important weakness of the solution with *mix zones*. They demonstrate that the initial assumption that, if two individuals cannot be tracked when they enter a *mix zone* from opposite directions and then reappear from them with different pseudonyms does not hold well if an adversary uses a statistical analysis. The experimental results show that the probability that each individual will return to the same direction from which he came is 1%, so if an adversary assumes that a user who entered from one direction and the user who left the mix zone at the other end are one person, regardless of IDs, he will be correct in 99% of cases.

There are certainly applications that do not conform completely to the system architecture from Figure 31.2. The possible differences are the expansion of the trusted zone to include an LBS, in which case a location server is not needed. The example of this type of application is provided in Priyantha et al. [64]:

indoor user location detection system. The beacons embedded in the building transmit the information about their locations. A device carried by a user detects the signals from beacons, and then determines the location of a user from multilateration of distances acquired from the signal strengths of beacons' signals. In this case, the information about the location is kept on the user's personal device, so privacy is not a concern. However, this simple case has a downside because the user must perform additional work in order to match his location to the location of an interesting object or service. Such a solution is possible only for applications in which results are stored at a device controlled by a user.

Finally, in some applications, it is necessary to maintain relationships between an individual and his profile at a data collection point on an LBS. An example of such an application is Networkcar service [65]. A network of sensors in a car checks the state of the engine and other functioning units in a car. Each car has a built-in gateway that connects the car with a mobile telephony network. At the same time, the gateway uses a GPS client to determine the position of the car. All this information is stored on a Web server. A user of the service (an owner or an authorized car mechanic) can log on to the service through the Web and examine the current location of the car, the conditions of its engine, and other information. The owner of the car receives a message if it has been stolen and taken out of a certain area. This type of application cannot use the proposed techniques in which the precision of location information is reduced because the continuous connection between user and location information sent to the database must be maintained.

31.5 Conclusion

For many military and civilian applications of wireless sensor networks, security and privacy protection protocols and algorithms are an indispensable part of the system architecture. Because of their unique properties, most notably limited resources and physical exposure of sensor nodes, sensor networks require a new type of security protocols. These protocols are tailored to the underlying system architecture, patterns of network traffic, and specific security requirements so that security-related resource consumption is minimized. Physical exposure of nodes, as well as the threat that their cryptographic secrets are potentially available to an adversary, demands that security protocols in sensor networks protect the integrity of the network even if cryptographic secrets are compromised.

Privacy protection is especially important for certain commercial applications of sensor networks. Users who are monitored by sensor networks expect their private information not to be publicly available. However, sensor networks need services from other entities that may not have satisfactory privacy protection mechanisms. In cases in which applications require less precise data than are available, a certain level of privacy protection can be achieved by decreasing precision of the data; therefore, the data cannot be easily matched to any particular individual.

References

1. Industrial automation, Ember Corp. (2003). Retrieved August 20, 2003, from http://www.ember.com/products/solutions/industrialauto.html.
2. Dynamic sensor networks. Retrieved September 1, 2003, from http://dsn.east.isi.edu/.
3. Anderson, R. and Kuhn, M., Tamper resistance — a cautionary note, in *Proc. 2nd USENIX Workshop Electron. Commerce*, USENIX, Berkeley, CA, 1996, 1.
4. Stajano, F. and Anderson, R., The resurrecting duckling: security issues for ad-hoc wireless networks, in *Proc. 7th Int. Workshop Security Protocols*, Christianson, B., Crispo, B., and Roe, M., Eds., Springer-Verlag, Heidelberg, 1999, 172.
5. Meguerdichian, S. et al., Localized algorithms in wireless ad-hoc networks: location discovery and sensor exposure, in *Proc. 2nd ACM Symp. Mobile Ad Hoc Networking and Computing* (MobiHOC), ACM Press, New York, 2001, 106.

6. Krishnamachari, B., Estrin, D., and Wicker, S., The impact of data aggregation in wireless sensor networks, in *Proc. Int. Workshop Distributed Event-Based Systems*, IEEE Computer Society, Los Alamitos, CA, 575.
7. Cheswick, B., An evening with Berferd in which a cracker is lured, endured, and studied, in *Proc. of Winter USENIX Conf.*, USENIX, Berkeley, CA, 1992, 163.
8. Stoll, C., *The Cuckoo's Egg*, Doubleday, New York, 1989.
9. Bellovin, S., Leech, M., and Taylor, T., ICMP traceback messages, Internet draft (work-in-progress), IETF, 2003.
10. Savage, S. et al., Practical network support for IP traceback, in *Proc. ACM SIGCOMM Conf. Applications, Technol., Architectures, Protocols Computer Commun.*, ACM Press, New York, 2000, 295.
11. Snoeren, A. et al., Single packet IP traceback, *IEEE/ACM Trans. Networking*, 10(6), 1, 2002.
12. Song, D. and Perrig, A., Advanced and authenticated marking schemes for IP traceback, in *Proc. 20th Joint Conf. IEEE Computer Commun. Soc.* (INFOCOM), IEEE, 2001, 878.
13. Boulis, A. and Srivastava, M.B., A framework for efficient and programmable sensor networks, in *Proc. 5th IEEE Conf. Open Architectures Network Program.* (OPENARCH 2002), New York, June 2002.
14. Cohen, F., Computer viruses, theory and experiments, *Computers Security*, 6(1), 22, 1987.
15. Kephart, J.O. and White, S.R., Measuring and modeling computer virus prevalence, in *Proc. IEEE Computer Soc. Symp. Res. Security Privacy*, IEEE Computer Society, Los Alamitos, CA, 1993, 2.
16. Spafford, E.H., Computer viruses — a form of artificial life? in *Artificial Life II*, Langton, C.G. et al., Eds., Addison–Wesley, Redwood City, CA, 1992, 727.
17. Denning, D.E., *Cryptography and Data Security*, Addison–Wesley, Redwood City, CA, 1982.
18. Neumann, P., *Computer-Related Risks*, Addison–Wesley, Redwood City, CA, 1995.
19. Cowan, C. et al., Buffer overflows: attacks and defenses for the vulnerability of the decade, *DARPA Inf. Survivability Conf. Exposition*, IEEE Computer Society, Los Alamitos, CA, 1999, 1119.
20. Wagner, M.G., Robust watermarking of polygonal meshes, in *Geometric Modeling Process.*, IEEE Computer Society, Los Alamitos, CA, 2000, 201.
21. Lampson, B.W., A note on the confinement problem, *Commun. ACM*, 16(10), 613, 1973.
22. Gold, B.D. et al., A security retrofit of VM/370, in *AFIPS Conf. Proc.*, AFIPS Press, 1979, 335.
23. Golomb, S.W., The limiting behavior of the Z-channel, *Trans. Inf. Theory*, 26(3), 372, 1980.
24. Hu, W., Reducing timing channels with fuzzy time, in *Proc. IEEE Computer Soc. Symp. Res. Security Privacy*, IEEE Computer Society, Los Alamitos, CA, 1991, 8.
25. Simmons, G.J., The history of subliminal channels, *IEEE J. Selected Areas Commun.*, 16(4), 452, 1998.
26. Rubin, A.D. and Geer, D.E., Mobile code security, *IEEE J. Internet Computing*, 2, 30, 1998.
27. Gong, L. et al., Going beyond the sandbox: an overview of the new security architecture in the Java development kit 1.2, in *Proc. USENIX Symp. Internet Technol. Syst.*, USENIX, 1997, 103.
28. Brigner, P., Creating signed, persistent Java applets, *Dr. Dobb's J.*, 24, 82, 1999.
29. Bowen, R. et al., Building survivable systems: an integrated approach based on intrusion detection and confinement, in *Proc. DARPA Inf. Security Symp.*, 2000, 1084.
30. Colby, C. et al., A certifying compiler for Java, *SIGPLAN Notices*, 35, 95, 2000.
31. Necula, G.C., Proof-carrying code, in *Proc. 24th ACM SIGPLAN-SIGACT Symp. Principles Programming Languages*, ACM Press, 1997, 106.
32. Lofstrom, K., Daasch, W.R., and Taylor, D., IC identification circuits using device mismatch, in *Proc. Int. Solid-State Circuits Conf.*, IEEE, 2000, 372.
33. Pitkow, J., In search of reliable usage data on the WWW, *Comp. Networks ISDN Syst.*, 29, 1343, 1997.
34. Franklin, M.K. and Malkhi, D., Auditable metering with lightweight security, *J. Comp. Security*, 6(4), 236, 1998.
35. Naor, M. and Pinkas, B., Secure accounting and auditing on the Web, *Comp. Networks ISDN Syst.*, 30, 541, 1998.

36. Rivest, R.L., Electronic lottery tickets as micropayments, in *Proc. Int. Conf. Financial Cryptography*, Hirschfeld, R., Ed., Springer–Verlag, Heidelberg, 1997, 307.
37. Findley, R. and Dixon, R., Dual smart card access control electronic data storage and retrieval system and methods, U.S. patent 5629508, 1997.
38. Menezes, A.J., Oorschot, P.C., and Vanstone, S.A., *Handbook of Applied Cryptography*, CRC Press, Boca Raton, FL, 1997.
39. Ross, C.D. et al., Method and apparatus for electronic licensing, U.S. patent 5553143, 1996.
40. Aura, T. and Gollmann, D., Software license management with smart cards, in *Proc. USENIX Workshop Smartcard Technol.*, USENIX Association, Berkeley, CA, 1999, 75.
41. Thomas, D.C., Method and apparatus for providing security or computer software, U.S. patent 4446519, 1984.
42. Law, Y.W., Etalle, S., and Hartel, P.H., Assessing security-critical energy-efficient sensor networks, in *Proc. 18th IFIP TC11 Int. Conf. Inf. Security Privacy Age Uncertainty* (SEC), Gritzalis, D. et al., Eds., Kluwer Academic Publishers, Boston, MA, 2003, 459.
43. Papadimitratos, P. and Haas, Z.J., Securing mobile ad hoc networks, in *Handbook of Ad Hoc Wireless Networks*, Ilyas, M., Ed., CRC Press, Boca Ration, FL, 2002.
44. Perrig, A. et al., SPINS: security protocols for sensor networks, in *Proc. 7th Int. Conf. Mobile Computing Networking* (MOBICOM), ACM Press, New York, 2001, 189.
45. Hill, J. et al., System architecture directions for network sensors, in *Proc. 9th Int. Conf. Architectural Support Programming Languages Operating Syst.* (ASPLOS), ACM Press, New York, 2000, 93.
46. Rivest, R.L., The RC5 encryption algorithm, in *Proc. 2nd Workshop Fast Software Encryption*, Preneel, B., Ed., Springer–Verlag, Heidelberg, 1995, 86.
47. Liu, D. and Ning, P., Efficient distribution of key chain commitments for broadcast authentication in distributed sensor networks, in *Proc. 10th Symp. Network Distributed Syst. Security*, Internet Society, Reston, VA, 2003, 263.
48. Slijepcevic, S. et al., On communication security in wireless ad-hoc sensor networks, in *Proc. 11th IEEE Int. Workshops Enabling Technol.: Infrastructure Collaborative Enterprises* (WETICE), IEEE Computer Society, Los Alamitos, CA, 2002, 139.
49. Hubaux, J.P., Buttyan, L., and Capkun, S., The quest for security in mobile ad hoc networks, in *Proc. 2nd ACM Symp. Mobile Ad Hoc Networking Computing* (MobiHOC), ACM Press, New York, 2001, 146.
50. Carman, D.W., Matt, B.J., and Cirincione, G.H., Energy-efficient and low-latency key management for sensor networks, in *Network Assoc. Labs Advanced Security Res. J.*, 5(1), 2003, 31.
51. Eschenauer, L. and Gligor, V.D., A key management scheme for distributed sensor networks, in *Proc. 9th ACM Conf. Computer Commun. Security*, ACM, New York, 2002, 41.
52. Yuan, L. and Qu, G., Design space exploration for energy-efficient secure sensor networks, in *Proc. IEEE Int. Conf. Application Specific Syst., Architectures, Processors* (ASAP), IEEE Computer Society, Los Alamitos, CA, 2002, 88.
53. Spencer, J.H., *The Strange Logic of Random Graphs*, 1st ed., Springer-Verlag, Heidelberg, 2000.
54. Chan, H., Perrig, A., and Song, D., Random key predistribution schemes for sensor networks, in *Proc. 2003 IEEE Symp. Security Privacy*, IEEE Computer Society, Los Alamitos, CA, 2003, 197.
55. Samarati, P. and Sweeney, L. Protecting privacy when disclosing information: k-anonymity and its enforcement through generalization and suppression, technical report, SRI International, 1998.
56. Gruteser, M. and Grunwald, D., Anonymous usage of location-based services through spatial and temporal cloaking, in *Proc. ACM/USENIX Int. Conf. Mobile Syst., Applications, Services* (MOBISYS), USENIX, Berkeley, CA, 2003.
57. Navigation Technologies, Navtech. Retrieved September 1, 2003, from http://www.navtech.com.
58. Avis Assist, Avis, Inc. Retrieved September 1, 2003, from http://www.avis.com/AvisWeb/JSP/US/en/deals/us_assist.jsp.
59. Autodesk Location Services, Autodesk (2003). Retrieved August 9, 2003, from http://locationservices.autodesk.com/.

60. Beresford, A. and Stajano, F., Location privacy in pervasive computing, *IEEE Pervasive Computing*, 2 (1), 46, 2003.
61. Gruteser, M. et al., Privacy-aware location sensor networks, in *Proc. 9th USENIX Workshop Hot Topics Operating Syst.* (HotOS), USENIX, Berkeley, CA, 2003.
62. Chaum, D., Security without identification: transaction systems to make Big Brother obsolete, *Commun. ACM*, 28 (10), 1030, 1985.
63. Pfitzmann, A. and Koehntopp, M., Anonymity, unobservability, and pseudonymity — a proposal for terminology, in *Proc. Workshop Design Issues Anonymity Unobservability*, Federrath, H., Ed., Springer-Verlag, Heidelberg, 2001, 1.
64. Priyantha, N.B., Chakraborty, A., and Balakrishnan, H., The Cricket location support system, in *Proc. 6th Int. Conf. Mobile Computing Networking* (MOBICOM), ACM Press, New York, 2000, 32.
65. Networkcar technology, Networkcar, Inc. Retrieved August 21, 2003, from http://www.networkcar.com.

32
A Taxonomy for Denial-of-Service Attacks in Wireless Sensor Networks

Anthony D. Wood
University of Virginia

John A. Stankovic
University of Virginia

32.1 Introduction 32-1
   A Note on Terminology • Denial of Service • Relation to Other Fields • Sensor Network Vulnerability
32.2 Attack Taxonomy 32-5
   Attacker • Capability • Target • Vulnerability • Result
32.3 Vulnerabilities and Defenses 32-10
   Jamming • Tampering • Collisions • Exhaustion and Interrogation • Selective Forwarding • Misdirection • Sinkholes • Wormholes • Sybil Attack • Flooding • HELLO Floods • Algorithmic Complexity Attack
32.4 Related Work 32-16
32.5 Conclusion 32-17

32.1 Introduction

Interest in wireless sensor networks (WSNs) continues to build momentum, with research results and technology beginning to transition to real-world applications. The application possibilities include sensing and actuating in many types of environments, from monitoring remote environmental sites or hostile battlefields to controlling the modern comfort of indoor health-care facilities [7, 27, 52]. Some networks may be used in places unsafe or undesirable for humans; others may be unobtrusively integrated into dense urban environments.

The WSN design space centers around small, wireless devices with magnetic, acoustic, optical, chemical, or other sensors on board [24]. Each is limited in resources, so nodes must communicate and coordinate to enact aggregate behaviors. Sensing nodes may not be serviceable once deployed, so low-power operation is essential for network longevity. Low per-unit cost and effective distributed algorithms will enable deployment at a large scale, where individual sensors are no longer identifiable or important. Failures are masked by robust aggregation, redundancy, and adaptive reconfiguration.

All the promises of robustness, fault tolerance, and cost-effective operation hinge on overcoming the inherent resource limitations of sensor nodes. Memory and processing cycles are limited, as they are in many embedded devices. Because the energy budget is small, wasted cycles and, especially, wasted radio transmissions are not wise. Without diligent consideration of these constraints, WSN protocols and architectures may never make it from the research lab to the field.

Security is another important, and frequently overlooked, requirement in many of these systems. Sensitive data generated by the network should be protected from unauthorized disclosure. Data and control mechanisms likewise should provide integrity and authenticity guarantees. These are all the more critical when the network can affect the environment using actuators or automatic responses.

Even with confidentiality and integrity, the WSN is not achieving its objectives if its services are not available to authorized users when they need it. In networks with such scarce resources, their improper consumption or destruction is a big concern. In addition to being a security problem, an inability of the network to perform its task may be a safety hazard, depending on the system being monitored or controlled.

This chapter explores denial-of-service vulnerabilities in WSNs. It begins by examining terminology, the definition of denial-of-service, and why it is a potential problem for WSNs. Section 32.2 presents a taxonomy for classifying attacks. A key part of the taxonomy — vulnerabilities — is described in Section 32.3, along with possible defenses. The chapter concludes with an overview of related work in denial-of-service research and some final comments.

32.1.1 A Note on Terminology

Some of the terms used in this chapter vary in their popular definition, depending on the author. For clarity and consistency, the following definitions have been adapted from the National Information Systems security glossary [11]:

Attack: attempt to gain unauthorized access to a service, resource, or information, or the attempt to compromise integrity, availability, or confidentiality. Note that success is not necessary.

Attacker, intruder, adversary: used synonymously to mean the originator of an attack.

Vulnerability, flaw: weakness in system security design, implementation, or configuration that could be exploited.

Threat: any circumstance or event (such as the existence of an attacker and vulnerabilities) with the potential to impact a system adversely through a security breach.

Risk: probability that an attacker will exploit a particular vulnerability, causing harm to a system asset.

32.1.2 Denial of Service

Classically, the definition of denial of service (DoS) involves three components: *authorized users*, a *shared service*, and a *maximum waiting time* [19]. Authorized users are said to deny service to other authorized users when they prevent access to or use of a shared service for longer than some maximum waiting time. Though this definition appears straightforward, its application raises several important questions.

Who are the users? Certainly, the users of a WSN as a whole are humans. Traditionally, processes acting on behalf of those humans are also considered users of the system. In a distributed general-purpose computing network, this is likely the end of the matter.

In a WSN, however, general interactive computing facilities are not likely to exist in any of the devices. The large number of devices, their relative inaccessibility, and low energy supplies are more manageable when localized autonomy and coordination are present. In such an environment, in-network services (such as localization, routing, and power management) are of more direct benefit to the sensor nodes that interact with them than to the human operators. Rather than treat only the human deployers, owners, or monitors of the network as users, it may be more useful to consider individual sensor nodes as users with respect to in-network services.

Distinctions such as insider vs. outsider threats can then apply at two levels. At the highest level, insiders are those who have an employee–employer or other trusted relationship with the WSN owners. On a lower level, individual, authentic sensor nodes are insiders. If their identities or secrets are stolen, say by physical tampering, an intruder is afforded the access, opportunity, and credentials to appear as an insider to the network. This possibility is affirmed by Anderson [4], who suggests that "software agent/mobile code technology may represent a nonhuman variant of the 'insider threat.'"

A particular difficulty on the Internet is that the network is vast and open, so it is currently not possible to authenticate every user. For some services, then, there are effectively no unauthorized users because users are not strongly authenticated at all. A WSN is a more closed and controlled system, but even here authentication is not without difficulty. Large-scale networks, as well as those in which periodic replenishment is expected after initial deployment, pose a variety of trust and identity management issues. These will not be explored further here.

What are the shared services? Depending on who the users are, the WSN may provide a few aggregation services, such as monitoring or control of an area. Shared access to these types of services could take the form of multiple battlefield commanders querying a WSN for the location of chemical or biological hazards. In-network services, such as routing, localization, and time synchronization, are also used by multiple sensor nodes concurrently. For this reason, it may be useful to view individual sensor devices as users, as discussed earlier.

How long is too long to wait? Many systems may only have loosely specified deadlines for service fulfillment. Strictly interpreting the DoS definition to preclude the possibility of DoS attacks is not useful in practice. Even discretionary services such as the World Wide Web have limits imposed by cognitive perceptions of action and response. Passive WSNs may be able to tolerate widely varying service fulfillment times, as long as the waits are "reasonable," perhaps probabilistically. Networks that can control their environment are more likely to require real-time properties consistent with precise specification of maximum waiting times.

Finally, modification or corruption of the intended and advertised service may also be considered denial of service [19]. In this case it is insufficient to deliver the appearance of service within some bounded time; only the specified service will do. Detection of this type of DoS is not as simple as setting a timer and may be impossible. Although modification of the service is a DoS to its legitimate users, it is perhaps more appropriately the concern of ensuring the accuracy and correctness of system function. Therefore, a broader definition, one inclusive of the above possibilities, is adapted from the National Information assurance glossary [11]:

Denial of service: the result of any action that prevents any part of a WSN from functioning correctly or in a timely manner. Even if a more precise definition is elusive, a DoS attack usually has the following properties:

- *Malicious.* The act is performed intentionally, not accidentally. Accidental failures are the domain of fault tolerance and reliability engineering. Because such failures can potentially produce equally disruptive results as DoS attacks, these fields have important contributions to make to the robustness of WSNs. They are not considered DoS, however, due to the lack of malice.
- *Disruptive.* A "successful" DoS attack degrades or disrupts some capability or service in the WSN. If the effect is not measurable, for example, if it is prevented, one may still say that an attack has occurred, but DoS has not. Note that disrupting the affected service may not be the end goal of the attacker.
- *Asymmetric.* Often the effect of an attack is much greater than the effort required to mount it. For example, sending a forged packet that overflows a remote buffer takes little effort, but may crash the server until an operator intervenes. Even in distributed denial-of-service (DDoS) attacks, the effort to "recruit" zombies and issue an order to flood a victim is small compared to the flood of traffic that reaches the target. This kind of asymmetry is not necessary, but makes an attack easier and more economical for the perpetrator.
- *Remote.* Especially in distributed systems, an attacker usually can (and wishes to) carry out an attack over the network. Often this is by unauthenticated or lightly authenticated users (such as those possessing a valid return IP address, in the case of TCP-SYN floods [44]). The high profile of many types of DoS attacks would make physical presence uncomfortable for the attacker.

With this understanding of DoS, its relation to other fields of study can be explored.

32.1.3 Relation to Other Fields

Denial of service is affected by other fields, such as security, reliability, performance, and software engineering. In some cases, transfer advances in these areas to gains in DoS resistance.

Security. As described earlier, DoS is a breach of the security characteristic of *availability*. Along with availability, *confidentiality* and *integrity* are the primary concerns of security. Integrity failures can cause DoS; for example, clients are denied service if the server's replies are always corrupted. However, ensuring integrity is not sufficient to prevent DoS. Protection models generally express what users may access — not that they will be able to regardless of what other users do [19]. This is the domain of availability and fairness.

Reliability. Results from the areas of reliability and fault tolerance may help a WSN be more robust to DoS attack. Loss-concealing and self-correcting coding schemes and protocols may increase the work an attacker must do to cause DoS. Reliability considers faults to be random and independent, however, in contrast to the intentional failures induced by a DoS attack.

Performance. Many DoS attacks consist of simple flooding, that is, overwhelming link bandwidth with large or numerous messages. Because large commercial sites must process many requests or messages in a timely fashion as part of routine business, they are designed with fast, efficient network stacks that minimize interrupts and data copying. High-performance systems, such as Web servers, will therefore naturally tolerate larger-scale DoS attacks based on flooding [28]. Although the resource-impoverished sensor devices in WSNs are in many ways the opposite of these high-performance server farms, some of the efficiency techniques used may be applicable. Of course, efficiency is not sufficient to thwart every DoS attack, even those that involve flooding.

Software engineering. DoS made possible by implementation flaws in programs continues to be underrepresented in the literature [3]. This is despite the fact that exploit programs are frequently published on the Web that allow intrusion or DoS of vulnerable computers on the Internet. According to NIST's ICAT vulnerability search engine, 335 remotely exploitable availability-impacting vulnerabilities were published in 2002 [25]. The large number of implementation-related flaws highlights the continuing need for research in software engineering because it can impact the process of writing secure programs.

32.1.4 Sensor Network Vulnerability

Although WSNs can be a low-cost, low-overhead way to monitor an environment in real time, some of their attributes make them even more susceptible to DoS attack or damage. The attribute that dominates all others has already been discussed: limited resources. A device with scarce resources is at risk of resource consumption under normal circumstances. When an adversary is actively attempting to consume or destroy its resources improperly, the situation is worse. Other attributes are:

- *Remote location.* Networks that are distant or unmonitored have a greater response time if manual (physical) intervention is required.
- *Large scale.* Due to the large number of devices likely to be deployed, manual intervention on each device is not feasible within cost constraints. If a vulnerability is discovered and exploited in the sensor program code, it will be no small matter to collect, reprogram, and redeploy each device physically. Wireless code download is a possible remedy, but is not without its own security concerns.
- *Cost-sensitive applications.* For a large-scale deployment to be cost-effective, the per-unit cost of sensor devices must be low. This applies added pressure to control hardware and software development costs. In the real world, this translates to hasty designs and numerous implementation errors.
- *Application specificity.* Resource constraints may dictate that well-defined and uncoupled network layers are compressed or merged, reducing code modularity. Unforeseen interactions between network layers and services may give rise to new vulnerabilities.

- *Attractive target.* The systems monitored or controlled by the WSN may be safety critical or highly visible, with significant consequences for failure. Depending on the motivation of the attacker, this may be precisely the profile of an attractive target.
- *Uncontrolled access.* Ubiquitous, wide-scale, and replenishable deployments may require relatively unfettered physical access to nodes. Odds of casual tampering or vandalism increase.
- *Middleware services.* As services are distributed among all or most nodes of the network, every device is a potential target for attack.

Effective mitigation of these inherent vulnerabilities requires careful consideration of the design and implementation of all protocols and code. To achieve a better understanding of the risks faced by the network, a taxonomy of DoS attacks against WSNs will now be described.

32.2 Attack Taxonomy

A taxonomy allows one to reason about attacks at a level higher than a simple list of vulnerabilities. It provides a classification system that ideally suggests ways to mitigate attacks by prevention, detection, and recovery. It can aid risk management by identifying vulnerabilities and making attacker characteristics explicit. Ideally, its insights can predict future attacks by exposing unguarded areas.

Every DoS attack is perpetrated by someone. The attacker has an identity and a motive and is able to do certain things in or to the WSN. An attack targets some service or layer, exploiting a vulnerability. An attack may be thwarted, or it may succeed with varying results. Each of these elements is necessary to understand the whole process of a DoS attack. Therefore, a useful and intuitive taxonomy should answer the following questions:

- Who is the attacker?
- What is the attacker capable of?
- What is the target?
- How is it attacked?
- What are the results?

The attack taxonomy presented here comprises the answers to each question in turn. Taken together, the attacker, capability, target, vulnerability, and results describe a DoS attack against a WSN. A summary of the taxonomy is shown in Figure 32.1. Each part will now be described in detail.

32.2.1 Attacker

It is valuable to know who the attacker is in so far as it implies what he is likely to do and how well he may be able to do it. The following list characterizes an attacker according to four dimensions: *motive*, *determination*, *knowledge*, and *resources*. It bears some similarity to that presented by Howard [22].

- Passerby — motivated by spontaneity; not determined; very little knowledge; few resources
- Vandal — desires to inflict damage, perhaps visibility; moderately determined; little knowledge; few resources necessary
- Hacker — desires access, motivated by curiosity and interest; highly determined; highly knowledgeable; moderate resources
- Raider — driven by personal or organizational monetary and/or political gain; highly determined; moderately to highly knowledgeable; moderate resources
- Terrorist or foreign power — causes real-world damage by compromise of critical systems, motivated by enmity; very determined; highly knowledgeable; very well resourced with time, money, and manpower.

This list is not meant to be exhaustive, but rather to cover a common and interesting range of attacker attributes. Those given represent a range of possibilities in each of the dimensions. Motivations vary, but

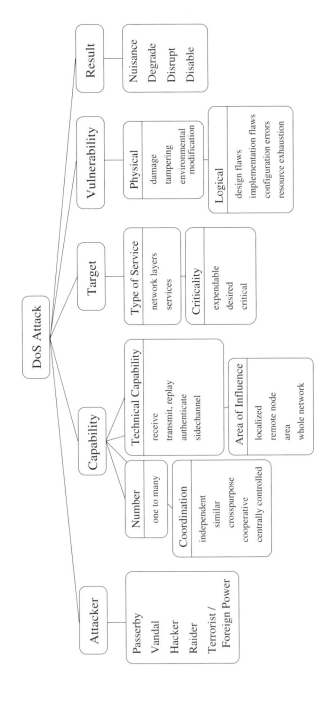

FIGURE 32.1 Taxonomy for denial-of-service attacks in wireless sensor networks.

the objective, for this discussion, is always to cause a DoS. In general, the more determined, knowledgeable, and well-resourced the attacker is, the more difficult is the defense.

An alternate dimension is the insider vs. outsider distinction, which is somewhat orthogonal to the above list because any of the attackers could be an insider or collude with one. As discussed before, the term "insider" is not without ambiguity. It could refer only to the owner, operators, controllers, or monitors of the WSN. It could also refer to all processes executed on behalf of these users by sensor devices.

If commands are not guaranteed to be authentic by the security architecture of the WSN, these processes may include execution of malicious code by legitimate or subverted nodes on behalf of an adversary. This is similar to a payroll clerk executing an email attachment: some aspects of the attack profile should be based on who executes the attachment, rather than on who sent it.

A "trust continuum" between insider and outsider is seen, instead of a bifurcation. Trust is concretely manifested by such aspects as knowledge, opportunity, authorization, affiliation, and access. Any moderately sized organization will have multiple levels between the extreme insider (owner) and outsider (competitor). Not everyone has the combination to the safe or the ability to change payroll.

As discussed by Anderson [4], insiders may be further divided into permanent, part-time, and temporary staff, contractors, developers, and others. Each has a distinct profile that can inform an analysis of the identity of an attacker.

32.2.2 Capability

Knowing what an attacker is capable of is important for efficiently defending a WSN. Realistically, all possible vulnerabilities cannot be eliminated within resource, cost, or usability constraints. System designs are targeted to address the threats most likely to be seen. Determining this real-world risk and selecting mitigation strategies depends partially on an enumeration of attacker capabilities, which will be presented. Capabilities in several dimensions will be considered, each of which is described in detail next: number of attackers, coordination, technical capabilities, and area of influence.

32.2.2.1 Number of Attackers

Clearly, attacks may be mounted by one or many attackers. Systems or particular defense strategies are often classified by their resistance to compromise. They may tolerate one compromise per system, one per neighborhood, or one per time epoch. Some solutions that work against one attacker may fail if, for example, enough attackers are available to partition the network.

32.2.2.2 Coordination of Attackers

When multiple attackers are present, their coordination may vary. Each implies unique approaches to defense. The attackers may be independent, each attempting to cause DoS according to his individual motivations. If the attacks are similar, they may be aggregated and considered as N separate instances of the same attack. Conversely, attackers may be working at cross-purposes, in which case there are N different attacks, possibly interfering with each other. For example, each may be attempting to selfishly gain priority in a routing service.

Autonomous attackers who cooperate to reach a common goal are generally more difficult to defend against. They may include a colluding authorized user (that is, an insider). A wormhole attack [23], in which two geographically separated nodes relay messages out of band and replay them (see Subsection 32.3.8), is an example of this type of coordination.

Multiple attackers may also be centrally controlled, as in many Internet DDoS attacks. An adversary instructs many penetrated systems ("zombies"), sometimes through an indirect channel, to commence a DDoS attack against a victim. This kind of attack may be easiest to address by detecting and disrupting the control channel the attacker uses to disseminate instructions.

32.2.2.3 Technical Capability

Another aspect of an attacker model is the set of particular technical capabilities available. These capabilities are given here in increasing order of power or complexity.

- An attacker may only be able to *receive* wireless radio transmissions. Limited to passive eavesdropping, the attacker cannot perform a DoS unless through other channels. For example, after homing in on a transmitting WSN node, the attacker may find and physically damage it. Listening to transmissions may precede other more complex attacks.
- Along with receiving, if an attacker can *transmit* in the wireless channel used by the WSN, it can interact with sensor devices. It may be able to impersonate a legitimate node by replaying an old message, unless the network guarantees message freshness.
- An attacker that can falsely *authenticate* itself to the WSN poses an even greater hazard. This may be possible through forgery or theft of legitimate credentials, or insider collusion. Depending on the authorization of the impersonated node, the attacker may have full access to all WSN services.
- A *side channel* may be available to the attacker for communication and coordination with other adversaries. Wired networks, other wireless channels, and optical communication allow the attacker to, for example, coordinate attacks despite disruption in the WSN's routing.
- An attacker may have a *more powerful class* of devices than exist in the WSN. These include higher bandwidth links, side channels, superior computational facilities, wireline electric power, and mobility. It will be most difficult to overcome this kind of capability asymmetry.

32.2.2.4 Area of Influence

The area of an attacker's influence varies depending on the WSN and on the attacker's capabilities. An individual adversary possessing power and radio resources similar to a WSN node may only be able to affect a localized region. Nodes outside this local communication range are unaffected. Access to packet routing in the network gives the adversary the ability to affect remote nodes. Widely distributed base stations may be reachable from anywhere in the network.

Due to spatial properties of physical environments monitored by sensor networks, attackers may be able to affect an arbitrary area of the network. An area-multicast service may be abused to flood many nodes with each message sent by an adversary. As a special case of the previous capability, an attacker may be able to affect the whole network, perhaps using an unauthenticated broadcast facility.

32.2.3 Target

Next the attacker's target will be considered. The type of target and its importance to the network are factors that affect the overall risk and constrain solutions. To avoid speculation about the ultimate objective of the attacker, the target of a DoS attack is defined to be the service that is being denied.

32.2.3.1 Type of Service

As described in Wood and Stankovic [49], attacks may be mounted predominantly against one layer or service in the network. They may also exploit service interactions. In a layered architecture, disrupting the lower layers is most advantageous for an attacker because most or all of the upper layers depend on it to function. Thus, an attack may affect multiple services by targeting their common underlying dependency.

Services include the typical layers found in a network stack: physical, link-layer, network and routing, transport, and application. Other WSN-specific services are also possible targets, such as: localization; time synchronization; group management; directory services; entity tracking; power management; event detection; topography discovery; code download; and aggregation. Any application-specific service may be targeted.

In a heterogeneous network, attacks may also be targeted at particular devices. However, these attacks must take the form of an attempted exploitation of a vulnerability in some service provided by the node. Thus, it is more useful even in this case to consider the type of service being attacked, rather than the specific identity of the victim.

32.2.3.2 Criticality of Target

The criticality of the target partially determines the risk associated with exploitation of related vulnerabilities. Critical services should be well protected against all forms of security violations, including DoS.

Some targeted services may be expendable or optional, such as the sensing coverage provided by a small number of nodes in a large WSN. The DoS at a few individuals may have no deleterious effect on the function of the network as a whole. Other services may be desired, but noncritical. The network can achieve most or all of its purposes, although at a degraded level. Optimal function of the network is inhibited. For example, complete sensing coverage or equalized power consumption may be unattainable. Critical services are those without which the WSN cannot function adequately. The loss of key services such as routing, directory services, or event detection may disrupt the entire operation of the network.

32.2.4 Vulnerability

After determining who an attacker is, what his capabilities are, and what the target is, the vulnerabilities of the WSN are considered. Vulnerabilities are weaknesses in the network, through which an attacker may gain or unduly exercise privilege. They arise due to conflicting requirements, cost constraints, designer shortcuts, and other reasons explored next. When attacker ability and intention meet vulnerability, risk is created.

32.2.4.1 Physical

Most computer science literature considers only the damage that a malicious program or user can do using computer or network processes. Because one desires to protect WSNs against all possible DoS attacks, it is necessary to consider those perpetrated by low-tech physical means also.

Low-cost packaging is unlikely to be very resistant to intentional physical damage. Only physical presence is required to mount the attack — a low threshold for a network that may be deployed ubiquitously. Part or all of the WSN node may be damaged. Although overt damage is easily detectable, more subtle physical tampering may go unnoticed. Here the node is physically modified in some way, perhaps to subvert it by replacing the program code, or secret keys may be read from its memory and used in another attack.

Sensor networks function by being partially embedded in the real world. This creates a unique vulnerability to environmental modification not found in traditional computer networks. An attacker who can systematically modify or falsify local sensor values in an area of the WSN may be able to mislead monitors or trigger improper control actions.

32.2.4.2 Logical

Logical vulnerabilities exist in computer programs and protocols. Four categories of these vulnerabilities can be identified:

- *Design flaws* allow the use of a protocol that violates assumptions about the environment or manner of use, while conforming to the protocol specification. This may be caused by ambiguity in the specification or by implementation variance among multivendor components. For example, a lack of authentication in a power management protocol may allow any node to put any other to sleep repeatedly.
- *Implementation flaws* are errors in hardware construction or software coding. An example is inadequate boundary checking, which may result in a buffer overflowing with attacker-controlled contents. This may result in an access violation and crash, the execution of arbitrary code, or the destruction of a shared resource. Improper loop logic that allows a node to enter an infinite loop is another example of an implementation flaw that would cause DoS.
- *Configuration errors* are the result of enabling improper settings for the particular environment or threat model appropriate to the WSN. Debug facilities, in particular, are often the source for such vulnerabilities.

- *Resource exhaustion* is possible even if designs, implementations, and configurations are correct. An attacker that can generate large amounts of traffic can flood a victim's network link. Poorly authenticated memory allocation or code execution may also enable an attacker to consume these resources and cause DoS.

32.2.5 Result

The final piece of the attack chain is the result of the DoS attempt. An attack may only be a nuisance if the targeted service is not harmed, due to preventative mechanisms. Depending on service availability requirements, if the duration of the attack is short and the service quickly recovers, the DoS attack may not be successful. Nuisance attacks may result in the masking of other attacks, however.

Network performance may be degraded during the attack, but not stopped. Services may continue to function in the network as a whole, but are temporarily unavailable in certain regions of the network. The attack may be of short duration. In a more serious vein, a service may be disrupted entirely for the duration of the attack plus some finite recovery time. The attack may last long enough to cause cascading failures in services dependent on the target.

A severe attack may result in a disabled service: the target ceases to function even after the attack stops. Such permanent disruption could be caused, for example, by physical damage, erased or corrupted memory, or crash failures without recovery. The attack or its consequences may last long enough to cause economic or physical damage.

32.3 Vulnerabilities and Defenses

Vulnerabilities are only one part of an attack, but it is useful to examine them in more detail. Without vulnerabilities the attack chain is broken and the WSN is perfectly secure (though not necessarily usable). Along with the available services that may be targeted, vulnerabilities are the only part of an attack under the control of the WSN designer.

Offering fewer services should be considered, if possible, because complexity is often a cause of insecurity. Reducing or streamlining services should reduce the opportunity for flaws to be created in design and implementation of the system. Smaller systems are easier to understand, and make an exhaustive analysis of attack possibilities more tractable. For example, broadening the scope of an existing remote configuration facility may be safer, from a security standpoint, than adding a more general code-download facility.

System requirements dictate many of the services in a WSN, but designers should always minimize the risk from vulnerabilities. Depending on the purpose and environment of the WSN and the threat model assumed, some vulnerabilities may be tolerated. For example, a network deployed in the interior of a guarded and physically secure military base may not require expensive tamper-proofing on every sensor device — unless the insider threat is significant.

Eliminating all vulnerabilities would prevent DoS attacks, but it is not possible in practice. Yu and Gligor [51] argue that preventing DoS is only possible with the enforcement of agreements between users, external to the services being shared. Although the focus is therefore on detection and recovery from attacks, one should also seek to limit DoS vulnerabilities wherever possible. Clearly, some designs are more resilient in the face of attack than others. Prevention is most coveted, but when necessary one settles for detection and recovery [16].

The assumption is that the attacker's objective, whatever his motivation, is to cause DoS against one or more of the services in the network (see Subsection 32.2.3.1). He may use any available vulnerability or — more problematically — a combination of them to achieve this. Flaws in the system cannot be viewed only individually; the security risk depends on their aggregation. The DoS attack taxonomy already presented tends to encourage this big-picture view.

In the following sections, some vulnerabilities of WSNs and possible defenses are described.

32.3.1 Jamming

Jamming is deliberate interference with radio reception to deny the target's use of a communication channel. For single-frequency networks, it is simple and effective, rendering the jammed node unable to communicate or coordinate with others in the network.

Constant transmission of a jamming signal is an expensive use of energy. An attacker limited in energy, as the WSN devices are, may use sporadic or burst jamming instead. In this attack, the attacker jams only when detecting radio transmissions in the area of the victim, which requires that he be nearby. Assuming that the attacker can detect all such transmissions and can react to them fast enough to jam before they finish, complete wireless channel denial is possible.

Defense. The most common defense against jamming attacks is the use of spread-spectrum communication [5, 39]. In frequency hopping, a device transmits a signal on a frequency for a short period of time, changes to a different frequency, and repeats. The transmitter and receiver must be coordinated. Direct sequence spreads the signal over a wide band, using a pseudorandom bit stream. A receiver must know the spreading code to distinguish the signal from noise.

Frequency-hopping schemes are somewhat resistant to interference from an attacker who does not know the hopping sequence. However, the attacker may be able to jam a wide band of the spectrum or even follow the hopping sequence by scanning for the next transmission and quickly tuning the transmitter. Direct-sequence spread spectrum is more effective at defeating the jammer, although its processing-gain advantage may be defeated by a high-power wideband jamming signal [47]. Due to synchronization and cost requirements, low-cost sensor devices may be limited to using the more susceptible single-frequency communication.

If jamming cannot be prevented, it may instead be detected and mapped by surrounding nodes [9, 50]. A description of the region may then be reported back to network monitors, who can use conventional means to remove the attacker. In-network knowledge of the extent of the jammed region may also allow for automatic routing avoidance or mobile jammer tracking.

Jamming always occurs at the receiver, that is, it does not interfere with transmissions. A sensor device with important data may temporarily overcome localized jamming by sending a high-power transmission to an unaffected node [49]. This node can then relay the message on behalf of the jammed node. Such a scheme must be used sparingly, however, because a high-power transmission will prematurely drain the device's energy, resulting in a permanent DoS.

If alternate modes of communication are available, such as acoustic, infrared, or optical [7], a node may switch to one of these schemes when the radio is jammed. Of course, these other channels may be jammed as well by a determined attacker.

32.3.2 Tampering

Large-scale WSNs may be deployed in densely populated areas, where physical access to individual nodes is impossible to prevent. Even casual passersby may be able to damage, destroy, or *tamper* with sensor devices. Destruction of the node could cause gaps in sensor or communication coverage. More well-equipped attackers can interrogate a device's memory, stealing its data or cryptographic keys. Its code can be replaced with a malicious program — one potentially undetectable to neighboring nodes. The capability profile of the subverted node is now that of a fully authorized insider.

Defense. Like jamming, an attack at this low level can render the entire sensor network useless. One intrinsic deterrent to tampering is the physical distribution of the network, that is, the geographic separation of individual nodes. Most low-cost attacks require at least brief physical presence near the targeted nodes. Protection measures outside the scope of the WSN may be sufficient to discourage tampering attacks. For example, upon detection, tracking, and reporting of human intruders within the compass of the network, guards or operators could physically intervene.

An obvious, albeit difficult, defense against tampering is to construct sensor devices with tamper-resistant packaging [2]. This is generally an expensive and imperfect approach, and one that must be narrowly tailored to an accurate threat model. Defending against accidental damage or curious mishandling is much easier than thwarting a determined and well-funded attack in an electronics lab. Another approach is to prevent detection of the nodes. Camouflaging the packaging, hiding the device, and using low probability of intercept (LPI) radio techniques are among the possibilities. Again, these may all increase the cost and complexity of WSN design.

32.3.3 Collisions

Similarly to jamming the physical radio channel, an attacker can willfully cause collisions or corruption at the link layer. By detecting and parsing radio transmissions near the victim, the attacker can disrupt key elements of packets, such as fields that contribute to checksums or the checksums themselves. With little effort or duration of transmission, the attacker may be able to cause the victim to discard a much longer packet, thus wasting the channel access as well as the transmitter's energy. Such disruption may trigger back-offs in contention-control mechanisms that delay other messages.

Defense. Unfortunately, in wireless networks detection of a collision with one's own transmission is difficult. Standard collision avoidance mechanisms do not help because they are cooperative by nature. An attacker simply ignores the avoidance protocol and transmits at the same time as the victim. Error-correcting codes can be used to provide some protection against corruption of message data. They are well suited for covering random transmission errors in which independent bits of a message may be flipped. An attacker, though, can always corrupt more data than the code can correct. The codes also cost additional processing and transmission overhead because the message contains greater redundancy.

32.3.4 Exhaustion and Interrogation

An attacker may be able to inflict DoS on a network by inducing repeated retransmission attempts. Even in the absence of high-rate traffic, if a node must continually retransmit due to collisions, as described earlier, eventually its energy may be exhausted. The attacker need only corrupt a small part of a much longer message or perhaps jam an acknowledgment from a neighbor. Randomizing back-offs is not a defense because the attacker can easily listen and wait for the next attempt. As long as he can react before the entire packet is transmitted, the cycle can continue.

At various layers in the network, small messages (such as queries) may elicit much larger responses. For example, an attacker may be able to replay a broadcast initialization command, causing nodes throughout the network to perform localization or time synchronization procedures. Such unauthenticated or stale messages provide an easy avenue for traffic amplification. This repeated solicitation of energy-draining responses is called *interrogation*.

Defense. Services can require that requests be authenticated, otherwise refusing to answer with even a negative acknowledgment. This may cause confusion with legitimate clients, however, who receive no indication of whether the service provider has failed or is ignoring them. Omitting extraneous messages from protocols improves message efficiency, but may result in brittle state synchronization between the endpoints.

Another defense is to rate-limit response even to properly authenticated nodes. Excessive requests will be queued or ignored without sending expensive radio transmissions. The rate must be high enough to provide sufficient bandwidth and timeliness for authorized users. A node detecting an interrogation attack may also notify upper layers of the condition. This is similar to the congestion or delivery-failure feedback that some link-layers provide to the network layer. Rate limitations may be adapted based on the recent history of request traffic.

32.3.5 Selective Forwarding

WSNs usually depend on every node to take part in routing for its neighbors if it can provide a desirable forwarding path. Various *selective forwarding* attacks can exploit this dependence to cause DoS via routing.

A subverted sensor device can simply neglect to forward certain messages. A random dropping policy raises the local loss rates and may induce costly end-to-end recovery mechanisms. An attacker may also drop messages to or from certain victims, such as base stations or other servers. At an extreme, a node not only could refuse to forward any packets, but could also advertise a desirable path to its neighbors, creating a routing *blackhole*. Any messages passing nearby will be diverted to the adversary, where they are silently dropped. In addition to causing a DoS to the senders of the messages, neighbors of the adversary suffer from increased contention due to the above-normal levels of traffic.

Defense. Using multiple disjoint routing paths [17] and diversity coding [1] can mitigate the effect of the attack. These defenses lessen the probability that a message will encounter an adversary along all routes to the destination. Diversity coding sends encoded messages along multiple paths so that the originals can be reconstructed to conceal message loss, without the cost of full duplication.

To counter these defenses, an attacker must subvert additional nodes along the disjoint paths or choose an important source to which to move closer, where jamming will be effective. Nodes can also monitor their neighbors to gain probabilistic assurance that messages are correctly forwarded. A node relays a message to its neighbor and then listens to the wireless channel, noting whether it overhears the neighbor's subsequent broadcast [33] of the same message. Although collisions, collusion, and asymmetric communication links limit the sender's ability to monitor every packet, the forwarding ratio can be used to inform a quality-rating mechanism. This mechanism is responsible for choosing a next-hop neighbor that has a high probability of properly forwarding subsequent messages.

Periodic end-to-end probing [10] can also alert a node to troublesome network paths, whether from congestion or malicious neglect. If an adversary can distinguish the probes from normal traffic, however, he can properly forward them so as not to arouse suspicion.

32.3.6 Misdirection

By forwarding messages along wrong paths, an attacker *misdirects* them, perhaps by advertising false routing updates. An attacker could inflict DoS on a particular sender by diverting only traffic originating from the victim node. A receiver could likewise be denied service if the attacker diverts traffic away from the node. This may be possible by rewriting the downstream path in routing algorithms that embed source routes in each packet. An attacker can also forge a source address when sending a request, so that the response will return to the victim. This could be done to confuse the victim or to flood it, if a service provides a mechanism for traffic amplification.

Defense. Routing updates should be authenticated to prevent malicious modification by untrusted adversaries [53]. A freshness mechanism can protect against replay attacks, while cryptographic integrity checks protect against unauthorized modification of a message while in transit. In WSNs that use a hierarchical structure for routing, egress filtering may be appropriate. Nodes that serve as collection points for subordinates' traffic may examine each message before forwarding it. Messages with source addresses that could not legitimately originate at lower levels of the hierarchy are discarded.

32.3.7 Sinkholes

Karlof and Wagner describe *sinkhole* attacks [29], in which an attempt is made to lure traffic from the sensor network to pass through an adversary. Low-cost routes may be erroneously flooded to lure the traffic, or a wormhole attack (see below) could be mounted actually to provide a low-cost route. In either case, the objective is for the attacker to be positioned so that other selective forwarding attacks, or merely eavesdropping, are easier to do.

Defense. One approach to avoiding sinkholes is to use routing algorithms resistant to arbitrary configurations, such as geographic forwarding [15, 29]. Because each node makes an independent forwarding decision based on the location of its neighbors, it is not as easy to attract routing to an attacker. Communicating parties may also use end-to-end verification of advertised latency or quality to detect when a path may contain an unwarranted diversion [29]. Upon detecting a problem, nodes may attempt systematic rerouting to avoid the malicious node [42].

32.3.8 Wormholes

In a *wormhole* attack, adversaries cooperate to provide a low-latency side channel for communication [23]. For example, two attackers may possess a second radio for communicating over a higher power, long-range link. Messages received at one attacker are relayed to the other using the side channel, where they are transmitted as if only one hop away from the original source. This ability to understate one's distance from another node may cause neighboring nodes to favor the attacker for routing — another example of a sinkhole. As long as the side channel exists, service may actually be enhanced, instead of denied. However, when the attacker moves or ceases to tunnel messages, the network may be left in an inconsistent state that requires reinitialization of some services to restore proper function.

Defense. As described for sinkholes, geographic forwarding is a tamper-resistant routing protocol (however, see the Sybil attack next). Each message is forwarded individually, choosing the next-hop node to be the neighbor closest to the ultimate destination. Such a scheme would not favor a wormhole in the network, although it may coincidentally use it. Hu et al. describe a defense based on packet leashes, in which the distance that a message may travel in a single hop is limited [23]. Each message includes a timestamp and the location of the sender. The receiver compares these with its location and time to determine if the maximum transmission range has been exceeded. The solution requires clock synchronization and accurate location verification, which may limit its applicability to WSNs.

32.3.9 Sybil Attack

Most protocols assume that nodes present a single unique identity. In a *Sybil attack*, an attacker presents multiple identities [14]. Coupled with insecure location claims, this means an attacker can appear to be in multiple places at the same time. By creating fake identities of nodes located at the edge of communication range all around a victim, chances are high that the attacker will be chosen as the next hop in geographic forwarding. The attack can also degrade any guarantees made by a multipath routing scheme, making selective forwarding easy.

Defense. Because identity fraud is central to the Sybil attack, proper authentication is a key defense [29]. A trusted key server or base station may be used to authenticate nodes to each other and bootstrap a shared session key for encrypted communications, as in SPINS [40]. This requires that every node share a secret key with the key server. If a single network key is used, compromise of any node in the WSN would defeat all authentication.

Another defense is location verification. Sastry et al. describe a simple protocol that uses the difference in time of flight of radio and sound waves to verify location claims securely [46]. The combination of these two defenses, verifying identities and locations, would prevent Sybil-based DoS attacks.

32.3.10 Flooding

A *flooding* attack overwhelms a victim's limited resources, whether memory, processing cycles, or bandwidth. In a homogeneous network, a subverted node may have the same resource limitations as a victim, making an attack relatively expensive to mount. However, if an attacker possesses or subverts a more powerful device, such as a base station or laptop, the cost of the flooding attack relative to the result is much lower. Poorly designed protocols can provide a stage for traffic amplification attacks, such as the Internet Smurf [8] attack. Here an attacker forges the victim's address as the source of a widely and easily distributed request. All nodes receiving the request reply to the victim, flooding its link. Thus, a few messages from an attacker result in a storm of responses to the victim.

Defense. To combat the consumption of memory resources at servers, Aura and Nikander describe principles for stateless connection management [54]. This approach securely stores the server's state in all messages, requiring the client to return it with every future response. The server need not store the state associated with the connection, which could lead to memory exhaustion if connections are not strongly authenticated (as is the case with TCP SYN-floods [44]).

Even without the overhead of ferrying server state in all messages, protocols can avoid allocating resources for unauthenticated traffic. The need for protocols that create traffic asymmetries, such as area multicast, should be carefully weighed against their potential to allow traffic amplification attacks.

Another approach to curtailing flooding is to require the clients of services to commit significant resources before connections are established. Client puzzles are one such method: servers dispense cryptographic puzzles that must be solved by brute force before connection-related resources on the server are allocated [6, 26]. The difficulty of the puzzles is scalable, so the server can increase the requirements when it believes it is under attack. In a WSN, this could adversely affect the many legitimate sensor devices, each of which has limited resources to commit.

A final strategy is to provide a way to detect the source of the flooding using a traceback mechanism. Existing schemes are IP-based and are appropriate for the Internet's scale and structure [13, 45, 48]. The goal, which could be ported to a WSN environment, is to allow even spoofed messages to be traced back to their actual source, using in-network auditing, periodic return-path messages, and other mechanisms.

32.3.11 HELLO Floods

Slightly different from conventional flooding, a *HELLO flood* is a single broadcast by a powerful adversary to many members of the WSN, announcing false neighbor status [29]. Many protocols use the exchange of HELLO messages to establish local neighborhood tables. The result of a HELLO flood is that every node thinks the attacker is within one-hop radio communication range. If the attacker subsequently advertises low-cost routes, nodes will attempt to forward their messages to the attacker. Retransmission attempts to the absent attacker cause traffic congestion and confusion in the entire routing system.

Defense. Verifying the bidirectionality of local links before using them is effective if the attacker possesses the same reception capabilities as the sensor devices [29]. However, if the attacker can use a sensitive receiver, it can eventually convince nodes in the network of its legitimacy. Authentication, as described for the Sybil defense, is also a possible solution. Nodes can use a trusted third party to verify the authenticity of each of their neighbors before forwarding messages to them.

32.3.12 Algorithmic Complexity Attack

An *algorithmic complexity attack* [12] requires several features to be present to be successful. First, a service must use some algorithm and data structure with super-linear, worst-case behavior, such as a hash table that inserts colliding elements into a hash chain. The input to the algorithm must be controllable by the attacker. The attacker must be able to compute values that will evoke worst-case performance and be able to deliver these values to the service. Once the data structure is primed, the attacker rapidly sends data that take a long time to process due to the inefficiency of the algorithm. During this time, processor cycles at the victim are consumed in a DoS attack. In the hash table example, this happens when all the inserted values hash to the same bucket and are appended to the chain after examining every other element — an $O(n)$ operation.

Defense. Removing any of the preceding requirements will disrupt the attack. When possible, services should use algorithms with efficient worst-case performance. The service can also remove determinism (from the point of view of the attacker) from the algorithm execution. In the hash table example, this can mean using a keyed cryptographic hash function so that the sequence of hash values computed by the program is unpredictable for the attacker. Another simple approach is to limit the size of data structures so that even in the worst case, less efficient algorithms cannot consume too much processing time.

The smaller memory sizes and lower-rate communication patterns that would ordinarily mitigate this kind of attack may be counteracted by the relatively slow microprocessors that may exist in WSN devices.

32.4 Related Work

This section briefly describes related work in two primary areas: development of DoS as a security concern and classification of DoS attacks.

- Gligor presents an early definition of DoS, as already discussed in Subsection 32.1.2, in terms of users, services, and maximum waiting times [19]. He argues that current operating system protection mechanisms are insufficient to address DoS. Interuser dependencies are identified as a critical element of DoS. Gligor claims that any dependency between users of a service that do not explicitly share objects in the service is undesirable and must be eliminated. He gives examples of DoS in operating systems due to bad sharing policies, mechanisms, and the interactions between them.
- Subsequent work [20, 21] refines the necessary and sufficient conditions for DoS and discusses the guises of DoS in computer networks. Gligor classifies attacks into three categories. Service-oriented attacks are local and are due to sharing and user dependencies, as in Gligor [19]. Intruder-oriented attacks involve the deletion or delay of packets in a network (modification and replay are handled by integrity models). Finally, performance fluctuations are due to congestion and flow control.
- Gligor claims that most intruder-based attacks cannot be prevented, and emphasizes the need to detect and recover from attacks. This idea was further developed by Yu and Gligor, who use temporal logic to express sharing agreements between users [51]. A key claim is that user agreements are external constraints and cannot be internalized and enforced by the shared service. A full grasp of the implications of this was slow in coming [16]: prevention of DoS is impossible except where it can be reduced to a "liveness" or safety problem.
- Millen describes a trusted service called the denial-of-service protection base (DPB), which enforces user agreements for resource allocation [34]. He suggests that probabilistic waiting-time policies are possible, although none are given. The DPB must be tamperproof, not prevented from running; implement all dependent services; and be able to revoke resources. These limitations present challenges for application of the DPB.
- Needham briefly describes application-layer attacks against a server, network, and client, including generating false alarms [37, 38]. He argues that, in some circumstances, availability is more important than confidentiality or authenticity. Needham suggests that the best protection against DoS is for a monitor service to depend on the same services that it is attempting to protect.
- Meadows addresses the problem that DoS attacks on the Internet often occur before parties are authenticated [32]. She presents a specification method for describing the incremental trust and capabilities that communicating parties have after performing each step of a protocol. Unlike authentication, DoS is a matter of degree and no protocol is fully immune.
- More recent work has focused on aspects of DoS attacks unique to wireless ad hoc networks [18] and WSNs [49]. In the former, Geng et al. propose using dynamic usage-based pricing to provide an economic disincentive to potential DoS attackers. In the latter, Wood and Stankovic survey attacks and defenses against a WSN and examine the weaknesses of two protocols that did not include security in their original designs.
- Other recent works include discussions of DoS possibilities alongside other concerns, such as secure routing in WSNs [29].

Various authors have presented taxonomies of attacks or vulnerabilities tailored to particular environments.

- Howard employs a systematic, process-based taxonomy to categorize computer and network attacks on the Internet, including DoS [22, §6.4]. Rather than listing particular vulnerabilities, he focuses on the attack as a whole, from an operational perspective. According to this taxonomy, an attacker uses tools to gain unauthorized access (through exploiting vulnerabilities), causing certain results consistent with the attacker's ultimate objectives. (This chapter explains attacks based (in part) on the vulnerabilities exploited, rather than the tools used. The programs and scripts used to exploit vulnerabilities become more automatic, complex, and widely applicable over time. Also, this chapter considers peculiarities of WSNs appropriate to the wired Internet.)

- Leiwo et al. classify Internet DoS attacks as tolerable or fatal [30]. Attack methods are classified based on the allowable deviations from a protocol: none, sequence, syntax, and semantics. They suggest that server resources only be allocated after client authentication and that the workload of clients should be higher than that of servers.
- Moore et al. approach DoS attacks empirically, using backscatter analysis [36]. They loosely classify attacks by examining the victims' responses sent to spoofed addresses that happened to reside in a large network they were monitoring. During their 3-week period of study, they recorded 12,805 attacks on more than 5000 hosts.
- Mirkovic et al. present a taxonomy of DDoS attacks and defenses [35]. Many of the attack characterizations are specific to distributed, centrally coordinated attacks on IP networks. They differentiate between protocol and brute-force (flooding) attacks on systems, noting that modifying protocols to protect against protocol attacks pushes the attack into the brute-force category. One part of the attack taxonomy, "degree of automation," is similar to Howard's tool-based approach.
- Rice and Davis take a different approach to categorizing attacks, performing a postmortem analysis of attack tools to build a genealogical attack tree [41]. This has the advantage of making relationships between vulnerabilities explicit and may also suggest future attack possibilities.
- Shields classifies the results of a network denial-of-service attack as a four-tuple: the protocol or layer used; effect on the limiting resource; resource being consumed or corrupted; and location of the device attacked [43].
- Lough discusses and reviews prior work on the properties of good taxonomies and presents VERDICT, a new taxonomy for computer security flaws [31]. VERDICT attributes the causes of vulnerabilities to improper verification, exposure, randomness, or deallocation. Lough applies the taxonomy to wireless networks, but only briefly considers DoS.

32.5 Conclusion

Wireless sensor network research is transitioning to interesting real-world applications. As the networks become more pervasive and accessible, they will face some of the same problems with which wired Internet and wireless ad hoc networks already struggle.

One such problem is denial of service. DoS attacks are commonplace on the Internet, from background noise to front-page headlines. WSNs, with all their inherent resource limitations, are particularly susceptible to consumption or destruction of these scarce resources. Protecting the availability of sensor networks that monitor and control critical systems, while remaining low cost and flexible, is a principal challenge that remains.

Every DoS attack originates from someone, uses various methods, and produces results. An attack taxonomy to facilitate higher level reasoning and analysis of risk in WSNs has been presented. It describes aspects of the entire process of a DoS attack:

- The attacker's identity, including motivation and resources
- What the attacker is capable of doing in the WSN
- Type and criticality of services targeted by the attack
- Physical and logical vulnerabilities that may be exploited
- The results of the attack, from nuisance to catastrophe

Because vulnerabilities are among the few parts of an attack under the control of the WSN designer, a variety of them have been reviewed, along with possible defenses. Many solutions given are not without cost; but they may be a necessary price for increased robustness against DoS attacks. One should seek to prevent DoS where possible, and detect, tolerate, and recover from the rest.

References

1. E. Ayanoglu, I. Chih-Lin, R.D. Gitlin, and J.E. Mazo. Diversity coding for self-healing and fault-tolerant communication networks. *IEEE Trans. Commun.* COM-41, 1677–1686, November 1993.
2. R. Anderson and M. Kuhn. Tamper resistance — a cautionary note. In *Proc. 2nd USENIX Workshop Electron. Commerce*, 1–11, Oakland, CA, November 1996.
3. R. Anderson. Why cryptosystems fail. In *Proc. 1st ACM Conf. Computer Commun. Security*, 215–227, Fairfax, VA, November 1993, ACM Press.
4. R. Anderson. Research and development initiatives focused on preventing, detecting and responding to insider misuse of critical defense information systems, CF-151-OSD. In workshop cosponsored by Rand's NSRD, 1999.
5. R. Anderson. *Security Engineering: A Guide to Building Dependable Distributed Systems*. Wiley Computer Publishing, New York, 2001.
6. T. Aura, P. Nikander, and J. Leiwo. DOS-resistant authentication with client puzzles. *Lect. Notes Computer Sci.*, 2133, 170–177, 2001.
7. I.F. Akyildiz, W. Su, Y. Sankarasubramaniam, E. Cayirci. Wireless sensor networks: a survey. *Computer Networks* (Amsterdam, Netherlands: 1999), 38(4), 393–422, March 2002.
8. CERT Coordination Center. Smurf IP denial-of-service attacks. Technical report, CERT Advisory CA-98:01, November, 1998.
9. K. Chintalapudi and R. Govindan. Localized edge detection in wireless sensor networks. In *Proc. IEE ICC Workshop Sensor Network Protocols Applications (SNPA)*, May 2003, 59–70.
10. S. Cheung and K. Levitt. Protecting routing infrastructures from denial of service using cooperative intrusion detection. In *New Security Paradigms Workshop*, Cumbria, U.K., September 1997, 94–106.
11. Committee on National Security Systems (CNSS). National information assurance glossary, NSTISSI No. 4009, May 2003.
12. S.A. Crosby and D.S. Wallach. Denial of service via algorithmic complexity attacks. In *Proc. USENIX Security 2003*, August 2003, 29–44.
13. D.Dean, M. Franklin, and A. Stubblefield. An algebraic approach to IP traceback. In *Proc. Network Distributed Syst. Security Symp. (NDSS)*, 3–12, February 2001.
14. J.R. Douceur. The sybil attack. In *In IPTPS*, 215–260, 2002.
15. G.G. Finn. Routing and addressing problems in large metropolitan-scale Internetworks. Technical report ISI/RR-87-180, ISI, March 1987.
16. V.D. Gligor, M. Blaze, and J. Ioannidis. Denial of service — panel discussion. April 2000. Security Protocols Workshop, Cambridge, U.K., LNCS 2133, Springer-Verlag, 194–203.
17. D. Ganesan, R. Govindan, S. Shenker, and D. Estrin. Highly resilient, energy-efficient multipath routing in wireless sensor networks. *Mobile Computing Commun. Rev.*, 4(5), October 2001.
18. X. Geng, Y. Huang, and A.B. Whinston. Defending wireless infrastructure against the challenge of DDoS attacks. *Mobile Networks Applications*, 7(3), 213–223, 2002.
19. V.D. Gligor. A note on the denial-of-service problem. In *Proc. IEEE Symp. Security Privacy*, 139–149, 1983.
20. V.D. Gligor. A note on the denial-of-service in operating systems. *IEEE Trans. Software Eng.*, SE-10(3), 320–324, May 1984.
21. V.D. Gligor. On denial of service in computer networks. In *Proc. Int. Conf. Data Eng.*, 608–617, IEEE, 1986.
22. J.D. Howard. An analysis of security incidents on the Internet 1989–1995. Ph.D. thesis, Department of engineering and Public Policy, Carnegie Mellon University, April 1997.

23. Y.-C. Hu, A. Perrig, and D.B. Johnson. Packet leashes: a defense against wormhole attacks in wireless networks. In *Proc. IEEE Infocom 2003*, 3, 1976–1986, April 2003.
24. J. Hill, R. Szewczyk, A. Woo, S. Hollar, D.E. Culler, and K.S.J. Pister. System architecture directions for networked sensors. In *Architectural Support Programming Languages Operating Syst.*, 93–104, 2000.
25. ICAT metabase. A CVE-based vulnerability database. NIST. http://icat.nist.gov/icat.cfm, visited August 2003.
26. A. Juels and J. Brainard. Client puzzles: a cryptographic defense against connection depletion attacks. In S. Kent, Ed., *Proc. NDSS '99 (Networks Distributed Security Syst.)*, 151–165, 1999.
27. J. Kahn, R. Katz, and K. Pister. Emerging challenges: mobile networking for "smart dust." *J. Commun. Networks*, 2(3), 188–196, September 2000.
28. F. Kargl, J. Maier, and M. Weber. Protecting Web servers from distributed denial-of-service attacks. In *Proc. 10th Int. WWW Conf.*, 514–524, 2001.
29. C. Karlof and D. Wagner. Secure routing in wireless sensor networks: attacks and countermeasures. In *1st IEEE Int. Workshop Sensor Network Protocols Applications*, 2003, 1–15.
30. J. Leiwo, T. Aura, and P. Nikander. Towards network denial-of-service resistant protocols. In *Proc. 15th Int. Inf. Security Conf. (IFIP/SEC 200)*, 301–310, Kluwer, Dordrecht, August 2000.
31. D.L. Lough. A taxonomy of computer attacks with applications to wireless networks. Ph.D. thesis, Virginia Tech. Computer Engineering Dept., April 2001.
32. C. Meadows. A formal framework and evaluation method for network denial of service. In *Proc. 12th Computer Security Foundations Workshop*, 4–13, IEEE Computer Society Press, June 1990.
33. S. Marti, T. Giuli, K. Lai, and M. Baker. Mitigating routing misbehavior in mobile ad hoc networks. In *Proc. 6th Annu. Int. Conf. Mobile Computing Networking (MOBICOM-00)*, 255–265, New York, August 2000, ACM Press.
34. J.K. Millen. A resource allocation model for denial of service. In *Proc. 1992 IEEE Computer Soc. Symp. Res. Security Privacy*, 137–147, IEEE Computer Society, May 1992.
35. J. Mirkovic, J. Martin, and P. Reiher. A taxonomy of DDoS attacks and DDoS defense mechanisms. Technical report CSD-TR-020018, Computer Science Dept., University of California, Los Angeles, 2002.
36. D. Moore, G. Voelker, and S. Savage. Inferring Internet denial-of-service activity. In *Proc. 10th USENIX Security Symp.*, 9–22, USENIX, August 2001.
37. R.M. Needham. Denial of service. In *Proc. 1st ACM Conf. Computer Commun. Security*, 151–153, ACM Press, 1993.
38. R.M. Needham. Denial of service: an example. *Commun. ACM*, 37(11), 42–46, 1994.
39. R.L. Pickholtz, D.L. Schilling, and L.B. Milstein. Theory of spread spectrum communications — a tutorial. *IEEE Trans. Commun.*, 20(5), 855–884, May 1982.
40. A. Perrig, R. Szewczyk, V. Wen, D. Culler, and J.D. Tygar. SPINs: security protocols for sensor networks. In *Proc. 7th Annu. Int. Conf. Mobile Computing Networks MOBICOM 2001*, 189–199, July 2001.
41. G. Rice and J.A. Davis. A genealogical approach to analyzing postmortem denial-of-service attacks. In *Secure Dependable Syst. Forensics Workshop*, University of Idaho, September 2002.
42. J. Staddon, D. Balfanz, and G. Durfee. Efficient tracing of failed nodes in sensor networks. In *Proc. 1st ACM Int. Workshop Wireless Sensor Networks Applications (WSNA)*, 122–130, ACM Press, 2002.
43. C. Shields. What do we mean by network denial of service? In *Proc. 2002 IEEE Workshop Inf. Assurance Security*, West Point, NY, June 2002, 1–8.
44. C.L. Schuba, I.V. Krsul, M.G. Kuhn, E.H. Spafford, A. Sundaram, and D. Zamboni. Analysis of a denial-of-service attack on TCP. In *Proc. 1997 Conf. Security Privacy (S&P-97)*, 208–223, Los Alamitos, May 1997, IEEE Press.

45. A.C. Snoeren, C. Partridge, L.A. Sanchez, C.E. Jones, F. Tchakountio, B. Schwartz, S.T. Kent, and W.T. Strayer. Single-packet IP traceback *IEEE/ACM Transactions on Networking*, 10(6), 721–734, 2002.
46. N. Sastry, U. Shankar, and D. Wagner. Secure verification of location claims. In *ACM Workshop Wireless Security (WiSe 2003)*, September 2003, 1–10.
47. M. Stahlbert. Radio jamming attacks against two popular mobile networks, 2000. Helsinki University of Technology, Tik-110.501 Seminar on Network Security.
48. S. Savage, D. Wetherall, A.R. Karlin, and T. Anderson. Practical network support for IP traceback. In *Proc. ACM SIGCOMM*, 295–306, August 2000.
49. A.D. Wood and J.A. Stankovic. Denial of service in sensor networks. *IEEE Computer*, 35(10), 54–62, October 2002.
50. A.D. Wood, J.A. Stankovic, and S.H. Son. JAM: a jammed-area mapping service for sensor networks. In *Proc. Real-Time Syst. Symp.*, Cancun, 2003, 286–297.
51. C.F. Yu and V.D. Gligor. A specification and verification method for preventing denial of service. *IEEE Trans. Software Eng.*, 16(6), 581–592, June 1990.
52. X. Yang, K.G. Ong, W.R. Dreschel, K. Zeng, C.S. Mungle, and C.A. Grimes. Design of a wireless sensor network for long-term, in-situ monitoring of an aqueous environment. *Sensors*, 2, 455–472, 2002.
53. L. Zhou and Z.J. Haas. Securing ad hoc networks. *IEEE Network*, 13(6), 24–30, 1999.
54. T. Aura and P. Nikander. Stateless connections. In *Proc. Int. Conf. Information and Communications Security (ICICS '97)*, LNCS 1334, 87–97, Springer-Verlag, 1997.

33
Reliability Support in Sensor Networks

33.1 Introduction **33**-1
33.2 Reliability Problems in Sensor Networks **33**-2
33.3 Existing Work on Reliability Support **33**-2
33.4 Supporting Reliability with Distributed Services **33**-3
33.5 Architecture of a Distributed Sensor System **33**-3
33.6 Directed Diffusion Network **33**-4
33.7 Distributed Services **33**-5
  Reconfigurable Smart Nodes • Distributed Lookup Server • Compositional Server • Adaptation Server
33.8 Mechanisms and Tools **33**-8
  Reliable Remote Service Execution • Connectors • Sensor Task Structures
33.9 Dynamic Adaptation of Distributed Sensor Applications **33**-10
  Consistent Schedule • Runtime Adaptation Control • Recovery from Sensor Failure and Degradation

Alvin S. Lim
Auburn University

33.10 Conclusions **33**-11

33.1 Introduction

Wireless sensor networks are becoming increasingly common due to the proliferation of very small, integrated low-powered sensors and mobile devices [1–3] capable of gathering critical, real-time information for remote surveillance, continual monitoring, and distributed target tracking. These mobile and miniaturized information devices are increasingly more powerful. In the current state of the art, they are equipped with small and embedded processors, wireless communication circuitry, information storage capability, smart sensors, and actuators. The embedded smart sensors and information storage enable the sensor nodes to pass real-world information into the digital system in the far end of the digital divide. Information is then fused and propagated through the sensor network to the intended end users or applications.

With increasing computing and wireless communication capabilities, the role of these sensor nodes in the distributed sensor network will expand from mere information dissemination to more demanding tasks of in-network processing and other distributed computation, such as sensor fusion, classification, and collaborative target tracking. These collaborative tasks are performed on the fly as the information is being propagated for specialized distributed embedded application. Thus, abundant processing powers in the numerous sensors can be exploited to process the sensor information before and as it is propagated in the sensor network.

33.2 Reliability Problems in Sensor Networks

The drawback in relying on these sensor nodes for in-network processing is that they and the sensor network infrastructure are prone to failure, insufficient energy supply, high error rate, and disconnection. Furthermore, sensor devices are mobile, deployed spontaneously, and frequently replaced and repositioned to provide more accurate surveillance and targeting information for dynamic control of the enterprise. Aggravating this problem is the ad hoc nature of sensor networks that are typically implemented with little or no fixed network support. Despite these dynamic changes in configuration of the sensor network, distributed in-network processing of sensor data must be performed reliably to ensure the correctness and accuracy of the results of embedded applications. Critical real-time information must still be disseminated dynamically from mobile sensor data sources through the self-organizing network infrastructure. Information fusion must be reliable in order for components to maintain correct control of dynamic replanning and reoptimization of the theater of operation, based on newly available information.

33.3 Existing Work on Reliability Support

The approaches used in most current work on reliable sensor applications are based on various algorithms for tolerating faults of sensors when fusing data from a set of sensors to generate a result with a certain level of accuracy and precision. The early work by Dolev [4] provides a typical solution to the Byzantine general's problem [5]. Other algorithms extend this basic idea for fault-tolerant sensor fusion and enhance the accuracy of the results [6], while still others improve the precision [7]. The Brooks–Iyengar hybrid algorithm [8] improves the precision and accuracy of the results in the presence of faulty sensors.

These works provide an important basis for this chapter's approach in that they develop the algorithms for identifying faulty sensors. The work described here provides the system mechanisms for replacing faulty sensors with other available sensors in the distributed sensor network. It is assumed that sensor networks are dense, i.e., there are many redundant sensors, because these miniaturized sensors [1–3] are small and inexpensive and can be deployed in very large numbers. The main thrust of this work is to provide the mechanisms for reliably replacing faulty sensors with reliable ones not only for sensor fusion algorithms, but also for more general distributed sensor algorithms, such as distributed target tracking and distributed query processing.

Other distributed environments have been developed for adapting distributed applications. Although these mechanisms are relevant, they may not provide good support for distributed sensor applications. These systems include Darwin [9]; ILI [10]; and Polylith [11]. They allow structural adaptation, but do not enforce consistency and correctness policies for adaptation. These responsibilities are left to the sensor application designers. Conic [12] and Argus [13] use transactions for preserving consistency, but restrict complex interaction and operations of enterprises.

Many reliable distributed systems are based on the distributed transaction model [14]. The authors' facilities provide a general mechanism for recovering from sensor failure that does not depend on the transaction model and does not have the restrictions of transactions. However, it may also support basic transactions as well as newer extended transactions, mobile transactions and nontransactional mechanisms. These flexible interaction capabilities may be required to support large and complex enterprises. The facilities allow new adaptation techniques based on transactions, although some may be restrictive, particularly for large distributed sensor applications with complex interactions.

Extended transactions, such as atomic abstract data types [15] and ACTA [16], have limitations in distributed sensor applications in that they require transactions to be serializable after nondependent operations are commuted. The burden of analyzing commutativity (or dependency) between operations is placed on the programmers. Optimistic approaches [17] also require committed transactions to be serializable. Cooperative transactions [18] extend basic nested transactions, but still require the partial order of lower level nested transactions to be equivalent to a total order of operations invoked by subtransactions of the cooperating transaction. The benefit of this approach is that during normal distributed sensor operations, these techniques may not require serializability and permit complex noncommutable interactions. During adaptation of the distributed sensor application, sensor nodes will be automatically restored to globally consistent states using analyses of dependency constraints.

Reliability issues have also been addressed at other layers of the sensor network protocol stack. Reliable multisegment transport (RMST) [19] implements a reliable transport protocol over a directed diffusion network. It is based on the selective repeat protocol. Ad hoc transport protocol (ATP) [20] also uses selective repeat protocol with a modification that intermediate nodes and not just the receiver may return data transmission and path failure information. Pump slowly fetch quickly (PSFQ) [21] uses negative acknowledgment methods for reliable transmission and is based on the assumption that data transmission in sensor networks is usually slow. It also uses intermediate nodes to recover lost data. In these protocols, reliability is maintained by the network protocol stack to allow reliable transmission of data. Although reliable data transmission simplifies the development, other support mechanisms are also required to maintain other aspects of reliable distributed sensor applications.

33.4 Supporting Reliability with Distributed Services

The reliability problems of sensor networks can be better overcome through the appropriate system mechanism for supporting reliability and reconfiguration of distributed sensor fusion applications. The three main distributed mechanisms for reliability and reconfiguration are service lookup, sensor node composition, and dynamic adaptation. Based on these services, another system mechanism, called connectors, is developed that supports reliable and reconfigurable communication between sensor nodes. Using these distributed lookup servers, composition servers, adaptation servers, and connectors, remote surveillance and target tracking systems may adapt these services to device failure and degradation, movement of sensor nodes, and changes in task and network requirements. Furthermore, new application-specific services may be deployed reliably to support existing distributed sensor applications while they are executing.

These mechanisms also enable sensor nodes to have capabilities for self-assembling impromptu networks that are incrementally extensible, sensor node mobility, and changes in task and network requirements. Nodes are aware of their own capabilities and those of other nodes around them that may provide the networking and system services or resources that they need. Although nodes are autonomous, they may cooperate with one another to disseminate information or assist each other in adapting to changes in the network configuration and failure or degradation of some sensor nodes.

33.5 Architecture of a Distributed Sensor System

The architecture consists of three key system layers (Figure 33.1):

- Self-organizing *application systems*, e.g., sensor information-processing layer and collaborative signal processing
- *Configurable distributed systems* that provide distributed services to the application systems
- *Robust sensor networking and physical devices layer* that routes messages through the ad hoc sensor network

At the physical device layer, different physical sensor and mobile devices may be assembled impromptu and reconfigured dynamically in an ad hoc wireless network. Each sensor node contains battery power source, wireless communications, multiple sensing modality, computation unit, and limited memory. Three common sensing modalities are supported by acoustic sensing using commercial microphones, seismic vibration using geophones, and motion detection using two-pixel infrared imagers. The wireless transceiver in the nodes provides communications between nodes, using time division multiplexing and frequency hopping spread spectrum. Each node contains a global positioning system (GPS) receiver that allows it to determine its current location and time. As described later, message routing and query processing use this location information.

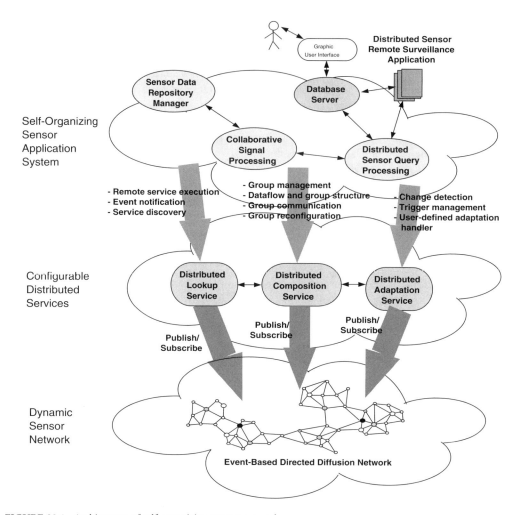

FIGURE 33.1 Architecture of self-organizing sensor networks.

At the networking layer, ad hoc routing protocols allow messages to be forwarded through multiple physical clusters of sensor nodes. Directed diffusion routing is used because of its ability to adapt to changes in sensor network topology dynamically and its energy-efficient localized algorithms. Distributed services support adaptation of applications systems in spite of dynamic changes in the sensor network. Application and system programs may also use simpler communication interfaces and abstraction than the raw network communication interface and metaphor (e.g., subscribe/publish used in diffusion routing). Furthermore, these distributed services may enhance overall performance, such as throughput and delay.

At the application system layer, distributed query processing and collaborative signal processing modules communicate with each other to support the surveillance and dynamic tracking functions of the enterprise.

33.6 Directed Diffusion Network

Directed diffusion protocol [22] is used for implementing all the distributed services and for retrieving data through dynamically changing ad hoc sensor networks. Diffusion routing converges quickly to network topological changes, conserves mobile sensor energy, and reduces the network bandwidth overhead because routing information is not periodically advertised.

Directed diffusion is a data-centric protocol, i.e., nodes are not addressed by IP addresses but by the data they generate. Data generated by a node are named by its *attribute–valu* pair. A sink node requests certain data by broadcasting an *interest* for the named data in the sensor network. The interest and gradient is established at intermediate nodes for this request throughout the sensor network. When a source node has a datum that matches the interest, it will be drawn down toward that sink node using this interest gradient that was established. Intermediate nodes may cache, transform data, or direct interests based on previously cached data. The sink node can determine if a neighbor node is in the shortest path whenever it receives a new datum earliest from that node. The sink node will reinforce this shortest path by sending a reinforcement packet with a higher data rate to this neighbor node, which forwards it to all the nodes in the shortest path.

Distributed services and applications use the publish and subscribe API provided by directed diffusion. Through the subscribe function, an application declares an interest that consists of a list of attribute–value pairs. The subscription is then diffused through the sensor network. A source node may indicate the type of data it offers through the publish function. It then sends the actual data through the handle returned from the publish function. The sink node then receives the data that have propagated through the sensor network using a recv function call with the handle returned from the subscribe call.

33.7 Distributed Services

Self-organizing networks may be built from reconfigurable smart sensor nodes that may be developed independently but may interact with other smart sensor nodes. Some smart sensor nodes may execute autonomously to provide networking and system services or control information retrieval and dissemination in the dynamically changing sensor network. To enhance the ability to reconfigure their networking, configuration, and adaptation functionalities, smart sensor nodes may make use of three main classes of distributed services: lookup, composition and adaptation services (Figure 33.1). These distributed services simplify sensor application development and also permit applications to execute more efficiently over a diffusion network by alleviating some of the problems of network traffic: communication delays, weak connectivity, mobility, disconnection, dynamic reconfiguration, and limited power.

33.7.1 Reconfigurable Smart Nodes

By exploiting these distributed services, sensor nodes can be self-aware, self-reconfigurable, and autonomous. These sensor nodes, known as reconfigurable smart nodes (in this chapter, referred to as smart nodes or sensor nodes), can be used to build scalable and self-organizing sensor networks. Smart nodes may represent sensor nodes, other types of mobile nodes, fixed nodes, or a cluster of these nodes. They may simultaneously be service providers for other smart nodes and clients of services that other smart nodes provide. Smart nodes may be dynamically composed into impromptu networks of clustered smart nodes that work together to provide abstract services for the agile sensor network. They may also adapt rapidly to abrupt changes in the sensors' capabilities, events, and new real-time information. Very large networks with hundreds of thousands of sensor nodes can be built by hierarchically composing reconfigurable smart nodes.

Smart sensor nodes may consist of hardware devices and software for interacting with the real-world systems. The hardware may contain computational, memory, wireless communication, and sensing devices. Smart nodes may contain control software for monitoring information from real-world devices, such as simple sensors, engaging in distributed signal processing and generating appropriate control signals to produce a desired result in the real-world system. The control software takes advantage of the functionalities provided by the networking and system software.

Smart nodes interact with other smart nodes through well-defined interfaces (for networking and systems operations) that also maintain interaction states to allow nodes to be reconfigured dynamically. These explicit interaction states and behavior information allow localized algorithms with the adaptation servers to maintain consistency when autonomous nodes and clusters are reconfigured dynamically, move

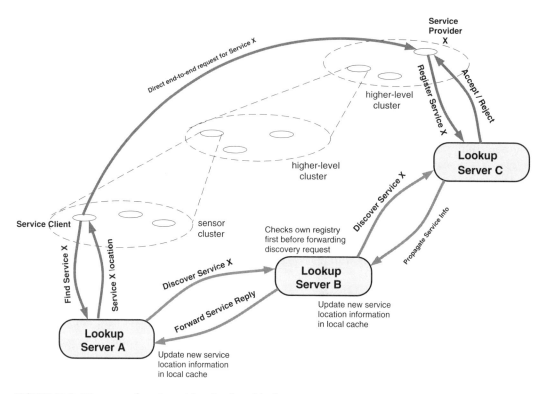

FIGURE 33.2 Discovery of services with a distributed lookup server.

around, or recover from failure. Smart node implementation and data (software and hardware) are encapsulated (hidden) from other nodes.

When new smart nodes are added to the sensor network, they register their services with a lookup server (Figure 33.2). Other nodes that require a service will discover the services available in a cluster through the lookup servers that return the location and interface of the service nodes. This is similar to Jini [14], which manages system-level services based on Java code executing in IP-based networks. On the other hand, reconfigurable smart nodes may provide lower level networking services using generic mobile codes executing in data-centric sensor networks. Client nodes then interact directly with the service node. Smart nodes are aware of their location, configuration, and services that they perform.

33.7.2 Distributed Lookup Server

New network and system services may be introduced by a sensor node for use by other nodes as the sensor network self-organizes. A sensor node that provides a service is called a service provider and a node that uses the service is called a service client. A sensor node may register a resource that it maintains or service that it can perform with a lookup server (Figure 33.2). A lookup server may contain information on services or resources at multiple clusters. Other nodes requiring the service may request it through a lookup server. If the service is recorded in the lookup server, it will return the location to the requesting node. Otherwise, a discovery protocol is used to locate the service through other lookup servers.

Applications use the lookup service through the following API for registering, finding, and calling a service. Sensor nodes that receive the service provider information can make service calls through a reliable remote execution method. The following describes the purposes and side effects of these lookup service function calls.

- `service_register()`. This function allows a service provider to register its service with a lookup server in the region. Services will remain in the lookup server for the lifetime specified by the service provider. The information supplied by the service provider ó service name, service type, location or address, interface definition, and mobile code ó is stored in the lookup server. Lookup servers for different regions may coordinate with each other to update their lists of service information.
- `service_deregister()`. This function allows a service provider to remove its service from the lookup server registry.
- `lookup_service()`. This function allows a service client to find the location or address of a service provider and/or the interface for using the service. Service lookup can also be based on cluster or predicate matching.
- `service_exec()`. This function allows a sensor node to request the service and obtain the results from the service provider. The service provider performs the requested service or remote procedure call and returns the results.
- `service_call()`. This function allows a service client to find and make a call for a service when the service client does not know the location or address of the service provider and/or the interface for using the service. It is implemented as a combination of the `lookup_service()` and `service_exec()` calls described next.

To search multiple lookup servers, a request message is propagated to all the lookup servers; the server that contains the service registration information will return the reply with the service location. It may also return the cluster name of that service. The lookup server that made the request will then cache that service location and cluster name information in its local registration cache. At regular frequency, service and resource registration information may be disseminated from one lookup server to others in the agile sensor network.

33.7.3 Compositional Server

The compositional server manages clusters of sensor nodes by allowing various smart nodes that may be added to or removed. It also manages network abstractions (or group behavior) and hierarchical composition of clusters. The compositional server simplifies dynamic reconfiguration of services provided by each smart node or cluster. The advantage of forming clusters is that failure and recovery of sensor nodes can be contained within a cluster where the scope of the effects of failure recovery is limited to only sensors in the cluster. This simplifies the development of a large self-organizing sensor network by allowing individual nodes and clusters to be specified and designed independently.

Compositional servers enhance clustering abstraction in sensor networks. A cluster of sensors may also provide distributed services by coordinating the tasks among the sensors, such as aggregating summary information. A head smart node in the cluster is responsible for control of cluster and inter-cluster communications and networking functions. Group communication to nodes in a cluster can be efficiently implemented. Synchronization constraints associated with network protocols and system services among smart nodes may be specified in clustered smart nodes. The capability to specify hierarchical composite clusters enables designers to build large and complex sensor networks by composing smaller sensor devices or clusters together at each level.

33.7.4 Adaptation Server

Adaptation servers utilize information from the compositional server, lookup server, and analytical tools to control smart nodes during dynamic reconfiguration and failure recovery. Adaptation servers monitor clusters of smart nodes during normal execution by probing the smart nodes, spontaneous signal from the sensors, or explicit network management directives for reconfiguration and failure recovery. Monitoring failures in the sensor nodes can be implemented through collaboration with the distributed lookup service or through an additional monitoring facility.

When a runtime reconfiguration is requested or triggered by a failure event, the adaptation server will generate the correct schedule of reconfiguration and recovery operations for recovering from the failure and ensure that the reconfigured and affected sensor nodes are globally consistent. To ensure correct adaptation and maintain consistency, the adaptation server makes use of analytical tools for dependency analysis and relevant information from compositional and lookup servers. When smart nodes are added or removed from the agile sensor network, a suite of analytical tools may be utilized to ensure that the sensor network still maintains its safety and liveness properties. Smart nodes (or clusters of smart nodes) may be specified and analyzed independently.

Although generic automatic recovery and reconfiguration procedures may be implemented, some sensor applications may have application-specific recovery requirements. Developers may define their own detection algorithms for other changes, including available services in the sensor network, migration of sensor nodes, and changes in task and network requirements. The procedures for maintaining the level of reliability required for the application may be specified by application programmers.

33.8 Mechanisms and Tools

To simplify failure recovery and reconfiguration of distributed sensor applications, the following mechanisms and tools, which utilize the preceding distributed system services, are listed here.

33.8.1 Reliable Remote Service Execution

Because sensor networks are ad hoc with dynamic routing protocols, data propagation is unreliable. Sensor applications will require reliable communication support over these unreliable sensor network layers. The following mechanism allows sensors to locate the appropriate service provider and execute the service remotely and reliably.

When a node wants to execute a service from a service provider, it searches its local service table to get the information on the service and the service provider. If it cannot find the relevant information, it will call the service lookup function to update its local service table. Three types of service interfaces including its service interface, which may be one of the three types, may be specified by the service provider's information: location or address of the service provider with known interface; interface definition of the service; or mobile code for the interface protocol.

Consider an example in which the service client uses remote procedure calls (RPCs) to request remote services from the service provider. RPCs are implemented as follows. The call to service_exec will first send an interest to the service provider through the subscribe function. The service provider then sends a datum to the client containing the permission to send the request. The client then sends a request and all the input data. If the input data are large, several packets may be sent reliably through an automatic repeat request protocol with retransmission. The service provider will then process the service remotely. The client will request the result of the service through another subscription. The provider then returns the result in a datum in response to the interest.

33.8.2 Connectors

Sensor nodes interact with each other through connectors, which encapsulate the properties and states of the interaction and contain specification of the communication methods and the interfaces of its endpoints and the attached sensor nodes. The runtime mechanisms enable active connectors to adapt dynamically as sensor nodes are being recovered, replaced, deployed, or removed. Separating the specification of well-defined sensor node applications from the node interaction and their composition behavior allows designers to implement easily replaceable, reconfigurable smart sensors and changeable interaction behavior. This abstraction and the supporting mechanism transparently handle the problems of maintaining recovery correctness and adaptation consistency independent of application developers.

This simplifies adaptation of sensor applications in response to failure and degradation of sensor nodes. It also simplifies the development of large-scale adaptable applications and hides implementation details. The interface of the components will be attached and matched with the interface of the endpoints of the connector. Connectors enable sensor nodes to be composed using a *service-oriented model*, i.e., sensors interact with each other based on the service provided and consumed, rather than the port number, host name, and IP address of the processes. The composition server maps the sensor nodes and connectors to the lower level processes, object entities, and communication facilities.

Connectors are units of adaptable communication. Sensor nodes use a connector without being aware of how and when changes are made in the communication methods or the sensor nodes that are being connected. For example, a sensor node at one end of the connector could fail and be replaced with another; the connector will manage the replacement transparently so that the sensor node at the other end may not be aware of the replacement. In another example, when a large volume of data is transmitted, the communication method may be automatically changed from RPCs to stream transmission.

The runtime mechanisms enable active connectors to adapt dynamically as sensor nodes are deployed, replaced, or removed. To enable adaptation of a connector, the stub library code of the connector contains functions for detecting requirements for change, managing the sequence of changes, and communicating with the composition server to adapt the cluster of sensor nodes. The connector will communicate with the composition server and adaptation server to determine the correct adaptation operation sequence that maintains the constraints and requirements for the sensor group application. The adaptation operations will then be executed to adapt the sensor node by replacing the failed sensor with an equivalent functional one. After replacement, the states of the sensor nodes will be restored to a globally consistent state. The community of sensor nodes in the distributed sensor application will then continue operation transparently as if the replacement had not occurred.

33.8.3 Sensor Task Structures

Data flow and task structures of the sensor nodes may be specified using connectors (Figure 33.3a) for the purpose of facilitating communication and reconfiguration among the group of sensors performing some distributed sensor tasks. The connector facility allows applications to assemble incrementally and reconfigure sensor nodes to form task groups. The advantage of specifying these task structures is that communication is more efficient and the system mechanisms automatically maintain the structure and reconfigure whenever problems such as failure occur in the nodes. The group structures can also be changed on the fly in response to changes in the global task requirements.

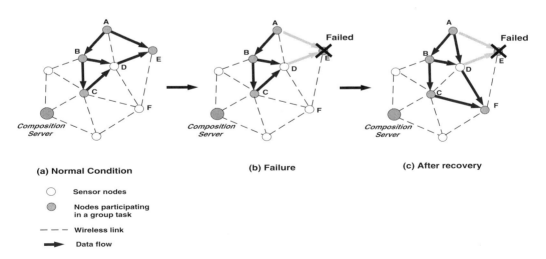

FIGURE 33.3 Automatic recovery from sensor failure.

During failure recovery and reconfiguration, the connector, task structure, and state information stored in the composition server allow the sensor nodes to be replaced and reconfigured automatically by the adaptation and composition services. Adaptation is thus performed transparently to the sensor application. Sensor applications need only deal with the simple communication interface provided by the connectors and do not need to be aware of the communications between the adaptation and composition server for performing recovery from failure. This simplifies distributed sensor application development and maintenance.

Depending on the level of consistency and accuracy that is important for the sensor application, the adaptation server will analyze the execution of the distributed sensor application to recover the application to a globally consistent state. Next, discussion focuses on the tools and methods for checking state consistency and correct recovery methods. Discussion of this example will continue in Subsection 33.9.3.

33.9 Dynamic Adaptation of Distributed Sensor Applications

Distributed sensor applications may adapt to external and internal change events by adapting sensor nodes and functionalities dynamically. Sensor node failure could trigger distributed sensor applications to adapt to recover from the failure. Dynamic adaptation may be initiated in two ways: (1) explicit request by the human operator of the sensor application; and (2) triggered by sensor values above a threshold or other changes in sensors or distributed environments. Making or triggering adaptation requests of the adaptation server specify, first, required adaptation operations of the sensor application and, second, necessary adaptation constraints.

Adaptation operations may include methods for modifying services, replacing sensor nodes or introducing new sensor services. These adaptation operations may occur concurrently with other normal operations of the distributed sensor applications. Adaptation constraints that define interaction and synchronization constraints may need to be specified to ensure correct and continuous execution of distributed sensor operations while adaptation is carried out.

From the set of operations to adapt sensor nodes and services, the adaptation server automatically generates a schedule of adaptation operations that preserves global consistency. Adaptation of higher level sensor clusters is performed in similar ways; adaptation operations are applied to group tasks and abstractions provided by clusters of sensor nodes. This will involve calls to the compositional servers that maintain the abstract behavior of the clusters. The compositional server managing the clusters of sensor nodes will propagate the adaptation operations to the individual sensor group successively at each level.

33.9.1 Consistent Schedule

Dynamic adaptation of distributed sensor applications requires appropriate schedules to ensure that adaptation operations maintain global consistency at the appropriate level required by the sensor application. The analysis for correct adaptation schedules may involve clusters of interacting sensor nodes and higher levels of sensor clusters with abstract services. By analyzing operations within sensor clusters, the size of the problem space used in the analysis is reduced, thus making it scalable to very large distributed sensor applications. Dynamic adaptation may cause inconsistency in a distributed sensor application.

The analyzer determines a correct adaptation schedule from the synchronization constraints and dependency information stored at the compositional server. From the analysis, sensor nodes not directly involved in the adaptation operations may still be affected by the adaptation. These affected sensor nodes may be required to perform adaptive operations to restore consistency in the sensor application after the adaptation is completed. Although recovery and adaptation operations may cause inconsistency in the current behavior, consistency may be restored by further recovery in the new configuration. These policies may be supplied by the designer and stored in the compositional server. The framework provides mechanisms that can be applied to any application-specific policies for enforcing correct adaptation of distributed sensor applications.

33.9.2 Runtime Adaptation Control

During the adaptation, synchronization constraints are satisfied; however, some sensor nodes may be temporarily inconsistent. Consistency will be restored at the end of the adaptation. The mechanisms allow different policies to be used for enforcing consistency in the distributed sensor application. The adaptation server controls synchronization and adaptation by receiving adaptation requests from the sensor nodes and initiating adaptation operations in the affected sensors according to the adaptation schedules. The adaptation server will use information from the compositional and lookup servers to ensure that adaptation operations will not lead to synchronization problems or eventual inconsistency in the affected sensor nodes.

33.9.3 Recovery from Sensor Failure and Degradation

To motivate this discussion, first consider a sensor failure that occurs in the example in Figure 33.3. When a failure occurs in *node E* (Figure 33.3b), the adaptation server will detect the failure using its monitoring facility. Through the trigger mechanism, it will then automatically determine the sequence of procedures for initiating recovery of affected sensor nodes and connectors. The adaptation server will communicate with the composition server to determine the task structures and states of the connectors in order to recover the sensor group into an equivalent group task structure. A new sensor node, F, with similar services as node E will be found through the lookup service and activated to replace node E (Figure 3.33c). Some connectors will be redirected or created to preserve the task structure specified for this sensor task group.

Recovery from sensor failure and degradation is a special case of dynamic adaptation of some or all of the services of a sensor node. Recovery operations may be used to restore those services or individual sensors. However, sensor nodes may interact with other sensors to perform coordinated services when the failure occurs. The distributed sensor application may be inconsistent when failure occurs or when restoring an intermediate sensor after a failure. The adaptation server analyzes the failure and determines the schedule of recovery operations that will restore global consistency of the service. The adaptation server makes use of the results of dependency analysis and correct recovery analysis from the analytical tools. The recovery schedule analysis also requires information from the composition server to determine the cluster composition and the lookup server for services involved in the cluster.

The algorithm for a consistent recovery schedule depends on the level of consistency and the recovery efficiency required by the application. When efficiency is not a major concern, recovery block or transactional method may be used when services can be reinitiated from the start of the recovery block, after the component is restored, or when a replacement sensor is deployed. When a high level of consistency and efficiency is required, the adaptation server may use an application-specific algorithm based on information from dependency analysis of the model of the services and the interaction behavior.

Sensor failure and recovery operations may cause operations of other sensor nodes that depend on them to become invalid and need to be recovered. The algorithm recursively checks the service models for a schedule of recovery operations and other adaptation operations that will transitively recover all dependent sensor operations. This can be computed rapidly at runtime if the necessary information is preprocessed and stored in the adaptation server. At runtime, the adaptation server will use the recovery schedule to execute the recovery operations on the failed and affected sensor nodes.

33.10 Conclusions

Future wireless sensor networks will support more general distributed sensor applications, such as sensor fusion, target detection, classification, collaborative target tracking, and distributed query processing. The key to supporting reliability in these distributed sensor applications is through appropriate system mechanisms that can recover automatically from sensor node failures, insufficient energy supply, high error rate, or mobility.

These mechanisms are based on three main distributed services for maintaining reliability in a distributed sensor network ó distributed lookup services, composition service, and dynamic adaptation service. Failures in sensor nodes and other changes in the sensor network applications can be monitored using known algorithms and registered with the distributed lookup service. Using a correct sequence of adaptation procedures, the adaptation server will use this change information to recover from sensor failures.

Connectors also use these services to support efficient replacement of faulty sensor nodes with minimal disruption in the continuous interaction among distributed sensors. The sequence of recovery procedures will preserve the level of reliability required for the application as specified by the application programmers.

Acknowledgments

This research is supported in part by the Space and Naval Warfare Systems Center (SPAWAR), San Diego, California, and the DARPA SensIT program.

References

1. Kahn, J.M. et al., Next century challenges: mobile networking for Smart Dust, *ACM Mobicom*, 271–278, 1999.
2. Pottie, G.J. and Kaiser W.J., Wireless integrated network sensors, *Commun. ACM*, 43(5), 51–58, May 2000.
3. The ultra low power wireless sensors project. http://www-mtl.mit.edu/~jimg/project_top.html
4. Dolev, D. et al., Reaching approximate agreement in the presence of faults, *J. ACM*, 499–516, July 1996.
5. Lamport, L. et al., The Byzantine generals problem, *ACM Trans. Programming Languages Syst.*, 382–401, July 1982.
6. Marzullo, K., Tolerating failures of continuous-valued sensors, *ACM Trans. Computer Syst.*, 284–304, November 1990.
7. Mahaney, S. and Schneider, F., Inexact agreement: accuracy, precision, and graceful degradation, *Proc. 4th ACM Symp. Principles Distributed Computing*, 237–249, 1995.
8. Brooks, R.R. and Iyengar, S., Robust distributed computing and sensing algorithm, *IEEE Computer*, 29(6), 53–60, June 1996.
9. Magee, J., Tseng, A., and Kramer, J., Composing distributed objects in Corba, *IEEE Int. Symp. Autonomous Decentralized Syst.*, 257–262, April 1997.
10. Martin, V. and Schwan, K., ILI: an adaptive infrastructure for dynamic interactive distributed applications, *IEEE Int. Conf. Configurable Distributed Syst.*, 1998.
11. Purtilo, J., The polylith software bus, *ACM TOPLAS*, January 1994.
12. Kramer, J. and Magee, J., The evolving philosoperís problem: dynamic change management, *IEEE Trans. Software Eng.*, 16(11), 1293–1306, November 1990.
13. Bloom, T., Dynamic module replacement in a distributed system, Technical Report MIT/LCS/TR-303, *MIT Lab. Computer Sci.*, March 1983.
14. Arnold, K. et al., *The Jini Specification*, Addison–Wesley, Reading, MA, 1999.
15. Weihl, W.E., Commutativity-based concurrency control for abstract data types, *IEEE Trans. Computers*, C-37(12), 1488–1505, December 1988.
16. Chrysanthis, P.K. and Ramamritham, K., ACTA: a framework for specifying and reasoning about transaction structure and behavior, *ACM SIGMOD*, 194–203, May 1990.
17. Herlihy, M., Apologizing versus asking permission: optimistic concurrency control for abstract data types, *ACM Trans. Database Syst.*, 15(11), 96–124, March 1988.
18. Bancilhon, F., Kim, W., and Korth, H., A model of CAD transactions, *VLDB*, 25–33, August 1985.
19. Stann, F. and Heidemann, J., Reliable data transport in sensor network, *IEEE Int. Workshop Sensor Network Protocols Applications (SNPA)*, Anchorage, AK, May 11, 102–112, 2003.

20. Sundaresan, K. et al., ATP: a reliable transport protocol for ad-hoc network, *ACM Int. Symp. Mobile Ad Hoc Networking Computing,* Annapolis, MD, June 1–3, 2003.
21. Wan, C.Y., Campbell, A.T., and Krishnamurthy, L., PSFQ: a reliable transport protocol for wireless sensor networks, *Proc. ACM Int. Workshop Wireless Sensor Networks Applications (WSNA),* Atlanta, 1–11, 2002.
22. Intanagonwiwat, C., Govindan, R., and Estrin, D., Directed diffusion: a scalable and robust communication paradigm for sensor networks, *ACM Mobicom,* Boston, 56–57, 2000.

34
Reliable Energy-Constrained Routing in Sensor Networks

Rajgopal Kannan
Louisiana State University

Lydia Ray
Louisiana State University

S. Sitharama Iyengar
Louisiana State University

Ram Kalidindi
Louisiana State University

34.1 Introduction 34-1
34.2 Game-Theoretic Models of Reliable and Length Energy-Constrained Routing 34-2
   Reliable Routing in Geographically Routed Sensor Networks • Distributed Implementation of Length-Constrained RQR
34.3 Distributed Length Energy-Constrained (LEC) Routing Protocol 34-5
   Data Transmission Phase • Path Determination Phase • Selection of β • Selection of Energy Depletion Indicator
34.4 Performance Evaluation 34-7
   Experimental Setup • Results and Analysis

34.1 Introduction

A wireless sensor network is an autonomous system of numerous tiny sensor nodes equipped with integrated sensing and data processing capabilities [1]. Sensor networks are distinguished from other wireless networks by the fundamental constraints under which they operate: (1) sensor nodes are untethered; and (2) sensor nodes are unattended. These constraints imply that network lifetime, i.e., the time during which the network can accomplish its tasks, is finite. Therefore, sensors must utilize their limited and unreplenishable energy as efficiently as possible.

The energy efficiency of routes is an important parameter; however, maximizing network information utility and lifetime implies that the *reliability* of a data transfer path from reporting to querying sensor is also a critical metric. This is especially true given the susceptibility of sensor nodes to denial-of-service (DoS) attacks and intrusion by adversaries that can destroy or steal node data [11]. The possibility of sensor node failure due to operation in hazardous environments cannot be discounted, especially for environmental monitoring and battlefield sensor network applications. For such networks to carry out their tasks meaningfully, sensors must route strategic and time-critical information via the most reliable paths available. Thus, an additional constraint on sensor operations can be introduced: sensor s_i can fail with probability $q_i = 1 - p_i$.

The primary issue addressed in this chapter is reliable energy-constrained intercluster routing within the framework of hierarchical cluster-based sensornet architectures. In a hierarchical architecture [3], nodes in close proximity form clusters, with one node in each cluster designated or elected as the cluster head with special responsibilities. Traffic between different clusters is routed through their corresponding cluster heads. Most hierarchical architectures are based on the assumption that cluster heads can communicate directly with each other. Here, discussion concerns a more realistic two-level hierarchical architecture in which cluster heads called *leader nodes* must use the underlying network infrastructure for communication, i.e., leader–leader and leader–sink routing.

Network partition is expedited by uneven energy distribution across sensors, resulting from improperly chosen routes. Ideally, data should be routed over a path in which participating nodes have higher energy levels relative to other nonparticipating nodes. Network operability will be prolonged if a critically energy-deficient node can survive longer by abstaining from a route rather than taking part in a route for a small gain in overall latency. Similarly, routing over less reliable paths increases energy depletion due to retransmissions. Therefore, *path length*, *path reliability*, and *path energy cost* are critical metrics affecting sensor lifetime.

Recent research in the literature has begun to consider these aspects. For example, Shah and Rabaey [4] describe a probabilistic routing protocol in which non least-energy cost paths are chosen periodically. In Yu et al. [5], a node attempts to balance energy across all its neighbors while finding shortest paths to the sink. However, no unified analytical model explicitly considers routing under the constraints of energy efficiency, path length, and path reliability. The choices of sensor nodes under these constraints are a natural fit for a game-theoretic framework.

This chapter describes a game-theoretic paradigm for solving the problem of finding reliable energy-optimal routing paths with bounded path length and defines two routing games in which sensors obtain benefits by linking to healthy and reliable nodes while paying a portion of path length costs. Thus, sensor nodes modeled as intelligent agents cooperate to find optimal routes. This model has the following benefits:

- Each sensor will tend to link to more reliable and healthier nodes; thus, network partition will be delayed.
- Because each node shares the path length cost, path lengths will tend to be as small as possible; therefore, delay is restricted in this model. Also, shorter path lengths will prevent too many nodes from taking part in a route, thus reducing overall energy consumption.

The Nash equilibria of these routing games define optimal reliable length energy-constrained paths. Computing optimal paths is NP-hard in arbitrary sensor networks, but it can be found in polynomial time (in a distributed manner) in sensor networks operating under a geographic routing regime. The following sections describe fully distributed, scalable, nearly stateless and easily implementable protocols for reliable and length energy-constrained intercluster routing.

34.2 Game-Theoretic Models of Reliable and Length Energy-Constrained Routing

Let $S = \{s_1, s_2, ..., s_n\}$ be the set of sensors in the sensor network participating in the routing game. Let $s_r = L_1$ and $s_q = L_2$ be a pair of leader nodes. Data packets are to be routed from L_1 to L_2 through an optimally chosen set, $S' \subset S$, of intermediate nodes by forming communication links.[*] Let $v_i \geq 0$ denote the value of information at sensor s_i to be routed to L_2, with $q_i = 1 - p_i$ the probability of sensor failure. Note that multicast communication between sets of leader nodes is not considered.

Strategies. Each node's strategy is a binary vector $I_i = (I_{i1}, I_{i2}, ..., I_{ii-1}, I_{ii+1}, ..., I_{in})$, where $I_{ij} = 1$ ($I_{ij} = 0$) represents sensor s_i's choice of sending/not sending a data packet to sensor s_j. Because a sensor typically relays a received data packet to one neighbor only, it is assumed that a node forms only one link for a given source and destination pair of leader nodes. In general, a sensor node can be modeled as having a mixed strategy [10], i.e., the I_{ij}'s are chosen from some probability distribution. However, in this chapter, the strategy space of sensors is restricted to pure strategies only. Furthermore, in order to eliminate some trivial equilibria, each sensor's strategy is nonempty and strategies resulting in a node linking to its

---

[*] In general, sensors in S will be simultaneously participating in routing paths between several such pairs.

ancestors (i.e., routing loops) are disallowed. Consequently, the strategy space of each sensor s_i is such that prob. $[l_{ij} = 1] = 1$ for exactly one sensor s_j and prob. $[l_{ij} = 1] = 0$ for all other sensors, such that no routing loops are formed [7].

Payoffs. Let $l = l_1 \times l_2 \times \ldots \times l_n$ be a strategy in the routing game resulting in a route P from source to destination leader node. Each sensor on P derives a payoff from participating in this route.

Reliable query routing (RQR) payoff model. Every sensor that receives data has an incentive for reaching the destination leader node s_q; thus, the benefit to any sensor s_i on P must be a function of the path reliability from s_i onwards. Because the network is unreliable, the benefit to player s_i should also be a function of the expected value of information arriving at s_i. Therefore, the payoff at s_i on linking to node s_j in P can be written as:

$$\prod_i(l) = \begin{cases} V_i R_i - c_{ij} & \text{if } s_i \in P \\ 0 & \text{otherwise} \end{cases}$$

where R_i denotes the path reliability from s_i onwards to s_q and $V_i = v_i + p_k V_k$ denotes the expected value of the data at node i with parent s_k in P.

Length energy-constrained routing (LEC) payoff model. In addition to reliability, leader–leader routing protocols must be designed to dissipate energy equitably over sensors. One possible approach is to prevent low-energy nodes from taking part in a route as long as they are energy-deficient relative to their neighbors. However, a route that focuses only on energy efficiency may be undesirably long because the lowest energy-cost path need not be the shortest. Conversely, longer paths will result in energy depletion at more sensors while also increasing delay.

Under this model, the payoff of sensor s_i on linking to s_j in P is defined as:

$$\pi_i(l) = E_j - \xi L(P) \tag{34.1}$$

where E_j is the residual energy level of node s_j and $L(P)$ the length of routing path P. E_j represents a benefit to s_i, thus inducing it to forward data packets to higher energy neighbors. The parameter ξ represents the proportion of path length costs borne by sensor s_i. Choosing ξ as a positive constant or proportional to path length will inhibit formation of longer routing paths. Conversely, setting ξ at zero or inversely proportional to path lengths will favor the formation of paths through high-energy nodes. Zeta is chosen as a nonzero positive constant for this routing game; thus, each sensor will forward packets to its maximal energy neighbor in such a way that the length of the path formed is bounded. This model encapsulates the process of decentralized route formation by making sensor nodes cooperate to achieve a joint goal (shorter routing paths) while optimizing their individual benefits.

A Nash equilibrium of this game under both payoff models corresponds to the path in which all participating sensors have chosen their best-response strategy, i.e., the one that yields the highest possible payoff given the strategies of other nodes. This equilibrium is the optimal reliable energy-constrained (RQR) and optimal length energy-constrained (LEC) path in the sensor network for the given leader pair. Note that the process of determining the optimal path requires each node to determine the optimal paths formed by each of its possible successors on receiving its data. The node then selects as next neighbor that node, the optimal path through which gives the highest payoff.

34.2.1 Reliable Routing in Geographically Routed Sensor Networks

Consider the reliable routing problem for sensor networks in which sensors are restricted to following a *geographic routing* regime. In other words, the strategy space of each sensor in the RQR game includes only neighbors geographically closer to the destination than it is. Routing paths under this regime are implicitly length constrained. For each sensor, it is assumed that the set of downstream neighbor nodes to a given destination can be found using a global positioning system (GPS) or some other localization protocol.

Let G be an arbitrary sensor network following geographic routing, with sensor success probabilities P, communication energy costs C, and data of value v_r to be routed from leader node s_r to the sink/leader node s_q, where $v_i = 0 \forall i \neq r$. Although the RQR problem is NP-hard for general sensor networks, it becomes surprisingly easy when the additional constraint of path length [7] is added.

Lemma 34.1. Let L_i be the longest geographically routed path from s_i to s_q in G. Then, s_i can determine its optimal RQR neighbor under the reliability payoff model in $|L_i|$ steps.

PROOF. The following simple observation is noted first: in a geographically routed network, all feasible routing paths from s_r to any node s_i and from s_i to the sink s_q intersect only at s_i. If any other such node existed it would need to be geographically closer than s_i to s_r (because it is on a feasible path from s_r to s_i) as well as to s_q (because it is on a feasible path from s_i to s_q), which is impossible.

Let $R(P_i(v_i))$ represent the reliability of the optimal RQR path P_i from s_j to s_q, transmitting information of value v_i. From the preceding observation, s_{ij} merely needs to know optimal values to s_q from each of its downstream neighbors. Let D_i represent this set; then the optimal neighbor for s_i is

$$N_{opt}(v_i) = \arg\max_{s_j \in D_i} \left\{ v_i p_i R(P_j(p_i v_i)) - c_{ij} \right\} \tag{34.2}$$

where v_i is the expected value of information received at s_i from a given upstream neighbor. The number of such values is proportional to the number of paths from s_r to s_i, which can be exponentially large. However, these values can be divided into disjoint, contiguous intervals in $(0..v_r]$, which makes next-hop selection much easier.

The lemma can now be formally proved by induction. Consider node s_i, whose longest path to the destination is of length one. It will link directly to s_q for all values v_i; $p_i p_q v_i > c_{iq}$. s_q is unreachable for smaller values of v_i. Thus, at node s_i, the optimal choices are divided into tuples consisting of (two) value intervals and optimal path reliabilities corresponding to each interval.

During the kth step of the algorithm, all nodes with $|L_i| = k$ follow the same reasoning, based on the optimal choices of downstream nodes in step $k - 1$. Each node has multiple optimal neighbors, based on a division of the incoming information value into disjoint intervals in $(0..v_r]$. These intervals are polynomial in number and calculated at each node on the basis of intersections of value intervals and optimal reliabilities from its downstream neighbors.

34.2.2 Distributed Implementation of Length-Constrained RQR

The results presented Kannan et al. [7] are now summarized. Let $D_i = \{s_{i_1}, s_{i_2}, \ldots, s_{i_j}\}$ be the set of downstream next-hop neighbors of s_j. For each node s_{ij} in this set, let the expected values of incoming information be divided into N_{ij} disjoint consecutive intervals $I_1^{ij}, I_2^{ij}, \ldots, I_{N_{ij}}^{ij}$, where $\cup_t I_t^{ij} = (0, v_r]$. Let $B(I_t^{ij})$ and $E(I_t^{ij})$ denote the (open) left and (closed) right endpoints and let $R(I_t^{ij})$ be the optimal path reliability from s_{ij} onwards for information of expected value in the given interval I_t^{ij}. When information of expected value v_i arrives at s_i and is forwarded, the expected value of information at s_{ij} is $p_i v_i$. Therefore, each value interval at s_{ij} corresponds to an equivalent "stretched" interval at s_i with left endpoint $B()/p_i$ and right endpoint $\max(v_r, E()/p_i)$. Henceforth, the notation I_t^{ij} refers to the stretched interval at s_i rather than the actual interval at s_{ij}.

Let $\pi_i(i_j, v_i, I_t^{ij})$ represent the payoff to sensor s_i on sending information of value $v_i \in I_t^{ij}$ to downstream neighbor s_{ij}. Note that the payoff function is continous and increasing through the entire range of v_i (as v_i increases, the payoff can only increase). It can therefore be assumed that all intervals give a positive payoff because intervals with negative or zero payoff can be identified and removed. The following lemma shows that the payoff optimality of two intersecting intervals at different neighbors s_{ij} and s_{ik} can be determined using a single fixed point.

Lemma 34.2. If $\pi_i(i_j, v_i, I_t^{i_j}) < \pi_i(i_k, v_i, I_u^{i_k})$ for $v_i = \text{Inf}[I_t^{i_j} \cap I_u^{i_k}]$, then $\pi_i(i_j, v_i, I_t^{i_j}) < \pi_i(i_k, v_i, I_u^{i_k})$ for all $v_i \in [I_t^{i_j} \cap I_u^{i_k}]$. If the two payoffs are equal at the fixed point, then $\pi_i(i_j, v_i, I_t^{i_j}) \leq \pi_i(i_k, v_i, I_u^{i_k})$ throughout the intersection if $R(I_t^{i_j}) \leq R(I_u^{i_k})$.

The lemma follows by definition of the payoff function in Equation 34.2. Thus, to compare two different intervals, it is necessary only to evaluate their payoff at the smallest intersecting point. Lemma 34.2 can be used to compute value ranges and corresponding optimal next neighbors (i.e., those that maximize payoff in the given value range) at each node, provided the optimal solutions are available at nodes one hop away. This can be achieved using reverse directional flooding of control packets [7] from the sink to the source.

Theorem 34.1. The optimal length-constrained RQR path in a sensor network with geographic routing can be computed in a distributed manner using reverse directional flooding with $O(|E|)$ total messages, where E is the number of edges in the sensor network. Optimal neighbors at each node can be found in $O(N_T \log|D_i|)$ time [7].

34.3 Distributed Length Energy-Constrained (LEC) Routing Protocol

A distributed implementation of the LEC routing protocol is described in terms of a simplified "team" version of the routing game. The protocol is derived from work presented in *Lecture Notes in Computer Science* [8]. This team-game routing protocol can be easily modified to obtain optimal LEC paths as well. In the team LEC game, each node on a path shares the payoff of the worst-off node on it. Formally, let L be the set of all distinct paths from a particular source and destination leader pair. Let $E_{min}(P)$ be the smallest residual energy value on path P. Then the equilibrium path of the team LEC game is defined as:

$$\hat{P} = \text{argmax}_{P \in L} (E_{min}(P) - \xi |P|) \quad (34.3)$$

For simplicity in the protocol description later, ξ is set to zero. However, the protocol can be easily modified for nonzero ξ as well as for computing optimal LEC paths in the original LEC game. The optimal path under this condition is interpreted as follows: given any path P, the durability of the path is inversely proportional to $E_{min}(P)$. A path with lower average energy but higher minimum energy should last longer than a route with the opposite attributes because the least energy node is the first to terminate and make that route obsolete. Thus, the inverse of the minimum node energy on a given path reflects the energy weakness of the path. The proposed protocol will select an optimal path of bounded length with the least energy weakness. Node energy levels are changing continously in a sensor network due to sensing, processing, and routing operations, so the optimal path needs to be recomputed periodically. Therefore, the proposed protocol operates in two different phases: data transmission and path determination as described next.

34.3.1 Data Transmission Phase

During this phase, data packets are transmitted from one leader node to the other through the optimal path (with least energy weakness). Each data packet also potentially collects information about the energy consumption en route by keeping track of residual energy levels of nodes on the path. When energy levels of a given critical number of nodes fall below a certain threshold, the data transmission phase ends and the new optimal path determination phase begins. The fundamental steps of the data transmission phase are:

- Each data packet is marked by the source leader node with the geographical position of the destination node and with a threshold value *th*; each packet contains a special *n*-bit energy depletion indicator (EDI) field, where *n* << packet size.
- Each sensor node receiving a data packet determines whether its energy level has fallen below the threshold *th*. If so and if the EDI field in the data packet is not exhausted, the node sets a single bit in the EDI field. Then it forwards the packet to the best next-hop neighbor according to its routing table. It is assumed that before the network starts any activity, all ordinary nonleader sensor nodes have the same energy level. Therefore, during the first data transmission phase, the best next-hop neighbor of a node is the one geographically nearest to the destination leader node. In all other phases, the routing table is updated according to the optimal LEC path calculation.
- If the receiver leader node gets a data packet with all *n* bits in the EDI field set to 1, it triggers a new optimal path selection procedure.

34.3.1.1 Calculation of the Threshold Value

The threshold value *th* plays a very important role in the data transmission phase because it is used to provide an approximate indication that the current optimal path has become obsolete. Intuitively, *th* must be a function of the current residual node energy levels in the network. In this chapter, the following function is used:

$$th = \beta E_{min} \tag{34.4}$$

where $0 < \beta < 1$ and E_{min} is the minimum energy level in the current optimal path. Because E_{min} changes with time, the threshold is recalculated in each path determination phase, consistent with current energy distribution across the network.

34.3.2 Path Determination Phase

This phase begins when the destination leader node receives critical EDI information and ends when the sending leader node has updated its routing table and recalculated the threshold value. The principal steps are:

- The destination leader node L_2 triggers this phase by flooding the network with control packets along the geographic direction of the source leader node L_1. Note that this *reverse directional flooding* occurs in a direction opposite that of data transfer.
- Each node forwards control packet to all its neighbors in the geographic direction of L_1. Each control packet contains a field EM_p that indicates the maximum of the minimum energy levels of all partial paths converging at the given node, i.e., the inverse of the energy weakness of the strongest partial path.
- On receiving the first control packet, each node sets a timer for a prefixed interval *T*. This time period should be large enough for the node to receive future control packets from most of its neighbors (corresponding to different partial paths from the leader node terminating at this node), but not large enough to cause high delays. With each arriving control packet, the node updates and stores the highest EM_p value seen so far. However, if its own energy level E_i is lower than all these values, it stores E_i. With each control packet, it also updates its routing table for destination L_2 to point to the node from which it has received the highest energy control packet. Note that this part of the protocol can be easily modified to incorporate path lengths in addition to the preceding minimum energy computations.
- When the timer expires, this node forwards a new control packet with EM_p field set to the stored energy value to all its neighbors in the geographic direction of L_1. Control packets arriving after the timer expires are discarded.
- Eventually, L_1 begins receiving control packets and sets its timer. Its value of *T* can be determined in many ways, depending on the specific requirements of applications. In this chapter, *T* is

calculated to ensure that most of the paths from L_1 to L_2 are included in the optimality calculations. If (D_{max}) is the maximum transmission delay between two nodes, the value of T is determined as ($MINHOP * D_{max}$), where $MINHOP$ is an estimate of the shortest path from L_1 to L_2. This value can be estimated *a priori* using GPSR routing [6] before the first data transmission phase. Note that the given value of T allows control packets from paths up to twice the length of the shortest path to be forwarded to $L - 1$. Also note that D_{max} is a function of the specific MAC-layer protocol implemented in the sensor network. Finally, when the timer expires at L_1, it selects the final E_{min} value as the highest EM_p value received, calculates the new value of *th*, and sets its routing table accordingly. The next data transmission phase can now begin.

34.3.3 Selection of β

Data transmission in the proposed protocol ends when residual energy levels of at least *n* nodes on the current path fall below threshold *th*. With high β, the smaller the threshold value is, the larger the useful data transmission phase. If the traffic is fairly bursty, *th* should be large so that data packets in the burst can be transmitted. Beta is an empirical value and can be modified based on previous observations. A useful rule of thumb to set the value of β for the current period is as follows:

$$\beta_{current} = \alpha \beta_{prev1} + (1-\alpha) \beta_{prev2} \tag{34.5}$$

where $0 \leq \alpha \leq 1$, β_{prev1} and β_{prev2} and are the previous and previous-to-previous values of β, respectively. Alpha should be chosen according to the specific requirement, i.e., whether the current value of β should be increased or decreased and to what extent.

34.3.4 Selection of Energy Depletion Indicator

Energy depletion indicator is an integer that indicates the maximum number of critical nodes allowed during a data transmission period. The main contribution of an energy depletion indicator is to regulate the duration of the data transmission phase. The higher the value of this parameter is, the longer the period of data transmission. Like β, this parameter is also empirical and can be modified based on previous observations. A rule of thumb similar to that of β can be used to modify the value of this parameter. Thus,

$$EDI_{current} = \gamma EDI_{prev1} + (1-\gamma) EDI_{prev2} \tag{34.6}$$

where $0 \leq \gamma \leq 1$, EDI_{prev1} and EDI_{prev2} are the previous and previous-to-previous values of *EDI*, respectively. Gamma should be chosen according to the specific requirement, i.e., whether the duration of the data transmission phase should be increased or decreased and to what extent.

34.4 Performance Evaluation

The main objective of the protocol is gradually to balance energy consumption across the network. To evaluate performance of this protocol, the following metrics, which reflect dispersion or concentration of energy consumption across a network, are used.

- *Variance of energy level.* The variance of the energy levels of all the nodes is the primary measure of dispersion. A high variance indicates higher energy consumption at some of the nodes compared to others.
- *Range of energy level.* This metric measures the difference between the energy levels of the maximum energy node and the minimum energy node over the entire network. A large value for this range is a result of unfair distribution of routing load among the nodes.

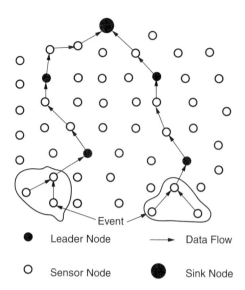

FIGURE 34.1 Mesh topology used for protocol evaluation.

34.4.1 Experimental Setup

In the simulation, 100 nodes are in a 1000- × 1000-m area, with one node at each of the positions of the 10 × 10 square grid. Figure 34.1 represents the mesh topology used for evaluating the protocol. The entire network is divided into five clusters. Two sensing areas are in the regions under clusters A and B. Sensor data packets are generated from these sensing areas at a uniform rate. The leader nodes in each of the clusters A and B collect these packets and send them to the leader nodes of clusters C and D, respectively, via intermediate sensor nodes. Leader nodes C and D forward these packets to the sink node in cluster E.

Each leader node selects the leader node that is geographically nearest to the sink for transmitting its received/sensed data. Leader–leader communication is accomplished through ordinary sensors. Reverse directional flooding is initiated when a leader node receives a sensor data packet indicating that at least *three* sensor nodes are close to the threshold *th*. A sender leader node sets *th* to the new βE_{min} obtained from the reverse flooding phase. The simulation is run for 900 sec with two leader–leader routing protocols: shortest path routing and the proposed team LEC protocol. These experiments are carried out on a simulation test bed that is an extension of Sensorsim [9].

34.4.2 Results and Analysis

It is assumed that before the network starts any activity, all ordinary sensor nodes have the same energy level; therefore, in the beginning, energy distribution is uniform across the network. When a network becomes active, the energy distribution across it gradually becomes nonuniform because nodes participating in a route inevitably consume more energy than other nodes do. A protocol that uses a fixed route until one node in the route is completely drained out of its energy ends up producing an energy distribution with high dispersion of energy levels. On the other hand, the proposed protocol tries to adapt to the dynamically changing energy distribution and gradually makes the initially uneven energy distribution uniform. Therefore, it is expected that the difference between the dispersion measures produced by this protocol and those produced by any protocol with fixed routing will increase with an increasing rate with time. This chapter compared the performance of this protocol with that of a protocol using a fixed shortest path for leader–leader communication.

Results of the simulation comparing performance of this protocol with that of shortest path routing reflect the outcome as expected. *Range* (the difference between the maximum and minimum value of a distribution) and *variance* were used as measures of dispersion of energy distribution to evaluate the protocol. Figure

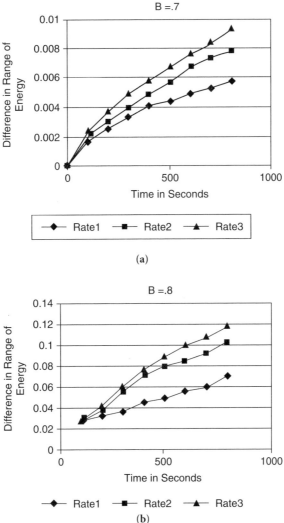

FIGURE 34.2 Difference in ranges of node energy distributions across the network over time under the two protocols for two different β values: (a) β = 0.7; (b) β = 0.8.

34.2(a) and (b) presents the difference in the ranges of node energy distributions across the network over time under the two protocols with two different values of β and three different traffic rates. In both figures, the difference of the ranges rises very sharply, indicating that this protocol yields a lower range of energy distribution compared to that produced by the fixed route protocol as time proceeds.

Moreover, with an increased traffic rate, this protocol produces a much better result compared to that of shortest path routing. This indicates that, with heavy traffic, the energy distribution across the network becomes more uneven in a fixed route protocol because the load is heavier on a particular route. In this case, frequent change of routing path is very useful in bringing uniformity to overall energy consumption. Note that routes changed more frequently with a higher value of β, performance of the proposed protocol is better when β = 0.8 than that when β = 0.7.

Figure 34.3(a) and (b) shows how the difference between the minimum energy level produced by this protocol and that by shortest path routing changes with time for two different β values, respectively. In both cases, the difference rises very sharply as time proceeds. With a higher value of β, the rise is sharper because change of route is accomplished more frequently; therefore, consumption of energy is more uniform under this protocol.

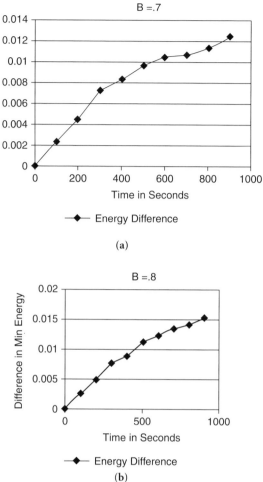

FIGURE 34.3 Difference between the minimum energy level produced by protocol and that by shortest path routing for two different β values: (a) β = 0.7; (b) β = 0.8.

Figure 34.4 represents the variance of residual energy distribution produced by this protocol with two different β values and that by the shortest path routing. The energy range metric does not measure the number of sensor nodes that are treated unfairly. The high variance of the shortest path routing indicates that a significant number of sensor nodes are treated unfairly, with network traffic concentrated at fewer nodes. This might expedite partition of the network due to energy depletion at critical nodes. With a higher value of β, the proposed protocol produces lower variance because of more frequent route changes.

Acknowledgments

This work was done in part with support from NSF under grants IIS-0312632 and IIS-0329738 and DARPA/AFRL under grant # F30602-02-1-0198.

References

1. I. Akyildiz, W. Su, Y. Sanakarasubramaniam and E. Cayirci, Wireless sensor networks: a survey, *Computer Networks J.*, 38(4), 393–422, 2002.

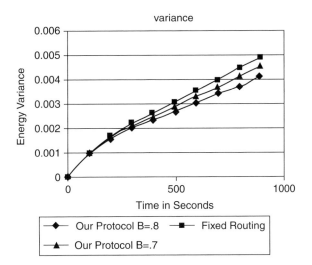

FIGURE 34.4 Variance of residual energy distribution produced by the protocol with two different β values and that by shortest path routing.

2. C. Intanagonwiwat, R. Govindan and D. Estrin, Directed diffusion: a scalable and robust communication paradigm for sensor networks, *Proc. 6th Ann. Int. Conf. Mobile Computing Networks (MobiCOM 2000)*, August 2000, Boston, MA.
3. M.J. Handy, M. Haase and D. Timmermann, Low energy adaptive clustering hierarchy with deterministic cluster-head selection, *4th IEEE Int. Conf. Mobile Wireless Commun. Networks*, Stockholm, 2002.
4. R.C. Shah and J.M. Rabaey, Energy aware routing for low energy ad hoc sensor networks, *Proc. IEEE WCNC'02*, March 2002.
5. Y. Yu, R. Govindan and D. Estrin, Geographical and energy aware routing: a recursive data dissemination protocol for wireless sensor networks, UCLA Computer Science Department technical report UCLA/CSD-TR-01-0023, May 2001.
6. B. Karp and H.T. Kung, GPSR: greedy perimeter stateless routing for wireless networks, *Proc. ACM/IEEE MobiCom*, August 2000.
7. R. Kannan et al., Game-theoretic models for reliable path-length and energy-constrained routing with data aggregation in wireless sensor networks, to appear in *IEEE J. Selected Areas Commun.* (also as LSU Computer Science technical report LSU/CSC-TR03-05).
8. Max–min length-energy-constrained routing in sensor networks, in *Lecture Notes Computer Sci.* LNCS 2920, EWSN 2004, Berlin, Germany, January 19–21 2004. (submitted to IEEE JSAC).
9. S. Park, A. Savvides and M.B. Srivastava, SensorSim: a simulation framework for sensor networks, *Proc. MSWiM 2000*, Boston, MA, August 11, 2000.
10. D. Fudenberg and J. Tirole, *Game Theory*, MIT Press, 1991.
11. L. Zhou and Z. Haas, Securing ad-hoc networks, *IEEE Network*, 13(6), 24–30, 1999.

35
Fault-Tolerant Interval Estimation in Sensor Networks

Yunmin Zhu
Sichuan University

Baohua Li
Sichuan University

35.1	Introduction	35-1
35.2	Sensor Network Formulation	35-2
	Sensor Outputs • Two Optimization Criteria	
35.3	Fault-Tolerant Interval Estimation without Knowledge of Confidence Degrees	35-4
35.4	Combination Rule and Optimal Fusion for Sensor Output	35-5
	Combined Intervals • Combined Confidence Degrees	
35.5	Fault-Tolerant Interval Estimation with Knowledge of Confidence Degrees	35-9
35.6	Extension to Sensor Estimate with Multiple Output Intervals	35-11
35.7	Robust Fault-Tolerant Interval Estimation	35-11
	Stability • Sensor Interval Endpoint Tolerance	
35.8	Conclusion	35-16

35.1 Introduction

Many advanced systems now use a large number of sensors in practical applications ranging from aerospace and defense, robotics and automation systems, to monitoring and control of process generation plants. Estimation fusion, or data fusion for estimation, is the problem of how to best utilize useful information contained in multiple sets of data for the purpose of estimating an unknown quantity — a parameter θ or process θ_t. These data sets are usually obtained from multiple sources (e.g., multiple sensors).

Evidently, estimation fusion has wide-spread applications because many practical problems involve multiple sensors — for instance, target tracking or state estimate for a dynamical process, including filtering, prediction, and smoothing. Estimation fusion has been investigated for more than two decades and many results have been obtained [1–6]. Although these results represent major progress on the point estimate fusion, relatively few results on the interval estimate fusion exist ([7–12] among others). In many practical applications, however, one may be more interested in finding an interval covering θ with a required confidence degree (coverage probability) than a guess of a single value as the value of θ. For example, it is not necessary to guarantee that a missile hit its target exactly; it is enough for the effective explosion region of the missile to cover the target with an allowable minimum confidence degree.

Fault tolerance is an important issue in network design because sensor networks work usually in a dynamic, uncertain situation; therefore, it is impossible to avoid faulty outputs from some sensors. When one considers fault-tolerant interval estimation in sensor networks, two optimization criteria — minimizing interval length with an allowable minimum confidence degree or maximizing confidence degree with an allowable maximum interval length for the interval estimation — are suggested. In terms of the two optimization criteria, several fault-tolerant interval estimation fusion methods are proposed.

When one knows how many sensors at most are faulty and the fused interval is requested to cover the true value with confidence degree one, Marzullo's method [9] gives the shortest interval even without prior knowledge of sensor confidence degrees. However, this method has an obvious drawback: the interval length may be very large even if every sensor output interval is short. If the final interval is too long, it will have little value for application. For example, such an interval (region) estimate has little value for weapon control if it exceeds the kill zone of a weapon very much, regardless of its correctness.

Prasad and colleagues [10] and Iyengar and Prasad [11] have presented a fault-tolerant interval integration function on the basis of Marzullo function. It can reduce the length of the output interval of Marzullo function under the assumption that all false intervals are located close to the unknown true value θ. Their output does not guarantee to cover θ and a reliable belief level of the output has not been shown.

In many practical situations, an interval estimate with confidence degree less than one is enough to use. If the prior knowledge of sensor confidence degrees can be known, the interval estimation fusion with the required confidence degree is developed, and various fault-tolerant interval integration results are shown. They, of course, depend on prior information of sensor fault and the optimization criteria. When the required confidence degrees are less than one, these results certainly give a smaller output interval than that given by Marzullo's method. If θ is a vector, the interval estimate can be extended to the set estimate. Without loss of generality, the interval estimate is mainly considered in this chapter.

Also, it is desirable that a fault-tolerant interval fusion algorithm be robust in the sense that a slight change in the input results has only a slight change in the output, in view of the unavoidable error in information acquisition and processing. Unfortunately, Marzullo's method is not robust in the able sense — it exhibits irregular behavior in that a slight difference in the input may produce a vastly different output.

To address this unstable behavior in sensor fusion process, Schmid and Schossmaier [12] introduced a fault-tolerant interval integration function called the Schmid–Schossmaier function, which makes its algorithm globally stable. However, the output of Marzullo's method is, in general, a real subset of the output of the Schmid–Schossmaier function; thus, its output interval is even larger than Marzullo's. A brief discussion on the robustness of the interval estimation fusion will be presented later.

In Section 35.2 of this chapter, the formulation for the multisensor interval estimation is presented. Section 35.3 presents the fault-tolerant interval estimation without knowledge of sensor confidence degrees. Then, based on outputs of sensors with their confidence degrees, the combination rule and the optimal fusion method for outputs of sensors are proposed in Section 35.4. Section 35.5 is dedicated to the fault-tolerant interval estimation fusion with knowledge of sensor confidence degrees. In Section 35.6, the results in the two preceding sections are extended to a case in which every sensor outputs multiple intervals. In Section 35.7, the Schmid–Schossmaier fusion method and a brief discussion on the robustness of the interval estimation fusion are presented. Finally, concluding remarks are given in Section 35.8.

35.2 Sensor Network Formulation

Consider the following case: l sensors are available to estimate an unknown parameter θ. When local sensors do not want to share all intimate details of their systems, or if the types of multiple sensor messages, such as image message, voice message, language message, and digital message, are too different from each other to obtain the joint distribution of all sensor data, the available information for the fusion center is only the local interval estimates and their confidence degrees given by all sensors.

35.2.1 Sensor Outputs

Each sensor yields an estimated closed interval of θ and its confidence degrees of covering θ, denoted as

$$I_i = [a_i, b_i], \; a_i \le b_i, \; i \le l, \quad (35.1)$$

and

$$\alpha_i = P_i(\theta \in I_i), \; i \le l. \quad (35.2)$$

*From the preceding message, it is known that the true θ is covered by either of I_i and its complementary set $I_i^c \stackrel{\text{def}}{=} (-\infty, a_i) \cup (b_i, \infty)$, which are still viewed as *an interval*, and confidence degree of I_i^c for covering θ is

$$P_i(\theta \in I_i^c) = 1 - \alpha_i, \; i \le l. \quad (35.3)$$

In practical applications, the two ends $-\infty$ and ∞ of I_i^c could be finite real numbers L_i and U_i, i.e., $I_i^c \stackrel{\text{def}}{=} [L_i, a_i) \cup (b_i, U_i]$. Even so, the two intervals $[L_i, a_i)$ and $(b_i, U_i]$ cannot be used individually because there is no information on their confidence degrees. Thus, each sensor actually outputs two intervals

$$I_i^1 \stackrel{\text{def}}{=} I_i, \; I_i^0 \stackrel{\text{def}}{=} I_i^c \quad (35.4\text{a})$$

and their confidence degrees

$$\alpha_i^1 \stackrel{\text{def}}{=} \alpha_i, \; \alpha_i^0 \stackrel{\text{def}}{=} 1 - \alpha_i. \quad (35.4\text{b})$$

Then, an entire local message received by the fusion center is an *interval list* of the output of all sensors

$$\{I_1^{r_1}, \cdots, I_l^{r_l}\}, \; r_i = 0, 1 \quad (35.5\text{a})$$

and the corresponding *confidence degree list* of the outputs of all sensors

$$\{\alpha_1^{r_1}, \cdots, \alpha_l^{r_l}\}, \; r_i = 0, 1. \quad (35.5\text{b})$$

These lists are called the *sensor outputs*.

35.2.2 Two Optimization Criteria

According to various practical requirements, there are corresponding interval estimate optimization criteria. The two most popular criteria in practice are:

A. Minimizing interval length under confidence degree constraint
B. Maximizing confidence degree under interval length constraint

Next, several fault-tolerant interval estimation fusion methods will be given in terms of the two criteria, without and with knowledge of confidence degrees.

---

*More generally, the interval I_i could be $(a_i, b_i]$, $[a_i, b_i)$.

35.3 Fault-Tolerant Interval Estimation without Knowledge of Confidence Degrees

In this section, all sensors deliver the fusion center their interval estimates but no knowledge of confidence degrees. Obviously, only criterion (A) with confidence degree one can be considered in this case. Also, from previous statistical data or another information resource, the fusion center knows in advance that, at most, f sensor estimates in the l sensor interval estimates do not cover θ, but is not sure which sensor estimates are faulty. Thus, the fusion center must tolerate possible faults and give a result that is as feasible as possible.

Suppose that the fusion center receives a list $\{I_1, I_2, \ldots, I_l\}$ of sensor interval estimates, where $I_i = [a_i, b_i]$, $a_i \leq b_i$, and, at most, f out of the l ($0 \leq f < l$) interval estimates are assumed to be faulty; the fusion center takes the list $\{I_1, I_2, \ldots, I_l\}$ as an input and outputs a closed interval, $I = [a,b]$, $a \leq b$, representing the final estimate of θ.

Let

$$I_{\{i_1, i_2, \ldots, i_{(l-f)}\}} = \bigcap_{j=1}^{l-f} I_{i_j} \neq \emptyset \tag{35.6}$$

and the list

$$\left\{ I_{\{i_1, i_2, \ldots, i_{(l-f)}\}} : \{i_1, i_2, \ldots, i_{(l-f)}\} \subseteq \{1, 2, \ldots, l\} \right\} \tag{35.7}$$

represent all the nonempty intersections of $(l - f)$ closed intervals in $\{I_1, I_2, \ldots, I_l\}$. The final interval $I = [a,b]$ given by Marzullo's method is that which is a shortest close interval containing the union of $I_{\{i_1, i_2, \ldots, i_{(l-f)}\}}$. Marzullo's method is given a functional representation, called Marzullo function, denoted by

$$I = M(\{I_1, I_2, \ldots, I_l\}) = \overline{\bigcup I_{\{i_1, i_2, \ldots, i_{(l-f)}\}}}, \tag{35.8}$$

where superscript "—" over a set on the straight line stands for convex closure of this set.

Obviously, Marzullo function gives the shortest interval that is guaranteed to contain θ. However, this method has two obvious drawbacks. A gap, if any, between all connected intervals $I_{\{i_1, i_2, \ldots, i_{(l-f)}\}}$ obviously cannot contain θ but is maintained as a subinterval in the output interval I. If such gaps are too many or too large, the output interval I will be very large even if every $I_{\{i_1, i_2, \ldots, i_{(l-f)}\}}$ is short. Another drawback is that, in many applications, 100% confidence degree is not necessary and an overly high confidence degree requirement may yield an overly long output interval — for instance, while a long interval has little confidence degree. If the output interval I were too long, it would have little value for application. For example, such an interval (region) estimate has little value for weapon control if it exceeds the kill zone of a weapon very much, regardless of its correctness. Now consider the following examples.

Example 35.1. The confidence degrees are unknown for the following sensor interval estimates:

$$\{[8, 11], [9, 12], [10, 13]\}.$$

TABLE 35.1 Fused Interval Outputs of Example 35.1

f	0	1	2	3
I	[10, 11]	[9, 12]	[9, 13]	∅
Confidence degree	1	1	1	1

TABLE 35.2 Fused Interval Outputs of Example 35.2

f	0	1	2	3
I	∅	[9, 11]	[8, 23]	∅
Confidence degree	1	1	1	1

Using the preceding, the output intervals of Marzullo function are given in Table 35.1. Reasonably, the length of the fused interval I becomes larger as f is increasing. The following example shows the drawback of the overlong fused interval of Marzullo function.

Example 35.2. The confidence degrees are unknown for the following sensor interval estimates:

$$\{[8, 11], [9, 12], [20, 23]\}.$$

Apparently, it is impossible that $f = 0$ because the intersection of the above three intervals is empty. Using the preceding, the output intervals of Marzullo function are given in Table 35.2. In this table, the fused interval [8, 23] for $f = 2$ is quite long because of the big gap between intervals [9, 12] and [20, 23].

Prasad and colleagues [10] and Iyengar and Prasad [11] presented a fault-tolerant interval integration function on the basis of Marzullo function. It can reduce the length of the output interval of Marzullo function under the assumption that all false intervals are located close to the unknown true value θ. However, because the meaning of "close" is ambiguous, their output does not guarantee to cover θ with confidence degree one as well as no confidence degree of the output can be shown.

If the fusion center knows in advance that at least f sensor estimates in the l sensor interval estimates do not cover θ, it is not guaranteed to exist at least an interval covering θ with confidence degree one. Therefore, this fault-tolerant problem is senseless. However, if all sensor confidence degrees are given and the final output interval could be a confidence degree smaller than one, the fault-tolerant question is meaningful and will be considered in Section 35.5.

35.4 Combination Rule and Optimal Fusion for Sensor Output

In this section, under the assumption of independent estimates across sensors, this section will develop a combination method for the fusion center optimally to combine sensor interval estimates and their confidence degrees. Using the sensor outputs, the following combination rule is proposed to combine sensor intervals and their confidence degrees.

35.4.1 Combined Intervals

The combined intervals at the fusion center now are all possible intersections of the sensor output intervals and their all possible connected unions. First, consider the following nonempty intersections of all possible sensor output intervals

$$I_{\{r_1,\ldots,r_l\}} = \bigcap_{i=1}^{l} I_i^{r_i}, \quad r_i = 0, 1. \tag{35.9}$$

Then, various unions of these nonempty intersections yield more intervals.

$$I_{\cup_{r=1}^{N}\{r_1,\ldots,r_l\}} = \bigcup_{r=1}^{N}\bigcap_{i=1}^{l} I_i^{r_i}, \quad r_i = 0,1, \quad N = 1,\ldots,2^l. \tag{35.10}$$

The union of all these nonempty intersections is the entire straight line $\Re$.

Because the two ends of l sensor interval outputs partition the straight line, at most, into $2l$ intervals (view the unconnected $I_{\{1^0,\ldots,l^0\}}$ as an interval), the upper bound of possible numbers of all nonempty intersections of sensor intervals is $2l$. That is to say, the number of all possible nonempty intersections of sensor intervals grows polynomially as the number of sensors increases.

35.4.2 Combined Confidence Degrees

Suppose that all the sensor estimates are mutually independent. By the laws of probability, the combined confidence degrees of the nonempty intervals at the fusion center are given by

$$P\left(\theta \in I_{\{r_1,\ldots,r_l\}} \middle| C\right) = \frac{1}{c}\prod_{i=1}^{l}\alpha_i^{r_i}, \quad I_{\{r_1,\ldots,r_l\}} \neq \emptyset, r_i = 0,1, \tag{35.11}$$

where the symbol C is defined as the following set of intervals

$$C = \left\{I_{\{r_1,\ldots,r_l\}} : I_{\{r_1,\ldots,r_l\}} \neq \emptyset\right\}, \tag{35.12}$$

and the parameter c is given by

$$c = \sum_{I_{\{r_1,\ldots,r_l\}}\neq\emptyset}\prod_{i=1}^{l}\alpha_i^{r_i}, \tag{35.13}$$

Clearly, the union of all of the nonempty intersections is the entire straight line $\Re$, and

$$P\left(\theta \in I_{\{r_1,\ldots,r_l\}} \middle| C\right) = 0 \quad \text{if } I_{\{r_1,\ldots,r_l\}} = \emptyset.; \tag{35.14}$$

Remark 35.1. The assumption of independent sensor estimates is necessary for the derivation of the combined confidence degrees given in Eq. 35.11. In practice, when θ is deterministic, the sensor estimates could be independent of each other when sensor noises are independent of each other; if θ is random, the sensor estimates are usually dependent on each other. When sensor estimations are significantly correlated, to get the confidence degrees of the combined interval, the joint probabilities

$$P(\theta \in I_1^{r_1},\cdots,\theta \in I_l^{r_l}), \quad r_i = 0,1,$$

of all possible sensor interval outputs are required, i.e.,

$$P\left(\theta \in I_{\{r_1,\ldots,r_l\}}\right) = P\left(\theta \in I_1^{r_1},\cdots,\theta \in I_l^{r_l}\right), \quad r_i = 0,1, \tag{35.15}$$

where all intervals $I_i^{r_i}$ are fixed. In some practical applications, they can be known from a specific experiment or historical data.

Remark 35.2. A special case of the combination rule is $\alpha_i^1 = 1$ for all $i \leq l$. This interval estimation fusion was well known to be the intersection of all sensor output intervals.

Now an example is given to show how to apply the preceding combination rule given in Sections 35.4.1 and 35.4.2.

Example 35.3. Consider the following sensor interval estimates and their confidence degrees

$$\{[8, 11], [9, 12], [10, 13]\}, \{0.8, 0.83, 0.85\}.$$

Therefore, all possible sensor outputs are

$$I_1^1 = [8, 11], I_1^0 = (-\infty, 8) \cup (11, \infty), I_2^1 = [9, 12],$$

$$I_2^0 = (-\infty, 9) \cup (12, \infty), I_3^1 = [10, 13], I_3^0 = (-\infty, 10) \cup (13, \infty)$$

and

$$\alpha_1^1 = 0.8, \ \alpha_1^0 = 0.2; \ \alpha_2^1 = 0.83, \ \alpha_2^0 = 0.17; \ \alpha_3^1 = 0.85, \ \alpha_3^0 = 0.15.$$

Using the preceding combination rule (Eqs. 35.9 to 35.13), the six fused nonempty intervals and their confidence degrees are given in Table 35.3.

Because the intervals and their confidence degrees are summable, more intervals and their confidence degrees can be obtained from the preceding intervals. For example, consider the original intervals estimated by three sensors and their current confidence degrees:

$$[8, 11]: 0.7963; \ [9, 12]: 0.9368; \ [10, 13]: 0.8545.$$

Comparing them with single sensor outputs, the modification of confidence degrees given by the combination rule is reasonable intuitively. The confidence degree of [9, 12] is improved significantly because its length, overlapped by [8, 11] and [10, 13], totals 4; however, the corresponding lengths for the other two intervals total 3. In other words, the estimate of the second sensor receives the most support from the other two sensors. Because the confidence degree of [8, 11] is the least in the three intervals and the significant improvement of the confidence degree of [9, 12] must have a bad impact on other intervals, the confidence degree of [8, 11] becomes a little smaller than that in single sensor case. Similarly, it can be explained why the confidence degree of [10, 13] is improved a little.

It is quite possible that the intersection $I_{\{1,\ldots,1\}}$ of all sensor intervals is empty. In this case, some (generalized intervals) of $I_{\{r_i,\ldots,r_l\}}$ may consist of several unconnected intervals (see the following example).

Example 35.4. The following are sensor interval estimates and their confidence degrees:

$$\{[8, 11]: 0.8; \ [12, 14]: 0.83; \ [13, 19]: 0.85\}.$$

TABLE 35.3 Fused Interval Outputs of Example 35.3

Intervals	$(-\infty, 8) \cup (13, \infty)$	[8,9)	[9,10)	[10,11]	(11,12]	(12,13]
Confidence degree	0.0059	0.0237	0.1159	0.6567	0.1642	0.0336

Therefore, all possible sensor outputs are

$$I_1^1 = [8, 11], I_1^0 = (-\infty, 8) \cup (11, \infty), I_2^1 = [12, 14],$$

$$I_2^0 = (-\infty, 12) \cup (14, \infty), I_3^1 = [13, 19], I_3^0 = (-\infty, 13) \cup (19, \infty)$$

and

$$\alpha_1^1 = 0.8, \ \alpha_1^0 = 0.2; \ \alpha_2^1 = 0.83, \ \alpha_2^0 = 0.17; \ \alpha_3^1 = 0.85, \ \alpha_3^0 = 0.15.$$

Using the preceding combination rule (Eqs. 35.9 to 35.13), the five fused nonempty intervals and their confidence degrees are given in Table 35.4. Then, the confidence degrees of the original three intervals estimated by sensors now become:

$$[8, 11]: 0.0926; \quad [12, 14]: 0.7532; \quad [13, 19]: 0.7713.$$

In addition, any two ends of sensor intervals can be used to yield more connected intervals to estimate θ. Although, precise confidence degrees may not be available for some of them, at least their lower bounds are known. For example,

$$[8, 14]: \geq 0.8458; \quad [8, 19]: \geq 0.9769.$$

These intervals with their imprecise confidence degrees can still be useful for the optimal interval estimation fusion (see Examples 35.5 and 35.6).

Thus, the combined outputs at the fusion center have three advantages:

- Deriving more intervals with their confidence degrees, i.e., higher resolution rate of intervals
- Deriving more reasonable coverage probability distribution over the entire real number space because one properly takes advantage of more information coming from multiple sensors
- Easy extension to higher dimensional set estimation fusion problems

The following examples show how to output the optimal interval estimation fusion.

Example 35.5. The sensor outputs are the same as in Example 35.3. In terms of criteria (A) and (B), the optimal interval estimation fusion is given in Table 35.5.

Example 35.6. The sensor outputs are the same as in Example 35.4. In terms of criteria (A) and (B), the optimal interval estimation fusion is given in Table 35.6.

TABLE 35.4 Fused Interval Outputs of Example 35.4.

Intervals	$(-\infty, 8) \cup (11, 12) \cup (13, \infty)$	[8,11)	[12,13)	[13,14)	(14,19)
Confidence degree	0.0231	0.0926	0.1130	0.6402	0.1311

TABLE 35.5 Optimal Fusion under Criteria (A) and (B) in Example 35.5

Confidence degree constraint	$\geq .6$	$\geq .8$	$\geq .9$
Interval length constraint	≤ 1	≤ 2	≤ 3
Optimal interval	[10, 11]	[10, 12]	[9, 12]
Confidence degree	0.6567	0.8209	0.9368
Interval length	1	2	3

TABLE 35.6 Optimal Fusion under Criteria (A) and (B) in Example 35.6

Confidence degree constraint	≥.6	≥.8	≥.95
Interval length constraint	≤1	≤7	≤11
Optimal interval	[13, 14]	[8, 14]	[8, 19]
Confidence degree	0.6567	0.8458	0.9769
Interval length	1	6	11

35.5 Fault-Tolerant Interval Estimation with Knowledge of Confidence Degrees

When every sensor delivers its interval estimate with confidence degree to the fusion center and the fusion center also knows that, at most, f sensor interval estimates are faulty, the method given in the last section can be extended to obtain corresponding conditional combined interval outputs and their confidence degrees. Furthermore, to obtain an optimal interval estimation fusion in terms of criterion (A) or (B) can be obtained.

Continue to assume that all sensor estimates are independent of each other. At most, f sensor interval estimates are faulty, but it is not known which of them are faulty. However, at least $l - f$ sensor interval estimates cover the true value θ. Of course, it is still not certain which sensor estimates they are. Therefore, by the definition of r_i given in 35.4 and 35.5, it is only necessary to consider the following set of intersections of interval outputs:

$$C = \left\{ I_{\{r_1,\ldots,r_l\}} : I_{\{r_1,\ldots,r_l\}} \neq \emptyset, \sum_{i=1}^{l} r_i \geq l - f \right\}. \tag{35.16}$$

Clearly, when $f = l$, the condition $\sum_{i=1}^{l} r_i \geq l - f = 0$ vanishes and the condition set C becomes 35.12, which was discussed in Section 35.4.

Thus, the conditional confidence degrees $I_{\{r_1,\ldots,r_l\}}$ for all intervals in C are defined as

$$P(\theta \in I_{\{r_1,\ldots,r_l\}} | C) = P(\theta \in I_{\{r_1,\ldots,r_l\}}; C) / P(C) = \frac{1}{c_0} \prod_{i=1}^{l} \alpha_i^{r_i}, \; r_i = 0, 1, \tag{35.17}$$

where the parameter c_0 is given by

$$c_0 = \sum_{\substack{I_{\{r_1,\ldots,r_l\}} \neq \emptyset \\ \sum_{i=1}^{l} r_i \geq l-f}} \prod_{i=1}^{l} a_i^{r_i}, \tag{35.18}$$

and

$$P(\theta \in I_{\{r_1,\ldots,r_l\}} | C) = 0 \text{ if } I_{\{r_1,\ldots,r_l\}} = \emptyset, \text{ or } \sum_{i=1}^{l} r_i < l - f. \tag{35.19}$$

Example 35.7. The sensor outputs are the same as those given in Example 35.3, and the fusion center knows that, at most, f sensors are faulty. The fused interval outputs are given in Table 35.7.

TABLE 35.7 Fused Fault-Tolerant Interval Outputs with, at most, f Faulty Sensors in Example 35.7

f	$(-\infty, 8) \cup (13, \infty)$	$[8, 9)$	$[9, 10)$	$[10, 11]$	$(11, 12]$	$(12, 13]$
3	0.0059	0.0237	0.1159	0.6567	0.1642	0.0336
2	0	0.0239	0.1166	0.6606	0.1651	0.0338
1	0	0	0.1237	0.7010	0.1753	0

TABLE 35.8 Optimal Fusion with $f = 1$ under Criteria (A) and (B) in Example 35.7

Confidence degree constraint	$\geq.7$	$\geq.8$	$\geq.9$
Interval length constraint	≤ 1	≤ 2	≤ 3
Optimal interval	$[10, 11]$	$[10, 12]$	$[9, 12]$
Confidence degree	0.7010	0.8763	1
Interval length	1	2	3

TABLE 35.9 Optimal Fusion with $f = 2$ under Criteria (A) and (B) in Example 35.7

Confidence degree constraint	$\geq.6$	$\geq.8$	$\geq.9$
Interval length constraint	≤ 1	≤ 2	≤ 3
Optimal interval	$[10, 11]$	$[10, 12]$	$[9, 12]$
Confidence degree	0.6606	0.8257	0.9423
Interval length	1	2	3

It can be seen from Table 35.7 and the following tables that the less f is, the more the confidence degrees are concentrated on the less intervals and the more reliable these intervals are. This makes sense because the less f is, the more reliable sensor estimates are. In terms of criteria (A) and (B), the optimal interval fusion is given in Table 35.8 and Table 35.9.

Remark 35.3. When all sensor confidence degrees α_i^1 are unknown for any $i \leq l$, and the criterion for the optimal interval estimation fusion is criterion (A) with the constraint of confidence degree being one, it is easy to verify that the optimal fault-tolerant interval fusion developed here becomes Marzullo's method. In other words, Marzullo's method is a special case of this method.

Remark 35.4. In fact, extra information may not be "at most, f sensor interval estimates are faulty." It could be others. When one knows all sensor output intervals and their confidence degrees, using the method proposed in Section 35.2, a corresponding conditional combination rule can be done. This issue can be viewed as a conditional confidence degree problem. So-called fault-tolerant interval fusion is just a specific case in the conditional confidence degree problem.

For example, if the fusion center knows by its experience that at least f sensor interval estimates are faulty, similar to Eqs. 35.16 to 35.19, new conditional confidence degrees can be derived; the only difference is that $\sum_{i=1}^{l} r_i \geq l - f$ in Eqs. 35.16, 35.18, and 35.19 now should be replaced by $\sum_{i=1}^{l} r_i \leq l - f$. Because this extra information implies obviously that sensor output intervals are less reliable than that without this extra information, and that the larger f is, the less reliable sensor output intervals are, it can be expected (see Table 35.8) that confidence degrees of the fused intervals will be assigned more to those intersections $I_{\{r_i,\ldots,r_l\}}$ in which more r_i of $\{r_1, \ldots, r_l\}$ are equal to 0 (see Table 35.8) as f becomes large.

Example 35.8. The sensor outputs are the same as those given in Example 35.3, and the fusion center knows that at least f sensors are faulty. The fused interval outputs are given in Table 35.10.

TABLE 35.10 Fused Fault-Tolerant Interval Outputs with at Least f Faulty Sensors in Example 35.8

f	$(-\infty, 8) \cup (13, \infty)$	[8, 9)	[9, 10)	[10, 11]	(11, 12]	(12, 13]
0	0.0059	0.0237	0.1159	0.6567	0.1642	0.0336
1	0.0173	0.0691	0.3374	0	0.4781	0.0980
2	0.0938	0.3750	0	0	0	0.5312

35.6 Extension to Sensor Estimate with Multiple Output Intervals

When sensors can output multiple interval estimates with their confidence degrees, namely, the ith sensor outputs intervals $I_i^1, \ldots, I_i^{S_i}$ with their confidence degrees $\alpha_i^1, \ldots, \alpha_i^{S_i}$, the results given in the previous sections can be extended to this more general case. Continue to assume the estimates $I_i^1, \ldots, I_i^{S_i}$ are independent of each other. Without loss of generality, intervals $I_i^1, \ldots, I_i^{S_i}$ are disjoint. If, originally, intervals I_i^j and I_i^k are joint, then define a new interval $I_i^{S_i+1} = I_i^j \cap I_i^k$ with the confidence degree $\{\alpha_i^{S_i+1} = \alpha_i^j \alpha_i^k\}$. By the summability of the confidence degrees, redefine

$$I_i^j := I_i^j - I_i^{r_i+1} \text{ with the confidence degree } \alpha_i^j := \alpha_i^j - \alpha_i^{S_i+1}$$

and

$$I_i^k := I_i^k - I_i^{r_i+1} \text{ with the confidence degree } \alpha_i^k := \alpha_i^k - \alpha_i^{S_i+1}.$$

Thus, I_i^j, I_i^k and $I_i^{S_i+1}$ become disjoint. Similarly, define a generalized interval

$$I_i^0 = \left(\bigcup_{j=1}^{S_i} I_i^j\right)^c \text{ with the confidence degree } \alpha_i^0 = 1 - \sum_{j=1}^{S_i} \alpha_i^j.$$

Then, the corresponding combination rule, its properties, and conditional combination rules can be presented.

35.7 Robust Fault-Tolerant Interval Estimation

It is better that a fault-tolerant interval fusion algorithm should be robust. The robustness here includes two aspects: stability and sensor-interval endpoint tolerance.

35.7.1 Stability

Stability means that a slight change in the input results has an only slight change in the output, in view of the unavoidable error in information acquisition and processing. Unfortunately, Marzullo's method is not stable — it exhibits irregular behavior in that a slight difference in the input may produce a vastly different output. See the following example.

Example 35.9. Their confidence degrees are unknown for following sensor interval estimates:

$$\{[8, 11], [9, 20], [20, 23]\}.$$

Apparently, when $f = 1$, using the preceding (35.6 to 35.8) the output interval of the Marzullo function is [9, 20]. However, if output interval [9, 20] of the second sensor changes to [9, 20 − δ] for any small

$\delta > 0$, the output interval of Marzullo function becomes [9, 11]. In fact, if a combined interval output is derived based on intersection of sensor interval outputs, this combined output is unstable, usually as the output interval of Marzullo function.

To address this unstable behavior in sensor fusion process, Schmid and Schossmaier [16] introduced a fault-tolerant interval integration function called the Schmid–Schossmaier function, which avoided intersection of sensor interval outputs, represented as

$$S(\{I_1, I_2, ..., I_l\}) = [\alpha, \beta], \quad (35.20)$$

where α is the $(f + 1)$-th largest left endpoint of l sensor intervals, and β is the $(f + 1)$-th smallest right endpoint of l sensor intervals. For Example 35.9, when three sensor intervals are

$$\{[8, 11], [9, 20], [20, 23]\}$$

$S(\{I_1, I_2, I_3\}) = [9, 20]$. If sensor output [9, 20] of the second sensor changes to $[9, 20 - \delta]$ for any small $\delta > 0$, $S(\{I_1, I_2, I_3\}) = [9, 20 - \delta]$, which also changes a small δ.

The main distinctive feature of the Schmid–Schossmaier function is that its algorithm is globally stable. However, $M(\{I_1, I_2, I_l\}) \subseteq S(\{I_1, I_2, I_l\})$ and, under some conditions, $M(\{I_1, I_2, I_l\})$ is a real subset of $S(\{I_1, I_2, I_l\})$. In this sense, $S(\{I_1, I_2, I_l\})$ is suboptimal. As a result, although the Schmid–Schossmaier algorithm has good stability, it suffers from the other drawback of the Marzullo function mentioned earlier: it includes gaps between connected intervals $I_{\{i_1, i_2, ..., i_{(l-f)}\}}$ and its output interval is large whenever Marzullo's output is large.

35.7.2 Sensor Interval Endpoint Tolerance

In practice, some errors exist in sensor output interval endpoints. For example, in practice, very often people know ends of sensor output intervals with error as follows:

$$I_i = [a_i^0 \pm \delta_i, b_i^0 \pm \varepsilon_i], \quad i \leq l, \quad (35.21)$$

where δ_i and ε_i are known. In other words, the true interval endpoints of $[a_i, b_i]$ are known imprecisely, but they satisfy the following inequalities

$$a_i^0 - \delta_i \leq a_i \leq a_i^0 + \delta_i, \quad b_i^0 - \varepsilon_i \leq b_i \leq b_i^0 + \varepsilon_i \, (i \leq l). \quad (35.22)$$

It is also assumed that their confidence degrees, $\alpha_i = P_i(\theta \in I_i)$, $i \leq l$ are known precisely. This implies that every sensor, in fact, outputs a set of intervals. Instead of the previous stability, this is another framework to consider robustness of the interval estimation fusion. Thus, a robust fault-tolerant interval estimation is proposed under the assumption that, at most, f intervals in the sensor interval list $\{I_1^1, \cdots, I_l^1\}$ are faulty no matter what the true sensor intervals are. In the following, only the case for optimization criterion (A) — minimizing interval length under confidence degree constraint α_0 — will be given. The method in this subsection can be used similarly in terms of criterion (B).

Naturally, to reach the goal, the min–max criterion is an appropriate candidate, i.e., first, find combined intervals that satisfy confidence degree constraint α_0 for all possible (worst case) true sensor interval outputs; then, choose the smallest interval from the obtained combined intervals. This method is divided into three steps:

1. Although all possible true sensor output interval lists $\{I_1^{r_1}, \cdots, I_l^{r_l}\}$ are countless, it is still possible to classify them into finite groups by the different kinds of orders of all the endpoints $\{\alpha_i, b_i : i \leq l\}$.
2. Considering each group, for local optimization, take all possible different fused intervals that are as short as possible under confidence degree constraint α_0.

3. Considering all those groups, for global optimization, obtain all possible convex closures of unions, which unite one of all those fused intervals of each group, and then take as the ultimate fused output the shortest interval in those convex closures of unions with confidence degrees greater than or equal to α_0.

Therefore, the ultimate fused output is the shortest interval estimate covering θ with the confidence degree of at least α_0, regarding all countless sensor output interval lists.

It has been known that for any given sensor output interval list $\{I_1^{r_1},\cdots,I_l^{r_l}\}$ ($r_i = 0,1$), all the endpoints $\{a_i, b_i : i \leq l\}$ rank on the straight line $\Re$ by increasing value and they partition $\Re$, at most, to 2 ± 1 disjoint intervals. Let q be the actual number of those intervals; clearly, $q \leq 2l + 1$. Each of all nonempty intersections of the sensor output intervals

$$I_{\{I_1^{r_1},\ldots,I_l^{r_l}\}} = \bigcap_{i=1}^{l} I_i^{r_i}, \; r_i = 0,1 \quad (35.23)$$

is equal to one (when all $r_i = 1$) or a union of some of those q intervals (when some of $r_i = 0$, i.e., I_i^0 consists of two separate parts). Thus, the upper bound of possible numbers of all nonempty intersections is q, and their conditional confidence degrees can be calculated by Eqs. 35.17 to 35.19. Using all nonempty intersections and their conditional confidence degrees, the optimal fused output can be taken very easily.

Consider first the definition — the kinds of orders of all the endpoints; some notations will be introduced in order to classify all sensor output interval lists by the different kinds of orders of all the endpoints $\{a_i, b_i : i \leq l\}$.

For example, suppose $1 \leq i \neq j \leq l$. Two left endpoints a_i and a_j have three possible kinds of orders, namely $a_i < a_j$, $a_i = a_j$, and $a_j < a_i$. Two right endpoints, b_i and b_j, have also the same possible kinds of orders, namely, $b_i < b_j$, $b_i = b_j$, and $b_j < b_i$. One left endpoint, a_i, and one right endpoint, b_j, have only two possible kinds of orders, namely, $a_i \leq b_j$ and $b_j < a_i$. Obviously, when $i = j$, $a_i \leq b_j$. Ranking endpoints $\{a_i, b_i : i \leq l\}$ by the preceding nine possible kinds of orders between two endpoints is defined as a kind of orders of them.

Clearly, all the endpoints of finite sensor output intervals can only yield at most finite possible kinds of orders. For all sensor output interval lists $\{I_1^{r_1},\cdots,I_l^{r_l}\}$, let N be the total number of different kinds of orders of all the endpoints $\{a_i, b_i : i \leq l\}$ and the nth kind denoted as $O^n (n \leq N)$. Let

$$O^n \stackrel{\text{def}}{=} c_0^n < c_1^n \sim \cdots \sim c_i^n \sim \cdots \sim c_{2l}^n < c_{2l+1}^n. \quad (35.24)$$

Here, $c_0^n = -\infty$; $c_{2l+1}^n = \infty$; c_i^n ($1 \leq i \leq 2l$) represents a certain left endpoint a_j or a certain right endpoint b_k and its corresponding fluctuant rang is denoted as $(c_i^n)^0 - \eta_i^n \leq c_i^n \leq (c_i^n)^0 + \eta_i^n$ with $(c_i^n)^0 = a_j^0$, $\eta_i^n = \delta_j$ or $(c_i^n)^0 = b_k^0$, $\eta_i^n = \varepsilon_k$; the symbol "$\sim$" represents "$<$," "$=$," or "$\leq$."

All sensor output interval lists can be classified into N groups by $0 \leq x \leq n - 1$. The nth $\leq$ group consists of all the senor output interval lists whose endpoints rank on $\Re$ as O^n. Clearly, all groups do not intersect mutually and their union just contains all sensor output interval lists.

The classification rule by $\{O^n : n \leq N\}$ has two advantages:

- Easily deriving all nonempty intersections $I_{\{I_1^{r_1},\ldots,I_l^{r_l}\}}$ with their conditional confidence degrees of each group from O^n ($n \leq N$), where $I_{\{I_1^{r_1},\ldots,I_l^{r_l}\}} = \langle c_i^n, c_{i+1}^n \rangle$ or $I_{\{I_1^{r_1},\ldots,I_l^{r_l}\}} = \bigcup_j \langle c_{i_j}^n, c_{i_j+1}^n \rangle$ with the symbol "$\langle$" representing "[" or "(" and the symbol "$\rangle$" representing "]" or ")"
- Easily calculating the optimal fused interval estimation output of each group by O^n ($n \leq N$) (see Eqs. 35.25 to 35.27)

Then, the optimal fused interval estimation outputs of each group will be taken.

For any sensor output interval list $\{I_1^{r_1}, \cdots, I_l^{r_l}\}$ in the nth group, let

$$F^n(I_1^{r_1}, \cdots, I_l^{r_l}) = \left\{ A_1^n, \cdots, A_{m_n}^n : A_i^n = \overline{\bigcup_{\{r_i, \cdots, r_l\}} I_{\{r_i, \cdots, r_l\}}} = [c_{i_1}^n, c_{i_2}^n], P(\theta \in A_i^n | C) \geq \alpha_0, \right.$$

any proper closed subset of A_i^n cannot be guaranteed that its confidence degree is (35.25)

greater than or equal to $\alpha_0, i \leq m_n \}$.

$F^n(\{I_1^{r_1}, \cdots, I_l^{r_l}\})$ shows all possible convex closures of unions of the nonempty intersections $I_{\{r_i, \cdots, r_l\}}$, which are as short as possible under the confidence degree constraint α_0.

Naturally, for the nth group, let

$$F^n(O^n) = \{B_1^n, \cdots, B_{m_n}^n\}, \tag{35.26}$$

where

$$B_i^n = \overline{\bigcup_{\{I_1^{r_1}, \cdots, I_l^{r_l}\}} A_i^n}$$

$$= \left[\inf_{\substack{(c_i^n)^0 - \eta_i^n \leq c_i^n \leq (c_i^n)^0 + \eta_i^n \\ 1 \leq i \leq 2l \\ O^n}} \{c_{i_1}^n\}, \sup_{\substack{(c_i^n)^0 - \eta_i^n \leq c_i^n \leq (c_i^n)^0 + \eta_i^n \\ 1 \leq i \leq 2l \\ O^n}} \{c_{i_1}^n\} \right] \tag{35.27}$$

The nth group contains countless interval lists because all of the endpoints $\{a_i, b_i : i \leq l\}$ change continuously in their ranges; however, using O^n and the given fluctuant ranges of all the endpoints, it is very easy to take B^n_i. Its left/right endpoint is actually equal to a certain $a_i^0 - \delta_i$, $a_i^0 + \delta_i$, $b_i^0 - \varepsilon_i$, or $b_i^0 + \varepsilon_i$, (see Example 35.10).

Finally, the convex closure of union of the shortest interval B_{mi}^n of each group cannot be directly taken as the final fused output interval. For global optimization, considering all those groups, one obtains all possible convex closures of unions that unite one of m_n intervals of each group, denoted as

$$G = \left\{ \overline{\bigcup_{\substack{n=1 \\ i_j \leq m_n}}^{N} B_{i_j}^n} : B_{i_j}^n \in F^n(O^n) \right\}. \tag{35.28}$$

In G there are $\prod_{n=1}^{N} m_n$ intervals and all their confidence degrees are greater than or equal to α_0. The shortest one in G is taken as the final fused output. Clearly, its confidence degree is not smaller than α_0.

Remark 35.5. The preceding method can be easily extended to the case with extra information, for example, "at least f intervals of any sensor output interval list are assumed to be fault," as in Remark 35.3. Of course, this method can be used without any prior information on sensor coverage faulty. The following example is given to show how to apply the preceding method.

Example 35.10. The following are sensor interval estimates with interval–end–error by the definitions 35.21 and 35.22 and their own confidence degrees:

$$\{I_1 = [8 \pm 0, 10 \pm 1]{:}0.75;\ I_2 = [9 \pm 0.5, 12 \pm 0]{:}0.83;\ I_3 = [10.75 \pm 0.75, 13 \pm 0]{:}0.91\}.$$

At most, $f = 1$ sensor output interval is faulty and the confidence degree constraint α_0 is 0.76.

For all the preceding sensor output interval lists, six endpoints have only three kinds of orders:

$$O^1 \stackrel{\text{def}}{=} -\infty < a_1 < b_1 < a_2 < a_3 < b_2 < b_3 < \infty;$$

$$O^2 \stackrel{\text{def}}{=} -\infty < a_1 < a_2 \leq b_1 < a_3 < b_2 < b_3 < \infty;$$

$$O^3 \stackrel{\text{def}}{=} -\infty < a_1 < a_2 < a_3 \leq b_1 < b_2 < b_3 < \infty.$$

Thus, all the sensor output interval lists are classified into three groups.

For the first group, using O^1 and the conditional combination rule of confidence degrees (35.17 to 35.19), the fused nonempty intervals and their confidence degrees are given in Table 35.11. Then,

$$F^1\left(\{I_1^{r_1}, I_2^{r_2}, I_3^{r_3}\}\right) = \{A_1^1 = I_{\{0,1,1\}} = [a_3, b_2]\},$$

$$F^1(O^1) = \{B_1^1 = [10, 12]\}. \tag{35.29}$$

For the second group, using O^2 and the conditional combination rule of confidence degrees, the fused nonempty intervals and their confidence degrees are given in Table 35.12. Then,

$$F^2\left(\{I_1^{r_1}, I_2^{r_2}, I_3^{r_3}\}\right) = \{A_1^2 = I_{\{0,1,1\}} = [a_3, b_2]\},$$

$$F^2(O^2) = \{B_1^2 = [10, 12]\}. \tag{35.30}$$

For the third group, using O^3 and the conditional combination rule of confidence degrees, the fused nonempty intervals and their confidence degrees are given in Table 35.13. Then,

$$F^3\left(\{I_1^{r_1}, I_2^{r_2}, I_3^{r_3}\}\right) = \{A_1^3 = \overline{I_{\{1,1,0\}} \cup I_{\{1,1,1\}}} = [a_2, b_1],\ A_2^3 = \overline{I_{\{0,1,1\}} \cup I_{\{1,1,1\}}} = [a_3, b_2]\},$$

$$F^3(O^3) = \{B_1^3 = [8.5, 11],\ B_2^3 = [10, 12]\}. \tag{35.31}$$

Therefore, $G = \{\overline{B_1^1 \cup B_1^2 \cup B_1^3} = [8.5, 12],\ \overline{B_1^1 \cup B_1^2 \cup B_2^3} = [10, 12]\}$ and the final fused output is $[10, 12]$, which is shorter than the convex closures of the second and third sensor interval outputs: $\overline{I_2} = [8.5, 12]$ and $\overline{I_3} = [10, 13]$.

TABLE 35.11 Fused Fault-Tolerant Interval Outputs and Their Confidence Degrees of First Group

Intervals	$I_{\{10,20,30\}}$		$I_{\{0,0,0\}}$	$I_{\{0,1,0\}}$	$I_{\{0,1,1\}}$	$I_{\{0,0,1\}}$
Intervals	$(-\infty, a_1) \cup (b_1, a_2) \cup (b_3, +\infty)$		$[a_1, b_1]$	$[a_2, a_3)$	$[a_3, b_2]$	(b_2, b_3)
Confidence degree	0		0	0	1	0

TABLE 35.12 Fused Fault-Tolerant Interval Outputs and Their Confidence Degrees of Second Group

Intervals	$I_{\{0,0,0\}}$	$I_{\{1,0,0\}}$	$I_{\{1,1,0\}}$	$I_{\{0,1,0\}}$	$I_{\{0,1,1\}}$	$I_{\{0,0,1\}}$
Intervals	$(-\infty, a_1) \cup (b_3, +\infty)$	$[a_1, a_2]$	$[a_2, b_1]$	(b_1, a_3)	$[a_3, b_2]$	$(b_2, b_3]$
Confidence degree	0	0	0.2288	0	0.7712	0

TABLE 35.13 Fused Fault-Tolerant Interval Outputs and Their Confidence Degrees of Third Group

Intervals	$I_{\{0,0,0\}}$	$I_{\{1,0,0\}}$	$I_{\{1,1,0\}}$	$I_{\{0,1,0\}}$	$I_{\{0,1,1\}}$	$I_{\{0,0,1\}}$
Intervals	$(-\infty, a_1) \cup (b_3, +\infty)$	$[a_1, a_2]$	$[a_2, a_3]$	$[a_3, b_1]$	$(b_1, b_2]$	$(b_2, b_3]$
Confidence degree	0	0	0.06905	0.6982	0.23275	0

35.8 Conclusion

Interval estimation fusion based on sensor interval estimates and their confidence degrees has been developed. When sensor estimates are independent of each other, a combination rule has been proposed to merge sensor estimates and their confidence degrees. Moreover, two popular optimization criteria have been suggested: (1) minimizing interval length with an allowable minimum confidence degree, or (2) maximizing confidence degree with an allowable maximum interval length for the interval estimation. In terms of the two criteria, an optimal interval estimation fusion can be obtained based on the combined intervals and their confidence degrees. When the fusion center receives interval estimate with confidence degree to the fusion center and also knows that, at most, f sensor interval estimates are faulty, results on the combined interval outputs and their confidence degrees can be extended to obtain a conditional combination rule and the corresponding optimal fault-tolerant interval estimation fusion in terms of the two criteria.

It is easy to see that Marzullo's fault-tolerant interval estimation fusion [9] is a special case in which the allowable minimum confidence degree is one. More generally, for any extra information, the corresponding conditional combination rule for the interval estimation fusion can be derived. When sensor estimates are dependent on each other, a similar interval estimation fusion method cannot be derived unless the joint probabilities of all possible sensor interval output lists can be known. Finally, the stability of the interval estimation fusion was briefly discussed and, using min–max criterion, an interval endpoint's tolerance fusion was derived.

Acknowledgment

The work of Y. Zhu is supported by NSF of China.

References

1. Bar-Shalom, Y. and Li, X.R., *Multitarget–Multisensor Tracking: Principles and Techniques*, YBS Publishing, 1995.
2. Bar-Shalom, Y., On the track-to-track correlation problem, *IEEE Trans. Automatic Control*, 26, 571, 1981.
3. Chong, C.Y., Mori, S., and Chang, K.C. Distributed multitarget multisensor tracking, in *Multitarget–Multisensor Tracking: Advanced Applications*, vol. 1, Bar-Shalom, Y., Ed., Artech House, Norwood, MA, 1990.
4. Bar-Shalom, Y., Ed., *Multitarget–Multisensor Tracking: Advanced Applications*, vol. 1 and 2, Atech House, Norwood, MA, 1990 and 1992.

5. Li, X.R., Zhu, Y.M., and Han, C.Z. Unified optimal linear estimation fusion, in *Proc. 2000 Int. Inf. Fusion Conf.*, Paris, July 2000.
6. Zhu Y.M. and Li, X.R., Best linear unbiased estimation fusion, in *Proc. 2nd Int. Inf. Fusion Conf.*, 2, 1054, ISIF, Sunnyvale, CA, July 1999.
7. Zhu, Y.M. and Li, B.H., Optimal interval estimation fusion based on sensor interval estimates and confidence degrees, in *Proc. 2003 Int. Conf. SPIE*, 5099, 268, Orlando, FL, April 2003.
8. Zhu, Y.M., Yu, G., and Li, X.R., Multisensor statistical interval estimation fusion, in *Proc. 2002 Int. Conf. SPIE*, 4731, 269, Orlando, FL, April 2002.
9. Marzullo, K., Tolerating failures of continuous-valued sensors, *ACM Trans. Computer Syst.*, 8, 284, 1990.
10. Prasad, L. et al., Functional characterization of fault-tolerant integration in distributed sensor networks, *IEEE Trans. Syst., Man, Cybernetics*, 21, 1082, 1991.
11. Iyengar, S.S. and Prasad, L., A general computational framework for distributed sensing and fault-tolerant sensor integration, *IEEE Trans. Syst., Man, Cybernetics*, 25, 643, 1995.
12. Schmid, U. and Schossmaier, K., How to reconcile fault-tolerant interval intersection with the Lipschiz condition, *Distributed Computing*, 14, 101, 2001.

36
Fault Tolerance in Wireless Sensor Networks

Farinaz Koushanfar
University of California at Berkeley

Miodrag Potkonjak
University of California at Los Angeles

Alberto Sangiovanni-Vincentelli
University of California at Berkeley

36.1	Introduction	36-1
	Motivation • Objectives	
36.2	Preliminaries	36-3
	Sensor Network	
36.3	Example of Fault Tolerance in a Sensor Network System	36-3
36.4	Classical Fault Tolerance	36-4
36.5	Fault Tolerance at Different Sensor Network Levels	36-5
	Physical Layer • Hardware • System Software • Middleware • Application	
36.6	Case Studies	36-8
	Heterogeneous Fault Detection • Discrepancy-Based Fault Detection and Correction	
36.7	Future Research Directions	36-12
36.8	Conclusion	36-13

36.1 Introduction

36.1.1 Motivation

The reliability of computer, communication, and storage devices was recognized early as one of the key issues in computer systems. Since the 1950s, techniques that enhance the reliability of computer and communication systems have been developed in academia and industry. It has been recognized that as complexity of computing and communication devices increases, fault tolerance will gain more importance. Surprisingly, it has never been the major design objective, mainly because reliability of individual components has been increasing at a much more rapid pace than was expected. In addition, creative packaging and cooling schemes have tremendously reduced the stress factor on computation and communication systems.

The only component of fault tolerance that has received a great deal of attention in industry is offline testing. The modern testers are $10+ million systems that are contributing increasingly to the cost of modern microprocessors. In addition, the percentage of logic that supports testing has been rapidly increasing in the last 10 years, from less than 1% to more than 5% of the total transistor count.

The rapid growth of the Internet was the first major facilitator of renewed interest in fault tolerance and related techniques such as self-repair [8, 9, 10, 13]. Because the Internet requires a constant mode of operation, a special effort has been made to develop fault-tolerant data canters. The emergence of wireless sensor networks will further increase the importance of fault tolerance while at the same time imposing a number of unique new conceptual and technical challenges to fault tolerance researchers.

At least three major groups of reasons support research in fault-tolerant sensor networks receiving significant attention. The first one is related to the technology and implementation aspects. At least two components of a sensor node, sensors and actuators, will directly interact with the environment and be subjected to a variety of physical, chemical, and biological forces. Therefore, they will have significantly lower intrinsic reliability than integrated circuits in fully enclosed packaging. In addition, wireless sensor nodes are exceptionally complex systems in which a variety of components interact in a complex way.

Furthermore, hundreds or maybe thousands of these nodes will form a distributed embedded network system that will handle a variety of sensing, actuating, communicating, signal processing, computation, and communication tasks. Wireless sensor networks will often be deployed as consumer electronic devices that will put significant constraints on cost and therefore quality of used components. More importantly, nodes operate under strict energy constraints that limit energy budgets dedicated to testing and fault tolerance.

The second reason is that applications will be equally as complex as the involved technology and architectures. More importantly, sensor networks will often operate in an autonomous mode without a human in the loop; security and privacy concerns will often prevent extensive testing procedures. This will adversely affect not only testing and fault tolerance but also related tasks such as debugging, in which reproduction of specific conditions during which a fault occurred will be difficult. Also, applications will require that sensor nodes are often deployed in uncontrolled and sometimes even hostile environments. Finally, and maybe most importantly, many applications of sensor networks will be safety critical and could have an adverse impact on humans and the environment, particularly when actuators are used.

The final reason is that, because wireless sensor networks are a new scientific and engineering field, the best way to address a particular problem is not quiet clear. In this situation, it is also difficult to predict accurately the best way to treat fault tolerance within a particular wireless sensor network approach. In addition, technology and envisioned applications for wireless sensor networks are changing at a rapid pace. For example, if one considers power consumption, each particular scheme will depend significantly on the relative power consumption of different approaches. Specifically, if communication energy is significantly higher than computation energy, it is important to develop localized algorithms that will require only a limited amount of communication.

Therefore, with respect to fault tolerance, it is important to consider schemes that conduct error detection using only local information. If one wants to ensure fault tolerance during the sensor fusion, the goal is to design fault-tolerant techniques that do not significantly increase the communication overhead. On the other hand, if the computation energy is significantly higher than the communication requirements, it is a good idea to support communication resources at one node with computation resources at other nodes. It is preferable to develop fault-tolerant sensor fusion approaches that require little additional computation regardless of any additional communication requirements.

36.1.2 Objectives

The primary goal of this chapter is to survey the field of fault tolerance in sensor networks. Fault tolerance is considered at four different levels of abstraction, starting from hardware and system software and going to the middleware and application layers. Fault tolerance is examined at each level of six individual components of a node: computing engine; communication and storage subsystems; energy supply; sensors; and actuators. It is also considered at the level of a node, as well as at the network level. Finally, resiliency against errors, where wireless sensor networks are treated as embedded distributed systems, is discussed.

Three aspects of fault tolerance are considered: fault models, error detection and diagnosis techniques, and resiliency mechanisms. In order to provide in-depth treatment of specific approaches, two case studies will be presented: one on error detection in sensor networks and one on heterogeneous built-in self-repair (BISR)-based fault tolerance. Finally, in order to provide a global vision, discussion will center on the relationship to sensor networks and traditional fault tolerance techniques as well as a set of predictions of future research directions in this field.

The next two sections provide relevant preliminary information. After that, fault tolerance is discussed at the node and network levels. Next, two case studies address in a more comprehensive way several technical details with respect to fault tolerance in sensor networks. Finally, future directions along the three dimensions of fault tolerance are suggested.

36.2 Preliminaries

36.2.1 Sensor Network

A wireless sensor network is a system of small, wirelessly communicating nodes in which each node is equipped with multiple components. In particular, each node has a computation engine; communication and storage subsystems; a battery supply; and sensing and, in some cases, actuating devices. Such a network is envisioned to integrate the physical world with the Internet and computations. The power supply on each node is relatively limited, and frequent replacement of the batteries is often not practical because of the large number of nodes in the network. Therefore, energy is the most constraining factor on the functionality of these networks. In order to save energy, nodes only use short-range communications, which have been proven to consume much less energy than long-range communications [44]. Short-range communication between the nodes implies localized interaction in the network.

There is a need to model the different components of a sensor network. Sensor networks are often abstracted and mapped into a graph in which each vertex corresponds to a wireless node and an edge corresponds to the communication between two nodes. If the communication between the nodes is bidirectional, the mapped graph of the network will be nondirected. However, if this communication is asymmetric, then the mapped graph becomes directed. The communication model between the nodes can be one to one, or one to many. In the one-to-one model, each node sends and receives messages from only one of the communication edges. In the one-to-many model, each message sent out by a node can be heard by all of its neighbors. Because of the great variety of different sensors, in terms of their functionality and in terms of their underlying technologies, providing a reasonable and practical model for sensors and actuators is a very complex task.

Many potential applications are envisioned for sensor networks. For example, they can be used in a battlefield, where they can detect and spy on enemies or support the positive forces. Also, they can be used in intelligent security systems in buildings and security-critical applications. They can be used for habitat-monitoring applications in which they can monitor and study changes in phenomena for a long time. A number of comprehensive surveys on sensor networks have been conducted [1, 24, 45].

36.3 Example of Fault Tolerance in a Sensor Network System

The problem of fault-tolerant multimodal sensor fusion for digital binary sensors can be informally introduced using the example shown in Figure 36.1(a) and (b). A sensor network recognition system is deployed in an office to identify people in that office as they walk in through the main door. Six people, named A, B, C, D, E, and F, work in the office. The system consists of two different types of sensors: (1) a height sensor, which is a set of light sensors in series; and (2) a voice recognition sensory system that requires everybody entering the room to speak a given pass phrase into a microphone. Figure 36.1(a) shows the selected identification characteristics of people in the office. Figure 36.1(b) shows the same characteristics mapped to a two-dimensional plot.

It is easy to see that the system can distinguish between two persons, P_1 and P_2, if they fall into different squares, when mapped to the chart shown in Figure 36.1(b). If all of the sensors work properly, each person will naturally fit into a different square, according to the figure. For most of the cases, even if one of the height sensors or voice sensors fails, the recognition of the right person is still possible. This is

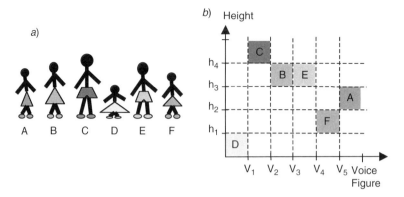

FIGURE 36.1 Example of a multimodal sensor network system. (From Koushanfar, F. et al., *IEEE Sensors*, 2, 1491–1496, June 2002. With permission.)

accomplished using heterogeneous fault tolerance, in which a failed sensor of one type can be replaced by the functionality of a sensor of another type. However, for the case of persons B and E, who are the same height, voice figure is the only way to distinguish the two persons, so the system does not have any fault tolerance to the failure of sensor v_3 that distinguishes between the objects B and E. If the office had only five people (excluding either B or E), then it would be completely fault tolerant. Complex sensor network systems can be designed in such a way that the heterogeneous fault tolerance scheme can make the system resilient to the failure of a specified number of sensors of specified modality.

36.4 Classical Fault Tolerance

Fault tolerance emerged early as a major concern during design of digital computing systems. For example, the Bell Lab's relay-based computer was performing multiple computations for the same inputs using the same program, in order to compare results and detect potential temporary malfunctioning. Also, UNIVAC 1, the first computer commercially available in 1951, utilized parity checking and arithmetic unit replication to enhance its reliability. At the same time, fault tolerance received research attention in the mid 1950s. Moore and Shannon [36] and von Neumann [42] conducted studies on how to design systems that preserve functionality after a subset of components experiences failures. Also, embedded computing systems such as Bell Lab's electronic switching system had extensive support for fault tolerance, mainly through enhanced serviceability features [16]. Furthermore, Apollo's mission to the Moon used triplicated computers in order to maximize the chances of success [46].

During the 1970s, research on fault tolerance started to diverge along several lines. Initially, the two most important and influential were fault tolerance in VLSI-based systems and fault tolerance in distributed systems [30, 31, 38, 40]. More recently, fault tolerance in data bases [7, 22], the Internet (e.g., reliable multicast and reliable distributed storage, peer–peer), and self-repair have been topics of major importance. In VLSI systems, fault tolerance has been addressed at all levels of abstraction, including circuit, logic, register, transfer, program, and system levels. It is common that reliable system design is discussed within three stages of life of a product: design, manufacturing, and operational. Before going into more technical details, it is important to define the basic entities.

Error is the manifestation of a fault inside a program. It is important to note that error can occur not only at the fault site, but also at some distance. Fault is an incorrect state of hardware or a program as a consequence of a component's failure. Permanent faults are continuous and stable in time. For example, permanent hardware faults are consequences of irreversible physical alteration within a component. An intermittent fault is one that has only occasional manifestation due to unstable characteristics of the hardware, or as a consequence of a program being in a particular subset of space. Finally, a transient fault is the consequence of temporary environmental impact on otherwise correct hardware. For example, often the impact of cosmic radiation may be transient.

Fault tolerance considers three main types of concerns: fault models; fault detection and diagnosis; and resiliency mechanisms. Each level of abstraction has its own types of faults [47]. For example, at the gate level, several fault models have been successfully used in the testing phase. One model is "stuck" where the logical value on interconnect gate or pin is permanently set to a value stuck at one or stuck at zero. Another model is the "bridging" fault model, in which two or more neighboring signal lines are physically connected introducing wired AND or wired OR functions, depending on the logic family used. Shorts and opens are another class of faults and correspond to missing or additionally introduced connections, respectively. Finally, in unidirectional faults, an error occurs in the same logical direction when a certain single failure occurs. Typical examples are an open circuit in a memory-select line resulting in a particular word incorrectly read as all ones, instead of all zeros. It is interesting and important to note that almost all testing approaches assume a single fault model, regardless of which type of fault is considered.

Reliability techniques compromise the following phases:

- Fault confinement establishes limits of fault effects over a particular area; therefore, contamination of other areas is prevented.
- Fault detection is a phase in which it is recognized that an unexpected event has occurred.
- Fault latency is the time that passes between the fault occurrence and the moment when the fault is detected.

Traditionally, fault detection techniques are classified into offline and online detections. Most often, for offline detection, special diagnostic programs are employed during idle periods of time or using multiplexing with a regular mode of operation. Online detection targets real-time fault identification and is performed simultaneously with a real work load. Typical online detection techniques include parity checking, duplication, and triplication.

- Diagnosis is a stage at which the exact occurrence of a fault is attributed to a specific atomic piece of hardware.
- Reconfiguration is the stage entered after diagnosis at which the system is restructured in such a way that faults do not have an impact on the correct output. Graceful degradation is a reconfiguration technique in which performance of the system is reduced, but the correct functionality is preserved.
- Recovery is a stage at which an attempt is made to eliminate the effects of faults. The two most widely used recovery techniques are fault masking and retry. The fault masking approach uses redundant correct information to eliminate the impact of incorrect information. In retry, after the fault is detected, a new attempt to execute a piece of a program is made in the hope that the fault is transient.
- Restart is the stage invoked after the recovery of correct, undamaged information. In cold restart, a complete resetting of the system is conducted.
- Repair is the stage during which the failed component is substituted with the operational component.

A number of excellent surveys, special issues, and textbooks on fault tolerance are available [25, 32].

36.5 Fault Tolerance at Different Sensor Network Levels

36.5.1 Physical Layer

The physical layer is responsible for establishing communication in a given medium between two nodes. Typical tasks at this level include modulation–demodulation and encoding–decoding. Traditionally, fully hardwired solutions have been used in order to minimize cost and maximize energy efficiency. A software radio is a wireless communication device in which parts or all of the physical layer functions are realized in software [33].

Software radios are a way to extend programmability into the physical layer and to enable adaptation to channel conditions. It has been demonstrated that adaptive link layer techniques can significantly improve the performance of wireless networks [17–19, 21]. The first commercially available radio was the speakeasy device [29]. Comprehensive surveys about software radios and related technologies have been conducted [34, 35, 41]. Currently, the major commercial application driver is incompatibility between the number of cellular and PCS communication standards. The primary reason for deployment of hardware and software radios has been to solve interoperability problems and to enhance performances in noisy media; however, they are also ideally suited for realization of a variety of fault tolerance techniques at the physical layer. For example, if some components of the software radio are not functional, one can switch to modulation and encoding schemes that can be realized with the still operational hardware resources. Adaptation to noise characteristics can also be considered from the fault tolerance point of view.

36.5.2 Hardware

At the hardware level, components can be divided into two groups. The first consists of a computation engine, storage subsystem, and power supply infrastructure that are very reliable. Exceptionally reliable systems are available that incorporate sophisticated fault tolerance techniques; even off-the-shelf microprocessors and DSP processors and controllers are very reliable devices with very low rates of malfunctioning.

At least three main reasons indicate why this does not necessarily imply that computational subsystems of sensor nodes will be exceptionally reliable. The first is that sensor nodes are very cost sensitive and therefore will not always be able to design using the highest quality components. The second is that strict energy constraints imply that repeated computations are often not realistic options. The third is that these systems are often deployed in much harsher environments than those in which today's computers function. Although programmability and flexibility are of high importance in sensor networks, strict energy constraints will result in extensive use of application-specific designs that can have up to two orders of magnitude less energy consumption for the same functionality. For these subsystems, heterogeneous BISR fault-tolerant schemes will simultaneously provide the targeted level of fault tolerance and low energy consumption [48].

With respect to storage components, several options exist. Memories can be divided into volatile and nonvolatile. Volatile memories are used for short-term storage; the best known and most widely used are SRAM and DRAM. Due to their relatively simple and regular structure, both types are very reliable, in particular DRAM. Starting with 4-Mb components, BISR has been often used to enhance the yield of DRAM memories. Note that memories are designed using the standard semiconductor processes. Nonvolatile memories such as flash and MRAM are also very reliable. Flash has the restriction that one cannot write too many times at the same location, so it will be used mainly for storage of system programs. Therefore, it can be concluded that storage systems are generally fault resilient and will very rarely incur a fault tolerance bottleneck.

With respect to energy supply, two parts of the energy supply subsystem can be distinguished. Traditional energy sources are rechargeable batteries and fault tolerance is achieved by providing a back-up battery. The first commercial fuel cells and subsystems that leverage on energy scavenging have recently been demonstrated. Although it appears that fuel cells will be very reliable, some energy-scavenging subsystems, such as the ones converting light into energy, can have very volatile performance. For energy distribution, the standard solution to enhance fault tolerance is to deploy multiple distribution networks.

Another component that greatly depends on the surrounding environment is the wireless radio. The standard way to enhance the performance of radios is to use aggressive error correction schemes and retransmission. These two schemes are examples of time redundancy. In addition, several schemes have been proposed in which two or more radios are used. Although the primary goal of these approaches is to save energy, they can also be used to enhance fault tolerance. In addition to the schemes that operate on the physical and link layer, it is important to mention techniques that operate at the network layer. However, these schemes are more naturally considered and implemented at the system software level.

As mentioned previously, sensors and actuators are the subsystems most prone to malfunctioning. In the case of sensors, three types of faults can be distinguished: (1) calibration systematic error; (2) random noise error; and (3) complete malfunctioning. The first two can be addressed through time redundancy; however, the last one is enhanced using hardware redundancy. Depending on the type of data and information processing, an effective fault tolerance scheme for sensors can be accomplished using heterogeneous BISR techniques [26]. Currently, no scheme other than hardware redundancy is envisioned for actuators.

36.5.3 System Software

System software consists of the operating system (OS) and utility programs. Fault tolerance at the system software level can be addressed in several ways with respect to the computational subsystem [4–6]. Probably the most promising is through software diversity: each program is implemented in n different versions in hope that different versions will not have identical bugs [2, 3]. The subsystem that can most benefit from fault tolerance realized at the system software level is the communication unit. For example, one can reroute messages using different paths in the multihop network. With respect to sensors and actuators, the most important piece of system software is the one related to calibration. Recently, a number of schemes have been proposed for this task [12, 43]. A very important component of system software is the one that supports distributed and simultaneous execution of localized algorithms. For example, in the case of energy minimization under functionality constraint requirements, several protocols have been developed for the coordination of distributed actions [14, 28]. It is important to note that when communication protocols are considered, there is a clear trade-off between complexity and effectiveness.

36.5.4 Middleware

At the system software level, in addition to the OS of the individual nodes, networking (communication) plays the most dominant role. Starting with the middleware level, emphasis is shifted toward data aggregation, data filtering, and sensor fusion. These are tasks mainly related to sensor readings. Because it is difficult to provide fault tolerance in an economic way at the level of a single sensor, numerous fault tolerant approaches for this task will appear at the middleware level. Although currently the majority of applications are very simple, in order to address real-life applications, it is necessary to develop much more complex middleware. In order to combat software faults, n-versioning is one of the options [2, 3].

In particular, heterogeneous approaches that can substitute the readings of one type of sensor with the readings of another type are very important because of their low overhead. Another important issue solely related to middleware is how many sensors of each type should be placed on which positions on a particular node. If error resiliency of communication is much higher than the error resiliency of sensors, solutions in which sensors of the same node are placed on the same node will be favored.

36.5.5 Application

Finally, fault tolerance can be addressed at the application level. For example, to identify a particular person, one can use sensors try to measure a variety of biometric features of that person. Each feature and possibly a combination of features will be sufficient to identify that person. Addressing fault tolerance at the application level may be very efficient; unfortunately, any given application will require a customized way in which to address the issue. On the other hand, an additional advantage of application-level fault tolerance is that it can be used to address faults in essentially any type of resource.

36.6 Case Studies

36.6.1 Heterogeneous Fault Detection

Before this case study is described, key facts and assumptions about fault models, fault detection, and embedded sensor networks will be briefly outlined.

Each sensor node has five components: computation; communication; storage; sensors; and, often, actuators. Widely accepted fault and error models for processors; FPGA-based components; SRAM and DRAM; nonvolatile memory and disks; and communication systems are readily available. However, the situation for actuators and sensors is very different. Both types of resources are conceptually more complex and intrinsically more diverse to allow for simple, yet realistic and widely applicable fault and error models.

Koushanfar et al. [26] restricted attention on faults in sensors. They adopted two fault models, the first of which is related to sensors that produce binary outputs. In this case, obviously, one can envision a number of applicable fault models. For example, one model can capture probability or statistics of erroneous reported results. Nevertheless, it appears that the most logical model with potentially largest applicability range is the permanent fault model in which the only possible outcomes are that the sensor is functional or not. For this fault model, the fault detection procedure is often straightforward — usually, just observing the output of the sensors.

The second fault model is related to sensors with continuous (analog) or multilevel digital outputs. The fault models for this type of sensor are even additionally more complex and diverse. They propose to measure the level of discrepancy of the output of individual sensors with the multimodal model used for fusion as the indication of the level of error in that sensor.

The approach has two key advantages. The first is that fault tolerance approaches are such that the developed technique is applicable to a great variety of fault models. This approach is particularly well suited for addressing transient errors and errors in measurements. The second advantage is that the approach simultaneously addresses fault detection and correction. Overall, Koushanfar and colleagues made only mild assumptions; the main one was that the majority of sensors were functioning correctly.

The sensor resource assignment (SRA) problem can be formulated in the following way:

INSTANCE: Set A_1 of points p_i $(x_{i1}, ..., x_{im})$ in m-dimensional space where $1 \leq i \leq n$, a positive integer J_1, set H that consists of $m(n-1)$ [$m-1$]-dimensional hyperplanes that are perpendicular to one of the m axes, such that each hyperplane is separating two points p_i and p_j that have the closest coordinates along the axis to which the hyperplane is perpendicular.

QUESTION: Find a subset of selected hyperplanes H, such that any two points p_i and p_j are separated by at least one of the selected hyperplanes and also the cardinality of H is, at most, J_1.

CLAIM: Sensor resource allocation (SRA) is NP-complete.

Next, these authors present their approach and algorithms for fault-tolerant sensor assignment. It is easy to envision a monolithic solution that simultaneously considers fault tolerance requirements and sensor allocation and assignment problems; however, following principles of separation of concerns and orthogonality, they designed a fully modular system with separate optimization mechanisms for the subtask: sensor assignment, sensor allocation, and fault tolerance. These three steps are addressed in the following way.

Koushanfar and coworkers employed two different algorithmic engines to the SRA problem: ILP based and simulated annealing based. The rationale behind the integer linear programming (ILP) approach is that although ILP solvers are often not fast, they are attractive because they guarantee an optimal solution. In addition, many smaller instances of practical importance can be solved using this approach. The point is that they must find the solution to the SRA problem before the deployment so that it is a one-time expense in computational time on the workstation and may be acceptable. In cases when ILP is not applicable, they provide the option of using simulated annealing as the optimization mechanism.

The ILP formulation for the SRA problem can be stated in the following way.

INPUTS: set of n, m-dimensional points $p_i(x_{i1}, x_{i2}, \ldots, x_{im})$, $1 \leq i \leq n$; set of all possible tests T, with elements t_k $(1 \leq t_k \leq m(n-1))$, where the $(l(n-1)+1)$ to $(l+1)(n-1)$ tests are in dimension l, $1 \leq l \leq m$, each separating two closest points in that dimension. The cost of each test t_k is c_k.
Variable X_k: $X_k = 1$ if test t_k is selected and $X_k = 0$ otherwise.
Objective function: to minimize the total cost of all of the selected tests. In other words:
OF:

$$\sum_{k=1}^{m(n-1)} x_k \cdot c_k$$

The constraint of the problem is that for each pair of points p_i and p_j, at least one test has a different outcome when applied to these two points. These authors define an auxiliary matrix $A[n \times k(m-1)]$, with constant elements a_{ik}, as $a_{ik} = 1$ if the test t_k produces 1 on point p_i and $a_{ik} = 0$ otherwise.

Using the matrix A and the variables, Koushanfar et al. find a linear expression that produces zero, if a test produces similar results on the two points p_i and p_j, and one otherwise. One such expression is $X_k \times (a_{ik} + a_{jk}) \times (1 - a_{ik} \times a_{jk})$, which has the required property. Therefore, to have a different test result on each set of two points p_i and p_j, they write the following constraints.

CONSTRAINTS: For each pair of points p_i and p_j,

$$\sum_{k=1}^{m(n-1)} x_k \cdot (a_{ik} + a_{jk}) \cdot (1 - a_{ik} a_{jk}) \geq 2$$

A standard simulated annealing code was used. The four components of simulated annealing (moves — neighborhood structure; objective function; cooling schedule; and stopping criteria) are defined in the following way. Move is the replacement of one sensor with another sensor of the same type; the goal is to maximize objective function. The standard geometric cooling schedule was used. Finally, as stopping criteria, the user-specified number of steps in which the improvement did not occur were used.

The resource allocation is conducted in the following way. The number of sensors that is lower bound on the potential solution is proposed as the initial solution. The bound is calculated assuming that all dimensions have the same number of sensors and each n-dimensional compartment will eventually contain one point. After that, the simulated annealing RSA algorithm is run. During this running process, the move is modified so that one type of sensor can be replaced with another type. Statistics about which type of sensor helps the most to improve objective function after each move are accumulated and this information is used to decide which type of sensor to add or remove.

For fault tolerance, one can envision three different mechanisms:

- The first is to specify in the ILP formulation or in the simulated annealing code that each two points must be separated by at least r hyperplanes. Because this approach essentially doubles the redundancy, this alternative was not accepted.
- The second alternative is to add exactly one extra sensor of each type to the solution generated by the sensor resource allocation problem. When a large number of sensor nodes of each type is used, the overhead is relatively low. Also, in this case, the need for storing or communicating more than one resource assignment solution is eliminated. Therefore, if moderate levels of fault tolerance are needed, this can be an attractive alternative.
- The final and most attractive alternative in terms of overhead is to leverage on heterogeneous back-up of sensors of different modality. Here, allocation is generated in the following way. First, the cost of an overall solution is calculated for each type of sensor for all allocations k, from 1 to smaller than the number allocated in the best resource allocation solutions. Then the cost of all these solutions is plotted on the y-axis on the graph, where the x-axis is the number of allocated sensors of analyzed type. In such a way, m graphs, where m is the number of sensors of different

modality, are obtained. Obviously, now it is necessary to use the RSA algorithm to analyze only allocations worse in terms of cost than the optimal solution and better than the solution from the second alternative. This analysis is conducted in the order dictated by increasing cost of the proposed solution.

The applicability of the preceding technique can be generalized, and therefore enhanced, in a number of ways. One possibility is to characterize objects using statistical data and to build a statistical model for decision making using data from sensors. Another, equally important and with equally large application, domain option is to conduct multimodal sensor fusion in order to support the decision process. As a matter of fact, multimodal, multilevel sensor fusion has emerged as one of the canonical problems in sensor networks. Informally, it can be defined in the following way: a number of sensors, some of them with different modalities, are given; the goal is to extract the information requested by a user as accurately as possible from noisy measurements.

Although the problem seems too general to be solved efficiently using a single approach, it can be addressed in a systematic way. It is necessary to develop, or even better to find, some already developed analytic models related to the measured quantities. Once the equations of an analytical model are assembled, the intriguing and important question is to try to figure out which measurements are faulty or have a high degree of noise. One way to answer this question is to try to find a subset of measurements that produces a consistent set of analytic models. Using this set of equations, the value for all quantities of interest can be calculated. Therefore, the key to providing fault-tolerant multimodal sensor fusion is to generate a model of the physical world that is rich enough to ensure that the system is solvable even when some of the equations are not used. The main difficulty is that the systems of equations are often nonlinear and therefore it is very difficult to say in advance when the system is well defined in a sense that it can be uniquely solved.

Probably the best way to clarify the introduced approach is to take a closer look at an example. For this purpose the scenario illustrated in Figure 36.2 will be used. An object O moving along its trajectory, which includes points p_i in an embedded sensor network, consists of a number of nodes, each represented by a shaded circle n_i. Four types of sensors — RSSI-based distance discovery; speedometer; accelerometer; and compass — are used to measure the angle in two-dimensional physical space. Three RSSI-based measurements can be used to locate the object O in any particular moment. Euclidian space, Newton mechanics, and trigonometry laws can be used to establish relationships between measurements.

Specifically, Equation 36.1 through Equation 36.9 are trilateration equations; Equation 36.10 through Equation 36.13 are Newton law equations and Equation 36.14 and Equation 36.15 are trigonometry laws. The key observation is that more equations (15) than variables (12) may have errors. Thus, if one sensor is not functioning, it can be calculated from the established system of equations. Also, for each variable, one can find how much it must be altered in order to make the whole system of equations maximally consistent; variables that must be altered most are most likely measured by faulty sensors. Therefore, one

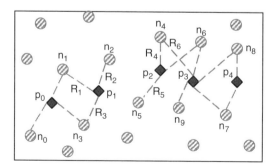

FIGURE 36.2 Sensors tracking an object. (From Koushanfar, F. et al., *IEEE Sensors*, 2003. With permission.)

way to identify and correct sensor measurements is to try all scenarios in which exactly one type of sensor measurement is not taken into account and compare the maximal error in the system. Another very important observation is that, by sampling all operational sensors more often, one can compensate for faulty sensors.

$$(x_1 - s_1)^2 + (y_1 - t_1)^2 = R_1^2 \tag{36.1}$$

$$(x_1 - s_2)^2 + (y_1 - t_2)^2 = R_2^2 \tag{36.2}$$

$$(x_1 - s_3)^2 + (y_1 - t_3)^2 = R_3^2 \tag{36.3}$$

$$(x_2 - s_4)^2 + (y_2 - t_4)^2 = R_4^2 \tag{36.4}$$

$$(x_2 - s_5)^2 + (y_2 - t_5)^2 = R_5^2 \tag{36.5}$$

$$(x_2 - s_6)^2 + (y_2 - t_6)^2 = R_6^2 \tag{36.6}$$

$$(x_3 - s_7)^2 + (y_3 - t_7)^2 = R_7^2 \tag{36.7}$$

$$(x_3 - s_8)^2 + (y_3 - t_8)^2 = R_8^2 \tag{36.8}$$

$$(x_3 - s_9)^2 + (y_3 - t_9)^2 = R_9^2 \tag{36.9}$$

$$\sqrt{(x_1 - x_2)^2 - (y_1 - y_2)^2} = \tfrac{1}{2} \cdot a_1 (\Delta t)^2 + v_1 (\Delta t) \tag{36.10}$$

$$\sqrt{(x_3 - x_2)^2 - (y_3 - y_2)^2} = \tfrac{1}{2} \cdot a_2 (\Delta t)^2 + v_2 (\Delta t) \tag{36.11}$$

$$a_1 . \Delta t = v_1 - v_0 \tag{36.12}$$

$$a_2 . \Delta t = v_2 - v_1 \tag{36.13}$$

$$\alpha_1 = \tan^{-1}\left(\frac{y_2 - y_1}{x_2 - x_1} \right) \tag{36.14}$$

$$\alpha_2 = \tan^{-1}\left(\frac{y_3 - y_2}{x_3 - x_2} \right) \tag{36.15}$$

36.6.2 Discrepancy-Based Fault Detection and Correction

The work by Koushanfar et al. [27] introduces a cross-validation-based technique for online detection of sensor faults — an approach that can be applied to a broad set of fault models. These authors define a fault as an arbitrary type of inconsistent measurement by a sensor that cannot be compensated systematically. In particular, they consider faults associated with incorrect measurements that cannot be corrected using calibration techniques. The approach is based on two ideas:

- Comparing the results of multisensor fusion with and without each of the sensors involved
- Using nonparametric statistical techniques to identify measurements that are not correctable, regardless of the mapping function used between the measured and accepted values

Sensor measurements are inevitably subject to errors of two types: (1) random fluctuations in data due to a noise in a sensor or in a sensed phenomenon; or (2) gross errors — faults. A practical method to distinguish a random noise is to run maximum likelihood or Bayesian approach on the multisensor fusion measurements. A random noise would exist if running these procedures improves the accuracy of final results of multisensor fusion. Although several efforts have attempted to minimize random errors, very little has been done for fault detection. In multisensor fusion, measurements from different sensors are combined in a model for consistent mapping of the sensed phenomena. Although the new fault detection technique is generic and can be applied to an arbitrary system of sensors that uses an arbitrary type of data fusion, for the sake of brevity and clarity they focus on equation-based sensor fusion [26].

Assume a set of sensors s_i ($0 \leq I \leq n$), each measuring a value x_i at a time t. The multimodal sensor fusion model equations are $f_1, \ldots, f_p$ and are typically nonlinear functions with the following forms: $f_j(x_1, x_2, \ldots, x_n) = 0$, ($0 \leq j \leq p$). The system of equations is overconstraint and solves the system $n + 1$ times. First, they solve with all the equations in the original format; then they ignore each variable and solve a least-constrained system with $n - 1$ variables (n times). They compare the values for each variable x_n in all $n + 1$ scenarios. In order to improve accuracy of fault detection, the system can be solved for m measurements by each sensor. At last, they conduct statistical analysis on the data for each sensor. If the obtained values for a sensor are not consistent within a confidence interval calculated by the percentile method [20], that sensor is considered faulty.

36.7 Future Research Directions

It is well known that it is very difficult to predict anything; nevertheless, certain directions will inevitably attract a higher level of research interest because of their intrinsic importance. Future research directions can be classified into four groups. The first two are related to fault models and testing. The third is related to resiliency mechanisms and the last to analogies between fault tolerance and other domains such as power minimization and security.

The development of theoretically attractable and realistic fault models is one of the key prerequisites for development of sound and real-life relevant fault tolerance techniques for sensor networks. Apart from fault tolerance models for components such as computation, storage and, to serious extent, communication are available; however very little has been published in terms of fault models for sensors and actuators. At the same time, these components are the most important for overall system fault tolerance. The development of fault models for sensors will be particularly difficult due to the great variety of their types, environments in which they will be deployed, and requirements in terms of fault tolerance of various applications. For example, it is clear that electromagnetic and mechanically based sensors will have fault characteristics very different from those of biological and chemical sensors.

Also, sensors deployed in harsh environments, such as nuclear plants, will have very different characteristics from those of sensors deployed in friendly environments, such as offices and residential areas. An intrinsic trade-off takes place between more complex fault models and relatively simple ones. In the VLSI domain, only the simplest fault models, such as those stuck at one and those stuck at zero, have been extensively used. However, in addition to these models, more complex models will be required for sensors. In particular, it will be interesting to see the kinds of fault models developed for the sensors common in a number of biological and chemical sensors that can be used only once for reading. In the VLSI domain, testing received significantly greater attention and application range than fault tolerance did. Testing needs to be addressed not only on component and individual node levels, but also at network and distributed system levels.

Closely related to testing is calibration, which can be defined as the process of mapping row sensor data to a new set of data that is, according to some statistical measure, more accurate than the initial readings. Calibration can be done offline and online. In the former case, the emphasis will be on the accuracy and strict interval of confidence; in the latter, the focus will be on the localized mode of operation. Also, calibration — not only of sensor readings but also of other parameters relevant to operation of sensor networks, including timing and the available energy level at each node — should be conducted.

At the application level, fault resiliency mechanisms required for common applications such as sensor fusion, data filtering, and data aggregation will be of primary importance. It is important to observe that, for each of these applications, a significant variety of approaches will be used. For example, in addition to equation-based sensor fusion, sensor fusion based on graphs, statistics, and stochastics will be possible. Each of these techniques has a number of unique peculiarities. Although specific fault resiliency techniques will not be developed for each of them, the primary emphasis will be on fault tolerance techniques that can be applied to multiple classes of approaches.

An interesting relationship exists between fault tolerance and several other fields. One of the main constraints in the deployment and operation of wireless sensor networks is energy. The most effective way to prolong the lifetime of the network is to place a subset of nodes in sleep mode. For example, power consumption of the software radio in three operational modes (transmission, receiving, and idle) rarely differs for more than a factor of two. At the same time, energy consumption in sleep mode is most often lower by two orders of magnitude (sometimes even more). A simple and powerful observation is that a node in sleeping mode can be treated as faulty and vice versa. It will be possible to retarget theoretical results and algorithms and even software for one objective to the other relatively easily.

Security and privacy are a major concern. For example, a key question concerns the extent to which one can trust results obtained using sensor fusion or data aggregation in a particular scenario in a particular sensor network, assuming that one or more nodes are compromised. In order to study this question, it will be necessary to develop a threat model and models of attacks in one or more nodes in a sensor network. These attacks can be modeled as worst-case fault models. Another interesting and important observation is that any technique that is resilient against nonintentional faults could also be retargeted to intentional faults.

36.8 Conclusion

Because of potential deployment in uncontrolled and harsh environments and due to the complex arch, wireless sensor networks are and will be prone to a variety of malfunctioning. The goal in this chapter was to identify the most important types of faults, as well as techniques for their detection and diagnosis, and to summarize the first techniques for ensuring efficiency of fault resiliency mechanisms. In addition to a comprehensive overview of fault tolerance techniques in general, and in particular in sensor networks, techniques were discussed that ensure fault resiliency during sensor fusion as well as the approach for heterogeneous built-in self-repair fault tolerance. The chapter concluded by outlining potential future research directions along several dimensions.

Acknowledgment

This material is based upon work supported in part by the National Science Foundation under Grant No. ANI-0085773 and NSF CENS Grant.

References

1. I.F. Akyildiz, W. Su, Y. Sankarasubramaniam, and E. Cyirci, Wireless sensor networks: a survey. *Computer Networks*, 38(4), 393–422, 2002.
2. A. Avizienis and J.P.J. Kelly, Fault tolerance by design diversity: concepts and experiments, *IEEE Computer*, 17(8), 67–80, August 1984.

3. A. Avizienis, The *n*-version approach to fault tolerant software, *IEEE Trans. Soft. Eng.*, SE-11(12), 1491–1501, 1985.
4. K. Birman. Replication and fault-tolerance in the ISIS system. In *Proc. 10th ACM Symp. Operating Syst. Principles*, 79–86, 1985.
5. K. Birman and T. Joseph. Reliable communication in the presence of failures. *ACM Trans. Computer Syst.*, 5, 47—76, February 1987.
6. K. Birman and R.V. Renesse, *Reliable Distributed Computing with the ISIS Toolkit*, IEEE Computer Society Press, 1994.
7. D. Bitton and J. Gray. Disk shadowing. *VLDB*, 331–338, 1988.
8. A. Brown and D.A. Patterson. Embracing failure: a case for recovery-oriented computing (ROC). *2001 High Performance Trans. Process. Symp.*, Asilomar, CA, October 2001.
9. A. Brown and D.A. Patterson. To err is human. *Proc. 1st Workshop Evaluating Architecting Syst. Dependability (EASY '01)*, Göteborg, Sweden, July 2001.
10. A. Brown and D.A. Patterson. Rewind, repair, replay: three *r*'s to dependability. *10th ACM SIGOPS Eur. Workshop*, Saint–Emilion, France, September 2002.
11. R.R. Brooks and S.S. Iyengar. Robust distributed computing and sensing algorithm. *IEEE Computer*, 25(6), 53–60, June 1996.
12. V. Bychkovskiy, S. Megerian, D. Estrin, and M. Potkonjak. Colibration: a collaborative approach to in-place sensor calibration. *2nd Int. Workshop Inf. Process. Sensor Networks (IPSN'03)*, 301–316, April 2003.
13. G. Candea, J. Cutler, A. Fox, R. Doshi, P. Garg, and R. Gowda. Reducing recovery time in a small recursively restartable system. *Proc. Int. Conf. Dependable Syst. Networks* (DSN-2002), Washington, D.C., June 2002.
14. B. Chen, K. Jamieson, H. Balakrishnan, and R. Morris. Span: an energy-efficient coordination algorithm for topology maintenance in ad-hoc wireless networks. In *ACM MobiCom*, July 2001.
15. S. Chessa and P. Santi, Crash faults identification in wireless sensor networks, *Computer Commun.*, 25(14), 1273–1282, Sept. 2002.
16. R.W. Downing, J.S. Nowak, and L.S. Tuomenoksa, No 1 ESS maintenance plan. *Bell Sys. Tech. J.* 43(5), pt. 1, 1961–2019, September 1964.
17. D. Eckhardt and P. Steenkiste, Measurement and analysis of the error characteristics of an in building wireless networks. In *Proc. SIGCOMM*, 243–254, 1996.
18. D. Eckhardt and P. Steenkiste, A trace-based evaluation of adaptive error correction for a wireless local area network. *J. Special Topics Mobile Networking Applications (MONET)*, 1998.
19. D. Eckhardt and P. Steenkiste, Improving wireless LAN performance via adaptive local error control. *INCP*, 1998.
20. B. Efron, *The Jackknife, The Bootstrap, and Other Resampling Plans*. S.I.A.M., Philadelphia, 1982.
21. C. Fragouli, P. Lettieri, and M. Srivastava, Low power error control for wireless links. *ACM/IEEE MobiCom*, 139–150, 1997.
22. J. Gray and A. Reuter. *Transaction Processing: Concepts and Techniques*. Morgan Kaufmann, San Mateo, California, 1993.
23. L. Guerra, M. Potkonjak, and J.M. Rabaey, High level synthesis techniques for efficient built-in-self-repair. *Int. Workshop DFT VLSI Syst.*, 41–48, 1993.
24. C. Intanagonwiwat, R. Govindan, and D. Estrin, Directed diffusion: a scalable and robust communication paradigm for sensor networks. *ACM/IEEE MobiCom*, 56–67, 2000.
25. P. Jalote. *Fault Tolerance in Distributed Systems*. P.T.R Prentice Hall, Englewood Cliffs, NJ, 1994.
26. F. Koushanfar, M. Potkonjak, and A. Sangiovanni–Vincentelli, Fault tolerance in wireless ad-hoc sensor networks. *IEEE Sensors*, 2, 1491–1496, June 2002.
27. F. Koushanfar, M. Potkonjak, and A. Sangiovanni–Vincentelli, Online fault detection in wireless sensor networks. *IEEE Sensors*, 2003, to appear.
28. F. Koushanfar, A. Davare, D.T. Nguyen, M. Potkonjak, and A. Sangiovanni–Vincentelli, Low power coordination in wireless ad-hoc networks. ISLPED, Aug 2003.

29. R. Lackey and D. Upmal, Speakeasy: the military software radio. *IEEE Commun. Mag.*, 33(5), 56–61, May 1995.
30. L. Lamport, R. Shostak, and M. Pease. The Byzantine general's problem. *ACM Trans. Programming Languages Syst.*, 4(3), 382–401, July 1982.
31. L. Lamport, The weak Byzantine generals' problem, J. ACM, 30, 668—676, July 1983.
32. P.A. Lee and T. Anderson. *Fault Tolerance: Principles and Practice*, 2nd ed., Springer–Verlag, Heidelberg, 1990.
33. N. Mandayam, A software radio architecture for linear multiuser detection. *IEEE JSAC*, 17(5), 814–823, May 1999.
34. J. Mitola, The software radio architecture. *IEEE Commun. Mag.*, May 1995.
35. J. Mitola, Technical challenges in the globalization of software radio. *IEEE Commun. Mag.*, February 1999.
36. E.F. Moore and C.E. Shannon, Reliable circuits using less reliable relays. J. Franklin Institute, 262, 191–208, September 1956.
37. V.D. Park and M.S. Caron, A highly adaptive distributed routing algorithm for mobile wireless networks. *Infocom 1997*.
38. M. Pease, R. Shostak, and L. Lamport. Reaching agreement in the presence of faults. *J. ACM*, 27(2), 228–234, 1980.
39. L. Prasad, S.S. Iyengar, R.L. Kashayap, and R.N. Madan, Functional characterization of fault tolerant interaction in distributed sensor network. *Phys. Rev. E*, 49(2), April 1994.
40. F.B. Schneider, Byzantine generals in action: implementing fail-stop processors, *ACM Trans. Computer Syst.*, 2(2), 145–154, May 1984.
41. W. H.W. Tuttlebee, Software-defined radio: facets of a developing technology, *IEEE Personal Commun. Mag.*, 6(2), 38–44, April 1999.
42. J. von Neumann, Probabilistic logics and the synthesis of reliable organisms from unreliable components. In C.E. Shannon and J. McCarthy, Eds. *Automata Studies*, Princeton University Press, 1956, 43–98.
43. K. Whitehouse and D. Culler, Calibration as parameter estimation in sensor networks. ACM WSNA, 2002.

IX

Performance and Design Aspects

37
Low-Power Design for Smart Dust Networks

Zdravko Karakehayov
Technical University of Sofia

37.1 Introduction 37-1
37.2 Location 37-1
37.3 Sensing 37-2
37.4 Computation 37-2
   Asynchronous Processors • Variable-Frequency Processors • Variable-Voltage Processors
37.5 Hardware–Software Interaction 37-5
37.6 Communication 37-7
   Mote-to-Mote Communication • Mote-to-Central Station Communication
37.7 Orientation 37-10
37.8 Conclusion 37-10

37.1 Introduction

Distributed sensor networks (DSNs) are composed of numerous small, low-cost, randomly located nodes. The network can be scalable to thousands of nodes that cooperatively perform complex tasks such as intelligent measurement. The network must be able to self-organize, adapt to random node spacing, execute algorithms for signal processing, and operate as power efficiently as possible. The major applications of DSNs are for monitoring environmental conditions, tracking the movements of birds and small animals, monitoring product quality, and building automation and defense networks. Smart Dust is a term recently coined at the University of California, Berkeley, to describe massively distributed sensor networks consisting of cubic-millimeter sized motes [1, 2]. The small size and anticipated low cost of the motes will help to collect information cost-effectively and less intrusively.

Each mote depends on low-capacity batteries as energy sources. Practically, the chance for battery replacement is nonexistent. As a result, every aspect of the Smart Dust networks, from mote location through computing and communication, is viewed from the low-power perspective.

37.2 Location

A deployment may leave numerous motes located in different areas of a large geographical region. The location of the motes affects energy efficiency in a number of ways. Sensor readings are of interest if only bound to a known location. Interrogation of motes before a location procedure would be a loss of energy. The global positioning system (GPS) is able to locate network nodes in outdoor environments. However, cost, power consumption, and size of the currently available GPS receivers are prohibitive for Smart Dust motes. Optical communication emerges as the most efficient method if a central station may be harnessed to provide energy for location tasks. Because motes may move, some applications would demand updating the positions regularly. Also, radio-frequency (RF) communication can be used by motes to locate themselves via beacon signals from reference points [3, 4].

As soon as the location procedure has been completed, some nodes will be actively involved in sensing, while others will wait for events and can be turned off to save energy. An event tracking, such as following light shadow edges over a sensor field, can be organized in two ways:

- All motes deactivate all subsystems except sensors that can obtain relevant data. If the sensors provide binary readings, they can be used to awake the motes in case of events.
- A more sophisticated power reduction approach will turn off all motes, except motes in the close vicinity of the event, completely. However, in case of a dense deployment the distance alone is not sufficient as a criterion.

Liu et al. [5] have developed a method for event tracking. The method identifies motes that will not be immediately approached by the event and can be turned off to save energy. The method is based on dual space transformation [6]. Figure 37.1 shows an example for event tracking.

With no loss of generality, it can be assumed that the event is a moving light shadow edge. The edge is presented in the primal space as the E line and the motes' locations are indicated as points. The line is uniquely defined in the primal space by the p slope and the y-intercept q. The line is transformed into the e point in the dual space; in turn, the points from the primal space are transformed into lines in the dual space. As a result, the dual space is partitioned into cells. The e point is contained in the shaded cell. Because the e point cannot intersect the m2 line before it crosses one of the cell boundaries, the M2 mote can stay turned off as long as none of M1, M3, and M4 sense a transition.

37.3 Sensing

The mote's sensors vary from application to application: temperature, light, magnetic field, vibration, and acoustic. Recent advances in technology have made it possible for these sensors to be released in ultralow sizes and power versions [2, 7].

Sensors convert physical variables into electrical signals. Typically, the signals are in the microvolt or millivolt range. An input signal conditioner is used to filter and amplify the signals. Energy is consumed in the sensor, amplifier, and analog-to-digital converter (ADC). The power consumption can be reduced with appropriate power management. The ADC's resolution has a significant impact on the energy budget. For instance, if the ADC's resolution is increased from 15 to 16 bits while keeping the other parameters unchanged, the power consumption is increased from 100 to 400 mW [8].

A common method for analog-to-digital conversion is the successive approximation [9]. Because the ADC determines one bit of the result in each cycle, it would be possible to apply selective resolution. Consequently, different samples will have different numbers of bits and different energy costs. Finally, one may only want to test if the input value belongs to a certain range. In this case, a microcontroller with an on-chip analog comparator can be a power-efficient solution. Microcontrollers such as the Atmel ATmega161L are capable of turning off the comparator to reduce the power consumption [10].

37.4 Computation

Motes incorporate a processor to carry out computations locally. Functionality typically requires the processor to run in outbursts separated by idle periods. Within the idle period, the processor may enter a power reduction mode to save energy [9]. The battery lifetime is influenced by the power efficiency of a running processor and the balance between active and idle periods.

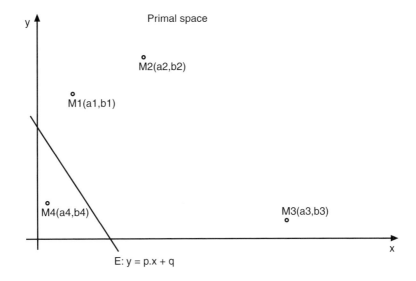

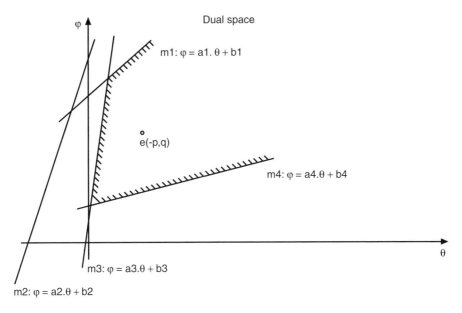

FIGURE 37.1 Primal-to-dual space transformation indicates the sequence of transitions.

37.4.1 Asynchronous Processors

The synchronous processor's clock distribution network is characterized by significant power consumption. Moreover, synchronous systems tend to maximize supply current transients. Smart Dust motes may have analog subsystems that are influenced by the electromagnetic radiation. Asynchronous designs promise to overcome the clock-related problems. In particular, a class of asynchronous implementations, termed self-timed systems, is capable of operating as fast as circumstances allow.

37.4.2 Variable-Frequency Processors

Using variable-frequency processors, power consumption can be gradually controlled by scaling the clock frequency. Typically, a phase lock loop (PLL) circuitry can multiply the oscillator frequency and an adjustable prescaler can divide the oscillator frequency. Based on the current task's deadline, the clock frequency may decline as much as possible. However, if the processor completes the task ahead of the deadline and enters a power-saving mode, the energy could be minimized [11]. In this case, the task's deadline period, T_{DL}, accommodates the active period, T_{ACT}, and the power-saving period, T_{PS}:

$$T_{DL} = T_{ACT} + T_{PS} \tag{37.1}$$

Assume that the power consumption scales linearly with the clock frequency:

$$P_{ACT} = k_{ACT} \times f_{CLK} + n_{ACT} \tag{37.2}$$

$$P_{PS} = k_{PS} \times f_{CLK} + n_{PS} \tag{37.3}$$

If the task's functionality requires NC processor clocks, the energy per task,

$$E_T = P \times T_{DL} = k_{PS} \times T_{DL} \times f_{CLK} + (n_{ACT} - n_{PS})\frac{NC}{f_{CLK}} + (k_{ACT} - k_{PS})NC + n_{PS} \times T_{DL} \tag{37.4}$$

Take the first derivative and calculate the critical numbers

$$f_{CLK} = \pm \sqrt{\frac{(n_{ACT} - n_{PS})NC}{k_{PS} \times T_{DL}}} \tag{37.5}$$

Consider two cases for the positive value:

- Let $n_{ACT} > n_{PS}$. Based on the second derivative test, the energy per task has a minimum for

$$f_{CLK,OPT} = \sqrt{\frac{n_{ACT} - n_{PS}}{k_{PS}}} \sqrt{\frac{NC}{T_{DL}}} \tag{37.6}$$

- If $n_{ACT} \leq n_{PS}$, the clock frequency must be selected as low as possible. The power-saving mode is not used.

$$f_{CLK,OPT} = \frac{NC}{T_{DL}} \tag{37.7}$$

Equation 37.6 does not guarantee that the deadline will be met. In some cases, the calculated clock frequency must be increased to meet the deadline.

Figure 37.2 shows an example mesh plot for the clock frequency. Assume that the processor is characterized by $n_{ACT} = 1$ mW; $n_{PS} = 0.1$ mW; and $k_{PS} = 1$ mW/MHz. The example is based on 256 combinations of deadline periods and cycles per task. For two combinations, the optimal clock frequencies have been replaced by higher values.

Actual tasks, which require replacement of the optimal clock frequency, can be viewed as targets for further improvement. Optimization of the code or relaxing the timing constraints would be an appropriate course of action.

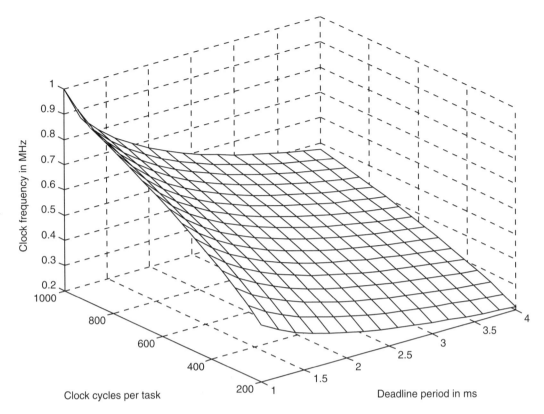

FIGURE 37.2 Mesh plot for the clock frequency.

37.4.3 Variable-Voltage Processors

Variable-voltage processors are capable of operating over a wide voltage range. Allocating such a processor for the network nodes allows power reduction by dynamically varying the supply voltage [12–15]. The method is often termed dynamic voltage scaling (DVS). DVS is an efficient method for power reduction; however, it imposes some limitations for the system:

- The system components must be capable of operating over a wide voltage range.
- A voltage converter loop hardware must be available.

Hong et al. [16] developed a design methodology for DVS. Figure 37.3 illustrates how to tune the voltages for extra power reduction. The tasks are specified by their arrival times, deadlines, and execution times at a nominal voltage. The schedule is viewed as a first iteration. It would be beneficial to extend the T2 task and reduce the V2 voltage. T1 is scheduled for V2 to shrink the execution time. The new border between T1 and T2 is placed just in the middle of the interval indicated by an arrow; no conflict takes place with the arrival time and the change is accepted.

Similarly, T3 is scheduled for V2 to allow extension for T2. The intention is to place the new border just in the middle of the interval marked by an arrow; however, T2 fails to meet the deadline and the border is aligned with the deadline. If the new schedule is more energy efficient, it is accepted.

37.5 Hardware–Software Interaction

A mote includes a CPU, memory, and peripherals. As a rule, peripherals possess three types of registers: data, control, and status. Data registers are employed as buffers between the CPU and peripherals, while control registers are used to adjust the I/O device functionality for a specific application. Status registers

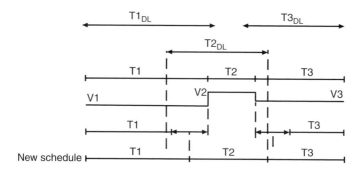

FIGURE 37.3 A schedule is modified for extra power reduction.

are read by the processor to check whether a specific operation is done. In spite of the huge variety of peripherals, the communication between the CPU and the I/O devices remains routine and easy to define.

Modifying one or more bits in a register, the CPU must keep the rest of the pattern unchanged. A common way to implement bit manipulation is to read a register, modify bits, and write the result back. The two memory accesses make the read–modify–write instructions power inefficient. In an attempt to improve the situation, Atmel has taken another approach with the AT91 microcontroller [17]. Instead of one control register, the microcontroller employs three registers mapped into three consecutive memory locations. The first register is used to set individual bits, the second to clear bits, and the third to obtain the current pattern. To set or reset a bit, a high bit is written to the corresponding position at the set or reset register.

In the AT91 microcontroller, a PLL circuitry and a programmable prescaler complement the ARM7TDMI core to a variable-frequency processor. The PLL circuitry multiplies the oscillator frequency; the highest multiplication factor is 64. As a result, the oscillator may run at a frequency 64 times lower than the actual clock and thus the oscillator saves energy. The programmable prescaler with a division factor of 64 allows the AT91 clock frequency to go down to 512 Hz. The CPU and embedded peripherals can be individually enabled and disabled. The ARM processor clock is enabled from the next interrupt or reset. The on-chip RAM reduces external memory accesses and allows further power reduction. Finally, the processor may switch to the 16-bit instruction set and benefit from a narrower memory.

Similarly to the analog-to-digital conversion, the measurement of time intervals also falls under the accuracy–power trade-off. Figure 37.4 shows how a counter/timer determines a time interval using different clock rates. The highest possible frequency provides the highest accuracy. If a Smart Dust application is based on the AT91 microcontroller, the number of counter transitions for a 50-ms period may vary with the frequency up to 50,000.

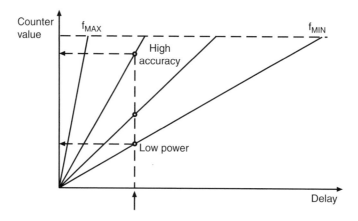

FIGURE 37.4 Using different frequencies to measure a time interval.

37.6 Communication

In a wireless sensor network, communication is the major consumer of energy. Smart Dust networks have two recognized communication styles: (1) RF is characterized with power consumption in the milliwatt range; (2) optical communication is associated with a lower energy cost but requires accurate pointing. Consequently, optical communication is more suitable for interaction between network motes and a central station. The RF approach is very common for communication between motes [8, 18, 19].

37.6.1 Mote-to-Mote Communication

The procedures for establishing and operating a network require the motes to communicate with one another. The task of routing packets from a source to a destination can be broken down into discovering the position of the destination and the actual forwarding of packets [20]. Furthermore, channel access can be implemented by two different methods: contention or explicit organization [21]. The contention-based approach is not suitable for DSNs because of its requirement to monitor the channel for a long span of time. Because the reception and transmission have almost the same energy cost, the organized channel access is characterized with better energy efficiency. At the same time, the process of establishing time division multiple access (TDMA) slots or frequency bands also consumes energy. In an attempt to alleviate this problem, some protocols employ a hierarchical structure that requires partitioning the network.

The two basic schemes to limit the mote's RF transmission power are: (1) a transmitter can vary its power to cover different distances under different environmental conditions; or (2) the link can be partitioned into several short intermediate hops and use constant transmission power. Any DSN with a sufficient density of nodes can benefit from multihop communication.

The energy used to send a bit over a distance d may be written as

$$E = A \times d^n \quad (37.8)$$

where A is a proportionality constant and n depends on the environment [18, 22]. The greater-than-linear relationship between energy and distance promises to reduce the energy cost when the link is partitioned.

Rewrite Equation 37.8 for NH number of hops. Also, include the energy for receiving E_R and energy for computation E_C:

$$E = A\left(\frac{d}{NH}D\right)^n NH + (E_R + E_C)NH \quad (37.9)$$

Assume equal distances for each hop. $D > 1$ is introduced to take into account the longer path inevitably associated with multihop communication. The energy has a minimum for

$$NH_{OPT} = (d \times D) \sqrt[n]{\frac{A(n-1)}{E_R + E_C}} \quad (37.10)$$

Figure 37.5 shows a plot for the energy per bit using different numbers of hops. The distance $d = 50$ m; $n = 4$; $A = 0.2$ fJ/m^4; $D = 1.2$; and $E_R + E_C = 30$ pJ. The energy per bit has a minimum for four hops.

A subtle effect of multihop communication is that energy consumption is distributed over the motes fairly. If the motes consume energy at about the same rate, the system lifetime is increased. Chen et al. [23] developed a coordination algorithm to increase the energy efficiency further. The algorithm is based on an assumption that when a wireless network has an ample density of nodes, only a small number of them need to be active to forward traffic. A distinctive feature of the method, named SPAN, is that the

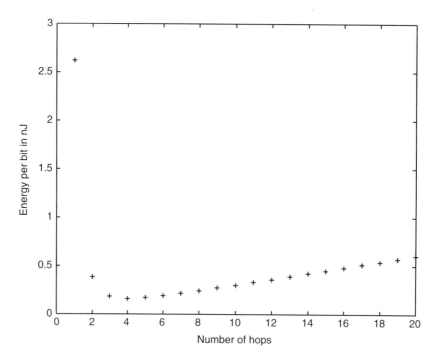

FIGURE 37.5 Energy-per-bit scales with the number of hops.

motes make a decision whether to sleep or be active based not only on the topology of the network, but also on the amount of energy available in the battery.

All motes of the network are dynamically split into two sets: motes that sleep and motes that stay awake to participate in the forwarding backbone topology. According to SPAN's terminology, the active motes are named coordinators. Each mote of the network makes periodic, local decisions on whether to sleep or become a coordinator. Coordinators are elected to achieve two goals: improved connectivity of the network and equal levels of energy remaining at each mote. All noncoordinator motes periodically participate in an election procedure to become coordinators; in parallel, all coordinators periodically pass through a withdrawal procedure to switch back to a sleep state. Figure 37.6 shows this election–withdrawal cycle. A mote becomes a coordinator to link two neighbor motes that cannot communicate directly or via one or two coordinators. Because several motes can run an election procedure simultaneously, there might be an overlap in the connectivity they introduce. The method attempts to minimize the number of coordinators to save energy.

To resolve contention, the election procedure is extended with a variable delay. As soon as the delay period is over, a coordinator announcement is sent out. If, at the end of the delay the mote receives other announcements for new coordinators, it reconsiders the need to become a coordinator. The election procedure distinguishes between two cases:

- All applicants for coordinators have equal energy left in their batteries. In this case, the more pairs of motes the applicant connects, the shorter is the delay. Also, to rotate coordinators with time, a random value influences the delay.
- The participating motes have unequal energy available in their batteries. In this case, the delay period is calculated on the base of the connection improvement and the amount of energy scaled to the maximum amount of energy the mote can have. The random factor is still included.

Each coordinator periodically runs a withdrawal procedure. A coordinator can go back to sleep if every pair of its neighbors can reach each other directly or via some other coordinators. Initially, the mote will stay as a coordinator if its withdrawal affects the network connectivity. However, after some time it will

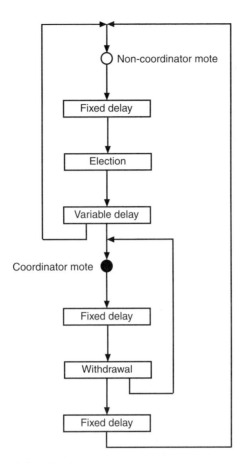

FIGURE 37.6 Span's election withdrawal cycle.

switch to noncoordinator state to give other neighbors a chance to become coordinators. As shown in Figure 37.6, a mote continues to serve as a coordinator for a fixed period of time after its withdrawal announcement is sent out. Thus, the routing protocol can use the old coordinator until a new coordinator is elected.

37.6.2 Mote-to-Central Station Communication

When one or more central stations communicate with a field of dust motes, optical systems are characterized by the lowest energy budget. Two methods can be used to apply optical communication for Smart Dust: passive reflective systems and active-steered laser systems [2]. Figure 37.7 shows an example of a passive reflective device, a corner-cube retroreflector (CCR). A CCR reflects the light via three mutually orthogonal mirrors. When a light beam enters the CCR, it bounces off the mirrors and is reflected back parallel to the direction from which it entered. Because one of the mirrors is mounted on a spring at an angle slightly askew from perpendicularity to the other mirrors, in this state little light returns to the remote receiver. No reflection of the light is considered a low logic level. To return the light to its source, high logic level, the mirror is shifted to a position perpendicular to other mirrors. The low-to-high transition consumes less than a nanojoule [2]. The high-to-low transition requires almost no energy.

Active-steered laser systems are suitable for mote-to-mote and mote-to-central station communications. The device consists of a semiconductor diode laser, collimating lens, and a two-degree-of-freedom micromirror [2]. Central stations can use imaging receivers to process transmissions from different angles.

FIGURE 37.7 Microfabricated corner-cube retroreflector. (From Hsu, V., Kahn, J.M., and Pister, K.S.J., www-ee.stanford.edu/~jmk/pubs/hsu.ms.11.99.pdf. With permission.)

This approach of separating transmissions according to their originating location is termed a space division multiple access (SDMA).

37.7 Orientation

Many applications will deploy motes in random orientation. Consequently, it will not be possible for all CCRs to return light to the central station. A CCR quadruplet is a solution that improves the accessibility of the motes. At the same time, some directions may be characterized with noise emissions and should be avoided. Furthermore, applications may require the motes to be invisible from a certain area.

It is proposed that the motes be magnetized and the CCR oriented to a predefined direction. When the motes fall through the air after being deployed, they will orient themselves. If the network has a sufficient density of motes, it may not need the motes, which change orientation upon landing. This approach for zero-power orientation is even more efficient for motes floating on the water. They could freely rotate to orient themselves. Figure 37.8 shows a deployment of two types of motes that differ in their CCR orientation; two central stations interrogate the motes. The DSNs community is growing and projects that simultaneously employ a single field can benefit from SDMA.

37.8 Conclusion

The low-power design of Smart Dust networks has a lot in common with many other computer applications. By allocating variable-frequency processors for the Smart Dust motes, clock frequency scaling can be applied to decrease power consumption. It is necessary to distinguish between two types of processors in order to decide whether it is more power efficient to operate quickly and then wait quietly, or just operate at the minimum speed possible. For the first case, the optimal clock frequency is calculated based on the required number of clock cycles and a deadline period. This approach also allows identifying tasks that require replacement of the optimal clock frequency. Thus, a set of tasks emerges as a target for further improvement. Variable-voltage processors could combine voltage scaling with frequency scaling if the hardware overhead is not prohibitive for a cubic-millimeter sized mote.

Hardware–software interaction also provides ample reserve for power reduction. Scaling down the theme of variable frequency from processors to counters, motes could measure time intervals, trading accuracy against number of transitions. The hardware–software interaction and the sensing show that redundant accuracy wastes energy in the same way as redundant computation speed.

The energy spent for communication is crucial for the success of wireless networks such as Smart Dust. Multihop communication can help power consumption to decline significantly and avoid obstacles for RF and optical systems. As an additional benefit, multihop transmissions distribute power consumption over the motes fairly and increase system lifetime.

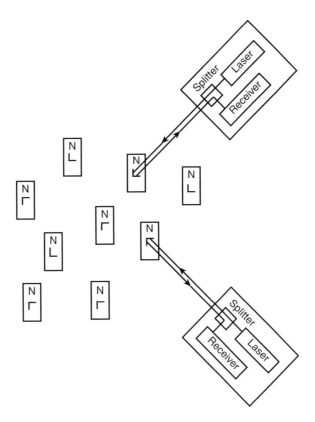

FIGURE 37.8 Each central station interrogates its own motes.

The location of the motes is a process specific to the network operation. Some applications may require only relative positions. Relative positions can be used to turn off motes, especially in case of event tracking. Finally, optical communication is associated with pointing and orientation. By using the Earth's magnetic field, zero-power orientation of the motes can be implemented for SDMA.

Acknowledgment

The author is grateful to Brett Warneke for important advice and to Victor Hsu for the CCR photograph.

References

1. Kahn, J.M., Katz, R.H., and Pister, K.S.J. Next century challenges: mobile networking for "Smart Dust," *J. Commun. Networks*, 2(3), 188, September 2000.
2. Warneke, B., Last, M., Liebowitz, B., and Pister, K.S.J. Smart Dust: communicating with a cubic-millimeter computer, *Computer*, 34, 44, January 2001.
3. Bulusu, N., Heidemann, J., and Estrin, D. GPS-less low-cost outdoor localization for very small devices, *IEEE Personal Commun.*, 7, 28, 2000.
4. Hightower, J. and Borriello, G. Location systems for ubiquitous computing, *IEEE Computer*, 34, 57, 2001.
5. Liu, J., Cheung, P., Guibas, L., and Zhao, F. A dual-space approach to tracking and sensor management in wireless sensor networks, Palo Alto Research Center technical report P2002-10077, 2002, available at www2. parc.com/spl/projects/cosense/pub/dualspace.pdf.
6. Berg, M., Kreveld, M., Overmars, M., and Schwarzkopf, O. *Computational Geometry: Algorithms and Applications*, Springer–Verlag, Berlin, 1997.

7. Gardner, J.W. *Microsensors: Principles and Applications*, John Wiley & Sons, New York, 1994.
8. Doherty, L., Warneke, B.A., Boser, B.E., and Pister, K.S.J. Energy and performance considerations for smart dust, *Int. J. Parallel Distributed Syst. Networks*, 4, 121, 2001.
9. Karakehayov, Z., Christensen K.S., and Winther, O. *Embedded Systems Design with 8051 Microcontrollers*, Marcel Dekker, New York, 1999.
10. Atmel Corporation, *AVR RISC Microcontroller, Data Book*, 1999.
11. Karakehayov, Z. Zero-power design for Smart Dust networks, *Proc. 1st IEEE Int. Conf. Intelligent Syst.*, Varna, 1, 302, 2002.
12. Macken, P., Degrauwe M., Paemel V., and Oguey H. A voltage reduction technique for digital systems, *Digest of Technical Papers, 37th IEEE Int. Solid-State Circuits Conf.*, 238, 1990.
13. Mudge, T., Power: a first-class architectural design constraint, *IEEE Computer*, 34, 52, 2001.
14. Burd, T. *Energy-Efficient Processor System Design*, Ph.D. thesis, University of California, Berkeley, 2001.
15. Sinha, A. and Chandrakasan, A. Dynamic power management in wireless sensor networks, *IEEE Design Test Computers*, 18, 62, 2001.
16. Hong, I. et al. Power optimization of variable-voltage core-based systems, *IEEE Trans. Computer-Aided Design Integrated Circuits Syst.*, 18(12), 1702, 1999.
17. Atmel Corporation, *AT91 ARM Thumb Microcontrollers, AT91M55800A*, 2002. Available at www.atmel.com.
18. Rabaey, J.M. et al. PicoRadio supports ad hoc ultra-low power wireless networking, *Computer*, 33, 42, July 2000.
19. Hill, J.L. and Culler, D.E. Mica: a wireless platform for deeply embedded networks, *IEEE MICRO*, 22, 12, November–December 2002.
20. Mauve, M. and Widmer, J. A survey on position-based routing in mobile ad hoc networks, *IEEE Network*, 15, 30, 2001.
21. Sohrabi, K. et al. Protocols for self-organization of a wireless sensor network, *IEEE Personal Commun.*, 7, 16, 2000.
22. Akyildiz, I.F. et al. A survey on sensor networks, *IEEE Commun. Mag.*, 40, 102, 2002.
23. Chen, B. et al. Span: An energy-efficient coordination algorithm for topology maintenance in ad hoc wireless networks, *ACM Wireless Networks J.*, 8, 481, 2002.

38
Energy-Efficient Design of Distributed Sensor Networks

Lin Yuan
University of Maryland

Gang Qu
University of Maryland

38.1 Introduction 38-1
  A Motivational Example
38.2 Background 38-4
  On Distributed Sensor Networks • On Dynamic Voltage Scaling
38.3 Preliminaries 38-6
  Model of Sensor Node and Energy Consumption in DSNs •
  Processor with Multiple Supply Voltages • The Message Header
38.4 DVS with Message Header 38-9
  Requirements and Constraints of Secure DSNs • Energy
  Consumption for Data Encryption and Decryption • Dynamic
  Voltage Scaling on Sensor Nodes
38.5 Simulation 38-11
  Simulation Platform • Simulation Results
38.6 Conclusions 38-17

38.1 Introduction

Distributed sensor networks (DSNs) produce high-quality information for civil and military applications with a large number of physical sensors (e.g., acoustic, seismic, visual) communicating to each other via ad hoc wireless networks. Advances in digital circuitry, wireless communications, battery technology, and microelectromechanical systems (MEMS) have resulted in smaller, less expensive, more versatile, and more reliable sensors that have longer durability. On many occasions, the durability of the DSN is defined as the sensor's limited battery capacity because of the difficulty of power-source maintenance for the sensors, in particular in hostile environments such as battlefields.

Furthermore, most DSNs use multihop communication to avoid energy-expensive long-distance transmission. In such communication, each sensor node communicates directly only to its neighbor nodes. Messages to a geographically distant node will be relayed by the sender node's neighbors and the neighbors' neighbors, and so on. Energy shortage on one node will cause its neighbor nodes to communicate more and thus quickly run out of energy, eventually disabling the entire DSN. Therefore, energy-efficient DSN design becomes one of the most interesting challenges and each individual sensor node must also be designed to be energy efficient and take precautions to conserve its energy [9, 16].

A sensor node consumes energy during communication and computation. Communication energy is the dominant factor and has attracted a lot of research attention. Energy-efficient algorithms and protocols have been proposed at network levels to balance available energy and thus extend the lifetime of each node throughout the network. These algorithms and protocols are quite effective in building energy-efficient paths in the multihop communication DSN. Computational energy, on the other hand, is consumed by the microprocessors in sensor nodes when data processing is performed locally. This energy is application dependent and, in circumstances in which the computation load is heavy, can take up to 30% of the total energy consumption [1]. However, compared to research on communication energy, little work has been reported on reducing the computational energy in DSN designs. In practice, one usually picks a microprocessor suitable for the desired data/signal processing of the DSN, without paying much attention to its energy efficiency.

Only recently have well-studied, low-power techniques such as dynamic power management (DPM) and dynamic voltage scaling (DVS) been applied to use in DSN microprocessors [27]. In DPM, the operating system puts the sensor nodes into idle states when no computation is required. This could yield substantial energy savings because the turned-off components in idle system states consume little or no energy. DPM targets the energy dissipation during a system's idle period; techniques like DVS provide additional energy savings in the active state without sacrificing sensor node performance.

DVS is a technique that varies the supply voltage and clock frequency, based on the computation load, to provide desired performance with the minimal amount of energy consumption. It has been demonstrated to be one of the most effective low-power system design techniques, particularly for real-time embedded systems. The key feature of DVS is the microprocessor's capability of operating at different voltages, which are normally provided by on-chip DC–DC converters. Many commercial high-performance microprocessors support DVS for energy and power efficiency. Examples include Transmeta's Crusoe; AMD's K-6; Intel's XScale and Pentium III and IV; and some DSPs developed in Bell Labs [32, 37–40].

This chapter discusses how to apply the DVS technique to design of energy-efficient DSNs. In particular, one can take advantage of the multiple voltage design methodology to reduce (computational) energy consumption in the sensor network based on two observations:

- *Three-stage processing in sensor nodes*. Message processing in sensor nodes can be modeled as three sequential stages: preprocessing, data processing, and postprocessing. In the preprocessing stage, the microprocessor takes raw data received from other sensors and/or its own sensing devices and performs the necessary operations (e.g., decryption and filtering, etc.) to obtain the required data format for data processing. During data processing, most of the calculations on data are performed. Finally, the microprocessor enters the postprocessing stage for data compression (if the data are stored locally) or encryption (if the data need to be sent out under security requirements). These three stages are sequential and dependent.
- *Large variety of data processing requirements*. Because of the wide deployment of sensor nodes and the ad hoc nature of DSNs, the computation load on sensor nodes can be largely unbalanced. For example, in some information-intensive areas, a sensor node may receive a large amount of data from its sensing device or other sensors. To extract useful information from the raw data, it must perform many computations locally. On the other hand, a sensor on the communication path of two other sensors may merely act as a messenger that only needs to forward messages to the next node in the path without performing any real computation on the data.

Multiple-voltage DVS systems capable of switching operating voltage among several simultaneously available levels will be considered. Such systems have been well studied in the VLSI design automation and real-time operating systems communities and are available commercially. This chapter's approach starts with characterizing the typical activity of the sensor nodes, particularly the data processing and communication requirements of the messages in the network. The operating systems embedded in the microprocess will then scale the operating voltage to the most energy-efficient level according to the importance of the incoming message and the current workload of the sensor node.

To take full advantage of a DVS system's energy efficiency, the concept of *message header*, which contains a small amount of additional information about the message and is inserted at the beginning of each message, is proposed. Such information includes the length and type of message; expected data processing time; length of the anticipated result; and the data processing deadline, among others. This information will help the sensor node to select the most energy-efficient operating voltage.

38.1.1 A Motivational Example

A secure DSN is used as an example to explain how DVS, with the help of a message header, can reduce energy consumption. Each sensor receives an encrypted message from other nodes every 5 s. The sensor must decrypt the message, process the data, and encrypt and send out the result before the arrival of the next message. A message contains a certain number of packets of fixed size. Suppose the RSA algorithm is used as the encryption function, which requires 110 and 5 ms to decrypt and encrypt a single packet, respectively.*

Now consider two messages, τ_1 and τ_2, both with 10 packets. Assume that τ_1 requires 2 s for data processing and needs 20 packets for the (encrypted) processing result, and that τ_2 demands a forward; therefore no data processing is needed and the encryption results in a 10-packet message. The microprocessor will be on for data decryption/encryption and processing with a power consumption of 230 mW at the 3.3-V reference voltage. It stays in the idle state from the completion of encryption to the arrival of the next message. For message τ_1, 110 ms × 10 = 1.1 s is necessary for decryption, 2 s for data processing, and 5 ms × 20 = 0.1 s for result encryption. For message τ_2, these numbers are 1.1, 0, and 0.05 s, respectively. This gives a total execution time of 4.35 s. If power consumption when the system is idle is ignored,** the energy consumption will be 230 mW × 4.35 s = 1 J (Figure 38.1).

If one implements the sensor using a DVS processor core with multiple voltages, the energy to process the same messages can be dramatically reduced. Table 38.1 gives the system's clock frequency and power consumption, as well as time and energy consumption for computing a 128-b multiplication (the basic function for the public key algorithm) under three different supply voltages.

In this energy-driven approach, a message header is added to the first packet of every message. A message header gives the receiver sensor information about the current message such as length of the message, expected processing time, and length of the result. Therefore, after encrypting the first packet, the sensor will be able to get the approximate computation load and to select the lowest voltage level accordingly, so that the required data processing and result encryption can be completed with the least amount of energy.

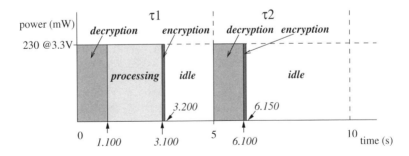

FIGURE 38.1 Sensor's energy consumption (1000 mJ) at a fixed 3.3-V supply voltage. The *x*-axis shows the starting time and ending time for each decryption, data processing, and encryption.

---

*The real times for RSA decryption and encryption on an 80-MHz MIPS R4000 processor are 72.7 and 3.5 ms, respectively [4]. Here, these numbers are scaled up for simplicity.

**Sensor energy dissipation at idle state is extremely low to extend the sensor's lifetime. For example, the Intel StrongARM 1100 processor used in wireless integrated networked sensors (WINSs) consumes less than 0.8 mW in its sleep mode.

TABLE 38.1 Processor's Performance at Different v_{dd}

v_{dd} (V)	Clock (MHz)	Power (mW)	128-b Multiplication	
			Time (µs)	Energy (nJ)
3.3	80	230	0.50	115
2.4	54	82	0.75	61.5
1.2	20	7.5	2.00	15

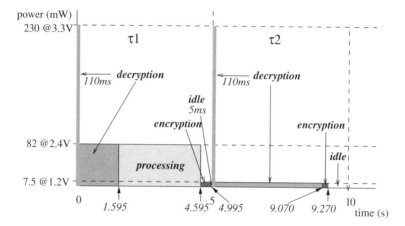

FIGURE 38.2 Sensor's power consumption (452 mJ) with a multiple voltage core.

As illustrated in Figure 38.2, the sensor decrypts the first packet of a message in the first 110 ms, then selects proper voltages for processing and encryption and eventually stays idle waiting for the next message. Notice that a lower voltage slows the clock frequency and requires longer execution time to perform the same amount of computation (Table 38.1). For instance, at 2.4 V, 3 s is needed to process τ_1; this can be done in 2 s under 3.3 V. In this example, it takes (230 mW × 0.11 s + 82 mW × 4.485 s + 7.5 mW × 0.4 s) = 396 mJ to execute τ_1, and (230 mW × 110 ms + 7.5 mW × 4.16 s) = 56.5 mJ for τ_2. The total energy consumption is 452.5 mJ, a savings of 54% over the 1 J consumed by the fixed 3.3-V system.

The next section surveys previous work on DSNs and dynamic voltage scaling. In Section 38.3, the basic model of sensor node in DSNs is explained, the energy–voltage relation for multiple voltage processors given, and the concept of message header introduced. Then, the authors' energy-efficient design approach for secure DSNs is demonstrated in Section 38.4. Section 38.5 reports all the simulation-related details and the chapter concludes in Section 38.6.

38.2 Background

38.2.1 On Distributed Sensor Networks

A DSN is composed of a large number of low-cost, compact, lightweight, disposable, and densely deployed sensor nodes. The primary mission of a DSN is to detect and report events occurring within the sensing range of the sensor network. Although each individual sensor node in the network generally has crude sensing functions (e.g., seismic, magnetic), the DSN can perform more reliable and sophisticated sensing functions through the cooperation of all the nodes in the sensor network. DSN has found numerous applications in various fields (e.g., health, military, home) and has attracted a lot of attention in recent years. Research in many areas (MEMS, wireless communication, network, cryptography, etc.) has addressed design and implementation issues of DSN from their points of view.

The wireless integrated network sensors (WINS) project at UCLA [33] and Rockwell Science Center [34] aims to develop low-power, low-cost, wireless MEMS-based microsensors that can sense, actuate, and communicate. Power efficiency is provided by power management over the network, low-power mixed signal circuits, and low-power radio frequency (RF) receivers [2]. Berkeley's Smart Dust project [35] uses optical, instead of RF, transmission techniques to make communication with reference to energy inexpensive [16]. More recently, researchers in MIT launched the Ultra Low-Power Wireless Sensor project, which targets design and fabrication of sensor systems capable of wirelessly transmitting data at 1 b/s — 1 Mb/s with average transmission power of 10 µW to 10 mW [36]. They have focused on developing energy-efficient communication protocols, in particular, energy-scalable algorithms [10, 29].

At a network level, Estrin et al. [9] discuss the scalable coordination problem and argue that a localized algorithm, in which sensors only interact with other sensors in a restricted vicinity, is promising and may be necessary for sensor network coordination. Over time, nodes that are a focal point for network traffic will lose energy more quickly than nodes at the edges of the network. For this reason, energy-aware routing protocols [18, 26, 28] are proposed in DSNs to balance the energy consumption in each sensor node. In the clustering protocols described by Mills [18] and Wang et al. [29], sensors periodically reorganize themselves to balance energy dissipation in the network and extend the overall lifetime of the network.

At the data-link layer, Ye and colleagues [31] propose a new MAC protocol, S-MAC, for sensor networks. S-MAC makes trade-offs between energy and latency according to traffic conditions and has better energy conserving properties than the traditional MAC protocol IEEE 802.11. Carman et al. [4] address the energy consumption issues in the domain of secure DSNs and develop novel key management protocols specifically designed for secure DSNs to reduce sensor nodes' energy consumption without sacrificing network security.

In order to reduce microprocessor's energy consumption, embedded processors typically have low-power modes that slow or halt the processor clock and place the device in a state that consumes less energy. For example, the Motorola DragonBall has sleep, doze, and run modes. Sinha and coworkers [27] propose an operating system-directed power management technique in DSN to shut down devices when they are not needed and wake them when necessary. This method can effectively extend a DSN's lifetime when sensor nodes do not execute instructions frequently. It is based on the fact that devices will consume little or no power when they are turned off. For example, the StrongARM microprocessor in Rockwell's sensor nodes consumes less than 1 mW in its sleep mode [1].

38.2.2 On Dynamic Voltage Scaling

Dynamically adapting voltage, and therefore clock frequency, to operate at the point of lowest power consumption for given temperature and process parameters was first proposed in the early 1990s [15, 17]. Later, Horowitz [12] described implementation of several digital power supply controllers based on this idea; Nielsen and colleagues [20] extended the dynamic voltage adaptation idea to take into account data-dependent computation times in self-timed circuits, Namgoong et al. [19] developed efficient DC–DC converters that allow output voltage to be rapidly changed (the time taken to reach steady state at the new voltage is less than 5 ms/V) under external control. More recently, Burd and colleagues [3] built a DVS-capable low-power ARM (lpARM) processor that can operate at 1.1 to 3.3 V, resulting in speeds between 10 and 100 MHz and power consumption between 18 and 220 mW. Hong et al. [11] developed a design methodology for the low-power, core-based, real-time system on chip using variable voltage hardware.

The preceding technology gives a DVS system the flexibility of operating at different voltages and clock frequencies to conserve energy. At the system level, there has been research on task-scheduling strategies for adjusting CPU speed so as to reduce energy consumption of DVS systems, particularly from the real-time system and operating system societies. Most of this work, based on a scheduling model suggested by Yao and coworkers [30], assumes that the CPU speed can be changed arbitrarily as a result of voltage scaling.

The feasible DVS system model by Qu [22] considers physical constraints on voltage scaling, such as how high, how low, and how fast voltage can be changed. DVS techniques can also be applied at a gate level, where power reduction is achieved by operating different gates at different supply and/or threshold voltage. The basic idea is to use high voltage along the critical path to keep the system's performance and low voltage off the critical path to reduce power.

Most research (particularly early research) work on DNS is on multiple DVS systems in which multiple voltages are simultaneously available on the chip. For practical reasons, this is also the DVS system that will be used in this chapter for the energy-efficient DSN design. The most important and relevant energy-reduction techniques on such multiple voltage DVS systems will now be surveyed.

Voltage scheduling can be performed at a behavioral level, typically on data flow graphs to exploit the parallelism among all operations. Specifically, operations on the critical path are conducted at the reference voltage to keep the required throughput, but operations off the critical path will be executed at reduced voltages to save power and energy. Raje and Sarrafzadeh [25] first proposed a multiple voltage scheduling algorithm to assign voltage level to each operation in a data flow graph to minimize power consumption with a given computation time constraint. Dual voltage (5.0 and 3.0 V) and three voltage (5.0, 3.0, and 2.4 V) were used for experimental purposes.

Chang and Pedram [6] presented a dynamic programming-based algorithm extending this to more general cases (such as cyclic graphs and throughput constraints) with four voltages (5.0, 3.3, 2.4, and 1.5 V) used in the simulation. Chen and Sarrafzadeh [8] related the DVS power minimization problem on a dual-voltage system to the maximal weighted independent set problem, which is polynomially solvable on a transitive graph. Then they developed a provably good algorithm to reduce a system's power consumption. In their simulation, 5.0 V was used as the high voltage while different voltages from 2.0 to 4.2 V were used as the low voltage.

The study of a multiple DVS system at a high level focuses on how to assign voltage to individual tasks (or jobs) in order to reduce energy consumption. Ishihara and Yasuura [14] showed that energy is minimized only when, at most, two voltages are applied to a single task. They formulated the voltage scheduling problem as an integer linear programming problem and relied on solving this problem to obtain the voltages for each task. Quan and Hu [23] studied the problem of determining the optimal voltage schedule for a real-time system with fixed-priority jobs. Their approach was based on an integer programming formulation, which could be efficiently solved, of the problem. Hua and Qu [13] proposed the voltage set-up problem, which targets how many levels and which values should be implemented for a multiple DVS system to achieve maximum energy savings. They derived analytical solutions for dual-voltage systems and gave efficient numerical methods to solve the general case. Their simulations suggested that a multiple DVS system can achieve energy savings very close to the ideal DVS model by Yao et al. if the voltages are selected properly.

38.3 Preliminaries

38.3.1 Model of Sensor Node and Energy Consumption in DSNs

A DSN consists of a collection of communicating sensor nodes in which each node (1) incorporates with one or more sensors to monitor the environment; (2) has limited processing capability to process the collected data into "high-value" information and to accomplish local control; and (3) is equipped with a radio transceiver to transmit information to or from neighbor nodes and, eventually, external users [34]. In DSNs, sensor nodes obtain information locally from target detection or environment monitoring. However, the major information source is the communication channel in the network. Because the energy required to transmit a bit can be much greater than the cost to process it [4] internally, raw data will typically be processed locally and the results exchanged within the network with fewer transmitted bits and less energy consumed.

Data packets from neighboring sensor nodes are received by the reception electronics and passed to the microprocessor. On receiving a new packet, preprocessing on the data is first conducted. This can be

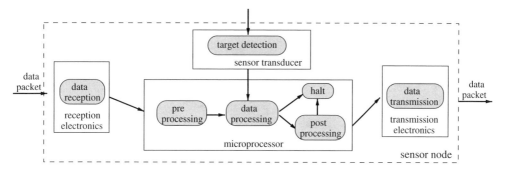

FIGURE 38.3 Data packet transmission and processing in a sensor node.

data unpackaging, decryption or verification, etc., depending on the applications of the DSN. Data processing is performed afterwards to obtain useful information and produce the required results. If the results need to be sent out to other sensor nodes, the microprocessor will compress or encrypt the results in the postprocessing step, and then send it to the transmission devices. Otherwise, when this sensor node is the final recipient of the data and does not need to cooperate with other nodes, the microprocessor will enter an idle state immediately after data processing. The transmission devices, normally consisting of transmission electronics and amplifiers, will send the data packet out. Figure 38.3 illustrates the mission of a sensor node in the network.

Energy dissipation through the sensor network is the sum of the energy consumed by all the sensor nodes in the network. This includes three parts: (1) energy dissipation on the sensor transducer; (2) energy dissipation for communication among sensor nodes; and (3) energy consumed by the microprocessor in computation. The amount of energy consumed by the sensor transducers depends on the sensitivity of the sensor and is normally a minor part of the entire sensor node's energy consumption.[*]

Radio transmission may contribute more than half of the peak power. It consumes more power in transmit mode than in receive mode because the transmit amplifier must be active at the transmitter's end. Radio transceivers are relatively complex circuits and it is difficult to reduce the communication energy consumption. A simplified model for communication energy can be found in Heinzelman et al. [10] and techniques to minimize this part of the energy consumption have been developed [2, 16, 33, 36]; however, this is beyond the scope of this chapter.

38.3.2 Processor with Multiple Supply Voltages

Reducing supply voltage can result in substantial reduction on switching power (also known as dynamic power), the dominant source of power dissipation in a CMOS circuit. Dynamic power is proportional to $\alpha C_L v_{dd}^2 f_{clock}$, where αC_L is the effective switched capacitance; v_{dd} s the supply voltage, and f_{clock} is the system clock frequency.

Roughly speaking, a system's power dissipation is halved if it is operating at a supply voltage 30% lower without changing any other system parameters. However, this power/energy savings comes at the cost of reduced throughput, slower system clock frequency, and longer gate delay. The gate delay is proportional to $\dfrac{v_{dd}}{(v_{dd} - v_t)^\beta}$, where v_t is the threshold voltage and $\beta \in (1.0, 2.0)$ is a technology-dependent constant. Naturally, we the power and delay trade off: on one hand, it is desirable to scale voltage as low as possible to reduce energy; on the other hand, a system operating at a low supply voltage may fail to complete the required computation.

---

[*]For example, in the AWAIRS I sensor developed at UCLA/Rockwell Science Center, the entire sensor node consumes a peak of 1 W of power. The processor consumes 300 mW; the radio consumes 600 mW at transmit mode and 300 mW in receive mode. The sensor transducers consumes less than 100 mW [1].

Assume that a set of discrete voltages is available at the same time and the system can operate at any voltage level. This system is referred to as a multiple DVS system and has been studied for years due to its easy implementation. With different supply voltages, the processor is able to operate at different speeds and therefore the time and power used to accomplish the same task (or same amount of computation) will also be different. Let $v_1 < v_2 < \ldots < v_n$ be the different voltages. Suppose that the processor finishes a task in time T_{ref} with power dissipation P_{ref} at the reference voltage v_{ref} and threshold voltage v_t. Then, at supply voltage v_i, to finish the same task, the processing time $T(v_i)$, the power dissipation $P(v_i)$, and the energy dissipation to complete this task $E(v_i)$ are given as follows:

$$T(v_i) = \frac{v_i}{(v_i - v_t)^\beta} \frac{(v_{ref} - v_t)^\beta}{v_{ref}} T(v_{ref}) \tag{38.1}$$

$$P(v_i) = \frac{v_i (v_i - v_t)^\beta}{v_{ref}(v_{ref} - v_t)^\beta} P(v_{ref}) \tag{38.2}$$

$$E(v_i) = P(v_i) t(v_i) = \frac{v_i^2}{v_{ref}^2} P(v_{ref}) T(v_{ref}) \tag{38.3}$$

38.3.3 The Message Header

To enable power control for energy efficiency, a sensor node at the receiving end must have additional information about the upcoming message to avoid selecting inappropriate voltages.* The earlier this information is available, the better decision can be made and the more energy can be saved by switching to the proper supply voltage. The complete knowledge of the message will not be available to the receiver sensor until it is completely revealed at the end of preprocessing. However, this may already be late for energy reduction because preprocessing can consume a nontrivial amount of energy in certain cases — for example, in some secure DSNs the asymmetric public key cryptographic algorithms (such as RSA) used to provide security consume a significant amount of energy in decryption.

On the other hand, the sender sensor knows the message better than the receiver before preprocessing the message. Therefore, it is proposed that the sender sensor collect additional information about the message and send it out as the *message header*. As depicted in Figure 38.4, the message header is embedded in the first data packet of a message and contains:

- Sender's information, such as the sensor's ID that the receiver sensor can verify to avoid attacks
- Receiver's information so that a receive sensor can tell whether it is the desired receiver of the message (for multicast)
- Message information, which is the key part of the message header and includes: size of the message in bits or in number of packets; time at which the message is generated and its latency requirement, both specified in a global clock; estimated computation load for data processing in CPU time at a reference voltage; predicted destiny of the message (e.g., being forwarded to other sensor nodes or not, being sent feedback to senders or not, etc.)
- (Part of the) data to be sent if space is left in the data packet after holding all the preceding information

---

*If the selected voltage is higher than necessary, further energy reduction is still possible; if the selected voltage is lower, then the sensor is in danger of missing the deadline or must raise the voltage level later to catch it, which will increase overall energy consumption.

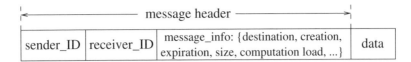

FIGURE 38.4 Content of the first data packet in a message.

Because most of this information, including the message header, requires little space,* the message will introduce one extra packet in the worst case. The energy consumed on processing the message header and transmitting this extra packet will negate some of the energy savings obtained by this proposed technique. However, it will be shown that this is insignificant.

To keep this discussion concrete, how to achieve energy efficiency will be demonstrated in the example of secure DSN design by DVS, with the help of a message header.

38.4 DVS with Message Header

38.4.1 Requirements and Constraints of Secure DSNs

Consider a DSN in a hostile environment in which communication among sensor nodes must be secure. Security is provided by standard data encryption protocols with pre-established keys. Requirements and constraints to be addressed during the design of secure DSNs include: embedded processor selection and memory design in the architecture layer; routing, scalability, and robustness in the network layer; real-time operating systems and power-aware software in the application layer; transmission technologies; sensor fabrication; and deployment issues.

In addition, a secure DSN must support the following features for security concerns:

- *Low energy consumption.* Once deployed in enemy territory, sensor nodes usually will be left unattended. Most sensor network designs are greatly influenced by finite battery limitations.
- *Confidentiality.* Sensors information, such as identity, private key, and public key, should be protected from disclosure to unauthorized parties and against traffic analysis.
- *Authenticity.* A sensor node should be able to verify the sender of a data packet on its arrival.
- *Unidirectional communication.* A sensor node should have the capability of receiving-only (to avoid detection) or transmitting-only (to avoid adverse jamming) communication.
- *Tamper resistance.* A sensor node should support tamper protection mechanisms, whether active or passive, to protect its data from being compromised.

38.4.2 Energy Consumption for Data Encryption and Decryption

Data encryption and decryption play vital roles in secure DSNs; they are performed in pre- and post-processing stages in a sensor node, respectively. Therefore it is useful to measure energy consumption for different data encryption/decryption algorithms on different embedded microprocessors. Assume a scenario in which the security requirement is high, messages are in small size, and public key algorithms are used to encrypt and decrypt messages. Table 38.2 gives information for computing a 128-b multiply function, the basic building block for most public key algorithms, at the reference 3.3 V. P represents power consumption; f represents clock frequency; t represents CPU time for the execution; and E represents energy consumption.

A processor's energy consumption for encryption and decryption operations is directly related to the costs of performing the basic modular arithmetic functions widely used in security protocols. For example, in RSA encryption, the modular operation (M^e mod n) must be computed for message M with

---

*For example, a sender can use 2 b to encode whether the message needs to be forwarded: 00 do not forward; 11 forward; 10/01 not sure, depends on the processing result.

TABLE 38.2 Energy Consumption for a 128-b Multiply Function

Processor	P(mW)	f(MHz)	t(μs)	E(mJ)
MIPS R4000	230	80	0.50	115
SA-1110 StrongARM	240	133	0.45	108
MC68328 DragonBal"	52	16	120	6200
MMC2001 M-Core	81	33	12.6	1020
ARC 3	2	40	4.2	8.4

a public key e such as 65537 and a modulus n of at least 1024-b size. For $e = 65537 = 2^{16} + 1$ and an n of 1024 b, using the Montgomery multiplication method, the number of 128-b operations needed to compute $M^e \bmod n$ is: 16 (number of 128-b multiply operations for $a^2 \bmod n$) + (number of 128-b multiply operations for $a \cdot b \bmod n$) = $16 \cdot \left[1.5 \cdot \left(\frac{1024}{64} \right)^2 + 1.5 \cdot \left(\frac{1024}{64} \right) \right] + \left[2 \cdot \left(\frac{1024}{64} \right)^2 + \frac{1024}{64} \right] = 7056$.

RSA decryption requires the computation of ($Md \bmod n$) with a 1024-b private key d. We use the Chinese Remainder Theorem and the 4-b exponent scanning technique for the calculation. The number of 128-b operations for this is:

$$2 \cdot \left\{ \left[\log_2 \frac{d}{2} \right] \cdot \left[1.5 \cdot \left(\frac{1024}{2 \cdot 64} \right)^2 + 1.5 \cdot \left(\frac{1024}{2 \cdot 64} \right) \right] + \frac{1}{4} \cdot \left[\log_2 \frac{d}{2} \right] \cdot \left[2 \cdot \left(\frac{1024}{2 \cdot 64} \right)^2 + \frac{1024}{2 \cdot 64} \right] \right\} = 145408.$$

Similarly, the number of 128-b operations needed for other popular cryptographic algorithms, such as DSA (digital signature algorithm) data signing/verification and ElGamal encryption/decryption, can be computed. Table 38.3 gives computational energy costs for these operations on different processors.

38.4.3 Dynamic Voltage Scaling on Sensor Nodes

Figure 38.5 gives an overview of how a microprocessor achieves energy efficiency by switching supply voltages. The encrypted data packets received by the radio transceivers are passed to the microprocessor, which decrypts and authenticates the data at the current voltage.* The microprocessor then checks whether the packet contains a message header; if not, it continues message decryption for the following packets. Otherwise, the microprocessor obtains information about the size of the message, estimated processing load, and size of the result from the message header.

TABLE 38.3 Energy Consumption for Public Key Algorithms at 3.3 V

	Computational Energy Consumption (mJ)					
	RSA		DSA		ElGamal	
Processor	Decryption	Encryption	Verification	Data Signing	Decryption	Encryption
MIPS R4000	16.7	0.81	9.9	20.0	9.94	134
SA-1110 StrongARM	15.0	0.74	9.1	18.2	9.1	123
MC68328 DragonBall	840	42	520	1040	520	7000
MMC2001 M-Core	137	6.9	85	169	85	1140
ARC 3	1.13	0.06	0.70	1.40	0.70	9.4

---

*If the microprocessor is in sleep mode, an interrupt will wake it up and it will set voltage at the default level.

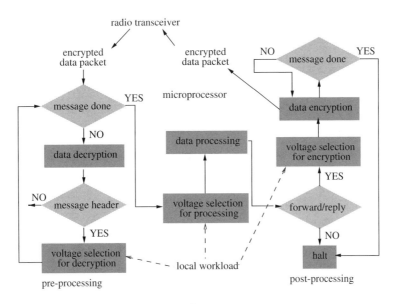

FIGURE 38.5 Dynamic voltage scaling on a sensor node.

Let W_{de}, W_{en}, W_{pr}, and W_{lo} be the computation workloads for decrypting one packet, encrypting one packet, and processing the current message (estimated by the sender sensor) and the jobs currently running on the receiver sensor node. Suppose the message has k packets and an estimated encrypted result of $\bar{k}$ packets to be completed by time t_d, the total workload that needs to be finished by t_d will be $k \cdot W_{de} + W_{pr} + \delta \cdot \bar{k} \cdot W_{en} + W_{lo}$, where $\delta = 0$ if there is no need to forward the result; otherwise $\delta = 1$.

The goal of preprocessing is to decrypt the message using the most energy-efficient voltage and determine the voltage for data processing. The decrypting voltage is decided based on the information provided by the message header; the microprocessor will decrypt the remaining packets of the message at this voltage. Once data decryption is done, the processor gains complete knowledge of the data and can update the voltage for data processing in a similar fashion. This completes the preprocessing and the microprocessor will start processing the data with the selected voltage.

After data processing, the microprocessor will halt (shift to idle state) if it is not necessary to forward the result to other sensor nodes. Otherwise, it will construct the message header, re-evaluate the size of the encrypted result $\bar{k}$, and select a proper voltage for data encryption. The encrypted data goes to the radio transceiver and will be sent out; this is referred to as postprocessing. The pseudocode is shown in Figure 38.6.

38.5 Simulation

This section explains the simulation platform and reports the results of 10 1-h simulations of sensor nodes in the secure DSN. The system is also configured by selecting different public key algorithms with different microprocessors; the most energy-efficient system configuration in the model is reported.

38.5.1 Simulation Platform

The behavior of a sensor node in a DSN is simulated assuming that the interarrival time of messages follows exponential distribution with parameter μ. With probability α, the sensor only needs to forward the message without any data processing; the (predicted) nonzero processing requirement is uniform between e_1 and e_2; the length of the message is uniform between a_1 and a_2; and the length of the (predicted) result is uniform between b_1 and b_2. Each message is nonpreemptive and has a deadline. In this simulation, 1.5 times the time from the arrival of one message until the start of the next is used as message deadline; this is equivalent to requiring a maximum of one message buffer. When a message misses its deadline, it must be dropped.

Dynamic voltage scaling on sensor node in DSN:

forever
{
  while (new message arrives)
  {
    $V_{dd} = V_{default}$; decrypt first packet in the message;
    extract information from message header of message size k, estimated computation load W_{pr} in data processing, and estimated size of encrypted result $\bar{k}$.
    if (unidentified message)
    {
      reject_message();
      /* go to idle state and wait for the next message */
    }
    else
    {
      if (forward only) /* no data processing needed */
      {
        select V_{dd} for decryption;
        /* select the lowest possible V_{dd} such that all jobs could be finished before deadline. */
        decrypt the whole message;
        re-evaluate message size for encryption, update $\bar{k}$;
        select V_{dd} for encryption;
        encrypt_message();
        send encrypted message to transceiver;
      }
      else /* message needs local data processing */
      {
        select V_{dd} for decryption;
        decrypt the whole message;
        re-evaluate computation load, update W_{pr};
        select V_{dd} for computation;
        data_processing ();
        if (forward needed) /* need to forward result to other sensors */
        {
          re-evaluate message size for encryption, update $\bar{k}$;
          select V_{dd} for encryption;
          encrypt_message();
          send encrypted message to transceiver;
        }
      }
    }
  }
}

FIGURE 38.6 Pseudocode of dynamic voltage scaling in DSN with message header.

A sequence of messages is generated using the preceding parameters and two simulations are conducted for each type of processor reported in Table 38.2. This takes place first on the traditional processor operating at 3.3 V and then on the same processor with three voltages 3.3, 2.4, and 1.2 V. The traditional processor will decrypt the message, process the data, encrypt the result, and send it out if necessary. It remains idle and does not consume any energy when there is no message to process.

For the same sequence of messages, assume that the sender sensor has constructed a 256-b message header for each message and built the first packet by encrypting this message header and the first 768 b of the message. The rest of the message is encrypted to 1024-b packets as before. Clearly, an overhead no worse than one packet per message may be experienced. The multiple voltage processor will process the messages with headers as described in Figure 38.5. Remember that a 256-b message header will be added to the result when "forward" or "reply" is required. This may also introduce a one-packet overhead.

After decrypting the message header, the receiver sensor may take one of the following actions:

- *Rejection.* The receiver is not the designated receiving nodes or the message is obsolete. The receiver will simply drop the rest of the message without further decryption.*
- *Forwarding.* The receiver is asked to forward the message, which happens in a multihop network. The receiver decrypts the entire message and encrypts it for forwarding. The decryption is based on the key agreement with the sender sensor, and the encryption is based on the key agreement with the sensor that will receive the forwarded message. The processing time in this case is 0, as mentioned earlier.
- *Acceptance.* The receiver decrypts the message, does the required processing, and makes decisions without sending any messages to other nodes, including the sender of the current message.
- *Proceeding.* This is a combination of acceptance and forwarding. The processing result needs to be encrypted and sent back to the sender or forwarded to other nodes.

The system's energy consumption on message decryption/verification, data processing, and data encryption/signing is monitored for both simulations. In the multiple voltage case, the total time during which the system is operating at different voltages is also tracked. Finally, this is repeated for each different combination of microprocessor and public key algorithms.

38.5.2 Simulation Results

First, the messages that the sensor receives in the simulated communications will be described. Then, for different microprocessors coupled with different public key algorithms, the average energy consumed by the traditional fixed voltage processor and the multiple voltage processor for the same sets of messages will be reported. To analyze where the energy saving comes from better, detailed time and energy data are given for the case of an MIPS R4000 processor with RSA. Several other case studies with questions such as the role of different public key algorithms and, eventually, how to guide system configurations are also conducted.

38.5.2.1 Messages Generated from Simulation

Behavior of one sensor node is simulated in the DSN; this sensor receives messages with the following parameters: interarrival rate of messages $\mu = 0.125$; forwarding-only probability $\alpha = 0.5$; range of nonzero processing time = [500 ms, 4000 ms]; original message and processing result = size [200 b, 20,000 b] (i.e., 1 to 20 packets). As discussed earlier, a message may be rejected by the receiver, so the rejection rate is set to be 0.2.

Figure 38.7 reports the total number of messages and the numbers of each different type of messages in 10 1-h simulations. The overhead (number of extra packets) of the messages and results caused by message headers in each simulation is also reported.

On average, the sensor node receives 464 messages in 1 h. Among them, 97 (21.0%) are rejected; 229 (49.4%) are forwarded without processing; and the remaining 137 (29.6%) require data processing. The average overhead on the received message is 118 packets — about 25.5% of the total messages, as expected.** The average overhead on the result is 92 packets, 24.9% of the messages that need to be

---

*For a fair comparison, assume that, on the traditional fixed voltage processor, the rejection can also be detected after the decryption of the first packet. This is true because the nondesignated receivers will not be able to decrypt the packet and information such as expiration time is widely used in most DSNs.

**The size of the message header is 256 b and a packet is of a fixed 1024-b size. The one-packet overhead occurs if the last packet contains more than 576 b of data and thus not sufficient space for the message header. Statistically, this happens 25% of the time.

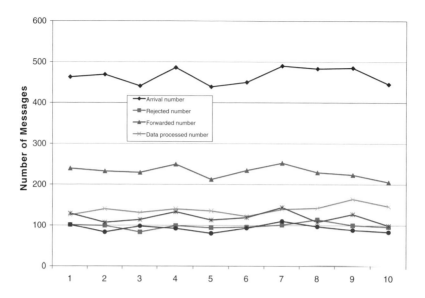

FIGURE 38.7 Messages created in 10 1-h simulations.

forwarded or replied to. Notice that each message carries, at most, one extra packet overhead because of the message header.

38.5.2.2 Simulation on Different System Configurations

For one set of messages generated from a 1-h simulation (described previously), further simulations are conducted on different combinations of microprocessors and public key algorithms. For each system, simulations are performed on the fixed voltage (3.3 V) core and the core with multiple voltages (3.3, 2.4, and 1.2 V). Figure 38.8 and Figure 38.9 report energy consumption and nonidle time for five representative systems (the ElGamal algorithm is excluded because of its high encryption cost):

- MIPS R4000 with RSA
- MIPS R4000 with DSA
- StrongARM with RSA
- StrongARM with DSA
- M-Core with RSA

Figure 38.8 indicates significant energy reduction in all systems, from 58% in the M-Core system to 73% in the StrongARM core, with an average of 64% energy savings. Considering that the microprocessor is responsible for 30% of the sensor node's total energy consumption, this means an energy savings of 20%, which is still significant. Although a constant 55% energy reduction from data processing occurs, energy savings from data decryption and data encryption are very different. For RSA, in which decryption is 20 times more expensive than encryption, the amount of energy savings from decryption is much more significant than that from encryption. The same result can be observed for the verification-expensive DSA. However, this is not as significant as in RSA because the verification in DSA is only twice as expensive as data signing. For systems with the same public key algorithm but different microprocessors, energy savings depend on the system's power/frequency performance. For a slow core such as M-Core, energy savings for all three phases of message processing are not as dramatic as those for fast-core StrongARM because the speed on slow cores cannot be reduced due to message density and data processing requirements.

Similar analysis holds for the system's nonidle time, as illustrated in Figure 38.9. Multiple voltage microprocessors have longer running time than the traditional fixed cores. This suggests that, to complete the same workload, it is better to run for a longer time at lower voltage, which is a well-known fact in the dynamic voltage scaling literature (see, for example, Ishihara and Yasuura [14] and Yao et al. [30]).

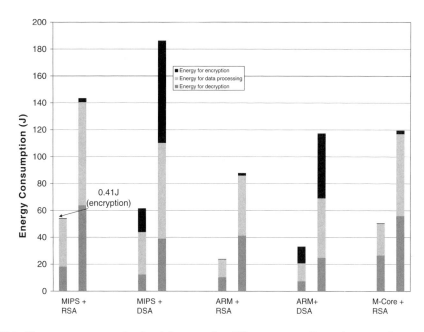

FIGURE 38.8 The energy consumption breakdown on five different systems. For each system, the one on the left represents the multiple voltage core and the other the traditional fixed-voltage core. The data unit on the y-axis is in joules.

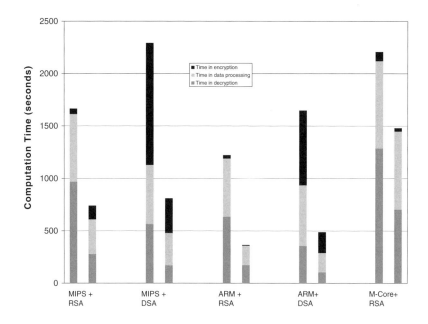

FIGURE 38.9 The nonidle time breakdown on five different systems. For each system, the one on the left represents the multiple voltage core and the other the traditional fixed-voltage core. The data unit on the y-axis is in seconds.

38.5.2.3 From Where Do Energy Savings Come?

Ten simulations on the MIPS R4000 using RSA as the public key algorithm are reported in detail for the analysis of the energy savings. As shown in Figure 38.7, during the 10 simulations, the sensor node

TABLE 38.4 Average Time and Energy Consumption for Traditional Fixed-Voltage and New Multiple Voltage Processor in 10 Simulations

Activity	Time		Energy	
	Fixed (s)	DVS (%)	Fixed (J)	DVS (%)
Decryption	276.11	350	63.51	28.8
Encryption	13.11	413	3.02	13.7
Processing	343.19	188	78.93	45.0
Total	632.41	264	145.45	37.3

received 464 messages on average. In the traditional approach, decryption was necessary for the first packet of the 97 rejected messages (refer to the section on simulation platform for the reasons for doing so); for the remaining 367 messages, 137 messages were processed and the results encrypted (as well as the 229 forward-only messages) before they were sent out. In the traditional approach, the processor is on for 632 s (276 s for decryption, 343 s for data processing, and 13 s for encryption) at the fixed 3.3 V; thus, it consumes 145 J of energy (see Table 38.4 for details).

Notice that data decryption and processing take about 44 and 54% of the total nonidle time and energy because: (1) RSA encryption is much less computation intensive compared to decryption (Table 38.3); (2) result encryption is necessary only for messages that need to be forwarded or replied to; and (3) for rejected messages, the first packet needs to be decrypted before the rejection.

In this new approach with the multiple voltage processor and message header, the insertion of additional information introduces one extra packet for 92 of the messages. The multiple voltage processor decrypts the message header at the default 3.3 V and then selects the proper voltage for the rest of the decryption. It will also drop the 97 rejected messages at this phase without any further decryption. Once the decryption is complete, the processor updates the voltage for data processing and/or result encryption. The total nonidle time is 1667 s, with an energy consumption of 54.2 J — a 62.7% energy savings over the traditional processor.

Table 38.5 gives the breakdown of run-time and energy consumption at different voltages and different processing phases (data decryption, data encryption, and data processing). Table 38.4 compares this with the fixed 3.3-V processor. Clearly, the multiple voltage processor spends most of the time at the low 1.2 V (about 72%) and 2.4 V (more than 25%) to save energy. It operates at the high 3.3 V only when necessary for the decryption of all first packets and the decryption and processing of messages with a high computation load that cannot be accomplished at a lower voltage. It never undertakes encryption at full speed* because the RSA encryption scheme is cheap, as witnessed from Table 38.3 and Table 38.4. However, it is not difficult to imagine that, in a secure DSN with DSA scheme, encryption will contribute more to energy savings.

38.5.2.4 Energy-Driven System Configuration

For a given set of messages with certain statistical information, it will be interesting to see what can guide selection of the right combination of microprocessor and public key algorithm to implement the secure DSN in the most energy-efficient way. For this purpose, the following simulations in which, for each different setting of messages, the energy consumption was stimulated on all possible system configurations, were conducted. The interarrival rate μ takes value from the set of $\{0.125, 0.1, 0.05, 0.025, 0.01\}$; the message size will be within one of the following ranges: $\{[200, 20000]; [200, 4000]; [200, 10000];$ or $[10000, 20000]\}$. The processing time falls into one of the following: $\{[500, 4000]$ or $[100, 1000],$

---

*This is a coincidence, however; it is very rare to encrypt at the highest voltage. This happens only when the processing load is so high that, after decryption and data processing, not enough time is left for encryption to be performed at a lower voltage. Considerng that encryption of one packet only needs 3.5 ms at 3.3 V, it is not surprising that this does not happen.

TABLE 38.5 Run-Time and Energy Consumption Breakdown for Simulation on MIPS R4000 with Multiple Voltages

Activity	3.3 V		2.4 V		1.2 V		Total	
	Time (s)	Energy (J)	Time (s)	Energy (J)	Time (s)	Energy (J)	Time (s)	Energy (J)
Decryption	34.30	7.89	44.91	3.68	888.26	6.69	967.47	18.26
Encryption	0	0	0.084	0.007	54.02	0.41	54.11	0.413
Processing	8.45	1.94	386.83	31.71	250.07	1.88	645.35	35.53
Total	42.75	9.83	431.82	35.39	1192.35	8.97	1666.93	54.2

[2000,10000]}. Finally, one of the following three sets is used for rejection rate, forward-only rate, and processing-demand rate: {(0.2,0.5,0.3); (0.1,0.3,0.6); or (0.1,0.6,0.3)}.

The preliminary results elicit several interesting observations. For example, if the microprocessor is not fast enough to keep pace with the message arrival rate, then it must drop some messages because of the deadline requirement. In traditional system settings in which no multiple voltages and no message header are present, only a few system configurations consistently finish all the processing without any message drop (e.g., MIPS/SA-1110) for the highest interarrival rate, $\mu = 0.125$, with moderate message size and data processing requirement. As the message arrival rate decreases, more and more system configurations can handle the messages, although the combination of SA-1110 and RSA remains the most energy efficient. When the interarrival rate reaches 0.05, the most power-efficient microprocessor ARC3 is able to handle the messages and becomes the best choice for DSN implementation.

Similar behavior can be found for rejection rate, message size, processing requirements, etc. Basically, larger rejection rates, fewer processing demands, smaller message sizes, fewer message drops, and more choices for system combinations occur when the sensor node receives messages less frequently. In general, MIPS or SA-1110 with RSA is a good combination when drop rate is high and ARC3 is the choice when it can handle the messages without any significant drops.

38.6 Conclusions

This chapter discussed how to apply the DVS technique in the design of DSNs to reduce sensor nodes' energy consumption. First, it was observed that computation workloads during data processing are not balanced among sensor nodes; even in the same sensor node, computation workload can be very different at different stages of data processing. This provides an opportunity to apply the DVS technique to reduce sensor nodes' computational energy. However, another observation — that a message's computation requirement may not be proportional to the length of the message — posts the challenge of how to scale voltage efficiently to reach the best energy efficiency.

The chapter introduced the concept of a message header to store message information. The message header is inserted at the beginning of each message and will be processed first by the receiver node. This gives the receiver a better understanding of the entire message and helps it to make proper selection of supply voltages to reduce the energy consumption. The approach was validated by applying it in a specific, secure DSN. It was simulated over a wide range of microprocessors and several different public key algorithms. Simulation results demonstrate the effectiveness of this approach: computational energy savings of about 60% were achievable despite the overhead of embedding extra information into the message header.

Acknowledgments

This work is in part supported by Minta Martin Research Fund. The authors thank David W. Carman, Peter S. Kruus, and Brain J. Matt from NAI Labs for their help.

References

1. J.R. Agre, L.P. Clare, G.J. Pottie, and N.P. Romanov, Development platform for self-organizing wireless sensor networks, *Proc. Int. Soc. Optical Eng.*, 257–268, April 1999.
2. K. Bult et al., Low power systems for wireless microsensors, *Proc. 1996 Int. Symp. Low Power Electron. Design*, 17–22, August 1996.
3. T.D. Burd, T. Pering, A. Stratakos, and R. Brodersen, A dynamic voltage-scaled microprocessor system, *IEEE Int. Solid-State Circuits Conf.*, 294–295, 466, February 2000.
4. D.W. Carman, P.S. Kruus, and B.J. Matt, Constraints and approaches for distributed sensor network security, NAI Labs technical report #00-10, September 2000.
5. A.P. Chandrakasan, S. Sheng, and R.W. Brodersen, Low-power CMOS digital design, *IEEE J. Solid-State Circuits*, 27(4), 473–484, 1992.
6. A.P. Chandrakasan, V. Gutnik., and T. Xanthopoulos, Data driven signal processing: an approach for energy efficient computing, *Int. Symp. Low Power Electron. Design*, 347–352, 1996.
7. J. Chang and M. Pedram, Energy Minimization using multiple supply voltages, *IEEE Trans. Very Large Scale Integration Syst.*, 5(4), 436–443, December 1997.
8. C. Chen and M. Sarrafzadeh, Provably good algorithm for low power consumption with dual supply voltages, *ICCAD'99: IEEE/ACM Int. Conf. Computer-Aided Design*, 76–79, November 1999.
9. D. Estrin, R. Govindan, J. Heidemann, and S. Kumar, Next century challenges: scalable coordination in sensor networks, *ACM/IEEE Int. Conf. Mobile Computing Networking*, 263–270, August 1999.
10. W.R. Heinzelman, A. Sinha, A. Wang, and A.P. Chandrakasan, Energy-scalable algorithms and protocols for wireless microsensor networks, *IEEE Int. Conf. Acoustics, Speech, Signal Process.*, 3722–3725, June 2000.
11. I. Hong et al., Power optimization of variable-voltage core-based systems, *IEEE Trans. Computer-Aided Design Integrated Circuits Syst.*, 18(12), 1702–1714, December 1999.
12. M. Horowitz, Low power processor design using self-clocking, *Workshop Low-Power Electron.*, August 1993.
13. S. Hua and G. Qu, Approaching the maximum energy saving on embedded systems with multiple voltages, *ICCAD'03: IEEE/ACM Int. Conf. Computer-Aided Design*, November 2003, 26–29.
14. T. Ishihara and H. Yasuura, Voltage scheduling problem for dynamically variable voltage processors, *Int. Symp. Low Power Electron. Design*, 197–202, August 1998.
15. V. Von Kaenel, P. Macken, M.G.R. Degrauwe, A voltage reduction technique for battery-operated systems, *IEEE J. Solid-State Circuits*, 25(5), 1136–1140, October 1990.
16. J.M. Kahn, R.H. Katz, and K.S.J. Pister, Next century challenges: mobile networking for Smart Dust, *ACM/IEEE Int. Conf. Mobile Computing Networking*, 271–278, August 1999.
17. P. Macken, M. Degrauwe, M. Van Paemel, and H. Oguey, A voltage reduction technique for digital systems, *IEEE Int. Solid-State Circuits Conf. Dig. Tech. Papers*, 238–239, February 1990.
18. D. Mills, Low energy communications and routing for microsensor networks, *Proc. ARL Fed. Lab. 4th Annu. Symp.*, March 2000.
19. W. Namgoong, M. Yu, and T. Meng, A high-efficiency variable-voltage CMOS dynamic DC-DC switching regulator, *IEEE Int. Solid-State Circuits Conf. Dig. Tech. Papers*, 380–381, February 1997.
20. L.S. Nielsen, C. Niessen, J. Sparso, and K. van Berkel, Low-power operation using self-timed circuits and adaptive scaling of the supply voltage, *IEEE Trans. Very Large Scale Integration Syst.*, 2(4), 391–397, December 1994.
21. T. Pering, T.D. Burd, and R.W. Brodersen. Voltage scheduling in the IpARM microprocessor system, *ISLPED'00: Int. Symp. Low Power Electron. Design*, 96–101, July 2000.
22. G. Qu, What is the limit of energy saving by dynamic voltage scaling? *IEEE/ACM Int. Conf. Computer Aided Design*, November 2001, 560–563.
23. G. Quan and X. Hu, Energy efficient fixed-priority scheduling for real-time systems on variable voltage processors, *DAC'01: IEEE/ACM Design Automation Conf.*, 828–833, June 2001.

24. J.M. Rabaey, Wireless beyond the third generation — facing the energy challenge, *Int. Symp. Low Power Electron. Design,* 1–3, August 2001.
25. S. Raje and M. Sarrafzadeh, Variable voltage scheduling, *ISLPED'95: Int. Symp. Low Power Electron. Design,* 9–14, April 1995.
26. R.C. Shah and J.M. Rabaey, Energy aware routing for low energy ad hoc sensor networks, *IEEE Wireless Commun. Networking Conf. (WCNC),* March 2002, 350–355.
27. A. Sinha and A.P. Chandrakasan, Dynamic power management in wireless sensor network, *IEEE Design Test Computers,* 18(2), 62–74, April 2001.
28. L. Tassiulas and J. Chang, Maximum lifetime routing in wireless sensor networks, *Proc. ARL Fed. Lab. 4th Annu. Symp.,* March 2000.
29. A. Wang, W.R. Heinzelman, and A.P. Chandrakasan, Energy-scalable protocols for battery-operated microsensor networks, *IEEE Workshop Signal Process. Syst.,* 483–492, October 1999.
30. F. Yao, A. Demers, and S. Shenker, A scheduling model for reduced cpu energy, *FOCS'95: IEEE Annu. Found. Computer Sci.,* 374–382, 1995.
31. W. Ye, J. Heidemann, and D. Estrin, An energy-efficient mac protocol for wireless sensor networks, *Proc. IEEE 21st Annu. Joint Conf. IEEE Computer Commun. Soc., INFOCOM 2002,* 3, 23–27, June 2002.
32. AMD Athlon 4 Processors, data sheet reference 24319. Advanced Micro Devices, Inc., 2001.
33. http://www.janet.ucla.edu/WINS.
34. http://wins.rsc.rockwell.com/.
35. http://robotics.eecs.berkeley.edu/~pister/Smart-Dust/.
36. http://www-mtl.mit.edu/~jimg/project_top.html.
37. http://www.intel.com/products/desk_lap/processors/laptop/pentium4-m.
38. http://www.chips.ibm.com:80/products/powerpc/chips.
39. http://www.transmeta.com.
40. http://developer.intel.com/design/iio/manual3/273411.htm.
41. http://www.sensorweb.com/.

39
Wireless Sensor Networks and Computational Geometry

Xiang-Yang Li
Illinois Institute of Technology

Yu Wang
Illinois Institute of Technology

39.1 Introduction 39-1
  Wireless Sensor Networks • Computational Geometry • Networking and Routing • Topology Control • Routing • Organization
39.2 Preliminaries 39-4
  Unit Disk Graph • Power-Attenuation Model • Spanners • Low-Weight Structures • Geometric Structures • Localized Algorithms
39.3 Topology Control 39-10
  Bounded Degree Structures • Planar Structures • Bounded Degree, Planar Structures • Bounded Degree, Planar, Low-Weight Structures • Fault Tolerance • Interference • Transmission Power Control • Clustering, Virtual Backbone
39.4 Localized Routing 39-33
  Simple Heuristics • Right-Hand Rule and Face Routing • Combining Face Routing with Greedy Routing • Routing on Delaunay Triangulation
39.5 Broadcasting 39-38
  Centralized Methods • Localized Methods
39.6 Summary and Open Questions............................ 39-42

39.1 Introduction

39.1.1 Wireless Sensor Networks

Due to potential applications in various situations such as battlefield, emergency relief, environmental monitoring, and so on, wireless sensor networks [50, 75, 118, 130] have recently emerged as a premier research topic. Sensor networks consist of a set of sensor nodes spread over a geographical area. These nodes are able to perform processing as well as sensing and are additionally capable of communicating with each other by means of a wireless ad hoc network. With coordination among these sensor nodes, the network will achieve a larger sensing task in urban environments as well as inhospitable terrain. The sheer numbers of these sensors and the expected dynamics in these environments present unique challenges in the design of wireless sensor networks. Many excellent research projects have been conducted to study problems in this new field [50, 69, 75, 118, 119, 130].

This chapter considers a wireless sensor network consisting of a set V of n wireless sensor nodes distributed in a two-dimensional plane. Each wireless sensor node has an omnidirectional antenna. This is attractive because a single transmission of a node, assumed to be a disk centered at the node, can be received by many nodes within its vicinity. The radius of this disk is called the *transmission range* of this sensor node. In other words, node v can receive the signal from node u if node v is within the transmission range of the sender u. Otherwise, two nodes communicate through multihop wireless links by using intermediate nodes to relay the message. Consequently, each node in the sensor network also acts as a router, forwarding data packets for other nodes. By a proper scaling, it is assumed that all nodes have the maximum transmission range equal to one unit. These wireless sensor nodes define a *unit disk graph* (UDG) (V) in which an edge is between two nodes if and only if their Euclidean distance is at most one.

In addition, each node is assumed to have a low-power global position system (GPS) receiver, which provides the position information of the node. If GPS is not available, the distance between neighboring nodes can be estimated on the basis of incoming signal strengths. Relative coordinates of neighboring nodes can be obtained by exchanging such information between neighbors [26]. With position information, computational geometry techniques can be applied to solve some challenging questions in sensor networks.

39.1.2 Computational Geometry

Computational geometry emerged from the field of algorithm design and analysis in the late 1970s. It studies various problems [49, 58, 131] from computer graphics, geographic information systems, robotics, scientific computing, wireless networks (recently), and others, in which geometric algorithms could play fundamental roles. Most geometric algorithms are designed for studying the structural properties, searching, inclusion or exclusion relations, of a set of points, a set of hyperplanes, or both. For example, the structural properties include convex hull, intersections, hyperplane arrangement, triangulation (Delaunay, regular, etc.), Voronoi diagram, and so on. The query operations often include point location, range searching (orthogonal, unbounded, or some variations) and so on. In this chapter, the concentration is on how to apply some structural properties of a point set for wireless sensor networks as wireless sensor devices are treated as two-dimensional points.

39.1.3 Networking and Routing

It is common to separate the network design problem from management and control of the network in the communication network literature. The separation is very convenient and helps to simplify these two tasks, which are already very complex, significantly. Nevertheless, a price is paid for this modularity because decisions made at the network design phase may strongly affect the network management and control phase. In particular, if the issue of designing efficient routing schemes is not taken into account by the network designers, then the constructed network might not be suited for supporting a good routing scheme.

A wireless sensor network needs special treatment because it intrinsically has its own special characteristics and some unavoidable limitations compared with traditional wired networks. Wireless sensor nodes are often powered only by batteries and often have limited memories. Therefore, it is more challenging to design a network topology suitable for designing an efficient routing scheme to save energy and storage memory consumption for wireless sensor networks, rather than the traditional wired networks. Also, because several thousand sensors may move often, construction of network topology should be easy to operate and updated in a dynamic way.

In technical terms, the question is therefore, "Is it possible (if possible, then how) to design a network that is a subgraph of the unit disk graph so that it can be constructed or updated efficiently and ensures attractive network features such as bounded node degree, low-stretch factor, and linear number of links, as well as attractive routing schemes such as localized routing with guaranteed performances?"

39.1.4 Topology Control

The size of the unit disk graph could be as large as the square order of the number of network nodes. Thus, a subgraph of the unit disk graph UDG (*V*), which is sparse, can be constructed locally in an efficient way and is still relatively good compared with the original unit disk graph for routes' quality.

Unlike the wired networks that typically have fixed network topologies, each node in a sensor network can potentially change the network topology by adjusting its transmission range and/or selecting specific nodes to forward its messages, thus controlling its set of neighbors. The primary goal of topology control in wireless sensor networks is to maintain network connectivity, optimize network lifetime and throughput, and make it possible to design power-efficient routing. Not every connected subgraph of the unit disk graph plays the same important role in network designing. One of the perceptible requirements of topology control is to construct a subgraph so that the shortest path connecting any two nodes in the subgraph is not much longer than the shortest path connecting them in the original unit disk graph. This aspect of path quality is captured by the *stretch factor* of the subgraph. A subgraph with constant stretch factor is often called a *spanner* and a spanner is called a *sparse spanner* if it has only a linear number of links. This chapter reviews and studies how to construct a spanner (a sparse network topology) efficiently for a set of static sensor nodes.

The other imperative requirement for network topology control in sensor networks is fault tolerance. To guarantee a good fault tolerance, the underlying network structure must be k-connected for some $k > 1$, i.e., given any pair of wireless sensor nodes, there need to be at least k disjoint paths to connect them.

Restricting the size of the network has been found to be extremely important in reducing the amount of routing information. The notion of establishing a subset of nodes that perform the routing has been proposed in many routing algorithms [44, 140, 146, 163]. These methods often construct a virtual backbone by using the connected dominating set [4, 150, 154], which is often constructed from a dominating set or maximal independent set.

39.1.5 Routing

Many routing algorithms have been proposed recently for wireless ad hoc networks; most of them can be used in wireless sensor networks. The routing protocols proposed may be categorized as table-driven protocols or demand-driven protocols. A good survey may be found in Royer and Toh [136].

Table-driven routing protocols maintain up-to-date routing information between every pair of nodes. The changes to the topology are maintained by propagating updates of the topology throughout the network. Destination-sequenced distance vector routing (DSDV) [127] and zone routing protocol (ZRP) [74, 166] are two table-driven protocols proposed recently. The mobile nature of the wireless sensor networks prevents these table-driven routing protocols from being widely used in large-scale wireless ad hoc networks. Thus, on-demand routing protocols are preferred.

Source-initiated, on-demand routing creates routes only when desired by the source node. Proposed methodologies include ad hoc on-demand distance vector routing (AODV) [128]; dynamic source routing (DSR) [23]; and the temporarily ordered routing algorithm (TORA) [122]. In addition, associativity based routing (ABR) [147] and signal stability routing (SSR) use various criteria for selecting routes.

Introducing a hierarchical structure into routing has also been used in many protocols, such as clusterhead gateway switch routing (CGSR) [33]; fisheye routing [123, 124]; and hierarchical state routing [70]. Dominating set-based methods have also been adopted by several researchers [44, 146, 163]. To facilitate this, several methods [4, 111, 150, 161] have been proposed to approximate the minimum dominating-set or minimum connected dominating-set problems in centralized and/or distributed ways.

Route discovery can be very expensive in communication costs, thus reducing response time of the network. On the other hand, explicit route maintenance can be even more costly in the explicit communication of substantial routing information and use of scarcity memory of wireless sensor nodes. The geometric nature of the multihop wireless sensor networks allows a promising idea: localized routing protocols.

Localized routing does not require nodes to maintain routing tables, a distinct advantage given the scarce storage resources and relatively low computational power available to wireless nodes. More importantly, given the numerous changes in topology expected in sensor networks, no recomputation of the routing tables is needed and therefore a significant reduction in overhead is expected. Thus, localized routing is scalable and also uniform, in the sense that all the nodes execute the same protocol when deciding to which node to forward a packet.

However, localized routing is challenging to design because even guaranteeing successful arrival at the packet's destination is a nontrivial task. This task was successfully solved by Bose et al. [20] (see also Karp and Kung [78]), thus opening the way for a second stage of research focusing on improving the *efficiency* of localized routings. Localized routing also does not have a built-in mechanism to avoid congestion by overloading nodes. Mauve et al. [117] conducted an excellent survey of position-based localized routing protocols.

39.1.6 Organization

The rest of this chapter is organized as follows. In Section 39.2, definitions necessary for more detailed review of current progress in applying computational geometry techniques to wireless sensor networks is applied. Specifically, the way in which the sensor network is modeled in this chapter is detailed; some geometry structures reviewed; graph spanners defined; and the localized algorithm concept introduced. Section 39.3 provides a detailed review of geometry structures suitable for topology control in wireless sensor networks, especially the structures with bounded stretch factor or with bounded node degree, or planar structures. Then a brief discussion of fault-tolerant and interference problems in topology control follows. This section also reviews the current status of controlling transmission power so that the total or maximum transmission power is minimized without sacrificing network connectivity. The state of the art of constructing virtual backbone for wireless networks is reviewed. Because many heuristics are proposed in this area, the focus is on the ones that have theoretic performance guarantees or are popular. After a review of geometric structures, the so-called localized routing methods are reviewed in Section 39.4. Many routing algorithms have been proposed in the literature; here, concentration is on localized routing protocols as they utilize the geometric nature of wireless sensor networks. Location service protocols are also discussed. Section 39.5 reviews broadcasting protocols that apply the geometric nature to guarantee performance. Finally, the chapter concludes in Section 39.6 by pointing out some possible future research questions.

39.2 Preliminaries

This section reviews definitions and concepts necessary for later discussion. It specifies how the sensor network is modeled in a geometric view, reviews some well-known geometry structures, and defines spanners and low-weight graphs.

39.2.1 Unit Disk Graph

Consider a sensor network consisting of a set V of sensor nodes distributed in a two-dimensional plane. By a proper scaling, it is assumed that all nodes have the maximum transmission ranges equal to one unit. These sensor nodes define a *unit disk graph* $UDG(V)$ in which an edge is between two nodes if and only if their Euclidean distance is, at most, one (see Figure 39.1a. Hereafter, it is always assumed that $UDG(V)$ is a connected graph. Given a set of points uniformly and randomly distributed in an area, if the transmission range satisfies some value, then the $UDG(V)$ is connected with high probability [63, 134, 137]. All nodes within a constant k hops of a node u in the unit disk graph $UDG(V)$ are called the *k-local nodes* of u and are denoted by $N_k(V)$. Usually, the constant k here is 1 or 2. The size of the unit disk graph could be as large as the square order of the number of sensor nodes, such as shown in Figure 39.1(b). Thus, in topology control (discussed in next section), the attempt is to construct a subgraph

Wireless Sensor Networks and Computational Geometry

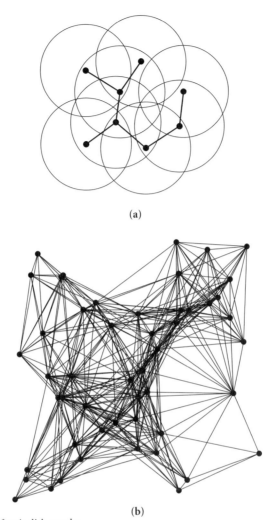

FIGURE 39.1 Examples of unit disk graphs.

(spanner) for the unit disk graph UDG(V), and the spanner is sparse and can be constructed locally in an efficient way.

39.2.2 Power-Attenuation Model

Energy conservation is a critical issue in sensor networks for the node and network life because the nodes are powered by batteries only. Each sensor node typically has a portable set with transmission and reception processing capabilities. To transmit a signal from a node to another node, the power consumed by these two nodes consists of the following three parts. First, the source node needs to consume some power to prepare the signal. Second, in the most common power-attenuation model, the power needed to support a link uv is $\|uv\|^\beta$, where $\|uv\|$ is the Euclidean distance between u and v, and β is a real constant between 2 and 5 dependent on the transmission environment. This power consumption is typically called *path loss*. Finally, when a node receives the signal, it needs to consume some power to receive, store, and then process that signal. For simplicity, this overhead cost can be integrated into one cost, which is almost the same for all nodes. Thus, c will be used to denote this constant overhead. In most results surveyed here, it is assumed that $c = 0$, i.e., the path loss is the major part of power consumption to transmit signals. The power cost $p(e)$ of a link $e = uv$ is then defined as the power consumed for transmitting signal from u to node v.

39.2.3 Spanners

Spanners have been studied intensively in recent years [7–9, 17, 28, 45, 76, 89, 164]. Let $G = (V,E)$ be an n vertex-connected weighted graph. The distance in G between two vertices $u,v \in V$ is the total weight (length) of the shortest path between u and v and is denoted by $d_G(u,v)$. A subgraph, $H = (V,E')$, where $E' \subseteq E$, is a t-spanner of G, if, for every $u,v \in V, d_H(u,v) \le t \cdot d_G(u,v)$. The value of t is called the *stretch factor*.

Spanners for Euclidean graphs are called *geometric spanners* or *Euclidean spanners*. This means the distance $d_G(u,v)$ in graph G between u and v is the Euclidean distance between vertices u and v. All previous algorithms that construct a t-spanner of the Euclidean complete graph $K(V)$ in computational geometry are centralized methods. The rapid development of wireless communication presents a new challenge for algorithm designing and analysis. Distributed algorithms are favored rather than the more traditional centralized algorithms.

Consider any unicast path $\Pi(u,v)$ in G (could be directed) from a node $u \in V$ to another node $v \in V$:

$$\Pi(u, v) = v_0 v_1 \ldots v_{h-1} v_h, \text{ where } u = v_0, v = v_h.$$

Here, h is the number of hops of the path Π. The total *transmission power*, $p(\Pi)$, consumed by this path, Π, is defined as

$$p(\Pi) = \sum_{i=1}^{h} \|v_{i-1} v_i\|^\beta$$

Let $p_G(u,v)$ be the least energy consumed by all paths connecting nodes u and v in G. Let H be a subgraph of G. The *power stretch factor* of the graph H with respect to G is then defined as

$$\rho_H(G) = \max_{u,v \in V} \frac{p_H(u,v)}{p_G(u,v)}$$

If G is a unit disk graph, use $\rho_H(V)$ instead of $\rho_H(G)$. For any positive integer n, let

$$\rho_H(n) = \sup_{|V|=n} \rho_H(V).$$

Similarly, define the length stretch factors $\ell_H(G)$ and $\ell_H(n)$. When the graph H is clear from the context, it is dropped from notations.

Li and colleagues [102] proved that, for a constant δ, $\rho_H(G) \le \delta$, if and only if, for any link $v_i v_j$ in graph G but not in H, $p_H(v_i,v_j) \le \delta \|v_i,v_j\|^\beta$. It is then sufficient to analyze the power stretch factor of H for each link in G but not in H. It is not difficult to show that, for any $H \subseteq G$ with a length stretch factor δ, its power stretch factor is, at most, δ^β for any graph G. In particular, a graph with a constant bounded length stretch factor must also have a constant bounded power stretch factor, although the reverse is not true. Finally, the power stretch factor has the following monotonic property: If $H_1 \subset H_2 \subset G$, then the power stretch factors of H_1 and H_2 satisfy $\rho_{H_1}(G) \ge \rho_{H_2}(G)$.

39.2.4 Low-Weight Structures

The power stretch factor previously discussed is defined for unicasting communications. However, in practice, it is also necessary to consider broadcast or multicast communications. Wan et al. [151] showed that the minimum energy cost of broadcasting or multicasting is related to the total energy cost of all links in the Euclidean minimum spanning tree, *MST*. They proved that a broadcasting method based on the Euclidean minimum spanning tree rooted at the sender uses energy no more than 12 times the minimum energy cost of any broadcasting scheme. Therefore, the network topology needs to be a low-weight structure. Given a structure G over a set of points, let $\omega(G)$ be the total length of the links in G and $\omega_\beta(G)$ be the total power needed to support all links in G, i.e., $\omega_\beta(G) = \sum_{uv \in G} \|uv\|^\beta$. Then, a structure G is called *low weight* if $\omega(G)$ is within a constant factor of $\omega(MST)$.

39.2.5 Geometric Structures

Several geometric structures have been studied recently by computational geometry scientists and network engineers. Here, we review the definitions of some of them THAT could be used in wireless sensor networking applications are reviewed. Let $G = (V,E)$ be a geometric graph defined on vertex set V with edge set E.

39.2.5.1 Minimum Spanning Tree, Relative Neighborhood Graph and Gabriel Graph

The *minimum spanning tree* of G, denoted by MST(G), is the tree belonging to E that connects all nodes and whose total edge length is minimized. MST(G) is obviously one of the sparsest connected subgraphs, but its stretch factor can be as large as $n - 1$.

The *relative neighborhood graph*, denoted by RNG(G), is a geometric concept proposed by Toussaint [148]. It consists of all edges $uv \in E$ such that there is no point $w \in V$ with edges uw and wv in E satisfying $\|uw\| < \|uv\|$ and $\|wv\| < \|uv\|$. See Figure 39.2(a) for an illustration. Notice that, if G is a directed graph, then edges uw and wv also are directed in the preceding definition, i.e., we have $\overrightarrow{uw}$ and $\overrightarrow{wv}$ instead of uw and wv.

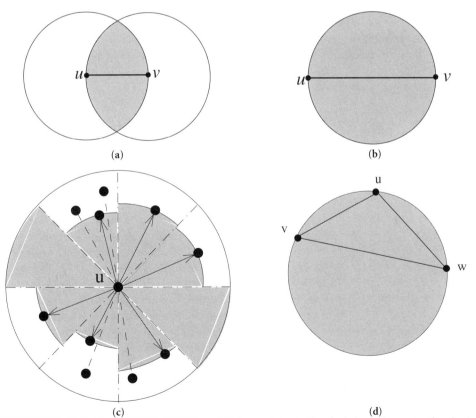

FIGURE 39.2 Definitions of RNG, GG, YG, and Del on point set. The shaded area is empty of nodes inside. (a) The lune using uv is empty for RNG; (b) the diametric circle using uv is empty for GG; (c) the shortest edge in each cone is added as a neighbor of u for Yao; and (d) the circumcircle of uvw is empty for Del.

Let *disk(u,v)* be the disk with diameter *uv*. Then, the *Gabriel graph* [53] GG(*G*) contains an edge *uv* from *G* if and only if *disk(u,v)* contains no other vertex $w \in V$ such that edges *uw* and *wv* from *G* exist. See Figure 39.2(b) for an illustration. The same holds true to the definition of RNG(*G*); if *G* is a directed graph, then edges *uw* and *wv* also are directed in the preceding definition of GG(*G*). GG(*G*) is a planar graph (that is, no two edges cross each other) if *G* is the complete graph or UDG. It is easy to show that RNG(*G*) is a subgraph of the Gabriel graph GG(*G*). For an undirected and connected graph *G*, GG(*G*) and RNG(*G*) are connected and contain the minimum spanning tree of *G*.

The Gabriel graph was used as a planar subgraph in the Face routing protocol [20, 47, 143] and the GPSR routing protocol [78], which guarantee packet delivery. Relative neighborhood graph RNG was used for efficient broadcasting (minimizing the number of retransmissions) in a one-to-one broadcasting model in [139].

39.2.5.2 Yao Graph and θ-Graph

The *Yao graph* [164], with an integer parameter k ≥ 6 denoted by $\overrightarrow{YG_k}(G)$, is defined as follows. At each node *u*, any *k* equally separated rays originated at *u* define *k* cones. In each cone, choose the shortest edge *uv* among all edges from *u*, if any there, and add a directed link $\overrightarrow{uv}$. Ties are broken arbitrarily. The resulting directed graph is called the Yao graph; see Figure 39.2(c) for an illustration. Let YG$_k$(*G*) be the undirected graph by ignoring the direction of each link in $\overrightarrow{YG_k}(G)$. If the link $\overrightarrow{vu}$ is added instead of the link $\overrightarrow{uv}$, the graph is denoted by $\overleftarrow{YG_k(G)}$, which is called the *reverse* of the Yao graph.

Some researchers have used a similar construction named the θ-graph [80, 81, 114]; the difference is that, in each cone, the edge with the shortest projection on the axis of the cone is chosen instead of the shortest edge (see Figure 39.3). Here, the axis of a cone is the angular bisector of the cone. More details are available in other texts [80, 81, 114].

All these definitions are exactly the conventional definitions [52, 79, 98, 164] when graph *G* is the completed Euclidean graph *K(V)*. MST(*V*), RNG(*V*), GG(*V*), and Yao(*V*) will be used to denote the corresponding resulting graph if *G* is the complete graph *K(V)*.

39.2.5.3 Delaunay Triangulation and Voronoi Diagram

Assume that no four vertices of *V* are cocircular. A triangulation of *V* is a *Delaunay triangulation*, denoted by *Del(V)*, if the circumcircle of each of its triangles does not contain any other vertices of *V* in its interior. A triangle is called the *Delaunay triangle* if its circumcircle is empty of vertices of *V*; see Figure 39.2(d) for an illustration. The *Voronoi region*, denoted by *Vor(p)*, of a vertex $p \in V$ is a collection of two

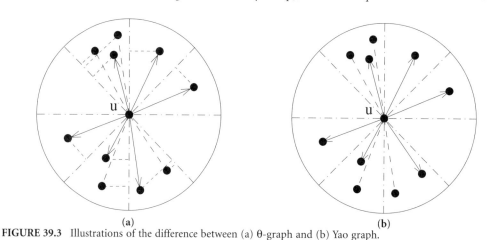

FIGURE 39.3 Illustrations of the difference between (a) θ-graph and (b) Yao graph.

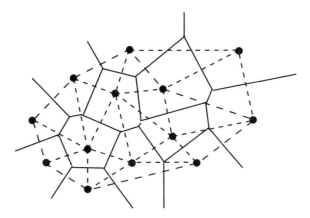

FIGURE 39.4 The Voronoi diagram and Delaunay triangulation of a set of two-dimensional nodes. The Delaunay triangulation is represented by dashed lines.

dimensional points such that every point is closer to p than to any other vertex of V. The *Voronoi diagram* for V is the union of all Voronoi regions $Vor(p)$, where $p \in V$.

The Delaunay triangulation $Del(V)$ is also the dual of the Voronoi diagram: two vertices p and q are connected in $Del(V)$ if and only if $Vor(p)$ and $Vor(q)$ share a common boundary. The shared boundary of two Voronoi regions, $Vor(p)$ and $Vor(q)$ is on the perpendicular bisector line of segment pq. The boundary segment of a Voronoi region is called the *Voronoi edge*; the intersection point of two Voronoi edges is called the *Voronoi vertex*. The Voronoi vertex is the circumcenter of some Delaunay triangle. See Figure 39.4 for an illustration of the relation between $Vor(V)$ and $Del(V)$.

39.2.5.4 Connected Dominating Set and Independent Set

In addition to these geometric structures, some graph notations will also be used in this chapter. A subset S of V is a *dominating set* if each node u in V is in S or is adjacent to some node v in S. Nodes from S are called dominators, while nodes not in S are called dominatees. A subset C of V is a *connected dominating set* (CDS) if C is a dominating set and induces a connected subgraph. Consequently, the nodes in C can communicate with each other without using nodes in $V - C$. A dominating set with minimum cardinality is called *minimum dominating set*, denoted by MDS. A connected dominating set with minimum cardinality is denoted by *minimum connected dominating set* (MCDS).

A subset of vertices in a graph G is an *independent set* if, for any pair of vertices, no edge is between them. It is a *maximal independent set* if no more vertices can be added to it to generate a larger independent set. It is a *maximum independent set* (MIS) if no other independent set has more vertices.

39.2.6 Localized Algorithms

The large numbers of sensors and the expected dynamics in sensor networks present unique challenges in network design. There are significant robustness and scalability advantages to designing applications using *localized algorithms* in which sensors only interact with other sensors in a restricted vicinity, but nevertheless collectively achieve a desired global objective (such as spanner or low weight). Estrin and colleagues [50] gave two attractive properties of localized algorithms: (1) because each node communicates only with other nodes in some neighborhood, the communication overhead *scales* well with increase in network size; and (2) for a similar reason these algorithms are *robust* to network partitions and node failures.

Specifically for topology control, it is preferred that the underlying network topology be constructed in a localized manner. Here, a distributed algorithm constructing a graph G is a *localized algorithm* if every node u can exactly decide all edges incident on u based only on the information of all nodes within constant hops of u (plus a constant number of additional nodes' information if necessary). It is easy to see that the Yao graph $YG(V)$, the relative neighborhood graph $RNG(V)$, and the Gabriel graph $GG(V)$

can be constructed locally. However, the Euclidean minimum spanning tree MST(V) and the Delaunay triangulation Del(V) cannot be constructed by any localized algorithm. The next section concerns localized algorithms that construct sparse and power-efficient network topologies.

39.3 Topology Control

Here, the power stretch factor of several sparse geometric structures for unit disk graphs, which can be used as the network topology of sensor networks, will be studied first. Notice that a trade-off can be made between sparseness of the topology and power efficiency. The power efficiency of any spanner is measured by its power stretch factor. Some geometric results on topology issues in sensor networks, such as fault tolerance and interference, will then be reviewed.

39.3.1 Bounded Degree Structures

In addition to the sparseness and spanner properties, it is also desirable that the node degree in the constructed topology be small and bounded from above by a constant. A small node degree reduces the MAC-level contention and interference, as well as help to mitigate the well-known hidden and exposed terminal problems. Therefore, this subsection reviews some bounded degree spanners.

39.3.1.1 Yao Structure

Applying Yao structure to bound node degree is a very natural idea. The Yao graph $YG_k(V)$ has length stretch factor $\dfrac{1}{1-2\sin\dfrac{\pi}{k}}$. Thus, its power stretch factor is no more than $\left(\dfrac{1}{1-2\sin\dfrac{\pi}{k}}\right)^{\beta}$. Li et al. [102] have proved a stronger result; see their work for a detailed proof of the following theorem.

Theorem 39.1. The power stretch factor of the Yao graph $YG_k(V)$ is, at most, $\dfrac{1}{1-\left(2\sin\dfrac{\pi}{k}\right)^{\beta}}$ [102].

Li and colleagues [103] also proposed to apply the Yao structure on top of the Gabriel graph structure (the resulting graph is denoted by $\overrightarrow{YGG_k}(V)$), and apply the Gabriel graph structure on top of the Yao structure (the resulting graph is denoted by $\overrightarrow{GYG_k}(V)$ $O(\log n)$. These structures are sparser than the Yao and Gabriel graph structures and they still have a constant bounded power stretch factor. The two structures are connected graphs if the UDG is connected, which can be proved by showing that RNG is a subgraph of both structures.

The two-phased approach by Wattenhofer et al. [156] consists of a variation of the Yao graph followed by a variation of the Gabriel graph. They tried to prove that the constructed spanner has a constant power stretch factor and the node degree is bounded by a constant. Unfortunately, their proof of the constant power stretch factor has some discrepancies and their result is erroneous, which was discussed in detail in Li et al. [102].

Li and coworkers [93] have proposed a structure similar to the Yao structure for topology control. Each node u finds a power $p_{u,\alpha}$ such that, in every cone of degree α surrounding u, there is some node that u can reach with power $p_{u,\alpha}$. Here, nevertheless, it is assumed that a node is reachable from u by the maximum power in that cone. Then the graph G_α contains all edges uv such that u can communicate with v using power $p_{u,\alpha}$. These authors proved that, if $\alpha \leq 5\pi/6$ and the UDG is connected, then graph G_α is a connected graph. On the other hand, if $\alpha > 5\pi/6$, they showed that the connectivity of G_α is not guaranteed by giving a counterexample [93].

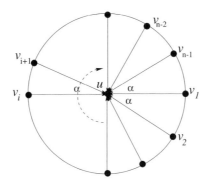

FIGURE 39.5 Node u has degree (or in-degree) $n-1$.

Although the directed graphs, $\overrightarrow{YG_k}(V)$, $\overrightarrow{GYG_k}(V)$ and $\overrightarrow{YGG_k}(V)$ have a bounded power stretch factor and a bounded out-degree k for each node, some nodes may have very large in-degrees. The node configuration given in Figure 39.5 will result in a very large in-degree for node u. Bounded out-degree offers advantages when several routing algorithms are applied. However, unbounded in-degree at node u will often cause large overhead at u. Therefore, it is often imperative to construct a sparse network topology such that the in-degree and the out-degree are bounded by a constant while still power efficient.

39.3.1.2 Sink Structure

Arya and colleagues [7] gave an ingenious technique to generate a bounded degree graph with constant length stretch factor. Li et al. [102] applied the same technique to construct a sparse network topology with a bounded degree and a bounded power stretch factor from YG(V). The technique is to replace the directed star consisting of all links toward a node u by a directed tree $T(u)$ of a bounded degree with u as the sink; tree $T(u)$ is constructed recursively. The algorithm is as follows.

Algorithm 39.1. Constructing YG*:

1. Construct the graph $\overrightarrow{YG_k}(V)$. Each node u will have a set of in-coming nodes $I(u) = \{v \mid \overrightarrow{vu} \in \overrightarrow{YG_k}(V)\}$.
2. For each node u, use the following algorithm tree($u,I(u)$) to build tree $T(u)$.

Algorithm 39.2. Constructing T(u) tree(u,I(u)):

1. To partition the unit disk centered at u, choose k equal-sized cones centered at u: $C_1(u)$, $C_2(u)$, ..., $C_k(u)$.
2. Node u finds the nearest node $y_i \in I(u)$ in $C_i(u)$, for $1 \le i \le k$, if there is any. Link $\overrightarrow{y_i u}$ is added to $T(u)$ and y_i is removed from $I(u)$. For each cone $C_i(u)$, if $I(u) \cap C_i(u)$ is not empty, call tree(y_i, $I(u) \cap C_i(u)$) and add the created edges to $T(u)$.

Figure 39.6(a) illustrates a directed star centered at u and Figure 39.6(b) shows the directed tree $T(u)$ constructed to replace the star with $k = 8$. The union of all trees $T(u)$ is called the *sink structure* $\overrightarrow{YG_k^*}(V)$.

Node u constructs the tree $T(u)$ and then broadcasts the structure of $T(u)$ to all nodes in $T(u)$. Because the total number of edges in the Yao structure is, at most, $k \cdot n$, where k is the number of cones divided, the total number of edges of $T(u)$ of all nodes u is also, at most, $k \cdot n$. Thus, the total communication cost of broadcasting the $T(u)$ to all its neighbors is still, at most, $k \cdot n$. Recall that k is a small constant.

The algorithm uses a directed tree $T(u)$ to replace the directed star for each node u. Therefore, if nodes u and v are connected by a path in $\overrightarrow{YG_k}$, they are also connected by a path in $\overrightarrow{YG_k^*}$. It is already known that $\overrightarrow{YG_k}$ is strongly connected if UDG(V) is connected, so does $\overrightarrow{YG_k^*}$.

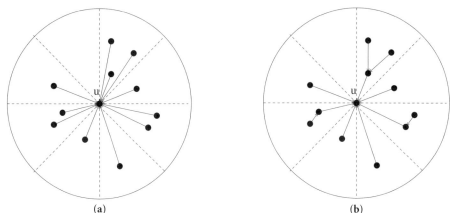

FIGURE 39.6 (a) Star formed by links to u; (b) directed tree $T(u)$ sunk at u.

Theorem 39.2. The power stretch factor of the graph $\overrightarrow{YG}^*_k(V)$ is at most $\left(\dfrac{1}{1-\left(2\sin\frac{\pi}{k}\right)^\beta}\right)^2$. The maximum degree of the graph $\overrightarrow{YG}^*_k(V)$ is, at most, $(k+1)^2 - 1$. The maximum out-degree is k [102].

The sink structure and the Yao graph structure need not have the same number of cones, and the cones do not need to be aligned. For setting up a power-efficient wireless networking, each node u finds all its neighbors in $YG_k(V)$, which can be done in linear time proportional to the number of nodes within its transmission range.

39.3.1.3 YaoYao Structure

Li and colleagues [103] have proposed an algorithm that constructs a sparse and power-efficient topology. Assume that each node v_i of V has a unique identification number $ID(v_i) = i$. The identity of a directed link $\overrightarrow{uv}$ is defined as $ID(\overrightarrow{uv}) = (\|uv\|, ID(u), ID(v))$.

Node u chooses a node v from each cone, if any exists, so the directed link $\overrightarrow{vu}$ has the smallest $ID(\overrightarrow{vu})$ among all directed links $\overrightarrow{wu}$ in $\overrightarrow{YG}_k(V)$ in that cone. The union of all chosen directed links is the final network topology, denoted by $\overrightarrow{YY}_k(V)$. If the directions of all links are ignored, the graph is denoted as $YY_k(V)$.

Theorem 39.3. Graph $\overrightarrow{YY}_k(V)$ is strongly connected if $UDG(V)$ is connected and $k > 6$ [103].

Wang and Li [153] proved that $\overrightarrow{YY}_k(V)$ is a spanner in a civilized graph. Here, a unit disk graph is a civilized graph if the distance between any two nodes in this graph is larger than a positive constant λ. Hunt and coworkers [68] called the civilized unit disk graph the λ-precision unit disk graph. Notice the sensor devices in wireless sensor networks cannot be too close or overlapped. Thus, it is reasonable to model the wireless sensor networks as a civilized unit disk graph.

Theorem 39.4. The power stretch factor of the directed topology $\overrightarrow{YY}_k(V)$ is bounded by a constant in a civilized graph [153].

Li and colleagues' [103] experimental results showed that this sparse topology has a small power stretch factor in practice. Li et al. [103] and Wang and Li [153] conjectured that $\overrightarrow{YY}_k(V)$ also has constants bounded length spanning ratio and power stretch factor, theoretically in any unit disk graph. Recently,

Jia et al. [72] claimed to have proved that $\overrightarrow{YY}_k(V)$ also has a constant bounded power stretch factor, theoretically in general graphs.

39.3.1.4 Symmetric Yao Graph

Li and Stojmenovic [99] and Li et al. [100] have also considered another undirected structure, called *symmetric Yao graph* $YS_k(V)$, which guarantees that the node degree is, at most, k. Each node u divides the region into k equal angular regions centered at the node and chooses the closest node in each region, if any. An edge uv is selected to graph $YS_k(V)$ if and only if directed edges $\overrightarrow{uv}$ and $\overrightarrow{vu}$ are in the Yao graph $\overrightarrow{YG}_k(V)$. Then it is obvious that the maximum node degree is k.

Theorem 39.5. The graph $YS_k(V)$ is strongly connected if $UDG(V)$ is connected and $k \geq 6$ [99, 153].

The experiment by Li et al. also showed that it has a small power stretch factor in practice. However, Grunewald and coworkers [59] recently showed that $YS_k(V)$ is not a spanner theoretically. See Li [96] for more detail.

39.3.2 Planar Structures

Some routing algorithms ask the network topology be planar, such as right-hand routing: *greedy perimeter stateless routing* (GPSR) [78]; *greedy face routing* (GFR) [20]; *adaptive face routing* (AFR) [87]; and *greedy other adaptive face routing* (GOAFR) [88]. Therefore, it is necessary for the constructed topology to be a planar graph, i.e., no two edges cross each other in the graph. In this subsection, we study three planar structures that can be used in sensor networks are studied.

39.3.2.1 RNG and GG

Remember that the relative neighborhood graph and Gabriel graph are planar graphs, and they can be constructed easily using localized methods. Because the relative neighborhood graph has a length stretch factor as large as $n-1$, obviously its power stretch factor is at most $(n-1)^\beta$. Li et al. [102] have showed that it is actually $n-1$. The Gabriel graph has a length stretch factor between $\frac{\sqrt{n}}{2}$ and $\frac{4\pi\sqrt{2n-4}}{3}$ [17]. Wang and colleagues [152] have showed that it is exactly $\sqrt{n-1}$; then its power stretch factor is at most $\left(\frac{4\pi\sqrt{2n-4}}{3}\right)^\beta$. Li et al. [102] have also proved that the power stretch factor of any Gabriel graph is one.

RNG and GG do not have bounded node degree. The node configuration given in Figure 39.5 will also result in a very large degree for node u.

39.3.2.2 Localized Delaunay Triangulation

Given a set of nodes V, it is well known that the Delaunay triangulation $Del(V)$ is a planar t-spanner of the completed graph $K(V)$. However, it is not appropriate to require the construction of the Delaunay triangulation in the wireless communication environment because of the possibly massive communications it requires. Given a set of points V, let $UDel(V)$ be the graph of removing all edges of $Del(V)$ that are longer than one unit, i.e., $UDel(V) = Del(V) \cap UDG(V)$. Li and colleagues [98] have considered the *unit Delaunay triangulation* $UDel(V)$ for planar spanner of UDG, which is a subset of the Delaunay triangulation. They have proved that $UDel(V)$ is a t-spanner of the unit disk graph $UDG(V)$.

Theorem 39.6. For any two vertices u and v of V, $\|\Pi_{UDel(V)}(u,v)\| \leq \frac{1+\sqrt{5}}{2}\pi \cdot \|\Pi_{UDG(V)}(u,v)\|$ [98].

Keil and Gutwin [80] have showed that the Delaunay triangulation is a t-spanner for a constant $t \approx 2.42$. This was proved by induction on the order of the lengths of all pair of nodes (from the shortest to the longest). It can be shown that the path connecting nodes u and v constructed by their method also satisfies that all edges of that path are shorter than $\|uv\|$. Consequently, the unit Delaunay triangulation UDel(V) is a $\frac{4\sqrt{3}}{9}\pi$-spanner of the unit disk graph UDG(V).

Li et al. [98] have given a localized algorithm that constructs a sequence of graphs, called *localized Delaunay* $LDel^{(k)}(V)$, that are supergraphs of UDel(V). Some definitions are necessary before the algorithm is presented. Triangle $\triangle uvw$ is called a *k-localized Delaunay triangle* if the interior of the circumcircle of $\triangle uvw$, denoted by $disk(u,v,w)$ does not contain any vertex of V that is a k-neighbor of u, v, or w, and all edges of the triangle $\triangle uvw$ have length no more than one unit. The *k-localized Delaunay graph* over a vertex set V, denoted by $LDel^{(k)}(V)$, has exactly all Gabriel edges in UDG and edges of all k-localized Delaunay triangles.

When it is clear from the context, the integer k will be omitted in the notation of $LDel^{(k)}(V)$. As shown in [98], the graph $LDel^{(1)}(V)$ may contain some edges intersecting. On the other hand, $LDel^{(2)}(V)$ is a planar graph.

Theorem 39.7. $LDel^{(k)}(V)$ *is a planar graph for any* $k \geq 2$ [98].

Notice that, although $LDel^{(1)}(V)$ is not a planar graph, the following theorem, proved by Li and colleagues [98], guarantees that it is sparse.

Theorem 39.8. Graph $LDel^{(1)}(V)$ *has thickness 2.*

Although the graph UDel(V) is a t-spanner for UDG(V), how to construct it locally is not known. One can construct $LDel^{(2)}(V)$, which is guaranteed to be a planar spanner of UDel(V), but a total communication cost of a simple approach is $O(m \log n)$ bits, where m is the number of edges in UDG(V) and could be as large as $O(n^2)$. In order to reduce the total communication cost to $O(n \log n)$ bits, $LDel^{(2)}(V)$ is not constructed; instead, a planar graph PLDel(V) is extracted from $LDel^{(1)}(V)$. Li and colleagues provided a novel algorithm to construct $LDel^{(1)}(V)$ by using linear communications and then make it planar in linear communication cost. The final graph still contains UDel(V) as a subgraph. Thus, it is a t-spanner of the unit-disk graph UDG(V). In the following, the order of three nodes in a triangle is immaterial.

Algorithm 39.3 Localized unit delaunay triangulation:

1. Each wireless node u broadcasts its identity and location and listens to messages from other nodes.
2. Assume that node u gathers the location information of $N_1(u)$. It computes the Delaunay triangulation Del($N_1(u)$) of its 1-neighbors $N_1(u)$, including u.
3. For each edge, uv, of Del($N_1(u)$), let $\triangle uvw$ and $\triangle uvz$ be two triangles incident on uv. Edge uv is a Gabriel edge if angles $\angle uvw$ and $\angle uzv$ are less than $\pi/2$. Node u marks all Gabriel edges uv, which will never be deleted.
4. Each node u finds all triangles $\triangle uvw$ from Del($N_1(u)$) such that all three edges of $\triangle uvw$ have length of, at most, one unit. If angle $\angle wuv \geq \frac{\pi}{3}$, node u broadcasts a message "proposal (u,v,w)" to form a 1-localized Delaunay triangle $\triangle uvw$ in $LDel^{(1)}(V)$, and listens to the messages from other nodes.
5. When a node u receives a message "proposal (u,v,w)," u accepts the proposal of constructing $\triangle uvw$ if $\triangle uvw$ belongs to the Delaunay triangulation Del($N_1(u)$) by broadcasting message "accept (u,v,w); otherwise, it rejects the proposal by broadcasting message" reject (u,v,w).
6. A node u adds the edges uv and uw to its set of incident edges if the triangle $\triangle uvw$ is in the Delaunay triangulation Del($N_1(u)$) and v and w have sent "accept (u,v,w)" or "proposal (u,v,w)."

It has been proved that the graph constructed by the preceding algorithm is $LDel^{(1)}(V)$. Indeed, for each triangle $\triangle uvw$ of $LDel^{(1)}(V)$, one of its interior angles is at least $\pi/3$ and $\triangle uvw$ is in Del($N_1(u)$), Del($N_1(v)$), and Del($N_1(w)$). Thus, one of the nodes among $\{u,v,w\}$ will broadcast the message proposal(u,v,w) to form a 1-localized Delaunay triangle $\triangle uvw$.

Because Del($N_1(u)$) is a planar graph, and a proposal is made only if $\angle wuv \geq \frac{\pi}{3}$, node u broadcasts at most six proposals; at most two nodes reply to each proposal. Therefore, the total communication cost is $O(n \log n)$ bits. The preceding algorithm also shows that $LDel^{(1)}(V)$ has $O(n)$ edges, which is knows from Theorem 39.8. Putting together the preceding arguments yields [98]:

Theorem 39.9. Algorithm 39.3 constructs $LDel^{(1)}(V)$ with total communication cost $O(n \log n)$ bits.

The algorithm to extract from $LDel^{(1)}(V)$, a planar subgraph, is then reviewed.
Algorithm 39.4. Planarize $LDel^{(1)}(V)$:

1. Each wireless node u broadcasts the Gabriel edges incident on u and the triangles Δuvw of $LDel^{(1)}(V)$ and listens to the messages from other nodes.
2. Assume node u gathered the Gabriel edge and 1-local Delaunay triangle information of all nodes from $N_1(u)$. For two intersected triangles, Δuvw and Δxyz, known by u, node u removes the triangle Δuvw if its circumcircle contains a node from $\{x,y,z\}$.
3. Each wireless node u broadcasts all the triangles incident on u that it has not removed in the previous step and listens to broadcasting by other nodes.
4. Node u keeps the edge uv in its set of incident edges if it is a Gabriel edge, or if there is a triangle Δuvw such that u, v, and w have all announced that they have not removed the triangle Δuvw in step 2.

The graph extracted by the preceding algorithm is denoted by $PLDel(V)$. Note that any triangle of $LDel^{(1)}(V)$ not kept in the last step of the planarization algorithm is not a triangle of $LDel^{(2)}(V)$, and therefore $PLDel(V)$ is a supergraph of $LDel^{(2)}(V)$. Thus,

$$UDel(V) \subseteq LDel^{(2)}(V) \subseteq PLDel(V) \subseteq LDel^{(1)}(V)$$

Similar to the proof that $LDel^{(2)}(V)$ is a planar graph, it has been shown that the algorithm generates a planar graph. The total communication cost to construct the graph $PLDel(V)$ is a $O(\log n)$ times the number of edges of the graph $LDel^{(1)}(V)$, which by Theorem 39.8 is $O(n)$. Putting all the preceding arguments and Theorem 39.6 together yields:

Theorem 39.10. $PLDel(V)$ is planar $\frac{4\sqrt{3}}{9}\pi$-spanner of $UPG(V)$, and can be constructed with total communication cost $O(n \log n)$ bits.

Li and colleagues [98] cannot construct $LDel^{(2)}$ in $O(n)$ messages due to the difficulty of collecting the two-hop neighbors for every node in $O(n)$ messages. Computing the two-hop neighborhood is not trivial because the UDG can be dense. The broadcast nature of the communication in ad hoc wireless networks is, however, very useful when computing local information. Recently, Gruia [25] proposed an approach (using $O(n)$ messages total) based on the specific connected dominating set introduced by Alzoubi, Wan, and Frieder [2]. This connected dominating set is based on a maximal independent set (MIS). In the algorithm, each node uses its adjacent nodes in the MIS to broadcast relevant information over a larger area. Listening to the information about other nodes broadcast by the MIS nodes enables a node to compute its two-hop neighborhood. For detailed algorithms and proofs, see Calinescu [25]. Using this approach, one can build $LDel^{(2)}$ in $O(n)$ messages; such an algorithm proposed by Wang and Li [155].

39.3.2.3 Partial Delaunay Triangulation

Stojmenovic and Li [99] have also proposed a geometry structure, namely, the partial Delaunay triangulation (PDT), which can be constructed in a localized manner. Partial Delaunay triangulation contains a Gabriel graph as its subgraph, and itself is a subgraph of the Delaunay triangulation — more precisely, the subgraph of the unit Delaunay triangulation $UDel(V)$. The algorithm for the construction of PDT goes as follows.

Let u and v be two neighboring nodes in the network. Edge uv belongs to $Del(V)$ if and only if a disk exists, with u and v on its boundary, that does not contain any other point from the set V. First, test whether $disk(u,v)$ contains any other node from the network. If it does not, the edge belongs to GG and therefore to PDT. If it does, check whether nodes exist on both sides of line uv or on only one side. If both sides of the line contain nodes from the set inside $disk(u,v)$, then uv does not belong to $Del(V)$.

Suppose now that only one side of line *uv* contains nodes inside the circle *disk(u,v)*, and let *w* be one such point that maximizes the angle ∠*uwv*. Let α = ∠*uwv*. Consider now the largest angle ∠*uxv* on the other side of the mentioned circle *disk(u,v)*, where *x* is a node from the set *S*. If ∠*uwv* + ∠*uxv* > π, then edge *uv* is definitely not in the Delaunay triangulation *Del(V)*.

The search can be restricted to common neighbors of *u* and *v*, if only one-hop neighbor information is available, or to neighbors of only one of the nodes if two-hop information (or exchange of the information for the purpose of creating PDT is allowed) is available. Then, whether edge *uv* is added to PDT is based on the following procedure.

Assume only $N_1(u)$ is known to *u* and that one node *w* from $N_1(u)$ is inside *disk(u,v)* with the largest angle ∠*uwv*. Edge *uv* is added to PDT if the following conditions hold: (1) no node from $N_1(u)$ lies on the different side of *uv* with *w* and inside the circumcircle passing through *u*, *v*, and *w*; and (2) sin α > $\frac{d}{R}$, where *R* is the transmission radius of each wireless node; *d* is the diameter of the circumcircle *disk(u,v,w)*; and α = ∠*uwv* (here α ≥ $\frac{\pi}{2}$).

Assume only one-hop neighbors are known to *u* and *v*, and one node *w* from $N_1(u) \cup N_1(v)$ is inside *disk(u,v)* with the largest angle ∠*uwv*. Edge *uv* is added to PDT if the following conditions hold: (1) no node from $N_1(u) \cup N_1(v)$ lies on the different side of *uv* with *w* and inside the circumcircle passing *u*, *v*, and *w*; and (2) sin α > $2\frac{d}{R}$, where *R* is the transmission radius of each wireless node and α = ∠*uwv*.

Obviously, PDT is a subgraph of UDel(*V*). It is not difficult to construct an example so that the spanning ratio of the partial Delaunay triangulation could be very large.

39.3.2.4 Restricted Delaunay Graph

Gao et al. [55] also proposed another structure, called *restricted Delaunay graph* (RDG) and showed that it has good spanning ratio properties and can be maintained locally. A restricted Delaunay graph of a set of points in the plane is a planar graph and contains all the Delaunay edges with length of, at most, one. In other words, these authors call *any* planar graph containing UDel(*V*) a restricted Delaunay graph. They described a distributed algorithm to construct a RDG such that, at the end of the algorithm, each node *u* maintains a set of edges *E(u)* incident to *u*. Those edges *E(u)* satisfy that

- Each edge in *E(u)* has length of, at most, one unit.
- The edges are consistent, i.e., an edge *uv* ∈ *E(u)* if and only if *uv* ∈ *E(v)*.
- The graph obtained is planar.
- The graph UDel(*V*) is in the union of all edges *E(u)*.

The algorithm works as follows. First, each node *u* acquires the position of its one-hop neighbors $N_1(u)$ and computes the Delaunay triangulation Del($N_1(u)$) on $N_1(u)$, including *u* itself. In the second step, each node *u* sends Del($N_1(u)$) to all of its neighbors. Let $E(u) = \{uv | uv \in \text{Del}(N_1(u))\}$. For each edge *uv* ∈ *E(u)*, and for each *w* ∈ $N_1(u)$, if *u* and *v* are in $N_1(w)$ and *uv* ∉ Del($N_1(u)$), then node *u* deletes edge *uv* from *E(u)*.

When these steps are finished, the resulting edges *E(u)* satisfy the four properties listed earlier. However, unlike the local Delaunay triangulation, the computation cost and communication cost of each node needed to obtain *E(u)* are not optimal within a small constant factor.

39.3.3 Bounded Degree, Planar Structures

The structures discussed so far have bounded degree, are planar, or are spanners, but none has all these three properties together. One recent result [155] can construct a bounded degree planar spanner in a localized manner (total communication cost is $O(n \log n)$ bits). No localized method is known before this result for constructing a planar spanner with bounded node degree. This method rigorously combines (localized) Delaunay triangulation LDel$^{(2)}(V)$ and the ordered Yao structure [21, 164].

Algorithm 39.5. Localized construction of planar spanner with bounded degree for UDG(V):

1. Compute the planar localized Delaunay triangulation $LDel^{(2)}(V)$ (using the method in Calinescu [25] to collect the location information of $N_2(u)$), so that every node u knows all its neighbors $N_{LDel^{(2)}}(u)$ and its node degree $d(u)$ in $LDel^{(2)}(V)$. Assume a synchronized method is used to collect $N_{LDel^{(2)}}(u)$ for every node u.
2. Build a local order π of V as follows: (Every node u initializes $\pi_u = 0$, i.e., unordered.)
   (a) If node u has $\pi_u = 0$ and $d(u) \leq 5$, then u queries* each node v, from its unordered neighbors, for the current degree $d(v)$. If node u has the smallest ID among all unordered neighbors v with $d(v) \leq 5$, node u sets

   $$\pi_u = \max\{\pi_v \mid v \in N_{LDel^{(2)}}(u)\} + 1,$$

   and broadcasts π_u to its neighbors $N_{LDel^{(2)}}(u)$.
   (b) If node u receives a message from its neighbor v saying that $\pi_v = k$, it updates its $d(u) = d(u) - 1$ and also updates the order π_v stored locally. So $d(u)$ represents how many neighbors are not ordered so far. If node u finds that $d(u) \leq 5$ and $\pi_u = 0$, it goes to step 2 (a). When node u finds that $d(u) = 0$ and $\pi_u > 0$, it can go to step 3.
3. Build structures based on local order π as follows (initialize all nodes unprocessed):
   (a) If an unprocessed node u has the highest local order in its unprocessed neighbors N_u in $LDel^{(2)}(V)$, let k be the number of processed neighbors** of u in $LDel^{(2)}(V)$. Assume that $v_1, v_2, \ldots, v_k$ is the processed neighbors of u in $LDel^{(2)}(V)$ (see Figure 39.7). Node u divides its transmission range into k *open* sectors cut by the rays from u to these processed neighbors. Then divide each sector into a minimum number of *open* cones of degree at most α with $\alpha \leq \pi/3$. For each cone, let $s_1, s_2, \ldots, s_m$ be the ordered unprocessed neighbors of u in $N_{LDel^{(2)}}(u)$. For this cone, node u first adds an edge us_i, where s_i is the nearest neighbor among $s_1, s_2, \ldots, s_m$. Node u then tells $s_1, s_2, \ldots, s_m$ to add all the edges $s_j s_{j+1}$, $1 \leq j \leq m$. Node u marks itself processed and tells all nodes in $N_{LDel^{(2)}}(u)$ that it is processed.
   (b) If an unprocessed node v receives a message for adding edge vv' from its neighbor u, it adds edge vv'.

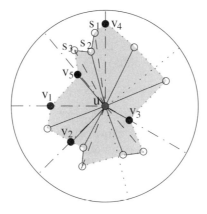

FIGURE 39.7 Constructing planar spanner with bounded degree for UDG(V): process node u.

---

*If some unordered neighbor with $d(v) \leq 5$ has smaller ID, that query round is called a *failed round*. Node u performs a new round of queries only if it finds that the number of its unordered neighbors has been reduced ($d(u)$ has reduced in step 2 (b)). Thus, there are at most five rounds of queries.

**There are, at most, five processed neighbors because graph $LDel^{(2)}(V)$ is planar.

4. When all nodes are processed, the final network topology is denoted by BPS(V).

Notice that *open* sectors are used in the algorithm, which means that one does not consider adding the edges on the boundaries (any edge involved previously processed neighbors). For example, in Figure 39.7, the cones do not include any edges uv_i. This guarantees the algorithm does not add any edges to node v_i after v_i has been processed. This approach bounds the node degree.

Theorem 39.11. The maximum node degree of the graph BPS(V) is, at most $19 + \left\lceil \frac{2\pi}{\alpha} \right\rceil$ [155].

For example, when $\alpha \leq \pi/3$, the maximum node degree is, at most, 25. Notice that the ordering computed by this method is not a total ordering. Some nodes may have the same order. However, no two neighboring nodes in $LDel^{(2)}(V)$ receive the same order. Thus, after all nodes are ordered, the algorithm will process all nodes. Observe that the algorithm does not process two neighboring nodes at the same time.

Assume that two nodes, say u and v, are processed at the same time. Remember that a node is processed only if it has the highest ordering among its unprocessed neighbors. Thus, nodes u and v must receive the same order, i.e., $\pi_u = \pi_v$, which is impossible in the ordering method.

The algorithms in Prosenjit et al. [22] and Li and Wang [105] always add the edges in the Delaunay triangulation to construct a bounded degree planar spanner for a set of points. Thus, the planarity of the final structure is straightforward. However, Algorithm 39.5 may add some edges (such as edges $s_i s_{i+1}$ added in step 4(b)) that do not belong to the UDel(V). Therefore, the proof of the following theorem is more complex.

Theorem 39.12. Graph BPS(V) is a planar graph [155].

Then Li and Wang prove that the structure is also a spanner.

Theorem 39.13. Graph BPS(V) is a t-spanner, where [155]

$$t = \max\left\{\frac{\pi}{2}, \pi \sin\left(\frac{\alpha}{2} + 1\right)\right\} \cdot C_{del}.$$

Here C_{del} is the spanning ratio of Delaunay triangulation.

For example, when $\alpha = \pi/3$, the spanning ratio is at most $\left(\frac{\pi}{2} + 1\right) \cdot C_{del}$; when $\alpha = 2\arcsin\left(\frac{1}{2} - \frac{1}{\pi}\right) \approx 20.9°$, the spanning ratio is at most $\frac{\pi}{2} \cdot C_{del}$. One expects to improve the bound on the spanning ratio further by using the following property: all such Delaunay neighbors s_i are inside the circumcircle of the triangle uvv'.

Theorem 39.14. Algorithm 39.5 uses, at most, O(n) messages, where each message has O(log n) bits [155].

PROOF. Calinescu has shown [25] that the two-hop neighbor information can be collected for all nodes using total $O(n)$ messages. The communication cost of building $LDel^{(2)}$ is $O(n)$ because every node only needs to propose, at most, six triangles and each propose is replied by two nodes.

The second step (local ordering) takes $O(n)$ messages because every node only queries, at most, five rounds and at the ith round of query the node sends at most $6-i$ query messages. For each query, only the queried node replies. After it has ordered, it broadcasts once to inform its neighbors.

The third step (bounded degree) also takes $O(n)$ messages because every node only broadcasts twice: (1) to tell its neighbors to add some edges; and (2) to claim that it is processed. The total messages of telling neighbors to add some edges is $O(n)$ because the total of added edges is $O(n)$ from the planar property of the final topology. Thus, the total communication cost is bounded by $O(n)$.

It is easy to show that the computation cost of each node is, at most, $O(d_2 \log d_2)$, where d_2 is the number of its two-hop neighbors in UDG. This can be improved to $O(d_1 \log d_1 + d_2)$, where d_1 is the

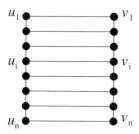

FIGURE 39.8 An instance of wireless sensor nodes for which every network structure described previously (except MST) has an arbitrarily large total weight.

number of its one-hop neighbors in UDG. The improvement is based on the fact that one only needs the triangles $\triangle wuv$ in $LDel^{(2)}(V)$ that have angle $\angle wuv \geq \pi/3$. All such triangles are definitely in $LDel^{(1)}(V)$. Thus, the Delaunay triangulation $Del(N_1(u))$ can be constructed instead. Then, check each candidate triangle $\triangle wuv$ from $LDel^{(1)}(V)$ to see if it contains any node from $N_2(u)$ inside its circumcircle. If it does not, then it belongs to $Del(N_2(u))$.

Observe that, after each node u collects the two-hop neighbors $N_2(u)$, the algorithms can be performed asynchronously. However, collecting $N_2(u)$ needs synchronized communication because, otherwise, a node cannot determine if it has already collected $N_2(u)$.

39.3.4 Bounded Degree, Planar, Low-Weight Structures

Remember that low weight is also a desirable feature for network topology in sensor networks. However, the total weight of any graph structures mentioned earlier (except MST) could be arbitrarily larger than MST theoretically [103, 104]. Figure 39.8 gives such an example of wireless sensor nodes. Here, $\|u_iv_i\| = 1$ and $\|u_iu_{i+1}\| = \|v_iv_{i+1}\| = \varepsilon$ for a very small positive real number ε. The graph shown in the example is the relative neighborhood graph $RNG(V)$. It is easy to show that

$$\frac{\sum_{e \in RNG}\|e\|^\beta}{\sum_{e \in MST}\|e\|^\beta} = \frac{n+2(n-1)\varepsilon^\beta}{1+2(n-1)\varepsilon^\beta} \to 0$$

when $\varepsilon \to 0$. Notice that all other graph structures (except MST) contain RNG as a subgraph for this node configuration. It then implies the previous claim.

This section discusses how to design algorithms achieving low weight (possibly) in addition to some other properties such as spanner, bounded degree, and planar. Unfortunately, until now, no efficient localized algorithm could achieve all three properties. Arya et al. [7, 8] gave a centralized algorithm to construct a spanner with bounded node degree; the total edge length is no more than a constant factor of that of $MST(V)$. However, it is very complicated to transform their algorithms to a distributed algorithm and the spanner is not guaranteed to be a planar graph.

39.3.4.1 Centralized Low-Weight Bounded Degree Planar Spanners

Recently, Bose et al. [22] proposed an algorithm that constructs a bounded degree and planar spanner for a given points set V. They showed that the length stretch factor of the final graph is $\frac{(\pi+1)2\pi}{((3\cos\pi)/6)(1+\varepsilon)}$ and node degree is at most 27. The running time of their algorithm is $O(n \log n)$. However, it is impossible for their method to have a localized evenly distributed version because they use BFS and many operations on polygons (such as degree-3 partitions). Notice that breadth-first search may take $O(n^2)$ communications.

Borrowing some ideas from their method, Li and Wang [105] proposed another method for constructing a low-weight bounded degree planar spanner.

1. First, compute the Delaunay triangulation of a set V of n nodes, $Del(V)$. Let $N_{Del}(u)$ be the neighbors of node u in the Delaunay triangulation $Del(V)$, and d_u be the degree of node u in $Del(V)$. By proper data structure, $N_{Del}(u)$ and du can be achieved in time $O(n)$.
2. Find an order π of V as follows. Let $G_1 = Del(V)$ and $d_{G,u}$ be the node degree of u in graph G. Find the node u_1 with the smallest value of $(d_{G_1,u}, ID(u))$; let $\pi_{u_1} = n$. Then, remove u_1 and its adjacent edges from G_1; the remaining graph G_2 is still a planar graph. Find the node u_2 with smallest value of $(d_{G_2,u}, ID(u))$; let . Repeat this procedure until G_n only has one node u_n; let $\pi_{u_n} = 1$. Let P_v denote the predecessors of v in π, i.e., $P_v = \{u \in V : \pi_u < \pi_v\}$. Because G_i is always a planar graph, the smallest value of $d_{G_i,u}$ is at most 5. Then, in ordering π, node u at most has five edges to its predecessors P_u in $Del(V)$.
3. Let E be the edge set of $Del(V)$ and E' be the edge set of the desired spanner. Initialize E' to be the empty set and all nodes in V are unprocessed. Then, for each node u in V, following the increasing order π, run the following steps to add some edges from E to E' (only the Delaunay neighbors $N_{Del}(u)$ of u are considered):
   (a) Use $v_1, v_2, \ldots, v_k$ to denote the predecessors of node u. Notice that u can have, at most, five edges to its predecessors (processed Delaunay neighbors) in E, i.e., $k \leq 5$. Then, $k \leq 5$ *open* sectors are at node u whose boundaries are rays emanated from u to the processed neighbors v_i of u in $Del(V)$. For each such sector at u, divide it into a minimum number of *open* cones of degree at most α, where $\alpha \leq \pi/2$ is a parameter.
   (b) For each such cone, let $s_1, s_2, \ldots, s_m$ be the geometrically ordered neighborhood $N_{Del}(u)$ of u in this cone. That is, $s_1, s_2, \ldots, s_m$ are all *unprocessed* nodes connected by some edges of E to u in this cone. For this cone, first add the shortest edge in E that is connected to u to the edge set E', then add to E' all the edges (s_j, s_{j+1}), $1 \leq j \leq m$.
   (c) Mark node u processed.
4. Repeat this procedure in the increasing order of π, until all nodes are processed. The final graph is formed by all edges in E'.
5. Run the greedy spanner algorithm [60] to bound the weight of the graph.

The following theorem was proved:

Theorem 39.15. Given a set V *of* n *points in a two-dimensional plane, the above* $O(n \log n)$*-time algorithm constructs a graph* [105]

- That is planar
- That is a t-*spanner, for* $t = \max\left\{\frac{\pi}{2}, \pi\sin\frac{\alpha}{2}+1\right\} \cdot C_{del}(1+\varepsilon)$
- In which each point of V has degree, at most, of $19 + \left\lceil\frac{2\pi}{\alpha}\right\rceil$
- *Whose total edge weight is bounded from above by a constant factor of the weight of the Euclidean minimum spanning tree of* V. *Here the constant factor depends on* ε.

Here $0 < \alpha < \pi/2$ is an adjustable parameter.

One can build Delaunay triangulation in $O(n \log n)$ and do ordering in time $O(n \log n)$ (using heap for the ordering based on degrees), and Yao structure in $O(n)$ (each edge is processed at most constant times and there are $O(n)$ edges to be processed). Thus, the time complexity of the algorithm is $O(n \log n)$, the same as the method by Bose et al. [22]. However, this algorithm has a smaller bounded node degree, and (more importantly) has potential to become a localized version for wireless sensor network application. The only problem here is the last step: greedy method cannot be performed in a local way.

39.3.4.2 Localized Low-Weight Bounded Degree Planar Structures

Recently, Li et al. [97, 107] proposed three localized structures that are low weight, planar, and have bounded node degree. However, they are not spanners.

39.3.4.2.1 Structure Based on RNG′

Li [97] gave the first localized method to construct a structure H with weight $O(\omega(MST))$ using total $O(n)$ local-broadcast messages. The method is based on a modified relative neighborhood graph. Notice that, traditionally, the relative neighborhood graph will always select an edge uv even if some node is on the boundary of $lune(u,v)$. Thus, RNG may have unbounded node degree, e.g., considering $n - 1$ points equally distributed on the circle centered at the nth point v, the degree of v is $n - 1$. For the sake of lowering the weight of a structure, the structure should contain as few edges as possible without breaking the connectivity. Li then naturally extended the traditional definition of RNG as follows.

The *modified relative neighborhood graph* consists of all edges uv such that

- The *interior* of $lune(u,v)$ contains no point $w \in V$.
- No point $w \in V$ with $ID(w) < ID(v)$ is on the boundary of $lune(u,v)$ and $\|wv\| < \|uv\|$
- No point $w \in V$ with $ID(w) < ID(u)$ is on the boundary of $lune(u,v)$ and $\|wu\| < \|uv\|$
- No point $w \in V$ is on the boundary of $lune(u,v)$ with $ID(w) < ID(u)$, $ID(w) < ID(v)$, and $\|wu\| < \|uv\|$.

Figure 39.9 illustrates when an edge uv is *not* included in the modified relative neighborhood graph. Li called this structure RNG′. Obviously, RNG′ is a subgraph of traditional RNG. Li proved [97] that RNG′ has a maximum node degree 6 and still contains an MST as a subgraph. However, RNG′ is still not a low-weight structure.

Obviously, graph RNG′ still can be constructed using n messages. Each node first locally broadcasts its location and ID to its one-hop neighbors. Then, every node decides which edge to keep solely based on the one-hop neighbors' location information collected. Because the definition is still symmetric, the edges constructed by different nodes are consistent, i.e., an edge uv is kept by a node u if it is also kept by node v. The computational cost of a node u is still $O(d \log d)$, where d is its degree in UDG. A simple edge-by-edge testing method has time complexity $O(d^2)$.

It is well known that the communication complexity of constructing a minimum spanning tree of an n-vertex graph G with m edges is $O(m + n \log n)$; the communication complexity of constructing MST for UDG is $O(n \log n)$ even under the local broadcasting communication model in wireless networks.

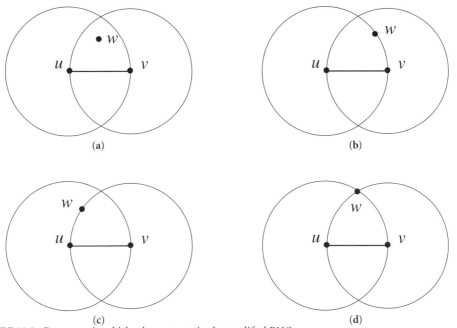

FIGURE 39.9 Four cases in which edges are not in the modified RNG.

Li showed [97] that it is *impossible* to construct a low-weighted structure using only one-hop neighbor information. The localized algorithm given by Li, which constructs a low-weighted structure using only some two-hop information, is as follows.

Algorithm 39.6. Construct low-weight structure H:

1. All nodes together construct the graph RNG′ in a localized manner.
2. Each node u locally broadcasts its incident edges in RNG′ to its one-hop neighbors. Node u listens to the messages from its one-hop neighbors.
3. Assume node u received a message revealing existence of edge $xy \in$ RNG′ from its neighbor x. For each edge $uv \in$ RNG′, if uv is the longest among uv, xy, ux, and vy, node u removes edge uv. Ties are broken by the label of the edges. Here, it is assumed that $uvyx$ is the convex hull of u, v, x, and y.
4. Let H be the final structure formed by all remaining edges in RNG′.

Obviously, if an edge uv is kept by node u, then it is also kept by node v.

Theorem 39.16. The total edge weight of H *is within a constant factor of that of the minimum spanning tree* [97].

This was proved by showing that the edges in H satisfy the *isolation property* (defined in Das et al. [46]). Li [97] also showed that the final structure contains *MST* of UDG as a subgraph.

Clearly, the communication cost of Algorithm 39.6 is, at most, $7n$: initially each node spends one message to tell its one-hop neighbors its position information, then each node uv tells its one-hop neighbors all its incident edges $uv \in$ RNG′ (there are, at most, $6n$ such messages because RNG′ has, at most, $3n$ edges). The computational cost of Algorithm 39.6 could be high because, for each link $uv \in$ RNG′, node u must test whether an edge $xy \in$ RNG′ and $x \in N_1(u)$ exists such that uv is the longest among uv, xy, ux, and vy. Next, some new algorithms that improve the computational complexity of each node while still maintaining low communication costs are presented.

39.3.4.2.2 Structure Based on LMST

The first new method [107] uses a structure called *local minimum spanning tree*, but it is first necessary to review its definition, proposed by Li and colleagues [95]. Each node u first collects its one-hop neighbors $N_1(u)$. Node u then computes the minimum spanning tree $MST(N_1(u))$ of the induced unit disk graph on its one-hop neighbors $N_1(u)$. Node u keeps a directed edge uv if and only if uv is an edge in $MST(N_1(u))$. The union of all directed edges of all nodes is called the *local minimum spanning tree*, denoted by $LMST_1$. If only symmetric edges are kept, the graph is called $LMST_1^-$, i.e., it has an edge uv if both directed edge uv and directed edge vu exist. If the directions of the edges in $LMST_1$ are ignored, the graph is called $LMST_1^+$, i.e., it has an edge uv if either directed edge uv or directed edge vu exists. Li and colleagues have proved that the graph is connected and has bounded degree 6 [107]; they have also showed that graph $LMST_1^-$ and $LMST_1^+$ are actually planar. Then, they extend the definition to k-hop neighbors. The union of all edges of all minimum spanning tree $MST(N_k(u))$ is the *k local minimum spanning tree*, denoted by $LMST_k$. For example, the two local minimum spanning tree can be constructed by the following algorithm.

Algorithm 39.7. Construct low-weight structure $LMST_2$ *by two-hop neighbors*:

1. Each node u collects its two-hop neighbors' information $N_2(u)$ using a communication-efficient protocol described in Calinescu [25].
2. Each node u computes the Euclidean minimum spanning tree $MST(N_2(u))$ of all nodes $N_2(u)$, including u itself.
3. For each edge $uv \in MST(N_2(u))$, node u tells node v about this directed edge.
4. Node u keeps an edge uv if $uv \in MST(N_2(u))$ or $vu \in MST(N_2(v))$. Let $LMST_2^+$ be the final structure formed by all edges kept.*

Li et al. [107] have proved that structures $LMST_2$ ($LMST_2^+$ and $LMST_2^-$) are connected, planar, low weighted, and have bounded node degree of, at most, 6. In general, the following theorem can be proved.

Theorem 39.17. Structure $LMST_k$ is connected, planar graph, and with bounded node degree of, at most, 6 for all $k \leq 1$. Structure $LMST_k$ is low weighted for all $k \geq 2$ [107].

Specifically, MST is a subgraph of $LMST_k$, $LMST_k \subseteq RNG'$.

Although the constructed structure, $LMST_2$, has several nice properties such as being bounded degree, planar, and low weighted, the communication cost of Algorithm 39.7 could be very large to save the computational cost of each node. The large communication costs are from collecting the two-hop neighbors' information $N_2(u)$ for each node u. Although the total communication of the protocol described by Calinescu [25] is $O(n)$, the hidden constant is large.

39.3.4.2.3 Structure Based on Combining RNG′ and LMST

The communication cost of collecting $N_2(u)$ can be improved by using a subset of two-hop information without sacrificing any properties. Define $N_2^{RNG'}(u) = \{w \mid vw \in RNG' \text{ and } v \in N_1(u)\} \cup N_1(u)$. The modified algorithm is described as follows.

Algorithm 39.8. Construct low weight structure IMRG by two-hop neighbors in RNG′:

1. Each node u tells its position information to its one-hop neighbors $N_1(u)$ using a local broadcast model. All nodes together construct the graph RNG′ in a localized manner.
2. Each node u locally broadcasts its incident edges in RNG′ to its one-hop neighbors. Node u listens to the messages from its one-hop neighbors.
3. Each node u computes the Euclidean minimum spanning tree $MST(N_2^{RNG'}(u))$ of all nodes $N_2^{RNG'}(u)$, including u itself.
4. For each edge $uv \in MST(N_2^{RNG'}(u))$, node u tells node v about this directed edge.
5. Node u keeps an edge uv if $uv \in MST(N_2^{RNG'}(u))$ or $vu \in MST(N_2^{RNG'}(v))$. Let $IMRG^+$ be the final structure formed by all edges kept. Similarly, the final structure is called $IMRG^-$ when edge $uv \in RNG'$ is kept if $uv \in MST(N_2^{RNG'}(u))$ and $uv \in MST(N_2^{RNG'}(v))$. Here, IMRG is the abbreviation of *incident MST and RNG graph*.

In the algorithm, node u constructs the local minimum spanning tree $MST(N_2^{RNG'}(u))$ based on the induced UDG of the point sets $N_2^{RNG'}(u)$. It is obvious that the communication cost of Algorithm 39.8 is, at most, $7n$.

Structures $IMRG^+$ and $IMRG^-$ are still connected, planar, bounded degree, and low weighted. They are obviously planar, and with bounded degree because both structures are still subgraphs of the modified relative neighborhood graph RNG′. Clearly, the constructed structures are a supergraph of the previous structures, i.e., $LSMT_2^+ \subseteq IMRG^+$ and $LSMT_2^- \subseteq IMRG^-$, because Algorithm 39.8 uses less information than Algorithm 39.7 in constructing the local minimum spanning tree. Thus, the two structures IMRG+ and IMRG- are still connected.

Theorem 39.18. Structures $IMRG^-$ and $IMRG^+$ are still low weighted [107].

Theorem 39.19. Algorithm 39.8 constructs structures $IMRG^-$ or $IMRG^+$ using at most $7n$ messages. The structures $IMRG^-$ or $IMRG^+$ are connected, planar, bounded degree, and low weighted. $IMRG^-$ and $IMRG^+$ have node degree of, at most, 6 [107].

Recall that, until now, no efficient localized algorithm could achieve all of the following desirable features: bounded degree, planar, low weight, and spanner. This is still an open problem. Some concrete examples of the geometry structures introduced in the previous sections are given in Figure 39.10.

---

*It keeps an edge if node u or node v wants to keep it. Another option is to keep an edge only if both nodes want to keep it. Let $LMST_2^-$ be the structure formed by such edges.

39.3.5 Fault Tolerance

Fault tolerance is one of the central challenges in designing wireless sensor networks. Sensor nodes may be battery constrained or subject to hostile environments, so individual *node failure* will be a regular or common event. To make fault tolerance possible, first of all, the underlying network topology must have multiple disjoint paths to connect any two given wireless sensor devices. Here the path could be vertex disjoint or edge disjoint. Vertex disjoint multiple paths are used in this chapter because of the communication nature of wireless sensor networks.

By setting the transmission range sufficiently large, the induced unit disk graph will be k-connected without doubt. However, because energy conservation is important to increase the life of the wireless sensor device, the question is how to find the minimum transmission range so that the induced unit disk graph is multiply connected. Recently, applying stochastic geometry, Penrose [125, 126], Bettstetter [15], and Li et al. [108] studied how to set the transmission radius to achieve the k-connectivity with certain probability for a network when wireless nodes or sensors are uniformly and randomly distributed over a two-dimensional region. Due to space limit, results are not presented here, but rather only the topology control problem.

Remember that, in topology control, one tries to maintain only a linear number of links using a localized construction method. However, this sparseness of the constructed network topology should not compromise on the fault tolerance and compromise too much on the power consumptions for communications. Therefore, this section studies a localized method to control the network topology, given a k-fault-tolerant deployment of wireless sensor nodes so that the resulting topology is still fault tolerant, but with much fewer communication links maintained. It will be shown that the constructed topology has only linear numbers of links and is a length spanner.

Levcopoulos et al. [90] proposed some algorithms for constructing fault-tolerant geometric spanners. Their algorithm can construct a spanner of degree $O(c^k)$, whose total edge length is bounded by $O(c^k)$ times the weight of an MST and resilient to k edge or vertex faults. However, their algorithms are too complex to have a localized version.

Lukovszki [114] gave a method to construct a spanner that can sustain k-node or link failures for complete graph. The topology control method [108] is based on this method and the Yao structure [164]. It is obvious that the Yao structure does not sustain k faults in a neighborhood of any node because each node only has, at most, p neighbors and one neighbor is selected in each cone, at most. However, the Yao structure can be modified as follows so that the structure is k-fault tolerant.

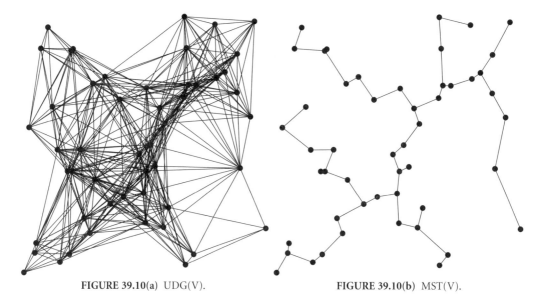

FIGURE 39.10(a) UDG(V).   **FIGURE 39.10(b)** MST(V).

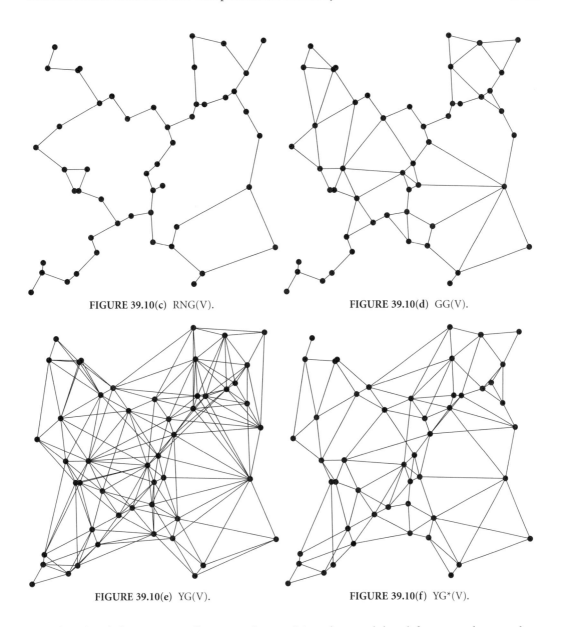

FIGURE 39.10(c) RNG(V).

FIGURE 39.10(d) GG(V).

FIGURE 39.10(e) YG(V).

FIGURE 39.10(f) YG*(V).

Each node u defines any p equally separated rays originated at u and thus defines p equal cones, where $p > 6$. In each cone, node u chooses the $k + 1$ closest nodes in that cone, if any are present, and adds directed links from u to these nodes. Ties are broken arbitrarily. Let $YG_{p,k+1}$ be the final topology formed by all nodes. Obviously, the following theorem holds.

Theorem 39.20. The structure $YG_{p,k+1}$ can sustain k node faults if the original unit disk graph is k-node fault tolerant [108].

The preceding structure approximates the original unit disk graph well. More specifically, it is a spanner even with k fault nodes.

Theorem 39.21. The structure $YG_{p,k+1}$ is a length spanner even with k node faults [108].

Due to limited power and resources of wireless sensor nodes, wireless topologies always prefer to have bounded node degree so that every wireless sensor node only keeps constant neighbors. The node degree of the structure $YG_{p,k+1}$ is, at most, $p(k + 1)$, where $p \geq 6$.

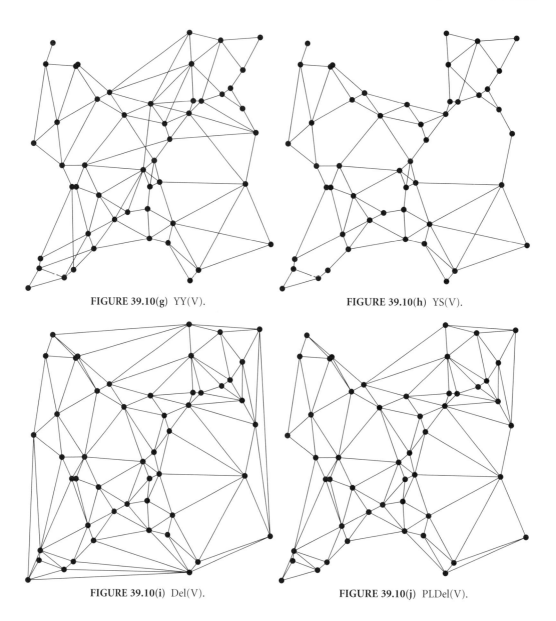

FIGURE 39.10(g) YY(V).

FIGURE 39.10(h) YS(V).

FIGURE 39.10(i) Del(V).

FIGURE 39.10(j) PLDel(V).

Another related problem is how to find small transmission range (power) for each node so that the resulting communication graph is k-connected. Hajiaghayi and colleagues [65] called this *power-optimal k-fault tolerance*. This problem is known to be NP-hard and related to the problem of *transmission power control*, which will be discussed in Subsection 39.3.7. Some heuristics [10, 134] for this problem have been proposed. Ramanathan and Rosales–Hain [134] consider the special case of two-fault tolerance and provide a centralized spanning tree heuristic for minimizing the maximum transmit power.

Recently, Bahramgiri et al. [10] generalized the cone-based local heuristic of Wattenhofer and colleagues [93, 156] to solve the k-fault tolerance. It can be proved that their resulting graph is also a length spanner even with k node faults (the proof is similar to that of Li et al. [108]). However, their method does not bound the node degree. Figure 39.11(a) shows an example in which node u can have as many as neighbors even after applying their method. Then, a careful enhancement of their protocol to bound the node degree will be given.

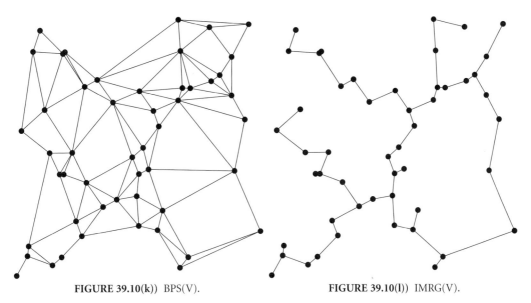

FIGURE 39.10(k)) BPS(V).   **FIGURE 39.10(l))** IMRG(V).

FIGURE 39.10 Different topologies from UDG(V). (a) UDG(V); (b) MST(V); (c) RNG(V); (d) GG(V); (e) YG(V); (f) YG*(V); (g) YY(V); (h) YS(V); (i) Del(V); (j) PLDel(V); (k) BPS(V); (l) IMRG(V).

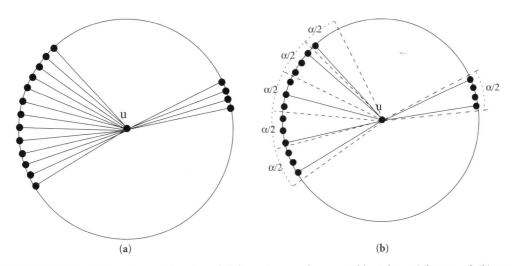

FIGURE 39.11 (a) Node u does not have bounded degree in a graph generated by Bahramgiri's protocol; (b) new method to bound node degree for Bahramgiri's protocol.

In Bahramgiri's method, the power is increased step by step until no gap is greater than α between the successive neighbors or the power reaches the maximum power. It was proved that if $\alpha \leq \frac{2\pi}{3k}$ then the resulting graph is k-connected. After applying their method, some links can be removed by the following method. For a node u, divide its transmission range into $\frac{4\pi}{\alpha}$ equal cones (each cone has an angle $\alpha/2$). Select only one neighbor in each cone c if any exist; delete all other links. However, if, for a cone c, one of its adjacent cones, say b, does not have any neighbors of u, select the boundary neighbor v so that vu forms the smallest angle with cone b; if both adjacent cones of c are empty, select *two* neighbors in c (close to the two boundaries of cone c respectively). If c does not have empty adjacent cones, one can select any one of the neighbors. See Figure 39.11(b) for illustration.

Because the gap between any two successive remaining neighbors is still not greater than α (except the empty cones), it is easy to show that the constructed graph is still *k*-connected if $\alpha \leq \frac{2\pi}{3k}$. The node degree is bounded by $\frac{2\pi}{\frac{\alpha}{2}} = \frac{4\pi}{\alpha}$. When $\alpha = \frac{2\pi}{3k}$, the node degree is bounded by 6*k*, which is almost the same as Li and colleagues' [108].

Both heuristics [10, 134] do not have provable bounds on the solution cost for power-optimal *k*-fault tolerance problems. Haijaghayi et al. [65] showed examples for which these heuristics perform arbitrarily worse than the optimal solution. Recently Lloyd and coworkers [113] presented a result that they prove gives an eight approximation for two-fault tolerance. Haijaghayi et al. then presented a more general result that some algorithms minimize power while maintaining *k*-fault tolerance with guaranteed approximation factors [65]. These will be reviewed them in Subsection 39.3.5.

39.3.6 Interference

In addition to spanner (which means connectivity and energy efficiency) and other properties discussed previously, it is desirable to have a topology with high capacity or throughput, so that it can route *as much traffic as* in the topology. One of the important issues affecting the throughput is interference. Modeling interference in a wireless environment is a complex task. The wireless medium is susceptible to path loss, noise, interference, and blockages due to physical obstructions. Rajaraman [133] reviewed several models from path loss, bit-error rate to interference. Gupta and Kumar [62] analyzed the throughput of ad hoc networks under the physical and protocol models of interference. For detailed definitions of these models, refer to Gupta and Kumar [62] and Rajaraman [133].

In Rajaraman's review, he claimed the throughput of a topology depends on, among other factors, the level of interference inherent to the topology. Define the *interference number* of an edge *e* in a graph *G* to be the maximum number of other edges in *G* that interfere with *e*, in the sense of the interference model. Define the interference number of the topology to be the maximum interference number of all edges in *G*. A plausible goal then is to seek a topology with a small interference number. The particular interference number achievable, however, depends on the relative positions of the wireless nodes and their transmission radii.

Most of the proposed topology control protocols did not study the interference number of their topology theoretically; instead, some of them showed simulation results on network throughput. Recently, Jia and colleagues [72] showed the interference analysis of YaoYao structure ($YY_k(V)$). They used the protocol model from Gupta and Kumar [62] as the interference model. First, they proved the following theorem to show the throughput achievable on $YY_k(V)$ is essentially limited only by its interference number, when compared with an optimal schedule on UDG(*V*).

Theorem 39.22. Let I be the interference number of $YY_k(V)$. Let W denote a set of packets that are successfully delivered by an arbitrary schedule of packet transmissions in UDG(V) in t steps. Then a schedule of transmissions exists in $YY_k(V)$ that delivers W in $O(tI + n^2)$ steps. Thus, for sufficiently large t and W, the throughput achievable on $YY_k(V)$ is an $\Omega(1/I)$ fraction of the optimal [72].

Then they established an upper bound on the interference number of $YY_k(V)$ for a random node distribution.

Theorem 39.23. If the n nodes are placed independently and uniformly at random in a unit square, the interference number of $YY_k(V)$ is $O(\log n)$ whp [72].

For other topologies, the interference analyses are still open problems.

39.3.7 Transmission Power Control

In the previous sections, it has been assumed that the transmission power of every node is equal and is normalized to one unit. This assumption is relaxed for a moment in this section. In other words, assume that each node can adjust its transmission power according to its neighbors' positions. A natural question is then how to assign the transmission power for each node so that the wireless sensor network is connected with optimization criteria minimizing the maximum (or total) transmission power assigned.

A transmission power assignment on the vertices in V is a function f from V into real numbers. The *communication graph*, denoted by G_f, associated with a transmission power assignment f, is a directed graph with V as its vertices and has a directed edge $\overrightarrow{v_i v_j}$ if and only if $\|v_i v_j\|^\beta \leq f(v_i)$. A transmission power assignment f is called *complete* if the communication graph G_f is strongly connected. Recall that a directed graph is strongly connected if, for any given pair of ordered nodes s and t, there is a directed path from s to t.

The *maximum cost* of a transmission power assignment f is defined as $mc(f) = \max_{v_i \in V} f(v_i)$. The total cost of a transmission power assignment f is defined as $sc(f) = \Sigma_{v_i \in V} f(v_i)$. The min–max assignment problem is then to find a complete transmission power assignment f whose cost $mc(f)$ is the least among all complete assignments. The min-total assignment problem is to find a complete transmission power assignment f whose cost $sc(f)$ is the least among all complete assignments.

Given a graph H, the power assignment f is induced by H if

$$f(v) = \max_{(v,u) \in E} \|vu\|^\beta,$$

where E is the set of edges of H. In other words, the power assigned to a node v is the largest power needed to reach all neighbors of v in H.

Transmission power control has been well studied by peer researchers in recent years. Monks et al. [120] conducted simulations that showed that implementing power control in a multiple access environment can improve the throughput performance of the non-power-controlled IEEE 802.11 by a factor of two. Therefore, it provides a compelling reason for adopting the power-controlled MAC protocol in wireless network.

The min–max assignment problem has been studied by several researchers [134, 137]. Let MST(V) be the Euclidean minimum spanning tree over a point set V. Ramanathan and Rosales–Hein [134] and Sanchez and colleagues [137] use the power assignment induced by MST(V). The correctness of using minimum spanning tree is proved in Ramanathan and Rosales–Hein [134]. Both algorithms compute the minimum spanning tree from the fully connected graph.

Notice that Kruskal's or Prim's minimum spanning tree algorithm has time complexity $O(m + n \log n)$, where m is the number of edges of the graph. Thus, the approach by these researchers [134, 137] has time complexity $O(n^2)$ in the worst case. In addition, different distributed implementation of this algorithm is not feasible because of the information that each node must store and process. In contrast, a simple $O(n \log n)$ time complexity centralized algorithm can be given to construct MST from RNG, which can also be implemented efficiently for distributed computation.

For an optimum transmission power assignment f_{opt}, call a link uv the *critical link* if $\|uv\|^\beta = mc(f_{opt})$. It has been proved [134] that the longest edge of the Euclidean minimum spanning tree MST(V) is always the critical link.

The best distributed algorithm [51, 54, 56] can compute the minimum spanning tree in $O(n)$ rounds using $O(m + n \log n)$ communications for a general graph with m edges and n nodes. The relative neighborhood graph, the Gabriel graph, and the Yao graph have $O(n)$ edges and contain the Euclidean minimum spanning tree. This implies the following theorem.

Theorem 39.24. The distributed min–max assignment problem can be solved in $O(n)$ *rounds using* $O(n \log n)$ *communications.*

The min-total assignment problem was studied by Kiroustis et al. [82] and by Clementi et al. [37–39]. Kiroustis and colleagues first proved that the min-total assignment problem is *NP-hard* when the mobile nodes are deployed in a three-dimensional space. A simple two-approximation algorithm based on the Euclidean minimum spanning tree was also given by these authors [82]. The algorithm guarantees the same approximation ratio in any dimensions. Then Clementi et al. [37–39] proved that the min-total assignment problem is still NP-hard when the mobile nodes are deployed in a two-dimensional space.

Recently, Calinescu et al. gave a method that achieves better approximation ratio than the approach by the minimum spanning tree by using an idea from the minimum Steiner tree. A natural generalization of the connectivity requirement is k-fault tolerance or k-connectivity. As mentioned in Subsection 39.3.5, some researchers studied the power assignments of wireless nodes that minimize power while maintaining k-fault tolerance. As power-optimal connectivity is NP-hard, power-optimal k-fault tolerance is NP-hard as well.

Many of the best known approximation algorithms (such as Cheriyan et al. [31]) are based on linear programming (LP) approaches. However, Haijaghayi et al. [65] showed that for the min-total k-connectivity assignment problem, the natural integer LP formulation has an integrality gap of $\Omega\left(\frac{n}{k}\right)$, implying that no approximation algorithm is based on LP with an approximation factor better than $\Omega\left(\frac{n}{k}\right)$.

Some heuristics [10, 134] are proposed. Bahramgiri et al. [10] showed that the cone-based topology control (CBTC) algorithm of Wattenhofer and colleagues [93, 156] can be extended to solve the k-fault tolerance. Examples have also been constructed that demonstrate that the approximation factor for CBTC algorithm is at least $\Omega\left(\frac{n}{k}\right)$ [65].

Recently, Lloyd and coworkers [113] presented a result that gives a centralized eight-approximation for min-total two-fault tolerance assignment. Haijaghayi et al. [65] then presented a more general result: three algorithms minimize power while maintaining k-fault tolerance. The first algorithm gives an $O(k\alpha)$ approximation where α is the best approximation factor for the related problem in wired networks (the best α so far is in $O(\log k)$ [31]). The second algorithm is based on an approximation algorithm introduced by Kortsarz and Nutov [84]. It is more complicated and achieves $O(k)$ approximation for general graphs. Their first two algorithms are centralized algorithms and then they present two distributed approximation algorithms for the cases of two- and three-connectivity in geometric graphs with approximation factors $2(4 \times 2^{\beta-1} + 1)$ and $2(1 + 7 \times 2^{\beta-1} + 12 \times 4^{\beta-1})$. Both these algorithms use the distributed minimum spanning tree algorithm. In addition, they demonstrate how to generalize these algorithms for k-connectivity in geometric graphs. However, their methods do not work for unit disk graphs, i.e., then the node transmission radius is bounded from above by a constant. It is still an open problem to achieve k-connectivity for UDG with an objective of minimizing the total edge length.

39.3.8 Clustering, Virtual Backbone

Although all the structures discussed so far are flat structures, another set of structures, called hierarchical structures, are used in wireless networks. Instead of all nodes involved in relaying packets for other nodes, the hierarchical routing protocols pick a subset of nodes that serve as the routers, forwarding packets for other nodes. The structure used to build this virtual backbone is usually the connected dominating set.

39.3.8.1 Centralized Methods

Guha and Khuller [61] studied the approximation of the connected dominating set problem for general graphs. They gave two different approaches, both of which guarantee approximation ratio of $\Theta(H(\Delta))$, where H is the harmonic function and Δ is the maximum node degree. Their approaches are for general graphs and thus do not utilize the geometric structure if applied to the wireless ad hoc networks.

One approach is to grow a spanning tree that includes all nodes. The internal nodes of the spanning tree are selected as the final connected dominating set. This approach has approximation ratio $2(H(\Delta) + 1)$. The other approach is first approximating the dominating set and then connecting the dominating set to a connected dominating set. Guha and Khuller [61] proved that this approach has approximation ratio $\ln\Delta + 3$.

One can also use the Steiner tree algorithm to connect the dominators. This straightforward method gives approximation ratio $c(H(\Delta) + 1)$, where c is the approximation ratio for the unweighted Steiner tree problem. Currently, the best ratio is $1 + \dfrac{\ln 3}{2} \simeq 1.55$, due to Robins and Zelikovsky [135].

By definition, any algorithm generating a maximal independent set is a clustering method. Methods that approximate the maximum independent set, the minimum dominating set, and the minimum connected dominating set will now be reviewed.

Hunt et al. [68] and Marathe et al. [115] also studied the approximation of the maximum independent set and the minimum dominating set for unit disk graphs. They gave the first PTASs for MDS in UDG. The method is based on the following observations: a maximal independent set is always a dominating set; given a square Ω with a fixed area, the size of any maximal dominating set is bounded by a constant C. Assume that n nodes are in Ω. Then, one can enumerate all sets with size, at most, C in time $\Theta(n^C)$. Among these enumerated sets, the smallest dominating set is the minimum dominating set. Then, using the shifting strategy proposed by Hochbaum [67], a PTAS was derived for the minimum dominating set problem [68, 115].

Because PTAS is for minimum dominating set and the graph *VirtG* connecting every pair of dominators within, at most, three hops is connected [154], one has an approximation algorithm (constructing a minimum spanning tree *VirtG*) for MCDS with approximation ratio $3 + \varepsilon$. Berman and colleagues [14] gave a $\frac{4}{3}$ approximation method to connect a dominating set and Robins et al. [135] gave a $\frac{4}{3}$ approximation method to connect an independent set. Thus, one can easily have an $\frac{8}{3}$ approximation algorithm for MCDS, which was reported in Alzoubi [3]. Recently, Cheng et al. [30] designed a PTAS for MCDS in UDG. However, it is difficult to efficiently make their method a distributed one.

39.3.8.2 Distributed Methods

Many distributed clustering (or dominating set) algorithms have been proposed in the literature [4–6, 34, 110, 111]. All algorithms assume that the nodes have distinctive identities (denoted by ID hereafter). In this subsection the terms *cluster head* and *dominator* are interchanged. The node that is not a cluster head is also called *dominatee*. A node is called *white* if its status is yet to be decided by the clustering algorithm; initially, all nodes are white. After the clustering method finishes, the status of a node could be *dominator* with color *black* or *dominatee* with color *gray*. The rest of this section concentrates on distributed methods that approximate the minimum dominating set and the minimum connected dominating set for unit disk graphs.

39.3.8.2.1 Clustering without Geometric Property

For general graphs, Jia et al. [73] described and analyzed randomized distributed algorithms for the minimum dominating set problem that run in polylogarithmic time, independent of the diameter of the network, and return a dominating set of size within a logarithmic factor from the optimum with high probability. Their best algorithm runs in $O(\log n \log \Delta)$ rounds with high probability, and every pair of neighbors exchanges a constant number of messages in each round. The computed dominating set is within $O(\log \Delta)$ in expectation and within $O(\log n)$ with high probability. Their algorithm works for weighted dominating sets also.

The method proposed by Das and colleagues [44, 141] contains three stages: approximating the minimum dominating set; constructing a spanning forest of stars; and expanding the spanning forest to a spanning tree. Here the *stars* are formed by connecting each dominatee node to one of its dominators.

The approximation method of MDS is essentially a distributed variation of the centralized Chvatal's greedy algorithm [35] for set cover. Notice that the dominating set problem is essentially the set cover problem, which has been well studied. Thus, it is no surprise that the method by Das et al. [44, 141] guarantees an $H(\Delta)$ for the MDS problem, where H is the harmonic function and Δ is the maximum node degree.

Although the algorithm proposed by Das et al. finds a dominating set and then grows it to a connecting dominating set, the algorithm proposed by Wu and Li [162, 163] takes an opposite approach. They first find a connecting dominating set and then prune out certain redundant nodes from the CDS. The initial CDS C contains all nodes that have at least two nonadjacent neighbors. A node u is said to be *locally redundant* if it has a neighbor in C with larger ID that dominates all other neighbors of u, or two adjacent neighbors with larger ID that, together, dominate all other neighbors of u. Their algorithm then keeps removing all locally redundant nodes from C. These authors showed that this algorithm works well in practice when the nodes are distributed uniformly and randomly, although they offer no theoretical analysis for the worst case as well as for the average approximation ratio. However, Alzoubi et al. [4] showed that the approximation ratio of this algorithm could be as large as $\frac{n}{2}$.

Stojmenovic and coworkers [146] proposed several synchronized distributed constructions of connecting dominating sets. In their algorithms, the connecting dominating set consists of two types of nodes: cluster head and border nodes (also called gateways or connectors). The cluster-head nodes are a maximal independent set, which is constructed as follows. At each step, all white nodes that have the lowest *rank* among all white neighbors are colored black, and the white neighbors are colored gray. The ranks of the white nodes are updated if necessary. Here, the following rankings of a node are used in various methods: the ID only [34, 110]; the ordered pair of degree and ID [29]; and an ordered pair of degree and location [146]. After the cluster-head nodes are selected, border nodes are selected to connect them. A node is a border-node if it is not a cluster head and at least two cluster heads are within its two-hop neighborhood. Alzoubi and colleagues [4] showed that the worst-case approximation ratio of this method is also $\frac{n}{2}$, although it works well in practice.

Basagni [11] and Basagni et al. [12] studied how to maintain the clustering in mobile wireless ad hoc networks. They use a general *weight* as a criterion for selecting the node as the cluster head, where the weight could be any criterion used before.

39.3.8.2.2 Clustering with Geometric Property

None of the preceding algorithm utilizes the geometric property of the underlying unit disk graph. Recently, several algorithms have been proposed with a constant worst-case approximation ratio by taking advantage of the geometric properties of the underlying graph. These methods typically use two messages similar to IamDominator and IamDominatee and have the following procedure: a white node claims to be a dominator if it has the smallest ID among all of its white neighbors, if any exist, and broadcasts IamDominator to its one-hop neighbors. A white node receiving an IamDominator message marks itself as dominatee and broadcasts IamDominatee to its one-hop neighbors.

The set of dominators generated by this method is actually a maximal independent set. Here, it is assumed that each node knows the IDs of all its one-hop neighbors; this can be achieved by asking each node to broadcast its ID to its one-hop neighbors initially. This approach of constructing MIS is well known. For example, Stojmenovic and colleagues [146] also used this method to compute the MIS.

The second step of backbone formation is to find some *connectors* (also called *gateways*) among all the dominatees to connect the dominators. Then, the connectors and the dominators form a *connected dominating set*. Recently, Wan et al. [150] proposed a communication-efficient algorithm to find connectors based on the fact that only a constant number of dominators are within k-hops of any node. The following observation is a basis of several algorithms for CDS. After clustering, one dominator node can be connected to many dominatees. However, it is well known that a dominatee node can only be connected to, at most, *five* dominators in the unit disk graph model. Generally, Wan et al. [150] and Wang and Li 154] showed that, for each node (dominator or dominatee), there are, at most, a constant number of dominators at most k units away.

Lemma 39.25. For every node v, the number of dominators inside the disk centered at v with radius k units is bounded by a constant $\ell_k < (2k + 1)^2$.

Lemma 39.26. Given a dominating set S, let VirtG be the graph connecting all pairs of dominators u and v if a path in UDG connects them with, at most, three hops. VirtG is connected.

It is natural to form a connected dominating set by finding connectors to connect any pair of dominators u and v if they are connected in *VirtG*. This strategy is also adopted by Wan et al. [150]. In their approach, Stojmenovic and colleagues [146] set any dominatee node as the connector if two dominators were within its two-hop neighborhood. This approach is very pessimistic and results in a very large number of connectors in the worst case [4]. Instead, Wan and colleagues [2] suggested finding only one unique shortest path to connect any two dominators that are, at most, three hops away. Wang and Li [154] and Alzoubi et al. [150] discussed in detail some approaches to optimize the communication cost and the memory cost. These authors proved the following theorem.

Theorem 39.27. The number of connectors found by this algorithm is at most ℓ_3 times the minimum. The size of the connected dominating set found by this algorithm is within a small constant factor of the minimum.

The graph constructed by this algorithm is called a CDS graph (or *backbone* of the network). If all edges that connect all dominatees to their dominators are added, the graph is called extended CDS, denoted by CDS'. It has been shown that the CDS' graph is a sparse spanner in terms of hops and length with factors 3 and 6 [2, 154]; meanwhile, CDS has a bounded node degree $\max(\ell_3, 5+ \ell_2)$. See Wang and Li [154] for detailed proofs.

Several routing algorithms require that underlying topology be planar. Notice that, in the formation algorithm of CDS, no geometric information is used. The resulting CDS may be a nonplanar graph. Even using some geometric information, the CDS still is not guaranteed to be a planar graph. Then Wang and Li [154] proposed a method to make the graph CDS planar without losing the spanner property of the backbone. Their method applies the localized Delaunay triangulation [98] on top of the induced graph from CDS, denoted by ICDS. It has been proved [98] that LDel(G) is a spanner if G is a unit disk graph. Notice that ICDS is a unit disk graph defined over all dominators and connectors. Consequently, LDel(*ICDS*) is a spanner in terms of length.

39.4 Localized Routing

The geometric nature of the multihop wireless sensor networks allows a promising idea: localized geometric routing (or localized routing) protocols. A routing protocol is *localized* if the decision on to which node to forward a packet is based only on:

- The information in the header of the packet. This information includes the source and destination of the packet, but more data could be included if its total length is bounded.
- The local information gathered by the node from a small neighborhood. This information includes the set of one-hop neighbors of the node, but a larger neighborhood set could be used if it could be collected efficiently.

Randomization is also used in designing the protocols. A routing is said to be *memoryless* if the decision on to which node to forward a packet is solely based on the destination, current node, and its neighbors within some constant hops. Localized routing is sometimes called in the literature *stateless* [77, 78], *online* [16, 18], or *distributed* [145].

In order to make the localized geometric routing work, the source node must learn the current (or approximately current) location of the destination node. For sensor networks collecting data, the destination node is often fixed; thus, location service is not needed in these applications. However, the help of a *location service* is needed in most application scenarios. Mobile nodes register their locations to the location service. When a source node does not know the position of the destination node, it queries the location service to get that information. In cellular networks, there are dedicated position servers.

It would be difficult to implement the centralized approach of location services in wireless sensor networks. First, for a centralized approach, each node must know the position of the node that provides the location services, which is a chicken-and-egg problem. Second, the dynamic nature of the wireless sensor networks makes it very unlikely that at least one location server will be available for each node. Algorithms for distributed location services have been studied recently [13, 64, 92, 142]. Due to space limits, the location service problem is omitted here. See Li [96] for detailed review.

39.4.1 Simple Heuristics

This subsection summarizes some localized geometric routing protocols proposed in the networking and computational geometry literature (see also Figure 39.12).

- *Compass routing*. Let t be the destination node. Current node u finds the next relay node v such that the angle $\angle vut$ is the smallest among all neighbors of u in a given topology [85].
- *Random compass routing*. Let u be the current node and t be the destination node. Let v_1 be the node above line ut such that $\angle v_1 ut$ is the smallest among all such neighbors of u. Similarly, define v_2 to be nodes below line ut that minimize the angle $\angle v_2 ut$. Then node u randomly chooses v_1 or v_2 to forward the packet [85].

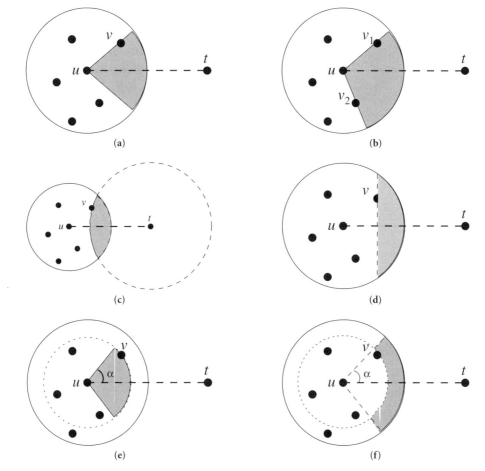

FIGURE 39.12 Various localized routing methods. The shaded area is empty and v is the next node. (a) compass; (b) random compass; (c) greedy; (d) most forwarding; (e) nearest neighbor; (f) farthest neighbor.

- *Greedy routing.* Let *t* be the destination node. Current node *u* finds the next relay node *v* such that the distance $\|vt\|$ is the smallest among all neighbors of *u* in a given topology [20].
- *Most forwarding routing (MFR).* Current node *u* finds the next relay node *v* such that $\|v't\|$ is the smallest among all neighbors of *u* in a given topology, where v' is the projection of *v* on segment *ut* [145].
- *Nearest neighbor routing (NN).* Given a parameter angle α, node *u* finds the nearest node *v* as forwarding node among all neighbors of *u* in a given topology such that $\angle vut \leq \alpha$.
- *Farthest neighbor routing (FN).* Given a parameter angle α, node *u* finds the farthest node *v* as forwarding node among all neighbors of *u* in a given topology such that $\angle vut \leq \alpha$.
- *Greedy compass.* Current node *u* first finds the neighbors v_1 and v_2 such that v_1 forms the smallest counterclockwise angle $\angle tuv_1$ and v_2 forms the smallest clockwise angle $\angle tuv_2$ among all neighbors of *u* with the segment *ut*. The packet is forwarded to the node of $\{v_1, v_2\}$ with minimum distance to *t* [18, 121]

It has been shown that the compass routing, random compass routing, and greedy routings guarantee to deliver packets from the source to the destination if Delaunay triangulation is used as network topology [20, 85]. This was proved by showing that the distance from the selected forwarding node *v* to the destination node *t* is less than the distance from current node *u* to *t*. However, the same proof cannot be carried over when the network topology is Yao graph, Gabriel graph, relative neighborhood graph, and the localized Delaunay triangulation. When the underlying network topology is a planar graph, the right-hand rule or face routing is often used to guarantee packet delivery after simple localized routing heuristics fail [20, 78, 86, 88, 145]. We will discuss them in next section.

Theorem 39.28. The greedy routing guarantees delivery of packets if the Delaunay triangulation is used as the underlying structure. The compass routing guarantees the delivery of the packets if the regular triangulation is used as the underlying structure. There are triangulations (not Delaunay) that defeat these two schemes. The greedy-compass routing works for all triangulations, i.e., it guarantees delivery of packets as long as a triangulation is used as the underlying structure. Every oblivious routing method is defeated by some convex subdivisions [121].

Here, a triangulation is *regular triangulation* if it is the projection of the lower convex hull of some three-dimensional polytopes *P* into the *X–Y* plane. Delaunay triangulation is a special regular triangulation in which all the vertices of *P* are on a paraboloid $z^2 = x^2 + y^2$. Another interesting triangulation is *greedy triangulation*, which is constructed by adding edges in the increasing order of their lengths to avoid crossing edges. Localized routing for greedy triangulation has also been studied [121] because the greedy triangulation cannot be constructed locally or very efficiently in a distributed manner. This part is omitted in this book. It is easy to see that no memoryless routing method works in the unit disk graph.

39.4.2 Right-Hand Rule and Face Routing

The *right hand rule* is a method long known for traversing a graph (in analogy to following the right-hand wall in a maze); it has been used in some wireless routing protocols [20, 77, 78, 145]. The rule states that when arriving at node *x* from node *y*, the next edge traversed is the next one sequentially counterclockwise about *x* from edge *xy*. In the example shown in Figure 39.13, *x* will forward the packet to *z*, following the right-hand rule, thus traversing face *P*. It is known that the right-hand rule traverses the interior of a closed polygonal region (a face) in clockwise edge order; it traverses an exterior region in counterclockwise edge order. In general, the right-hand rule is applied in planar graphs (in which no edges intersect each other). Karp [77] has given a *no-crossing heuristic* to deal with a case in which edges cross.

Applying the right-hand rule in planar graphs, a routing protocol called *face routing* has been proposed [85] (the algorithm is called *compass routing II*). Consider a planar graph *G*. The nodes and edges of graph *G* partition the Euclidean plane into contiguous regions called the *faces* of *G*. The main idea of the face routing is to walk along the faces intersected by the line segment *st* between the source *s* and the

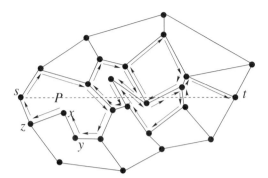

FIGURE 39.13 An illustration of the face routing algorithm.

destination t. In each face, the algorithm uses the right-hand rule to explore the boundaries. On its way around a face, the algorithm keeps track of the points at which it crosses the line st. Having completely surrounded a face, the algorithm returns to one of these intersections lying closest to t, where it proceeds by exploring the next face close to t. Figure 39.13 gives an illustration see Kranakis et al. [85] and Kuhn et al. [87] for detailed algorithms. They also proved that the face routing algorithm guarantees to reach the destination t after traversing, at most, $O(n)$ edges, where n is the number of nodes.

Although face routing terminates in linear time, it is not satisfactory because already a very simple flooding algorithm will terminate in $O(n)$ steps. Kuhn and colleagues [87] proposed a new method called *adaptive face routing* (AFR), in which restricted search areas are used to avoid exploring the complete boundary of faces. The exploration of faces is restricted to an ellipse area, the size of which is set to an initial estimate of the optimal path length. If face routing fails to reach the destination (when it reaches the ellipse, it must turn back), the algorithm will restart with a bounding ellipse of doubled size. These authors proved that the algorithm will finally find a path to t if s and t are connected. Also, the number of steps of AFR is bounded by $O(c^2(p^\star))$, where $p^\star$ is an optimal path and $c(p^\star)$ is the cost of that path. In their proof, Kuhn et al. assumed the unit disk graph to be a civilized graph. Finally, they offered a tight lower bound by showing that *any localized* geometric routing algorithm has worst-case cost $O(c^2(p^\star))$.

Recently, Kuhn et al. [88] extend *adaptive face routing* to a routing algorithm called *other adaptive face routing* (OAFR). Instead of changing to the next face at the "best" intersection of the face boundary with st, OAFR returns to the boundary point closest to the destination. These authors proved that the cost of OAFR is also bounded by $O(c^2(p^\star))$, which is asymptotically optimal.

39.4.3 Combining Face Routing with Greedy Routing

Greedy routing was used in early routing protocols for wireless networks. However, it is easy to construct a simple example to show that greedy algorithm will not succeed to reach the destination but fall into a local minimum, a node without any "better" neighbors. A natural approach to improve the potential of greedy routing for practical purposes is to combine greedy routing and face routing (or the right-hand rule) to recover the routing after simple greedy routing fails in local minimum. Many wireless protocols have used this approach [20, 78, 86, 88, 145, 165].

Greedy perimeter stateless routing (GPRS) [77, 78] is a famous routing protocol for wireless networks. It uses RNG or GG as the planar routing topology, then combines greedy and right-hand rule to forward packets in the network. When a node receives a greed-mode packet, it searches its neighbor table for a neighbor closer to the destination t. If one exists, it will forward the packet to that neighbor. When no neighbor is closer, the node marks the packet into perimeter mode. GPSR forwards perimeter-mode packets using a simple planar graph traversal (right-hand rule). When a packet enters perimeter mode, GPSR records in the packet the location L_p. Then, when receiving a perimeter-mode packet, GPSR will first compare it with forwarding node's location. GPRS returns a packet to greedy mode if the distance from the forwarding node to t is less than that from Lp to t. For more detail, see Karp [77] and Karp and Kung [78]. GPRS can guarantee delivery of packets when the underlying network topology is a planar graph.

Recently, Kuhn et al. [88] proposed a new algorithm to combine greedy routing with their OAFR, calling the new method greedy other adaptive face routing (GOAFR). The idea is similar to GPSR. When greedy method falls in local minimum, GOAFR uses OAFR to recover the routing. These authors proved the cost of GOAFR is bounded by $O(c^2(p^*))$, which is asymptotically optimal. In addition, they show that the algorithm is also average-case efficient through extensive simulations. Kuhn and colleagues showed simulations of a variety of face routing algorithms and their combinations with a greedy approach. Unlike GPSR, when face routing is performed in GOAFR, it does not return to greedy method until OAFR completely finishes exploration of the face. This may affect the efficiency of routing.

Kuhn et al. [86] used an "early fallback" technique to return to greedy routing as soon as possible; their new algorithm is called GOAFR+. It employs two counters p and q to keep track of how many of the nodes visited during the current face routing phase are located closer (counted by p) and how many are not closer (counted by q) to the destination than the starting point of the current face routing phase. When a certain fallback condition holds, GOAFR+ directly falls back to greedy mode. This modification makes an obvious improvement for the average case performance. Their theoretical analysis also proves that GOAFR+ is asymptotically optimal in the worst case.

39.4.4 Routing on Delaunay Triangulation

With respect to localized routing, there are several ways to measure the quality of the protocol. In Kuhn's analysis, they used the number of steps (hops) in a path to measure the quality of their routing methods. Given the scarcity of power resources in wireless sensor networks, minimizing the total power used is imperative. A stronger condition is to minimize the total Euclidean distance traversed by the packet.

Bose and Morin [18] and Morin [121] also studied the performance ratio of previously studied localized routing methods. They proved that none of the previously proposed heuristics guarantees a constant ratio of the traveled distance of a packet compared with the minimum. They gave the first localized routing algorithm such that the traveled distance of a packet from u to v is, at most, a constant factor of $\|uv\|$ when the Delaunay triangulation is used as the underlying structure.

Their algorithm is based on the proof of the spanner property of the Delaunay triangulation [48]. Without loss of generality, let $b_0 = u, b_1, b_2, \ldots, b_{m-1}, b_m = v$ be the vertices corresponding to the sequence of Voronoi regions traversed by walking from u to v along the segment uv. If a Voronoi edge or a Voronoi vertex happens to lie on the segment uv, then choose the Voronoi region lying above uv (see Figure 39.14). Given two nodes u and v, $tunnel(u,v)$ is defined as the collection of triangles that intersect the segment uv. The sequence of nodes b_i, $0 \leq i \leq m$, defines a path from u to v. In general, Dobkin and colleagues [48] refer to the path constructed this way between some nodes u and v as the *direct DT path* from u to v.

Assume that line uv is the x-axis. The path constructed by Dobkin et al. uses the direct DT path as long as it is above the x-axis. Assume that the path constructed so far has led to some node b_i such that b_i is above uv, and b_{i+1} is below uv. Let j be the least integer larger than i such that b_j is above uv. Here, j exists because $b_m = v$ is on uv. Then the path constructed by Dobkin et al. uses the direct DT path to b_j or takes a *shortcut*. These authors [48] offer more detail about when to choose the direct DT path from b_i to b_j; when to choose the shortcut path from b_i to b_j; and how the shortcut path is defined.

Bose and Morin basically use a type of binary search method to find which path is better (see Morin [121] for more details of finding the path). However, their algorithm needs the Delaunay triangulation as the underlying structure, which is expensive to construct in wireless ad hoc networks. They further extend their method to any triangulations satisfying the diamond property [19]. Here, a triangulation satisfying the diamond property if, for every edge uv in the triangulation, either Δuvw_1 or Δuvw_2 is empty of other vertices, where w_i satisfying $\angle w_i uv = \angle w_i vu = \frac{\pi}{6}$, for $i = 1,2$.

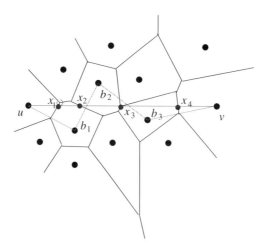

FIGURE 39.14 There is a good approximation path using the edge of tunnel (u,v).

Li and Wang [106] showed that the local Delaunay triangulation, PLDel, can be used to approximate the Delaunay triangulation, Del, almost always when the network is connected and the sensor nodes are randomly deployed. Consequently, the method in Morin [121] can be used on local Delaunay triangulation almost always.

Localized routing protocols support mobility by eliminating the communication-intensive task of updating the routing tables. However, mobility can affect localized routing protocols in the performance and the guarantee of delivery. Thus far, no work has taken place to design protocols with guaranteed delivery when the network topology changes during the routing.

39.5 Broadcasting

Before this section, only unicast routing protocols have been considered; however, in wireless networks, broadcast is a very important operation because it provides an efficient way of communication that does not require global information and functions well in the case of changing topologies. Although many broadcast/multicast algorithms [32, 71, 112, 129, 149, 159, 160] have been proposed for wireless ad hoc networks, most of them are not power aware. Not until recently have research efforts been made to devise power-efficient multicast/broadcast algorithms for wireless networks. Li and Hou [94] provide a detailed taxonomy of existing work. Here, existing work is simply put into two groups: centralized methods and localized methods.

39.5.1 Centralized Methods

39.5.1.1 Assumptions

Minimum-energy broadcast/multicast routing in a simple ad hoc networking environment has been addressed by pioneering work [36, 39, 83, 157]. To assess the complexities one at a time, the nodes in the network are assumed to be distributed randomly in a two-dimensional plane and there is no mobility. Nevertheless, as Wieselthier and colleagues argued [157], the impact of mobility can be incorporated into this static model because the transmitting power can be adjusted to accommodate the new locations of the nodes as necessary. In other words, the capability to adjust the transmission power provides considerable "elasticity" to the topological connectivity and thus may reduce the need for hand-offs and tracking. In addition, as these authors assumed [157], bandwidth and transceiver resources are sufficient. Under these assumptions, centralized (as opposed to distributed) algorithms have been presented for minimum-energy broadcast/multicast routing [42, 91, 109, 157]. In this simple networking environment, these centralized algorithms are expected to serve as the basis for further studies on distributed algorithms in a more practical network environment, with limited bandwidth and transceiver resources, as well as the node mobility.

39.5.1.2 Centralized Methods

Some centralized methods are based on optimization. The scheme proposed by Marks and coworkers [116] is built upon an alternate search-based paradigm in which the minimum-cost broadcast/multicast tree is constructed by a search process. Two procedures are devised to check the viability of a solution in the search space. Preliminary experimental results show that this method renders better solutions than BIP, though at a higher computational cost. Liang [109] showed that the minimum-energy broadcast tree problem is NP-complete and proposed an approximate algorithm to provide a bounded performance guarantee for the problem in the general setting. Essentially, Liang reduced the minimum-energy broadcast tree problem to an optimization problem on an auxiliary weighted graph and solved the optimization problem so as to give an approximate solution for the original problem. Liang also proposed another algorithm that yields better performance under a special case.

Das et al. [42] proposed an evolutionary approach using genetic algorithms. The same authors also presented three different integer programming models that can be used to find the solutions to the minimum-energy broadcast/multicast problem [43]. The major drawback of optimization-based schemes is, however, that they are centralized and require the availability of global topological information.

Some centralized methods are based on greedy heuristics. Three greedy heuristics have been proposed for the minimum-energy broadcast routing problem: MST (minimum spanning tree); SPT (shortest-path tree); and BIP (broadcasting incremental power) [157]. The MST heuristic first applies the Prim's algorithm to obtain an MST and then orients it as an arborescence rooted at the source node. The SPT heuristic applies the Dijkstra's algorithm to obtain an SPT rooted at the source node. The BIP heuristic is the node version of Dijkstra's algorithm for SPT. It maintains, throughout its execution, a single arborescence rooted at the source node. The arborescence starts from the source node, and new nodes are added to it, one at a time, on the minimum incremental cost basis until all nodes are included in the arborescence. The incremental cost of adding a new node to the arborescence is the minimum additional power increased by some node in the current arborescence to reach this new node.

The implementation of BIP is based on the standard Dijkstra's algorithm, with one fundamental difference on the operation whenever a new node q is added. Whereas Dijkstra's algorithm updates the node weights (representing the currently known distances to the source node), BIP updates the cost of each link (representing the incremental power to reach the head node of the directed link). This update is performed by subtracting the cost of the added link pq from the cost of every link qr that starts from q to a node r not in the new arborescence. They have been evaluated through simulations [157], but little is known about their analytical performances in terms of the approximation ratio. Here, the approximation ratio of a heuristic is the maximum ratio of the energy needed to broadcast a message based on the arborescence generated by this heuristic to the least necessary energy by any arborescence for any set of points.

For a pure illustration purpose, another slight variation of BIP has been discussed in detail [151]. This greedy heuristic is similar to the Chvatal's algorithm [35] for the set cover problem and is a variation of BIP. Like BIP, an arborescence, which starts with the source node, is maintained throughout the execution of the algorithm. However, unlike BIP, many new nodes can be added, one at a time. Similar to the Chvatal's algorithm [35], the new nodes added are chosen to have the minimal *average* incremental cost, which is defined as the ratio of the minimum additional power increased by some node in the current arborescence to reach these new nodes to the number of these new nodes. This heuristic is called the broadcast average incremental power (BAIP). In contrast to the $1 + \log m$ approximation ratio of the Chvatal's algorithm [35], where m is the largest set size in the set cover problem, Wan and colleagues showed that the approximation ratio of BAIP is at least $\frac{4n}{\ln n} - o(1)$, where n is the number of receiving nodes.

Wan et al. [151] showed that the approximation ratios of MST and BIP are between 6 and 12 and between $\frac{13}{3}$ and 12, respectively; on the other hand, the approximation ratios of SPT and BAIP are at least $\frac{n}{2}$ and $\frac{4\pi}{\ln n}$ $-o(1)$, respectively, where n is the number of nodes. Their proof techniques are detailed in the next subsection.

The Iterative maximum-branch minimization (IMBM) algorithm was another effort [91] to construct power-efficient broadcast trees. It begins with a basic broadcast tree in which the source directly transmits to all other nodes. Then it attempts to approximate the minimum-energy broadcast tree by iteratively replacing the maximum branch with lower power, more-hop alternatives.

BIP and IMBM operate under the assumption that the transmission power of each node is unconstrained, i.e., every node can reach every other node. Both algorithms are centralized in the sense that they require that: (1) the source node know the position/distance of every other node; and (2) each node know its downstream, on-tree neighbors so as to propagate broadcast messages. As a result, it may be difficult to extend both algorithms into distributed versions because a significant amount of information must be exchanged among nodes.

39.5.1.3 Theoretical Analysis of Minimum-Energy Broadcast

Any broadcast routing is viewed as an arborescence (a directed tree) T, rooted at the source node of the broadcasting, that spans all nodes. Let $f_T(p)$ denote the transmission power of the node p required by T. For any leaf node p of T, $f_T(p) = 0$. For any internal node p of T,

$$f_T(p) = \max_{pq \in T} \|pq\|^\beta,$$

in other words, the βth power of the longest distance between p and its children in T. The total energy required by T is $\Sigma_{p \in p} f_T(p)$. Thus, the minimum-energy broadcast routing problem is different from the conventional link-based minimum spanning tree (MST) problem.

Indeed, although the MST can be solved in polynomial time by algorithms such as Prim's algorithm and Kruskal's algorithm [41], it is still unknown whether the minimum-energy broadcast routing problem can be solved in polynomial time. In its general graph version, minimum-energy broadcast routing can be shown to be NP-hard [57]; even worse, it cannot be approximated within a factor of $(1 - \varepsilon)\log\Delta$, unless NP $\subseteq$ DTIME $[n^{O(\log \log n)}]$, where Δ is the maximal degree and ε is any arbitrary small positive constant. However, this intractability of its general graph version does not necessarily imply the same hardness of its geometric version. In fact, as shown later, its geometric version can be approximated within a constant factor. Nevertheless, this suggests that the minimum-energy broadcast routing problem is considerably more difficult than the MST problem.

Recently, Clementi et al. [36] proved that the minimum-energy broadcast routing problem is an NP-hard problem and obtained a parallel but weaker result to those of Wan and colleagues [151], who gave some lower bounds on the approximation ratios of MST and BIP by studying some special instances. Their deriving of the upper bounds relies extensively on the geometric structures of Euclidean MSTs.

A key result is an upper bound on the parameter $\Sigma_{e \in mst(P)} \|e\|^2$ for any finite point set P of radius one. Note that the supreme of the total edge lengths of $mst(P)$, $\Sigma_{e \in mst(P)} \|e\|^2$, over all point sets P of radius one is infinity. However, the parameter $\Sigma_{e \in mst(P)} \|e\|^2$ is bounded from above by a constant for any point set P of radius one. Wan and colleagues used c to denote the supreme of $\Sigma_{e \in mst(P)} \|e\|^2$ over all point sets P of radius one. The constant c is, at most, 12 [151].

Theorem 39.29. $6 \leq c \leq 12$ [151].

The proof of this theorem involves complicated geometric arguments; see Wan et al. [151] for more detail. Note that, for any point set P of radius one, the length of each edge in $mst(P)$ is, at most, one. Therefore, Theorem 39.29 implies that, for any point set P of radius one and any real number $\beta \geq 2$,

$$\sum_{e\in mst(P)} \|e\|^\beta \le \sum_{e\in mst(P)} \|e\|^2 \le c \le 12.$$

The next theorem proved Wan and colleagues explores a relation between the minimum energy required by a broadcasting and the energy required by the Euclidean MST of the corresponding point set [151].

Lemma 39.30. For any point set P in the plane, the total energy required by any broadcasting among P is at least $\frac{1}{c}\Sigma_{e\in mst(P)} \|e\|^\beta$ *[151].*

Consider any point set P in a two-dimensional plane. Let T be an arborescence oriented from some $mst(P)$. Then the total energy required by T is at most $\Sigma_{e\in T_p} \|e\|^\beta$. From Lemma 39.30, this total energy is at most c times the optimum cost. Thus, the approximation ratio of the link-based MST heuristic is at most c. Together with Theorem 39.29, this observation leads to the following theorem.

Theorem 39.31. The approximation ratio of the link-based MST heuristic is, at most, c and therefore is, at most, 12 [151].

In addition, they derived an upper bound on the approximation ratio of the BIP heuristic. Once again, the Euclidean MST plays an important role.

Lemma 39.32. For any broadcasting among a point set P in a two-dimensional plane, the total energy required by the arborescence generated by the BIP algorithm is, at most, $\Sigma_{e\in mst(P)} \|e\|^\beta$ *[151].*

39.5.2 Localized Methods

The centralized algorithms do not consider computational and message overheads incurred in collecting global information. Several of them also assume that the network topology does not change between two runs of information exchange. These assumptions may not hold in practice, because the network topology may change from time to time, and the computational and energy overheads incurred in collecting global information may not be negligible. This is especially true for large-scale sensor networks in which the topology is changing dynamically due to the changes of position, energy availability, environmental interference, and failures. This implies that centralized algorithms that require global topological information may not be practical.

Santivanez et al. [138] show that flooding is a good solution for the sake of scalability and simplicity. Several flooding techniques for wireless networks have been proposed [66, 132, 144], each with respect to a certain optimization criterion. However, none of them takes advantage of the feature that the transmission power of a node can be adjusted.

Some distributed heuristics have been proposed [1, 24, 158]. Most of them are based on the distributed MST method. A possible drawback of this method is that it may not perform well under frequent topological changes because it relies on information multiple hops away to construct the MST (see Li and Hou [94] for more detail). The relative neighborhood graph, the Gabriel graph and the Yao graph have $O(n)$ edges and contain the Euclidean minimum spanning tree. This implies that one can construct the minimum spanning tree using $O(n \log n)$ messages.

Cartigny et al. [27] proposed a localized algorithm, called RBOP built upon the notion of relative neighborhood graph. In RBOP, the broadcast is initiated at the source and propagated, following the rules of neighbor elimination, on the topology represented by RNG. Simulation results show that the performance degradation could be as high as 100% compared to BIP. Li and Hou [94] proposed another localized algorithm, called BLMST, which basically uses LMST as the broadcast topology. Their simulations show the performance of BLMST is much better than that of BROP and comparable to that of BIP.

However, as shown in Li [97] and Subsection 39.3.4 (by Figure 39.8), the total weights of RNG and LMST could still be as large as $O(n)$ times the total weight of MST. Given a graph G, let $\omega_b(G) = \Sigma_{e\in G} \|e\|^\beta$. Then $\omega_1(\text{RNG}) = \Theta(n) \cdot \omega_1(\text{MST})$ and $\omega_1(\text{LMST}) = \Theta(n) \cdot \omega_1(\text{MST})$. Subsection 39.3.4.2 describes three low-weight planar graphs: H, LSMT$_2$ and IMRG. All of the three low-weight planar structures can be constructed by localized methods, and the total communication costs are $O(n)$. It is easy to show that the energy consumption using those structures is within $O(n^{\beta-1})$ of the optimum, i.e., $\omega_\beta(H) = O(n^{\beta-1}) \cdot \omega_\beta(MST)$, $\omega_\beta(LSMT_2) = O(n^{\beta-1}) \cdot \omega_\beta(MST)$, $\omega_\beta(IMRG) = O(n^{\beta-1}) \cdot \omega_\beta(MST)$ for any $\beta \geq 1$. This improves the previously known "lightest" structure RNG by $O(n)$ factor since in the worst case $\omega_1(RNG) = \Theta(n) \cdot \omega_1(MST)$ and $\omega_\beta(RNG) = \Theta(n^\beta) \cdot \omega_\beta(MST)$.

39.6 Summary and Open Questions

Wireless sensor networks have attracted considerable attention recently due to potential wide applications in various areas and the ubiquitous computing. Much excellent research has been conducted to study the electronic and the networking parts of wireless sensor networks. Networking also has many interesting topics, such as topology control; routing; energy conservation; QoS; mobility management; and so on. This chapter presented an overview of recent progress in applying computational geometry techniques to solve questions such as topology construction and localized routing in wireless sensor networks.

Nevertheless, many excellent results were not covered in this chapter due to space limits. For example, Meguerdichian et al. [119] and Li et al. [101] solved the coverage problems in sensor networks, combining computational geometry and graph theoretic techniques. Their algorithms rely heavily on geometrical structures such as Delaunay triangulation, Voronoi diagram, and relative neighbor graph. With more research work in sensor networks, field computational geometry techniques can help solve more questions and play an important role.

Interesting open questions for topology control and localized routing in wireless sensor networks abound:

- Is the YaoYao structure $YY_k(V)$ a length spanner for general graphs?
- Can a localized structure be designed that achieves all desirable features such as bounded degree, planar, low weight, and spanner?
- If interference and fault tolerance are considered, how can the network topology be designed/
- When the overhead cost c of signal transmission is not negligible, are the structures reviewed here still power spanners?
- How can the network topology be controlled when different nodes have different transmission ranges so that the topology has some nice properties?
- Can a localized routing protocol be designed that achieves constant ratio of the length of the found path to the minimum? (The answer is probably negative; see Kuhn et al. [87].)

References

1. A. Ahluwalia, E. Modiano, and L. Shu. On the complexity and distributed construction of energy-efficient broadcast trees in static ad hoc wireless networks. In *Proc. 36th Annu. Conf. Inf. Sci. Syst. (CISS)*, Princeton, 2002.
2. K. Alzoubi, P.-J. Wan, and O. Frieder. Message-optimal connected-dominating-set construction for routing in mobile ad hoc networks. In *Proc. 3rd ACM Int. Symp. Mobile Ad Hoc Networking Computing (MobiHoc'02)*, Lausanne, 2002.
3. K.M. Alzoubi. Virtual backbone in wireless ad hoc networks. Ph.D. thesis, Illinois Institute of Technology, 2002.
4. K.M. Alzoubi, P.-J. Wan, and O. Frieder. New distributed algorithm for connected dominating set in wireless ad hoc networks. In *Proc. IEEE HICSS-35*, Hawaii, 2002.
5. A.D. Amis and R. Prakash. Load-balancing clusters in wireless ad hoc networks. In *Proc. 3rd IEEE Symp. Application-Specific Syst. Software Eng. Technol.*, 2000.

6. A.D. Amis, R. Prakash, D. Huynh, and T. Vuong. Max–min d-cluster formation in wireless ad hoc networks. In *Proc. 19th Annu. Joint Conf. IEEE Computer Commun. Soc. INFOCOM*, 1, 32–41, 2000.
7. S. Arya, G. Das, D. Mount, J. Salowe, and M. Smid. Euclidean spanners: short, thin, and lanky. In *Proc. 27th ACM STOC*, Las Vegas, 1995.
8. S. Arya and M. Smid. Efficient construction of a bounded degree spanner with low weight. In *Proc. 2nd Annu. Eur. Symp. Algorithms (ESA)*, vol. 855, *Lecture Notes Computer Sci.*, 48–59, 1994.
9. S. Arya, G. Das, D. M. Mount, J.S. Salowe, and M. Smid. Euclidean spanners. In *Proc. 27th Annu. ACM Symp. Theory Computing*, 1995.
10. M. Bahramgiri, M.T. Hajiaghayi, and V.S. Mirrokni. Fault-tolerant and 3-dimensional distributed topology control algorithms in wireless multi-hop networks. In *Proc. 11th Annu. IEEE Int. Conf. Computer Commun. Networks (ICCCN)*, 2002.
11. S. Basagni. Distributed clustering for ad hoc networks. In *Proc. IEEE Int. Symp. Parallel Architectures, Algorithms, Networks (I-SPAN)*, 1999.
12. S. Basagni, I. Chlamtac, and A. Farago. A generalized clustering algorithm for peer-to-peer networks. In *Proc. Workshop Algorithmic Aspects Commun.*, 1997.
13. S. Basagni, I. Chlamtac, V.R. Syrotiuk, and B.A. Woodward. A distance routing effect algorithm for mobility (dream). In *Proc. ACM/IEEE MobiCom'98*, 1998.
14. P. Berman, M. Furer, and A. Zelikovsky. Applications of matroid parity problem to approximating steiner trees. Technical report 980021, Computer science, UCLA, 1998.
15. C. Bettstetter. On the minimum node degree and connectivity of a wireless multihop network. In *3rd ACM Int. Symp. Mobile Ad Hoc Networking Computing (MobiHoc'02)*, June 2002.
16. P. Bose, A. Brodnik, S Carlsson, E.D. Demaine, R. Fleischer, A. Lopez-Ortiz, P. Morin, and J.I. Munro. Online routing in convex subdivisions. In *Proc. Int. Symp. Algorithms Computation*, 2000.
17. P. Bose, L. Devroye, W. Evans, and D. Kirkpatrick. On the spanning ratio of gabriel graphs and beta-skeletons. In *Proc. Latin Am. Theor. Infocomatics (LATIN)*, 2002.
18. P. Bose and P. Morin. Online routing in triangulations. In Proc. 10th Annu. Int. Symp. Algorithms Computation ISAAC, 1999.
19. P. Bose and P. Morin. Competitive online routing in geometric graphs. In Proc. VIII Int. Colloquium Structural Inf. Commun. Complexity (SIROCCO 2001), 2001.
20. P. Bose, P. Morin, I. Stojmenovic, and J. Urrutia. Routing with guaranteed delivery in ad hoc wireless networks. *ACM/Kluwer Wireless Networks*, 7(6), 609–616, 2001.
21. Prosenjit Bose, Joachin Gudmundsson, and Pat Morin. Ordered theta graphs. In *Proc. Can. Conf. Computational Geometry (CCCG)*, 2002.
22. P. Bose, J. Gudmundsson, and M. Smid. Constructing plane spanners of bounded degree and low weight. In *Proc. Eur. Symp. Algorithms (ESA)*, 2002.
23. J. Broch, D. Johnson, and D. Maltz. The dynamic source routing protocol for mobile ad hoc networks, 1998.
24. M. Cagalj, J.-P. Hubaux, and C. Enz. Minimum-energy broadcast in all-wireless networks: Np-completeness and distribution issues. In *Proc. ACM MOBICOM 02*, 2002.
25. G. Clinescu. Computing 2-hop neighborhoods in ad hoc wireless networks. In *Proc. Ad Hoc-Now 03*, 2003.
26. S. Capkun, M. Hamdi, and J.P. Hubaux. Gps-free positioning in mobile ad-hoc networks. In *Proc. Hawaii Int. Conf. Syst. Sci.*, 2001.
27. J. Cartigny, D. Simplot, and I. Stojmenvic. Localized minimum-energy broadcasting in ad-hoc networks. In *Proc. IEEE INFOCOM 2003*, 2003.
28. B. Chandra, G. Das, G. Narasimhan, and J. Soares. New sparseness results on graph spanners. In *Proc. 8th Annu. Symp. Computational Geometry*, 1992.
29. G. Chen, F. Garcia, J. Solano, and I. Stojmenovic. Connectivity based k-hop clustering in wireless networks. In *Proc. IEEE Hawaii Int. Conf. Syst. Sci.*, 2002.

30. X. Cheng et al. Polynomial-time approximation scheme for minimum connected dominating set in ad hoc wireless networks. *Network*, 42(4), 202–208, 2003.
31. J. Cheriyan, S. Vempala, and A. Vetta. Approximation algorithms for minimum-cost k-vertex connected subgraphs. In *Proc. 34th Annu. ACM Symp. Theory Computing*, 306–312. ACM Press, 2002.
32. C.C. Chiang and M. Gerla. On-demand multicast in mobile wireless networks. In *Proc. IEEE INCP*, 1998.
33. C.C. Chiang. Routing in clustered multihop, mobile wireless networks with fading channel. In *Proc. IEEE SICON'97*, 197–211, Apr. 1997.
34. I. Chlamtac and A. Farago. A new approach to design and analysis of peer to peer mobile networks. *Wireless Networks*, 5, 149–156, 1999.
35. V. Chvátal. A greedy heuristic for the set-covering problem. *Math. Operations Res.*, 4(3), 233–235, 1979.
36. A. Clementi, P. Crescenzi, P. Penna, G. Rossi, and P. Vocca. On the complexity of computing minimum energy consumption broadcast subgraphs. In *Proc. 18th Annu. Symp. Theor. Aspects Computer Sci., LNCS 2010*, 2001.
37. A. Clementi, P. Penna, and R. Silvestri. Hardness results for the power range assignment problem in packet radio networks. In *Proc. II Int. Workshop Approximation Algorithms Combinatorial Optimization Problems (RANDOM/APPROX'99), LNCS* (1671), 197–208, 1999.
38. A. Clementi, P. Penna, and R. Silvestri. The power range assignment problem in radio networks on the plane. In *Proc. XVII Symp. Theor. Aspects Computer Sci. (STACS'00), LNCS*(1770), 651–660, 2000.
39. A. Clementi, P. Penna, and R. Silvestri. On the power assignment problem in radio networks. *Mobile Networks and Application*, 9(2), 125–140, 2004.
40. A.E.F. Clementi, P. Penna, and R. Silvestri. On the power assignment problem in radio networks, 2000.
41. T.J. Cormen, C.E. Leiserson, and R.L. Rivest. *Introduction to Algorithms*. MIT Press and McGraw–Hill, 1990.
42. A.K. Das, R.J. Marks, M. El-Sharkawi, P. Arabshahi, and A. Gray. Minimum power broadcast trees for wireless networks: an ant colony systems approach. In *Proc. IEEE Int. Symp. Circuits Syst.*, 2002.
43. A.K. Das, R.J. Marks, M. El-Sharkawi, P. Arabshahi, and A. Gray. Minimum power broadcast trees for wireless networks: integer programming formulations. In *Proc. IEEE INFOCOM 2003*, 2003.
44. B. Das and V. Bharghavan. Routing in ad-hoc networks using minimum connected dominating sets. In *Proc. IEEE Int. Conf. Commun. (ICC'97)*, 1, 376–380, 1997.
45. G. Das and G. Narasimhan. A fast algorithm for constructing sparse euclidean spanners. In *Proc. 10th Annu. Symp. Computational Geometry*, 1994.
46. G. Das, G. Narasimhan, and J. Salowe. A new way to weigh malnourished euclidean graphs. In *Proc. ACM Symp. Discrete Algorithms*, 215–222, 1995.
47. S. Datta, I. Stojmenovic, and J. Wu. Internal node and shortcut based routing with guaranteed delivery in wireless networks. *Cluster Computing*, 5(2), 169–178, 2002.
48. D.P. Dobkin, S.J. Friedman, and K.J. Supowit. Delaunay graphs are almost as good as complete graphs. *Discrete and Computational Geometry*, 5, 399–407, 1990.
49. H. Edelsbrunner. *Algorithms in Combinatorial Geometry*. Springer–Verlag, 1987.
50. D. Estrin, R. Govindan, J.S. Heidemann, and S. Kumar. Next century challenges: scalable coordination in sensor networks. In *Proc. Mobile Computing Networking (Mobicom)*, 1999.
51. M. Faloutsos and M. Molle. Creating optimal distributed algorithms for minimum spanning trees. Technical report CSRI-327 (also submitted in WDAG'95), 1995.
52. H.N. Gabow, J.L. Bently, and R.E. Tarjan. Scaling and related techniques for geometry problems. In *Proc. ACM Symp. Theory Computing*, 1984.
53. K.R. Gabriel and R.R. Sokal. A new statistical approach to geographic variation analysis. *Systematic Zoology*, 18, 259–278, 1969.

54. R. Gallager, P. Humblet, and P. Spira. A distributed algorithm for minimum weight spanning trees. *ACM Trans. Programming Languages Syst.*, 5(1), 66–77, 1983.
55. J. Gao, L.J. Guibas, J. Hershburger, L. Zhang, and A. Zhu. Geometric spanner for routing in mobile networks. In *Proc. 2nd ACM Symp. Mobile Ad Hoc Networking Computing (MobiHoc 01)*, 2001.
56. J.A. Garay, S. Kutten, and D. Peleg. A sub-linear time distributed algorithms for minimum-weight spanning trees. In *Proc. Symp. Theory Computing*, 1993.
57. M.R. Garey and D.S. Johnson. *Computers and Intractability*. W.H. Freeman and Co., New York, 1979.
58. P.-L. George and H. Borouchaki. *Delaunay Triangulations and Meshing*. HERMES, Paris, 1998.
59. M. Grünewald, T. Lukovszki, C. Schindelhauer, and K. Volbert. Distributed maintenance of resource efficient wireless network topologies, 2002. *Proc. 8th Eur. Conf. Parallel Computing* (Euro-Par'02).
60. J. Gudmundsson, C. Levcopoulos, and G. Narasimhan. Improved greedy algorithms for constructing sparse geometric spanners. In *Proc. Scand. Workshop Algorithm Theory*, 2000.
61. S. Guha and S. Khuller. Approximation algorithms for connected dominating sets. In *Proc. Eur. Symp. Algorithms*, 1996.
62. P. Gupta and P. Kumar. Capacity of wireless networks. Technical report, University of Illinois, Urbana–Champaign, 1999.
63. P. Gupta and P.R. Kumar. Critical power for asymptotic connectivity in wireless networks. *Stochastic Analysis, Control, Optimization and Applications: A Volume in Honor of W.H. Fleming*, W.M. McEneaney, G. Yin, and Q. Zhang (Eds.), Birkhauser, Boston, 1998.
64. Z. Haas and B. Liang. Ad-hoc mobility management with uniform quorum systems. *IEEE/ACM Trans. Networking*, 7(2), 228–240, 1999.
65. M.T. Hajiaghayi, N. Immorlica, and V.S. Mirrokni. Power optimization in fault-tolerant topology control algorithms for wireless multi-hop networks. In *Proc. ACM MOBICOM 03*, 2003.
66. C. Ho, K. Obraczka, G. Tsudik, and K. Viswanath. Flooding for reliable multicast in multi-hop ad hoc networks. In *Proc. 3rd Int. Workshop Discrete Algorithms Methods Mobile Computing Commun.*, Seattle, WA, 1999.
67. D.S. Hochbaum and W. Maass. Approximation schemes for covering and packing problems in image processing and VLSI. *J. ACM*, 32, 130–136, 1985.
68. H.B. Hunt III, M.V. Marathe, V. Radhakrishnan, S.S. Ravi, D.J. Rosenkrantz, and R.E. Stearns. NC-approximation schemes for NP- and PSPACE -hard problems for geometric graphs. *J. Algorithms*, 26(2), 238–274, 1998.
69. C. Intanagonwiwat, R. Govindan, and D. Estrin. Directed diffusion: a scalable and robust Commun. paradigm for sensor networks. In *Proc. Mobile Computing Networking (Mobicom)*, 2000.
70. A. Iwata, C.-C. Chiang, G. Pei, M. Gerla, and T.-W. Chen. Scalable routing strategies for ad hoc wireless networks. *JSAC99, IEEE J. Selected Areas Commun.*, 17(8), 1369–1379, August 1999.
71. J.G. Jetcheva, Y.-C. Hu, D.A. Maltz, and D.B. Johnson. A simple protocol for multicast and broadcast in mobile ad hoc networks, 2001.
72. L. Jia, R. Rajaraman, and C. Scheideler. On local algorithms for topology control and routing in ad hoc networks. In *Proc. 15th Annu. ACM Symp. Parallel Algorithms Architectures*, 2003.
73. L. Jia, R. Rajaraman, and T. Suel. An efficient distributed algorithm for constructing small dominating sets. In *Proc. ACM PODC*, 2000.
74. M. Joa-Ng and I-T. Lu. A peer-to-peer zone-based two-level link state routing for mobile ad hoc networks. *IEEE J. Selected Areas Commun.*, 17(8), 1415–1425, August 1999.
75. J.M. Kahn, R.H. Katz, and K.S.J. Pister. Next century challenges: mobile networking for "smart dust." In *Proc. Int. Conf. Mobile Computing Networking (MOBICOM)*, 1999.
76. M.I. Karavelas and L.J. Guibas. Static and kinetic geometric spanners with applications. In *Proc.12th Annu. Symp. Discrete Algorithms*, 168–176, 2001.
77. B. Karp. Geographic routing for wireless networks. Ph.D. thesis, Harvard University, 2000.

78. B. Karp and H.T. Kung. GPSR: greedy perimeter stateless routing for wireless networks. In *Proc. ACM/IEEE Int. Conf. Mobile Computing . Networking (MobiCom)*, 2000.
79. J. Katajainen. The region approach for computing relative neighborhood graphs in the lp metric. *Computing*, 40, 147–161, 1988.
80. J.M. Keil and C.A. Gutwin. Classes of graphs which approximate the complete Euclidean graph. *Discrete and Computational Geometry*, 7(1), 13–28, 1992.
81. J.M. Keil. Approximating the complete Euclidean graph. In *Proc. SWAT 88: 1st Scand. Workshop Algorithm Theory*, 1988.
82. L. Kirousis, E. Kranakis, D. Krizanc, and A. Pelc. Power consumption in packet radio networks. In *Proc. Symp. Theor. Aspects Computer Sci. (STACS)'97.*, 1997.
83. L.M. Kirousis, E. Kranakis, D. Krizanc, and A. Pelc. Power consumption in packet radio networks. *Theor. Computer Sci.*, 243, 289–305, 2000.
84. G. Kortsarz and Z. Nutov. Approximating node connectivity problems via set covers. In *Proc. 3rd Int. Workshop Approximation Algorithms Combinatorial Optimization (APPROX 2000)*, 2000.
85. E. Kranakis, H. Singh, and J. Urrutia. Compass routing on geometric networks. In *Proc. 11th Can. Conf. Computational Geometry*, 1999.
86. F. Kuhn, R. Wattenhofer, Y. Zhang, and A. Zollinger. Geometric ad-hoc routing: of theory and practice. In *Proc. 22nd ACM Int. Symp. Principles Distributed Computing (PODC)*, 2003.
87. F. Kuhn, R. Wattenhofer, and A. Zollinger. Asymptotically optimal geometric mobile ad-hoc routing. In *Proc. 6th Int. Workshop Discrete Algorithms Methods Mobile Computing Commun. (Dial-M)*, 24–33. ACM Press, 2002.
88. F. Kuhn, R. Wattenhofer, and A. Zollinger. Worst-case optimal and average-case efficient geometric ad-hoc routing. In *Proc. 4th ACM Int. Symp. Mobile Ad-Hoc Networking Computing (MobiHoc)*, 2003.
89. C. Levcopoulos, G. Narasimhan, and M. Smid. Efficient algorithms for constructing fault-tolerant geometric spanners. In *Proc. 30th Annu. ACM Symp. Theory Computing*, 1998.
90. C. Levcopoulos, G. Narasimhan, and M. Smid. Improved algorithms for constructing fault tolerant geometric spanners. *Algorithmica*, 32(1), 144–156, 2002.
91. F. Li and I. Nikolaidis. On minimum-energy broadcasting in all-wireless networks. In *Proc. IEEE 26th Annu. IEEE Conf. Local Computer Networks (LCN'01)*, 2001.
92. J. Li, J. Jannotti, D. De Couto, D. Karger, and R. Morris. A scalable location service for geographic ad-hoc routing. In *Proc. 6th ACM Int. Conf. Mobile Computing Networking (MobiCom'00)*, 2000.
93. L. Li, J.Y. Halpern, P. Bahl, Y.-M. Wang, and R. Wattenhofer. Analysis of a cone-based distributed topology control algorithms for wireless multi-hop networks. In *Proc. ACM Symp. . Principles Distributed Computing (PODC)*, 2001.
94. N. Li and J.C. Hou. BLMST: a scalable, power efficient broadcast algorithm for wireless sensor networks, UIUC Computer Science Department Technical Report, 2003.
95. N. Li, J.C. Hou, and L. Sha. Design and analysis of a mst-based topology control algorithm. In *Proc. IEEE INFOCOM 2003*, 2003.
96. X.-Y. Li. Topology control in wireless ad hoc networks. In *Ad Hoc Networking*, IEEE Press, 2003.
97. X.-Y. Li. Approximate MST for UDG locally. In *Proc. 9th Annu. Int. Computing Combinatorics Conf. COCOON 2003*, 2003.
98. X.-Y. Li, G. Calinescu, and P.-J. Wan. Distributed construction of planar spanner and routing for ad hoc wireless networks. In *Proc. 21st Annu. Joint Conf. IEEE Computer Commun. Soc. (INFOCOM)*, 3, 2002.
99. X.-Y. Li and I. Stojmenovic. Partial Delaunay triangulation and degree limited localized bluetooth scatternet formation. In *Proc. Int. Conf. Ad Hoc Networks*.
100. X.-Y. Li, I. Stojmenovic, and Y. Wang. Partial delaunay triangulation and degree limited localized bluetooth multihop scatternet formation, *IEEE Trans. Parallel Distributed Systems*, 15(4), 350–361, 2004.

101. X.-Y. Li, P.-J. Wan, and O. Frieder. Coverage in wireless ad-hoc sensor networks. In *Proc. IEEE Int. Conf. Commun. (ICC)*, 2002.
102. X.-Y. Li, P.-J. Wan, and Y. Wang. Power efficient and sparse spanner for wireless ad hoc networks. In *Proc. IEEE Int. Conf. Computer Commun. Networks (ICCCN01)*, 2001.
103. X.-Y. Li, P.-J. Wan, Y. Wang, and O. Frieder. Sparse power efficient topology for wireless networks. In *Proc. IEEE Hawaii Int. Conf. Syst. Sci. (HICSS)*, 2002.
104. X.-Y. Li, P.-J. Wan, Y. Wang, and O. Frieder. Sparse power efficient topology for wireless networks. *J. Parallel Distributed Computing*, 2002. Accepted for publication. Short version in IEEE ICCCN 2001.
105. X.-Y. Li and Y. Wang. Efficient construction of bounded degree planar spanner. In *Proc. 9th Annu. Int. Computing Combinatorics Conf. (COCOON2003)*, 2003.
106. X.-Y. Li and Y. Wang. Localized routing for wireless ad hoc networks. In *IEEE Int. Conf. Commun. (ICC 2003)*, 2003.
107. X.-Y. Li, Y. Wang, P.-J. Wan, Wen-Zhan Song, and O. Frieder. Localized low weight graph and its applications in wireless ad hoc networks, 2003. in *Proc. IEEE INFOCOM*, 2004.
108. X.-Y. Li, Y. Wang, P.-J. Wang, and C.-W. Yi. Fault tolerant deployment and topology control for wireless ad hoc networks. In *Proc. 4th ACM Int. Symp. Mobile Ad Hoc Networking Computing (MobiHoc2003)*, 2003.
109. W. Liang. Constructing minimum-energy broadcast trees in wireless ad hoc networks. In *Proc. ACM Int. Symp. Mobile Ad Hoc Networking Computing (MOBIHOC)*, 2002.
110. C.R. Lin and M. Gerla. Adaptive clustering for mobile wireless networks. *IEEE J. Selected Areas Commun.*, 15(7), 1265–1275, 1997.
111. J.-C. Lin, S.-N. Yang, and M.-S. Chern. An efficient distributed algorithm for minimal connected dominating set problem. In *Proc. 10th Annu. Int. Phoenix Conf. Computers Commun. 1991*, 1991.
112. M. Liu, R. Talpade, and A. McAuley. Amroute: Adhoc multicast routing protocol, 1998.
113. L. Lloyd, R. Liu, M.V. Marathe, R. Ramanathan, and S.S. Ravi. Algorithmic aspects of topology control problems for ad hoc networks. In *IEEE MOBIHOC*, 2002.
114. T. Lukovszki. New results on geometric spanners and their applications. Ph.D. thesis, University of Paderborn, 1999.
115. M.V. Marathe, H. Breu, H.B. Hunt III, S..S. Ravi, and D.J. Rosenkrantz. Simple heuristics for unit disk graphs. *Networks*, 25, 59–68, 1995.
116. R.J. Marks, A.K. Das, M. El-Sharkawi, P. Arabshahi, and A. Gray. Minimum power broadcast trees for wireless networks: optimizing using the viability lemma. In *Proc. IEEE Int. Symp. Circuits Syst.*, 245–248, 2002.
117. M. Mauve, J. Widmer, and H. Harenstein. A survey on position-based routing in mobile ad hoc networks. *IEEE Network Mag.*, 15(6), 30–39, 2001.
118. S. Megerian and M. Potkonjak. Wireless sensor networks. Book chapter in *Wiley Encyclopedia of TeleCommunications*, J.G. Proakis, Ed., 2002.
119. S. Meguerdichian, F. Koushanfar, M. Potkonjak, and M. Srivastava. Coverage problems in wireless ad-hoc sensor network. In *Proc. IEEE INFOCOM'01*, 2001.
120. J. Monks, V. Bharghavan, and W.-M. Hwu. Transmission power control for multiple access wireless packet networks. In *Proc. IEEE Conf. Local Computer Networks (LCN)*, 2000.
121. P. Morin. Online routing in geometric graphs. Ph.D. thesis, Carleton University School of Computer Science, 2001.
122. V.D. Park and M.S. Corson. A highly adaptive distributed routing algorithm for mobile wireless networks. In *Proc. IEEE INFOCOM'97, Kobe, Japan*, April 1997.
123. G. Pei, M. Gerla, and T.-W. Chen. Fisheye state routing: a routing scheme for ad hoc wireless networks. In *Proc. ICC 2000, New Orleans, LA*, 2000.
124. G. Pei, M. Gerla, and T.-W. Chen. Fisheye state routing in mobile ad hoc networks. In *Proc. Workshop Wireless Networks Mobile Computing, Taipei, Taiwan, Apr*, 2000.

125. M. Penrose. The longest edge of the random minimal spanning tree. *Ann. Appl. Probability*, 7, 340–361, 1997.
126. M. Penrose. On *k*-connectivity for a geometric random graph. *Random Structures Algorithms*, 15, 145–164, 1999.
127. C.E. Perkins and P. Bhagwat. Highly dynamic destination-sequenced distance-vector routing (DSDV) for mobile computers. *Computer Commun. Rev.*, 234–244, October 1994.
128. C.E. Perkins and E.M.Royer. Ad-hoc on demand distance vector routing. In *Proc. 2nd IEEE Workshop Mobile Computing Syst. Applications, New Orleans, LA*, 90–100, February 1999.
129. C.E. Perkins, E.M. Belding–Royer, and S. Das. Ad hoc on demand distance vector (AODV) routing, 2003.
130. G.J. Pottie and W.J. Kaiser. Wireless integrated network sensors. *Commun. ACM*, 43(5), 551–558, 2000.
131. F.P. Preparata and M.Ian Shamos. *Computational Geometry: an Introduction*. Springer–Verlag, 1985.
132. A. Qayyum, L. Viennot, and A. Laouiti. Multipoint relaying: an efficient technique for flooding in mobile wireless networks. Technical research report RR-3898, INRIA, February 2000. Conf. version in *Proc. IEEE HICSS'35*, 2001.
133. R. Rajaraman. Topology control and routing in ad hoc networks: a survey. In *SIGACT News*, 33, 60–73, 2002.
134. R. Ramanathan and R. Rosales–Hain. Topology control of multihop wireless networks using transmit power adjustment. In *Proc. IEEE INFOCOM*, 2000.
135. G. Robins and A. Zelikovsky. Improved steiner tree approximation in graphs. In *Proc. ACM/SIAM Symp. Discrete Algorithms*, 2000.
136. E. Royer and C. Toh. A review of current routing protocols for ad-hoc mobile wireless networks. i.e.,*EE Personal Commun.*, Apr. 1999.
137. M. Sanchez, P. Manzoni, and Z. Haas. Determination of critical transmission range in ad-hoc networks. In *Proc. Multiaccess, Mobility Teletraffic Wireless Commun. (MMT'99)*, 1999.
138. C. Santivanez, B. McDonald, I. Stavrakakis, and R. Ramanathan. The scalability of ad hoc routing protocols. In *Proc. IEEE INFOCOM 2002*, 2002.
139. M. Seddigh, J. Solano Gonzalez, and I. Stojmenovic. RNG and internal node based broadcasting algorithms for wireless one-to-one networks. *ACM Mobile Computing. Commun. Rev.*, 5(2), 37–44, 2002.
140. P. Sinha, R. Sivakumar, and V. Bharghavan. Cedar: Core extraction distributed ad hoc routing, In *Proc. IEEE INFOCOM*, 1999.
141. R. Sivakumar, B. Das, and V. Bharghavan. An improved spine-based infrastructure for routing in ad hoc networks. In *IEEE Symp. Computers Commun.*, Athens, Greece, June 1998.
142. I. Stojmenovic. A routing strategy and quorum based location update scheme for ad hoc wireless networks. Technical report TR-99-09, Computer Science, SITE, University of Ottawa, 1999.
143. I. Stojmenovic and S. Datta. Power and cost aware localized routing with guaranteed delivery in wireless networks. In *Proc. 7th IEEE Symp. Computers Commun. ISCC*, 2002.
144. I. Stojmenovic, M. Seddigh, and J. Zunic. Dominating sets and neighbor elimination based broadcasting algorithms in wireless networks. *IEEE Trans. Parallel Distributed Syst.*, 13(1), 14–25, 2002.
145. I. Stojmenovic and X. Lin. Loop-free hybrid single-path/flooding routing algorithms with guaranteed delivery for wireless networks. *IEEE Trans. Parallel Distributed Syst.*, 12(10), 2001.
146. I. Stojmenovic, M. Seddigh, and J. Zunic. Dominating sets and neighbor elimination based broadcasting algorithms in wireless networks. *IEEE Trans. Parallel Distributed Syst.*, 13(1), 14–25, 2002.
147. C-K. Toh. A novel distributed routing protocol to support ad-hoc mobile networks. In *Proc 1996 IEEE 15th Annu. Int. Phoenix Conf. Comp. Commun.*, Mar 1996.
148. G.T. Toussaint. The relative neighborhood graph of a finite planar set. *Pattern Recognition*, 12(4), 261–268, 1980.
149. Y.-C. Tseng, S.-Y. Ni, Y.-S. Chen, and J.-P. Sheu. The broadcast storm problem in a mobile ad hoc network. *Wireless Networks*, 8, 153–167, 2002. Short version in MOBICOM 99.

150. P.-J. Wan, K.M. Alzoubi, and O. Frieder. Distributed construction of connected dominating set in wireless ad hoc networks. In *INFOCOM*, 2002.
151. P.-J. Wan, G. Calinescu, X.-Y. Li, and O. Frieder. Minimum-energy broadcast routing in static ad hoc wireless networks. *ACM Wireless Networks*, 2002. Preliminary version appeared in IEEE INFOCOM 2000.
152. W.Z. Wang, X.-Y. Li, K. Moaveninejad, Y. Wang, and W.-Z. Song. The spanning ratios of beta-skeleton. In *Proc. Canadian Conf. Computational Geometry (CCCG)*, 2003.
153. Y. Wang and X.-Y. Li. Distributed spanner with bounded degree for wireless ad hoc networks. In *Proc. Int. Parallel Distributed Process. Symp.: Parallel Distributed Computing Issues Wireless Networks Mobile Computing*, April 2002.
154. Y. Wang and X.-Y. Li. Geometric spanners for wireless ad hoc networks. In Proc. 22nd IEEE Int. Conf. Distributed Computing Syst. (ICDCS), 2002.
155. Y. Wang and X.-Y. Li. Localized construction of bounded degree planar spanner for wireless networks. In *Proc. ACM DIALM-POMC Joint Workshop Foundations Mobile Computing*, 2003.
156. R. Wattenhofer, L. Li, P. Bahl, and Y.-M. Wang. Distributed topology control for wireless multihop ad-hoc networks. In *Proc. IEEE INFOCOM'01*, 2001.
157. J. Wieselthier, G. Nguyen, and A. Ephremides. On the construction of energy-efficient broadcast and multicast trees in wireless networks. In *Proc. IEEE INFOCOM 2000*, 2000.
158. J.E. Wieselthier, G.D. Nguyen, and A. Ephremides. The energy efficiency of distributed algorithms for broadcasting in ad hoc networks. In *Proc. IEEE 5th Int. Symp. Wireless Personal Multimedia Commun. (WPMC)*, 2002.
159. B. Williams and T. Camp. Comparison of broadcasting techniques for mobile ad hoc networks. In *Proc. ACM Int. Symp. Mobile Ad Hoc Networking Computing (MOBIHOC)*, 194–205, 2002.
160. C.W. Wu, Y.C. Tay, and C.-K. Toh. Ad hoc multicast routing protocol utilizing increasing ID-numbers (AMRIS) functional specification, 1998.
161. J. Wu, F. Dai, M. Gao, and I. Stojmenovic. On calculating power-aware connected dominating sets for efficient routing in ad hoc wireless networks. *IEEE/KICS J. Commun. Networks*, 4(1), 59–70, 2002.
162. J. Wu and H. Li. Domination and its applications in ad hoc wireless networks with unidirectional links. In *Proc. Int. Conf. Parallel Process. 2000*, 2000.
163. J. Wu and H. Li. A dominating-set-based routing scheme in ad hoc wireless networks. *Special Issue Wireless Networks TeleCommun. Systs J.*, 3, 63–84, 2001.
164. A.C.-C. Yao. On constructing minimum spanning trees in k-dimensional spaces and related problems. *SIAM J. Computing*, 11, 721–736, 1982.
165. Y. Yu, R. Govindan, and D. Estrin. Geographical and energy-aware routing: a recursive data dissemination protocol for wireless sensor networks. Technical report, UCLA Computer Science Department Technical Report UCLA/CSD-TR-01-0023, 2001.
166. Z.J. Haas and M.R. Pearlman. The zone routing protocol (ZRP) for ad hoc networks. In Internet draft — mobile ad hoc networking (MANET), Working Group of the Internet Engineering Task Force (IETF), November 1997.

40
Localized Algorithms for Sensor Networks

Jessica Feng
University of California at Los Angeles

Farinaz Koushanfar
University of California at Berkeley

Miodrag Potkonjak
University of California at Los Angeles

40.1 Introduction 40-1
    Motivation • Chapter Organization
40.2 Models and Abstractions 40-2
40.3 Centralized Algorithm 40-4
40.4 Case Studies 40-8
    Energy Management and Topology Maintenance • $(MI)^2$ • Solving ILP Problems by $(MI)^2$-Based Paradigm • GPSR
40.5 Analysis 40-12
40.6 Protocols and Distributed Localized Algorithms 40-13
40.7 Pending Challenges 40-15

40.1 Introduction

40.1.1 Motivation

Recently, wireless multihop networks (WMNs) have emerged as a promising architecture for realization of a various embedded distributed networked systems. WMNs can be used for a variety of tasks, including human communication and Internet-like data distribution. The most exciting application of wireless ad hoc networks is probably serving as the building platform for wireless sensor networks. In wireless sensor networks, each node is equipped with a certain amount of communication, computing, storage, sensing, and, in some scenarios, actuating resources. Wireless ad hoc sensor networks have the potential to bridge the gap between the Internet and the physical world. Numerous applications in the military environment as well as in personal and industrial tasks have been envisioned.

At the same time, wireless ad hoc sensor networks pose a number of new technological and optimization challenges. It is apparent that in order to address these challenges, sensor networks must operate in autonomous mode. In addition, in order to address low-energy, privacy, security, and scalability issues better, wireless sensor networks will require new types of algorithms that will use minimal amounts of communication. The goal of this chapter is to discuss the state of the art of algorithms commonly known as localized algorithms.

It is interesting to compare localized algorithms to other types of algorithms that have been excessively studied in computer science and related areas. In theoretical computer science and operational research, a great variety of algorithms has been developed for a wide range of combinatorial problems. These algorithms are developed under the following set of assumptions. The first is that constraints are on only two types of resources: storage and speed of computation. A number of models have been developed under this assumption, such as the Turing machine, Post's model, and the universal register machine. It has been demonstrated that these models are essentially equivalent. The inputs for the algorithm are specified at the beginning of its execution; run time and storage requirements serve as measurements of the quality of the solutions and algorithms. It is customary to consider algorithms that have run-time as polynomial functions with respect to the length of the input expressed in bits as efficient and the ones that require exponential time as inefficient.

On a more practical note, a number of paradigms that can be used to develop efficient algorithms have been identified, including divide and conquer; branch and bound; dynamic programming; and reduce and conquer. The key observation is that algorithms are designed and analyzed mainly based on how well they scale as the size of the input increases asymptotically. In addition, algorithms that guarantee optimal solutions and approaches guaranteeing that obtained solutions are within a certain vicinity of the optimal solution are widely studied (e.g., approximation algorithms), as well as algorithms that provide heuristic solutions when the problem is computationally intractable [3, 6, 7].

Although localized algorithms and even sensor networks have only been attracting research and development attention recently, already a wide literature and great variety of proposed approaches regarding the topic exist. It is already impossible to provide a comprehensive survey of all proposed algorithms for all wireless ad hoc sensor network tasks. The main objective in this chapter is to identify the most suitable abstractions and the most efficient techniques as foundations for developing localized algorithms. In addition, special emphasis is placed in summarizing how to analyze and evaluate localized algorithms. The goal is to cover all the most important developments as well as provide insights on why these algorithms are effective. In addition to presentation of already published results, several new algorithms that are optimal or superior to the published ones in terms of performance are proposed.

40.1.2 Chapter Organization

Section 40.2 summarizes all the proposed models, abstractions, and foundations for designing and analyzing localized algorithms in wireless sensor networks. In the next section, centralized algorithms that provide a comparison metric to localized algorithms are discussed. Section 40.4 presents several case studies for canonical problems in wireless sensor networks, as well as the existing algorithms, approaches and general paradigms. A number of widely applied analysis metrics and standards are presented in Section 40.5. In order to enable distributed localized algorithms, the different protocols in Section 40.6 can be applied in developing them; proposed techniques and algorithms for distributed localized algorithms are also discussed. Finally, Section 40.7 states some of the future conceptual, technological, and theoretical challenges related to localized algorithms.

40.2 Models and Abstractions

This section summarizes information about relevant models and abstractions required to specify and analyze localized algorithms. Much diversity is present among potential combinations of properties of models that can be used for this task. Many of them are interesting because they provide favorable trade-offs between their capability of capturing real-life sensor networks and their suitability for analysis and development of a variety of optimization techniques. Attention is focused on two groups: (1) those mainly related to widely used models in the literature; and (2) models favored by current and expected technology trends.

Currently, only static networks are considered when one studies models related to network topology. However, in the near future, a variety of models for mobile networks will appear. In order to ensure connectivity of all nodes, the standard assumption is that all nodes, when viewed at the graph level, form a single connected component. In addition, the edge between two nodes can be unidirectional or bidirectional. The first option is used when all nodes are equipped with identical radio transmitters and receivers. The second indicates situations in which node A can hear node B, but not vice versa. In addition, sometimes one or more nodes have special positions as gateways to the Internet or as base stations. The most important assumption about the network is related to the question of how much each node knows about the locations and connectivity of all other nodes.

The current standard assumption is that each node is only aware of its own neighborhood, i.e., nodes to which it can directly communicate. Sometimes this definition is enhanced to k-hop neighbors. In the future, schemes that explicitly state what is stored at each node will emerge. Essentially, as data structures play a crucial rule in the development of standard computer algorithms, data placement plays a crucial rule in localized algorithms. It is also important to note that as storage technology rapidly emerges, assuming that each node has only information about its own neighborhood is unrealistic. However, although information in static networks can be easily stored in each node, it would be expensive for each node to inform too many nodes about its status when the network is mobile or when an energy-saving procedure is conducted using sleeping mode.

Currently, it is most often assumed that nodes in the network are randomly deployed with uniform distribution in unit square areas. The assumption is justified in some scenarios, for example, when nodes are dropped from airplanes. However, it is obvious that new methodologies and approaches for WSNs with very different structural properties will emerge in order to address the needs of specific applications. In these networks, sensor placement will affect performance of localized algorithms in a very profound way.

Another aspect that is rarely discussed but crucially important is related to space topology and obstacles. For example, in environmental monitoring, simply ignoring trees and physical obstacles would inevitably result in incorrect conclusions. Finally, note that three-dimensional tasks are commonly significantly more difficult than two-dimensional tasks.

Currently, the standard assumption is that all nodes are equipped with identical transceivers and identical omnidirectional antennas. This assumption has the direct ramification that all two-communicating parties have the same transmission and reception strength. However, the communication range can be modeled in various ways depending on radios used. Four of the most intuitive options include:

- In the unit disk model, all nodes in the network have identical radio range.
- A generalization of the unit disk model is the arbitrary disk model, in which each node has an arbitrary radio range and is uniform along all directions. In this case, situations exist in which node A can hear node B, but node B cannot necessarily hear node A. Therefore, the arbitrary disk model requires directed graph for representation of the network connectivity.
- Another communication model relinquishes assumption of the uniformity of signal propagation along all directions and captures the statistical behavior of propagation signal as a probabilistic function of distance between the communicating node pair. Probability is different along different directions, but is a monotonically nonincreasing function along any given direction. Examples of the function that may be applied include the distance formula and the square of distance.
- Another option aims to incorporate complete arbitrariness in communication patterns. It assumes that communication between any two nodes, regardless of their positions, is established with a certain user-defined probability.

In addition to communication range, assumptions on the structure of transmitted data also play an important role in designing localized algorithms and evaluating their performance. The most widely adopted schemes are: (1) number of bits sent; (2) number of packets with no packet size restrictions; and (3) number of packets in which each packet has limited size.

The first option does not involve the concept of packet. Information is measured in terms of number of bits sent and received between two nodes that can communicate directly. The second scheme adopts the notion of packet, but packets are of a relatively large size relative to the information that must be sent so that they can be considered unlimited size packets. The last option imposes an upper limit of information that each packet can contain. Depending on the adopted communication models and the packet structure models, relative performances across different algorithms may significantly differ. Therefore, constructing algorithms most suited for the particular set-up so that they maximize the advantages of the assumptions is of great importance.

A number of energy consumption models exist. A specific example of an energy consumption model for wireless radio is given by Digitan. Assume 2 Mb/s 802.11; transmission takes 1.9 W of energy; reception takes roughly 1.5 W; idle/listening takes 0.75 W; and sleeping consumes only 0.025 W. The main observation is that unless the node is in the sleeping mode, no significant amount of energy can be conserved even if the node is in idle mode. The conclusion is simple and with strong ramifications: often it is more important to design localized algorithm that can be executed while a large percentage of nodes is in the sleeping mode.

Storage models can be categorized in two classes: direct and indirect storage. Direct storage implies that all the information each node stores is kept physically within the node. In indirect storage, data used by a node during execution of the localized algorithm are stored somewhere else — at some other node or possibly a separate gateway storage device. Therefore, this scheme requires an explicit step of referencing and communication in order to gain access to the information. Clearly, direct storage has advantages over indirect storage in terms of access time, flexibility, and communication cost. On the other hand, indirect storage can enable significantly better sharing of data as well as significant storage capacity enhancement.

Fault models are a well-studied topic and have been discussed comprehensively in VLSI and computer architecture literature. However, fault tolerance and therefore fault models have never been one of the dominating concerns and objectives for VLSI designs. The reason is that the properties of VLSI technology and design styles facilitate strong resiliency against faults naturally. However, wireless ad hoc sensor networks are vulnerable against faults (also equivalent attacks and data skewing) because of their wireless communication and localized mode of operation.

Furthermore, use of such networks also enhances the importance and the need for privacy and security. In addition, the observed physical world is full of obstacles that interfere with communication and sensing tasks. Sensor networks are often deployed in the physical world where the environment is complex or even hostile. For example, consider a habitat-monitoring sensor network deployed in a forest. Simply ignoring the existence of trees, plants, and other obstacles will lead to incorrect conclusions. Currently, fault tolerance is rarely addressed in sensor networks and the development of a fault-tolerant localized algorithm still must be addressed.

Sensing models capture sensitivity of a sensor as a function of parameters such as distances, properties of the environment, and position. For example, one can assume all sensors have only two sensitivity modes: detecting or not detecting an event. A widely used model for sensitivity is one in which the accuracy of sensing decreases according to a certain function of distance between the sensor and the target object. Linear and quadratic functions are often used.

40.3 Centralized Algorithm

This section discusses centralized algorithms for sensor networks. After the definition of centralized algorithms, their major advantages and disadvantages are briefly outlined. After that, several different scenarios in which centralized algorithms can be specified and analyzed are summarized. Special emphasis is placed on two phases: data collection and result dissemination. Several optimal centralized algorithms for common tasks in wireless sensor networks are presented.

Centralized algorithms in wireless ad hoc sensor network are procedures in which all information from all nodes in the network is first collected at a single, usually predefined, node. The problem is solved at this node and consequently the results of the optimization are disseminated to all nodes that requested this information. Therefore, three phases of centralized algorithms can be identified:

- Information collection in which readings of all sensors from all nodes are collected to a single computational point
- Optimization mechanism execution on that node
- Results of the optimization sent to all other nodes using multihop communication

One must study centralized algorithms for a given problem in which the primary goal is to develop the localized algorithm for several reasons. The first reason is that the centralized algorithm provides an upper bound of what is achievable with respect to the quality of the solution. At the same time, this algorithm also provides an upper bound of expected communication cost with respect to the corresponding localized algorithm. Note that both of the previous bounds are not actually guaranteed. For example, in the case of upper bound of the quality of the solution, if the problem is computationally intractable, it may happen that the localized algorithm "gets lucky" and produces a better solution than the centralized algorithm. In the case of communication cost, the centralized algorithm may get unlucky and some nodes are visited several times; therefore, energy consumption higher than the corresponding localized algorithm is the result.

There is a wide consensus that localized algorithms are the correct alternative for wireless ad hoc sensor networks. In a number of situations, centralized algorithms are obviously competitive if not better. For example, if the network is reasonably small and one must conduct several optimization problems at the same time, centralized algorithms are certainly attractive options to consider. Also, centralized algorithms are particularly well suited for a mapping problem in which each node must get a specific set of attributes.

It is important and interesting to consider relative advantages and disadvantages of centralized algorithms with respect to corresponding localized algorithms. In a number of aspects, centralized algorithms have significant advantages over localized algorithms. For example, the main logistic advantage is that optimization mechanisms do not need to be customized as in the case of localized algorithms. In addition, absolutely the same data collection and distribution algorithm and software can be applied to all problems. Furthermore, synthesis and analysis of centralized algorithms are significantly simpler conceptually and logistically than in the case of localized algorithms. For mapping problems, centralized algorithms are often competitive in terms of the communication cost. Finally, performance and cost of centralized algorithms most often have significantly lower variance in terms of quality of solution and communication cost than those of localized algorithms.

Nevertheless, localized algorithms have significant advantages in many situations that often greatly outweigh their limitations. For example, if size of the network increases, localized algorithms inevitably become the only realistic option. In particular, they show great advantages when search problems are addressed. Furthermore, localized algorithms provide strong advantages in terms of fault tolerance, security, and privacy. Finally, localized algorithms are much better suited for customization with respect to specific optimization mechanisms and communication models.

The advantages and disadvantages of centralized algorithms will be illustrated using several different abstractions and modeling scenarios. Three scenarios in which a centralized node is the Internet gateway that contains unlimited computation, storage, and energy supply resources will be considered. Note that, in this case, the centralized node has enough storage to contain all information regarding all nodes and their connectivity.

First consider a case in which the communication cost is measured in terms of transmitted data bits. This problem can be solved optimally. All that is required is that each node send its information using the shortest path to the centralized node. Dijkstra's algorithm can provide the solution in linear time in terms of number of edges in the graph. Notice that, because each node is sending information using the most sufficient route, the optimality of the algorithm is guaranteed.

In the second scenario, the communication cost is measured in terms of the total number of packets transmitted. In this case, the assumption that a packet has unlimited size is adopted; this is reasonable when the network is relatively small and the packet size limit is relatively large. In this case, the problem can also be optimally solved. The solution is based on the observation that each node must send its information at least once to some other node. Therefore, if the algorithm only requires each node to send its information once, the optimality is automatically achieved. The first step of the algorithm is to conduct breath-first search (BFS) in order to find the distance in terms of hops of each node from the centralized node. After that, each node in the network is scheduled to transmit its data or the data that it has received in a decreasing order according to its distance from the centralized node.

The third scenario is the situation in which the packet size is fixed to a certain amount and the goal is again to transmit the minimal number of packets. Unfortunately, the problem is now computationally intractable. Still, it can be solved optimally using integer linear programming (ILP)-based approaches. Note that, in many situations — particularly when the network is relatively small and sparse — this is attractive because it must be solved only once per lifetime for the network. The following variables are introduced:

$$X_{ij} = \begin{cases} m & \text{node } i \text{ sends } m \text{ bits to node } j \\ 0 & o/w \end{cases} \quad (40.1)$$

$$X_i = \begin{cases} l & \text{node } i \text{ sends } l \text{ outgoing bits} \\ 0 & o/w \end{cases} \quad (40.2)$$

$$Y_{ij} = \begin{cases} k & \text{node } i \text{ sends } k \text{ packets to node } j \\ 0 & o/w \end{cases} \quad (40.3)$$

There are two types of constraints. First, for each and every node, the outgoing number of bits that it sends out must equal the sum of the received bits plus the number of bits recorded. The second constraint ensures that the number of packets is sufficient to transfer the number of bits that need to be transmitted:

$$\left(\sum_{i=1}^{n} x_{ij}\right) + (R_i) = \forall i \quad (40.4)$$

$$y_{ij} > \frac{x_{ij}}{P} \quad (40.5)$$

where
R_i = number of bits that node i has recorded
P = packet size limit in terms of bits
n = total number of nodes in the network

The objective function is to minimize the number of total packets sent; therefore:

$$\min: \sum_{i=1}^{n} \sum_{j=1}^{n} y_{ij} \quad (40.6)$$

Now consider the same three scenarios when there is no explicitly predefined centralized node. If the assumption is that each and every node is aware of the situation of the entire network, only minor modifications to the existing approaches would be sufficient. In the first scenario in which communication cost is measured in terms of bits, conduct the 1-to-n shortest path using Dijkstra's algorithm at each and every node, and select as the centralized node the one node with the smallest sum of shortest paths to all other nodes. In the case of the second scenario, in which the packet size is large enough to be considered unlimited size, all nodes have the same quality to be the centralized node; therefore, any arbitrary node can be served as the centralized node. In the case of the third scenario, in which the packet size is limited, one arbitrary node solves the system using the same ILP formulation with the assumption of a different centralized node, selects the node that provides the best objective function value when it is assumed to be the centralized node, and notifies this node to continue the procedure.

If the assumption is that each node only knows its limited neighborhood information, the problem becomes more complicated. In this case, the "spiral" algorithm [10] is proposed. Starting from an arbitrarily selected node, the goal is to minimize the number of times each node is visited in order to

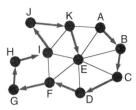

FIGURE 40.1 Example topology.

collect all the information in the network. The algorithm can be best understood in a geometric context. Consider the following illustrative example:

Figure 40.1 presents a network with 11 nodes; each is only aware of its own one-hop neighbors. Let node A be the arbitrary starting point; using the clockwise "sweeping" search technique, A finds the first occurrence of a nonvisited node, i.e., node B in this case. Therefore node A sends all its information to B. Now B applies the same technique to find the next first occurrence of a nonvisited node; this is node C. Node B forwards what node A has sent and node B's own data recorded to node C. This procedure continues until node E, which "sweeps" 360°. However, all the nodes encountered have been visited, so node E concludes that it has all the information in the network and announces that it is the centralized node.

Once all the information is present at the centralized node, it can apply various optimization techniques to obtain solutions. Focus on the last phase of the centralized algorithm — solution dissemination. The problem is equivalent to the broadcasting problem, which can be again addressed using ILP. Define the following variables:

$$X_i = \begin{cases} 1 & \text{node } i \text{ broadcasts} \\ 0 & o/w \end{cases} \quad (40.7)$$

$$X_{ij} = \begin{cases} 1 & \text{node } i \text{ sends message to } j \\ 0 & o/w \end{cases} \quad (40.8)$$

Using these specified variables, the following three constraints are enforced. First, each node must receive the information from some other node in the network. The second type of constraint ensures that only nodes within communication range of each other can communicate. The third type of constraint ensures that the broadcasting node is only charged once no matter how many nodes have received messages from it.

$$\sum_{\substack{i=1 \\ i \neq j}}^{n} x_{ij} \geq 1 \quad j = 1,\ldots,n \quad (40.9)$$

where n = total number of nodes in the network.

$$x_{ij} \quad \text{if } E_{ij} \neq 1 \quad (40.10)$$

$$x_{ij} \leq x_i \quad i = 1,\ldots n; \quad j = 1,\ldots n; \quad i \neq j \quad (40.11)$$

The objective is again to minimize the number of packets sent, i.e., minimize the number of nodes that broadcast:

$$\min: \sum_{i=1}^{n} x_i \tag{40.12}$$

40.4 Case Studies

40.4.1 Energy Management and Topology Maintenance

A number of alternative power minimization methods act above the MAC layer powering off redundant nodes' radios in order to expand the battery lifetimes. For example, AFECA [19] trades off energy consumption and the quality of the message delivery services based on the application requirements. GAF [20] is another power-saving scheme that saves energy by powering off the redundant nodes. GAF identifies the redundant nodes by using the geographic location and a conservative estimate of the radio ranges. It superimposes a virtual grid proportional to the communication radius of the nodes onto the network. Because the nodes in one grid are equal from the routing perspective, the radios of the redundant nodes within a grid can be turned off. The nodes awake within a grid rotate to balance their energy.

One of the main advantages of GAF is that it is completely static and localized. All nodes are capable of estimating virtual grids and determining equivalent nodes. In addition to saving 40 to 60% of the energy compared to an unmodified ad hoc routing protocol, GAF also suggests that network lifetime increases proportionally to node density. On the other hand, a significant performance bottleneck can be easily created by grids that contain very limited number of nodes. Moreover, sometimes it is acceptable to let all nodes in some grids sleep (e.g., the boundary nodes) in order to reduce energy further. However, this situation cannot be recognized by GAF.

SPAN is a power-saving, distributed, randomized coordination approach [1] that preserves connectivity in wireless networks. The work presented in Koushanfar and colleagues [9] has proved the necessary and sufficient conditions for putting the radios in the sleep mode, while still guaranteeing connectivity. A major advantage of this scheme is that all the decisions are made locally and individually. Therefore, it is much more robust, flexible, and scalable than the centralized schemes. In addition, according to the condition of the network, coordinator nodes are adjusted and re-elected locally as well. However, SPAN shares some similar limitations with GAF, in particular with respect to energy savings. For example, in some situations not all coordinator nodes need to be awake.

There are also a number of research efforts that trade off between latency and energy consumption. The power management approach presented in Kravets and Krishnan [12] selectively chooses short periods of time to suspend and shut down the communication unit; they queue the data before suspending the communication. STEM is a power-saving strategy [17] that does not try to preserve the capacity of the network. STEM works by putting an increasing number of nodes into sleep mode, and then encountering the latency to set up a multihop path. Nodes in STEM must have an extra low power radio (paging channel) that does not go into sleeping mode and constantly monitors the network to wake up the node in case of an interesting event.

40.4.2 $(MI)^2$

In traditional computer science, backbones for designing efficient algorithms are optimization paradigms such as branch-and-bound, dynamic programming, divide-and-conquer, and iterative improvement. This section introduces the maximally informed maximally informing $(MI)^2$ paradigm — the first systematic approach for the design of localized algorithms. In order to make the presentation self-contained, key assumptions are first summarized and typical sensor network optimization problems that will serve as illustrative examples briefly described. After that, an explanation is offered on how to apply the $(MI)^2$ strategy in a systematic way during each of the four phases of a localized algorithm: information gathering, system structuring, optimization mechanism, and result dissemination. Key insights and key trade-offs in designing localized algorithms are described. The realization of such algorithms on a number of typical sensor network tasks, such as routing and minimum spanning tree, is illustrated.

Given a network, assume that each node has minimal state information about the network and is only aware of nodes within its communication range. This is so because: (1) it is necessary to minimize storage requirements at each node; (2) nodes go to sleeping mode from time to time in order to minimize the energy consumption [2, 16]; and (3) updating the routing tables might not be possible as a result of nodes' high mobility.

The goal of the shortest path problem is to find a path between S and D such that the path has the smallest cardinality (i.e., the smallest number of nodes on the path). The MST problem asks to find the minimum spanning tree for a subset of nodes in the network. The connected dominating set problem addresses selecting a subset of nodes of minimal cardinality in such a way that each node is in the subset or has a neighbor in the subset. The importance of the selected problems for wireless ad hoc networks is self-evident. For example, the connected dominating set ensures that information can be efficiently collected or distributed from the nodes in the dominating set to all other nodes [18].

Although previous research in this area has implicitly specified the four phases in the design of localized algorithms, the phases are explicitly identified and formalized here for the first time. More importantly, the novelty of this approach is that insights and systematic generic methods to leverage the $(MI)^2$ paradigm have been developed in each step. This results in efficient localized algorithms on a great variety of problems.

40.4.2.1 Phase 1: Information Gathering

The information gathering (IG) phase is where the inputs to the procedure are prepared. If the information from multiple nodes is needed, routing between the nodes and the order in which nodes are visited and information is gathered will have a large effect on the amount of energy consumed. According to the maximally informing paradigm, each step of the IG phase must select the next node to be visited or contacted in such a way that the maximal amount of relevant information required for the application of the optimization mechanisms is acquired. The maximally informing principle can be realized in several ways, depending on the considered scope and objective function of the optimization problem. When considering the scope, one can take a greedy local view in which one considers which nodes can be contacted in a few hops if a particular node is visited next.

When considering the objective function, one can contact a node that will expose the largest number of constraints itself, or contact a node that has neighbor nodes that will reveal the largest number of constraints. For example, one alternative is to select a node that is likely to have many unvisited neighbors as the next node. In this case, the amount of obtained information is maximized. Another alternative is to visit a node that has a large unexplored area within its communication range with a high likelihood of containing nodes relevant for optimization.

For example, in shortest path routing, one can always contact the node closest to the destination node in a greedy way. An alternative is to contact the node with the largest area in its communication range, with a large percentage of points that are closer to the destination.

The final important observation related to the IG phase is that, in certain situations, visiting some nodes is perhaps more important than visiting others. One such situation is when the goal is to find the connected dominating set for all nodes in a geographic region. In this situation, it is crucial to visit all nodes on the outer perimeter of the network because their information could guarantee that all of the relevant nodes are considered. Therefore, in this situation, the $(MI)^2$ paradigm indicates that these nodes should be visited first.

40.4.2.2 Phase 2: System Structuring

Every node in the system has some amount of processing capability. However, not all of the system nodes need to compute the optimization procedure all the time. In the system-structuring phase, the decision about when and where to conduct optimization mechanism computations is made.

According to the $(MI)^2$ paradigm, two principles for selection of computation centers are followed. The first is to assemble enough information initially to conduct at least part of the computation meaningfully as soon as possible. This point is particularly well illustrated on MRA [9] and exposure tasks [13]. The second principle is always to conduct computations at the boundary of an already visited region in order to reduce the requirements for obtaining additional information. This point is clearly illustrated with the exposure task.

Finally, note that different optimization mechanisms dictate different system structuring phases. In some tasks, such as MRA, the local information is sufficient to guarantee the optimum solution. However, in computationally intractable optimization problems in which the interaction between all of the nodes in the system defines the output, the quality of the solution may be seriously hampered using only localized scopes. In such situations, it is necessary to obtain information about a large neighborhood for each node before the optimization mechanism is started.

40.4.2.3 Phase 3: Optimization Mechanism

Once the needed input is at a computation center, the optimization procedure is executed. The separation-of-concerns principle suggests that the phases should be as independent as possible. However, in the majority of problems, strong interdependence exists between the information-gathering phase and the optimization mechanism (OM) phase because, based upon the specific needs for executing an optimization mechanism, the relevant information must be acquired.

The first observation is that constructive and deterministic algorithms are strongly preferred to iterative improvement and probabilistic algorithms. This is because the former algorithms require only one pass through all inputs, but the latter require multiple passes. For example, for the MST problem, Prim's algorithm is much better suited for implementation as a localized algorithm than Kruskal's algorithms. This is the case because Prim's MST algorithm starts from an arbitrary node and at each step selects the shortest edge incident to one of the nodes already visited and does not form a cycle with the edges in the existing partial MST. This edge is then added to the partially built MST. Therefore, the algorithm uses only information about nodes that are already visited and their neighbors. On the other hand, Kruskal's algorithm requires one to consider all edges in the graph at each step and select the globally shortest edge. Therefore, before starting the execution of Kruskal's algorithm, it is necessary to obtain information about the whole graph.

The $(MC)^2$ optimization paradigm is well suited for use in conjunction with the $(MI)^2$ paradigm. In order to gain maximal benefit from the merged $(MC)^2(MI)^2$ paradigm, it is often advantageous to consider variants of $(MC)^2$ that only consider nodes adjacent to already explored nodes. This must be done in such a way that communication requirements are reduced. Other optimization paradigms naturally well suited for design of localized algorithms and, in particular with the $(MI)^2$ paradigm, are branch and bound and dynamic programming-based algorithms. Finally, note that in some cases, such as exposure calculations, one can directly use the available optimization mechanism. In others, such as the MRA problem, in order to design an efficient localized algorithm, one must develop a new optimization mechanism and, therefore, a new centralized algorithm that operates locally and with the partial information.

40.4.2.4 Phase 4: Information Dissemination

The information dissemination phase is the step in which the output of the optimization procedure is sent to the nodes requiring that information. The maximally informed paradigm states that one should disseminate information about the output of the optimization node to a particular node while close to that node. In the ideal case of balanced optimization and information distribution phases, all information that some node requires should be sent when visiting the last of its neighbors.

40.4.3 Solving ILP Problems by $(Mi)^2$-Based Paradigm

To demonstrate the wide application range of a paradigm for designing localized algorithms, apply it to a set of specific problems that can be specified and solved using a particular optimization solving strategy.

This subsection presents an (MI)²-based approach for solving an instance of a problem specified as an integer linear program. ILP is a widely used procedure for specifying and solving combinatorial optimization problems. ILP formulations are readily available for a large variety of combinatorial optimization problems, or they are easy to develop [14]. In particular, in a special case of ILP, called 0-1 ILP, all variables must be assigned to one of two binary values [14]. For example, all problems discussed in this chapter can be easily specified and solved using a 0-1 ILP formulation.

ILP formulation has three different components: variables, objective function, and constraints. Variables can take only integer values; the objective function and constraints must be linear. Note that, if the requirement that variable must be integers is removed, ILP reduces to a linear program (LP) that also has a wide range of applications [15]. An ILP defined over a set of variables x_i has the following standard form:

$$\text{Max } E^T.\mathbf{X} \tag{40.13}$$

$$\text{such that: } A^T.\mathbf{X} \text{ } B, \text{ } C^T.\mathbf{X} = D \tag{40.14}$$

where A, B, C, D, and E are matrices composed of real constants, and $\mathbf{X}$ is vector consisting of variables x_i. The first clause is the objective function (OF), while the equations on the second line are the constraints.

Assume that each node has information about one or more coefficients from matrices A, B, C, D, and E. Furthermore, the node has a list of its neighbors and a list of information of each neighbor, but does not necessarily have all the information that each neighbor has. The reason for this assumption is that, for many optimization parameters (such as energy level, sleep state, occupancy of buffers), collected sensor information is transmitted only on demand in a sensor network in order to reduce power consumption. Finally, each node must be informed about the value of all variables x_i important to it.

The (MI)²-based approach for locally solving an ILP instance is based on the following observation and intuition: a particular value can be assigned to a particular variable x_i, only after information is obtained about all constraints that contain x_i. Furthermore, it is advantageous first to resolve variables that are components of the most difficult (strict) constraints. Also, it is important that at the time of assigning a particular value to a variable, as much information as possible is available about all constraints that contain the variables contained by the constraint under consideration. In order to maximize the objective function, it is important to assign high values to variables with high coefficients and to keep estimating the highest possible value of the OF in view of the already observed constraints.

One can envision two approaches with respect to the relationship between the OF and constraints: optimistic (in which it is preferable to maximize the objective function at potential danger that later some constraints will become unable to be satisfied), and pessimistic (in which constraints are favored at the expense of the OF). The (MI)²-based localized ILP procedure is summarized using the pseudocode presented in Figure 40.2.

The IF is weighted sum of resolving power of the information available at the node and resolving power of its neighbors. The weighs of neighbors to a node are scaled by the average number of neighbors

---

Procedure (Localized (MI)²-based ILP Procedure)

Initialization;
while (*termination criteria* ==No) {
    Contact a neighbor that has highest information function (IF);
if (there are neighbors that do not have unvisited neighbors) {
    execute the optimization mechanism and communicate
    assigned values of the assigned variables to them } }

FIGURE 40.2 Pseudocode for (MI)²-based localized ILP procedure.

from already visited nodes. Resolving power is proportional to reduction in information uncertainty, according to the classical information theoretical definition.

The optimization mechanism used is based on the maximally constrained, minimally constraining principle. Essentially, one tries to assign each variable in such a way that it resolves a maximal number of constraints or increases the chance that they are later satisfied. The optimization function is treated as a constraint that is dynamically updated. The initial value is provided by a simple probabilistic analysis and consequently the value is updated by extrapolating the values obtained from the already visited nodes.

40.4.4 GPSR

Routing is one of the fundamental tasks in wireless networks. Although one can envision a number of different types of routing, the focus here is on a case in which a single message must be sent from a node to another node. Only one localized routing algorithm will be considered so that it can be described and analyzed in sufficient detail.

Karp and Kung [8] have developed a stateless routing protocol for wireless networks: greedy perimeter stateless routing (GPSR). The development of GPSR is based on two main assumptions. First, it assumes that each node (router) in the network is aware of its geographic location and the geographic locations of all its direct (one-hop) neighbors. Second, it assumes that the geographic location of the destination is also known. GPSR abandons traditional routing concepts that require continual distribution of the current map of the entire network's topology to all nodes. The packet forwarding decisions are made based only on the positions and knowledge of local nodes and the final destination location. More specifically, each node considers the locations of all neighbors, and makes a greedy decision to forward the data packet to the node closest to the destination. Therefore, GPSR is stateless in the sense that it does not keep additional information about the rest of the network beyond its neighborhood. As a consequence, GPSR scales better than traditional routing protocols and is much more adaptive to mobility.

GPSR protocol has two phases: greedy forwarding and perimeter forwarding. Greedy forwarding refers to the phase in which a series of nodes follow the same rule and each node makes a greedy decision of forwarding the data packet to the one neighbor that the current node believes is the closest to the destination. However, greedy forwarding would fail in a situation in which a node is the local minimum in terms of its geographic distance to the destination, i.e., when all its neighbors have longer distances to the destination than it does. In this case, control is switched to perimeter forwarding mode from greedy forwarding in order to escape the deadlock. Perimeter forwarding essentially follows the right-hand rule, which seeks to find an alternative route around and eventually converges to the destination.

In addition to being stateless and having exceptional scalability, GPSR has a number of other noble properties. It is efficient in the sense that it often selects the optimal or near-optimal path when the network is dense. It is also conceptually (and from implementation point of view) very simple and clean. However, it has a number of limitations. For example, if the network is not very dense, it is easy to show that the greedy approach is not the best choice because the scope of the problem considered is limited with respect to available information. In addition, there is no guarantee that GPSR will eventually converge to the destination. Situations exist in which forwarding phase and perimeter phase oscillate within a set of nodes and never converge on the correct destination. It is also difficult, if not impossible, to see how to generalize the approach when additional information is available to a three-dimensional case, or in the presence of obstacles.

40.5 Analysis

Creation of algorithms has two interdependent phases: synthesis and analysis. Although synthesis of localized algorithms is widely considered a difficult and demanding task, analysis often does not receive the proper attention and treatment. In this section, the most important issues related to analysis of localized algorithms are discussed.

Analysis of localized algorithms can be defined as a process of characterizing the effectiveness of a proposed localized algorithm for a given problem. It is a complex and often cumbersome task for several reasons. First, it is not easy to identify which properties of the algorithms are interesting and important. Even when these properties are identified, it is often unclear how to define each of them exactly. In addition, it is often difficult to calculate or measure these properties. For example, some are associated with solving computationally intractable problems.

The next layer of complexity comes from a need to consider more than one property simultaneously. Furthermore, it is not clear *a priori* what should be the representative and realistic properties of instances. Finally, one can consider localized algorithms as generalizations of on-line algorithms in which the designer has an impact on information that will be obtained next. Therefore, unpredictability often results in randomness of characteristics of a particular algorithm.

The primary goal of localized algorithms is to minimize the amount of energy spent on communication. This does not necessarily mean minimization of the number of packets. The current technology indicates that the most effective way of saving energy is through placing the radios of as many nodes as possible into sleeping mode. Also, note that in future applications, energy minimization will not be necessarily equivalent in the first approximation the minimization of energy devoted to communication. Depending on the technology, and even more on the targeted applications, computation or some other components may have the dominant role.

The primary constraint is to achieve the user-requested level of optimality and/or accuracy. Because of complex error propagation through the sensor fusion phase, it is sometimes difficult to select the most appropriate definition of accuracy.

Historically, the performance of algorithms has been evaluated as the size of their input asymptotically increases. Also, in traditional networking research, one of the key issues is scaling the protocols as the size of the network increases. Although many wireless sensor networks will be of limited size, scaling localized algorithms is already widely studied. A better way to evaluate localized algorithms for limited-sized networks is probably the development of benchmarks. Unfortunately, of the very few benchmarks available at present, all are synthetic.

In addition to these three metrics, amenability to provide fault tolerance, satisfy real-time constraints (such as throughput and latency), maintain privacy and security, and facilitate mobility will also be of prime importance for evaluation of localized algorithms.

One can envision many ways to combine two or more metrics. For example, in operation research literature, it is common to derive a set of solutions that form a Pareto optimal curve. In the computer science literature, it is more common to take one metric as the optimization goal and others as constraints.

40.6 Protocols and Distributed Localized Algorithms

This section briefly discusses the distributed localized algorithms in which more than one thread of computation is executed at the same time. Distributed localized algorithms have a number of advantages in terms of their ability to respond faster to changes in the environment and the network, fault tolerance, and their resiliency against security attacks. First the desiderata for protocols that govern the execution of distributed localized algorithms are stated. After that, one generic approach is presented for development of protocols for distributed localized algorithms [9].

Proper computation and synchronization strategy should have the following characteristics:

- *Concurrency.* The computation (decision making) should take place at as many places in the network as possible. In particular, nodes should be constantly updating their resources to cope with the dynamics in the network.
- *Synchronization (avoiding deadlocks).* The computing nodes should not have a conflict on the resources they use. For example, assume that a node $v1$ finds a node $v2$ redundant in terms of a specific functionality. At the same time, $v2$ also finds $v1$ redundant. If $v1$ and $v2$ decide to go to

sleep (using each other as a back up), a deadlock will occur. A good synchronization strategy must avoid deadlock situations like this.
- *Overhead.* The computation and synchronization strategy should add an overhead as low as possible to the network, especially in terms of its power consumption and communication overhead.
- *Latency.* Higher latency in putting a node into sleeping mode implies more idle energy consumption. Also, nodes should be updated for changes in the network to adapt to the network dynamics.
- *Fault tolerance.* Fault is inevitable in sensor networks. The computation and synchronization strategy should be designed so that the faults in any number of nodes cannot corrupt its functionality.

Koushanfar et al [9] have developed an approach termed "distributed token mechanism" that attempts to fulfill the stated requirements. A token indicates that the node has control of the local flow of the sleeping procedure. At each point of time, more than one token can be present in the network to comply with the concurrency requirements. A token is generated by an awakened node that needs to check the eligibility of the nodes within its local scope to enter the sleep state. The token is eliminated as soon as it examines the functionality of its local scope of the network and selects the nodes for sleeping. The node with the token locks its local area of consideration so that no other nodes use the same resources and the nodes acting on the mutual resources are synchronized. To lock a node means to consider it only for one token at each point of time.

The localized and distributed nature of the token generation makes it very tolerant to faults at the individual nodes. The pseudocode for the distributed token mechanism procedure is shown in Figure 40.3; a node v_i that is not already locked by any other nodes considers running the sleeping procedure (steps 1 through 4) and therefore generates a token. A node that has slept before generates the token at a random time r_i within the interval ($0 < r_i < r$max) (steps 5 through 9); a node that has already changed its state into sleep at least once generates a token as soon as it wakes up (steps 10 and 11). A node with the token locks all of the unlocked nodes within its local scope of consideration (step 12). This node then runs the sleeping procedure, which decides which of the locked nodes can enter the sleep state (step 13) and for how long (step 14). The token node then announces the decision to its neighborhood (step 15) and unlocks the locked nodes (step 16).

The random initiation time (r_i) assigned to each node in the beginning of the procedure serves the purpose of avoiding simultaneous requests on the use of mutual resources. Because the sleep intervals are assigned independently to nodes depending on the power and topology of the neighborhood, the

```
Procedure Distributed Token mechanism
1.   at ∀ node v_i,
2.   {
3.     while (node v_i is not locked by another node)
4.     {
5.       if (never have slept before)
6.       {
7.         set a random initiation time r_i (0 < r_i < r_max);
8.         generate a token at the time r_i;
9.       }
10.      else {
11.        generate a token as soon as v_i wakes up;
12.        lock all the unlocked nodes in the v_i's scope;
13.        select the best node to sleep;
14.        select the sleep interval for the sleeping node;
15.        announce the decision in the neighborhood;
16.        unlock the locked nodes;
17.      }
18.   }
```

FIGURE 40.3 Pseudocode for the distributed token mechanism procedure.

wake-up times are different. Therefore, after a node wakes up, it can immediately start another round of sleeping strategy without having too many locked nodes in its neighborhood. It is reasonable not to be concerned about the collisions in the network because they rely on the network's MAC layer to resolve any such conflicts.

40.7 Pending Challenges

This section outlines some of the potential trends for developing localized algorithms. It is always dangerous to make predictions, in particular when the topic is broad and application dependent; nevertheless, one can identify some major trends. Future research directions are classified into two broad categories related to: (1) the conceptual novelties for developing localized algorithms; and (2) optimization and algorithmic techniques. Due to space limitations, many important directions, such as interaction of localized algorithms with privacy and security; mobility; fault tolerance; applications within real-time systems; and use for actuator-based system, are omitted.

It is well known that mandatory prerequisites for developing high-quality algorithms are sound theoretical foundations. In some cases, one can develop such foundations, for example, PRAM, URM and the Von Neumann models of computation. When it is difficult to define a single widely applicable model, such as in parallel computing, progress is much slower. Currently, several models have been proposed for wireless ad hoc networks, including that of Zonoozi and Dassanayake [21]. However, it seems that the completely random nature of these models makes them of relatively limited practical relevance. Several other fields have also developed theoretical models. For example, in VLSI computations, the standard model is the one that assumes planarity and finite feature size of transistors and interconnects. The development of sound foundations for wireless sensor networks is a complex and difficult task because one must model at least four aspects of the systems: computation, communication, storage, and sensing.

Future algorithmic techniques can be naturally classified into two groups: one is related to design and the other is related to the analysis of localized algorithms. Design-related issues include the development of new paradigms that will facilitate systematic creation of localized algorithms, in particularly data collection and dissemination. An example of this is the maximally informing and maximally informed paradigm [11].

Currently, although a number of localized algorithms have been published, relatively little is known about their optimality in terms of quality of solution and expected energy cost. Several approaches have been proposed for this purpose, including the development of low bounds and probabilistic analysis. This trend will continue and will include new hard bound techniques as well as statistical guarantees.

Obviously, a strong correlation exists between how nodes are deployed and performances of localized algorithms. It is easy to see that different localized algorithms are best suited for different wireless sensor network organizations. Interestingly, this topic has not been addressed. In particular, sensor networks with regular structure such as grid can facilitate the development of fast and efficient localized algorithms. Another important issue with respect to localized algorithms for sensor networks is the development of optimization mechanisms that are resilient against unavoidable errors in sensor measurements. Finally, there will be a particular need to develop comprehensive approaches that combine continuous, discrete, and statistical techniques in order to obtain efficient localized algorithms. An example of this is exposure coverage [13].

Another side of the coin for localized algorithm development is the analysis of localized algorithms. Soon many activities will be conducted to define and develop scalable algorithms that are scalable not only with respect to the size of the network, but also with respect to the intensity of errors and the quality of solutions. Localized algorithms are, in a sense, the generalization of the concept of on-line algorithms in which one can decide which piece of information to obtain next. Competitive analysis of on-line algorithms has been a widely studied topic; it will be important for localized algorithms as well.

From a practical point of view, the most urgent issue is to develop standard benchmark examples that can properly capture the properties of real-life applications. Once the benchmarks are available, it would be important to analyze localized algorithms using statistical and perturbation analysis [4, 5].

A new network (distributed systems) architecture will appear, and it will be well suited for specific classes of tasks and applications. In addition, there is an urgent need for rapid prototyping and simulation platforms on which performances of localized algorithm can be accurately observed and quantified. Another important research direction is the development of design patterns for common localized algorithms. Design patterns changed the way in which software development is conducted and it will have a high impact in sensor networks.

Acknowledgment

This material is based upon work supported in part by the National Science Foundation under Grant No. ANI-0085773 and NSF CENS Grant.

References

1. Chen, B. et al. Span: an energy-efficient coordination algorithm for topology maintenance in ad hoc wireless networks, *Int. Conf. Mobile Computing Networking (MOBICOM)*, 85, 2001.
2. Estrin D. et al. Next century challenges: scalable coordination in sensor networks, *Int. Conf. Mobile Computing Networking (MOBICOM)*, 263, 1999.
3. Goemans, M.X. and Williamson, D.P. Improved approximation algorithms for maximum cut and satisfiability problems using semidefinite programming, *J. Assoc. Computing Machinery*, 42, 1115, 1995.
4. Grossglauser, M., and Tse, D.N.C. Mobility increases the capacity of ad hoc wireless networks, *IEEE/ACM Trans. Networking*, 10, 477, 2002.
5. Grossglauser, M., and Vetterli, M. Locating nodes with ease: mobility diffusion of last encounters in ad hoc networks, *Annu. Joint Conf. IEEE Computer Commun. Soc. (INFOCOM)*, 1954, 2003.
6. Hochbaum, D., Ed. *Approximation Algorithms for NP-hard Problems*, PWS Publishing Company, 1997.
7. Johnson, D. Approximation algorithms for combinatorial problems, *J. Computer Syst. Sci.*, 9, 256, 1974.
8. Karp, B. and Kung, H.T. GPSR: greedy perimeter stateless routing for wireless networks, *Int. Conf. Mobile Computing Networking (MOBICOM)*, 243, 2000.
9. Koushanfar, F. et al. Low power coordination in wireless ad hoc networks, *Int. Symp. Low Power Electron. Design*, 475, 2002.
10. Koushanfar, F. et al. Algorithms for resource discovery in wireless networks, unpublished manuscript, 2003.
11. Koushanfar, F. et al. Maximally-informing and maximally-informed algorithms for wireless networks, unpublished manuscript, 2003.
12. Kravets, P. and Krishnan, P. Application-driven power management for mobile communication, *Wireless Networks*, 6, 263, 2000.
13. Meguerdichian, S. et al. Exposure in wireless ad hoc sensor networks, *Int. Conf. Mobile Computing Networking (MOBICOM)*, 139, 2001.
14. Nemhauser, G.I. and Wolsey, L.A. *Integer and Combinatorial Optimization*, John Wiley & Sons, 1988.
15. Papadimitriou, C. and Steiglitz, K. *Combinatorial Optimization: Algorithms and Complexity*, Prentice Hall, Englewood Cliffs, NY, 1982.
16. Rabaey J.M. et al. PicoRadio supports ad hoc ultra-low power wireless networking, *Computer Mag.*, 42, 2000.

17. Schurgers, C., Tsiatsis, V., and Srivastava, M. STEM: topology management for energy-efficient sensor networks, *IEEE Aerospace Conf.*, 78, 2002.
18. Stojmenovic, I., Seddigh, M., and Zunic, J. Dominating sets and neighbor elimination based broadcasting algorithms in wireless networks, *IEEE Trans. Parallel Distributed Syst.*, 13, 14, 2002.
19. Xu, Y., Heidemann, J., and Estrin, D. Adaptive energy-conserving routing for multihop ad hoc networks, technical report *527*, USC/Information Science Institute, October, 2000.
20. Xu, Y., Heidemann, J., and Estrin, D. Geography-informed energy conservation for ad hoc routing, *Int. Conf. Mobile Computing Networking (MOBICOM)*, 70, 2001.
21. Zonoozi, M. and Dassanayake, P. User mobility modeling and characterization of mobility patterns, *IEEE J. Selected Areas Commun.*, 15, 1239, 1997.

Index

Index

(MC)² optimization paradigm, **40**-10
(MC)²(MI)² paradigm, **40**-10
(MI)², **40**-8
10-wire transducer independent interface. *See* TII
9-b interprocessor protocol. *See* NBIP
915 MHz ISM band, **3**-25

A

A-GPS, **21**-1
A/D converters, **2**-3
ABC, **20**-17
ABR, **39**-3
Absolute location, **20**-2
Absolute positioning, **20**-4
Abstract sensors, **24**-1
Abstract syntax notation.1. *See* ASN.1
Abstraction level, **31**-4
Accelerometers, **5**-4–**5**-5, **5**-14, **12**-5, **36**-10
Accelerometers, packaging, **5**-12–**5**-13
Access control, **30**-3
Access control lists. *See* ACL
Access point, **3**-20
Access protocols, designing, **14**-2
Accounting management, **3**-14–**3**-15
Accuracy, **1**-5, **3**-11, **9**-10, **15**-8, **18**-3
Accuracy, sensing and, **13**-9
Accuracy, WSN, **9**-2
Acknowledgment mechanisms, **16**-9
ACL, **2**-7
Acoustic, beamforming, **13**-7
Acoustic, sensors, **9**-6, **13**-9, **32**-1, **38**-1
Acoustic, spectra, ambient, **11**-3
ACPI, **27**-2
ACPI, standard, **29**-3
Acquisition, **1**-9
ACTA, **33**-2
Active Badge system, **21**-10
Active Bat system, **20**-9, **21**-10
Active networks, **4**-5–**4**-6, **26**-17
Active perception, **7**-6–**7**-7
Active power management, **27**-5
Active sensor model of system-level programmability, **4**-4
Active-steered laser communication, **5**-10, **37**-9
Activity-based routing. *See* ABR
Actuation, **1**-6, **20**-10, **32**-1
Actuator modules, **12**-4, **36**-2
Actuator modules, local control of, **16**-2, **16**-7
Actuator modules, operating in conjunction with sensor modules, **8**-2
Actuator-based system, **40**-15
Actuator-sensor-interface standard. *See* ASI standard

Actuators, **36**-7
Ad hoc architecture, **9**-7, **9**-9
Ad hoc architecture, multihop, **12**-1
Ad hoc deployment, **6**-4
Ad hoc deployment, wireless sensor networks, **13**-10
Ad hoc localization system. *See* AHLoS
Ad hoc multihop routing protocols, **30**-1
Ad hoc network management protocol. *See* ANMP
Ad hoc networks, **1**-4
Ad hoc networks, comparison with sensor networks, **16**-1
Ad hoc networks, MAC protocols for, **28**-4–**28**-10
Ad hoc on-demand distance vector. *See* AODV
Ad hoc radio network, single-hop. *See* ARN
Ad hoc request/respond mode, **15**-9
Ad hoc routing techniques, **16**-9
Ad hoc sensor networks, **31**-8
Ad hoc threshold-based mode, **15**-9
Ad hoc transport protocol. *See* ATP
Ad hoc wireless networks, **17**-1, **18**-2
Ad hoc wireless networks, scaling, **13**-4
Adaptability, **9**-10, **20**-10, **30**-4
Adaptation, **3**-12, **33**-10
Adaptation, server, **33**-7
Adaptive clustering, **6**-11
Adaptive election algorithm. *See* AEA
Adaptive face routing. *See* AFR
Adaptive listening, **28**-16
Adaptive localized algorithms, achieving desired local behavior with, **6**-22
Adaptive periodic threshold-sensitive energy-efficient sensor network. *See* APTEEN
Adaptive probabilistic response. *See* APR
Adaptive reconfiguration, **32**-1
Adaptive routing, **6**-18
ADC, **12**-10
ADC, Maxim MAX186, **12**-12
ADC, resolution, **37**-2
Addition, **25**-7
Address centric, **6**-7
Address-based routing, **20**-2
Addressing, **17**-2
Addressing, hierarchical, **13**-16–**13**-18
Addressing, IP, **26**-2
Addressing, network layer, **30**-3
Addressing, scalable, **15**-8
Administrative state attribute, **3**-25
Admission manager, smart messages and, **26**-4
ADV, **6**-19
Advanced configuration and power management interface. *See* ACPI
Advanced signal processing, **20**-10
Adversary, **32**-2
AEA, **28**-19

I-3

Aerospace applications, **35**-1
AFECA, **40**-8
Affect space mapping, **7**-8
AFR, **39**-36
Agent Tcl, **4**-5
Agents in wireless sensor networks, **3**-17–**3**-18
Agents in wireless sensor networks, cloning of, **3**-25
Agents in wireless sensor networks, migration of, **3**-24
Aggregating WSNs, **9**-6
Aggregation, **3**-6, **3**-25, **6**-3, **6**-16, **8**-2, **8**-12, **9**-3, **9**-19, **13**-2, **16**-10, **18**-5, **18**-10, **32**-1
Aggregation, diffusion limited, **25**-15
Aggregation, directed diffusion methods of, **6**-9
Aggregation, filters, **4**-6–**4**-7
Aggregation, greedy, **9**-20, **18**-16
Aggregation, hierarchical clustering and, **8**-3
Aggregation, periodic, algorithms for, **4**-3
Agilent, **10**-11
Agriculture, use of sensor networks in, **1**-3
AHLoS, **16**-4
Aircraft, **22**-2
Aircraft drag reduction, **1**-3
Algorithmic complexity attack, **32**-15
Algorithms, **4**-3, **9**-13–**9**-20, **12**-2, **22**-3, **40**-1
Algorithms, backoff, **6**-10
Algorithms, Bellman-Ford, **10**-23
Algorithms, Brooks-Iyengar hybrid, **25**-4
Algorithms
    centralized, **19**-2
Algorithms, clustering, **8**-3, **8**-8, **13**-11, **13**-14
Algorithms
    combinatorial optimization, **2**-15
Algorithms, consistent recovery schedule, **33**-11
Algorithms, control, **9**-7
Algorithms, coordination, **37**-7
Algorithms, data aggregation, **9**-19
Algorithms, data integration, **25**-4
Algorithms, development paradigms, **40**-2
Algorithms, Dijkstra, **6**-17
Algorithms
    distributed, **19**-2
    distributed estimation, **4**-11
Algorithms, distributed locationing, **30**-3
Algorithms, distribution, **31**-4
Algorithms, dual network clustering, **13**-13
Algorithms
    DV-hop, **20**-19
    EAR, **28**-16
    edge detection, **4**-3
Algorithms, encryption, **38**-10
Algorithms, energy efficient, **27**-1
Algorithms, energy-efficient hierarchical clustering, **9**-18
Algorithms, energy-scalable, **27**-1
Algorithms
    fusion, **4**-11
Algorithms, GAF, **18**-12
Algorithms, genetic, **25**-3
Algorithms, geometric, **39**-2
Algorithms, graph search, **9**-13
Algorithms, greedy, **27**-2
Algorithms
    Hop-TERRAIN, **20**-18
Algorithms, Internet multicast, **31**-9
Algorithms, interval fusion, **35**-2
Algorithms, LEDES, **29**-13
Algorithms, LEDF, **29**-4
Algorithms, linked-cluster, **13**-12
Algorithms
    localized, **19**-2
Algorithms, localized, **39**-9
Algorithms, max-min zPmin, **6**-17
Algorithms, minimum cost, **6**-10
Algorithms, multidimensional scaling, **15**-10
Algorithms, multiple winner, **6**-11
Algorithms, nearest neighbor, **21**-6
Algorithms, optimal polynomial-time, **9**-13
Algorithms, performance of, **40**-13
Algorithms
    periodic aggregation, **4**-3
Algorithms, polling, **18**-7
Algorithms, polynomial-time, **9**-14, **9**-20
Algorithms, position and location tracking, **21**-4
Algorithms, practical virtual force, **9**-14
Algorithms, probability-based, **21**-7
Algorithms, real-time message scheduling, **17**-2
Algorithms
    relaying, **10**-16
Algorithms, resource-adaptive, **6**-18
Algorithms, routing, **10**-16
Algorithms
    routing, **17**-2
Algorithms, sensor-signal processing, **23**-1
Algorithms
    single winner, **6**-11
Algorithms, single-hop cluster head formation, **9**-17
Algorithms, single-level multihop clustering, **9**-18
Algorithms
    SMACS, **28**-16
    target tracking, **4**-3
    TORA, **1**-7
Algorithms, Viterbi-like tracking, **21**-8
All-or-nothing model, **1**-9
Allocation, **12**-8, **13**-8–**13**-10
Alternating pulse modulation. *See* APM
Altruistic routing, **13**-19
Ambient heat, **12**-5
AMD, K-6, **38**-2
AMD, PowerNow!, **29**-5
Ammunition, monitoring, **16**-2
Amplifiers, **38**-7
Analog Devices, **10**-11
Analog-to-digital conversion, **18**-4
Analog-to-digital converter modules. *See* ADC
Analog-to-digital converters. *See* A/D converters
Analysis, **40**-12
Analytical modeling, **9**-22, **36**-10
Anchor nodes, **20**-4, **20**-7
Anchor nodes, dense populations of, **20**-6
Angle of arrival. *See* AOA
Angular constraints, **20**-7
Angular rates, **22**-13
Animal tagging, **11**-2

Index

ANMP, **3**-24, **11**-3
Annealing code, **36**-9
Anonymity, **31**-13
ANSER project, **22**-1
ANSER project, picture compliant system, **22**-14
ANSER project, vision-sensing node in, **22**-10
Ant pheromone model, **25**-15
Antennae, multiple sectored, **20**-6
Antennas, **1**-4, **5**-5, **13**-10
Antennas, multiple input, multiple output (MIMO), **2**-8
Anytime deployment, **9**-3
AOA, **16**-4, **20**-5, **21**-1
AOA, positioning with, **21**-2
AODV, **1**-7, **13**-11, **13**-15, **39**-3
AODV, routing, **30**-4
APIs, **4**-11, **9**-21
APIs, cooperative computing model, **26**-7
APIs, smart message, **26**-8
APM, **10**-3
Application communication, **3**-10–**3**-11
Application layer, **3**-24, **30**-2
Application layer, protocols, **16**-3
Application parallelism, **2**-11
Application systems, self-organizing, **33**-3
Application-specific design, **9**-10, **31**-2
Application-specific queries, **16**-9
Application-specific sensor designs, **31**-4
Applications, **10**-3, **16**-2
Applications, distributed, **4**-6
Applications, efficient sensor networks, **4**-2–**4**-3
Applications, fault tolerance, **36**-7
Applications, management, **3**-19
Applications, querying, **8**-2
Applications, smart message, **26**-12–**26**-14
Applications, targeted, **12**-3
Applications, tasking, **8**-2
APR, **8**-5
APTEEN, **6**-13–**6**-15, **18**-8, **18**-14
Arborescence, **39**-39
Architectures, energy efficient, **27**-1
Architectures, Flynn classes for, **24**-3
Architectures, implementation, **8**-4–**8**-12
Architectures, interconnecting, **10**-15
Architectures, internet coupling, **10**-14
Architectures, network, traditional, **25**-2
Architectures, state of the art, **12**-2
Area coverage, **19**-4–**19**-8
ARGOS, **11**-2
Arithmetic unit replication, **36**-4
ARM THUMB processor, **12**-10, **13**-6
ARN, single-hop, **23**-12
Arousal, **7**-8
ARQ, **16**-13
Art Gallery problem, **19**-4
Artificial neural networks, **7**-4
ASI standard, **10**-16
ASI-ASI coupler, **10**-18
ASIC units, **12**-4
Asleep state, **23**-2
ASN.1, **3**-21
Assembly line sensing, **27**-1

Asset monitoring and management, **1**-3
Asset tracking, **5**-5
Assisted global positioning system. *See* A-GPS
Associative spreadsheet, **8**-5, **8**-8
Assumption-based coordinates. *See* ABC
Asymmetric delay, effects of on synchronization, **16**-5
Asymmetric denial of service attack, **32**-3
Asynchronous processors, **37**-3
AT91 microcontroller, **37**-6
ATLAS, **24**-18
ATmega161L, **37**-2
Atmel, **37**-2
Atmel, AVR microcontroller, **9**-11
Atomic abstract data types, **33**-2
Atomic multilateration, **21**-5
ATP, **33**-3
ATRV mobile robots, use of CTT protocol for, **25**-14
Attack, **32**-2
Attack, taxonomy, **32**-5
Attacker, **32**-2
Attacker, capability, **32**-7
Attacker, types of, **32**-5
Attention control, **7**-4
Attribute-based addressing, **18**-10
Attribute-based naming, **4**-6–**4**-7, **6**-3, **6**-9, **8**-4, **8**-8, **16**-10
Attributed change value record, **3**-21
Attributes, **6**-14, **16**-10
Audio perception, **7**-4
Audio sensors, **12**-5
Auditing, **31**-4
Authenticity, **38**-9
Authorized users, **32**-2
Automatic channel handoff procedures, **20**-2
Automatic organization, tiered architectures, **13**-10
Automatic recovery, **33**-8
Automatic repeat request. *See* ARQ
Automatic warehouse inventory tracking, **9**-6
Automation systems, **35**-1
Automobile industry, WSNs in, **10**-21
Automotive telematics, **1**-2
Automotive telemetric applications, **31**-1
Autonomic systems, **3**-3
Autonomous proactive computing, **4**-2
Autonomous sensor nodes, **3**-9–**3**-10
Availability, **3**-11, **32**-4
AVR microcontroller, **9**-11
AWAIR I, **13**-7
Awake state, **23**-2
Awareness horizon, **17**-5

B

B, selection of, **34**-7
Back-off counter, **28**-4
Back-off scheme, **19**-6
Back-off window, **17**-3
Backbone, **10**-16
Backbone, communication, **16**-10
Backbone, connected, **13**-12
Backbone, creation protocols, **13**-15, **13**-19
Backoff algorithm, **6**-10

Badges, **12**-8
Bahramgiri's protocol, **39**-27
BAIP, **39**-39
Balanced design, **12**-2
Balanced k-clustering, **9**-17
Bandwidth, **24**-7
Bandwidth, communication, **22**-4
Bandwidth, conservation, **21**-9
Bandwidth, efficient radios, **12**-6
Bandwidth, flat ad hoc network, **13**-4
Bandwidth, requirements, **3**-24
Bandwidth, reusable, **12**-16
Bandwidth, savings with active networks, **4**-5
Bandwidth, use of in wireless networks, **2**-7
Bargaining, **3**-11
Barrier coverage, **19**-4, **19**-9
Base stations, **13**-4, **23**-3
Base stations, interoperation of multiple, **20**-9
Base stations, placement of, **14**-2
Base stations, security advantages of, **31**-5
Basic devices, **10**-8
Batch transfer, **5**-13
Battery, low-capacity, **37**-1
Battery, models, **15**-9
Battery, technology, **38**-1
Battery cells, **13**-4
Battery cells, high-density, **12**-5
Battle damage assessment, **16**-2
Battle-space monitoring, **1**-4
Battlefield surveillance, **16**-2, **32**-1, **39**-1
Battlefield surveillance, target tracking, **31**-1
Bayes rule, **21**-7
Bayesian fuser, **25**-4
Bayesian inference, **7**-4
Bayesian update, **22**-3
Beacon-based localization, **16**-4, **37**-2
Beaked whales, **11**-2
Bearing measurements, **22**-13
Bearing-only sensors, **22**-9
Bell Lab's, **36**-4
Bell Labs, **38**-2
Bellman-Ford algorithm, **10**-23
Beluga whales, **11**-2
Benchmarks, **12**-2, **40**-13
Berkeley BWRC research group, **12**-4
Berkeley Mote node. *See* Mote node
Betavoltaics, **5**-12
BFS, **40**-5
Biased distribution, **9**-13
Biasing route selection, **13**-11
BiCMOS, **5**-4
Binary exponential backoff (BEB) scheme, **28**-8
Binary multilevel de Bruijn network. *See* BMD
Binary operators, **23**-5
Biological attack detection, **16**-2
Biological sensors, **5**-4–**5**-5
Biomedical applications, **9**-4–**9**-5, **15**-2
BIP, **39**-39
BISR-based fault tolerance, **36**-2, **36**-6
Bitbus, **10**-7
Bitbus, -NBIP coupler, **10**-19–**10**-21

Black burst, **18**-9
Blackboard-based model, **4**-6
Blanket coverage, **19**-4
Blind points, **19**-6
BLMST, **39**-41
Block and base classes, **10**-13
Block functions, **10**-15
Blockages, **39**-28
Blue whales, **11**-2
BlueRings, **10**-23
Bluetooth, **1**-6, **5**-5, **10**-13, **10**-22, **10**-24, **12**-16, **16**-12, **28**-2
Bluetooth, operation, **28**-6
Bluetooth, power consumption modes, **27**-2
Bluetrees, **10**-23
BMD, **25**-3
Boeing, **10**-11
Border effects, **20**-15
Border nodes, **30**-11
Border zone, **6**-15
Bosch, **10**-5
Bottom-up method of active perception, **7**-7
Bounded degree structures, **39**-10, **39**-16
    low-weight, **39**-19
Bounded latency, **18**-6
Bowhead whales, **11**-2
Boxes, tracking of, **1**-3
Braided multipath routing, **6**-21
Branch and bound, **40**-2
Breath-first search. *See* BFS
Bricks, **26**-5
Bridges, **10**-8, **10**-16
Bridges, bibus-NBIP coupler, **10**-19
Bridging fault model, **36**-5
Bridging nodes, **10**-23
Broadcast average incremental power. *See* BAIP
Broadcast communication, **28**-13, **39**-38–**39**-42
Broadcast communication, authentication of, **31**-8
Broadcast communication, short-range, **1**-1
Broadcast property, omnidirectional antennas, **1**-6
Broadcast topology, **22**-8
Broadcasting incremental power. *See* BIP
Brooks-Iyengar hybrid algorithm, **25**-4, **33**-2
Brüel & Kjaer, **10**-11
Brumby MKIII aircraft, **22**-2
Buffer overflow, **31**-3
Bulk machining, **5**-2
Bus control, **10**-2
Bus coupler, **10**-16
Business 2.0, **1**-2
Business management, **3**-3, **3**-5
Byzantine general's problem, **33**-2

C

C4 computing, **1**-3
C4 computing, ISRT systems, **9**-2, **16**-2, **18**-1
Caching, **6**-9
CAD, **12**-8
Calibration, **12**-5, **12**-8, **36**-12
Calibration, systematic error, **36**-7
Cameras, image capture, **26**-11

Cameras, intelligent, **26**-1
CAN, **10**-5, **10**-15
CAN bus, use of by ANSER system, **22**-14
Cantilevers, **5**-2
CAOC, **2**-3
Capacitors, voltage-tunable high-Q, **5**-5
Capacity, **1**-6–**1**-7
Carbon nanotubes, **5**-5
Cardiograms, **11**-3
Carrier boards, **13**-7
Carrier frequency generation, **3**-25
Carrier sense multiple access. *See* CSMA
Carrier sense multiple access with collision avoidance. *See* CSMA/CA
Carrier-sense multiple access. *See* CSMA-based medium access
CCF, **27**-6
CCLK, **27**-6
CCP, **19**-7
CCR, **5**-9, **37**-9
CCR, packaging, **5**-12–**5**-13
CDL, **2**-5
CDMA, **18**-7, **28**-4, **28**-11, **28**-16
CDS, **19**-7, **39**-9, **39**-30, **39**-32
CEC, **2**-3, **13**-12
CEC, detection reports from sensors, **2**-6
CEDAR, **13**-12
Cell-based positioning, **20**-9
Cell-based WSNs, **31**-5
Cellular systems, **12**-13
Center of gravity method of defuzzification, **7**-12
Central node election, **6**-10
Centralized algorithms, **19**-2, **40**-4–**40**-7
Centralized control, lack of, **28**-2
Centralized data fusion, **15**-6
Centralized data gathering, **28**-12
Centralized low-weight bounded degree planar structures, **39**-19
Centralized management of WSNs, **3**-19
Centralized methods of broadcasting, **39**-38
Cerfcube, **13**-7
CERN, **24**-18
CGSR, **39**-3
Chain-based architecture, hierarchical, **18**-11
Channel access, **1**-7–**1**-8, **28**-4
Channel filter, **22**-4, **22**-8
Channel partitioning schemes, **18**-7
Characteristics linearization, **10**-1
Chemical attack detection, **16**-2
Chemical energy storage, **5**-12
Chemical sensors, **5**-4–**5**-5, **32**-1
Childhood learning, **9**-5
ChipCon CC1000, **11**-6
Choleski factorization, **20**-8
Chronos periods, **11**-4
Chvatal's algorithm, **39**-39
CI, **22**-4
CI, broadcast with update, **22**-6
Circuits, energy efficiency of, **27**-1
Civil engineering, **1**-3
Civilian applications of sensor networks, **25**-4, **27**-1, **38**-1

Class-based addressing, **30**-3
Classification of network layer protocols, **18**-10
Client puzzles, **32**-15
Client/server model, **4**-6
Climate control, local, **9**-3
Clock, synchronization and security, **31**-6
Clock drift, **28**-15
Clock glitches, effects of on synchronization, **16**-5
Clock synchronization, **9**-22, **18**-7, **28**-6
Cloning of agents, **3**-25
Close formations, **10**-22
Cluster heads, **18**-10, **18**-15, **34**-1, **39**-31
Cluster heads, border nodes and, **30**-13
Cluster heads, energy aware communication and, **29**-20
Cluster-based architecture, **18**-11, **28**-23
Cluster-based multihop architecture, **15**-5
Cluster-head node, **3**-18, **4**-4, **13**-11
Cluster-head node, rotation of, **14**-2
Clusterhead gateway switch routing. *See* CGSR
Clustering, **1**-6, **3**-10, **18**-14
Clustering, adaptive, **6**-11
Clustering, algorithm, **8**-3, **8**-8
Clustering, architectures, **9**-16
Clustering, data, **24**-16
Clustering, distributed, **39**-31
Clustering, dual network, **13**-13
Clustering geometric property, **39**-32
Clustering, hierarchical, **4**-8, **8**-3
Clustering, load-balanced approach, **9**-17
Clustering, logical, **13**-16
Clustering, mechanisms of, **13**-11–**13**-15
Clustering, random competition-based, **13**-13
Clustering, virtual backbone, **39**-30
Clustering without geometric property, **39**-31
Clustering-based heuristic for maximum lifetime data aggregation. *See* CMLDA
Clustering-based hierarchies, **18**-5
Clusters, **13**-10
Clusters, roaming, **13**-18
CMIP, **3**-24
CMLDA, **18**-16
CMOS, **5**-2
CMOS, energy consumed by, **9**-12
CMOS, high aspect ratio micromachining, **5**-4
CMOS, micromachining and, **5**-3–**5**-4
CMOS, power consumption of, **29**-3
CMOS, power dissipation in, **38**-7
CMS, **24**-18
CMSA/CA, **10**-5
CNC, **10**-7
Code bricks, **26**-5
Code cache, smart messages and, **26**-4
Code division multiplexing access. *See* CDMA
Code identification, **9**-13
Code signing, **31**-3
Code updates, **31**-3
Coding schemes, **32**-4
Coherent processing, **6**-10–**6**-11, **18**-16
Collaboration, **3**-6, **8**-2

Collaborative computation, efficient, **23**-15
Collaborative multilateration, **21**-6
Collaborative signal and information processing. *See* CSIP
Collimating lenses, **5**-10
Collision, **18**-8, **28**-11, **32**-12
Collision, analysis of results, **24**-18
Collisionless region, **1**-8
Combination rule, **35**-5
Combinatorial optimization algorithms, **2**-15
Combined air operations center. *See* CAOC
Combined intervals, **35**-5
Commercial applications of sensor networks, **16**-2
Commercial systems, security and privacy protection of, **31**-11
Commitment distribution mechanism, **31**-7
Committees, **25**-2
Common data link. *See* CDL
Common management information protocol. *See* CMIP
Common nodes, **3**-17
Common object request broker architecture. *See* CORBA
Communication delays, **33**-5
Communication energy, **38**-1, **38**-7
Communication graph, **39**-29
Communications, **1**-2, **2**-5, **3**-4, **3**-10–**3**-11, **5**-5–**5**-10, **9**-14, **10**-14, **13**-9, **15**-3, **20**-10, **28**-10, **29**-1, **37**-7–**37**-10
Communications, backbone, **13**-2
Communications cost reduction, **13**-9
Communications, coverage area map, **3**-25
Communications, delayed, **22**-3
Communications, determination of costs, **3**-15
Communications, energy, **18**-4, **36**-2
Communications, energy, consumption, **18**-5
Communications, energy-efficient, overview, **29**-4
Communications, integration of sensing with, **15**-1
Communications, layer, **15**-4
Communications, limits of, **6**-2
Communications, link failures, **15**-3
Communications, localization of, **4**-2
Communications, maintenance, **10**-22
Communications, management, **22**-4
Communications, multihop, **16**-9
Communications, polling mode, **10**-17
Communications, protocols, **16**-1, **17**-1, **36**-7
Communications, protocols, industrial sensor fitting, **10**-2–**10**-10
Communications, radio subsystems, **12**-4
Communications, range of, **6**-4, **19**-2, **19**-6
Communications, redundant paths, **14**-2
Communications, safe, **9**-10
Communications, steered agile laser, **5**-13
Communications, strategies for, **23**-15
Communications, subsystems, **36**-2
Communications, systems for industrial automation, **10**-2
Communications, technology, **17**-1
Communicator programs, **1**-2
Comparison metrics, **28**-24
Compass, **36**-10
Compass, routing, **39**-34
Complex sensor measurement, flexible support for, **11**-5

Complexity of dynamic WSNs, **15**-8
Composite channel model, **1**-7
Compositional server, **33**-7
Compression, **12**-8
Computability, **12**-3
Computation, mote, **37**-2–**37**-5
Computation, requirements, **38**-16
Computation engine, **36**-6
Computation resources, **2**-5, **6**-4, **9**-14, **13**-1, **36**-2
Computation resources, cost of, **23**-1
Computation resources, efficiency of collaborative, **23**-15
Computation resources, guaranteeing, **2**-8–**2**-10
Computation resources, limited supply of in WSNs, **6**-2
Computational complexities, **25**-5
Computational energy, **38**-2
Computational load, time-varying, **27**-2, **27**-5
Computational module, **3**-7
Computer numeric controller. *See* CNC
Computer viruses, **31**-3
Computing energy, **18**-4
Computing engine, **36**-2
Computing power, limitations of, **1**-1
Concurrency, **40**-13
Concurrency-intensive operation, **12**-3
Condorcet Jury models, **25**-4
Conference sessions on sensor networks, **1**-2
Confidence degrees, combined, **35**-6
Confidence degrees, fault-tolerant interval estimation with knowledge of, **35**-9–**35**-11
Confidence degrees, fault-tolerant interval estimation without knowledge of, **35**-4–**35**-8
Confidence levels, **20**-19
Confidentiality, **32**-4, **38**-9
Configurable distributed systems, **33**-3
Configuration, **3**-4, **3**-8–**3**-9
Configuration, errors, **32**-9
Configuration, management, **3**-12
Congestion, **32**-12
Congestion control, **3**-6, **16**-7, **16**-11, **17**-4
Connected dominating set. *See* CDS
Connected random coverage, **19**-6
Connection reestablishment procedure, **28**-5
Connections, **3**-23, **10**-16
Connections, end-points, **17**-3
Connections, state of, **17**-4
Connectivity, **1**-10, **3**-6, **6**-5, **19**-2
Connectivity, degree of, **20**-5
Connectivity, detection, **25**-11
Connectivity, maintaining, **1**-7
Connectivity, modeling, **15**-10
Connectivity, rules, **10**-24
Connectors, **33**-8, **39**-32
Constant velocity, time-correlated, **22**-5
Consumer electronic devices, **36**-2
Consumption, **29**-1
Containers, tracking of, **1**-3
Contaminant transport, **1**-3
Contention-based protocols, **28**-4, **28**-11, **37**-7
Contention-control mechanisms, back-offs in, **32**-12
Context declarations, **17**-6
Context labels, **17**-5, **17**-7

Index

Continental United States. *See* CONUS
Continuous monitoring, **3**-20
Continuous monitoring, sensor networks, **28**-13
Continuous phenomenon, **3**-10
Continuous sensing, **16**-2
Continuum theory, **15**-10
Control, **15**-4
Control, messages, **18**-15
Control, messages, overhead due to, **18**-7
Control, registers, **37**-5
Control algorithms for dynamic systems, **9**-7
Control overhead, **6**-5
Control packet overhead, **28**-11
Controlled multiplexing, **18**-7
Controller area network. *See* CAN
Controllers, cost to reach field, **30**-11
Controls, **28**-6
CONUS, **2**-3
Convergence, **20**-16
Convex hull, **39**-2
Cooling schemes, **36**-1
Cooltown, **17**-7
Cooperational model, **3**-24
Cooperative computing applications, **26**-2, **26**-3
Cooperative computing applications, programming interface for, **26**-7
Cooperative effort, **1**-1
Cooperative engagement capability. *See* CAC
Cooperative mobile robots, **10**-21
Cooperative processing, **6**-10
Cooperative ranging, **20**-16, **20**-19
Cooperative transactions, **33**-2
Coordinated policy, **12**-6
Coordinated sleeping, **28**-14
Coordinated vehicle tracking, **8**-9
Coordination algorithm, **37**-7
Coordinator, motes, **37**-8
CORBA, **2**-11, **17**-6
CORBA, sensor interfaces and, **2**-15
Core clock. *See* CCLK
Core clock configuration register. *See* CCF
Core voltage, **27**-7
Corner cube reflectors, **23**-3
Corner cube retroreflector. *See* CCR
Cost, **15**-8, **20**-9
Cost, computational, **23**-1
Cost, deployment, **9**-3
Cost, effectiveness, tiered architectures, **13**-3–**13**-4
Cost, map, **3**-24
Cost, smart message migration, **26**-9
Cost, WSN, **9**-2
Cost reduction factor, **13**-11
COTS processor systems, **2**-9
COUGAR, **17**-7
Cougar, **4**-7
Count time, **6**-12
Counters, **31**-6
Counting, **3**-15
Coupler, **10**-16
Covariance intersection. *See* CI
Coverage, **3**-11, **3**-14

barrier, **19**-9
Coverage, monitoring, **15**-4
Coverage
    point, **19**-8
Coverage, problems, **19**-4
Coverage configuration protocol. *See* CCP
Covert communication channels, **31**-3
CPU power savings, **29**-7
CPU speed, adjusting, **38**-5
CPU-centric DPM, **29**-2
CR-LDP, **2**-6
Cramer-Rao bound of sensor location accuracy, **16**-4
Creative packaging, **36**-1
Cricket, **17**-7, **20**-9, **21**-11
Cricket Compass, **12**-5
Critical mass, **17**-6
Critical nodes, **1**-5
Cross-layer communication protocol stack, **18**-5
Cross-layer protocol, **1**-5, **9**-20
Crossbow Technology, **1**-11, **10**-11, **13**-5
Crossover, **25**-7
Crusoe, **38**-2
Cryptographic algorithm, **38**-10
Cryptographic hash function, **32**-15
Cryptographic tools, **31**-3
Crystalline semiconductors, **5**-2
CSIE/NCTU, **21**-11
CSIP, **9**-19, **18**-5, **18**-16
CSMA, **18**-8, **28**-4
CSMA, PEDAMACS and, **28**-21
CSMA-based medium access, **16**-13
CSMA/CA-based schemes, **18**-8, **28**-4
CTT protocol, **25**-14
Current data object, **3**-22
Current remaining energy level summary control, **3**-21
Cyclic graphs, **38**-6

D

Dallas Semiconductor, **10**-10, **10**-13
DAP, **10**-23
DARPA, **1**-2
DATA, **6**-19
Data
    -centric routing protocol, **16**-10
Data, -link layer, **3**-25, **10**-16
Data, acquisition, **20**-16
Data, aggregation, **3**-25, **9**-19, **13**-2
Data
    aggregation, **16**-10
Data, aggregation, **18**-5
Data
    aggregation, **18**-10
Data, aggregation, **18**-16, **20**-10, **36**-13. *See also* aggregation
Data
    aggregation (*See also* aggregation)
Data, aggregation, by funneling, **30**-10
Data, atomic abstract, **33**-2
Data, bricks, **26**-5
Data, bricks, serialization and deserialization of, **26**-9
Data, centric, **6**-7

Data, clustering, **24**-16
Data, collaborative processing, **13**-9
Data, collection, **6**-10
Data, communication, **30**-1
Data, compression, routing and, **1**-7
Data, convergence time, **26**-14
Data, delivery, **3**-6, **13**-1
Data delivery, protocols, **18**-11
Data, dissemination, **8**-5, **8**-12, **26**-12
Data, dissemination, architectures, **15**-3
Data, encryption, **3**-25
Data, encryption, energy consumption for, **38**-9
Data, exchange, **20**-16
Data exchange, smart message, **26**-5
Data, filtering, **36**-13
Data, flow, specification of, **33**-9
Data forwarding, **18**-11
Data, frame transmission and reception, **3**-25
Data, funneling, **30**-10–**30**-13
Data, fusion, **2**-1, **7**-3–**7**-4, **9**-9, **14**-2, **16**-10
Data fusion, **18**-10
Data, fusion, **25**-4, **35**-1
Data, fusion, approaches, **9**-19
Data, fusion, architectures, **15**-6
Data, fusion, layer, **15**-4
Data, fusion, multimodality, **2**-8
Data, fusion, multiple air vehicles, **22**-13
Data, fusion, real-time distributed network computing goals for, **2**-5
Data, fusion, scalable methods of, **15**-8
Data high bandwidth, **10**-22
Data, integration, **25**-3
Data, integration, methods, **25**-4
Data, link layer protocols, **16**-11–**16**-13, **18**-6, **30**-3
Data low bandwidth, **10**-22
Data, migration, **26**-2
Data, negotiation, **6**-18
Data, processing, **18**-16, **28**-10
Data, processing, coherent and noncoherent, **6**-10–**6**-11
Data, processing, leveraging, **6**-22
Data, processing, requirements, **38**-2
Data, propagation phase of energy-aware routing, **30**-6
Data, quality, **3**-14
Data, redundancy, **6**-3, **9**-6
Data, registers, **37**-5
Data, reporting of a parameter, **3**-15
Data, searching and profiling, **12**-8
Data, sensing and reporting, **6**-1
Data sharing, smart message, **26**-5
Data, signing/verification, **38**-10
Data storage, smart message, **26**-5
Data, traffic, integration of positioning with, **20**-12
Data, transfer path, reliability of, **34**-1

Data, transmission, **1**-4, **1**-6
Data, transmission, phase of length-energy constrained routing, **34**-5
Data, transport, **30**-2
Data, unpackaging, **38**-7
Data flow graph, **38**-6
Database model of system-level programmability, **4**-4
Datagram receiving/sending, **25**-11
DataSpace, **17**-7
DBTMA, **18**-9
DCF, **18**-8, **28**-4
DDF-based tracking systems, **22**-1
DDF-based tracking systems, algorithms for, **22**-14
Deadlocks, **40**-13
Debugging, **12**-8
Decentralized computing, **4**-2
Decentralized data fusion-based tracking systems. *See* DDF-based tracking systems
Decentralized tracking, **22**-2
Decision making, **7**-2, **24**-3
Decision making, cycle of fusion model, **7**-4
Decomposition of tasks, **13**-8–**13**-10
Decryption, **38**-7
Decryption, energy consumption for, **38**-9
Dedicated topology, **10**-23
Deep embedment, **20**-1
Deep reactive ion etch. *See* DRIE
Defenses against attack, **32**-10
Defuzzification process, **7**-12
Degradation, recovery from, **33**-11
Delaunay graph, restricted, **39**-16
Delaunay triangulation, **39**-8, **39**-13. *See also* triangulation
Delaunay triangulation, lines, **19**-9
Delaunay triangulation, partial, **39**-15
Delaunay triangulation, routing on, **39**-37
Delay, **1**-7, **6**-7, **11**-6, **15**-3
Delay, -constrained systems, **1**-7
Delay, balancing, **1**-7
Delay, simulating, **11**-9–**11**-11
Delayed communication, **22**-3
Deletion, **25**-7
Delivery, **18**-6
Delivery-failure feedback, **32**-12
Demand-based MAC, **16**-12
Dempster-Shafer, **7**-4
Dempster-Shafer theory of evidence, **25**-4
Denial of service. *See* DoS
Dense deployment, **1**-1, **13**-8
Density, **3**-12
Dependence model, **3**-24
Deploy-and-leave, **3**-1
Deployment, **3**-11, **9**-8, **15**-2
Deployment, optimization, **15**-12
Deployment, sensor, **25**-5–**25**-8
Deployment, strategies, **9**-13–**9**-14
Dept. of Computer Science and Information Engineering, National Chiao Tung Univ. *See* CSIE/NCTU
Design, application-specific, **9**-10
Design, architecture, **15**-2
Design balanced, **12**-2

Design, diversity, **12**-3
Design, flaws, **32**-9
Design, flow, **12**-8
Design, patterns, **40**-16
Design, principle, **16**-9
Design, security, **9**-10
Design
    techniques for, **12**-2
Destination-sequenced distance vector. *See* DSDV
Detection, **32**-10, **36**-2
Detection, prevention, **32**-12
Detection accuracy, **25**-8
Detection probability table, **29**-20
Determination to attack, **32**-5
Determinism, **32**-15
Deterministic coverage, **15**-12, **19**-7
Deterministic coverage, point, **19**-8
Deterministic decision-making, **24**-3
Deterministic deployment, **19**-2
Deterministic sensor deployment, **19**-2
Device buses, **10**-2
Device power states, switching between, **29**-3
Device scheduling, MUSCLE algorithm for, **29**-14
Device scheduling, two-state I/O devices, **29**-11
Devices, **4**-9–**4**-11
DFuse, **4**-11
Diagnosis techniques, **36**-2, **36**-5
Different sensors, **24**-2
Diffused computational operation, **8**-6
Diffusion limited aggregation. *See* DLA
Diffusion routing, **30**-8, **33**-4
DIFS counter, **17**-3, **28**-4
Digital binary sensors, **36**-3
Digital circuitry, **38**-1
Digital communication and TEDS for multidrop systems (IEEE P1451.3), **10**-12
Digital power supply controllers, **38**-5
Digital signal processors. *See* DSPs
Digital signature algorithm. *See* DSA
Dijkstra algorithm, **6**-17, **39**-39, **40**-5
DIN mess bus, **10**-4–**10**-5
Direct connected architecture, **15**-5
Direct connections, **10**-16
Direct perception, **7**-4
Direct sensor networking, **10**-2
Direct storage, **40**-4
Direct-conversion transceivers, **5**-8
Directed diffusion, **6**-7, **6**-20, **9**-19, **16**-10, **17**-2, **17**-7, **18**-16, **30**-4
Directed diffusion, network, **33**-4
Directed diffusion, query-based routing and, **6**-21
Directed diffusion, using smart messages, **26**-13
Directed diffusion, with in-network processing, **4**-6–**4**-7
Directional antennas, in engineered networks, **13**-10
Directional flooding phase, **30**-11
Directional source-aware routing protocol. *See* DSAP
Directional value, **14**-13
Directly addressed packets. *See* DAP
Directly attached storage, **2**-8
Disaster, detection, **1**-3
Disaster, prevention and relief, **9**-4, **18**-2

Disaster, recovery, **1**-3, **31**-8
Disaster, wireless sensor networks and areas of, **3**-1
Discharge rate-dependent model, **15**-10
Disconnection, **33**-5
Discovery, **3**-15
Discretization, **21**-8
Disjoint, **19**-5
Disjoint, routing paths, **32**-13
Disjoint, sets, **19**-8
Disk model, **1**-8
Dispersion measures, **34**-8
Disruptive denial of service attack, **32**-3
Dissemination, **3**-6, **3**-10–**3**-11, **8**-5, **8**-12
Dissemination, architecture, **15**-6
Dissemination, directed diffusion and, **6**-8
Dissemination, type, **3**-20
Distinguished nodes, **8**-9
Distributed algorithms, **19**-2
Distributed clustering, **39**-31
Distributed computing, **6**-1, **20**-10
Distributed coordination, **8**-2
Distributed coordination function. *See* DCF
Distributed data collection, **8**-2, **28**-12
Distributed embedded network system, **36**-2
Distributed estimation algorithm, **4**-11
Distributed events, **17**-1
Distributed localized algorithms, protocols and, **40**-13
Distributed locationing algorithms, **30**-3
Distributed management, MANNA architecture, **3**-16. *See also* MANNA
Distributed network computing, sensor data fusion goals for real-time, **2**-5
Distributed periphery. *See* DP
Distributed programming, **4**-4
Distributed programming, paradigms, **17**-6
Distributed routing, **39**-33
Distributed sensor applications, dynamic adaptation of, **33**-10
Distributed sensor network. *See* DSN
Distributed sensor processing, **24**-3
Distributed services, supporting reliability with, **33**-3
Distributed signal processing, **1**-11
Distributed systems, **22**-2
Distributed systems, fault tolerance in, **36**-4
Distributed token mechanism, **40**-14
Distributed transaction model, **33**-2
Distribution, **20**-5
Distribution algorithms, **31**-4
Diversity coding, **32**-13
Divide and conquer, **40**-2
DLA, **25**-15
DLL, **30**-3
DNC, **13**-13
Domains, **3**-17
Domains, terminology, **10**-16
Dominating sets, **39**-9
Dominator, **39**-31
Dorigo's telecom routing work, **25**-15
DoS, **3**-5, **32**-2
    asymmetric, **32**-3
    classification of attacks, **32**-16

development of as security concern, **32**-16
   disruptive, **32**-3
   remote, **32**-3
DoS, results of, **32**-10
DoS, security management and, **3**-14
DoS, targets, **32**-8
Dos, malicious, **32**-3
Downlink wireless communication channels, **20**-2
DP, **10**-9
DPM, **9**-14, **27**-2, **29**-2, **38**-2
DPM, CPU-centric, **29**-2
DPM, I/O-centric, **29**-3
DPN, **2**-7
DR-TDMA, **18**-8, **18**-9
Drag reduction, **1**-3
DragonBall, **38**-5
DRAM, **36**-6
DREAM, **17**-2
DRIE, **5**-2
DRIE, SOI/CMOS, **5**-15
DSA, **38**-10
DSAP, **14**-10, **14**-13–**14**-15
   analysis of, **14**-15–**14**-19
   three-dimension analysis of, **14**-18
DSDV, **1**-7, **10**-23, **13**-16, **30**-4, **39**-3
DSN, **15**-2, **24**-3, **37**-1, **38**-4
   ant pheromone model for, **25**-15
DSN, architecture of, **33**-3
DSN, communication between sensors in, **25**-10
DSN
   definition of, **25**-1
DSN, design, **38**-1
DSN
   foundational aspects of, **25**-2–**25**-5
DSN, mobile-agent based, **25**-3
DSN, model of sensor node and energy consumption in, **38**-6
DSN
   multifractal model, **25**-15
DSN, routing paradigms for, **25**-8–**25**-16
DSN, secure, requirements and constraints of, **38**-9
DSN
   spin glass model, **25**-14
DSPs, **15**-8, **27**-1, **38**-2
DSR, **1**-7
DSR, routing, **30**-4
Dual busy tone multiple access. *See* DBTMA
Dual network clustering. *See* DNC
Dual space transformation, **37**-2
Dual-port RAM connections, **10**-16
Duplicate data removal, **13**-1
Duplicate suppression, **6**-21, **18**-16
Duplication, **36**-5
Duty cycle assignment, **8**-12
DV-hop, **20**-18
DVMRP, **10**-14
DVS, **9**-15, **18**-4, **29**-2, **37**-5, **38**-2
DVS, background, **38**-5
DVS, on sensor nodes, **38**-10
DVS, power savings due to, **27**-12
DVS, with message header, **38**-9–**38**-11

DVS circuit, **27**-6
Dynamic adaptation, **33**-11
Dynamic information, **3**-23–**3**-24
Dynamic network topology, **22**-6
Dynamic Power Management. *See* DPM
Dynamic power management, **27**-2
Dynamic power management concepts, **27**-1
Dynamic power optimization, **9**-14–**9**-16
Dynamic private networks. *See* DPN
Dynamic programmability, **4**-3
Dynamic programming, **40**-2
Dynamic reconfiguration, **33**-5
Dynamic reservation TDMA. *See* DR-TDMA
Dynamic sensor networks, modeling of, **15**-7–**15**-13
Dynamic slot assignment, **18**-9
Dynamic source routing. *See* DSR
Dynamic topologies, **9**-9, **22**-3
Dynamic transmission range, WSNs with, **23**-10
Dynamic velocity monotonic scheduling, **17**-3
Dynamic voltage scaling. *See* DVS
Dynamic voltage scheduling, **27**-5
Dynamic WSN_SUM protocol, **23**-10

E

E^2. *See* energy efficiency
EAP muscles, **7**-10
EAR, **13**-11
EAR algorithm, **28**-18
EAR algorithm, advantages and disadvantages of, **28**-18
EAR+A, **13**-11
Earliest deadline first. *See* EDF
Early fallback, **39**-37
Earth monitoring, **16**-2
Eaton, **10**-11
Eavesdrop and register algorithms. *See* EAR algorithm
EC-MAC, **28**-4
EC-MAC, advantages and disadvantages of, **28**-8
EC-MAC
   frame structure, **28**-7
   protocol for wireless ATM networks, **28**-7
Economy, **10**-2
Ecosystem monitoring, **9**-3
EDC, **10**-11
EDF, **1**-7, **29**-4
Edge detection algorithms, **4**-3
Edge routing, **14**-10
Edge routing, comparison with interior routing, **14**-11
EDI, **34**-6
EDI, selection of, **34**-7
EDP, **5**-2
EEPROM, **12**-9
Eight-neighbors WSN, **14**-6
Elasticity, **39**-38
Election procedure, **37**-8
Electric sensors, **24**-1
Electro-optical or infrared images. *See* EO/IR images
Electromagnetic sensors, **24**-1
Electromechanical sensors, **24**-1
Electronic head, **7**-8–**7**-11
Electronic nose, **10**-21

Electronic sensors, **24**-1
Electronic switching system, **36**-4
Electronics, low duty cycle, **9**-12–**9**-13
Electrophysiological signals, **11**-3
Electroplating, **5**-2
Elemental structures, micromachines, **5**-2
ElGamal encryption/decryption, **38**-10
Elimination yield non-preemptive priority multiple access.
    See EY-NPMA
EM direct sequence radios, **13**-6
Embedded commitments, **31**-7
Embedded computing, **26**-17, **27**-2
Embedded computing, cost of, **26**-1
Embedded computing, mission-critical, **29**-12
Embedded PCs, **13**-7
Embedded programming, **4**-4
Embedded sensor nodes, due to drop in operating speed, **29**-3
Embedded signal processors, **2**-8–**2**-9
Embedded systems, WSNs as, **12**-13
Ember, **1**-11, **13**-6
Emergency relief, **39**-1
Emergent protocols, adaptive routing using, **25**-14
Emotion tag, **7**-10
Empiric decision-making, **24**-3
Encoding schemes, **36**-6
Encrypted communications, **32**-14
End-to-end probing, **32**-13
End-to-end retransmissions, **16**-9
End-to-end verification of advertised latency, **32**-13
Endevco, **10**-11
Endpoint tolerance, **35**-12
Energy, **3**-2
Energy, -efficient hierarchical clustering algorithm, **9**-18
Energy, availability, **39**-41
Energy, balancing, **1**-7
Energy, conservation, **39**-5
Energy, constrained WSNs, **3**-20
Energy, constraints, **1**-4, **28**-10
Energy, consumption, **1**-9–**1**-10, **3**-15, **6**-4, **9**-14, **9**-16, **29**-1
Energy, consumption, low, **38**-9
Energy, consumption, minimization of, **25**-8
Energy, consumption, reduction of by energy-aware routing, **30**-8
Energy, cost, **23**-1
Energy, design of efficient protocols, **1**-6
Energy, dissipation, **15**-9, **23**-2, **38**-7
Energy, efficiency, **1**-5, **1**-7, **14**-14, **18**-3, **18**-10, **19**-2
Energy, efficiency, dynamic WSNs, **15**-7
Energy, efficiency, information dissemination protocols, **18**-15
Energy, efficient protocols for MAC, **18**-6–**18**-10
Energy, expenditure, **18**-4
Energy, harvesting, **5**-12
Energy, level severity assignment profile, **3**-21
Energy, limitations of, **1**-1, **16**-11
Energy, limited supply of in WSNs, **6**-2
Energy, management, **40**-8
Energy, management, node-level I/O-device-oriented, **29**-11–**29**-19

Energy, management, node-level processor-oriented, **29**-4–**29**-10
Energy, map, **3**-25
Energy, measure of usage, **6**-7
Energy, minimization, **29**-2
Energy, models, **15**-9
Energy, range and variance of levels, **34**-7
Energy, refined evaluation model, **29**-26
Energy, residual, **3**-24
Energy, storage, **5**-10–**5**-12
Energy, supply, **12**-5, **36**-2
Energy depletion indicator. *See* EDI
Energy scavenging, **36**-6
Energy-aware communication, **29**-19–**29**-32
Energy-aware routing, **30**-2, **30**-4, **30**-5–**30**-7. *See also* EAR
Energy-conserving MAC. *See* EC-MAC
Energy-efficient communication, overview of, **29**-4
Engineered networks, **13**-10
Engineering, civil, sensor network applications in, **1**-3
Engineering, general, applications for sensor networks, **1**-2–**1**-3
Enhanced 911 mandate, **20**-9, **21**-1
Enhanced serviceability, **36**-4
Entity followers, **17**-5
Entity leader, **17**-4
Entity members, **17**-4
Entity-aware transport, **17**-4
Environment, **3**-9, **3**-23
Environment, detection and monitoring, **9**-3–**9**-4
Environment, sensing, **8**-2
Environmental, control applications, **8**-2, **16**-2
Environmental, interference, **39**-41
Environmental, monitoring, **16**-2, **18**-1, **39**-1
Environmental, monitoring, remote sites, **32**-1
Environmental, monitoring, use of sensor networks in, **1**-3
Environmental attributes, **17**-2
EO/IR images, **2**-2
Equipment, **3**-22
Equipment, monitoring, **16**-2
Error, **36**-4
Error, -prone wireless medium, **9**-9
Error, control, **3**-25, **12**-5, **12**-8, **16**-11, **16**-13
Error, correction schemes, **36**-6
Error, detection, **18**-6, **36**-2
Error, handling, **7**-6
Error-correcting codes, **32**-12
ESRT, **16**-8
Estimation fusion, **35**-1
Etching, **5**-2
Ethernet, use of by ANSER system, **22**-14
Ethernet-fieldbus coupler, **10**-18
Ethylene diamine pyrochatechol. *See* EDP
Euclidean distance, **21**-7
Euclidean distance, spanners, **39**-6
Euclidean space, **36**-10
European Laboratory for Particle Physics, **24**-18
European workshop on wireless sensor networks.
    See EWSN
Event, -driven sensor networks, **6**-1, **28**-13
Event, -driving monitoring, **3**-20
Event, -to-sink transport, **16**-7

Event, detection, **16**-2
Event, driven reporting, **19**-1
Event
    forwarding discriminator, **3**-21
Event, identification, **16**-2
Event, radius model, **6**-9
Event
    record, **3**-21
Event, tracking, **37**-2
Event-based network behavior, **28**-10
Event-driven computation, **27**-8
Event-driven messaging system, **21**-11
Event-to-sink reliable transport protocol. *See* ESRT
EWSN, **1**-2
Execution migration, **26**-2
Exhaustion, **32**-12
Explicit organization, **37**-7
Exposed terminal, **18**-8
Exposure, **3**-14
Exposure, coverage, **12**-8
Exposure-based model of coverage, **19**-9
Extended coverage, **15**-2
Extended transactions, **33**-2
Extensibility, **3**-24
Extensible markup language. *See* XML
EY-NPMA, **18**-9

F

Fabrication processes for micromachining, **5**-2–**5**-3
Face routing, **39**-35
Facial expression, modeling, **7**-10
Facial expression driver. *See* FED
Factory instrumentation, **9**-6
Fading, **1**-1
Fading channels, **1**-8
Fading zone, **23**-3
Fail-over strategies, **2**-10
Failure detection, **3**-20
Failure recovery, **33**-8
Failures, **39**-41
Fairness, **1**-7
False alarms, **29**-31
False-alarm probability, **15**-8
Farthest neighbor routing. *See* FN
Fast Fourier transform, **13**-7
Fault detection and diagnosis, **36**-5
Fault detection and diagnosis, discrepancy-based, **36**-11
Fault discover, **12**-8
Fault management, **3**-13
Fault models, **36**-2, **36**-5, **36**-12
Fault resiliency mechanisms, **36**-13
Fault tolerance, **1**-5, **2**-5, **2**-8, **3**-1, **6**-4, **9**-2, **9**-10, **12**-5,
    12-16, **15**-2, **18**-2, **18**-3, **19**-2, **21**-9, **25**-4,
    35-1, **36**-1, **39**-24, **40**-13, **40**-14
Fault tolerance, BISR-based, **36**-2
Fault tolerance, bottleneck, **36**-6
Fault tolerance, classical, **36**-4
Fault tolerance, example in sensor network system, **36**-3
Fault tolerance, NRM, **2**-16
Fault tolerance, robust interval estimation, **35**-11–**35**-16

Fault tolerance, VLSI-based systems, **36**-4
Fault-location protocol, **23**-12
Fault-tolerant energy-efficient summing protocol, **23**-9
Fault-tolerant WSN_SUM protocol, **23**-9
Faulty node location, **23**-2, **23**-11–**23**-15
FBAR, **5**-8
FCC, **22**-13
FDMA, **18**-7, **28**-4, **28**-11, **28**-16
Feature fusion, **7**-3–**7**-4, **7**-8
FEC, **16**-13
Fechner, G.T., **7**-2
FED, **7**-8
FH-CDMA, **28**-6
Fidelity, **3**-12
Fieldbus, **10**-2, **10**-15
Fieldbus, foundation, **10**-8
FIFO buffer, scheduling and, **1**-8
File table maintenance, **25**-11
Film bulk acoustic resonators. *See* FBAR
Filtering, **8**-2, **10**-1, **20**-15, **35**-1
Filtering, motes, **13**-9
Finback whales, **11**-2
Fingerprinting, **21**-6
Fingertip accelerometer virtual keyboards, **1**-2
Finite state model, **27**-2
Fire detection, **9**-4, **13**-9, **15**-8, **18**-15, **31**-1
Fire indication, **7**-11–**7**-12
Fire monitoring, **19**-1
Firewalls, **31**-3
Fisheye state routing. *See* FSR
Five-neighbors WSN, **14**-4
Fixed allocation, **16**-12
Fixed naming schemes, **26**-2
Fixed sensor nodes, **15**-3
Fixed topology, **14**-2
Fixed topology, impact on low-power requirements, **14**-3
Fixed zoning, **6**-15
Fixed-size cluster routing, **6**-15–**6**-16
Flash memory, **12**-4, **36**-6
Flat ad hoc network, **13**-13, **15**-5
Flat ad hoc network, bandwidth, **13**-4
Flat ad hoc network, cluster size in, **13**-14
Flat architecture, **13**-1, **18**-11
Flat multihop protocols, **18**-12
Flat routing, **6**-6–**6**-11
Flat wireless sensor networks, **3**-17–**3**-18
Flaw, **32**-2
Flexibility, **4**-5, **10**-2, **12**-3, **15**-2, **18**-3, **36**-6
Flexible architectures, **9**-9
Flexible logging, **11**-3
Flight control computer. *See* FCC
Flip-chip bonding, **5**-13
Floating-point operations per second. *See* FLOPS
Flooding, **3**-11, **6**-18, **6**-21, **14**-2, **26**-14, **39**-41
Flooding, attack, **32**-14
Flooding, DoS attacks based on, **32**-4
Flooding, reverse directional, **34**-6
Flooding-based routing protocol, **18**-11
FLOPS, **2**-5, **2**-9
Flow control mechanisms, **2**-7, **18**-6
Flow sensors, **5**-4–**5**-5

Index

Fluidic microassembly, **5**-13
Flynn classes for computer architecture, **24**-3
FN, **39**-35
Foreign power, **32**-5
Forest fire detection, **31**-1
Forward error correction. *See* FEC
Forward path, **16**-7
Foundation fieldbus, **10**-8
Four-dimensional fix, **20**-3
Four-neighbors WSN, **14**-4
Fragmenting gateway, **10**-18
Frame synchronization message. *See* FSM
Framing, **18**-6
Free node, **20**-3
Free-space optical transmission, **1**-6, **5**-8
Frequency, reduction, **27**-7
Frequency division multiplexing access. *See* FDMA
Frequency hopping code division multiple access. *See* FH-CDMA
Frequency noise, effects of on synchronization, **16**-5
Frequency selection, **3**-25
Frequency shift keying. *See* FSK
Frequency-hopping schemes, **32**-11
Freshness constraints, **17**-6, **31**-6, **32**-13
Freshwater quality monitoring, **1**-3
Friendly forces, monitoring, **16**-2
Frontend node, **8**-5
FSK, **10**-3
FSK, telemetry, **11**-5
FSM, **28**-7
FSR, **13**-17, **39**-3
Fuel cells, **18**-17, **36**-6
Full cells, **12**-5
Functional architecture, **3**-17–**3**-21
Functional decomposition, sensor network, **13**-2
Fuser, **24**-7
Fuser, Bayesian, **25**-4
Fuser, merit factor, **24**-14
Fusion, **9**-6, **18**-10
Fusion, optimal, **35**-5
Fusion algorithms, **4**-11
Fusion element, **24**-7
Fusion of data, **2**-1
Fuzzy inference, **7**-12
Fuzzy logic, **7**-4

G

GA, **25**-3, **39**-39
GA, computation of, **25**-7
GA, mobile agent routing using, **25**-8
GA, optimal sensor deployment using, **25**-6
Gabriel graph, **39**-8, **39**-13
GAF, **9**-16, **13**-13, **40**-8
GAF, algorithm, **18**-12
Gateways, **10**-15, **13**-1, **16**-10, **39**-32
Gateways, fragmenting, **10**-18
Gateways, nodes, **18**-15. *See also* cluster heads
Gateways, TCP/IP node, **13**-7
Gateways, tunneling, **10**-18
Gaussian node placement, **15**-13

GCF, **25**-3, **25**-10
GDOP, **20**-15
General engineering, applications for sensor networks, **1**-2
General interORB protocol. *See* GIOP
Genetic algorithm. *See* GA
Geocast, **21**-2
Geographic forwarding, **32**-13
Geographic routing protocol, **4**-3, **13**-15, **20**-2, **30**-4, **34**-3
Geographical adaptive fidelity. *See* GAF
Geolocation, **2**-2–**2**-3
Geometric algorithms, **39**-2
Geometric dilution of precision. *See* GDOP
Geometric routing, **20**-2
Geometric spanners, **39**-6
Geometric structures, **39**-7
Geophysical monitoring, **1**-3
GFR, **39**-13
Gimbals, **22**-11
GIOP, **2**-11
Global closest first. *See* GCF
Global Hawk, **2**-3
Global positioning system. *See* GPS
Global queue, **11**-14
Global task requirements, **33**-9
Globalfuture, **1**-2
Globecom, **1**-2
GloMoSim simulator, **8**-9
GMTI, **2**-2
GOAFR, **39**-13, **39**-37
Goodput, **3**-14
Gossiping, **3**-11
GPIO, **12**-10
GPS, **8**-3, **12**-10, **13**-2, **20**-4, **21**-1, **30**-3, **34**-3, **37**-1, **39**-2
GPS, accuracy of, **20**-15
GPS, integration, **20**-9
GPS, location query, **26**-11
GPS, Nymph and, **13**-5
GPS, PC/104s and, **13**-9
GPS, radio, **12**-6
GPS, trilateration of receivers, **20**-8
GPSR, **39**-13, **39**-36, **40**-12
Gradient paths, **6**-9
Graph generator, **2**-12–**2**-13
Graph search, **2**-14–**2**-15
Graph search, algorithm, **9**-13
Graph theory, **20**-10, **31**-10
Great Duck Island, **9**-3
Greedy aggregation, **9**-20, **18**-16
Greedy algorithm, **27**-2
Greedy compass routing, **39**-35
Greedy face routing. *See* GFR
Greedy forwarding, **40**-12
Greedy heuristics, **25**-10, **39**-39
Greedy other adaptive face routing. *See* GOAFR
Greedy perimeter stateless routing. *See* GPSR
Greedy routing, **39**-35
Green power, **18**-17
Grid-based sensor deployment, **15**-12
Grids, as analytically tractable models, **1**-10
Ground moving target indication. *See* GMTI
Group communication, **17**-6

Group membership management, 17-6
Guaranteed zone, 6-15
Gyroscopes, 5-4–5-5

H

H-AODV, 13-15
Habitat monitoring, 1-3, 13-4, 15-2, 16-2
Hacker, 32-5
Half gateway, 10-15
Handoff, 20-2
Hard threshold, 6-14
Hardware, -software interaction, 37-5
Hardware, architectures of sensor nodes, 31-4
Hardware, compact, 28-10
Hardware, constraints, 6-4, 28-24
Hardware, fault tolerance, 36-6
Hardware, miniaturization, 17-1
Hardware, operation, 1-4
Hardware, operation, impact on optimum protocol design, 1-11
Hardware, redundancy, 9-14, 36-7
Hardware, sensor network, 13-5–13-8
Hardware, techniques, 9-11–9-13
HART, 10-3
Hash table, 32-15
HDVI, 2-3
Health Improvements through Space Technologies and Resources. See HI-STAR
Health monitoring, sensor network applications in, 1-4
Heartbeats, 17-5
HELLO floods, 32-15
Heterogeneous built-in self-repair-based fault tolerance. See BISR-based fault tolerance
Heterogeneous fault detection, 36-7
Heterogeneous wireless sensor networks, 3-18, 9-22, 13-1, 13-19, 15-5
Heuristics, 20-15, 30-4, 39-34
Heuristics, greedy, 25-10
HI-STAR, 9-4
Hidden terminal, 18-8
Hierarchical addressing, 13-16
    mapping unique IDs to, 13-17
Hierarchical architecture, 18-11, 25-2
Hierarchical architecture, evaluation of merit factor for, 24-8
Hierarchical architecture, parameterization of, 24-6
Hierarchical clustering, 4-8, 8-3
Hierarchical clustering, architectures, 9-16
Hierarchical clustering, scalability of, 15-8
Hierarchical communication, forming, 13-13
Hierarchical energy efficient protocols, 18-14
Hierarchical power-aware routing, 6-17
Hierarchical routing, 6-11–6-18
Hierarchical state routing, 39-3. See also HSR
Hierarchical state routing, mapping in, 13-18
Hierarchical systems, 24-3, 24-5
Hierarchical wireless sensor networks, 3-18
Hierarchy, routing in, 13-15
High bandwidth data, 10-22
High data rate, problems with, 24-5

High-definition vector imaging. See HDVI
High-density battery cells, 12-5
High-energy physics, architecture evaluation in, 24-18
High-Q capacitors, voltage-tunable, 5-5
High-range resolution profiles. See HRR profiles
High-resolution image processing, 24-5
Highly integrated processes, 5-3–5-4
Highway addressable remote transducer. See HART
HIPERLAN, 18-9
Histogram method, implementation of likelihood function with, 21-8
History data object, 3-22
Home automation, 18-2
Home intelligence, 9-5
Home phoneline networking alliance. See HPNA
HomeRF, 12-16
Homogeneous wireless sensor networks, 3-17, 13-1, 15-5
Honeypots, 31-3
Hop count, 20-18
Hop metric, 13-11
Hop-TERRAIN, 20-18
Hopping sequence, determination of, 28-6
Host system interaction, smart messages and, 26-5
Hostile environment, 31-2
Hot spots, 13-18
Hot swapping, 10-13
Householder transform, 20-8
HPNA, 10-12
HRR profiles, 2-3
HSR, protocol, 13-16
Human perception, 7-2
Human-computer interaction, 7-7–7-8
Human-computer interface, 7-2
Humidity monitoring, 9-2, 18-1
Humpback whales, 11-2, 11-8
Hybrid CI/IF update, broadcast with, 22-6
Hybrid collect model, 3-9–3-10
Hybrid data fusion architecture, 15-6
Hybrid laser/vision sensors, 22-13
Hybrid method of facial expression modeling, 7-10
Hybrid protocols, 18-7
Hybrid TDMA/FDMA-based medium access, 16-12
Hyperbolic trilateration, 20-5
Hyperplane arrangement, 39-2

I

I/O devices, incorrect switching of, 29-12
I/O-centric DPM, 29-3
I/O-device-oriented energy management, node-level, 29-11–29-19
ICC, 1-2
ICNs, 10-2
ICNs, interconnection of, 10-16
ID assignment, 30-3
Identification, 20-16
Identity fraud, 32-14
Idle channel sensing, 18-7
Idle listening, 28-11
Idle power management, 27-2
Idle power management, hooks, 27-8

IEC 1158-2, **10**-8
IEEE 1118, **10**-7
IEEE 1451 family, **10**-11–**10**-13
IEEE 802.11, **18**-8, **21**-11, **28**-2, **28**-4
IEEE 802.11, power saving mode in, **28**-5
IEEE 802.11b, **12**-16
IEEE Proceedings, **1**-2
IGMP, **10**-14
Illumination, changes in, **22**-10
ILP, **9**-13, **36**-8, **40**-5
ILP, solving by (MI)² paradigm, **40**-10
Image processing, **22**-15, **24**-12
Image processing, high-resolution, **24**-5
Image-based modeling, **7**-10
Imote, **13**-6
Implementation, **1**-11
Implementation, flaws, **32**-9
Implication refinement, **7**-4
Implosion problem, **3**-19, **6**-21, **14**-2, **16**-10
Implosion problem, response, **8**-5
In-network processing, **31**-2
In-network processing, directed diffusion with, **4**-6–**4**-7
In-theatre sensors, **2**-8
Incorrect switching of I/O devices, **29**-12
Independent range estimates, **20**-7
Independent routing, **13**-16
Independent set, **39**-9
Indirect connections, through FIFO queue, **10**-16
Indirect storage, **40**-4
Individual sensor nodes, **12**-8
Indoor tour guide, **21**-11
Indoor user location detection system, **31**-15
Induction, **1**-3
Industrial automation, communication systems for, **10**-2
Industrial communication networks. *See* ICNs
Industrial plants, sensing and maintenance in, **1**-2–**1**-3
Industry, wireless sensor networks in, **10**-21–**10**-24
INFOCOM, **1**-2
Information abstraction, SINA systems, **8**-5
Information accuracy, **15**-3
Information aggregation logic, **8**-6
Information architecture, **3**-21–**3**-24
Information collection protocols, energy efficient, **18**-12
Information dissemination protocols, **40**-10
Information dissemination protocols, energy efficient, **18**-15
Information filter, **22**-1, **22**-3
Information fusion, **7**-3–**7**-4, **33**-2
Information gathering methods, **18**-12, **23**-2, **23**-5, **40**-9
Information gathering methods, SINA systems, **8**-5–**8**-6
Information log, **3**-21
Information processing in sensor networks. *See* IPSN
Information propagation model, **11**-7–**11**-9
Information quality, **15**-3
Information sharing, **22**-2
Information summarization, **18**-10
Information-driven schemes, **9**-19
Infostations, **11**-1, **11**-3
Infrared sensors, **5**-5
Infrastructure communication, **3**-10–**3**-11
Initialization task, **23**-12

Input interface, **7**-2
Insider *vs.* outside threats, **32**-2
Installation properties, **10**-3
Integer linear programming. *See* ILP
Integrated circuit sensors, **5**-5
Integrated machining processes, **5**-3–**5**-4
Integrated Ornstein-Uhlenbeck. *See* IOU
Integrated sensing and processing. *See* ISP
Integrity, **32**-4
Intel, **10**-5, **10**-17, **13**-6, **27**-2, **38**-2
Intel, Strong ARM microprocessor, **9**-11
Intellectual property tagging, **31**-4
Intelligence, surveillance and reconnaissance sensors. *See* ISR
Intelligent agents, **3**-24, **10**-1
Intelligent cameras, **26**-1
Intelligent data collection, **8**-2
Intelligent patient monitoring, **27**-1
Intelligent scheduling algorithms, **28**-6
Intelligent sensors, **12**-1, **24**-2
Intensity zone, **23**-3
Intentional naming system, **17**-2, **17**-7
Interactive surroundings, **9**-5, **18**-2
Interbus, **10**-4
Intercluster routing, reliable energy-constrained, **34**-1
Interconnecting architectures, **10**-15
Interest, **6**-8
Interest, packets, **30**-10
Interest, propagation in energy-aware routing, **30**-6
Interfaces, serial, **10**-17
Interference, **39**-28
Interference, analysis, **1**-9
Interference, loss, **1**-8
Interference, number, **39**-28
Interior routing, **14**-10
Interior routing, comparison with edge routing, **14**-11
Intermediate system, **10**-16
Intermittent faults, **36**-4
International standards, sensor networking and, **10**-2
Internet, multicast algorithms, **31**-9
Internet coupling architectures, **10**-14
Internet group management protocol. *See* IGMP
Internet protocol. *See* IP technologies
Interoperable System Project. *See* ISP
Interplatform communication, **22**-14
Interprocess communication, **22**-14
Interrogation, **32**-12
Interruption of processors, **2**-9
Intersections, **39**-2
Intertask time, **29**-15
Interval estimate, **35**-2, **35**-16
Interval estimate, fault-tolerant, with knowledge of confidence degrees, **35**-9–**35**-11
Interval estimate, fault-tolerant, without knowledge of confidence degrees, **35**-4–**35**-8
Intervals, combined, **35**-5
Intrapiconet scheduling, **10**-23
Intruder, **32**-2
Intrusion detection, **6**-1, **31**-3
Inventory control, **6**-1, **16**-2, **19**-1
Inversion, **25**-7

IOU, process model, **22**-5
IP addressing, **26**-2
IP technologies, military applications for sensor networks and, **2**-2
IPAQ, **13**-4, **13**-7, **26**-2
IPSN, **1**-2
Ising model, **25**-14
ISIT, **1**-2
ISO open systems interconnection vocabulary, **10**-16
ISO/OSI communication model, **10**-8
Isolation property, **39**-22
ISP, **1**-2, **10**-8
ISR, **2**-1
Iterations, **20**-15
Iterative multilateration, **21**-6

J

Jamming signal, **18**-9, **32**-11
Java KVM, **26**-2, **26**-8
JIK network, **25**-3
Joint directors of laboratories model, **7**-4
Joint distributions, **25**-4

K

K-6, **38**-2
K-anonymity, **31**-13
K-clustering, **9**-17
K-coverage, **19**-7
K-fault tolerance, **39**-26
Kalman filters, **7**-4, **22**-3
Kernel method, implementation of likelihood function by, **21**-8
Key chain commitments, **31**-7
Key distribution schemes, **31**-9
Key establishment, **31**-4
Key rings, **31**-9
Keyboards, virtual, fingertip accelerometer, **1**-2
Keyframing, **7**-10
Knapsack problem, **25**-5
Knowledge base, **7**-7
Knowledge to attack, **32**-5
Kruskal's algorithm, **39**-29, **39**-40

L

LAN, **2**-8, **10**-2
  NRMs, **2**-16
LAN, wireless coverage, **20**-9
Landmark routing, **13**-16
Landmark routing, mapping in, **13**-18
LANMAR, **13**-17
LAR, **17**-2
Large nodes, **13**-9
LAS, **10**-8
Laser-based sensors, **22**-13
Late binding, **8**-8
Latency, **1**-5, **9**-10, **15**-8, **18**-3, **18**-15, **22**-15, **28**-8, **40**-14
Latency, bounded, **18**-6

Latency, bounds, **17**-4
Latency, end-to-end verification of, **32**-13
Law enforcement applications, **31**-1, **31**-8
Layer-by-layer protocol, **1**-5
LBS, **31**-12
LCA, **13**-12
LCF, **25**-3, **25**-10
LCS, **20**-17
LEACH, **6**-11–**6**-13, **6**-20, **13**-10, **14**-2, **18**-8, **19**-6
LEACH, distributed, **18**-14
LEACH, operational phases of, **6**-11
Leach's Storm Petrel, **9**-3
Leader handoff, **17**-6
Leakage energy, **18**-4
Leakage power, **9**-13
Leakage power, constant, **15**-9
LEC payoff model, **34**-2
LEC payoff model, distributed, **34**-5
LEDES, **29**-11, **29**-17, **29**-33
LEDF algorithm, **29**-4
Length-constrained RQR, distributed implementation of, **34**-4
Length-energy constrained routing, game-theoretic models of, **34**-2
Level of abstraction, **31**-4
LFP, **10**-23
License key, **31**-4
Licensing, network usage, **31**-4
Lifetime, dynamic WSNs, **15**-7
Light compass, **12**-11
Light detection, **6**-1, **9**-2, **11**-3
Light sensor, **26**-11
Light-weight protocol stack, **3**-24
Lighthouse location system, **20**-9
Limited energy provision of sensor networks, **1**-4
Limited power, **33**-5
Limited resources, **31**-2
Lincoln Space Surveillance Complex, **2**-1
Linda-like model, **4**-6
Linear acceleration, **22**-13
Linear mobility, **11**-9
Linear model, **15**-10
Linear program routing, **30**-8
Linear programming, **30**-5
Linear programs, **20**-6
Linear topology, node overlay in, **13**-4
Link access, **18**-6
Link active scheduler. *See* LAS
Link layer, **30**-3
Link master device. *See* LMD
Link utilization, **2**-13
Linked-cluster algorithm. *See* LCA
Lipschitz condition, **25**-4
Listeners, **21**-11
Listening, **1**-9
Lithium batteries, **5**-10
Little's formula, **11**-12
Living with errors, **20**-12
LLA, **3**-3
LMD, **10**-8
LMST, structure based on, **39**-22

Index

Local aggregators, **6**-16
Local area networks. *See* LAN
Local closest first. *See* LCF
Local coordinate system. *See* LCS
Local deadline misses, **17**-4
Local information processing, **29**-1
Local minimum spanning tree. *See* LMST
Local miss ratio, **17**-4
Localization, **1**-11, **13**-2, **13**-7, **16**-4, **32**-2
Localization, communications, **4**-2
Localization, target approach, **29**-19
Localization, techniques, **20**-10
Localized algorithms, **19**-2, **39**-9, **40**-1
Localized algorithms, analysis of, **40**-13
Localized data fusion architecture, **15**-6
Localized low-weight bounded degree planar structures, **39**-20
Localized methods of broadcasting, **39**-41
Localized processing, **9**-9
Localized routing, **39**-4, **39**-33–**39**-38
Location, **3**-12, **30**-3, **37**-1
Location, abstractions of, **20**-2
Location, awareness, **8**-3, **18**-10
Location, computation to derive, **20**-16
Location, navigation techniques to derive, **20**-5
Location, privacy of information, **31**-12
Location, service, **39**-33
Location, systems, experimental, **21**-10
Location, techniques, **20**-12
Location, technology, **20**-2
Location, tracking, **21**-1, **21**-8
Location, verification, **32**-14
Location sensing, **16**-2
Location-assisted routing protocols, **17**-2
Location-aware network protocols, **21**-2
Location-based access control, **21**-11
Location-based routing, **20**-2
Location-based services, **31**-12
Location-based services, commercial, **20**-9
Log, **3**-21
Logging, flexible, **11**-3
Logical areas, **21**-11
Logical clustering, **13**-16
Logical layered architecture. *See* LLA
Logical roles, **13**-2
Logical sensor, **24**-1
Logical vulnerability, **32**-9
Longevity, tiered architectures, **13**-4
LonWorks, **10**-6
Look-ahead, **29**-16
Lookup server, distributed, **33**-6
Loops, **6**-8
Loops, routing, **10**-23
Loose formations, **10**-22
Loss-concealing coding schemes, **32**-4
Lost flooded packet. *See* LFP
Low bandwidth data, **10**-22
Low data-rate links, **14**-2
Low duty cycle electronics, **9**-12–**9**-13, **20**-10
Low energy adaptive clustering hierarchy. *See* LEACH
Low-energy device scheduler. *See* LEDES
Low-power dissipation, **14**-2
Low-power electronics, **12**-1
Low-weight structures, **39**-6
LRG, **20**-18
LVS, **20**-18

M

M-Core system, **38**-14
MAC layer, **13**-9, **16**-12, **17**-3
MAC layer, adaptability of schemes for, **28**-23
MAC layer, components, **12**-16
MAC layer, design issues, **28**-3
MAC layer, energy-efficient protocols for, **18**-6–**18**-10
MAC layer, need for protocols, **28**-12
MAC layer, optimizing schemes, **28**-10
MAC layer, power-saving mode on, **9**-10
MAC layer, protocols for, **3**-25, **11**-6
MAC layer, protocols for, classification and comparison of, **18**-7
MAC layer, protocols for, distributed, **18**-8
MAC layer, protocols for, hybrid, **18**-9
MAC layer, QoS, **3**-6
MAC layer, scheduling and, **1**-8
MAC layer, schemes, **28**-2
MAC layer, sharing of media by, **12**-6
MAC layer, sources of energy consumption at, **18**-7
MACA, **28**-4
MACAW, **28**-4
Machining, **5**-2
Macroprogramming, **4**-12
Macrosensors, **9**-1
MADSN, **25**-3, **25**-8
Magnetic sensors, **5**-4–**5**-5, **24**-1, **32**-1
MagnetOS, **4**-11, **17**-7
Maia processor, **12**-4
Maintenance, **3**-4, **3**-11–**3**-12, **8**-7
Maintenance, communication, **10**-22
Maintenance, ID, **6**-3
Maintenance, properties, **10**-3
Malfunctioning, **36**-7
Malicious denial of service attack, **32**-3
MAMPS, **9**-12, **13**-7, **15**-3, **27**-1, **27**-6
Managed element, **3**-22
Managed object class, **3**-21
Management architecture, **3**-19
Management by delegation. *See* MbD
Management dimensions, **3**-3–**3**-15
Management functional areas, **3**-3, **3**-12–**3**-15, **3**-16
Management information base. *See* MIB
Management levels, **3**-3, **3**-5–**3**-8
Management log, **3**-21
Management operation schedule, **3**-21
Management services, **3**-15–**3**-17
MANET, **6**-2, **16**-12, **20**-10
MANNA, **3**-2, **3**-15–**3**-25
MANTIS project, **13**-5
Manual organization, tiered architectures, **13**-10
Manufacturing, cost of, **16**-1
Mapping, **2**-12, **18**-1
Mapping, database, **2**-15

Mapping, hierarchical addressing, **13**-16
Maps, **31**-13
Marine mammals, tagging, **11**-2
Markov chain, **11**-16
Mars, WSNs on the surface of, **9**-5
Marzullo's method, **25**-4, **35**-2, **35**-4
Massive deployment, **9**-8
Master agent, **21**-9
Master aggregator node, **6**-11, **6**-16
Master key, **31**-5
Master node control, **28**-6
Master transmission, **28**-6
Master-slaves configuration, **10**-16
Master-slaves configuration, NBIP protocol for, **10**-17
Maté, **4**-9, **17**-7, **26**-17
Materialization, **8**-7
MAV, **5**-1
Max-min path, **6**-17
Max-min zPmin algorithm, **6**-17
MAX1717 step-down controller, **27**-6
Maxim MAX 186 ADC, **12**-12
Maximal breach path. *See* MBP
Maximal independent set, **39**-9
Maximal support path. *See* MSP
Maximum independent set. *See* MIS
Maximum information gain, **22**-4
Maximum lifetime data aggregation. *See* MLDA
Maximum waiting time, **32**-2
MbD, **3**-24
MBP, **19**-9
MCDS, **39**-9
MCFA, **6**-10
MCLK, **27**-6
MCU, **9**-11
MCU, computational workload of, **9**-15
Measured power consumption, **27**-12
Measurement, **20**-16
Measurement bus, **10**-4–**10**-5
Mechanical sensors, **24**-1
Medical care, **9**-4–**9**-5, **18**-2
Medical records, depersonalization of, **31**-11
Medical sensing, **1**-4
Medium access, **1**-8, **10**-2
Medium access, control, **3**-25
Medium access, prioritization, **17**-3
Medium access control layer. *See* MAC layer
Medusa MK-2 node, **12**-9, **13**-6
Meeting-oriented model, **4**-6
Membership declaration, **6**-10
Membranes, **5**-2
Memory, **7**-7
Memory, capabilities, **7**-5
Memory, limitations of, **1**-1
Memory clock. *See* MCLK
MEMS, **1**-5, **3**-1, **5**-1, **9**-1, **9**-12, **12**-6, **18**-17
MEMS, -based microprocessors, **38**-5
MEMS, advances in, **23**-1, **38**-1
MEMS, basics of, **5**-2–**5**-4
MEMS, communication, **5**-5–**5**-10
MEMS, design, **18**-1
MEMS, fusion of technology with electronics, **1**-11

MEMS, sensor network technology and, **1**-2
MEMS, sensor selection criteria, **5**-4–**5**-5
MEMS, systems, **5**-13–**5**-15
Merit factor, **24**-5
Merit factor, evaluation for parallel and hierarchical
   systems, **24**-8–**24**-11
Mesh networking, **1**-1
Mesh topology, **14**-2
Message authentication code (MAC), **31**-6
Message communication, **17**-2
Message generation rate, **14**-3
Message header, **38**-2, **38**-8
Message looping, **8**-8
Message passing, **26**-2
Message pattern, **14**-3
Message routing, **33**-3
Message velocity, **17**-3
Message-passing approach, **22**-15
Messaging system, event-driven, **21**-11
Metadata, **18**-15
Metering, **31**-4
Metering, remote, **9**-5
Metrics object. *See* MO
MFR, **39**-35
MH-TRACE, **18**-10
MIB, **3**-17
MIB, implementation and usage, **3**-21
Mica, **13**-5
Mica, two tier network with, **13**-9
Mica, **4**-9
Mica2 mote, **13**-5, **23**-2
Micro-surgery, **1**-4
Microadaptive multidomain power-aware sensors.
   See μAMPS
Microaerial vehicles. *See* MAV
Microassembly, **5**-13
Microbatteries, **5**-10–**5**-12
Microcontroller unit, **10**-17, **37**-2. *See also* MCU
Microelectromechanical devices. *See* MEMS
MicroLan protocol, **10**-13
Micromachine, fabrication techniques, **5**-2–**5**-3
Microphones, **5**-4–**5**-5
Micropower sources, **5**-10–**5**-12
Microprocessors, programmable, **15**-8
Microsensing, **16**-2
Microsensor devices, **9**-9
Microsoft, **27**-2
Microsystems, **5**-13–**5**-15
Microwire, **10**-10
Middleware, **2**-6, **2**-11, **9**-21, **12**-8, **17**-6
Middleware, fault tolerance, **36**-7
Middleware, SINA systems, **8**-4
Migration, smart message, **26**-6
Military applications of sensor networks, **1**-3–**1**-4, **9**-2–**9**-3,
   9-14, **16**-2, **17**-1, **19**-8, **25**-4, **27**-1, **31**-1, **31**-8,
   35-1, **38**-1
Millenial, **1**-11
MIMD, **24**-4
MIMO systems, **2**-8
Mini ATRV mobile robots, use of CTT protocol for, **25**-14
Miniaturization, **1**-5, **1**-11

Index

Miniaturization, MEMS and, **5**-4–**5**-5
Minimal generalization, principle of, **31**-11
Minimal interference, **9**-3
Minimum connected dominating set. *See* MCDS
Minimum cost forwarding algorithm. *See* MCFA
Minimum coverage, **25**-6
Minimum spanning tree. *See* MST
Minimum-energy broadcast/multicast routing, **39**-38
Minimum-energy broadcast/multicast routing, theoretical analysis of, **39**-40
MIPS processors, **38**-13
Mirror scanner, **22**-11
MIS, **39**-9
MISD, **24**-4
Misdirection, **32**-13
Missing-detection probability, **15**-8
MIT Lincoln Laboratory, **2**-1, **2**-16, **13**-7, **15**-3
Mix zones, **31**-14
Mixed mode communication protocols and TEDS formats (IEEE P1451.4), **10**-12
Mixed-mode interface for analog transducers, **10**-12
Mixed-mode sensors, **10**-13
MLDA, **9**-20, **18**-16
MMWN, **13**-18
MO, **2**-13–**2**-14
MobiCom, **1**-2
MobiHoc, **1**-2
Mobile ad hoc network. *See* MANET
Mobile agent-based distributed sensor network. *See* MADSN
Mobile agents, **3**-24, **25**-3
Mobile agents, active networks, **4**-5–**4**-6
Mobile agents, clustering schemes, **9**-19
Mobile agents, comparison with smart messages, **26**-16
Mobile agents, routing using GA, **25**-8
Mobile code, **31**-3
Mobile computing, **1**-2
Mobile computing, active networks and, **4**-5
Mobile networks, **14**-3
Mobile node wireless communications, **10**-21
Mobile nodes, **1**-4
Mobile nodes, EAR algorithm in, **28**-18
Mobile robot sensors, **24**-2
Mobile sensor nodes, **15**-3
Mobile targets, identification of, **2**-2–**2**-3
Mobile wireless networks, connectivity through time for, **25**-10
Mobile wireless nodes, **12**-13
Mobility, **1**-10, **15**-3, **17**-6, **33**-5, **40**-13, **40**-15
Mobility, degree of, **20**-5
Mobility, management, **28**-23
Mobility, patterns of whales, **11**-9
Mobility management, **3**-7
Model-specific register. *See* MSR
Modeling, **1**-8–**1**-10
Modeling, architecture, **15**-2
Modeling, connectivity, **15**-10
Modeling, deployment and sensing coverage, **15**-12
Modeling, techniques for sensor networks, **12**-2
Modulation, **3**-25
Molybdenum oxide batteries, **5**-11

Monitored object, **3**-22
Monitoring applications, **3**-25, **15**-3, **27**-1, **31**-1, **32**-3
Monitoring applications, remote object, **23**-1
Monitoring station, **19**-1. *See also* sink node
Moore's law, **12**-3, **20**-1
MOSFETs, **5**-5
MOSFETs, addressable, **31**-4
MOSPF, **10**-14
Most forwarding routing. *See* MFR
Mote node, **12**-8, **13**-5
Mote-to-central station communication, **37**-9
Mote-to-mote communication, **37**-7
Mote/TinyOS development team, **12**-8
Motes, **4**-9, **31**-5, **37**-1
Motes, orientation of, **37**-10
Motive for attack, **32**-5
Motorola, **38**-5
Mounting, errors, **22**-10
Movement detection, **6**-1
MRA, **40**-10
MRAM, **36**-6
MRAM, nanoelectronics-based, **12**-4
MRI algorithm, **25**-3
MSET, **1**-2
MSP, **19**-9
MSP430 processor, **11**-3
MSR, **29**-6
MST, **39**-6, **39**-39, **39**-41
MST, algorithm, **39**-29
MTESLA, **31**-6
Multicast, advantage, **1**-6
Multicast, communication, **10**-14
Multicast, messages, **28**-13
Multicast, routing protocol, **18**-11
Multicast advantage, **3**-11
Multidimensional scaling algorithm, **15**-10
Multifractal model, **25**-15
Multihop ad hoc networks, **12**-14–**12**-16
Multihop ad hoc networks, power conservation for large-scale, **15**-11
Multihop clusters, **9**-17
Multihop communication, **16**-9, **18**-5, **37**-7, **38**-1
Multihop mode, **15**-6
Multihop network topology, **20**-10
Multihop routing, **1**-1, **1**-4, **1**-7, **9**-6, **12**-16
Multihop routing, energy efficiency achieved through, **9**-8
Multihop self-organizing sensor network, **16**-12
Multihop TRACE. *See* MH-TRACE
Multihop wireless capabilities, **3**-11
Multihop wireless sensor networks, geometric nature of, **39**-3
Multihop WSNs, **23**-3
Multihop WSNs, locating fault sensors in, **23**-12
Multihop WSNs, sum problem in, **23**-6
Multilateration, **20**-6, **21**-5, **31**-15
Multimodal sensor fusion, fault tolerant, **36**-3, **36**-10
Multipath routing, **6**-6, **6**-20–**6**-21, **30**-1
Multipath routing, braided, **6**-21
Multiple input, multiple output antenna systems. *See* MIMO systems
Multiple output intervals, sensor estimate, **35**-11

Multiple shutdown states, **27**-2
Multiple winner algorithm. *See* MWA
Multiple-packet storage methods, **11**-14
Multiple-voltage DVS systems, **38**-2
Multiplexing, **10**-2
Multiplexing, controlled, **18**-7
Multiplexing, of data streams, **3**-25
Multiresolution integration algorithm. *See* MRI algorithm
Multisensor exploitation. *See* MSET
Multisensor fusion, **36**-12
Multisensor integration, **24**-3
Multistate constrained low-energy scheduler. *See* MUSCLES
Multistate I/O devices, low-energy device scheduling, **29**-14
MUNIN, **17**-6
MUSCLES, **29**-14, **29**-17, **29**-33
Mutation, **25**-7
MWA, **6**-11
Myograms, **11**-3

N

Name-based shared memory, **26**-2
Naming, fixed schemes for, **26**-2
Naming, NES, **26**-2
Naming, scalable, **15**-8
Naming, smart message, **26**-5
Nanoelectronics-based MRAM, **12**-4
Nanosensors, **5**-5
Nanotechnology, sensor network technology and, **1**-2
Nash equilibria, **34**-2
National Instruments, **10**-11
National Science Foundation. *See* NSF
Natural disaster, sensor networks at site of, **17**-1
Naval exercises, effect on marine mammals, **11**-2
Navigation filter, **22**-13
Navigation solutions, classification of, **20**-5
Navigation solutions, trilateration, **20**-7
NBIP, **10**-17
NCAP, **10**-11
NCAP, object-oriented definition of, **10**-13
NCR protocol, **28**-20
Nearest neighbor algorithms, **21**-7
Nearest neighbor in signal space. *See* NNSS
Nearest neighbor routing. *See* NN
Negotiation-based protocols, **6**-21, **18**-5
Neighbors, discovery of, **26**-11
Neighbors, list management, **30**-3
Neighbors, number of, **14**-1
Neighbors, protocol, **28**-19
Neilsen Media Research, **31**-4
NES, **26**-1
NES, cooperative computing model, **26**-3, **26**-17
NES, deployment of, **26**-2
NEST, **1**-2
Netcard II, **13**-7
NETEX, **1**-2
NetRatings, Inc., **31**-4
Network, **3**-22
Network, -attached storage, **2**-8

Network, active, **4**-5–**4**-6
Network, architectures, **25**-2
Network, block functions, **10**-15
Network, capable application processor information model (IEEE1451.1), **10**-11
Network, capacity, **13**-4
Network, connectivity, **1**-10
Network, connectivity, clustering and, **13**-12
Network, cost, **2**-13
Network, daemons, **25**-8
Network, design, **18**-5
Network, element management, **3**-7
Network element management, management functions for, **3**-12
Network, engineered, **13**-10
Network, implosion, **14**-2
Network, interconnections, industrial, **10**-15–**10**-21
Network, inventory, **3**-15
Network, layer, **3**-25, **30**-3
Network, layer, protocols, **16**-9–**16**-11, **18**-10–**18**-17
Network level configuration management functions, **3**-12
Network, lifetime, **19**-4
Network, lifetime, maximizing, **19**-2
Network, management, **3**-3, **3**-6–**3**-8
Network, management, challenges of, **3**-2–**3**-3
Network, management, domains in, **3**-17
Network, modeling, **1**-8–**1**-10
Network, operating system (NOS), **12**-16
Network, resource management, **2**-11–**2**-16
Network, resource manager. *See* NRM
Network, resources, guaranteeing, **2**-6
Network, retrieving state, **8**-12
Network, routing, **10**-23
Network, routing, efficiency of, **13**-15
Network, scalability, **3**-24
Network, structuring protocol, **10**-24
Network terrestrial, **2**-6
Network, topology, **10**-24, **40**-2
Network, topology, determining and detecting changes in, **1**-7
Network, topology, map, **3**-24
Network, topology
  three-dimensional analysis of, **14**-12
  two-dimensional analysis of, **14**-10
Network, traffic, **10**-22
Network, volatility, resilience of smart messages to, **26**-3
Network wireless, **2**-7
Network capable application processor. *See* NCAP
Network time protocol. *See* NTP
Network-based tracking, **21**-9
Networked embedded software technology. *See* NEST
Networking, **39**-2
Networking, active, **4**-4
Networking, convergence of with real-time computing, **2**-6–**2**-10
Networking, IEEE 1451.1, **10**-13
Networking, in extreme environments. *See* NETEX
Networks of embedded systems. *See* NES

Index

Neural networks, **7**-4
NEURON microcontroller, **10**-6
Newton mechanics, **36**-10
NG node, **15**-3
Nibble, **21**-11
Nickel-zinc batteries, **5**-11
Nine-bit interprocessor protocol. *See* NBIP
NIST, Manufacturing Engineering Laboratory, **10**-11
NN, **39**-35
NNSS, **21**-7
NNSS-AVG, **21**-7
No-crossing heuristic, **39**-35
Node coverage, **19**-7
Node failure, **39**-24
Node state transition, **17**-5
Node-level energy management, overview of, **29**-2–**29**-4
Node-level energy minimization techniques, **29**-1
Node-level operating systems, **4**-3
Nodes, **3**-17. *See also* specific nodes
Nodes, -level scheduling, **1**-7–**1**-8
Nodes, administrative state control function, **3**-25
Nodes, anchor, **20**-6
Nodes, application-aware, **6**-8
Nodes, architecture, **12**-2
Nodes, breakdown of, **10**-22
Nodes, bridging, **10**-23
Nodes, characteristics of, **20**-10
Nodes, classification of, **13**-7
Nodes, communication between, **14**-2
Nodes, communication between, small and large, **13**-8
Nodes, compromised, **31**-10
Nodes, cooperative computing, **26**-4
Nodes, densities, **20**-5
Nodes, detection prevention, **32**-12
Nodes, distinguished, **8**-9
Nodes, distribution, **1**-10
Nodes, failures, **1**-1
Nodes
    fixed, **15**-3
Nodes, frontend, **8**-5
Nodes, gateway, **9**-17, **19**-8
Nodes, ill-connected, **20**-19
Nodes, lifetime, **28**-3
Nodes, local deadline misses, **17**-4
Nodes, low-power, **13**-4
Nodes, master aggregator, **6**-11
Nodes, Medusa MK-2, **12**-9
Nodes, mobile, **10**-21
Nodes
    mobile, **15**-3
Nodes, mobile, **20**-10
Nodes, mobile, wireless, **12**-13
Nodes, mote, **12**-8–**12**-9
Nodes
    packet-forwarding load of, **13**-19
Nodes, placement, **20**-10
Nodes, role of in fixed-size cluster routing, **6**-15
Nodes, scheduling coverage, **19**-5
Nodes, scope of, **20**-3
Nodes, sensor, **4**-4, **13**-9
Nodes, sensor, components of, **5**-1, **12**-4

Nodes, sensor, miniature, **5**-10
Nodes, sensor, network, **12**-8–**12**-13
Nodes, sensor, resource limitations of, **32**-1
Nodes, sensor, small, **13**-5
Nodes, sink, **6**-7
Nodes, small-sized sensor, **23**-1
Nodes, snooper, **22**-2
Nodes, source, **6**-7
Nodes, spatial programming and, **4**-8–**4**-9
Nodes, specialized, **14**-2
Nodes
    subtype, **30**-3
    type, **30**-3
Nodes, velocity, **13**-13
Nodes
    wall-powered, **13**-19
Noise characteristics, adaptation to, **36**-6
Noise error, **36**-7
Noise level monitoring, **9**-2
Noise-analysis, **1**-9
Non-volatile memories, **36**-6
Nonaggregating WSNs, **9**-6
Noncoherent processing, **6**-10–**6**-11
Noncoordinator motes, **37**-8
Noninvasive deployment, **9**-3
Nonparametric statistical techniques, **36**-12
Nonpreemptive scheduling, **26**-6
Nonpropagating WSNs, **9**-6
Nonvolatile memory, **12**-4
Northern right whales, **11**-2
NRM, **2**-6, **2**-12
NRM, agents, **2**-15
NRM, experimental results, **2**-16–**2**-19
NRM
    fault tolerance, **2**-16
    federation, **2**-16
NSF, research projects on sensor networks funded by, **1**-2
NTP, **16**-5, **22**-14
Nuclear attack detection, **16**-2
Nuclear plants, **36**-12
Nuclear power plant monitoring, **15**-8
Nuisance attacks, **32**-10
Nymph, **13**-5

O

OASIS, **2**-17
Object classes, **3**-21
Object detection, **6**-1, **9**-2
Object motion detection, **9**-2
Object tracking, **26**-1, **26**-3, **27**-1
Objective function, **40**-9
Observation uncertainty, **22**-12
Observe, orient, decide, and act model. *See* OODA model
Observer, **3**-9
Obstacles, **40**-3
Ocean area coverage problem, **19**-4
Off-duty eligibility rule, **19**-5
Off-line phase of pattern matching, **21**-6
Offline testing, **36**-1
OGDC, **19**-6

Olfactory sense, active perception nature of, 7-6
Omnibus model, 7-4
Omnidirectional antennas, in engineered networks, 13-10
Omnidirectional links, 5-5
On-demand ad hoc routing protocols, 13-11
On-demand monitoring, 3-20
On-demand position updates, 20-19
On-demand reporting, 19-1
On-demand routing protocols, 39-3
Online algorithms, 40-13
Online device scheduler, 29-11
Online fault detection techniques, 36-5
Online routing, 39-33
OODA model, 7-4
Open systems interconnection vocabulary, 10-16
Opens, 36-5
Operating frequency, 27-2, 27-5
Operating system-directed dynamic power management. *See* OS-DPM
Operational binaries, 16-9
Operator assisted integrated systems. *See* OASIS
Opnet, 30-7
Optical communication, 5-8, 37-1, 37-7
Optical receiver photodiode, packaging, 5-12–5-13
Optical sensors, 32-1
Optical transmission techniques, 1-6
Optimal fusion for sensor output, 35-5
Optimal geographical density control. *See* OGDC
Optimal polynomial-time algorithm, 9-13
Optimistic approaches, 33-2
Optimization criteria, 35-3
Optimization mechanisms, 40-10
Optimum protocol design, 1-11
Ordinary vehicle tracking method, 8-9
OS-DPM, 27-2
OSPF, 2-7
Out-of-band signaling, 28-8
Overdetermined topologies, 20-13
Overhead, 13-19, 18-7, 25-4, 28-11, 40-14
Overhearing, 18-7, 28-11
Overlap, 14-2, 16-10
Oxidation, 5-2

P

PA, 10-9
PA-VBS, 9-17, 18-15
Packaging, 5-12–5-13
Packet network, throughput in, 1-6–1-7
Packet transmission, 28-3
Packets, 10-16, 28-8, 34-5, 40-3
Packets, aggregation of, 30-10
Packets, collision of, 23-3
Packets, direct transfer, 30-7
Packets, leashes, 32-14
Packets, losses, 10-23, 25-11
Packets, payloads, 30-2
PAMAS, 28-4, 28-8
PAMAS, advantages and disadvantages of, 28-9
Parallel architectures, evaluation of merit factor for, 24-8
Parallel architectures, parameterization of, 24-6

Parallel connections, 10-16
Parallel processing, 2-11
Parallel system, merit factor in, 24-5
Parallel vector, signal, and image processing library for C++. *See* parallel VSIPL++
Parallel vector library. *See* PVL
Parallel VSIPL++, 2-11
Pareto optimal curve, 40-13
Parity checking, 36-4, 36-5
Particle filtering, 16-5
Partitioning, 9-19
Passerby, 32-5
Passive perception, 7-6–7-7
Passive reflective systems, 37-9
Path determination phase of LEC routing, 34-6
Path energy cost, metrics affecting, 34-2
Path loss, 39-5, 39-28
Path loss exponential functions, 15-9
Path loss exponential functions, minimization of, 25-8
Patient tracking and monitoring, 9-5, 27-1
Pattern matching, 21-6
PBNM, 3-4
PC/104, 13-7
PCB, 10-11
PCS cellular networks, positioning using, 21-1
PDAs, 13-7
PEDAMACS, 28-14, 28-20
PEDAMACS, advantages and disadvantages of, 28-22
Peer-to-peer communication, 5-9–5-10, 10-16
Peer-to-peer multihop architecture, 15-5
PEGASIS, 6-13, 9-18, 18-14
Pentium architecture, 38-2
Perception, 7-2
Perceptive localization framework. *See* PLF
Perceptual reasoning machine. *See* PRM
Perceptual systems, modeling of, 7-3–7-8
Performability, 3-11
Performance, 32-4
Performance, analyses, 1-9
Performance, management, 3-13–3-14
Performance, metrics, 1-5
Performance, parameters, formulation of, 2-13–2-14
Perimeter forwarding, 40-12
Periodic aggregation algorithms, 4-3
Permanent anonymous IDs, 31-14
Permanent faults, 36-4
PFU Systems, Inc., 13-7
Phase noise, effects of on synchronization, 16-5
Phase of arrival. *See* PoA
Phase-locked loop. *See* PLL
Phenomenon, 3-9
Photolithography, 5-2
Photorealism, 7-10
Photovoltaics, 12-12
Physical architecture, 3-24–3-25
Physical architecture, components of, 12-16
Physical devices layer, 33-3
Physical layer, 3-25
Physical layer, fault tolerance at, 36-5
Physical layer, information, 10-13
Physical location, 20-2

Physical model, **1**-8
Physical sensor, **24**-1
Physical signal sampling, **18**-4
Physical size, **12**-3
Physical vulnerability, **32**-9
Pick-and-place methods of microassembly, **5**-13
Piconet, communication inside, **10**-22
Piconet, traffic in, **28**-6
PicoNode, **12**-11
Piconodes, **9**-12
PicoRadio, **5**-15, **9**-12, **30**-2, **30**-10
PIM, **10**-14
Pixels, variance in, **22**-10
Planar micromachining, **5**-2
Planar structures, **39**-13
Planetary exploration, **1**-3, **16**-2
Planning, **3**-11
Plates, **5**-2
Platform orientation uncertainty, **22**-12
Platform position uncertainty, **22**-12
Pleasantness, **7**-10
Pleasure, **7**-8
PLF, **16**-5
PLL, **27**-6
PLL, circuitry, **37**-4
Plug-and-play integration, **2**-11
Plug-N-Run, **13**-7
PMTU, **12**-10
PoA, **21**-1
Point coverage, **19**-8
Point location, **39**-2
Point solutions, **2**-12
Point-to-point chain topology, **10**-3, **20**-2
Poisson, node placement, **15**-13
Poisson, process, **15**-9
Poisson, queue, **11**-14
Policy-based network management. See PBNM
Polling algorithms, **18**-7
Polling mode communication, **10**-17
Polling mode communication, centralized, **10**-20
Polynomial-time algorithm, **9**-14
Position, changes of, **39**-41
Position awareness, **6**-4
Position fix, **20**-3
Position updates, on-demand precision, **20**-19
Positioning, **12**-8, **20**-16, **21**-1
Positioning, key components of, **20**-16
Positioning, metrics of systems for, **20**-5
Positioning, techniques, **20**-12
Positioning, using network topology for, **20**-12
Post's model, **40**-1
Postprocessing, **38**-11
Power, -aware routing, **14**-10, **14**-14
Power, awareness, **28**-10
Power, conservation, **9**-14, **14**-2
Power, conservation, policies, **15**-8
Power, consumption, **12**-3, **15**-3, **20**-10
Power, consumption, measured, **27**-12
Power, control, components, **12**-16
Power, control capability, **9**-15
Power, duty cycles, ultralow, **20**-12

Power, efficiency, **38**-5
Power, management, **3**-7, **32**-2, **37**-2
Power, management, active, **27**-5
Power, management, aggressive, **14**-2
Power, management, idle, **27**-2
Power, requirements of sensor networks, **1**-4
Power, saving, **21**-9
Power, stretch factor, **39**-6
Power, supply, **3**-7, **10**-3, **12**-2, **12**-4
Power, supply, digital controllers, **38**-5
Power, supply, infrastructure, **36**-6
Power, supply, sensor networks, **1**-5–**1**-6
Power, usage, analysis of, **14**-10
Power, usage, calculation of, **14**-9
Power, usage, number of neighbors and, **14**-1
Power aware multiple access. See PAMAS
Power dissipation, **23**-2
Power management and tracking unit. See PMTU
Power-attenuation model, **39**-5
Power-aware sensor model, **27**-4
Power-aware virtual base stations. See PA-VBS
Power-efficient and delay-aware medium access protocol for sensor networks. See PEDAMACS
Power-efficient gathering in sensor information systems. See PEGASIS
Power-optimal k-fault tolerance, **39**-26
PowerDevice, **27**-3
PowerNow!, **29**-5
PPS, **21**-1
Practical virtual force algorithm, **9**-14
PRAM, **40**-15
Precise positioning service. See PPS
Precision agriculture, **1**-3
Precision on-demand position updates, **20**-16
Predetermined sensor deployment, **9**-13
Predictive schemes, **29**-3, **35**-1
Preprocessing, **6**-10, **38**-6, **38**-11
Preprocessing, elements of, **24**-12
Pressure detection, **9**-2, **18**-1
Pressure sensors, **5**-4–**5**-5
Prevention, **32**-10
Prim's MST algorithm, **39**-29, **39**-40, **40**-10
Principal component analysis, **7**-4
Priority aloha protocol, **18**-9
Priority scheduling, **1**-7
Priority scheduling, consistency in, **17**-3
Privacy, **9**-9, **12**-3, **36**-13, **40**-13, **40**-15
Privacy, protection, **31**-1
Privacy protection, **31**-11–**31**-15
PRM, **7**-6
Proactive computing, **4**-2
Proactive routing, **10**-23, **30**-4
Probabilistic algorithms, **40**-10
Probabilistic deployment, **25**-5
Probabilistic key distribution, **31**-9
Probabilistic localization approach, **29**-25
Probability table, **29**-21
Process automation. See PA
Process fieldbus. See Profibus
Process generation plants, **35**-1
Process instrumentation, **10**-1

Process refinement, 7-4
Processing, 1-6, 3-4, 3-10, 7-4, 13-9, 15-3
Processing, directed diffusion and, 6-8
Processing, limitations of, 16-11
Processing, parallel, 2-11
Processing, power, 24-10
Processing, QoS, 3-6
Processing, structure, forming, 13-13
Processor interruption, avoiding, 2-9–2-10
Processor power modes, 27-9
Processor-oriented energy management, node level, 29-4–29-10
Processors, 12-4
Processors, asynchronous, 37-3
Processors, multiple supply voltages, 38-7
Processors, variable-frequency, 37-4
Processors, variable-voltage, 37-5
Product quality monitoring, 16-2
Production, map, 3-24
Profibus, 10-9, 10-15
Profibus PA, 10-9
Programmability, 36-6
Programmability, sensor networks, 4-3
Programmable microprocessors, 15-8
Programming interface for cooperative computing model, 26-7
Proof-carrying code, 31-3
Propagation, speed of, 1-10
Proportional to absolute temperature circuits. *See* PTAT circuits
Protection, 1-4
Protocol design, impact of hardware on, 1-11
Protocol design, network structuring, 10-24
Protocol design, stack, 30-2–30-4
Protocols, 9-13–9-20. *See also* specific protocols
Protocols, dynamic WSN_SUM, 23-10
Protocols, efficiency of, 23-2
Protocols, emergent, adaptive routing using, 25-14
Protocols, energy efficient, 27-1
Protocols, fault-location, 23-13
Protocols, fault-tolerant WSN_SUM, 23-9
Protocols, scaling, 40-13
Protocols
    summing, 23-6
    WSN_SUM, 23-7
Pruning rule k, 19-8
PSFQ, 16-9, 33-3
PTAT circuits, 5-5
Public key cryptography, 31-9
Pull model of communication, 30-1
Pump slow, fetch quickly. *See* PSFQ
Pure sensing, 1-2
Puzzles, 32-15
PVL, 2-11

Q

Θ-graph, 39-8
QoS, 1-10–1-11, 2-6, 6-5, 19-1
QoS, concerns, 9-9
QoS, definition of end-to-end parameters, 2-11
QoS, design with resource constraints, 9-10
QoS, energy efficient constraints, 18-3
QoS, routing, 1-7
QoS, service management and, 3-5–3-6
QoS, soft, 18-6
Quality, 15-8
Quality of service. *See* QoS
Quantity of service, 1-10
Quartz, 5-2
Queries, 3-9
Query operations, 39-2
Query processing, 33-2, 33-3
Query-based routing, 6-21
Querying, 8-1
Querying, applications, 8-2

R

RADAR, 20-9, 21-11
Radar applications, sensor networks in, 2-1
Radar sensors, 22-11
Radial constraints, 20-7
Radial lens distortion, 22-10
Radiation detectors, 5-4–5-5
Radio frequency, 1-6. *See also* RF
Radio jamming, 32-11
Radio signal propagation, 20-10
Radio tags, 11-4
Radio transmission, 38-7
Radioactive isotopes, storing energy with, 5-12
Radios, 12-6
Radios, direct sequence spread spectrum, 13-6
Radios, fault tolerance, 36-6
Radios, in ad hoc multihop architecture, 12-16
Radios, network connectivity through, 25-11
Radios, off-the-shelf, 13-7
Radios, power consumption of, 28-10
Radios, short-range wireless communication, 12-1
Radios, small sensor node, 13-5
Radios, wireless, 36-6
RAID systems, 2-10
Raider, 32-5
Random access, 16-12
Random compass routing, 39-34
Random competition-based clustering. *See* RCC
Random coverage, connected, 19-6
Random coverage, energy efficient, 19-4
Random deployment, 9-8, 9-13
Random noise error, 36-7
Random point coverage, 19-8
Random sensor deployment, 19-2
Random source model, 6-9
Randomization, 39-33
Range, 9-13, 34-8
Range, searching, 39-2
Range returns, 22-13
Range-only sensors, 22-9
Ranging, 20-16
RAP, 17-3
Rate-limit response, 32-12
Rayleigh fading, 1-9

Index

RBOP, **39**-41
RBS, **16**-5
RC5 block cipher, **31**-6
RCA, **13**-14
RCC, **13**-13
Reactive routing, **13**-16, **30**-4
Real time properties, **10**-2
Real-time data acquisition, **10**-4
Real-time distance-aware scheduling, **17**-3
Real-time distributed network computing, sensor data fusion goals for, **2**-5
Real-time message scheduling algorithms, **17**-2
Real-time operating systems. *See* RTOS
Real-time OS platforms, **27**-12
Real-time phase of pattern matching, **21**-6
Real-time research, **17**-2
Reasoning, **7**-4, **7**-6
Rebinding of tasks, **2**-10
Received signal strength. *See* RSS
Received signal strength indicator. *See* RSSI
Receivers, **1**-4
Reception, **1**-9
Rechargeable batteries, **36**-6
Reconfiguration, **6**-22, **36**-5
Reconfiguration, runtime, **33**-8
Reconnaissance, **16**-2
Recovery, **32**-10, **36**-5
Recovery, operations, **33**-11
Red Hat eCos, **12**-10, **27**-11
Reduce and conquer, **40**-2
Redundancy, **28**-10, **32**-1
Redundancy, exploitation of, **6**-22
Redundant array of independent disks. *See* RAID systems
Redundant neighboring nodes, **28**-3
Redundant nodes, selection of, **28**-5
Reference-broadcast synchronization. *See* RBS
Refined energy evaluation model, **29**-26
Refinement, **20**-19
Region of interest. *See* RoI
Relation-based schemes, **9**-19
Relative location, **20**-2
Relative location-based localization, **16**-4
Relative neighborhood graph. *See* RNG
Relative positioning, **20**-3
Relaxation model, **15**-10
Relay organization, **13**-12
Relay traffic, **1**-8
Relay-based computer, **36**-4
Reliability, **3**-11, **9**-10, **15**-2, **32**-4, **36**-1
Reliability, support, **33**-2
Reliable multisegment transport. *See* RMST
Reliable query routing payoff model. *See* RQR payoff model
Reliable sensor applications, **33**-2
Remapping of tasks, **2**-10
Remote denial of service attack, **32**-3
Remote metering, **9**-5
Remote method invocation, **17**-5
Remote object monitoring, **23**-1
Remote procedure call framework. *See* RPC framework
Remote service execution, **33**-8
Remote virus sensing, **9**-4
Rendezvous clustering algorithm. *See* RCA
Rene2, **4**-9
ReOrg, **13**-12
Repair, **36**-5
Replay attacks, **32**-13
Replied sensors, **24**-2
Report once strategy, **29**-31
REQ, **6**-19
Request-response model, **30**-1
Residual energy, **3**-24
Resiliency mechanisms, **36**-2, **36**-5
Resonators, packaging, **5**-13
Resource
    adaptation, **14**-2
    blindness, **14**-2
Resource, consumption, **20**-10
Resource, efficiency design, **9**-9
Resource, efficiency design, necessity of, **18**-2
Resource, exhaustion, **32**-10
Resource, limitations, **9**-8
Resource, management, **15**-4
Resource, utilization, **2**-15
Resource-adaptive algorithms, **6**-18
Resource-biased path selection, **13**-11
Resources to attack, **32**-5
Response implosion problem, **3**-20, **8**-5
Response-frame format, **10**-17
Restart, **36**-5
Restoration scheme, local path, **6**-7
Result encryption, **38**-16
Retail stores, tracking of goods, **1**-3
Retasking/reprogramming, **16**-9
Retransmissions, **16**-9, **18**-6, **36**-6
Reverse directional flooding, **34**-6
Reverse key chain, **31**-6
Reverse multicast tree, **16**-10
Reverse path, **16**-7
RF, characterization of, **37**-7
RF, circuits, **27**-1
RF, combined, **20**-9
RF, communication, **5**-5–**5**-8, **37**-2
RF, design, **9**-1, **18**-1
RF, identification systems (RF-ID), **20**-9
RF, Monolithic 916.50 transceiver, **12**-9
RF, Monolithics TR1000, **11**-6
RF, propagation techniques, range estimation based on, **20**-13
RF, transmit power, **15**-9
RFC 2488, **2**-8
RFC3433, **3**-22
Right hand rule, **39**-35
Risk, **32**-2
Risk maps, **9**-4
RMST, **33**-3
RNG, **39**-7, **39**-13, **39**-21
Road condition information, **31**-13
Roaming cluster, **13**-18
Robotics, **35**-1
Robots, control of, **9**-6
Robots, mobile, **10**-21

Robots, sensor, **1**-2
Robust sensor networking, **33**-3
Robust start-up, **20**-16
Robustness, **1**-7, **3**-1, **10**-3, **12**-3, **15**-8, **20**-1, **22**-3
Rockwell Science Center, **13**-7, **38**-5
RoI, **24**-12, **24**-19
RoI, builders, **24**-12
RoI, clustering and, **24**-16
Room occupation monitoring, **31**-1
Round-trip approach, **22**-14
Route, aggregation, **13**-2
Route, biasing, **13**-11
Route, discovery, **13**-16, **39**-3
Route, maintenance of energy-aware routing, **30**-6
Routers, **10**-16
Routers, bibus-NBIP coupler, **10**-19
Routing, **1**-4, **1**-7, **4**-6–**4**-7, **18**-10, **32**-2, **39**-2. *See also* specific routing types; specific techniques
Routing, ad hoc protocols, **1**-7, **16**-9
Routing, ad hoc protocols, multihop, **30**-1
Routing, adaptive, **6**-18
Routing, adaptive, using emergent protocols, **25**-14
Routing, address-based, **20**-2
Routing, alternatives, **15**-3
Routing, altruistic, **13**-19
Routing
    analysis of power usage, **14**-10
    cluster-based, **6**-11–**6**-18
Routing, data-centric protocol, **16**-10
Routing
    diffusion, **30**-8
Routing, diffusion, **33**-4
Routing
    edge, **14**-10
Routing, energy efficiency of, **34**-1
Routing, energy-aware, **13**-11
Routing, fisheye state, **13**-17
Routing, flat, **6**-6–**6**-11
Routing, freedom, **14**-12
Routing
    game-theoretic models of reliable and length energy-constrained, **34**-2
    games, **34**-2
Routing, geographic protocol, **4**-3
Routing, geometric, **20**-2
Routing
    hierarchical, **6**-11–**6**-18
Routing, hierarchical, **13**-15
Routing, hierarchical, power-aware, **6**-17–**6**-18
Routing, independent, **13**-16
Routing
    interior, **14**-10
Routing, landmark, **13**-16
Routing
    linear program, **30**-8
Routing, localized, **39**-33–**39**-38
Routing, location assisted protocols, **17**-2
Routing, location-based, **20**-2
Routing, maximization of lifetimes, **30**-5
Routing, mechanisms, **13**-11
Routing, message, **33**-3

Routing, mobile agent, **25**-3
Routing, mobile wireless sensor networks, **2**-7
Routing
    motivation and design issues, **6**-3–**6**-4
Routing, multicast protocols, **10**-14
Routing, multihop, **1**-1, **12**-16
Routing, multipath, **6**-6, **6**-20–**6**-21
Routing, network, **10**-23
Routing, network layer, **30**-3
Routing, paradigms for DSNs, **25**-8–**25**-16
Routing
    power-aware, **14**-10
Routing, proactive, **10**-23
Routing, protocols, **18**-11, **26**-17, **31**-4
Routing, protocols, characteristics, **30**-4
Routing, protocols, table-driven, **39**-3
Routing, protocols, tamper-resistant, **32**-14
Routing, QoS, **3**-6
Routing, query-based, **6**-21
Routing, reactive, **13**-16
Routing, reduction of overhead with location information, **21**-2
Routing
    requirements of WSNs and MANETs, **6**-2
Routing, scalable, **15**-8
Routing, shortest-path, **14**-10, **40**-9
Routing, smart messages and, **4**-8, **26**-5
Routing, table construction, **25**-11
Routing, trees, **31**-5
Routing, VGA, **6**-16
RPC framework, **10**-14
RQR payoff model, **34**-2
RSA, **38**-8
RSA, decryption, **38**-10
RSA, encryption, **38**-16
RSS, **16**-4, **21**-1, **21**-3
RSSI, **20**-5
RSSI, errors induced by measurements, **20**-19
RSSI-based distance discovery, **36**-10
RSVP-TE, **2**-6
RT-Linux, **29**-6
RTOS, **12**-8
RTOS, interruption of, **2**-10
RTS-CTS control messages, **18**-9, **28**-8
RTS-CTS-DATA-ACK exchange, **13**-12, **28**-14
Runtime adaptation control, **33**-11
Runtime reconfiguration, **33**-8

S

S-MAC, **18**-9
Safety, **3**-11, **9**-9
Safety, mechanism of transmission, **10**-2
Samplecast, **8**-5
Sampling, **1**-6, **31**-4
Sampling, operation, **8**-5
Sampling, sensing and, **13**-9
Sandboxing, **31**-3
SAR, **2**-2, **6**-6–**6**-7
SARS, **9**-5
Satellite systems, **12**-13, **20**-8

Scalability, 1-5, 3-14, 3-24, 6-4, 15-8, 18-3, 20-10, 28-23, 30-4
Scalability, energy quality, 27-1
Scalability, tiered architectures, 13-4
Scalable architectures, 9-9
Scalable coordination problem, 38-5
Scalar weights, 2-14
Scaling laws, 1-7
Scanners, 3-22
Schedule, 6-14
Schedule, consistency, 33-10
Schedule change protocol. *See* SCP
Scheduled response, 8-5
Scheduling, 1-7–1-8, 11-4, 29-12
Scheduling, MUSCLES algorithm for multistate I/O devices, 29-14
Scheduling online, 29-12
Scheduling, smart message, 26-6
Scheduling, valid instant, 29-15
Scheduling-based protocols, 28-4, 28-11
Schmid-Schossmaier function, 25-4, 35-2, 35-12
Scientific applications, lifetime requirements of, 13-4
Scientific exploration, 9-5, 18-2
Score-based ranking, 29-21
SCP, 28-19
Script interpreters, 4-4
SDLC frame, 10-20
SDMA, 37-10
Security, 1-3, 1-11, 3-11, 6-1, 9-22, 12-3, 18-3, 31-1, 32-2, 32-4, 36-13, 40-13, 40-15
Security, -related properties, 31-2
Security, architectures, 31-4–31-10
Security, design, 9-10
Security, management, 3-14
SEE, 8-7
Segment length, 10-2
Sei whales, 11-2
Seismic, sensors, 38-1
Selective forwarding, 32-12
Self-configuration, 6-22, 20-10
Self-containment, 20-1
Self-correcting coding schemes, 32-4
Self-healing minefields, 1-4
Self-management, 3-16
Self-orchestrated operation, 8-5
Self-organizing capabilities, 1-1, 3-12, 6-3, 15-2, 20-1
Self-organizing medium access control for sensor networks. *See* SMACS
Self-regulated sensor deployment, 9-13
Self-repair, 36-1, 36-4
Self-routing, smart message, 26-7, 26-17
Self-timed systems, 37-3
Semantic security, 31-6
Semiconductors, 5-2, 12-5
Semiconductors, tracking, 31-4
Sense-think-act paradigm, 7-3, 7-7
Sensed value, 6-14
Sensing, 3-4, 3-9–3-10, 9-14, 13-1, 13-8, 15-3, 16-2, 18-5, 28-10, 29-1, 32-1
Sensing, accuracy, 27-1

Sensing, coverage area map, 3-24, 3-25
Sensing, energy, 18-4
Sensing, horizon, 17-4
Sensing, integration of with signal processing and communications, 15-1
Sensing, layer, 15-4
Sensing, location-dependent carrier, 18-8
Sensing, models, 40-4
Sensing, motes, 37-2
Sensing, power dissipation due to, 23-2
Sensing, ranges, 19-2
Sensing task queries, 6-21
SensIT, 1-2
Sensor, 12-4
Sensor, -centric design, 12-11
Sensor, acquisition, 1-6
Sensor, actuator networks, 10-2
Sensor, algorithms, 33-2
Sensor, collection, 24-3
Sensor, communication between, 25-10
Sensor, cost, 3-14
Sensor, coverage problem, 19-1
Sensor, data, filtering, 12-8
Sensor, data, fusion, 12-8
Sensor, data, perception of, 7-1
Sensor, data fusion, real-time distributed network computing goals for, 2-5
Sensor, deployment strategies, 9-13–9-14, 19-2, 25-4–25-8
Sensor, deployment strategies, using GA, 25-6
Sensor, design, 22-9–22-13
Sensor, detection requirements, 15-1
Sensor, element, 3-8
Sensor, fusion, 2-11, 7-3–7-4, 35-2, 36-13
Sensor, fusion, fault tolerance during, 36-2
Sensor, fusion, interval, 25-4
Sensor, fusion, methods, 7-4
Sensor, fusion, OASIS application, 2-17
Sensor, grid-based deployment, 15-12
Sensor, information technology. *See* SensIT
Sensor, integrated circuit, 5-5
Sensor, integration strategy, 11-3
Sensor, integration strategy, architectures for, 24-3–24-18
Sensor, interface, 2-15
Sensor, interval endpoint tolerance, 35-12
Sensor, lifetime, metrics affecting, 34-2
Sensor, measurement, complex, 11-5
Sensor, measurement, errors in, 36-12
Sensor, modules, operating in conjunction with actuator modules, 8-2
Sensor, monitoring requirements, 15-1
Sensor, networking, 10-1, 10-9
Sensor, networking, internet-based, 10-13–10-15
Sensor, networks, 36-3. *See also* WSN; specific network types
Sensor, networks, advantages of, 1-2
Sensor, networks, application of GPS to, 20-9
Sensor, networks, applications, 1-2–1-4, 1-11, 4-2–4-3, 8-1–8-2, 16-2
Sensor, networks, architecture, 15-3
Sensor, networks, basic features of, 1-1
Sensor, networks, collaboration of nodes in, 13-8, 16-7

Sensor, networks
    communication and application types in, **28**-12
Sensor, networks, communication mode-based
    classification, **15**-5
Sensor, networks, communication protocols for, **16**-1
Sensor, networks, comparison with ad hoc networks, **16**-1
Sensor, networks, context aware services, **20**-2
Sensor, networks, deployment of, **6**-1
Sensor, networks
    design methodology, **12**-2
Sensor, networks, diagnosis of, **8**-9
Sensor, networks, differences from traditional data
    networks, **4**-2
Sensor, networks
    dimensions for management of, **3**-4
Sensor, networks, embedded systems and, **26**-17
Sensor, networks
    embedded wireless, **12**-1
Sensor, networks, error detection in, **36**-2
Sensor, networks, fault tolerance at different levels,
    36-5–**36**-7
Sensor
    networks, formulation, **35**-2
Sensor, networks, functional architecture for, **8**-3–**8**-4
Sensor, networks, functional decomposition, **13**-2
Sensor, networks, hardware spectrum, **13**-5–**13**-8
Sensor, networks, modeling, **15**-9
Sensor, networks, mote nodes, **12**-8
Sensor, networks, multihop self-organizing, **16**-12
Sensor, networks, nodes, **12**-8–**12**-13
Sensor, networks, objectives of, **16**-7
Sensor, networks, organization and processing, **24**-1–**24**-3
Sensor, networks, performance metrics, **1**-5
Sensor, networks, programmability of, **4**-3
Sensor, networks, programming languages, **8**-6–**8**-7
Sensor, networks, programming model, **17**-5
Sensor, networks, protocol suite for, **17**-2–**17**-5
Sensor, networks, protocols and applications. *See* SNPA
Sensor, networks, requirements, **12**-2
Sensor
    networks, requirements, **12**-3
Sensor, networks, resource limitations of, **26**-1
Sensor, networks, sparse, **13**-10
Sensor, networks, technical challenges, **1**-4-**1**-11
Sensor, networks, ubiquitous, **30**-1
Sensor, nodes, **3**-7, **3**-17–**3**-18, **3**-20, **9**-10, **19**-2, **38**-6
Sensor
    nodes (*See also* nodes)
Sensor, nodes, application-aware, **6**-8
Sensor, nodes, architecture, **27**-3
Sensor, nodes, channel status of, **23**-4
Sensor, nodes, communication-centered design, **12**-11
Sensor, nodes, components of, **15**-9
Sensor, nodes, deployment optimization of, **15**-12
Sensor, nodes, large, **13**-7
Sensor, nodes, lifetime of, **27**-1
Sensor, nodes, resource limitations of, **32**-1
Sensor, nodes, three-stage processing, **38**-2
Sensor
    output, **35**-3
Sensor, processing systems, **24**-3

Sensor, readings, **37**-1
Sensor, robots, **1**-2
Sensor, sampling, **1**-6
Sensor, selection criteria, **5**-4–**5**-5
Sensor, signal processing algorithms, **23**-1
Sensor, signals, **2**-5
Sensor, states, **19**-2
Sensor, subsystems, modularity of, **22**-3
Sensor, systems, evolution of, **24**-2
Sensor execution environment
Sensor information networking architecture. *See* SINA
Sensor interval estimates, **35**-16
Sensor management protocol. *See* SMP
Sensor network architecture. *See* SNA
Sensor network encryption protocol. *See* SNEP
Sensor node, **33**-6
Sensor protocols for information via negotiation. *See* SPIN
Sensor query and data dissemination protocol. *See* SQDDP
Sensor querying and tasking language. *See* SQTL
Sensor querying language. *See* SQL
Sensor resource assignment problem. *See* SRA problem
Sensor Synergy, **10**-11
Sensor-MAC. *See* S-MAC
Sensor-specific application programming interface
    extensions, **27**-12
Sensoria, **1**-11
Sensorim, **34**-8
Sensors, **36**-2, **36**-7. *See also* specific sensors
Sensors, failure, recovery from, **33**-11
Sensors, selection of, **29**-24
Sensors, task structures, **33**-9
SensorWare, **4**-9–**4**-11, **26**-17
Sensory perception, **7**-5, **7**-8
SenSys, **1**-2
Sentient Computing, **17**-7
Sequential assignment routing. *See* SAR
Sercos, **10**-7
Serial bus connections, **10**-1
Serial communication processors, NBIP protocol for, **10**-17
Serial connections, **10**-16
Serial real time communication system. *See* Sercos
Serial system, merit factor in, **24**-6
Service management, **3**-3, **3**-5–**3**-6
Service-oriented model, **33**-9
Session key, **31**-6
Settled nodes, **20**-7
Setup phase of energy-aware routing, **30**-6
Seven-neighbors WSN, **14**-6
Severe acute respiratory syndrome. *See* SARS
SF_6 plasma, **5**-2
SGTC, **29**-6
Shared memory, name-based, **26**-2
Shared service, **32**-2
Shared wireless infostation model. *See* SWIM
Short interframe space. *See* SIFS
Short-range broadcast communication, **1**-1, **13**-10
Short-range wireless communication radios, **12**-1
Shortest path problem, **40**-9
Shortest-path tree. *See* SPT
Shorts, **36**-5
Shutdown policy, **27**-2

Index

Si4112 phase-locked loop (PLL) RF synthesizer, **11**-5
SIFS, **28**-4
Signal, analysis, **7**-2
Signal, attenuation, **25**-11
Signal, collaborative processing, **13**-9
Signal, conditioning, **10**-1, **18**-4
Signal, detection, **3**-25
Signal, fading, **20**-13
Signal, modulation, **10**-2
Signal, physical sampling, **18**-4
Signal, processing, **1**-4, **9**-3, **18**-16, **20**-10
Signal, processing, distributed, **1**-11
Signal, processing, integration with sensing, **15**-1
Signal, propagation, **18**-4
Signal, sources, **16**-4
Signal, strength, positioning by, **21**-3
Signal propagation properties, **20**-5
Signal stability routing. *See* SSR
Signal-to-noise ratio. *See* SNR
Signal-to-noise-and-interference ratio. *See* SNIR
SiidTech, Inc., **31**-4
Silicon Labs, **11**-5
Silicon semiconductors, **5**-2
SIMD, **24**-4
Simple network management protocol. *See* SNMP
Simple protocols, **30**-4
SimpleFace, **7**-10
SINA, **4**-7–**4**-8, **8**-1, **8**-4–**8**-9, **17**-7
SINA, architectural view of, **8**-8
Single hop WSNs, **23**-3
Single object refinement, **7**-4
Single winner algorithm. *See* SWE
Single-hop clusters, **9**-17
Single-hop WSNs, **9**-6
Single-level multihop clustering algorithm, **9**-18
Single-packet storage methods, **11**-13
Single-route failure, **6**-6
Single-sensor systems, **24**-2
Sink node, **3**-18, **6**-7, **16**-7, **19**-1, **33**-5
Sink structure, **39**-11
Sink-to-sensors transport, **16**-8
Sinkholes, **32**-13
SISD, **24**-4
Situation refinement, **7**-4
Six-neighbors WSN, **14**-5
Six-neighbors WSN, three dimensions, **14**-6
Size restriction, **39**-3
SLAM, **22**-14
Slave agent, **21**-9
Sleep modes, **1**-6, **19**-4, **40**-3
Sleep modes, transitioning to, **27**-2, **27**-4
Sleep time thresholds, **27**-4
Sleeping, **1**-9
Slot assignments, **18**-9
SM, **4**-8–**4**-9
SMACS, **18**-8, **28**-13, **38**-5
SMACS, advantages and disadvantages of, **28**-16
SMACS, EAR algorithms and, **28**-16
SMACS, neighbor discover in, **28**-15
SMACS, operation, **28**-14

Small minimum energy communication network. *See* SMECN
Small nodes, **13**-9
Small sensor nodes, **13**-5
Smart agents, **10**-1
Smart Dust, **9**-12, **13**-5, **15**-3, **20**-1, **20**-9, **20**-14, **37**-1
Smart Dust, networks, communication in, **37**-7
Smart Dust, optical communication for, **37**-9
Smart kindergartens, **9**-5
Smart messages, **26**-2, **26**-5–**26**-7. *See also* SM
Smart messages, cost of migration, **26**-8
Smart messages, execution of, **26**-3
Smart messages, lifecycle of, **26**-6
Smart messages, self-routing of, **26**-7
Smart messages, tracking, **26**-4
Smart messages, transfer, **26**-9
Smart nodes, reconfigurable, **33**-5
Smart office spaces, **1**-3
Smart office spaces, constructing, **16**-2
Smart roads, **9**-6
Smart space, **9**-5
Smart transducer interface module. *See* STIM
Smart transducer interface standards, **10**-11–**10**-13
SMECN, **6**-15, **16**-10
SMFAs, **3**-3
Smoke sensor nodes, **9**-6
Smoothing, **35**-1
SMP, **16**-3
SNA, **8**-1
SNEP, **31**-5
SNIR, **1**-8
SNMP, **3**-24
Snooper nodes, **22**-2
SNPA, **1**-2
SNR, **13**-8
Social studies, **1**-3
Socket-style connections, **17**-2
Soft threshold, **6**-14
Software, **3**-8
Software, development, **9**-20–**9**-22
Software, engineering, DoS due to, **32**-4
Soil composition detection, **9**-2
Solar cells, **12**-5, **18**-17
Solar cells, energy efficiency of, **1**-6
Solar radiation, **5**-12
Solution dissemination, **40**-7
Sound detection, **6**-1
Source nodes, **6**-7
Source routing strategy, **10**-16, **31**-5
Southern right whales, **11**-2
Space, constraints, **17**-2
Space, division, **18**-14
Space, location in, **20**-2, **20**-10
Space, monitoring, **15**-2
Space, topology, **40**-3
Space division multiple access. *See* SDMA
SPAN, **18**-12, **37**-7, **40**-8
Spanner, **39**-3, **39**-6
Sparse anchor problem, **20**-16
Sparse sensor network, **13**-10
Sparse spanner, **39**-3

Sparse topology and energy management. *See* STEM
Spatial concurrency, **13**-19
Spatial density, **20**-10
Spatial diversity, exploitation of, **6**-22
Spatial location, **9**-13
Spatial programming, **4**-8–**4**-9
Spatial reuse gain, **1**-8
Spatially distributed sensors, **24**-2
Spec mote, **13**-5
Special probe packet, **28**-8
Species, tracking movement of, **16**-2
Specific management functional areas. *See* SMFAs
SPEED, **17**-4
Speed of propagation, **1**-10
Speedometer, **36**-10
Sperm whales, **11**-2
SPIN, **6**-18, **9**-19, **14**-2, **18**-15, **26**-14, **31**-5, **32**-14
SPIN, protocols, **6**-19
SPIN, using smart messages, **26**-12
Spin glass model, **25**-14
Spiral algorithm, **40**-6
Spreadsheets, associative, **8**-5, **8**-8
SPS, **21**-1
SPT, **39**-39
SQDDP, **16**-3
SQL, **4**-7, **8**-7
SQTL, **4**-7–**4**-8, **8**-6, **16**-3
SRA problem, **36**-8
SRAM, **12**-9, **36**-6
SSL, **31**-3
SSR, **39**-3
Stability, **3**-25, **35**-11
Stack layers, **16**-9
Stance, **7**-8, **7**-10
Stand positioning service. *See* SPS
Star topology, **10**-3, **20**-2
Stargate board, **13**-7
Start-up positioning scheme, **20**-17
Start-up transient behavior, effect on energy dissipation, **15**-9
State change record, **3**-21
State estimate, **35**-1
State management, **3**-7
State transitions, **9**-15
Stateless connection management, **32**-14
Stateless routing, **39**-33
Static information, **3**-21–**3**-23
Static networks, **3**-11
Static velocity-monotonic scheduling, **17**-3
Statistical optimization framework, **9**-14
Status registers, **37**-5
Stayton board, **13**-7
Steady mode, **15**-9
STEM, **18**-12, **40**-8
Stickiness probabilities, **25**-15
STIM, **10**-11
Stochastic coverage, **15**-12
Stochastic decision-making, **24**-3
Stochastic geometry, **39**-24
Stochastic models, **29**-3
Stop-grant timeout count. *See* SGTC

Storage, **12**-2, **20**-10
Storage, calculating requirements, **11**-12–**11**-17
Storage, models, **40**-4
Storage, multiple packet methods, **11**-14
Storage, requirements, **11**-1
Storage, sensor data, **12**-8
Storage, single-packet methods, **11**-13
Storage, subsystems, **36**-2, **36**-6
Storage, technology, **40**-3
Storage, units, **12**-4
Storage area networks, **2**-8
Storage buffer, **2**-5
Storage buffer, guaranteeing resources, **2**-8
Storing, **3**-15
Stove-piped stand-alone systems, **2**-5
Strategy space, **34**-3
Stretch factor, **39**-3
Strip topology, node overlay in, **13**-4
Strong ARM microprocessor, **9**-11, **13**-7, **27**-2, **38**-3, **38**-14
Strong ARM microprocessor, SA-1110, **27**-6
Strong ARM microprocessor, SA-1110, power consumption of, **27**-10
Structural model, **3**-24
Structural properties, **39**-2
Structure optimizing rules, **10**-24
Structures, monitoring of with sensor networks, **1**-3
Stuck fault model, **36**-5
Subgraphs, **39**-4
Sublimated XeF_2, **5**-2
Subnetworks, **10**-16
Subsampling transceivers, **5**-8
Subscribe function, **33**-5
Successive approximation, **37**-2
Sum problem, **23**-6
Summing protocol, energy efficient, **23**-6
Supply chain monitoring, **5**-5
Support object class, **3**-21
Surface micromachining, **5**-2
Surgery, sensor network applications in, **1**-4
Surveillance, **1**-4, **6**-1, **9**-6, **16**-2
Surveillance, finite region, **25**-6
SWE, **6**-11
Sweep coverage, **19**-4
SWIM, **11**-1, **11**-3
SWIM, networks, **11**-6–**11**-7
SWIM, placement of stations, **11**-11
Switched-capacitor DC-DC converter, **12**-5
Switching energy, **18**-4
Switching power, **9**-13
Switching power, consumption, **27**-5
Sybil attack, **32**-14
Symbolic location, **20**-2
Symmetric cryptography, **31**-8
Symmetric keys, **31**-9
Symmetric Yao graph, **39**-13
Synchronization, **1**-11, **40**-13
Synchronization, clock, **9**-22
Synchronization, clock, security and, **31**-6
Synchronization, constraints associated with network protocols, **33**-7
Synchronization, SMAC, **28**-15

Index

Synchronization, smart messages and, **26**-5
Synchronization, time, **12**-5, **13**-2, **22**-14
Synchronization, time, effect of, **28**-23
Synchronization, time, location and, **6**-22
Synchronous systems, **37**-3
Synthesis techniques, **40**-12
Synthesis techniques, state of the art, **12**-2
Synthetic aperture radar. *See* SAR
System, **3**-23
System, -level security, **31**-3
System, configuration, energy-driven, **38**-16
System, design, algorithmic, **22**-3–**22**-9
System, lifetime, **18**-3
System, monitoring, **15**-2
System, partitioning, **9**-19, **18**-4
System, software, fault tolerance of, **36**-7
System, status inquiry, **26**-11
System, structuring, **40**-9
System architecture, **9**-13–**9**-20
System configuration, **3**-2
System faults, working through, **2**-10
System lifetime, **1**-5
System power model, **27**-4
System task processes, **2**-9
System-level programmability, **4**-3–**4**-6
System-level programmability, frameworks for, **4**-6–**4**-11
System-level scheduling, **1**-7–**1**-8
System-on-modules, **13**-7
SystemState, **27**-3

T

Tactical surveillance, **6**-1
TADAP, **16**-3
Tag space, **4**-8, **26**-2, **26**-5
Tag space, cost of operation, **26**-11
Tagging of marine mammals, **11**-2
Tagging with RF-ID, **20**-9
Tamper resistance, **38**-9
Tamper-resistant routing protocol, **32**-14
Tampering, **32**-11
Tangential lens distortion, **22**-10
Target detection, **6**-10, **15**-2
Target field imaging, **6**-1
Target identification, **2**-8
Target localization, **7**-4
Target localization, energy-reduction method, **29**-19
Target modeling, **22**-5
Target of DoS attack, **32**-8
Target reporting, **29**-21
Target tracking algorithms, **4**-3, **22**-2, **33**-2, **35**-1
Targeted applications, **12**-3
Targeting, **16**-2
Task assignment and data advertisement protocol. *See* TADAP
Task decomposition, **13**-8–**13**-10
Task management, **3**-7
Task structures, specification of, **33**-9
Task-application mappings, **2**-15
Task-based network behavior, **28**-10
Tasking, **8**-1

Tasking applications, **8**-2
Taylor series, **20**-8
TBIMs, **10**-12
TCP, **16**-7, **25**-11
TCP, end-to-end semantics, **16**-8
TCP, SYN-floods, **32**-15
TCP/IP, **10**-2, **10**-18
TCP/IP, gateway note, **13**-7
TDMA protocol, **13**-10, **18**-7, **18**-14, **28**-4, **28**-6, **28**-11, **28**-16, **29**-31, **37**-7
TDMA protocol, advantages and disadvantages of, **28**-7
TDMA protocol, medium access, **16**-12
TDMA/CDMA MAC, **6**-11
TDoA, **20**-5
TDoA, positioning with, **21**-2
TDP, **16**-5
Technological trends, **12**-2
Technology Review, **1**-2
TEDS, **10**-11
TEEN, **6**-13–**6**-15, **18**-14
Telemonitor, **10**-11
Temperature, **18**-1
Temperature, effects of on synchronization, **16**-5
Temperature, measuring, **31**-1
Temperature, readings, **20**-10
Temperature detection, **6**-1, **8**-2, **9**-2, **11**-3, **13**-9
Temporal geocast, **21**-2
Temporally-ordered routing algorithm. *See* TORA
Temporary anonymous identifications, **31**-13
Terrestrial networks, **2**-6–**2**-7
Terrorist, **32**-5
Testability, **3**-11
Testing, **12**-8, **36**-12
Tetramethylammonium hydroxide. *See* TMAH
Texas Instruments, **11**-3
Thermal sensors, **5**-4–**5**-5
Thin-film lithium energy cells, **5**-11
Thin-film vanadium oxide batteries, **5**-11
Threat, **32**-2
Three-dimensional networks, analysis of, **14**-12
Three-neighbors WSN, **14**-4
Threshold based strategy, **21**-9
Threshold data object, **3**-22
Threshold value, calculation of, **34**-6
Threshold-sensitive energy-efficient sensor network. *See* TEEN
Thresholds, **6**-14
Throughput, **1**-5, **1**-6–**1**-7, **15**-8
Throughput, constraints, **38**-6
THUMB processor, **12**-10
Tick-driven scheduling, **29**-12
Tiered architectures, **6**-22, **13**-1, **13**-3–**13**-5
  cost effectiveness of, **13**-3–**13**-4
Tiered architectures, drawbacks of, **13**-18
Tiered architectures, forming, **13**-10–**13**-15
Tiered architectures
  longevity of, **13**-4
Tiered architectures, routing and addressing in, **13**-15–**13**-18
Tiered architectures
  scalability, **13**-4

TII, **10**-11
Time, concept, **7**-4–**7**-7
Time, location in, **20**-2, **20**-10
Time, redundancy, **36**-6
Time, synchronization, **12**-5, **13**-2, **22**-14, **32**-3
Time, synchronization, protocols, **16**-5–**16**-7
Time difference of arrival. *See* TDOA
Time division multiplexing access. *See* TDMA
Time of arrival. *See* TOA
Time reservation using adaptive control for energy efficiency. *See* TRACE
Time-diffusion synchronization protocol. *See* TDP
Time-distance of arrival. *See* TDoA
Time-driven sensor networks, **6**-1
Time-of-flight measurement approach, **22**-13
Time-stamp counter. *See* TSC
Time-varying computational load, **27**-5
Timekeeping system, **11**-4
Timeout timer, **17**-5
TinyDB, **4**-7
TinyOS, **4**-3, **11**-6, **12**-2, **12**-16, **13**-6, **17**-7
TinyOS, Maté and, **4**-9
TMAH, **5**-2
TOA, **16**-4, **20**-5, **21**-1
TOA, positioning with, **21**-2
Top-down method of active perception, **7**-7
TopDisc, **8**-1, **8**-9–**8**-12
TopDisc, architectural view of, **8**-11–**8**-12
Topology, **3**-11, **10**-2, **10**-22, **12**-16, **14**-1
Topology, broadcast, **22**-8
Topology
    control, **39**-3
Topology, control, **39**-10–**39**-33
Topology, database, **2**-15–**2**-16
Topology, dedicated, **10**-23
Topology, determining and detecting changes in, **1**-7
Topology, discovery, **20**-17
Topology, dynamic, **9**-9, **22**-3, **22**-6
Topology, fixed, **14**-2
Topology, flat network, cluster size in, **13**-14
Topology, forwarding backbone, **37**-8
Topology, hierarchical overlay network, **13**-2
Topology, issues for design, **14**-3–**14**-8
Topology, large-scale, **15**-10
Topology, maintenance, **40**-8
Topology, management, **9**-16
Topology, map, **3**-25
Topology, mesh, **14**-2
Topology, monitoring, **15**-4
Topology, multihop network, **20**-10
Topology, network, **10**-24
Topology
    network, **39**-2
Topology, node overlay in, **13**-4
Topology, optimization, **15**-10
Topology, overdetermined, **20**-13
Topology
    point-to-point chain, **10**-3
Topology, relative, **20**-3
Topology, self-configuration and, **9**-10
Topology
    star, **10**-3
Topology, TopDisc, **8**-9
Topology, tree network, **10**-3, **22**-8
Topology, using for positioning, **20**-12
Topology discovery for sensor networks. *See* TopDisc
TORA, **1**-7, **39**-3
Toshiba, **27**-2
Toxic zones, monitoring of, **27**-1
Toxic zones, wireless sensor networks and, **3**-1
TRACE, **18**-10
Trace detectors, **24**-5
Traceback mechanisms, **32**-15
Tracking, **12**-8, **15**-2, **23**-1
Tracking, smart message, **26**-4
Trade-off region, **1**-8
Traditional data networks, differences from sensor networks, **4**-2
Traffic, **1**-10, **34**-1
Traffic, checking, **3**-15
Traffic, concentration problem, **3** 19
Traffic, distribution, **9**-19
Traffic, models, **15**-9
Traffic, monitoring, **9**-6
Traffic, pattern, **10**-22
Traffic adaptive medium access protocol. *See* TRAMA
TRAMA, **28**-13
TRAMA, advantages and disadvantages of, **28**-20
TRAMA, operation, **28**-18
Transceiver, **3**-8
Transducer bus interface modules. *See* TBIMs
Transducer electronic data sheet. *See* TEDS
Transducer to microprocessor communication protocol and TEDS formats (IEEE 1451.2), **10**-11
Transducers, **12**-5
Transformation, **31**-13
Transient fault, **36**-4
Transitioning, **27**-2
Translocation, **25**-7
Transmeta, **38**-2
Transmission, **1**-9
Transmission, distance, **15**-9
Transmission, effect of fixed number of, **14**-11
Transmission, electronics, **38**-7
Transmission, energy-expensive long-distance, **38**-1
Transmission, low range, **28**-10
Transmission, media, **6**-5
Transmission, number of, **14**-9
Transmission, overlapping ranges, **14**-2, **23**-4
Transmission, power, **9**-15
Transmission, power, control, **39**-26, **39**-29
Transmission, power, limitations of, **1**-1
Transmission, range, **39**-2
Transmission, range limitations, **20**-10, **23**-10
Transmission, rate and timing, **10**-2
Transmission, reduction of overhead for, **30**-2
Transmitters, **1**-4
Transmitters, electronics energy dissipation from, **15**-9
Transport capacity, **1**-5, **1**-6–**1**-7
    event-to-sink, **16**-7
    sink-to-sensors, **16**-8
Transport control module, **25**-11

Index

Transport control protocol. *See* TCP
Transport layer, **3**-25
Transport layer, mechanisms, **16**-11
Transport layer, protocols, **16**-7–**16**-9
Transport protocol, **17**-4
Transport-layer, abstraction, **17**-2
Transport-layer, address space, **17**-3
TreC, **8**-11
Tree network topology, **10**-3, **22**-8, **24**-3
Tree network topology, dynamic, **22**-7
Tree network topology
    joining, **22**-8
Tree-based architecture, **18**-11
Triangulation, **20**-5, **39**-2
Trigonometry laws, **36**-10
Trilateration, **20**-5, **20**-13, **21**-4
Trilateration, navigation solutions using, **20**-7
Triplication, **36**-5
Trojan horses, **31**-3
Trust continuum, **32**-7
Trusted key server, **32**-14
TSC, **29**-6
Tunneling, **10**-16
Tunneling, gateway, ethernet-fieldbus coupler, **10**-18
Tuple space model, **20**-12
Turing machine, **40**-1
Two-dimensional networks, **13**-4
Two-dimensional networks, analysis of, **14**-10
Two-entity model, **4**-2
Two-fault tolerance, **39**-28
Two-state I/O devices, device scheduling for, **29**-11

U

UAV, **2**-3
UAV, applications, **22**-13
UBP, **10**-23
UCLA, **13**-7, **15**-3
UCLA Medusa MK-2 node. *See* Medusa MK-2 node
UCLinux, **12**-10
UDG, **39**-3, **39**-4
UDP, **25**-11
UGS, **1**-2
Ultra Low-Power Wireless Sensor, **38**-5
Ultra wide-band. *See* UWB
Ultracapacitor, **5**-10
Ultralow power radio, **12**-6
Ultrasound, **20**-5
Ultrasound, ranging, **20**-9
Unattended ground sensors. *See* UGS
Unattended operation, **9**-9
Uncertainty problem, **3**-21, **22**-12
Underwater exploration, **9**-5
Unicast, **3**-11
Unicast, messages, **28**-12
Unicast, routing protocol, **18**-11
Unidirectional communication, **38**-9
Unit disk graph. *See* UDG
Universal register machine, **40**-1
University of Colorado, **13**-6
UNIX, **10**-14, **13**-6

Unmanned aerial vehicle. *See* UAV
Unrechargeable energy provision of sensor networks, **1**-4
Update broadcast packets. *See* UBP
Updating, **20**-16
Uplink wireless communication channels, **20**-2
Urban, planning, **1**-3
Urban, warfare, **1**-4
URM, **40**-15
Usage, diversity, **12**-3
Usage standard map, **3**-24
User access, **31**-4
User datagram protocol. *See* UDP
User layer, **15**-4
UWB, **20**-5

V

Valence, **7**-8
Valid scheduling instant, **29**-15
Vandal, **32**-5
Variable voltage hardware, **38**-5
Variable voltage processing, **27**-5, **37**-5
Variable-frequency processors, **37**-4
Variance, **34**-8
Vehicle tracking, **8**-9
Velocity constraints, enforcement of, **17**-3
Velocity monotonic scheduling. *See* VMS
Verification, **12**-8, **38**-7
Verification, architecture, **15**-2
VGA routing, **6**-16
Vibration, **5**-12
Video image processing, **24**-12
Video radar data, **2**-5
Video rewrite, **7**-10
View composition, **8**-7
Virtual grid architecture routing. *See* VGA routing
Virtual hosts, **17**-4
Virtual keyboards, **16**-2
Virtual keyboards, fingertip accelerometer, **1**-2
Virtual machines, **4**-4, **26**-2
Virtual machines, motes and, **4**-9
Virtual machines, smart messages and, **26**-4
Virtual private networks, **31**-3
Virus monitoring, **9**-4
Viruses, **31**-3
Visible light pulses, **20**-5
Vision, active perception nature of, **7**-6
Vision, sensors, **22**-9
Visual perception, **7**-4
Visual sensors, **38**-1
Visual system, perception by, **7**-3
Viterbe algorithm, **21**-8
Viterbi-like tracking algorithm, **21**-8
VLSI, **1**-11
VMS, **17**-3
Voice communication, mesh networks and, **1**-1
Volatile memories, **36**-6
Voltage, core, **27**-7
Voltage, scheduling, **38**-6
Voltage processing, variable, **27**-5
Voltage-tunable high-Q capacitors, **5**-5

Von Neumann models of computation, **40**-15
Voronoi diagrams, **9**-13, **39**-2, **39**-8
Voronoi diagrams, lines of, **19**-9
VTC, **1**-2
Vulnerabilities, **32**-2, **32**-4
Vulnerabilities, minimization of risks from, **32**-10

W

Wafer bonding, **5**-2, **5**-13
WAHN, **28**-2
WAHN, architectures, **31**-4
WAHN, MAC protocols for, **28**-4–**28**-10
WAHN, multihop, **28**-5
Wake-up protocol, **4**-3, **28**-5
WAN, NRMs, **2**-16
Warehouse inventory tracking, **9**-6
WAS, **2**-3
WASN, **19**-2, **40**-1
WASN, centralized algorithms for, **40**-4
Waterfall method, **7**-4
WBM, **3**-24
Weak connectivity, **33**-5
Weapon control, **35**-2
Weather monitoring, **6**-1
Web-based management protocol. See WBM
Weber's law, **7**-2
Websites, ARGOS, **11**-2
Websites, ASI, **10**-4
Websites, bitbus, **10**-7
Websites, Business 2.0, **1**-2
Websites, CAN, **10**-5
Websites, foundation fieldbus, **10**-8
Websites, Globalfuture, **1**-2
Websites, Great Duck Island, **1**-3
Websites, HART, **10**-3
Websites, LonWorks, **10**-7
Websites, Microwire, **10**-10
Websites, Profibus, **10**-9
Websites, Sercos, **10**-7
Websites, smart message, **26**-8
Websites, WINLAB, **11**-3
Weighted averages, **7**-4
Whales, tagging, **11**-2
Whales, tagging, packet propagation, **11**-8
Wide area networks. See WAN
Wide area surveillance. See WAS
Wild land fire detection, **9**-4
WINLAB, **11**-3
WINS, **5**-15, **9**-12, **15**-3, **19**-2, **38**-5
WINS, NG 2.0, **13**-7
WINS, power consumption for, **15**-9
Wireless, ad hoc networks, **1**-1, **9**-7
Wireless, communications, **38**-1
Wireless, error prone medium, **9**-9
Wireless, Ethernet, use of by ANSER system, **22**-14
Wireless, front ends, **20**-12
Wireless, Internet access, mesh networks and, **1**-1
Wireless, link, **1**-8–**1**-9
Wireless, multicast advantage, **1**-6
Wireless, multihop networks. See WMNs

Wireless, networking, operational characteristics of, **25**-11
Wireless, networking, topology effect on, **14**-3
Wireless, networks, **2**-7–**2**-8
Wireless, networks, packet losses in, **25**-11
Wireless, radio, **36**-6
Wireless, reprogramming, **1**-11
Wireless, sensor networks. See WSN
Wireless, sensor networks and applications. See WSNA
Wireless, telemetry, **11**-2
Wireless ad hoc networks. See WAHN
Wireless ad hoc sensor networks. See WASN
Wireless communication protocols (IEEE P1451.5), **10**-13
Wireless Integrated Network Sensor. See WINS
Wireless local area network. See WLAN
Wireless transceivers, **19**-2. See also sensor nodes
Withdrawal procedure, **37**-8
WLAN, **12**-14, **28**-2
WLAN, transceivers, **12**-6
WMNs, **40**-1
WNS, location in, **20**-10–**20**-21
Workload prediction strategy, **9**-15
WorldFIP initiatives, **10**-8, **10**-15
Wormholes, **32**-14
Worst-case fault models, **36**-13
WSN_SUM protocol, **23**-7
WSNA, **1**-2
WSNs, **3**-1, **28**-1, **39**-1. See also sensor
  ad hoc, **31**-4
WSNs, advantages and disadvantages of EC-MAC for, **28**-8
WSNs, advantages and disadvantages of IEEE 802.11 for, **28**-5
WSNs
  advantages and disadvantages of PAMAS for, **28**-9
WSNs, advantages and disadvantages of TDMA schemes for, **28**-7
WSNs
  agents, **3**-17
WSNs, applications of, **27**-1. See also specific applications
WSNs
  cell-based, **31**-4
WSNs, characteristics, **9**-7–**9**-8, **15**-2, **28**-2
WSNs, cheap, compact, low-power, **9**-11–**9**-12
WSNs, classification of, **9**-6–**9**-7
WSNs, configuration, **3**-8–**3**-9
WSNs, coverage, **9**-1, **9**-13
WSNs, cross-layer communication protocol stack for, **18**-5
WSNs, definition of, **23**-3
WSNs, definition of topology in, **14**-1
WSNs, deployment and operation of, **36**-13
WSNs
  design challenges, **28**-10
WSNs, design objectives, **9**-9–**9**-11
WSNs, differences from traditional wireless networks, **6**-2
WSNs, dimensions for management, **3**-4–**3**-5
WSNs, dynamic, performance metrics of, **15**-7–**15**-13
WSNs, dynamic transmission range, **23**-10
WSNs
  eight-neighbors, **14**-6
WSNs, embedded systems, **12**-13
WSNs, energy constrained, **3**-20
WSNs, energy consumption in, **18**-3, **28**-10, **29**-1, **29**-19

WSNs, energy evaluation model for target localization, **29**-24
WSNs
 five-neighbors, **14**-4
WSNs, flat, **3**-17
WSNs
 four-neighbors, **14**-4
WSNs, functional layers of, **15**-4
WSNs, functionalities, **3**-8–**3**-12
WSNs, future design issues, **9**-22
WSNs
 goals, **3**-23
WSNs, high node density, **28**-3
WSNs, identifying faulty nodes in, **23**-11–**23**-15
WSNs, in industry, **10**-21–**10**-24
WSNs, information gathering in, **23**-5
WSNs, limitations of, **6**-2
WSNs, locationing characteristics, **20**-12
WSNs, locationing schemes for, **20**-16–**20**-20
WSNs, management challenges, **3**-2–**3**-3
WSNs
 management context, **3**-23
 manager, **3**-17
WSNs, MANNA management models, **3**-15–**3**-17
WSNs, medium access protocols for, **28**-13
WSNs, monitoring capabilities of, **9**-2
WSNs, multihop, **23**-3
WSNs, multiplexing protocols for, **18**-7
WSNs, network vulnerability, **32**-4
WSNs, nodes, **28**-3
WSNs, nodes, characteristics of, **20**-10
WSNs
 observer, **3**-23
WSNs, optimal MAC layer schemes for, **28**-23
WSNs, passive, **32**-3
WSNs, power-constrained, **14**-2
WSNs, protocols for large number of sensors in, **23**-2
WSNs, reliability problems in, **33**-2
WSNs, reliable monitoring paradigms, **9**-1
WSNs, routing, challenges in, **6**-4–**6**-6
WSNs, routing, motivation and design issues, **6**-3–**6**-4

WSNs, routing, protocols, **6**-6–**6**-21
WSNs, security challenges and enabling mechanisms, **31**-2–**31**-4
WSNs, self-regulating, **9**-5
WSNs
 seven-neighbors, **14**-6
WSNs, single-hop, **23**-3
WSNs, six-neighbors, **14**-5
WSNs, spatial density of, **20**-10
WSNs, technical approaches to implementation and design, **9**-11–**9**-22
WSNs, technical challenges and requirements of, **9**-8
WSNs, three dimensions, **14**-6
WSNs
 three-neighbors, **14**-4
WSNs, workload, **9**-14
WTLS, **31**-3

X

XML, **8**-7, **10**-12
XScale microarchitecture, **13**-7, **38**-2

Y

Yao graph, **39**-8
Yao graph, symmetric, **39**-13
Yao structure, **39**-10
YaoYao structure, **39**-12

Z

Zebranet, **1**-3
Zero-power orientation, **37**-10
Zn-air cell batteries, **5**-11
Zone routing, **6**-18, **39**-3
Zoning, fixed, **6**-15
ZRP, **39**-3

DISCARDED
CONCORDIA UNIV. LIBRARY